Precision Medicine for Investigators, Practitioners and Providers

Precision Medicine for Investigators, Practitioners and Providers

Edited by

Joel Faintuch
Department of Gastroenterology
São Paulo University Medical School
São Paulo, Sao Paulo, Brazil

Salomao Faintuch
Department of Radiology
Harvard Medical School
Boston, United States

ELSEVIER
ACADEMIC PRESS
An imprint of Elsevier

Academic Press is an imprint of Elsevier
125 London Wall, London EC2Y 5AS, United Kingdom
525 B Street, Suite 1650, San Diego, CA 92101, United States
50 Hampshire Street, 5th Floor, Cambridge, MA 02139, United States
The Boulevard, Langford Lane, Kidlington, Oxford OX5 1GB, United Kingdom

Notices

Knowledge and best practice in this field are constantly changing. As new research and experience broaden our understanding, changes in research methods, professional practices, or medical treatment may become necessary.

Practitioners and researchers must always rely on their own experience and knowledge in evaluating and using any information, methods, compounds, or experiments described herein. In using such information or methods they should be mindful of their own safety and the safety of others, including parties for whom they have a professional responsibility.

To the fullest extent of the law, neither the Publisher nor the authors, contributors, or editors, assume any liability for any injury and/or damage to persons or property as a matter of products liability, negligence or otherwise, or from any use or operation of any methods, products, instructions, or ideas contained in the material herein.

Library of Congress Cataloging-in-Publication Data
A catalog record for this book is available from the Library of Congress

British Library Cataloguing-in-Publication Data
A catalogue record for this book is available from the British Library

ISBN: 978-0-12-819178-1

For information on all Academic Press publications visit our website at https://www.elsevier.com/books-and-journals

Publisher: Stacy Masucci
Acquisition Editor: Elizabeth Brown
Editorial Project Manager: Pat Gonzalez
Production Project Manager: Maria Bernadette Vidhya
Cover Designer: Miles Hitchen

Typeset by TNQ Technologies

Contents

31. Decision support systems in breast cancer

Alexandros Vamvakas, Katerina Vassiou, Dimitra Tsivaka and Ioannis Tsougos

D—Big data, artificial intelligence, and deep phenotyping

32. Precision medicine in the assessment of suicide risk

Thiago Henrique Roza, Lucas Mohr Patusco, Aline Zimerman, Pedro Ballester and Ives Cavalcante Passos

33. Artificial intelligence in gastroenterology

James K. Ruffle, Adam D. Farmer and Qasim Aziz

34. Algorithms for epilepsy monitoring

Giampaolo Vatti, Paolo Detti, Zabalo Manrique de Lara, Lucia Billeci, Alessandro Tonacci and Maurizio Varanini

35. Precision medicine in ophthalmology: an evolving revolution in diagnostic and therapeutic tools

Marina Roizenblatt, Jaime Roizenblatt, Kim Jiramongkolchai, Peter Louis Gehlbach, Rodrigo Brant, Mauricio Maia and Rubens Belfort

36. Phenotyping COPD

José Luis López-Campos

E—Lifestyle, age-related, and environmental agents

37. Lifestyle medicine

Patrice Couzigou

Contributors

Jessica Almqvist, Associate Professor of Public International Law. Department of Public Law and Legal Philosophy, Autónoma University in Madrid, Madrid, Spain

Amir Asadi, Engineering Technology & Industrial Distribution, College of Engineering, Texas A&M University, College Station, TX, United States

Qasim Aziz, Centre for Neuroscience and Trauma, Blizard Institute, Wingate Institute of Neurogastroenterology, Barts and the London School of Medicine & Dentistry, Queen Mary University of London, London, United Kingdom

Joana Bisol Balardin, Instituto do Cérebro, Hospital Israelita Albert Einstein, São Paulo, São Paulo, Brazil

Pedro Ballester, Machine Intelligence Research Group, PUCRS, Porto Alegre, Brazil; Programa de Pós-Graduação em Ciência da Computação, Pontifícia Universidade Católica do Rio Grande do Sul, Porto Alegre, Brazil

Hermes Vieira Barbeiro, Emergências Clínicas/LIM 51, Faculdade de Medicina da Universidade de São Paulo, São Paulo, Brazil

Denise Frediani Barbeiro, Emergências Clínicas/LIM 51, Faculdade de Medicina da Universidade de São Paulo, São Paulo, Brazil

Soumeya Bekri, Department of Metabolic Biochemistry, Rouen University Hospital, Rouen, France; Normandie Univ, UNIROUEN, CHU Rouen, INSERM U1245, Rouen, France

Rubens Belfort, Jr., Department of Ophthalmology, Federal University of São Paulo, São Paulo, Brazil; Vision Institute, IPEPO, Paulista Medical School, Federal University of São Paulo, São Paulo, Brazil

Miguel A. Bergero, Urology, Sanatorio Privado San Geronimo, Santa Fe, Argentina

Adri Bester, Bowels and Brains Lab, School of Applied Science, London South Bank University, London, United Kingdom

Gargi Bhattacharjee, School of Biological Sciences and Biotechnology, Institute of Advanced Research, Koba Institutional Area, Gandhinagar, Gujarat, India

Claudinei Eduardo Biazoli, Jr., Universidade Federal do ABC, Center for Mathematics, Computing and Cognition, Santo André, SP, Brazil

Lucia Billeci, Institute of Clinical Physiology, National Research Council of Italy (IFC-CNR), Pisa, Italy

Graziela Biude da Silva Duarte, Department of Food and Experimental Nutrition, Faculty of Pharmaceutical Science, University of São Paulo, São Paulo, Brazil

Alex B. Blair, Department of Surgery, Johns Hopkins Hospital, Baltimore, MD, United States

Darren Braddick, Department of R&D, Cementic S.A.S., Paris, France

Rodrigo Brant, Department of Ophthalmology, Federal University of São Paulo, São Paulo, Brazil

Robert A. Britton, Center for Metagenomics and Microbiome Research, Department of Molecular Virology and Microbiology, Baylor College of Medicine, Houston, TX, United States

Richard A. Burkhart, Department of Surgery, Johns Hopkins Hospital, Baltimore, MD, United States

J.A. Byrne, Children's Cancer Research Unit, Kids Research and Discipline of Child and Adolescent Health, Faculty of Medicine and Health, University of Sydney, Westmead, NSW, Australia

Joaquim M.S. Cabral, iBB – Institute for Bioengineering and Biosciences and Department of Bioengineering, Instituto Superior Técnico, Universidade de Lisboa, Lisbon, Portugal; The Discoveries Centre for Regenerative and Precision Medicine, Lisbon Campus, Instituto Superior Técnico, Universidade de Lisboa, Lisbon, Portugal

Thiago Cabral, Department of Ophthalmology, Federal University of São Paulo (UNIFESP), São Paulo, Brazil; Adjunct Professor in Ophthalmology, Department of Specialized Medicine, CCS – Federal University of Espirito Santo (UFES), Vitoria, Espirito Santo, Brazil; Vision Center Unit, Ophthalmology, Empresa Brasileira de Servicos Hospitalares (EBSERH), HUCAM-UFES, Vitoria, Espirito Santo, Brazil

José S. Câmara, CQM – Centro de Química da Madeira, Universidade da Madeira, Campus da Penteada, Funchal, Portugal; Faculdade de Ciências Exatas e da Engenharia, Universidade da Madeira, Campus da Penteada, Funchal, Portugal

Carlos Campillo-Artero, Balearic Health Service, Majorca, Balearic Islands, Spain; and Center for Research in Health and Economics, Barcelona School of Management, Universitat Pompeu Fabra, Barcelona, Catalonia, Spain

Grover Enrique Castro Guzman, University of São Paulo, Department of Computer Science, São Paulo, São Paulo, Brazil

Tina Catela Ivkovic, Masaryk University, Brno, Czech Republic

Juan P. Cayún, Laboratory of Chemical Carcinogenesis and Pharmacogenetics (CQF), Department of Basic and Clinical Oncology, Faculty of Medicine, University of Chile, Latin American Society of Pharmacogenomics and Personalized Medicine (SOLFAGEM) & Latin American Network for Implementation and Validation of Pharmacogenomic Clinical Guidelines (RELIVAF), Quinta Normal, Santiago, Chile

Naseem A. Charoo, Zeino Pharma FZ LLC, Dubai Science Park, Dubai, United Arab Emirates

Y. Chen, Children's Cancer Research Unit, Kids Research and Discipline of Child and Adolescent Health, Faculty of Medicine and Health, University of Sydney, Westmead, NSW, Australia

Marina Codari, Department of Electronics, Informatics and Bioengineering, Politecnico di Milano, Milan, Italy

Graciely G. Correa, São Paulo State University (UNESP), Graduate Program Bioscience and Biotechnology Applied to Pharmacy, Araraquara, Brazil

Mairene Coto-Llerena, Institute of Pathology, University Hospital Basel, Basel, Switzerland

Patrice Couzigou, Professor Emeritus of Medicine, University of Bordeaux, Bordeaux, France

Daniel A. Cozetto, São Paulo State University (UNESP), School of Pharmaceutical Sciences, Department of Bioprocess and Biotechnology, Araraquara, Brazil

Gemma Currie, Institute of Cardiovascular and Medical Sciences, University of Glasgow, Glasgow, United Kingdom

L. Dalla-Pozza, The Cancer Centre for Children, The Children's Hospital at Westmead, Westmead, NSW, Australia

Victor N. de Jesus, São Paulo State University (UNESP), School of Pharmaceutical Sciences, Department of Bioprocess and Biotechnology, Araraquara, Brazil

Zabalo Manrique de Lara, Department of Information Engineering and Mathematics, University of Siena, Siena, Italy

Christian Delles, Institute of Cardiovascular and Medical Sciences, University of Glasgow, Glasgow, United Kingdom

Iñigo de Miguel Beriain, Chair in Law and the Human Genome Research Group, Department of Public Law, University of the Basque Country, UPV/EHU, Bizkaia, Spain; IKERBASQUE, Basque Foundation for Science, Bilbao, Spain

Juliana de Moura, Department of Bioprocesses and Biotechnology Engineering, Federal University of Paraná, Curitiba, Paraná, Brazil; Department of Basic Pathology, Federal University of Paraná, Curitiba, Paraná, Brazil

Cintia S. de Paiva, Department of Ophtalmology, São Paulo University Medical School, São Paulo, Brazil; Department of Ophthalmology, Baylor College of Medicine, Houston, TX, United States

Rodrigo G. de Souza, Department of Ophtalmology, São Paulo University Medical School, São Paulo, Brazil; Department of Ophthalmology, Baylor College of Medicine, Houston, TX, United States

Paolo Detti, Department of Information Engineering and Mathematics, University of Siena, Siena, Italy

Romina Díaz, Department of Chemical and Bioprocess Engineering, School of Engineering. Pontificia Universidad Catolica de Chile, Santiago, Chile

Jesse M. Ehrenfeld, Vanderbilt University Medical Center, Nashville, TN, United States

Bluma Linkowski Faintuch, Radiopharmacy Center, Institute of Energy and Nuclear Research, São Paulo, São Paulo, Brazil

Joel Faintuch, Department of Gastroenterology, São Paulo University Medical School, São Paulo, São Paulo, Brazil

Jacob J. Faintuch, Department of Internal Medicine, Hospital das Clinicas, São Paulo, São Paulo, Brazil

Salomao Faintuch, Department of Radiology, Harvard Medical School, Boston, United States

Telma A. Faraldo Corrêa, Food Research Center (FoRC), CEPID-FAPESP, Research Innovation and Dissemination Centers São Paulo Research Foundation, São Paulo, Brazil; Department of Food and Experimental Nutrition, Faculty of Pharmaceutical Science, University of São Paulo, São Paulo, Brazil

Adam D. Farmer, Centre for Neuroscience and Trauma, Blizard Institute, Wingate Institute of Neurogastroenterology, Barts and the London School of Medicine & Dentistry, Queen Mary University of London, London, United Kingdom; Department of Gastroenterology, University Hospitals of North Midlands NHS Trust, Stoke on Trent, United Kingdom

Paulo J.C. Freire, São Paulo State University (UNESP), School of Pharmaceutical Sciences, Department of Bioprocess and Biotechnology, Araraquara, Brazil

Andre Fujita, University of São Paulo, Department of Computer Science, São Paulo, São Paulo, Brazil

Daniel Garrido, Department of Chemical and Bioprocess Engineering, School of Engineering. Pontificia Universidad Catolica de Chile, Santiago, Chile

Athalye-Jape Gayatri, Neonatal Directorate, Perth Children's Hospital, Perth, WA, Australia; Neonatal Directorate, King Edward Memorial Hospital for Women, Perth, WA, Australia; School of Medicine, University of Western Australia, Perth, WA, Australia

Nisarg Gohil, School of Biological Sciences and Biotechnology, Institute of Advanced Research, Koba Institutional Area, Gandhinagar, Gujarat, India

Marta Gómez de Cedrón, Precision Nutrition and Cancer Program. Molecular Oncology and Nutritional Genomics of Cancer Group. IMDEA Food Institute, CEI UAM + CSIC, Madrid, Spain

Dolores Gonzalez Moron, Neurogenetics Clinic, Hospital JM Ramos Mejia, Buenos Aires, Argentina

Tetsuya Ishii, Office of Health and Safety, Hokkaido University, Sapporo, Hokkaido, Japan

Claude J. Pirtle, Vanderbilt University Medical Center, Nashville, TN, United States

Abhishek Jain, Department of Biomedical Engineering, College of Engineering, Texas A&M University, College Station, TX, United States

R.V. Jamieson, The Children's Hospital at Westmead & Children's Medical Research Institute, Eye & Developmental Genetics Research Group, Westmead, NSW, Australia

Kim Jiramongkolchai, Department of Ophthalmology, Wilmer Eye Institute, Johns Hopkins University School of Medicine, Baltimore, MD, United States

Thomas Kaiser, Department of Surgery, University of Minnesota, Minneapolis, MN, United States; Bio-Technology Institute, University of Minnesota, St. Paul, MN, United States

Maged N. Kamel Boulos, Sun Yat-sen University, Guangzhou, Guangdong, China

Sri Harsha Kanuri, Department of Clinical Pharmacology, Institute of Personalized Medicine (IIPM), IU School of Medicine, Indianapolis, IN, United States

Marcelo A. Kauffman, Neurogenetics Clinic, Hospital JM Ramos Mejia, Buenos Aires, Argentina

Khushal Khambhati, School of Biological Sciences and Biotechnology, Institute of Advanced Research, Koba Institutional Area, Gandhinagar, Gujarat, India

Mansoor A. Khan, Irma Lerma Rangel College of Pharmacy, Texas A&M Health Science Center, Texas A&M University, College Station, TX, United States

Rolf P. Kreutz, Krannert Institute of Cardiology, Indiana University School of Medicine, Indianapolis, IN, United States

Mathew Kuttolamadom, Engineering Technology & Industrial Distribution, College of Engineering, Texas A&M University, College Station, TX, United States

Hitesh Lal, Sports Injury Centre, Safdarjung Hospital and Vardhman Mahavir Medical College, New Delhi, India

Jose Ronaldo Lima de Carvalho, Jr., Department of Ophthalmology, Columbia University, New York, NY, United States; Jonas Children's Vision Care and Bernard & Shirlee Brown Glaucoma Laboratory, New York, NY, United States; Department of Ophthalmology, Hospital das Clinicas de Pernambuco (HCPE) – Empresa Brasileira de Servicos Hospitalares (EBSERH), Federal University of Pernambuco (UFPE), Recife, Pernambuco, Brazil; Department of Ophthalmology, Federal University of São Paulo (UNIFESP), São Paulo, Brazil

Milca R.C.R. Lins, São Paulo State University (UNESP), Graduate Program Bioscience and Biotechnology Applied to Pharmacy, Araraquara, Brazil

José Luis López-Campos, Medical-Surgical Unit of Respiratory Diseases, Biomedicine Institute of Sevilla, (IBIS), University Hospital Virgen del Rocío, University of Sevilla, Sevilla, Spain; Research Center in Respiratory Diseases Net (CIBERES), Carlos III Health Institute (ISCIII), Madrid, Spain

Peter Louis Gehlbach, Department of Ophthalmology, Wilmer Eye Institute, Johns Hopkins University School of Medicine, Baltimore, MD, United States

Blanca Lumbreras, Department of Public Health, University Miguel Hernández, Alicante, the Valencian Community, Spain; and CIBER of Epidemiology and Public health (CIBERESP)

Vinit B. Mahajan, Omics Laboratory, Byers Eye Institute, Department of Ophthalmology, Stanford University, Palo Alto, CA, United States; Veterans Affairs Palo Alto Health Care System, Palo Alto, CA, United States

Mauricio Maia, Department of Ophthalmology, Federal University of São Paulo, São Paulo, Brazil; Vision Institute, IPEPO, Paulista Medical School, Federal University of São Paulo, São Paulo, Brazil

Indra Mani, Department of Microbiology, Gargi College, University of Delhi, New Delhi, Delhi, India

J. Alfredo Martinez, Department of Nutrition, Food Science and Physiology, University of Navarra, and Center for Nutrition Research, University of Navarra, Pamplona, Spain; CIBERobn, Physiopathology of Obesity, Carlos III Institute, Madrid, Spain; Navarra Institute for Health Research (IdiSNA), Pamplona, Spain; Madrid Institute of Advanced Studies (IMDEA Food), Madrid, Spain

Pablo F. Martinez, Urology, Hospital Italiano de Buenos Aires, Buenos Aires, Argentina

Sheon Mary, Institute of Cardiovascular and Medical Sciences, University of Glasgow, Glasgow, United Kingdom

Tanmay Mathur, Department of Biomedical Engineering, College of Engineering, Texas A&M University, College Station, TX, United States

Cláudia C. Miranda, iBB – Institute for Bioengineering and Biosciences and Department of Bioengineering, Instituto Superior Técnico, Universidade de Lisboa, Lisbon, Portugal; The Discoveries Centre for Regenerative and Precision Medicine, Lisbon Campus, Instituto Superior Técnico, Universidade de Lisboa, Lisbon, Portugal

Reza Mirnezami, Colorectal Surgeon and Honorary Lecturer in Surgery, Department of Surgery & Cancer, Imperial College London, London, United Kingdom

Charlotte K.Y. Ng, Department of BioMedical Research, University of Bern, Bern, Switzerland

Francesc Palau, Department of Genetic Medicine and Pediatric Institute of Rare diseases, and Director, Sant Joan de Déu Research Institute, Sant Joan de Déu Children's Hospital, Barcelona, Spain; Institute of Medicine and Dermatology, Hospital Clínic, Barcelona, Spain; CSIC Research Professor and Adjunct Professor of Pediatrics, University of Barcelona School of Medicine and Health Sciences, Barcelona, Spain; Group Leader, Neurogenetics and Molecular Medicine Group, CIBERER, Barcelona, Spain

Happy Panchasara, School of Biological Sciences and Biotechnology, Institute of Advanced Research, Koba Institutional Area, Gandhinagar, Gujarat, India

Navaneeth K.R. Pandian, Department of Biomedical Engineering, College of Engineering, Texas A&M University, College Station, TX, United States

Karen Sophia Park, Department of Ophthalmology, Columbia University, New York, NY, United States; Jonas Children's Vision Care and Bernard & Shirlee Brown Glaucoma Laboratory, New York, NY, United States

Ives Cavalcante Passos, Molecular Psychiatry Laboratory, Hospital de Clínicas de Porto Alegre, Porto Alegre, Brazil; Programa de Pós-Graduação em Psiquiatria e Ciências do Comportamento, Universidade Federal do Rio Grande do Sul, Porto Alegre, Brazil

Maria Pastor-Valero, Department of Public Health, University Miguel Hernández, Alicante, the Valencian Community, Spain; and CIBER of Epidemiology and Public health (CIBERESP)

Francesca Patella, Radiology Unit, ASST Santi Paolo e Carlo, Milan, Italy

Mohit Kumar Patralekh, Central Institute of Orthopaedics, Safdarjung Hospital and Vardhman Mahavir Medical College, New Delhi, India

Lucas Mohr Patusco, Molecular Psychiatry Laboratory, Hospital de Clínicas de Porto Alegre, Porto Alegre, Brazil; Programa de Pós-Graduação em Psiquiatria e Ciências do Comportamento, Universidade Federal do Rio Grande do Sul, Porto Alegre, Brazil

Danielle B. Pedrolli, São Paulo State University (UNESP), School of Pharmaceutical Sciences, Department of Bioprocess and Biotechnology, Araraquara, Brazil

Jorge A.M. Pereira, CQM – Centro de Química da Madeira, Universidade da Madeira, Campus da Penteada, Funchal, Portugal

Filippo Pesapane, Postgraduate School in Radiodiagnostics, Università degli Studi di Milano, Milan, Italy

João Vitor Pincelli, Laboratory of Special Techniques, Department of Pathology and Clinical Pathology, Hospital Israelita Albert Einstein, São Paulo, São Paulo, Brazil

Salvatore Piscuoglio, Institute of Pathology, University Hospital Basel, Basel, Switzerland; Visceral Surgery Research Laboratory, Clarunis, Department of Biomedicine, University of Basel, Basel, Switzerland

Jose J. Ponce-Lorenzo, Department of Medical Oncology, University General Hospital of Alicante, Alicante, Spain

Priscilla Porto-Figueira, CQM – Centro de Química da Madeira, Universidade da Madeira, Campus da Penteada, Funchal, Portugal

V.S. Priyadharshini, Instituto Nacional de Enfermedades Respiratorias, Delegación Tlalpan, Mexico; Escuela Superior de Medicina del Instituto Politecnico Nacional, Plan de San Luis y Díaz Miron, Mexico

Peter Natesan Pushparaj, Center of Excellence in Genomic Medicine Research, Faculty of Applied Medical Sciences, King Abdulaziz University, Jeddah, Mecca Province, Kingdom of Saudi Arabia; Department of Medical Laboratory Technology, Faculty of Applied Medical Sciences, Jeddah, Mecca Province, Kingdom of Saudi Arabia

Luis A. Quiñones, Laboratory of Chemical Carcinogenesis and Pharmacogenetics (CQF), Department of Basic and Clinical Oncology, Faculty of Medicine, University of Chile, Latin American Society of Pharmacogenomics and Personalized Medicine (SOLFAGEM) & Latin American Network for Implementation and Validation of Pharmacogenomic Clinical Guidelines (RELIVAF), Quinta Normal, Santiago, Chile

Bruna Jardim Quintanilha, Department of Nutrition, School of Public Health, University of São Paulo, São Paulo, Brazil; Food Research Center (FoRC), CEPID-FAPESP, Research Innovation and Dissemination Centers São Paulo Research Foundation, São Paulo, Brazil

Ziyaur Rahman, Irma Lerma Rangel College of Pharmacy, Texas A&M Health Science Center, Texas A&M University, College Station, TX, United States

Ana Ramírez de Molina, Precision Nutrition and Cancer Program. Molecular Oncology and Nutritional Genomics of Cancer Group. IMDEA Food Institute, CEI UAM + CSIC, Madrid, Spain

Omar Ramos-Lopez, Department of Nutrition, Food Science and Physiology, University of Navarra, and Center for Nutrition Research, University of Navarra, Pamplona, Spain; Faculty of Medicine and Psychology, Autonomous University of Baja California, Tijuana, BC, Mexico

Kenneth S. Ramos, University of Arizona Health Sciences, Office of the Senior Vice President Health Sciences, Tucson, AZ, United States; University of Arizona College of Medicine-Phoenix, Tucson, AZ, United States; University of Arizona College of Medicine-Tucson, Tucson, AZ, United States; University of Arizona Center for Applied Genetics and Genomic Medicine, Tucson, AZ, United States

Srikanth Rapole, Proteomics Lab, National Centre for Cell Science (NCCS), Ganeshkhind, SPPU Campus, Pune, Maharashtra, India

João Renato Rebello Pinho, Laboratory of Special Techniques, Department of Pathology and Clinical Pathology, Hospital Israelita Albert Einstein, São Paulo, São Paulo, Brazil; LIM 03/LIM 07 – Departments of Gastroenterology and Pathology, São Paulo University Medical School, São Paulo, São Paulo, Brazil

Bruna Zavarize Reis, Department of Food and Experimental Nutrition, Faculty of Pharmaceutical Science, University of São Paulo, São Paulo, Brazil

Juan Pablo Rey-Lopez, University of Sydney, School of Public Health, Sydney, NSW, Australia

Nathan V. Ribeiro, São Paulo State University (UNESP), School of Pharmaceutical Sciences, Department of Bioprocess and Biotechnology, Araraquara, Brazil

Marcelo Macedo Rogero, Department of Nutrition, School of Public Health, University of São Paulo, São Paulo, Brazil; Food Research Center (FoRC), CEPID-FAPESP, Research Innovation and Dissemination Centers São Paulo Research Foundation, São Paulo, Brazil

Marina Roizenblatt, Department of Ophthalmology, Wilmer Eye Institute, Johns Hopkins University School of Medicine, Baltimore, MD, United States; Department of Ophthalmology, Federal University of São Paulo, São Paulo, Brazil; Vision Institute, IPEPO, Paulista Medical School, Federal University of São Paulo, São Paulo, Brazil

Jaime Roizenblatt, Division of Ophthalmology, São Paulo University School of Medicine, São Paulo, Brazil

Thiago Henrique Roza, Molecular Psychiatry Laboratory, Hospital de Clínicas de Porto Alegre, Porto Alegre, Brazil; Programa de Pós-Graduação em Psiquiatria e Ciências do Comportamento, Universidade Federal do Rio Grande do Sul, Porto Alegre, Brazil

Noah S. Rozich, Department of Surgery, Johns Hopkins Hospital, Baltimore, MD, United States

James K. Ruffle, Centre for Neuroscience and Trauma, Blizard Institute, Wingate Institute of Neurogastroenterology, Barts and the London School of Medicine & Dentistry, Queen Mary University of London, London, United Kingdom; Department of Radiology, University College London NHS Foundation Trust, London, United Kingdom

Patole Sanjay, Neonatal Directorate, King Edward Memorial Hospital for Women, Perth, WA, Australia; School of Medicine, University of Western Australia, Perth, WA, Australia

Fábio P. Saraiva, Adjunct Professor in Ophthalmology, Department of Specialized Medicine, CCS – Federal University of Espirito Santo (UFES), Vitoria, Espirito Santo, Brazil; Vision Center Unit, Ophthalmology, Empresa Brasileira de Servicos Hospitalares (EBSERH), HUCAM-UFES, Vitoria, Espirito Santo, Brazil

Francesco Sardanelli, Unit of Radiology, IRCCS Policlinico San Donato, San Donato Milanese, Italy; Department of Biomedical Sciences for Health, Università degli Studi di Milano, San Donato Milanese, Italy

João Ricardo Sato, Universidade Federal do ABC, Center for Mathematics, Computing and Cognition, Santo André, SP, Brazil

Aletta E. Schutte, Hypertension in Africa Research Team (HART), MRC Unit for Hypertension and Cardiovascular Disease, North-West University, Potchefstroom, South Africa

Luke A. Schwerdtfeger, Department of Biomedical Sciences, Colorado State University, Fort Collins, CO, United States

Amirali Selahi, Department of Biomedical Engineering, College of Engineering, Texas A&M University, College Station, TX, United States

Prakash Chand Sharma, University School of Biotechnology, Guru Gobind Singh Indraprastha University, Dwarka, New Delhi, India

Rao Shripada, Neonatal Directorate, Perth Children's Hospital, Perth, WA, Australia; School of Medicine, University of Western Australia, Perth, WA, Australia

Patrick J. Silva, University of Arizona Health Sciences, Office of the Senior Vice President Health Sciences, Tucson, AZ, United States

Vijai Singh, School of Biological Sciences and Biotechnology, Institute of Advanced Research, Koba Institutional Area, Gandhinagar, Gujarat, India; Present address: Department of Biosciences, School of Sciences, Indrashil University, Rajpur, Gujarat, India

Ondrej Slaby, Masaryk University, Brno, Czech Republic; Masaryk Memorial Cancer Institute, Brno, Czech Republic

Bruno Araujo Soares, Department of Bioprocesses and Biotechnology Engineering, Federal University of Paraná, Curitiba, Paraná, Brazil; Department of Basic Pathology, Federal University of Paraná, Curitiba, Paraná, Brazil

Francisco Garcia Soriano, Emergências Clínicas/LIM 51, Faculdade de Medicina da Universidade de São Paulo, São Paulo, Brazil

Kamila Souckova, Masaryk University, Brno, Czech Republic

Patrick N. Squizato, São Paulo State University (UNESP), School of Pharmaceutical Sciences, Department of Bioprocess and Biotechnology, Araraquara, Brazil

Nickolas Stabellini, Laboratory of Special Techniques, Department of Pathology and Clinical Pathology, Hospital Israelita Albert Einstein, São Paulo, São Paulo, Brazil

Christopher Staley, Department of Surgery, University of Minnesota, Minneapolis, MN, United States; BioTechnology Institute, University of Minnesota, St. Paul, MN, United States

João Paulo Stanke Scandelari, Department of Basic Pathology, Federal University of Paraná, Curitiba, Paraná, Brazil

Matteo B. Suter, Medical Oncology Unit, ASST Sette Laghi, Varese, Italy

D.E. Sylvester, Children's Cancer Research Unit, Kids Research and Discipline of Child and Adolescent Health, Faculty of Medicine and Health, University of Sydney, Westmead, NSW, Australia

Ravindra Taware, Proteomics Lab, National Centre for Cell Science (NCCS), Ganeshkhind, SPPU Campus, Pune, Maharashtra, India

Abdellah Tebani, Department of Metabolic Biochemistry, Rouen University Hospital, Rouen, France

Luis M. Teran, Instituto Nacional de Enfermedades Respiratorias, Delegación Tlalpan, Mexico

Luigi M. Terracciano, Institute of Pathology, University Hospital Basel, Basel, Switzerland

Taleb Ba Tis, Department of Materials Science and Engineering, North Carolina State University, Raleigh, NC, United States

Stuart A. Tobet, Department of Biomedical Sciences, Colorado State University, Fort Collins, CO, United States; School of Biomedical Engineering, Colorado State University, Fort Collins, CO, United States

Alessandro Tonacci, Institute of Clinical Physiology, National Research Council of Italy (IFC-CNR), Pisa, Italy

Miguel Toribio-Mateas, Bowels and Brains Lab, School of Applied Science, London South Bank University, London, United Kingdom; School of Health and Education, Faculty of Transdisciplinary Practice, Middlesex University, London, United Kingdom

Stephen H. Tsang, Department of Ophthalmology, Columbia University, New York, NY, United States; Jonas Children's Vision Care and Bernard & Shirlee Brown Glaucoma Laboratory, New York, NY, United States; Department of Pathology & Cell Biology, Stem Cell Initiative (CSCI), Institute of Human Nutrition, Vagelos College of Physicians and Surgeons, New York, NY, United States

Dimitra Tsivaka, Medical Physics Department, Medical School, University of Thessaly, Larisa, Greece; Neuroimaging Department, IoPPN, King's College London, London, United Kingdom

Ioannis Tsougos, Medical Physics Department, Medical School, University of Thessaly, Larisa, Greece; Neuroimaging Department, IoPPN, King's College London, London, United Kingdom

Alexandros Vamvakas, Medical Physics Department, Medical School, University of Thessaly, Larisa, Greece

Maurizio Varanini, Institute of Clinical Physiology, National Research Council of Italy (IFC-CNR), Pisa, Italy

Katerina Vassiou, Radiology and Anatomy Department, Medical School, University of Thessaly, Larisa, Greece

Giampaolo Vatti, Department of Neurological and Sensory Sciencese, Azienda Ospedaliera Universitaria Senese, Siena, Italy

Renu Verma, University School of Biotechnology, Guru Gobind Singh Indraprastha University, Dwarka, New Delhi, India

Kalyani Verma, Defence Institute of Physiology and Allied Sciences, Defence Research and Development Organisation, Timarpur, Delhi, India

Luiz Otávio Vittorelli, Laboratory of Special Techniques, Department of Pathology and Clinical Pathology, Hospital Israelita Albert Einstein, São Paulo, São Paulo, Brazil

Caterina Volonté, Independent Researcher, London, United Kingdom

Markus von Flüe, Visceral Surgery Research Laboratory, Clarunis, Department of Biomedicine, University of Basel, Basel, Switzerland

Arsalan Wafi, Clinical Research Fellow, Department of Cardiovascular Surgery, St George's Hospital, University of London, London, United Kingdom

Bruna Mayumi Wagatuma Bottolo, Department of Basic Pathology, Federal University of Paraná, Curitiba, Paraná, Brazil

Qingshan Wei, Department of Chemical and Biomolecular Engineering, North Carolina State University, Raleigh, NC, United States; Emerging Plant Disease and Global Food Security Cluster, North Carolina State University, Raleigh, NC, United States

Peng Zhang, Vanderbilt University Medical Center, Nashville, TN, United States

Shengwei Zhang, Department of Chemical and Biomolecular Engineering, North Carolina State University, Raleigh, NC, United States

Zhigang Zhu, Department of Surgery, University of Minnesota, Minneapolis, MN, United States; BioTechnology Institute, University of Minnesota, St. Paul, MN, United States

Aline Zimerman, Molecular Psychiatry Laboratory, Hospital de Clínicas de Porto Alegre, Porto Alegre, Brazil; Programa de Pós-Graduação em Psiquiatria e Ciências do Comportamento, Universidade Federal do Rio Grande do Sul, Porto Alegre, Brazil

Preface

Many healthcare professionals seem puzzled and discomfited about precision medicine: A bright future or a false start? Better health for all or a splurge of precious resources for the benefit of nobody? "Cui bono," who is behind it, for whom will it be profitable? All medical specialties? Only oncology subareas, genetic diseases, and a few other niches? Or not even those, just manufacturers of reagents, laboratory machinery, and electronic equipment? After a reasonable time, robust investments, and a few thousand scientific articles, are there tangible medical achievements or just hot air? Are morbidity and mortality rates falling somewhere, on account of precision medicine progress?

Editorials in respected journals have not hesitated to employ expressions like "hype," "hoax," and "hunting elephants" [1–4]. Skepticism abounds, along with large servings of cautionary words like daunting task [3], distraction, misapplication, and iatrogenic interventions [4].

Of course in the past it was much worse. Sharp technical and conceptual innovations like Jenner's smallpox vaccination elicited not only controversy but even chicanery [5].

One does not deny that any novel proposal must prove its worth, especially when expensive and rather complicated technology is involved. Even seasoned investigators and laboratories do not take lightly the handling of omics, artificial intelligence, large-scale biobanks, digital health, big data, and bioinformatics. Guidelines are scarce or nonexistent, and emerging procedures and algorithms need to stand the test of time. Conspicuous epidemiological results, like decline in morbidity and mortality indexes, can take decades to materialize.

Yet the Pandora box of precision medicine is by now wide open, and few worms and ghosts materialized. Even the social and ethical mayhem, of mishandling of sensitive clinical and genetic information, has remained a mere threat, not a fact. Millions of personal genomes are now safely stored in public and private databanks, without breaches or inappropriate manipulation. Science has a way of weeding out its own excesses and false steps, and of concentrating on robust, well-trodden paths.

This book is not a festschrift about the wonders of genomics, proteomics, robotics, organoids, wearables, deep learning, or point of care cutting edge resources. Of course all these topics are covered, and many more, however with a grain of salt, whenever appropriate. The emphasis was real experts presenting real-world applications, not experimental theories or outlandish hypotheses. This book was fortunate to count with world-class collaborators, who indeed originate from many latitudes and continents, sparing no time or effort to present the most complete and up-to-date information, yet in handy and practical chapters. The editors are seriously indebted to all of them.

Joel Faintuch
Salomao Faintuch

References

[1] Mennel RG. Precision medicine: hype or hoax? Proc (Bayl Univ Med Cent). 2015;28(3):397–400.

[2] Pitt GS. Cardiovascular precision medicine: hope or hype? Eur Heart J 2015;36(29):1842–3.

[3] Joyner MJ. Precision medicine, cardiovascular disease and hunting elephants. Prog Cardiovasc Dis 2016;58(6):651–60.

[4] Lourenco AP, Leite-Moreira AF. Cardiovascular precision medicine: Bad news from the front? Porto Biomed J 2017;2(4):97–132.

[5] Riedel S. Edward Jenner and the history of smallpox and vaccination. Proc (Bayl Univ Med Cent) 2005;18(1):21–5.

Section I

Tools for investigators

Chapter 1

Introduction

Joel Faintuch[1] and Salamao Faintuch[2]

[1]*Department of Gastroenterology, São Paulo University Medical School, São Paulo, São Paulo, Brazil;* [2]*Department of Radiology, Harvard Medical School, Boston, United States*

History

The roots

The origins of medicine are intertwined with those of religion. Hippocrates (460–370 BCE) was acclaimed as Father of Medicine because he was able to disentangle the two fields, albeit not completely. Moreover, he was a superb observer, and a master of empiricism, to the point that a few of his aphorisms are still cited today.

Some claim that the Egyptians should get the laurels. The Edwin Smith papyrus (3000 BCE?), not only reports diseases and injuries, but also teaches the doctor how to perform a physical examination, and to logically and deductively interpret the findings [1].

The branches

Some degree if not of magic and mysticism, at least of incomplete knowledge and shaky scientific foundations, survived till quite recent times. Charles Sidney Burwell, dean of Harvard Medical School (1935–49), addressed new students with the message: "Half of what we are going to teach you is wrong, and half of it is right. Our problem is that we don't know which half is which."

The fruits

Randomized controlled trials started in the 1960s and meta-analysis in the 1970s. Yet until the systematization of evidence-based medicine (EBM), in the 1990s [2], expert opinion played a significant role in science. "Magister dixit, ergo verum est": the master told it; therefore, it is true. The physicist Max Planck (1858–1947) had already alerted, that new scientific truth has to wait for funerals, even with a raft of evidence. Academics often defend their beliefs for life, rarely reforming their ideas. Of course, this was more relevant in the XIX century, when authorities abhorred criticism.

Fundamental and paradigm-changing as it was, EBM was destined to be promptly outshined by further developments. Within a mere decade or two genomic medicine, personalized medicine, structured medicine, and precision medicine started successfully sharing the limelight.

These models do not simply embody a patient-centered approach, with new "omic" tools for molecular disease characterization, and innovative biomarker-validated results. They represent a quantum leap from old hypothesis-based medicine, in which an investigator devised a theory based on his or her intuition, and looked for corroborating evidence.

Currently, systems biology seeks to mathematically model complex biological phenomenons. The overarching paradigms are not theoretical or pathophysiology-related assumptions, but data-driven architectures of diagnosis and treatment. Paraphrasing Lord Kelvin (1824–1907), "when you can … express it in numbers, you know something about it; but when you cannot measure it … your knowledge is of a meager and unsatisfactory kind."

State of the art

Individual versus society

The rationale of personalized medicine, which has been considerably expanded by precision medicine, is "the right drug for the right patient based on genetic data" [3]. Indeed, a large share of medical prescriptions is believed to be useless, or worse, just a trigger of side effects, because of mismatch with patient's genetics, enzymes, lifestyle, and environmental factors [4]. Many have argued that this emphasis on the individual, will be conducted at the expense of community, depriving the population of the much-needed funds and healthcare resources.

A recent editorial [5] stopped short of declaring patient-centered precision medicine, as incompatible with public health concerns. This is obviously pessimistic.

Precision Medicine for Investigators, Practitioners and Providers. https://doi.org/10.1016/B978-0-12-819178-1.00001-0

As highlighted by others [4], precision medicine is steadily benefitting all stakeholders. Analogously to the advances of analytical chemistry in the 18th and 19th century, only a handful of pioneers predicted the potential range and scope of such initiatives or gave them sufficient attention.

Clinical chemistry

When Antoine Lavoisier (1743–94) described and measured oxygen intake and carbon dioxide production by living tissue (respiration), this was considered as outlandish and devoid of interest, as the most obscure omics in our time. Nowadays, few seriously ill patients would do without a pulse oximeter. Sugar in the urine of diabetics was first regularly diagnosed by the yeast test of Francis Home (1719–1813) before a preliminary chemical detection procedure was devised by Karl August Trommer (1806–79). Trommer never received much recognition. However, Home eventually became famous, on account of a brilliant career, not because of the test. Chemistry was deemed secondary, in comparison to clinical acumen, even in the setting of a life-threatening condition such as diabetes.

Substantial expenses

Cost-effectiveness in scientific progress has always been a primary concern. In the times of Lionel Smith Beale (1828–1906), professor at Kings College, clinical laboratories did not exist in London. In order to entice institutions to create investigation laboratories, his proposal was to attract talented, university-affiliated young physicians and surgeons. They should be paid 100 pounds a year, "just sufficient to provide the necessaries of existence." He was afraid that hospitals would be reticent to invest in new technologies, as indeed happens all the time [6].

An exaggeration of disciplines and technologies

Precision medicine was launched in quite a modest packaging: "prevention and treatment strategies that take individual variability into account" [7]. Yet it was obvious since its inception, that genomics would not make it alone, and that together with other omics, would act as a magnet for myriads of parallel techniques and developments.

This is the very essence of scientific breakthroughs. Rarely does one deal with a single question and the corresponding straightforward answer. Typically, each stride opens up a Pandora box of both evils and marvels, promises and challenges.

In the book Elementa Medicinae [8], John Brown (1735–88) was able to define the components of a sort of "precision medicine" of his time: "Chemistry, Statistics, the Microscope, the Stethoscope, and all new helps and methods." It certainly raised as many eyebrows in the 18th century, as actual precision medicine does in the 21st century. High dimensional datasets, complex algorithms, digital health, and omics, are just a few of the paths pursued worldwide [9].

P4 medicine

Proposed in 2011 in the cancer setting [10], this modality of personalized medicine has disseminated to a number of arenas, including nutrition and lifestyle medicine [11]. The four tenets are predictive, personalized, preventive, and participatory. How do they translate to precision medicine? Omics investigations may be included; however, often, they are not. Novel biomarkers are definitely the goal, and large datasets are built, typically automated, and web-based, with input from the interested individuals. This would fulfill the personalized-participatory features.

A major advantage concerning conventional trials, is the ability to seamlessly follow the population for a long period, something desirable in order to track the natural history of chronic diseases. However, protocols are still costly and cumbersome, even if patients upload their personal information at home, thus avoiding time-consuming visits to the office or hospital. Permanent incentive and guidance to the individuals must be provided, to inhibit drop-out, and to assure that timely, and structured or semi-structured information is introduced in the system, which is easier to process. The next steps are analogous to other studies: warehousing, biostatistical analysis, interpretation, and validation.

Pitfalls

Roadmap to the future, or U-turn to the past?

According to detractors, the fact that precision medicine is individual-centered, instead of community-driven, is the original sin. Costly and scarce resources are directed toward genetical diseases, pharmacogenomical testing of drugs, microbiomic explorations, and search for big-data or molecular targets and biomarkers, to the detriment of the real pressing needs of the society. These encompass mushrooming, exorbitant, and disabling non-communicable diseases, perpetuated by poor lifestyle and environmental perils including air, soil, and water pollution, along with socioeconomic disparities concerning housing, education, and access to care.

Classic preventive medicine and public health efforts would, therefore, better serve the population, with less expenditures and a more predictable outcome. As further proof of misguided efforts, nearly 2 decades of precision medicine had no measurable effect on population morbidity and mortality [12].

Yet precision medicine and public health are synergistic, not antagonistic. The growth of one does not endanger the other, as progress will be advantageous to all. It could take a generation to fully materialize; however, it is emerging much sooner, as success is mounting. What is expected is a sensible framework, with defined strategies and feasible goals, highlighting cost-effectiveness and applicability in daily practice. A few priorities are already available [13].

Primary tasks

Disease classification

Genotypic, phenotypic, and subphenotype benchmarks for disease classification and handling have already been introduced, and the momentum will certainly grow. One example is diabetes, in which, besides classic types 1, 2 and gestational, the concepts of Latent Autoimmune Diabetes of Adulthood (LADA), Maturity Onset Diabetes of the Young (MODY), and Neonatal Diabetes Mellitus (NDM) have been advocated [14,15], on the basis of big data of electronic health records feedback.

Another classification admits five types: severe autoimmune diabetes (SAID), severe insulin-deficient diabetes (SIDD), severe insulin-resistant diabetes (SIRD), mild obesity-related diabetes (MOD), and mild age-related diabetes (MARD) [16]. It is worth underlining that these proposals are not hypothesis-driven, or built on traditional consensus or expert opinion. They were deducted after large scale genotypic, phenotypic, and outcome data analysis, which portends reliable applications and measurable responses.

Molecular biomarkers

Many research protocols, therapeutical trials, and ordinary treatments are based on traditional clinical and biochemical outcomes. These are not necessarily wrong, as the principal end-point for a painkiller course must be pain relief, for an antidiarrheal drug should be normal stools, and for an antidiabetic agent, restoration of glucose homeostasis.

Yet gene expression, microbiome shifts, and profile of crucial molecules and pathways lend themselves to more consistent pathophysiological insights and mechanistic confirmations. They more effectively neutralize placebo effect and observer bias, which interfere with a clinical appraisal.

Decision support systems

Algorithms in mathematics are known since Euclid devised one that carries his name, about 300 BCE. Medical applications were announced much more recently, in the middle of last century, but in connection with administrative systems and machines, not as a treatment aid or bedside tool [17]. Only around the 1970s did the first therapeutic algorithms enter medical practice.

Algorithm steps and decision support tools are as good as the information which underlies their construction. With weak variables, solid results cannot be accomplished. This leads back to square one, namely massive data collection, and deep genotyping and phenotyping, with the help of multiomics. The key steps here are extracting value from big data and empowering clinical decision making.

It has been argued that most electronic health records and follow-up notes are confusing, if not biased. Denoising algorithms for artificial intelligence protocols exist, yet no universal design was proven adequate. Custom design is still required in most circumstances, increasing demands of time and expenditures. Nevertheless, potential rewards are commensurate.

The linking of lifelong personal records with genomic markers has the potential of yielding as rich and trustworthy information, as in the most accurate gene knockout laboratory models. Indeed it has been likened to human knockout experiments, and pilot investigations are being carried out at the Broad Institute (Massachusetts Institute of Technology and Harvard Medical School). The Human knockout Project aims, among others, to conduct genotype base recall and deep phenotyping of humans with loss-of-function gene variants.

Social, behavioral, psychological, and environmental circumstances

Some will criticize that this item undermines all previous arguments. If precision data has to do with endogenous omics and massive data-driven strategies, social science would mean engaging reverse gear, and relying on supposedly less cutting-edge social and environmental information, part of which can be qualitative and subjective. Nevertheless, these exogenous and sometimes fuzzy influences are steadfastly defended as the missing "omes" (or omics), of social background (philome) and environmental impact (aerome, hydrome, terrome, nutriome, and biome) [18].

One ingenuous example of the complex interfaces between the two universes was recently provided in a simple treadmill protocol. Participants were preliminarily tested for a gene variant, involved in exercise performance. Then the information was passed to them, not necessarily correctly, and the outcome was measured. A significant difference occurred, in agreement with the forwarded information, independent of its veracity [19]. The psychological impact of the test was rapid and meaningful, whereas the value of the genetics remained unproven in those circumstances.

Additional priorities

Cancer diagnosis, management, and prognosis

This is already a priority in many centers. Indeed tumor genotyping and cancer chemotherapy pharmacogenomics, have been going on for decades, long before these terms were coined and became mainstream.

Cancer risk assessment

This is not a novelty either; however, a clear horizon is not available yet. As addressed in this book, targeted versus standard genetic screening, commercial versus tailor-made gene panels, and handling of borderline or atypical sequencing results are much debated. Yet this is a good fight, a healthy anxiety that with accumulating evidence, will certainly be crowned with successful guidelines and routines.

Chronic conditions and inherited diseases

These have not been overlooked either, and the book brings many examples. The same is true for pharmacogenetics and pharmacogenomics, prenatal and population screening, as well as whole genome sequencing.

Outcomes

This detail cannot be overemphasized. Precision medicine should use the clinical benefit as the main yardstick, not surrogate chemical or genetic variables only. Biomarkers are obviously important, as alluded to, and they are, however, necessary to confirm and enhance, not to replace the clinical arena.

Value

Such an item has similarly been addressed before and deserves to be focused again. Manufacturers are fully aware of its importance, as are policy-makers, along with public and private healthcare managers. Although market wisdom alone cannot be trusted, risky, or ambiguous investments will be weeded out by vigilant professional and community organizations.

Ongoing studies

Liquid biopsy and cancer screening

This is an example of how reality moves forward, and goals are achieved before the ink of roadmaps and future perspectives, becomes dry. When genomic studies came to the bedside, a couple of decades ago, paving the way for the first steps in precision medicine, circulating DNA was below the radar. Indeed not many professionals, even in the cancer field, have ever sequenced such material. Yet tumor cells have been recognized in the bloodstream for 150 years [20], and also free DNA had been identified quite a long time ago [21]. This includes circulating free DNA (cfDNA), circulating tumor nucleic acids (ctNA), and other subclasses.

Different areas were already availing themselves of molecular information provided by such samples, such as prenatal screening, atherosclerosis, cardiovascular diseases, and diabetes.

Genotyping of DNA for universal cancer diagnosis

Compared to straightforward surgical biopsies, this is still a tricky and expensive material to process, given the often incongruent techniques and gene panels applied to cancer cells, tumor-derived cell-free nucleic acids, exosomes, and tumor-educated platelets in the plasma. Hence, the clinical value of commercial liquid biopsy, as the method is usually known, has been much debated, particularly considering that the price tag may be in the range of 4000 US dollars [22].

Yet important breakthroughs are mounting, and the technique was listed in 2015 among the top 10 advances in the MIT Technological Review [23]. Plasma can be easily collected, and consequently, if cost-effectiveness is proven, the method is amenable to mass use.

One cunning assay [24] takes advantage of epigenetic reprogramming, and methylation landscape in many cancers, called the *Methylscape*. With the help of a highly sensitive electrochemical potentiostat, which analyzes circulating free DNA in 10 min, the DNA methylation pattern was demonstrated to be highly diagnostic for breast and colorectal cancer. Results with other locations are encouraging, and the authors predict that it could become a universal cancer biomarker.

The approach adopted by others [25] relies on a multianalytical approach, simultaneously measuring circulating proteins, and mutations in cell-free DNA, along with an algorithm for result interpretation. Findings in over 1000 patients, suffering from clinically detected, nonmetastatic disease representing 60% of cancer deaths in the USA, exhibited over 70% sensitivity, with less than 1% false positives. Moreover, cancer location was partially detected (two possible organs) in 83% and fully unveiled (correct site) in 68%.

This is not a negligible feat, given the fact that plasma measurements are nonspecific, and DNA could originate from an anatomical region. Despite the intricate methodology, the authors anticipate a cost of 500 US dollars per test, much more affordable than current alternatives.

Polygenic risk score and genome-wide score

Inherited diseases, both benign and malignant, are suspected since antiquity. Hippocrates (460–377 BCE), Aristotle (384–322 BCE) and Epicurus (341–270 BCE), already preached about the familial origin of physical traits and defects. Prenatal screening is substantially more recent, having started in the 1960s. In the 1970s, the World Health Organization published the first clinical guidelines for genetic disorders [26].

A recent development was the popularization of direct-to-consumer genetic tests, in some parts of the world. Although often distributed by reputable companies, and adopting gene panels spotlighting relevant mutations, these monogenic approaches raise more questions than answers. Indeed, a few common diseases are related to a single gene, penetrance can be variable, and environmental factors should not be neglected. Consequently, actionable results are more controversial than one could wish for. Negative screenings are reassuring for mostly improbable and unusual circumstances, whereas positive ones may trigger severe stress and anxiety, that are not always justified [27].

Disorders like type 2 diabetes may suffer the impact of 400 or more point mutations or other genetic variants in the DNA. Shifts with similar orders of magnitude could underlie common forms of obesity, dyslipidemia, coronary artery complications, Alzheimer's, and cancer modalities. Polygenic scores, or better still genome-wide scores, are demanded. And answers are steadily arriving, in the laboratory and also in the industry.

Much overlap is inevitable when one deals with scores of genes so that the same analysis can indicate risk factors for certain cancers, metabolic illnesses, and neurodegenerative diseases. This is not a problem, as such scores should be used as guidelines to increase surveillance and screening procedures, improve lifestyle, and combat additional risks such as alcohol, tobacco, and sedentarism.

What about people who already comply with advice for a healthy life? It is a minority that adopts a balanced diet, never fails to exercise, avoids tobacco and alcohol, and follows recommendations by all scientific societies, concerning periodical clinical, biochemical, and imaging tests for early disease diagnosis. Consequently, there is much to gain from an algorithm-based report, which timely indicates major vulnerabilities in the genome.

Studies can entail searching data banks with millions of enrolled people [28], and up to 6.6 million positions in the individual DNA [29]. Indeed there are no less than 3.1 billion base pairs in the total human genome. Big datasets indeed exist in several countries, and in the USA it is estimated that over 10 million people have supplied DNA information to such sites as 23andMe and Ancestry.com. Commercial providers like Geisinger Health System in Pennsylvania, USA, are already offering a restricted number of free genetic screenings for patients [30].

Clinical trials and molecular biomarkers

New disease classifications and biomarker-proven diagnosis and therapy are at the core of all to-do lists in precision medicine. Yet the debate continues on how to improve efficiency, expand reliability, and reproducibility, and in the meantime diminish the staggering costs of clinical investigation.

One of the first attempts was the Clinical Trials Transformation Initiative, which preceded the concept of precision medicine. Nevertheless, along with administrative, ethical, technical, financial, logistical, and public health concerns, it did not miss precision medicine priorities, such as novel endpoints, mobile technology as an adjunct to information collecting, and integration of health data sources, with the help of advances in data sciences [31]. Surrogate endpoints, molecular, genomic or otherwise, are one of the currently envisaged options, to save time when clinical endpoints require a very long follow up, as well as to bypass ethical and social bottlenecks, for instance, when gender or ethnicity is addressed.

Extant protocol designs are deemed heavy and unwieldy for many purposes, particularly multidrug and multidisease contexts. One group has devised an umbrella, basket, or platform possibilities. All consider a defined master plan, with multiple arms addressing different biomarkers, illnesses, and drug prescriptions. The advantage, with regard to conventional independent trials, is that all arms are operative at the same time, within a key framework, eventually using the same populations and even the same set of biomarkers. All therapeutic options may thus be assessed in a short sequence of steps, saving time and expenditure [32].

The opposite hurdle may emerge for specific drugs or interventions, in the course of single chronic evolving illnesses. As the disease trajectories vary, so may baseline findings, as well as therapeutic results, even if prescription and population are nominally the same. The N-of-1 model aims to circumvent such limitations [33]. Without abdicating the classic platform of randomized controlled trials, the trick is to enroll the same individuals in multiple crossover trials, at successive times, and not just a single parallel-arm or crossover model as usual. Instead of a snapshot of the therapy, which can be misleading, one will count with solid, repeated outcomes.

The demise of the 3-phase clinical investigation is also defended in some quarters, as all jobs, from study design to recruitment to drug administration and data analysis, must be conducted in triplicate. By means of a single, seamless process, the same population would undergo all three phases without interruptions [34]. Criticisms notwithstanding, especially from regulatory authorities who fear that certain safeguards will not be fulfilled, the unified model has been operative for a few drugs.

Artificial intelligence

Artificial intelligence (AI) is nowhere to be found, among the original tenets of precision medicine. Yet big data analysis and decision support systems, which are fundamental to the advent and implementation of this modality, would be seriously hampered, without AI methods.

It is true that big data and AI do not strictly link to healthcare and clinical decisions. Humankind has never been so electronically connected, with 3.7 billion internet users in 2017, and 4.1 billion expected for 2021. By the same token, astronomical numbers of emails are daily exchanged (269 billion in 2017, 320 billion for 2021) [35]. Nevertheless, the healthcare system is becoming so reliant upon big data, that it was bestowed with the lion's share of investments. Big data in healthcare will represent a market of 68.8 billion US dollars by 2025, with just the blockchain sector reaching 5.6 billion dollars [36,37].

One is not dreaming of utopia, of a brave new world. Apparatus-assisted diagnosis is a reality since 1954, when Nash devised a simple and hardly successful mechanical device, processing basic clinical information [38]. The outlook started to improve with computer-assisted protocols like MYCIN developed at Stanford University in the 1970s, and DXplain, launched by Massachusetts General Hospital in the 1980s. This last program is still operative and highly regarded [39].

What is the contribution of AI to such endeavors? Instead of dealing with a few hundred or thousand pieces of information, it can handle exponential amounts, and also store them in the cloud. One can upload more clinical experience than a specialist would accumulate in a lifetime, or in countless lifetimes. The process usually encompasses machine learning and deep learning, to cope with structured data, including coded language, images, electrodiagnostic procedures, and genomic sequencing results. Machine learning systems become more efficient with use, literally learning from each case, and applying the information in subsequent cases. Of course, algorithms and statistical methods cannot be omitted from such programs.

Unstructured information, such as natural language retrieved from electronic medical records and surgical notes, or directly submitted by the patient through his cellphone, can also be handled. However, they require previous conversion to structured data, with the help of natural language processing.

Will AI replace the doctor? In some areas, machines are already better than humans, including detection of skin cancer, diabetic eye disease, and heart arrhythmias. Extensive use in diagnostic imaging portends a brilliant future, although according to the Google X-ray project, 10–20 years might elapse before it becomes routine [40].

Smartphone assistance for medical emergencies has been successfully started in parts of London, UK, by the National Health Service. Chatbots are the employed interface, allowing screening and guidance to patients, without the participation of healthcare personnel. Over one million people are already covered, and the whole nation should be able to use it in the future [41].

One should be careful to accumulate only such data that directly or indirectly links to, or could act as confounders, regarding the objectives of the analysis or intervention. Any irrelevant numbers in a dataset, introduced for administrative or other extraneous purposes, will inflate the volume without increasing the efficiency of algorithms. On the contrary, data processing and decision making will be slowed and hampered in such circumstances. The name "fat data" has been coined, to distinguish it from proper big data.

Machine learning has actually been implied in a recent scientific meeting [42], as a potential trigger of a crisis in science. One of the most convenient abilities of machine learning, obviously, concerns the interpretation of massive datasets, accumulated during the years by universities and research facilities. Thanks to innovative and flexible algorithms, along with superb statistical power, this technique is able to extract tendencies, associations, and patterns, which could robustly underlie clinical decisions.

The caveat is that acceptance of a new methodology requires at least two populations or two independent data banks. In the first one, primary information is collated, screened, and processed, significant associations are deducted and highlighted, and equations or statistical models are assembled. The reproducibility and reliability of the results then need to be tested in a new real-world setting in order to validate the original findings.

Massive scientific information, patiently and seriously accumulated along the years, is notably scarce. Moreover, machine learning utilization, especially in the case of unstructured information or "fat data," which are common settings, is cumbersome and time-consuming. It can also be fully misleading if the wrong mining tools and filters are applied. Many protocols rely on a single vast bank, and results typically garner worldwide acclaim, on the basis of the huge numbers involved. However, they are rarely proven wrong, if and when a second major dataset is similarly investigated, and hence, the potential crisis.

Of course, conflict and controversy are nothing new in science. Sir Isaac Newton (1643–1727), one of the most eminent presidents of the Royal Society (Britain), and Robert Hooke (1635–1703) a brilliant contemporary mind, engaged in bitter and prolonged acrimony, concerning the nature of light, as particle (Newton) versus energy (Hooke), as well as other topics. Growth pains are inevitable for all modalities of artificial intelligence, especially in the tricky and multifactorial world of biology and medicine.

Digital imaging for surgery, robotics, and anticancer drug design

Virtual reality and augmented reality along with 3-D (three dimensions) experiences have a long cinematographic history, before progressing toward resources for product advertisement, ground, and air navigation, along with other commercial applications. "The great train robbery" (Edwin S Porter, USA, 1903), already tried to introduce very basic stereoscopic vision into the movie. Beyond business and entertainment, modern options are being devised to assist healthcare professionals, in jobs that have been barely envisaged yet, such as 3-D disease visualization, therapy planning, and management simulation at multiple levels, not only anatomical and physiological, but cellular, genomic, metabolomic, and immunological.

If that brings to mind incredibly powerful "google earth" devices for scanning tumors and other lesions, aims are not far from that [43]. Lord Kelvin wanted to express knowledge in numbers. Available virtual reality protocols are able to convert numerical information into digital images. And if histological, cellular, and omics information are amalgamated, the imaging protocol can encompass as many layers and angles as desired.

Virtual reality in liver surgery? Potentially involving automated robotics [44]? Virtual microdissection of head and neck cancer for microenvironment exploration [45]? For immune molecular subgroup identification [46]? All of these are on the table, based on evolving datasets and techniques.

Wearable, 3-D printable, embedded in cell phones or household appliances, and delivered via nanoparticles or drug carrier vectors, were not devices, tools or concepts espoused by the pioneers of the human genome project, or early precision medicine. Indeed they are peripheral to the core principles, however handy, smart, and fully consistent with the same doctrines. They promise not only to bridge the gap between heavily equipped laboratories and the bedside, or even the roadside, but also to create their own theoretical frameworks, pathways, and algorithms, toward legitimate, effective, and customized patient care (Fig. 1.1).

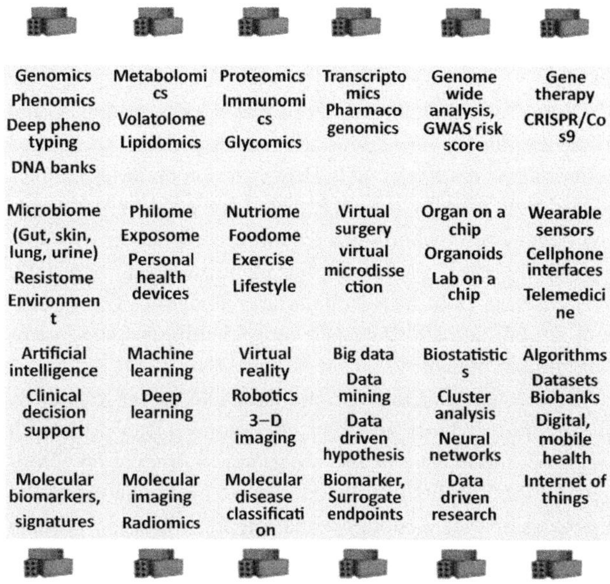

FIGURE 1.1 Representative building blocks for precision medicine.

Genomic medicine in the real world

As alluded to, precision medicine is dichotomizing and subbranching into multiple directions, some of them not strictly related to the original omics framework. This means that certain endeavors, especially in the bioengineering, bioinformatics, computer science, electronics, and artificial intelligence realm, are more straightforward and advance quite rapidly. In contrast, primordial clinical genomics can still hit the wall, the remarkable success of next-generation sequencing of genomes and exomes notwithstanding.

There is not much mystery behind such difficulties. When one peruses thousands of genes, it is not improbable to find dozens of somatic and germline changes. Some could be true mutations and gene defects, whereas others could result from technical failure, tissue heterogeneity, measurement ambiguity, or unreported patterns. Reference datasets for the fields of genomics and other omics are already available, as pointed out in specific chapters; however, they do not cover all possible natural variants, let alone the unknown or artifactual ones.

Even if a mutation is identified, there could be insufficient information to characterize it as either driver (protagonist), or passenger (innocent bystander). In other words, actionable and druggable defects, for which treatment is formally indicated, are not always distinguished from others, undeserving of any intervention. Of course, information is accumulating, and operational guidelines should soon emerge [46].

Geospatial correlations of disease

Environmental factors, more specifically, geography and climate, have always played a relevant role in human disease. To a large extent, these effects were prominent for infectious and parasitic entities. On account of specific vectors (insects and other arthropods), certain illness distributions paralleled the habitat of such invertebrate hosts. Diarrhea and hepatitis spread by contaminated drinking water, are also typical of underdeveloped, tropical countries.

Malaria alone has been deemed responsible for the fate of so many generals and armies, that it is said to have shaped human history along millennia [47]. Among others, it is believed that Alexander the Great (356–323 BCE) succumbed to the parasite, and the catastrophic invasions of Attila the Hun (406–453 CE) and Genghis Khan (1162–1227 CE), were eventually derailed in Europe by the same agent. As recently as World War II, malaria risk was a primary concern for armies on both sides of the conflict.

Nutritional deficiencies like kwashiorkor, pellagra, and hypovitaminosis D, have always displayed a marked geographical pattern as well. However, the major epidemics of the XXI century are represented by noncommunicable and degenerative conditions encompassing cancer, obesity, type 2 diabetes, cardiovascular abnormalities, and fatty liver disease, both nonalcoholic and alcoholic. How pertinent are latitude and longitude to epidemiology, diagnosis, and clinical course?

Few conclusive answers are available so far; however, investigational tools are fully operational. They link with precision medicine via the environment, including geography, along with lifestyle, as essential modulators of the repercussions of genomics and other omics, on human health. Incidentally, primary measuring devices are also seasoned protagonists in this context, namely GPS-equipped smartphones, tablets, and even watches and shoes.

By virtue of geographic information systems, GPS, and remote sensing, it has never been so easy and reliable to track human beings all the time. Will paradigm-changing concepts emerge from such investigations? At least in the field of obesity, spatial technologies are proving worthwhile [48].

Final considerations

There is little danger that doctors, nurses, dietitians, pharmacists, and allied professionals, will follow the course of alchemists, phrenologists, or leech collectors, any time soon. On the other hand, radical shifts in the way patients are screened, diagnosed, treated, and followed are imminent, and many can already be witnessed in our days. Healthcare professions will have a role in society for the foreseeable future, even if they need to dialogue as much with computers and robots, as with real people. Yet, in a sense, this has been occurring in medical science since the days of Hippocrates and Galen, although at an extraordinarily slower pace. After all, is it not a well-founded change in the quintessence of progress?

'Tis not the many oaths that makes the truth,

But the plain single vow that is vow'd true.

Shakespeare W. All's Well That Ends Well. Act 4. Scene 2.

References

[1] Stiefel M, Shaner A, Schaefer SD. The Edwin Smith Papyrus: the birth of analytical thinking in medicine and otolaryngology. Laryngoscope 2006;116(2):182–8.

[2] Evidence-based medicine. A new approach to teaching the practice of medicine. J. Am. Med. Assoc. 1992;268(17):2420–5.

[3] Phillips KA, Deverka PA, Sox HC, Khoury MJ, Sandy LG, Ginsburg GS, Tunis SR, Orlando LA, Douglas MP. Making genomic medicine evidence-based and patient-centered: a structured review and landscape analysis of comparative effectiveness research. Genet. Med. 2017;19(10):1081–91.

[4] Vogenberg FR, Barash CI, Pursel M. Personalized medicine. Part 1: evolution and development into theranostics. PT 2010;35(10):560–76.

[5] Chowkwanyun M, Bayer R, Galea S. "Precision" public health — between novelty and hype. NEJM 2018;379(15):1398–400.

[6] Beale LS. The microscope in medicine. 4th ed. London, UK: J and A Churchill; 1878.

[7] Collins FS, Varmus H. A new initiative on precision medicine. NEJM 2015;372(9):793–5.

[8] Brown J, Moscatti P. Elementa Medicinae, 1794, I.G.Hanisch, Venice, Italy.

[9] Mathé E, Hays JL, Stover DG, Chen JL. The omics revolution continues: the maturation of high-throughput biological data sources. Yearb. Med. Inform 2018;27(1):211–22.

[10] Hood L, Friend SH. Predictive, personalized, preventive, participatory (P4) cancer medicine. Nat. Rev. Clin. Oncol. 2011;8(3):184–7.

[11] Westerman K, Reaver A, Roy C, Ploch M, Sharoni E, Nogal B, Sinclair DA, Katz DL, Blumberg JB, Blander G. Longitudinal analysis of biomarker data from a personalized nutrition platform in healthy subjects. Sci. Rep. 2018;8(1):14685.

[12] Khoury MJ, Galea S. Will precision medicine improve population health? J. Am. Med. Assoc. 2016;316(13):1357–8.

[13] Divaris K. Fundamentals of precision medicine. Compend. Contin. Educ. Dent. 2017;38(8 Suppl):30–3.

[14] Hosszúfalusi N, Vatay A, Rajczy K, et al. Similar genetic features and different islet cell autoantibody pattern of latent autoimmune diabetes in adults (LADA) compared with adult-onset type 1 diabetes with rapid progression. Diabetes Care 2003;26(2):452–7.

[15] Niddk. www.niddk.nih.gov/health-information/diabetes/overview; 2018.

[16] Ahlqvist E, Storm P, Käräjämäki A, Martinell M, Dorkhan M, Carlsson A, Vikman P, Prasad RB, Aly DM, Almgren P, Wessman Y, Shaat N, Spégel P, Mulder H, Lindholm E, Melander O, Hansson O, Malmqvist U, Lernmark Å, Lahti K, Forsén T, Tuomi T, Rosengren AH, Groop L. Novel subgroups of adult-onset diabetes and their association with outcomes: a data-driven cluster analysis of six variables. Lancet Diabetes Endocrinol 2018;6(5):361–9.

[17] Fletcher KH. Matter with mind: a neurological research robot. Research 1951;4(7):305–7.

[18] Davis MM, Shanley TP. The missing -omes: proposing social and environmental nomenclature in precision medicine. Clin. Transl. Sci 2017;10(2):64–6.

[19] Science Mag. www.sciencemag.org/news/2018/12/just-thinking-you-have-poor-endurance-genes-changes-your-body; 2018.

[20] Ashworth T. A case of cancer in which cells similar to those in the tumors were seen in the blood after death. Aust. Med. J. 1869;14:146–9.

[21] Leon SA, Shapiro B, Sklaroff DM, Yaros MJ. Free DNA in the serum of cancer patients and the effect of therapy. Cancer Res. 1977;37(3):646–50.

[22] Torga G, Pienta KJ. Patient-paired sample congruence between 2 commercial liquid biopsy tests. JAMA Oncol. 2018;4(6):868–70.

[23] Technology Review. www.technologyreview.com/s/544996/10-breakthrough-technologies-of-2015-where-are-they-now; 2018.

[24] Sina AA, Carrascosa LG, Liang Z, Grewal YS, Wardiana A, Shiddiky MJA, Gardiner RA, Samaratunga H, Gandhi MK, Scott RJ, Korbie D, Trau M. Epigenetically reprogrammed methylation landscape drives the DNA self-assembly and serves as a universal cancer biomarker. Nat. Commun. 2018;9(1):4915.

[25] Cohen JD, Li L, Wang Y, Thoburn C, Afsari B, Danilova L, Douville C, Javed AA, Wong F, Mattox A, Hruban RH, Wolfgang CL, Goggins MG, Dal Molin M, Wang TL, Roden R, Klein AP, Ptak J, Dobbyn L, Schaefer J, Silliman N, Popoli M, Vogelstein JT, Browne JD, Schoen RE, Brand RE, Tie J, Gibbs P, Wong HL, Mansfield AS, Jen J, Hanash SM, Falconi M, Allen PJ, Zhou S, Bettegowda C, Diaz Jr LA, Tomasetti C, Kinzler KW, Vogelstein B, Lennon AM, Papadopoulos N. Detection and localization of surgically resectable cancers with a multi-analyte blood test. Science 2018;359(6378):926–30.

[26] Anonymous. Genetic disorders: prevention, treatment and rehabilitation. Report of a WHO scientific group. World Health Organ. Tech. Rep. Ser. 1972;497:1–46.

[27] Holland CMA, Arbe-Barnes EH, McGivern EJ, Forgan RMC. The 10th Oxbridge varsity medical ethics debate-should we fear the rise of direct-to-consumer genetic testing? Philos. Ethics Humanit. Med. 2018;13(1):14.

[28] Jansen PR, Watanabe K, Stringer S, Skene N, Bryois J, Hammerschlag AR, de Leeuw CA, Benjamins J, Munoz-Manchado AB, Nagel M, Savage JE, Tiemeier H, White T, Tung JY, Hinds DA, Vacic V, Sullivan PF, van der Sluis S, TJC Polderman JC, Smit AB, Hjerling-Leffler J, van Someren EJW, Posthuma D. Genome-wide analysis of insomnia (N=1,331,010) identifies novel loci and functional pathways. bioRxiv 2018:39439556. www.biorxiv.org/content/biorxiv/early/2018/02/01/214973.full.pdf.

[29] Khera AV, Emdin CA, Drake I, Natarajan P, Bick AG, Cook NR, Chasman DI, Baber U, Mehran R, Rader DJ, Fuster V, Boerwinkle E, Melander O, Orho-Melander M, Ridker PM, Kathiresan S. Genetic risk, adherence to a healthy lifestyle, and coronary disease. N. Engl. J. Med. 2016;375(24):2349–58.

[30] Medscape. www.medscape.com/viewarticle/902769; 2018.

[31] Tenaerts P, Madre L, Landray M. A decade of the clinical trials transformation initiative: what have we accomplished? What have we learned? Clin. Trials 2018;15(1_Suppl.):5–12.

[32] Woodcock J, LaVange LM. Master protocols to study multiple therapies, multiple diseases, or both. N. Engl. J. Med. 2017;377(1):62–70.

[33] Senn S. Statistical pitfalls of personalized medicine. Nature 2018;563(7733):619–21.

[34] Prowell TM, Theoret MR, Pazdur R. Seamless oncology-drug development. N. Engl. J. Med. 2016;374(21):2001–3.

[35] Radicati. www.radicati.com/wp/wp-content/uploads/2017/01/Email-Statistics-Report-2017-2021-Executive-Summary.pdf; 2018.

[36] PR Newswire. www.prnewswire.com/.news-releases../global-big-data-in-healthcare-market-to-reach-6875-billion-by-2025-reports-bis-research-678151823.html; 2018.

[37] PR Newswire. www.prnewswire.com/.news-releases/global-blockchain-in-healthcare-market-to-reach-561-billion-by-2025-reports-bis-research-680230953.html; 2018.

[38] Nash FA. Differential diagnosis, an apparatus to assist the logical faculties. Lancet 1954;266(6817):874–5.

[39] Harvard. http://lcs.mgh.harvard.edu/projects/dxplain.html; 2018.

[40] Li Z, Wang C, Han M, Xue Y, Wei W, Li LJ, Fei-Fei L, Syracuse University, PingAn Technology, US Research Lab, Google Inc. Thoracic disease identification and localization with limited supervision. arXiv 2018. 1711.06373v6, https://arxiv.org/pdf/1711.06373.pdf.

[41] MyHealth. www.myhealth.london.nhs.uk/111babylon; 2018.

[42] Allen G. Machine learning: the view from statistics. 2019. https://aaas.confex.com/aaas/2019/meetingapp.cgi/Session/21598.

[43] (a) ScienceBlog. https://scienceblog.cancerresearchuk.org/2017/02/10/3-of-the-toughest-questions-in-cancer-and-more-than-70-million-to-solve-them/; 2019.(b) Quero G, Lapergola A, Soler L, Shabaz M, Hostettler A, Collins T, Marescaux J, Mutter D, Diana M, pessaux P. Virtual and augmented reality in oncologic liver surgery. Surg. Oncol. Clin. 2019;28(1):31–44.

[44] Ma BBY. Virtual microdissection in the molecular subtyping of head and neck squamous carcinoma – a "virtual reality" of the tumor microenvironment? Ann. Oncol. November 23, 2019;30(1):8–10.

[45] Chen YP, Wang YQ, Lv JW, Li YQ, Chua MLK, Le QT, Lee N, Colevas AD, Seiwert T, Hayes DN, Riaz N, Vermorken JB, O'Sullivan B, He QM, Yang XJ, Tang LL, Mao YP, Sun Y, Liu N, Ma J. Identification and validation of novel microenvironment-based immune molecular subgroups of head and neck squamous cell carcinoma: implications for immunotherapy. Ann. Oncol. 2019;30(1):68–75.

[46] McGraw SA, Garber J, Jänne PA, Lindeman N, Oliver N, Sholl LM, Van Allen EM, Wagle N, Garraway LA, Joffe S, Gray SW. The fuzzy world of precision medicine: deliberations of a precision medicine tumor board. Per. Med. 2017;14(1):37–50.

[47] Malaria Site. www.malariasite.com/wars-victims; 2019.

[48] Jia P, Xue H, Yin L, Stein A, Wang M, Wang Y. Spatial technologies in obesity research: current applications and future promise. Trends Endocrinol. Metab. 2019;30(3):211–23.

Chapter 2

The role of the microbiome in precision medicine

João Vitor Pincelli[1], Luiz Otávio Vittorelli[1], Nickolas Stabellini[1] and João Renato Rebello Pinho[1,2]

[1]*Laboratory of Special Techniques, Department of Pathology and Clinical Pathology, Hospital Israelita Albert Einstein, São Paulo, São Paulo, Brazil;* [2]*LIM 03/LIM 07 — Departments of Gastroenterology and Pathology, São Paulo University Medical School, São Paulo, São Paulo, Brazil*

Introduction

One gram of stool contains about 10^{11} organisms, composed by up to thousands of different bacterial species, most of them anaerobic [1]. Compared to the number of genes present in the human host, the collective genome of these microorganisms can be 150-fold larger [2], showing the potential influence of these bacterial groups on host environmental settings and homeostasis. The majority of them is anaerobic, a factor that makes culture difficult and more expensive, resulting in a complex and limiting process to characterize the microbiota composition by usual methods. That explains why the evolution of studies in this field is highly correlated to the development of nucleic acid sequencing.

These symbiotic organisms evolved alongside mammal and nonmammal hosts, providing physiological benefits while being protected and gaining easy access to nutrients. Some examples of benefits offered to human hosts are digestion of certain types of carbohydrates, biotransformation of bile acids, vitamin synthesis, urea hydrolysis, and immune system regulation [3].

Health, disease, nutrition and other lifestyle repercussions

The intestinal microbiota formation starts at childhood, with fast growth, both in size and diversity, mainly during the first 5 years of life. During adolescence, the number of bacteria colonizing the gut is similar to the adult but vastly differs in composition [4]. In healthy adults, the gut microbiota is composed mainly of two phyla: Bacteroidetes and Firmicutes [5]. The exact composition in each person is unique; however, the main proportions tend to stay constant in the population [6]. In older people, microbiotic composition usually drops both in number and diversity of bacteria.

Compositional trends can be associated with certain obesity profiles. The ease of gain or loss of weight can exhibit a correlation with the species of bacteria that predominates in the gut. Further associations with a large number of diseases are being traced encompassing neurology, psychiatry, cardiology, immunology, and oncology [5].

Composition of the microbiota is directly connected to endogenous and exogenous characteristics of the host, such as diet, genetics, lifestyle, and some kinds of medications. The diet is deeply related to the host microbiome [7]. Someone with a tendency to high carbohydrate and lipid ingestion usually exhibits higher Firmicutes concentration, while a hypocaloric diet moves the trend to a higher Bacteroidetes proportion [8]. Genetics actuates not only by creating a tendency for some kind of diet but also modulates the immune system, carbohydrate digestion, and even the physical architecture of the bowel [9,10]. The lifestyle is deeply associated with dietary and exercise habits, both of which are powerful determinants of microbiome composition. Finally, antibiotics and immunosuppressants also correlate with the composition of an individual microbiome [11].

The period between birth and early childhood is known to be crucial for the acquisition and development of the microbiome [12]. In general, the prenatal and early childhood environments have a decisive role in the development of the immune system and other stress response systems, both implying microbiome composition [13,14]. In this way, exposure to factors like malnutrition, insufficient or inadequate breastfeeding, and other harmful factors, can alter the formation and diversification of the pool of bacteria in the gut, potentially leading to health damages in the

Precision Medicine for Investigators, Practitioners and Providers. https://doi.org/10.1016/B978-0-12-819178-1.00002-2

future [15]. In the adult, the microbiome of people that share the same environment and habits are different from those out of this social circle [16,17]. The specific factors of the surroundings of a patient that play significant roles in compositional and numerical changes are still undefined.

Obesity

Since its correlation is partly understood, and prevalence is high worldwide, obesity is one of the targets in the field. The objective is to trace better correlations, find specific targets, and assess the efficacy of interventions in this symbiotic relationship.

Inflammatory bowel disease

In inflammatory bowel disease, an alteration in the interaction of the microbiota with the immunological system is admitted, and a very different composition is featured [18,19]. There are two major subtypes of this condition: ulcerative colitis and Crohn's disease [20], both presenting alterations in the microbiome when compared to the regional standard. Other diseases that have a promising future in research are cardiovascular conditions, type II diabetes, oncologic diseases, and systemic lupus erythematosus [21].

Malignancies

Cancer should not be overlooked. A 20% prevalence of microbe-induced tumors is anticipated [22], even in body sites considered sterile, in which microbial DNA was recently found [23]. Large bowel was the first investigated site, on account of high-level colonization. Tests in mice reveal that the intestinal microbiome interferes with tumoral development through immunological modulation. Microbiota transplantation from cancerous mice to a germ-free model, thus conventionalizing such animals with a tumor-bearing model, confirmed faster and more aggressive tumorigenesis [24].

Other studies demonstrate an increased remission rate in some patients treated with antibiotics, achieving even complete remission [23]. Alongside with these findings, there are many centers looking to elucidate the specific molecules involved in these phenomena, especially those involved in immunomodulatory effects, as some of these microbes can synthetize genotoxic and tumorigenic molecules [23]. The most conspicuous example of a bacterium causing a carcinogenic lesion is *Helicobacter pylori* infection and gastric adenocarcinoma [25]. It involves immunological modulation, causing downregulation of antitumorigenic agents, as well as proliferative signaling and inflammation [23].

Fecal microbiota transplantation (FMT)

FMT arises as a possibility for treating some gastrointestinal diseases, for instance, recurrent infections caused by *Clostridium difficile* [26,27]. For inflammatory bowel disease, despite promising results, there are not enough data to support clinical implementation [28,29]. Metabolic syndrome is another potential candidate; however, safety and efficacy are incompletely known.

Multiomics, specialized equipment, techniques, and diagnostic implications

Preparing the biological sample to proceed with DNA sequencing analysis is a complex process, which demands trained professionals, capable of dealing with genetic material purification, preparation, amplification, and sequencing. Fortunately, there are complete protocols to guide each kind of analysis, from many types of samples [30]. These protocols are free to consult on the Internet, and represent a landmark toward standardization of microbiota analysis, which has the potential of making different studies comparable. Greater power for meta-analysis could also emerge, as acquisition and preparation of the sample can drastically alter the results.

Next generation sequencing

Next Generation Sequencing (NGS), is a relatively new method of analyzing genetic material, capable of dealing fairly fast output with very large amounts of data [31]. This kind of analysis needs specialized professionals, including bioinformatics assessment, as well as next-generation nucleic acid sequences, linked to a large amount of computing power. Small nucleotide pieces which are initially generated, have to be aligned as different units, in a complete nucleic acid sequence.

Usually, analysis is done by the amplification of the 16S region of the bacterial ribosomal RNA [32], which is basically constant throughout bacterial species, making it a specially interesting target for amplification. This procedure makes the taxonomical classification possible and easily executable, by most of the available software. However, only proportions of each phylum become available. The final interpretation of results needs a specialized team of microbiologists, pathologists and/or geneticists, to avoid fallacious results, like some famous examples that erroneously linked microbiome and autism years ago [33].

Metabolomics and other omics

Chemical mediators of the effects of microorganisms and their interaction can be unearthed, with the help of additional techniques [34]. The transcriptome can inform the level of expression of a set of genes, for example, measuring mRNA. Several techniques can be employed, a modified NGS system being the most used nowadays.

Proteomics, on the other hand, requires more complex procedures. One way of doing so is sorting out proteins by an electrical charge, and then analyzing through mass spectrometric identification. Metabolomics, a measurement of small molecules that are active in intermediate tasks inside the cell metabolism, will enrich the picture. It does not keep a direct correlation with transcription intermediates, as proteins do. It can be estimated by spectrographic methods, using time-of-flight mass spectrometry, for instance.

Therapeutic protocols

The first generation of therapy was composed by natural probiotics and prebiotics, which act against pathogens occupying spaces and binding sites, altering local pH, and secreting substances toxic to certain bacteria. A new generation is emerging, based on genetic engineering advances and involves recombinant probiotics, microbial associations, and selective antimicrobials (developed with the aid of CRISP-CAS systems) [35] (Fig. 2.1).

The therapies are classified as additive, subtractive, or modulatory [35]. The additive model consists of supplementing the host microbial ecosystem with individual strains or associations (consortia) of natural and engineered microorganisms. The subtractive model consists in the elimination of deleterious microorganisms by means of antibiotics, peptides, chemicals or bacteriophages. The modulatory model involves administration of nonliving or prebiotic agents, to modulate the composition or activity of the endogenous microbiome. The future trend is the combined use of additive and subtractive models (Table 2.1).

FMT is an additive technique consisting of the transfer of healthy microbiome from a donor to a receptor, using oral (capsule) or endoscopic ways, for the treatment of recurrent infections by *C. difficile*. This technique already has an effectiveness of over 90% and is approved by the medical community [36]. Also within the additive framework, studies demonstrate the functionality of certain strains of *Escherichia coli* for the treatment of cholera and

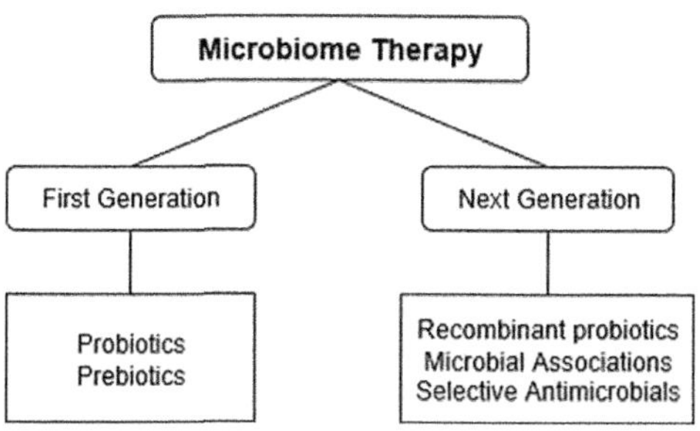

FIGURE 2.1 Generations of interventions in microbiome therapy.

TABLE 2.1 Types of microbiome therapy.

Therapy	Methods
Additive	Supplementation of the host microbial ecosystem with single strains or consortia of natural or bioengineered microorganisms
Subtractive	Elimination of deleterious species
Modulatory	Administration of nonliving prebiotics (paraprobiotics)

obesity, *Lactobacillus jensenii* for simian immunodeficiency virus, *Lactobacillus gasseri* for diabetes, *Lactococcus lactis* for colitis, and FMT for certain modalities of inflammatory bowel disease [35].

Subtractive therapies always relied on antibiotics as a standard, but the undesirable effects of killing nontarget microbes, and the creation of superbugs, have partly moved the focus to bacteriophages, which have become a great expectation for the future.

Another field of activity is in anticancer therapies, with a modulatory role. During treatment with oxaliplatin, for example, the intestinal microbiome stimulates the production of ROS, leading to tumor regression. In contrast, cyclophosphamide impairs the intestinal epithelial barrier, leading to the passage of gram-positive bacteria into the bloodstream, a factor that elevates the defense cells, leading to immune-mediated tumor regression [37]. The intestinal microbiome plays an important role in the pharmacokinetics of drugs, concerning biotransformation and effectiveness. There are also descriptions of microbiome markers of drug efficacy.

Importance for health care providers and institutions

Advances in genetic engineering and personalized medicine, coupled with individual variability of the microbiome, place it as a center of great expectation for the future. The profiling and manipulation of the human microbiome can provide substantial opportunities for diagnosis, intervention, risk management, and risk stratification [38].

Personalized medicine (PM) seeks to provide effective and adapted therapeutic strategies based on an individual's genomic, epigenomic, and proteomic profile, providing not only treatment but also prevention. Its insertion in health care alters the approach to diagnosis and treatment, in addition to leading to greater participation of the patient during and after treatment [39]. The adoption of this strategy can reduce costs and optimize time, avoiding unnecessary changes and adverse effects of drug prescriptions, and at this point lies the importance of the study and use of the microbiome.

This is underscored by the creation of projects like the NIH Human Microbiome Project/HMP [40] and MetaHIT (European Commission project). Antibiotics are now seen in a new light. They provide a good cost-benefit ratio [41]; however, the exacerbated use has led to the phenomenon of antimicrobial resistance (AMR), a fact that raises costs, length of stay in hospitals, morbidity, and mortality [42]. As alluded to, microbiome-friendly alternatives are being searched. *C. difficile* infection is the main cause of nosocomial infection, and hospital costs associated with it in the United States exceed $3.2 billion per year. FMT has been shown to be more effective, with comparable or even lower costs than standard management, for recurrent *C. difficile* infection [43].

Ongoing lines of investigation and research opportunities in the field

Between 2013 and 2017, publications about genetic sequencing of the human microbiota have reached 12.900 papers, which represent around 80% of 40 years of publications about this topic [44]. Most are trying to find a direct quantitative relationship between bacteria and host since the culture of these organisms is expensive and difficult to conduct [45,46]. For this reason, the NIH invested 173 million dollars in the development of HMP, a massive free access database that contains the sequencing of thousands of healthy subjects, making comparative studies easier [47].

The most common method of sequencing used in research so far is NGS [48]. Many technological developments are making this tool more available, particularly to clarify how the interaction between microorganisms and host work [49]. With faster and cheaper sequencing and analysis platforms, using the microbiome as a biomarker for some kinds of diseases is becoming a possibility [50].

Personalization of synthetic microbial communities for clinical practice is another goal. With the development of Metagenomic Systems Biology, the building, organization, and activity of the microbiota can be understood more easily [51]. Bonding mathematical and computational models has been the first choice to understand the more complex steps in microbiota interaction. A very promising pathway seems to be the development of mechanistic and phenomenological models, that could be capable of predicting function and intrinsic characteristics of a hypothetical microbiota [52]. However, those advances are extremely demanding, complex, and expensive.

Probiotics are an example of how microbiome development is conducted today. These are ingestible microorganisms that, at the right dose, are beneficial to the patient [53]. Some of the advantages of probiotic use are a reduction of lactose intolerance, gastric discomfort, and diarrhea risk during prolonged antibiotic therapy. The most commonly used bacteria are *Lactobacillus* and *Bifidobacterium* taxa [54]. Similar treatment options should be aimed at obesity control. The large majority of studies have been conducted in animal models, with diet-induced obesity. Some human studies are still going on [54]. *L. gasseri* SBT2055 treatment in Japanese adults elicited a large reduction in abdominal fat, and also a minor reduction in body weight, BMI, hip and waist diameter, however without statistical significance [54–56].

Taxonomic signatures as biomarkers for some conditions are becoming more robust. In obesity, the accuracy of a profile for obesity prediction was 70% [57]. The microbiome can be relevant for patients with melanoma undergoing immune checkpoint blockade. Microorganisms in the intestine, which signal colitis induced by blockade treatment, can be fingerprinted [58].

A good example of the utilization of metabolomics is the demonstration that butyrate, produced by some species of *Clostridium*, interacts with G protein receptors, modulating the activity of Treg cells [51,52,59]. Means to modulate which microorganism and in what proportion should be present in a given environment, for a better health outcome, is another priority.

References

[1] Uhlig HH, Powrie F. Dendritic cells and the intestinal bacterial flora: a role for localized mucosal immune responses [Internet] J. Clin. Investig. 2003;112:648–51. Available from: https://www.jci.org/articles/view/19545.

[2] Xu J, Gordon JI. Honor thy symbionts. Proc. Natl. Acad. Sci. September 2, 2003;100(18):10452–9 [Internet]. Available from: https://www.pnas.org/content/100/18/10452.

[3] Baetge EE. Next-generation nutritional biomarkers to guide better health care. In: 84th nestlé nutrition institute workshop, Lausanne, September 2014 [internet]; 2016. Available from: https://books.google.com.br/books?hl=pt-BR&lr=&id=Osh2CwAAQBAJ&oi=fnd&pg=PP1&dq=3-%09Baetge+EE,+Dhawan+A,+Prentice+AM+(eds):+Next-Generation+Nutritional+Biomarkers+to+Guide+Better+Health+Care.+Nestlé+Nutr+Inst+Workshop+Ser.+Nestec+Ltd.+Vevey/S.+Karger+AG+.

[4] Cheng J, Ringel-Kulka T, Heikamp-De Jong I, Ringel Y, Carroll I, De Vos WM, et al. Discordant temporal development of bacterial phyla and the emergence of core in the fecal microbiota of young children. ISME J. 2016;10(4):1002–14 [Internet]. Available from: https://www.nature.com/articles/ismej2015177.

[5] Lynch SV, Pedersen O. The human intestinal microbiome in health and disease. Phimister EG, editor N. Engl. J. Med. December 15, 2016;375(24):2369–79 [Internet]. Available from: http://www.nejm.org/doi/10.1056/NEJMra1600266.

[6] Consortium THMP, Huttenhower C, Gevers D, Knight R, Abubucker S, Badger JH, et al. Structure, function and diversity of the healthy human microbiome. Nature June 13, 2012;486:207 [Internet]. Available from: https://doi.org/10.1038/nature11234.

[7] Yatsunenko T, Rey FE, Manary MJ, Trehan I, Dominguez-Bello MG, Contreras M, et al. Human gut microbiome viewed across age and geography [Internet] Nature 2012;486:222–7. Nature Publishing Group; Available from: http://www.nature.com/articles/nature11053.
[8] Ley RE, Turnbaugh PJ, Klein S, Gordon JI. Microbial ecology: human gut microbes associated with obesity. Nature 2006;444(7122):1022–3 [Internet]. Available from: https://www.nature.com/articles/4441022a.
[9] Hall AB, Tolonen AC, Xavier RJ. Human genetic variation and the gut microbiome in disease. Nat. Rev. Genet. 2017;18(11):690–9 [Internet]. Available from: https://www.nature.com/nrg/journal/v18/n11/abs/nrg.2017.63.html.
[10] Goodrich J, Davenport E, et al. Genetic determinants of the gut microbiome in UK twins. Cell Host Microbe 2016. Elsevier [Internet]. Available from: https://www.sciencedirect.com/science/article/pii/S1931312816301536.
[11] Knights D, Silverberg MS, Weersma RK, Gevers D, Dijkstra G, Huang H, et al. Complex host genetics influence the microbiome in inflammatory bowel disease. Genome. Med. December 2, 2014;6(12):107 [Internet], http://genomemedicine.biomedcentral.com/articles/10.1186/s13073-014-0107-1.
[12] Herd P, Palloni A, Rey F, Dowd JB. Social and population health science approaches to understand the human microbiome. Nat. Human Behav 2018. nature.com [Internet]. Available from: https://www.nature.com/articles/s41562-018-0452-y.
[13] McDade TW. The ecologies of human immune function. Annu. Rev. Anthropol. October 2005;34(1):495–521 [Internet]. Available from: http://www.annualreviews.org/doi/10.1146/annurev.anthro.34.081804.120348.
[14] Fagundes CP, Glaser R, Kiecolt-glaser JK. Brain, behavior, and immunity stressful early life experiences and immune dysregulation across the lifespan. Behav. Med. 2012:1–5 [Internet]. Available from: https://www.sciencedirect.com/science/article/pii/S0889159112001821.
[15] Codagnone MG, Spichak S, O'Mahony SM, O'Leary OF, Clarke G, Stanton C, et al. Programming bugs: microbiota and the developmental origins of brain health and disease. Biol. Psychiatry January 2018:150–63 [Internet]. Available from: https://linkinghub.elsevier.com/retrieve/pii/S0006322318316056.
[16] Dominguez-Bello MG, Costello EK, Contreras M, Magris M, Hidalgo G, Fierer N, et al. Delivery mode shapes the acquisition and structure of the initial microbiota across multiple body habitats in newborns. Proc. Natl. Acad. Sci. 2010;107(26):11971–5 [Internet]. Available from: http://www.pnas.org/content/107/26/11971.short.
[17] Mueller NT, Whyatt R, Hoepner L, Oberfield S, Dominguez-Bello MG, Widen EM, et al. Prenatal exposure to antibiotics, cesarean section and risk of childhood obesity. Int. J. Obes. 2015;39(4):665–70 [Internet]. Available from: https://www.nature.com/articles/ijo2014180.
[18] Wlodarska M, Kostic AD, Xavier RJ. An integrative view of microbiome-host interactions in inflammatory bowel diseases [Internet] Cell Host Microbe 2015;17:577–91. Available from: https://www.sciencedirect.com/science/article/pii/S1931312815001663.
[19] Imhann F, Vich Vila A, Bonder MJ, Fu J, Gevers D, Visschedijk MC, et al. Interplay of host genetics and gut microbiota underlying the onset and clinical presentation of inflammatory bowel disease. Gut 2018;67(1):108–19 [Internet]. Available from: https://gut.bmj.com/content/67/1/108.abstract.
[20] Huttenhower C, Kostic AD, Xavier RJ. Inflammatory bowel disease as a model for translating the microbiome [Internet] Immunity 2014;40:843–54. Available from: https://www.sciencedirect.com/science/article/pii/S1074761314001939.
[21] Zhernakova DV, Le TH, Kurilshikov A, Atanasovska B, Bonder MJ, Sanna S, et al. Individual variations in cardiovascular-disease-related protein levels are driven by genetics and gut microbiome. Nat. Genet. 2018;50(11):1524–32 [Internet]. Available from: https://www.nature.com/articles/s41588-018-0224-7.
[22] De Martel C, Ferlay J, Franceschi S, Vignat J, Bray F, Forman D, et al. Global burden of cancers attributable to infections in 2008: a review and synthetic analysis. Lancet Oncol. June 1, 2012;13(6):607–15 [Internet]. Available from: https://www.sciencedirect.com/science/article/pii/S1470204512701377.
[23] Goodman B, Gardner H. The microbiome and cancer [Internet] J. Pathol. 2018;244:667–76. Available from: http://www.ncbi.nlm.nih.gov/pubmed/29377130.
[24] Zackular JP, Baxter NT, Iverson KD, Sadler WD, Petrosino JF, Chen GY, et al. The gut microbiome modulates colon tumorigenesis. MBio November 5, 2013;4(6) [Internet]. e00692-13. Available from: http://www.ncbi.nlm.nih.gov/pubmed/24194538.
[25] Peek RM, Blaser MJ. *Helicobacter pylori* and gastrointestinal tract adenocarcinomas. Nat. Rev. Cancer January 2002;2(1):28–37 [Internet]. Available from: http://www.ncbi.nlm.nih.gov/pubmed/11902583.
[26] van Nood E, Vrieze A, Nieuwdorp M, Fuentes S, Zoetendal EG, de Vos WM, et al. Duodenal infusion of donor feces for recurrent *Clostridium difficile*. N. Engl. J. Med. January 31, 2013;368(5):407–15 [Internet]. Available from: http://www.nejm.org/doi/10.1056/NEJMoa1205037.
[27] Kelly CR, Kahn S, Kashyap P, Laine L, Rubin D, Atreja A, et al. Update on fecal microbiota transplantation 2015: indications, methodologies, mechanisms, and outlook. Gastroenterology 2015;149(1):223–37 [Internet]. Available from: https://www.sciencedirect.com/science/article/pii/S0016508515006800.
[28] A.S. B, C.R. K. Fecal transplant in inflammatory bowel disease. Gastroenterol. Clin. North Am. 2017;46(4):825–37 [Internet]. Available from: https://www.gastro.theclinics.com/article/S0889-8553(17)30086-9/abstract.
[29] Shi Y, Dong Y, Huang W, Zhu D, Mao H, Su P. Fecal microbiota transplantation for ulcerative colitis: a systematic review and meta-analysis. Singh UP, editor PLoS One June 13, 2016;11(6):e0157259 [Internet]. Available from: http://dx.plos.org/10.1371/journal.pone.0157259.
[30] Versalovic J, Keitel W, Watson M, Dunne M. HMP initiative 1: core microbiome sampling protocol a human microbiome project – core microbiome sampling protocol A. 2009. Available from: http://www.fda.gov/cder/guidance/959fnl.pdf.
[31] Harismendy O, Ng PC, Strausberg RL, Wang X, Stockwell TB, Beeson KY, et al. Evaluation of next generation sequencing platforms for population targeted sequencing studies. Genome Biol. March 27, 2009;10(3):R32 [Internet]. Available from: http://genomebiology.biomedcentral.com/articles/10.1186/gb-2009-10-3-r32.
[32] Weisburg WG, Barns SM, Pelletier DA, Lane DJ. 16S ribosomal DNA amplification for phylogenetic study. J. Bacteriol. January 1, 1991;173(2):697–703 [Internet]. Available from: http://www.ncbi.nlm.nih.gov/pubmed/1987160.

[33] Reardon S. Bacterium can reverse autism-like behaviour in mice. Nature December 5, 2013 [Internet]. Available from: http://www.nature.com/doifinder/10.1038/nature.2013.14308.
[34] Zhang W, Li F, Nie L. Integrating multiple "omics" analysis for microbial biology: application and methodologies. Microbiology February 1, 2010;156(2):287–301 [Internet]. Available from: http://mic.microbiologyresearch.org/content/journal/micro/10.1099/mic.0.034793-0.
[35] Mimee M, Citorik R, Lu TK. Microbiome therapeutics—advances and challenges. Elsevier [Internet] Adv. Drug Deliv. Rev 2016. Available from: https://www.sciencedirect.com/science/article/pii/S0169409X16301429.
[36] Hongyu Zhang DJ. Manipulation of microbiome, a promising therapy for inflammatory bowel diseases. J. Clin. Cell Immunol. 2014;05(04) [Internet]. Available from: https://pdfs.semanticscholar.org/2f1e/8625cfda846f097d689894f2f719278c6a08.pdf.
[37] Bashiardes S, Tuganbaev T, Federici S, Elinav E. The microbiome in anti-cancer therapy [Internet] Semin. Immunol. 2017;32:74–81. Available from: https://www.sciencedirect.com/science/article/pii/S1044532316300914.
[38] Zmora N, Zeevi D, Korem T, Segal E, Elinav E. Taking it personally: personalized utilization of the human microbiome in health and disease [internet] Cell Host Microbe 2016;19:12–20. Available from: https://linkinghub.elsevier.com/retrieve/pii/S1931312815005089.
[39] Mathur S, Sutton J. Personalized medicine could transform healthcare (review). Biomed. Rep. July 2017;7(1):3–5 [Internet]. Available from: https://www.spandidos-publications.com/10.3892/br.2017.922.
[40] Peterson J, Garges S, Giovanni M, McInnes P, Wang L, Schloss JA, et al. The NIH human microbiome project. Genome Res. December 1, 2009;19(12):2317–23 [Internet]. Available from: http://genome.cshlp.org/cgi/doi/10.1101/gr.096651.109.
[41] Simoens S. Health economics of antibiotics. Pharmaceuticals April 29, 2010;3(5):1348–59 [Internet]. Available from: http://www.mdpi.com/1424-8247/3/5/1348.
[42] Founou RC, Founou LL, Essack SY. Clinical and economic impact of antibiotic resistance in developing countries: a systematic review and meta-analysis [Internet]. Butaye P, editor PLoS One 2017;12:e0189621. Available from: http://dx.plos.org/10.1371/journal.pone.0189621.
[43] Arbel LT, Hsu E, McNally K. Cost-effectiveness of fecal microbiota transplantation in the treatment of recurrent clostridium difficile infection: a literature review. Cureus 2017 [Internet]. Available from: https://www.ncbi.nlm.nih.gov/pmc/articles/PMC5652885/.
[44] Cani PD. Human gut microbiome: hopes, threats and promises [Internet] Gut. BMJ Publishing Group 2018;67:1716–25. Available from: http://www.ncbi.nlm.nih.gov/pubmed/29934437.
[45] Browne HP, Forster SC, Anonye BO, Kumar N, Neville BA, Stares MD, et al. Culturing of "unculturable" human microbiota reveals novel taxa and extensive sporulation. Nature May 4, 2016;533(7604):543–6 [Internet]. Available from: http://www.nature.com/articles/nature17645.
[46] Guilhot E, Khelaifia S, La Scola B, Raoult D, Dubourg G. Methods for culturing anaerobes from human specimen [Internet]. Future Medicine Ltd London, UK Future Microbiol. 2018;13:369–81. Available from: https://www.futuremedicine.com/doi/10.2217/fmb-2017-0170.
[47] McDonald D, Birmingham A, Knight R. Context and the human microbiome [Internet] Microbiome. BioMed. Central 2015;3:52. Available from: http://www.microbiomejournal.com/content/3/1/52.
[48] Blaut M, Collins MD, Welling GW, Doré J, van Loo J, de Vos W. Molecular biological methods for studying the gut microbiota: the EU human gut flora project. Br. J. Nutr. May 9, 2002;87(6):203–11 [Internet]. Available from: http://www.journals.cambridge.org/abstract_S0007114502000971.
[49] McNulty NP, Yatsunenko T, Hsiao A, Faith JJ, Muegge BD, Goodman AL, et al. The impact of a consortium of fermented milk strains on the gut microbiome of gnotobiotic mice and monozygotic twins. Sci. Transl. Med. October 26, 2011;3(106):106ra106 [Internet]. Available from: http://www.ncbi.nlm.nih.gov/pubmed/22030749.
[50] Arnold JW, Roach J, Azcarate-Peril MA. Emerging technologies for gut microbiome research [internet]. Elsevier Current Trends Trends Microbiol. 2016;24:887–901. Available from: https://www.sciencedirect.com/science/article/pii/S0966842X16300713.
[51] Waldor MK, Tyson G, Borenstein E, Ochman H, Moeller A, Finlay BB, et al. Where next for microbiome research? PLoS Biol. January 20, 2015;13(1):e1002050 [Internet]. Available from: http://dx.plos.org/10.1371/journal.pbio.1002050.
[52] Manor O, Levy R, Borenstein E. Mapping the inner workings of the microbiome: genomic- and metagenomic-based study of metabolism and metabolic interactions in the human microbiome [Internet]. Cell Press Cell Metabol. 2014;20:742–52. Available from: https://www.sciencedirect.com/science/article/pii/S1550413114003295.
[53] The World Health Organization. Health and nutritional properties of probiotics in food including powder milk with live lactic acid bacteria. Fao Who October 2001:1–34 [Internet]. Available from: https://ci.nii.ac.jp/naid/10030377877/.
[54] Nova E, Pérez De Heredia F, Gómez-Martínez S, Marcos A. The role of probiotics on the microbiota: effect on obesity [internet]. John Wiley & Sons, Ltd Nutr. Clin. Pract. 2015;31:387–400. Available from: http://doi.wiley.com/10.1177/0884533615620350.
[55] Kadooka Y, Sato M, Imaizumi K, Ogawa A, Ikuyama K, Akai Y, et al. Regulation of abdominal adiposity by probiotics (*Lactobacillus gasseri* SBT2055) in adults with obese tendencies in a randomized controlled trial. Eur. J. Clin. Nutr. June 10, 2010;64(6):636–43 [Internet]. Available from: http://www.nature.com/articles/ejcn201019.
[56] Kadooka Y, Sato M, Ogawa A, Miyoshi M, Uenishi H, Ogawa H, et al. Effect of *Lactobacillus gasseri* SBT2055 in fermented milk on abdominal adiposity in adults in a randomised controlled trial. Br. J. Nutr. November 25, 2013;110(9):1696–703 [Internet]. Available from: http://www.journals.cambridge.org/abstract_S0007114513001037.
[57] Peters BA, Shapiro JA, Church TR, Miller G, Trinh-Shevrin C, Yuen E, Friedlander C, Hayes RB, Ahn J. A taxonomic signature of obesity in a large study of American adults. Sci. Rep. June 27, 2018;8(1):9749.
[58] Dubin K, Callahan MK, Ren B, Khanin R, Viale A, Ling L, et al. Intestinal microbiome analyses identify melanoma patients at risk for checkpoint-blockade-induced colitis. Nat. Commun. December 2, 2016;7(1):10391 [Internet]. Available from: http://www.nature.com/articles/ncomms10391.
[59] Arpaia N, Campbell C, Fan X, Dikiy S, Van Der Veeken J, Deroos P, et al. Metabolites produced by commensal bacteria promote peripheral regulatory T-cell generation. Nature December 13, 2013;504(7480):451–5 [Internet]. Available from: http://www.nature.com/articles/nature12726.

Chapter 3

High-throughput omics in the precision medicine ecosystem

Abdellah Tebani[1] and Soumeya Bekri[1,2]

[1]*Department of Metabolic Biochemistry, Rouen University Hospital, Rouen, France;* [2]*Normandie Univ, UNIROUEN, CHU Rouen, INSERM U1245, Rouen, France*

Introduction: toward high-resolution medicine

The new era of biomedicine has been set by technological advances that enable the assessment and management of human health, at an amazing resolution scale. This includes the different biological information layers of genome, epigenome, transcriptome, metabolome, microbiome, phenotype, lifestyle, and behavioral attributes. This sets the foundation of a new data-driven medical practice that drives the precision medicine (PM) era [1]. PM may be defined as the longitudinal assessment, understanding, and management of an individual's health, based on the measurement of its basic components at the individual level, and derived from population knowledge. This will enable disease prevention and dynamic intervention, compared to conventional strategies that mainly rely on static and fragmented pictures of health, in a very sparse timeframe.

PM strategies aim to set an integrated framework for health and disease management. The high molecular and physiological data granularity aims to define the individual baseline of health. The dynamic assessment of this baseline allows early detection of deviations, to set preventive intervention. It also allows tailored treatment and monitoring when a disease is detected. Finally, a population of finely tuned individuals presents an amazing resource to set public health policies and assess disease and health trajectories throughout the population at different scales: health institution, group of patients, country and global population level, for large epidemiological perspectives.

This very promising data-driven medicine is due to an impressive convergence of different disciplines, mainly biomedical sciences, engineering, and data sciences. This convergence led to the generation of a large amount of clinical and biological data, which requires big data strategies to digest, process, and navigate through, for effective clinical actionability. Different international PM initiatives have been launched [2] such as "Precision Medicine Initiative" (USA), "The precision medicine initiative for Alzheimer's disease," the "100 k Wellness Project" (USA), the UK Biobank, The China Kadoorie Biobank, and the Estonian biobank of the Estonian Genome Center. These PM endeavors will allow a more systemic definition of health and disease [3].

For ages, biomedicine addressed the different parts of biological systems separately, and so physicians handled and treated each disease. Currently, customized contextualization of disease states leads to better diagnosis and treatment. Indeed, this integrative network vision of biological systems is highly effective to finely describe the structure, organization, and function of the system's components, at both the individual and population levels. The structure involves basic biomolecules (genes, gene expression products, proteins, metabolites). The topological connections between these components define the organization, whereas the function reflects how the system behaves, regarding internal and environmental stimuli [4].

These system medicine concepts [3] set the foundation for P4 medicine, which aims to be predictive, preemptive, participatory, and personalized [5]. For achieving this goal, omics technologies play a key role. The potential of multiomics data integration strategies within the clinical context is discussed, as well as the clinical actionability of omics-based biomarkers.

Omics is reshaping medical practice

Parsing genome complexity has gone far, since the discovery of the DNA structure [6]. The milestone was the whole human genome sequencing [7,8]. High-throughput

Precision Medicine for Investigators, Practitioners and Providers. https://doi.org/10.1016/B978-0-12-819178-1.00003-4

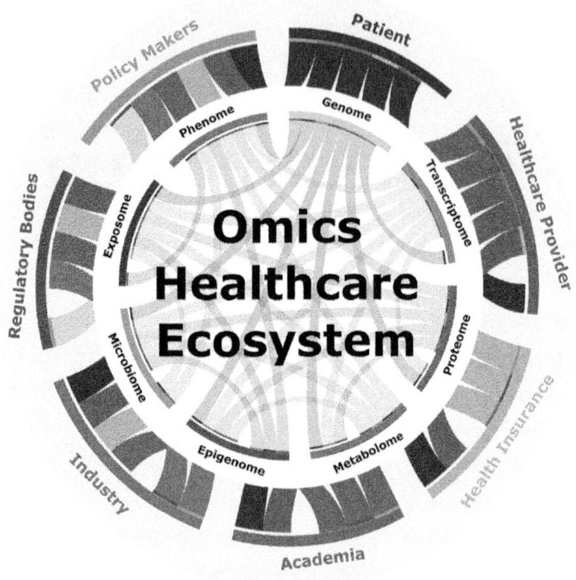

FIGURE 3.1 Omics and main stakeholders in the precision medicine healthcare ecosystem.

omics technologies simultaneously characterize a large number of biomarkers, including genes or their functional and structural patterns (i.e., methylation), transcripts, proteins, and metabolites or bacteria of the microbiome. The biological samples are often derived from human tissues, blood, urine, or other biofluids. Biomarkers can be used for disease diagnosis, response to treatment prediction, or for survival prognostics. Key biological information layers, such as genome, epigenome, transcriptome, proteome, metabolome, exposome, and microbiome, can be considered. These levels are interconnected through information transfer channels, including transcription, translation, posttranslational protein modification, and metabolic flux [9]. Fig. 3.1 shows the different omics layers of the biological information. Omics technologies exhibit an uneven maturity regarding their use in a clinical setting. Genomics with next-generation sequencing (NGS)-based technology, seems to be the closest to make it into the clinic, compared to transcriptomics and epigenomics. Metabolomics is closer than proteomics, since targeted metabolomics analyses using mainly mass spectrometry, is already being widely used for drug monitoring, metabolic investigations, and newborn screening.

High-throughput sequencing (HTS) technologies

The terms next-generation sequencing (NGS), massively parallel sequencing, or high-throughput sequencing (HTS), refer to innovative technologies that allow the sequencing reactions, without any physical separation of individual reactions. Instead, these sequencing reactions happen simultaneously on a solid surface, such as glass or beads, which allow billions of sequencing reactions to occur in parallel. This impressively enhances the throughput compared to classic Sanger sequencing. HTS are classified according to their applications for investigating genome, epigenome, or transcriptome, and the size of the interrogated sequences defines its medical diagnostics scope.

These strategies include capturing few targeted protein-coding regions of a selected panel of genes (Targeted Sequencing TS), sequencing of the entire genetic code of a person, which is called *whole-genome sequencing* (WGS), and sequencing parts of the genome that contain only exonic regions, which is called *whole-exome sequencing* (WES). WGS and WES are used to discover variants associated with a cell function or a disease [10]. NGS-based transcriptome analysis (RNA-seq) [11] entails for quantitative gene expression profiling, whereas epigenome methods focus on chromatin structure [12].

Genomics

By using current technology, it is possible to process a genome sequence within hours [13], with much higher throughput compared to the conventional Sanger sequencing [14]. Instrumental development has made a profound contribution to downsize platforms and lower the HTS costs, while fostering performance gains, including smoothening the computational burden. The sequencing cost per genome reduction has surpassed Moore's law, with the ongoing sub-$1000 USD genome era [13,15,16]. The high-throughput and cost-effectiveness of current HTS allows to sequence a panel of genes, WGS or WES in a matter of hours or days, depending on the technology and implemented protocols. HTS technologies and analysis workflow condition the time frame, and also the clinical accuracy and precision of the sequencing, including variant calling, as well as the type of detectable variants, such as single-nucleotide polymorphisms (SNPs) [17], indels [18], copy number variations (CNVs) [19], and fusion genes [20].

Targeted sequencing

For targeted sequencing (TS), a list of genes of interest is sequenced simultaneously. It is a very cost-effective first-tier test, with a high read depth of the targeted regions. This approach is compatible with most benchtop short-read sequencers. Thus, TS enables sequence variant detection in samples with very low non-reference allele frequencies [21,22]. The widespread availability of disease-specific gene panels lower incidental findings by facilitating variant interpretation [23]. One of the drawbacks of TS is the difficulty of detecting clinically relevant variants, especially

CNVs, in regions with an insufficient number of reads, leading to discontinuous coverage. This coverage lack may result from poor enrichment of GC-rich regions, and the absence of enrichment probes [24]. Furthermore, gene panels require continuous updates, given the changing pattern of gene-association studies. Thus, TS negative results might be nonconclusive, requiring additional investigation by second-tier sequencing WES and/or WGS.

The exome that represents around 2% of the human genome, however, includes a large proportion of the disease-causing variants [25]. The clinical exome covers only ~20% of the entire exome (around 5000 genes), and thus, requires further investigation when variants are not detected [26]. An advantage of WES is that it might lead to novel gene-disease associations, in a rather cost-effective fashion [27,28]. Furthermore, WES facilitates trio analyses (including parents), leading to better diagnostic yield [29]. Given the flexibility of data analysis, a gene panel can be selected in the WES in silico data mining, and subsequently expanded if needed [30]. Although WES has better coverage than TS, it has similar limitations and might fail to cover poorly enriched parts of the exome [24]. WES coverage can be enhanced by combining enrichment kits, or a higher concentration of capture probes [31,32].

Whole genome sequencing and genome-wide association studies

WGS exhibits the broadest coverage and variant detection [31,32]. WGS makes no assumptions about the exome and the human genome that is not yet fully characterized [33,34]. A recent deep transcriptional study, identified over 2000 unannotated isoforms of protein-coding mRNAs [35]. WGS enables the interrogation of both the coding and noncoding regions of the genome. This sets WGS apart in detecting non-exonic variants [24], and enhancing CNV detection [36,37]. Similar to WES, WGS can also be used in silico gene panels selection. So far, interpreting deep intronic, intergenic, and regulatory sequence variants has been difficult, by relying only on DNA level [38]. WGS offers a more comprehensive picture of the genome and can be mined over and over throughout an individual's lifetime. It is a valuable tool to parse the relationships between genomic and complex phenotypic traits. However, this will mainly depend on HTS technologies performances in terms of reading length, mapping uncertainty, haplotype phasing, and assessment of epigenetic markers.

Clinical translation of genomic findings

Clinical history, family history, and physical examination are mandatory, for effective sequence variant interpretation. Still, clinical validity is the most challenging aspect of NGS. From a clinical perspective, comparing different diagnostic strategies is of paramount interest; however, it requires standardized and adopted metrics [34]. Sanger sequencing remains the gold standard for sequencing technologies and allows confident calling of genotypes. Since noninferiority is a basic prerequisite for any clinical adoption in the medical innovation ecosystem, Goldfeder et al. recently suggested an appealing metric, to quantify the clinical grade for sequencing technologies reporting standard [33].

Epigenomics and environmental influences

Epigenomics is the chemical modification of the DNA, histones, nonhistone chromatin proteins, or nuclear RNA, through internal or external modifiers [39]. This leads to the modification of the gene expression pattern (extinction/activation), without changing the sequence [40]. These molecular modifications are related to environmental exposures at different stages throughout the life span [41]. The four main actors of epigenetic machinery include DNA methylation, histone modification, micro RNA (miRNA) expression and processing, and chromatin condensation [42,43]. Thus, epigenetic information is under the control of genome sequence, environmental exposure, and stochasticity. Epigenetic markers present several potentials for clinical translation, such as their technical stability, especially DNA methylation [44], and their early detectability over the whole genome, and not only in coding regions [45].

Main technologies still use bisulfite conversion to facilitate the detection of methylated cytosines [12]. Epigenome methods generally focus on chromatin structure and include histone modification ChIP-seq (Chromatin ImmunoPrecipitation sequencing), which is a method to identify DNA-associated protein-binding sites [46]. It allows precise characterization of transcription factor binding sites and patterns of histone modification. DNase-seq is a method in which DNase I digestion of chromatin, is combined with next-generation sequencing, to identify regulatory regions of the genome, including enhancers and promoters [47], DNA methylation [48], and ATAC-seq, which stands for Assay for Transposase-Accessible Chromatin sequencing. This method combines next-generation sequencing, with in vitro transposition of sequencing adapters into native chromatin [49,50].

Clinical impact

Epigenetic alterations are already getting to the biomarker ecosystem. Moreover, their reversible nature offers a promising therapeutic target to suppress disease using epigenetic-based drugs. Epigenetics is being implemented in patient management in oncology [51–54]. It is also actively investigated in neurological, infectious, and immune disorders [45].

Transcriptomics

The transcriptome refers to all RNA found in a cell or a given biological sample and reflects its functional state. It includes ribosomal RNA (rRNA), messenger RNA (mRNA), transfer RNA (tRNA), micro RNA (miRNA), and other noncoding RNA (ncRNA). Transcript abundances can be quantified using either microarrays (Chips) or RNA-sequencing (RNAseq). Microarrays are based on oligonucleotide probes that hybridize to specific RNA transcripts, whereas RNAseq uses direct sequencing of RNAs. Thus, RNA-seq allows a better mapping of all transcripts, at both qualitative and quantitative levels [11]. Furthermore, HTS has dramatically widened the transcriptional scope, using low quantities of RNA [55]. Transcriptomics has been used for molecular subtyping, of drug response and cancer [56]. Based on RNA patterns, Parker et al. identified four breast cancer subtypes associated with chemotherapy patient response [57]. RNA-Seq led to the characterization of novel molecular subgroups, related to treatment response and/or survival in various cancer studies, including pancreatic [58], esophageal [59], prostate [60], and cholangiocarcinoma [61]. miRNA has also been shown to be important in disease development and progression [62,63].

Proteomics

All the previous layers (genome, epigenome, and transcriptome) feed the proteome, which is the complement of proteins in a given organism [64]. The sequence, structure, and expression levels of proteins are encoded by the genome. However, the proteins can be changed at the translational and posttranslational steps [65]. The proteins exhibit various spatial configurations, intracellular localization, and intermolecular interactions, which confer them a highly versatile functionality. However, this raises big challenges for proteomics-based test development. Proteomics is the study of proteins and their structural and functional status, localization, and interactions.

The proteome can mainly be interrogated, using mass spectrometry (MS) and protein microarrays. MS and protein separation now allows rapid and accurate detection, of hundreds of human proteins and peptides, from a small amount of body fluid or tissue [66,67]. Array-based proteomic assays are typically dependent on an antibody for a specific protein. The most commonly used techniques for multiplexed assays are reverse phase protein arrays (RPPA), multiplexed immunofluorescence, and antibody-based chips or beads. These techniques provide a quantitative assay that analyzes simultaneously up to hundreds of proteins on low amounts of sample. These assays are powerful in identifying and validating cellular targets. One of the limits of array-based techniques is the inherent reliance on quality antibodies or prior knowledge of substrates.

Proteomic atlas

Recent initiatives demonstrated the potential of proteomics in translation, through a tissue-based map of the Human Proteome Atlas [68] and The Cancer Protein Atlas [69]. Most direct protein measurement techniques are MS-based. Several methods have been developed, including stable isotope labeling (SILAC), tandem mass tags (TMT), and isobaric tags (iTRAQ). So far, their main limitation is the challenge of protein absolute quantification. Given the importance and ubiquitousness of posttranslational modifications, the development of techniques that quantify these changes is of paramount importance. Compared to array-based techniques, single reaction monitoring (SRM) MS-based methods, can accurately measure multiple peptides from a single protein. Hoofnagle et al. reported a strong correlation between SRM and an immunoassay-based platform [70]. Mundt et al. recently reported an MS-based proteomics study, on patient-derived xenografts, to identify potential mechanisms of resistance to phosphoinositide 3-kinase inhibitors, with a potential to clinical actionability [71]. As for other omics, much of the clinical utility of translational proteomics will be conditioned by sample accessibility, quantity, and quality, along with correlation to accurate and relevant clinical data.

Metabolomics

The metabolome defines the metabolites present in a given organism [72–74]. Metabolites are small organic molecules involved in enzymatic reactions. They can be endogenous, naturally produced by the host or cells under study, or exogenous, including drugs and nutrients. Thus, metabolomics is a technology that aims to biochemically characterize a metabolome, and its changes regarding genetic, environmental, drug, or dietary factors [75]. Hence, metabolomics is an appealing tool to define metabolic phenotypes (metabotypes), that could be used for individual stratification. Metabolomics aims to measure small molecules (less than 1500 Da), in blood, tissue, urine, tissues, or even breath [76].

Various strategies can be combined, such as extraction, metabolite enrichment, and analytical techniques. Metabolic profiles are a rich functional readout of the biological state and represent a valuable resource for defining phenotypes. Since small molecules are key information carriers, they lie at the intersection of the different omics; genome, epigenome, proteome, and also environment (exposome and microbiome). Thus, metabolomics is increasingly used along with other omics to characterize clinical samples [77–79], and in biomarker discovery [80–83]. The ultimate goal is to measure all metabolites in a given sample; however, the current analytical scope is far from covering the whole metabolome [84]. Unlike HTS technologies, which can measure genome-wide features, such as gene expression or sequence variants with one assay, metabolomics requires multiple analytical techniques and instrumentation for broader metabolome coverage, given the quantitative and qualitative chemical space.

Metabolomic analysis

In practice, a specific combination of sample preparation and analytical techniques, is often used for a certain class of metabolites [76]. Two main analytical strategies can be used: nuclear magnetic resonance (NMR) [85] and mass spectrometry (MS) [86]. MS-based techniques are often combined with a separation technique, such as liquid (LC) or gas (GC) chromatography [87], capillary electrophoresis (CE) [88], or ion mobility (IMS) [89]. While NMR is considered robust, quantitative, and the gold standard for compound identification, MS-based methods are by far more sensitive, with higher coverage.

Targeted approaches

In targeted studies, a small set of metabolites with known chemical characteristics are accurately quantified. For untargeted metabolomics, the aim is to detect as many metabolites as possible, within a given sample. Untargeted approaches yield semi-quantitative measurements of thousands of signals, including known and unknown chemical entities. Annotation of the unknown signals is one of the biggest bottlenecks. However, despite the high level of technical and biological noise, and the increased complexity in data analysis, untargeted approaches are favorable for discovering novel biomarkers, or for data-driven hypothesis generation [90]. Metabolomics strategies are increasingly being adopted in translational and clinical research, thanks to the advances in automation, and improved quantification both NMR- and MS-based.

Reference datasets

Challenges regard database curation, and data analysis workflow development, to handle the data deluge. So far, Metabolomics Workbench [91] and MetaboLights [92], are the main metabolomics repositories. Integration of metabolomics data with other omics is increasingly performed [93,94]. The multiplication of large cohort studies using metabolomics, and the complexity of harmonizing data, and incorporating clinical and environmental attributes, require further standardization and informatics infrastructure. Of special interest, the impact of alterations in the microbiome on host metabolism is relevant, since these relationships can be indicative of various human diseases [95,96], including obesity and diabetes [97], cardiovascular diseases [98], inflammatory diseases [99], and cancer [100]. Different studies reported metabolite regulation of epigenetic events [101,102].

Phenomics

The term "phenome" refers to any morphological, biochemical, physiological, or behavioral attributes of an organism. Hence, phenotypes exhibit interindividual variability, that is tracked by phenomic strategies. Parsing the phenome requires determining the phenotypic features and their correlations to define the traits. Phenomics relies on both deep phenotyping (DP) and phenomics analysis (PA). DP refers to a comprehensive approach to data acquisition, at different time and space scales that include clinical assessment, laboratory analyses, pathology, imaging, environment, and lifestyle. PA involves the study of relationship patterns between individuals with related phenotypes, and/or between genotype–phenotype associations. PA relies on both clinical data and high-dimensional data integration [9], analysis, and visualization [103–107].

Electronic health records and phenome data mining

Electronic health records (EHRs) proved a valuable resource for analyzing phenotypic traits and developing phenome-wide association studies (PheWASs). PheWASs are designed to study phenotype association with a given genotype [108]. The Human Phenotype Ontology (HPO) provides the most comprehensive resource for deep phenotyping, for translational and clinical research [109]. Several HPO-based algorithms have been developed to sustain phenotype-driven genomic diagnosis. Usage of HPO includes analysis of clinical WES/WGS data [110–113], as well as integrative data analysis [114–116]. Using EHR-derived codes, mapped to HPO terms, generated EHR-based phenotypes leading to better variant pathogenicity periodization in Mendelian diseases [117].

Opportunities and stumbling blocks

Recently, Paik et al. created an impressive dynamic longitudinal representation of an individual's health, to identify novel disease associations for the risk stratification of patients. They created a dynamic large-scale disease network to map disease trajectories by using over

10.4 million patients [118]. State-of-the-art phenome-wide association studies have been recently reviewed [119]. The first barrier for achieving full clinical actionability of phenomics are the issues of data sharing, the condfidentiality of medical records, and privacy concerns. The second barrier is the cost and effort of handling EHRs. Indeed, manual curation of structured data by clinicians is not scalable. The third barrier is a lack of comparability and consistency among data and knowledge resources, which requires data standardization and harmonization for smoother interoperability.

Omics data analysis and the curse of dimensionality

What is an omics-based biomarker?

A biomarker is a characteristic that is objectively measured and evaluated, as an indicator of normal biological processes, pathogenic states, or pharmacologic response to a therapeutic intervention [120]. According to the FDA, biomarkers are measurable endpoints. They may also include imaging investigations, clinical laboratory testing, or their combination, such as radiomics [121]. The Institute of Medicine Committee on the Review of Omics-Based Tests for Predicting Patient Outcomes in Clinical Trials defines "omics" as the study of related sets of biological molecules, in a comprehensive fashion. Omics-based tests are defined as "an assay composed of, or derived from, multiple molecular measurements, and interpreted by a fully specified computational model, to produce a clinically actionable result" [122].

Criteria for omics-based biomarkers, and machine learning

Omics-based test development entails several crucial steps: analytical development, computational modeling of the predictor, then its assessment, and finally the validation of its clinical utility. The high-dimensionality and complexity of omics data require mathematical modeling to digest it for effective medical decision-making through machine learning (ML) methods [123]. Two main ML methods are used: unsupervised and supervised methods. Unsupervised methods track patterns in the data and draw the inference. Labeling the data is not necessary. They are widely used to identify novel phenotypic groups related to common underlying features [124,125]. Commonly used techniques are principal component analysis, cluster analysis, self-organizing maps, and hierarchical clustering [126].

Unsupervised and supervised machine learning

In unsupervised ML, no predictive model is built. For assessing new data, two options are possible; either mapping into the clustered space or the clustering or dimension reduction is performed with all of the data once again. Unsupervised ML is very interesting when sample labeling is missing or incorrect. Furthermore, it is a very effective visualization tool for high-dimensional data. Supervised ML methods are applied to labeled data. Labels are used to train the ML model to recognize feature patterns. Thus, new data can be assigned to a specific group based on the built predictive model [126]. Of note, the output of unsupervised methods can be used as input to supervised techniques.

ML methods are context-specific and require careful experimental design. The performance of ML methods can be affected by different factors such as feature selection, user-defined parameters, and the selection of the classifiers. The performance of predictive models can be assessed with independent labeled datasets (external validation) or using n-fold cross-validation (internal validation). In this case, a dataset is divided into n folds, n-1 folds are used for training, whereas the remaining fold is used for testing purposes. This process is iterated n times. For more details about ML, the reader may refer to recent reviews [123,126].

Omics-based biomarkers are already tested within clinical trial contexts, such as the NCI-Match, to help determine drug repurposing options for cancer patients [127], and the Michigan Oncology Sequencing Project (Mi-ONCOSEQ) uses an NGS approach, to stratify clinical trial-eligible patients [128]. Diagnostic acceptance of HTS by regulatory bodies is crucial, for clinical adoption of the technology. There are commercial breast cancer predictive transcriptome-based tests, such as PAM50 (Prosigna, NanoString Technologies, Seattle, WA, USA) [129] and MammaPrint (MammaPrint BluePrint, Agendia NV, Amsterdam, The Netherlands) [130]. Although there is an active endeavor to develop omics-based biomarkers, there is still some caution regarding the massive use of HTS in the clinical setting [131,132].

Omics challenges in a clinical and translational context

A very heterogeneous background of regulatory approval and clinical acceptance entails omics-based biomarkers. Key challenges may include sample collection [133–135], quality assessment [136], platform choice [13], data analysis, validation, and management [136,137].

Sample heterogeneity and biological noise

Tissue and cell-type specificity define two main challenges in omics: tissue and cell type selection, and tissue heterogeneity. The most used human sample is peripheral blood, including its derivatives plasma, serum, and leukocytes. Using blood-based specimens is convenient. However, using them as a surrogate tissue requires cautious validation and interpretation [138–140]. Depending on the location of a tissue sample or the individual physiological condition, the proportions of different cell types can change substantially. In silico optimization may be considered for cell type deconvolution [133–135]. Cell-specific HTS technologies are another alternative. Single-cell RNA-Seq (scRNA-Seq) has been successful in predicting treatment response. Standard processing methods used with "bulk" or multi-cell data are not always appropriate for both. The ultimate goal is to measure each omic in each purified cell type, for consistent and deep inference on the molecular mechanism of a disease [141,142].

Analytical noise and data interpretation

Consistent results require high reproducibility and repeatability [143]. These are often affected by "batch effects." Moreover, used technologies depend on the choice of the platform, often with different protocols, which lead to heterogeneous outputs [50]. This missing standardization presents a big challenge regarding the quality and accuracy of metrics comparison. For proteomics and metabolomics, technologies have different sensitivities and molecular coverage [144]. These heterogeneities hamper omics meta-analysis. Using harmonized Standard Operating Procedures (SOPs) may reduce heterogeneity [145–148]. Furthermore, standard quality control and metrics [149], along with appropriate computational tools, can address some issues of this technical variation.

In a clinical setting, accuracy, reproducibility, and standardization of NGS outputs can be enhanced through reference standards [150]. Regardless of the used technology, the initial base-calling is usually performed using proprietary software; however, novel methods are available [137]. For postalignment, selecting appropriate data analysis pipeline is crucial. Of note, there is a huge consensus lack between manufacturers, given the relative immaturity of the market [13,136]. For MS-based omics, standardization of protocols for downstream data analyses, including quality control, transformation, normalization, and differential analysis, are also difficult to establish, due to differences in experimental study design and data acquisition. Skilled workforce recruitment, including bioinformaticians, clinicians, and data scientists, is crucial to develop and manage the most appropriate tools with such a deeply multidisciplinary context. Thus, training the next-generation workforce is of paramount importance [151].

Clinical relevance and accuracy

The genome functional and structural complexity presents challenges for NGS sequencing strategies regarding the accuracy, which is the mandatory prerequisite for clinical decision-making. For example, a gene may have one to hundreds of exons, and the guanine-cytosine richness is also very challenging for capture chemistry-based targeted sequencing. Furthermore, simple repeats that are shorter than the read could be analyzed by short-reads sequencing strategies. However, when the read length is shorter than the repeat stretch, it is very hard to quantify the size of the repeated region [152]. Moreover, using short reads lacks phase information, which is very important from a clinical perspective.

Computational solutions have been proposed to handle such issues [153]. Obviously, long-read sequencing strategies, using either longer molecule barcoding fragments, combined with short-read sequencing and in silico assembly [154], or direct sequencing of longer molecules may help [155]. Chaisson et al. provided seminal evidence for the utility of long-read sequencing by closing euchromatic gaps in the GRCh38 human reference genome [155]. The clinical application requires rapid, timely answers; however, NGS strategies often need time-consuming library preparation. More automation would enhance turnover, and make a step forward to clinical adoption of NGS. Both a high standard of accuracy and rapid reporting of results are mandatory. Regarding high-throughput MS-based omics, such as proteomics and metabolomics, big challenges still lie ahead. For untargeted metabolomics, metabolite identification is a very limiting step [156].

For proteomic analysis, splice variants, post-translational modifications, and low-abundance protein detection are still technically challenging [157]. Direct measurements using methods analyzing their unique chemical attributes, such as structure, mass, and the charge is a possible way to go [158]. Immunocapture enrichment of low-abundance proteins may enhance sensitivity [159], although this might affect the throughput. The omics community is actively working to overcome these limitations [148,160]. The lack of reference metabolomes causes many data analysis issues, particularly for untargeted metabolomics studies, where the identification of metabolites is difficult.

Data integration

Multiomics integration and analysis are very appealing to get systematic biological insights. However, the pipeline is not easily streamlined, given the heterogeneity and complexity of biomedical data, including biological features (biomolecules, diseases), and related metadata (sampling metadata and clinical data). Data integration mainly includes omics data collection, processing, and quality assessment at different omics layers. Then, the processed data can be combined either at the data level using concatenation, at the model level, or using prior knowledge by mapping omics data to biological databases. The latter is done through the graph and network theory, such as genome-scale modeling, coexpression networks, protein-protein interaction networks, and regulatory networks [9,123,126]. These tools can holistically analyze multiple data types, to provide systems-level biological insights. However, the lack of metadata standards hampers efficient data integration, and structured approach is needed to effectively use training and test datasets.

Data management and governance

Data governance concerns storage, sharing, and privacy. Challenges for omics integration in clinical environments include data acquisition, validation, analytics, storage, reporting, and interoperability with already implemented laboratory systems and infrastructure. Important downstream offline steps should be consistently tracked, such as sample preparation, extraction, enrichment, along with upstream steps, such as bioinformatics analysis, quality assurance, data interpretation, and reporting. All these steps add complexity layers and potential error sources. The ultimate informatics solution would be fully and smoothly integrated with the laboratory information system (LIS), to track samples from order receipt to reporting [136]. Furthermore, data storage, sharing, navigation, and processing resources have to be considered [161].

Using sequence comparison algorithms is time-consuming and computationally intensive. Compression techniques offer another effective storage solution [162]. Besides the computational burden of the omics bioinformatics pipeline, one should add background data handling related to wet laboratory steps, sample metadata, sample processing, and tracking, reports, and quality control data. With such high-dimensional data management issues, omics clinical implementation should be approached with big data analytic solutions, and high-performance computing [161]. From an informatics perspective, not all omics are equally ready for routine applications. NGS seems to be much closer to get into clinical practice.

Data integrity, sharing, and ethical issues

In the digital era, cybersecurity measures are needed for encryption, authentication, access authorization management [163], along with donor anonymity protection, in publicly available repositories. Good practice guidelines and standards related to genomic data governance are issued while keeping translational research vivid [164]. For data sharing, meta-data standardization is essential. However, such standards are inconsistently used across omics, as demonstrated in Minimum Information About a Microarray Experiment (MIAME) and Minimum Information about a high-throughput nucleotide SEQuencing Experiment (MINSEQE) guidelines [165]. To reach a further level, global harmonization, along with an adapted regulation ecosystem for omics strategies, are urgently needed. Finally, data sharing within the scientific community raises controversial legal and ethical concerns [166].

Toward a data-driven healthcare workforce

Healthcare educational curriculum should be reviewed and upgraded, including health informatics, computer sciences, statistics, and bioinformatics. Indeed, clinicians should be aware of the basics of predictive data analytics and their recommendations, to be able to understand their limitations, given the upcoming deluge of such technologies [167]. New career pathways need to be designed to fill this talent and training gap. Policy makers need a better understanding of these disruptive technologies, to get a clearer picture of clinical decision-making, in this challenging data-driven healthcare system (Fig. 3.2).

Conclusion

In order to achieve omics full integration in clinical frameworks, the computational resources required for processing must be accessible, cost-effective, efficient, and with low entry-barrier skills. Standardization of analytical and computational pipelines will streamline clinical validation, with emphasis on data interoperability. Novel regulatory approval frameworks, reimbursement policies, and transdisciplinary training of the health care workforce with overlapping expertise are among the highest priorities.

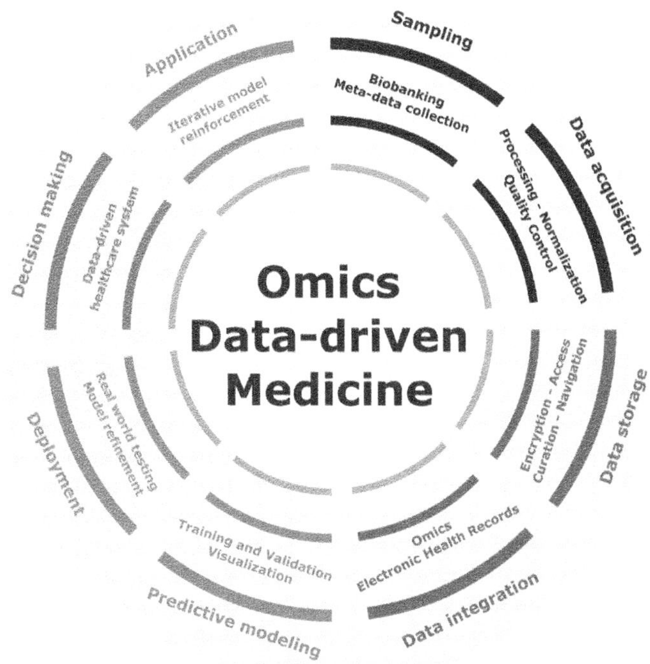

FIGURE 3.2 General workflow of omics profiling in precision medicine. Major steps include; (i) sampling and metadata collection, through effective biobanking infrastructure and resources, (ii) data acquisition and processing using relevant omics technologies, along with high standards of quality assurance and quality control. (iii) data governance, including data storage, access, sharing, and handling. (iv) data integration in the clinical informatics framework, with healthcare records and other relevant data. (v) data modeling and predictive model building, assessment, tuning, and validation in the real world. (vi) building data-driven healthcare ecosystems, based on model predictions and their continuous and real-time refinement. This pipeline is generic, and other versions may be adapted, depending on the requirements.

References

[1] Greene JA, Loscalzo J. Putting the patient back together — social medicine, network medicine, and the limits of reductionism. N. Engl. J. Med. 2017;377(25):2493–9.

[2] Vegter MW. Towards precision medicine; a new biomedical cosmology. Medicine. Health Care and Philosophy; 2018.

[3] Berlin R, Gruen R, Best J. Systems medicine—complexity within, simplicity without. J. Healthcare Inform. Res. 2017;1(1):119–37.

[4] Berlin R, Gruen R, Best J. Systems medicine disease: disease classification and scalability beyond networks and boundary conditions. Frontiers Bioeng. Biotechnol. 2018;6(112).

[5] Sagner M, et al. The P4 health spectrum – a predictive, preventive, personalized and participatory continuum for promoting healthspan. Prog. Cardiovasc. Dis. 2017;59.

[6] Watson JD, Crick FH. The structure of DNA. Cold Spring Harbor Symp. Quant. Biol. 1953;18:123–31.

[7] Venter JC, et al. The sequence of the human genome. Science 2001;291(5507):1304–51.

[8] Lander ES, et al. Initial sequencing and analysis of the human genome. Nature 2001;409(6822):860–921.

[9] Ritchie MD, et al. Methods of integrating data to uncover genotype-phenotype interactions. Nat. Rev. Genet. 2015;16(2):85–97.

[10] Schwarze K, et al. Are whole-exome and whole-genome sequencing approaches cost-effective? A systematic review of the literature. Genet. Med. 2018;20(10):1122–30.

[11] Mortazavi A, et al. Mapping and quantifying mammalian transcriptomes by RNA-Seq. Nat. Methods 2008;5(7):621–8.

[12] Mensaert K, et al. Next-generation technologies and data analytical approaches for epigenomics. Environ. Mol. Mutagen. 2014;55(3):155–70.

[13] Goodwin S, McPherson JD, McCombie WR. Coming of age: ten years of next-generation sequencing technologies. Nat. Rev. Genet. 2016;17(6):333–51.

[14] Baker M. Next-generation sequencing: adjusting to data overload. Nature Publishing Group; 2010.

[15] Loman NJ, et al. Performance comparison of benchtop high-throughput sequencing platforms. Nat. Biotechnol. 2012;30(5):434.

[16] Mardis ER. DNA sequencing technologies: 2006–2016. Nat. Protoc. 2017;12(2):213.

[17] Dworkis DA, et al. Severe sickle cell anemia is associated with increased plasma levels of TNF-R1 and VCAM-1. Am. J. Hematol. 2011;86(2):220–3.

[18] White MB, et al. A frame-shift mutation in the cystic fibrosis gene. Nature 1990;344(6267):665.

[19] Shlien A, Malkin D. Copy number variations and cancer. Genome Med. 2009;1(6):62.

[20] Vihinen M. Systematics for types and effects of DNA variations. BMC Genom. 2018;19(1):974.

[21] Feliubadaló L, et al. Benchmarking of whole exome sequencing and ad hoc designed panels for genetic testing of hereditary cancer. Sci. Rep. 2017;7:37984.

[22] García-García G, et al. Assessment of the latest NGS enrichment capture methods in clinical context. Sci. Rep. 2016;6:20948.

[23] Hehir-Kwa JY, et al. Towards a European consensus for reporting incidental findings during clinical NGS testing. Nature Publishing Group; 2015.

[24] Meienberg J, et al. Clinical sequencing: is WGS the better WES? Hum. Genet. 2016;135(3):359–62.

[25] Choi M, et al. Genetic diagnosis by whole exome capture and massively parallel DNA sequencing. Proc. Natl. Acad. Sci. U.S.A. 2009;106(45):19096–101.

[26] Eldomery MK, et al. Lessons learned from additional research analyses of unsolved clinical exome cases. Genome Med. 2017;9(1):26.

[27] Alazami AM, et al. Accelerating novel candidate gene discovery in neurogenetic disorders via whole-exome sequencing of prescreened multiplex consanguineous families. Cell Rep. 2015;10(2):148–61.

[28] Trujillano D, et al. Clinical exome sequencing: results from 2819 samples reflecting 1000 families. Eur. J. Hum. Genet. 2017;25(2):176.

[29] Farwell KD, et al. Enhanced utility of family-centered diagnostic exome sequencing with inheritance model–based analysis: results from 500 unselected families with undiagnosed genetic conditions. Genet. Med. 2015;17(7):578.

[30] Kalia SS, et al. Recommendations for reporting of secondary findings in clinical exome and genome sequencing, 2016 update (ACMG SF v2. 0): a policy statement of the American College of Medical Genetics and Genomics. Genet. Med. 2017;19(2):249.

[31] Mattick JS, et al. Whole genome sequencing provides better diagnostic yield and future value than whole exome sequencing. Med. J. Aust. 2018;209(5):197–9.

[32] Farooqi MS, et al. High concentration capture probes enhance massively parallel sequencing assays. Clin. Chem. 2016;62(7):1032–4.

[33] Goldfeder RL, et al. Human genome sequencing at the population scale: a primer on high-throughput DNA sequencing and analysis. Am. J. Epidemiol. 2017;186(8):1000.

[34] Lionel AC, et al. Improved diagnostic yield compared with targeted gene sequencing panels suggests a role for whole-genome sequencing as a first-tier genetic test. Genet. Med. 2018;20(4):435.

[35] Mercer TR, et al. Targeted RNA sequencing reveals the deep complexity of the human transcriptome. Nat. Biotechnol. 2012;30(1):99.

[36] Deveson IW, et al. Universal alternative splicing of noncoding exons. Cell Syst. 2018;6(2):245–255. e5.

[37] Tattini L, D'Aurizio R, Magi A. Detection of genomic structural variants from next-generation sequencing data. Frontiers Bioeng. Biotechnol. 2015;3:92.

[38] Kremer LS, et al. Genetic diagnosis of Mendelian disorders via RNA sequencing. Nat. Commun. 2017;8:15824.

[39] Feinberg AP. The key role of epigenetics in human disease prevention and mitigation. N. Engl. J. Med. 2018;378(14):1323–34.

[40] Skvortsova K, Iovino N, Bogdanović O. Functions and mechanisms of epigenetic inheritance in animals. Nat. Rev. Mol. Cell Biol. 2018:1.

[41] Huang B, Jiang C, Zhang R. Epigenetics: the language of the cell? Epigenomics 2014;6(1):73–88.

[42] Sadakierska-Chudy A, Filip M. A comprehensive view of the epigenetic landscape. Part II: histone post-translational modification, nucleosome level, and chromatin regulation by ncRNAs. Neurotox. Res. 2015;27(2):172–97.

[43] Sadakierska-Chudy A, Kostrzewa RM, Filip M. A comprehensive view of the epigenetic landscape part I: DNA methylation, passive and active DNA demethylation pathways and histone variants. Neurotox. Res. 2015;27(1):84–97.

[44] García-Giménez JL, et al. Challenges in the analysis of epigenetic biomarkers in clinical samples. Clin. Chem. Lab. Med. 2017;55(10):1474–7.

[45] Berdasco M, Esteller M. Clinical epigenetics: seizing opportunities for translation. Nat. Rev. Genet. 2019 Feb;20(2):109–27.

[46] Barski A, et al. High-resolution profiling of histone methylations in the human genome. Cell 2007;129(4):823–37.

[47] Yaragatti M, Basilico C, Dailey L. Identification of active transcriptional regulatory modules by the functional assay of DNA from nucleosome-free regions. Genome Res. 2008;18(6):930–8.

[48] Lister R, et al. Highly integrated single-base resolution maps of the epigenome in Arabidopsis. Cell 2008;133(3):523–36.

[49] Buenrostro JD, et al. ATAC-seq: a method for assaying chromatin accessibility genome-wide. Curr. Protoc. Mol. Biol. 2015;109: 21.29.1–9.

[50] Meyer CA, Liu XS. Identifying and mitigating bias in next-generation sequencing methods for chromatin biology. Nat. Rev. Genet. 2014;15(11):709–21.

[51] Imperiale TF, et al. Multitarget stool DNA testing for colorectal-cancer screening. N. Engl. J. Med. 2014;370(14):1287–97.

[52] Lamb YN, Dhillon S. Epi proColon® 2.0 CE: a blood-based screening test for colorectal cancer. Mol. Diagn. Ther. 2017;21(2):225–32.

[53] Ned RM, Melillo S, Marrone M. Fecal DNA testing for colorectal cancer screening: the ColoSure™ test. PLoS Curr. 2011;3.

[54] Moran S, et al. Epigenetic profiling to classify cancer of unknown primary: a multicentre, retrospective analysis. Lancet Oncol. 2016;17(10):1386–95.

[55] Wang Z, Gerstein M, Snyder M. RNA-Seq: a revolutionary tool for transcriptomics. Nat. Rev. Genet. 2009;10(1):57–63.

[56] Taylor BS, et al. Integrative genomic profiling of human prostate cancer. Cancer Cell 2010;18(1):11–22.

[57] Parker JS, et al. Supervised risk predictor of breast cancer based on intrinsic subtypes. J. Clin. Oncol. 2009;27(8):1160.

[58] Raphael BJ, et al. Integrated genomic characterization of pancreatic ductal adenocarcinoma. Cancer Cell 2017;32(2):185–203. e13.

[59] Network CGAR. Integrated genomic characterization of oesophageal carcinoma. Nature 2017;541(7636):169.

[60] Abeshouse A, et al. The molecular taxonomy of primary prostate cancer. Cell 2015;163(4):1011–25.

[61] Farshidfar F, et al. Integrative genomic analysis of cholangiocarcinoma identifies distinct IDH-mutant molecular profiles. Cell Rep. 2017;18(11):2780–94.

[62] Keller A, et al. Multiple sclerosis: microRNA expression profiles accurately differentiate patients with relapsing-remitting disease from healthy controls. PLoS One 2009;4(10):e7440.

[63] Daugaard I, et al. Small RNA sequencing reveals metastasis-related microRNAs in lung adenocarcinoma. Oncotarget 2017;8(16): 27047.

[64] James P. Protein identification in the post-genome era: the rapid rise of proteomics. Q. Rev. Biophys. 1997;30(4):279–331.

[65] Khoury GA, Baliban RC, Floudas CA. Proteome-wide post-translational modification statistics: frequency analysis and curation of the swiss-prot database. Sci. Rep. 2011;1.

[66] Betzen C, et al. Clinical proteomics: promises, challenges and limitations of affinity arrays. Proteonomics Clin. Appl. 2015;9(3–4):342–7.

[67] Sabbagh B, et al. Clinical applications of MS-based protein quantification. Proteonomics Clin. Appl. 2016;10(4):323–45.

[68] Uhlén M, et al. Tissue-based map of the human proteome. Science 2015;347(6220):1260419.

[69] Li J, et al. Explore, visualize, and analyze functional cancer proteomic data using the cancer proteome Atlas. Cancer Res. 2017;77(21):e51–4.

[70] Hoofnagle AN, et al. Multiple-reaction monitoring-mass spectrometric assays can accurately measure the relative protein abundance in complex mixtures. Clin. Chem. 2012:173856. clinchem. 2011.

[71] Mundt F, et al. Mass spectrometry–based proteomics reveals potential roles of NEK9 and MAP2K4 in resistance to PI3K inhibition in triple-negative breast cancers. Cancer Research; 2018.

[72] Oliver SG, et al. Systematic functional analysis of the yeast genome. Trends Biotechnol. 1998;16(9):373–8.

[73] Nicholson JK, Lindon JC, Holmes E. "Metabonomics": understanding the metabolic responses of living systems to pathophysiological stimuli via multivariate statistical analysis of biological NMR spectroscopic data. Xenobiotica 1999;29(11):1181–9.

[74] Fiehn O. Metabolomics—the link between genotypes and phenotypes. Plant Mol. Biol. 2002;48(1–2):155–71.

[75] Beger RD, et al. Metabolomics enables precision medicine:"a white paper, community perspective". Metabolomics 2016;12(9):149.

[76] Tebani A, Afonso C, Bekri S. Advances in metabolome information retrieval: turning chemistry into biology. Part I: analytical chemistry of the metabolome. J. Inherit. Metab. Dis. 2018 May;41(3):379–91.

[77] Zhang J, et al. Nondestructive tissue analysis for ex vivo and in vivo cancer diagnosis using a handheld mass spectrometry system. Sci. Transl. Med. 2017;9(406).

[78] St John E, et al. Real time intraoperative classification of breast tissue with the intelligent knife. Eur. J. Surg. Oncol. 2016;42(5):S25.

[79] He J, et al. Ambient mass spectrometry imaging metabolomics method provides novel insights into the action mechanism of drug candidates. Anal. Chem. 2015;87(10):5372–9.

[80] Sandlers Y. The future perspective: metabolomics in laboratory medicine for inborn errors of metabolism. Transl. Res. 2017;189:65–75.

[81] Tebani A, et al. Clinical metabolomics: the new metabolic window for inborn errors of metabolism investigations in the post-genomic era. Int. J. Mol. Sci. 2016;17(7).

[82] Bekri S. The role of metabolomics in precision medicine. Expert Rev. Pre. Med Drug Dev. 2016;1(6):517–32.

[83] Collino S, Martin FP, Rezzi S. Clinical metabolomics paves the way towards future healthcare strategies. Br. J. Clin. Pharmacol. 2013;75(3):619–29.

[84] Wishart DS, et al. HMDB 3.0—the human metabolome database in 2013. Nucleic Acids Res. 2013;41:D801.

[85] Nagana Gowda GA, Raftery D. Recent advances in NMR-based metabolomics. Anal. Chem. 2017;89(1):490–510.

[86] Kohler I, et al. Analytical pitfalls and challenges in clinical metabolomics. Bioanalysis; 2016.

[87] Haggarty J, Burgess KEV. Recent advances in liquid and gas chromatography methodology for extending coverage of the metabolome. Curr. Opin. Biotechnol. 2017;43:77–85.

[88] Rodrigues KT, et al. Strategies involving mass spectrometry combined with capillary electrophoresis in metabolomics. In: Metabolomics: from fundamentals to clinical applications. Springer; 2017. p. 99–141.

[89] Tebani A, et al. Optimization of a liquid chromatography ion mobility-mass spectrometry method for untargeted metabolomics using experimental design and multivariate data analysis. Anal. Chim. Acta 2016;913:55–62.

[90] Wishart DS. Emerging applications of metabolomics in drug discovery and precision medicine. Nat. Rev. Drug Discov. 2016;15(7):473–84. advance online publication, Jul.

[91] Sud M, et al. Metabolomics Workbench: an international repository for metabolomics data and metadata, metabolite standards, protocols, tutorials and training, and analysis tools. Nucleic Acids Res. 2015;44(D1):D463–70.

[92] Haug K, et al. MetaboLights—an open-access general-purpose repository for metabolomics studies and associated meta-data. Nucleic Acids Res. 2012;41(D1):D781–6.

[93] Menni C, et al. Mixing omics: combining genetics and metabolomics to study rheumatic diseases. Nat. Rev. Rheumatol. 2017;13(3):174.

[94] Patel S, Ahmed S. Emerging field of metabolomics: big promise for cancer biomarker identification and drug discovery. J. Pharm. Biomed. Anal. 2015;107:63–74.

[95] Shaffer M, et al. Microbiome and metabolome data integration provides insight into health and disease. Transl. Res. 2017;189:51–64.

[96] Garg N, et al. Natural products as mediators of disease. Nat. Prod. Rep. 2017;34(2):194–219.

[97] Boulangé CL, et al. Impact of the gut microbiota on inflammation, obesity, and metabolic disease. Genome Med. 2016;8(1):42.

[98] Feng Q, et al. Integrated metabolomics and metagenomics analysis of plasma and urine identified microbial metabolites associated with coronary heart disease. Sci. Rep. 2016;6:22525.

[99] Clemente JC, Manasson J, Scher JU. The role of the gut microbiome in systemic inflammatory disease. BMJ 2018;360:j5145.

[100] Ma Y, Li X. Integrated microbiome and metabolome analysis to reveal the interplay between commensal bacteria and metabolites in colorectal cancer. J. Clin. Oncol. 2018;36(15_Suppl.):e15696. e15696.

[101] Lai CQ, et al. Epigenomics and metabolomics reveal the mechanism of the APOA2-saturated fat intake interaction affecting obesity. Am. J. Clin. Nutr. 2018;108(1):188–200.

[102] Sharma U, Rando OJ. Metabolic inputs into the epigenome. Cell Metabol. 2017;25(3):544–58.

[103] Bilder RM, et al. Phenomics: the systematic study of phenotypes on a genome-wide scale. Neuroscience 2009;164(1):30–42.

[104] Freimer N, Sabatti C. The human phenome project. Nat. Genet. 2003;34(1):15–21.

[105] Gerlai R. Phenomics: fiction or the future? Trends Neurosci. 2002;25(10):506–9.

[106] Oetting WS, et al. Getting ready for the human phenome project: the 2012 forum of the human variome project. Hum. Mutat. 2013;34(4):661–6.
[107] Groza T, et al. The human phenotype ontology: semantic unification of common and rare disease. Am. J. Hum. Genet. 2015;97(1):111–24.
[108] Denny JC, Bastarache L, Roden DM. Phenome-wide association studies as a tool to advance precision medicine. Annu. Rev. Genom. Hum. Genet. 2016;17:353–73.
[109] Köhler S, et al. The human phenotype ontology project: linking molecular biology and disease through phenotype data. Nucleic Acids Res. 2013;42(D1):D966–74.
[110] Stokman MF, et al. Clinical and genetic analyses of a Dutch cohort of 40 patients with a nephronophthisis-related ciliopathy. Pediatr. Nephrol. 2018;33(10):1701–12.
[111] Zhu Q, et al. EHR based genetic testing knowledge base (iGTKB) development. BMC Med. Inf. Decis. Mak. 2015;15(4):S3.
[112] Tumienė B, et al. Diagnostic exome sequencing of syndromic epilepsy patients in clinical practice. Clin. Genet. 2018;93(5):1057–62.
[113] Short PJ, et al. De novo mutations in regulatory elements in neurodevelopmental disorders. Nature 2018;555(7698):611.
[114] Haghighi A, et al. An integrated clinical program and crowdsourcing strategy for genomic sequencing and Mendelian disease gene discovery. NPJ Genom. Med. 2018;3(1):21.
[115] Rao A, et al. Phenotype-driven gene prioritization for rare diseases using graph convolution on heterogeneous networks. BMC Med. Genom. 2018;11(1):57.
[116] Adler A, et al. PhenoDis: a comprehensive database for phenotypic characterization of rare cardiac diseases. Orphanet J. Rare Dis. 2018;13(1):22.
[117] Bastarache L, et al. Phenotype risk scores identify patients with unrecognized Mendelian disease patterns. Science 2018;359(6381):1233–9.
[118] Paik H et al. Tracing diagnosis trajectories over millions of inpatients reveal an unexpected association between schizophrenia and rhabdomyolysis. www.biorxiv.org/content/10.1101/473082v1
[119] Bush WS, Oetjens MT, Crawford DC. Unravelling the human genome-phenome relationship using phenome-wide association studies. Nat. Rev. Genet. 2016;17(3):129–45.
[120] Atkinson AJ, Colburn WA, DeGruttola VG, DeMets DL, Downing GJ, Hoth DF, Oates JA, Peck CC, Schooley RT, Spilker BA, Woodcock J, Zeger SL. Biomarkers and surrogate endpoints: preferred definitions and conceptual framework. Clin. Pharmacol. Ther. 2001;69(3):89–95.
[121] Halim A-B. Biomarkers in drug development: a useful tool but discrepant results may have a major impact. INTECH Open Access Publisher; 2011.
[122] Micheel CM, Nass SJ, Omenn GS. Evolution of translational omics: lessons learned and the path forward. National Academies Press; 2012.
[123] Zitnik M, et al. Machine learning for integrating data in biology and medicine: principles, practice, and opportunities. Inf. Fusion 2019;50:71–91.
[124] Miotto R, et al. Deep patient: an unsupervised representation to predict the future of patients from the electronic health records. Sci. Rep. 2016;6:26094.
[125] Glicksberg BS, et al. Automated disease cohort selection using word embeddings from electronic health records. Pac. Symp. Biocomput. 2018;23:145–56.
[126] Camacho DM, et al. Next-generation machine learning for biological networks. Cell Jun 14 2018;173(7):1581–92.
[127] Barroilhet L, Matulonis U. The NCI-MATCH trial and precision medicine in gynecologic cancers. Gynecol. Oncol. 2018;148(3):585–90.
[128] Roychowdhury S, et al. Personalized oncology through integrative high-throughput sequencing: a pilot study. Sci. Transl. Med. 2011;3(111):111ra121. 111ra121.
[129] Wallden B, et al. Development and verification of the PAM50-based Prosigna breast cancer gene signature assay. BMC Med. Genom. 2015;8(1):54.
[130] Van De Vijver MJ, et al. A gene-expression signature as a predictor of survival in breast cancer. N. Engl. J. Med. 2002;347(25):1999–2009.
[131] Prasad V. Perspective: the precision-oncology illusion. Nature 2016;537(7619):S63. S63.
[132] Prasad V. Why the US Centers for Medicare and Medicaid Services (CMS) should have required a randomized trial of Foundation Medicine (F1CDx) before paying for it. Oxford University Press; 2017.
[133] McGregor K, et al. An evaluation of methods correcting for cell-type heterogeneity in DNA methylation studies. Genome Biol. 2016;17(1):1–17.
[134] Buettner F, et al. Computational analysis of cell-to-cell heterogeneity in single-cell RNA-sequencing data reveals hidden subpopulations of cells. Nat. Biotechnol. 2015;33(2):155–60.
[135] Houseman EA, Molitor J, Marsit CJ. Reference-free cell mixture adjustments in analysis of DNA methylation data. Bioinformatics 2014;30(10):1431–9.
[136] Roy S, et al. Standards and guidelines for validating next-generation sequencing bioinformatics pipelines: a joint recommendation of the Association for Molecular Pathology and the College of American Pathologists. J. Mol. Diagn. 2018;20(1):4–27.
[137] Crockett DK, Voelkerding KV. Bioinformatics tools in clinical genomics. In: Netto GJ, Schrijver I, editors. Genomic applications in pathology. New York, NY: Springer New York; 2015. p. 177–96.
[138] GTEx Consortium AK, Deluca DS, Segrè AV, Sullivan TJ, Young TR, Gelfand ET, Trowbridge CA, Maller JB, Tukiainen T, Lek M, Ward LD, Kheradpour P, Iriarte B, Meng Y, Palmer CD, Esko T, Winckler W, Hirschhorn JN, Kellis M, MacArthur DG, Getz G, Shabalin AA, Li G, Zhou YH, Nobel AB, Rusyn I, Wright FA, Lappalainen T, Ferreira PG, Ongen H, Rivas MA, Battle A, Mostafavi S, Monlong J, Sammeth M, Melé M, Reverter F, Goldmann JM, Koller D, Guigó R, McCarthy MI, Dermitzakis ET, Gamazon ER, Im HK, Konkashbaev A, Nicolae DL, Cox NJ, Flutre T, Wen X, Stephens M, Pritchard JK, Tu Z, Zhang B, Huang T, Long Q, Lin L, Yang J, Zhu J, Liu J, Brown A, Mestichelli B, Tidwell D, Lo E, Salvatore M, Shad S, Thomas JA, Lonsdale JT, Moser MT, Gillard BM, Karasik E, Ramsey K, Choi C, Foster BA, Syron J, Fleming J, Magazine H, Hasz R, Walters GD, Bridge JP, Miklos M, Sullivan S, Barker LK, Traino HM, Mosavel M, Siminoff LA, Valley DR, Rohrer DC, Jewell SD, Branton PA, Sobin LH, Barcus M, Qi L, McLean J, Hariharan P, Um KS, Wu S, Tabor D, Shive C, Smith AM, Buia SA, Undale AH, Robinson KL, Roche N, Valentino KM, Britton A, Burges R, Bradbury D, Hambright KW, Seleski J, Korzeniewski GE, Erickson K, Marcus Y, Tejada J, Taherian M, Lu C, Basile M, Mash DC, Volpi S, Struewing JP, Temple GF, Boyer J, Colantuoni D, Little R, Koester S, Carithers LJ,

Moore HM, Guan P, Compton C, Sawyer SJ, Demchok JP, Vaught JB, Rabiner CA, Lockhart NC, Ardlie KG, Getz G, Wright FA, Kellis M, Volpi S, Dermitzakis ET. Human genomics. The Genotype-Tissue Expression (GTEx) pilot analysis: multitissue gene regulation in humans. Science 2015;348(6235):648–60.

[139] Torell F, et al. Multi-organ contribution to the metabolic plasma profile using hierarchical modelling. PLoS One 2015;10(6): e0129260.

[140] Do KT, et al. Network-based approach for analyzing intra- and interfluid metabolite associations in human blood, urine, and saliva. J. Proteome Res. 2015;14(2):1183–94.

[141] Bock C, Farlik M, Sheffield NC. Multi-omics of single cells: strategies and applications. Trends Biotechnol. 2016;34(8):605–8.

[142] Leonavicius K, et al. Multi-omics at single-cell resolution: comparison of experimental and data fusion approaches. Curr. Opin. Biotechnol. 2019;55:159–66.

[143] Ioannidis JP, Khoury MJ. Improving validation practices in "omics" research. Science 2011;334(6060):1230–2.

[144] May JC, McLean JA. Advanced multidimensional separations in mass spectrometry: navigating the big data deluge. Annu. Rev. Anal. Chem. 2016;9(1):387–409.

[145] Kolker E, et al. Toward more transparent and reproducible omics studies through a common metadata checklist and data publications. Omics 2014;18(1):10–4.

[146] Tenenbaum JD, Sansone SA, Haendel M. A sea of standards for omics data: sink or swim? J. Am. Med. Inform. Assoc. 2014;21(2):200–3.

[147] Chitayat S, Rudan JF. Chapter 10 – phenome centers and global harmonization. In: Metabolic phenotyping in personalized and public healthcare. Boston: Academic Press; 2016. p. 291–315.

[148] Rocca-Serra P, et al. Data standards can boost metabolomics research, and if there is a will, there is a way. Metabolomics 2016;12(1):14.

[149] Dunn WB, et al. The importance of experimental design and QC samples in large-scale and MS-driven untargeted metabolomic studies of humans. Bioanalysis 2012;4(18):2249–64.

[150] Hardwick SA, Deveson IW, Mercer TR. Reference standards for next-generation sequencing. Nat. Rev. Genet. 2017;18(8):473.

[151] Schiffthaler B, et al. Training in high-throughput sequencing: common guidelines to enable material sharing, dissemination, and reusability. PLoS Comput. Biol. 2016;12(6):e1004937.

[152] Ashley EA. Towards precision medicine. Nat. Rev. Genet. 2016;17(9):507–22.

[153] Tewhey R, et al. The importance of phase information for human genomics. Nat. Rev. Genet. 2011;12(3):215–23.

[154] Zheng GX, et al. Haplotyping germline and cancer genomes with high-throughput linked-read sequencing. Nat. Biotechnol. 2016;34(3):303–11.

[155] Chaisson MJ, et al. Resolving the complexity of the human genome using single-molecule sequencing. Nature 2015;517(7536): 608–11.

[156] Vinaixa M, et al. Mass spectral databases for LC/MS- and GC/MS-based metabolomics: state of the field and future prospects. Trac. Trends Anal. Chem. 2016;78:23–35.

[157] Wu L, et al. Variation and genetic control of protein abundance in humans. Nature 2013;499(7456):79–82.

[158] Deutsch EW, et al. Human proteome project mass spectrometry data interpretation guidelines 2.1. J. Proteome Res. 2016;15(11):3961–70.

[159] Whiteaker JR, et al. An automated and multiplexed method for high throughput peptide immunoaffinity enrichment and multiple reaction monitoring mass spectrometry-based quantification of protein biomarkers. Mol. Cell. Proteom. 2010;9(1):184–96.

[160] Fehniger TE, et al. Four areas of engagement requiring strengthening in modern proteomics today. J. Proteome Res. 2014;13(12):5310–8.

[161] Ma T, Zhang A. Omics informatics: from scattered individual software tools to integrated workflow management systems. IEEE ACM Trans. Comput. Biol. Bioinform 2017;14(4):926–46.

[162] Biji C, Nair AS. Benchmark dataset for whole genome sequence compression. IEEE ACM Trans. Comput. Biol. Bioinform 2017;14(6):1228–36.

[163] Langmead B, Nellore A. Cloud computing for genomic data analysis and collaboration. Nat. Rev. Genet. 2018;19(4):208.

[164] Mittos A, Malin B, De Cristofaro E. Systematizing genome privacy research: a privacy-enhancing technologies perspective. Proc. Privacy Enhanc. Technol. 2019;1:87–107.

[165] McQuilton P, et al. BioSharing: curated and crowd-sourced metadata standards, databases and data policies in the life sciences. Database; 2016.

[166] Jiang X, et al. A community assessment of privacy preserving techniques for human genomes. BMC Med. Inform. Decis. Mak. 2014;14(1):S1.

[167] Fridsma DB. Health informatics: a required skill for 21st century clinicians. BMJ 2018;362:k3043.

Chapter 4

Recent advances in the infant gut microbiome and health

Romina Díaz and Daniel Garrido
Department of Chemical and Bioprocess Engineering, School of Engineering. Pontificia Universidad Catolica de Chile, Santiago, Chile

Introduction

Microbial colonization occurring during birth has been increasingly shown to play a fundamental role in the health of the newborn [1,2]. The early gut microbiome assembles in a period of over 2 years [2], and the consequences of how this process develops, could also last in the long term. This natural process is considered to create a critical window, where interruptions, for example, by antibiotics or C-section, could result in aberrant microbiome compositions. This aberrant state (dysbiosis) has been shown to contribute to certain diseases later in life, for example, asthma [3,4], type 2 diabetes [5], or food allergies [6].

Next-generation sequencing technologies are providing a better knowledge of the colonization process [7,8]. The gut microbiome composition dynamically changes during the first year of life (9), and it is modulated by different factors including gestational age [10], delivery type [11], antibiotic use [12], and infant feeding [9]. The infant gut microbiome composition is distinguishable from the adult [13], and it transitions to a mature state after the introduction of solid food [14].

Breast milk is the natural and ideal food for the newborn, providing energy and nutrients that are essential during the first months of life [15]. The composition of breast milk is unique in that it contains a wide array of bioactive molecules, especially free human milk oligosaccharides (HMO) [16]. Breastfed infants have a lower incidence of diarrhea, allergies, and inflammatory diseases [17]. Some of the benefits can last beyond childhood [16].

Computational models can be designed, focusing on gut microbiome composition associated with infant feeding [18]. These models are based on metabolic cross-feeding interactions between gut microbes, and while still at early stages, hold promise regarding predictive capabilities [19,20].

Factors influencing gut microbiome assembly and development

Gut colonization could start in the prenatal stage, indicating that the fetus does not develop in a sterile environment [21,22]. Viable microorganisms have been isolated from the umbilical cord, but not other sources, such as amniotic fluid [23]. Several of these studies report circulating bacterial DNA in the amniotic fluid, not necessarily representing viable microorganisms [24]. Therefore, most of the newborn exposures to microorganisms occur during and after birth.

The first crucial microbial exposure occurs with the rupture of the amniotic membrane. Dynamics of early colonization are influenced by delivery type, gestational age, antibiotic use, and infant feeding.

Delivery type

Vaginally delivered infants are usually colonized by a different set of gut microbes compared to C-section. Vaginal delivery exposes the newborn to the vaginal and gastrointestinal microbiome of the mother, as well as the mother's skin. The vertical mother-infant microbial transmission was shown to be essential for infant health in the short term [25]. There are implications also, in the long term [26]. For identifying vertical mother-infant transmission events, it is necessary to isolate the same strain variant within mother-infant pairs.

Ferretti et al. used high-resolution shotgun metagenomics with strain-level computational profiling, to characterize the transfer of microbes from 25 mother-infant pairs during the first 4 months of life [25]. Multiple maternal body sites contribute to the developing infant microbiome, with gut strains providing the most significant amount [25]. Similarly, Yassour et al. used longitudinal

Precision Medicine for Investigators, Practitioners and Providers. https://doi.org/10.1016/B978-0-12-819178-1.00004-6

metagenomic sequencing in 44 mother-infant pairs. The authors identified dominant and secondary strain inheritance events, in which genetic differences between strains may provide selective advantages in infant gut colonization [26].

During C-section, microbes associated with the mother's skin and surrounding environment are prominent, nominally *Staphylococcus*, and *Streptococcus* [27]. More heterogeneity occurs compared to those born vaginally, resulting in a delay of colonization by essential gut microbes, and increased risk of infections [28]. Neonates can be inoculated with vaginal secretions from their mothers, in order to make their microbiome more similar to infants born vaginally [11]. While promising, this procedure still puts the infant at risk for certain pathogens [29]. These differences are observed for 3 months and tend to disappear after 6 months [30]. Introduction of solid foods also abrogates microbiome imbalances [31].

Antibiotics

The administration of antibiotics is frequent in the newborn and has been associated with an increased risk of asthma [3,4], type 2 diabetes [5], or food allergies [6]. A study of Korpela et al. reported that macrolide use in 2–7 years old Finnish children, was associated with marked changes in the gut microbiota composition and metabolism, predisposing to weight gain and asthma in later childhood [32]. Risk of cow's milk allergy and cephalosporin use in children was also reported [6].

Changes depend on treatment duration and dose [33,34]. Cefepime and ceftazidime are broad-spectrum cephalosporins for bacterial infections [35]. Quinolones (ciprofloxacin) are similarly broad-spectrum bactericidal agents, active against many Gram-positive and Gram-negative bacteria [33]. Penicillin (amoxicillin) is another broad-spectrum beta-lactam antibiotic, that interferes with the synthesis of the bacterial cell wall peptidoglycan [33]. Penicillins, cephalosporins, and second-generation macrolides are usually prescribed to treat pharyngitis, respiratory infections, or pneumonia [36]. Overuse of antibiotics significantly diminishes microbiome numbers and diversity [37].

Another concern is antimicrobial resistance in infants [38]. Bokulich et al. studied microbial development during the first 2 years of life, of 43 infants who received antibiotics. Decreasing numbers of *Bacteroides* were prominent. Less bacterial diversity, with reduced abundance of *Bifidobacterium* and higher colonization by *Enterococcus*, is also observed [39].

The microbiome recovers after antibiotic therapies; however, a delay in microbiome maturation is noticed [40]. Methods to circumvent such drawbacks include antiquorum sensing molecules and bacteriophages [34,41,42]. Phages programmed to deliver the CRISPR-Cas system were employed against antibiotic-resistant bacteria [43].

Breastfeeding by antibiotic-taking mothers could also alter the milk microbiome, or transfer the drug directly to the infant intestine [44,45]. Antibiotics should, therefore, be avoided when there is no clear indication of use. If required, spectrum and treatment time should be optimized [45].

Feeding type

Breast milk is a complex fluid that meets all the nutritional requirements of the newborn [46]. It contains high concentrations of lactose and lipids, in addition to proteins and peptides, for example, lactoferrin, caseins, and immunoglobulins. Human milk also contains a wide range of bioactive molecules [47], which promote proper health, growth, and protection of the newborn [48,49]. The third most abundant components in breast milk are HMO, which have a prebiotic role, being selectively consumed by beneficial members of the infant gut microbiome [50].

HMO as prebiotics is almost exclusively fermented by *Bifidobacterium* species [16]. HMO reach high concentrations in mature milk (10–15 g/L) and especially in the colostrum (15–23 g/L) [51]. Structurally, HMO represents a pool of more than 45 different linear or branched oligosaccharides, composed of 3–15 carbohydrate units, including glucose, galactose (Gal), N-acetylglucosamine, fucose (Fuc) and sialic acid. More complexity is added, considering that these monosaccharides are arranged in multiple glycosidic linkages [16].

One of the most important factors influencing HMO composition is host genetics [52]. Secretor status is a factor that genetically varies among mothers, influencing the assembly of infant intestinal microbiome [53]. Secretory milk is characterized by glycoconjugates containing the Fucα1-2Gal determinant, in which FUT2 fucosyltransferase is responsible for the synthesis of these conjugates. FUT2 is encoded by the secretory locus in chromosome 19 [16]. The resulting genotype can be phenotypically confirmed by composition of HMO in breast milk.

Nonsecretory mothers lacking the enzyme FUT2 do not produce α1-2 fucosylated oligosaccharides, such as 2′-fucosyllactose (2FL) or lacto-N-fucopentose (LNFP). This genotype is found in 30% of women worldwide [51,53,54]. FUT3 is another fucosyltransferase with different specificity, which is associated with the Lewis gene and produces glycoconjugates containing the Fucα1-3Gal determinant. Allelic variations between secretor and Lewis genes result in milk with fucosylated, nonfucosylated HMOs, or combinations thereof [55]. Content of fucose in breast milk could be associated with a higher risk of certain diseases, with secretory milk protecting against diarrhea caused by *Campylobacter*, and enterotoxigenic *E. coli* [51,53].

Infants fed by nonsecretor mothers showed a delay in the establishment of *Bifidobacterium* species [53]. Lacto-N-neotetraose (LNnT) and Lacto-N-tetraose (LNT) are coregulated with FUT2-dependent 2FL concentration. Interestingly, LNnT showed a positive and LNT a negative relation with 2FL [56].

Selective effects of HMO in the gut microbiome

Infant formulas are being increasingly supplemented with prebiotics, in part to replicate the bifidogenic effect of breast milk [16,57]. Commonly used prebiotics include fructooligosaccharides (FOS) and galactooligosaccharides (GOS), which are added in a ratio 9:1. FOS are a mixture of linear polymers of fructose in β2-1 linkage, and a degree of polymerization (DP) of 3–6 [58], obtained from chicory and onions. Instead, GOS are enzymatically produced by transglycosylation reactions using lactose as substrate, resulting in oligosaccharides with DP of 3–15 [59]. FOS and GOS are linear polymers, much simpler structures compared to HMO.

HMO can be industrially produced by de novo synthesis using microbial, enzymatic, or chemical methods [60]. Adding HMO to infant formula fills a gap, and brings them one step closer to matching breast milk complexity [61]. Recently, LNnT has been synthetized chemically (>97% purity), and evaluated for safety in rats [62]. In addition, 2FL (>99% purity) has been chemically synthetized from fucose and lactose [60]. Currently, the oligosaccharides 2FL and LNnT are commercially available [61].

Formula supplemented with 2′FL (1 g/L) and LNnT (0.5 g/L) is characterized by an intestinal microbiome dominated by *Bifidobacterium* species, resembling breastfed infants [15,61]. Stool concentration of short-chain fatty acids (SCFA) such as acetate, propionate, and butyrate, were similar among the different forms of infant feeding [61]. The European Union considers these two HMO as novel foods (Commission Implemented Regulation 2017/2470). In addition, the FDA has approved three HMO in the category Generally Regarded as Safe (GRAS no 650), including HMO 3-sialyllactose.

HMO promote the growth of *Bifidobacterium* species. This genus includes saccharolytic species that are particularly dominant in the infant gut microbiome, but not significantly in the adult. *B. breve, B. bifidum, and B. longum* subsp. *infantis (B. infantis),* are commonly found in the infant gut [63]. Among these, *B. infantis* is able to ferment all types of HMO, including fucosylated and sialylated molecules. Strategies for *B. infantis* consumption are based on the bacterial import of intact HMO inside the bacterial cell, mediated by several Solute Binding Proteins (SBPs) [16], in addition to a complete array of glycolytic enzymes [64]. *B. bifidum* relies on a set of diverse membrane-associated extracellular glycosyl hydrolase families (GHs), with similar enzymatic affinities for HMO [65].

Twelve vaginally born infants were also breastfed by secretor mothers. An increased abundance of *Bifidobacterium* followed. Genomic analysis of bifidobacterial isolates indicated that not all of them were capable of utilizing HMO efficiently, indicating that HMO utilization is rather strain-dependent [66]. This observation has also been obtained in in vitro studies [67,68].

Predictive models and simulation of the infant gut microbiome

A few groups have optimized and validated in vitro fermentation models of the infant intestinal microbiota, using bioreactors [69]. Exploration using these systems is almost unlimited, allowing studies on microbial metabolism of drugs and xenobiotics, and the impact of nutrients on the gut microbiome. A simple approach is batch fermentation, where a fecal suspension is used as inoculum and added to a closed bioreactor. This model is especially useful to test fermentation of novel poly- and oligosaccharides, and the concomitant production of SCFA [70]. A low carbohydrate medium with autoclaved fecal supernatants leads to stable coculturing of a complex microbiota [71]. Unfortunately, batch bioreactors do not allow the control of pH, and accumulation of metabolites can exert toxic activities on certain microbes [72]. The effect of *Salmonella* infection in the child gut [73], the impact of antibiotic treatment on gut microbes [74], and iron bioavailability [75] are some of the questions addressed so far.

Continuous bioreactors are more suitable and representative of the large intestine, as retention time reflects intestinal transit. Appropriate addition of nutrients and adjustment of environmental conditions leads to stable mixed bacterial cultures [76]. More complex systems allow selective absorption of water, salts, and SCFA from the bioreactor, in addition to a mucosal layer for biofilm formation [77,78].

In vitro systems use complex and sometimes undefined fecal samples in the bioreactors, with significant loss of reproducibility. We recently used a continuous bioreactor inoculated with a simple microbial community in order to simulate how infant gut microbes adapt to dietary changes [79]. We simulated the transition of a microbiome receiving infant formula supplemented with fructooligosaccharides (FOS), to breast milk, by adding 2-fucosyllactose as carbon source. Since FOS are more frequently utilized by gut microbes, the system reached a higher cell density on this substrate, compared to 2FL. Microbial abundance in the bioreactor matched the ability of the microorganisms to consume FOS (*Lactobacillus acidophilus*), or 2FL (*Bifidobacterium infantis*) more quickly.

SCFA production also correlated with the metabolism of the dominant gut microbe, on each phase. During the

FOS phase, carbohydrate left in the supernatants was negligible, indicating active consumption. After the switch, 2FL initially accumulated in the bioreactor, and only after a defined amount of time, its utilization was higher than accumulation. After validation, these systems could be used to study other dietary interventions, or the impact of antibiotics on specific gut microbes, avoiding the ethical implications of clinical studies.

Simple, pairwise coculture experiments between gut microbes, are sufficient to predict the composition of larger gut microbial communities [80]. This indicates that cross-feeding of intermediate metabolites, or macromolecular degradation products, are a strong force shaping gut microbiome composition, and therefore, the chemical nature of the substrates arriving at the colon is important. For example, *B. bifidum* has extracellular activity on several HMO, releasing fucose, sialic acid, and galactose to the medium. These metabolites could be used by *B. breve* species [81,82].

The fucose moiety on 2FL, as well as fucose part of mucin oligosaccharides, could also be cross-fed between infant bifidobacteria and the butyrate-producer *Eubacterium rectale* [83,84].

The metabolic interaction data generated in these experiments could be useful to develop predictive, mechanistic models during the consumption of certain dietary substrates. Kinetic models can be used to investigate the temporal changes in community structure, and the tools from dynamic systems theory can be used to analyze system properties, including stability and parameter sensitivity [85].

The Lotka-Volterra equation determines an interaction coefficient between two microorganisms. This variable could be obtained either experimentally from cocultures, or estimated from co-occurrence networks from large-scale microbiome data. In a gut microbiome synthetic community of 12 microorganisms, pairwise microbial interactions based on metabolite cross-feeding were the major factor explaining community abundance and structure, contributing to system stability [20]. A mechanistic model used FOS, as substrate. This approach incorporated metabolic cross-feeding interaction terms to classic microbial growth laws [86]. Mono and coculture experimental data, concerning microbial abundance, prebiotic consumption, and acid production, were used to calibrate and later validate a mathematical model, based on ordinary differential equations of microbial growth. In general, there was good agreement between experimental coculture results and modeled data, indicating that the behavior of the microbial system could be explained by these equations [86].

References

[1] Korpela K, Costea P, Coelho LP, Kandels-Lewis S, Willemsen G, Boomsma DI, et al. Selective maternal seeding and environment shape the human gut microbiome. Genome Res. 2018;28:561–8.

[2] Stiemsma LT, Turvey SE. Asthma and the microbiome : defining the critical window in early life. Allergy Asthma Clin. Immunol. 2017;13:1–9.

[3] Iizumi T, Battaglia T, Ruiz V, Perez GIP. Gut microbiome and antibiotics. Arch. Med. Res. 2017;48:727–34.

[4] Kozyrskyj AL, Ernst P, Becker AB. Increased risk of childhood asthma from antibiotic use in early life *. Chest. 2007;131:1753–9.

[5] Mikkelsen KH, Knop FK, Frost M, Hallas J, Pottegard A. Use of antibiotics and risk of type 2 diabetes: a population-based case-control study. J. Clin. Endocrinol. Metab. 2015;100:3633–40.

[6] Metsälä J, Lundqvist A, Virta LJ, Kaila M, Gissler M, Virtanen SM. Mother's and offspring's use of antibiotics and infant allergy to cow's milk. Epidemiology. 2013;24:303–9.

[7] Tamburini S, Shen N, Wu HC, Clemente JC. The microbiome in early life : implications for health outcomes. Nat. Publ. Gr. 2016;22:713–22.

[8] Aagaard K, Riehle K, Ma J, Segata N, Mistretta TA, Coarfa C, et al. A metagenomic approach to characterization of the vaginal microbiome signature in pregnancy. PLoS One. 2012;7:1–15.

[9] Bäckhed F, Roswall J, Peng Y, Feng Q, Jia H, Kovatcheva-Datchary P, et al. Dynamics and stabilization of the human gut microbiome during the first year of life. Cell Host Microbe. 2015;17:690–703.

[10] La Rosa PS, Warner BB, Zhou Y, Weinstock GM, Sodergren E, Hall-Moore CM, et al. Patterned progression of bacterial populations in the premature infant gut. Proc. Natl. Acad. Sci. 2014;111:12522–7.

[11] Dominguez-Bello MG, De Jesus-Laboy KM, Shen N, Cox LM, Amir A, Gonzalez A, et al. Partial restoration of the microbiota of cesarean-born infants via vaginal microbial transfer. Nat. Med. 2016;22:250–3.

[12] Dreyer JL, Liebl AL. Early colonization of the gut microbiome and its relationship with obesity. Hum. Microbiome J. 2018;10:1–5.

[13] Heintz C, Mair W. You are what you host: microbiome modulation of the aging process. Cell. 2014;156:408–11.

[14] Tanaka M, Nakayama J. Development of the gut microbiota in infancy and its impact on health in later life. Allergol. Int. 2017;66:515–22.

[15] Vandenplas Y, Berger B, Carnielli VP, Ksiazyk J, Lagström H, Luna MS, et al. Human milk oligosaccharides: 2'-fucosyllactose (2'-FL) and lacto-n-neotetraose (LNnT) in infant formula. Nutrients. 2018;10:1–12.

[16] Thomson P, Medina DA, Garrido D. Human milk oligosaccharides and infant gut bifidobacteria: molecular strategies for their utilization. Food Microbiol. 2017;75:37–46.

[17] Horta BL, Loret De Mola C, Victora CG. Long-term consequences of breastfeeding on cholesterol, obesity, systolic blood pressure and type 2 diabetes: a systematic review and meta-analysis. Acta Paediatr. Int. J. Paediatr. 2015;104:30–7.

[18] Zmora N, Suez J, Elinav E. You are what you eat: diet, health and the gut microbiota. Nat. Rev. Gastroenterol. Hepatol. 2019;16:35–56.

[19] Kettle H, Louis P, Holtrop G, Duncan SH, Flint HJ. Modelling the emergent dynamics and major metabolites of the human colonic microbiota. Environ. Microbiol. 2015;17:1615–30.

[20] Venturelli OS, Carr AC, Fisher G, Hsu RH, Lau R, Bowen BP, et al. Deciphering microbial interactions in synthetic human gut microbiome communities. Mol. Syst. Biol. 2018;14:1–19.

[21] Collado MC, Rautava S, Aakko J, Isolauri E, Salminen S. Human gut colonisation may be initiated in utero by distinct microbial communities in the placenta and amniotic fluid. Nat. Publ. Gr. 2016;6:1–13.

[22] Underwood MA, Sohn K. The microbiota of the extremely preterm infant. Clin. Perinatol. 2017;44:407–27.

[23] Jiménez E, Fernández L, Marín ML, Martín R, Odriozola JM, Nueno-Palop C, et al. Isolation of commensal bacteria from umbilical cord blood of healthy neonates born by cesarean section. Curr. Microbiol. 2005;51:270–4.

[24] Perez-Muñoz ME, Arrieta MC, Ramer-Tait AE, Walter J. A critical assessment of the "sterile womb" and "in utero colonization" hypotheses: implications for research on the pioneer infant microbiome. Microbiome. 2017;5:1–19.

[25] Ferretti P, Pasolli E, Tett A, Asnicar F, Gorfer V, Fedi S, et al. Mother-to-Infant microbial transmission from different body sites shapes the developing infant gut microbiome. Cell Host Microbe. 2018;24:133–45.

[26] Yassour M, Jason E, Hogstrom LJ, Arthur TD, Tripathi S, Siljander H, et al. Strain-level analysis of mother-to-child bacterial transmission during the first few months of life. Cell Host Microbe. 2018;24:146–54.

[27] Karlström A, Lindgren H, Hildingsson I. Maternal and infant outcome after caesarean section without recorded medical indication: findings from a Swedish case-control study. BJOG An Int. J. Obstet. Gynaecol. 2013;120:479–86.

[28] Moore HC, De Klerk N, Holt P, Richmond PC, Lehmann D. Hospitalisation for bronchiolitis in infants is more common after elective caesarean delivery. Arch. Dis. Child. 2012;97:410–4.

[29] Amenyogbe N, Kollmann TR, Ben-Othman R. Early-life host–microbiome interphase: the key frontier for immune development. Front. Pediatr. 2017;5:1–12.

[30] Rutayisire E, Huang K, Liu Y, Tao F. The mode of delivery affects the diversity and colonization pattern of the gut microbiota during the first year of infants' life: a systematic review. BMC Gastroenterol. 2016;16:1–12.

[31] Stinson LF, Payne MS, Keelan JA. Planting the seed: origins, composition, and postnatal health significance of the fetal gastrointestinal microbiota. Crit. Rev. Microbiol. 2017;43:352–69.

[32] Korpela K, Salonen A, Virta LJ, Kekkonen RA, Forslund K, Bork P, et al. Intestinal microbiome is related to lifetime antibiotic use in Finnish pre-school children. Nat. Commun. 2016;7:1–8.

[33] Dominguez-Bello MG, De Jesus-Laboy KM, Shen N, Cox LM, Amir A, Gonzalez A, et al. Same exposure but two radically different responses to antibiotics: resilience of the salivary microbiome versus long-term microbial shifts in feces. MBio. 2015;6:1–11.

[34] Langdon A, Crook N, Dantas G. The effects of antibiotics on the microbiome throughout development and alternative approaches for therapeutic modulation. Genome Med. 2016;8:1–16.

[35] Arnold CJ, Ericson J, Cho N, Tian J, Wilson S, Chu VH, et al. Cefepime and ceftazidime safety in hospitalized infants. Pediatr. Infect. Dis. J. 2015;34:964–8.

[36] Vaz LE, Kleinman KP, Raebel MA, Nordin JD, Lakoma MD, Dutta-Linn MM, et al. Recent trends in outpatient antibiotic use in children. Pediatrics. 2014;133:375–85.

[37] Brunser O, Gotteland M, Cruchet S, Figueroa G, Garrido D, Steenhout P. Effect of a milk formula with prebiotics on the intestinal microbiota of infants after an antibiotic treatment. Pediatr. Res. 2006;59:451–6.

[38] Duranti S, Lugli GA, Mancabelli L, Turroni F, Milani C, Mangifesta M, et al. Prevalence of antibiotic resistance genes among human gut-derived bifidobacteria. Appl. Environ. Microbiol. 2017;83:1–14.

[39] Tanaka S, Kobayashi T, Songjinda P, Tateyama A, Tsubouchi M, Kiyohara C, et al. Influence of antibiotic exposure in the early postnatal period on the development of intestinal microbiota. FEMS Immunol. Med. Microbiol. 2009;56:80–7.

[40] Bokulich NA, Chung J, Battaglia T, Henderson N, Jay M, Li H, et al. Supplementary materials for antibiotics, birth mode, and diet shape microbiome maturation during early life. Sci. Transl. Med. 2016;8:343–82.

[41] Rémy B, Mion S, Plener L, Elias M, Chabrière E, Daudé D. Interference in bacterial quorum sensing: a biopharmaceutical perspective. Front. Pharmacol. 2018;6:1–17.

[42] Piewngam P, Zheng Y, Nguyen TH, Dickey SW, Joo HS, Villaruz AE, et al. Pathogen elimination by probiotic *Bacillus* via signalling interference. Nature. 2018;562:532–7.

[43] Yosef I, Manor M, Kiro R, Qimron U. Temperate and lytic bacteriophages programmed to sensitize and kill antibiotic-resistant bacteria. J. Reprod. Med. 2015;54:661–8.

[44] Mathew JL. Effect of maternal antibiotics on breast feeding infants. Postgrad. Med. 2004;80:196–200.

[45] Nogacka AM, Salazar N, Arboleya S, Suárez M, Fernández N, Solís G, et al. Early microbiota , antibiotics and health. Cell. Mol. Life Sci. 2017;75:83–91.

[46] Smilowitz JT, O'Sullivan A, Barile D, German JB, Lonnerdal B, Slupsky CM. The human milk metabolome reveals diverse oligosaccharide profiles. J. Nutr. 2013;143:1709–18.

[47] Hennet T, Borsig L. Breastfed at tiffany's. Trends Biochem. Sci. 2016;41:508–18.

[48] Morrow AL, Ruiz-Palacios GM, Jiang X, Newburg DS. Human-milk glycans that inhibit pathogen binding protect breast-feeding infants against infectious diarrhea. J. Nutr. 2005;135:1304–7.

[49] Brandtzaeg P. The mucosal immune system and its integration with the mammary glands. J. Pediatr. 2010;156:S8–15.

[50] Garrido D, Kim JH, German JB, Raybould HE, Mills DA. Oligosaccharide binding proteins from *Bifidobacterium longum* subsp. infantis reveal a preference for host glycans. PLoS One. 2011;6:1–13.

[51] Smilowitz JT, Lebrilla CB, Mills DA, German JB, Freeman SL. Breast milk oligosaccharides: structure-function relationships in the neonate. Annu. Rev. Nutr. 2014;34:143–69.

[52] Andreas NJ, Hyde MJ, Gomez-Romero M, Lopez-Gonzalvez MA, Villaseñor A, Wijeyesekera A, et al. Multiplatform characterization of dynamic changes in breast milk during lactation. Electrophoresis. 2015;36:2269–85.

[53] Lewis ZT, Totten SM, Smilowitz JT, Popovic M, Parker E, Lemay DG, et al. Maternal fucosyltransferase 2 status affects the gut bifidobacterial communities of breastfed infants. Microbiome. 2015;3:15–7.

[54] Kunz C, Meyer C, Collado MC, Geiger L, García-Mantrana I, Bertua-Ríos B, et al. Influence of gestational age, secretor, and Lewis blood group status on the oligosaccharide content of human milk. J. Pediatr. Gastroenterol. Nutr. 2017;64:789–98.

[55] Totten SM, Zivkovic AM, Wu S, Ngyuen U, Freeman SL, Ruhaak LR, et al. Comprehensive profiles of human milk oligosaccharides yield highly sensitive and specific markers for determining secretor status in lactating mothers. J. Proteome Res. 2012;11:6124–33.

[56] Sprenger N, Lee LY, De Castro CA, Steenhout P, Thakkar SK. Longitudinal change of selected human milk oligosaccharides and association to infants' growth, an observatory, single center, longitudinal cohort study. PLoS One. 2017;12:1–15.

[57] Martin CR, Ling P-R, Blackburn GL. Review of infant feeding: key features of breast milk and infant formula. Nutrients. 2016;8:1–11.

[58] Roberfroid M, Gibson GR, Hoyles L, McCartney AL, Rastall R, Rowland I, et al. Prebiotic effects: metabolic and health benefits. Br. J. Nutr. 2010;104:S1–63.

[59] Barboza M, Sela DA, Pirim C, LoCascio RG, Freeman SL, German JB, et al. Glycoprofiling bifidobacterial consumption of galacto-oligosaccharides by mass spectrometry reveals strain-specific, preferential consumption of glycans. Appl. Environ. Microbiol. 2009;75:7319–25.

[60] Coulet M, Phothirath P, Allais L, Schilter B. Pre-clinical safety evaluation of the synthetic human milk, nature-identical, oligosaccharide 2'-O-Fucosyllactose (2'FL). Regul. Toxicol. Pharmacol. 2014;68:59–69.

[61] Donovan SM, Comstock SS. The immune benefit of breastfeeding has been attributed in part to the diverse bioactive components in human milk human milk oligosaccharides influence neonatal mucosal and systemic immunity. Nutr. Metab. 2017;69:42–51.

[62] Coulet M, Phothirath P, Constable A, Marsden E, Schilter B. Pre-clinical safety assessment of the synthetic human milk, nature-identical, oligosaccharide Lacto-N-neotetraose (LNnT). Food Chem. Toxicol. 2013;62:528–37.

[63] Medina DA, Pinto F, Ovalle A, Thomson P, Garrido D. Prebiotics mediate microbial interactions in a consortium of the infant gut microbiome. Int. J. Mol. Sci. 2017;18:1–16.

[64] Garrido D, Dallas DC, Mills DA. Consumption of human milk glycoconjugates by infant-associated *Bifidobacteria*: mechanisms and implications. An. Microbiol. 2013;159:649–64.

[65] Kitaoka M. Bifidobacterial enzymes involved in the metabolism of human milk oligosaccharides. Adv Nutr. 2012;3:422S–9S.

[66] Matsuki T, Yahagi K, Mori H, Matsumoto H, Hara T, Tajima S, et al. A key genetic factor for fucosyllactose utilization affects infant gut microbiota development. Nat. Commun. 2016;7:1–12.

[67] Ruiz-Moyano S, Totten SM, Garrido DA, Smilowitz JT, Bruce German J, Lebrilla CB, et al. Variation in consumption of human milk oligosaccharides by infant gut-associated strains of *Bifidobacterium breve*. Appl. Environ. Microbiol. 2013;79:6040–9.

[68] Garrido D, Ruiz-Moyano S, Kirmiz N, Davis JC, Totten SM, Lemay DG, et al. A novel gene cluster allows preferential utilization of fucosylated milk oligosaccharides in *Bifidobacterium longum* subsp. longum SC596. Sci. Rep. 2016;6:1–18.

[69] Payne AN, Chassard C, Banz Y, Lacroix C. The composition and metabolic activity of child gut microbiota demonstrate differential adaptation to varied nutrient loads in an in vitro model of colonic fermentation. FEMS Microbiol. Ecol. 2012;80:608–23.

[70] Pompei A, Cordisco L, Raimondi S, Amaretti A, Pagnoni UM, Matteuzzi D, et al. In vitro comparison of the prebiotic effects of two inulin-type fructans. Anaerobe. 2008;14:280–6.

[71] Kim BS, Kim JN, Cerniglia CE. In vitro culture conditions for maintaining a complex population of human gastrointestinal tract microbiota. J. Biomed. Biotechnol. 2011;2011:15–7.

[72] Gumienna M, Lasik M, Czarnecki Z. Bioconversion of grape and chokeberry wine polyphenols during simulated gastrointestinal in vitro digestion. Int. J. Food Sci. Nutr. 2011;62:226–33.

[73] Le Blay G, Rytka J, Zihler A, Lacroix C. New in vitro colonic fermentation model for Salmonella infection in the child gut. FEMS Microbiol. Ecol. 2009;67:198–207.

[74] Maccaferri S, Vitali B, Klinder A, Kolida S, Ndagijimana M, Laghi L, et al. Rifaximin modulates the colonic microbiota of patients with Crohn's disease: an in vitro approach using a continuous culture colonic model system. J. Antimicrob. Chemother. 2010;65:2556–65.

[75] Dostal A, Fehlbaum S, Chassard C, Zimmermann MB, Lacroix C. Low iron availability in continuous in vitro colonic fermentations induces strong dysbiosis of the child gut microbial consortium and a decrease in main metabolites. FEMS Microbiol. Ecol. 2013;83:161–75.

[76] Duncan SH, Louis P, Thomson JM, Flint HJ. The role of pH in determining the species composition of the human colonic microbiota. Environ. Microbiol. 2009;11:2112–22.

[77] Van Den Abbeele P, Marzorati M, Derde M, De Weirdt R, Joan V, Possemiers S, et al. Arabinoxylans, inulin and *Lactobacillus reuteri* 1063 repress the adherent-invasive *Escherichia coli* from mucus in a mucosa-comprising gut model. NPJ Biofilms Microbiomes. 2016;2:1–8.

[78] De Paepe K, Verspreet J, Verbeke K, Raes J, Courtin CM, Van de Wiele T. Introducing insoluble wheat bran as a gut microbiota niche in an in vitro dynamic gut model stimulates propionate and butyrate production and induces colon region specific shifts in the luminal and mucosal microbial community. Environ. Microbiol. 2018;20:3406–26.

[79] Medina DA, Pinto F, Ortuzar MV, Garrido D. Simulation and modeling of dietary changes in the infant gut microbiome. FEMS Microbiol. Ecol. 2018;94:1–11.

[80] D'hoe K, Vet S, Faust K, Moens F, Falony G, Gonze D, et al. Integrated culturing, modeling and transcriptomics uncovers complex interactions and emergent behavior in a three-species synthetic gut community. Elife. 2018;7:1–30.

[81] Egan M, O'Connell Motherway M, Kilcoyne M, Kane M, Joshi L, Ventura M, et al. Cross-feeding by *Bifidobacterium breve* UCC2003 during co-cultivation with *Bifidobacterium bifidum* PRL2010 in a mucin-based medium. BMC Microbiol. 2014;14:1–14.

[82] Centanni M, Ferguson SA, Sims IM, Biswas A, Tannock GW. *Bifidobacterium bifidum* ATCC 15696 and *Bifidobacterium breve* 24b metabolic interaction based on 2′-O-fucosyl-lactose studied in steady-state cultures in a Freter-style chemostat [Internet] Appl. Environ. Microbiol. 2019;85:1–44.

[83] Schwab C, Ruscheweyh HJ, Bunesova V, Pham VT, Beerenwinkel N, Lacroix C. Trophic interactions of infant *Bifidobacteria* and *Eubacterium hallii* during L-fucose and fucosyllactose degradation. Front. Microbiol. 2017;8:1–14.

[84] Bunesova V, Lacroix C, Schwab C. Mucin cross-feeding of infant *Bifidobacteria* and *Eubacterium hallii*. Microb. Ecol. 2018;75:228–38.

[85] Aström KJ, Murray RM. Feedback systems: an introduction for scientists and engineers. September 2006. p. 1–9.

[86] Pinto F, Medina DA, Pérez-Correa JR, Garrido D. Modeling metabolic interactions in a consortium of the infant gut microbiome. Front. Microbiol. 2017;8:1–12.

Chapter 5

Paraprobiotics

Rao Shripada[1,3], Athalye-Jape Gayatri[1,2,3] and Patole Sanjay[2,3]

[1]Neonatal Directorate, Perth Children's Hospital, Perth, WA, Australia; [2]Neonatal Directorate, King Edward Memorial Hospital for Women, Perth, WA, Australia; [3]School of Medicine, University of Western Australia, Perth, WA, Australia

Introduction

Probiotics are live organisms that when administered, confer health benefits to the host [1]. Studies have shown that probiotics have the potential to improve the outcomes of patients with ulcerative colitis [2,3], *Clostridium difficile*-associated diarrhea [4], antibiotic-associated diarrhea [5], traveler's diarrhea [6] and irritable bowel syndrome [7]. Probiotics have also been shown to be beneficial in the prevention of necrotizing enterocolitis (NEC) [8,9] and late-onset sepsis in preterm infants [10], ventilator-associated pneumonia [11], postoperative infections [12] and many other conditions [13].

In spite of the encouraging results, there is hesitancy among healthcare professionals to use probiotics, considering the occasional risk of serious infections due to the administered probiotic organism [14–22]. The other concerns are altered long-term immune responses and the possibility of development and spread of antibiotic resistance. Paraprobiotics on the other hand are inactivated organisms and hence unlikely to be harmful, while retaining the beneficial effects of live probiotics. They are incapable of growing in vitro, which can be verified by plating in adequate culture media [23–25]. Without the need for maintaining cold chain to preserve viability of the probiotic organisms, the costs associated with paraprobiotics are also expected to be lower.

Nomenclature

Paraprobiotics are also known as inactivated probiotics, nonviable probiotics, killed probiotics, ghost probiotics, modified probiotics, sterilized probiotics, and postbiotics. Taverniti et al. from Italy first proposed the term "paraprobiotic" for these agents [25]. The prefix "para" (from the ancient Greek, παρά) was chosen because of its meaning of "alongside of" or "atypical," which can simultaneously indicate similarity to and the difference from the conventional definition of a probiotic [25].

Methods of inactivation

Inactivation could be achieved using physical or chemical strategies, including heat treatment, γ or UV rays, chemical or mechanical disruption, pressure, lyophilization, or acid deactivation [25]. The excellent review by de Almada et al., covers the various methods for inactivation of probiotics [23] (Table 1).

Randomized controlled trials (RCTs)

Till date, nearly 60 RCTs have been conducted evaluating the efficacy and safety of paraprobiotics in various conditions in humans (Table 2).

Atopic dermatitis and allergic diseases

Harima-Mizusawa et al. examined whether citrus juice supplemented with heat-killed *Lactobacillus* (*L.*) *plantarum* YIT 0132 (LP0132) could alleviate the symptoms of perennial allergic rhinitis [26]. Patients consumed LP0132-fermented juice ($n = 17$) or unfermented citrus juice (placebo: $n = 16$) once a day for 8 weeks. The participants in the LP0132 group showed a significant reduction in the nasal symptom score and stuffy nose score. Significant attenuation of type 2 helper T cells (Th2 cells)/helper T cells, serum total immunoglobulin E (IgE), and eosinophil cationic protein (ECP), and augmentation of type 1 helper T cells (Th1 cells)/Th2 cells at 8 weeks of intervention was reported as well.

In the study by Inoue et al., 49 atopic dermatitis patients aged $\geq$16 years used heat-killed *L. acidophilus* L-92 [27]. Skin lesions were assessed using the SCORAD index. The L-92 group had significantly lower SCORAD scores

Precision Medicine for Investigators, Practitioners and Providers. https://doi.org/10.1016/B978-0-12-819178-1.00005-8

TABLE 1 Methods of inactivation of probiotics [a].

Method	Key mechanisms	Key features
1.Thermal treatment		
a. Pasteurization: <100°C **b.** Sterilization: ≥100°C	Damage to cell membrane, protein denaturation, solute losses, enzyme inactivation, reduction of intracellular pH, conformational changes in ribosomes and nucleoids, disintegration of RNA filaments	Range of temperature and duration in various studies: 60–121°C for 15–60 min)
2.Nonthermal treatments		
a. Ionizing radiation	Damage to nucleic acids	Gamma irradiation; Cobalt 60 source for 20 h at 8.05 Gy/minute
b. Ultraviolet (UV) rays	Denaturation of proteins, formation of DNA photoproducts	UV details not specified, UV for 30 min, UV for 20 min, UV for 5 min under 39 W germicidal lamp
c. High pressure	Damage to cell membranes	400 or 600 MPa at 37°C for 10 min
d. Sonication	Rupture of cell wall, damage to cell membrane, DNA damage, activation of intracellular esterases, inhibition of cell metabolism	Low-frequency ultrasound (20 kHz) generated by submerging 10 mm diameter probe in 30 mL probiotic cell suspension. Sonic power: 300 W/cm^2. Ultrasonic power and irradiation time: 60 W/cm^2 and 0–20 min. Suspension temperature maintained at 20 ± 1°C in ice water bath to prevent lethal thermal effect, Probiotic culture centrifuged at 13,000 rpm for 10 min and then sonicated for 5 min.
e. Pulsed Electric Field (PEF) technology	Electroporation (cell membrane damage)	Pulsed high voltage (40 kV/cm); applied to probiotic bacteria containing food products placed between two electrodes
f. Ohmic or Joule or resistive heating	Thermal induced damage due to alternating electric current and electroporation.	Based on Joule's law
g. Supercritical CO_2 technology	Induction of intracellular acidosis, cell membrane damage and depletion of vital constituents, enzyme inactivation, CO2 and HCO3 induced direct damage, intracellular electrolyte imbalance	Use of CO_2 at temperatures and pressures above their critical point values (Tc:31.1°C; Pc: 7.38 MPa)
h. Dehydration	Damage to cell membrane, conformational changes to proteins, nucleic acids and ribosomes	Lyophilization (Freeze drying): Water in the material to be dried is frozen first and then sublimed under vacuum conditions. Spray drying: Formation of powder by using an atomizer for pulverization of liquids and formation of small droplets in contact with hot air.
i. Modification of pH	Damage to cell membranes, chemical denaturation of DNA and ATP and enzyme inactivation	Lowering pH of the medium to 3.0–4.to was effective in inactivating probiotics; duration of exposure to low pH not specified
3.Combination treatments		
UV-C light and heat treatment	Denaturation of proteins, formation of DNA photoproducts, cell membrane damage, enzymatic denaturation	UV-C light (0-10.6 kJ/m^2) with heating at 50°C

[a] *Based on de Almada et al. [23].*

TABLE 2 **Paraprobiotics in various conditions and populations**

1. Acute diarrhea (Prevention in children, treatment in adults and children)
2. Antibiotic associated diarrhea (Prevention in adults and children)
3. Atopic dermatitis and allergic diseases (Prevention and treatment in adults, children and neonates)
4. Chronic diarrhea (Treatment in adults)
5. Constipation (Treatment in female adults)
6. Dental diseases (Treatment in adults and prevention in children)
7. General Immunity (Treatment in healthy and elderly adults)
8. Helicobacter pylori infection (Treatment in children and adults)
9. Infantile colic (Treatment in infants)
10. Intestinal permeability (Prevention in healthy adults)
11. Irritable bowel syndrome (Treatment in adults)
12. Lactose malabsorption (Prevention in children)
13. Malignancies (As an adjuvant treatment in adults)
14. Multiple organ dysfunction (Prevention in adult intensive care unit patients)
15. Necrotizing enterocolitis (Prevention in preterm infants)
16. Neonatal sepsis (Prevention in preterm infants)
17. Obesity (Treatment in adults)
18. Ocular diseases (Prevention in healthy adults)
19. Pancreatitis (Treatment in adults)
20. Prevention of postoperative infections (Adults undergoing major abdominal surgery)
21. Sleep disturbances (Treatment in adults)
22. Traveler's diarrhea (Prevention in adult travelers)
23. Viral Infections (Prevention in adults and children)

($p = 0.002$), decreased ratios of change for eosinophil count ($p = 0.03$), and increased ratios of change for serum TGF-β ($p = 0.03$).

Morisset et al. evaluated the impact of an infant formula containing heat-killed *Bifidobacterium* (*B.*) *breve* C50 and *Streptococcus thermophilus* 065 (HKBBST) in 129 infants at high risk of atopy [28]. Infants were given HKBBST or a standard infant formula since birth until one year of age, and were followed at 4, 12, and 24 months after birth. The use of HKBBST milk did not alter the proportion of cow's milk protein allergy (CMPA), but decreased the proportion of positive skin prick test to cow's milk (1.7% vs. 12.5%, $p = 0.03$), and the incidence of digestive (395% vs. 63%, $p = 0.01$) and respiratory "potentially allergic adverse events" (PAAE) (7% vs. 21%, $p = 0.03$) at 12 months, and that of respiratory PAAEs at 24 months (13% vs. 35%, $p = 0.01$).

In the investigation of Moroi et al., the clinical effect of a supplementary diet containing heat-killed lactic acid bacterium (LAB) *L. paracasei* K71 (LAB diet), was assessed in 34 adults with atopic dermatitis [29] who were being treated with conventional topical corticosteroids and tacrolimus. The skin severity scores decreased significantly from baseline at week 8 ($p < 0.05$) and at week 12 ($p < 0.01$). The consumption of topical therapeutics in the placebo group was 1.9-times greater compared to the LAB group, but the difference was not statistically significant.

Gotoh et al. conducted an RCT to determine whether the administration of heat-killed *L. gasseri* OLL2809 would benefit patients with Japanese cedar pollinosis (JCP) [30]. Participants were randomly allocated to receive oral placebo, or tablets containing 100 mg of *L. gasseri* OLL2809 per day for 8 weeks during the pollen season. The results showed no obvious differences between the groups. Subgroup analysis revealed that the OLL2809 subgroups with CAP-RAST scores of 4 or 5, exhibited improvement in nasal symptoms scores and serum allergy-related items, including Japanese cedar pollen-specific IgE levels [30].

Peng et al. evaluated the efficacy of heat-killed *L. paracasei* 33 (LP33) for treating allergic rhinitis induced by house-dust-mite [31]. Ninety patients were assigned to one of the three treatment groups. Patients in groups A and B received two capsules per day of live or heat-killed LAB (5×10^9 colony forming units (CFU)/capsule), respectively, over a period of 30 days while Group C received a placebo. The overall quality-of-life score was significantly improved in groups A and B [31].

Negative results

Kirjavainen et al. [32] assessed the efficacy of oral supplementation with a viable and heat-inactivated probiotic (*L. rhamnosus GG*), in 35 infants with atopic eczema and CMPA. Treatment with heat-inactivated LGG was associated with adverse gastrointestinal symptoms and diarrhea; hence, the recruitment of patients was stopped early. Atopic eczema and subjective symptoms were significantly alleviated in all the intervention groups. The SCORAD scores decreased from 13 (IQR 4–29) to 8 (0–29) units in the placebo group, from 19 (4–47) to 5 (0–18) units in the viable LGG group, and from 15 (0–29) to 7 (0–26) units in the heat-inactivated LGG group. The decrease in the SCORAD scores within the viable LGG group tended to be greater than within the placebo group. Treatments did not appear to affect the gut microbiota. The authors concluded that atopic eczema and CMPA could be potentially managed by supplementation of infant formulas with viable, but not heat-inactivated LGG [32].

Additional studies with paraprobiotic agents for prevention [33] or treatment of atopic dermatitis, with improvement in symptoms, can be found in the literature [34–36].

Irritable bowel syndrome

Symptoms (abdominal pain, bloating, number of stools per day, and stool consistency), impact on health-related quality of life (HRQOL), and consequences on anal continence, were evaluated in 297 patients with irritable bowel syndrome (IBS) [37] before and after 1 month of probiotic treatment with Lacteol (inactivated *Lactobacillus* LB plus fermented culture medium). The pain scores decreased from 4.46 ± 0.15 to 2.8 ± 0.14 ($p < 0.0001$). Bloating decreased from 4.49 ± 0.18 to 2.5 ± 0.15 ($p < 0.0001$). The HRQOL score, which is inversely correlated with quality of life, decreased from 5.99 ± 0.14 to 3.92 ± 0.16 ($p < 0.0001$). The mean number of stools per week decreased from 17.59 to 12.83 ($p < 0.0001$). Before treatment, 54% of patients had watery stools, and 46% had smooth stools; after completing the treatment, only 18.5% of patients had watery stools, and 34% had normal stools [37].

Guyonnet et al. [38], and Halpern et al. [39] have also reported symptomatic improvement with inactivated probiotics in adult patients with IBS.

Antibiotic-associated diarrhea

Aiming at prevention of antibiotic-associated diarrhea (AAD) [40], 87 hospitalized patients were randomized to receive a commercial probiotic milk drink fermented with LGG, *L. acidophilus* La-5 and *Bifidobacterium* Bb-12, or a milk drink with heat-killed bacteria as a placebo for 14 days. Sixty-three patients completed the study; two patients (5.9%) in the live probiotic group and eight (27.6%) in the heat-killed lactobacillus group developed AAD ($p = 0.035$). These results suggest that fermented milk with multistrain live probiotics, not the heat-killed supplement, may prevent AAD in hospitalized adult patients.

Merenstein et al. [41] reported no difference in the incidence of AAD (21.9% vs. 18%) in 125 children randomized to prophylactic modified or living probiotic.

Infections

Cow's milk (group A) or rice (group B) fermented with heat-inactivated *L. paracasei* CBA L74, or placebo (group C), were compared in children for 3 months during the winter season. The proportion of those who experienced at least one common infectious disease (CID) was lower in group A (51.8%) and B (65.9%), versus group C (80.3%). This included upper respiratory tract infections and acute gastroenteritis. A net increase of all fecal biomarkers of innate and acquired immunity was observed for groups A and B. The investigators concluded that dietary supplementation with cow's milk or rice supplemented with heat-killed *L. paracasei* CBA L74 prevents CIDs in children attending daycare, by enhancing their innate and acquired immunity [42]. In a substudy of 20 infants from the original RCT, Berni Canani [43] reported that the heat-killed lactobacillus group had higher fecal butyrate levels [43].

Salazar-Lindo et al. [44] (80 outpatients) and Boulloche et al. [45] (103 inpatients) reported the reduced mean duration of acute diarrhea in children, after treatment with inactivated probiotic. However, Khanna et al. [46] did not observe such a benefit in 98 in-patient children. Sharieff et al. [47] noticed no difference in the prevalence of diarrhea, in 50 children given prophylactic inactivated probiotic compared with placebo.

Experimental human rhinovirus infection (HRV) was induced by intranasal HRV A39 instillation in 59 adults, who had consumed juice enriched with live or heat-inactivated *Lactobacillus* LGG, or control juice in the 3 preceding weeks. There was a tendency toward the lowest HRV loads (log10 copies/mL) in the LGG groups: 6.20 in live, 6.30 in inactivated LGG, and 7.25 in placebo group, $p = 0.57$ [48].

In contrast, Kumpu et al. [49] reported no significant differences in the incidence of rhinovirus infection in 60 adults given prophylactic live (74%) or inactivated LGG (90%), or placebo (90%).

Heat-killed *L. pentosus* b240 was administered to elderly adults [50]. Participants in the low-dose and high-dose b240 groups were given tablets containing 2×10^9 or 2×10^{10} cells, respectively, of heat-killed b240. The cumulative incidence rate of the common cold was 47.3%, 34.8% and 29.0% for the placebo, low-dose and high-dose b240 groups, respectively ($p = 0.012$). General health perception, as determined by the SF-36, increased in the b240 groups in a dose-dependent manner ($p < 0.025$) [50].

Hirose et al. [51] reported significant reduction in respiratory tract infections, in 78 adults with high levels of stress, after administration of inactivated *L. plantarum* L137 versus placebo (48.7% vs. 61.5%, $p < 0.05$).

Paraprobiotics plus culture medium

In a study, heat-killed *L. acidophilus* LB (10 billion CFU), plus 160 mg of spent culture medium was administered to infants with nonrotavirus diarrhea. Shortened recovery time by 1 day was documented. In conjunction with the simultaneously performed in vitro study, it was concluded that heat-killed *L. acidophilus* LB plus its culture medium, antagonizes the C1845-induced increase in paracellular permeability in intestinal Caco-2/TC7 cells, and produces a clinically significant benefit in children with nonrotavirus diarrhea [52].

Mitra et al. reported no significant difference in the median duration of Vibrio cholerae or E-coli-induced diarrhea, in 197 adults given inactivated or live *Streptococcus faecium* SF 68 [53].

One hundred and thirty-seven adult patients with chronic diarrhea were randomly allocated to receive either a 4-week course of Lacteol Fort (heat-killed *L. acidophilus* LB) twice a day or Lacidophilin (living lactobacilli) thrice daily. At the second and fourth week of treatment, mean bowel frequency was significantly lower in the Lacteol group ($p < 0.05$) [54].

As an adjunct to oral rehydration therapy in children (3–24 months) with acute diarrhea [55], the same Lacteol Fort (heat-killed *L. acidophilus* LB) or placebo was offered on admission and every 12-hourly for five doses. After 24 h of treatment, the number of rotavirus-positive children with watery stools was significantly lower ($p = 0.012$) in the *L. acidophilus* LB group, along with the mean duration of diarrhea [55].

Intestinal dysbiosis

Inactivated L. *Acidophilus* was prescribed to 63 children (<4 years), with infective diarrhea or extraintestinal infections. The patients were divided into three groups: the first and the second included children with acute diarrhea; the third included children with extraintestinal infections. The first group received inactivated *L. acidophilus* LB, the second (Control) group did not receive a probiotic, whereas the third group received antibiotic treatment plus inactivated *L. acidophilus*. Statistically significant clinical improvement was noted in the first, inactivated *Lactobacillus* group, while in the third group, antibiotic-induced gut dysbiosis was prevented [56].

In infants with rotavirus diarrhea, IgA enzyme immunoassay antibody responses were higher in infants treated with viable *L. casei* GG, versus inactivated *L. casei* GG. There was a significant difference at convalescence, with rotavirus-specific IgA secreting cells found in 10/12 infants receiving viable, but in only 2/13 infants receiving inactivated *L. casei* GG. The authors concluded that viable *L. casei* GG stimulated rotavirus-specific IgA antibody responses, which played an important role in the prevention of reinfections [57].

On the other hand, Briand et al. [58] reported no difference in the incidence of traveler's diarrhea, in 348 adult travelers given inactivated prophylactic *L. acidophilus* versus placebo.

Immune functions

Elderly participants in two nursing homes were randomized to receive a jelly containing 10 billion heat-killed *L. paracasei* MCC1849 cells (LP group) or placebo, for 6 weeks [59]. Three weeks after starting the supplements, all participants were given an influenza vaccine. There were no significant differences in immune parameters between the groups, including in antibody responses against the vaccination. However, in the subgroup of those ≥85 years of age, the antibody responses to the A/H1N1 and B antigens were improved in the LP group [59].

Heat-killed *Lactobacillus gasseri* TMC0356 or placebo was orally administered to 28 healthy subjects aged 50–70 years for 4 weeks, at a dosage of 1.0×10^9 cfu/day. The number of CD8 (+) T cells significantly increased in the TMC0356 group ($p < 0.05$). The population of CD8 (+) CD28 (+) T cells and the amount of lymphocyte transformation both significantly decreased in the placebo group ($p < 0.05$), but not in the subjects who received TMC0356, indicating enhanced immunity in elderly people [60].

In healthy adults, the influence of heat-killed *L. plantarum* (HK-LP) on immune function and quality of life (QOL) was tested. Among the measured biomarkers, the percentage change in Concanavalin A-induced proliferation ($p = 0.036$) and the Th1:Th2 ratio ($p = 0.002$) in the HK-LP group was greater. The degree of improvement in QOL was higher in the HK-LP group at week 8 ($p = 0.049$), and tended to be higher at week 12 ($p = 0.092$) [34].

Infantile colic

Vandenplas et al. [61] studied the efficacy and safety of APT198K (xyloglucan plus heat-killed *L. reuteri* SGL01 and *B. brevis* SGB01), versus a lactase dietary supplement, for infantile colic. On day 8, the mean duration of crying per episode was significantly shorter in the APT198K group (9.14 ± 5.34 vs. 13.22 ± 5.29 min; $P = 0.014$), and remained so until the end of the observations, on day 11 [61].

Dental diseases

Thirty-nine patients undergoing supportive periodontal therapy (SPT) were randomly assigned to receive a capsule containing 10 mg of heat-killed *L. plantarum* HK L-137 or a placebo, daily for 12 weeks. The bleeding on probing, and number of teeth or sites with probing depth (PD) ≥4 mm were significantly reduced in both groups, while there was significantly greater PD reduction ($p < 0.05$) at site(s) with PD ≥ 4 mm at baseline, in the experimental group at week 12 [62].

Incidence of dental caries was the end point in 245 seven-year-old treated children [63,64]. They were randomized to chewable tablets containing pyridoxine and heat-killed lactic acid bacteria (LAB), or placebo for 16 weeks. A consistent reduction in the incidence of dental caries was observed in the heat killed lactic acid bacteria group. After 2 years of follow up, a 42% reduction in the incidence of dental caries was confirmed [63,64].

Helicobacter pylori infection

Mehling et al. conducted an RCT in 22 *Helicobacter pylori* positive, asymptomatic adults, with nonviable *L. reuteri* DSMZ 17648 (Pylopass). A significant reduction of *H. pylori* was observed in the nonviable *Lactobacillus* group [65]. *H. pylori*-positive patients (n = 120) were randomly assigned in another protocol, to a 7-day triple therapy based on rabeprazole, clarithromycin, and amoxicillin (RCA: n = 60), or to RCA regimen supplemented with a lyophilized and inactivated culture of *L. acidophilus* (RCAL: n = 60). Eradication was successful in 42/60 (70%) of patients in the RCA versus 52/60 (87%) patients in the RCAL group ($p = 0.02$) [66].

Negative results

Gotteland et al. conducted a multicentre RCT in 295 asymptomatic children (6–16 y) who tested positive for *H. pylori* by (13) C-urea breath test (UBT) [67]. Participants were allocated to four groups: cranberry juice/ *L. johnsonii* La1 (CB/La1), placebo juice/La1 (La1), cranberry juice/heat-killed La1 (CB), and placebo juice/ heat-killed La1 (control). The *H. pylori* eradication rates were 1.5% in the control group versus 14.9%, 16.9%, and 22.9% in the La1, CB, and CB/La1 groups, respectively ($p < 0.01$) [67]. The control group, i.e., heat-killed La1 plus placebo juice, had the lowest clearance rates, suggesting that heat-killed La1 was not beneficial.

A cohort of 326 asymptomatic children (9.7 ± 2.6 y) was screened for *H. pylori* by the (13) C-UBT, and *H. pylori*-colonized children were randomized to receive the following: live La1 or ST11 (groups 1 and 3), heat-killed La1 or ST11 (groups 2 and 4), and placebo (group 5) once a day for 4 weeks. A moderate but significant difference was detected in children receiving live La1 (−7.64 per 1000; 95% confidence interval: −14.23 to −1.03), whereas other groups showed no differences [68].

Preterm infants

Infant formula containing heat-inactivated *B. breve* C50 and *Streptococcus thermophilus* 065[69], was assessed in 58 preterm infants with gestation 30–35 weeks during their hospital stay. Fecal calprotectin was significantly lower in infants fed fermented formula for 2 weeks, and secretory IgA increased with both mother's milk and the fermented formula. The study formula was well tolerated and did not significantly modulate the bacterial colonization, but had benefits on inflammatory and immune markers [69].

In a neonatal unit, 60 infants received LP (*L. acidophilus*), 60 received KP (killed probiotic), and 30 received a placebo on day one of life. Both LP and KP reduced the risk of NEC/necrotizing enterocolitis (absolute risk reduction (ARR): 16%, 15%, respectively), and LOS/ length of stay (ARR: 18%) [70].

Lactose malabsorption

Thirty-nine children were randomized to receive live probiotic and another 40 to killed probiotic. The mean hydrogen levels on BHT/breath hydrogen test decreased from baseline, after administration of live probiotic as well as killed probiotic ($p < 0.001$). The mean hydrogen levels at 120 min was similar between the live and killed probiotic groups ($p = 0.453$) [71].

Surgical conditions

In an RCT by Rayes et al. [72–74], the incidence of bacterial sepsis was compared in 172 patients following major abdominal surgery or liver transplantation. The interventions included: (a) conventional parenteral or enteral nutrition, (b) enteral nutrition with fiber and live *L. plantarum* 299, or (c) enteral nutrition with fiber and heat-inactivated lactobacilli. The incidence of postoperative infections was 31% in the conventional group compared to 4% in the lactobacillus-group, and 13% in the heat inactivated group. In the liver transplant recipients (n = 95), 13% of group-b patients developed infections compared to 48% of group-a, and 34% of group-c patients. The duration of antibiotic therapy was significantly shorter in both live as well as heat inactivated *Lactobacillus*-groups [72–74].

Acute pancreatitis

Kecskés et al. [75] and Olah et al. [76] studied the effect of *L. plantarum* 299 in reducing acute pancreatitis related morbidity and mortality. Patients with acute pancreatitis were randomized to receive a freeze-dried preparation containing 10^9 live *L. plantarum* 299 (N = 22) or heat-inactivated *L. plantarum* 299. Supplementation was continued for a week two-times-a-day. Infected necrosis and abscesses occurred in 1/22 (4.5%) in the live L. plantarum versus 7/23 (30%) in the heat inactivated group ($p = 0.023$). The length of stay was not significantly different in the treatment versus the placebo group (13.7 vs. 21.4 days, respectively) [75].

Malignancies

LC9018 (heat-killed *L. casei* YIT9018) was employed in 228 patients with Stage IIIB cervical cancer undergoing radiotherapy [77,78]. It enhanced tumor regression ($p < 0.1$) by radiation. The combination therapy also prolonged survival and the relapse-free interval ($p < 0.05$), compared to radiation alone. Radiation-induced leukopenia was less severe ($p < 0.05$) in the LC9018 group [77,78].

Ocular diseases

The effects of heat-killed *L. paracasei* KW 3110 in ocular disorders and symptoms of eye fatigue, were investigated among healthy subjects [79]. The Visual Display Terminal (VDT) load-induced reduction of critical flicker frequency, tended to be milder in the *L. paracasei* KW3110 versus the placebo group, during the fourth week. Subgroup analysis showed a significant benefit of *L. paracasei* KW3110 supplementation, in subjects with high-level eye fatigue ($p = 0.020$). The in vitro study found that *L. paracasei* KW3110 suppressed blue light-induced retinal pigment epithelial cell death [79].

Sleep disturbances

Seventeen male volunteers (41–69y) with insomnia (Athens Insomnia Scale (AIS) score≥6) were given placebo or heat-killed *L. brevis* SBC8803 capsules for 10 days. The sleep journals revealed an improvement in "waking" ($p = 0.047$) SBC8803 group, and there was a marginally significant effect on "drowsiness during the following day" ($p = 0.067$). Effects on the EEG delta power value ($\mu V(2)$/min) were revealed by a stratified analysis based on age, AIS, and the Beck Depression Inventory (BDI). Specifically, beneficial effects were found among subjects in their 40s who consumed the SBC8803 capsules ($p = 0.049$) and ($p = 0.045$) [80]. The authors concluded that a beneficial effect on sleep due to consumption of heat-killed *L. brevis* SBC8803 was found in subjects with slightly challenged sleep.

Other conditions

Forty-four female students with a tendency for constipation (20.2 ± 3.3 yr) were asked to consume 30 g test pickles daily for 2 weeks. They were divided into 3 groups: viable-cell group (n = 14, viable LAB), dead-cell group (n = 15, heat sterilized LAB), and placebo groups (n = 15) [81]. Marked enhancement of NK-cell activity and improved bowel symptoms were observed in subjects consuming pickles containing dead LAB cells.

Gotteland et al. [82] evaluated the ability of probiotic LGG to protect the gastrointestinal mucosa against indomethacin-induced alterations of permeability. They carried out four gastrointestinal permeability tests in 16 healthy volunteers: (i) basal; (ii) after indomethacin; (iii) after 5 days of living LGG ingestion before indomethacin administration; (iv) after 5 days of heat-killed LGG ingestion before indomethacin administration. Heat-killed LGG did not modify the indomethacin-induced increase of gastrointestinal permeability, while live bacteria significantly reduced the alteration of gastric but not indomethacin-induced intestinal permeability [82].

Higashikawa et al. [83] reported significant reduction in body mass index, in 62 obese adults treated with inactivated or live *P. pentosaceus* LP 28, or placebo [-0.06 (±0.12) versus 0.07 (±0.1) versus 0.4(±0.15), $p < 0.05$]. Alberda et al. [84] reported no difference in multi-organ dysfunction scale, in 28 adult ICU patients given prophylactic inactivated or living VSL#3 compared to placebo.

Systematic review

Zorzela et al. have reported a systematic review evaluating the efficacy and safety of modified (heat-killed or sonicated) probiotics [85]. Modified probiotics were not significantly more or less effective than the living probiotic in 86% of the preventive trials, and 69% of the treatment trials. Modified probiotic strains were significantly more effective in 15% of the treatment trials [85].

Ongoing trials

Hammerman et al. (NCT02796703) from Israel plan to recruit 450 preterm neonates of very low birth weight (<1500 gm), to receive daily supplementation with a heat-inactivated probiotic or placebo, starting with the initiation of feeds [9]. Supplements will be continued until the infant tolerates enteral feeds of 100 mL/kg/day, or reaches

35 weeks postconceptional age. Biotikid, a probiotic mixture (*L. rhamnosus, L. casei, B. infantis, B. bifidum, B. longum, L, acidophilus*, and *S. thermophilus*) will be heated to 100°C for 10 min. The hypothesis is that prophylaxis with heat-inactivated probiotics will reduce the incidence of NEC in preterm infants. The investigators will also measure intestinal barrier integrity, as reflected by levels of I-FABP in urine and fecal Calprotectin. It is not clear if the trial has begun recruiting participants as yet.

Jape et al. [86] from Western Australia are conducting a pilot RCT ($n = 70$, ACTRN12618000489291p) comparing the efficacy and safety of paraprobiotics in preterm infants with gestation <32 weeks. The intervention arm will receive a heat-inactivated mixture of 3 bifidobacteria strains (*B. breve* M-16V, *B. longum* subsp. *infantis* M-63 and *B. longum* subsp. *longum* BB536). The control arm will receive mixture of the three live bifidobacteria. The supplement will be commenced with enteral feeds and continued till at least 37 weeks corrected gestational age. Primary outcomes include safety and tolerance of the heat-inactivated preparation and fecal calprotectin levels. Secondary outcomes include fecal 16srRNA analysis, fecal cytokine, and short chain fatty acid levels, as well as the assessment of gut barrier integrity and intestinal transit time. Clinical outcomes such as NEC, all-cause mortality, time to full enteral feeds, duration of parenteral nutrition, and hospital stay will be secondary outcomes of interest.

The future of paraprobiotics

Adequately powered RCTs evaluating the efficacy and safety of paraprobiotics are essential to explore this new frontier [87]. The potential benefits of paraprobiotics with regards to probiotic sepsis, antibiotic resistance, better shelf life, and possibly lesser costs, are important, but consideration of their safety must be taken into consideration. Heat-killed probiotics are able to act as biological response modifiers. While moderate stimulation of proinflammatory cytokines could be beneficial in maintaining a good immunological balance and increasing resistance to infections, excessive stimulation is undesirable. Experts have warned that dead probiotic cells have the potential to cause harmful effects, and hence, careful screening and selection of strains as well as safety monitoring is highly essential [24].

The RCTs in this field should evaluate paraprobiotics against a placebo, in conditions where probiotic therapy is not a standard of care. Assuring follow up of participants in such trials is important to assess the long term safety and efficacy of paraprobiotics. Paraprobiotics could also be studied in conditions where probiotics are contraindicated. Last but not least, assessing the cost-benefit ratio of different methods of inactivating probiotics is essential.

References

[1] Hill C, Guarner F, Reid G, et al. Expert consensus document. The international scientific association for probiotics and prebiotics consensus statement on the scope and appropriate use of the term probiotic. Nat. Rev. Gastroenterol. Hepatol 2014;11(8):506–14.

[2] Chibbar R, Dieleman LA. Probiotics in the management of ulcerative colitis. J. Clin. Gastroenterol 2015;49(Suppl. 1):S50–55.

[3] Ganji-Arjenaki M, Rafieian-Kopaei M. Probiotics are a good choice in remission of inflammatory bowel diseases: a meta analysis and systematic review. J. Cell. Physiol 2018;233(3):2091–103.

[4] Goldenberg JZ, Ma SS, Saxton JD, et al. Probiotics for the prevention of *Clostridium* difficile-associated diarrhea in adults and children. Cochrane Database Syst. Rev 2013;(5):Cd006095.

[5] Hayes SR, Vargas AJ. Probiotics for the prevention of pediatric antibiotic-associated diarrhea. Explore 2016;12(6):463–6.

[6] McFarland LV, Goh S. Are probiotics and prebiotics effective in the prevention of travellers' diarrhea: a systematic review and meta-analysis. Trav. Med. Infect. Dis 2019;27:11–9.

[7] Barbara G, Cremon C, Azpiroz F. Probiotics in irritable bowel syndrome: where are we? Neuro Gastroenterol. Motil 2018;30(12):e13513.

[8] Deshpande G, Rao S, Patole S, Bulsara M. Updated meta-analysis of probiotics for preventing necrotizing enterocolitis in preterm neonates. Pediatrics 2010;125(5):921–30.

[9] Dermyshi E, Wang Y, Yan C, et al. The "golden age" of probiotics: a systematic review and meta-analysis of randomized and observational studies in preterm infants. Neonatology 2017;112(1):9–23.

[10] Rao SC, Athalye-Jape GK, Deshpande GC, Simmer KN, Patole SK. Probiotic supplementation and late-onset sepsis in preterm infants: a meta-analysis. Pediatrics 2016;137(3):e20153684.

[11] Weng H, Li JG, Mao Z, et al. Probiotics for preventing ventilator-associated pneumonia in mechanically ventilated patients: a meta-analysis with trial sequential analysis. Front. Pharmacol 2017;8:717.

[12] Arumugam S, Lau CS, Chamberlain RS. Probiotics and synbiotics decrease postoperative sepsis in elective gastrointestinal surgical patients: a meta-analysis. J. Gastrointest. Surg. 2016;20(6):1123–31.

[13] Wilkins T, Sequoia J. Probiotics for gastrointestinal conditions: a summary of the evidence. Am. Fam. Physician 2017;96(3):170–8.

[14] Kothari D, Patel S, Kim SK. Probiotic supplements might not be universally-effective and safe: a review. Biomed. Pharmacother 2019;111:537–47.

[15] Didari T, Solki S, Mozaffari S, Nikfar S, Abdollahi M. A systematic review of the safety of probiotics. Expert Opin. Drug Safety 2014;13(2):227–39.

[16] Esaiassen E, Hjerde E, Cavanagh JP, Simonsen GS, Klingenberg C. *Bifidobacterium* bacteremia: clinical characteristics and a genomic approach to assess pathogenicity. J. Clin. Microbiol 2017;55(7):2234–48.

[17] Bertelli C, Pillonel T, Torregrossa A, et al. *Bifidobacterium longum* bacteremia in preterm infants receiving probiotics. Clin. Infect. Dis. 2015;60(6):924–7.

[18] Esaiassen E, Cavanagh P, Hjerde E, Simonsen GS, Stoen R, Klingenberg C. *Bifidobacterium longum* subspecies infantis bacteremia in 3 extremely preterm infants receiving probiotics. Emerg. Infect. Dis 2016;22(9):1664–6.

[19] Tena D, Losa C, Medina MJ, Saez-Nieto JA. Peritonitis caused by *Bifidobacterium longum*: case report and literature review. Anaerobe 2014;27:27–30.

[20] Suwantarat N, Romagnoli M, Wakefield T, Carroll KC. Ventriculoperitoneal shunt infection caused by *Bifidobacterium breve*. Anaerobe 2014;28:1–3.

[21] Ohishi A, Takahashi S, Ito Y, et al. *Bifidobacterium* septicemia associated with postoperative probiotic therapy in a neonate with omphalocele. J. Pediatr 2010;156(4):679–81.

[22] Van den Nieuwboer M, Brummer RJ, Guarner F, Morelli L, Cabana M, Claasen E. The administration of probiotics and synbiotics in immune compromised adults: is it safe? Benef. Microbes 2015;6(1):3–17.

[23] de Almada CN, Almada CN, Martinez RC, Sant'Ana AS. Paraprobiotics: evidences on their ability to modify biological responses, inactivation methods and perspectives on their application in foods. Trends Food Sci. Technol 2016;58:96–114.

[24] Adams CA. The probiotic paradox: live and dead cells are biological response modifiers. Nutr. Res. Rev 2010;23(1):37–46.

[25] Taverniti V, Guglielmetti S. The immunomodulatory properties of probiotic microorganisms beyond their viability (ghost probiotics: proposal of paraprobiotic concept). Genes Nutr 2011;6(3):261–74.

[26] Harima-Mizusawa N, Kano M, Nozaki D, Nonaka C, Miyazaki K, Enomoto T. Citrus juice fermented with *Lactobacillus plantarum* YIT 0132 alleviates symptoms of perennial allergic rhinitis in a double-blind, placebo-controlled trial. Benef. Microbes 2016;7(5):649–58.

[27] Inoue Y, Kambara T, Murata N, et al. Effects of oral administration of *Lactobacillus acidophilus* L-92 on the symptoms and serum cytokines of atopic dermatitis in Japanese adults: a double-blind, randomized, clinical trial. Int. Arch. Allergy Immunol 2014;165(4):247–54.

[28] Morisset M, Aubert-Jacquin C, Soulaines P, Moneret-Vautrin DA, Dupont C. A non-hydrolyzed, fermented milk formula reduces digestive and respiratory events in infants at high risk of allergy. Eur. J. Clin. Nutr 2011;65(2):175–83.

[29] Moroi M, Uchi S, Nakamura K, et al. Beneficial effect of a diet containing heat-killed *Lactobacillus paracasei* K71 on adult type atopic dermatitis. J. Dermatol 2011;38(2):131–9.

[30] Gotoh M, Sashihara T, Ikegami S, et al. Efficacy of oral administration of a heat-killed *Lactobacillus gasseri* OLL2809 on patients of Japanese cedar pollinosis with high Japanese-cedar pollen-specific IgE. Biosci. Biotechnol. Biochem 2009;73(9):1971–7.

[31] Peng GC, Hsu CH. The efficacy and safety of heat-killed *Lactobacillus paracasei* for treatment of perennial allergic rhinitis induced by house-dust mite. Pediatr. Allergy Immunol 2005;16(5):433–8.

[32] Kirjavainen PV, Salminen SJ, Isolauri E. Probiotic bacteria in the management of atopic disease: underscoring the importance of viability. J. Pediatr. Gastroenterol. Nutr 2003;36(2):223–7.

[33] Lau S, Gerhold K, Zimmermann K, et al. Oral application of bacterial lysate in infancy decreases the risk of atopic dermatitis in children with 1 atopic parent in a randomized, placebo-controlled trial. J. Allergy Clin. Immunol 2012;129(4):1040–7.

[34] Hirose Y, Murosaki S, Yamamoto Y, Yoshikai Y, Tsuru T. Daily intake of heat-killed *Lactobacillus plantarum* L-137 augments acquired immunity in healthy adults. J. Nutr 2006;136(12):3069–73.

[35] Arkwright PD, David TJ. Effect of *Mycobacterium vaccae* on atopic dermatitis in children of different ages. Br. J. Dermatol 2003;149(5):1029–34.

[36] Brothers S, Asher MI, Jaksic M, Stewart AW. Effect of a *Mycobacterium vaccae* derivative on paediatric atopic dermatitis: a randomized, controlled trial. Clin. Exp. Dermatol 2009;34(7):770–5.

[37] Tarrerias AL, Costil V, Vicari F, et al. The effect of inactivated *Lactobacillus* LB fermented culture medium on symptom severity: observational investigation in 297 patients with diarrhea-predominant irritable bowel syndrome. Dig. Dis 2011;29(6):588–91.

[38] Guyonnet D, Chassany O, Ducrotte P, et al. Effect of a fermented milk containing *Bifidobacterium animalis* DN-173 010 on the health-related quality of life and symptoms in irritable bowel syndrome in adults in primary care: a multicentre, randomized, double-blind, controlled trial. Aliment. Pharmacol. Ther 2007;26(3):475–86.

[39] Halpern GM, Prindiville T, Blankenburg M, Hsia T, Gershwin ME. Treatment of irritable bowel syndrome with Lacteol Fort: a randomized, double-blind, cross-over trial. Am. J. Gastroenterol 1996;91(8):1579–85.

[40] Wenus C, Goll R, Loken EB, Biong AS, Halvorsen DS, Florholmen J. Prevention of antibiotic-associated diarrhoea by a fermented probiotic milk drink. Eur. J. Clin. Nutr 2008;62(2):299–301.

[41] Merenstein DJ, Foster J, D'Amico F. A randomized clinical trial measuring the influence of kefir on antibiotic-associated diarrhea: the measuring the influence of Kefir (MILK) study. Arch. Pediatr. Adolesc. Med 2009;163(8):750–4.

[42] Nocerino R, Paparo L, Terrin G, et al. Cow's milk and rice fermented with *Lactobacillus paracasei* CBA L74 prevent infectious diseases in children: a randomized controlled trial. Clin. Nutr 2017;36(1):118–25.

[43] Berni Canani R, De Filippis F, Nocerino R, et al. Specific signatures of the gut microbiota and increased levels of butyrate in children treated with fermented cow's milk containing heat-killed *Lactobacillus paracasei* CBA L74. Appl. Environ. Microbiol 2017;83(19).

[44] Salazar-Lindo E, Figueroa-Quintanilla D, Caciano MI, Reto-Valiente V, Chauviere G, Colin P. Effectiveness and safety of *Lactobacillus* LB in the treatment of mild acute diarrhea in children. J. Pediatr. Gastroenterol. Nutr 2007;44(5):571–6.

[45] Boulloche J, Mouterde O, Mallet E. Management of acute diarrhea in infants and toddlers-controlled-study of the antidiarrheal efficacy of killed *Lactobacillus*-acidophilus (lb strain) versus a placebo and a reference agent (loperamide). In: Paper presented at: annales de pediatrie; 1994.

[46] Khanna V, Alam S, Malik A, Malik A. Efficacy of tyndalized *Lactobacillus acidophilus* in acute diarrhea. Indian J. Pediatr 2005;72(11):935–8.

[47] Sharieff W, Bhutta Z, Schauer C, Tomlinson G, Zlotkin S. Micronutrients (including zinc) reduce diarrhoea in children: the Pakistan Sprinkles Diarrhoea Study. Arch. Dis. Child 2006;91(7):573–9.

[48] Tapiovaara L, Kumpu M, Makivuokko H, et al. Human rhinovirus in experimental infection after peroral *Lactobacillus rhamnosus* GG consumption, a pilot study. Int. Allergy Rhinol. 2016;6(8):848–53.

[49] Kumpu M, Kekkonen RA, Korpela R, et al. Effect of live and inactivated *Lactobacillus rhamnosus* GG on experimentally induced rhinovirus colds: randomised, double blind, placebo-controlled pilot trial. Benef. Microbes 2015;6(5):631–9.

[50] Shinkai S, Toba M, Saito T, et al. Immunoprotective effects of oral intake of heat-killed *Lactobacillus pentosus* strain b240 in elderly adults: a randomised, double-blind, placebo-controlled trial. Br. J. Nutr 2013;109(10):1856–65.

[51] Hirose Y, Yamamoto Y, Yoshikai Y, Murosaki S. Oral intake of heat-killed *Lactobacillus plantarum* L-137 decreases the incidence of upper respiratory tract infection in healthy subjects with high levels of psychological stress. J. Nutr. Sci 2013;2:e39.

[52] Lievin-Le Moal V, Sarrazin-Davila LE, Servin AL. An experimental study and a randomized, double-blind, placebo-controlled clinical trial to evaluate the antisecretory activity of *Lactobacillus acidophilus* strain LB against nonrotavirus diarrhea. Pediatrics 2007;120(4):e795–803.

[53] Mitra AK, Rabbani GH. A double-blind, controlled trial of bioflorin (*Streptococcus faecium* SF68) in adults with acute diarrhea due to *Vibrio cholerae* and enterotoxigenic *Escherichia coli*. Gastroenterology 1990;99(4):1149–52.

[54] Xiao SD, Zhang DZ, Lu H, et al. Multicenter, randomized, controlled trial of heat-killed *Lactobacillus acidophilus* LB in patients with chronic diarrhea. Adv. Ther 2003;20(5):253–60.

[55] Simakachorn N, Pichaipat V, Rithipornpaisarn P, Kongkaew C, Tongpradit P, Varavithya W. Clinical evaluation of the addition of lyophilized, heat-killed *Lactobacillus acidophilus* LB to oral rehydration therapy in the treatment of acute diarrhea in children. J. Pediatr. Gastroenterol. Nutr 2000;30(1):68–72.

[56] Michielutti F, Bertini M, Presciuttini B, Andreotti G. [Clinical assessment of a new oral bacterial treatment for children with acute diarrhea]. Minerva Medica 1996;87(11):545–50.

[57] Kaila M, Isolauri E, Saxelin M, Arvilommi H, Vesikari T. Viable versus inactivated *Lactobacillus* strain GG in acute rotavirus diarrhoea. Arch. Dis. Child 1995;72(1):51–3.

[58] Briand V, Buffet P, Genty S, et al. Absence of efficacy of nonviable *Lactobacillus acidophilus* for the prevention of traveler's diarrhea: a randomized, double-blind, controlled study. Clin. Infect. Dis 2006;43(9):1170–5.

[59] Maruyama M, Abe R, Shimono T, Iwabuchi N, Abe F, Xiao JZ. The effects of non-viable *Lactobacillus* on immune function in the elderly: a randomised, double-blind, placebo-controlled study. Int. J. Food Sci. Nutr 2016;67(1):67–73.

[60] Miyazawa K, Kawase M, Kubota A, et al. Heat-killed *Lactobacillus gasseri* can enhance immunity in the elderly in a double-blind, placebo-controlled clinical study. Benef. Microbes 2015;6(4):441–9.

[61] Vandenplas Y, Bacarea A, Marusteri M, Bacarea V, Constantin M, Manolache M. Efficacy and safety of APT198K for the treatment of infantile colic: a pilot study. J. Comp. Eff. Res 2017;6(2):137–44.

[62] Iwasaki K, Maeda K, Hidaka K, Nemoto K, Hirose Y, Deguchi S. Daily intake of heat-killed *Lactobacillus plantarum* L-137 decreases the probing depth in patients undergoing supportive periodontal therapy. Oral Health Prev. Dent 2016;14(3):207–14.

[63] Bayona-Gonzalez A, Lopez-Camara V, Gomez-Castellanos A. Final results of a dental caries clinical trial using heat killed lactic bacteria (*Streptococci* and *Lactobacilli*) orally. Pract. Odontol 1990;11(6):41–7.

[64] Bayona Gonzalez A, Lopez Camara V, Gomez Castellanos A. [Prevention of caries with *Lactobacillus* (final results of a clinical trial on dental caries with killed *Lactobacillus* [*streptococcus* and *Lactobacillus*] given orally)]. Pract. Odontol 1990;11(7):37–9. 42-33, 45-36.

[65] Mehling H, Busjahn A. Non-viable *Lactobacillus reuteri* DSMZ 17648 (Pylopass) as a new approach to *Helicobacter pylori* control in humans. Nutrients 2013;5(8):3062–73.

[66] Canducci F, Armuzzi A, Cremonini F, et al. A lyophilized and inactivated culture of *Lactobacillus acidophilus* increases *Helicobacter pylori* eradication rates. Aliment. Pharmacol. Ther 2000;14(12):1625–9.

[67] Gotteland M, Andrews M, Toledo M, et al. Modulation of *Helicobacter pylori* colonization with cranberry juice and *Lactobacillus johnsonii* La1 in children. Nutrition 2008;24(5):421–6.

[68] Cruchet S, Obregon MC, Salazar G, Diaz E, Gotteland M. Effect of the ingestion of a dietary product containing *Lactobacillus johnsonii* La1 on *Helicobacter pylori* colonization in children. Nutrition 2003;19(9):716–21.

[69] Campeotto F, Suau A, Kapel N, et al. A fermented formula in preterm infants: clinical tolerance, gut microbiota, down-regulation of faecal calprotectin and up-regulation of faecal secretory IgA. Br. J. Nutr 2011;105(12):1843–51.

[70] Awad H, Mokhtar H, Imam SS, Gad GI, Hafez H, Aboushady N. Comparison between killed and living probiotic usage versus placebo for the prevention of necrotizing enterocolitis and sepsis in neonates. Pakistan J. Biol. Sci 2010;13(6):253–62.

[71] Rampengan NH, Manoppo J, Warouw SM. Comparison of efficacies between live and killed probiotics in children with lactose malabsorption. Southeast Asian J. Trop. Med. Publ. Health 2010;41(2):474–81.

[72] Rayes N, Seehofer D, Muller AR, Hansen S, Bengmark S, Neuhaus P. [Influence of probiotics and fibre on the incidence of bacterial infections following major abdominal surgery – results of a prospective trial]. Zeitschrift fur Gastroenterol 2002;40(10):869–76.

[73] Rayes N, Seehofer D, Hansen S, et al. Early enteral supply of *Lactobacillus* and fiber versus selective bowel decontamination: a controlled trial in liver transplant recipients. Transplantation 2002;74(1):123–7.

[74] Rayes N, Hansen S, Seehofer D, et al. Early enteral supply of fiber and *Lactobacilli* versus conventional nutrition: a controlled trial in patients with major abdominal surgery. Nutrition 2002;18(7–8):609–15.

[75] Kecskes G, Belagyi T, Olah A. Early jejunal nutrition with combined pre- and probiotics in acute pancreatitis—prospective, randomized, double-blind investigations. Magy. Sebeszet 2003;56(1):3–8.

[76] Olah A, Belagyi T, Issekutz A, Gamal ME, Bengmark S. Randomized clinical trial of specific *Lactobacillus* and fibre supplement to early enteral nutrition in patients with acute pancreatitis. Br. J. Surg 2002;89(9):1103–7.

[77] Okawa T, Niibe H, Arai T, et al. Effect of LC9018 combined with radiation therapy on carcinoma of the uterine cervix. A phase III, multicenter, randomized, controlled study. Cancer 1993;72(6):1949–54.

[78] Okawa T, Kita M, Arai T, et al. Phase II randomized clinical trial of LC9018 concurrently used with radiation in the treatment of carcinoma of the uterine cervix. Its effect on tumor reduction and histology. Cancer 1989;64(9):1769–76.

[79] Morita Y, Jounai K, Miyake M, Inaba M, Kanauchi O. Effect of heat-killed *Lactobacillus paracasei* KW3110 ingestion on ocular disorders caused by Visual Display Terminal (VDT) loads: a randomized, double-blind, placebo-controlled parallel-group study. Nutrients 2018;10(8).

[80] Nakakita Y, Tsuchimoto N, Takata Y, Nakamura T. Effect of dietary heat-killed *Lactobacillus brevis* SBC8803 (SBL88) on sleep: a non-

randomised, double blind, placebo-controlled, and crossover pilot study. Benef. Microbes 2016;7(4):501–9.

[81] Takii Y, Nishimura S, Yoshida-Yamamoto S, Kobayashi Y, Nagayoshi E. Effects of intake of pickles containing *Lactobacillus brevis* on immune activity and bowel symptoms in female students. J. Nutr. Sci. Vitaminol 2013;59(5):402–11.

[82] Gotteland M, Cruchet S, Verbeke S. Effect of *Lactobacillus* ingestion on the gastrointestinal mucosal barrier alterations induced by indometacin in humans. Aliment. Pharmacol. Ther 2001;15(1):11–7.

[83] Higashikawa F, Noda M, Awaya T, et al. Antiobesity effect of *Pediococcus pentosaceus* LP28 on overweight subjects: a randomized, double-blind, placebo-controlled clinical trial. Eur. J. Clin. Nutr 2016;70(5):582–7.

[84] Alberda C, Gramlich L, Meddings J, et al. Effects of probiotic therapy in critically ill patients: a randomized, double-blind, placebo-controlled trial. Am. J. Clin. Nutr 2007;85(3):816–23.

[85] Zorzela L, Ardestani SK, McFarland LV, Vohra S. Is there a role for modified probiotics as beneficial microbes: a systematic review of the literature. Benef. Microbes 2017;8(5):739–54.

[86] Jape G. Safety and efficacy of live versus heat-inactivated probiotic strains in preterm infants: a pilot randomized controlled trial. 2018. ACTRN12618000489291p.

[87] Deshpande G, Athalye-Jape G, Patole S. Para-probiotics for preterm neonates-the next frontier. Nutrients 2018;10(7).

Chapter 6

Fecal material transplant and ocular surface diseases

Rodrigo G. de Souza[1,2], Robert A. Britton[3] and Cintia S. de Paiva[1,2]
[1]*Department of Ophtalmology, São Paulo University Medical School, São Paulo, São Paulo, Brazil;* [2]*Department of Ophthalmology, Baylor College of Medicine, Houston, TX, United States;* [3]*Center for Metagenomics and Microbiome Research, Department of Molecular Virology and Microbiology, Baylor College of Medicine, Houston, TX, United States*

Introduction

Several studies have contributed to show the relationship between microbiome and diseases such as cancer, asthma, obesity, metabolic syndrome, atherosclerosis, inflammatory bowel disease, diabetes, and other autoimmune disorders [1–7]. The role of the microbiome in etiology and pathogenesis has been addressed in germ-free animal models and laboratories that perform fecal transplant experiments [6–13].

The highest concentration of bacteria can be found in the gastrointestinal tract [14,15]. It is also known that microbiome composition varies according to the places in the body, such as the skin, urinary tract, and oral mucosal surface [16].

Ocular microbiome

It was long believed that the ocular surface was a sterile site because of the difficulty in obtaining viable microbes, using traditional culture-based techniques [17–20], and also because of the potent antimicrobial compounds that are present in tears, such as lysozyme and antimicrobial peptides [21–24]. However, the ocular surface is not a sterile site, but paucibacterial compared to other parts of the body [25–27]. Its microbiome is mainly composed of bacteria, although fungi and viruses are also present [13,28–30]. The cornea seems to be a more sterile environment, while the conjunctiva is more permissive to microbes, although still usually in a low number [31,32].

In agreement with the idea that there is a bona fide ocular microbiome, studies have shown important differences between the set of bacteria of the ocular surface and those on the surrounding facial skin [26,33]. Healthy conjunctiva, tears, and eyelid margins had less microbial species than elsewhere in the body [20,34,35]. Without cultivation or fluorescence in situ hybridization flow cytometry showing communities on the eye surface and given the low bacterial load, it is challenging to say there is a true eye community. It could either be sequencing contamination or bacteria coming into contact with the eye but not forming a stable community.

Nevertheless, studies with conjunctival swabs showed that the ocular environment might contain many genera of bacteria, including "commensal" bacteria such as *Corynebacterium* and coagulase-negative *Staphylococcus*, as well as "pathogenic" bacteria such as *Pseudomonas*. These genera can comprise 30%–70% of the ocular surface bacteria. They are influenced by potentially adverse situations, such as the use of contact lenses, and certain geographic locations and ethnicities [17,25,26,33,34,36–39].

Pathogens and commensals

The species usually found on conjunctival surfaces are Gram-positive bacteria, especially *Staphylococcus* sp. coagulase-negative, including *Staphylococcus epidermidis*, followed by other microbial species in less abundance such as *Propionibacterium* sp. (*Propionibacterium acnes*), *Corynebacterium* sp., *Staphylococcus aureus*, *Streptococcus* sp., *Micrococcus* sp., *Bacillus* sp., and *Lactobacillus* sp. Gram-negatives are less frequent on the healthy ocular surface, although *Pseudomonas aeruginosa*, *Enterobacter* sp., *Escherichia coli*, *Proteus* sp., and *Acinetobacter* sp. can also be found [20,31,40–44]. In the experience of Shin and colleagues, commensal conjunctival diversity is higher than on the skin below the eye, with an abundance of *Haemophilus*, *Streptococcus*, *Staphylococcus*, and *Corynebacterium* sp. [33].

Precision Medicine for Investigators, Practitioners and Providers. https://doi.org/10.1016/B978-0-12-819178-1.00006-X

Microbiomic methods and functional classification

The techniques that are used to identify the microbiota can also influence on the numbers and findings [25–27,45]. Deep sequencing found 12 genera that could constitute the ocular microbiota, although the percentage among the species was very variable [34,46]. There is a fine line separating beneficial and pathogenic ocular bacteria, and the list of either subtype changes, as our knowledge about a particular bacterial species increases [47]. For example, while *Corynebacterium* spp. has the potential to cause serious eye infections, in the elderly and immunosuppressed subjects, other studies have identified *Corynebacterium mastitidis* as commensal bacteria, improving the immune response of the ocular surface and protecting against infections. These data suggest that *Corynebacterium* spp. and probably any other bacteria can act as commensal in the steady state and become pathogenic under situations where the immune system is suppressed or overactive [13,37].

Demographic effects

Few studies have analyzed the association between microbiome with gender and aging. Although most reported no changes with age, some of them showed alterations in the diversity of the species of children (younger than 10 years) and adults, which may be related to maturation and progressive reduction of immune competence with aging [33,38,45,46,48,49].

Immunologic repercussions of the microbiome on the ocular surface

Commensal bacteria may play an essential role in the modulation and immunity of eye diseases. Caspi and colleagues [29] noticed a relationship between intestinal microbiome and autoimmune disease of the retina. The link is consistent with observations in mice with multiple sclerosis and autoimmune diabetes [50,51] and in humans involving intestinal bacteria and autoimmune diseases [52,53]. The relationship between the microbiome and immunological mechanisms involved in glaucoma and retinal diseases, such as age-related macular degeneration, requires functional studies for a better understanding of how this association modulates evolution of the diseases [13,54].

Shared antigens

Major histocompatibility molecules have been associated with systemic and ocular diseases [55], such as HLA-A29–associated chorioretinopathy, HLA-DQw3–associated Vogt–Koyanagi–Harada disease, and acute anterior uveitis–associated with HLA-B27 [56,57]. HLA-B27 is also associated with other autoimmune diseases, in addition to the influence on the constitution of intestinal microbiome [58].

Metabolic alterations and induction of regulatory T cells can also lead to alterations in the intestinal microbiota [59,60]. Some common bacteria found on ocular surfaces (such as *P. acnes* and coagulase-negative *Staphylococcus*) can stimulate T cells, antigen-presenting cells, eosinophils, and basophils, among others [13,61].

Immune globulin modulation

The primary immunoglobulin on the ocular surface is IgA [62,63], important not only for maintaining homeostasis and neutralizing toxins, viruses, and bacteria but also for exerting antiinflammatory activities and dendritic cell maturation [64–67]. Many studies have shown evidence that IgA levels are directly related to the commensal microbiome [68–72].

Evidence of a gut–eye–lacrimal gland axis and ocular changes in germ-free mice

In transgenic mice that express the retinoid-binding protein in $CD4^+$ T cells, these are activated in the gut. T cells then migrate to the eye and are capable of inducing uveitis, an autoimmune/inflammatory process inside the eye. When these transgenic mice were given antibiotics, there was an improvement of the autoimmune phenotype [29]. The commensal bacteria *C. mastitidis*, isolated from C57BL/6J animals reared at the NIH facility (but not from commercial vendors), was shown to modulate *Candida albicans and P. aeruginosa* keratitis, through the production of IL-17A from $\gamma\delta$ T cells [37].

Germ-free Swiss Webster mice were more susceptible to *P. aeruginosa* keratitis than conventional Swiss Webster, and this effect was mediated by both ocular and intestinal microbes [31].

The germ-free model

The first studies using germ-free mice started about 50 years ago [73,74]. Mice that are born in a sterile environment are viable and can reproduce; however, many alterations have been noted [75]. They tend to be smaller, with smaller litters and decreased number of goblet cells in the intestine [76]. Goblet cells are capable of secreting mucins and immunoregulatory factors, which inhabit not only intestinal tract and vagina but also the conjunctiva and ocular mucosa.

Microbiome inhibition with antibiotics

The effect of the microbiome on the ocular surface was first investigated using a cocktail of oral antibiotics [25]. When C57BL/6J mice were subjected to experimental dry eye for about 5–10 days, with a cocktail of antibiotics in the drinking water for 24 days, there was a worsening of the dry eye phenotype. This was observed as greater corneal barrier disruption, decreased numbers of goblet cells, increased $CD4^+$ T cell infiltration in the conjunctiva, and production of cytokines [25]. 16S analysis of stool samples from these mice showed a significant decrease in alpha diversity after antibiotic treatment and desiccating stress and clustering of operational taxonomic units (OTUs) in beta diversity analysis.

This was accompanied by significant changes in microbial composition, with a decrease of commensal bacteria genera such as *Lactobacillus* and *Bacteroides* and an increase in potentially pathogenic bacteria such as *E. coli*, *Shigella*, *Enterobacter*, and *Pseudomonas*. Evaluation of the ocular surface and lacrimal glands of C57BL/6J mice showed that naïve 8-week-old germ-free mice also exhibited a spontaneous dry eye phenotype, which mimicked the results obtained with oral antibiotics and desiccating stress. The production of cytokines and the levels of corneal barrier disruption were worse in germ-free females than males, recapitulating the female predisposition seen in Sjögren syndrome patients [6].

Dacryoadenitis model

Similar results were observed in the CD25 knockout (KO) mouse strain. CD25 is the IL-2 receptor alpha chain, and it is the binding arm of the heterodimeric CD25 receptor [77,78]. Without CD25, mice spontaneously develop splenomegaly, colitis, hemolytic anemia, and systemic autoimmunity, inclusive of T cell infiltration in lacrimal and salivary glands. These effects are thought to be due to loss of IL-2 signaling, spontaneous T cell activation, and the inability of generating regulatory T cells [78–81].

Many laboratories including ours have characterized the dacryoadenitis (lacrimal gland inflammation) and ocular surface alterations in these mice and observed a progressive worsening in T cell infiltration and production of IFN-γ, IL-21, and IL-12; loss of goblet cells; and $CD4^+$ T cell infiltration in conjunctiva, as these mice aged from 8 to 16 weeks of age [82–85]. IFN-γ has been shown to play a pathogenic role in Sjögren syndrome [86–88], as genetic deletion of IFN-γ in CD25KO mice improved dacryoadenitis, decreased glandular apoptosis, and improved goblet cell density [89,90].

Germ-free synergic effects

When CD25KO mice were born and raised in a germ-free environment, they displayed a more severe and accelerated dry eye phenotype. This effect was mediated by $CD4^+$ T cells, as adoptive transfer of germ-free CD4 T cells was capable of recapitulating the dry eye phenotype, in immunodeficient mice [7]. Taken together, the findings in these two very distinct models (one nonautoimmune and an autoimmune) point out for a protective role of commensal bacteria in dry eye and Sjögren syndrome.

Fecal transplantation and eye disease

As our understanding of the role of the microbiome in ocular surface diseases increases, it has become evident that manipulation of either the ocular microbiome or the fecal microbiome could be used as potential new therapies for eye diseases. It has been shown that the presence of the commensal *C. mastitidis* in the conjunctiva of C57BL/6J mice from NIH facility can protect from *Candida* and *Pseudomonas* corneal infections, potentially blinding conditions [37]. Antibiotic administration (either as eye drops or systemic) can turn Swiss Webster mice from resistant to susceptible to *P. aeruginosa* infections [31].

Experimental investigations

Fecal material transplanted into germ-free C57BL/6 or CD25KO mice improved goblet cell density, ameliorated corneal barrier disruption, and decreased dacryoadenitis and pathogenicity of $CD4^+$ T cells [6,7]. Because CD25KO mice are unable to generate regulatory T cells after fecal microbiome transfer (FMT), this effect is thought to be independent of the intestinal generation of regulatory T cells, a previously suggested mechanism [91]. It is possible that bacteria themselves or microbial metabolites modulate the beneficial effects observed after FMT. In support of this idea, probiotic *Lactobacillus reuteri* DSM17938 was shown to modulate autoimmunity in scurfy mice, a mouse strain with a Foxp3 mutation, which is also devoid of regulatory T cells. When *L. reuteri* was given as oral gavage from day 8 to day 21 to scurfy mice, it greatly ameliorated the disease by significantly decreasing Th1 cells and diminishing T cell infiltration in the liver and lungs [92].

Human studies

In a study of patients with Sjögren syndrome, decreased intestinal microbiome diversity was observed, and clustering of microbial communities by principal component analysis showed segregation of normal subjects differently from Sjögren syndrome patients [25]. The number of OTUs inversely correlated with systemic severity score; however, future investigations are needed to delineate if these alterations are cause or consequence of the disease. Small pilot studies have shown eye beneficial effects of oral supplementation of potential commensal bacteria [93,94], but total FMT in patients with ocular surface diseases has not been attempted yet. It is essential to move away from FMT to defined bacterial therapeutics.

Conclusions

16S and whole-genome bacterial sequencing remain expensive to perform in all patients as a diagnostic tool. Yet, as our knowledge increases and more potential "good bacteria" are identified, it is likely that FMT and other modalities of bacterial therapy may be used as a novel therapy to treat certain chronic eye diseases.

Acknowledgments

This work was supported by the NIH EY026893 (C.S.D.P.), NIH EY-002520 (core grant for Vision Research Department of Ophthalmology), NIH NCI P30CA125123 (Pathology and Histology Core), Research to Prevent Blindness (Department of Ophthalmology), Research to Prevent Blindness Stein Innovation Award and NIH grants R33AI121522 and R01AI123278 (RAB), The Oshman Foundation, William Stamps Farish Fund, The Hamill Foundation, and the Sid Richardson Foundation.

References

[1] Mullaney JA, Stephens JE, Costello ME, Fong C, Geeling BE, Gavin PG, Wright CM, Spector TD, Brown MA, Hamilton-Williams EE. Type 1 diabetes susceptibility alleles are associated with distinct alterations in the gut microbiota. Microbiome 2018;6:35.

[2] Ohtani N. Microbiome and cancer. Semin. Immunopathol. 2015;37:65–72.

[3] Sanmiguel A, Grice EA. Interactions between host factors and the skin microbiome. Cell. Mol. Life Sci. 2015;72:1499–515.

[4] Sanz Y, Olivares M, Moya-Perez A, Agostoni C. Understanding the role of gut microbiome in metabolic disease risk. Pediatr. Res. 2015;77:236–44.

[5] Viaud S, Daillere R, Boneca IG, Lepage P, Langella P, Chamaillard M, Pittet MJ, Ghiringhelli F, Trinchieri G, Goldszmid R, Zitvogel L. Gut microbiome and anticancer immune response: really hot Sh*t! Cell Death Differ. 2015;22:199–214.

[6] Wang C, Zaheer M, Bian F, Quach D, Swennes AG, Britton RA, Pflugfelder SC, de Paiva CS. Sjogren-like lacrimal keratoconjunctivitis in germ-free mice. Int. J. Mol. Sci. 2018;19.

[7] Zaheer M, Wang C, Bian F, Yu Z, Hernandez H, De Souza RG, Simmons KT, Schady D, Swennes AG, Pflugfelder SC, Britton RA, De Paiva CS. Protective role of commensal bacteria in Sjogren syndrome. J. Autoimmun. 2018:30179–83. pii: S0896-8411.

[8] Atarashi K, Tanoue T, Shima T, Imaoka A, Kuwahara T, Momose Y, Cheng G, Yamasaki S, Saito T, Ohba Y, Taniguchi T, Takeda K, Hori S, Ivanov Ii, Umesaki Y, Itoh K, Honda K. Induction of colonic regulatory T cells by indigenous *Clostridium* species. Science 2011;331:337–41.

[9] Markle JG, Frank DN, Mortin-Toth S, Robertson CE, Feazel LM, Rolle-Kampczyk U, Von Bergen M, Mccoy KD, Macpherson AJ, Danska JS. Sex differences in the gut microbiome drive hormone-dependent regulation of autoimmunity. Science 2013;339:1084–8.

[10] Nell S, Suerbaum S, Josenhans C. The impact of the microbiota on the pathogenesis of IBD: lessons from mouse infection models. Nat. Rev. Microbiol. 2010;8:564–77.

[11] Rabot S, Membrez M, Bruneau A, Gerard P, Harach T, Moser M, Raymond F, Mansourian R, Chou CJ. Germ-free C57BL/6J mice are resistant to high-fat-diet-induced insulin resistance and have altered cholesterol metabolism. FASEB J. 2010;24:4948–59.

[12] Roy U, Galvez EJC, Iljazovic A, Lesker TR, Blazejewski AJ, Pils MC, Heise U, Huber S, Flavell RA, Strowig T. Distinct microbial communities trigger colitis development upon intestinal barrier damage via innate or adaptive immune cells. Cell Rep. 2017;21:994–1008.

[13] St Leger AJ, Caspi RR. Visions of eye commensals: the known and the unknown about how the microbiome affects eye disease. Bioessays 2018;40:e1800046.

[14] Belkaid Y, Harrison OJ. Homeostatic immunity and the microbiota. Immunity 2017;46:562–76.

[15] He J, Li Y, Cao Y, Xue J, Zhou X. The oral microbiome diversity and its relation to human diseases. Folia Microbiol. 2015;60:69–80.

[16] Whiteside SA, Razvi H, Dave S, Reid G, Burton JP. The microbiome of the urinary tract—a role beyond infection. Nat. Rev. Urol. 2015;12:81–90.

[17] Huang Y, Yang B, Li W. Defining the normal "core microbiome" of conjunctival microbial communities. Clin. Microbiol. Infect. 2016, Jul;22(7):643.e7–643.e12.

[18] Moeller CT, Branco BC, Yu MC, Farah ME, Santos MA, Hofling-Lima AL. Evaluation of normal ocular bacterial flora with two different culture media. Can. J. Ophthalmol. 2005;40:448–53.

[19] Nolan J. Evaluation of conjunctival and nasal bacterial cultures before intra-ocular operations. Br. J. Ophthalmol. 1967;51:483–5.

[20] Willcox MD. Characterization of the normal microbiota of the ocular surface. Exp. Eye Res. 2013;117:99–105.

[21] Caffery B, Joyce E, Boone A, Slomovic A, Simpson T, Jones L, Senchyna M. Tear lipocalin and lysozyme in Sjogren and non-Sjogren dry eye. Optom. Vis. Sci. 2008;85:661–7.

[22] Careba I, Chiva A, Totir M, Ungureanu E, Gradinaru S. Tear lipocalin, lysozyme and lactoferrin concentrations in postmenopausal women. J. Med. Life 2015;8:94–8. Spec Issue.

[23] Kolar SS, Mcdermott AM. Role of host-defence peptides in eye diseases. Cell. Mol. Life Sci. 2011;68:2201–13.

[24] Mcdermott AM. Antimicrobial compounds in tears. Exp. Eye Res. 2013;117:53–61.

[25] de Paiva CS, Jones DB, Stern ME, Bian F, Moore QL, Corbiere S, Streckfus CF, Hutchinson DS, Ajami NJ, Petrosino JF, Pflugfelder SC. Altered mucosal microbiome diversity and disease severity in Sjogren syndrome. Sci. Rep. 2016;6:23561–71.

[26] Doan T, Akileswaran L, Andersen D, Johnson B, Ko N, Shrestha A, Shestopalov V, Lee CS, Lee AY, VAN Gelder RN. Paucibacterial microbiome and resident DNA virome of the healthy conjunctiva. Invest. Ophthalmol. Vis. Sci. 2016;57:5116–26.
[27] Doan T, Akileswaran L, Andersen D, Ko N, Shrestha A, Lee CS, Lee A, Van Gelder R. Human ocular surface microbiome composition revealed by next-generation sequencing. Investig. Ophthalmol. Visual Sci. 2015;56. 4067–4067.
[28] Ge C, Wei C, Yang BX, Cheng J, Huang YS. Conjunctival microbiome changes associated with fungal keratitis: metagenomic analysis. Int. J. Ophthalmol. 2019;12:194–200.
[29] Horai R, Zarate-Blades CR, Dillenburg-Pilla P, Chen J, Kielczewski JL, Silver PB, Jittayasothorn Y, Chan CC, Yamane H, Honda K, Caspi RR. Microbiota-dependent activation of an autoreactive T cell receptor provokes autoimmunity in an immunologically privileged site. Immunity 2015;43:343–53.
[30] Shivaji S, Jayasudha R, Sai Prashanthi G, Kalyana Chakravarthy S, Sharma S. The human ocular surface fungal microbiome. Invest. Ophthalmol. Vis. Sci. 2019;60:451–9.
[31] Kugadas A, Christiansen SH, Sankaranarayanan S, Surana NK, Gauguet S, Kunz R, Fichorova R, Vorup-Jensen T, Gadjeva M. Impact of microbiota on resistance to ocular *Pseudomonas aeruginosa*-induced keratitis. PLoS Pathog. 2016;12:e1005855.
[32] Taravati P, Lam D, Van Gelder RN. Role of molecular diagnostics in ocular microbiology. Curr. Ophthalmol. Rep. 2013;1.
[33] Shin H, Price K, Albert L, Dodick J, Park L, Dominguez-Bello MG. Changes in the eye microbiota associated with contact lens wearing. MBio 2016;7:e00198.
[34] Dong Q,BJ, Iovieno A, Bates B, Garoutte A, Miller D, Revanna KV, Gao X, Antonopoulos DA, Slepak VZ, Shestopalov VI. Diversity of bacteria at healthy human conjunctiva. Invest. Ophthalmol. Vis. Sci 2011;52:5.
[35] Ozkan J, Willcox MD. The ocular microbiome: molecular characterization of a unique and low microbial environment. Curr. Eye Res. 2013 Dec, 117, 99-105.
[36] Ozkan J, Nielsen S, Diez-Vives C, Coroneo M, Thomas T, Willcox M. Temporal stability and composition of the ocular surface microbiome. Sci. Rep. 2017;7:9880.
[37] St Leger AJ, Desai JV, Drummond RA, Kugadas A, Almaghrabi F, Silver P, Raychaudhuri K, Gadjeva M, Iwakura Y, Lionakis MS, Caspi RR. An ocular commensal protects against corneal infection by driving an interleukin-17 response from mucosal gammadelta T cells. Immunity 2017;47:148–158.e5.
[38] Wen X, Miao L, Deng Y, Bible PW, Hu X, Zou Y, Liu Y, Guo S, Liang J, Chen T, Peng G-H, Chen W, Liang L, Wei L. The influence of age and sex on ocular surface microbiota in healthy adults. Investig. Ophthalmol. Visual Sci. 2017;58:6030–7.
[39] Zegans ME, Van Gelder RN. Considerations in understanding the ocular surface microbiome. Am. J. Ophthalmol. 2014;158:420–2.
[40] Campos MS, Campos E Silva Lde Q, Rehder JR, Lee MB, O'brien T, Mcdonnell PJ. Anaerobic flora of the conjunctival sac in patients with AIDS and with anophthalmia compared with normal eyes. Acta Ophthalmol. 1994;72:241–5.
[41] Elander TR, Goldberg MA, Salinger CL, Tan JR, Levy B, Abbott RL. Microbial changes in the ocular environment with contact lens wear. CLAO J. 1992;18:53–5.
[42] Graham JE, Moore JE, Jiru X, Moore JE, Goodall EA, Dooley JS, Hayes VE, Dartt DA, Downes CS, Moore TC. Ocular pathogen or commensal: a PCR-based study of surface bacterial flora in normal and dry eyes. Invest. Ophthalmol. Vis. Sci. 2007;48:5616–23.
[43] Hsu HY, Lind JT, Tseng L, Miller D. Ocular flora and their antibiotic resistance patterns in the midwest: a prospective study of patients undergoing cataract surgery. Am. J. Ophthalmol. 2013;155:36–44.e2.
[44] Ozkan J, Zhu H, Gabriel M, Holden BA, Willcox MD. Effect of prophylactic antibiotic drops on ocular microbiota and physiology during silicone hydrogel lens wear. Optom. Vis. Sci. 2012;89:326–35.
[45] Cavuoto KM, Banerjee S, Miller D, Galor A. Composition and comparison of the ocular surface microbiome in infants and older children. Transl. Vision Sci. Technol. 2018;7. 16–16.
[46] Zhou Y, Holland MJ, Makalo P, Joof H, Roberts CH, Mabey DC, Bailey RL, Burton MJ, Weinstock GM, Burr SE. The conjunctival microbiome in health and trachomatous disease: a case control study. Genome Med. 2014;6:99.
[47] Teweldemedhin M, Gebreyesus H, Atsbaha AH, Asgedom SW, Saravanan M. Bacterial profile of ocular infections: a systematic review. BMC Ophthalmol. 2017;17:212. 212.
[48] O'sullivan O, Coakley M, Lakshminarayanan B, Conde S, Claesson MJ, Cusack S, Fitzgerald AP, O'toole PW, Stanton C, Ross RP. Alterations in intestinal microbiota of elderly Irish subjects post-antibiotic therapy. J. Antimicrob. Chemother. 2013;68:214–21.
[49] Terzulli M, Contreras-Ruiz L, Kugadas A, Masli S, Gadjeva M. TSP-1 deficiency alters ocular microbiota: implications for Sjogren's syndrome pathogenesis. J. Ocul. Pharmacol. Ther. 2015;31:413–8.
[50] Berer K, Mues M, Koutrolos M, Rasbi ZA, Boziki M, Johner C, Wekerle H, Krishnamoorthy G. Commensal microbiota and myelin autoantigen cooperate to trigger autoimmune demyelination. Nature 2011;479:538–41.
[51] Tai N, Peng J, Liu F, Gulden E, Hu Y, Zhang X, Chen L, Wong FS, Wen L. Microbial antigen mimics activate diabetogenic CD8 T cells in NOD mice. J. Exp. Med. 2016;213:2129–46.
[52] Obando-Pereda GA. GAKG-RGEKG an epitope that provokes immune cross-reactivity between *Prevotella* sp. and human collagen: evidence of molecular mimicry in chronic periodontitis. Autoimmune Dis. 2016:5472320.
[53] Pianta A, Arvikar SL, Strle K, Drouin EE, Wang Q, Costello CE, Steere AC. Two rheumatoid arthritis-specific autoantigens correlate microbial immunity with autoimmune responses in joints. J. Clin. Investig. 2017;127:2946–56.
[54] Rowan S, Jiang S, Korem T, Szymanski J, Chang ML, Szelog J, Cassalman C, Dasuri K, Mcguire C, Nagai R, Du XL, Brownlee M, Rabbani N, Thornalley PJ, Baleja JD, Deik AA, Pierce KA, Scott JM, Clish CB, Smith DE, Weinberger A, Avnit-Sagi T, Lotan-Pompan M, Segal E, Taylor A. Involvement of a gut-retina axis in protection against dietary glycemia-induced age-related macular degeneration. Proc. Natl. Acad. Sci. U. S. A. 2017;114:E4472–e4481.
[55] Asquith M, Rosenbaum JT. The interaction between host genetics and the microbiome in the pathogenesis of spondyloarthropathies. Curr. Opin. Rheumatol. 2016;28:405–12.
[56] Davis JL, Mittal KK, Freidlin V, Mellow SR, Optican DC, Palestine AG, Nussenblatt RB. HLA associations and ancestry in Vogt-Koyanagi-Harada disease and sympathetic ophthalmia. Ophthalmology 1990;97:1137–42.
[57] Nussenblatt RB, Mittal KK, Ryan S, Green WR, Maumenee AE. Birdshot retinochoroidopathy associated with HLA-A29 antigen and immune responsiveness to retinal S-antigen. Am. J. Ophthalmol. 1982;94:147–58.
[58] Rosenbaum JT, Lin P, Asquith M. The microbiome, HLA, and the pathogenesis of uveitis. Jpn. J. Ophthalmol. 2016;60:1–6.

[59] Nakamura YK, Janowitz C, Metea C, Asquith M, Karstens L, Rosenbaum JT, Lin P. Short chain fatty acids ameliorate immune-mediated uveitis partially by altering migration of lymphocytes from the intestine. Sci. Rep. 2017;7:11745.
[60] Nakamura YK, Metea C, Karstens L, Asquith M, Gruner H, Moscibrocki C, Lee I, Brislawn CJ, Jansson JK, Rosenbaum JT, Lin P. Gut microbial alterations associated with protection from autoimmune uveitis. Invest. Ophthalmol. Vis. Sci. 2016;57:3747–58.
[61] Zhang X, Volpe EA, Gandhi NB, Schaumburg CS, Siemasko KF, Pangelinan SB, Kelly SD, Hayday AC, Li DQ, Stern ME, Niederkorn JY, Pflugfelder SC, de Paiva CS. NK cells promote Th-17 mediated corneal barrier disruption in dry eye. PLoS One 2012;7:e36822.
[62] Macpherson AJ, Koller Y, Mccoy KD. The bilateral responsiveness between intestinal microbes and IgA. Trends Immunol. 2015;36:460–70.
[63] Wells PA, Hazlett LD. Immunocytochemical localization of immunoglobulins at the corneal surface of the mouse. Exp. Eye Res. 1985;40:779–96.
[64] Cerutti A. The regulation of IgA class switching. Nat. Rev. Immunol. 2008;8:421–34.
[65] Mazanec MB, Nedrud JG, Kaetzel CS, Lamm ME. A three-tiered view of the role of IgA in mucosal defense. Immunol. Today 1993;14:430–5.
[66] Peterson DA, Mcnulty NP, Guruge JL, Gordon JI. IgA response to symbiotic bacteria as a mediator of gut homeostasis. Cell Host Microbe 2007;2:328–39.
[67] Pilette C, Detry B, Guisset A, Gabriels J, Sibille Y. Induction of interleukin-10 expression through Fcalpha receptor in human monocytes and monocyte-derived dendritic cells: role of p38 MAPKinase. Immunol. Cell Biol. 2010;88:486–93.
[68] Allansmith MR, Gudmundsson OG, Hann LE, Keys C, Bloch KJ, Taubman MA, Sullivan DA. The immune response of the lacrimal gland to antigenic exposure. Curr. Eye Res. 1987;6:921–7.
[69] Casola S, Otipoby KL, Alimzhanov M, Humme S, Uyttersprot N, Kutok JL, Carroll MC, Rajewsky K. B cell receptor signal strength determines B cell fate. Nat. Immunol. 2004;5:317–27.
[70] Chaudhuri J, Alt FW. Class-switch recombination: interplay of transcription, DNA deamination and DNA repair. Nat. Rev. Immunol. 2004;4:541–52.
[71] Dullaers M, Li D, Xue Y, Ni L, Gayet I, Morita R, Ueno H, Palucka KA, Banchereau J, Oh S. A T cell-dependent mechanism for the induction of human mucosal homing immunoglobulin A-secreting plasmablasts. Immunity 2009;30:120–9.
[72] Gill RF, Pirockinaite G, O'sullivan NL, Montgomery PC. Nasal-associated lymphoid tissue is not an absolute requirement for the induction of rat tear IgA antibody responses. Curr. Eye Res. 2010;35:1–8.
[73] Gudmundsson OG, Cohen EJ, Greiner JV, Taubman MA, Allansmith MR. Mononuclear and IgA-containing cells in the lacrimal gland of germ-free and conventional rats. Exp. Eye Res. 1984;39:575–81.
[74] Tazume S, Umehara K, Matsuzawa H, Yoshida T, Hashimoto K, Sasaki S. Immunological function of food-restricted germfree and specific pathogen-free mice. Jikken Dobutsu 1991;40:523–8.
[75] Al-Asmakh M, Zadjali F. Use of germ-free animal models in microbiota-related research. J. Microbiol. Biotechnol. 2015;25:1583–8.
[76] Kandori H, Hirayama K, Takeda M, Doi K. Histochemical, lectin-histochemical and morphometrical characteristics of intestinal goblet cells of germfree and conventional mice. Exp. Anim. 1996;45:155–60.
[77] Taniguchi T, Minami Y. The IL-2/IL-2 receptor system: a current overview. Cell 1993;73:5–8.
[78] Willerford DM, Chen J, Ferry JA, Davidson L, Ma A, Alt FW. Interleukin-2 receptor alpha chain regulates the size and content of the peripheral lymphoid compartment. Immunity 1995;3:521–30.
[79] Cobbold SP, Nolan KF, Graca L, Castejon R, Le MA, Frewin M, Humm S, Adams E, Thompson S, Zelenika D, Paterson A, Yates S, Fairchild PJ, Waldmann H. Regulatory T cells and dendritic cells in transplantation tolerance: molecular markers and mechanisms. Immunol. Rev. 2003;196:109–24.
[80] Junghans RP, Waldmann TA. Metabolism of Tac (IL2Ralpha): physiology of cell surface shedding and renal catabolism, and suppression of catabolism by antibody binding. J. Exp. Med. 1996;183:1587–602.
[81] Sharma R, Bagavant H, Jarjour WN, Sung SS, Ju ST. The role of Fas in the immune system biology of IL-2R alpha knockout mice: interplay among regulatory T cells, inflammation, hemopoiesis, and apoptosis. J. Immunol. 2005;175:1965–73.
[82] de Paiva CS, Hwang CS, Pitcher 3rd JD, Pangelinan SB, Rahimy E, Chen W, Yoon KC, Farley WJ, Niederkorn JY, Stern ME, Li DQ, Pflugfelder SC. Age-related T-cell cytokine profile parallels corneal disease severity in Sjogren's syndrome-like keratoconjunctivitis sicca in CD25KO mice. Rheumatology 2010;49:246–58.
[83] Rahimy E, Pitcher 3rd JD, Pangelinan SB, Chen W, Farley WJ, Niederkorn JY, Stern ME, Li DQ, Pflugfelder SC, De Paiva CS. Spontaneous autoimmune dacryoadenitis in aged CD25KO mice. Am. J. Pathol. 2010;177:744–53.
[84] Sharma R, Zheng L, Guo X, Fu SM, Ju ST, Jarjour WN. Novel animal models for Sjogren's syndrome: expression and transfer of salivary gland dysfunction from regulatory T cell-deficient mice. J. Autoimmun. 2006;27:289–96.
[85] Zheng L, Sharma R, Gaskin F, Fu SM, Ju ST. A novel role of IL-2 in organ-specific autoimmune inflammation beyond regulatory T cell checkpoint: both IL-2 knockout and Fas mutation prolong lifespan of Scurfy mice but by different mechanisms. J. Immunol. 2007;179:8035–41.
[86] Cha S, Brayer J, Gao J, Brown V, Killedar S, Yasunari U, Peck AB. A dual role for interferon-gamma in the pathogenesis of Sjogren's syndrome-like autoimmune exocrinopathy in the nonobese diabetic mouse. Scand. J. Immunol. 2004;60:552–65.
[87] Saito I, Terauchi K, Shimuta M, Nishiimura S, Yoshino K, Takeuchi T, Tsubota K, Miyasaka N. Expression of cell adhesion molecules in the salivary and lacrimal glands of Sjogren's syndrome. J. Clin. Lab. Anal. 1993;7:180–7.
[88] Tsubota K, Fukagawa K, Fujihara T, Shimmura S, Saito I, Saito K, Takeuchi T. Regulation of human leukocyte antigen expression in human conjunctival epithelium. Investig. Ophthalmol. Vis. Sci. 1999;40:28–34.
[89] de Paiva CS, Rocha EM. Sjogren syndrome: what and where are we looking for? Curr. Opin. Ophthalmol. 2015;26:517–25.

[90] Pelegrino FS, Volpe EA, Gandhi NB, Li DQ, Pflugfelder SC, De Paiva CS. Deletion of interferon-gamma delays onset and severity of dacryoadenitis in CD25KO mice. Arthritis Res. Ther. 2012;14:R234.
[91] Atarashi K, Tanoue T, Oshima K, Suda W, Nagano Y, Nishikawa H, Fukuda S, Saito T, Narushima S, Hase K, Kim S, Fritz JV, Wilmes P, Ueha S, Matsushima K, Ohno H, Olle B, Sakaguchi S, Taniguchi T, Morita H, Hattori M, Honda K. Treg induction by a rationally selected mixture of Clostridia strains from the human microbiota. Nature 2013;500:232–6.
[92] He B, Hoang TK, Wang T, Ferris M, Taylor CM, Tian X, Luo M, Tran DQ, Zhou J, Tatevian N, Luo F, Molina JG, Blackburn MR, Gomez TH, Roos S, Rhoads JM, Liu Y. Resetting microbiota by *Lactobacillus reuteri* inhibits T reg deficiency-induced autoimmunity via adenosine A2A receptors. J. Exp. Med. 2017;214:107–23.
[93] Kawashima M, Nakamura S, Izuta Y, Inoue S, Tsubota K. Dietary supplementation with a combination of lactoferrin, fish oil, and *Enterococcus faecium* WB2000 for treating dry eye: a rat model and human clinical study. Ocul. Surf. 2016;14:255–63.
[94] Kim J, Choi SH, Kim YJ, Jeong HJ, Ryu JS, Lee HJ, Kim TW, Im SH, Oh JY, Kim MK. Clinical effect of IRT-5 probiotics on immune modulation of autoimmunity or alloimmunity in the eye. Nutrients 2017;9.

Further reading

[1] Kugadas A, Gadjeva M. Impact of microbiome on ocular health. Ocul. Surf. 2016;14:342–9.

Chapter 7

CRISPR technology for genome editing

Gargi Bhattacharjee[1], Indra Mani[2], Nisarg Gohil[1], Khushal Khambhati[1], Darren Braddick[3], Happy Panchasara[1] and Vijai Singh[1,4]

[1]*School of Biological Sciences and Biotechnology, Institute of Advanced Research, Koba Institutional Area, Gandhinagar, Gujarat, India;* [2]*Department of Microbiology, Gargi College, University of Delhi, New Delhi, Delhi, India;* [3]*Department of R&D, Cementic S.A.S., Paris, France;* [4]*Present address: Department of Biosciences, School of Sciences, Indrashil University, Rajpur, Gujarat, India*

Introduction

Synthetic biology is a rapidly growing field that deals with design and construction of synthetic biological parts, devices, and circuits, and it holds great potential for its use in therapeutic, biomedical, industrial, and biotechnological applications [1]. It was developed to ensure tight and regulatory control over gene expression and to cure wide range of diseases resulting from single mutation, or defect in genes that may get easily transferred from generation to generation, such as in case of sickle cell anemia [2], hemophilia [3], Duchenne muscular dystrophy [4], certain modalities of cataract [5], and cystic fibrosis [6].

Genome engineering is a key tool for the correction or repair of genetic defects either to cure the disease or to use it for creating animal disease models. It has been developed for accelerating genomics and genome editing [7]. In the past decade, genome editing tools such as Zinc-finger nucleases (ZFNs) and transcription activator-like effector nucleases (TALENs), both involving artificial endonucleases, have allowed to perform targeted genome editing [8]. However, these techniques have suffered from complexity, expense, and time issues.

Clustered regularly interspaced short palindromic repeats (CRISPRs) and CRISPR-associated proteins (Cas) are parts of an RNA-guided adaptive system in prokaryotes that confers protection against phages and virulent promiscuous plasmids. They are categorized into three major types, namely Type I, II, and III [9]. Type II CRISPR-Cas is one of the simplest systems requiring only a multidomain single Cas9 endonuclease. It is rapid, cost-effective, specific, and an efficient genome editing toolbox used in a wide range of organisms and cell types [10–13].

CRISPR-Cas9 has been extended for correction of genetic defects [14], removal of bacterial pathogens [15,16], and eradication of viruses such as HIV-1 [17], human papillomavirus [18], hepatitis B virus [19], Epstein-Barr virus from human cells [20], and many more.

Biogenesis and mechanism of CRISPR-Cas systems

CRISPR systems found within wild type organisms endow them an ability to distinguish between self and nonself DNA, by memorizing foreign DNA sequences and guiding endonuclease to lyze or destroy such DNA when encountered within the cell. For simplicity, the events performed during CRISPR-Cas mediated immune response are divided into three categories, commonly called adaptation, biogenesis, and interference. During adaptation, Cas proteins of CRISPR system distinguish DNA as self or nonself by detection of the protospacer adjacent motif (PAM). PAMs are three to six bp nucleotide motif which are recognized by the Cas proteins of CRISPR system. Once the PAM site is detected (usually found in the foreign genome), it is possible to integrate the sequence of a few base pairs adjacent to the PAM motif into CRISPR loci found within the host organism. This would integrate the sequence of few base pair adjacent to PAM motif into the CRISPR loci found within the host organism. This would instill a memory of the foreign DNA in the host. During biogenesis, the integrated foreign sequence would be expressed in the form of RNA, which on combining with tracrRNA (trans-activating CRISPR RNA) would form premature crRNA. Once the premature crRNA converts into mature crRNA, it is ready to guide the Cas endonuclease to make double-stranded breaks into the foreign DNA sequence.

TracrRNA holds an important role in mature crRNA, as it helps to bind Cas endonuclease to the "expressed foreign RNA sequence." Interference is the step in which mature crRNA would guide Cas endonuclease to target the foreign

Precision Medicine for Investigators, Practitioners and Providers. https://doi.org/10.1016/B978-0-12-819178-1.00007-1

sequence upon entering the cell, and to subsequently make double-stranded breaks into the foreign DNA. The foreign DNA complementary to the mature crRNA (other than the tracrRNA region) and adjacent to the PAM motif would be subjected to these double-strand breaks. Without crRNA, the Cas endonuclease would remain inactive in the cell [12,13].

In order to create a genome editing technology, the chosen cell should express Cas9 endonuclease and single guide RNA (sgRNA) that form a complex in the target region in the presence of PAM, creating a double-stranded break (DSB), which may be repaired by nonhomologous end joining (NHEJ) or homology directed repair (HDR) mechanism (Fig. 7.1). This allows to create gene knockout, knock-in, addition, deletion, and many more kinds of modifications [12,13]. It is currently used in animals, plants, insects, and microorganisms.

Therapeutic use of CRISPR-Cas9 in humans and mammals

CRISPR-Cas9 was used to eliminate the gene that encodes C-C chemokine receptor 5 (CCR5) in human embryos, with possibility of rendering the offspring resistant to HIV, smallpox, and cholera [21]. The most commonly targeted genes that are used for editing by CRISPR-Cas9, in human and mammalian diseases, are listed in Table 7.1.

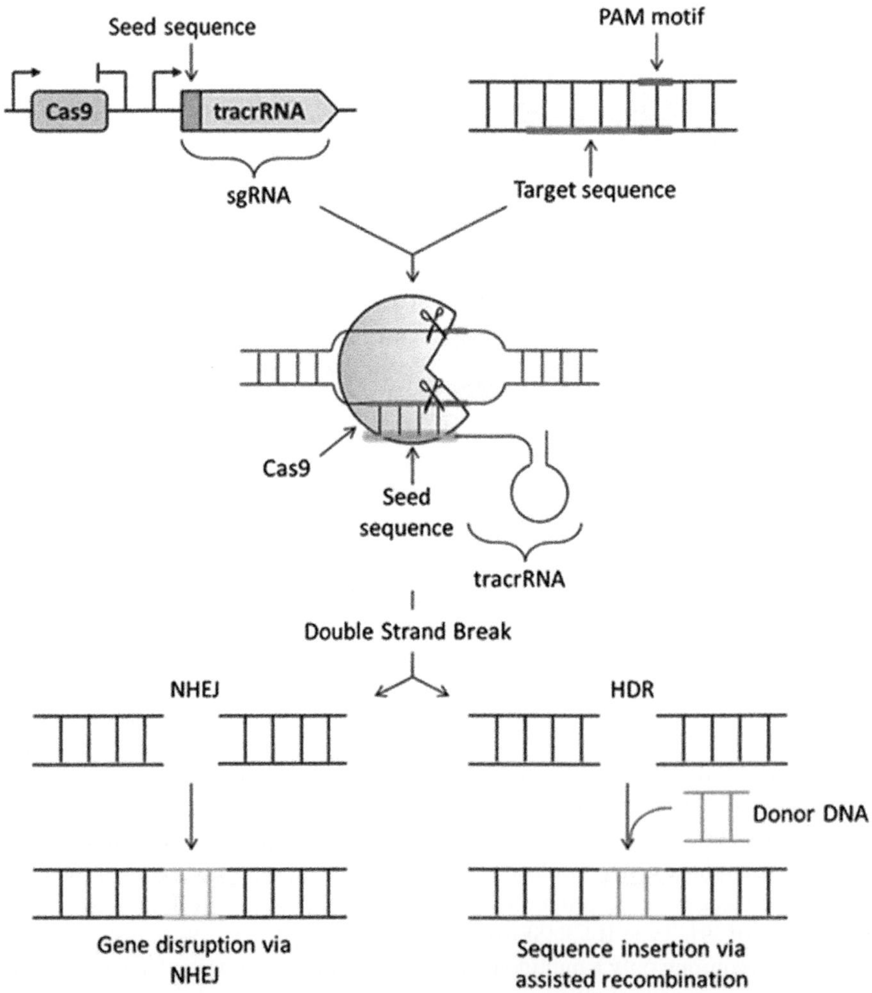

FIGURE 7.1 Schematic representation of targeted genome editing through CRISPR-Cas9 system. Cas9 and sgRNA form a complex which recognizes the PAM site, a trinucleotide sequence lying adjacent to the target sequence. Cas9 being an endonuclease, it introduces a DSB in a sequence-specific manner. DSB is later repaired through two different DNA mechanisms, including nonhomologous end joining (NHEJ) and homology directed repair (HDR). NHEJ creates error prone deletion or insertion of nucleotide, within the targeted region of DNA, leading to premature stop codon. In HDR a short DNA template, with sequence homologous to the strands flanking the target site, is introduced at the point of repair or gene insertion (*blue* [light gray in print version]) [10,12].

TABLE 7.1 A list genes edited through CRISPR-Cas9 for inherited disease therapy.

Disease	Gene target	References
Cancer	Chimeric antigen receptor (CAR)	[26]
Cataract	*Crygc* gene	[5]
Duchenne muscular dystrophy	Dystrophin gene (DMD)	[29]
Tyrosinemia	Fumarylacetoacetate hydrolase (FAH)	[31]
Cystic fibrosis	*CFTR* locus	[6]
Beta-thalassemia	Human hemoglobin beta (*HBB*) gene	[42]
Urea cycle disorder	Hypoxanthine-guanine phosphoribosyl transferase	[38]
Sickle cell anemia	*HBB* locus	[47]
Blindness	*Nrl* gene	[49]
Cardiovascular disease	*Gper1*	[52]
Neurological disease	*IGF*-1 gene	[56]
Huntington disease	*mHTT*	[61]

Cancer

Cancer is the biggest reason for disease-associated mortality, with ever-increasing incidences globally [22]. Small molecules and antibodies can be of therapeutic interest, when they interfere with proteins of oncogenic signaling routes. Breakpoint cluster region protein-abelson murine leukemia viral oncogene homolog 1 (BCR-ABL) is targeted by imatinib in chronic myeloid leukemia [23] and epidermal growth factor receptor (EGFR), by specific antibodies in colorectal cancer [24] are prominent examples. In cancer immunotherapy, engineered T-cells could locate and kill cancer cells. The FDA (United States Food and Drug Administration) has approved two drugs for so-called chimeric antigen receptor (CAR) immunotherapy.

CRISPR-Cas9 gene knockouts in T-cells combined with the expression of CAR are an attractive model. Most approaches involve the inactivation of the programmed death-1 (PD-1) receptor in T-cells [25,26]. PD-1 regulates anticancer cell immune response, and is inhibited by receptor blocking ligands that are produced by cancer cells to evade immune response. Targeting PD-1 with dedicated RNA-guided nucleases (RGNs), which do not involve CRISPR, significantly increases abundance of antitumor cytotoxic T-cells (CTLs) in both solid and hematologic cancers [25,26]. Possibilities with CRISPR techniques are envisaged as well.

Cataract

In humans, genetics can underlie certain forms of congenital cataract and eventually age-related forms as well. A mouse model has been used to study the dominant cataract disorder, caused by a defined mutation in the *Crygc* gene. These mice carried a 1 bp deletion in exon 3 of *Crygc*, stopping codon at the 76th amino acid, thus preventing the production of truncated γC-crystallin in nuclear cataracts, in both homozygous and heterozygous mutant mice [27]. A dominant mutation in the *Crygc* gene could also be saved through coinjection of Cas9 mRNA and sgRNA into zygotes. The resulting mice were fertile, and able to transmit the corrected allele to offspring [5].

Duchenne muscular dystrophy

The prevalence of Duchenne muscular dystrophy (DMD) is 1 in approximately 5000 men worldwide and it is a lethal X-linked recessive neuromuscular disorder [28]. Due to mutations, there is disruption of the reading frame in the *dystrophin* gene, triggering the diseased state. *The dystrophin* gene (*DMD*) was elucidated by a loss of the dystrophin protein that leads to decrease of muscle membrane integrity and susceptibility to degeneration. CRISPR-Cas9 permanent exon skipping can restore the disrupted *DMD* reading frame, leading to dystrophin normalization [29]. Difficulties include offtarget effects, activation of the immune response, and suboptimal *in vivo* delivery. Another hurdle for clinical applicability is the wide spectrum of mutations [30].

Tyrosinemia

Hereditary tyrosinemia Type 1 (HT1) is an autosomal recessive metabolic disorder caused by mutations in the gene encoding fumaryl acetoacetate hydrolase (FAH) [31,32]. Due to the accumulation of toxic metabolites, HTI causes severe liver cirrhosis, liver failure and even hepatocellular carcinoma. Several strategies have been devised to improve HTI symptoms in animal models [31]. *Ex vivo* hepatocyte-directed gene editing using CRISPR-Cas9 could correct a mouse model with HT1, in which a single point mutation results in the loss of FAH function [32]. A *Fah* mutant rat model was developed, to investigate whether genome editing with Cas9 nickase (Cas9n) could correct the mutation and whether Cas9n could induce indels in both on-target and off-target sites in cell lines. Gene therapy stopped progression of liver cirrhosis in the *Fah* mutant rat, a measurable phenotype, which could not be achieved in the HTI mouse model [31].

Cystic fibrosis

Cystic fibrosis (CF) occurs in 1 of 2500 live births with a preponderance in people of northern European ancestry [33]. Early destruction of the pancreas, intestinal dysfunction and recurrent lung infection are major features [34]. CRISPR-Cas9 method introduced mutations in cystic fibrosis transmembrane conductance regulator (CFTR) locus of ovine fetal fibroblasts, followed by somatic cell nuclear transfer (SCNT), which gave rise to $CFTR^{-/-}$ and $CFTR^{+/-}$ lambs, consistent with the pathology of human CF [6]. Therapeutic attempts are going on in various centers.

Urea cycle disorder

Urea cycle disorders (UCDs) are infrequent enzymopathies with an incidence of 1 in 35,000 births, resulting in ~113 new cases per year in the United States [35]. These disorders are a leading cause of inherited hyperammonemia with potential loss of psychomotor function, intellectual disabilities, seizures, and death [36]. UCDs result from a deficiency in one of six hepatic enzymes, or two mitochondrial transporters, that regulate nitrogen metabolism and urea production [37]. A CRISPR-Cas9-based strategy to genetically modify the exon-1 of the hypoxanthine-guanine phosphoribosyltransferase locus could successfully re-establish arginase activity [38]. Patient-specific human induced pluripotent stem cells (iPSCs) involving genetically modified cell therapy could assist in the handling of urea cycle disorders.

Beta-thalassemia

β-thalassemia is caused by mutations of the (*HBB*) gene, leading to deficient β-globin expression [39]. Induced pluripotent stem cells (iPSCs) have been used as a promising tool for understanding disease pathophysiology and evaluating the effect of drugs and gene therapeutics [40]. CRISPR-Cas9 system is an option to edit the gene that encodes HBB [41]. CRISPR-Cas9, TALENs, and ZFNs were compared for their targeting efficiency on the HBB gene locus [42]. Indel frequency of CRISPR-Cas9 was superior to TALENs and ZFNs.

Sickle cell anemia

Sickle cell disease (SCD), a common occurrence in developing countries [43], is due to a single-nucleotide mutation of the human hemoglobin beta (*HBB)* gene. In deoxygenated situations, ischemia, anemia, hemolysis and multiorgan injury may ensue [44]. Correction of the SCD mutation in patient-derived iPSCs has been attempted through TALEN-induced HDR [45] and ZFN-based strategies [46]. CRISPR-Cas9 system has similarly been utilized resulting in improved efficacy, as compared to previous nucleases [47].

Genetic retinopathies

Hereditary forms of retinal degeneration such as retinitis pigmentosa often develop in early childhood [48]. CRISPR-Cas9 mediated *Nrl* gene (encoding for a neural retina-specific leucine zipper protein) disruption in rod cells is a promising option [49].

Cardiovascular disease and gut microbiota

Familial hypercholesterolemia (FH) affects more than 30 million people, with prevalence rates of 1/200–1/500 for heterozygous FH (HeFH) and 1/160,000–1/300,000 for homozygous FH (HoFH) [50]. CRISPR-Cas9 genome-editing technology addressed a three-base pair homozygous deletion in the low-density lipoprotein cholesterol receptor (LDLR) exon 4 in iPSCs derived from a patient with HoFH [51].

G-protein–coupled estrogen receptor (*Gper1*) is known to be involved in cardiovascular disease, but its systematic function in blood pressure control remains unclear. Genetically salt-sensitive hypertensive rats with edited *Gper1* exhibited lower blood pressure accompanied by altered microbiota and other changes. Microbiota transplantation from hypertensive $Gper1^{+/+}$ rats reversed the cardiovascular protective effect exerted by genomic deletion of *Gper1* revealing the role of *Gper1* in facilitating microbiota alterations, that in turn contribute to cardiovascular pathology [52,53].

Neurological disease

Amyotrophic lateral sclerosis (ALS) is a comparatively common motor neuron (MN) illness, and many patients die from respiratory failure [54]. About 10% are familial-based (fALS), and mutations in superoxide dismutase 1 (*SOD1*) gene account for a further 20%. Other possibilities are mutations in transactive response DNA-binding protein 43 and in chromosome 9 open reading frame 72 (C9ORF72) [55]. Deletion of insulinlike growth factor-I (IGF-1) gene in ALS mice achieved approximately 50% reduction in the levels of the IGF-1 mRNA and protein. Knocking out IGF-1 increased the level of Bax, decreasing Bcl-xl and Bcl-2 levels in mitochondria, and inducing release of CytC into the cytosol, suggesting that apoptosis was activated after IGF-1 deletion in this study [56].

Huntington disease

Huntington disease (HD) is an autosomal dominant inherited neurodegenerative disease believed to be primarily caused by toxic gains of function [57]. Genetic testing and intervention in the long premanifestation phase could postpone or prevent later symptom onset [58]. Extensions of a CAG repeat in the huntington *HTT* gene result in the ubiquitous expression of mutant huntington protein (mHTT), thought to be the predominant toxic species [59], and accordingly the reduction of mHTT production would improve pathogenic outlook.

There is some evidence that loss of function through mHTT contributes to HD, which could be exacerbated by reducing HTT production [60]. There are reports about CRISPR-Cas9 use to permanently deactivate the mutant *HTT* allele in patient-derived fibroblasts, using two constructs to excise a large region of *HTT* DNA resulting in a near-total reduction in both the mHTT protein and RNA [61], similarly to the HD rodent model [62].

Antimicrobial interventions

Multidrug bacterial resistance (MDR) is an emerging hospital scourge, and discovery of new antimicrobial agents has come to a standstill in recent years [63]. Transfer of MDR plasmids by bacterial conjugation is a well-known mechanism for dissemination of the trouble. CRISPR-Cas9 engineered plasmids could remove resistance-harboring plasmids from bacteria, thus resensitizing them to antibiotics.

Genome engineering of the gut microbiota

As alluded to, virulence-causing and antimicrobial resistant genes from a wide range of the bacteria can be eradicated [15,16,64]. CRISPR-Cas9 could selectively eliminate a particular bacterial strain out of a mixed population of *Escherichia coli*, with plasmids expressing Cas9 and sgRNA unique to each genotype [64]. Phagemid-based transduction of CRISPR-Cas, with sequence-specific lethality in virulence-gene carrying *S. aureus*, successfully removed plasmids encoding AMR genes [15]. The efficiency of these CRISPR-Cas9 phagemids was tested on mice skin colonized with *S. aureus*. In *E. coli*, selective elimination of AMR genes from plasmids using programmed CRISPR-Cas9 constructs was accomplished, and *in vivo* efficiency was examined on *Galleria mellonella* larvae, exposed to enterohemorrhagic *E. coli*. [16].

A pronounced shortfall that was discovered was the fact that the re-sensitized cells lacked a selective benefit over the resistant ones, which allowed the untamed resistant bacteria to perpetuate indefinitely within the target population. Using temperate and lytic phage to resensitize bacteria to β-lactam antibiotics seems possible. Temperate phage-mediated delivery extended resistance to even the lytic phage, thereby increasing selectivity for challenged bacteria [65].

The complication of manipulating highly diverse sequence variants of β-lactamase genes was overcome by implementing CRISPR-Cas9 against broad-spectrum β-lactamase gene classes in *E. coli* that shared target sequences across more than 200 mutational variants [66].

Human immunodeficiency virus

HIV-1/AIDS remains a major public health threat, affecting more than 35 million individuals worldwide [67]. Combination antiretroviral therapy (ART) has proven efficacious; however, the virus resides latently in the infected cells, including resting CD4+ T cells, macrophages, and follicular dendritic cells, in some circumstances resuming replication cycle shortly after termination of therapy [68].

Available antiviral strategies lack the ability to clear the integrated proviral genome, leaving chances for viral rebound [69]. Sideeffects of ART, escape by untreated mutants, and morbidities in aging populations are other concerns [70].

With CRISPR-Cas9 technology, inactivation of viral gene replication and expression in latently infected cells such as microglia, T-cells and promonocytes becomes feasible, paving the way for vanquishing latent viruses that are either unapproachable or escape existing therapies. Whole-genome sequencing analysis and SURVEYOR assays reported no off-target editing or genotoxicity, arising from CRISPR [71]. CRISPR tested at 10 sites of HIV-1 DNA, in latently infected JLat10.6 cells, revealed mutations at each of the target sites. The second exon of Rev displayed the highest degree of mutation, thereby reducing HIV-1 gene expression and viral load by 20-fold [17].

Human iPSCs, engineered to express the HIV-targeted CRISPR-Cas9 system could be conferred resistance against HIV-1 by functionally differentiating into HIV reservoir cell types [72]. Genes expressing crucial viral proteins such as CCR5, a coreceptor that facilitates the entry of HIV-1, are other alternatives to ward off HIV infection [73].

Transduction of lentiviral vectors expressing CCR5-specific gRNAs and Cas9 into HIV-1 susceptible human CD4+ cells, presented a high degree of CCR5 gene disruption. This also resulted in resistance against R5-tropic HIV-1 with better results than cells with undisrupted CCR5 during R5-tropic HIV-1 infection [20]. Several other groups have attempted to target the latently infected cell pool [74,75].

Epstein-Barr virus

Epstein-Barr virus (EBV) is a human lymphocryptovirus associated with Burkitt lymphoma, mononucleosis, nasopharyngeal carcinoma and gastric cancer. The genetic material is comprised of linear double-stranded DNA sized about 168−184 kbp, with 85 genes, an internal repeat region, and terminal repeat regions [76]. The primary targets of EBV are the B-lymphocytes and epithelial cells where it may reside latently for several years. During latency, the genome circularizes and exists within the host cell as an episome, maintaining a steady copy number [77].

During latency, only the genes of utmost importance are expressed. These include EBNA-1 and EBNA-2 that are involved in the process of viral DNA replication and cellular transformation. EBNA-2 in turn controls the expression of a few other important latency-associated genes such as LMP-1 and LMP-2 which send out active signals for the proliferation of infected B-cells. LMP-1 and EBNA proteins, being highly antigenic, are ideally destroyed by cytotoxic-T cells (CTLs) in healthy individuals, although in immunosuppressed backgrounds, CTLs typically fail to clear the cells [78]. LMP-1 in association with many other latency-associated proteins including EBV-encoded small RNA-1 and RNA-2 (EBER1 and EBER2), and BamHI rightward transcripts (BARTs) can regulate cellular transformation in Burkitt lymphoma and nasopharyngeal carcinoma [79].

CRISPR-Cas9 mediated targeting of essential viral sequences such as EBNA1, EBNA3C and LMP1 induced marked declines in viral proliferation and helped to restore apoptotic pathways in Raji (Burkitt) cells [20]. A targeted deletion of 558bp from the promoter locus of BART (BamHI A rightward transcript) using two gRNAs (i.e., dual targeted gRNA) led to loss of function in BART expression with absolutely no off-target cleavage [80]. EBVs can very easily transform primary B-cells into immortalized lymphoblastoid cells (LCLs), a basis for some vital genetic informations.

A limitation of CRISPR editing in LCLs is low lentiviral transduction frequency that often hinders outcomes, demanding coexpression of multiple additional components. This barrier could be circumvented through LCL nucleofection of Cas9/sgRNA ribonucleoprotein complexes (RNPs) which have already been shown to ensure a high degree of reproducibility [81].

Human papillomavirus

Human papillomaviruses (HPVs) are small DNA viruses with a genome size of around 8 kb. They infect cutaneous or mucosal epithelial cells, genital tissues and the upper respiratory tract. Out of the 200 genetically discrete subtypes of HPV, HPV16 and HPV18 account for almost 70% of cervical malignancies, while HPV16 alone has been reported to be responsible for >80% of HPV^+ anogenital and head-and-neck cancers [82]. HPV-induced tumors basically occur as a function of two HPV-encoded oncogenes, E6 and E7, the chief targets of which are host tumor suppressor proteins.

E6 binds and blocks p53, while E7 forms a complex with the retinoblastoma family of tumor suppressor proteins (pRb, p107 and p130) which helps the infected cells to escape cell-cycle checkpoints, promoting progression [83]. Targeting E6 and E7 genes through CRISPR could help eliminate HPV induced cancers, restoring natural cell cycle arrest and apoptosis.

Indel mutations in E6 and E7 genes of HeLa and SiHa cervical carcinoma cell lines integrated with HPV18 or HPV16, respectively, activated p53 or pRb and restored cell-cycle checkpoints upon introduction of E6- or E7-specific gRNAs and Cas9 into the cell [84,85].

In vitro analysis confirmed that HPV16-E7-specific sgRNA could bring about nicks at only desired sites of HPV16-E7 DNA with no effect on HPV negative C33A and HEK293 cells [71]. Disruption of E7 DNA led to downregulation of E7 protein and upregulation of tumor suppressor protein pRb [71]. CRISPR-Cas9 mediated editing of the promoter as well as the transcribing region of HPV16 E6 and E7, followed by its transduction into the SiHa cells led to the accumulation of p53 and p21 proteins while rate of cell proliferation gradually dwindled.

Inhibition of tumorigenesis was observed in nude mice [86]. CRISPR also allowed HPV16 genome editing in primary human anal cancer cells *in vivo*. AAV (Adeno-associated virus) vectors used for delivery of *S. aureus* Cas9 gene and cognate sgRNAs into tumor cells resulted in diminished tumor volume [87]. Three high-risk HPV-positive cervical cancer cell lines (HeLa, HCS-2, and SKG-I) were transfected along with the constitutively expressed Cas9 gene. Multiple mutations were reported after transduction with AAV-sgE6. The generated indels unveiled reduction in E6 expression with subsequent increase in p53 expression and apoptosis. Intratumoral administration of AAV-sgE6 repressed subcutaneous tumor growth while no adverse effects were observed [88].

Hepatitis B virus

There are over 350 million chronic HBV (hepatitis B virus) carriers worldwide [89]. HBV is a small, circular, hepadnavirus with a partially double-stranded DNA genome and may lead to hepatic cirrhosis, liver failure and hepatocellular carcinoma (HCC). The HBV genome persists as an episomal covalently closed circular DNA (cccDNA) that is highly stable within CHB patients and assists viral mRNA and pregenomic RNA synthesis by serving as a template. Once HBV DNA becomes integrated into host genome, it may cause the alteration or disruption of several key host genes associated with proliferative signaling and survival [90]. Nucleotide analogs (NAs) or interferons restrict viral reverse transcriptase activity and repress HBV replication; however, these alone have limited chance to fully abolish remaining replicative viruses with inherent cccDNA [91] or integrated HBV DNA [92].

CRISPR-Cas9 editing in Huh-7 hepatocyte-derived cellular carcinoma cells and mouse models transfected with HBV-expression vector disrupted HBV templates and intrahepatic HBV genome-containing plasmids, promoting their clearance, both *in vitro* and *in vivo.* [93] In HepG2 hepatoma cells exhibiting HBV receptors, Cas9 mediated cleavage in cccDNA followed by NHEJ DNA repair mechanism resulted in 8-fold inhibition of HBV infections [94].

Inhibition against HBV DNA production was also observed in *in vitro* models following lentiviral transduction of Cas9 and HBV-specific gRNAs. The result was 1000-fold reduction of total viral load with a coincident 10-fold decrease observed for cccDNA levels [95]. Excision of 3175-bp integrated HBV DNA fragment along with disruption of HBV cccDNA in a stable HBV cell line resulted in large reduction of HBV cccDNA, HBV DNA, HBeAg and HBsAg which remained below the negative critical values for more than 10 months [90].

Importance for healthcare providers and institutions

Editas Medicine (USA) is one of the pharmaceutical companies seriously investing in many human diseases that are expected to be treated through use of CRISPR-Cas9 system. CRISPR-Cas9 has also been successfully used as an antimicrobial agent for treating human pathogenic bacteria such as *S. aureus*, with a focus on eliminating strains that present rampant drug resistance. Bacteriophage-based delivery systems that can allow for specific treatment of a harmful pathogen without killing the background gut microbiota (i.e., the beneficial microorganisms) are an exciting prospect. CRISPR pills could treat multidrug-resistant pathogens, taken similarly to other orally available medicines. A number of human viruses (HIV, HBV and HPV) has been inhibited and eradicated from *in vitro* and *in vivo* experimental backgrounds, providing preclinical information for human trials.

CRISPR has enabled ultrasensitive (atto-molar detection limit) diagnostic tools, for pathogens and otherwise undetectable diseases, by companies like Mammoth Biosciences (USA). Knockout cell lines, reporter cell lines, and plasmid reagents for nucleic acid transfection and gene editing, all boasting CRISPR technology, are routinely commercialized in the international market, at moderate prices.

Research opportunities in diagnostics

Cas versions have been identified and modified for gene regulation, loci imaging, epigenetic analysis and diagnostics. One of the mutant versions of Cas9 is referred to as dead Cas9 (dCas9), a catalytically inactive form that retains the ability to recognize the particular targeted sequence via sgRNA, as Cas9 does, however without generating double-strand breaks. This version has been used for CRISPR interference (CRISPRi) technology, which has facilitated developments in gene regulation, loci imaging and epigenetics, A test kit joining dCas9 with a magnetic nanoparticle (dCas9-MN), along with its corresponding gRNA binds with the target sequence of interest, if it is present in the sample. The magnetic nanoparticles would help to separate the entire bound complex (dCas9-MN) together with the target sequence from components through magnetic separation. In the absence of the target sequence in the sample, only the dCas9-MN guide-RNA complex would be retrieved. Addition of SYBR stain would give a fluorescence signal for the target sequence which could be recognized as a positive signal and measured by spectrofluorimetry. This concept has been successfully demonstrated by the sensitive detection of *mecA* gene of methicillin-resistant *S. aureus* (MRSA) within a concentration of 10 CFU/mL of the bacteria [96].

Cas13a is an endonuclease, similar to better known Cas9, found in type IV Class 2 systems which has shown great potential as point-of-care diagnostics [97]. A nonspecific RNase, upon activation through target recognition, it can be harnessed to detect nucleic acids in given samples at attomolar-level concentrations [97]. RNA or DNA recognition in samples doesn't demand expensive instrument usage and combined with lateral flow assays is well-suited for pathogen detection in field trials [98]. Zika and HIV viruses, *K. pneumoniae* and *E. coli*, as well as SNPs and various cancers can be listed as successful examples [97].

Essentially, a gRNA corresponding to Cas13a is designed as a complex and Cas13a would engage its nuclease activity on target recognition (i.e., an RNA molecule complementary to the seed sequence of gRNA), eliciting its nonspecific RNase activity. The target RNA molecule from genomic dsDNA would be formed with the help of RPA (recombinase polymerase amplification) following T7 transcription.

If the sample contains nucleic acid in the form of RNA then it would undergo RT-RPA (reverse transcriptase recombinase polymerase amplification) for sequence amplification which would then be followed by T7 transcription [97,99]. With the probe RNA molecule already present in the solution, the presence of a quencher molecule on one end and a fluorophore molecule on another end would aid detection as the nonspecific RNase activity of Cas13a would cleave the RNA probe molecule and liberate the fluorophore molecule, expressing fluorescence and undergoing detection.

The protocol (commonly referred to as Specific High-sensitivity Enzymatic Reporter unlocking-SHERLOCK) necessitates standard instrumentation for signal detection, which can be circumvented with lateral flow assays (SHERLOCK v2) [98]. The whole cost for detection can thus be minimized to as little as USD 0.61$ per test. Variants of the Cas13 endonuclease and their assorted targeting preferences to different RNA probe sequences enable detection of different nucleic acid sequences in a single reaction mixture, using specific fluorescence signals. The drawback is that a fluorescence detector system is indispensable for multiplex nucleic acid detection in a single reaction [98]. Systems such as SHERLOCK v2 of multiplexed and portable nucleic acid detection platforms, with Cas13, Cas12a and Csm6 could be helpful in epidemic outbreaks for rapid and sensitive pathogen detection, as well as for easy and cheap genotyping in other fields.

Conclusions and future perspectives

CRISPR-Cas applications in humans can raise ethical issues, demanding proper legislation, decent discussions and debates, and the balancing of the benefit of the technology for improving human life against the threats of nefarious misinterpretation and misappropriation.

Further research needs to be done in order to use CRISPR more precisely and effectively in therapeutics, biomedical, industrial and biotechnological applications in order to deliver its large potential.

Abbreviations

ALL Acute lymphoblastic leukemia
BCR Breakpoint cluster region protein
CAR Chimeric antigen receptor
Cas9 CRISPR associated protein 9
CCR5 C-C chemokine receptor 5
CRISPR Clustered regularly interspaced short palindromic repeats
crRNAs CRISPR RNAs
DSBs Double-stranded breaks,
HDR Homology directed repair
HEK Human embryonic kidney cells
NHEJ Nonhomologous end joining
PAM Protospacer adjacent motif
sgRNA Single guide RNA
tracrRNA Transactivating CRISPR RNA

Acknowledgment

This work was supported by The Puri Foundation for Education in India.

References

[1] Singh V. Recent advancements in synthetic biology: current status and challenges. Gene 2014;535(1):1–11.
[2] Rees DC, Williams TN, Gladwin MT. Sickle-cell disease. Lancet 2010;376(9757):2018–31.
[3] Manco-Johnson MJ, Abshire TC, Shapiro AD, et al. Prophylaxis versus episodic treatment to prevent joint disease in boys with severe hemophilia. N. Engl. J. Med. 2007;357(6):535–44.
[4] Long C, McAnally JR, Shelton JM, Mireault AA, Bassel-Duby R, Olson EN. Prevention of muscular dystrophy in mice by CRISPR/Cas9–mediated editing of germline DNA. Science 2014;345(6201):1184–8.
[5] Wu Y, Liang D, Wang Y, et al. Correction of a genetic disease in mouse via use of CRISPR-cas9. Cell Stem Cell 2013;13:659–62.
[6] Fan Z, Perisse IV, Cotton CU, et al. A sheep model of cystic fibrosis generated by CRISPR/Cas9 disruption of the *CFTR* gene. JCI Insight 2018;3(19):e123529.
[7] Gibson TJ, Johnson CB, Murrell JC, Chambers JP, Stafford KJ, Mellor DJ. Components of electroencephalographic responses to slaughter in halothane-anaesthetised calves: effects of cutting neck tissues compared with major blood vessels. N.Z. Vet. J. 2009;57(2):84–9.
[8] Gaj T, Gersbach CA, Barbas III CF. ZFN, TALEN, and CRISPR/Cas-based methods for genome engineering. Trends Biotechnol. 2013;31(7):397–405.
[9] Horvath P, Barrangou R. CRISPR/Cas, the immune system of bacteria and archaea. Science 2010;327(5962):167–70.
[10] Cong L, Ran FA, Cox D, et al. Multiplex genome engineering using CRISPR/Cas systems. Science 2013;339(6121):819–23.
[11] Jiang W, Bikard D, Cox D, Zhang F, Marraffini LA. RNA-guided editing of bacterial genomes using CRISPR-Cas systems. Nat. Biotechnol. 2013;31(3):233–9.
[12] Singh V, Braddick D, Dhar PK. Exploring the potential of genome editing CRISPR-Cas9 technology. Gene 2017;599:1–18.
[13] Singh V, Gohil N, Ramírez García R, Braddick D, Fofié CK. Recent advances in CRISPR-Cas9 genome editing technology for biological and biomedical investigations. J. Cell. Biochem. 2018;119(1):81–94.
[14] Nelson CE, Hakim CH, Ousterout DG, et al. *In vivo* genome editing improves muscle function in a mouse model of Duchenne muscular dystrophy. Science 2016;351(6271):403–7.
[15] Bikard D, Euler CW, Jiang W, et al. Exploiting CRISPR-Cas nucleases to produce sequence-specific antimicrobials. Nat. Biotechnol. 2014;32(11):1146–50.

[16] Citorik RJ, Mimee M, Lu TK. Sequence-specific antimicrobials using efficiently delivered RNA-guided nucleases. Nat. Biotechnol. 2014;32(11):1141–5.

[17] Zhu W, Lei R, Le Duff Y, et al. The CRISPR/Cas9 system inactivates latent HIV-1 proviral DNA. Retrovirology 2015;12:22.

[18] Kennedy EM, Kornepati AV, Goldstein M, et al. Inactivation of the human papillomavirus E6 or E7 gene in cervical carcinoma cells using a bacterial CRISPR/Cas RNA-guided endonuclease. J. Virol. 2014;88(20):11965–72.

[19] Zhen S, Hua L, Liu YH, et al. Harnessing the clustered regularly interspaced short palindromic repeat (CRISPR)/CRISPR-associated Cas9 system to disrupt the hepatitis B virus. Gene Ther. 2015;22(5):404–12.

[20] Wang J, Quake SR. RNA-guided endonuclease provides a therapeutic strategy to cure latent herpesviridae infection. Proc. Natl. Acad. Sci. U.S.A. 2014;111(36):13157–62.

[21] Normile D. CRISPR bombshell: Chinese researcher claims to have created gene-edited twins. Science 2018. https://doi.org/10.1126/science.aaw1839.

[22] Torre LA, Bray F, Siegel RL, Ferlay J, Lortet-Tieulent J, Jemal A. Global cancer statistics, 2012. CA Cancer J. Clin. 2015;65:87–108.

[23] Kantarjian H, Sawyers C, Hochhaus A, et al. Hematologic and cytogenetic responses to imatinib mesylate in chronic myelogenous leukemia. N. Engl. J. Med. 2002;346(9):645–52.

[24] Cunningham D, Humblet Y, Siena S, et al. Cetuximab monotherapy and cetuximab plus irinotecan in irinotecan-refractory metastatic colorectal cancer. N. Engl. J. Med. 2004;351:337–45.

[25] Shao J, Xu Q, Su S, et al. Engineered cells for costimulatory enhancement combined with IL-21 enhance the generation of PD-1-disrupted CTLs for adoptive immunotherapy. Cell. Immunol. 2017;320:38–45.

[26] Zhao Z, Shi L, Zhang W, Han J, Zhang S, Fu Z. CRISPR knock out of programmed cell death protein 1 enhances anti-tumor activity of cytotoxic T lymphocytes. Oncotarget 2018;9(4):5208–15.

[27] Zhao L, Li K, Bao S, et al. A 1-bp deletion in the gamma C-crystallin leads to dominant cataracts in mice. Mamm. Genome 2010;21(7–8):361–9.

[28] Emery AE. Population frequencies of inherited neuromuscular diseases—a world survey. Neuromuscul. Disord. 1991;1:19–29.

[29] Shimo T, Hosoki K, Nakatsuji Y, Yokota T, Obika S. A novel human muscle cell model of duchenne
muscular dystrophy created by CRISPR/Cas9 and evaluation of antisense-mediated exon skipping. J. Hum. Genet. 2018;63:365–75.

[30] Lim KRQ, Yoon C, Yokota T. Applications of CRISPR/Cas9 for the treatment of duchenne muscular dystrophy. J. Personalized Med. 2018;8(4):E38.

[31] Shao Y, Wang L, Guo N, et al. Cas9-nickase-mediated genome editing corrects hereditary tyrosinemia in rats. J. Biol. Chem. 2018;293(18):6883–92.

[32] VanLith C, Guthman R, Nicolas CT, et al. Curative *Ex vivo* hepatocyte-directed gene editing in a mouse model of hereditary tyrosinemia type 1. Hum. Gene Ther. 2018;29(11):1315–26.

[33] Welsh M, Ramsey B, Accurso F, Cutting GR. Cystic fibrosis. In: Scriver CR, Beaudet AL, Valle D, editors. The metabolic and molecular bases of inherited disease. Columbus (OH): McGraw-Hill; 2001. p. 5121–89.

[34] Ratjen F, Grasemann H. New therapies in cystic fibrosis. Curr. Pharm. Des. 2012;18(5):614–27.

[35] Summar ML, Koelker S, Freedenberg D, et al. The incidence of urea cycle disorders. Mol. Genet. Metab. 2016;110(1–2):179–80.

[36] Foschi FG, Morelli MC, Savini S, et al. Urea cycle disorders: a case report of a successful treatment with liver transplant and a literature review. World J. Gastroenterol. 2015;21:4063–8.

[37] Uchino T, Snyderman SE, Lambert M, et al. Molecular basis of phenotypic variation in patients with argininemia. Hum. Genet. 1995;96(3):255–60.

[38] Lee PC, Truong B, Vega-Crespo A, Gilmore WB, Hermann K. Restoring ureagenesis in hepatocytes by CRISPR/Cas9-mediated genomic addition to arginase-deficient induced pluripotent stem cells. Mol. Ther. Nucleic Acids 2016;5(11):e394.

[39] Cao A, Galanello R. Beta-thalassemia. Genet. Med. 2010;12(2):61–76.

[40] Cai B, Sun S, Li Z, et al. Application of CRISPR/Cas9 technologies combined with iPSCs in the study and treatment of retinal degenerative diseases. Hum. Genet. 2018;137(9):679–88.

[41] Cyranoski D, Reardon S. Chinese scientists genetically modify human embryos. Nature News 2015. https://doi.org/10.1038/nature.2015.17378.

[42] Antony JS, Latifi N, Haque AKMA, et al. Gene correction of *HBB* mutations in $CD34^+$ hematopoietic stem cells using *Cas9* mRNA and ssODN donors. Mol. Cell. Pediatr. 2018;5(1):9.

[43] Weatherall DJ. The inherited diseases of hemoglobin are an emerging global health burden. Blood 2010;115(22):4331–6.

[44] Hoban MD, Orkin SH, Bauer DE. Genetic treatment of a molecular disorder: gene therapy approaches to sickle cell disease. Blood 2016;127:839–48.

[45] Sun N, Zhao H. Seamless correction of the sickle cell disease mutation of the HBB gene in human induced pluripotent stem cells using TALENs. Biotechnol. Bioeng. 2014;111(5):1048–53.

[46] Zou J, Mali P, Huang X, Dowey SN, Cheng L. Site-specific gene correction of a point mutation in human iPS cells derived from an adult patient with sickle cell disease. Blood 2011;118(17):4599–608.

[47] Huang X, Wang Y, Yan W, et al. Production of gene-corrected adult beta globin protein in human erythrocytes differentiated from patient iPSCs after genome editing of the sickle point mutation. Stem Cells 2015;33(5):1470–9.

[48] Holmes D. Retinal repair: visions of the future. Nature 2018;561(7721):S1.

[49] Yu W, Mookherjee S, Chaitankar V, et al. *Nrl* knockdown by AAV-delivered CRISPR/Cas9 prevents retinal degeneration in mice. Nat. Commun. 2017;8:14716.

[50] Cuchel M, Bruckert E, Ginsberg HN, Raal FJ, Santos RD. European atherosclerosis society consensus panel on familial H: homozygous familial hypercholesterolaemia: new insights and guidance for clinicians to improve detection and clinical management. A position paper from the consensus panel on familial hypercholesterolaemia of the European atherosclerosis society. Eur. Heart J. 2014;35:2146–57.

[51] Omer L, Hudson EA, Zheng S, Hoying JB, Shan Y, Boyd NL. CRISPR correction of a homozygous low-density lipoprotein receptor mutation in familial hypercholesterolemia induced pluripotent stem cells. Hepatol. Commun. 2017;1(9):886–98.

[52] Waghulde H, Cheng X, Galla S, et al. Attenuation of microbiota dysbiosis and hypertension in a CRISPR/Cas9 gene ablation rat model of GPER1. Hypertension 2018;72(5):1125–32.

[53] Musunuru K. How genome editing could be used in the treatment of cardiovascular diseases. Per. Med. 2018;15(2):67−9.

[54] Thomsen GM, Gowing G, Svendsen S, Svendsen CN. The past, present and future of stem cell clinical trials for ALS. Exp. Neurol. 2014;262 Pt B:127−37.

[55] Ajroud-Driss S, Siddique T. Sporadic and hereditary amyotrophic lateral sclerosis (ALS). Biochim. Biophys. Acta 2015;1852: 679−84.

[56] Wen D, Cui C, Duan W, et al. The role of insulin-like growth factor 1 in ALS cell and mouse models: a mitochondrial protector. Brain Res. Bull. 2019;144:1−13.

[57] Lee JM, Ramos EM, Lee JH, et al. CAG repeat expansion in Huntington disease determines age at onset in a fully dominant fashion. Neurology 2012;78(10):690−5.

[58] Wild EJ, Tabrizi SJ. Premanifest and early Huntington's disease. In: Bates G, Jones L, Tabrizi SJ, editors. Huntington's disease. Oxford: Oxford University Press; 2014.

[59] Ross CA, Tabrizi SJ. Huntington's disease: from molecular pathogenesis to clinical treatment. Lancet Neurol. 2011;10(1):83−98.

[60] Saudou F, Humbert S. The biology of huntingtin. Neuron 2016;89(5):910−26.

[61] Shin JW, Kim KH, Chao MJ, et al. Permanent inactivation of Huntington's disease mutation by personalized allele-specific CRISPR/Cas9. Hum. Mol. Genet. 2016;25(20):4566−76.

[62] Yang S, Chang R, Yang H, et al. CRISPR/Cas9-mediated gene editing ameliorates neurotoxicity in mouse model of Huntington's disease. J. Clin. Investig. 2017;127:2719−24.

[63] Bugg TD, Braddick D, Dowson C, Roper DI. Bacterial cell wall assembly: still an attractive antibacterial target. Trends Biotechnol. 2011;29(4):167−73.

[64] Gomaa AA, Klumpe HE, Luo ML, Selle K, Barrangou R, Beisel CL. Programmable removal of bacterial strains by use of genome-targeting CRISPR-Cas systems. mBio 2014;5(1). e00928-13.

[65] Yosef I, Manor M, Kiro R, Qimron U. Temperate and lytic bacteriophages programmed to sensitize and kill antibiotic-resistant bacteria. Proc. Natl. Acad. Sci. U.S.A. 2015;112(23):7267−72.

[66] Kim YJ, Woo JH, Kim MJ, et al. Opportunistic diseases among HIV-infected patients: a multicenter-nationwide Korean HIV/AIDS cohort study, 2006 to 2013. Korean J. Intern. Med. 2016;31(5):953−60.

[67] UNAIDS. Global AIDS update 2016. 2016. http://www.unaids.org/sites/default/files/media_asset/global-AIDS-update-2016_en.pdf.

[68] Schrager LK, D'souza MP. Cellular and anatomical reservoirs of HIV-1 in patients receiving potent antiretroviral combination therapy. JAMA 1998;280(1):67−71.

[69] Deeks SG, Autran B, Berkhout B, et al. Towards an HIV cure: a global scientific strategy. Nat. Rev. Immunol. 2012;12(8): 607−14.

[70] Sterne JA, Hernán MA, Ledergerber B, et al. Long-term effectiveness of potent antiretroviral therapy in preventing AIDS and death: a prospective cohort study. Lancet 2005;366(9483):378−84.

[71] Hu Z, Yu L, Zhu D, et al. Disruption of HPV16-E7 by CRISPR/Cas system induces apoptosis and growth inhibition in HPV16 positive human cervical cancer cells. BioMed Res. Int. 2014:612823.

[72] Liao HK, Gu Y, Diaz A, et al. Use of the CRISPR/Cas9 system as an intracellular defense against HIV-1 infection in human cells. Nat. Commun. 2015;6:6413.

[73] Stone D, Kiem HP, Jerome KR. Targeted gene disruption to cure HIV. Curr. Opin. HIV AIDS 2013;8(3):217−23.

[74] Bella R, Kaminski R, Mancuso P, et al. Removal of HIV DNA by CRISPR from patient blood engrafts in humanized mice. Mol. Ther. Nucleic Acids 2018;12:275−82.

[75] Dampier W, Sullivan NT, Mell JC, et al. Broad-spectrum and personalized guide RNAs for CRISPR/Cas9 HIV-1 therapeutics. AIDS Res. Hum. Retrovir. 2018;34(11):950−60.

[76] Baer R, Bankier AT, Biggin MD, et al. DNA sequence and expression of the B95-8 Epstein—Barr virus genome. Nature 1984;310(5974):207−11.

[77] Adams A, Lindahl T. Epstein-Barr virus genomes with properties of circular DNA molecules in carrier cells. Proc. Natl. Acad. Sci. U.S.A. 1975;72(4):1477−81.

[78] Pannone G, Zamparese R, Pace M, et al. The role of EBV in the pathogenesis of Burkitt's Lymphoma: an Italian hospital based survey. Infect. Agents Cancer 2014;9(1):34.

[79] Marquitz AR, Raab-Traub N. The role of miRNAs and EBV BARTs in NPC. Semin. Cancer Biol. 2012;22(2):166−72. Academic Press.

[80] Yuen KS, Chan CP, Wong NHM, et al. CRISPR/Cas9-mediated genome editing of Epstein−Barr virus in human cells. J. Gen. Virol 2015;96(3):626−36.

[81] Jiang S, Wang LW, Walsh MJ, et al. CRISPR/Cas9-mediated genome editing in Epstein-Barr virus-transformed lymphoblastoid B-cell lines. Curr. Protoc. Mol. Biol. 2018;121:31.12.1−31.12.23.

[82] Zhen S, Li X. Oncogenic human papillomavirus: application of CRISPR/Cas9 therapeutic strategies for cervical cancer. Cell. Physiol. Biochem. 2017;44(6):2455−66.

[83] Mighty KK, Laimins LA. The role of human papillomaviruses in oncogenesis. Viruses and human cancer. Berlin, Heidelberg: Springer; 2014. p. 135−48.

[84] Kennedy EM, Cullen BR. Bacterial CRISPR/Cas DNA endonucleases: a revolutionary technology that could dramatically impact viral research and treatment. Virology 2015;479−480:213−20.

[85] Cheng YX, Chen GT, Yang X, Wang YQ, Hong L. Effects of HPV pseudotype virus in cutting E6 gene selectively in SiHa cells. Curr. Med. Sci. 2018;38(2):212−21.

[86] Zhen S, Hua L, Takahashi Y, Narita S, Liu YH, Li Y. *In vitro* and *in vivo* growth suppression of human papillomavirus 16-positive cervical cancer cells by CRISPR/Cas9. Biochem. Biophys. Res. Commun. 2014;450(4):1422−6.

[87] Hsu DS, Kornepati AV, Glover W, Kennedy EM, Cullen BR. Targeting HPV16 DNA using CRISPR/Cas inhibits anal cancer growth *in vivo*. Future Virol. 2018;13(7):475−82.

[88] Yoshiba T, Saga Y, Urabe M, et al. CRISPR/Cas9-mediated cervical cancer treatment targeting human papillomavirus E6. Oncol. Lett. 2018:2197−206.

[89] Seo Y, Yano Y. Short-and long-term outcome of interferon therapy for chronic hepatitis B infection. World J. Gastroenterol. 2014;20(37):13284−92.

[90] Li H, Sheng C, Wang S, et al. Removal of integrated hepatitis B virus DNA using CRISPR-Cas9. Front. Cell. Infect. Microbiol. 2017;7:91.

[91] Dandri M, Burda MR, Will H, Petersen J. Increased hepatocyte turnover and inhibition of woodchuck hepatitis B virus replication by adefovir in vitro do not lead to reduction of the closed circular DNA. Hepatology 2000;32(1):139−46.

[92] Zucman-Rossi J, Laurent-Puig P. Genetic diversity of hepatocellular carcinomas and its potential impact on targeted therapies. Pharmacogenomics 2007;8(8):997−1003.

[93] Lin SR, Yang HC, Kuo YT, et al. The CRISPR/Cas9 system facilitates clearance of the intrahepatic HBV templates *in vivo*. Mol. Ther. Nucleic Acids 2014;3:e186.

[94] Seeger C, Sohn JA. Targeting hepatitis B virus with CRISPR/Cas9. Mol. Ther. Nucleic Acids 2014;3:e216.

[95] Kennedy EM, Bassit LC, Mueller H, et al. Suppression of hepatitis B virus DNA accumulation in chronically infected cells using a bacterial CRISPR/Cas RNA-guided DNA endonuclease. Virology 2015;476:196–205.

[96] Guk K, Keem JO, Hwang SG, et al. A facile, rapid and sensitive detection of MRSA using a CRISPR-mediated DNA FISH method, antibody-like dCas9/sgRNA complex. Biosens. Bioelectron. 2017;95:67–71.

[97] Gootenberg JS, Abudayyeh OO, Lee JW, et al. Nucleic acid detection with CRISPR-Cas13a/C2c2. Science 2017;356(6336):438–42.

[98] Gootenberg JS, Abudayyeh OO, Kellner MJ, Joung J, Collins JJ, Zhang F. Multiplexed and portable nucleic acid detection platform with Cas13, Cas12a, and Csm6. Science 2018;360(6387):439–44.

[99] Khambhati K, Bhattacharjee G, Singh V. Current progress in CRISPR-based diagnostic platforms. J. Cell. Biochem. 2018. https://doi.org/10.1002/jcb.27690.

Chapter 8

Engineering microbial living therapeutics

Danielle B. Pedrolli[1], Nathan V. Ribeiro[1], Patrick N. Squizato[1], Daniel A. Cozetto[1], Victor N. de Jesus[1], Paulo J.C. Freire[1], Milca R.C.R. Lins[2] and Graciely G. Correa[2]

[1]*São Paulo State University (UNESP), School of Pharmaceutical Sciences, Department of Bioprocess and Biotechnology, Araraquara, Brazil;*
[2]*São Paulo State University (UNESP), Graduate Program Bioscience and Biotechnology Applied to Pharmacy, Araraquara, Brazil*

Context

Recent advances in understanding the relationship between the microbiome and the human host have revealed a microbe-host interconnectivity. Human-hosted microbes contribute to health or illness together with other factors such as genetic, environmental, and lifestyle that help define the human ecosystem [1]. The ability of such microorganisms to interact with the host can in principle allow for their use in the delivery of therapy in a targeted and controllable fashion. In that context, synthetic biology offers a means of engineering the microbiota as living therapeutics, for treating and preventing illnesses that conventional medicines have failed to cure [2,3].

The human microbiome as a therapeutic platform

A vast microbial community lives on and within the human body, including bacteria, fungi, and phages [4]. Bacteria seem to account for the majority of nonhuman cells inhabiting the human body. It is estimated that an adult body carries 3.8×10^{13} to 4.4×10^{13} bacterial cells, which is about the same amount of own cells (2.1×10^{13} to 3.0×10^{13}) [5]. The most fruitful body habitats for bacteria are the gastrointestinal, urogenital, oral, nasal, and skin areas. Each of these is characterized by one or a few signature taxa, in almost every healthy North American Adult (Human Microbiome Project), but no bacterial taxa are universally present among all body habitats and individuals. Therefore, the diversity of the human microbiome is unique to each individual, and strongly determined by body habitat [4].

Coevolution between host–microbe and microbe–microbe has shaped a range of metabolic functions. Unlike microbial taxa, several metabolic pathways seem to be ubiquitous among individuals and body habitats. One exception includes the Sec and Tat polypeptide pathways, for protein secretion across the bacterial cytoplasmic membrane, indicating a high degree of host–microbe and microbe–microbe interactions in the healthy human microbiota [4].

The metabolic pathways provided by the human microbiome maintain proper body health by providing nutritional and digestive activities, and resistance to colonization by dangerous pathogens [4]. Recent findings have shown that commensal microbes play an even larger set of active roles in complex body tasks, including energy homeostasis, fat deposition, immunity, inflammation, and neurological function. Perturbations of the host–microbiota balance, known as dysbiosis, have been associated with different diseases, including autoimmune and inflammatory diseases, dermatological illness, obesity, diabetes, neurodegenerative and mental illnesses [6–11]. Use of broad-spectrum antibiotics, for example, has been shown to destabilize the gut microbial community and lead to dysbiosis [12].

Relevant pathogens, commensals, and probiotics

The human microbiome is mostly constituted of commensals microbes, which benefit from the body environment and do no harm to it, and symbiotic microbes, which take benefit from and give benefit to the human body. Each habitat presents different selection pressures and, therefore, harbors different signature microbial taxa. The oral cavity is dominated by *Streptococcus*, the vaginal introitus by *Lactobacillus*, the skin is universally colonized by *Staphylococcus epidermis*, and *Bacteroides* are abundant in the gut microbiome [4].

Precision Medicine for Investigators, Practitioners and Providers. https://doi.org/10.1016/B978-0-12-819178-1.00008-3

Some gut symbiotic microbes are also classified as probiotic, which are live microorganisms that, when administered in adequate amounts, confer a health benefit on the host [13]. Health benefits have been demonstrated for specific probiotic strains of the genera *Lactobacillus*, *Bifidobacterium*, *Saccharomyces*, *Pediococcus*, *Leuconostoc*, *Bacillus*, and some strains of *Enterococcus*, *Streptococcus*, *Escherichia coli* [14].

The capacity of such microorganisms to harmonically inhabit human body niches can in principle allow for their use in the delivery of therapy in a targeted and controllable way [3,15].

The clinical potential of wild-type bacterial communities was demonstrated with fecal microbiota transplantation, in combating recurrent *Clostridium difficile* infections [16]. Formulations of probiotic strains have also been used to compete with diarrheagenic bacteria (e.g., *E. coli*, *Campylobacter jejuni*, and *Shigella* species). Commonly used probiotic against these pathogens include *Lactobacillus acidophilus*, *Lactobacillus rhamnosus*, *Saccharomyces boulardii*, *Bifidobacterium bifidum*, and *Bacillus coagulans* [17]. Moreover, the VL#3 formulation containing eight strains of *Bifidobacterium*, *Lactobacillus*, and *Streptococcus* has been prescribed for inflammatory bowel disease [3]. The therapeutic potential of microbes has been recognized for nearly two centuries [18].

The concept of bioengineered microorganisms

Bioengineered microorganisms are good at synthesizing biologically active molecules. Indeed, many therapeutic compounds are currently produced in industrial scale by bacteria and fungi. Biotech companies profitably commercialize hormones, antigens, enzymes, vitamins, antibiotics, among others, produced by genetically engineered microbes [15]. Traditionally, microbial production of therapeutics and its delivery are two separate processes. However, bioengineered microorganisms are on the way to change it.

Tiny living factories

Next-generation therapies will likely consist of tiny living factories able to autonomously self-replicate, detect an abnormal condition, produce and deliver the therapy inside the human body, and promote self-destruction after completing its healing task. In vivo production and delivery of therapy by microorganisms have several advantages over traditional systemic treatment. First, production costs would decrease considerably. As the living factories will produce the therapy directly in the human body, there will be no need for costly downstream processes. The cost to propagate large amounts of engineered bacteria is much lower than to produce pharmaceutical-grade biological molecules.

Second, reduced side effects are expected, as the administration route is noninvasive, and therapy is locally administered. Third, the required dosage is much lower, as the distance from factory to target is shortened [19], which is particularly relevant for unstable and acid-sensitive compounds. Finally, living organisms can produce more than one compound, and the timing can be controlled to be simultaneous or sequential [3,15].

Living therapy may be more efficient than traditional methods for treating challenging diseases, like chronic conditions, cancer, and antibiotic-resistant pathogens, providing uninterrupted treatment at (near) constant concentration. They can also provide in situ delivery of medicines, avoiding the damage to healthy cells caused by systemic treatment. Finally, living therapeutics may be armed to fight the increasing number of antibiotic-resistant pathogens [2,15,20]. Molecular tools for engineered bacteria enable the construction of sophisticated genetic circuits, to perform complex tasks [15].

Potential clinical targets

Microbes can be engineered to either diagnose or treat diseases and disorders. In some cases, they have been engineered to do both, and only deliver the treatment after a positive diagnosis. These have been called "smart probiotics" [21].

Potential targets for engineered microbial living therapeutics include metabolic disorders; tumors; infectious, neurodegenerative, and autoimmune diseases [3,20–23] (Fig. 8.1). Each of these conditions has singularities that make the design strategy unique.

Gastrointestinal disorders

Traditional treatment for IBD with antiinflammatory, immune system suppressors and antibiotics aims at reducing inflammation that triggers symptoms. However, treatment can cause systemic side effects and, in some cases, therapy fails to ameliorate symptoms [22,24]. Therefore, bioengineered microbes overexpressing molecules able to alleviate inflammation could be an attractive treatment. Agents encompass interleukin 10 and 27 (IL-10, IL-27) [25,26], transforming growth factor-β (TGF-β) [27], human keratinocyte growth factor-2 (KGF-2) [19], antitumor necrosis factor (TNF)-α nanobodies [28], and protease inhibitor (Elafin) [29]. All these proteins have been successfully expressed by engineered microorganisms, and ameliorated inflammation in IBD animal models.

Phenylketonuria and celiac disease are good candidates, as systemic conditions with important digestive system involvement. Phenylalanine metabolizing enzymes have been coded into bioengineered probiotics, to eliminate harmful phenylalanine amino acid in the intestine of phenylketonuria patients [30,31].

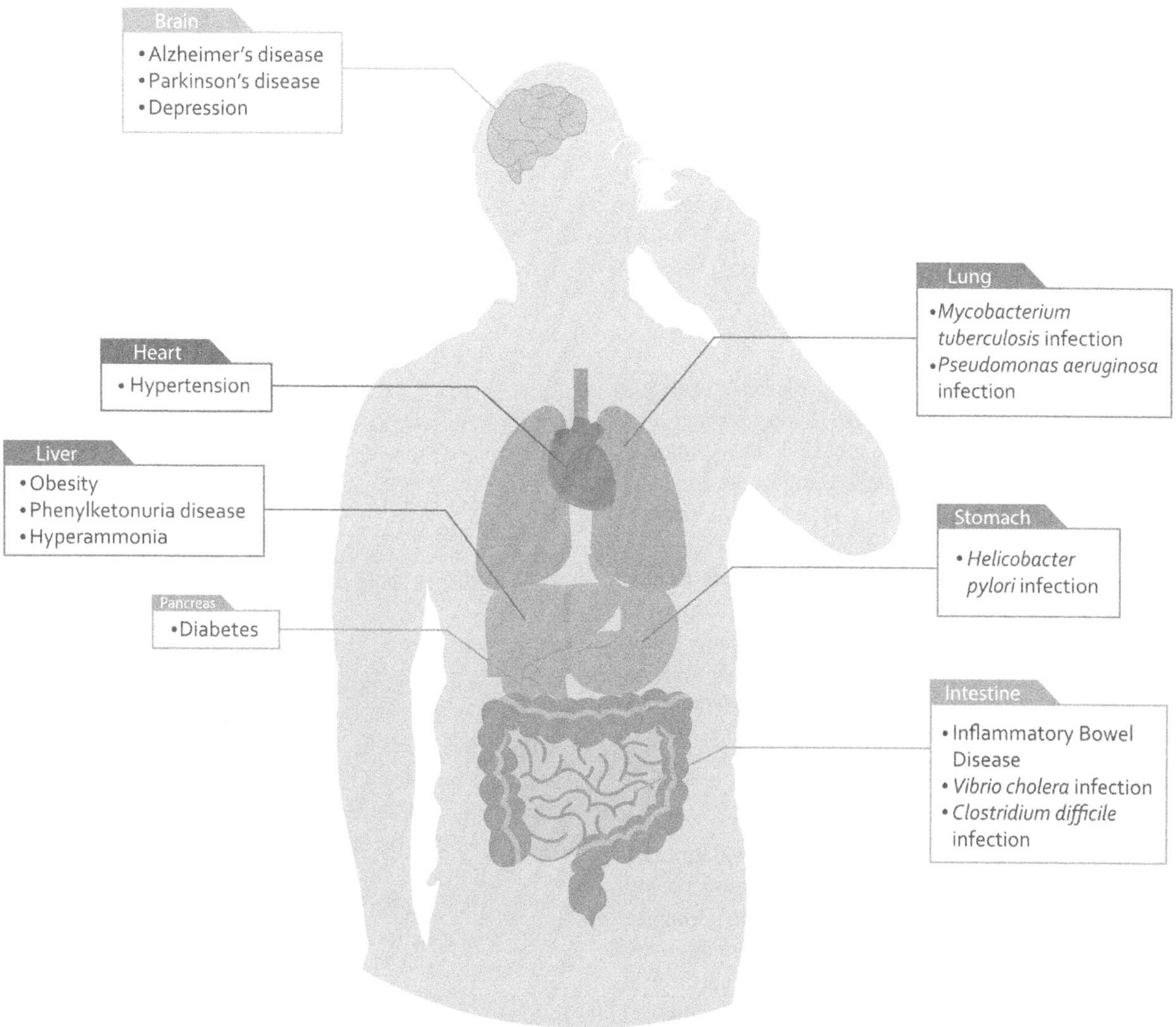

FIGURE 8.1 Potential clinical targets for engineered microbial therapeutics. Some relevant bacterial infections are exemplarily shown; however, any infectious agent can be targeted. Although not shown specifically affecting any body part, tumors are also potential targets for bioengineered microbes.

Type 1 and autoimmune diabetes

Two different approaches have been taken in diabetic models: (i) secretion of glucagon like peptide-1 (GLP-1), to reprogram intestinal cells into glucose-responsive insulin-secreting cells [32,33]; (ii) secretion of human proinsulin [34] or a synthetic single-chain insulin analog [35,36]. A combination therapy of anti-CD3 antibodies, with *Lactococcus lactis* secreting human proinsulin and interleukin-10, has been reported to cure 66% of mice with new-onset autoimmune diabetes [34].

Cancer

There is a long history of anaerobic bacteria proliferating in solid tumors, and resulting in antitumor effects [37]. For safety purposes, usually attenuated versions of pathogenic bacteria are employed. However, they cannot eradicate tumors alone. Therefore, scientists are upgrading these bacteria to produce anticancer compounds [37]. Anticancer engineered bacteria produce three kind of compounds: (i) Cytotoxic agents such as hemolysin [38] and cytolysin [39]; (ii) Prodrug-converting enzymes such as myrosinase, which converts glucosinolates from diet into anticancer sulforaphane for colorectal-cancer chemoprevention [40]; (iii) Immunomodulators such as interferon-gamma against melanoma [41], and flagellin B against murine colon carcinoma cells MC38 [42].

Resistant pathogenic microorganisms

Emerging multidrug resistant pathogens can also be targeted, to circumvent the depletion of antibiotic options [23]. Commensal *E. coli* reprogrammed to swim toward infection-promoting *Pseudomonas aeruginosa*, and release antimicrobial and antibiofilm proteins, was able to eliminate the pathogen from infected mice [43]. Bioengineered L. *lactis* efficiently inhibited growth of multidrug-resistant *Enterococcus faecalis* strains [44]. Fungal infections can also be fought by living therapeutics. *E. coli* was engineered to sense and respond to candidemia, by producing cis-2-dodecanoic acid, a compound able to inhibit filamentation and virulence gene expression in *Candida albicans*.

Mucosal vaccines

Another valid approach is engineering microbes to display an antigen and act like a vaccine [20]. Mucosal vaccines are a modern attempt to present antigens, and elicit protective immune responses, directly at the aerodigestive, intestinal, and urogenital mucous membranes [1]. Anti-*Helicobacter pylori* engineered probiotic vaccine elicited immune response, when tested in mice [45]. Engineered *Bacillus subtilis* spores expressing *H. pylori* urease B protein on the surface generated humoral response in mice, and significantly reduced *H. pylori* load [46]. Human vaginal *Lactobacillus* has been engineered to produce neutralizing antibody fragments against HIV [47].

Disease fingerprints

Microbes have been engineered to detect inflammation biomarkers such as nitrogen oxides [48], tetrathionate [49], and thiosulfate [50]. Bioengineered *E. coli* colonized mouse gut, retaining the ability to respond to inflammation for at least 6 months. Lactic acid bacteria reported the presence of pathogens such as *Staphylococcus aureus* [51] and *Vibrio cholerae* [52]. In both cases, the bioengineered microbe was able to detect a quorum-sensing molecule, which then triggers expression of a reporter gene. The test output is a color change in a catalyzed reaction.

Required synthetic biology

Engineering efficient and smart living therapeutics requires a robust chassis (carrier microorganism), and a big toolbox (manipulation of genome and gene expression) enabling input sensing, gene expression control, memory, production, and delivery of therapeutic molecules. Additionally, biocontainment enabling tools are desirable to address safety issues (Figure 8.2) [15].

Living therapeutics chassis

Choosing the chassis will most likely define microbial and pharmaceutical doses in treated individuals. Chassis including eukaryotic cells, bacteriophages, and bacteria have been used as drug delivery systems. So far, probiotic bacteria are the best option [53].

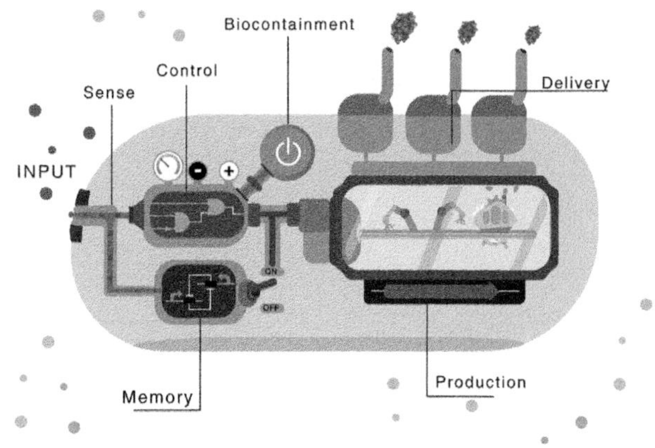

FIGURE 8.2 Living therapeutics modules. In order to accomplish a task, engineered living therapeutics must carry input sensors to detect external signals, control devices to integrate the inputs, and actuators to generate a response according to the decision made by the logic circuits that comprise the control module. Actuators are usually proteins that act on targets outside the cell and, after production, are secreted by the delivery system. The biocontainment device promotes cell death at nonpermissive condition [15].

Preferred gut chassis focus *E. coli*, *Bacteroides*, and lactic acid bacteria. Other body environments may be more suitable for some conditions. Microbial biogeographic preferences can be leveraged to treat diseases or maintain host health [54]. *Lactobacillus jensenii* [55] and *Staphylococcus epidermidis* [56] have been chosen for engineering living therapeutics for the vaginal tract and the skin, respectively, and *Salmonella typhimurium* for tumors [41].

E. coli is the most used chassis and a large collection of biological parts is currently available for engineering this species. It has the advantage of high prevalence in the human gut (76% of US stool samples) [4]. *L. lactis* has a long record as probiotic bacterium. It cannot invade mammalian tissues on ingestion, nor does it cause infection. Like *E. coli*, *L. lactis* is a noncolonizing bacterium that is cleared shortly after administration [57].

Bacteroides is the most abundant genus in the microbiome [4], stably and resiliently colonizing the human gut [58]. *Bacteroides thetaiotaomicron* and *Bacteroides fragilis* have emerged in the last 5 years as living delivery chassis. Colonizing bacteria, contrasting with transient ones, are likely to efficiently deliver medicine for long periods. However, they raise safety concerns, as they may persist and proliferate inside the human body even after the end of treatment. Noncolonizing bacteria can be useful to treat acute diseases, when multiple doses in a short period followed by natural clearance are an acceptable approach. On the other hand, colonizing bacteria can be useful to treat chronic diseases [15].

Microbial features such as amenability to DNA transformation, availability of molecular tools to modify the genome, and the range of **sensors** and **actuators** compatible with the chassis are also essential to consider [21].

Sense and control

Most of the **control** of gene expression in engineered microbes is performed at the transcription level. Promoters are the core of the "sense and control" drawer in the toolbox. Complex microbial behaviors can be engineered by combining regulated promoters, to create internal logic circuits for integration of different inputs [48]. Libraries of constitutive and regulated promoters spanning a broad range of gene expression strengths were organized for *B. thetaiotaomicron* and other *Bacteroides*. Constitutive promoters exhibit a range of gene expression that spans up to 30,000-fold, and predictably functions in a diverse range of species in the same genus.

Promoters may also be activated by rhamnose, chondroitin sulfate, arabinogalactan, isopropyl β-D-1-thiogalactopyranoside (IPTG), and anhydrotetracycline (aTc). Moreover, promoter parts can be combined with one synthetic ribosome binding site from the libraries, for fine-tuning gene expression [59,60]. A device for CRISP interference for gene repression (CRISPRi), activated by IPTG, is available in the *Bacteroides* toolbox. Although CRISPRi may be used to repress gene expression up to 45-fold, dCas9 protein, which is coupled with the CRISPR locus when gene editing is desired, does exhibit some toxicity to the cell [60].

For lactic acid bacteria, constitutive promoters span a range of fold change for *Lactococcus* and *Lactobacillus* [61,62], one of which was observed to be twice as strong (generating twice as much protein) as the widely used *L. lactis* P_{nisA}. A CRISPRi device was also developed for *L. lactis*, in which the nisin inducible promoter (P_{nisA}) drives dCas9 expression. However, dCas9 synthesis considerably decreased the growth rate of *L. lactis* [63].

Targeted delivery of molecules

Control of gene expression, necessary for condition- or site-specific agents, is preferably done with small molecules. Common compounds used for decades as inducers are IPTG, aTc, and nisin. Recently, a new tool entered the box: salicylate-inducible cascade expression for *Salmonella typhimurium*. Upon salicylate binding, protein NahR activates its own transcription, and the transcription of XylS2 through P_{sal} promoter. Then, XylS2 activates transcription driven by P_M promoter [64]. One advantage is safety as salicylate has been commercialized worldwide since 1899, under the name Aspirin.

Disease-specific promoters

Induction of gene expression by exogenous molecules is useful, as it can be triggered externally at any time, in any condition. However, those molecules can cause undesired effects and increase treatment cost. Promoters regulated by disease-specific conditions can be a solution, such as tetrathionate-sensing device composed by TtrR/TtrS two-component system, constructed in *E. coli* [49], and nitrite/nitrate-sensitive P_{YeaR} promoter used in *E. coli* and *B. subtilis*, both useful for inflammation dependent gene expression [48]. Other examples include stress-inducible controlled expression (SICE) for *L. lactis*, which responds to pH- and bile salt-stress [25]; and temperature-sensing switches for *E. coli*, which can respond to fever [65].

Quorum sensing molecules

Many bacteria, however not all of them, are endowed with a quorum-sensing (QS) system, able to secrete and detect signal molecules, tracking them down. This kind of "sense and control" device has been constructed in *E. coli*, *L. lactis*, and *Lactobacillus reuteri* to fight *P. aeruginosa* [43], *V. cholerae* [52,66], *C. albicans* [67], *Enterococcus faecalis* [44], and *S. aureus* [51]. The strategy consists in inserting the signal molecule-sensor device from the pathogen into the engineered chassis, placing the gene of interest under control of the QS-activated promoter [68]. An exception is *E. coli* W, which naturally has a sensor for *C. albicans* QS-molecule hydroxyphenylacetic acid [67].

QS devices can be engineered as information transfer systems for interspecies communication, enabling a programmable synthetic gut consortium. In that case, one can be engineered to act as the signaler, by carrying the acyl homoserine lactone signal synthesis protein (LuxI). Another cell can be engineered to act as the responder, detecting the signal through the regulatory protein (LuxR), and generating the output accordingly [69].

Memory

Some applications for living therapeutics require a continuous response to a transient signal. Bacteria can be engineered to report the diagnosis. In such case, they must record an input inside the body, and display the output after release [60,70]. Memory devices are also useful to decouple propagation from production. That means bacteria will not produce the therapeutic molecules during cell propagation in the industrial facility. However upon a transient induction event—before administration or inside the human body—the memory device will permanently turn on production of the therapy.

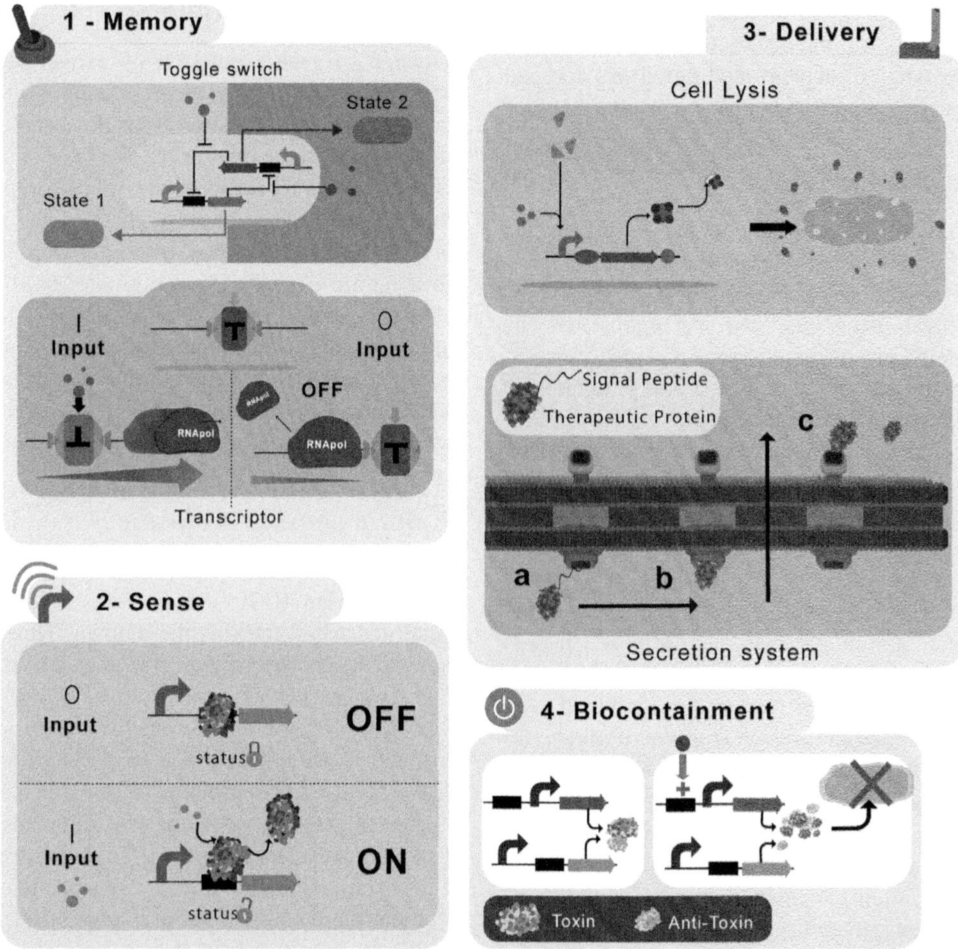

FIGURE 8.3 Successful actuator strategies used in engineered microbial living therapeutics. (1) Memory modules provide continuous response for transient inputs: Toggle switches have bistable gene-regulatory network constructed with double negative feedback circuits. Inputs activate one or the other state. Transcriptors control the flow of RNA polymerase through a transcription unit that has a unidirectional premature terminator, surrounded by recombination sites. The sites are recognized by a serine recombinase that flips the DNA sequence in between. In default state, the terminator prevents transcription. A specific input stimulates synthesis of serine recombinase, which flips terminator to reverse DNA strand and frees the way for full-length transcription. The terminator will remain flipped, no matter the input status. (2) Sensors are commonly made of proteins able to specifically bind to a small molecule, and regulate promoter activity turning gene expression ON or OFF. (3) Biocontainment based on toxin/antitoxin counteraction. In the default state, the antitoxin is constitutively produced to neutralize any toxin originated from promoter leakiness. In nonpermissive conditions, toxin synthesis is activated in high levels, to overcome antitoxin neutralizing capacity leading to cell death. (4) Delivery plays crucial role in therapeutics, ensuring that produced active compound reaches target. Cell lysis strategy consists of producing membrane-degrading proteins, in response to internal or external signals. Cell disruption frees cytoplasm content, releasing therapeutics in surrounding environment. The secretion process relies on membrane transporters that recognize signal peptides linked to therapeutic protein. During secretion process, signal tag is removed [15].

Toggle switches and serine integrases are depicted in Fig. 8.3. Toggle switches were constructed in *E. coli* and *S. typhimurium*, based on mutual repression activity of phage lambda cI/Cro genetic switch [49,69,70]. Serine integrases catalyze unidirectional inversion of DNA between two recognition sites, flipping the DNA sequence. They were employed to construct synthetic memory devices in *B. thetaiotaomicron* and *E. coli* [48,60]. To achieve a readable response, a unidirectional transcription terminator, surrounded by recombination sites, was placed between the promoter and the reporter gene in *E. coli.* In its initial configuration, transcription of the reporter gene prematurely terminates. Induced synthesis of BxB1 integrase causes the terminator to flip, allowing complete transcription of the reporter gene.

Production and delivery of therapeutic molecules

After sensing and processing external inputs, living therapeutics will execute the treatment protocol they have been programmed for. Therefore, the active compound must be synthesized and delivered. Proteins, peptides, and small metabolites are employed as therapeutic molecules in engineered microbes. Living therapeutics against pathogenic bacteria have been engineered to produce antimicrobial peptides only, or together with antibiofilm proteins [43,44,66,68].

Small metabolites may be able to freely cross the bacterial membrane. Proteins and peptides require exportation to the extracellular environment in most cases. Adding a secretion tag to the polypeptide chain is a common approach that causes no harm to the producer cell (Fig. 8.3). The secretion tag *usp45* is widely used in engineered *L. lactis* [35,36,44,71], while *yebF* tag has been used for *E. coli* [40,66,68]. Although secretion tags are commonly employed, membrane exporters have a limited capacity. Therefore, polypeptides that are synthesized at high rates tend to saturate the secretion capacity of the microbial cell and accumulate in the cytoplasm. Moreover, secretion efficiency seems to be affected by cargo polypeptide sequence. Hence, secretion devices with improved efficiency are needed. Ideally, they will be orthogonal, and work independently of the cargo sequence [15].

Alternatively, some bacteria have been engineered to lyse and release their contents to the environment (Fig. 8.3). To ensure proper growth, production and subsequent release of the cargo, living therapeutics carry sophisticated control devices, connecting the expression of phage lysins to an internal [43,66] or external input signal [64]. An ingenious oscillator was constructed in *S. typhimurium*: the LuxR/LuxI quorum-sensing system controls expression of φX174 E lysis gene, creating synchronized bacterial growth and lysis cycles [38].

Diffusing and making way to the target may limit range of action of the molecules. To overcome the limits of diffusion, motile bacteria can be reprogrammed to selectively swim toward the target. By controlling *cheZ* expression through QS-responsive P_{lasI} promoter, it is possible to reprogram chemotactic behavior of *E. coli*, to swim toward QS-signal AHL produced by *P. aeruginosa*. Further addition of degradation tag YbaQ to CheZ reduced basal activity and broadened responsive range of inducer concentration [68].

Another approach for delivering therapeutics in situ is to display proteins on the chassis surface to act anchored (such as antibody display [55]), or to promote adhesion or infiltration into the target. In *E. coli*, ice nuclease protein (INP) tag exports polypeptide cargos to the membrane. When fused to INP, histonelike protein A promotes microbial infiltration into tumor mass, for targeted delivery of anticancer compounds [40]. Synthetic modular adhesins are also available, at least for *E. coli*. Constructed by combining β-domain from intimin and nanobodies, synthetic adhesins accept modification of specificity by exchanging the variable domain in nanobodies. Consequently, adhesins can be engineered against virtually any antigen of interest, targeting a chosen surface or cell [72].

Bioengineering constraints

Heterologous gene expression can impose a metabolic burden to engineered microbes, reducing fitness and affecting natural capacity to confer a health benefit on the host. That is particularly relevant when several devices (sense, control, memory, production, etc.) are combined in one chassis, to create a sophisticated biological system. For example, natural production of lactic acid by *L. lactis* in the mouse gut inhibits *V. cholerae* colonization. The burden of heterologous expression reduced the acidifying capability and the viability of *L. lactis* in coculture with *V. cholerae* [52]. One needs to create the right balance between cellular basal metabolism and synthetic metabolism [15].

Genetic stability

Most engineered living therapeutics carry recombinant DNA integrated in the chromosome. Plasmids do not offer genetic stability, as they may be lost in the absence of a selection pressure, or be unequally segregated during cell division. Researchers may address the problem by deleting one essential gene in the host, and subsequently complementing with a plasmid copy of the same gene [40,43]. A more sophisticated approach was taken to engineer two devices in *E. coli* strain Nissle 1917. In the first device, a toxin (hok) encoded in the plasmid was counteracted by a short-lived antitoxin (sok), encoded in the host chromosome. Thus, the loss of the plasmid killed the cell. The second device was the alp7AR cassette encoded in the plasmid. Alp proteins produced filaments that dynamically pushed plasmids to the poles of the cell, ensuring equal segregation during cell division [73].

Biocontainment

Containment mechanisms efficiently prevent spreading of bacteria and recombinant DNA into the environment, potentially a source of ecological risks. Many strategies have been successfully applied [74]; however, very few have been specifically constructed for living therapeutics. Pioneers in the field introduced auxotrophy for engineered probiotics. The strategy is not optimal, as the microbe may still survive outside the human body, by obtaining essential metabolites generated by other organisms. Genetic switches are an interesting alternative. Ideal kill switches will be triggered by environmental inputs, and be highly lethal in nonpermissive conditions, yet stable enough in permissive conditions.

Genetic circuits constructed in *E. coli*, employ temperature-sensitive switches to control the expression of toxin-antitoxin system CcdB/CcdA (Fig. 8.3). Although ingenious, temperature-sensitive kill switches have escape frequencies of 10^{-4} to 10^{-5} [65,75], which are higher than recommended by the National Institutes of Health (10^{-8}) [76]. More effective kill switches—with escape frequencies $\leq 10^{-8}$—have been constructed [77,78].

A device called iDNA employs CRISPR-Cas9 system targeting host genome [77]. Degradation of host genome not only kills the cell but also inactivates recombinant genes. However, the device lacks an environmental trigger that would make it suitable for living therapeutics. Another highly effective containment strategy is bacterial addiction to noncanonical amino acids or synthetic nucleotides. Microbial genetic code is modified to confer metabolic dependence on nonstandard amino acids or nucleotides for survival. Altering the chemical language isolates engineered microbes from natural ecosystems, yielding escape frequencies as low as 10^{-11} [79,80]. Yet, as long as the analog compound has to be supplemented to the bacteria, addiction is not a suitable method, and further research is required [15].

Research opportunities in the field

Synthetic biology was built upon model microorganisms such as *E. coli* and *Saccharomyces cerevisiae*, which have decades of genetic studies and molecular biology tools development behind them. However, it is becoming increasingly clear that new chassis are required for synthetic biology capabilities, in order to advance in medicine and industry [81]. There is a growing need for genetic tools for manipulation of nonmodel microorganisms such as lactic acid bacteria, bifidobacteria, and *Bacteroides*. Opportunities include developing sensors and diagnostics, preventing or treating pathogen infection, boosting the host immune system, enhancing metabolism, altering microbiomes, and understanding structure–function relationships between microbiota and host [54].

In order to accomplish a task, engineered living therapeutics must carry input sensors to detect external signals, internal logic circuits for integration of inputs, and actuators to generate a response according to the decision made by the logic circuits (Fig. 8.2). Unlike electronic devices, biological parts can only be used to build one circuit per cell, to ensure orthogonality among circuits. Orthogonality entails parallel systems, independent of evolutionary constraints, and uncoupled from cell regulation, encompassing promoters, transcription regulators, or regulatory RNAs. The more regulators are available, the more complex genetic circuits can be built.

Development of new sensors for detecting nutrients, toxic compounds, disease-specific markers, and pathogens will support fine-tuning of production, delivery, and containment devices. Efficient, modular, and multilayered containment systems have been engineered recently [74], but they lack a signal-sensor to differentiate the environment inside and outside the body. Unlike sensors and regulators, the development of actuators is directly linked to the application pursued. They will be needed at the same rate as drugs for conventional therapies. So far, most available actuators are conventional drugs adapted for living therapeutics. Future efforts will probably add to the toolbox new compounds.

Heterologous gene expression can impose a metabolic burden. To overcome loss of fitness, a higher dose of bacteria is required. It may be helpful to attach to the target surface. *E. coli* carrying synthetic adhesins efficiently bind to the target cells enabling reduction of the bacterial doses needed, by two orders of magnitude [72]. Nevertheless, adhesion is not an option to eliminate toxic compounds from the gut (e.g., phenylalanine in phenylketonuria [30,31]) or deliver drugs to be absorbed by the epithelium (e.g., insulin in diabetes [35]). Engineering a fusion protein of insulin to a cell penetrating peptide, such as penetratin, may improve penetration through the gut epithelium [82].

Still, diffusion through the gut will limit the efficacy for both insulin and phenylalanine lyase, possibly requiring a very high dose of living therapeutics. Biological parts enabling to settle down in one, or all gut niches, would be ideal. Mice intestines host two distinct niches: the lumen and the mucous layer [83]. If that is also true for humans, living therapeutics programmed to colonize mucus could be more efficient in delivering drugs to the epithelium. Lumen could be a better choice to eliminate toxic compounds [15].

Early results

Today, the only FDA-approved microbiota-based therapy is fecal microbiota transplant, to treat *C. difficile* infections not responsive to standard therapies [84]. However, there are over 170 microbiome-based therapeutics (not only bioengineered), across various stages of development [1,85].

Companies such as Intrexon, Synlogic, Aduro Biotech, and Azitra have an array of bioengineered living therapeutics, most under development or preclinical stage; however, a few have entered phase 1 and 2 clinical trials.

SYNB1020, an *E. coli* engineered probiotic developed by Synlogic to treat hyperammonemia, was well tolerated by male and female volunteers in phase 1 clinical study. The study indicated in vivo SYNB1020 activity. Therapeutic concentrations reached steady state by the second day of dosing, and SYNB1020 was no longer detectable in feces 2 weeks after the last dose. SYNB1020 is currently being evaluated in a phase 2a study in patients with cirrhosis (ClinicalTrials.gov, identifier: NCT03447730) [86]. SYNB1618 for phenylketonuria has entered phase 1 trial (ClinicalTrials.gov, Identifier: NCT03516487) [30].

Intrexon has bioengineered *L. lactis* in phase 1 (ActoBiotics AGO19) and another in phase 2 (ActoBiotics AGO13). AGO19 was engineered to deliver autoantigen human proinsulin (hPINS), and tolerance-enhancing cytokine hIL-10 (ClinicalTrials.gov, Identifier: NCT03751007). AG013 was engineered to deliver human trefoil factor 1 to mucosal tissues of the oral cavity, to prevent oral mucositis in head and neck cancer patients receiving chemo-radiation, and to prevent relapse of oral mucositis in cancer patients receiving chemotherapy (ClinicalTrials.gov, Identifier: NCT03234465).

A couple of products already failed phase 2 trial. Engineered *L. lactis* from ActoBio Therapeutics (former ActoGenix) was considered inefficient in the treatment of Crohn disease, possibly due to a low production of interleukin-10 in the gut [22], and strain CRS-207 from Aduro Biotech was no more effective at fighting pancreatic cancer than standard chemotherapy [87].

Azitra, Inc., received 5.4 million USD in 2018 to further develop and start human studies for their platform technology for skin diseases, based on the bacterium *S. epidermis* [88].

Safety and regulatory concerns must be addressed, before such microbes can be commercially available.

Medical, environmental, and ethical challenges

The safety of patients and containment of recombinant genetic elements are of concern. Living therapeutics must be equipped with efficient biocontainment devices, to avoid spreading of bacteria and recombinant DNA into the environment during treatment. After treatment, a clearance protocol should be taken to eliminate therapeutic microbes from patients to prevent unintended consequences. Persistent therapeutic microbes might be unintentionally passed to others, including healthcare workers or family members, again leading to unintended consequences [89]. Even more challenging than biocontainment is to contain genetic material exchange between bioengineered microbes and the natural microbiota.

It is uncertain how well will the public accept such a disruptive treatment approach. People will have to be educated about it. Moreover, healthcare providers will have to be trained to deal with bioengineered microorganisms, and detailed protocols will have to be established. Technical issues essential for commercialization approval, regulatory and safety concerns regarding the therapeutic use of bioengineered microbes, and societal and ethical questions must be addressed before such agents can be effectively used in the clinic [48].

Glossary

5′’-UTR mRNA mRNA 5′’-untranslated end.

Actuator Biological part or device that generates an output (e.g., motility, death, antibiotic production).

Adhesin Cell-surface components of bacteria that promote adhesion to other cells or to surfaces.

AHL *Pseudomonas aeruginosa* quorum-sensing signal N-Acyl homoserine lactone.

Auxotrophy Inability of an organism to synthesize a compound needed for growth.

Biological part A sequence of DNA that encodes a biological function (e.g., promoter, RBS, gene).

Cell penetrating peptide (CPP) Peptides able to pass cross tissue and cell membranes via energy-dependent or − energy-independent mechanisms.

Control A collection of all synthetic regulatory parts (e.g., promoter, RBS, protein tags, and regulatory proteins and RNAs) that determines when and how many actuators will be produced by the cell. Control devices are directly connected to sensors.

dCas9 (dead Cas9) A catalytically inactive form of Cas9.

Delivery Biological part or device that promotes exportation or release of molecules to the extracellular environment.

Device A collection of biological parts that performs a higher order function.

Elafin: A natural protease inhibitor that is normally produced by the human intestine and protects the gut from insults. The capacity to produce Elafin is lost in patients suffering from inflammatory bowel diseases.

Intimin Enterohemorrhagic *E. coli* adhesin.

Memory An artificial genetic circuit that alters the system behavior upon transient exposition to a signal. Unlike a regular control device, a memory device will continue to exhibit the altered behavior even if the signal input is over.

Nanobodies Ig domains of high affinity, solubility, and stability based on the variable domain of heavy chain-only antibodies.

Quorum-sensing (QS) Cell-to-cell communication through small molecules.

Sensors Protein and/or nucleic acid device able to recognize a signal (usually a small molecule) and induce a cellular response (e.g., gene expression).

Toggle switch Bistable gene-regulatory network.

Transcriptor Transistor-like device that regulates the flow of RNA polymerase along DNA.

Acknowledgments

D.B.P is funded through the Young Investigator Award grant #2014/17564-7 from São Paulo Research Foundation (FAPESP), and by the National Council for Scientific and Technological Development (CNPq) grant #401990/2016-8.

References

[1] Mays ZJ, Nair NU. Synthetic biology in probiotic lactic acid bacteria: at the frontier of living therapeutics. Curr. Opin. Biotechnol. 2018;53:224–31.

[2] Singh B, Mal G, Marotta F. Designer probiotics: paving the way to living therapeutics. Trends Biotechnol. 2017;35(8):679–82.

[3] Ozdemir T, Fedorec AJH, Danino T, Barnes CP. Synthetic biology and engineered live biotherapeutics: toward increasing system complexity. Cell Syst. 2018;7(1):5–16.

[4] Human Microbiome Project C. Structure, function and diversity of the healthy human microbiome. Nature 2012;486(7402):207–14.

[5] Sender R, Fuchs S, Milo R. Revised estimates for the number of human and bacteria cells in the body. PLoS Biol. 2016;14(8):e1002533. e1002533.

[6] Amar J. Microbiota–host crosstalk: a bridge between cardiovascular risk factors, diet, and cardiovascular disease. Am. J. Hypertens. 2018;31(8):941–4.

[7] Jameson KG, Hsiao EY. Linking the gut microbiota to a brain neurotransmitter. Trends Neurosci. 2018;41(7):413–4.

[8] Marques FZ, Mackay CR, Kaye DM. Beyond gut feelings: how the gut microbiota regulates blood pressure. Nat. Rev. Cardiol. 2018;15(1):20–32.

[9] Lima-Ojeda JM, Rupprecht R, Baghai TC. "I Am I and my bacterial circumstances": linking gut microbiome, neurodevelopment, and depression. Front. Psychiatry 2017;8:153.

[10] Torres-Fuentes C, Schellekens H, Dinan TG, Cryan JF. The microbiota-gut-brain axis in obesity. Lancet Gastroenterol. Hepatol. 2017;2(10):747–56.

[11] Cho I, Blaser MJ. The human microbiome: at the interface of health and disease. Nat. Rev. Genet. 2012;13:260.

[12] Langdon A, Crook N, Dantas G. The effects of antibiotics on the microbiome throughout development and alternative approaches for therapeutic modulation. Genome Med. 2016;8(1):39. 39.

[13] Hill C, Guarner F, Reid G, Gibson GR, Merenstein DJ, Pot B, Morelli L, Canani RB, Flint HJ, Salminen S, Calder PC, Sanders ME. The International Scientific Association for Probiotics and Prebiotics consensus statement on the scope and appropriate use of the term probiotic. Nat. Rev. Gastroenterol. Hepatol. 2014;11:506.

[14] Fijan S. Microorganisms with claimed probiotic properties: an overview of recent literature. Int. J. Environ. Res. Public Health 2014;11(5):4745–67.

[15] Pedrolli DB, Ribeiro NV, Squizato PN, de Jesus VN, Cozetto DA, Tuma RB, Gracindo A, Cesar MB, Freire PJC, da Costa AFM, Lins MRCR, Correa GG, Cerri MO. Engineering microbial living therapeutics: the synthetic biology toolbox. Trends Biotechnol. 2019;37(1):100–15.

[16] van Nood E, Vrieze A, Nieuwdorp M, Fuentes S, Zoetendal EG, de Vos WM, Visser CE, Kuijper EJ, Bartelsman JF, Tijssen JG, Speelman P, Dijkgraaf MG, Keller JJ. Duodenal infusion of donor feces for recurrent *Clostridium difficile*. N. Engl. J. Med. 2013;368(5):407–15.

[17] Fijan S. Microorganisms with claimed probiotic properties: an overview of recent literature. Int. J. Environ. Res. Public Health 2014;11(5):4745–67.

[18] Paton AW, Morona R, Paton JC. Bioengineered microbes in disease therapy. Trends Mol. Med. 2012;18(7):417–25.

[19] Hamady ZZ, Scott N, Farrar MD, Lodge JP, Holland KT, Whitehead T, Carding SR. Xylan-regulated delivery of human keratinocyte growth factor-2 to the inflamed colon by the human anaerobic commensal bacterium *Bacteroides ovatus*. Gut 2010;59(4):461–9.

[20] Chua KJ, Kwok WC, Aggarwal N, Sun T, Chang MW. Designer probiotics for the prevention and treatment of human diseases. Curr. Opin. Chem. Biol. 2017;40:8–16.

[21] Landry BP, Tabor JJ. Engineering diagnostic and therapeutic gut bacteria. Microbiol. Spectr. 2017;5(5).

[22] Cano-Garrido O, Seras-Franzoso J, Garcia-Fruitós EJMCF. Lactic acid bacteria: reviewing the potential of a promising delivery live vector for biomedical purposes. Microb. Cell Fact 2015;14(1):137.

[23] Hwang IY, Koh E, Kim HR, Yew WS, Chang MW. Reprogrammable microbial cell-based therapeutics against antibiotic-resistant bacteria. Drug Resist. Updates 2016;27:59–71.

[24] Neurath MF. New targets for mucosal healing and therapy in inflammatory bowel diseases. Mucosal Immunol. 2013;7:6.

[25] Benbouziane B, Ribelles P, Aubry C, Martin R, Kharrat P, Riazi A, Langella P, Bermudez-Humaran LG. Development of a Stress-Inducible Controlled Expression (SICE) system in *Lactococcus lactis* for the production and delivery of therapeutic molecules at mucosal surfaces. J. Biotechnol. 2013;168(2):120–9.

[26] Hanson ML, Hixon JA, Li W, Felber BK, Anver MR, Stewart CA, Janelsins BM, Datta SK, Shen W, McLean MH, Durum SK. Oral delivery of IL-27 recombinant bacteria attenuates immune colitis in mice. Gastroenterology 2014;146(1):210–221.e13.

[27] Hamady ZZR, Scott N, Farrar MD, Wadhwa M, Dilger P, Whitehead TR, Thorpe R, Holland KT, Lodge PJA, Carding SR. Treatment of colitis with a commensal gut bacterium engineered to secrete human tgf-β1 under the control of dietary xylan. Inflamm. Bowel Dis. 2011;17(9):1925–35.

[28] Vandenbroucke K, de Haard H, Beirnaert E, Dreier T, Lauwereys M, Huyck L, Van Huysse J, Demetter P, Steidler L, Remaut E, Cuvelier C, Rottiers P. Orally administered L. lactis secreting an anti-TNF Nanobody demonstrate efficacy in chronic colitis. Mucosal Immunol. 2009;3:49.

[29] Motta J-P, Bermúdez-Humarán LG, Deraison C, Martin L, Rolland C, Rousset P, Boue J, Dietrich G, Chapman K, Kharrat P, Vinel J-P, Alric L, Mas E, Sallenave J-M, Langella P, Vergnolle N. Food-grade bacteria expressing Elafin protect against inflammation and restore colon homeostasis. Sci. Transl. Med. 2012;4(158):158ra144.

[30] Isabella VM, Ha BN, Castillo MJ, Lubkowicz DJ, Rowe SE, Millet YA, Anderson CL, Li N, Fisher AB, West KA, Reeder PJ, Momin MM, Bergeron CG, Guilmain SE, Miller PF, Kurtz CB, Falb D. Development of a synthetic live bacterial therapeutic for the human metabolic disease phenylketonuria. Nat. Biotechnol. 2018;36:857–64.

[31] Durrer KE, Allen MS, Hunt von Herbing I. Genetically engineered probiotic for the treatment of phenylketonuria (PKU); assessment of a novel treatment in vitro and in the PAHenu2 mouse model of PKU. PLoS One 2017;12(5):e0176286.

[32] Duan FF, Liu JH, March JC. Engineered commensal bacteria reprogram intestinal cells into glucose-responsive insulin-secreting cells for the treatment of diabetes. Diabetes 2015;64(5):1794.

[33] Agarwal P, Khatri P, Billack B, Low W-K, Shao J. Oral delivery of glucagon like peptide-1 by a recombinant Lactococcus lactis. Pharmaceut. Res. 2014;31(12):3404–14.

[34] Takiishi T, Cook DP, Korf H, Sebastiani G, Mancarella F, Cunha JP, Wasserfall C, Casares N, Lasarte JJ, Steidler L, Rottiers P, Dotta F, Gysemans C, Mathieu C. Reversal of diabetes in NOD mice by clinical-grade proinsulin and IL-10-secreting *Lactococcus lactis* in combination with low-dose anti-CD3 depends on the induction of Foxp3-positive T cells. Diabetes 2017;66(2):448–59.

[35] Mao R, Wu D, Hu S, Zhou K, Wang M, Wang Y. Secretory expression and surface display of a new and biologically active single-chain insulin (SCI-59) analog by lactic acid bacteria. Appl. Microbiol. Biotechnol. 2017;101(8):3259–71.

[36] Ng DT, Sarkar CA. Nisin-inducible secretion of a biologically active single-chain insulin analog by Lactococcus lactis NZ9000. Biotechnol. Bioeng. 2011;108(8):1987–96.

[37] Zhou S, Gravekamp C, Bermudes D, Liu K. Tumour-targeting bacteria engineered to fight cancer. Nat. Rev. Canc. 2018;18(12):727–43.

[38] Din MO, Danino T, Prindle A, Skalak M, Selimkhanov J, Allen K, Julio E, Atolia E, Tsimring LS, Bhatia SN, Hasty J. Synchronized cycles of bacterial lysis for in vivo delivery. Nature 2016;536(7614):81–5.

[39] Ryan RM, Green J, Williams PJ, Tazzyman S, Hunt S, Harmey JH, Kehoe SC, Lewis CE. Bacterial delivery of a novel cytolysin to hypoxic areas of solid tumors. Gene Ther. 2009;16:329.

[40] Ho CL, Tan HQ, Chua KJ, Kang A, Lim KH, Ling KL, Yew WS, Lee YS, Thiery JP, Chang MW. Engineered commensal microbes for diet-mediated colorectal-cancer chemoprevention. Nat. Biomed. Eng. 2018;2(1):27–37.

[41] Yoon W, Park YC, Kim J, Chae YS, Byeon JH, Min S-H, Park S, Yoo Y, Park YK, Kim BM. Application of genetically engineered *Salmonella typhimurium* for interferon-gamma–induced therapy against melanoma. Eur. J. Cancer 2017;70:48–61.

[42] Zheng JH, Nguyen VH, Jiang S-N, Park S-H, Tan W, Hong SH, Shin MG, Chung I-J, Hong Y, Bom H-S, Choy HE, Lee SE, Rhee JH, Min J-J. Two-step enhanced cancer immunotherapy with engineered *Salmonella typhimurium* secreting heterologous flagellin. Sci. Transl. Med. 2017;9(376). eaak9537.

[43] Hwang IY, Koh E, Wong A, March JC, Bentley WE, Lee YS, Chang MW. Engineered probiotic *Escherichia coli* can eliminate and prevent *Pseudomonas aeruginosa* gut infection in animal models. Nat. Commun. 2017;8:15028.

[44] Borrero J, Chen Y, Dunny GM, Kaznessis YN. Modified lactic acid bacteria detect and inhibit multiresistant enterococci. ACS Synth. Biol. 2015;4(3):299–306.

[45] Liu W, Tan Z, Xue J, Luo W, Song H, Lv X, Zheng T, Xi T, Xing Y. Therapeutic efficacy of oral immunization with a non-genetically modified Lactococcus lactis-based vaccine CUE-GEM induces local immunity against *Helicobacter pylori* infection. Appl. Microbiol. Biotechnol. 2016;100(14):6219–29.

[46] Zhou Z, Gong S, Li X-M, Yang Y, Guan R, Zhou S, Yao S, Xie Y, Ou Z, Zhao J, Liu Z. Expression of *Helicobacter pylori* urease B on the surface of Bacillus subtilis spores. J. Med. Microbiol. 2015;64(1):104–10.

[47] Marcobal A, Liu X, Zhang W, Dimitrov AS, Jia L, Lee PP, Fouts TR, Parks TP, Lagenaur LA. Expression of human immunodeficiency virus type 1 neutralizing antibody fragments using human vaginal Lactobacillus. AIDS Res. Hum. Retrovir. 2016;32(10–11):964–71.

[48] Courbet A, Endy D, Renard E, Molina F, Bonnet J. Detection of pathological biomarkers in human clinical samples via amplifying genetic switches and logic gates. Sci. Transl. Med. 2015;7(289):289ra83.

[49] Riglar DT, Giessen TW, Baym M, Kerns SJ, Niederhuber MJ, Bronson RT, Kotula JW, Gerber GK, Way JC, Silver PA. Engineered bacteria can function in the mammalian gut long-term as live diagnostics of inflammation. Nat. Biotechnol. 2017;35(7):653–8.

[50] Daeffler KNM, Galley JD, Sheth RU, Ortiz-Velez LC, Bibb CO, Shroyer NF, Britton RA, Tabor JJ. Engineering bacterial thiosulfate and tetrathionate sensors for detecting gut inflammation. Mol. Syst. Biol. 2017;13(4):923.

[51] Lubkowicz D, Ho CL, Hwang IY, Yew WS, Lee YS, Chang MW. Reprogramming probiotic Lactobacillus reuteri as a biosensor for *Staphylococcus aureus* derived AIP-I detection. ACS Synth. Biol. 2018;7(5):1229–37.

[52] Mao N, Cubillos-Ruiz A, Cameron DE, Collins JJ. Probiotic strains detect and suppress cholera in mice. Sci. Transl. Med. 2018;10(445).

[53] Claesen J, Fischbach MA. Synthetic microbes as drug delivery systems. ACS Synth. Biol. 2015;4(4):358–64.

[54] Bober JR, Beisel CL, Nair NU. Synthetic biology approaches to engineer probiotics and members of the human microbiota for biomedical applications. Annu. Rev. Biomed. Eng. 2018;20(1):277–300.

[55] Marcobal A, Liu X, Zhang W, Dimitrov AS, Jia L, Lee PP, Fouts TR, Parks TP, Lagenaur LA. Expression of human immunodeficiency virus type 1 neutralizing antibody fragments using human vaginal Lactobacillus. AIDS Res. Hum. Retrovir. 2016;32(10–11):964–71.

[56] Munivar AM, Whitfill TM. Therapeutic treatment of skin disease with recombinant commensal skin microorganisms. 2016.

[57] Song AA, In LLA, Lim SHE, Rahim RA. A review on *Lactococcus lactis*: from food to factory. Microb. Cell Factories 2017;16(1):55.

[58] Lee SM, Donaldson GP, Mikulski Z, Boyajian S, Ley K, Mazmanian SK. Bacterial colonization factors control specificity and stability of the gut microbiota. Nature 2013;501(7467):426–9.

[59] Whitaker WR, Shepherd ES, Sonnenburg JL. Tunable expression tools enable single-cell strain distinction in the gut microbiome. Cell 2017;169(3):538–546 e12.

[60] Mimee M, Tucker AC, Voigt CA, Lu TK. Programming a human commensal bacterium, *Bacteroides thetaiotaomicron*, to sense and respond to stimuli in the murine gut microbiota. Cell Syst. 2015;1(1):62–71.

[61] Zhu D, Liu F, Xu H, Bai Y, Zhang X, Saris PE, Qiao M. Isolation of strong constitutive promoters from *Lactococcus lactis* subsp. lactis N8. FEMS Microbiol. Lett. 2015;362(16).

[62] Tauer C, Heinl S, Egger E, Heiss S, Grabherr R. Tuning constitutive recombinant gene expression in *Lactobacillus plantarum*. Microb. Cell Factories 2014;13:150.

[63] Berlec A, Skrlec K, Kocjan J, Olenic M, Strukelj B. Single plasmid systems for inducible dual protein expression and for CRISPR-Cas9/CRISPRi gene regulation in lactic acid bacterium *Lactococcus lactis*. Sci. Rep. 2018;8(1):1009.

[64] Camacho EM, Mesa-Pereira B, Medina C, Flores A, Santero E. Engineering Salmonella as intracellular factory for effective killing of tumour cells. Sci. Rep. 2016;6:30591.

[65] Piraner DI, Abedi MH, Moser BA, Lee-Gosselin A, Shapiro MG. Tunable thermal bioswitches for *in vivo* control of microbial therapeutics. Nat. Chem. Biol. 2017;13(1):75–80.

[66] Jayaraman P, Holowko MB, Yeoh JW, Lim S, Poh CL. Repurposing a two-component system-based biosensor for the killing of *Vibrio cholerae*. ACS Synth. Biol. 2017;6(7):1403–15.

[67] Tscherner M, Giessen TW, Markey L, Kumamoto CA, Silver PA. A synthetic system that senses Candida albicans and inhibits virulence factors. ACS Synth. Biol. 2019;8:434–44.

[68] Hwang IY, Tan MH, Koh E, Ho CL, Poh CL, Chang MW. Reprogramming microbes to be pathogen-seeking killers. ACS Synth. Biol. 2014;3(4):228–37.

[69] Kim S, Kerns SJ, Ziesack M, Bry L, Gerber GK, Way JC, Silver PA. Quorum sensing can Be repurposed to promote information transfer between bacteria in the mammalian gut. ACS Synth. Biol. 2018;7(9):2270–81.

[70] Kotula JW, Kerns SJ, Shaket LA, Siraj L, Collins JJ, Way JC, Silver PA. Programmable bacteria detect and record an environmental signal in the mammalian gut. Proc. Natl. Acad. Sci. USA. 2014;111(13):4838–43.

[71] Shigemori S, Watanabe T, Kudoh K, Ihara M, Nigar S, Yamamoto Y, Suda Y, Sato T, Kitazawa H, Shimosato T. Oral delivery of *Lactococcus lactis* that secretes bioactive heme oxygenase-1 alleviates development of acute colitis in mice. Microb. Cell Factories 2015;14:189.

[72] Pinero-Lambea C, Bodelon G, Fernandez-Perianez R, Cuesta AM, Alvarez-Vallina L, Fernandez LA. Programming controlled adhesion of *E. coli* to target surfaces, cells, and tumors with synthetic adhesins. ACS Synth. Biol. 2015;4(4):463–73.

[73] Danino T, Prindle A, Kwong GA, Skalak M, Li H, Allen K, Hasty J, Bhatia SN. Programmable probiotics for detection of cancer in urine. Sci. Transl. Med. 2015;7(289):289ra84.

[74] Lee JW, Chan CTY, Slomovic S, Collins JJ. Next-generation biocontainment systems for engineered organisms. Nat. Chem. Biol. 2018;14:530–7.

[75] Stirling F, Bitzan L, O'Keefe S, Redfield E, Oliver JWK, Way J, Silver PA. Rational design of evolutionarily stable microbial kill switches. Mol. Cell 2017;68(4):686–697 e3.

[76] Wilson DJ. NIH guidelines for research involving recombinant DNA molecules. Account. Res. 1993;3(2–3):177–85.

[77] Caliando BJ, Voigt CA. Targeted DNA degradation using a CRISPR device stably carried in the host genome. Nat. Commun. 2015;6:6989.

[78] Gallagher RR, Patel JR, Interiano AL, Rovner AJ, Isaacs FJ. Multilayered genetic safeguards limit growth of microorganisms to defined environments. Nucleic Acids Res. 2015;43(3):1945–54.

[79] Mandell DJ, Lajoie MJ, Mee MT, Takeuchi R, Kuznetsov G, Norville JE, Gregg CJ, Stoddard BL, Church GM. Biocontainment of genetically modified organisms by synthetic protein design. Nature 2015;518(7537):55–60.

[80] Marliere P, Patrouix J, Doring V, Herdewijn P, Tricot S, Cruveiller S, Bouzon M, Mutzel R. Chemical evolution of a bacterium's genome. Angew Chem. Int. Ed. Engl. 2011;50(31):7109–14.

[81] Adams BL. The next generation of synthetic biology chassis: moving synthetic biology from the laboratory to the field. ACS Synth. Biol. 2016;5(12):1328–30.

[82] Kristensen M, Franzyk H, Klausen MT, Iversen A, Bahnsen JS, Skyggebjerg RB, Fodera V, Nielsen HM. Penetratin-Mediated transepithelial insulin permeation: importance of cationic residues and pH for complexation and permeation. AAPS J. 2015;17(5):1200–9.

[83] Li H, Limenitakis JP, Fuhrer T, Geuking MB, Lawson MA, Wyss M, Brugiroux S, Keller I, Macpherson JA, Rupp S, Stolp B, Stein JV, Stecher B, Sauer U, McCoy KD, Macpherson AJ. The outer mucus layer hosts a distinct intestinal microbial niche. Nat. Commun. 2015;6:8292.

[84] Tauxe WM, Dhere T, Ward A, Racsa LD, Varkey JB, Kraft CS. Fecal microbiota transplant protocol for Clostridium difficile infection. Lab. Med. 2015;46(1):e19–23.

[85] Microbiome therapeutics and diagnostics market. 2017. p. 2017–30.

[86] Kurtz CB, Millet YA, Puurunen MK, Perreault M, Charbonneau MR, Isabella VM, Kotula JW, Antipov E, Dagon Y, Denney WS, Wagner DA, West KA, Degar AJ, Brennan AM, Miller PF. An engineered *E. coli* Nissle improves hyperammonemia and survival in mice and shows dose-dependent exposure in healthy humans. Sci. Transl. Med. 2019;11(475). eaau7975.

[87] Maxmen A. Living therapeutics: scientists genetically modify bacteria to deliver drugs. Nat. Med. 2017;23(1):5–7.

[88] Azitra, Inc. https://azitrainc.com/. (accessed 18.01.2019).

[89] Weiman S. In: Fox J, editor. Harnessing the power of microbes as therapeutics: bugs as drugs: report on an American academy of microbiology colloquium held in San Diego, CA, in april 2014; 2015. Washington (DC).

Chapter 9

Organ-on-a-chip and 3D printing as preclinical models for medical research and practice

Abhishek Jain, Tanmay Mathur, Navaneeth K.R. Pandian and Amirali Selahi

Department of Biomedical Engineering, College of Engineering, Texas A&M University, College Station, TX, United States

Introduction

An ultimate quest of most biomedical research is to determine the basic communication between cells and molecules, which assemble to form human tissues and organs. The expectation is that with this knowledge in both health and disease, scientists can identify most potent targets, against which drugs can be discovered and tested. However, such knowledge cannot be attained directly from humans, and therefore models are used. Our understanding of the complex molecular signaling in disease modalities, and subsequent development of therapeutic targets, has been heavily influenced by animal models. In the last 4 decades, animal models have been extensively used, to develop our understanding of several lethal diseases, such as, cancer progression and metastasis [1], vascular disorders [2], diabetes mellitus [3], and cardiac complications [4].

Small versus large animals

Among animal models, small animals or rodent models have been extensively studied [5]. Small animal models offer the scope of genetic and molecular modification, allowing the development of more humanized variants [6]. However, a major drawback is the sheer difference in the anatomical sizes compared to humans. This limits their ability to precisely recapitulate human physiology in vivo. Large animal models based on porcine, canine, and equine biology more closely resemble human physiology, and thus have been used in extensive preclinical and clinical trials [7]. Large animals also have sufficient tissue content, making them ideal for immunohistopathological applications. However, compared to small animal models, genetic modification of large animals is very difficult and requires larger breeding times to see the desired effects [8].

Additionally, these models do not allow dissection of specific signaling pathways, and analyzing cell-cell interactions is very difficult. Primate models more closely mimic the human biological system; however, their use in research is extremely limited, due to increased ethical concerns [9]. Further, with stringent drug screening protocols, lesser drug candidates are emerging, and the average investment per drug has increased up to 150% of the estimated amounts a decade ago [10]. Adding to that, increased ethical concerns, and deeper understanding of the dissimilarities between humans and animals, have made it difficult to rely exclusively on animal models. Thus, the need of the hour is to develop "humanized" disease models that can recapitulate human pathophysiology with maximum detail.

In vitro models and cell cultures

To independently dissect the signaling that occurs on a cellular or tissue level, in vitro models have also been used (Fig. 9.1) [11]. Static, well plate cultures with monolayers of cells have been used to analyze vascular disorders. Due to their simplicity, these systems can be easily multiplexed. Hence these models have been increasingly used for high throughput drug screening applications [12]. However, due to their lower complexity, these models cannot recapitulate intracellular signaling, and have been shown to induce phenotypical changes in cells [13]. These models also do not incorporate the effects of fluid forces, on the cellular functioning and structure. Increasing an order of complexity by involving fluid forces and nonplanar cultures, researchers have developed flow chambers [14]. These modalities have been able to impart fluid mechanical forces to cell monolayers, at the millimeter scale, hence

Precision Medicine for Investigators, Practitioners and Providers. https://doi.org/10.1016/B978-0-12-819178-1.00009-5

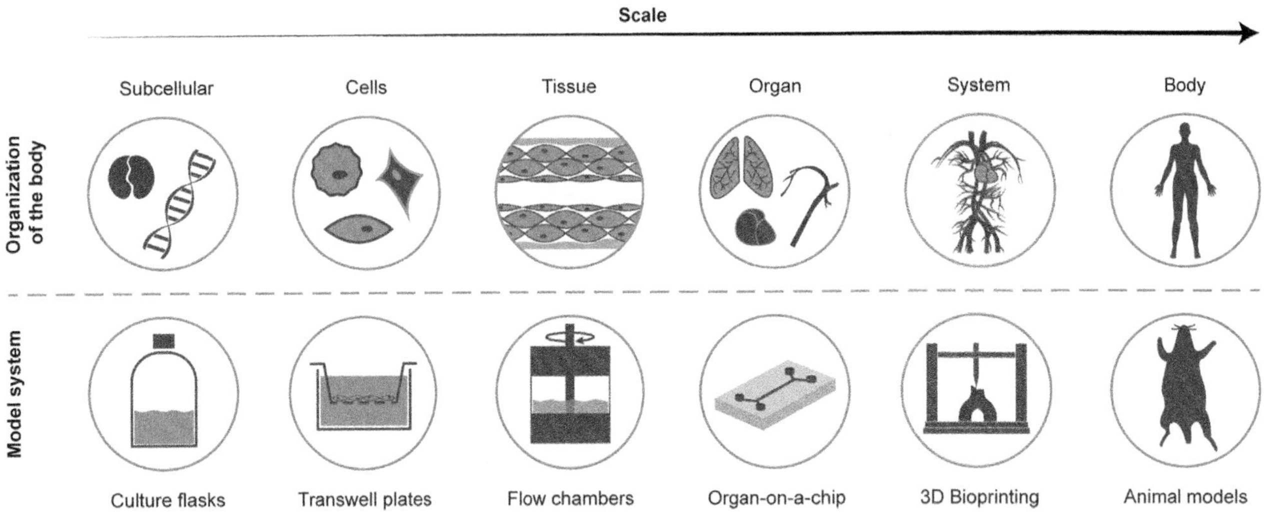

FIGURE 9.1 Overview of the in vitro and in vivo models to study cellular, tissue-level, or organ-level functioning. As the complexity of the disease pathophysiology increases, the requirements for a modeling modality that can more faithfully recapitulate the exact mechanistic events increase.

also attracting attention as "macro models" [15]. However, due to their large sizes, these models require large sample volumes, growth factors, and cells. Additionally, they cannot be reliably used for modeling small vessels like capillaries or arterioles.

3D structures

With the recent advancement in microfabrication and soft lithographic processes, researchers have engineered 3D in vitro systems that cannot only mimic the cellular/tissue level microphysiological environment but can also allow a systematic analysis of the key factors responsible for disease progression.

In vitro preclinical models: organ-on-a-chip

An organ-on-a-chip (OOC) is a multichannel, 3D microfluidic cell culture chip, that simulates the activities, mechanics, and physiological response of entire organs, or part of organ systems, representing a type of artificial organ. An advantage of organ-on-a-chip is to model the smallest functional unit of organs or tissues, to mimic the physiologic and pathophysiologic interactions in different locations inside the human body. Achieving the desired functionality of the tissues can be possible, by culturing the cells in microfluidic channels, and providing conditions similar to those that exist in vivo; including temperature, pressure, flow rate, pH, osmotic pressure, and nutrient content [16].

Lung-on-a-chip

Current lung models cannot reproduce organ-level complexity, and pathophysiological responses. Even though microfluidic devices were used to model a few tissue interfaces, the first pioneering design of organ-on-a-chip of the lung was by Huh et al. [17]. Their bioinspired microdevice reproduced complex integrated organ-level responses to bacteria and inflammatory cytokines, introduced into the alveolar space, by mimicking the cyclic mechanical strain caused by physiological breathing, using multichannel microfluidic devices separated by thin PDMS membranes [17]. After this first demonstration, they further advanced the platform to model the alveolocapillary interface (Fig. 9.2A).

This system contained two microfluidic channels separated by a thin porous PDMS membrane, and revealed that cyclic mechanical strain expresses the inflammatory responses of the lung to nanoparticles more prominently [18]. Later, the same platform was used to model small airways and viral infections [19]. Other groups have shown that the functionality of the alveolar barrier could be restored by coculturing epithelial and endothelial cells. Moreover, they could successfully enhance the cell culture efficacy, in this dynamical environment [20]. Incorporating smooth muscle cells in the lung-on-a-chip, to mimic their interactions with epithelial cells was achieved by Young's group, by culturing SMCs in collagen and Matrigel hydrogels that provide the necessary environment for cellular growth [21].

Jain et al. have successfully modeled inflammation-induced thrombosis, in their microfluidic alveolus-on-a-chip, by coculturing human primary alveolar epithelium and endothelium, in blood flow conditions [22]. Scientists could use the alveolus-on-a-chip to assess the therapeutic potentiality of a molecule, since it creates opportunities for future drug assays at lower cost, and extremely high speed. The door is also open for modeling of other pulmonary diseases by the way of the microchip techniques.

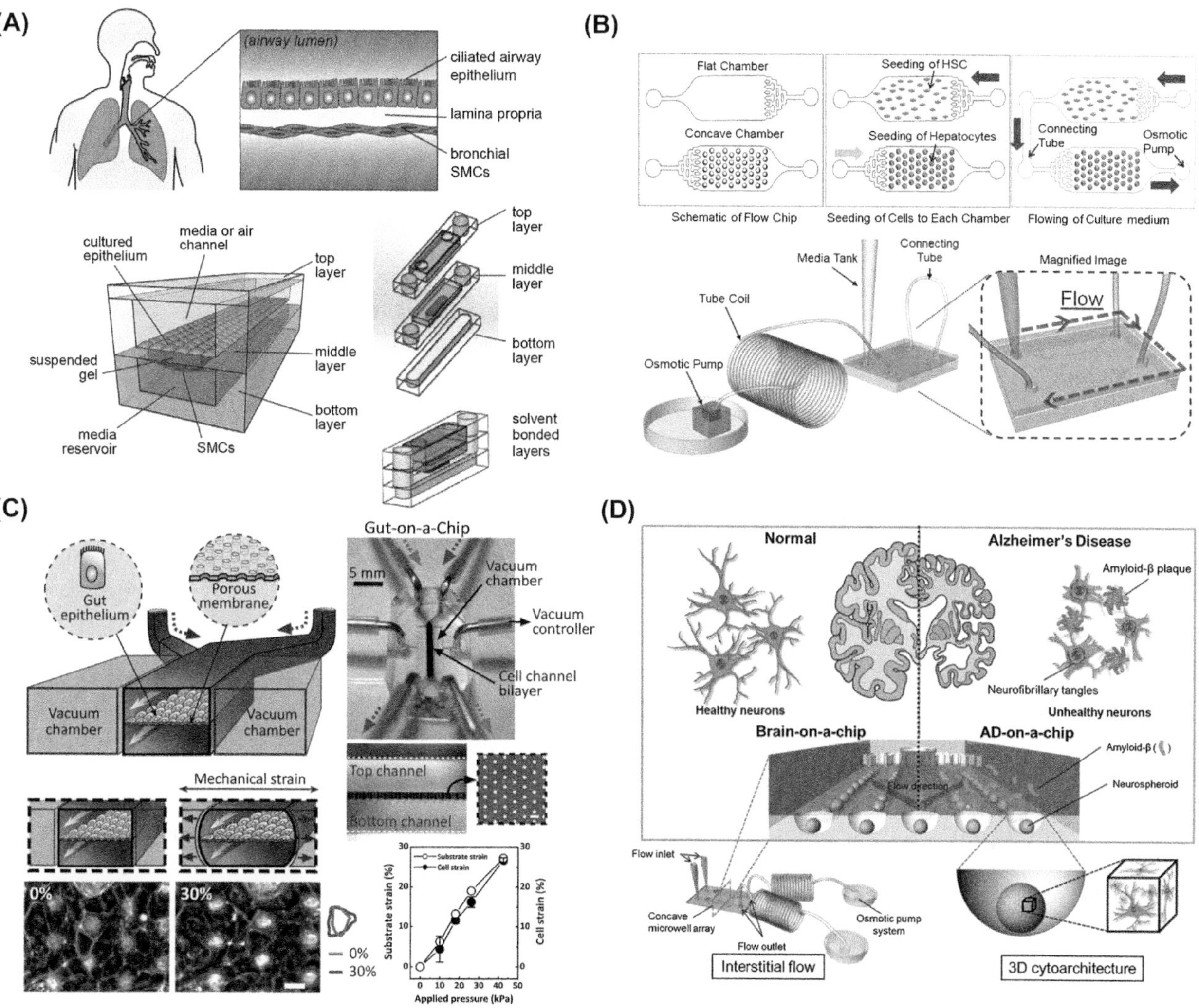

FIGURE 9.2 Overview of some organ-on-a-chip models. (A) Biologically inspired human lung-on-a-chip microdevice to form alveolar-capillary barrier on a thin, porous, flexible PDMS membrane coated with ECM. By applying vacuum to the side channels and a mechanical stretch, this device is capable of re-creating physiological breathing movements. Image taken from [17]. (B) The liver-on-a-chip cell seeding process, in which each cell suspension was introduced into the inlet port of the flow-chip, through microfluidic channels (*top figure*), and the experimental setup (*bottom figure*). Image taken from [23]. (C) A schematic of the gut-on-a-chip device, showing the flexible porous ECM-coated membrane lined by gut epithelial cells. In the *bottom left*, the phase contrast images of intestinal monolayers, in the absence (left) and presence (right) of mechanical strain exerted by applying suction to the vacuum chambers is shown. Image taken from [28]. (D) A three-dimensional illustration of the brain-on-a-chip with interstitial level of flow. The chip contains a concave microwell array for the formation of homogenous neurospheroids. *Image taken from Park J, et al. Three-dimensional brain-on-a-chip with an interstitial level of flow and its application as an in vitro model of Alzheimer's disease. Lab Chip 2015;15(1):141–150.*

Liver-on-a-chip

A three-dimensional liver-on-a-chip, to investigate the interactions of hepatocytes and hepatic stellate cells in which two types of cells are cocultured without direct cell-cell contact, has been designed by Lee et al. (Fig. 9.2B) [23]. As mentioned before, one of the main difficulties with in vivo studies is following and monitoring the steps of metabolic activities, as the experiment is running. Therefore, organs-on-a-chip integrated with sensors provide a functional platform to follow steps in a biological process. Bavli et al. presented a device capable of maintaining human tissue in vitro, over a course of a month, while taking advantage of proper sensors to track the dynamics of mitochondrial dysfunction [24].

Some improvements can be made, regarding the coculture, in which other cells involved in the liver, like Kupffer cells, could be added. Another way of improvement would be to re-create the bile ducts, which have not been investigated yet, despite their central position in the liver. Once these improvements have been made, complex diseases like the hepatitis B, may be modeled from this liver-on-chip.

Kidney-on-a-chip

One of the most difficult organs to model is kidney, since it is formed by several tissue layers, interacting extensively with each other. In one of the recent publications in this area, authors aimed to present a model of the kidney glomerulus, which is the primary site of blood filtration. With the kidney-on-a-chip, the authors could successfully demonstrate human kidney major cells differentiation, with high efficiency [25]. Similar works have aimed at mimicking proximal kidney tube formation in vitro, for testing a wide range of kidney-related diseases [26,27]. Future research in this area should focus on the optimal balance of a physiological relevant design (multi-compartmental 3D tubular structure of perfused human renal cells) that leads to a high predictive value and high-throughput applicability [27].

Gut-on-a-chip

Due to the complexities in fluid flow conditions and mechanical structures, traditional in vitro models were unable to represent an efficient model of the human gut. Peristaltic movement of the gut is the major factor leading to the complicated flow condition. In two recent publications, Kim et al. have been able to recapitulate multiple dynamic physical and functional features of the human intestine that provide a platform for drug toxicity tests, as well as intestinal disease modeling (Fig. 9.2C) [28,29]. Human acute radiation syndrome (ARS), resulting from partial-body or whole-body radiation, can cause severe gastrointestinal bleeding, sepsis, and death [30].

Gut-on-a-chip models have the potential to replace animal models, or even human tests for radiation study, mimicking tight junction disruption, as well as drug-related responses in the epithelial layer [31]. We could argue that these models are simplifications of the reality, but there is a possibility of progressively adding other features. For example, blood could be directly used as flowing fluid, instead of the current medium. Another opportunity would be the development of a gut-liver-on-a-chip; that could allow research on the first-pass effect, in the metabolism of drugs and nutrients [32].

Skin-on-a-chip

Scientists have tried to coculture the main three layers of skin including epidermal, dermal, and endodermal layers, to mimic inflammation and edema conditions, and test the devices with specific drugs to observe their functionality [33]. Lim et al. formed a 3D skin model, and then applied periodic displacement to stretch the skin for 7 days, forming a wrinkled skin-on-a-chip platform, to test antiwrinkle cosmetics and medicines [34]. Current studies focus on the tissue-tissue or organ-organ interactions, physically separated in vivo, but connected by metabolites in the medium [35,36].

Brain-on-a-chip

Of all other organs, human brain genetics and structure are most different from animal models. The blood-brain barrier (BBB) has been of significant interest, since it is an important mechanism for protecting the brain from fluctuations in plasma composition, and from circulating agents such as neurotransmitters and xenobiotics, capable of disturbing neural function [37,38]. Park et al. have developed a microfluidic chip, based on 3D neurospheroids, by introducing flow conditions similar to the interstitial space of the brain, and then studied the effect of flow on cell-cell interaction, as well as neural differentiation (Fig. 9.2D) [39].

Another group has attempted to dissect neural differentiation and maturation, using a brain-on-a-chip model that bridges the gap between cell culture and in vitro/in vivo conditions, through the recapitulation of self-organized neural differentiation. Neural progenitor cells expressed a significantly enhanced chemotactic response, toward the embryonic brain developing chemokines [40]. The BBB-on-chip models would be better, if adapted to human neuron cells. Induced pluripotent stem cells (iPSCs) have amazing potentiality to produce human neurons but, as explained by Dauth et al. [41], cannot yet make differentiation specific to a brain region.

Heart-on-a-chip

Conventional in vitro models to study cardiovascular diseases do not mimic the in vitro physiology, since they usually consist of a single cell culture layer under a static flow condition, in a considerable large geometry. Heart-on-a-chip platforms have systematically modified the in vitro models [42]. In one of the latest works, authors engineered a cardiomyocyte beating tissue using mechanical stimuli that result in a more efficient mechanical and electrical coupling [43]. In this work, the mechanical and electrical response showed a high coupling response. This platform also made it possible to provide mechanical stimuli, while culturing the cells, and also was capable of testing different concentrations of isoprenaline.

To investigate cardiac contractility, a muscle-inspired 3D chip was designed and tested by Ahn et al. [44]. They used gelatin as extracellular matrix, in addition to titanium oxide and silver nanoparticles. In this device the cardiotoxicity of nanoparticles alters the calcium signaling to sarcomere, providing a toolbox to measure the contractile effects [44]. In the future, researchers plan to link the brain and blood-brain barrier models with chips based on other human organs. Doing so will contribute to the knowledge about the "downstream" effects of experimental brain treatments on other organs.

Vessel-on-a-chip

Many diseases including diabetes and cancer are influenced by vascular system mechanics and physiology [45].

The model presented by Jain et al. uses interconnected channels covered with human endothelium cells, to mimic a network of stenosed arteriolar vessels, permitting evaluation of blood clotting using small sample volumes. By measuring the time needed to form blood clots inside the device, coagulation and platelet function can be dynamically followed in vitro for patient blood samples (Fig. 9.3) [46]. Another study presented a microfluidic platform to mimic vascular injury, by integrating a pneumatic valve into the chip and flowing blood [47].

Review articles provide more comprehensive details into the design and function of these devices [48–52]. A general commonality is that they are manufactured using techniques borrowed from the semiconductor microchip industry, and hence the term, "on-a-chip."

Equipment and techniques for microfabrication

Microfabrication techniques typically include replica molding [53], soft lithography [53], and microcontact printing [54], for organ-on-a-chip platforms [55]. Polydimethylsiloxane (PDMS) is a polymer widely used for the fabrication and prototyping of organ-on-a-chip microdevices. PDMS being transparent allows for high-quality imaging of cultured cells. Other prominent properties of PDMS that make it suitable for microfluidic applications are its flexibility, gas permeability, biocompatibility, low autofluorescence, and price affordability [56]. The PDMS stamp shown in Fig. 9.3A, is made using a mold which could be fabricated with different techniques, two of which are soft-lithography and three-dimensional (3D) printing.

In soft-lithography, the silicon wafer is coated with a photoresist layer, and then negatively or positively cross-linked using photomask and UV light source. Then, the uncross-linked photoresist is washed, using a developer to form the desired pattern on the silicon wafer that provides the mold. For microfluidic device fabrication, a cross-linking agent is typically mixed with liquid PDMS, after which the mixture is degassed to release air-bubbles, poured into the desired mold, and baked at 70–80°C for curing. Finally, the PDMS slab containing the device features can be removed from the mold, punched to make outlets for fluid flow, and finally can be bonded using oxygen plasma (Fig. 9.3B) [53,57,58].

Cell coculturing

Microfabrication techniques also provide methods to fabricate thin, flexible, and porous membranes, to form semipermeable surfaces for coculturing two different cell types, including epithelial and endothelial layers on either sides of the membrane [17]. By microcontact printing of proteins, the pattern of cultured cells on these membranes can be altered, based on the required physiology of the in vivo condition. In this method, the desired pattern is first shaped on a mold surface, followed by fabricating the stamp, by pouring and curing the liquid PDMS over the mold.

After removing from the mold, the stamp surface is covered with the target protein, which will get transferred to the membrane substrate upon contact. Since the cells are growing only on the area covered with protein, the form of the cell culture on the substrate can be patterned (Fig. 9.3C) [54]. Recently, 3D printing presents a promising alternative to traditional techniques such as soft-lithography, enabling rapid, single-step, and cost-effective production of highly complex organ-on-a-chip devices [59].

3D printing/additive manufacturing

One limitation of organ-on-a-chip platforms is that these models are not well-suited to incorporate the structural complexity and anatomy of the organs being modeled. Additionally, the scale and sizes of current microfluidic platforms limit the extent of modeling organ-level functioning. Although still in a nascent stage, attempts at implementing 3D bioprinting for modeling organ-level functioning have witnessed an exponential growth in the last 5 years, to the extent that the number of articles published on 3D printing and additive manufacturing have exceeded that of traditional organ-chips by nearly 3000 in 2017 [48].

3D printing for cardiac muscle and valves

3D printing has majorly focused on developing functional tissue and blood vessel models. Due to their ability to support cell growth, while mimicking the structural and biological integrity of native tissue, 3D printed constructs have been extensively utilized in cardiovascular applications. One exciting application is 3D printed heart valves. Current strategy to mitigate heart valve failure is either implanting mechanical constructs or utilizing valves produced from an allographic source [60].

However, these strategies either have a very low lifespan or require extensive use of anticoagulants to ensure proper in vivo functioning. To improve the hemocompatibility and success of the implants, researchers have employed 3D printed valve structures, using autologous cardiac muscle cells [61]. The resulting valves have the ability to grow/repair themselves along with the patients. Combined with novel imaging techniques, 3D printing also offers the advantage of incorporating the patient's native valve structure, and thus produces a patient-specific tissue that also takes spatial heterogeneity of the mechanical properties into account [62].

3D printing has been an alternative to fabricate thick, porous substrates that can allow stable cardiomyocyte culture, for myocardial tissue replacement. Due to the low success rates of cell therapy, researchers have used these

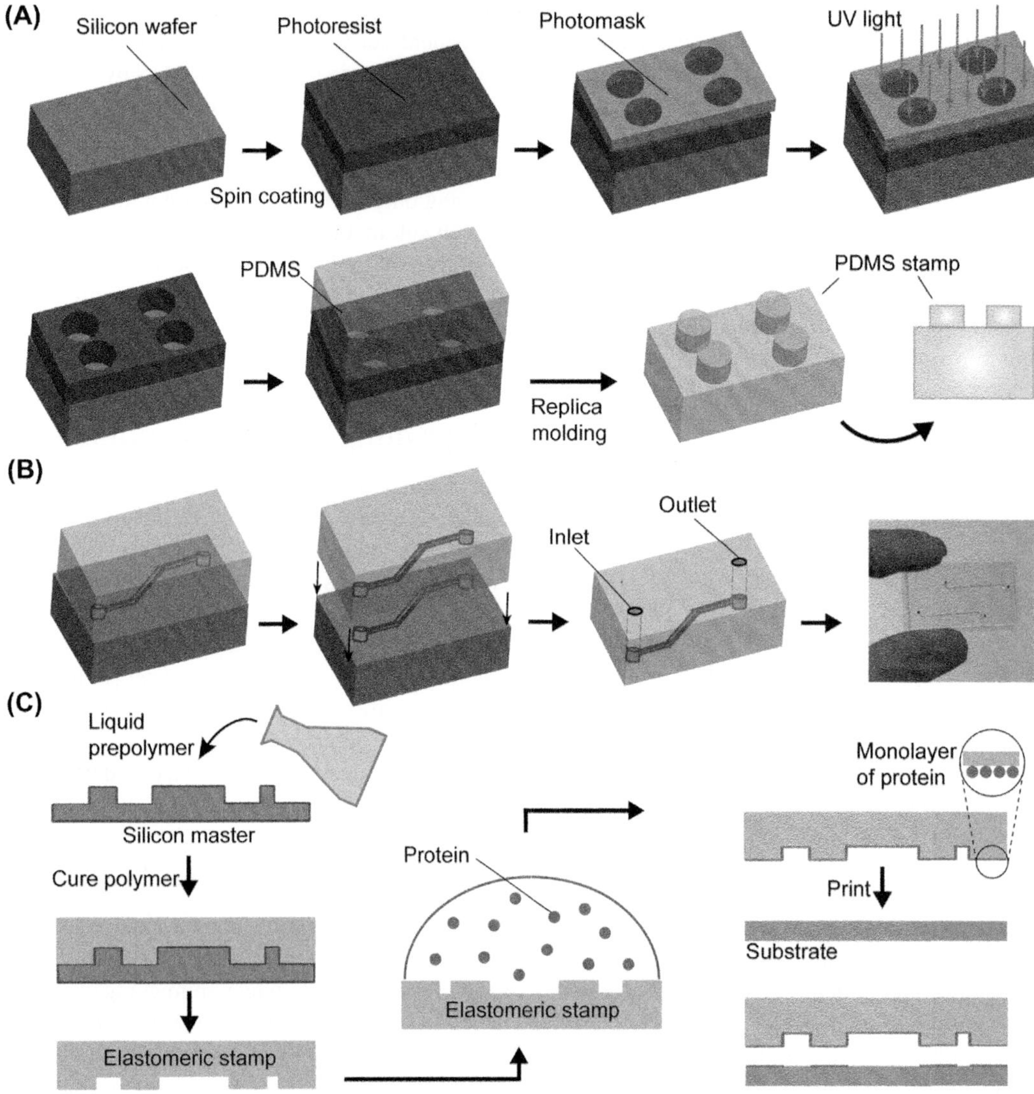

FIGURE 9.3 Microfabrication techniques and applications. (A) Use of replica molding to create PDMS stamps. The silicon wafer is first coated with a photoresist layer. Then, using a photomask, the photoresist is exposed to the UV light to form the desired pattern. Finally, the uncross-linked photoresist is washed, and the silicon mold is employed to make the PDMS stamp. (B) Use of soft lithography to generate a single PDMS channel. After the PDMS slab is cast out of the silicon mold, the inlet(s) and outlet(s) is punched, and the PDMS is bonded to the glass slide already covered with a thin layer of PDMS. (C) Microcontact-printed proteins used to pattern cells. The PDMS stamp is fabricated using the method shown in section A. Then, the stamp is covered with a solution of desired proteins which forms a monolayer of proteins on the PDMS. Finally, the desired pattern is printed on the substrate.

bioprinted tissues in tandem with cell grafts, to overcome the oxygen diffusion limitations that otherwise exist [63]. The production of viable tissue constructs, while maintaining uniformly dispersed cells, has been possible by 3D printing.

Synthesis of functionalized blood vessels

3D printing or additive manufacturing techniques offer the production of biocompatible vascular grafts. Current solutions to replace or repair damaged vessels, in patients suffering from coronary artery disease, are either through angioplasty or using bypass grafts [64]. The effectiveness and longevity of grafts are sometimes poor, when availability of adequate autologous vessel sections for angioplasty is lacking, or due to the low in vivo survival times of vascular grafts. 3D printed functional blood vessels overcome immunorejection, and have improved survival in vivo, overcoming thrombogenesis of current vascular grafts [65]. Functional blood vessel models for assessing pharmacological therapeutics have also been devised. Niklason et al. demonstrated the effect of prostaglandin, endothelin-1, and serotonin on vascular wall contraction [66]. Diebolt et al. have also assessed the effects of polyphenols on vascular contraction/dilation, through calcium-independent mechanisms, using a tissue engineered vein model [67]. The interactions of microparticles shed by lymphocytes/leukocytes with vascular components have been extensively studied using 3D printed vessel models [68].

Biological 3D printing methods

The four most commonly used principles that are suitable for producing biologically relevant models, are stereolithography (SLA), pressure-based extrusion (PBE), selective laser sintering (SLS), and fused deposition molding (FDM) [69]. Though the mechanism of action for each of these techniques is different, the fabrication of the complete structure is still accomplished, through depositing biomaterial in a layer-by-layer manner. Stereolithography techniques use photochemical methods, to cure light-sensitive materials. A laser/UV light signal is focused in a bath containing photocross-linkable resins [70].

Similarly, SLS techniques use a beam of focused lasers, to heat and fuse layers of thin powder, to form solid objects [71]. FDM utilizes a nozzle-based system, through which material is melted and deposited on a flat substrate [72]. PBE-based techniques are similar to FDM but instead push viscous biopolymers, using syringe pumps or pressure gradients [73]. Although SLS and SLA are fast and have a high throughput, they require sophisticated optical setups, and are only compatible with bioinks that can either withstand high temperatures or are light sensitive. On the other hand, FDM and PBE are fairly simple, and produce geometries with high resolution and varying densities [74].

Although limitation of thermally stable materials is still valid for FDM, PBE is the most widely used technique, as it offers the advantage of printing viscous hydrogels and biomaterials, and is appropriate for working with biological matter like cells. However, this makes PBE-based deposition techniques very time consuming. The choice of the 3D printing method depends on the size and tissue to be fabricated. Additionally, these techniques are limited by the type of bioink they can use.

3D bioinks

Natural polymers harnessed from sources like native extracellular matrix (ECM) components (collagen, fibronectin) or gelatin and alginate, which are polysaccharide-based polymers, have better cell adhesion and proliferative abilities, due to presence of natural binding domains [75]. However, these materials have poor printability, and often lack mechanical strength. Synthetic polymers designed to maintain structural integrity, mechanical and chemical stability, are often used to overcome these limitations, although these materials do not contain cell-specific binding domains, and thus have poor biocompatibility [76]. Hence, researchers have started modifying synthetic polymer backbones, with natural polymers or peptides, to ensure better functionality and bioactivity.

These copolymer formulations allow fine tuning of the structural and chemical properties, producing artificial bioinks that mimic native ECM conductivity, stiffness, and bioactivity [77]. Researchers have also incorporated nanocomposites in the bioink formulations, to improve print fidelity and conductivity [78].

Human cells in organ-on-a-chip and 3D printing

At present, the most easily available source of human cells is commercial cell lines. Difficulty in accessing certain human tissues to extract cells and difficulty in long term culture have limited the use of primary cells in organ-on-a-chip [79]. The drawback of cell lines is that they are mostly derived from cancer, and do not represent the exact physiology, genetic data, and phenotype of healthy organs [80]. Also, both primary cells and cell lines needed to make different organs-on-chip have to be extracted from different tissues.

The emergence of stem cells can overcome some of the limitations [81]. Stem cells extracted from healthy individuals or patients can potentially differentiate into different tissue cell types (cardiomyocytes, endothelial cells, epithelial cells), with same genetic material and thus, can capture the variability in diseases and drug response of human population [82–84]. The three main sources are embryonic stem cells (ESCs), induced pluripotent stem cells (iPSCs), and adult stem cells (ASCs) [85].

Stem cells derived from blastocysts or inner mass of human embryos are known as ESCs. ESCs are pluripotent and show consistency in expression of phenotype markers, across different labs. However their source, human embryos, is highly controversial and considered unethical [86]. Meanwhile, ASCs are easy to extract from the bone marrow or adipose tissues [87]. They only differentiate into mesodermal cells of bones, muscle, cartilage, and fat lineages [88]. ASCs are also not preferred in organ-on-a-chip due to their lack in consistency with the expression of phenotype markers.

Fibroblast cells derived from skin tissue or blood cells of adults, are transfected with certain transcription factors (e.g., Yamanaka factors), to convert them into stem cells called iPSCs [85,89]. IPSCs are easily derivable, differentiate into many cell types, noninvasive during coculture, and excellent for long-term culture. Though iPSCs have been successfully used in some labs for organ-on-a-chip, robust protocols need to be developed to obtain iPSCs that can differentiate homogenously into cells with mature phenotypes [90–92]. Because of these reasons, primary cells and cell lines are still preferred over stem cells in organ-on-a-chip. Different sources of cells and tissues are summarized in Table 9.1.

TABLE 9.1 Tissue sources for organ-on-chip, their advantages and disadvantages.

Cell Source	Advantages	Disadvantages
Primary cell & tissue biopsies	• Can maintain natural extracellular matrix • Patient specific • Derived from adult • Mature phenotype	• Inaccessibility of some tissue samples • Difficult to maintain long term cell culture • Different sources for each tissue type
Cell lines	• Readily available • Established protocols • Derived from adult • Mature phenotype • Noninvasive	• Not representative of normal physiology • Derived from cancerous cells • Different sources for each tissue type
Embryonic stem cells (ESCs)	• Long-term culture • Pluripotent differentiation • Consistent among protocols	• Derived from embryos • Use highly regulated • Immature phenotype • Limited differentiation protocols
Adult stem cells (ASCs)	• Patient specific • Derived from adult • Long-term culture • Multipotent differentiation • Easy to extract from biopsies	• Inconsistent derivation protocol • Heterogeneous population • Immature phenotype • Limited differentiation protocols
Induced pluripotent stem cells (iPSC)	• Patient specific • Derived from adult • Long-term culture • Pluripotent differentiation • Easy to extract from skin biopsies and blood draws • Noninvasive	• Immature phenotype • Limited differentiation protocols
Circulating progenitor cells	• Patient specific • Derived from adult • Mature phenotype • Very limited potency	• Large volumes of blood required • Difficult to maintain long-term cell culture • Not available for all tissues

Improvements in primary cells

There is also a need to diversify the primary cells used in organ-on-a-chip. Human umbilical vein endothelial cells (HUVECs) are widely used to model both arteries and veins on a chip, as these cells have shown good results and variation under arterial and venous microenvironment [93,94]. HUVECs are also readily available commercially. Primary cells derived from arterial tissues such as human coronary artery endothelial cells (HCAECs), and human pulmonary artery endothelial cells (HPAECs), along with those from venous tissues, represented by human saphenous vein endothelial cells (HSaVECs), are also commercialized, displaying phenotype difference with HUVECs [95–97].

Diversified phenotypes for human-on-a-chip

Use of these primary cells will create more physiologically relevant models of organ-on-a-chip, compared to using HUVECs. Tissue-specific primary cells can also be replaced by late progenitor cells, nominally blood outgrowth endothelial cells (BOECs), also known as endothelial colony forming cells (ECFCs) [98]. Long culture of these cells under arterial or venous microenvironment will result in cells that are phenotypically similar to primary cells derived from arteries or veins, respectively. Such different vasculatures-on-chip will play an important role in connecting different organs-on-chip, to form a human-on-a-chip (also known as body-on-a-chip).

Human-on-a-chip—interconnected multiple organs-on-chips

Though the term "human-on-a-chip" should contain all the organs that are present in a human body, at present researchers call chips that contain two to three organs-on-a-chip that are interconnected with each other mostly by vasculature-on-a-chip as human-on-a-chip [99] (Fig. 9.1A). Human-on-a-chip is a sought-after technology, as it helps in the study of pharmacokinetics (PC) and pharmacodynamics (PD) of drugs, in an efficient way during the preclinical trials. Most of the present clinical trials of drugs are conducted on young adult male volunteers [100]. Human-on-a-chip models made from either stem cells or primary cells from a diverse population will be able to include adult females, young children, and older people, thus testing the efficacy of drugs in a larger diversity of population.

"Human" preclinical assays

Human-on-chip models should ideally contain gut-on-a-chip, lung-on-a-chip or vasculature-on-a-chip, to incorporate the different drug administration pathways in humans like oral, inhaling, or intravenous. Further, these human-on-chip models should also contain liver-on-a-chip, as liver metabolizes and clears most drugs, partly by secretion of bile [101]. The third component of the system should be the target in which the response to the drug is of interest. Tumors can be the organ of interest, as PC and PD of a newly developed chemotherapy drug can be investigated (Fig. 9.1B).

Step-by-step global assessment

In a system with vasculature, liver, and tumor, the prescribed dose of drug will be administered to the vasculature-on-a-chip to mimic intravenous administration, including physiologic hemodilution. Metabolization in liver-on-a-chip could be the next step, depending on product pharmacology, before it reaches the tumor-on-a-chip and acts on tumor cells. The system will help us understand the effects of the drug on tumor cells, liver cells, and vascular cells, not overlooking secretions like cytokines, released into the system by these cells when in contact with the drug. A combination of gut, liver, vasculature, and kidney can be used to study overall drug absorption, distribution across different organs, metabolism, and excretion from Ref. [101] (Fig. 9.4).

Technical losses of the model

While studying drug absorption, distribution, and excretion, it is important to include the various losses that are inherently included in a human-on-a-chip. At present most of the organs-on-chip are fabricated using silicones (PDMS) that are porous and elastomeric in nature [101]. Though the porous nature is beneficial for oxygenation of the cells cultured in them, it can lead to loss of drugs by retention and permeation, known as package loss of the effluents. Empirical relations that relate the surface area of channels, volume of PDMS used to make the organ-on-a-chip, and package loss have to be developed, for in vivo drug dosage extrapolation.

Fine-tuning of the drug investigation model

These relations will be important in organs such as liver that have very low flow rate (~15 μL/h) [102]. When the flow rate becomes small, the mass transfer of drugs and other constituents in the system will be diffusion dominant, which will increase package losses. Use of newer optically

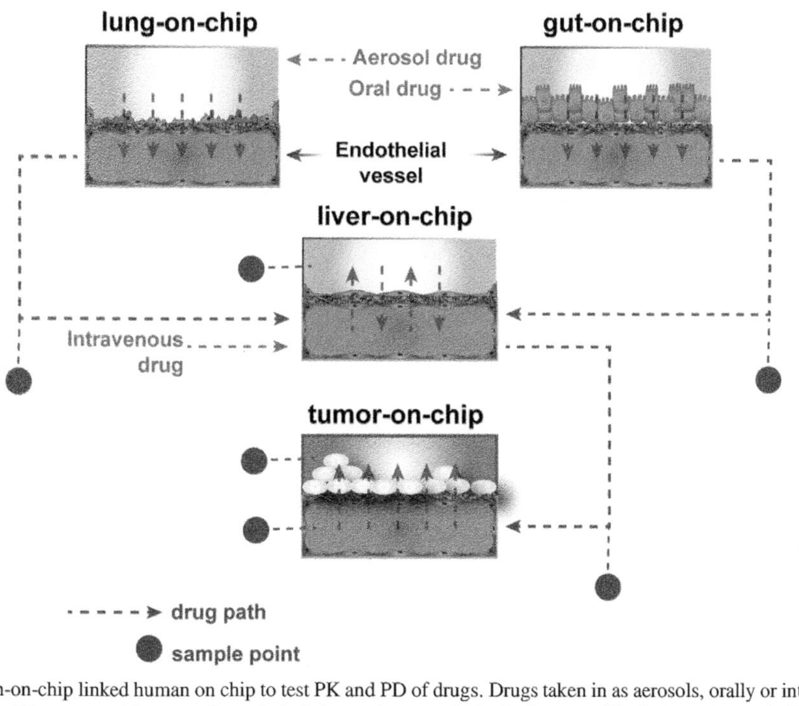

FIGURE 9.4 A multi organ-on-chip linked human on chip to test PK and PD of drugs. Drugs taken in as aerosols, orally or intravenously, can be input to the epithelial side of lung-on-chip, gut-on-chip, or to the endothelial vessel, respectively. In liver-on-chip the hepatic metabolism of the drug takes place. Finally, the drug enters the tumor-on-a-chip through the attached endothelial vessel, and its effects are observed and studied here.

clear and flexible hydrogels that are denser and allow less package losses needs to be explored.

Fine control of flow and shear rate have to be maintained when incorporating different organs-on-a-chip, to form a human-on-a-chip, as this will have an adverse effect on the phenotype profile of the cells. The dimensions of each organ-on-a-chip have to be designed, such that the shear rate in each of them is maintained at physiological rates, when the volume flow rate of media and drugs are equal in all of them. When the artery-on-a-chip and vein-on-a-chip are connected, the dimensions of the vein-on-a-chip have to be bigger, so as to maintain lower shear rate experienced by the venous endothelium, compared to arterial endothelium [103].

Multifunctional devices

Most organs-on-a-chip developed at present mimic one particular organ level function only. In line with human-on-a-chip requirements, where different organs-on-a-chip are connected, multiple organ-on-a-chip devices should be considered, to mimic different functions of the same organ. Multiple blood-brain barrier (BBB) constructs on-a-chip were used to assess the effectiveness of psychoactive drug methamphetamine [104].

Such models not only allow the assessment of drugs but also help discover unknown organ level interactions. In the above BBB-on-a-chip procedures, previously unknown metabolic coupling between BBB and neurons was discovered. Such complex yet dissectible models of brain, liver, and heart will be very useful, as these are highly affected by drugs in humans [99]. When studying PC and PD of a drug on heart-on-a-chip, instead of only depending on the image-based analysis, one can also use impedance-based biosensors embedded on the organ-on-a-chip. Such biosensors can give real-time continuous readout of the health of the system and the effectiveness of the drug.

Biosensors in organ-on-a-chip

Biosensors are analytical devices that combine a biologically sensitive recognition element, attached to a physicochemical transducer, to identify the presence of one or more specific biological elements [105]. Biologically sensitive elements sense the concentration or kinetics of specific elements, and emit physical signals like light, conformational change, or strain. These signals are then converted into optical or electrical signals by the transducer. A detector will in turn convert these signals into readouts. Biosensor-based detection techniques are more reliable, accurate, cost-effective, and easy to use, compared to the present microscopy-based detection techniques that are the "gold standard" in microfluidics [106].

Transducers use electrochemical, electrical, mechanical, and optical methods, to convert the signals from recognition elements. Transepithelial/transendothelial electrical resistance (TEER) is a widely used resistance/impedance-based method that detects the health of an organ-on-a-chip [107]. For example, TEER can detect in real time if a lumen is fully formed or healthy in a vasculature-on-a-chip. In one of the widely used TEER models, two electrodes (either metal wires or patterned thin-film electrodes) are embedded on the top and bottom sides of the microfluidic channel encasing the lumen. For the resistance-based method, voltage drop across the electrodes, for a low frequency square wave form alternating current (AC) is measured (as direct current can harm the cells).

For the impedance-based method, the electrodes are swept by different frequencies of a small amplitude AC, and the resulting amplitude and phase response of the current is measured. One of the main challenges with TEER is the difficulty in embedding the electrodes in an organ-on-a-chip under sterile conditions. Though embedding of electrodes is difficult, platinum electrodes can be used to sense pH and oxygen levels, in an organ-on-a-chip containing cell culture [108]. Oxygen levels play an important role in maintaining cell health and functioning, and hypoxia can make tumor cells drug resistant [109].

Oxygen sensing fluorophores can also be used in a tumor-on-a-chip platform, to sense the variations in oxygen. When the fluorophores are excited (e.g., 510 nm light), they emit energy at wavelengths detectable by photodiodes. The intensity of emitted light linearly reduces with an increase in oxygen attachment, and under hypoxic conditions the fluorophores will emit highest intensity of light possible. Hypoxia can also lead to morphologic change in cells, which can be detected using embedded strain gauges [110].

Large-scale physiological monitoring

A series of biosensors can also be used in tandem to get multiple readouts from organs-on-a-chip that contain multiple cell types. One can use a combination of TEER and microelectrode arrays (MEAs) in a heart-on-a-chip that contains endothelial lumen and cardiomyocytes, in a dual channel system separated by a porous membrane [111]. While TEER measures the barrier function and vascular permeability of the endothelial lumen, MEAs measure the electrical activity and cardiac function of the cardiomyocytes. MEAs can also be fabricated using ink jet printing, which is an additive manufacturing method similar to 3D bioprinting [112].

3D bioprinting

Photolithography and soft lithography used to fabricate organs-on-a-chip are not scalable, as they are made as 2D projections and require skilled individuals. This can be overcome by using 3D bioprinting, an exponentially growing field [48]. With 3D printing one can include the connecting tissues that are not present in organ-on-chips models at present, thereby addressing the complete biological output and physiology of the in vivo system. Vasculature-on-a-chip models at present only contain the extracellular matrix and the endothelial lumen, not fibroblasts, pericytes, or smooth muscle cells (SMCs). Also, cellular arrangements and phenotypes in most organs-on-a-chip may differ from in vivo, as those are built on rectangular channels.

The additive nature of 3D bioprinting will help in fabrication of cylindrical macroscale organs. Also, 3D printing with help of supplementary fabrication techniques like electrospinning can contribute to arrangement of cells and tissues as observed in vivo. For example, SMCs always align in a direction perpendicular to flow in vivo. This is not yet achieved in vitro, as SMCs align in the direction of flow. Electrospinning can provide radial grooves in the ECM, and thereby guide cells to align in a direction perpendicular to flow. Though limitation in bioinks that can truly mimic the mechanical and chemical properties of the ECM is a big factor, foundations for future research and development of such materials have already been laid. Also, 3D printers that are able to handle multiple materials, construct hierarchical structures, and multiple length scales need to be explored.

Conclusion

As the number of organs that have been successfully simulated increase, the recent focus is to use these technologies in translational research, perhaps for advanced point-of-care procedures. Organ-on-a-chip researchers have already begun incorporating patient-specific biomarkers, including blood and its flow rate, iPSCs derived from patients, and patient-specific atherosclerotic plaque geometry. Drug development and patient treatment expenses will eventually diminish. However, before implementing organ-on-a-chip and 3D bioprinting in precision medicine, it is crucial to verify that their readouts are able match in vivo data.

Access of developers to patient-specific information is required, which is sensitive, and ethical concerns could arise. However, long-term studies that correlate organ-chip and 3D printing data to patient outcomes may help in validating these technologies. Pharmaceutical companies, research labs, and clinics need to adopt innovations, and be willing to test these new models.

References

[1] Olive KP, et al. Inhibition of Hedgehog signaling enhances delivery of chemotherapy in a mouse model of pancreatic cancer. Science 2009;324(5933):1457–61.

[2] Beuzard Y. Mouse models of sickle cell disease. Transfus. Clin. Biol. 2008;15(1–2):7–11.

[3] Hughes D, et al. Adventitial transplantation of blood outgrowth endothelial cells in porcine haemodialysis grafts alleviates hypoxia and decreases neointimal proliferation through a matrix metalloproteinase-9-mediated pathway—a pilot study. Nephrol. Dial. Transplant. 2009;24(1):85–96.

[4] Savitz SI. A critical appraisal of the NXY-059 neuroprotection studies for acute stroke: a need for more rigorous testing of neuroprotective agents in animal models of stroke. Exp. Neurol. 2007;205(1):20–5.

[5] Perlman RL. Mouse models of human disease: an evolutionary perspective. Evol. Med. Public Health 2016;2016(1):170–6.

[6] Whitelaw CB, et al. Engineering large animal models of human disease. J. Pathol. 2016;238(2):247–56.

[7] Liao J, Huang W, Liu G. Animal models of coronary heart disease. J. Biomed. Res. 2015;30.

[8] Hasenfuss G. Animal models of human cardiovascular disease, heart failure and hypertrophy. Cardiovasc. Res. 1998;39(1):60–76.

[9] Hansen BC, Bodkin NL. Primary prevention of diabetes mellitus by prevention of obesity in monkeys. Diabetes 1993;42(12):1809–14.

[10] Mullin R. Cost to develop new pharmaceutical drug now exceeds $2.5B. 2014. Available from: https://cen.acs.org/articles/92/web/2014/11/Tufts-Study-Finds-Big-Rise.html.

[11] Carden MA, et al. Extracellular fluid tonicity impacts sickle red blood cell deformability and adhesion. Blood 2017;130(24):2654–63.

[12] Fillinger MF, et al. Coculture of endothelial cells and smooth muscle cells in bilayer and conditioned media models. J. Surg. Res. 1997;67(2):169–78.

[13] Ryan AJ, et al. Towards 3D in vitro models for the study of cardiovascular tissues and disease. Drug Discov. Today 2016;21(9):1437–45.

[14] Higgins JM, et al. Sickle cell vasoocclusion and rescue in a microfluidic device. Proc. Natl. Acad. Sci. U.S.A. 2007;104(51):20496–500.

[15] Young EW, Simmons CA. Macro- and microscale fluid flow systems for endothelial cell biology. Lab Chip 2010;10(2):143–60.

[16] Wikswo JP. The relevance and potential roles of microphysiological systems in biology and medicine. Exp. Biol. Med. 2014;239(9):1061–72.

[17] Huh D, et al. Reconstituting organ-level lung functions on a chip. Science 2010;328(5986):1662–8.

[18] Huh D, et al. A human disease model of drug toxicity–induced pulmonary edema in a lung-on-a-chip microdevice. Sci. Transl. Med. 2012;4(159):159ra147.

[19] Benam KH, et al. Small airway-on-a-chip enables analysis of human lung inflammation and drug responses in vitro. Nat. Methods 2016;13(2):151.

[20] Stucki AO, et al. A lung-on-a-chip array with an integrated bio-inspired respiration mechanism. Lab Chip 2015;15(5):1302–10.

[21] Humayun M, Chow C-W, Young EW. Microfluidic lung airway-on-a-chip with arrayable suspended gels for studying epithelial and smooth muscle cell interactions. Lab Chip 2018;18(9):1298–309.

[22] Jain A, et al. Primary human lung alveolus-on-a-chip model of intravascular thrombosis for assessment of therapeutics. Clin. Pharmacol. Ther. 2018;103(2):332–40.

[23] Lee S-A, et al. Spheroid-based three-dimensional liver-on-a-chip to investigate hepatocyte–hepatic stellate cell interactions and flow effects. Lab Chip 2013;13(18):3529–37.
[24] Bavli D, et al. Real-time monitoring of metabolic function in liver-on-chip microdevices tracks the dynamics of mitochondrial dysfunction. Proc. Natl. Acad. Sci. U.S.A. 2016;113(16):E2231–40.
[25] Musah S, et al. Mature induced-pluripotent-stem-cell-derived human podocytes reconstitute kidney glomerular-capillary-wall function on a chip. Nat. Biomed. Eng. 2017;1(5):0069.
[26] Jang K-J, et al. Human kidney proximal tubule-on-a-chip for drug transport and nephrotoxicity assessment. Integr. Biol. 2013;5(9):1119–29.
[27] Wilmer MJ, et al. Kidney-on-a-chip technology for drug-induced nephrotoxicity screening. Trends Biotechnol. 2016;34(2):156–70.
[28] Kim HJ, et al. Human gut-on-a-chip inhabited by microbial flora that experiences intestinal peristalsis-like motions and flow. Lab Chip 2012;12(12):2165–74.
[29] Kim HJ, Ingber DE. Gut-on-a-Chip microenvironment induces human intestinal cells to undergo villus differentiation. Integr. Biol. 2013;5(9):1130–40.
[30] Singh VK, Romaine PL, Newman VL. Biologics as countermeasures for acute radiation syndrome: where are we now? Taylor & Francis; 2015.
[31] Jalili-Firoozinezhad S, et al. Modeling radiation injury-induced cell death and countermeasure drug responses in a human Gut-on-a-Chip. Cell Death Disease 2018;9(2):223.
[32] Choe A, et al. Microfluidic Gut-liver chip for reproducing the first pass metabolism. Biomed. Microdevices 2017;19(1):4.
[33] Wufuer M, et al. Skin-on-a-chip model simulating inflammation, edema and drug-based treatment. Sci. Rep. 2016;6:37471.
[34] Lim HY, et al. Development of wrinkled skin-on-a-chip (WSOC) by cyclic uniaxial stretching. J. Ind. Eng. Chem. 2018;68:238–45.
[35] Sung JH, et al. Microfabricated mammalian organ systems and their integration into models of whole animals and humans. Lab Chip 2013;13(7):1201–12.
[36] Luni C, Serena E, Elvassore N. Human-on-chip for therapy development and fundamental science. Curr. Opin. Biotechnol. 2014;25:45–50.
[37] Shen HH. News feature: better models for brain disease. Proc. Natl. Acad. Sci. U.S.A. 2016;113(20):5461–4.
[38] Abbott NJ. Astrocyte–endothelial interactions and blood–brain barrier permeability. J. Anat. 2002;200(6):629–38.
[39] Park J, et al. Three-dimensional brain-on-a-chip with an interstitial level of flow and its application as an in vitro model of Alzheimer's disease. Lab Chip 2015;15(1):141–50.
[40] Kilic O, et al. Brain-on-a-chip model enables analysis of human neuronal differentiation and chemotaxis. Lab Chip 2016;16(21):4152–62.
[41] Dauth S, et al. Neurons derived from different brain regions are inherently different in vitro: a novel multiregional brain-on-a-chip. Journal of neurophysiology 2016;117(3):1320–41.
[42] Jastrzebska E, Tomecka E, Jesion I. Heart-on-a-chip based on stem cell biology. Biosens. Bioelectron. 2016;75:67–81.
[43] Marsano A, et al. Beating heart on a chip: a novel microfluidic platform to generate functional 3D cardiac microtissues. Lab Chip 2016;16(3):599–610.
[44] Ahn S, et al. Mussel-inspired 3D fiber scaffolds for heart-on-a-chip toxicity studies of engineered nanomaterials. Anal. Bioanal. Chem. 2018;410(24):6141–54.
[45] Cochrane A, et al. Advanced in vitro models of vascular biology: human induced pluripotent stem cells and organ-on-chip technology. Adv. Drug Deliv. Rev. 2018;13317.
[46] Jain A, et al. A shear gradient-activated microfluidic device for automated monitoring of whole blood haemostasis and platelet function. Nat. Commun. 2016;7:10176.
[47] Sakurai Y, et al. A microengineered vascularized bleeding model that integrates the principal components of hemostasis. Nat. Commun. 2018;9(1):509.
[48] Gold K, Gaharwar AK, Jain A. Emerging trends in multiscale modeling of vascular pathophysiology: organ-on-a-chip and 3D printing. Biomaterials; 2018.
[49] Huh D, et al. Microengineered physiological biomimicry: organs-on-chips. Lab Chip 2012;12(12):2156–64.
[50] Shuler ML. Advances in organ-, body-, and disease-on-a-chip systems. Lab Chip 2019;19(1):9–10.
[51] Huh D, Hamilton GA, Ingber DE. From 3D cell culture to organs-on-chips. Trends Cell Biol. 2011;21(12):745–54.
[52] Kimura H, Sakai Y, Fujii T. Organ/body-on-a-chip based on microfluidic technology for drug discovery. Drug Metab. Pharmacokinet. 2018;33(1):43–8.
[53] Bhatia SN, Ingber DE. Microfluidic organs-on-chips. Nat. Biotechnol. 2014;32(8):760.
[54] Bernard A, et al. Microcontact printing of proteins. Adv. Mater. 2000;12(14):1067–70.
[55] Jiang B, et al. Organs on microfluidic chips: a mini review. Sci. China Chem. 2014;57(3):356–64.
[56] Duffy DC, Duffy DC, McDonald JC, Schueller OJA, Whitesides GM. Anal. Chem. 1998;70:4974.
[57] Wang Z, et al. Organ-on-a-chip platforms for drug delivery and cell characterization: a review. Sensor. Mater. 2015;27(6):487–506.
[58] Shamloo A, Selahi A, Madadelahi M. Designing and modeling a centrifugal microfluidic device to separate target blood cells. J. Micromech. Microeng. 2016;26(3):035017.
[59] Amin R, et al. 3D-printed microfluidic devices. Biofabrication 2016;8(2):022001.
[60] Jana S, et al. Scaffolds for tissue engineering of cardiac valves. Acta Biomater. 2014;10(7):2877–93.
[61] Masoumi N, et al. Tri-layered elastomeric scaffolds for engineering heart valve leaflets. Biomaterials 2014;35(27):7774–85.
[62] Hockaday LA, et al. Rapid 3D printing of anatomically accurate and mechanically heterogeneous aortic valve hydrogel scaffolds. Biofabrication 2012;4(3):035005.
[63] Gaetani R, et al. Cardiac tissue engineering using tissue printing technology and human cardiac progenitor cells. Biomaterials 2012;33(6):1782–90.
[64] Hillis LD, et al. ACCF/AHA guideline for coronary artery bypass graft surgery: a report of the American college of cardiology foundation/American heart association task force on practice guidelines. Circulation 2011;124(23):e652–735.
[65] Wu W, DeConinck A, Lewis JA. Omnidirectional printing of 3D microvascular networks. Adv. Mater. 2011;23(24):H178–83.
[66] Niklason LE, et al. Functional arteries grown in vitro. Science 1999;284(5413):489–93.

[67] Diebolt M, et al. Polyphenols modulate calcium-independent mechanisms in human arterial tissue-engineered vascular media. J. Vasc. Surg. 2007;46(4):764–72.
[68] Pricci M, et al. Applications of human tissue-engineered blood vessel models to study the effects of shed membrane microparticles from T-lymphocytes on vascular function. Tissue Eng. A 2009;15(1):137–45.
[69] Mosadegh B, et al. Current progress in 3D printing for cardiovascular tissue engineering. Biomed. Mater. 2015;10(3):034002.
[70] Melchels FP, Feijen J, Grijpma DW. A review on stereolithography and its applications in biomedical engineering. Biomaterials 2010;31(24):6121–30.
[71] Yeong WY, et al. Porous polycaprolactone scaffold for cardiac tissue engineering fabricated by selective laser sintering. Acta Biomater. 2010;6(6):2028–34.
[72] Gu BK, et al. 3-dimensional bioprinting for tissue engineering applications. Biomater. Res. 2016;20:12.
[73] Li XM, et al. 3D-Printed biopolymers for tissue engineering application. Int. J. Polym. Sci. 2014; Article ID 829145, 13 pages.
[74] Richards DJ, et al. 3D printing for tissue engineering. Isr. J. Chem. 2013;53(9–10):805–14.
[75] Itoh M, et al. Scaffold-free tubular tissues created by a bio-3D printer undergo remodeling and endothelialization when implanted in rat aortae. PLoS One 2015;10(9).
[76] Tsai KJ, et al. Biomimetic heterogenous elastic tissue development. NPJ Regen. Med. 2017;2.
[77] Chimene D, et al. Nanoengineered Ionic-Covalent Entanglement (NICE) bioinks for 3D bioprinting. ACS Appl. Mater. Interfaces 2018;10(12):9957–68.
[78] Paul A, et al. Injectable graphene oxide/hydrogel-based angiogenic gene delivery system for vasculogenesis and cardiac repair. ACS Nano 2014;8(8):8050–62.
[79] Wnorowski A, Yang H, Wu JC. Progress, obstacles, and limitations in the use of stem cells in organ-on-a-chip models. Adv Drug Deliv Rev; 2018.
[80] Kaur G, Dufour JM. Cell lines: valuable tools or useless artifacts. Spermatogenesis 2012;2(1):1–5.
[81] Nguyen A, Pavlovich M. Engineering a strong bond between stem cells and biotechnology. Cell Stem Cell 2018;22(3):279–80.
[82] Geraili A, et al. Controlling differentiation of stem cells for developing personalized organ-on-chip platforms. Adv. Healthc. Mater. 2018;7(2).
[83] Liang P, et al. Drug screening using a library of human induced pluripotent stem cell-derived cardiomyocytes reveals disease-specific patterns of cardiotoxicity. Circulation 2013;127(16):1677–91.
[84] Sharma A, et al. High-throughput screening of tyrosine kinase inhibitor cardiotoxicity with human induced pluripotent stem cells. Sci. Transl. Med. 2017;9(377).
[85] Takahashi K, et al. Induction of pluripotent stem cells from adult human fibroblasts by defined factors. Cell 2007;131(5):861–72.
[86] Pera MF, Reubinoff B, Trounson A. Human embryonic stem cells. J. Cell Sci. 2000;113(Pt 1):5–10.
[87] Hassan HT, El-Sheemy M. Adult bone-marrow stem cells and their potential in medicine. J. R. Soc. Med. 2004;97(10):465–71.
[88] Zuk PA, et al. Human adipose tissue is a source of multipotent stem cells. Mol. Biol. Cell 2002;13(12):4279–95.
[89] Rufaihah AJ, et al. Human induced pluripotent stem cell-derived endothelial cells exhibit functional heterogeneity. Am. J. Transl. Res. 2013;5(1):21–35.
[90] Baxter M, et al. Phenotypic and functional analyses show stem cell-derived hepatocyte-like cells better mimic fetal rather than adult hepatocytes. J. Hepatol. 2015;62(3):581–9.
[91] Terryn J, et al. Recent advances in lineage differentiation from stem cells: hurdles and opportunities? F1000Res. 2018;7:220.
[92] Siller R, et al. Modelling human disease with pluripotent stem cells. Curr. Gene Ther. 2013;13(2):99–110.
[93] Costa PF, et al. Mimicking arterial thrombosis in a 3D-printed microfluidic in vitro vascular model based on computed tomography angiography data. Lab on a Chip 2017;17(16):2785–92.
[94] Urbaczek AC, et al. Endothelial cell culture under perfusion on a polyester-toner microfluidic device. Sci. Rep. 2017;7(1):10466.
[95] Lakota K, et al. Increased responsiveness of human coronary artery endothelial cells in inflammation and coagulation. Mediat. Inflamm. 2009;2009:146872.
[96] Meyrick B, Berry Jr LC, Christman BW. Response of cultured human pulmonary artery endothelial cells to endotoxin. Am. J. Physiol. 1995;268(2 Pt 1):L239–44.
[97] Schmiedl UP, Maravilla KR, Nelson JA. Improved method for in vivo magnetic resonance contrast media research. Investig. Radiol. 1991;26(1):65–70.
[98] Ecklu-Mensah G, et al. Blood outgrowth endothelial cells (BOECs) as a novel tool for studying adhesion of *Plasmodium falciparum*-infected erythrocytes. PLoS One 2018;13(10):e0204177.
[99] Lee SH, Sung JH. Organ-on-a-Chip technology for reproducing multiorgan physiology. Adv. Healthc. Mater. 2018;7(2).
[100] Glue P, et al. Ascending-dose study of noribogaine in healthy volunteers: pharmacokinetics, pharmacodynamics, safety, and tolerability. J. Clin. Pharmacol. 2015;55(2):189–94.
[101] Prantil-Baun R, et al. Physiologically based pharmacokinetic and pharmacodynamic analysis enabled by microfluidically linked organs-on-chips. Annu. Rev. Pharmacol. Toxicol. 2018;58:37–64.
[102] Schepers A, et al. Engineering a perfusable 3D human liver platform from iPS cells. Lab Chip 2016;16(14):2644–53.
[103] Diamond SL. Systems analysis of thrombus formation. Circ. Res. 2016;118(9):1348–62.
[104] Maoz BM, et al. A linked organ-on-chip model of the human neurovascular unit reveals the metabolic coupling of endothelial and neuronal cells. Nat. Biotechnol. 2018;36(9):865–74.
[105] Luka G, et al. Microfluidics integrated biosensors: a leading technology towards lab-on-a-chip and sensing applications. Sensors (Basel) 2015;15(12):30011–31.
[106] Bhalla N, et al. Introduction to biosensors. Essays Biochem. 2016;60(1):1–8.
[107] Srinivasan B, et al. TEER measurement techniques for in vitro barrier model systems. J. Lab. Autom. 2015;20(2):107–26.
[108] Bonk SM, et al. Design and characterization of a sensorized microfluidic cell-culture system with electro-thermal micro-pumps and sensors for cell adhesion, oxygen, and pH on a glass chip. Biosensors (Basel) 2015;5(3):513–36.
[109] Michiels C. Physiological and pathological responses to hypoxia. Am. J. Pathol. 2004;164(6):1875–82.
[110] Pandya HJ, Park K, Desai JP. Design and fabrication of a flexible MEMS-based electromechanical sensor array for breast cancer diagnosis. J. Micromech. Microeng. 2015;25(7).
[111] Maoz BM, et al. Organs-on-Chips with combined multi-electrode array and transepithelial electrical resistance measurement capabilities. Lab Chip 2017;17(13):2294–302.
[112] Yang H, et al. 3-D printed adjustable microelectrode arrays for electrochemical sensing and biosensing. Sensor. Actuator. B Chem. 2016;230:600–6.

Chapter 10

Designing multicellular intestinal systems

Luke A. Schwerdtfeger[1] and Stuart A. Tobet[1,2]

[1]*Department of Biomedical Sciences, Colorado State University, Fort Collins, CO, United States;* [2]*School of Biomedical Engineering, Colorado State University, Fort Collins, CO, United States*

Introduction

Modern organotypic methods for maintaining human colon outside of the body, in vitro, emerged in the 1970s [1]. A number of the early in vitro intestinal studies were performed using colon explants [2], enteric nervous system explants [3], or gut musculature cultures [4]. Some of these early explant systems housed the full cellular diversity of the in vivo intestinal mucosa, however, were rarely maintained beyond 72 h [2]. While some studies cultured explants for 20 days or more [1], the details of tissue health were minimal. The use of large quantities of serum (e.g., fetal bovine serum) and antibiotics in culture media is a persistent problem in intestinal in vitro systems, due to the variability of serum across sources, and the impacts of antibiotic treatment on the intestinal microbiome. Early explant systems are likely to have included some complement of microbiota; however, bacterial load or diversity was not analyzed.

Cancer cells

Moving away from organotypic systems during the later part of the 20th century, immortalized cell lines (e.g., Caco-2 cells) became common in studies of intestinal physiology. Caco-2 cells were originally derived from colorectal adenocarcinoma from a 72-year-old male [5], and have been used to study a variety of topics, ranging from gut epithelial barrier permeability [6] to applications in drug discovery [7]. Another line, HT-29 cells, was derived from colorectal adenocarcinoma from a 44-year-old female (ATCC). These cells display different characteristics, such as mucin secretory competence in a small portion of cells [8]. It is worth noting that a host of other immortalized cell lines has been employed as models of mammalian intestines in vitro. However most of these cell lines, including Caco-2 and HT-29, suffer a problem of “passage number,” which is created when phenotypes change, based on the number of cell divisions and the development of subclones [9].

Stem cells

Development of organoid systems using induced pluripotent stem cells [10], or patient isolated progenitors [11], has resulted in a powerful tool for studying intestinal epithelial cell biology, in a 3D structured system not previously attainable with immortalized cell cultures. These systems, however, fail to capture the unique physiology of the intestinal wall ex vivo, because the cells either dedifferentiate at some point to less than normal function, or they lack sufficient cellular diversity even when they redifferentiate. The ability to answer questions regarding cell-cell interactions is limited by the absence of neural, immune, and microbial components, relative to the original tissue.

Gut slices

Organotypic slice models provide an alternative to cell cultures or organoids, for capturing the complex cellular heterogeneity and microbial presence of the in vivo gut wall, yet in vitro. Precision cut intestinal slices [12] (PCIS) represented an early use of an organotypic slice model, for study of the intestinal capacity for metabolizing drugs [13]. PCIS models were unfortunately only viable in the first ~24 h of culture. Longer maintenance of mouse intestinal tissues ex vivo was achieved with a similar protocol to one used previously in brain [14]. This procedure allowed the maintenance of 250 μm thick slices, in serum and antibiotic free media, for up to 6 days ex vivo [15], and provided for the study of antibiotic effects on intestinal segmental contraction rates.

While the rate of segmental contraction ex vivo was representative of in vivo behavior, maintenance of organotypic slices beyond 48 h required the use of calcium ion channel blockers (e.g., nicardipine) to prevent long-term muscular contractions that otherwise damaged epithelia [15]. Advances in organotypic slice technology provide for maintenance of epithelial populations, neuronal and glial cells, immune responses to pathogen, and a subset of the

Precision Medicine for Investigators, Practitioners and Providers. https://doi.org/10.1016/B978-0-12-819178-1.00010-1

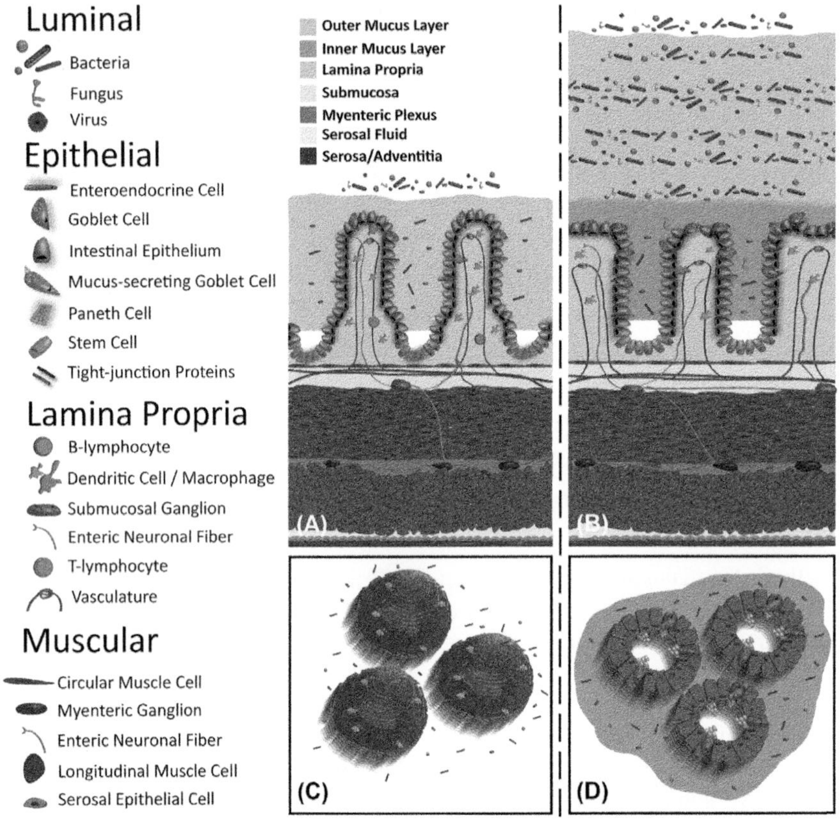

FIGURE 10.1 Schematic diagram illustrating the organization and complexity of the intestinal tripartite barrier system, with some of the extensive cellular heterogeneity. Luminal contents (e.g., microbiota) sit above and inside a mucus layer in the ileum (A), and in outer and inner mucus layers in the colon (B). Diverse epithelial elements comprise a barrier between gut luminal microbiota and internal *lamina propria* contents. Organization of this barrier varies across the 10 anatomical regions of the gut tract (stomach, duodenum, jejunum, ileum, caecum, ascending colon, transverse colon, descending colon, sigmoid colon, and rectum); however, a crypt-villus axis in the small intestine (e.g., ileum; A) and crypts in the colon (B) predominate. Internal to the *lamina propria*, two neuronal plexuses are visible. First, the submucosal plexus sits upon a circular smooth muscle layer. Second, the myenteric plexus is housed between circular and longitudinal smooth muscle layers. Finally, a layer of serosal fluid generated by serosal epithelium is visualized between the longitudinal musculature and a layer of adventitia composed of connective tissue. Schematics of the view looking inward from the lumen of both the ileum (C) and colon (D) demonstrate the varied organization of the apical epithelium. Cellular representations are keyed on the left, organized by barrier compartment (e.g., luminal, epithelial, *lamina propria*, muscular), and color codes for each region are provided.

native microbiome ex vivo [15]. Although organotypic methods maintain more cellular diversity than many in vitro systems, there are other issues that would be helpful to address. Principally, there is a steep oxygen gradient across the intestinal wall [16], one not reproduced in organotypic slice models, cell line cultures, or organoids. This gradient has a major impact on gut physiology, in the intestinal wall, and on luminal microbiome.

Gut physiological specifications

General considerations

Evolution has provided a number of barrier tissues within mammals that protect components of one compartment from the potential dangers of external environments. The function of such barriers is essential to maintaining healthy states, with an increasing number of gastrointestinal disorders attributed to barrier failures [17,18]. Barrier tissues automatically define multiple environments, "external" and "internal" to epithelial monolayers, which also have multiple constituent components (Fig. 10.1). This chapter focuses on mammalian intestines where the external space is the lumen, including microbial components, the epithelial lining includes multiple cell types and associated extended mucus layers, and the internal component is the tissue comprised of diverse cellular elements that include neural and immune cells. There are cellular interactions within each compartment, and significant molecular signaling between compartments. Mechanistic understanding of the interplay between the diverse cell types that form such a multicompartmental barrier structure is essential for developing treatments for barrier disorders.

Reconstituting the complex physiology of the intestinal wall ex vivo is a tall order. With no fewer than 15 cell types, a complex microbiome, a functional nervous system, and the body's largest immune system, the intestinal wall is one of the largest and most heterogeneous organs in any mammalian organism (Fig. 10.1). In vitro systems frequently take one of two divergent paths toward modeling in vivo physiology. One is to create reductionist systems that allow the dissection of unique roles of individual component parts. The other is to re-create the complexity of the organ in vitro, allowing control of the system, with responses to perturbation being potentially more accurate indicators of in vivo function.

There are pros and cons to both paths. For the former, component parts function differently alone than when together. For the latter, ex vivo methods only offer an approximation of in vivo physiology. It is a premise of this chapter that the latter has been overlooked in much of intestinal research, and yet provides an important component with translational potential. Distilling the complex physiology of the intestinal wall into three "compartments," namely luminal, epithelial, and tissue (Fig. 10.1), with individualized specifications for each, aids in guiding the design of more physiologically relevant culture systems.

Luminal

The contents of the intestinal lumen vary depending upon diet, gut segment, and local composition of microbiota. Home to thousands of species of bacteria, fungi, and archaea [19], the gut lumen is a complex environment. Understanding the interspecies interactions, quorum signaling, and secretions from these microorganisms, is an ongoing investigation for microbiology. Secretory products from luminal microbiota influence tissue physiology; and a growing number of functional pathways have been recently discovered, including direct metabolite–cell surface receptor signaling [20].

Many bacterial metabolite–host signaling pathways are still unknown. Environmental factors such as antibiotic treatment [21] and oxygen culture condition [22] influence the production of microbially derived metabolites. Designing culture methodologies that incorporate the native microbiome of the organism of interest is an important target to advance understanding of bacterial–host signaling mechanisms. Spiking monolayer cell cultures with specific bacteria to mimic a "microbiome" has been done in microfluidic devices [23,24]. However, given that only a small percentage of the native gut flora can be cultured under the conditions of most microfluidic devices, it is difficult to consider these models reflective of microbiomes in vivo.

Culture methods that maintain explant tissues directly from donors have potential to maintain larger subsets of the native microbiota, due to their presence in the lumen of the organ providing explants. Key design specifications for the luminal compartment can be thought of in two ways that perhaps match the way the microbial community is naturally organized. From one perspective, a defined microbiome could be "selected," as an engineered blend of anaerobic, facultative, and aerobic microbes. Alternatively, creating conditions that result in a specific metabolomic profile could influence microbial composition, as there are many ways to get to the same metabolic profile. In both cases, the in vitro luminal environment will be an end result of the initial conditions: source of microbial components and substrates for microbial energy utilization (e.g., glucose, oxygen, etc.), and bacterial survival skills (e.g., microbial interactions).

Epithelial

Epithelial cells of the mammalian intestine constitute a critical barrier between the luminal contents and gut tissue, and are an aspect of the intestine that has drawn significant attention for in vitro modeling. Separating the lumen from gut epithelia is a secreted layer of mucous that serves dual roles, as a pathogen barrier and as a source of nutrients for luminal microbiota. Goblet cells in the intestinal epithelial layer secrete large quantities of mucins, principally Muc2 in the intestine, with small quantities of other mucins (e.g., Muc5ac). These mucins provide protein cores for chains of O-linked glycans that are comprised of varied structures (e.g., galactose/galactosamine base, with variable inclusion of fucose or sialic acid) that also vary by gut region and across species [25].

Beneath this mucus layer, apical epithelia form a barrier that is interlinked with tight junction–associated proteins, such as ZO-1 (zona occludins) and claudins [26]. Molecular regulation of tight junctions is complex and has been reviewed elsewhere [27]. Interspersed among the epithelia are mucous secreting goblet cells, defensin secreting Paneth cells, and hormone producing enteroendocrine cells, among others, all derived from crypt $Lgr5^+$ stem cell populations. These stem cells are supported by mesenchymal cells, and surrounded by soluble factors and extracellular matrix [28]. Mesenchymal regulation of this stem cell niche may be important in vivo, or in response to tissue injury [29]; however, it is not required to generate crypt-villus axes in vitro [30].

Communication between luminal contents (i.e., bacterial components) and epithelial cells has been demonstrated, often involving the tolllike receptor family members [31]. Apical surfaces of at least one type of barrier cell (i.e., tuft cells) have been shown to possess receptors, and bind specific microbial secreted metabolites [32]. Nonetheless, specific -metabolite-tissue receptor signaling mechanisms are still poorly understood. Epithelial cells also communicate directly with cells in the tissue compartment, via dendritic cell–epithelial interactions [33], enteric neurons via gut neuroepithelial circuits [34], and enteric glia that lie beneath the epithelial layer [35].

Key design specifications for the barrier epithelial compartment should not only account for the cellular diversity of the epithelial monolayer, with the extended mucus layer, but also for cells that regularly traffic in, out, or through the layer. Intraepithelial lymphocytes (IELs) are a subset of T-lymphocytes that possess the unique ability to both present antigen and act as natural killer T-cells [36]. IELs have been shown to move in and out of the epithelial layer in response to pathogens; a characteristic that was recently described as epithelial "flossing" [37].

While flossing between epithelial cells is not unique to IELs, as lamina propria dendritic cells act similarly [38], the speed of their movements ($\sim$6–7 μm/min [39]) is astounding. Additionally, phenotypically distinct IELs inhabit the small intestine compared to the colon, and can regulate colonic barrier integrity [40]. Finally, design considerations for the epithelial compartment should also be able to account for relevant signaling molecules that are secreted both cellularly and via mucous inhabiting microbiota.

Tissue

Interspersed throughout the mucosa, the *lamina propria* is comprised of diverse immune cell populations, neuronal and glial cells and projections, significant extracellular matrix, and vascular and lymph networks (Fig. 10.1). Due to the complex anatomical arrangement of *lamina propria*, which penetrates into the cores of small intestinal villi, and surrounds colonic crypts, engineering cellular components to mimic this unique epithelial–*lamina propria* arrangement is difficult, and may require different designs depending upon gut region of interest (e.g., small vs. large intestine).

For example, gut associated lymphoid tissue is distributed differently in the ileum (Peyer patches) versus the colon where they appear in smaller isolated lymphoid follicles. Interactions between barrier cells and tissue/*lamina propria* via neural–epithelial signaling [34] are apparent. Tissue components also possess the ability to directly interact with the luminal contents. This is accomplished via multiple routes; first, as noted in the previous paragraph for both dendritic cells and IELs. Second, tuft cells can sense luminal metabolites such as succinate [32] activating type 2 mucosal immune responses [41]. Additionally, invasive bacteria have been shown to directly signal to sensory nerve endings in the *lamina propria* [42].

Many intestinal models have focused on building systems, from the simplest state of mostly single cell type monolayers, and even those derived from $Lgr5^+$ epithelial progenitors do not include the diverse *lamina propria* and submucosal cell populations seen in vivo (e.g., neural, immune, vascular). Organotypic slice methods maintain the neural, immune, and epithelial components of the tissue [15]. However, depending on the orientation of the slice preparation (e.g., perpendicular to the wall or other), there may be more or less potential to mimic the full "barrier function" of the intestinal wall. Key design specifications for the tissue compartment should include as many of the diverse *lamina propria* and submucosal cell populations in vitro as possible.

Cell-cell and microorganism-cell signaling

Signaling between intestinal neural components and microbial, epithelial, and immune cells are important areas for ongoing investigations. In the apical epithelium, direct synapses between a subset of enteroendocrine cells and neurons—both vagus and enteric in origin—have been shown anatomically [35] and functionally, with glutamate release from enteroendocrine cells stimulating nodose ganglion vagal afferents [43]. Innervation of other epithelial cell types has been observed, with tuft cells directly contacting enteric sensory neurons [44]. Moving inward from the apical epithelia, numerous components of the *lamina propria* and musculature regularly signal with enteric or vagal neurons.

Mast cells in the *lamina propria* have been shown to be in contact with both enteric and vagal neurons [45], and signal bidirectionally with these cells [46]. In the intestinal muscle layers, *muscularis* macrophages are also in contact with vagal afferents [47], and are activated in response to enteric sympathetic neuron signaling during bacterial infection [48]. *Lamina propria* dendritic cell and IEL interactions with vagal or enteric sensory neurons are yet unexplored areas of investigation that are necessary to be filled in.

Compartmental environment problems

In the case of oxygen concentration, the intestinal wall presents an extreme situation in that it maintains a steep oxygen gradient in vivo [49]. The gradient spans 0.1%–1% in the luminal compartment, to 5%–10% oxygen in the heavily vascularized submucosa and musculature layers [50]. Without an ability to mimic the anaerobic-aerobic interface in vitro, the nonphysiologic oxygen concentrations of many preparations will likely lead to both cellular and microbial signaling among the components of the gut wall that will fall short of modeling in vivo physiology.

Culturing organotypic slices in a static, fixed oxygen incubator creates a nonphysiologic oxygen environment that can influence tissue metabolism (Schwerdtfeger and Tobet, personal observation), and significantly impact the gut microbiome [51]. Similarly, nutrient provisions in the tissue compartment come from the arterial supply (e.g., glucose), whereas in the luminal compartment food consumption and metabolism play larger roles. Engineering

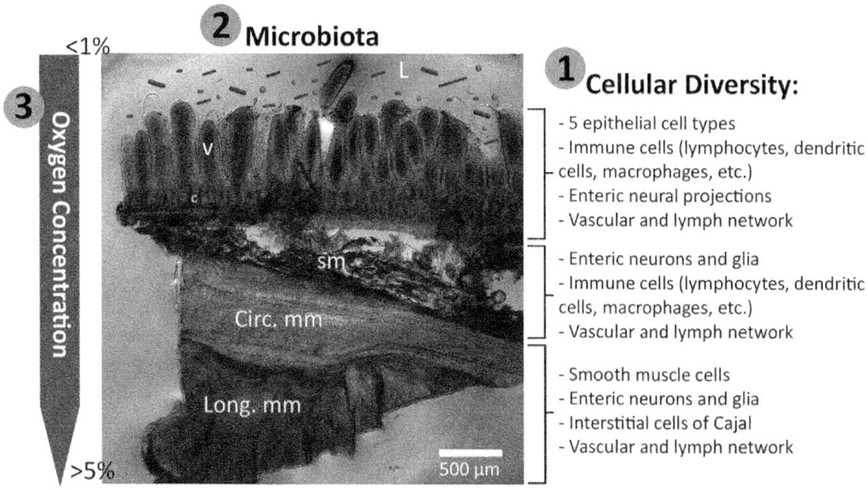

FIGURE 10.2 Graphical outline of three key components of intestinal tissues, superimposed on an image of a section of human gut wall (jejunum) that should be appreciated in designing next generation organotypic culture systems, to improve physiologic relevance. "L" signifies lumen, "V" a villus, "c" a crypt, "sm" submucosa, "Circ. mm" the circular muscle layer, "Long. mm" the longitudinal muscle layer. Scale bar is 500 μm.

a model that re-creates differential environments will require differential flow across tissues, in a multichannel microfluidic system.

Designing new organotypic systems

Harnessing the power of advancing engineering technologies like microfluidics allows for development of culture methods that incorporate media flow, making more accurate modeling of in vivo physiology possible. Current microfluidic systems provide a technical foundation for culturing intestinal cell lines and tissues, with continuously perfused media/nutrients, and clearance of waste products. Engineering advances of microfluidic culture systems have recently been reviewed [52–54]. These technologies have been used for culture of liver [55] and intestine [56] among others. Their use in tissues like the intestine, however, need to better account for the compartmental requirements unique to barrier tissues.

Recently, microfluidic technology was used for culture of mouse intestinal explants, and was able to show a direct bacterial-immune-neuronal signaling event ex vivo [57], providing evidence of the importance of maintaining cellular heterogeneity of the intestinal wall in vitro. While this model advanced the microfluidic gut-on-chip paradigm, it required the use of 95% oxygen, which would have likely altered the explanted microbiome. Furthermore, this method was only performed using juvenile (12–14 days old) mice, and had a limited survival time of 24 h. Given that the prepubertal gut wall is significantly different than after puberty, in terms of microbiome and immune system [58], there are many unanswered questions concerning the utility of this early system.

The thickness of the intestinal wall creates an inherent problem of nutrient and oxygen penetration. Human intestinal tissue is far thicker than cell monolayers and even organoids (Fig. 10.2), increasing to ~3 mm in the small intestine (Fig. 10.3), and significantly thicker in the colon. Such tissue thicknesses require novel approaches to

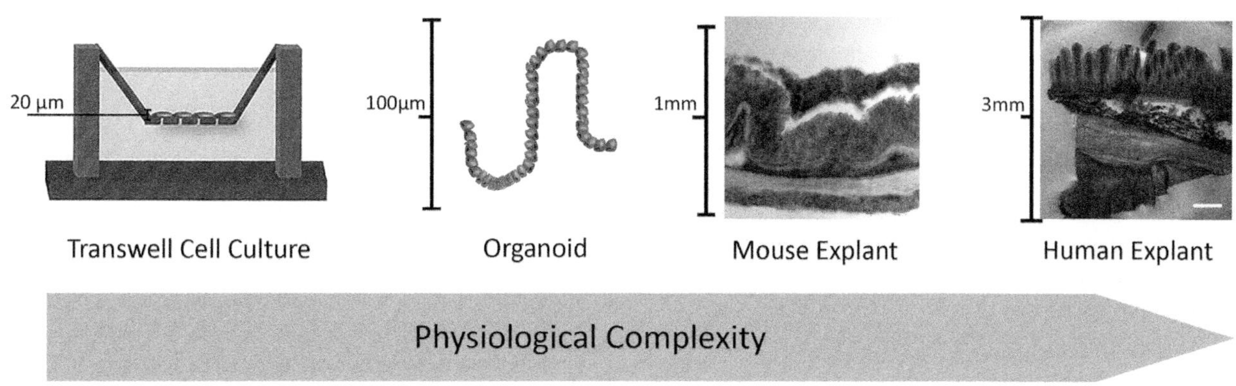

FIGURE 10.3 Schematic illustrating the progression in physiological complexity across different intestinal culture systems.

maintain tissue in a healthy state, due to oxygen and nutrient penetration into the gut wall.

Oxygen consumption by colonic epithelia, driven by oxidation of fatty acids such as butyrate, drives environmental oxygen down in the lumen, creating an ideal environment for anaerobic microbes [59]. Given the influence of oxygen conditions on intestinal microbiome [51,60], bacterial metabolism [61], and secretions [22], re-creating the anaerobic–aerobic interface in novel culture methods is of paramount importance, if the system is to faithfully mimic in vivo colonocyte metabolism. Many standard culture conditions (e.g., antibiotic use) alter the gut microbiome, which in turn alters microbially derived metabolites that are key regulators of intestinal function. Additionally, the impact of the mode, rate, and volume of delivery of resources like oxygen and glucose are profound. Media flow rate has been shown to influence tissue metabolism [62].

Impact of antibiotics

Antibiotic use is a standard tissue culture practice, to prevent "infection" of host cultures. Given that dense bacterial communities inhabit every barrier tissue of the mammalian body, understanding barrier function(s) requires understanding microbial-host relationships. For one, microbially derived metabolites are key regulators of intestinal function, and treatment with antibiotics has been shown to alter the metabolome of rats [21]. Second, gut bacteria are a major source of antimicrobial compounds [63], and host cells such as Paneth cells produce large quantities of antimicrobial compounds [64]. These considerations make it difficult to understand host–bacterial signaling when the investigator is adding extra antibiotics. Given the importance of the role of bacterial components in cell signaling and barrier function, exogenous antibiotic use in vitro should be either precise in targeting particular microbial elements or eliminated.

Conclusions

The design specifications discussed throughout this chapter offer a framework to guide development of novel intestinal—and barrier in general—culture methodologies that faithfully recapitulate in vivo physiology. Optimizing design strategies, particularly concerning tissue health validation and microfluidic flow paradigms, is critical to moving forward. It will be beneficial for future model systems to verify maintenance of components of tissue barriers and validate their ability to mimic selected aspects of in vivo physiology.

Acknowledgments

The authors wish to thank Drs. Zaid Abdo and David Dandy for their thoughtful comments on the manuscript.

Funding

Luke Schwerdtfeger was supported by an NSF NRT - GAUSSI fellowship; Results presented in this paper are based upon collaborative work supported by a National Science Foundation NRT Grant No. 1450032. Any opinions, findings, conclusions or recommendations expressed in this paper are those of the author(s) and do not necessarily reflect the views of the National Science Foundation.

References

[1] Autrup H, Barrett LA, Jackson FE, Jesudason ML, Stoner G, Phelps P, Trump BF, Harris CC. Explant culture of human colon. Gastroenterology 1978;74:1248–57.

[2] Senior PV, Pritchett CJ, Sunter JP, Appleton DR, Watson AJ. Crypt regeneration in adult human colonic mucosa during prolonged organ culture. J. Anat. 1982;134(3):459–69.

[3] Jessen KR, Saffrey MJ, Burnstock G. The enteric nervous system in tissue culture. I. Cell types and their interactions in explants of the myenteric and submucous plexuses from Guinea pig, rabbit and rat. Brain Res. 1983;262:17–35.

[4] Brookes SJH, Costa M. Identification of enteric motor neurons which innervate the circular muscle of the Guinea pig small intestine. Neurosci. Lett. 1990;118:227–30.

[5] Fogh J, Fogh J, Orfeo T. One hundred and twenty-seven cultured human tumor cell lines producing tumors in nude mice. J. Natl. Cancer Inst. 1977;59:221–6.

[6] Hidalgo IJ, Raub TJ, Borchardt RT. Characterization of the human colon carcinoma cell line (Caco-2) as a model system for intestinal epithelial permeability. Gastroenterology 1989;96:736–49.

[7] Peng Y, Yadava P, Heikkinen AT, Parrott N, Railkar A. Applications of the 7-day Caco-2 model in drug discovery and development. Eur. J. Pharm. Sci. 2014;56:120–30.

[8] Gagnon M, Zihler Berner A, Chervet N, Chassard C, Lacroix C. Comparison of the Caco-2, HT-29 and the mucus-secreting HT29-MTX intestinal cell models to investigate Salmonella adhesion and invasion. J. Microbiol. Methods 2013;94:274–9.

[9] Sambuy Y, De Angelis I, Ranaldi G, Scarino M, Stammati A, Zucco F. The Caco-2 cell line as a model of the intestinal barrier: influence of cell and culture-related factors on Caco-2 cell functional characteristics. Cell Biol. Toxicol. 2005;21:1–26.

[10] Rodansky ES, Johnson LA, Huang S, Spence JR, Higgins PDR. Intestinal organoids: a model of intestinal fibrosis for evaluating anti-fibrotic drugs. Exp. Mol. Pathol. 2015;98:346–51.

[11] Kasendra M, Tovaglieri A, Sontheimer-Phelps A, Jalili-Firoozinezhad S, Bein A, Chalkiadaki A, Scholl W, Zhang C, Rickner H, Richmond CA, Li H, Breault DT, Ingber DE. Development of a primary human small intestine-on-a-chip using biopsy-derived organoids. Sci. Rep. 2018;8:2871.

[12] De Kanter R, Tuin A, Van De Kerkhof E, Martignoni M, Draaisma AL, De Jager MH, De Graaf IAM, Meijer DKF, Groothuis GMM. A new technique for preparing precision-cut slices from small intestine and colon for drug biotransformation studies. J. Pharmacol. Toxicol. Methods 2005;51:65–72.

[13] De Graaf IAM, Olinga P, de Jager MH, Merema MT, de Kanter R, van de Kerkhof E, Groothuis GMM. Preparation and incubation of precision-cut liver and intestinal slices for application in drug metabolism and toxicity studies. Nat. Protoc. 2010;5(9):1540–51.

[14] Tobet SA, Walker HJ, Seney ML, Yu KW. Viewing cell movements in the developing neuroendocrine brain. Integr. Comp. Biol. 2003;43:794–801.

[15] Schwerdtfeger LA, Ryan EP, Tobet SA. An organotypic slice model for ex vivo study of neural, immune, and microbial interactions of mouse intestine. Am. J. Physiol. Gastrointest. Liver Physiol. 2016;310:G240–8.

[16] Zheng L, Kelly CJ, Colgan SP. Physiologic hypoxia and oxygen homeostasis in the healthy intestine. A review in the theme: cellular responses to hypoxia. Am. J. Physiol. Cell Physiol. 2015;309:C350–360.

[17] Arik YB, van der Helm MW, Odijk M, Segerink LI, Passier R, van den Berg A, van der Meer AD. Barriers-on-chips: measurement of barrier function of tissues in organs-on-chips. Biomicrofluidics 2018;12:042218.

[18] Bischoff SC, Barbara G, Buurman W, Ockhuizen T, Schulzke J, Serino M, Tilg H, Watson A, Wells J. Intestinal permeability – a new target for disease prevention and therapy. BMC Gastroenterol. 2014;13:25.

[19] Sekirov I, Russell SL, Antunes LC, Finlay BB. Gut microbiota in health and disease. Physiol. Rev. 2010;90:859–904.

[20] Schneider C, O'Leary CE, von Moltke J, Liang H, Ang QY, Turnbaugh PJ, Radhakrishnan S, Pellizzon M, Ma A, Locksley RM. A metabolite-triggered tuft cell-ILC2 circuit drives small intestinal remodeling. Cell 2018;174:271–84.

[21] Sun J, Schnackenberg LK, Khare S, Yang X, Greenhaw J, Salminen W, Mendrick DL, Beger RD. Evaluating effects of penicillin treatment on the metabolome of rats. J. Chromatogr. B 2013;932:134–43.

[22] Schertzer JW, Brown SA, Whiteley M. Oxygen levels rapidly modulate *Pseudomonas aeruginosa* social behaviours via substrate limitation of PqsH. Mol. Microbiol. 2010;77:1527–38.

[23] Kim HJ, Li H, Collins JJ, Ingber DE, Beebe DJ, Ismagilov RF. Contributions of microbiome and mechanical deformation to intestinal bacterial overgrowth and inflammation in a human gut-on-a-chip. Proc. Natl. Acad. Sci. U.S.A. 2016;113(1):E7–15.

[24] Shah P, Fritz JV, Glaab E, Desai MS, Greenhalgh K, Frachet A, Niegowska M, Estes M, Jager C, Seguin-Devaux C, Zenhausern F, Wilmes P. A microfluidics-based in vitro model of the gastrointestinal human-microbe interface. Nat. Commun. 2016;7:11535.

[25] Holmen Larsson JM, Thomsson KA, Rodriguez-Pineiro AM, Karlsson H, Hansson GC. Studies of mucus in mouse stomach, small intestine, and colon. III. Gastrointestinal Muc5ac and Muc2 mucin O-glycan patterns reveal a regiospecific distribution. Am. J. Physiol. Gastrointest. Liver Physiol. 2013;305:G357–363.

[26] Groschwitz KR, Hogan SP. Intestinal barrier function: molecular regulation and disease pathogenesis. J. Allergy Clin. Immunol. 2009;124:3–20.

[27] Buckley A, Turner JR. Cell biology of tight junction barrier regulation and mucosal disease. Cold Spring Harb. Perpsect. Biol. 2018;10(1). pii: a029314.

[28] Pastula A, Marcinkiewicz J. Cellular interactions in the intestinal stem cell niche. Arch. Immunol. Ther. Exp. 2018. https://doi.org/10.1007/s00005-018-0524-8.

[29] Miyoshi H, Ajima R, Luo CT, Yamaguchi TP, Stappenbeck TS. Wnt5a potentiates TGF-β signaling to promote colonic crypt regeneration after tissue injury. Science 2012;338:108–13.

[30] Sato T, Vries RG, Snippert HJ, van de Wetering M, Barker N, Stange DE, van Es JH, Abo A, Kujala P, Peters PJ, Clevers H. Single Lgr5 stem cells build crypt-villus structures *in vitro* without a mesenchymal niche. Nat. Lett. 2009;459:262–5.

[31] Carvalho FA, Aitken JD, Vijay-Kumar M, Gewirtz AT. Toll-like receptor-gut microbiota interactions: perturb at your own risk! Annu. Rev. Physiol. 2012;74:177–98.

[32] Nadjsombati MS, McGinty JW, Lyons-Cohen MR, Jaffe JB, DiPeso L, Schneider C, Miller CN, Pollack JL, Nagana Gowda GA, Fontana MF, Erle DJ, Anderson MS, Locksley RM, Raftery D, von Moltke J. Detection of succinate by intestinal tuft cells triggers a type 2 innate immune circuit. Immunity 2018;49:33–41.

[33] Jaensson E, Uronen-Hansson H, Pabst O, Eksteen B, Tian J, Coombes JL, Berg PL, Davidsson T, Powrie F, Johansson-Lindbom B, Agace WW. Small intestinal CD103+ dendritic cells display unique functional properties that are conserved between mice and humans. J. Exp. Med. 2008;205:2139–49.

[34] Bohorquez DV, Samsa LA, Roholt A, Medicetty S, Chandra R, Liddle RA. An enteroendocrine cell-enteric glia connection revealed by 3D electron microscopy. PLoS One 2014;9:e89881.

[35] Bohorquez DV, Shahid RA, Erdmann A, Kreger AM, Wang Y, Calakos N, Wang F, Liddle RA. Neuroepithelial circuit formed by innervation of sensory enteroendocrine cells. J. Clin. Investig. 2015;125:782–6.

[36] Hu MD, Jia L, Edelblum KL. Policing the intestinal epitheial barrier: innate immune functions of intraepithelial lymphocytes. Curr. Pathobiol. Rep. 2018;6(1):35–46.

[37] Hoytema van Konijnenburg DP, Reis BS, Pedicord VA, Farache J, Victora GD, Mucida D. Intestinal epithelial and intraepithelial T cell crosstalk mediates a dynamic response to infection. Cell 2017;171(4):783–94.

[38] Niess J, Brand S, Gu X, Landsman L, Jung S, McCormick B, Vyas J, Ploegh H, Fox J, Littman D, Reinecker H. CX3CR1-mediated dendritic cell access to the intestinal lumen and bacterial clearance. Science 2005;307:254–8.

[39] Edelblum KL, Shen L, Weber CR, Marchiando AM, Clay BS, Wang Y, Prinz I, Malissen B, Sperling AI, Turner JR. Dynamic migration of γδ intraepithelial lymphocytes requires occluding. Proc. Natl. Acad. Sci. U.S.A. 2012;109(18):7097–102.

[40] Kuhn KA, Schulz HM, Regner EH, Severs EL, Hendrickson JD, Mehta G, Whitney AK, Ir D, Ohri N, Robertson CE, Frank DN, Campbell EL, Colgan SP. Bacteroidales recruit IL-6-producing intraepithelial lymphocytes in the colon to promote barrier integrity. Mucosal Immunol. 2017;11(2):357–68.

[41] Gerbe F, Sidot E, Smyth DJ, Ohmoto M, Matsumoto I, Dardalhon V, Cesses P, Garnier L, Pouzolles M, Brulin B, Bruschi M, Harcus Y, Zimmermann VS, Taylor N, Maizels RM, Jay P. Intestinal epithelial tuft cells initiate type 2 mucosal immunity to helminth parasites. Nat. Lett. 2016;529:226–30.

[42] Pinho-Ribeiro FA, Baddal B, Haarsma R, O'Seaghdha M, Yang NJ, Blake KJ, Portley M, Verri WA, Dale JB, Wessels MR, Chiu IM. Blocking neuronal signaling to immune cells treats streptococcal invasive infection. Cell 2018;173:1083–97.

[43] Kaelberer MM, Buchanan KL, Klein ME, Barth BB, Montoya MM, Shen X, Bohorquez DV. A gut-brain neural circuit for nutrient sensory transduction. Science 2018;361:1219.

[44] Middelhoff M, Westphalen B, Hayakawa Y, Yan KS, Gershon MD, Wang TC, Quante M. Dclk1-expressing tuft cells: critical modulators of the intestinal niche? Am. J. Physiol. Gastrointest. Liver Physiol. 2017;313:G285–99.

[45] Sharkey KA, Kroese ABA. Consequences of intestinal inflammation on the enteric nervous system: neuronal activation induced by inflammatory mediators. Anat. Rec. 2001;262:79–90.

[46] Sharkey KA, Mawe GM. Neuroimmune and epithelial interactions in intestinal inflammation. Curr. Opin. Pharmacol. 2002;2. 669–667.

[47] Cailotto C, Gomez-Pinilla PJ, Costes LM, van der Vliet J, Giovangiulio MD, Nemethova A, Matteoli G, Boeckxstaens GE. Neuro-anatomical evidence indicating indirect modulation of macrophages by vagal efferents in the intestine but not in the spleen. PLoS One 2014;9(1):e87785.

[48] Gabanyi I, Muller PA, Feighery L, Oliveira TY, Costa-Pinto FA, Mucida D. Neuro-immune interactions drive tissue programming in intestinal macrophages. Cell 2016;164:378–91.

[49] Kelly CJ, Colgan SP. Breathless in the gut: implications of luminal O_2 for microbial pathogenicity. Cell Host Microbe 2016;19:427–8.

[50] Thermann M, Jostarndt L, Eberhard F, Richter H, Sass W. Oxygen supply of the human small intestine in mechanical ileus. Langenbecks Arch. Chir. 1985;363:179–84.

[51] Albenberg L, Esipova TV, Judge CP, Bittinger K, Chen J, Laughlin A, Grunberg S, Baldassano RN, Lewis JD, Li H, Thom SR, Bushman FD, Vinogradov SA, Wu GD. Correlation between intraluminal oxygen gradient and radial partitioning of intestinal microbiota. Gastroenterology 2014;147:1055–63.

[52] Bein A, Shin W, Jalili-Firoozinezhad S, Park MH, Sontheimer-Phelps A, Tovaglieri A, Chalkiadaki A, Kim HJ, Ingber DE. Microfluidic organ-on-a-chip models of human intestine. Cell. Mol. Gastroenterol. Hepatol. 2018;5:659–68.

[53] McLean IC, Schwerdtfeger LA, Tobet SA, Henry CS. Powering ex vivo tissue models in microfluidic systems. Lab Chip 2018;18:1399–410.

[54] van Duinen V, Trietsch SJ, Joore J, Vulto P, Hankemeier T. Microfluidic 3D cell culture: from tools to tissue models. Curr. Opin. Biotechnol. 2015;35:118–26.

[55] Prodanov L, Jindal R, Bale SS, Hegde M, William J, Golberg I, Bhushan A, Yarmush ML, Usta OB. Long term maintenance of a microfluidic 3-D human liver sinusoid. Biotechnol. Bioeng. 2016;113:241–6.

[56] van Midwoud PM, Merema MT, Verpoorte E, Groothuis GM. A microfluidic approach for in vitro assessment of interorgan interactions in drug metabolism using intestinal and liver slices. Lab Chip 2010;10:2778–86.

[57] Yissachar N, Zhou Y, Ung L, Lai NY, Mohan JF, Ehrlicher A, Weitz DA, Kasper DL, Chiu IM, Mathis D, Benoist C. An intestinal organ culture system uncovers a role for the nervous system in microbe-immune an intestinal organ culture system uncovers a role for the nervous system in microbe-immune crosstalk. Cell 2017;168:1135–48.

[58] Markle JGM, Frank DN, Mortin-Toth S, Robertson CE, Feazel LM, Rolle-Kampczyk U, von Bergen M, McCoy KD, Macpherson AJ, Danska JS. Sex differences in the gut microbiome drive hormone-dependent regulation of autoimmunity. Science 2013;339:1084–8.

[59] Litvak Y, Byndloss MX, Baumler AJ. Colonocyte metabolism shapes the gut microbiota. Science 2018;362:1017.

[60] von Martels JZH, Sadaghian Sadabad M, Bourgonje AR, Blokzijl T, Dijkstra G, Faber KN, Harmsen HJM. The role of gut microbiota in health and disease: in vitro modeling of host-microbe interactions at the aerobe-anaerobe interphase of the human gut. Anaerobe 2017;44:3–12.

[61] Wessel AK, Arshad TA, Fitzpatrick M, Connell JL, Bonnecaze RT, Shear JB, Whiteley M. Oxygen limitation within a bacterial aggregate. mBio 2014;5:e00992.

[62] Lee J, Choi JR, Ha SK, Choi I, Lee SH, Kim D, Choi N, Sung JH. A microfluidic device for evaluating the dynamics of the metabolism-dependent antioxidant activity of nutrients. Lab Chip 2014;14:2948–57.

[63] Garcia-Gutierrez E, Mayer MJ, Cotter PD, Narbad A. Gut microbiota as a source of novel antimicrobials. Gut Microbes 2018;0(0):1–21.

[64] Farin HF, Karthaus WR, Kujala P, Rakhshandehroo M, Schwank G, Vries RGJ, Kalkhoven E, Nieuwenhuis EES, Clevers H. Paneth cell extrusion and release of antimicrobial products is directly controlled by immune cell-derived IFN-γ. J. Exp. Med. 2014;211(7): 1393–405.

Chapter 11

Translational interest of immune profiling

Peter Natesan Pushparaj[1,2]
[1]*Center of Excellence in Genomic Medicine Research, Faculty of Applied Medical Sciences, King Abdulaziz University, Jeddah, Mecca Province, Kingdom of Saudi Arabia;* [2]*Department of Medical Laboratory Technology, Faculty of Applied Medical Sciences, Jeddah, Mecca Province, Kingdom of Saudi Arabia*

Introduction

Immunoprofiling is a process of deciphering antigen receptor (AR) heterogeneity, as well as antibody, cellular, and cytokine responses in health and disease [1,2]. Immunoprofiling is used to identify novel biomarkers that can be used to monitor disease severity, remission, and predict a patient's response to therapy. The quantitative measurement of antigen receptors, using immunoprofiling strategies such as antibodies and T-Cell Receptors (TCRs), aids in the development of new therapies and vaccines. Many cutting-edge technologies are used in immunoprofiling, such as flow cytometry, mass cytometry by time-of-flight (CyTOF), barcoding antibody-based protein arrays, next generation sequencing (NGS) technologies, Multicolor or Multiplex immunohistochemistry (mIHC), NanoString's nCounter technology, and Luminex multiple-analyte profiling technology (xMAP) [2,3]. Besides the development of numerous methods for bioinformatics, statistics-driven interpretation, and visualization of high dimensional data derived from these technologies, augment the reliability of immunoprofiling interpretation [4,5].

Growing worldwide impact

The bibliometric database Scopus [6] showed 1989 publications in the area of immunoprofiling until the end of 2018 (Fig. 11.1). The 2018 Nobel Prize for Physiology or Medicine was awarded to immunologists James Patrick Allison and Tasuku Honjo, for their pioneering work in the field of cancer immunotherapy. The data mining of National Institutes of Health Research Portfolio Online Reporting Tools (NIH RePORTER), showed that the funding for precision medicine had been increased from US$100 million for the year 2018, to US$186 million for the year 2019. Besides, NIH has thus far invested more than US$180 million in research projects involving immunoprofiling strategies. The University of California San Francisco (UCSF) has recently announced a new US$10 million "Immunoprofiler" initiative, for the advancement of drug discovery, and precision immunotherapy to better understand the response to immunotherapies for cancer, by partnering with AbbVie, Amgen, and Bristol-Myers Squibb. Many Biotech companies are offering specific immunoprofiling services or products, as listed in Table 11.1.

Immunoprofiling techniques in translational research

Immunoprofiling strategies rely on the latest technologies available in the market, such as Next-Generation Sequencing (NGS), Flow Cytometry or Fluorescent-Activated Cell Sorting (FACS), Mass Cytometry or cytometry by time-of-flight, barcoding antibody-based protein arrays, Multicolor or Multiplex Immunohistochemistry (mIHC), NanoString's nCounter technology, and Luminex Multiple-Analyte Profiling Technology (xMAP) [2,3].

Next-generation sequencing (NGS) methods in immunoprofiling

NGS is a powerful tool for the analyses of antigen specificity, as well as clonality and diversity of T lymphocytes. [7] has developed a new T-cell receptor (TCR) repertoire analysis method using 454 DNA sequencing technology,

Precision Medicine for Investigators, Practitioners and Providers. https://doi.org/10.1016/B978-0-12-819178-1.00011-3

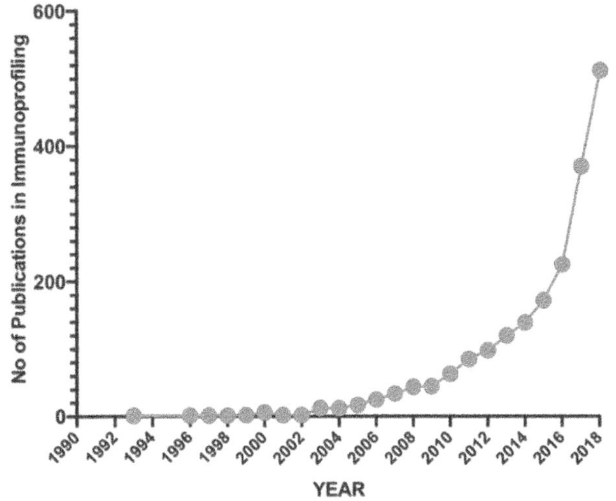

FIGURE 11.1 The Number of Publications in Immunoprofiling. The bibliometric analyses using the search terms "Immune Profiling Or Immunoprofiling" in Scopus showed that only one document was published that related to the Immunoprofiling in the year 1998. However, the number of publications related to Immunoprofiling has increased significantly in the past 20 years. In total, 1989 publications were published in the area of Immunoprofiling till the year 2018.

combined with an adaptor-ligation mediated polymerase chain reaction (PCR). Conventionally, mRNA or DNA are isolated from single cells, or in bulk, and prepared for sequencing by PCR using immune specific primers for complementarity-determining region 3 (CDR 3). This codes the antigen-binding domain of antigen receptor (AR), as well as subregions [variable (V), diversity (D), and joining (J)], and the subsequent sequencing provides clues about AR diversity, specificity, and clonality of the immune cells isolated from peripheral blood or tissue samples.

Importantly, high-throughput single-cell RNA sequencing (scRNA-Seq), has changed our perception of complex cell populations, but it does not give information about protein expression. To address this issue in scRNA-Seq, [8] used cellular indexing of transcriptomes and epitopes, by sequencing (CITE-seq) method that utilizes oligonucleotide-labeled antibodies to integrate cellular protein and transcriptome measurements in single cells. Latest cutting-edge NGS technologies such as the 10X Genomics platform can be used to characterize single cells from a variety of samples (Fig. 11.2) [9].

TABLE 11.1 List of Biotech companies that offer specific immunoprofiling services or products.

Company Name	Description
10X Genomics	Genomics, Haplotyping instrumentation platform
Adaptive Biotechnologies	T cell and B cell receptor analysis with clonoseq
Clontech (Takara Bio)	Wide array of molecular biology products
Cofactor Genomics	Genomic analysis
Core immunology Laboratory at SFGH	Immunology assays, expert consultation, and education to the UCSF and global research community
Definiens	Targeted drug development in immuno-oncology therapy including tissue Phenomics® technology
Enpicom	Bioinformatics Immunoprofiling
Fluidigm	Instrumentation for single-cell analyses
GENEWIZ	Sanger DNA sequencing, gene synthesis, molecular biology, high-throughput/next generation sequencing, bioinformatics, and GLP regulatory services
iRepertoire	Reagents and services for amplifying and sequencing the immune repertoire with next generation sequencing (NGS)
MIODx	ClonoMap platform TCR and BCR profiles single-cell sequencing
OncoCellMDx	Immunoprofiling technologies created at Harvard Medical School
Repertoire genesis	Next-Gen T-cell receptor (TCR)/B cell receptor (BCR) repertoire analysis
Texcell	Viral safety, immunology, and bioanalytical services

Adapted and modified from https://www.biotech-careers.org/company-coreactivity/immunoprofiling.

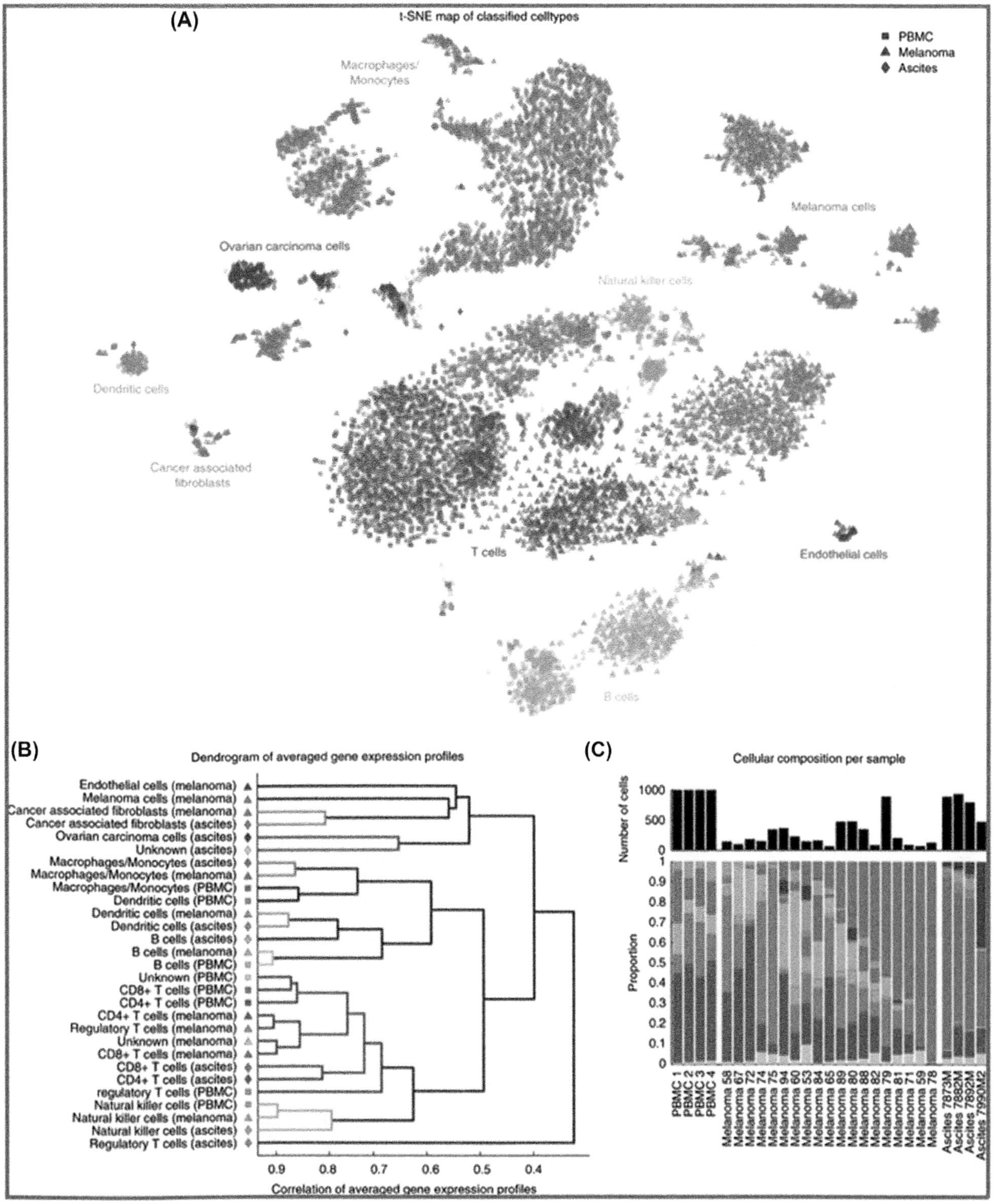

FIGURE 11.2 Comparison of gene expression profiles of single cells from different data sources. (A) Single cells were arranged in two dimensions based on similarity of their gene expression profiles by the dimensionality reduction technique t-SNE. The clusters that emerge spontaneously can be associated with cell types (colors) and data source (symbol types: squares for PBMC-data sets, triangles for the melanoma-data sets, and diamonds for ascites data-sets). (B) Pair-wise correlation of averaged gene expression profiles of clusters encoding cell type and origin as identified in a visualized as dendrogram. (C) Number of cells and cellular composition per sample. *Adapted from Schelker M, Feau S, Du J, Ranu N, Klipp E, MacBeath G, Schoeberl B, Raue A. Estimation of immune cell content in tumour tissue using single-cell RNA-seq data. Nat. Commun. December 11, 2017;8(1):2032. https://doi.org/10.1038/s41467-017-02289-3 and this work is licensed under a Creative Commons Attribution 4.0 Generic License.*

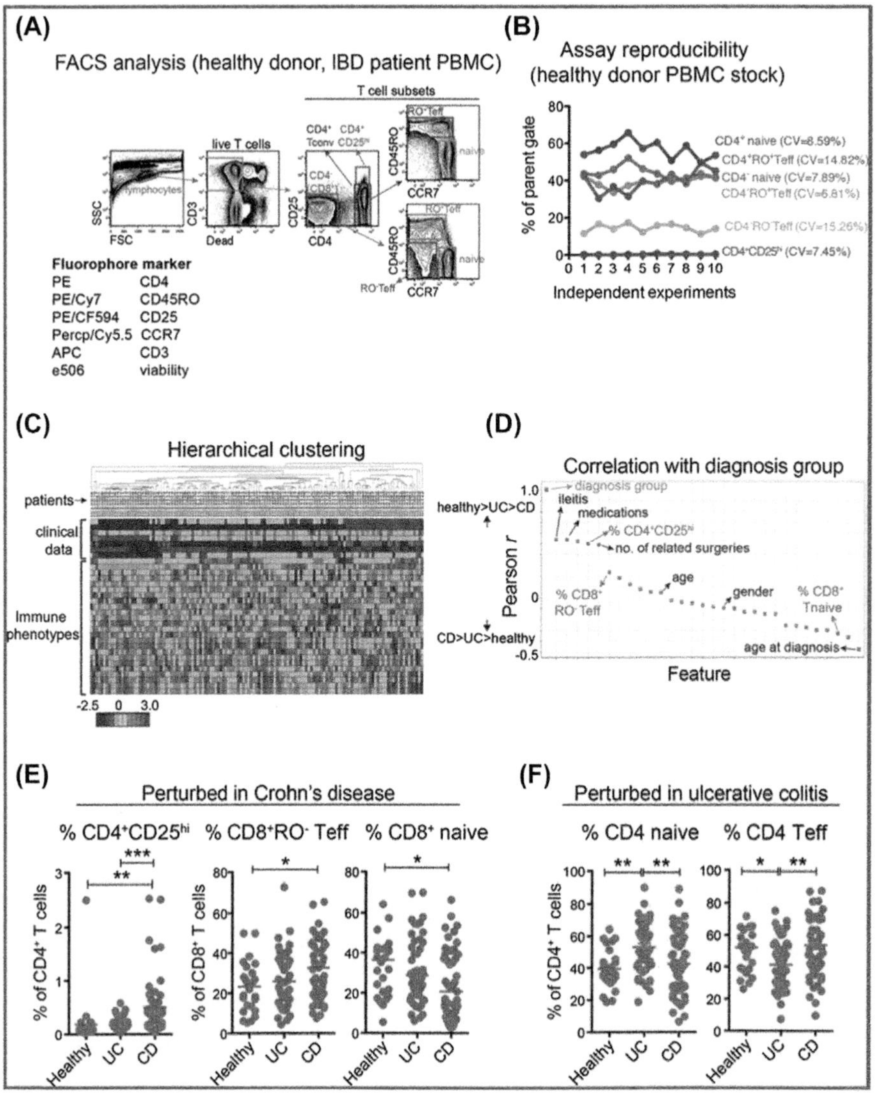

FIGURE 11.3 Informatics-based identification of immune dysregulation in clinical inflammatory bowel diseases. (A) Bottom left, 6-parameter FACS panel used for analyzing expression of surface antigens on peripheral blood mononuclear cells (PBMC) from healthy adult donors and adult IBD patients. Gating strategy for FACS analysis of human PBMC; immune subsets used in downstream analysis are indicated by gates and text. (B) Percentages of major T-cell subsets in a healthy control PBMC stock, determined by repeated FACS analysis as in (A), over 10 independent staining experiments. Each subset is quantified based on the percentages within relevant parent gates (as in (A)); coefficients of variation (CVs) are indicated for each subset by color-matched text. (C) Heat map showing hierarchical clustering of seven disease endpoints and 24 immune phenotypes in healthy adults (n = 26) and IBD patients ((ulcerative colitis (UC), n = 50; Crohn's disease (CD), n = 53). (D) Rank-ordered (Pearson r) correlation values of all disease endpoints and immune phenotypes relative to diagnosis group (i.e., healthy donors, group 1; UC patients, group 2; CD patients, group 3). Relevant disease endpoints and immune phenotypes are indicated by black and red (gray in print version) text, respectively; the correlation of the reference variable with itself (r = 1.0) is shown at top left in gray. (E) Immune cell subsets (CD4+CD25hi−left; CD8+RO- Teff−middle; CD8+ naïve−right) identified by hierarchical clustering and ranked Pearson coefficients (as in (C, D)) perturbed in CD patient PBMC. (F) Immune cell subsets (CD4+ naive−left; CD4+ Teff−right) identified by hierarchical clustering and ranked Pearson coefficients perturbed in UC PBMC. Red (gray in print version) lines indicate median values for each group. $*P < 0.05$, $**P < 0.01$, $***P < 0.001$, One-way ANOVA. Teff, effector/memory T cells. *Adapted from Delmas A, Oikonomopoulos A, Lacey PN, Fallahi M, Hommes DW, Sundrud MS. Informatics-based discovery of disease-associated immune profiles. PLoS One. September 26, 2016;11(9):e0163305. https://doi.org/10.1371/journal.pone.0163305 and this work is licensed under a Creative Commons Attribution 4.0 Generic License.*

Flow cytometry or fluorescence-activated cell sorting (FACS) in immunoprofiling

Flow cytometry is an important tool for immunoprofiling [10]. Latent tuberculosis infection (LTBI) is present in one-third of the world's population, and about 10% of the individuals with LTBI will get active tuberculosis (TB). It is difficult to diagnose individuals with LTBI, using currently available diagnostic methods. Hence, flow cytometry-based immunoprofiling methods have been established [11,12]. Robison et al. [13] measured cytokines secreted from antigen-challenged PBMCs, by silicon photonic sensor arrays, and used feature selection algorithms to deduce positively correlated multi-biomarker signatures for LTBI status and reactivation risk assessment.

Fonseka et al. [14] used mixed-effects modeling of associations of single cells (MASC), a reverse single-cell association strategy. In mass cytometry analyses of CD4+ T cells from the blood in rheumatoid arthritis (RA), a significantly increased population of CD4+ T cells, recognized as CD27- HLA-DR+ effector memory cells, occurred in RA patients [14]. Delmas et al. [15] have shown the application of FACS, combined with robust informatics approaches (Fig. 11.3), to delineate the immunoprofiling in inflammatory bowel diseases (IBD) [9]. estimated the cellular composition in three ovarian cancer ascites samples, using single-cell and bulk RNA sequencing, as well as FACS, and found that single-cell RNA-Seq was robust in the determination of immune cells compared to FACS method (Fig. 11.4).

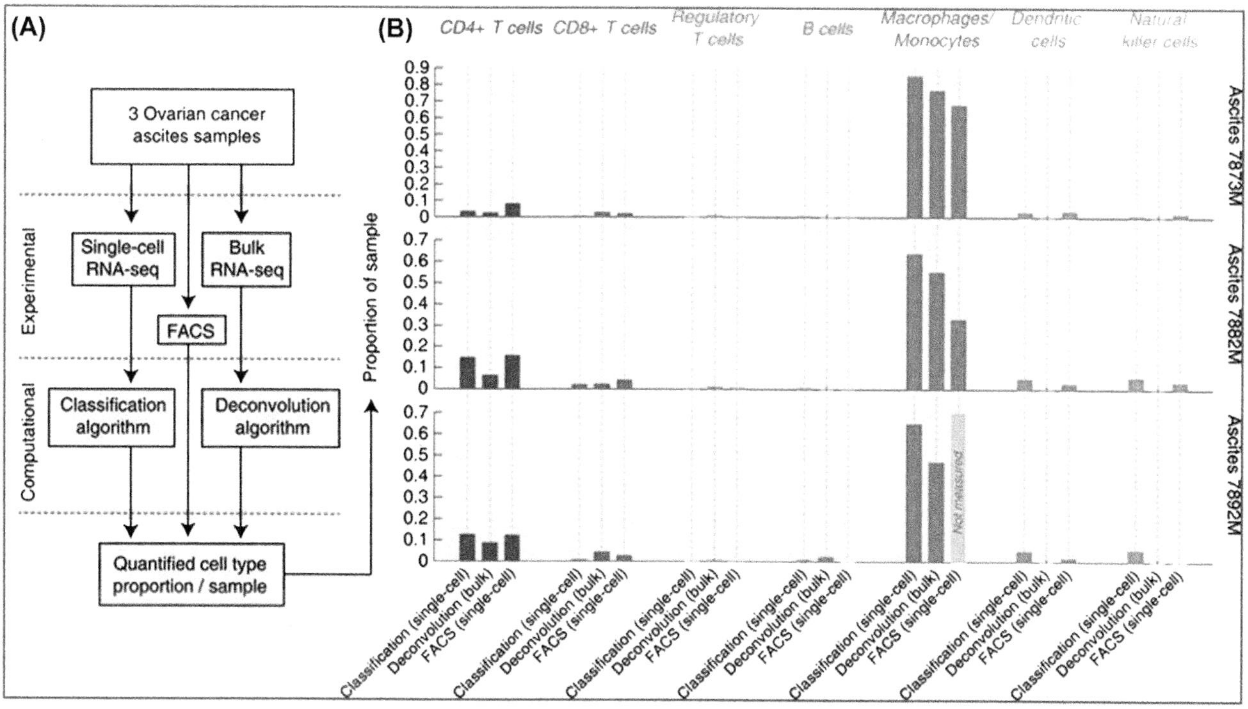

FIGURE 11.4 Estimates of cellular composition by three different methods. Three ovarian cancer ascites samples were profiled by single-cell and bulk RNA sequencing, as well as FACS. (A) Process diagram of data and results generation for three ovarian cancer ascites samples. (B) Estimates of the cellular composition are derived by: (1) classification based on single-cell RNA sequencing data; (2) computational deconvolution on the bulk RNA sequencing data using the single-cell RNA sequencing derived RGEP3; (3) quantification by FACS. For sample 7892, macrophages/monocytes quantification could not be determined by FACS. *Adapted from Schelker M, Feau S, Du J, Ranu N, Klipp E, MacBeath G, Schoeberl B, Raue A. Estimation of immune cell content in tumour tissue using single-cell RNA-seq data. Nat. Commun. December 11, 2017;8(1):2032. https://doi.org/10.1038/s41467-017-02289-3 and this work is licensed under a Creative Commons Attribution 4.0 Generic License.*

Mass cytometry or cytometry by time of flight (CyTOF) in immunoprofiling

Mass cytometry, or cytometry by time-of-flight (CyTOF), is a valuable tool in the high-dimensional analysis of single cells [16]. In CyTOF antibodies are labeled with isotopes, followed by the labeling of cells, before analysis using high-throughput spectrometry. Compared with FACS, CyTOF provides more parameters to quantify, which is limited by the overlap between the emission spectra of individual fluorophores [16,17]. CyTOF requires a starting cell number of 1000–1,000,000 for the single-cell analysis that can easily be derived from biopsies. CyTOF analysis generates various visualization plots to characterize immune cell populations [16,17]. Furthermore, spanning-tree progression analysis of density-normalized events (SPADE), hierarchically clusters single cells to decipher cell lineages, rare cell populations, and functional markers in a biopsy [16]. used linear discriminant analysis (LDA) classifier, to automatically categorize single-cell populations in CyTOF data. Table 11.2 lists antibody panels available for immunoprofiling assays using CyTOF [17].

TABLE 11.2 Panel of antibodies available for CyTOF[a].

Antigen	Isotope	Clone	Company
CD1c (BDCA-1)	166Er	AD5-8E7	Miltenyi biotec
CD3	154Sm	UCHT1	Fluidigm
CD4	145Nd	RPA-T4	Fluidigm
CD8a	146Nd	RPA-T8	Fluidigm
CD11b (Mac-1)	209Bi	ICRF44	Fluidigm
CD11c[a]	152Sm	Bu15	BioLegend
CD14	151Eu	M5E2	Fluidigm
CD15[a]	173Yb	MEM-158	Exbio
CD16	148Nd	3G8	Fluidigm
CD19	142Nd	HIB19	Fluidigm
CD25	169Tm	2A3	Fluidigm
CD39	160Gd	A1	Fluidigm
CD45	89Y or 156Gd	HI30	Fluidigm
CD45RA	155Gd	HI100	Fluidigm
CD45RO	164Dy	UCHL1	Fluidigm
CD56 (NCAM)	176Yb	N901	Fluidigm
CD62L (L-selectin)	153Eu	DREG-56	Fluidigm
CD68 [b]	171Yb	Y1/82A	Fluidigm
CD69	144Nd	FN50	Fluidigm
CD127 (IL-7Rα)	149Sm	A019D5	Fluidigm
CD141 (BDCA-3)[a]	165Ho	AD5-14H12	Miltenyi biotec
CD152 (CTLA-4) [b]	161Dy	14D3	Fluidigm
CD197 (CCR7)	159Tb	G043H7	Fluidigm
CD275 (ICOS-L)[a]	163Dy	2D3/B7-H2	BD Biosciences
CD278 (ICOS)	143Nd	DX29	Fluidigm
CD279 (PD-1)	175Lu	EH12.2H7	Fluidigm
CD303 (BDCA-2)	147Sm	201A	Fluidigm
CD366 (Tim-3)	150Nd	F38-2E2	BioLegend

Continued

TABLE 11.2 Panel of antibodies available for CyTOF[a].—cont'd

Antigen	Isotope	Clone	Company
Foxp3 [b]	162Dy	259D/C7	Fluidigm
HLA-DR	170Er	L243	Fluidigm
Ki-67 [b]	172Yb	B56	Fluidigm

[a]*Adapted and modified from Poláková I, Pelák O, Thürner D, Pokrývková B, Tachezy R, Kalina T, Šmahel M. Implementation of mass cytometry for immunoprofiling of patients with solid tumors. J. Immunol. Res. February 11, 2019;2019:6705949. https://doi.org/10.1155/2019/6705949 and this work is licensed under a Creative Commons Attribution 4.0 Generic License.*
[b]*All products for surface staining, except for those marked with a double asterisk (intracellular staining).*

Nanostring nCounter technology in immunoprofiling

NanoString's nCounter analysis system has advantages over conventional gene expression methods, such as direct measurement of mRNA levels without enzymatic reactions, and the digital output of data. NanoString's nCounter technology can be used to detect low abundance genes in formalin-fixed paraffin-embedded (FFPE) tissues in translational research, and also detects genes that are found to be absent by the microarray method. Importantly, the nCounter system has the ability to evaluate DNA, RNA, and protein targets in a sample at a particular point of time. This feature of nCounter technology gives the flexibility to measure immune cell profiling by both gene and protein levels and gives additional information about cellular functions.

Additionally, nCounter technology offers a variety of immune and immune oncology-related panels of genes related to response to therapy, disease-specific survival, and disease remission and relapse. For example, the nCounter platform offers a 53-gene signature panel for melanoma and a 50-gene signature panel for breast cancer assessments. Both immune cells and tumor tissues have been analyzed, for molecular expression profiles of genes and proteins that are specific to immune cells. Table 11.3_lists enriched

TABLE 11.3 List of enriched genes (>2 Fold) in immune cells[b].

Immune cells	Enriched genes
All T cells	Human: *CD3G, CD3D, TRA, CD6, CD5, NPDC1, CD28, CAMK4, GFI1, GATA3, SH2D1A, TRB, TNFRSF25, NK4, TACTILE, BCL11B, CD3E, INPP4B, MAL, NPDC1, ITM2A, ITK, LCK, NFATC3, RORA, MGC19764, TCF7, ZAP70, LEF1, SPOCK2, PRKCQ, SATB1, RASGRP1, LRIG1, DPP4, CD3Z, PDE4D, FYN, WWP1, LAT, DUSP16, KIAA0748, CDR2, STAT4, FLT3LG, IL6ST*
CD8+T cells	Human: *FCGBP, C1ORF21, PHEMX, KLRG1, ZNF145, ADRB2, DUSP2, IL2RB, CCL5, GZMC, TBX21, CCL4L, GZMH, PRF1, GNLY, CST7, GPR56, KLRC1, S100B, D12S2489E, CD8B1, CD8A*
T Regulatory cells	Human: *AKAP2, ANXA2, C8FW, CALM2, CDH13, ENTPD1, EPSTI1, F5, GBP2, GBP5, HS3ST3B1, HSPCA, ICA1, LAIR2, LGALS3, LOC51191, MKP-7, NINJ2, P53DINP1, PMAIP1, PTPLA, GPR2, HLA-DMA, HLA-DPA1, HLA-DPB1, HLA-DQB1, HLA-DRA, HLA-DRB4, HLA-DRB5, HPGD, PTT1, SAT, SHMT2, TMOD3, TNFRSF6, VIL2, LRRC32, CHAC1, EGR1, PYCR1, IL13, C3AR1, ZBED2, IL10, GAB3, NR4A1, IL1R2, ULBP1, IL17F, EIF4EBP1, RGS16, BCAT1, VEGFA, GIMAP1, GPR44, GPT2, PSPH, NLN, MS4A3, UCK2, TNIP3, TNFRSF9, IL7R, NR4A3, CSGALNA, CT1, SLC43A1, ZBTB32, SCFD2, HK2, C17ORF96, C1ORF186, LYZ, SFXN4, CYP1B1, SLC7A5, C17ORF58, SLC4A5, CD52, KLRB1, S100A4, POLA1, SLC7A1, ATPBD4, UBXN8, CASP8, FAIM3, IL12RB2, MPP7, ZNF831, BNIP3L, LRP8, UTRN, SOCS3, LY9, EVI2B, SFXN3, TNFRSF4, ING4, BLM, TNFSF11, SLFN5, CCR8, CD244, MYO1F, HOMER1, ATF3, AGPAT4, CTH, JAKMIP1, TFRC, RPL26L1, MICAL1, GPRASP1, TMX4, FYB, ZNF792, WDR4, HUWE1, CD83, TRIB2, LARP1B, SLC38A5, AUH, CTPS, CYTH1, APOL3, CHD6, IL1R1, LPCAT4, CITED2, DUSP4, SSBP2, LAYN, ZFAND2B, PIK3R1, GCNT4, PPP3CA, TIMP1, SIRPG, CCR3, PTP4A3, SERINC5, MAGOHB, GIMAP4, KLHL24, CCNG2, WDR70, MYBL1, CNPY4, TIMM8A, CCDC28A, MNAT1, MS4A2, GZMB, E2F5, HIBCH, C9ORF91, NHP2, CDC25B, VDR, CSF1, GPATCH4, NFKBID, TMCO7, EHBP1, FAM173B, FARS2, FLVCR2, NPM3, PNPO, P2RY8, TMEM63A, MXI1, MRTO4, SESN2, C1ORF38, STK38, SLC6A6, ALG14, LXN, TNFRSF18, ACSS1, UBA7, MTSS1, PSMA6, FAM8A1, RAB31, TNRC6B, SESN1, CEP68, LAPTM5, PLAGL2, LRRC33, EIF2B3, URB2, DDX21, ITGA4, RGS2, RORC, IFNAR2, CSDA, SORL1, LRRC37B, IL10RA, PRKACB, CBWD1, SMAD3, C1ORF96, FCER1A, HSP90AB1, EMG1*

Continued

TABLE 11.3 List of enriched genes (>2 Fold) in immune cells[b].—cont'd

Immune cells	Enriched genes
B Cells	Human: *PSEN2, FZD5, PLEKHF2, SIDT2, MTPN, TUBB6, FLJ21127, SMC6L1, KIAA1026, NUP88, GNA12, RNTRE, GSTZ1, RFX5, XYLT1, GL012, EPB41L2, LOC51760, SCRN1, PTD008, CYB561D2, SCAP2, MMP11, ERDJ5, TTC7A, PMAIP1, GARNL4, TMEPAI, SH3BP5, EDD, UCP4, EVI5L, PIK3C2B, SMAD3, LHFPL2, LOC57228, GM2A, HRK, ZNF207, LOC92497, F5, DMXL1, CLIC4, PRCP, DKFZP434C0328, STRBP, PHF16, JUP, TEAD2, PPP3CA, CNR2, ATP5B, ABCA1, BMF, ZCCHC7, KIAA0274, SLC22A3, SPI1, AMFR, ANXA4, EBF, ITPR1, HIST1H2BK, COPS3, COL14A1, SAV1, APOBEC3B, CCNG2, STAG3, LOC348938, POLD4, TBC1D1, MK2S4, APG4A, DAPP1, TNFRSF18, HERPUD1, POU2F2, ADK, CKIP-1, ACTA2, PARP14, MTSS1, DTX1, DDR1, RRAS2, RNF141, SYPL, SSPN, CORO1C, NFKBIE, CHD7, SP140, WDR34, BTNL9, ATP6V0A1, UROS, HLA-DOA, C22ORF13, IFI27, JMJD2B, WEE1, ODC1, KIAA0746, WDR11, SYNGR2, FOXP1, BLR1, MYO1E, MYBL2, GYLTL1B, SETBP1, KLF1, INPPL1, LOC54103, ZNF154, MHC2TA, CD24, ARHGEF3, BIRC3, TRIM56, HLA-DPB1, UVRAG, CD38, PEA15, FLJ10697, TLR10, ARHGAP10, HLA-DRA, SHMT2, LGALS9, FLJ10853, CEBPB, PRICKLE1, LCP21, CR1, KLHL14, TLR7, C20ORF72, HLA-DQB2, CORO1A, PRKCE HECW2, FLJ25604, TRIM26, FBXO41, GCNT1, LRMP, UBE2J1, IGKV3D, PCCA, GLDC, CYSLTR1, BTLA, NET5, HLA-DPA1, BLNK, CD79B, TM4SF8, TFEB, CD11ORF24,ADRBK2, CDKN2A, RIPK2, STX7, NCF4, SRGAP2, SLC2A5, CTSZ, AIM2, TCF4, TPD52, MAP3K8, MOBKL2B, HLA-DOB, IFIT3, MGC24039, BRD4, RALGPS2, SEMA4B, BSG,IGKC, HLA-DRB6, FUBP1, UNC84B, IFNGR2, HSPA5, HLA-DRB5, MARCKS, KYNU, PACAP, RHOBTB2, FLJ12363, TMED8, FGD2, FBXO10, IL4R, CD1C, MGC50844,MRPL49, CTSH, LYN, WASPIP, C3ORF6, EGR1, IGKV15, SPAP1, CHERP, IGHG3, ADAM28, HLA-DMA, HLA-DMB, PALM2-AKAP2, CD86, LAF4, LOC283663, FREB, DKFZP586A0522, UREB1, IGLL1, HLA-DQA2, KIAA1219, PLCG2, MARCH-1, BCNP1, PNOC, CD20, PSCD1, BCL11A, VPS28, SWAP70, SYK, BCL7A, NAP1L, TEM6, BTK,RAB30, FCGR2B, STAT6, BRDG1, LOC201895, ITGB1, CD74, CORO2B, VPREB3, HSPC182, MGC15619, RHOH, IGHA, CYBASC3, KIAA0125, RAM2, CD83, FCRH1, FLII,SNX10, IGHG1, EPHX1, FLJ10979, GTPAP, IGLL3, E2F5, TNFRSF17, BLK, FLJ00332, SNX2, CD200, HLA-DQB1, TRIO, CYBB, PIK3AP1, IRF4, CD22, BACE2, IGJ, PAX5,RGS13, TCL1A, MEF2C, C13ORF18, POU2AF1, HHEX, IRF8, BANK1, OSBPL10, SLC2A1, NCF1, IGL@, IGLC2, SPAP1, IGHG3, HLA-DMA, HLA-DMB, CD79A, HSPA6, IGHD, KIAA0476, SLC7A7, NKG7, CD19, SAMD9, LY86, SPIB, NAPSB, RNASE6, IGHM, LY64, CD72, IRTA2, CD1D, HLA-DQA1, SAS, CTSH, LYN, WASPIP, C3ORF6, EGR1, FREB, DKFZP586A0522*
Resting NK cells	Human: *IL18RAP, PRF1, GZMH, TKTL1*
Macrophages	[a]Murine: *Pecr, Tmem195, Ptplad2, 1810011H11Rik, Fert2, Tlr4, Pon3, Mr1, Arsg, Fcgr1, Camk1, Pld3, Tpp1, Ctsd, Pla2g15, Lamp2, Pla2g4a, MerTK, Tlr7, Cd14, Tbxas1, Fcgr3, Fgd4, Sqrdl, Csf3r, Plod1, Tom1, Myo7a, A930030A15Rik, Sepp1, Glul, Cd164, Tcn2, Dok3, Ctsl, Tspan14*
M1 macrophages	[a]Murine: *Cd38, Cfb, Slfn4, H2-Q6, Fpr1, Slfn1, Gpr18, Ccrl2, Fpr2, Cxcl10, Mpa2lOasl1, Tlr2, Ms4a4c, LOC100503664, Irak3, Hp, Itgal, Herc6, Cd300lf, Isf20, Pstpip2, Cp, Isg15, Herc6, Probe, 1452408_at, Ifi44, E030037K03, Rik, Saa3, Ifit1, Marco, F11r, Rsad2, Ddx60, Pilr1, Cpd, Gngt2, Mx1, Pyhin1, Epb4.1l3, Slfn8, Arh-gap24, Nfkbiz, Gbp6, Stat1, Zpb1, D14Erd668e, Ddx58, Tuba4a, H2-T10, Ebi3, Fam176b, Xaf1, Stat2, Sepx1, Ifit2*
M2 macrophages	[a]Murine: *Ptgs1, Egr2, Olfm1, Flrt2, P2ry1, Vwf, Amz1, Tmem158, Tiam1, Rhoj, Mmp9, Mrc1, Bcar3, Il6st, Tanc2, Mmp12, Tcfec, Clec71, Matk, Myc, Clec10a, Atp6v0a1, Lmna, Chst7, Atp6v0d2, Emp2, Socs6, Atp6v0a1, Plk2, Ptpla*
Classical monocytes	Human: *S100A12, ALOX5AP, PAD14, NRG1, MCEMP1, THBS1, CRISPLD2, F13A1, MOSC1, CYP27A1, CD163, QPCT, ADAM19, ASGR2, RNASE4, ALDH1A1, EDG3, PROK2, CLEC4D, S100A9, OSM, CSPG2, GALS2, ANG, VNN2, EBI2, UTS2, RPPH1, MGST1, IL8, CCR2, SELL, DYSF, MT1F, CD14, RANSE6, CSF3R, STAB1, CYP1B1, ADHFE1, PLA2G7, MMP25, ZNF395, SLC2A3, EGR1, CMTM2, CD9*
Intermediate monocytes	Human: *GFRA2, NKG7, PLVAP, PLAC8, MARCKSL1, E2F2, G8P4, CLEC10A, SCD, COTL1, SLC29A1, DDIT4, TGM2, LILRA3, ATF5, GPA33, C1QC, EVA1, MERTK, MGLL, DDEFL1, MARCO, NR1H3, FBP1, ACP2, GBP1, GPBAR1, SASH1, OLFM1, TIMP1, HLA-DOA, CAMK1, POUFUT1, EPB4IL3, H19, ZNF703, SNX5, CLEC10, AK-ALPHA-1, DPB1, DHRS9, MTHFD2, RGL1, PRDM1, FADS1, SLC2A8, CSK, ISOC2, CD300C, FGD2*
Nonclassical monocytes	Human: *C1QA, HES4, CDKN1C, C1QB, RHOC, CLEC4F, ADA, TAGLN, TCF7L2, CD798, CTSL, SFTPD, ABI3, SETBP1, FGFRL1, CD798, SPRED1, SNFT, EVL, MTSS1, SH2D1B, CYFIP2, NELF, STS-1, SIGLEC10, CKS1B, PTP4A3, LYPD2, SIDT2, INSIG1, LTB, PAPSS2, ABCC3, CASP5, VOM1, HSPB1, GNGT2, HMOX1, RRAS, RNF122, CDH23, FMNL2, RGS12, SGGB3A1, FER1L3, CALML4, IFITM3, CKB, CEACAM3, IFITM1*
Mature DC's	Human: *ACCL5, CXCL10, CCR7, IL15, IF127, IF144L, IFIH1, IFIT1, MX1, ISG15, ISG20, IRF7, GBP4, DUSP5, NFKB1A, ATF3, TNFSF10, IL6, IL8, IL7R, CCL4, TNFA1P6, IFIT3, OASL, GBP1, HES4, CYP27B1, RIPK2, TNFRSF9, SOD2, CD38, CD44, CD80, CD83, CD86, INDO MT2A, TRAF1, GADD45B, MT1M, MTIP2, BIRC3, USP18, TUBB2A, CCL8, EBI3, IFITM1, MT1B, MT1E, MT1G, MT1H, GADD45A, CD200, LAMP3, RGS1, SAT1*
Monocyte-derived DC's	Human: *CD1A, CD1B, CD1C, CD23A, MRC1, CD36, HLA-DQA, HLA-DP light chain, HLA-DG protein 41, RIL, RAP1GAP, CCND2, DUSP5, PPIC, STAC, PRKACB, TRIB2, SHB*

Continued

TABLE 11.3 List of enriched genes (>2 Fold) in immune cells[b].—cont'd

Immune cells	Enriched genes
Mature neutrophils	Human: *ALPL, IL8RB, FCGR3B, SEMA3C, HM74, SOD2, FCGR3A, IL-8, STHM, IL8RA, FCGR2A, CSF3R, NCF2, AOAH*
Immature neutrophils	Human: *AZU1, ELA2, BPI, LCN2, MPO, CTSG, MMP8, DEFA4, DEFA3, CAMP, X-CGD*

[a]*Genes unique to mice due to lack of human specific studies.*
[b]*Adapted from Lyons YA, Wu SY, Overwijk WW, Baggerly KA, Sood AK. Immune cell profiling in cancer: molecular approaches to cell-specific identification. NPJ Precis. Oncol. August 15, 2017;1(1):26. https://doi.org/10.1038/s41698-017-0031-0, and this work is licensed under a Creative Commons Attribution 4.0 Generic License.*

genes, or gene expression signatures, with differential expression of more than 2-fold with other immune cell types (Table 11.3) [3].

Application of immunoprofiling of human peripheral blood samples from an aging cohort identifies changes in the immune system that inform our understanding of age-associated complex diseases [18]. The immunoprofiling of specific subsets of immune cells is necessary, and Nanostring has gene panels to define specific subsets of immune cells in tissue biopsies (Table 11.4) [19].

TABLE 11.4 NanoString gene panels used to define specific immune cell subsets in tissue biopsies[a].

Immune cell type	NanoString gene panel
T Cells	CD2, CD3E, CD3G, CD6
Helper T cells	ANP32B, BATF, NUP107, CD28, ICOS
TH1	CD38, CSF2, IFNG, IL12RB2, LTA, CTLA4, TXB21, STAT4
TH2	CXCR6, GATA3, IL26, LAIR2, PMCH, SMAD2, STAT6
TH17	IL17A, IL17RA, RORC
Follicular helper T cells	CXCL13, MAF, PDCD1, BCL6
Memory T cells	
Central memory T cells	ATM, DOCK9, NEFL, REPS1, USP9Y
Effector memory T cells	AKT3, CCR2, EWSR1, LTK, NFATC4
Regulatory T cells	FOXP3
Cytotoxic CD8 T cells	CD8A, CD8B, FLT3LG, GZMM, PRF1
Gamma delta T cellls	CD160, FEZ1, TARP
B Cells	BLK, CD19, CR2, HLA-DOB, MS4A1, TNFRSF17
Natural killer cells	BCL2, FUT5, NCR1, ZNF205
CD56 high	FOXJ1, MPPED1, PLA2G6, RRAD
CD56 low	GTF3C1, GZMB, IL2IR
Dendritic cells	
Myeloid dendritic cells	CCL13, CCL17, CCL22, CD209, HSD11B1
Immature dendritic cells	CD1A, CD1B, CD1E, F13A1, SYT17
Activated dendritic cells	CCL1, EBI3, IDO1, LAMP3, OAS3
Plasmacytoid dendritic cells	IL3RA
Myeloid cells	
Macrophages	APOE, CCL7, CD68, CHIT1, CXCL5, MARCO, MSR1
Mast cells	CMA1, CTSG, KIT, MS4A2, PRG2, TPSAB1
Neutrophils	CSF3R, FPR2, MME
Eosinophils	CCR3, IL5RA, PTGDR2, SMPD3, THBS1

[a]*Adapted from Lim SY, Rizos H. Immune cell profiling in the age of immune checkpoint inhibitors: implications for biomarker discovery and understanding of resistance mechanisms. Mamm. Genome. December 2018;29(11–12):866–878. https://doi.org/10.1007/s00335-018-9757-4 and distributed under CC-BY 4.0 license.*

Multicolor multiplex immunohistochemistry(mIHC) in immunoprofiling

The immunoprofiling analysis of FFPE biopsies is essential to deduce the complexity of tumor immunology and identify novel predictive biomarkers for cancer immunotherapy. Immunoprofiling analysis of tissues requires the evaluation of combined markers, including inflammatory cell subpopulations and immune checkpoints, in the tumor microenvironment. The novel multiplex immunohistochemical methods like automated multiplex immunofluorescence (IF) method, for the multiparametric analyses in tissue specimens, is an essential tool in immunoprofiling of cancer [20]. Surface et al. [20] used unconjugated primary antibodies optimized for standard immunohistochemistry, on FFPE tissue samples, with a six-marker multiplex antibody panel comprising PD-L1, PD-1, CD68, CD8, Ki-67, and AE1/AE3 cytokeratins, and 4′,6-diamidino-2-phenylindole as a nuclear cell counterstain.

Parra et al. [21] validated multiplex immunofluorescence (mIF) panels, to apply to FFPE tissues using a set of immune marker antibodies, with the Opal 7-color Kit (PerkinElmer) for the characterization of the expression of PD-L1, PD-1, and subsets of tumor-associated immune cells. They quantified the expression of immune markers in mIHC, using the Vectra 3.0 multispectral microscopy and image analysis InForm 2.2.1 software (PerkinElmer), and compared with conventional IHC results obtained for each immune marker.

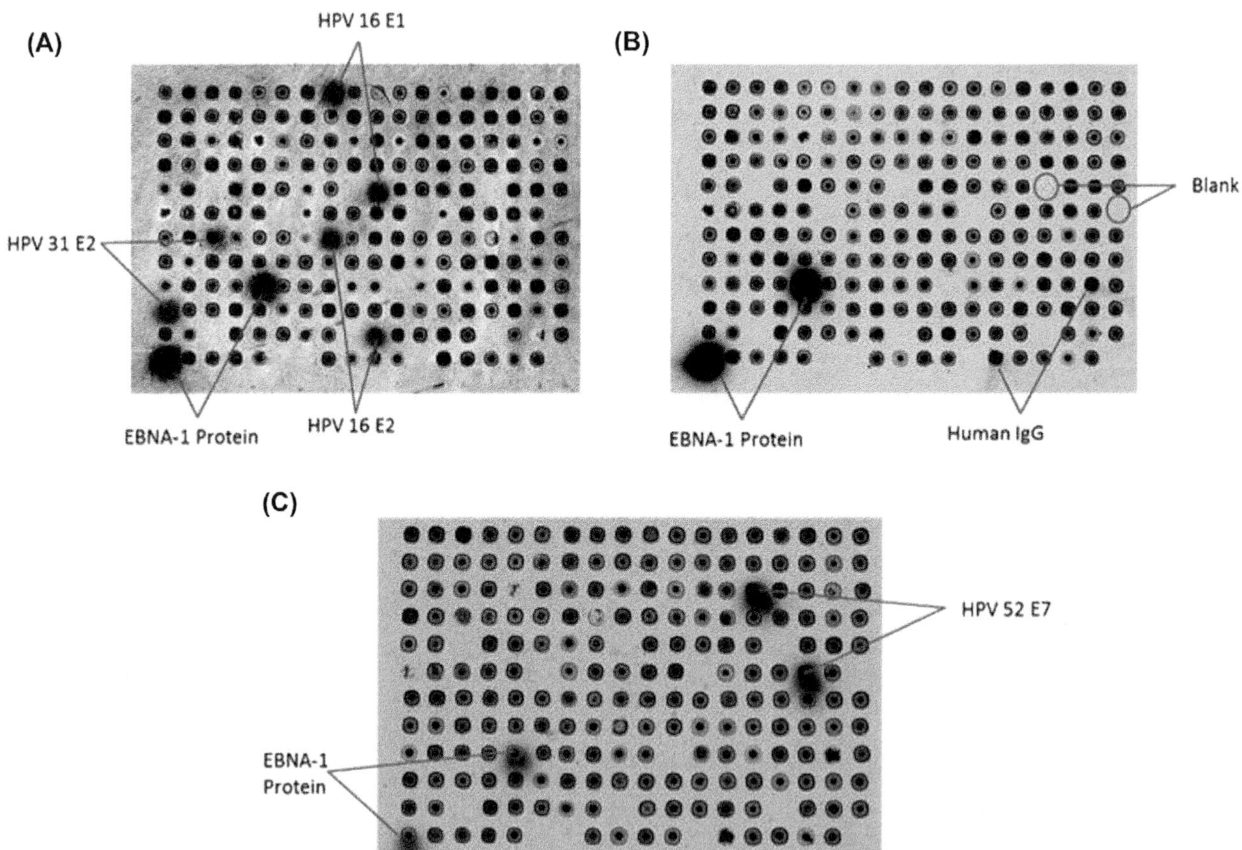

FIGURE 11.5 Detection of IgG in human serum using HPV protein arrays. (A) Detection of IgG Abs in serum from a patient with ICC. Immunoreactivity to the positive control EBV EBNA-1 protein, and HPV E4 protein from four different HPV types (16, 31, 35, and 45) are detected. (B) and (C) Detection of IgG Abs in sera from two women with CIN II/III. Immunoreactivity to HPV16 E4, HPV52 E4 (B) and HPV58E4 (C) as well as EBNA-1 protein, is shown. Dark spots represent the individual proteins (HPV Ags and non HPV-related controls in random order) displayed on the arrays after adjusting the raw images to extreme brightness and contrast. Positive spots (with diffused signal) are labeled. *Adapted from Ewaisha R, Panicker G, Maranian P, Unger ER, Anderson KS. Serum immune profiling for early detection of cervical disease. Theranostics August 23, 2017;7(16):3814–3823. https://doi.org/10.7150/thno.21098 and distributed under CC-BY 4.0 license.*

Protein arrays in immunoprofiling

Serum immunoprofiling was done for the detection of human papillomavirus (HPV) using nucleic acid programmable protein arrays (NAPPA), with the proteomes of two low-risk HPV types, and 10 oncogenic high-risk HPV types. NAPPA is accurate, reproducible, high-throughput, and flexible. It is an important tool for functional proteomics and protein-protein interaction studies [22]. The scale, specificity, and heterogeneity of the serologic response to HPV in cervical disease can be monitored by the HPV protein arrays (Fig. 11.5) [23]. Similarly, recombinant antibody microarray is a high-throughput proteomic technique that can detect multiplexed panels of both high- and low-abundant proteins in biofluids [24]. Diffuse large B-cell lymphoma (DLBCL) is a form of aggressive lymphoma, and is heterogeneous in terms of clinical progression and molecular findings, thus making it challenging to treat [24]. Immunoprofiles of DLBCL patients during the treatment period were assessed, using the recombinant antibody microarray (Figs. 11.6–11.8).

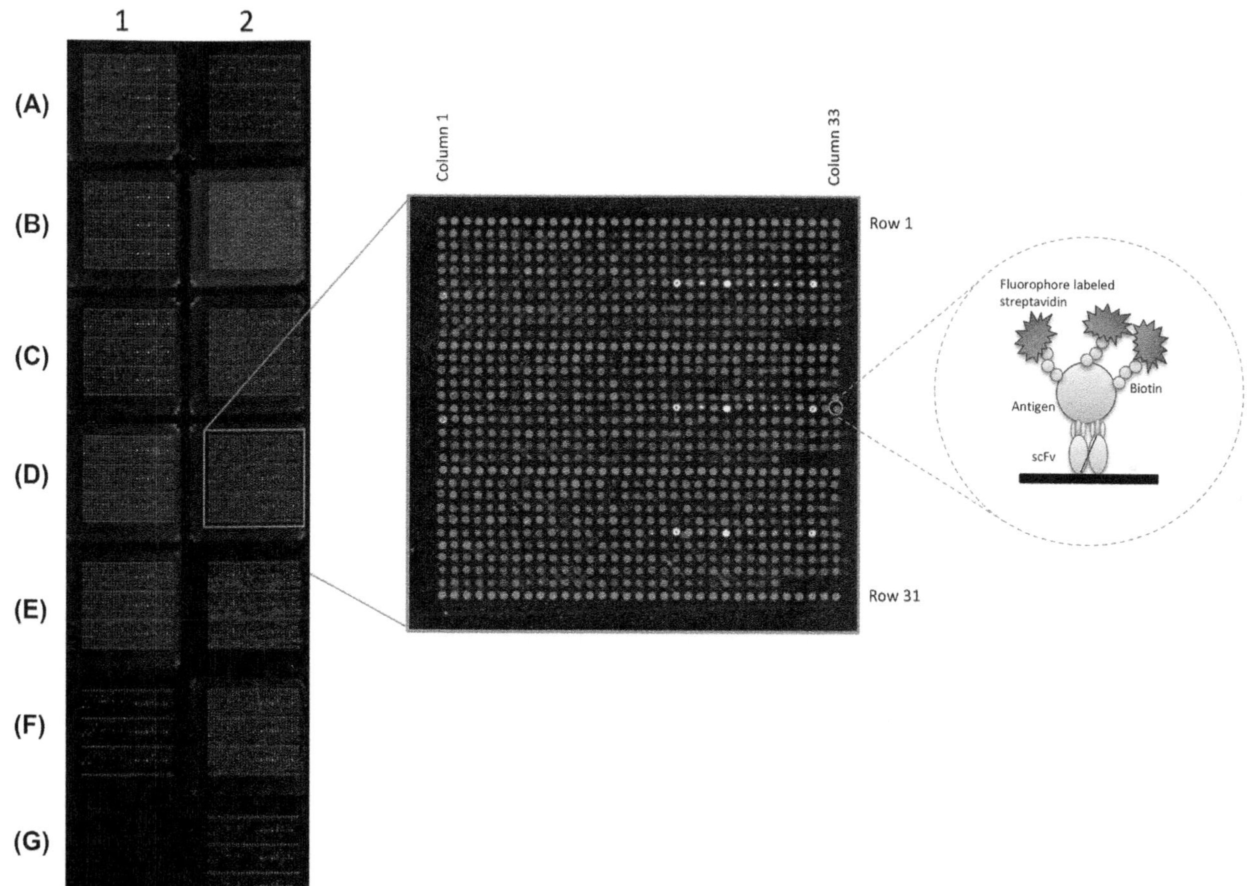

FIGURE 11.6 A representative scanned microarray image of a recombinant antibody microarray hybridized with plasma from a DLBCL patient. Thirteen identical subarrays denoted 1A-F and 2A-G were spotted. Subarray D2 is enlarged, showing the array layout with 33x31 spots. The arrays consist of three segments separated by printed rows of labeled BSA (row 1, 11, 21 and 31). Each antibody was printed in three replicate spots, one in each segment. *Adapted from Pauly F, Fjordén K, Leppä S, Holte H, Björkholm M, Fluge Ø, Møller Pedersen L, Eriksson M, Isinger-Ekstrand A, Borrebaeck CA, Jerkeman M, Wingren C. Plasma immunoprofiling of patients with high-risk diffuse large B-cell lymphoma: a Nordic Lymphoma Group study. Blood Cancer J. November 18, 2016;6(11):e501. https://doi.org/10.1038/bcj.2016.113 and distributed under CC-BY 4.0 license.*

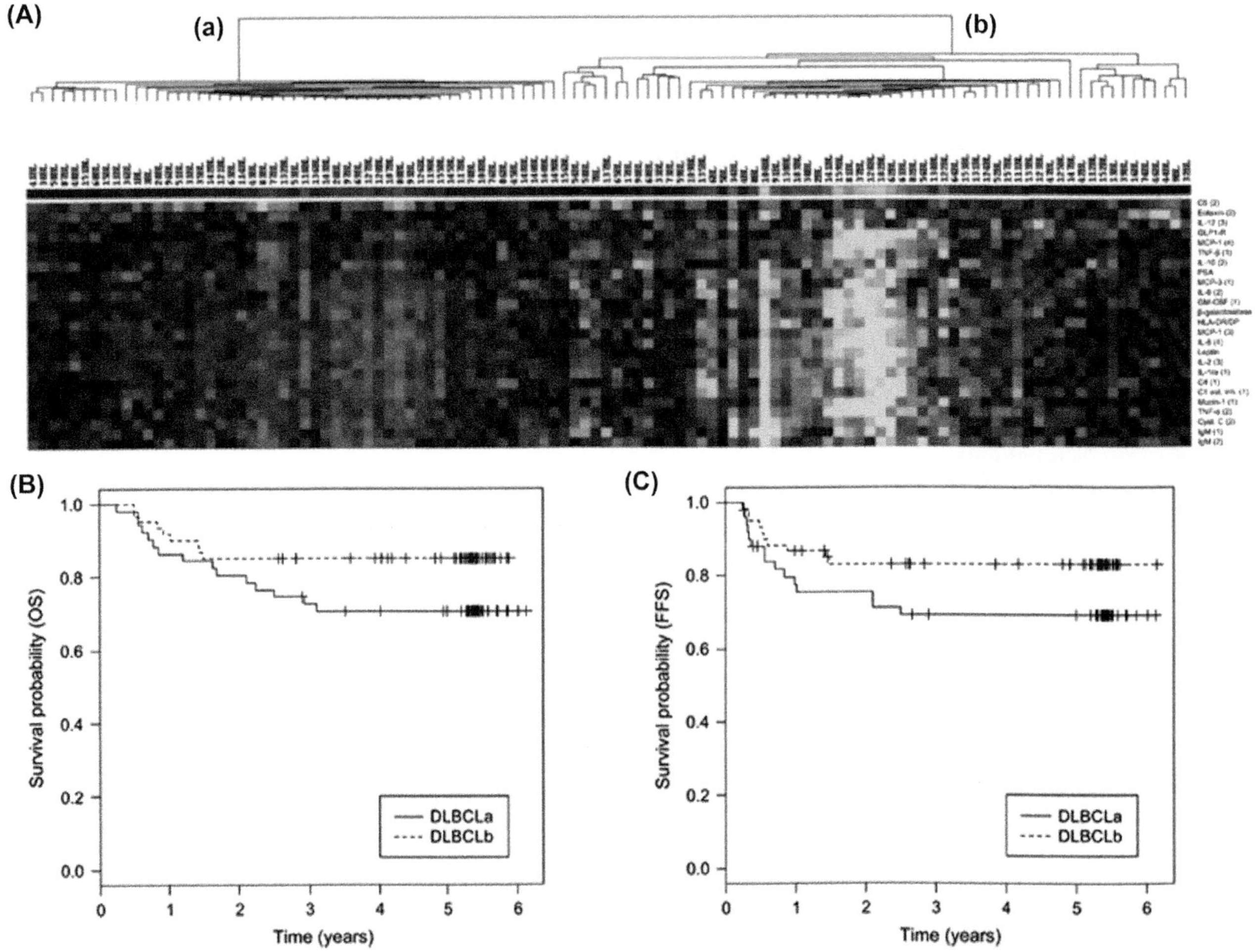

FIGURE 11.7 Stratification of DLBCL patients according to subgroup. (A) Subdivision of DLBCL patients into subgroups DLBCLa and DLBCLb based on a 23 biomarker signature identified in our previous study.1 (B) Kaplan–Meier plot demonstrating OS of DLBCLa and DLBCLb, log-rank P = 0.07 for OS. (C) Kaplan–Meier plot demonstrating FFS of DLBCLa and DLBCLb, log-rank P = 0.05 for FFS. *Adapted from Pauly F, Fjordén K, Leppä S, Holte H, Björkholm M, Fluge Ø, Møller Pedersen L, Eriksson M, Isinger-Ekstrand A, Borrebaeck CA, Jerkeman M, Wingren C. Plasma immunoprofiling of patients with high-risk diffuse large B-cell lymphoma: a Nordic Lymphoma Group study. Blood Cancer J. November 18, 2016;6(11):e501. https://doi.org/10.1038/bcj.2016.113 and distributed under CC-BY 4.0 license.*

Luminex xMAP technology in immunoprofiling

The xMAP (x = analyte or biomarker, MAP = multi-analyte profiling) technology was invented in the 1990s, by the scientists at Luminex Corporation (USA), for simultaneous detection of analytes in biological samples [25]. xMAP assay is a solid-phase isolation method that uses fluidics, optics, and digital signal analysis, with patented microsphere (bead)-based technology. xMAP technology is an open platform offered by Luminex to academia and industrial partners, to develop multiplex assays for a variety of applications [26].

Since the xMAP technology is easy, rapid, reproducible, high-throughput, cost-effective, and generates high-quality data, it is commonly utilized in pharmaceutical, clinical, and research laboratories [26,27]. The xMAP instruments currently available in the market encompass LUMINEX 200, FLEXMAP 3D, and MAGPIX [25]. The FLEXMAP 3D platform can be used to quantify up to 500 analytes simultaneously in a sample.

Recently, we describe the use of xMAP technology for the multiplex detection of an array of cytokines, chemokines, and growth factors, in the serum of patients suffering from autoimmune diseases, such as rheumatoid arthritis (RA), using microsphere-based multiplex immuno-assay formats (MBMI) [2]. More essentially, the immunoprofiling of serum or plasma autoantibodies can be cross validated (Fig. 11.9) using Luminex xMAP technology [28].

(A)

Baseline → Cycle 3

Protein	FC	*P*	Adjusted *P*
IL-10 (3)		4E-06	0.0007
IL-5 (2)		5E-06	0.0007
IL-9 (1)		2E-05	0.002
CIMS-7		3E-05	0.002
IL-3 (2)		4E-05	0.002
IL-11 (2)		7E-05	0.004
TNF-β (4)		9E-05	0.004
IL-9 (3)		0.0001	0.004
IFN-γ (1)		0.0001	0.004
MCP-1 (1)		0.0001	0.004
TGF-β1 (3)		0.0001	0.004
IL-3 (1)		0.0002	0.004
IL-8 (2)		0.0002	0.004
IL-10 (2)		0.0003	0.006
VEGF (2)		0.0003	0.006
TNF-α (3)		0.0004	0.006
VEGF (1)		0.0004	0.006
RANTES (1)		0.0004	0.006
IL-4 (2)		0.0004	0.006
IL-6 (2)		0.0005	0.008
IL-18 (1)		0.0006	0.008
TNF-β (2)		0.0006	0.008
IL-1α (2)		0.0007	0.008
TNF-α (1)		0.0009	0.01
C1 inhibitor (3)		0.0009	0.01

(B)

Cycle 3 → Cycle 8

Protein	FC	*P*	Adjusted *P*
C3 (5)		0.0002	0.05
ICAM-1		0.001	0.11
CIMS-27		0.002	0.11
CIMS-2		0.003	0.11
IL-6 (5)		0.003	0.11
OSTP (1)		0.003	0.11
IgM (3)		0.004	0.11
OSBPL3 (2)		0.004	0.11
MCP-1 (6)		0.005	0.11
Factor B (3)		0.005	0.11
CIMS-30		0.005	0.11
IL-3 (2)		0.005	0.11
C5 (3)		0.006	0.11
TNFRSF14 (2)		0.006	0.11
IL-6 (6)		0.006	0.11
PTK-6		0.006	0.11
C3 (4)		0.008	0.11
CIMS-11		0.008	0.11
CIMS-5		0.008	0.11
CIMS-17		0.008	0.11
C4 (4)		0.009	0.11
RPS6KA2 (2)		0.009	0.11
TNFRSF14 (2)		0.01	0.11
C1 inhibitor (2)		0.01	0.11
Keratin19 (1)		0.01	0.11

(C)

Baseline → Cycle 8

Protein	FC	*P*	Adjusted *P*
IL-3 (1)		0.0005	0.16
IL-9 (1)		0.002	0.31
HADH2 (2)		0.004	0.31
UBE2C (1)		0.004	0.31
VEGF (2)		0.007	0.36
IL-5 (2)		0.008	0.36
VEGF (1)		0.009	0.36
IL-1β (2)		0.01	0.38
CIMS-12		0.01	0.40
Cyst. C		0.02	0.44
MCP-3 (1)		0.02	0.44
BTK (4)		0.02	0.44
C4 (1)		0.02	0.44
C5 (2)		0.02	0.47
IL-8 (2)		0.03	0.50
IL-12 (2)		0.03	0.51
IgM (2)		0.03	0.51
GM-CSF (2)		0.03	0.53
APOA4 (3)		0.04	0.53
IL-10 (3)		0.04	0.53
FASN (2)		0.04	0.53
TNF-β (4)		0.04	0.53
IL-10 (1)		0.05	0.60

FIGURE 11.8 Immunoprofiles of DLBCL patients during the course of treatment. Top 25 significantly deregulated proteins ($P < 0.05$) as determined by paired t-tests between time points (A) BL and Cy3, (B) Cy3 and Cy8 and (C) BL and Cy8 (only 23 deregulated proteins). Fold changes (FC) are presented in heat maps; red (dark gray in print version) – upregulated, green (gray in print version) – down-regulated and black – equal levels. *Adapted from Pauly F, Fjordén K, Leppä S, Holte H, Björkholm M, Fluge Ø, Møller Pedersen L, Eriksson M, Isinger-Ekstrand A, Borrebaeck CA, Jerkeman M, Wingren C. Plasma immunoprofiling of patients with high-risk diffuse large B-cell lymphoma: a Nordic Lymphoma Group study. Blood Cancer J. November 18, 2016;6(11):e501. https://doi.org/10.1038/bcj.2016.113 and distributed under CC-BY 4.0 license.*

Miscellaneous technologies in immunoprofiling

Microengraving technique uses suspension of cells on top of an array of microwells, and invert onto a glass slide with capture reagent, followed by incubation and scanning using fluorescent microarray imager [3]. It is useful for the quantification of cytokines and antigen-specific immunoglobulins, functional responses of immune cells, and other secreted proteins. This technique requires only 50,000–100,000 cells, and the cells are viable after the analysis. It is not suitable for the measurement of cytosolic proteins [3]. The cytosolic and surface proteins of single cells can be measured using barcoded microchip system. The microchips have microchambers that hold about 10, 000 cells, and houses the barcode, or antibody array for capture, lysis, and detection of various proteins, such as cytokine production from macrophages and T cells, monitoring the therapeutic efficiency of immunotherapy, and functional responses of immune cells [3].

The latest technologies used in immunoprofiling, potential biomarkers screened, sample preparation methods, and the relevant bioinformatics tools, have been summarized in Table 11.5.

Future directions

Immunoprofiling is a way to measure the state of an individual's immune system, at a given point in health and disease. More essentially, the cutting-edge high-throughput

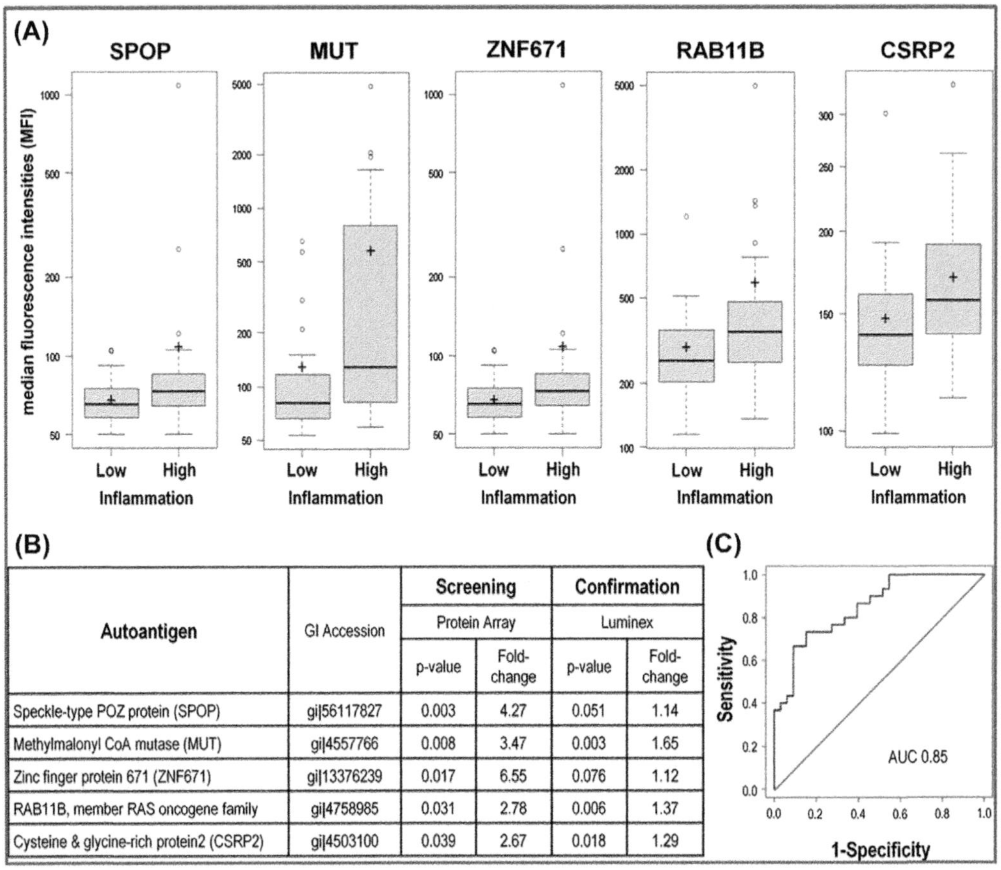

Autoantigen	GI Accession	Screening		Confirmation	
		Protein Array		Luminex	
		p-value	Fold-change	p-value	Fold-change
Speckle-type POZ protein (SPOP)	gi|56117827	0.003	4.27	0.051	1.14
Methylmalonyl CoA mutase (MUT)	gi|4557766	0.008	3.47	0.003	1.65
Zinc finger protein 671 (ZNF671)	gi|13376239	0.017	6.55	0.076	1.12
RAB11B, member RAS oncogene family	gi|4758985	0.031	2.78	0.006	1.37
Cysteine & glycine-rich protein2 (CSRP2)	gi|4503100	0.039	2.67	0.018	1.29

FIGURE 11.9 Cross-validation of identified autoantibody profile using a Luminex-beads protein array. The identified profile was validated in an independent set of prostate cancer patients ($n = 60$) using the bead-based Luminex technology to identify autoantibodies. A Box plots of mean fluorescent intensities (MFI) values for the five top autoantibody candidates significantly increased in the prostate high inflammation group in both screens. B Table displaying fold-change and *P*-values of the autoantibodies significantly upregulated in high-inflammation serum samples of the screening and validation patient cohorts. C ROC curve for the top five autoantibodies for the classification of samples of the validation set. The identified biomarker profile discriminates between high and low inflammation patients with an AUC of 0.85. *Adapted from Schlick B, Massoner P, Lueking A, Charoentong P, Blattner M, Schaefer G, Marquart K, Theek C, Amersdorfer P, Zielinski D, Kirchner M, Trajanoski Z, Rubin MA, Müllner S, Schulz-Knappe P, Klocker H. Serum autoantibodies in chronic prostate inflammation in prostate cancer patients. PLoS One February 10, 2016;11(2):e0147739. https://doi.org/10.1371/journal.pone.0147739 and this work is licensed under a Creative Commons Attribution 4.0 Generic License.*

TABLE 11.5 Summary of novel technologies used in immune biomarker profiling.

Technology	Suggestions and potential biomarkers	Sample preparation	Bioinformatic tools
Whole exome sequencing for neoantigen discovery	• Mutation load for CTLA-4 and PD-1 blockade therapy • Neoantigen-specific T-cell response	DNA from tumor and normal cells	EBcall, JointSNVMix, MuTect, SomaticSniper, Strelka, VarScan 2, BIMAS, RNAKPER SYFPEITHI, IDEB, NetMCHpan, TEPITOPEpan, PickPocket, Multipred2, MultiRTA
Gene signature and pattern	• MAGE-A3 gene signature • chemokine expression in melanoma • Neoantigen signature	DNA and RNA from tumor, lymph node and PBMCs	BRB-ArrayTools, LIMMA, SAM, PAM, Partek, Genomic Suite, GSEA, Ingenuity IPA
Epigenetic-differentiation based immune cell quantification	• immune cell lineage specific epigenetic modification • Leukocyte ratios in blood and tissue	Genomic DNA from fresh or frozen whole blood, PBMC, lymph node and fresh tissue or FFPE tissue and blood clots	HOMER package Motif Finder algorithm findMotifGenome.pl, MatInspector (Genomatix), Mendelian randomization

Continued

TABLE 11.5 Summary of novel technologies used in immune biomarker profiling.—cont'd

Technology	Suggestions and potential biomarkers	Sample preparation	Bioinformatic tools
Protein microarray (seromics)	• TAA antibody response • Broad antibody signature • new antigen discovery	Fresh or frozen serum and plasma	Prospector, LIMMA package, PAA package, Spotfire package
Flow cytometry and Mass cytometry	• Use best flow practices and recommended flow panels • Multimers for T-cell epitope screening • TAA-specific T-cell response for CTLA-4 blockade therapy • CD4+ICOS+ T cells for CTLA-4 blockade therapy • Baseline MDSC for CTLA-4 blockade therapy	Whole blood; Fresh or frozen PBMCs and TILs; Fresh or frozen cells from ascites or pleural effusion	Computational algorithm-driven analysis for MDSC, cytobank, FlowJo, SPADE, PhenoGraph, PCA, viSNE, citrus, ACCENSE, Isomap, 3D visualization
T And B cell receptor deep sequencing	• CD3 T-cell count • T cell clonotype stability for CTLA-4 blockade therapy • Baseline T-cell clonality in tumor in PD-1 blockade therapy	DNA from FFPE; Frozen cells from tumor, lymph node or PBMCs; Fresh or frozen cells from ascites or pleural effusion	Shannon Entropy, Morisita's distance, estimated TCR gene rearrangements per diploid genomes, clonality, ImmuneID, Adaptive ImmunoSeq software
Multicolor IHC staining	• CD3 immune score • CD8/FOXP3 ratio for tumor necrosis • PD-L1 expression on tumor in PD-1 blockade therapy	FFPE tissue; Fresh or frozen tissue	TissueGnostic system, PerkinElmer system

Adapted and modified from Yuan J, Hegde PS, Clynes R, Foukas PG, Harari A, Kleen TO, Kvistborg P, Maccalli C, Maecker HT, Page DB, Robins H, Song W, Stack EC, Wang E, Whiteside TL, Zhao Y, Zwierzina H, Butterfield LH, Fox BA. Novel technologies and emerging biomarkers for personalized cancer immunotherapy. J. Immunother. Cancer. January 19, 2016;4:3. https://doi.org/10.1186/s40425-016-0107-3 and this work is licensed under a Creative Commons Attribution 4.0 Generic License.

immunoprofiling strategies, have effectively been used to identify the predictive biomarkers and mechanism of resistance to immunotherapy. For example, the resistance to checkpoint inhibitor (CI) antibodies that are approved for treating many cancer types [19,30], and further classification of immunologically responsive and unresponsive tumor types, can be done by various immunoprofiling methods as pictorially depicted (Fig. 11.10) by Yuan et al. (2016). Hence, a combinatorial immunoprofiling approach in preclinical models is necessary to predict the aggressiveness of diseases such as cancer [31].

The immunoprofiling data, combined with data integration and machine learning, is important to gain valuable information about the impact of different adjuvant formulations on vaccine-induced immune responses [1]. Integrated bioinformatics is required for the combination of complex morphological phenotypes, with the "multiomics" datasets that drive precision medicine [32].

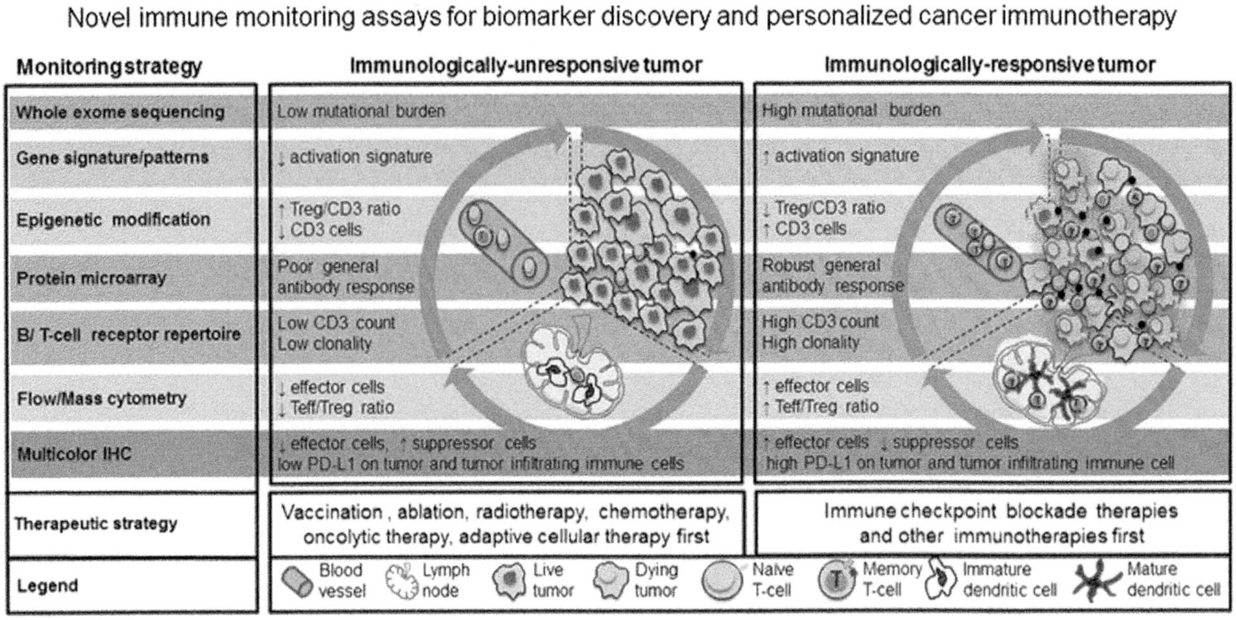

FIGURE 11.10 High-throughput immune assessment for biomarker discovery and personalized cancer immunotherapy. Immunologically ignorant and immunologically responsive tumors are classified by the presence of immune cells in the tumor microenvironment. Potential biomarkers identified from high-throughput technologies can further differentiate these tumors by the mutation load, gene/protein/antibody signature profile, phenotype and function of immune cells, and can also provide clinical strategies for personalized cancer immunotherapies. The new and innovative technologies that can be utilized to identify potential biomarkers include whole exome sequencing, gene signature, epigenetic modification, protein microarray, B/T-cell receptor repertoire, flow/mass cytometry and multicolor IHC. Arrows indicate a decrease (↓) or increase (↑). *Adapted from Yuan J, Hegde PS, Clynes R, Foukas PG, Harari A, Kleen TO, Kvistborg P, Maccalli C, Maecker HT, Page DB, Robins H, Song W, Stack EC, Wang E, Whiteside TL, Zhao Y, Zwierzina H, Butterfield LH, Fox BA. Novel technologies and emerging biomarkers for personalized cancer immunotherapy. J. Immunother. Cancer January 19, 2016;4:3. https://doi.org/10.1186/s40425-016-0107-3 and this work is licensed under a Creative Commons Attribution 4.0 Generic License.*

Acknowledgments

This work is funded by the National Plan for Science, Technology and Innovation (MAARIFAH)-King Abdulaziz City for Science and Technology-The Kingdom of Saudi Arabia-award number 12-BIO2267-03. The authors also acknowledge with thanks the Science and Technology Unit (STU), King Abdulaziz University for their excellent technical support.

References

[1] Chaudhury S, Duncan EH, Atre T, Storme CK, Beck K, Kaba SA, Lanar DE, Bergmann-Leitner ES. Identification of immune signatures of novel adjuvant formulations using machine learning. Sci. Rep. November 30, 2018;8(1):17508. https://doi.org/10.1038/s41598-018-35452-x. PubMed PMID: 30504893; PubMed Central PMCID: PMC6269591.

[2] Bahlas S, Damiati L, Dandachi N, Sait H, Alsefri M, Pushparaj PN. Rapid immunoprofiling of cytokines, chemokines and growth factors in patients with active rheumatoid arthritis using Luminex Multiple Analyte Profiling technology for precision medicine. Clin. Exp. Rheumatol. 2019 January–February;37(1):112–9. Epub 2018 Jun 14. PubMed PMID: 29998825.

[3] Lyons YA, Wu SY, Overwijk WW, Baggerly KA, Sood AK. Immune cell profiling in cancer: molecular approaches to cell-specific identification. NPJ Precis. Oncol. August 15, 2017;1(1):26. https://doi.org/10.1038/s41698-017-0031-0. eCollection 2017. Review. PubMed PMID: 29872708; PubMed Central PMCID: PMC5871917.

[4] Greiff V, Miho E, Menzel U, Reddy ST. Bioinformatic and statistical analysis of adaptive immune repertoires. Trends Immunol. November 2015;36(11):738–49. https://doi.org/10.1016/j.it.2015.09.006. Epub 2015 Oct 25. Review. PubMed PMID: 26508293.

[5] Miho E, Yermanos A, Weber CR, Berger CT, Reddy ST, Greiff V. Computational strategies for dissecting the high-dimensional complexity of adaptive immune repertoires. Front. Immunol. February 21, 2018;9:224. https://doi.org/10.3389/fimmu.2018.00224. eCollection 2018. Review. PubMed PMID: 29515569; PubMed Central PMCID: PMC5826328.

[6] Baskaran S, Agarwal A, Leisegang K, Pushparaj PN, Panner Selvam MK, Henkel R. An in-depth bibliometric analysis and current perspective on male infertility research. World J. Mens. Health February 20, 2019. https://doi.org/10.5534/wjmh.180114 [Epub ahead of print] PubMed PMID: 30799564.

[7] Kitaura K, Shini T, Matsutani T, Suzuki R. A new high-throughput sequencing method for determining diversity and similarity of T cell receptor (TCR) α and β repertoires and identifying potential new invariant TCR α chains. BMC Immunol. October 11, 2016;17(1):38. PubMed PMID: 27729009; PubMed Central PMCID: PMC5059964.

[8] Stoeckius M, Hafemeister C, Stephenson W, Houck-Loomis B, Chattopadhyay PK, Swerdlow H, Satija R, Smibert P. Simultaneous epitope and transcriptome measurement in single cells. Nat. Methods September 2017;14(9):865–8. https://doi.org/10.1038/nmeth.4380. Epub 2017 Jul 31. PubMed PMID: 28759029; PubMed Central PMCID: PMC5669064.

[9] Schelker M, Feau S, Du J, Ranu N, Klipp E, MacBeath G, Schoeberl B, Raue A. Estimation of immune cell content in tumour tissue using single-cell RNA-seq data. Nat. Commun. December 11, 2017;8(1):2032. https://doi.org/10.1038/s41467-017-02289-3. PubMed PMID: 29230012; PubMed Central PMCID: PMC5725570.

[10] Larsen FT, Bed'Hom B, Naghizadeh M, Kjærup RB, Zohari S, Dalgaard TS. Immunoprofiling of peripheral blood from infectious bronchitis virus vaccinated MHC-B chicken lines – monocyte MHC-II expression as a potential correlate of protection. Dev. Comp. Immunol. July 2019;96:93–102. https://doi.org/10.1016/j.dci.2019.02.004. Epub 2019 Feb 11. PubMed PMID: 30763593.

[11] Halliday A, Whitworth H, Kottoor SH, Niazi U, Menzies S, Kunst H, Bremang S, Badhan A, Beverley P, Kon OM, Lalvani A. Stratification of latent *Mycobacterium tuberculosis* infection by cellular immune profiling. J. Infect. Dis. May 1, 2017;215(9):1480–7. https://doi.org/10.1093/infdis/jix107. PubMed PMID: 28329119; PubMed Central PMCID: PMC5451604.

[12] Pennington K, Sasieta HC, Ramos GP, Erskine CL, Van Keulen VP, Peikert T, Escalante P. Flow cytometric immune profiling in infliximab-associated tuberculosis. Clin. Med. Insights Case Rep. August 24, 2017;10. https://doi.org/10.1177/1179547617724776. eCollection 2017. 1179547617724776. PubMed PMID: 28890660; PubMed Central PMCID: PMC5574471.

[13] Robison HM, Escalante P, Valera E, Erskine CL, Auvil L, Sasieta HC, Bushell C, Welge M, Bailey RC. Precision immunoprofiling to reveal diagnostic signatures for latent tuberculosis infection and reactivation risk stratification. Integr. Biol. (Camb.) February 4, 2019. https://doi.org/10.1093/intbio/zyz001 [Epub ahead of print] PubMed PMID: 30722034.

[14] Fonseka CY, Rao DA, Teslovich NC, Korsunsky I, Hannes SK, Slowikowski K, Gurish MF, Donlin LT, Lederer JA, Weinblatt ME, Massarotti EM, Coblyn JS, Helfgott SM, Todd DJ, Bykerk VP, Karlson EW, Ermann J, Lee YC, Brenner MB, Raychaudhuri S. Mixed-effects association of single cells identifies an expanded effector CD4(+) T cell subset in rheumatoid arthritis. Sci. Transl. Med. October 17, 2018;10(463). https://doi.org/10.1126/scitranslmed.aaq0305. pii: eaaq0305. PubMed PMID: 30333237; PubMed Central PMCID: PMC6448773.

[15] Delmas A, Oikonomopoulos A, Lacey PN, Fallahi M, Hommes DW, Sundrud MS. Informatics-based discovery of disease-associated immune profiles. PLoS One September 26, 2016;11(9):e0163305. https://doi.org/10.1371/journal.pone.0163305. eCollection 2016. PubMed PMID: 27669154; PubMed Central PMCID: PMC5036861.

[16] Abdelaal T, van Unen V, Höllt T, Koning F, Reinders MJT, Mahfouz A. Predicting cell populations in single cell mass cytometry data. Cytometry March 12, 2019. https://doi.org/10.1002/cyto.a.23738 [Epub ahead of print] PubMed PMID: 30861637.

[17] Poláková I, Pelák O, Thürner D, Pokrývková B, Tachezy R, Kalina T, Šmahel M. Implementation of mass cytometry for immunoprofiling of patients with solid tumors. J. Immunol. Res. February 11, 2019;2019:6705949. https://doi.org/10.1155/2019/6705949. eCollection 2019. PubMed PMID: 30886872; PubMed Central PMCID: PMC6388349.

[18] Raychaudhuri S, Gupta RM. Immunoprofiling comes of age. Nat. Med. March 2019;25(3):362–4. https://doi.org/10.1038/s41591-019-0387-5. PubMed PMID: 30842672.

[19] Lim SY, Rizos H. Immune cell profiling in the age of immune checkpoint inhibitors: implications for biomarker discovery and understanding of resistance mechanisms. Mamm. Genome December 2018;29(11–12):866–78. https://doi.org/10.1007/s00335-018-9757-4. Epub 2018 Jul 2. Review. PubMed PMID: 29968076; PubMed Central PMCID: PMC6267680.

[20] Surace M, DaCosta K, Huntley A, Zhao W, Bagnall C, Brown C, Wang C, Roman K, Cann J, Lewis A, Steele K, Rebelatto M, Parra ER, Hoyt CC, Rodriguez-Canales J. Automated multiplex immunofluorescence panel for immuno-oncology studies on formalin-fixed carcinoma tissue specimens. J. Vis. Exp. January 21, 2019;143. https://doi.org/10.3791/58390. PubMed PMID: 30735177.

[21] Parra ER, Uraoka N, Jiang M, Cook P, Gibbons D, Forget MA, Bernatchez C, Haymaker C, Wistuba II, Rodriguez-Canales J. Validation of multiplex immunofluorescence panels using multispectral microscopy for immune-profiling of formalin-fixed and paraffin-embedded human tumor tissues. Sci. Rep. October 17, 2017;7(1):13380. https://doi.org/10.1038/s41598-017-13942-8. PubMed PMID: 29042640; PubMed Central PMCID: PMC5645415.

[22] Manzano-Román R, Fuentes M. A decade of Nucleic Acid Programmable Protein Arrays (NAPPA) availability: news, actors, progress, prospects and access. J. Proteomics April 30, 2019;198:27–35. https://doi.org/10.1016/j.jprot.2018.12.007. Epub 2018 Dec 12. PubMed PMID: 30553075.

[23] Ewaisha R, Panicker G, Maranian P, Unger ER, Anderson KS. Serum immune profiling for early detection of cervical disease. Theranostics August 23, 2017;7(16):3814–23. https://doi.org/10.7150/thno.21098. eCollection 2017. PubMed PMID: 29109779; PubMed Central PMCID: PMC5667406.

[24] Pauly F, Fjordén K, Leppä S, Holte H, Björkholm M, Fluge Ø, Møller Pedersen L, Eriksson M, Isinger-Ekstrand A, Borrebaeck CA, Jerkeman M, Wingren C. Plasma immunoprofiling of patients with high-risk diffuse large B-cell lymphoma: a Nordic Lymphoma Group study. Blood Cancer J. November 18, 2016;6(11):e501. https://doi.org/10.1038/bcj.2016.113. PubMed PMID: 27858932; PubMed Central PMCID: PMC5148057.

[25] Angeloni S, Cordes R, Dunbar S, Garcia C, Gibson G, Martin C, et al. xMAP cookbook: a collection of methods and protocols for developing multiplex assays with xMAP technology. 2nd ed. Austin, TX: Luminex; 2014.

[26] Graham H, Chandler DJ, Dunbar SA. The genesis and evolution of bead-based multiplexing. Methods April 1, 2019;158:2–11.

[27] Kellar KL, Mahmutovic AJ, Bandyopadhyay K. Multiplexed microsphere-based flow cytometric immunoassays. Curr. Protoc. Cytom. February 2006 Chapter 13:Unit13.1.

[28] Schlick B, Massoner P, Lueking A, Charoentong P, Blattner M, Schaefer G, Marquart K, Theek C, Amersdorfer P, Zielinski D, Kirchner M, Trajanoski Z, Rubin MA, Müllner S, Schulz-Knappe P, Klocker H. Serum autoantibodies in chronic prostate inflammation in prostate cancer patients. PLoS One February 10, 2016;11(2):e0147739. https://doi.org/10.1371/journal.pone.0147739. eCollection 2016. PubMed PMID: 26863016; PubMed Central PMCID: PMC4749310.

[29] Yuan J, Hegde PS, Clynes R, Foukas PG, Harari A, Kleen TO, Kvistborg P, Maccalli C, Maecker HT, Page DB, Robins H, Song W, Stack EC, Wang E, Whiteside TL, Zhao Y, Zwierzina H, Butterfield LH, Fox BA. Novel technologies and emerging biomarkers for personalized cancer immunotherapy. J. Immunother. Cancer January 19, 2016;4:3. https://doi.org/10.1186/s40425-016-0107-3. eCollection 2016. Review. PubMed PMID: 26788324; PubMed Central PMCID: PMC4717548.

[30] Ward FJ, Dahal LN, Abu-Eid R. On the road to immunotherapy-prospects for treating head and neck cancers with checkpoint inhibitor antibodies. Front. Immunol. September 24, 2018;9:2182. https://doi.org/10.3389/fimmu.2018.02182. eCollection 2018. Review. PubMed PMID: 30319637; PubMed Central PMCID: PMC6165864.

[31] Fortis SP, Mahaira LG, Anastasopoulou EA, Voutsas IF, Perez SA, Baxevanis CN. Immune profiling of melanoma tumors reflecting aggressiveness in a preclinical model. Cancer Immunol. Immunother. December 2017;66(12):1631–42. https://doi.org/10.1007/s00262-017-2056-1. Epub 2017 Sep 4. PubMed PMID: 28871365.

[32] Koelzer VH, Sirinukunwattana K, Rittscher J, Mertz KD. Precision immunoprofiling by image analysis and artificial intelligence. Virchows Arch. April 2019;474(4):511–22. https://doi.org/10.1007/s00428-018-2485-z. Epub 2018 Nov 23. Review. PubMed PMID: 30470933.

Further reading

[1] Bethmann D, Feng Z, Fox BA. Immunoprofiling as a predictor of patient's response to cancer therapy-promises and challenges. Curr. Opin. Immunol. April 2017;45:60–72. https://doi.org/10.1016/j.coi.2017.01.005. Epub 2017 Feb 20. Review. PubMed PMID: 28222333.

[2] Frei AP, Bava FA, Zunder ER, Hsieh EW, Chen SY, Nolan GP, Gherardini PF. Highly multiplexed simultaneous detection of RNAs and proteins in single cells. Nat. Methods March 2016;13(3):269–75. https://doi.org/10.1038/nmeth.3742. Epub 2016 Jan 25. PubMed PMID: 26808670; PubMed Central PMCID: PMC4767631.

[3] Lee JK, Bangayan NJ, Chai T, Smith BA, Pariva TE, Yun S, Vashisht A, Zhang Q, Park JW, Corey E, Huang J, Graeber TG, Wohlschlegel J, Witte ON. Systemic surfaceome profiling identifies target antigens for immune-based therapy in subtypes of advanced prostate cancer. Proc. Natl. Acad. Sci. U.S.A. May 8, 2018;115(19):E4473–82. https://doi.org/10.1073/pnas.1802354115. Epub 2018 Apr 23. PubMed PMID: 29686080; PubMed Central PMCID: PMC5949005.

[4] Morin A, Kwan T, Ge B, Letourneau L, Ban M, Tandre K, Caron M, Sandling JK, Carlsson J, Bourque G, Laprise C, Montpetit A, Syvanen AC, Ronnblom L, Sawcer SJ, Lathrop MG, Pastinen T. Immunoseq: the identification of functionally relevant variants through targeted capture and sequencing of active regulatory regions in human immune cells. BMC Med. Genomics September 13, 2016;9(1):59. https://doi.org/10.1186/s12920-016-0220-7. PubMed PMID: 27624058; PubMed Central PMCID: PMC5022205.

[5] Mulder DT, Mahé ER, Dowar M, Hanna Y, Li T, Nguyen LT, Butler MO, Hirano N, Delabie J, Ohashi PS, Pugh TJ. CapTCR-seq: hybrid capture for T-cell receptor repertoire profiling. Blood Adv. December 11, 2018;2(23):3506–14. https://doi.org/10.1182/bloodadvances.2017014639. Erratum in: Blood Adv. 2019 Jan 22;3(2):121. PubMed PMID: 30530777; PubMed Central PMCID: PMC6290103.

[6] Peterson VM, Zhang KX, Kumar N, Wong J, Li L, Wilson DC, Moore R, McClanahan TK, Sadekova S, Klappenbach JA. Multiplexed quantification of proteins and transcripts in single cells. Nat. Biotechnol. October 2017;35(10):936–9. https://doi.org/10.1038/nbt.3973. Epub 2017 Aug 30. PubMed PMID: 28854175.

[7] Zemmour D, Zilionis R, Kiner E, Klein AM, Mathis D, Benoist C. Single-cell gene expression reveals a landscape of regulatory T cell phenotypes shaped by the TCR. Nat. Immunol. March 2018;19(3):291–301. https://doi.org/10.1038/s41590-018-0051-0. Epub 2018 Feb 12. Erratum in: Nat Immunol. 2018 May 3;:. PubMed PMID: 29434354; PubMed Central PMCID: PMC6069633.

Chapter 12

Organoids: a model for precision medicine

Noah S. Rozich, Alex B. Blair and Richard A. Burkhart
Department of Surgery, Johns Hopkins Hospital, Baltimore, MD, United States

Introduction

While two-dimensional (2D) cell lines have been widely used to study multiple diseases, they are subject to genetic drifts, are limited by 2D culture conditions, are difficult to quickly and reliably generate de novo from patient samples, and often lack cellular heterogeneity [1]. Xenograft models overcome some of these limitations, such as the monolayer environment of cell lines, by recruiting host tissue, including stromal components, immune cells, and vascular structures. Additionally, while certainly not one-to-one representations of humans, animal models can be manipulated to develop conditions that reliably re-create specific disease processes.

The analysis of such options can provide valuable insight into disease mechanism, and can assist in identifying drug targets, interactions, and potential harmful side effects [2]. However, like 2D cell lines, xenograft models are similarly plagued by many limitations, including the variable influence of the host system, and low concordance between test data and clinical utility [3,4]. In addition, the process of generating these models is both time and resource consuming, further limiting realistic everyday clinical applicability [5].

Organoid features

Ex vivo three dimensional (3D) culture systems of developing tissue, called organoids, have been developed in an attempt to overcome these limitations. Strictly defined, an organoid is a "collection of organ-specific cell types, that develops from stem cells or organ progenitors, and self-organizes through cell sorting and spatially restricted lineage commitment, in a manner similar to in vivo" [6]. Tissue-derived organoids retain the ability to self-organize into structures, similar to the tissue from which they are derived, and mimic some of the original function [7].

Unlike previous in vitro culture methods, such as 2D cell lines, organoid cultures are less subject to selection, and remain genetically stable throughout multiple passages [8,9]. Additionally, they resemble their organs of origin in both function and structure, and are composed of a mixture of cell types that differentiate from stem cells or progenitor cells in a manner similar to living tissue. Importantly, organoids retain the genetic signatures of the original tissue, and therefore provide a powerful tool for studying genetic processes [10].

Unlike many in vivo models, organoids can be expanded to create biobanks, cryopreserved, and tested with methods similar to those of traditional 2D culture. Furthermore, they can be generated from multiple types of tissue, and often require limited amounts of starting tissue to be established. Organoids also provide a population of high purity cells from a particular tissue lineage, with limited influence from other cell types such as immune cells, fibroblasts, etc., allowing for focused experiments that may not be possible with in vivo models.

The concept of organoids

The term "organoid" is literally defined as "from an organ,", and has origins that track as far back as 1907 [11]. However, organoids as they are understood and studied today have roots in studies from the 1970s by James Rheinwald and Howard Green, where human keratinocytes were cocultured with mouse 3T3 fibroblasts, to generate confluent sheets of epidermal skin cells [12]. Later, in Green's lab, these human epidermal "stem cells" were cultured, and used to grow sheets of epidermis to successfully treat burn patients [13].

While the term "stem cells" was not used to describe the cells cultured in these experiments, these studies laid the groundwork that led to the culture of human embryonic stem cells, and ultimately the generation of the first

Precision Medicine for Investigators, Practitioners and Providers. https://doi.org/10.1016/B978-0-12-819178-1.00012-5

organoids from both mice [14] and human [15] intestinal stem cells. Since that time, organoids have been generated from multiple organ systems, and the field has developed into a robust area of basic and translational research that relies upon an expanding array of technology.

Organoid requirements

First, multiple organ-specific cell types can be necessary to form an organoid. Second, organoids should, at least in part, mirror the original function of their derived organ, such as secretion, contraction, or neural conduction. Third, organoids should self-assemble into structures that resemble the organ from which it is derived, both in cellular composition and structural organization; thus, a 3D matrix is required for their growth.

Organoids can be derived from two different types of cells: pluripotent stem cells (PSCs), either embryonic stem cells or induced pluripotent stem cells, and organ-specific adult stem cells, often referred to as progenitor cells (PCs) [16]. However, there are examples of epithelial cells that lack stem cell markers, being capable of forming organ-resembling tissue in 3D culture [17–19]. These cells heavily rely on interactions with a 3D matrix, rich with extracellular matrix components that simulate a basement membrane. While these "organoids" aren't generated from PSCs or PCs, they are generally defined as such throughout the literature.

For organoids derived from PSCs, specific culture conditions are generally tailored to support the desired tissue type in vitro. In addition to standard nutrients, serum, and antibiotics, variables to support organoid establishment can include apoptotic inhibitors, and growth factors unique to the desired cell type. In contrast, organoids derived from PCs require somewhat less stringent media and culture conditions, similar to those present in their tissue of origin, where they normally reside and function to regenerate damaged tissue [20]. When these cells are harvested from normal or diseased tissue, and embedded in a 3D matrix with proper culture conditions, they congregate and grow into epithelial structures resembling their organotypic phenotype, are genetically stable, and are able to be expanded in culture through multiple passages [14,21–25].

Organoid models

Organoids have been developed from multiple organs systems, and generated from cells representing all three primordial germ layers. While examples exist of organoids generated from animal tissue, herein we will focus primarily on organoids generated from human tissue.

Gastrointestinal tract organoids

The human gastrointestinal (GI) tract encompasses everything, from the oropharynx to the anus, and develops primarily from endodermal tissue, differentiating into foregut, midgut, and hindgut structures [26]. Organoids derived from the human GI tract have been generated both from PSCs and PCs. Utilizing various signal transduction pathways and growth factors, including wingless/integrated pathway (Wnt), fibroblast growth factor (FGF), activin A, and prostaglandin E_2 (PGE_2), human PSCs can be driven to differentiate into the different tissue types of the human GI tract.

A laminin-rich 3D gel matrix, derived from the Engelbreth-Holm-Swarm tumor cell line, called Matrigel, has most frequently been described as the scaffold in which GI organoids are grown [6]. Matrigel is unique in that it is liquid at 0°C, but solid at 37°C, allowing for easy dissolution and passage of the growing cells within. Using Matrigel to re-create the extracellular matrix and the signaling principles described, the lab of Hans Clever developed intestinal organoids from both midgut and hindgut origin [14,27]. Organoids have since been generated using PSCs and PCs from multiple different organs that make up the GI tract, including the tongue, esophagus, and pyloric and chief cells from the stomach [28].

Abdominal and retroperitoneal organs

Using PCs generated from adult mouse bile duct tissue, researchers embedded these cells in Matrigel, and added media supplemented with epithelial grow factor (EGF), hepatocyte growth factor (HGF), R-spondin, FGF, and nicotinamide, generating cystic organoids expressing biliary ductal markers [29]. While human liver organoids have not been similarly generated from PCs, using induced human PSCs, researchers have successfully grown vascularized liver-buds in vitro in Matrigel [30]. In addition, human cholangiocytes have been generated from PSCs that express mature biliary markers, and retain epithelial function similar to mature cholangiocytes [31].

Normal human pancreatic organoids have been established, using digested ductal tissue plated in Matrigel with specific culture conditions, including EGF, R-spondin, Noggin, Wnt, FGF, nicotinamide, and PGE_2 [32]. Organoids have even been generated from such complex organ systems as the kidney, where human PSCs originated nephrons and a collecting duct network, surrounded by renal interstitial and endothelial cells [33].

Thoracic and neural organoids

Human PSCs, both embryonic and induced, have been used to generate human lung organoids. By manipulating culture conditions with factors including vascular endothelial growth factor (VEGF), sonic hedgehog molecule (SHH), and FGF, spherical lung organoids were created resembling proximal airway tissue, and expressing cell markers for basal cells, ciliated cells, and club cells [34]. Again, using human PSCs, multiple studies have shown the ability to generate brain organoids that organize into distinct cortical layers containing glial cells and functional neurons [35,36].

In fact, Lancaster et al. created cerebral organoids without utilizing growth factors, but instead embedded neuroectodermal tissue in Matrigel, and generated a heterogeneous population of "mini-brains" that resembles human fetal brain tissue [37].

Organoids and disease

Cystic fibrosis

Cystic Fibrosis (CF) is a rare but life shortening genetic disorder that results from various mutations in the transmembrane conductance regulator gene, *CFTR,* which regulates the absorption and secretion of salt and water in different organ systems [38]. Historically, treatments were primarily aimed at symptom relief, including pulmonary therapy, mucolytics, antibiotics, expectorants, and bronchodilators. However, the development of drugs targeting CFTR, has led to renewed interest in drug testing and development. Of note, treatment response heterogeneity, even for patients with the same *CFTR* mutations, underscores the importance of developing a precision approach for CF treatment.

A study by Dekkers et al. used human intestinal organoids, to create a model of CF that responded to CTFR-targeting drugs, with quantified response heterogeneity based on patient-specific *CTFR* mutations [39]. Efforts are currently underway, to create biobanks of colorectal organoids derived from CF patients, and use them to direct treatment with patient-specific pharmacotyping [40]. Furthermore, as CF affects multiple organ systems, there is potential to use organoids to elucidate organ-specific pathophysiology, and aid in the development of drugs that act on different target organs.

Inflammatory bowel disease

Inflammatory bowel disease (IBD) is a chronic inflammatory condition affecting the human GI tract in the form of ulcerative colitis, Crohn disease, or indeterminate colitis. The pathophysiology of this spectrum of disease is poorly understood, but is believed to involve interactions between host immune cells and microbiome with GI epithelial cells, via direct contact and secreted bacterial metabolites [41]. Intestinal organoids have been used to study association between bacteria associated with IBD, and the effect of their metabolites, such as butyrate, to induced changes in gene expression for transcription factors, that affect cellular metabolism and growth [42].

Utilizing organoids grown in varying culture conditions, researchers were able to show that exposure to butyrate, inhibited growth of intestinal stem cell organoids, but not of differentiated intestinal organoids, leading the authors to hypothesize that normal colonic architecture, protects intestinal PCs from exposure to damaging bacterial metabolites [43]. Ingestion of certain bacterial metabolites by human enterocytes has been shown to lead to hypoxic conditions, which can in turn induce expression of inflammatory cytokines, and protective factors known to be elevated in IBD [41]. Using human organoids grown in hypoxic conditions, researchers have demonstrated increased expression of inflammatory cytokines, such as TNFα, and further shown that treatment with pharmaceutical inhibitors has decreased expression of these inflammatory cytokines [44].

Several bacteria have been loosely implicated in the development of IBD [45]; however, little is understood about the mechanism driving these relationships. Recently, researchers have successfully injected intestinal organoids with *Bacteroides thetaiotaomicron* and *Clostridium difficile* [46,47], demonstrating the ability to coculture anaerobic bacteria in organoid culture, simulating infection.

Infectious disease

Organoids are particularly well suited as a model for the study of infectious disease, given that all cell types from an organ system can be generated in vitro, providing insight into the interactions between pathogens, and specific cells types that are required for host infection. For example, *C. difficile* infection is quite common among hospitalized patients, the complications of which range from intractable diarrhea and pain to septic shock, toxic megacolon, and death. After infecting human intestinal organoids with *C. difficile*, researchers found that the secreted bacterial toxin inhibited the barrier function of intestinal cells, and diminished expression of a Na^+/H^+ exchanger (NHE3), both contributing to the pathogenicity of *C. difficile* [46].

Examples exist that leverage organoids to model infection with *Helicobacter pylori, Cryptosporidium parvum, Salmonella enterica, Toxoplasma gondii,* Rotavirus, Norovirus, influenza, and Zika virus [6,41]. Qian et al. used induced human PSCs to generate forebrain-specific organoids that recapitulated key features of cerebral development [48]. These brain organoids were later infected with strains of Zika virus, where increased neuronal cell death and decreased neuronal layer thickness were noted, resembling microcephaly.

Organoids and cancer

Heterogeneity in genetic profile, epigenetic regulation, cellular metabolism, invasiveness, and tumor phenotypes are important challenges in oncology, which largely contribute to variable treatment response [49]. An improved understanding of tumor biology for different types of cancers requires the use of preclinical models, and organoids provide an elegant solution for many of the notable limitations of two-dimensional cancer cell lines and animal xenograft models. In addition, they provide a platform for therapeutic drug testing and development.

Modeling cancer

Cancer occurs when normal tissues acquire genetic and cellular changes that allow for uncontrolled proliferation, and loss of differentiation [50]. Normal human-derived organoids can be genetically manipulated to express tumor-related genes, such as *KRAS*, *TP53*, *APC*, *SMAD4*, and *CDKN2A*, in order to study their effects on "normal tissue,", and malignant progression in vitro and in vivo. Huang et al. generated normal pancreatic tissue organoids from PSCs, and induced mutations in *KRAS* or *TP53*, two frequent mutations in pancreatic cancer, via a lentivirus vector system [51]. Transplanting these mutated organoids into immunodeficient mice resulted in abnormal ductal architecture and changes in nuclear morphology, consistent with neoplastic transformation [40].

Other studies have used genome editing technology to mutate genes in colonic organoids, including *KRAS*, *APC*, *TP53*, and *SMAD4*. Drost et al. found that loss of *APC* and *P53* was sufficient to generate aneuploidy in intestinal organoids, signifying tumor progression [52]. Furthermore, Matano et al. added a *PIK3CA* mutation to normal colon organoids and found that, while these mutated organoids would indeed form tumors when transplanted into the capsule of mice kidneys, these driver mutations alone were not sufficient to cause metastasis, suggesting that additional mutations are required for more invasive tumor behavior [53].

Discovering biomarkers

Discovering tumor biomarkers entails analyzing genetic, molecular, and histological data gathered from patient tumors, which is often limited by the cellularity of the actual patient tumor specimen. Utilizing similar techniques for generating organoids from normal tissue, organoids can be established from malignant tissue, both from biopsies as well as from operatively resected tissue specimens, and studies suggest that these organoids accurately resemble their tumors of origin with regards to key histological, molecular, and genetic characteristics [8,32]. After several passages, the selective utilization of growth factors in media and incubation can preferentially select for various components of the tumor microenvironment (TME).

For example, removing tumor-associated fibroblasts, immune cells, and other stromal components, may allow a population of "pure" malignant cells to thrive. These cultures genetically resemble the original tumor, and may preserve the clonal heterogeneity present within most tumors [54]. These organoids can be used to create biobanks of tumor tissue, useful for genetic profiling and drug sensitivity testing. Van der Wetering et al. created a biobank of 20 consecutive colorectal cancer organoid lines derived from patients, and, after sequencing these samples, the authors were able to correlate genetic mutations with drug sensitivities in a multivariable model [55]. An organoid line devoid of *APC* or *CTNNB1* mutations that carried a mutation in the Wnt pathway regulator *RNF42* was extremely sensitive to a small-molecule inhibitor of Wnt secretion, while other lines were not [55].

Huang et al. tested an epigenetic drug that inhibits histone methyltransferase *EZH2*, on pancreatic cancer patient–derived organoids. They found that only a subset of organoids, carrying a particular epigenetic marker (H3K27me3), was significantly responsive to the drug, with potential clinical applications [51]. Also a biobank of gastric cancer organoids was created, sequenced, and used to identify genetic and molecular subtypes, to be used for therapeutic drug screening [56].

Drug sensitivity and pharmacotyping

Walsh et al. [57] made use of optical metabolic imaging (OMI), a tool used to assess changes in metabolism in response to drug therapy, as a way to predict drug sensitivity. Pancreatic cancer organoids were derived and exposed to therapeutics. While the authors did not correlate the OMI data with actual patient responses, this platform provides an attractive method for using patient-derived organoids to determine drug sensitivities for individual treatment planning. Similar studies have also been conducted in breast cancer [58] or with other technologies such as *DeathPro,* in ovarian cancer. *DeathPro* is a microscopy-based assay to assess drug-induced growth arrest and cell death [59].

Organoids are convenient as a high through-put method for precision drug pharmacotyping [60]. Organoids can be generated from individual patients, cultured to an appropriate biomass, and then dissociated into single cells and plated onto a multiwell plate. Therapies of varying doses can then be introduced into this single-cell plate suspension, and survival assays can be conducted, to generate specific drug sensitivity profiles for each organoid line. Preliminary work in endometrial [61], colorectal [62], and pancreatic cancer [60] shows great potential as a drug screening tool.

Drug development could be benefitted, testing the efficacy of new drugs on human cancer tissue in vitro. The organoids developed from patient tumor tissue provide a highly cellular repository that recapitulates key features of the original tumors. They are easily expanded, stable over many passages, and readily manipulated. OMI, *DeathPro,* and other evolving scientific techniques offer the opportunity to test cancer organoid drug sensitivity, like a bacterial culture in a clinical laboratory.

Immune therapy

Through passaging, organoids lack many stromal components of the tumor microenvironment (TME), including tumor-infiltrating lymphocytes (TIL) through which immune-based therapies function. However, techniques are being developed to overcome these limitations [63,64]. In a recent study by Dijkstra et al., the authors generated patient-derived organoids from patients with mismatch repair deficient (MMRd) colorectal cancer (CRC), and nonsmall cell lung cancer (NSCLC), and cocultured these with autologous immune cells isolated from peripheral blood mononuclear cells (PBMCs) [63]. By staining for interferon gamma (IFNγ) and degranulation marker CD107a, the authors not only demonstrated that $CD8^+$ T cells recognized tumor tissue, but also confirmed a lack of reactivity to normal tissue, and decreased survival of tumor organoids [63].

In another breakthrough study, Neal et al. used an air-liquid interface culture method that preserves aspects of the complex architecture of the TME, including stromal components and TILs, to generate patient-derived tumor organoids. The authors present data suggesting preserved endogenous tumor immune cells (T cells, B cells, macrophages, natural killer (NK) cells, activated TILs), and recapitulated the programmed cell death protein 1 (PD1) immune checkpoint, in a manner that was responsive to blockade [64].

Limitations

One major limitation of organoids is the inability to accurately recapitulate the native TME with current methods [28]. The TME is an integral part of tumor phenotype, function, and physiology [65]. In fact, the intricate relationship between malignant cells, cancer associated fibroblasts, infiltrating immune cells, stromal cells, and extracellular matrix proteins, plays an important role in tumor survival, progression, and protection from cytotoxic agents.

Efforts are underway using coculture, and air-liquid interface techniques, to simulate and retain aspects of the TME [62–64]. However, organoids lack innervation and vascularization that likely limit their growth potential, and the ability to recapitulate the full spectrum of disease. The transplantation of organoids into living hosts, such as a murine model, is another area of active study [66]. Finally, while organoids do not undergo clonal selection to the extent of 2D cancer cell lines, the ability to maintain the tumor heterogeneity of the original tumor remains an area of study.

Future directions

Currently, the organoid landscape is restricted primarily to the laboratory, and has not been successfully integrated into clinical decision making. However, as previously discussed, preliminary work using organoids for drug screening and pharmacotyping has been promising, and the shift into clinical practice is approaching. A recently opened multiinstitutional trial for pancreatic cancer, supported by the Stand Up To Cancer initiative (SU2C, USA), will grow patient-derived tumor organoids from pretreatment tumor biopsies, with borderline or locally advanced pancreatic cancer.

A FOLFIRINOX (5-fluorouracil, leucovorin, irinotecan, and oxaliplatin)-based neoadjuvant treatment regimen will be given to each patient and, in parallel, the organoids will be tested with the same regimen, in order to determine whether they accurately predict treatment response. A promising next step, following proof of concept and feasibility, would be to modify chemotherapy regimens based on organoid response, to optimize therapeutic benefit on an individual patient basis.

References

[1] Neal JT, Kuo CJ. Organoids as models for neoplastic transformation. Annu. Rev. Pathol. 2016;11:199–220.

[2] Bredenoord AL, Clevers H, Knoblich JA. Human tissues in a dish: the research and ethical implications of organoid technology. Science 2017;355(6322):eaaf9414.

[3] Mak IW, Evaniew N, Ghert M. Lost in translation: animal models and clinical trials in cancer treatment. Am. J. Tourism Res. 2014;6(2):114.

[4] Franco R, Cedazo-Minguez A. Successful therapies for Alzheimer's disease: why so many in animal models and none in humans? Front. Pharmacol. 2014;5:146.

[5] Hidalgo M, Amant F, Biankin AV, Budinská E, Byrne AT, Caldas C, et al. Patient-derived xenograft models: an emerging platform for translational cancer research. Cancer Discov 2014;4(9):998–1013.

[6] Lancaster MA, Knoblich JA. Organogenesis in a dish: modeling development and disease using organoid technologies. Science 2014;345(6194):1247125.

[7] Fang Y, Eglen RM. Three-dimensional cell cultures in drug discovery and development. Slas Discov 2017;22(5):456–72.

[8] Huch M, Gehart H, Van Boxtel R, Hamer K, Blokzijl F, Verstegen MM, et al. Long-term culture of genome-stable bipotent stem cells from adult human liver. Cell 2015;160(1):299–312.

[9] Sato T, Stange DE, Ferrante M, Vries RG, Van Es JH, Van Den Brink S, et al. Long-term expansion of epithelial organoids from human colon, adenoma, adenocarcinoma, and Barrett's epithelium. Gastroenterology 2011;141(5):1762–72.

[10] Clevers H. Modeling development and disease with organoids. Cell 2016;165(7):1586–97.

[11] Wilson HV. A new method by which sponges may be artificially reared. Science 1907;25(649):912–5.

[12] Rheinwatd JG, Green H. Seria cultivation of strains of human epidemal keratinocytes: the formation keratinizin colonies from single cell is. Cell 1975;6(3):331–43.

[13] O'Connor N, Mulliken J, Banks-Schlegel S, Kehinde O, Green H. Grafting of burns with cultured epithelium prepared from autologous epidermal cells. Lancet 1981;317(8211):75–8.

[14] Sato T, Vries RG, Snippert HJ, Van De Wetering M, Barker N, Stange DE, et al. Single Lgr5 stem cells build crypt−villus structures in vitro without a mesenchymal niche. Nature 2009;459(7244):262.
[15] Spence JR, Mayhew CN, Rankin SA, Kuhar MF, Vallance JE, Tolle K, et al. Directed differentiation of human pluripotent stem cells into intestinal tissue in vitro. Nature 2011;470(7332):105.
[16] Dutta D, Heo I, Clevers H. Disease modeling in stem cell-derived 3D organoid systems. Trends Mol. Med. 2017;23(5):393−410.
[17] Weaver VM, Bissell MJ. Functional culture models to study mechanisms governing apoptosis in normal and malignant mammary epithelial cells. J. Mammary Gland Biol. Neoplasia 1999;4(2):193−201.
[18] Debnath J, Brugge JS. Modelling glandular epithelial cancers in three-dimensional cultures. Nat. Rev. Cancer 2005;5(9):675.
[19] Montesano R, Schaller G, Orci L. Induction of epithelial tubular morphogenesis in vitro by fibroblast-derived soluble factors. Cell 1991;66(4):697−711.
[20] Bartfeld S, Clevers H. Stem cell-derived organoids and their application for medical research and patient treatment. J. Mol. Med. 2017;95(7):729−38.
[21] Sato T, Stange DE, Ferrante M, Vries RG, Van Es JH, Van Den Brink S, et al. Long-term expansion of epithelial organoids from human colon, adenoma, adenocarcinoma, and Barrett's epithelium. Gastroenterology 2011;141(5):1762−72.
[22] Jung P, Sato T, Merlos-Suárez A, Barriga FM, Iglesias M, Rossell D, et al. Isolation and in vitro expansion of human colonic stem cells. Nat. Med. 2011;17(10):1225.
[23] Huch M, Dorrell C, Boj SF, Van Es JH, Li VS, Van De Wetering M, et al. In vitro expansion of single Lgr5 liver stem cells induced by Wnt-driven regeneration. Nature 2013;494(7436):247.
[24] Huch M, Gehart H, Van Boxtel R, Hamer K, Blokzijl F, Verstegen MM, et al. Long-term culture of genome-stable bipotent stem cells from adult human liver. Cell 2015;160(1):299−312.
[25] Barker N, Huch M, Kujala P, van de Wetering M, Snippert HJ, van Es JH, et al. Lgr5 ve stem cells drive self-renewal in the stomach and build long-lived gastric units in vitro. Cell Stem Cell 2010;6(1):25−36.
[26] Heath JK. Transcriptional networks and signaling pathways that govern vertebrate intestinal development, vol. 90; 2010. p. 159−92.
[27] Spence JR, Mayhew CN, Rankin SA, Kuhar MF, Vallance JE, Tolle K, et al. Directed differentiation of human pluripotent stem cells into intestinal tissue in vitro. Nature 2011;470(7332):105.
[28] Fatehullah A, Tan SH, Barker N. Organoids as an in vitro model of human development and disease. Nat. Cell Biol. 2016;18(3):246.
[29] Huch M, Dorrell C, Boj SF, Van Es JH, Li VS, Van De Wetering M, et al. In vitro expansion of single Lgr5 liver stem cells induced by Wnt-driven regeneration. Nature 2013;494(7436):247.
[30] Takebe T, Zhang R, Koike H, Kimura M, Yoshizawa E, Enomura M, et al. Generation of a vascularized and functional human liver from an iPSC-derived organ bud transplant. Nat. Protoc. 2014;9(2):396.
[31] Ogawa M, Ogawa S, Bear CE, Ahmadi S, Chin S, Li B, et al. Directed differentiation of cholangiocytes from human pluripotent stem cells. Nat. Biotechnol. 2015;33(8):853.
[32] Boj SF, Hwang C, Baker LA, Chio IIC, Engle DD, Corbo V, et al. Organoid models of human and mouse ductal pancreatic cancer. Cell 2015;160(1):324−38.
[33] Takasato M, Pei XE, Chiu HS, Maier B, Baillie GJ, Ferguson C, et al. Kidney organoids from human iPS cells contain multiple lineages and model human nephrogenesis. Nature 2015;526(7574):564.
[34] Dye BR, Hill DR, Ferguson MA, Tsai Y, Nagy MS, Dyal R, et al. In vitro generation of human pluripotent stem cell derived lung organoids. Elife 2015;4:e05098.
[35] Lancaster MA, Renner M, Martin C, Wenzel D, Bicknell LS, Hurles ME, et al. Cerebral organoids model human brain development and microcephaly. Nature 2013;501(7467):373.
[36] Mariani J, Simonini MV, Palejev D, Tomasini L, Coppola G, Szekely AM, et al. Modeling human cortical development in vitro using induced pluripotent stem cells. Proc. Natl. Acad. Sci. U.S.A 2012;109(31):12770−5.
[37] Camp JG, Badsha F, Florio M, Kanton S, Gerber T, Wilsch-Bräuninger M, et al. Human cerebral organoids recapitulate gene expression programs of fetal neocortex development. Proc. Natl. Acad. Sci. U.S.A 2015;112(51):15672−7.
[38] Noordhoek J, Gulmans V, van der Ent K, Beekman JM. Intestinal organoids and personalized medicine in cystic fibrosis: a successful patient-oriented research collaboration. Curr. Opin. Pulm. Med. 2016;22(6):610−6.
[39] Dekkers JF, Wiegerinck CL, De Jonge HR, Bronsveld I, Janssens HM, De Winter-de Groot, Karin M, et al. A functional CFTR assay using primary cystic fibrosis intestinal organoids. Nat. Med. 2013;19(7):939.
[40] Dekkers JF, Berkers G, Kruisselbrink E, Vonk A, De Jonge HR, Janssens HM, et al. Characterizing responses to CFTR-modulating drugs using rectal organoids derived from subjects with cystic fibrosis. Sci. Transl. Med. 2016;8(344):344ra84.
[41] Bartfeld S. Modeling infectious diseases and host-microbe interactions in gastrointestinal organoids. Dev. Biol. 2016;420(2):262−70.
[42] Lukovac S, Belzer C, Pellis L, Keijser BJ, de Vos WM, Montijn RC, et al. Differential modulation by *Akkermansia muciniphila* and *Faecalibacterium prausnitzii* of host peripheral lipid metabolism and histone acetylation in mouse gut organoids. mBio 2014;5(4):1438.
[43] Kaiko GE, Ryu SH, Koues OI, Collins PL, Solnica-Krezel L, Pearce EJ, et al. The colonic crypt protects stem cells from microbiota-derived metabolites. Cell 2016;165(7):1708−20.
[44] Xue X, Ramakrishnan S, Anderson E, Taylor M, Zimmermann EM, Spence JR, et al. Endothelial PAS domain protein 1 activates the inflammatory response in the intestinal epithelium to promote colitis in mice. Gastroenterology 2013;145(4):831−41.
[45] Hanauer SB. Inflammatory bowel disease: epidemiology, pathogenesis, and therapeutic opportunities. Inflamm. Bowel Dis. 2006;12(Suppl. 1):S9.
[46] Leslie JL, Huang S, Opp JS, Nagy MS, Kobayashi M, Young VB, et al. Persistence and toxin production by *Clostridium difficile* within human intestinal organoids result in disruption of epithelial paracellular barrier function. Infect. Immun. 2015;83(1):138−45.
[47] Engevik MA, Aihara E, Montrose MH, Shull GE, Hassett DJ, Worrell RT. Loss of NHE3 alters gut microbiota composition and influences *Bacteroides thetaiotaomicron* growth. Am. J. Physiol. Gastrointest. Liver Physiol. 2013;305(10):G711.
[48] Qian X, Nguyen HN, Song MM, Hadiono C, Ogden SC, Hammack C, et al. Brain-region-specific organoids using mini-bioreactors for modeling ZIKV exposure. Cell 2016;165(5):1238−54.

[49] McGranahan N, Swanton C. Biological and therapeutic impact of intratumor heterogeneity in cancer evolution. Cancer Cell 2015;27(1):15–26.

[50] Hanahan D, Weinberg RA. Hallmarks of cancer: the next generation. Cell 2011;144(5):646–74.

[51] Huang L, Holtzinger A, Jagan I, BeGora M, Lohse I, Ngai N, et al. Ductal pancreatic cancer modeling and drug screening using human pluripotent stem cell–and patient-derived tumor organoids. Nat. Med. 2015;21(11):1364.

[52] Drost J, Van Jaarsveld RH, Ponsioen B, Zimberlin C, Van Boxtel R, Buijs A, et al. Sequential cancer mutations in cultured human intestinal stem cells. Nature 2015;521(7550):43.

[53] Matano M, Date S, Shimokawa M, Takano A, Fujii M, Ohta Y, et al. Modeling colorectal cancer using CRISPR-Cas9–mediated engineering of human intestinal organoids. Nat. Med. 2015;21(3):256.

[54] Sachs N, Clevers H. Organoid cultures for the analysis of cancer phenotypes. Curr. Opin. Genet. Dev. 2014;24:68–73.

[55] van de Wetering M, Francies HE, Francis JM, Bounova G, Iorio F, Pronk A, et al. Prospective derivation of a living organoid biobank of colorectal cancer patients. Cell 2015;161(4):933–45.

[56] Yan HH, Siu HC, Law S, Ho SL, Yue SS, Tsui WY, et al. A comprehensive human gastric cancer organoid biobank captures tumor subtype heterogeneity and enables therapeutic screening. Cell Stem Cell 2018;23(6):897. e11.

[57] Walsh AJ, Castellanos JA, Nagathihalli NS, Merchant NB, Skala MC. Optical imaging of drug-induced metabolism changes in murine and human pancreatic cancer organoids reveals heterogeneous drug response. Pancreas 2016;45(6):863.

[58] Walsh AJ, Cook RS, Sanders ME, Aurisicchio L, Ciliberto G, Arteaga CL, et al. Quantitative optical imaging of primary tumor organoid metabolism predicts drug response in breast cancer. Cancer Res. 2014, Sep 15;74(18):5184–94.

[59] Jabs J, Zickgraf FM, Park J, Wagner S, Jiang X, Jechow K, et al. Screening drug effects in patient-derived cancer cells links organoid responses to genome alterations. Mol. Syst. Biol. 2017;13(11):955.

[60] Aberle MR, Burkhart RA, Tiriac H, Olde Damink S, Dejong C, Tuveson DA, et al. Patient-derived organoid models help define personalized management of gastrointestinal cancer. Br. J. Surg. 2018;105(2):e60.

[61] Girda E, Huang EC, Leiserowitz GS, Smith LH. The use of endometrial cancer patient–derived organoid culture for drug sensitivity testing is feasible. Int. J. Gynecol. Cancer 2017;27(8):1701.

[62] Boehnke K, Iversen PW, Schumacher D, Lallena MJ, Haro R, Amat J, et al. Assay establishment and validation of a high-throughput screening platform for three-dimensional patient-derived colon cancer organoid cultures. J. Biomol. Screen 2016;21(9):931–41.

[63] Dijkstra KK, Cattaneo CM, Weeber F, Chalabi M, van de Haar J, Fanchi LF, et al. Generation of tumor-reactive T cells by co-culture of peripheral blood lymphocytes and tumor organoids. Cell 2018;174(6):1598. e12.

[64] Neal JT, Li X, Zhu J, Giangarra V, Grzeskowiak CL, Ju J, et al. Organoid modeling of the tumor immune microenvironment. Cell 2018;175(7):1988. e16.

[65] Mbeunkui F, Johann DJ. Cancer and the tumor microenvironment: a review of an essential relationship. Cancer Chemother. Pharmacol. 2009;63(4):571–82.

[66] Taguchi A, Kaku Y, Ohmori T, Sharmin S, Ogawa M, Sasaki H, et al. Redefining the in vivo origin of metanephric nephron progenitors enables generation of complex kidney structures from pluripotent stem cells. Cell Stem Cell 2014;14(1):53–67.

Section II

Precision medicine for practitioners

A—Genetics and genomics

Chapter 13

Modern applications of neurogenetics

Dolores Gonzalez Moron and Marcelo A. Kauffman

Neurogenetics Clinic, Hospital JM Ramos Mejia, Buenos Aires, Argentina

We have seen an exponential growth in the knowledge of the genetic basis of diseases during the past 50 years [1]. In about a third of monogenetic disorders the nervous system is compromised. Every tertiary hospital should consider incorporating neurogenetic clinics, in which multidisciplinary teams could offer a thorough assistance for patients affected by neurogenetic conditions [2].

The practice of neurogenetics

DNA sequencing technologies are advancing and therapeutics for some conditions are emerging [3]. A large number of conditions can be tested for, at reduced cost and with improved diagnostic accuracy [4]. Single genes, panels of genes, whole exomes, or even whole genomes can be addressed whenever necessary [5,6].

Indications

A positive family history or the onset of disorders at early age is important; however, they are not the only clues for neurogenetic conditions [7]. Almost 50% of neurogenetic conditions starting in infancy are caused by so called de novo mutations, presenting as sporadic cases [8]. On the other hand, not less than a third of neurogenetic disorders start at adult age [9]. Neurogenetic disorders should always be considered, especially when the observed phenotype is complex, or belongs to the so-called "neurogenetic niches," such as cerebral palsy, malformations of cortical development, intellectual disability, epilepsy, ataxia, movement disorders, early onset dementia, and neuromuscular disorders.

The neurogenetic niches

Cerebral palsy is an umbrella term usually applied for any patient exhibiting a congenital neurological deficit, with a very slow progressive course or not progressive at all. An acquired or traumatic etiology is frequently assumed, even in cases lacking clear history. De novo point mutations in KCNC3, ITPR1, and SPTBN2 genes were unearthed using next-generation sequencing assays, in a large cohort of individuals with ataxic cerebral palsy [10]. Exome sequencing also identified de novo mutations in different genes such as TUBA1A, SCN8A, and KDM5C [11]. In a cohort of patients suffering from hemiplegic cerebral palsy, Zarrei et al. detected de novo CNVs and/or sex chromosome abnormalities in 7.2% of probands, impacting important developmental genes such as GRIK2, LAMA1, DMD, PTPRM, and DIP2C [12].

Intellectual deficit

Historically, a specific diagnosis has been achieved in only a small minority of children suffering from intellectual disability; however, results are improving [13]. In almost half of the patients [14], an etiology can be found. The vast majority are de novo truncating mutations in any of the hundreds of genes recently implied in intellectual disabilities. Each of these genes accounts for less than 1% of cases, highlighting the extreme genetic heterogeneity of this population [15]. Health economic studies suggest that testing is most cost-effective when performed early in the patient's diagnostic odyssey [16]. Nevertheless, identifying the cause of these disorders is of paramount importance for genetic counseling and therapy, whenever it is available.

Epilepsy

Several genes are causes of monogenic epilepsies, and hundreds are risk factors for complex genetic epilepsies. Genetic testing plays a pivotal role [17]. In certain conditions, the likelihood of finding a genetic etiology may be higher, particularly in epileptic encephalopathies, in which analogously to intellectual disability, the main findings are de novo truncating mutations within extreme genetic heterogeneity architecture [18]. Furthermore, accurate genetic diagnosis may define specific treatments, such as in case of

Precision Medicine for Investigators, Practitioners and Providers. https://doi.org/10.1016/B978-0-12-819178-1.00013-7

Dravet Syndrome [19] and KCNT1-related epilepsy [20], among others.

The diagnostic yield in this field has been quite variable, ranging from about 10% in older reports to about 50% in recent ones [21]. In children with newly diagnosed epilepsy, an overall yield of 40.4% was accomplished with various genetic tests [22]. Exome sequencing and multigene panel testing are the most cost-effective procedures for epilepsy [23].

Movement disorders

Molecular genetics has allowed better classifications and definitions of different clinical syndromes [24], with translation into clinical practice [25]. Accurate genetic diagnoses in movement disorders often pave the way for specific and disease modifying treatments [26]. However, experienced professionals need to characterize the complex phenotypes shown by such patients, in order to improve diagnosis and clinical care [27].

Dementia

About 20% of the population aged 55 years and older has a family history of dementia [28]. For most, this is due to a genetically complex disease, where many variations of small effects interact to increase risk of dementia [29]. Early onset dementias are classified as rare diseases (less than 1%) [30]. Good assessment of the cognitive phenotype is of paramount importance, for obtaining better diagnostic yields and rational use of genetic tests. Phenotype-guided ordering of single gene, or small multigene panels, can result in high diagnostic yields [31].

Early onset Alzheimer disease is often caused by mutations in APP, PSEN1, and PSEN2 [32]. Frontotemporal dementia might be caused by an abnormal number of repeats of a pentanucleotide in C9orf72 [33] or point mutations in PGRN, MAPT, VCP, among other genes less frequently compromised [34]. In addition, more than 30 monogenic disorders present with or include dementia as a clinical symptom [35]. More comprehensive approaches are recommended, after excluding the more prevalent phenotypes [36].

Neuromuscular abnormalities

Hundreds of individual disorders can affect muscle, nerve, motor neuron, or neuromuscular junction. Their onset is any time, from in utero until old age. They are most often genetic, thus markedly benefiting from sequencing technologies [37]. Any type of mutation in human DNA can cause genetic neuromuscular disorders. Exome and multigene panel sequencing are recommended early in the evaluation of neuromuscular disorders, in many cases clarifying diagnosis and minimizing invasive investigation [38]. Diagnostic yields of 26%−65% emphasize the importance of shortening the diagnostic odyssey, minimizing unnecessary testing, and providing opportunities for clinical and investigational therapies in these heterogeneous group of patients [39,40].

Genetic counseling

Diagnosis should be as specific as possible [41]. It is not finalized until the causing molecular defect is individualized. Recurrence risk assessment also becomes an important part of genetic counseling [42]. Inheritance pattern associated with the genetic defect, penetrance, and age of onset, as well as de novo mutations inferred from the site and time of development of the anomaly, are all relevant. Prognosis, natural history, and referral to disease-specific support groups should not be neglected [43].

Neurogenetics on a personalized research-based clinic

We have implemented a clinic and a laboratory specialized in neurogenetics, which make use of their own resources, within a framework of research [2]. We demonstrated the clinical utility of exome sequencing in our patient cohort, obtaining a diagnostic yield of 40% among a diverse group of neurological disorders [44]. Furthermore, we were able to expand the phenotypic spectrum of known genes, and identify new pathogenic variants in several genes [45]. Preliminary cost-analysis lends support to the assertion that exome sequencing is more cost-effective than other molecular diagnostic approaches based on single- or panel-gene analysis. Our results were comparable with previous experiences reported by others [5], and highlight the advantages of working as a personalized research group, where phenotypic and genotypic information can be thoughtfully assessed, in contrast to commercial diagnostic laboratories that only have access to focused, heterogeneous, and often less informative clinical phenotypic reports, filled by the external ordering physician. This interdisciplinary work proved useful for reducing the long diagnostic delays, impacting medical management, and optimizing genetic counseling for these families.

We have specially addressed molecular diagnosis of ataxias [46,47] and malformations of cortical development [48]. Ataxias have a worldwide prevalence of about 3−5 cases per 100,000 [49]. More than 100 conditions can be classified as ataxic disorder. Pursuing diagnosis can require the use of different molecular genetic techniques. A great number of dominant ataxias and the most frequent recessive ataxia, Friedreich's ataxia, are caused by abnormally repetitive sequences of trinucleotides [50]. Thus, molecular diagnosis requires assays able to quantify the number of these repeats.

On the other hand, the rest of dominant and recessive ataxias are caused by point mutations or short indels. Its diagnosis is amenable to sequencing based-assays, such as multigene panel sequencing or exome sequencing. Applying both types of assays, we were able to identify the causing molecular defect in about a third of our cohort. The most common ataxias in our population were SCA type 2, SCA type 3 and Friedreich's ataxia. Exome sequencing led us to identify the genetic cause, in about a half of those cases negative for abnormal repeat trinucleotides. This figure is comparable to others [51].

The malformations of cortical development constitute another neurogenetic niche that specially captivated our attention in the past years. Human brain cortex development is a complex and highly regulated process that involves neural proliferation, differentiation, migration, and postmigrational development [52] (Table 13.1). Disruption in any of these steps can result in structural brain anomalies called Malformations of Cortical Development (MCDs), which are an important cause of epilepsy and neurodevelopment delay [53]. There are more than 30 types of MCDs classified in three major groups depending on the primary developmental step interrupted [54,55]. Although a few MCDs can be caused by environmental or acquired factors (e.g., CMV infection), most probably, the majority of the MCDs have a genetic origin. Historically linkage analysis, positional cloning, and gene sequencing have allowed to identify some genetic causes of MCDs; however, the genetic background of a very high proportion of MCDs remained elusive. The advent of high-throughput next-generation sequencing changed immensely our knowledge of the molecular basis of MCDs. The new technologies allowed to identify new candidate genes, to expand the phenotypic spectra of known genes, and to achieve a better understanding of the molecular pathways in brain development and pathologic processes underlying MCDs [56,57].

This notion can be illustrated with many types of MCDs, for example, the Lissencephaly (LIS) spectrum. LIS comprises a spectrum of malformations caused by a defect in neuronal migration that includes agyria, pachygyria, and subcortical band heterotopia (SBH) [58,59]. Of the 20 genes associated to LIS more than a half has been discovered in the past 6 years [60]. A targeted sequencing panel of 17 LIS-associated genes was applied by Di Donato et al. [60], recently, to identify a causal mutation in 34% of 216 children with unexplained LIS. These results added to historical molecular analysis (deletion 17p13.3 and DCX, LIS1, ARX sequencing) detected mutations in 81% of a cohort of 811 LIS patients and supplied relevant data regarding LIS-associated genes prevalence, phenotypic expression, and allowed a new biological network-based classification of LIS.

TABLE 13.1 Genetic etiologies of MCDS.

MCT type	Group	Causing genes	Associated pathways and etiology
Microcephaly	Group I	MCPH1, CENPJ, CDK5RAP2, WDR62, NDE1, NDE1, ASPM, CDK5RAP2, TUBA1A, TUBB2B, TUBB3, TUBG1, LIS1, DCX, DYNC1H, KIF5C, and NDE1	Neurogenesis and cell replication, tubulin, and microtubule-associated proteins (MAP)
Megalencephaly spectrum	Group I	WDR62, PIK3R2, PIK3CA, and AKT3	mTOR
FCD type IIA	Group I	MTOR, DEPDC5, PIK3CA	mTOR
FCD type IIB	Group I	MTOR, DEPDC5, NPRL3	mTOR
Tubulinopathies (lissencephaly, basal ganglia dysgenesia, cortical dysgryria)	Group II	TUBA1A, TUBB2B, TUBB3, TUBB, TUBA (,TUBG1, LIS1, DCX, DYNC1H, KIF5C, KIF2A, NDE1	Microtubule structure and function, tubulins, and centrosome expressed MAPs
Lissencephalies	Group II	ARX, RELN, VLDR, ACTB, ACTG1, CDK5	Reelin, forebrain transcriptional regulation
Gray matter heterotopia	Group II	FLNA, ARGEF2, ERMARD, FAT4, DCHS1, LRP2, C6orf70, NEDD4L	Actin filaments/Neuroepithelium/mTOR
Cobblestone malformations	Group II	GPR56, LAMB1; LAMB2, LAMC3, SRD5A3	Dystroglycanopathies
Polymicrogyria (PMG)	Group III	GPR56, TUBB2B, SRPX2, TBR2, PAX6, NDE1, WDR62, FH, OCLN, CHD7, RAB3GAP1, RAB18, NEDD4L, MTOR, PIK3R2	mTOR, microtubule structure and function

In our center, to contribute to phenotypic expansion and to collaborate to elucidate the pathological pathways in MCDs, we searched for germinal and somatic mutations in a cohort of 38 patients with neuronal migration disorders (Periventricular nodular heterotopia, Subcortical band heterotopia, and Lissencephaly). A conclusive genetic diagnosis was achieved in 14 patients.

Remarkably, we found a somatic mutation in four out of twelve patients in whom we applied a targeted high-coverage NGS (mean coverage of about 4000×). This technique allowed us to detect mosaic mutations with a very low alternate allele frequency (between 10% and 15%) from peripheral blood samples [48]. Thus, it provides a unique opportunity for the study of brain mosaic diseases overcoming two of its historical difficulties: the limited accessibility to brain tissue and Sanger low sensibility to detect mosaic variants (threshold of 15%–20%) [61] (Fig. 13.1). Other MCDs that have largely benefited from

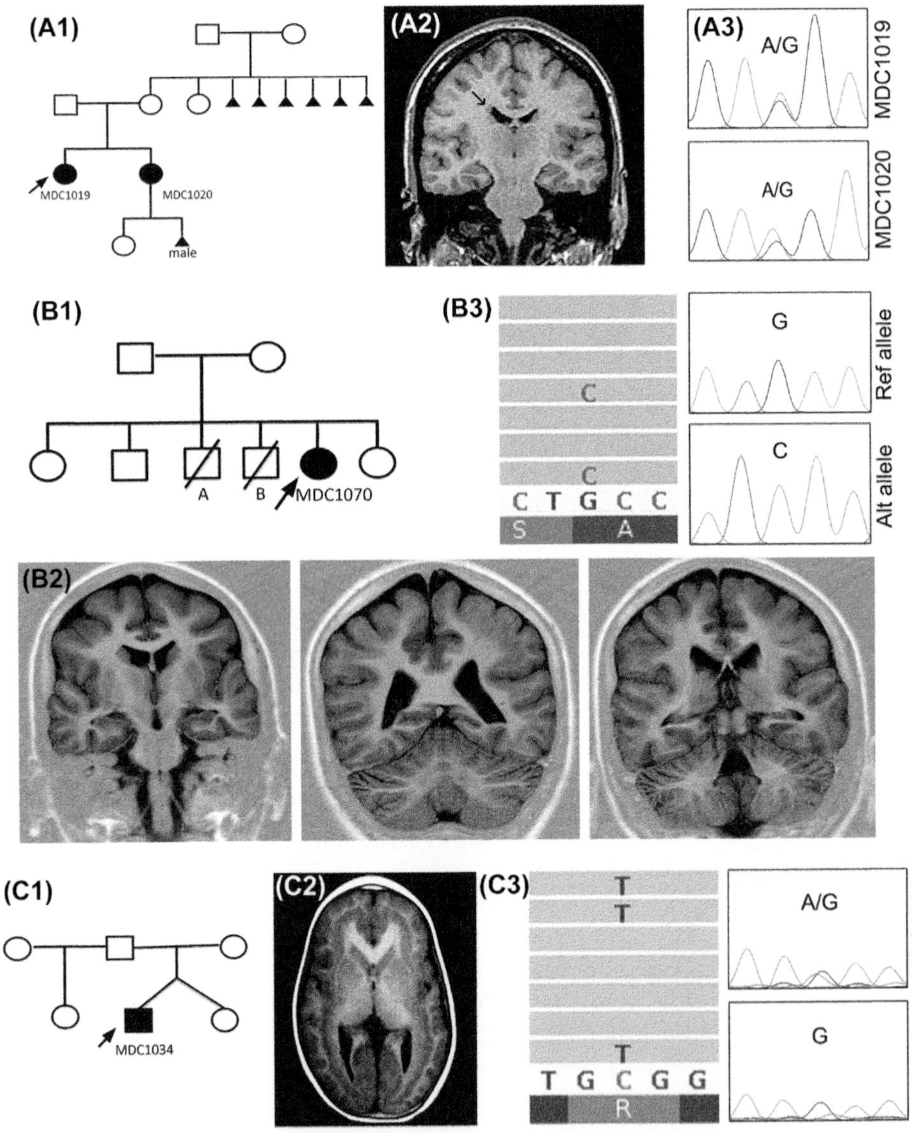

FIGURE 13.1 Familiar, radiological and molecular findings in four individuals of our MCD cohort (MDC1019, MDC1020, MDC1070, and MDC1034). A-Case MDC1019 and MDC1020 illustrate the expanded phenotype of FLNA mutations which ranged in this pair of siblings from diffuse bilateral heterotopic periventricular nodules responsible for refractory epilepsy (MDC 1019) to an isolated nodule in an apparently asymptomatic patient (MDC 1020). A1 Pedigree A2. Coronal T1 MRI from MDC 1020 participant. The image shows an isolated heterotopic nodule adjacent to the right lateral ventricle (*arrow*). A3. Sanger sequencing of *FLNA* gene showing the NM_001110556.1: c.4159G>A mutation. B-Case MDC1070 represents the Lissencephaly spectrum. In this case a mosaic PAFAH1B1 mutation resulted in a less severe phenotype of SBH with posterior predominance. B1 Pedigree B2. MDC 1070 MRI. Inversion-Recovery Coronal MRI images show a posterior (P > A) band of subcortical heterotopia as well as simplified gyri and a thin layer of white matter between the cortex and band. B3. NGS (left) and Sanger sequencing after subcloning (right) of *PAFAH1B1* gene showing the presence of the somatic mutation NM_000430: c.628G>C; p.A210P (alternate allele read frequency 14.99%). C-Case MDC1034 illustrates a somatic mutation of DCX. Similarly to case MDC 1070, this mosaic mutation, although present in an X-linked gene in a male patient, caused a less severe phenotype (HBS). C1 Pedigree C2. Coronal T1-WI shows a diffuse, thick (>12 mm) subcortical heterotopic band. C3. NGS (left) and Sanger (right) sequencing for *DCX* gene showing the NM_178152.1: c.176G>A mutation. Please consider that the Sanger sequencing was performed on the coding strand (reverse of reference).

new sequencing technologies are the megalencephaly, dysplasic megalencephaly, hemimegalencephaly, and focal cortical dysplasias. Until recently, these malformations were not well understood in pathogenic terms, specially the FCD. However, the focal nature of these lesions and the pathological resemblance to tubers in tuberous sclerosis led to the idea that somatic mosaic mutation in the mTOR pathway (that includes tuberous sclerosis associated genes: TSC1 and TSC2) could be the responsible. This hypothesis was in part confirmed by the identification of mosaic mutations in many mTOR genes (DEPDC5, AKT3, TSC1, PIK3CA, PIK3R2, mTOR, etc.) in hemimegalencephaly, megalencephaly, and FCD type 2 [62–64]. In our cohort of patients with MCDs, we identified a somatic mutation in the RHEB gene through high depth and ultrahigh depth next generation sequencing in a patient with hemimegalencephaly and drug resistant epilepsy. It was only present in the brain tissue at a mutant allele fraction of 21%, being undetectable by Sanger. The RHEB gene encodes a protein that has a key role in growth and cell cycle progression due to its action in regulation of the mTOR pathway. Hyperactivation of the mTORC1 was observed in dysmorphic neurons of our patient, in contrast to apparently normal adjacent neurons [65]. The acknowledgment of the mTOR pathway in the generation of cortical malformations has implications not only in a pathological, molecular, and diagnosis sphere but also provides a therapeutic window due to Rapamicyn's ability to inhibit this pathway [66].

The practice of neurogenetics is a work of multidisciplinary teams. The members of these teams should be proficient in the five main components that constitute a thorough evaluation and management of patient with neurogenetic diseases (*modified from* 7): (a) an adequate knowledge of the neurology of these disorders; (b) sufficient experience and training in the genetic of these disorders; (c) interpersonal skills required for genetic counseling; (d) have a family perspective in order to recognize those family members at risk that will necessitate of these teams' work as well and (e) comprehensive knowledge of state-of-the-art diagnostic technologies applicable in this field. It is our belief, that following these premises, we will continue to solve and shorten the many diagnostic odysseys historically suffered by these complex patients.

References

[1] Lee O, Porteous M. Genetic testing and reproductive choice in neurological disorders. Practical Neurol. 2017;17(4):275–81.

[2] Rodriguez-Quiroga SA, Cordoba M, Gonzalez-Moron D, Medina N, Vega P, Dusefante CV, et al. Neurogenetics in Argentina: diagnostic yield in a personalized research based clinic. Genet. Res. 2015;97:e10.

[3] van Karnebeek CD, Bowden K, Berry-Kravis E. Treatment of neurogenetic developmental conditions: from 2016 into the future. Pediatr. Neurol. 2016;65:1–13.

[4] Galatolo D, Tessa A, Filla A, Santorelli FM. Clinical application of next generation sequencing in hereditary spinocerebellar ataxia: increasing the diagnostic yield and broadening the ataxia-spasticity spectrum. A retrospective analysis. Neurogenetics 2018;19(1):1–8.

[5] Bardakjian TM, Helbig I, Quinn C, Elman LB, McCluskey LF, Scherer SS, et al. Genetic test utilization and diagnostic yield in adult patients with neurological disorders. Neurogenetics 2018;19(2):105–10.

[6] Stelten BML, Huidekoper HH, van de Warrenburg BPC, Brilstra EH, Hollak CEM, Haak HR, et al. Long-term treatment effect in cerebrotendinous xanthomatosis depends on age at treatment start. Neurology 2019;92(2):e83–95.

[7] Bird TD. Approaches to the patient with neurogenetic disease. Clin. Lab. Med. 2010;30(4):785–93.

[8] Clark MM, Stark Z, Farnaes L, Tan TY, White SM, Dimmock D, et al. Meta-analysis of the diagnostic and clinical utility of genome and exome sequencing and chromosomal microarray in children with suspected genetic diseases. NPJ Genom. Med 2018;3:16.

[9] Lynch DS, Wade C, Paiva ARB, John N, Kinsella JA, Merwick A, et al. Practical approach to the diagnosis of adult-onset leukodystrophies: an updated guide in the genomic era. J Neurol Neurosurg Psychiatry May 2019;90(5):543–4.

[10] Parolin Schnekenberg R, Perkins EM, Miller JW, Davies WI, D'Adamo MC, Pessia M, et al. De novo point mutations in patients diagnosed with ataxic cerebral palsy. Brain 2015;138(Pt 7):1817–32.

[11] McMichael G, Bainbridge MN, Haan E, Corbett M, Gardner A, Thompson S, et al. Whole-exome sequencing points to considerable genetic heterogeneity of cerebral palsy. Mol. Psychiatry 2015;20(2):176–82.

[12] Zarrei M, Fehlings DL, Mawjee K, Switzer L, Thiruvahindrapuram B, Walker S, et al. De novo and rare inherited copy-number variations in the hemiplegic form of cerebral palsy. Genet. Med. 2018;20(2):172–80.

[13] Pekeles H, Accogli A, Boudrahem-Addour N, Russell L, Parente F, Srour M. Diagnostic yield of intellectual disability gene panels. Pediatr Neurol 2019 Mar;92:32–6.

[14] Al-Nabhani M, Al-Rashdi S, Al-Murshedi F, Al-Kindi A, Al-Thihli K, Al-Saegh A, et al. Reanalysis of exome sequencing data of intellectual disability samples: yields and benefits. Clin. Genet. 2018;94(6):495–501.

[15] Chen C, Chen D, Xue H, Liu X, Zhang T, Tang S, et al. IDGenetics: a comprehensive database for genes and mutations of intellectual disability related disorders. Neurosci. Lett. 2018;685:96–101.

[16] Monroe GR, Frederix GW, Savelberg SM, de Vries TI, Duran KJ, van der Smagt JJ, et al. Effectiveness of whole-exome sequencing and costs of the traditional diagnostic trajectory in children with intellectual disability. Genet. Med. 2016;18(9):949–56.

[17] Weber YG, Biskup S, Helbig KL, Von Spiczak S, Lerche H. The role of genetic testing in epilepsy diagnosis and management. Expert Rev. Mol. Diagn 2017;17(8):739–50.

[18] Allen NM, Conroy J, Shahwan A, Lynch B, Correa RG, Pena SD, et al. Unexplained early onset epileptic encephalopathy: exome screening and phenotype expansion. Epilepsia 2016;57(1):e12–7.

[19] Ziobro J, Eschbach K, Sullivan JE, Knupp KG. Current treatment strategies and future treatment options for Dravet syndrome. Curr. Treat. Options Neurol. 2018;20(12):52.

[20] Dilena R, DiFrancesco JC, Soldovieri MV, Giacobbe A, Ambrosino P, Mosca I, et al. Early treatment with quinidine in 2 patients with epilepsy of infancy with migrating focal seizures (EIMFS) due to gain-of-function KCNT1 mutations: functional studies, clinical responses, and critical issues for personalized therapy. Neurotherapeutics 2018;15(4):1112−26.

[21] Snoeijen-Schouwenaars FM, van Ool JS, Verhoeven JS, van Mierlo P, Braakman HMH, Smeets EE, et al. Diagnostic exome sequencing in 100 consecutive patients with both epilepsy and intellectual disability. Epilepsia 2019;60(1):155−64.

[22] Berg AT, Coryell J, Saneto RP, Grinspan ZM, Alexander JJ, Kekis M, et al. Early-life epilepsies and the emerging role of genetic testing. JAMA Pediatr 2017;171(9):863−71.

[23] Sanchez Fernandez I, Loddenkemper T, Gainza-Lein M, Sheidley BR, Poduri A. Diagnostic yield of genetic tests in epilepsy: a meta-analysis and cost-effectiveness study. Neurology 2019. in press.

[24] Rossi M, Anheim M, Durr A, Klein C, Koenig M, Synofzik M, et al. The genetic nomenclature of recessive cerebellar ataxias. Mov. Disord. 2018;33(7):1056−76.

[25] Di Fonzo A, Monfrini E, Erro R. Genetics of movement disorders and the practicing clinician; who and what to test for? Curr. Neurol. Neurosci. Rep. 2018;18(7):37.

[26] Hogg E, During E, Tan E E, Athreya K, Eskenazi J, Wertheimer J, et al. Sustained quality-of-life improvements over 10 years after deep brain stimulation for dystonia. Mov. Disord. 2018;33(7):1160−7.

[27] Bentivoglio AR, Loi M, Valente EM, Ialongo T, Tonali P, Albanese A. Phenotypic variability of DYT1-PTD: does the clinical spectrum include psychogenic dystonia? Mov. Disord. 2002;17(5):1058−63.

[28] Wolters FJ, van der Lee SJ, Koudstaal PJ, van Duijn CM, Hofman A, Ikram MK, et al. Parental family history of dementia in relation to subclinical brain disease and dementia risk. Neurology 2017;88(17):1642−9.

[29] De Jager PL, Yang HS, Bennett DA. Deconstructing and targeting the genomic architecture of human neurodegeneration. Nat. Neurosci. 2018;21(10):1310−7.

[30] Bonvicini C, Scassellati C, Benussi L, Di Maria E, Maj C, Ciani M, et al. Next generation sequencing analysis in early onset dementia patients. J Alzheimers Dis 2019;67(1):243−56.

[31] Carmona S, Hardy J, Guerreiro R. The genetic landscape of Alzheimer disease. Handb. Clin. Neurol. 2018;148:395−408.

[32] Van Cauwenberghe C, Van Broeckhoven C, Sleegers K. The genetic landscape of Alzheimer disease: clinical implications and perspectives. Genet. Med. 2016;18(5):421−30.

[33] Vatsavayai SC, Nana AL, Yokoyama JS, Seeley WW. C9orf72-FTD/ALS pathogenesis: evidence from human neuropathological studies. Acta Neuropathol 2019 Jan;137(1):1−26.

[34] Karch CM, Wen N, Fan CC, Yokoyama JS, Kouri N, Ross OA, et al. Selective genetic overlap between amyotrophic lateral sclerosis and diseases of the frontotemporal dementia spectrum. JAMA Neurol 2018;75(7):860−75.

[35] Zalar B, Maver A, Kovanda A, Peterlin A, Peterlin B. Clinical exome sequencing in dementias: a preliminary study. Psychiatr. Danub. 2018;30(2):216−9.

[36] Xu Y, Liu X, Shen J, Tian W, Fang R, Li B, et al. The whole exome sequencing clarifies the genotype- phenotype correlations in patients with early-onset dementia. Aging Dis 2018;9(4):696−705.

[37] Chae JH, Vasta V, Cho A, Lim BC, Zhang Q, Eun SH, et al. Utility of next generation sequencing in genetic diagnosis of early onset neuromuscular disorders. J. Med. Genet. 2015;52(3):208−16.

[38] Angelini C, Giaretta L, Marozzo R. An update on diagnostic options and considerations in limb-girdle dystrophies. Expert Rev. Neurother. 2018;18(9):693−703.

[39] Haskell GT, Adams MC, Fan Z, Amin K, Guzman Badillo RJ, Zhou L, et al. Diagnostic utility of exome sequencing in the evaluation of neuromuscular disorders. Neurol. Genet 2018;4(1): e212.

[40] Schofield D, Alam K, Douglas L, Shrestha R, MacArthur DG, Davis M, et al. Cost-effectiveness of massively parallel sequencing for diagnosis of paediatric muscle diseases. NPJ Genom. Med 2017;2.

[41] Wofford S, Noblin S, Davis JM, Farach LS, Hashmi SS, Mancias P, et al. Genetic testing practices of genetic counselors, geneticists, and pediatric neurologists with regard to childhood-onset neurogenetic conditions. J. Child Neurol. 2019 883073818821036, in press.

[42] Uhlmann WR, Roberts JS. Ethical issues in neurogenetics. Handb. Clin. Neurol. 2018;147:23−36.

[43] Craufurd D, MacLeod R, Frontali M, Quarrell O, Bijlsma EK, Davis M, et al. Diagnostic genetic testing for Huntington's disease. Practical Neurol. 2015;15(1):80−4.

[44] Cordoba M, Rodriguez-Quiroga SA, Vega PA, Salinas V, Perez-Maturo J, Amartino H, et al. Whole exome sequencing in neurogenetic odysseys: an effective, cost- and time-saving diagnostic approach. PLoS One 2018;13(2). e0191228.

[45] Cordoba M, Rodriguez S, Gonzalez Moron D, Medina N, Kauffman MA. Expanding the spectrum of Grik2 mutations: intellectual disability, behavioural disorder, epilepsy and dystonia. Clin. Genet. 2015;87(3):293−5.

[46] Cordoba M, Rodriguez-Quiroga S, Gatto EM, Alurralde A, Kauffman MA. Ataxia plus myoclonus in a 23-year-old patient due to STUB1 mutations. Neurology 2014;83(3):287−8.

[47] Garcia AM, Abrevaya S, Kozono G, Cordero IG, Cordoba M, Kauffman MA, et al. The cerebellum and embodied semantics: evidence from a case of genetic ataxia due to STUB1 mutations. J. Med. Genet. 2017;54(2):114−24.

[48] Gonzalez-Moron D, Vishnopolska S, Consalvo D, Medina N, Marti M, Cordoba M, et al. Germline and somatic mutations in cortical malformations: molecular defects in Argentinean patients with neuronal migration disorders. PLoS One 2017;12(9). e0185103.

[49] Storey E. Genetic cerebellar ataxias. Semin. Neurol. 2014;34(3):280−92.

[50] Den Dunnen WFA. Trinucleotide repeat disorders. Handb. Clin. Neurol. 2017;145:383−91.

[51] Sun M, Johnson AK, Nelakuditi V, Guidugli L, Fischer D, Arndt K, et al. Targeted exome analysis identifies the genetic basis of disease in over 50% of patients with a wide range of ataxia-related phenotypes. Genet. Med. 2019;21(1):195−206.

[52] Guerrini R, Dobyns WB. Malformations of cortical development: clinical features and genetic causes. Lancet Neurol 2014 Jul;13(7):710−26.

[53] Desikan RS, Barkovich AJ. Malformations of cortical development. Ann. Neurol. 2016;80(6):797−810.

[54] Barkovich AJ, Kuzniecky RI, Jackson GD, Guerrini R, Dobyns WB. A developmental and genetic classification for malformations of cortical development. Neurology 2005;65(12):1873–87.

[55] Barkovich AJ, Guerrini R, Kuzniecky RI, Jackson GD, Dobyns WB. A developmental and genetic classification for malformations of cortical development: update 2012. Brain 2012;135 (Pt 5):1348–69.

[56] Parrini E, Conti V, Dobyns WB, Guerrini R. Genetic basis of brain malformations. Mol. Syndromol 2016;7(4):220–33.

[57] Manzini MC, Walsh CA. What disorders of cortical development tell us about the cortex: one plus one does not always make two. Curr. Opin. Genet. Dev. 2011;21(3):333–9.

[58] Bahi-Buisson N, Souville I, Fourniol FJ, Toussaint A, Moores CA, Houdusse A, et al. New insights into genotype-phenotype correlations for the doublecortin-related lissencephaly spectrum. Brain 2013;136(Pt 1):223–44.

[59] Di Donato N, Chiari S, Mirzaa GM, Aldinger K, Parrini E, Olds C, et al. Lissencephaly: expanded imaging and clinical classification. Am. J. Med. Genet. 2017;173(6):1473–88.

[60] Di Donato N, Timms AE, Aldinger KA, Mirzaa GM, Bennett JT, Collins S, et al. Analysis of 17 genes detects mutations in 81% of 811 patients with lissencephaly. Genet. Med. Off. J. Am. Coll. Med. Genet 2018;20(11):1354–64.

[61] Qin L, Wang J, Tian X, Yu H, Truong C, Mitchell JJ, et al. Detection and quantification of mosaic mutations in disease genes by next-generation sequencing. J. Mol. Diagn 2016;18(3):446–53.

[62] Lee JH, Huynh M, Silhavy JL, Kim S, Dixon-Salazar T, Heiberg A, et al. De novo somatic mutations in components of the PI3K-AKT3-mTOR pathway cause hemimegalencephaly. Nat. Genet. 2012;44(8):941–5.

[63] Leventer RJ, Scerri T, Marsh AP, Pope K, Gillies G, Maixner W, et al. Hemispheric cortical dysplasia secondary to a mosaic somatic mutation in MTOR. Neurology 2015;84(20):2029–32.

[64] Lim JS, Kim WI, Kang HC, Kim SH, Park AH, Park EK, et al. Brain somatic mutations in MTOR cause focal cortical dysplasia type II leading to intractable epilepsy. Nat. Med. 2015;21(4):395–400.

[65] Salinas V, Vega P, Piccirilli MV, Chicco C, Ciraolo C, Christiansen S, et al. Identification of a somatic mutation in the RHEB gene through high depth and ultra-high depth next generation sequencing in a patient with Hemimegalencephaly and drug resistant Epilepsy. Eur. J. Med. Genet. 2018 in press.

[66] Way SW, Rozas NS, Wu HC, McKenna 3rd J, Reith RM, Hashmi SS, et al. The differential effects of prenatal and/or postnatal rapamycin on neurodevelopmental defects and cognition in a neuroglial mouse model of tuberous sclerosis complex. Hum. Mol. Genet. 2012;21(14):3226–36.

Chapter 14

Pediatric genomics and precision medicine in childhood

Francesc Palau[1,2,3,4]

[1]Department of Genetic Medicine and Pediatric Institute of Rare diseases, and Director, Sant Joan de Déu Research Institute, Sant Joan de Déu Children's Hospital, Barcelona, Spain; [2]Institute of Medicine and Dermatology, Hospital Clínic, Barcelona, Spain; [3]CSIC Research Professor and Adjunct Professor of Pediatrics, University of Barcelona School of Medicine and Health Sciences, Barcelona, Spain; [4]Group Leader, Neurogenetics and Molecular Medicine Group, CIBERER, Barcelona, Spain

Introduction

The DNA molecule, the structure and regulation of the genome, and the variations of genome and epigenetic modification of gene expression, are the bases of individuality and inheritance of human beings. This complexity manifests itself early in embryonic and fetal development, and in the early stages of postnatal life. Understanding the biological importance of genetic variability has become a fundamental tool to know the pathophysiology of disease, the appropriate diagnosis, management, and therapeutic response in pediatrics [1,2].

Studies performed at the end of 20th century estimated the burden of genetic conditions in children and young adults as 5.4/1000. The genetic load of pediatric disease increased to 53/1000, by considering multifactorial disorders, and 79/1000 if congenital anomalies were also included. In addition, most of the admissions and chronic patients in referent pediatric hospitals have a genetic component either as primary cause or susceptibility factor [3,4]. In 2018, the census of children and adolescents with diagnosis of a rare disease (RD, most genetic), in the Sant Joan de Déu Children's Hospital, was 17,656 patients. The average annual cost per RD patient, compared with the rest of patients, has been multiplied by a factor of 5.2 (internal data).

Since the report of the first draft of the human genome in 2003, genomics is increasingly being part of the clinical medicine and research [5]. Next generation sequencing (NGS) allows to perform whole exome sequencing (WES) and whole genome sequencing (WGS), to diagnose single-gene disorders, and chromosomal microarrays (CMA), by which genomic rearrangement disorders and part of chromosomal abnormalities are detected, have become part of the routine. Genomics is also now more relevant to investigate the etiology of pediatric multifactorial diseases, because it allows to confirm the presence of variants of susceptibility of several genes, and to contrast the oligogenic hypothesis in some disorders.

DNA-based diagnostics of pediatric disease

Molecular diagnosis has moved from gene mapping by linkage analysis and positional cloning to parallel massive genome sequencing. Molecular genetic testing by Sanger sequencing, MLPA (multiplex ligation-dependent probe amplification), or TP-PCR (triplet repeat primed polymerase chain reaction) allows finding the mutation or pathogenic variant causative of a specific disorder, with no locus genetic heterogeneity. This is the case of cystic fibrosis by using Sanger sequencing of *CFTR* gene, to detect non-frequent mutations; the molecular diagnosis of Duchenne muscular dystrophy combines MLPA for deletions and duplications of exons, and Sanger sequencing for point mutations; in the case of muscular spinal atrophy (SMA), genetic testing is based on the analysis of exon 7 deletion in the *SMN1* gene, and copy number of the *SMN2* gene by MLPA; TP-PCR is the first test to investigate the trinucleotide repeat expansion in disorders such as X-fragile syndrome, myotonic dystrophy, and Friedreich ataxia.

There are, however, a number of clinical entities that are associated with several genes, because of locus heterogeneity. In such a case, gene-by-gene sequencing becomes more complicated, as it results in Charcot-Marie-Tooth

Precision Medicine for Investigators, Practitioners and Providers. https://doi.org/10.1016/B978-0-12-819178-1.00014-9

disease and related neuropathies, genetic retinal dystrophies, congenital deafness, or Joubert syndrome.

The genome as the diagnostic target

Low-resolution, whole-genome approach to genetic diagnosis has also changed from the classical karyotype used to diagnose chromosomal abnormalities (aneuploidies, translocations, inversions), and segmental aneuploidies with 5–7 Mb resolution, to molecular cytogenetics by using CMA with resolution of approximately 100 kb, which allow to detect microdeletions and microduplications that are associated with more than 50 developmental disorders. Thus, the availability to detect genomic dose changes in a specific chromosomal region depends on the size of the genome rearrangements.

Classical syndromes such as 5p- (cri-du-chat) or 4p- (Wolf-Hirschhorn) syndromes are diagnosed by traditional karyotype, but also by CMA. By contrast, proper diagnosis of most of genomic developmental disorders associated with segmental rearrangements requires CMA as a first instance. Examples are Williams syndrome associated with a 1.5 Mb microdeletion at chromosome 7q11.2, 22q11 deletion or duplication syndromes as a consequence of microdeletions or microduplication at chromosome 22q11.2 that are associated with several classical phenotypes such a DiGeorge syndrome, velo-cardio-facial syndrome and conotruncal heart disease, or rearrangements at 17p11.2 with microdeletion that is expressed as the Smith-Magenis syndrome, and microduplication that is associated with Potocki-Lupski syndrome [6].

Segmental versus panoramic profile

The resolution of the genome has changed from the microscopic vision of the chromosome through the analysis of the karyotype, to obtaining either the complete sequence (whole genome sequencing, WGS) or the coding of DNA sequence (whole exome sequencing, WES). An exponential increase allows obtaining useful genetic information for diagnosis and treatment opportunities [7,8], as well as new challenges with respect to some issues, such as the genotype-phenotype relationship and the bioethics aspects [9].

Phenotype-genotype correlations

NGS-based genetic diagnosis allows searching for three phenotype-genotype clinical situations, that is, (i) *one disease– one gene* univocal relationship, (ii) *one disease–several genes*, defined by locus genetic heterogeneity, and (iii) *several phenotypes/diseases– one gene* that represents the pleiotropic effects of one mutant gene. NGS genetic testing can be addressed by three main ways, two of them sequencing the encoding exons, that is, the sequence that is translated into amino acids, and one that is sequencing the whole genome: (i) customized gene panel related to specific and well-known clinical phenotypes, with a targeted symptom/sign or biological pathway; the number of genes may vary from few genes (i.e., rasopathies that include the NF1 gene of neurofibromatosis type 1, and genes related to Noonan syndrome among others), to a large number of genes related to major clinical signs as it is the case of epilepsy for epileptic syndromes and epileptic encephalopathies, muscle weakness for muscular dystrophies, congenital myopathies and peripheral neuropathies, or immune deficits for primary immunodeficiencies; as more genes are described, this approach requires updating.

Gene panels and sequencing output

An extended gene panel that covers a very large number of genes is the clinical exome sequencing (CES), covering between 5000 to almost 7000 genes associated with pathological entities; (ii) whole exome sequencing (WES) that involves the coding regions (~20,000 genes), which equates to approximate 1.5% of the entire human genome; and (iii) whole genome sequencing (WGS), which represents the sequence of the entire genome, including exons and introns of genes, and the intergenic regions that could be relevant in gene regulation and expression such as promoter, enhancer, and silencer sequences. The information about genetic variation depends on the quality of DNA obtained from human samples, the depth of sequencing (number of reads per nucleotide), coverage of genes (percentage of exons/gene sequenced), and bioinformatic tools applied to the sequence analysis.

Diagnostic methodological selection

Cost-benefit ratio and institution's budget are important, but also technological capacity of implementing targeted gene panels, CES, or WES [10–12]. WGS is still a tool for genetic research through discovery of mutant genes, modifier genes and genetic variation, in regulatory sequences that may explain phenotypic variation, such as clinical expressivity and penetrance. WGS will become a relevant tool in the near future, as cost reduction and biological interpretation in the clinical setting become a reality. WES is becoming the current standard test for patients. Depending on the NGS sequencer used, CES is also a very good option to apply for genetic testing. Running as single test on an affected individual, it is possible to detect the causative genetic variant in as many as 25% of patients.

These results may increase to 40% by trio-WES approach, as the familial segregation of the candidate variants can be used as a filter (inheritance pattern), in the bioinformatics pipeline. The number of applications of tests in genetic laboratories of teaching hospitals is usually very high. Thus, depending on the NGS sequencing facilities and costs, WES-trio is not the first operative option, so one working option is to use CES to investigate single patients and, in selected negative cases with well-defined phenotypes, apply WES-trio.

In 2018, at the Sant Joan de Déu Children's Hospital, 857 pediatric patients with suspected genetic disease were investigated through CES testing (6710 genes, TSO Expanded, Illumina). The diagnostic rate was 40% (341 patients). In addition, 23 undiagnosed patients with a neurodevelopmental disorder, were selected for WES-trio, and genetic diagnosis was obtained in 11 of them (48%). Such a CES-selected WES-trio protocol seems to be a good approach when using NGS in a clinical setting.

Whole exome sequencing

Exome sequencing is a powerful tool; however, there are a number of potential limitations (also CES). Some of them are solved or better addressed by WGS [13]. These include detection of copy number variants (CNVs) that reflect the presence of a deletion or duplication, either benign or pathogenic; detection of deep intronic sequences, regulatory sequences in 5′ promoters or enhancers/silencers, or microRNAs; detection of mitochondrial DNA mutations. Structural genomic variants such as chromosomal translocations, inversions, and aneuploidies will be better solved by WGS. Other limitations are mutation mechanisms that still require classical molecular approaches to be detected such as uniparental disomy, tandem repeat sequences (trinucleotide expansions), or recognition of a pseudogene or nonfunctional gene copies, like the *GBA* gene and the related pseudogene in Gaucher disease, or *SMN1* functional copy and *SMN2* nonfunctional copy in spinal muscular atrophy/SMA.

Integration and interpretation of genetic information

Within the 3000 Mb size of the human haploid genome, there are 5 million variants. The exome contains approximately 40,000 annotated variants. Thus, such background information requires the use of bioinformatics tools, to go from the variation pool of the individual to the gene variant that could be the pathogenic mutation related to the patient's disease. In a WES test filtering for variants frequency (excluding common variants) may reduce the focus to 500 rare, functional, missense variants; then, by applying biological consequences, the number could be reduced to 10–20 rare functional variants, in genes that may be relevant, and filtering by phenotype, using ontologies such the Human Phenotype Ontology or HPO [14], and inheritance of the total candidate variants may be reduced to one to four or none [15].

Team effort

Bioinformaticians have become part of the "genomic team," along with laboratory geneticists and technicians that have to collaborate closely with genetic counselors and clinicians. They design the analytical workflow for variant calling, annotation, and interpretation of the genomic sequence and variation. There are a lot of software programs, but the way in which the raw data is processed, the pathogenic variants are identified, and the clinical data are integrated into the workflow pipeline is still elaborated in a very local and specific way in each laboratory. Annotation of variants as part of the genome is a necessary process, before the sequence analysis.

Variant annotation

To carry out proper annotation, the allelic differences have to be identified by variant calling. There is specific software that allows this process to be carried out, such as SNPeff, which includes databases such as 1000 Genomes, dbSNP, ExAc, and ClinVar. The process is done by comparing the base composition, and position of an individual genome, to the equivalent position in a reference human genome. The final result used to be stored in a standardized variant calling format, the VCF file. Exclusion of variants is mainly based on biological functionality, frequency, and inheritance pattern. Thus, nonfunctional variants and common variants are not included in the pathogenicity analysis.

In addition, if the variant gene does not agree with the well-known inheritance pattern, it is also excluded. In case of doubt, raw data can be reanalyzed applying different filtering approaches, so a good candidate variant not previously recognized may emerge. Important issues to take into account is the distribution of variants in different geographical populations, or with different ethnicity, and the low representation of pathogenic variants associated with pediatric disease, in databases based on adult disease cohorts. Therefore, an important point that must still be taken into account is to address genomic diagnoses in the context of patient variability, the population stratified by age or ethnic origin, or environmental exposures that may affect the allelic frequency, of pathogenic and not pathogenic variants.

Clinical and experimental branches of modern genomics

Establishing the pathogenicity of a variant is not always an easy issue. Finding a variant previously reported in a well-established gene does not usually represent a problem. However, the NGS approach is increasing the phenotype-genotype interpretation scenario, in clinical pediatrics [16], which is variable depending on the test applied, that is, CES, WES, or WGS. The most frequent questions are as follows: (i) patients with one (autosomal dominant or X-linked) or two variants (autosomal recessive) of unknown significance, because the variant has not been reported, or it is a missense variant with not obvious biological effect, in a well-defined phenotype already associated with the gene (new *allelic heterogeneity*?); (ii) the patient's phenotype has not been associated with the candidate gene, and the variant is pathogenic, but the gene is related to a different disease (is there a *pleiotropic effect*?); (iii) in case of WES or WGS, the finding of one or two variants, with evident or likely pathogenic effect, in a gene that has not been previously associated with human disease (*new disorder*?).

To address these situations, the phenotype-genotype gap requires the confluence of two points. First, it is an important in-depth analysis of the phenotype beyond a work diagnosis, that is, to collect the clinical signs and symptoms, and apply the ontology phenotype criteria that delimit the clinical picture in a way to increase the congruence between the phenotype and the candidate variants, through the bioinformatics workflow. The current approach of genome-based diagnostics requires dialog within the triangle structure that includes pediatricians or pediatric subspecialists with clinical expertise, laboratory geneticists, and clinical bioinformaticians. There are also cases for which referring to a clinical genetics service may be appropriate.

Variant pathogenicity

Incorporation of functional studies to establish biochemical evidences, both in vitro and cellular or in animal models that confirm biological changes of the variant are becoming relevant approaches in clinical medicine. Nevertheless, this "functional genomics" approach is not easy to be established in a routine clinical setting. Most of functional predictions are based on *in silico* software that uses information about biological effects of the variant in gene expression and splicing, biochemical consequences on protein biology or structure, and evolutionary conservation.

There are a number of predictors such as SIFT, PolyPhen2, MutationTaster, and CADD, among others, which may indicate the pathogenicity of a variant. The American College of Medical Genetics and Genomics and the Association of Molecular Pathology (ACMG-AMP) guidelines are a very useful tool to classify variants [17]. However, additional research, biological knowledge, and data mining will be needed to improve prediction tools based just in the sequence and variation found in a patient.

Clinical management of genetic variation

NGS offers diagnosis, prognosis, and possible treatment. The physician has also to manage a huge amount of genome variation that is difficult to use in clinical practice. We can differentiate several aspects of the genetic variation found in the individual. First, technical aspects related to genetic testing (analytical validity, clinical validity, and clinical utility) have to be well established. Second, the biological significance of any gene and any variation is not completely understood in many cases. Thus, we are still far from understanding the effect of the whole genetic variation, in penetrance, clinical expressivity, and severity of the disease. Third, we still do not know the number of human genes associated with pathology.

The Online Mendelian Inheritance in Man (OMIM) catalog (updated March 22nd, 2019) has 16,054 gene entries (from ~20,000 genes), and description of 5498 phenotypes with known molecular basis. Although the number of disease genes and new clinical syndromes has increased exponentially in the last decade, it could be postulated that variation in any gene may be the primary cause of a human disease or condition.

Genetic counseling in the genomics era

The National Society of Genetic Counselors adopted the following definition [18]: "Genetic counseling is the process of helping people understand and adapt to the medical, psychological and familial implications of genetic contributions to disease. This process integrates the following: (i) interpretation of family and medical histories to assess the chance of disease occurrence or recurrence; (ii) education about inheritance, testing, management, prevention, resources and research; and (iii) counseling to promote informed choices, and adaptation to the risk or condition."

The new technological approaches to ask the genome for genetic variants, either pathogenic or benign (including rare or common polymorphisms), have increased the capacity to find the specific cause of the genetic rare disease or birth defect. Finding a genomic deletion or duplication by chromosomal microarrays, point or indel mutations by massive parallel NGS, intragenic exon deletion or duplication by MLPA, or microsatellite expansion by TP-PCR, allows defining the proper segregation patterns, classic Mendelian inheritance, and nontypical inheritance to understand clinical expressivity, penetrance, and anticipation within the family.

Specific attention

Having one or more genetic counseling sessions is especially relevant in a number of clinical situations: (i) prenatal diagnosis of genetic disease or congenital anomaly, because of familial history or maternal screening, and/or fetal imaging data (ultrasound and MRI findings); (ii) decision-making of parents with a newborn or infant that has a life-threatening disease or birth defect for which a genetic cause is suspected, and the child requires specific types of therapies or aggressive interventions; (iii) late-onset disease within the family, and presymptomatic or predisposition testing is requested, something that is especially relevant in healthy minors; and (iv) counseling prior to performing genome sequencing in a family involving the patient, parents, and other relatives, in association with informed consent, where specific options related to what they want to be reported back to them are indicated.

While the first three situations have been handled for a long time, when applying classical chromosomal or molecular genetic tests, the last indication is increasingly necessary, due to the systematic application of genomic sequencing tests, either WES or WGS.

Precision medicine in childhood and human development

The relevant aspects of precision medicine in pediatrics focus on the biological pathways and genes shared by different disorders, the genetic approach of treatments, preventive medicine and, again, to achieve diagnosis as the starting point when a patient consults for the first time.

Genes and pathophysiology: pathways in epilepsy

Neurodevelopment is a very complex process, and associated disorders are related to genetic etiology, including targets for searching of gene candidates and networks. There are more than 500 different genes associated with epilepsy [19], and the number is increasing over time. The number of entries at OMIM related to early infantile epileptic encephalopathies is 73 (https://www.omim.org/phenotypicSeries/PS308350). Thus, complexity defines epilepsy genomics. Epileptic genes encode proteins involved in membrane excitability, synaptic and neurotransmission functions, and neuronal firing and wiring, as compartments that are part of neuronal synchronization disorders [20].

Such complexity also includes the association of epilepsies with other neurological syndromes, cognitive and behavioral disorders, and autism. Along with genes such as the voltage-gated potassium channel KCNQ2, associated with either early infantile epileptic encephalopathy 7 (EEIE7) or benign familial neonatal seizures 1, sodium channel alpha subunit SCN1A associated with Dravet syndrome or EEIE6, or the glucose transporter gene *SLC2A1* related to GLUT1-deficiency syndrome, genomic microdeletions on 15q13.3, 15q11.2, and 16p13.11 chromosomes are well-established risk factors for genetic generalized epilepsy, and neurodevelopmental problems [21].

Other neurodevelopmental aberrations

Genome analysis by NGS or CMA allows investigating and recognizing pathogenic variants, in major neurodevelopmental phenotypes, such as intellectual disability (ID) and autism spectrum disorders (ASD). Twin studies have demonstrated high heritability rate in ASD [22]. Many patients express ASD but also ID, as the consequence of de novo pathogenic mutations, in either CNVs [23–25] or single nucleotide variants (SNVs) [26]. Contribution of genome variation in the development of autism is not only due to the rare variants but also to common genetic variants that have been estimated to be 40%–60% [27]. The SFARI Database currently includes 740 genes that have potential links to ASD with 2347 (83.9%) rare variants and 449 (16%) common variants (https://gene.sfari.org/about-human-gene/statistics/, updated March 31, 2019). Two major networks are neuron structure including synapse, neuronal projections, postsynaptic density, long-term potentiation and cytoskeleton, and chromatin organization and modeling. Research programs are showing new genes related to neurodevelopmental disorders, sometimes as allelic phenotypic variants in genes that are also associated with other neurological syndromes such as adult-onset ataxia [28].

Several well-defined syndromic disorders associated with ID also involve specific pathways. The above-mentioned rasopathies are one of the well-known group of disorders, related because causing genes are part of the RAS cascade, having ERK as final effector in the nucleus. Coffin-Siris syndrome is characterized by developmental or cognitive delay, facial features, central hypotonia, fifth-digit nail/distal phalanx hypoplasia, and hirsutism and hypertrichosis. Most of Coffin-Siris–related genes (eight OMIM entries: ARID1A, ARID1B, DPF2, ARID2, SMARCC2, SMARCE1, SMARCA4, and SMARCB1) are part of the BRG1-associated factor (BAF) chromatin remodeling complex (also known as SWI/SNF family) [29] that facilitates gene activation by assisting transcription machinery to gain access to gene targets. BAF complex contains a DNA-stimulated ATPase activity, capable of destabilizing histone-DNA interactions [30].

Germline versus somatic mutations

The advent of NGS and single-cell sequencing technologies shows that the inherited or de novo germline mutations is not the only mechanism in neurodevelopmental disorders. Somatic mutations are also important cause of neuronal migration, brain overgrowth disorders, epileptic encephalopathies, intellectual disability, and autism spectrum disorders [31].

Childhood cancer: developmental tumors

Genome profile in pediatric cancer offers several new options, to affected children and their families [32,33]. Germline mutation rate is established as 8.5%−14% in pediatric cancer patients, and there are a large number of cancer predisposing syndromes (e.g., Li-Fraumeni syndrome caused by mutations in the TP53 gene). Thus, germline molecular testing is part of the current clinical management in pediatric oncology practice. As other Mendelian disorders, genetic testing is currently performed by NGS. There is some discussion about the convenience to use multigene panels or WES (or WGS) approaches, because of the return of secondary findings, and the frequent finding of variant of uncertain clinical significance, and the ethical consideration of testing minors.

However, as it happens in other clinical conditions, it seems that WES may be not ethically disruptive. Pediatric cancer information on the genetic and genomic profile of somatic tumor cellscan be utilized for personalized treatment. Whole exome sequencing of tumor DNA and RNA can reveal actionable mutations that can change treatment and genetic counseling, and also address clinical trials in cohorts of specific patients [34]. In addition, investigation of the genomic profiles in pediatric tumors opens new perspectives to understand the developmental origin of cancer in children and adolescents. Although the number of targetable mutations is less in pediatric tumors than in adult cancer (because of biological differences), implementation of precision medicine is changing clinical pediatric oncology.

Currently, there are several aspects in the genetic and epigenetic profiling of leukemia and solid tumors that are relevant for precision approach [33,35,36]: (i) enrichments of targetable gene fusions or gene pathways; (ii) relative frequency of rare mutations in actionable genes, in unexpected tumor types; (iii) as mentioned above, the importance of abnormal germline mutation in pediatric gene predisposition cancer syndromes, but also in patients with no familial history [37].

Therapies and molecular therapeutic targets

Molecular targets, drugs, and advanced therapies are often based on genomic knowledge and individual genome profile of the patient. Starting with classical gene therapy by delivering a transgene [38], the number of new genetic-based treatment approaches is increasing.

In cystic fibrosis, patients carrying mutations that involve stabilization of misfolded cystic fibrosis transmembrane regulator/CFTR and trafficking to the epithelial cell membrane like F508del, or are not available to open the CFTR channel properly (e.g., G551D), can benefit from molecules such as ivacaftor and lumacaftor that act as potentiator or corrector of the molecular defect, respectively [39,40].

Spinal muscular atrophy represents a success story of a genetic rare disease, in which the major defect is the exon 7 deletion of the survival motor neuron 1 (*SMN1*) gene, on chromosome 5q13.2 [41]. Two gene therapy approaches, based on the knowledge of specific gene mutation and pathomechanisms of the disease, are now offering new important results in the disease treatment: expression modification of the endogenous gene by a molecular therapy, and expression of a transgene generated by genetic engineering.

Nusinersen (Spinraza, Biogen, Cambridge, MA, USA) is an antisense oligonucleotide (ASO) that increases the inclusion rate of exon 7 in the transcripts of the mRNA, of the survival motor neuron 2 gene (*SMN2*), by binding to an ISS-N1 site (silencer of the intron splicing process), located in the intron 7 of the messenger pre-mRNA of the *SMN2* gene. By joining, the ASO shifts the splicing factors, which normally suppress the splicing. Displacement of these factors results in the retention of exon 7 in the SMN2 mRNA and, consequently, when the SMN2 mRNA is produced, it can be translated into its full-length functional SMN protein. Both early-onset [42] and late-onset [43] forms of SMA are benefited by this gene therapy. There are still some gaps [44], and direct gene therapy is also a promising therapeutic strategy [45].

Gene and cell therapy

Gene silencing and gene editing are other therapeutic options, addressed to modify gene expression by RNA interference, or correcting the malfunction of the endogenous gene mutation in situ, at the DNA molecule of the patient, by precise targeted homologous recombination. A number of different tools for gene editing have been developed (zinc finger nucleases, TALENs, and CRISPR-Cas9) [46,47]. CRISPR has become the most reliable system for gene editing [47,48], and is applied to biological research, and the development of new molecular tools for somatic gene therapy, and the treatment of monogenic diseases.

Human pluripotent stem cells are actively investigated in therapeutics, especially by applying induced pluripotent stem (iPS) cells, generated from differentiated cells. An interesting case is the confluence of cellular, gene, and immune approaches in modern therapy, represented by the CAR-T (chimeric antigen receptor in T cells) system. CAR-T system is transforming the prognosis of children affected by severe or refractory forms of acute lymphoblastic leukemia, and diffuse large B cell lymphoma. In this antitumor immunotherapy, T lymphocytes of the patient are genetically manipulated to express a receptor, capable of recognizing a molecule in the cell membrane of the tumor and acting against it, and are then infused into the patient's blood [49]. In the case of B-cell ALL, CAR-T cells express the receptor against the B-cell specific CD19 antigen, and attack the neoplastic B-cells. The treatment is a personalized immunotherapy for the ill child based on cellular and gene expression knowledge [50].

Genetic screening

Classically, the diseases that have been good candidates for newborn screening/NBS are those with early onset, severity, complete penetrance in infancy and childhood, availability of effective treatments, confirmation tests, and opportunity of vigilance. NBS is mainly based on biochemical tests. These tests are in most cases limited to some conditions for which a good biomarker is available, such as metabolites and enzymatic activities for inherited metabolic diseases, thyroid hormones for congenital hypothyroidism, immunoreactive trypsinogen for cystic fibrosis, and isoelectric focusing/high performance liquid chromatography for sickle cell anemia.

Genome NGS allows us to approach early onset genetic diseases, through the analysis of genes related to illnesses that can benefit from NBS [51–53]. Needless to say, the ethical aspects of these incidental and secondary findings are of great importance for the future of the newborn and the relation to the genomic profile of the parents.

Rare and undiagnosed diseases: searching for a diagnosis via gene and genome variation

A rare disease (RD) is defined in the European Union as a condition that affects less than 5 people per 10,000 inhabitants, and in the United States, fewer than a total of 200,000 people. Other criteria are chronicity, disability, severity, early onset in pediatric life (but also in adults), and a genetic origin [54]. The number of different rare diseases is between 6000 and 7000. Many of them are really ultrarare. Symptoms and signs in patients with a rare disease, and sometimes also with nontypical common disease, are nonspecific, and diagnosis becomes difficult. This situation has generated the concept of undiagnosed disease [55,56], a working category that is telling us that we need to perform new actions to get the diagnosis, including in-depth phenotyping and genome analysis.

The evaluation of undiagnosed or rare diseases (URDs) is a challenge that requires a multidisciplinary team [57], and specific programs both hospital-based at local level (e.g., Sant Joan de Déu Children's Hospital, www.sjdhospitalbarcelona.org/en/children/rare-diseases), and networks or consortia (https://commonfund.nih.gov/diseases) [58]. Kliegman et al. [13] recognize several factors that can influence the evaluation of children with URD, and categorize them as those specific to the disease, the patient, the physician, and the diagnostic tests limits. Regarding the last point, genome sequencing and genetic tests are changing the vision we have had until recently, about our diagnostic capacity. However, we still need more knowledge to know how to interpret genetic variation in a personalized way, in many patients with URD [59].

How genomics is changing medical thinking

The vision of the International Rare Diseases Research Consortium (IRDiRC) for the period 2017–27, is to enable all people living with a rare disease to receive accurate diagnosis, care, and available therapy, within 1 year of coming to medical attention. Diagnosis depends on NGS-based genomic testing and gene discovery, and therapy requires innovative approaches [60].

Regarding medical thinking, there is another issue: how genomics is changing the way the physician generates the clinical hypothesis or uses information. In Fig. 14.1 we try to compare classical approach, and that generated by the capacity to ask for the individual's genome. Panel A represents the hypothesis-driven process, based on symptoms and syndromic diagnosis, suggesting which genetic studies related to the retina, brain, and heart disease should be performed. After ruling out Refsum disease, Charcot-Marie-Tooth (CMT) neuropathy genes, and cardiomyopathy genes, the patient is reevaluated and the diagnosis of Friedreich ataxia (FRDA) is proposed, which is confirmed after genetic analysis of the *FXN* gene that shows GAA expansion and a missense mutation.

One-way versus two-way road

In this model the pediatrician *proposes* the genetic test based on progressive differential diagnosis. On the contrary, the precise recognition of the phenotypic ontologies of the main symptoms and signs, and the genome or exome sequencing, allow reaching the genetic diagnosis without the need to propose a clinical hypothesi (Fig. 14.1B). In fact, what the pediatrician does is to *contrast* the clinical phenotype with the genetic variants obtained after sequencing of the genome. This new medical thought proposes to change the linear model (univocal), of the clinical and diagnostic phenotype toward the genotype, based on a clinical hypothesis to the bidirectional model (one-to-one), where the diagnosis is the result of the dialog between the phenotype and the genotype. Both approaches are useful, and depending on the clinical setting one of them becomes the first option.

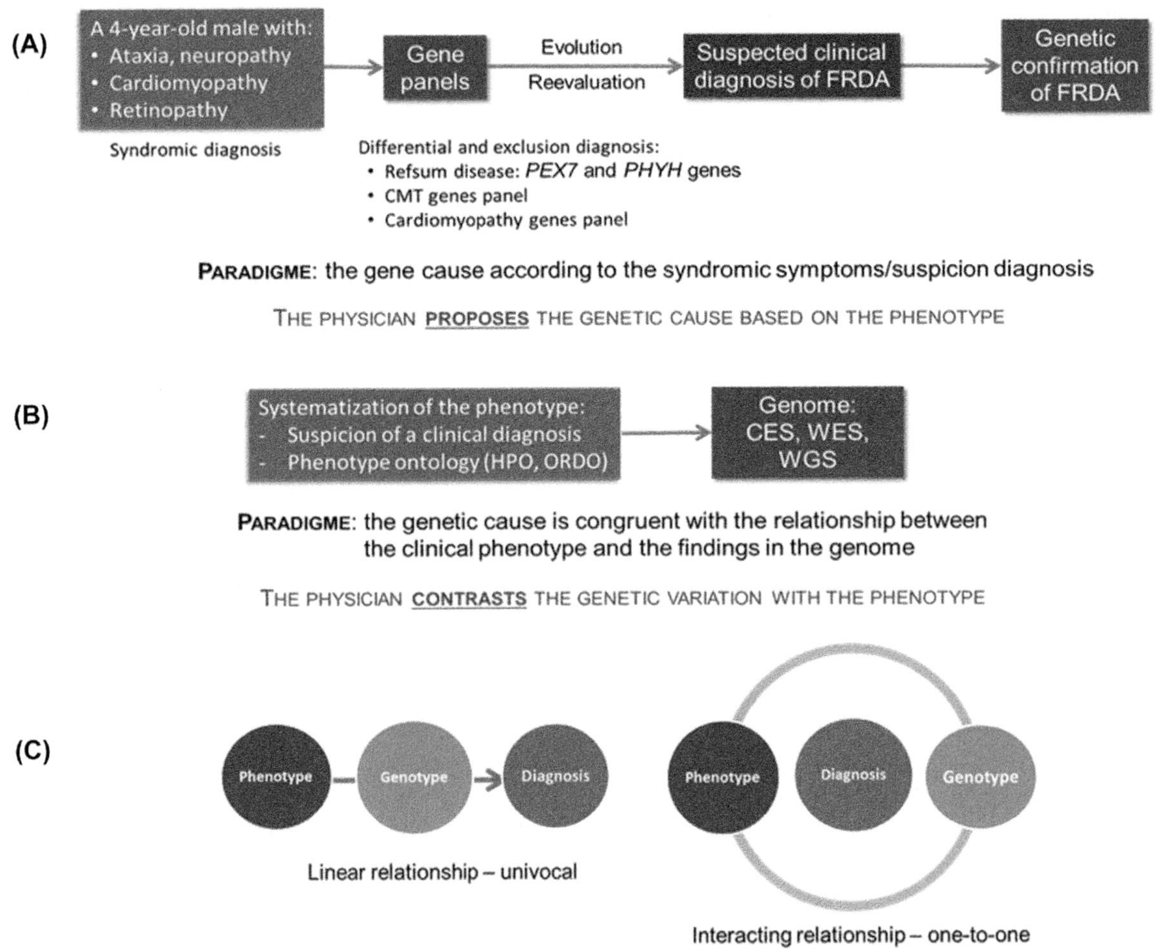

FIGURE 14.1 Models of medical thinking of disease. (A) The hypothesis-driven model, based on clinical suspicion and progressive approach to diagnosis, in a linear manner, and taking into account disease evolution; the starting point is the syndromic and clinical expression of the characteristics of the phenotype, within the context of differential and exclusion diagnosis. (B) Nondirective model based on in-depth phenotyping, and categorization by phenotype ontologies and WES/WGS. (C) Representation of the "linear or univocal thinking model," following the phenotype- genotype diagnosis sequence, and the "circular or bidirectional thinking model" getting the diagnosis as a dialog between the almost simultaneous clinical interpretation of phenotype and genotype data. *CES*, clinical exome sequencing; *CMT*, Charcot-Marie-Tooth disease; *FRDA*, Friedreich ataxia; *HPO*, Human Phenotype Ontology; *ORDO*, Orphanet Rare Disease Ontology; *WES*, whole exome sequencing; *WGS*, whole genome sequencing.

To understand the challenge of integral knowledge in pediatrics, moving beyond genomics is needed. We must combine different approaches to omics, related to epigenetic changes, proteins, metabolites, and cellular compartments, encompassing physiological pathways and networks, and their relation to environmental changes.

Funding

The Palau's research group is funded by grants from the Spanish Ministry of Science, Innovation and Universities, CIBERER and the Instituto de Salud Carlos III, Generalitat de Catalunya, European Commission DG SANCO, Ramon Areces Foundation, Isabel Gemio Foundation and Amigos de Nono Foundation.

References

[1] Lee B. Integration of genetics into pediatric practice. In: Kliegman RM, Stanton BF, St Geme III JW, Schor NF, editors. Nelson texbook of pediatrics. 20th ed. Philadelphia: Elsevier; 2016. p. 580–4.
[2] Smith M. Unravelling complexities in genetics and genomics: impact on diagnosis counseling and management. Singapore: World Scientific Publishing; 2017.
[3] Baird PA, Anderson TW, Newcombe HB, et al. Genetic disorders in children and young adults: a population study. Am. J. Hum. Genet. 1988;42:677–93.
[4] Yoon PW, Olney RS, Khoury MJ, Sappenfield WM, Chavez GF, Taylor D. Contribution of birth defects and genetic diseases to pediatric hospitalizations. A population-based study. Arch. Pediatr. Adolesc. Med. 1997;151:1096–103.

[5] Connolly JJ, Hakonarson H. The impact of genomics on pediatric research and medicine. Pediatrics 2012;129:1150–60.

[6] Bacino CA, Lee B. Cytogenetics. In: Kliegman RM, Stanton BF, St. Geme III JW, y Schor NF, editors. Nelson's texbook of pediatrics. 20th ed. Philadelphia: Elsevier; 2016. p. 604–26.

[7] Strande NT, Berg JS. Defining the clinical value of genomic diagnosis in the era of next-generation sequencing. Annu. Rev. Genom. Hum. Genet. 2016;17:303–32.

[8] Gay LM. Pediatrics: sequencing the next generation. Cell 2012;148:1073–4.

[9] Scott DA, Lee B. The genetic approach to pediatric medicine. In: Kliegman RM, Stanton BF, St. Geme JW III, y Schor NF, editors. Nelson's texbook of pediatrics. 20th ed. Philadelphia: Elsevier. p. 584–7.

[10] Yang Y, Muzny DM, Xia F, Niu Z, Person R, Ding Y, et al. Molecular findings among patients referred for clinical whole-exome sequencing. JAMA 2014;312:1870–9.

[11] Lee H, Deignan JL, Dorrani N, Strom SP, Kantarci S, Quintero-Rivera F, et al. Clinical exome sequencing for genetic identification of rare Mendelian disorders. JAMA 2014;312:1880–7.

[12] Iglesias A, Anyane-Yeboa K, Wynn J, Wilson A, Truitt Cho M, Guzman E, et al. The usefulness of whole-exome sequencing in routine clinical practice. Genet. Med. 2014;16:922–31.

[13] Kliegman RM, Bordini BJ, Basel D, Nocton JJ. How doctors think: common diagnostic errors in clinical judgment—lessons from an undiagnosed and rare disease program. Pediatr. Clin. 2017;64:1–15.

[14] Robinson PN, Mundlos S. The human phenotype ontology. Clin. Genet. 2010;77:525–34.

[15] Wright CF, FitzPatrick DR, Firth HV. Paediatric genomics: diagnosing rare disease in children. Nat. Rev. Genet. 2018;19:253–68.

[16] Bick D, Dimmock D. Whole exome and whole genome sequencing. Curr. Opin. Pediatr. 2011;23:594–600.

[17] Richards S, Aziz N, Bale S, Bick D, Das S, Gastier-Foster J, et al. Standards and guidelines for the interpretation of sequence variants: a joint consensus recommendation the American college of medical genetics and genomics and the association of molecular pathology. Genet. Med. 2015;17:405–24.

[18] Resta R, Biesecker BB, Bennett RL, Blum S, Hahn SE, Strecker MN, et al. A new definition of genetic counseling: national society of genetic counselors' task force report. J. Genet. Couns. 2006;15:77–83.

[19] Wang J, Lin ZJ, Liu L, Xu HQ, Shi YW, Yi YH, et al. Epilepsy-associated genes. Seizure 2017;44:11–20.

[20] Noebels J. Pathway-driven discovery of epilepsy genes. Nat. Neurosci. 2015;18:344–50.

[21] Mullen SA, Carvill GL, Bellows S, Bayly MA, Trucks H, Lal D, Sander T, Berkovic SF, Dibbens LM, Scheffer IE, Mefford HC. Copy number variants are frequent in genetic generalized epilepsy with intellectual disability. Neurology 2013;81:1507–14.

[22] Colvert E, Tick B, McEwen F, et al. Heretability of autism spectrum disorder in a UK population-based twin sample. JAMA Psychiatry 2015;72:415–23.

[23] De Rubeis S, Buxbaum JD. Genetics and genomics of autism spectrum: embracing complexity. Hum. Mol. Genet. 2015;24(R1):R24–31.

[24] Geschwind DH, Flint J. Genetics and genomics of psychiatric disease. Science 2015;349:1489–94.

[25] Sanders SJ, He X, Willsey AJ, et al. Insights into autism spectrum disorder genomic architecture and biology from 71 risk loci. Neuron 2015;87:1215–33.

[26] De Rubeis S, He X, Goldberg AP, et al. Synaptic, transcriptional and chromatin genes disrupted in autism. Nature 2014;515:209–15.

[27] Geschwind DH, State MW. Gene hunting in autism spectrum disorder: on the path to precision medicine. Lancet Neurol. 2015;14:1109–20.

[28] Gennarino VA, Palmer EE, McDonell LM, et al. A mild PUM1 mutation is associated with adut-onset ataxia, whereas haploinsufficiency causes developmental delay and seizures. Cell 2018;172:924–36.

[29] Santen GWE, Aten E, Vulto-van Silfhout AT, et al. Coffin-Siris syndrome and BAF complex: genotype-phenotype study in 63 patients. Hum. Mutat. 2013;34:1519–28.

[30] Ronan JL, Wu W, Crabtree GR. From neural development to cognition: unexpected roles for chromatin. Nat. Rev. Genet. 2013;14:347–59.

[31] D'Ganna AM, Walsh CA. Somatic mosaicism and neurodevelopmental disease. Nat. Neurosci. 2018;21:1504–14.

[32] Berger MF, Mardis ER. The emerging clinical relevance of genomics in cancer medicine. Nat. Rev. Clin. Oncol. 2018;15:353–65.

[33] Dean SJ, Farmer M. Pediatric cancer genetics. Curr. Opin. Pediatr. 2017;29:629–33.

[34] Forrest SJ, Geoerger B, Janeway KA. Precision medicine in pediatric oncology. Curr. Opin. Pediatr. 2018;30:17–24.

[35] Tasian SK, Hunger SP. Genomic characterization of paediatric acute lymphoblastic leukaemia: an opportunity for precision medicine therapeutics. Br. J. Haematol. 2017;176:867–82.

[36] Mody RJ, Prensner JR, Everett J, Pearsons DW, Chinnaiyan AM. Precision medicine in pediatric oncology: lessons learned and next steps. Pediatr. Blood Cancer 2017;64:e26288.

[37] Zhang J, Walsh MF, Wu G, et al. Germline mutations in predisposition genes in pediatric cancer. N. Engl. J. Med. 2015;373:2336–46.

[38] Verma IM. Gene therapy that works. Science 2013;341:853–5.

[39] Condren ME, Bradshaw MD. Ivacaftor: a novel gene-based therapeutic approach for cystic fibrosis. J. Pediatr. Pharmacol. Ther. 2013;18:8–13.

[40] Kuk K, Taylor-Cousar JL. Lumacaftor and ivacaftor in the management of patients with cystic fibrosis: current evidence and future prospects. Ther. Adv. Respir. Dis. 2015;9:313–26.

[41] Prakash V. Spinraza—a rare disease success story. Gene Ther. 2017;24:497.

[42] Finkel RS, Mercuri E, Darras BT, et al. Nusinersen versus sham control in infantile-onset spinal muscular atrophy. N. Engl. J. Med. 2017;377:1723–32.

[43] Mercuri E, Darras BT, Chiriboga CA, et al. Nusinersen versus sham control in later-onset spinal muscular atrophy. N. Engl. J. Med. 2018;378:625–35.

[44] Girard T, Servais L. Nusinersen treatment of spinal muscular atrophy: current knowledge and existing gaps. Dev. Med. Child Neurol. 2019;61:19–24.

[45] Mendell JR, Al-Zaidy S, Shell R, et al. Single-dose gene-replacement therapy for spinal muscular atrophy. N. Engl. J. Med. 2017;377:1713–22.

[46] Pennisi E. The CRISPR craze. Science 2013;341:833–6.

[47] Adli M. The CRISPR tool kit for genome editing and beyond. Nat. Commun. 2018;9:1911.

[48] Lander ES. The heroes of CRISPR. Cell 2016;164:18–28.

[49] Hartmann J, Schüßler-Lenz M, Bondanza A, Buchholz CJ. Clinical development of CAR T cells-challenges and opportunities in translating innovative treatment concepts. EMBO Mol. Med. 2017;9:1183–97.

[50] Srivastava S, Riddell SR. Engineering CAR-T cells: design concepts. Trends Immunol. 2015;36:494–502.

[51] Berg JS, Agrawal PB, Bailey DB, et al. Newborn sequencing in genomic medicine and public health. Pediatrics 2017;139. e20162252.

[52] Howard HC, Knoppers BM, Cornel MC, et al. Whole-genome sequencing in newborn screening? A statement on the continued importance of targeted approaches in newborn screening programmes. Eur. J. Hum. Genet. 2015;23:1593–600.

[53] Johnston J, Lantos JD, Goldenberg A, et al. Sequencing newborns: a call for nuanced use of genomics technologies. Hastings Cent. Rep. 2018;48(Suppl. S2):S2–51.

[54] Berman JJ. Rare disease and orphan drugs. London: Academic Press – Elsevier; 2014.

[55] Gahl WA, Adams DR, Markello TC, Boerkoel NF, Tifft CJ. Genetic approaches to rare and undiagnosed diseases. In: Kliegman RM, Stanton BF, St. Geme III JW, y Schor NF, editors. Nelson's texbook of pediatrics. 20th ed. Philadelphia: Elsevier; 2016. p. 629–33.

[56] Gahl WA, Boerkoel CF, Boehm M. The NIH undiagnosed diseases program: bonding scientists and clinicians. Dis. Mod. Mech. 2012;5:3–5.

[57] Kliegman RM, Ruggeri BE, Smith MM. The team-based approach to undiagnosed and rare diseases. Pediatr. Clin. 2017;64:17–26.

[58] Ramoni RB, Mulvihill JJ, Adams DR, et al. The undiagnosed diseases network: accelerating discovery about health and disease. Am. J. Hum. Genet. 2017;100:185–92.

[59] Palau F. Personalized medicine in rare diseases. Per. Med. 2012;9:137–41.

[60] Boycott KM, Ardigó D. Addressing challenges in the diagnosis and treatment of rare genetic diseases. Nat. Rev. Drug Discov. 2018;17:151–2.

Chapter 15

Molecular pathogenesis and precision medicine in gastric cancer

Renu Verma and Prakash Chand Sharma
University School of Biotechnology, Guru Gobind Singh Indraprastha University, Dwarka, New Delhi, India

Introduction

According to the World Health Organization (WHO), cancer-related health disorders caused 9.6 million deaths in 2018 and remained the second leading cause of deaths globally. The most common types of cancer in men include lung, prostate, colorectal, stomach and liver, while breast, colorectal, lung, cervix, and thyroid cancers are more common among women. Approximately 30%–50% of cancer-related deaths could be prevented by addressing key risk factors, like consumption of tobacco products and alcohol, minimizing infection-related factors, and by maintaining a healthy lifestyle. Cancer results from a large number of genetic and epigenetic changes in the genome that affect mismatch repair genes, tumor suppressor genes, and oncogenes. These alterations interrupt molecular pathways responsible for proper functioning and regulation of cell growth, apoptosis, and metastasis.

Worldwide, GC is the sixth most common cancer (1.03 million cases in 2018), and the third leading cause of cancer-related mortality (783,000 deaths in 2018) [1]. GC is more common in developing countries; however, it is relevant in all continents. The scarcity of biomarkers for early detection, classification, and prognosis, has been a barrier in the management of GC.

Next-generation sequencing (NGS) techniques

Illumina sequencing

Illumina utilizes the sequencing-by-synthesis approach, with a flow channel (8-channel sealed glass microfabricated device), which allows bridge amplification, namely amplification of fragments over a solid surface. For incorporation of nucleotides into the cluster fragments, DNA polymerase along with four 3′-OH blocked fluorescently labeled nucleotides, are simultaneously added to the flow channel. These fragments are primed with oligomeric units. After each incorporation event, remaining molecules are washed away. Next, the imaging step conducted on an optic instrument scans each lane of flow cells in 100-tile segments. Once it is done, chemicals which block the 3′-OH blocking groups, are added to flow cell, so that each strand is prepared for another round of incorporation. Poor quality sequences are removed, by a quality checking pipeline, and a base-calling algorithm assigns sequences and quality value to each read.

454 sequencing

Roche 454 sequencing can sequence much longer reads simultaneously, for the detection of minor variations. Also known as 454 FLX pyrosequencing, it was the first developed next-generation sequencing technique. The downstream reaction takes place with the release of pyrophosphate, after a DNA polymerase incorporates a nucleotide. It produces light with the help of the luciferase enzyme, which can be registered by a suitable detector. In the Roche approach, agarose beads, which carry oligonucleotides on their surface, are mixed with the fragment library to amplify single-stranded DNA copies. A fragment: bead complex mixture is formed, which is encapsulated into oil-water micelles containing PCR reactants. Clonal amplification takes place in aqueous micro-reactors. Each bead is decorated with one million copies of DNA fragment, which are then sequenced together. Substitution error is common in Roche sequencing because each nucleotide is incorporated specifically.

Precision Medicine for Investigators, Practitioners and Providers. https://doi.org/10.1016/B978-0-12-819178-1.00015-0

Ion Torrent semiconductor sequencing

Ion Torrent uses semiconductor-based technology. About one million DNA molecules are present on the surface of the semiconductor chip micro hole. For sequencing, this chip is passed through the flow of nucleotides, and complementary nucleotides are incorporated in the DNA, followed by the release of hydrogen ion, which is detected by a hypersensitive ion sensor. As it is a direct detection method, no scanning or light is required. The high concentration of H^+ ion causes a change in pH and produces a high electronic signal, which is converted into a digital signal. Although being a simple, less expensive, and reliable technique with a smaller machine set up, the technique may not be suitable for sequencing large genomes.

SOLiD sequencing

Applied Biosystems SOLiD sequencer is based on the principle of two base encoding. It uses a library consisting of adaptor-flanked fragments. Similar to other NGS-techniques, emulsion PCR is the approach to amplify DNA fragments on the surfaces of 1-μm magnetic beads, for a signal during a reaction. When these fragments are deposited on flow cell slide, primer is annealed to the adaptor sequences, followed by addition of DNA ligase and fluorescently labeled octamers, whose fourth and fifth bases are encoded by fluorescent labels. After fluorescence detection, labeled bases are removed from the ligated octamer, and then another round of hybridization and ligation takes place.

Other NGS platforms include pacific biosciences, sequel, and nanopore alternatives, with less read length and higher error rate.

Classification of gastric cancer

A significant advance has been the genomic and molecular classification of GC, provided by The Cancer Genome Atlas (TCGA), based on whole genome sequencing, whole exome sequencing, RNA sequencing, and microRNA sequencing. TCGA system categorizes GC into four subtypes, namely EBV positive, microsatellite unstable, genomically stable, and chromosomal instability [2].

Epstein–Barr virus (EBV) positive GC

This category is represented by 9% of gastric cancers and is characterized by CpG island methylation phenotype and high levels of DNA hypermethylation. Overexpression of programmed death ligand 1 and 2 (*PD-L1* and *PD-L2*) has also been associated and could be used for therapeutic purposes. There is a strong affinity of *PIK3CA* mutations with EBV positive gastric cancers.

Microsatellite unstable GC

This subtype has been observed in 22% of GC incidences, being characterized by microsatellite instability (MSI). CpG island methylation phenotype is registered, including hypermethylation of the MLH1 promoter. Mutational analysis of MSI samples has identified 37 significantly mutated genes, including *TP53*, *PIK3A*, *KRAS*, and *ARID1A*. Unlike colorectal cancer, *BRAF* and *V600E* mutations are not associated with microsatellite instable GCs.

Genomically stable GC

GS subgroup comprises 20% of cases with gastric cancer, exhibiting diffuse histology with *CDH1* mutations. Other features of genomically stable gastric cancers include the presence of mutations in the *RHOA* gene and overexpression of cell adhesion pathway genes. The fusion of CLDN18-ARHGAP26 has also been observed.

GC with chromosomal instability

It is noticed in the remaining 50% of gastric cancers, mainly with intestinal histology. The focus here is aneuploidy and amplifications of receptor tyrosine kinases (RTKs). The group displays a high propensity for *TP53* mutations. Based on recurrent amplifications of the *VEGFA* gene, angiogenesis has been predicted as an important feature of chromosomal unstable gastric cancers.

Older classifications of GC based on histology are given by Lauren and WHO. Lauren [3] subdivided GCs into intestinal type (54%), diffuse type (32%), and indeterminate type (15%). The WHO has classified GC into four histological subtypes viz. tubular, papillary, mucinous, and poorly cohesive [4] (Fig. 15.1).

Genomic, transcriptomic, microbiomic, metabolomic and proteomic studies in gastric cancer

The alluded to technologies or related ones, able to sequence a vast number of short reads of DNA and RNA, much more quickly and cheaply than the previously used Sanger method [5], have been widely applied for gastric cancer studies, within a number of omics settings (Fig. 15.2).

Genomics and transcriptomics

To ascertain the role of guanosine triphosphate-binding protein 4 (*GTPBP4)* in GC, transcriptome profiling using Illumina platform was performed in a cancer cell line MKN45, with and without *GTPBP4* knockdown. The expression of the tumor suppressor gene, *p53*, was found to

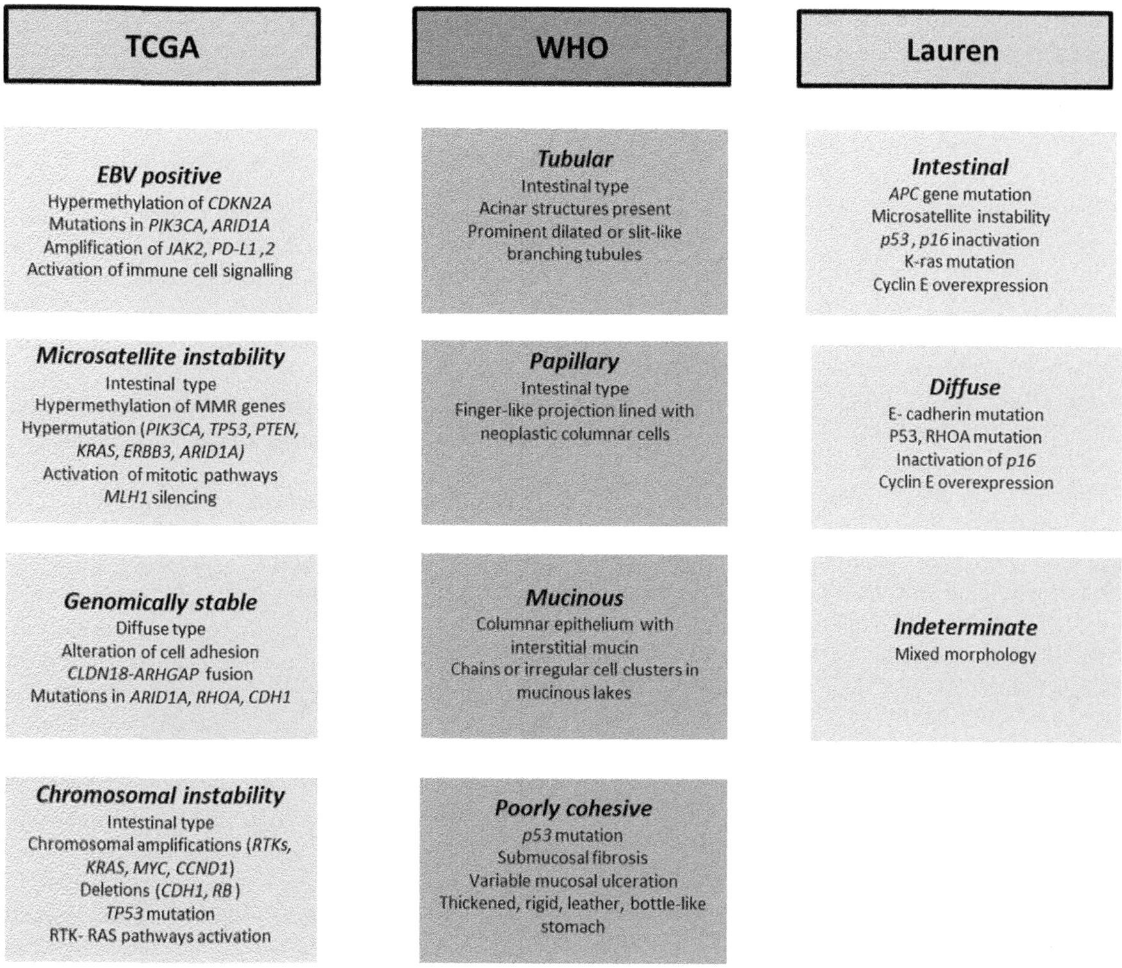

FIGURE 15.1 Molecular classification of gastric cancer (TCGA), compared to older clinical and histological guidelines (WHO, Lauren).

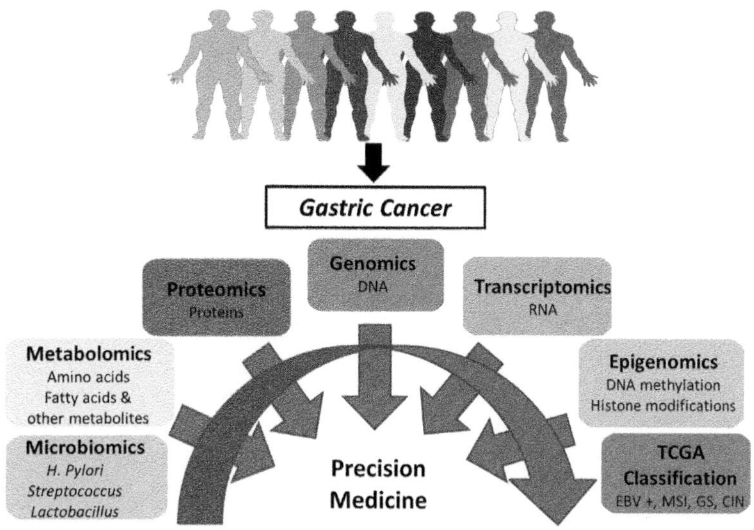

FIGURE 15.2 Relevant omics techniques for gastric cancer diagnosis, treatment, prognosis, and classification (TCGA).

be increased in knockdown mutants, while its negative effectors were downregulated [6]. A novel method "non-overlapping integrated reads (NOIR)," was introduced for detection and quantification of mutations in circulating DNA, using Ion Torrent sequencing. Frequency of mutation in a tumor suppressor gene, *TP53*, was determined at five different progression stages, showing an increase in mutation level with the progression of gastric cancer [7].

Exploiting a noninvasive approach for ion personal genome machine (PGM) based, targeted sequencing using stool specimens, five hotspots of mutations in *APC*, *CDKN2A*, and *EGFR* genes, and seven novel mutations in *APC, CDH1, DDR2, HRAS, NRAS, PTEN*, and *SMARCB1* were detected [8]. Differentially expressed 74 long non-coding RNAs, and 449 mRNAs, were identified in 3 GC samples through Illumina HiSeq sequencing. Genes *FEZF1-AS1, HOTAIR*, and *LINC01234* were perceived to have potential diagnostic value in gastric cancer [9]. Similarly, transcriptome sequencing revealed differentially expressed 1181 mRNAs and 390 long noncoding RNAs in GC, using Illumina platform. Also, the role of four lncRNAs, including AC016735.2, AP001626.1, RP11-400N13.3, and RP11-243M5.2, was recognized, as a source of potential biomarkers in GC [10].

Helicobacter pylori *and gastric microbiome*

Presence of *Helicobacter pylori* and Epstein Barr virus in the microbiome of endoscopic biopsies has been documented through whole-genome sequencing. The bacterial content of the gastric microbiome in actively infected *H. pylori*-positive individuals is increased. Whole genome sequencing performed on patients undergoing phase II pazopanib treatment, revealed a mutation in *BRAF* V600E, causing drug resistance, which could lead to metastasis [11]. A study on integrated transcriptome with exome sequencing analyzed 24 significantly mutated genes in microsatellite stable (MSS) tumors, and 16 in microsatellite unstable (MSI) tumors, along with splice site variants. An isoform of *ZAK* gene, TV1, was found to be upregulated, inducing robust transcriptional activation of several cancer-related signaling genes such as *AP1* and *NFkB*, known to be modulated by ZAK activity, while isoform TV2 displayed variable levels in GC [12].

A new approach using a green fluorescent protein (GFP) expressing attenuated adenovirus, wherein telomerase promoter regulates viral replication (TelomeScan, OBP-401), has been developed, to identify biologically malignant subpopulations in cytology-positive GC patients, Using a panel of target genes on MiSeq platform, peritoneal washes from positive TelomeScan patients revealed 774 genetic variants, including single-nucleotide polymorphisms (SNP), deletions, insertions, and point mutations [13]. A study on RNA-Seq and microarray data from TCGA, reported lncRNA as a key regulator of gastric cancer development and progression. Shorter survival and poorer prognosis occurred in patients with high HOXA11-AS expression [14].

Liquid biopsy

Circulating tumor DNA (ctDNA), one of the modalities of liquid biopsy, refers to DNA released from cancer cells into the bloodstream. Targeted deep sequencing has shown *TP53* mutation in primary GC tissues [15]. NGS-based genomic profiling has provided a better picture of amplification followed by base substitutions of activating mutations in *ERBB2* in tumor tissues. Patients with these mutations can be benefitted from approved targeted *ERBB2* inhibitor therapy [16].

Deep sequencing revealed an amplification of *FGFR2*, that was found exclusively in the primary lesion, and a deletion in the gene *TGFBR2* occurring exclusively during metastasis [17]. Whole exome and genome sequencing employed in autosomal-dominant cancer-predisposition syndrome GAPPS (Gastric adenocarcinoma and proximal polyposis of the stomach), could not detect causal point mutations, which were detected through Sanger sequencing, emphasizing a shortcoming of the NGS technique [18].

Transcription factors, splicing factors, tumor suppressor genes, and many other genes were looked upon, for splice variants in Epstein–Barr virus-associated gastric cancer through RNA sequencing. Various splice variants were found to be linked with EBV positive GC samples acquired from the TCGA database [19]. Targeted sequencing of 46 cancer-related genes, helped in identifying differences in mutation frequency pattern, in gastroesophageal junction and gastric carcinoma. *TP53* mutations were the most common in gastroesophageal junction, while mutations in *APC* and *CTNNB1* were prevalent in gastric carcinoma [20]. *CTNNB1* mutations were also detected in all the gastrointestinal tumor samples in another study using the same targeted multigene NGS approach [21].

A high proportion of 78% of 116 GC cases, harbored at least one clinically relevant genomic alteration in *KRAS, CDKN2A, CCND1, ERBB2, PIK3CA, MLL2, MET, PTEN, ATM, DNMT3A, NF1, NRAS*, and *MDM2*, and 116 cases had alterations in *TP53, ARID1A*, and *CDH1* [22]. A mutation common to cancers that activates the PI3/AKT signaling pathway, *PIK3CA* gene mutation, was quantified using pyrosequencing in GC patients, suggesting no prognostic relationship of the gene [23].

Transcriptome analysis

Transcriptome refers to the complete set of transcripts of a cell or population of cells. RNA-Seq approach has surpassed the well-known microarray technique, for the assessment of the level of gene expression. Unlike microarrays, RNA-Seq can be used for the analysis of expression of novel transcripts without using probes.

Transcriptome profiling using Illumina platform revealed a high number of expressed genes in tumor (13,228) and normal (13,674) tissues in GC patients. Also, Cadherin-1 gene (*CDH1*), with 309 fold upregulation (24), was highlighted in GC, while another study reported expression change to be 36 fold [24]. Dermatopontin gene (*DPT*) plays an important role in cell-matrix interactions and is a key gene in TGF-β signaling. *DPT* gene has been postulated to modify the behavior of *TGFBR2*, through interaction with decorin [25]. Low expression (~40 fold) of *DPT* was detected in a study on Chinese GC patients, along with downregulated *TGFBR2*. Other reports have recorded downregulation of these genes [24]' corroborating the low expression of *DPT* in oral cancer validated by qRT-PCR [26]. *TGFBR2* gene has also been linked with the microsatellite instability and is being explored as a potential biomarker in GC.

Length polymorphism at microsatellite loci, in coding regions of genes, affects their expression by the premature occurrence of a stop codon. We have also observed microsatellite instability in coding regions of some tumor suppressor and mismatch genes, which have led to the formation of truncated proteins in GC tissues (unpublished work). The findings emphasized the significance of the particular TCGA subgroup (MSI unstable). *TGFBR2* showed lack of expression in MSI-H samples. Genes having MSI in their untranslated regions displayed differential expression, as compared to genes without UTR mutations. Upregulated and downregulated genes (137 and 139, respectively), containing mutations at microsatellite loci were observed, and 96% of these mutations were present in the UTR regions. These observations suggest an influence of mutations in UTR on gene expression.

Transcriptome results validated by q-PCR revealed significantly downregulated expression of *MGLL*, *SORL1*, *C20orf194*, *WWC3*, and *PXDC1* genes in MSI-H cell lines. Mutations in 3′ UTR region of *MGLL* gene, resulted in 42.6% downregulation of recombinant luciferase, indicating the presence of aberrant gene products as a consequence of MSI [27].

Receptor tyrosine kinases

Receptor tyrosine kinases (RTKs) play a crucial role in the activation of various intracellular signaling pathways. Role of several RTKs inhibitors, in the antiproliferative activity, has been witnessed in clinical trials in target-specific therapy. Silencing and overexpression of the *ARID1A* gene led to both increased and decreased proliferation, respectively, in tissue culture. Silencing of the *ARID1A* gene also increases the level of E2F1 and cyclin E1 transcription factors. Long recurrence-free survival from mutation or deficiency of protein of *ARID1A* [28] has been predicted.

A tyrosine kinase receptor gene *EGFR* exhibited amplification and overexpression in GC [29]. Inhibitors of another gene of the RTK family, fibroblast growth factor receptor 2 (*FGFR2*), have shown some clinical efficacy in GC [30]. Ki23057, one of the FGFR inhibitors, along with 5-fluorouracil, has displayed synergistic antitumor effects for GC treatment [31]. Loss of function of the *SMAD4* gene helps in epithelial–mesenchymal transition, and its re-expression has been seen in reversing the process [32]. Expression of one of the important genes involved in breast cancer, *BRCA1*, is correlated with sensitivity to chemotherapeutics in gastric cancer [33].

Microbiomics

Helicobacter pylori, a gram-negative bacteria, has infected half of the world's human population, out of which 1%–3% develop GC [34]. Virulence factors affecting gastric cancer risk include *cag* and VacA pathogenicity. Although *H. pylori* has been defined as one of the strong risk factors for GC, other gastric microbes could also influence the development of the disease. Pyrosequencing of GC samples showed an abundance of *Bacilli* and members of the Streptococcaceae family when compared to samples of chronic gastritis and intestinal metaplasia [35]. Decreased acidity of the gastric lumen has been associated with the increased risk of *Clostridium difficile* infection [36].

Gastric microbiota was found to be abundantly represented by *H. pylori*, *Haemophilus*, *Serratia*, *Neisseria* and *Stenotrophomonas* using MiSeq platform, and an increased abundance was observed in the bacterial diversity after eradication of *H. pylori* [37]. Frequency of *H. pylori* significantly decreased in a tumoral microenvironment, as compared to normal and peritumoral microhabitats. *Prevotella copri, Bacteroides uniformis*, and *H. pylori* count decreased while *Prevotella melaninogenica*, *Streptococcus anginosus*, and *Propionibacterium acnes* increased in tumoral gastric microbiota. Overall, the enrichment of bacterial diversity decreased in tumoral and peritumoral microhabitat [38].

Metabolomics

Compared to human genome, epigenome, transcriptome, and proteome, the metabolome is not directly involved in the information flow of the central dogma, which encompasses the steps by which DNA instructions are converted in a functional product. However, metabolomics measures both upstream and downstream changes that are close to environmental exposures and phenotypic changes [39]. The two main techniques to explore the metabolomic status of the target tissue are nuclear magnetic resonance (NMR) and mass spectrometry (MS).

TABLE 15.1 Summary of metabolites found in gastric cancer.

S. No.	Sample Type	Technique	Metabolites	Expression	References
1.	Gastric Juice	SIFT-MS	Acetaldehyde, Acetone, Acetic acid, Hexanoic acid, Hydrogen cyanide, Hydrogen sulfide, Methanol, Methyl phenol	Upregulated	Kumar et al. [43]
			Formaldehyde	Downregulated	
2.	Serum	GC-MS	Hexadecanenitrile, Sarcosine, Valine	Upregulated	Song et al. [44]
			Cholesterol,	Downregulated	
			1,2,4,- Benzenetricarboxylic acid,		
			2-Amino-4-hydroxypteridinone,		
			9,12 Octadecadienoic acid,		
			9-Octadecenoic acid,		
			9-Octadecenoic acid,		
			Fumaric acid, Glutamine, Hexanedioic acid		
			Mesyl-arabinose, Benzeneacetonitrile, Nonahexacontanoic acid,		
			Trans-13- octadecenoic acid		
3.	Serum	GC-MS	3-Hydroxypropionic acid, 3-Hydroxyisobutyric acid	Upregulated	Ikeda et al. [45]
			Octanoic acid, Phosphoric acid, Pyruvic acid	Downregulated	
4.	Serum	GC-MS	11-Eicosenoic acid, 2-Hydroxybutyrate, Asparagine, Azelaic acid, Glutamic acid, Ornithine, Pyroglutamate, Urate, y-tocopherol	Upregulated	Yu et al. [46]
			Creatinine, Threonate	Downregulated	
5.	Tissue	HR-MAS-MRS	Alanine, Choline, Glycine, Triacylglycerides	Upregulated	Calabrese et al. [47]
6.	Tissue	GC-MS	1-Phenanthrene, a-Ketoglutaric acid, Benzenepropanoic acid, Carboxylic acid, Fumaric acid, Octadecanoic acid, Squalene, Valeric acid, Xylonic acid	Upregulated	Song et al. [48]
			3-Hydroxybutanoic acid, 9-Hexadecanoic acid, 9-Octadecenamide, Arachidonic acid, Cis-vaccenic acid, Hexadecanoic acid	Downregulated	
7.	Tissue	GC-MS	Acetamide, Butanetriol, Butenoic acid, Galactofuranoside, Glutamine, Hypoxanthine, Isoleucine, L-Cysteine, L-Tryosine, Naphtalene, Oxazolethione, Phenanthrenol, Serine, Valine	Upregulated	Wu et al. [49]
			D-Ribofuranose, L-Altrose, L-Mannofuranose, Phosphoserine	Downregulated	

Continued

TABLE 15.1 Summary of metabolites found in gastric cancer.—cont'd

S. No.	Sample Type	Technique	Metabolites	Expression	References
8.	Tissue	GC-MS	Fructose, Glyceraldehyde, Isocitric acid, Lactic acid, Pyruvic acid	Upregulated	Cai et al. [50]
			Fumaric acid	Downregulated	
9.	Gastric juice	HPLC	Phenylalanine, Tryptophan, Tyrosine	Upregulated	Deng et al. [51]
10.	Gastric juice	LC-MS	Anthranilic acid, Indole-3-lactic acid, Kynurenic acid, Kynurenine, Nicotinic acid, Tryptophan	Upregulated	Choi et al. [52]
11.	Tissue	HR-MAS-NMR	Alanine, Glutamate, Isoleucine, Lactate, Leucine, Lysine, Phenylalanine, Taurine, Valine	Upregulated	Jung et al. [41]
12.	Serum	GC-MS	b-Hydroxybutyrate, Citrate, Succinate, Docosahexaenoic acid, Fumurate, Glutamic acid, Hepatanoic acid, Hexadecenoic acid, Succinate	Upregulated	Aa et al. [53]
			Glucose	Downregulated	

GC-MS, Gas Chromatography Mass Spectrometry; *HPLC*, High Performance Liquid Chromatography; *HR-MAS-MRS*, High Resolution Magic Angle Spinning Magnetic Resonance Spectroscopy; *HR-MAS-NMR*, High Resolution Magic Angle Spinning Nuclear Magnetic Resonance Spectroscopy; *LC-MS*, Liquid Chromatography- Mass Spectrometry; *SIFT-MS*, Selected Ion Flow Tube Mass Spectrometry.

Historical vignettes

Metabolic reprogramming is a hallmark of cancer, linked to tumorigenesis. Otto Warburg (1883–1970) observed a characteristic metabolic pattern, of large glucose consumption for glycolysis in tumor cells even under conditions of sufficient oxygen (Warburg effect). Lactic acid concentration increases in urine and tissue samples in gastric cancer patients [40,41]. The utility of metabolomics in diagnosis and prognosis has been recognized [42]. A list of different metabolites in gas chromatography (GC) is given in Table 15.1.

Proteomics

Proteomics addresses virtually all proteins expressed in a cell, tissue, or organism [54]. Proteomics-related approaches have been used to identify differentially expressed proteins between normal and GC samples (Table 15.2). Enhanced coverage of protein sequences is required to detect low abundance proteins in proteomic studies [64]. The proteomic approaches use electrophoresis, mainly two-dimensional electrophoresis, liquid chromatography (LC), and mass spectrometry (MS) analysis for quantification and identification of expressed proteins [65]. MALDI-TOF, widely used in microbiology laboratories, as well as its variation SELDI-TOF, are two techniques of mass spectrometry used to identify proteins associated with gastric cancer. HSP27 has been found upregulated and downregulated, in gastric cancer indicating heterogeneity pattern [61,66]. Proteins enolase-alpha (ENOA), nicotinamide N-methyltransferase (NNMT), annexin 2 (ANXA2) and transgelin (TGLN), were found to be upregulated in GC samples. Gastrokine-1(GKN1) and carbonic anhydrase 2 (CA2), involved in energy metabolism, exhibit downregulation in GC samples [50,67].

Downregulation of lactate dehydrogenase (LDH) subunit LDHA and upregulation of pyruvate dehydrogenase (PDH) subunit PDHB has been observed to inhibit cell growth and cell migration [50]. Annexins are calcium-dependent and membrane-binding intracellular proteins. One such protein, ANXA2, has been reported to have an increased expression in GC [57]. Also, increased ANXA1 expression in a GC cell line with lymph node metastasis, compared with a GC cell line derived from a primary tumor, was observed [71].

Various proteins have been described in the new TCGA classification including caspase 7 (CASP7), proliferating cell nuclear antigen (PCNA), BCL2-associated X protein (BAX), spleen tyrosine kinase (SYK), Src family tyrosine kinase LCK (LCK), to have elevated expression in EBV positive subgroup, whereas upregulated expression of claudin 7 (CLDN7), von Hippel-Lindau tumor suppressor (VHL), and cyclin B1 (CCNB1) was detected in the microsatellite instability subtype. On the other hand, KIT proto-oncogene receptor tyrosine kinase (KIT), v-myc avian myelocytomatosis viral oncogene homolog (MYC), v-akt murine thymoma viral oncogene homolog (AKT), and protein kinase C alpha (PRKCA) expressions, were highly elevated in the genomically stable subtype [2].

TABLE 15.2 Details of proteomic studies in gastric cancer.

S.No.	Sample Size	No. of Differentially Expressed Proteins	Techniques	Important Protein(s)	References
1.	107	20↑	LC-MS/MS	EPHA2	Kikuchi et al. [55]
2.	9	15↑, 13↓	2DE, MS	S100A2	Liu et al. [56]
3.	12	15↑, 9↓	2DE, MS/MS	GAL4, HADHA, HADHB, HNRNPM	Kocevar et al. [57]
4.	15	42↑, 39↓	Nano-RPLC-MS/MS	ANXA1	Zhang et al. [58]
5.	12	19↑, 11↓	2DE, MS	SEPT2, UBE2N, TALDO1, GKN1, MRPL12, PACAP, GSTM3, TPT1	Kocevar et al. [59]
6.	8	26↑, 6↓	2DE, MALDI-TOF MS	ENOA, GDI2, GRP78, GRP94, PPIA, PRDX1, PTEN,	Bai et al. [60]
7.	3	7↑, 16↓	DIGE-MS, MS	HSP60, HSP27, ZNF160, SELENBP1, EEF1A1, mutant desmin, fibrinogen gamma, tubulin alpha 6, prostaglandin F synthase	Wu et al. [61]
8.	6	57↑, 50↓	2DE, MS/MS	HYOU1, TTHY, KPYM, GRP78, FUMH, ALDOA, LDHA	Liu et al. [62]
9.	3	12↑, 7↓	2DE, MS/MS	ANXA2, ANXA4	Lin et al. [63]

↑, denotes upregulation; ↓, denotes downregulation; *2DE,* Two Dimensional Gel Electrophoresis; *DIGE-MS,* Two Dimensional-Differential In Gel Electrophoresis Mass Spectrometry; *LC-MS/MS,* Liquid Chromatography Tandem Mass Spectrometry; *MALDI-TOF-MS,* Matrix-Assisted Laser Desorption Ionization Time of Flight Mass Spectrometry; *MS,* Mass Spectrometry; *Nano-RPLC-MS/MS,* Nanoliter Reverse-Phase Liquid Chromatography Mass/Mass Spectrometry

Epigenomic influences

Methylation across the genome is unraveled through whole-genome bisulfite sequencing, as well as targeted sequencing aiming to screen the specific desirable genomic regions of interest. An epigenetic trait has been defined as a "stably heritable phenotype resulting from changes in a chromosome, without alterations in the DNA sequence" [72]. Aberrant DNA methylation profiles and histone modifications are linked to developmental defects, obesity, asthma, and neurodegenerative disorders, besides cancer [73]. However, given the complexity of epigenetic mechanisms, which are influenced by aging, genetic variations such as polymorphisms, and environmental factors, deciphering epigenetic information is a challenge [74].

Epigenetic changes are somewhat similar to genetic mutations, that change the underlying structure of the DNA, contributing toward the initiation and progression of cancer [75]. For normal gene expression, epigenetic machinery responsible for DNA methylation, DNA hydroxymethylation, post-translational modifications (PTMs) of histone proteins, nucleosome remodeling, and regulation by noncoding RNAs, performs in harmony with *cis* and *trans* acting elements [76,77].

Tumor suppressor genes

Aberrant DNA methylation in the promoter region of genes, which leads to inactivation of tumor suppressor and other cancer-related genes, is the most well-defined epigenetic activity during gastric tumorigenesis. In mammalian cells, DNA methylation consists of covalent attachment of a methyl group, to the 5′ position of cytosine residues in CG dinucleotides. CG dinucleotides are not randomly distributed throughout the genome but tend to cluster in regions called *CpG islands*, mainly present in the promoter region of the genes [76,77]. An accepted definition of CpG islands describes them as DNA sequences, more than 200 base pair long, with CG content greater than 50%, and an observed/expected CpG ratio of more than 60% [76]. Methylation can also occur at nonpromoter CpG islands, defined as CpG shores, located in the vicinity of CpG islands up to 2 kb in length [78]. Methylation of CpG islands is typically associated with gene silencing, while demethylation of these sites enables transcription [76]. Various risk factors like age, diet, chronic inflammation, infection with *H. pylori*, and EBV, are also causative agents of aberrant gene methylation in GC [79].

The methylation status of *LPHN2* has been found to be a potential novel epigenetic biomarker, for cisplatin treatment in GC [80]. Defective DNA methylation in *CDH1*, *CHFR*, *DAPK*, *GSTP1*, *p15*, *p16*, *RARβ*, *RASSF1A*, *RUNX3*, and *TFPI2*, has been considered as a serum biomarker for the detection of GC [79]. A large number of genes have been identified to be methylated in the gastric mucosa of GC patients. Among them, *RASGRF1* methylation has been found to be significantly elevated, in mucosa from patients with either intestinal- or diffuse-type GC [81]. Silencing of miRNAs is also associated with hypermethylation of CpG islands. Methylation of miR34-b/c was ubiquitous in GC cell lines, but not in normal gastric mucosa from healthy *H. pylori*-negative individuals [82]. Aberrant DNA methylation in noncancerous gastric mucosa has been implicated in gastric carcinogenesis.

Pyrosequencing has been proved to be a more reliable method, in comparison to both methylation-specific polymerase chain reaction (MSP), and bisulfite sequencing [83]. In a comparative analysis, the frequency of promoter region methylation in the *TCF4* gene was reported to be higher, when analyzed by using pyrosequencing than MSP in advanced GC samples [84].

Hypermethylation in *GPX3* promoter region with a 10% cut off, was observed using pyrosequencing in 60% of the GC samples, and six out of nine cell lines [85]. Hypermethylation in the *EDNRB* gene in GC tissues has been observed and correlated with tumor infiltration. Similarly, loss of expression of the *FAT4* gene was observed in highly methylated GC cell lines, and removal of methylation by demethylating agent restored its expression. Methylation status of *FAT4* has also been associated with *H. pylori* infection in GC [86].

By analyzing 295 GC samples for CpG methylation level in 86 genes and 14 miRNAs, the Cancer Genome Atlas (TCGA) has grouped the hypermethylated genes into three categories: hypermethylated in EBV-positive subtype, hypermethylated in both EBV-positive and MSI-high subtypes, and other hypermethylated genes. Prominent methylation changes were observed in *RUNX1*, *ARHGDIB*, *PSME1*, *GZMB*, and *RBM5* genes, while *VAMP5* and *POLG* genes showed a marginal methylation difference between normal and GC cells.

Molecular biomarkers in gastric cancer

Current markers used for GC diagnosis in clinical use include CA 19-9, CA-50, and CA-72. They lack high sensitivity and specificity, which hampers their large-scale efficient and unambiguous use. Other molecular biomarkers can be classified into genetic, epigenetic, and protein markers.

Genetic markers of chemotherapy response

DPD and heparin-binding epidermal growth factor (HB-EGF) like genes are considered as related to 5FU resistance. Metallothionein-IG and HB-EGF are also potential molecular marker candidates for cisplatin resistance genes. Paclitaxel and cisplatin treatment have been predicted with *TP53* codon 72 SNP.

Epigenetic markers

Micro-RNA miR-21 was found linked to trastuzumab resistance in which miR-21 has been shown to have a regulatory effect on the treatment response. In blood and gastric secretions, long noncoding RNAs (lncRNAs) have been found to be potential biomarkers for GC. LncRNAs, such as H19, HOTAIR, and MEG3, have been suggested to have a functional role in tumorigenesis and tumor progression. Decreased methylation leads to increased expression of the secreted protein *BMP4*. Bcl-2/adenovirus E1B1 19 kDa interacting protein three and DAPK (death-associated protein kinase) methylation products, lead to lower response to fluoropyrimidine-based chemotherapy.

Protein markers

Thymidylate synthetase (TS) and DPD are indicative of 5FU tumor sensitivity. In serum, AMBP protein has been observed to correlate with chemotherapeutic response to paclitaxel and capecitabine. Another protein in serum, TUBB3, has been suggested to be involved in resistance to paclitaxel and capecitabine. FOXM1 protein in tissue predicts resistance to docetaxel. REG4 predicted resistance to docetaxel (Table 15.3).

Conclusions

Recent advances in medical science concerning prevention and treatment of GC have recorded significant success, yet the National Cancer Database (NCDB) indicates 5-year survival rate of 31% for GC, which is lower than for many tumors. Role of perioperative chemotherapy and/or radiotherapy in the improvement of overall survival (OS) has been recognized; however, the tendency to metastasis and recurrence still remains an area of concern. Adoption of precision medicine helps clinicians to customize treatment options according to patient needs, using various molecular diagnostic methods to design a better curative regimen.

TABLE 15.3 Details of drugs used for the treatment of metastatic gastric cancer.

Line of Treatment	Target	Drug	Phase	Median overall Survival (months)	References
First	EGFR	EOX ± panitumumab	III	11.3 versus 8.8	Langer et al. [87]
First		TCF ± panitumumab	III	11.7 versus 10	Tebbutt et al. [88]
First		CX ± cetuximab	III	9.4 versus 10.7	Ott et al. [89]
Second		Gefitinib versus placebo	III	3.73 versus 3.67	Langer et al. [90]
Second		Gefitinib	II	5.4	Janmaat et al. [91]
First	HER2	CX/CF ± trastuzumab	III	13.8 versus 11.1	Yoshikawa at al. [68]
First		OX ± lapatinib	III	12.2 10.5	Metzger et al. [92]
Second		Paclitaxel ± lapatinib	III	11.0 versus 8.9	Matsubara et al. [93]
Second		TDM-1	III	7.9 versus 8.6	Igney et al. [94]
Second	mTOR	Everolimus	II	8.3	Yoon et al. [95]
Second		Everolimus	II	10.1	Doi et al. [96]
Second	VEGF	CX ± bevacizumab	II	12.1 versus 10.1	Zhao et al. [97]
First	VEGF-2	FOLFOX ± ramucirumab	II	11.7 versus 11.5	Yoon et al. [98]
Second		Ramucirumab ± placebo	III	5.2 versus 3.8	Cunningham et al. [69]
Second		Ramucirumab ± paclitaxel	III	9.6 versus 7.4	Ajani et al. [70]
Second		Apatinib ± placebo	III	6.5 ± 4.7	Li et al. [99]

CF, Cisplatin 5-Fluorouracil; *CX*, Cisplatin Capecitabine; *EOX*, Epirubicin Oxaliplatin Capecitabine; *FOLFOX*, 5-Fluorouracil Leucovorin Oxaliplatin; *OX*, Oxaliplatin Capecitabine; *TCF*, Docetaxel Cisplatin 5-Fluorouracil; *TDM-1*, ado Trastuzumab Emtansine.

References

[1] Bray F, Ferlay J, Soerjomataram I, Siegel RL, Torre LA, Jemal A. Global cancer statistics 2018: GLOBOCAN estimates of incidence and mortality worldwide for 36 cancers in 185 countries. CA Cancer J. Clin. 2018;68(6):394−424.

[2] Cancer Genome Atlas Research N. Comprehensive molecular characterization of gastric adenocarcinoma. Nature 2014;513(7517):202−9.

[3] Lauren P. The two histological main types of gastric carcinoma: diffuse and so-called intestinal-type carcinoma. An attempt at a histo-clinical classification. Acta Pathol. Microbiol. Scand. 1965;64:31−49.

[4] Hu B, El Hajj N, Sittler S, Lammert N, Barnes R, Meloni-Ehrig A. Gastric cancer: classification, histology and application of molecular pathology. J. Gastrointest. Oncol. 2012;3(3):251−61.

[5] Shendure J, Ji H. Next-generation DNA sequencing. Nat. Biotechnol. 2008;26(10):1135−45.

[6] Li L, Pang X, Zhu Z, Lu L, Yang J, Cao J, et al. GTPBP4 promotes gastric cancer progression via regulating P53 activity. Cell. Physiol. Biochem. 2018;45(2):667−76.

[7] Kukita Y, Matoba R, Uchida J, Hamakawa T, Doki Y, Imamura F, et al. High-fidelity target sequencing of individual molecules identified using barcode sequences: de novo detection and absolute quantitation of mutations in plasma cell-free DNA from cancer patients. DNA Res 2015;22(4):269−77.

[8] Youssef O, Sarhadi V, Ehsan H, Bohling T, Carpelan-Holmstrom M, Koskensalo S, et al. Gene mutations in stool from gastric and colorectal neoplasia patients by next-generation sequencing. World J. Gastroenterol. 2017;23(47):8291−9.

[9] Gu J, Li Y, Fan L, Zhao Q, Tan B, Hua K, et al. Identification of aberrantly expressed long non-coding RNAs in stomach adenocarcinoma. Oncotarget 2017;8(30):49201−16.

[10] Wang Y, Zhang J. Identification of differential expression lncRNAs in gastric cancer using transcriptome sequencing and bioinformatics analyses. Mol. Med. Rep. 2018;17(6):8189−95.

[11] Park C, Ha SY, Kim ST, Kim HC, Heo JS, Park YS, et al. Identification of the BRAF V600E mutation in gastroenteropancreatic neuroendocrine tumors. Oncotarget 2016;7(4):4024−35.

[12] Liu J, McCleland M, Stawiski EW, Gnad F, Mayba O, Haverty PM, et al. Integrated exome and transcriptome sequencing reveals ZAK isoform usage in gastric cancer. Nat. Commun. 2014;5:3830.

[13] Watanabe M, Kagawa S, Kuwada K, Hashimoto Y, Shigeyasu K, Ishida M, et al. Integrated fluorescent cytology with nano-biologics in peritoneally disseminated gastric cancer. Cancer Sci. 2018;109(10):3263−71.

[14] Sun M, Nie F, Wang Y, Zhang Z, Hou J, He D, et al. LncRNA HOXA11-AS promotes proliferation and invasion of gastric cancer by scaffolding the chromatin modification factors PRC2, LSD1, and DNMT1. Cancer Res. 2016;76(21):6299−310.

[15] Hamakawa T, Kukita Y, Kurokawa Y, Miyazaki Y, Takahashi T, Yamasaki M, et al. Monitoring gastric cancer progression with circulating tumour DNA. Br. J. Canc. 2015;112(2):352–6.
[16] Chmielecki J, Ross JS, Wang K, Frampton GM, Palmer GA, Ali SM, et al. Oncogenic alterations in ERBB2/HER2 represent potential therapeutic targets across tumors from diverse anatomic sites of origin. Oncol. 2015;20(1):7–12.
[17] Alsinet C, Ranzani M, Adams DJ. One patient, two lesions, two oncogenic drivers of gastric cancer. Genome Biol. 2014;15(8):444.
[18] Li J, Woods SL, Healey S, Beesley J, Chen X, Lee JS, et al. Point mutations in exon 1B of APC reveal gastric adenocarcinoma and proximal polyposis of the stomach as a familial adenomatous polyposis variant. Am. J. Hum. Genet. 2016;98(5):830–42.
[19] Armero VES, Tremblay MP, Allaire A, Boudreault S, Martenon-Brodeur C, Duval C, et al. Transcriptome-wide analysis of alternative RNA splicing events in Epstein-Barr virus-associated gastric carcinomas. PLoS One 2017;12(5). e0176880.
[20] Li-Chang HH, Kasaian K, Ng Y, Lum A, Kong E, Lim H, et al. Retrospective review using targeted deep sequencing reveals mutational differences between gastroesophageal junction and gastric carcinomas. BMC Cancer 2015;15:32.
[21] Mafficini A, Amato E, Fassan M, Simbolo M, Antonello D, Vicentini C, et al. Reporting tumor molecular heterogeneity in histopathological diagnosis. PLoS One 2014;9(8). e104979.
[22] Ali SM, Sanford EM, Klempner SJ, Rubinson DA, Wang K, Palma NA, et al. Prospective comprehensive genomic profiling of advanced gastric carcinoma cases reveals frequent clinically relevant genomic alterations and new routes for targeted therapies. Oncologist 2015;20(5):499–507.
[23] Harada K, Baba Y, Shigaki H, Ishimoto T, Miyake K, Kosumi K, et al. Prognostic and clinical impact of PIK3CA mutation in gastric cancer: pyrosequencing technology and literature review. BMC Cancer 2016;16:400.
[24] Tian P, Liang C. Transcriptome profiling of cancer tissues in Chinese patients with gastric cancer by high-throughput sequencing. Oncol. Lett 2018;15(2):2057–64.
[25] Zhang FG, He ZY, Wang Q. Transcriptome profiling of the cancer and normal tissues from gastric cancer patients by deep sequencing. Tumour Biol. 2014;35(8):7423–7.
[26] Yamatoji M, Kasamatsu A, Kouzu Y, Koike H, Sakamoto Y, Ogawara K, et al. Dermatopontin: a potential predictor for metastasis of human oral cancer. Int. J. Cancer 2012;130(12):2903–11.
[27] Yoon K, Lee S, Han TS, Moon SY, Yun SM, Kong SH, et al. Comprehensive genome- and transcriptome-wide analyses of mutations associated with microsatellite instability in Korean gastric cancers. Genome Res. 2013;23(7):1109–17.
[28] Zang ZJ, Cutcutache I, Poon SL, Zhang SL, McPherson JR, Tao J, et al. Exome sequencing of gastric adenocarcinoma identifies recurrent somatic mutations in cell adhesion and chromatin remodeling genes. Nat. Genet. 2012;44(5):570–4.
[29] Ku GY, Ilson DH. Esophagogastric cancer: targeted agents. Cancer Treat Rev. 2010;36(3):235–48.
[30] Holbrook JD, Parker JS, Gallagher KT, Halsey WS, Hughes AM, Weigman VJ, et al. Deep sequencing of gastric carcinoma reveals somatic mutations relevant to personalized medicine. J. Transl. Med. 2011;9:119.
[31] Yashiro M, Shinto O, Nakamura K, Tendo M, Matsuoka T, Matsuzaki T, et al. Synergistic antitumor effects of FGFR2 inhibitor with 5-fluorouracil on scirrhous gastric carcinoma. Int. J. Cancer 2010;126(4):1004–16.
[32] Pohl M, Radacz Y, Pawlik N, Schoeneck A, Baldus SE, Munding J, et al. SMAD4 mediates mesenchymal-epithelial reversion in SW480 colon carcinoma cells. Anticancer Res. 2010;30(7):2603–13.
[33] Shim HJ, Yun JY, Hwang JE, Bae WK, Cho SH, Lee JH, et al. BRCA1 and XRCC1 polymorphisms associated with survival in advanced gastric cancer treated with taxane and cisplatin. Cancer Sci. 2010;101(5):1247–54.
[34] Amieva M, Peek Jr RM. Pathobiology of *Helicobacter pylori*-induced gastric cancer. Gastroenterol. 2016;150(1):64–78.
[35] Eun CS, Kim BK, Han DS, Kim SY, Kim KM, Choi BY, et al. Differences in gastric mucosal microbiota profiling in patients with chronic gastritis, intestinal metaplasia, and gastric cancer using pyrosequencing methods. Helicobacter 2014;19(6):407–16.
[36] Jump RL. *Clostridium difficile* infection in older adults. Aging Health 2013;9(4):403–14.
[37] Li TH, Qin Y, Sham PC, Lau KS, Chu KM, Leung WK. Alterations in gastric microbiota after *H. Pylori* eradication and in different histological stages of gastric carcinogenesis. Sci. Rep. 2017;7:44935.
[38] Liu X, Shao L, Liu X, Ji F, Mei Y, Cheng Y, et al. Alterations of gastric mucosal microbiota across different stomach microhabitats in a cohort of 276 patients with gastric cancer. EBioMedicine 2019:236–348.
[39] Koo I, Wei X, Zhang X. Analysis of metabolomic profiling data acquired on GC-MS. Methods Enzymol. 2014;543:315–24.
[40] Hu JD, Tang HQ, Zhang Q, Fan J, Hong J, Gu JZ, et al. Prediction of gastric cancer metastasis through urinary metabolomic investigation using GC/MS. World J. Gastroenterol. 2011;17(6):727–34.
[41] Jung J, Jung Y, Bang EJ, Cho SI, Jang YJ, Kwak JM, et al. Noninvasive diagnosis and evaluation of curative surgery for gastric cancer by using NMR-based metabolomic profiling. Ann. Surg. Oncol. 2014;21(Suppl. 4):S736–42.
[42] Xiao S, Zhou L. Gastric cancer: metabolic and metabolomics perspectives (review). Int. J. Oncol. 2017;51(1):5–17.
[43] Kumar S, Huang J, Cushnir JR, Spanel P, Smith D, Hanna GB. Selected ion flow tube-MS analysis of headspace vapor from gastric content for the diagnosis of gastro-esophageal cancer. Anal. Chem. 2012;84(21):9550–7.
[44] Song H, Peng JS, Dong-Sheng Y, Yang ZL, Liu HL, Zeng YK, et al. Serum metabolic profiling of human gastric cancer based on gas chromatography/mass spectrometry. Braz. J. Med. Biol. Res. 2012;45(1):78–85.
[45] Ikeda A, Nishiumi S, Shinohara M, Yoshie T, Hatano N, Okuno T, et al. Serum metabolomics as a novel diagnostic approach for gastrointestinal cancer. Biomed. Chromatogr. 2012;26(5):548–58.
[46] Yu L, Aa J, Xu J, Sun M, Qian S, Cheng L, et al. Metabolomic phenotype of gastric cancer and precancerous stages based on gas chromatography time-of-flight mass spectrometry. J. Gastroenterol. Hepatol. 2011;26(8):1290–7.
[47] Calabrese C, Pisi A, Di Febo G, Liguori G, Filippini G, Cervellera M, et al. Biochemical alterations from normal mucosa to gastric cancer by ex vivo magnetic resonance spectroscopy. Cancer Epidemiol. Biomark. Prev. 2008;17(6):1386–95.

[48] Song H, Wang L, Liu HL, Wu XB, Wang HS, Liu ZH, et al. Tissue metabolomic fingerprinting reveals metabolic disorders associated with human gastric cancer morbidity. Oncol. Rep. 2011;26(2):431–8.

[49] Wu H, Xue R, Tang Z, Deng C, Liu T, Zeng H, et al. Metabolomic investigation of gastric cancer tissue using gas chromatography/mass spectrometry. Anal. Bioanal. Chem. 2010;396(4):1385–95.

[50] Cai Z, Zhao JS, Li JJ, Peng DN, Wang XY, Chen TL, et al. A combined proteomics and metabolomics profiling of gastric cardia cancer reveals characteristic dysregulations in glucose metabolism. Mol. Cell Proteom 2010;9(12):2617–28.

[51] Deng K, Lin S, Zhou L, Li Y, Chen M, Wang Y, et al. High levels of aromatic amino acids in gastric juice during the early stages of gastric cancer progression. PLoS One 2012;7(11):e49434.

[52] Choi JM, Park WS, Song KY, Lee HJ, Jung BH. Development of simultaneous analysis of tryptophan metabolites in serum and gastric juice – an investigation towards establishing a biomarker test for gastric cancer diagnosis. Biomed. Chromatogr. 2016;30(12): 1963–74.

[53] Aa JY, Yu LZ, Sun M, Liu LS, Li MJ, Cao B, et al. Metabolic features of the tumor microenvironment of gastric cancer and the link to the systemic macroenvironment. Metabolomics 2012;8(1): 164–73.

[54] Leal MF, Wisnieski F, de Oliveira Gigek C, do Santos LC, Calcagno DQ, Burbano RR, et al. What gastric cancer proteomic studies show about gastric carcinogenesis? Tumour Biol 2016;37(8):9991–10010.

[55] Kikuchi S, Kaibe N, Morimoto K, Fukui H, Niwa H, Maeyama Y, et al. Overexpression of Ephrin A2 receptors in cancer stromal cells is a prognostic factor for the relapse of gastric cancer. Gastric Cancer 2015;18(3):485–94.

[56] Liu YF, Liu QQ, Wang X, Luo CH. Clinical significance of S100A2 expression in gastric cancer. Tumour Biol 2014;35(4):3731–41.

[57] Kocevar N, Grazio SF, Komel R. Two-dimensional gel electrophoresis of gastric tissue in an alkaline pH range. Proteomics 2014;14(2–3):311–21.

[58] Zhang ZQ, Li XJ, Liu GT, Xia Y, Zhang XY, Wen H. Identification of Annexin A1 protein expression in human gastric adenocarcinoma using proteomics and tissue microarray. World J. Gastroenterol. 2013;19(43):7795–803.

[59] Kocevar N, Odreman F, Vindigni A, Grazio SF, Komel R. Proteomic analysis of gastric cancer and immunoblot validation of potential biomarkers. World J. Gastroenterol. 2012;18(11):1216–28.

[60] Bai Z, Ye Y, Liang B, Xu F, Zhang H, Zhang Y, et al. Proteomics-based identification of a group of apoptosis-related proteins and biomarkers in gastric cancer. Int. J. Oncol. 2011;38(2):375–83.

[61] Wu C, Luo Z, Chen X, Wu C, Yao D, Zhao P, et al. Two-dimensional differential in-gel electrophoresis for identification of gastric cancer-specific protein markers. Oncol. Rep. 2009;21(6): 1429–37.

[62] Liu R, Li Z, Bai S, Zhang H, Tang M, Lei Y, et al. Mechanism of cancer cell adaptation to metabolic stress: proteomics identification of a novel thyroid hormone-mediated gastric carcinogenic signaling pathway. Mol. Cell Proteomics 2009;8(1):70–85.

[63] Lin LL, Chen CN, Lin WC, Lee PH, Chang KJ, Lai YP, et al. Annexin A4: a novel molecular marker for gastric cancer with *Helicobacter pylori* infection using proteomics approach. Proteomics Clin. Appl. 2008;2(4):619–34.

[64] Manadas B, Mendes VM, English J, Dunn MJ. Peptide fractionation in proteomics approaches. Expert Rev. Proteomics 2010;7(5):655–63.

[65] Scherl A. Clinical protein mass spectrometry. Methods 2015;81:3–14.

[66] Ryu JW, Kim HJ, Lee YS, Myong NH, Hwang CH, Lee GS, et al. The proteomics approach to find biomarkers in gastric cancer. J. Korean Med. Sci. 2003;18(4):505–9.

[67] Leal MF, Chung J, Calcagno DQ, Assumpcao PP, Demachki S, da Silva ID, et al. Differential proteomic analysis of noncardia gastric cancer from individuals of northern Brazil. PLoS One 2012;7(7):e42255.

[68] Yoshikawa T, Sasako M, Yamamoto S, Sano T, Imamura H, Fujitani K, et al. Phase II study of neoadjuvant chemotherapy and extended surgery for locally advanced gastric cancer. Br. J. Surg. 2009;96(9):1015–22.

[69] Cunningham D, Starling N, Rao S, Iveson T, Nicolson M, Coxon F, et al. Capecitabine and oxaliplatin for advanced esophagogastric cancer. N. Engl. J. Med. 2008;358(1):36–46.

[70] Ajani JA, Winter K, Okawara GS, Donohue JH, Pisters PW, Crane CH, et al. Phase II trial of preoperative chemoradiation in patients with localized gastric adenocarcinoma (RTOG 9904): quality of combined modality therapy and pathologic response. J. Clin. Oncol. 2006;24(24):3953–8.

[71] Hou Q, Tan HT, Lim KH, Lim TK, Khoo A, Tan IB, et al. Identification and functional validation of caldesmon as a potential gastric cancer metastasis-associated protein. J. Proteome Res. 2013;12(2):980–90.

[72] Berger SL, Kouzarides T, Shiekhattar R, Shilatifard A. An operational definition of epigenetics. Genes Dev 2009;23(7):781–3.

[73] Kang C, Song JJ, Lee J, Kim MY. Epigenetics: an emerging player in gastric cancer. World J. Gastroenterol. 2014;20(21):6433–47.

[74] Heyn H, Moran S, Hernando-Herraez I, Sayols S, Gomez A, Sandoval J, et al. DNA methylation contributes to natural human variation. Genome Res 2013;23(9):1363–72.

[75] You JS, Jones PA. Cancer genetics and epigenetics: two sides of the same coin? Cancer Cell 2012;22(1):9–20.

[76] Sandoval J, Esteller M. Cancer epigenomics: beyond genomics. Curr. Opin. Genet. Dev. 2012;22(1):50–5.

[77] Sandoval J, Peiro-Chova L, Pallardo FV, Garcia-Gimenez JL. Epigenetic biomarkers in laboratory diagnostics: emerging approaches and opportunities. Expert Rev. Mol. Diagn 2013;13(5):457–71.

[78] Doi A, Park IH, Wen B, Murakami P, Aryee MJ, Irizarry R, et al. Differential methylation of tissue- and cancer-specific CpG island shores distinguishes human induced pluripotent stem cells, embryonic stem cells and fibroblasts. Nat. Genet. 2009;41(12): 1350–3.

[79] Qu Y, Dang S, Hou P. Gene methylation in gastric cancer. Clin. Chim. Acta 2013;424:53–65.

[80] Jeon MS, Song SH, Yun J, Kang JY, Kim HP, Han SW, et al. Aberrant epigenetic modifications of LPHN2 function as a potential cisplatin-specific biomarker for human gastrointestinal cancer. Cancer Res. Treat 2016;48(2):676–86.

[81] Takamaru H, Yamamoto E, Suzuki H, Nojima M, Maruyama R, Yamano HO, et al. Aberrant methylation of RASGRF1 is associated with an epigenetic field defect and increased risk of gastric cancer. Cancer Prev. Res. 2012;5(10):1203–12.

[82] Suzuki H, Yamamoto E, Nojima M, Kai M, Yamano HO, Yoshikawa K, et al. Methylation-associated silencing of microRNA-34b/c in gastric cancer and its involvement in an epigenetic field defect. Carcinogenesis 2010;31(12):2066–73.

[83] Reed K, Poulin ML, Yan L, Parissenti AM. Comparison of bisulfite sequencing PCR with pyrosequencing for measuring differences in DNA methylation. Anal. Biochem. 2010;397(1):96–106.

[84] Joo JK, Kim SH, Kim HG, Kim DY, Ryu SY, Lee KH, et al. CpG methylation of transcription factor 4 in gastric carcinoma. Ann. Surg. Oncol. 2010;17(12):3344–53.

[85] Peng DF, Hu TL, Schneider BG, Chen Z, Xu ZK, El-Rifai W. Silencing of glutathione peroxidase 3 through DNA hypermethylation is associated with lymph node metastasis in gastric carcinomas. PLoS One 2012;7(10):e46214.

[86] Yoshida S, Yamashita S, Niwa T, Mori A, Ito S, Ichinose M, et al. Epigenetic inactivation of FAT4 contributes to gastric field cancerization. Gastric Cancer 2017;20(1):136–45.

[87] Langer R, Specht K, Becker K, Ewald P, Ott K, Lordick F, et al. Comparison of pretherapeutic and posttherapeutic expression levels of chemotherapy-associated genes in adenocarcinomas of the esophagus treated by 5-fluorouracil- and cisplatin-based neoadjuvant chemotherapy. Am. J. Clin. Pathol. 2007;128(2):191–7.

[88] Tebbutt NC, Price TJ, Ferraro DA, Wong N, Veillard AS, Hall M, et al. Panitumumab added to docetaxel, cisplatin and fluoropyrimidine in oesophagogastric cancer: ATTAX3 phase II trial. Br. J. Cancer 2016;114(5):505–9.

[89] Ott K, Vogelsang H, Marton N, Becker K, Lordick F, Kobl M, et al. The thymidylate synthase tandem repeat promoter polymorphism: a predictor for tumor-related survival in neoadjuvant treated locally advanced gastric cancer. Int. J. Cancer 2006;119(12):2885–94.

[90] Langer R, Specht K, Becker K, Ewald P, Bekesch M, Sarbia M, et al. Association of pretherapeutic expression of chemotherapy-related genes with response to neoadjuvant chemotherapy in Barrett carcinoma. Clin. Cancer Res. 2005;11(20):7462–9.

[91] Janmaat ML, Gallegos-Ruiz MI, Rodriguez JA, Meijer GA, Vervenne WL, Richel DJ, et al. Predictive factors for outcome in a phase II study of gefitinib in second-line treatment of advanced esophageal cancer patients. J. Clin. Oncol. 2006;24(10):1612–9.

[92] Metzger R, Danenberg K, Leichman CG, Salonga D, Schwartz EL, Wadler S, et al. High basal level gene expression of thymidine phosphorylase (platelet-derived endothelial cell growth factor) in colorectal tumors is associated with nonresponse to 5-fluorouracil. Clin. Cancer Res. 1998;4(10):2371–6.

[93] Matsubara J, Nishina T, Yamada Y, Moriwaki T, Shimoda T, Kajiwara T, et al. Impacts of excision repair cross-complementing gene 1 (ERCC1), dihydropyrimidine dehydrogenase, and epidermal growth factor receptor on the outcomes of patients with advanced gastric cancer. Br. J. Cancer 2008;98(4):832–9.

[94] Igney FH, Krammer PH. Death and anti-death: tumour resistance to apoptosis. Nat. Rev. Cancer 2002;2(4):277–88.

[95] Yoon DH, Ryu MH, Park YS, Lee HJ, Lee C, Ryoo BY, et al. Phase II study of everolimus with biomarker exploration in patients with advanced gastric cancer refractory to chemotherapy including fluoropyrimidine and platinum. Br. J. Cancer 2012;106(6):1039–44.

[96] Doi T, Muro K, Boku N, Yamada Y, Nishina T, Takiuchi H, et al. Multicenter phase II study of everolimus in patients with previously treated metastatic gastric cancer. J. Clin. Oncol. 2010;28(11):1904–10.

[97] Zhao L, Pan Y, Gang Y, Wang H, Jin H, Tie J, et al. Identification of GAS1 as an epirubicin resistance-related gene in human gastric cancer cells with a partially randomized small interfering RNA library. J. Biol. Chem. 2009;284(39):26273–85.

[98] Yoon HH, Bendell JC, Braiteh FS, Firdaus I, Philip PA, Cohn AL, et al. Ramucirumab combined with FOLFOX as front-line therapy for advanced esophageal, gastroesophageal junction, or gastric adenocarcinoma: a randomized, double-blind, multicenter Phase II trial. Ann. Oncol. 2016;27(12):2196–203.

[99] Li J, Qin S, Xu J, Xiong J, Wu C, Bai Y, et al. Randomized, double-blind, placebo-controlled phase III trial of Apatinib in patients with chemotherapy-refractory advanced or metastatic adenocarcinoma of the stomach or gastroesophageal junction. J. Clin. Oncol. 2016;34(13):1448–54.

Chapter 16

Molecular alterations and precision medicine in prostate cancer

Prakash Chand Sharma[1] and Kalyani Verma[2]

[1]*University School of Biotechnology, Guru Gobind Singh Indraprastha University, Dwarka, New Delhi, India;* [2]*Defence Institute of Physiology and Allied Sciences, Defence Research and Development Organisation, Timarpur, Delhi, India*

Introduction

Prostate cancer (PC) is the second most common cancer, and the fifth leading cause of cancer deaths in men. Incidence rates vary by more than 25-fold worldwide, being most frequently diagnosed in industrialized countries. However, the mortality rates are highest in less developed territories, because of late detection and inadequate treatment facilities. Disease risk and mortality rates are higher in Africans as compared to other human races, suggesting a role of ethnic and genetic diversity in the prevalence of prostate cancer [1]. In addition to race/ethnicity, family history and age are the other major risk factors [2]. Prostate cancer like some other cancers, is a heterogeneous disease, both in terms of pathological and clinical presentation. Multiple tumor foci are commonly detected within a single prostate gland, with varying degree of dysplasia and tissue disorganization [3].

Classification of prostate cancer

The World Health Organization (WHO) classification updated in 2016 [4] now considers changes, including (i) The Gleason grading system, modified to precisely represent clinical outcomes, (ii) Intraductal carcinoma of the prostate (IDC-P), and large-cell neuroendocrine carcinoma (LCNEC), as the newly identified subtypes of PC, (iii) The updated histological variants of acinar adenocarcinoma, and (iv) New immunohistochemical markers, which are useful for disease diagnosis. PC is histopathologically classified into (a) Glandular neoplasms (acinar adenocarcinoma, intraductal carcinoma, and ductal adenocarcinoma), (b) Urothelial carcinoma, (c) Squamous neoplasms (adenosquamous carcinoma and squamous cell carcinoma), (d) Basal cell carcinoma, and (e) Neuroendocrine tumors (adenocarcinoma with neuroendocrine differentiation, small-cell neuroendocrine carcinoma and large-cell neuroendocrine carcinoma).

However, molecular characterization of PC would enable the classification of PC into subgroups, considering their evolution from a poorly understood, heterogeneous group of diseases with variable clinical courses, to better defined molecular subtypes. The Cancer Genome Atlas (TCGA) research network, has identified genomic and other molecular alterations, to help in the identification of potential therapeutic targets and predict prognostic effects of disease treatment [4].

Among the 333 cases of primary prostatic cancers, based on the data on somatic mutations, gene fusions, somatic copy-number alterations, gene expression, and DNA methylation, a molecular taxonomy of the primary disease is available. Of these primary cancers, 75% were assigned to one of seven molecular classes, based on distinct oncogenic drivers viz. fusions involving (1) *ERG* 46%, (2) *ETV1* 8%, (3) *ETV4* 4%, and (4) *FLI1* 1% and mutations in (5) *SPOP* 11%, (6) *FOXA1* 3%, and (7) *IDH1* 1% [5] (Fig. 16.1).

TCGA further classifies primary prostatic cancers into two groups, one with rearrangements in *ETS* family transcription factors (*ERG*, *ETV1*, *ETV4*, and *FLI1*) (*ETS* positive), and the other without (*ETS* negative). *ETS*-positive prostatic cancers are further subclassified according to the specific *ETS*-fusion gene member involved: *ERG*, *ETV1*, *ETV4*, and *FLI1*. On the other hand, *ETS*-negative prostatic cancers are classified in accordance with mutations in *SPOP*, *FOXA1*, and *IDH1*.

ETS rearrangements and *TMPRSS2-ERG* fusions, have been detected in high-grade prostatic intraepithelial neoplasia (HGPIN), and low-grade prostatic cancer, suggesting that these changes appear to be an early event in

Precision Medicine for Investigators, Practitioners and Providers. https://doi.org/10.1016/B978-0-12-819178-1.00016-2

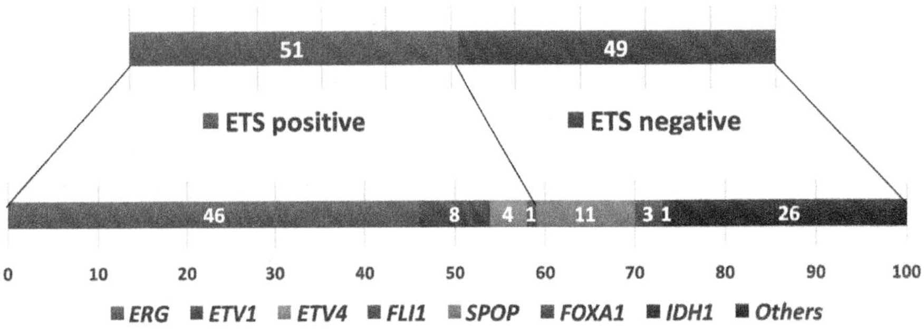

FIGURE 16.1 Molecular taxonomy of primary prostate cancer (The Cancer Genome Atlas/TCGA research network).

prostate carcinogenesis [4]. *ETS*-rearranged cancers are notably enriched in genomic alterations like *PTEN* deletions, *TP53* alterations, PIK3 pathway alterations, and specific amplification of 3p [6].

In the case of *ETS*-negative prostate cancers, *SPOP* mutations remain the most common point mutation (6%−15%), and are generally detected in HGPINs. These mutations presumably drive one of the early events in prostatic carcinogenesis [7]. On the other hand, *IDH1* mutations appear to represent a rare event in PC, with relatively few somatic copy number alterations (SCNAs), and high levels of genomic hypermethylation [8]. *FOXA1* is an androgen receptor transcription factor, associated with the highest level of AR transcriptional activity, that promotes prostatic cancer progression [8].

Genetic alterations

Cancer development results from progressive accumulation of multiple genetic events, with additional genetic alterations down-stream in the tumor microenvironment, responsible for the progression and migration of the disease to other parts of the body [9]. Mainly, two categories of molecular genetic alterations, single-nucleotide polymorphisms (SNPs) and microsatellite instability (MSI) are used in risk stratification.

Single-nucleotide polymorphisms (SNPs)

SNPs denote a DNA sequence variation, occurring when a single-nucleotide varies from the nucleotide normally present/expected. SNPs can occur both in coding and noncoding regions of the genome, being at a higher frequency in the latter. SNPs affect the mRNA stability, translational efficiency, structural changes, and activity of proteins. They can also alter chromatin organization, and influence the gene expression through alternative splicing and other mechanisms [10].

Over 100 SNPS can have implications in the occurrence of prostate cancer [11]. Genome-wide association study (GWAS) uses a hypothesis-neutral genome-based approach, and compares the DNA variations in a large number of targets and controls, to screen for genetic variants associated with the disease risk. Unlike the candidate gene approach, untargeted GWAS allows discovery of new variants in a nonselective (agnostic) way, with the ability to assay 300,000 to 5 million SNPs concurrently [12]. The genes showing prevalence of SNPs could prove important in clinical use. The first identified genomic region, 8q24, has been documented to have the highest number of independent genetic variants, associated with PC.

The 8q24 region is located near the *MYC* proto-oncogene, and may act as an enhancer by interfering with the chromatin conformation. Located on chromosome 10, another important SNP, rs10993994, is present 2 bp upstream of the transcription initiation site of microseminoprotein-β (*MSMB*). *MSMB* codes for a protein called PSP94, which is exclusively produced in the prostate and secreted in the semen. PSP94 protein is thought to be involved in the regulation of growth and apoptosis of PC cells, and can be measured in the plasma after release from these cells [11,12]. Other sites of potential clinical significance include SNP rs4245739 on chromosome 1 near the *MDM4* gene, which is a negative regulator of *TP53* and rs11568818, which shows linkage disequilibrium with the gene *MMP7*, encoding for a matrix metalloproteinase.

MMP7 has been reported to be associated with metastasis and poor prognosis [13]. These variants could perhaps play a role in the ability to differentiate, between low- and high-risk disease status [14]. Another common SNP, rs2735839, is suggested to have a specific function, as it lies between *KLK2* and *KLK3* genes, reported to influence PC risk. *KLK3* encodes prostate-specific antigen (PSA) protein [15], while *KLK2* encodes kallikrein-related peptidase 2 (hK2), whose expression is used to avoid unnecessary biopsy, in previously unscreened subjects with elevated total PSA [16]. Fine mapping results indicate that 10 SNPs on *HNF1B* gene were significantly associated with PC, the most significant association being that of SNP rs4430796. SNP rs1048656 seven on *JAZF1* was also reported to be associated with PC susceptibility [17].

GWAS can fail to develop a full understanding, because of a spectrum of genetic variations with high to low penetrance, and usually, low penetrance variations go unnoticed or undetected. GWAS overlooks rare variants, thus missing heritability and enrichment for SNPs in gene desert regions, compared to protein coding and promoter regions [18]. Some recent variants in the principal genes related to PC are briefly described in Table 16.1. Among several genetic variants identified to be associated with PC risk, common SNPs, which appear to be the genetic modifier, likely to have an important role in PC susceptibility, are the targets need to be focused on to develop a biomarker.

Microsatellite instability

Microsatellites, also known as Simple Sequence Repeats (SSRs), are tandemly repeated nucleotide motifs of one to six base pairs long, and are ubiquitous in both coding and noncoding regions of eukaryotic and prokaryotic genomes, being abundant in noncoding regions. On the basis of nucleotide arrangement within the repeat motifs, microsatellite sequences are classified as simple perfect (tandem array of a single repeat motif), imperfect (perfect repeats interrupted by nonrepeat units), and compound (two basic repeat motifs present together).

Microsatellite loci witness a very high mutation rate of about 10^{-6} to 10^{-2} per generation, owing to slipped-strand mispairing, and subsequent errors occurring during DNA replication, repair, and recombination processes [39]. Microsatellite-based molecular markers have emerged as markers of choice, because of their high polymorphic nature, abundance, codominant inheritance, and broad distribution in the genome. The limitations are relatively high development cost and technical challenge, of prior knowledge of sequence harboring a microsatellite. In humans, microsatellites have been recognized to play a major role in neurodegenerative disorders like Huntington's disease, fragile X syndrome, etc. and various cancers, including colon, gastric, and endometrial cancers [40].

Several studies have described microsatellite instability (MSI) in PC, and reported large variation in the frequency, ranging from 2% to 65%. There are some reports establishing the relationship between MSI and clinicopathological variables, suggesting that MSI plays an important role in the development and progression of PC, and can identify patients at high risk [41].

Androgen receptor

Androgen hormone plays a role in the malignant growth of prostate cells, via androgen receptor (AR). Androgen receptor binds to dihydrotestosterone (DHT), and stimulates the transcription of androgen responsive genes. Therefore, *AR* gene is proposed to be a strong candidate for assessing the risk and progression of PC. A single copy of the *AR* gene, 90 kb long, is present in the Xq11-12 chromosomal region. It harbors two polymorphic microsatellite repeats (CAG and GGC), separated by approximately 1.1 kb. The CAG (polyglutamine) repeats vary from 5 to 33, the average being 20, and the GGN tract (GGT_3 GGG GGT_2 GGC_n, polyglycine) varies from 9 to 24, wih an average of 16 repeats. These GGN repeats are commonly called GGC repeats, because in this tract, only the GGC repeat motif is variable, and remaining repeat motifs are consistent.

Transcriptional regulatory activity of AR correlates with CAG repeats, and may affect a variety of systems. These repeats show racial variation, as Asian descendants have the highest number of CAG repeats, whites have an intermediate number, and African-Americans have the lowest number. Interestingly, the risk of PC development is found to be in the reverse order [42].

CAG repeats and cancer biology

Severity not only increases with the length of the repeats, but there seems to be a critical threshold length, for development of the disease. The underlying molecular mechanisms of CAG associated PC development are not very clear. However, long back, the modulation of AR transcriptional potency, which inversely correlates with the CAG repeats, has been suggested [42]. An inverse relationship between CAG repeat length and the level of transactivation activity of androgen receptor is also anticipated. Shorter CAG alleles could cause more rapid growth of prostate cells, elevating the risk of PC [43]. Shorter CAG repeat length is associated with the development of PC in men at a younger age [44]. Furthermore, shorter CAG repeats are correlated with lymph node positive disease [45], biochemical failure [46], advanced stage [43], and high grade, that is, Gleason score >7 [43,44].

CAG and GGC repeats

A synergistic effect was seen toward an increased risk of PC [43,47] such that CAG and CAG + GGC repeats can be used to predict the PC risk, but GGC marker alone is not sufficient. MSI level at these repeats could also help in identifying persons at an advanced stage of the disease.

Steriod 5 α reductase type II (SRD5A2) gene

The active androgen DHT (dihydrotestosterone), encoded by *SRD5A2* gene, catalyzes the conversion of testosterone to DHT, with NADPH as the cofactor. This gene is expressed in genitals, skin, liver, and prostate gland. The gene harbors polymorphic dinucleotide repeats $(TA)_n$ in its 3′ UTR. Rajender et al. [48] have concluded that longer (TA) repeats, that is, $(TA)_9$, in contrast to $(TA)_0$ and $(TA)_{0/9}$, are associated with increased risk of PC in South Indian population. However, Ezzi et al. [49] are of the opinion that $(TA)_0$ allele is associated with increased risk of PC in Lebanese men.

TABLE 16.1 Role of SNPs in genes implicated in prostate cancer.

No.	Gene	SNP	Role in Prostate Cancer	References
1.	*HOXB13*	rs138213197 (G84E)	Increased risk of PC for men with family history and men diagnosed below the age of 55 years	[19]
2.	*XPD*	rs13181 (Lys751Gln)	Increased risk of PC	[20]
		rs238406 (Arg156Arg)	High-risk PC	[21]
3.	*hOGG1*	rs1052133 (Ser326Cys)	Increased risk of PC	[21]
			Lower risk of PC	[22]
4.	*XRCC1*	rs915927	PC metastasis	[23]
			Increased risk of PC	[22]
5.	*RNASEL*	rs12757998	Reduction in risk of recurrence	[24]
		rs635261	PC metastasis	[23]
6.	*IL6*	rs1800796 (-572C > G)	Increased risk of PC	[25,26]
		rs1800795	Increased risk of PC	[27]
7.	*AKT1*	rs3730358	Decreased risk of PC	[28]
		rs2494750	PC metastasis	[23,29]
8.	*SRD5A1*	rs3822430, rs1691053	High risk of presenting initial PSA levels >20 ng/mL	[30]
		rs3736316, rs3822430, rs1560149, rs248797 and rs472402	Risk of high-grade PC	[31]
9.	*SRD5A2*	rs2300700	Risk of high-grade PC	[31]
		rs508562	Frequent in patients with high PSA and high-grade PC	[32]
		rs11675297	Frequent in metastatic stage	
10.	*CYP17A1*	rs17115149	Aggressiveness according to Gleason score	[33]
		rs1004467	Elevated testosterone and dihydrotestosterone (DHT) levels	[34]
11.	*CYP1B1*	rs1056836	Associated with biochemical recurrence	[35]
		rs1056836	Decreased risk of PC	[35]
12.	*VDR*	rs731236, rs7975232	PC risk	[36]
		rs731236, rs1544410, rs3782905	Associated with high PSA level	
		rs1544410, rs2239185	Associated with high Gleason score	
		rs731236 (*TaqI*)	Age at diagnosis (>58 years old)	[37]
		rs1544410 (*Bsm*I)	Lower PSA levels (<10 ng/mL)	
		rs7975232 (*Apa*I)	Lower PSA levels	
13.	*CYP24A1*	rs2585428, rs2296241 and rs4809959	Inversely associated with PSA levels	[38]
		rs972650	Effect on elevated PSA level	
		rs2585428, rs2296241and rs2762941	Negative association with clinical stage	
		rs2244719	Associated with advanced pathological stage	
		rs2209314	Negative association with pathological stage and positive association with pathological Gleason score	
		rs6127119	Negative association with pathologic Gleason score	
		rs2248461, rs2248359, rs6022999, rs2585428 and rs4809959	PC risk	

CYPIIA1 gene

CYPIIA1 gene encodes for enzyme PY50scc, which mediates the first step in the synthesis of steroid sex hormone. This gene harbors $(TTTTA)_n$ polymorphic repeat in its promoter region. Kumazawa et al. [50] observed that PC patients lacking $(TTTTA)_4$ repeat motif, carried an increased risk of metastatic disease (Stage D) and high-grade (Gleason Score $\geq$8/poorly differentiated) cancer. Thus, the absence of $(TTTTA)_4$, that is, homozygosity for $(TTTTA)_6$ or longer allele, could be a useful predictive marker. According to Celhar et al. [51], there is significant association of $(TTTTA)_6$ allele with biochemical relapse after prostatectomy, in patients with organ confined PC, low grade PC, or both.

Epigenetic modifications

Epigenetics involves heritable phenotypic changes, which do not involve alterations in the DNA sequence, such as DNA methylation and histone modification. Epigenetic alterations are frequently involved in both initiation and progression in prostate cancer. They occur at a much higher frequency than genetic mutations, and are also common in premalignant stages of the disease.

DNA methylation

The phenomenon is nearly ubiquitous in multicellular organisms. DNA methylation most often occurs in the context of CpG dinucleotides, although non-CpG DNA methylation can also occur. It has been observed that CpG islands of a large number of genes, which are unmethylated in normal tissue, become methylated with varying degrees in human cancers [52]. As a tumor biomarker, DNA methylation is stable, and easily assayed in body fluids like blood, urine, and saliva. Moreover, DNA methylation constitutes a positive signal against negative background, reducing the chances of being masked by contaminated normal DNA. These are advantages in clinical samples, where DNA derived from neoplastic cells is often limited [53].

Single-gene evaluation

Gene specific hypermethylation refers to the increase in DNA methylation status of promoter region of a specific gene or intergenic region, causing inactivation of genes involved in the various processes like DNA repair, cell-cycle regulation, and apoptosis. The genes having a potential of being targets for discovery of biomarkers are summarized in Table 16.2.

Genome-wide methylation

Genome-wide methylation analysis screens many genes as potential biomarkers, in a single step, as a panel of markers (two or more genes) has increased sensitivity and specificity, for diagnostic or prognostic purposes. *AOX1, HIF3A, SCGB3A1,* and *SEPT9* are important, differentially methylated genes in PC [54]. Zhao et al. [55] validated eight differentially methylated CpGs (*ALKBH5, ATP11A, FHAD1, KLHL8, PI15,* and three intergenic regions), which distinguish patients with metastatic-lethal tumors from non-recurrent tumors. Wu et al. [56] identified 3567 differentially methylated regions (DMRs), between normal and cancer tissues, and 913 DMRs distinguishing low from high-grade prostate cancers. Most of these DMRs were located at CGI shores.

The top five candidate DMRs corresponding to low to high Gleason score include *OPCML, ELAVL2, EXT1, IRX5,* and *FLRT2*. Hence, a panel of selected DMRs may serve as novel surrogate biomarkers, for Gleason score in PC. Global DNA methylation differences between tumor subtypes (*TMPRSS2:ERG* (T2E) gene fusion, T2E-positive and T2E-negative) have also been documented [57]. A large number of differentially methylated CpG sites were identified and DNA methylation profiles accurately distinguished between tumor T2E subgroups. Top-ranked differentially methylated CpGs genes include C3orf14, *CACNA1D, GREM1, KLK10, NT5C, PDE4D, RAB40C, SEPT9,* and *TRIB2*, several of which had a corresponding alteration in mRNA expression.

These genes may have various roles in the pathogenesis of PC, and calcium-channel gene *CACNA1D* is a known ERG-target. Methylation status of 485,577 CpG sites from genomic regions, with a broad spectrum of CpG densities, interrogating both gene-associated and nonassociated regions, using the recently developed Illumina 450K methylation platform, revealed *BNC1, FZD1, RPL39L, SYN2, LMX1B, CXXC5, ZNF783,* and *CYB5R2*, as top candidate novel genes that are frequently methylated, and whose methylation was associated with the inactivation of gene expression in PC cell lines. Pathway analysis of genes with altered methylation patterns revealed involvement of a cancer-related network of genes, whose activity may be regulated by *TP53, MYC, TNF, IL1, IFN-γ,* and *FOS* in prostate cancer [58].

Histone modification

Histones, the DNA-packaging proteins, are subjected to a large number of post-translational modifications. These proteins possess N-terminal tails, that could be covalently or reversibly modified by the addition or deletion of acetyl, methyl, ubiquitin, sumoyl, and adenosine diphosphate-ribosyl groups. A unique complex modification pattern (histone code) exists in balance, and functions to regulate the transcriptional activity levels of genes. Histone modifications can be analyzed by chromatin immunoprecipitation (ChIP) method [59]. Histone modifications are less

TABLE 16.2 Epigenetic alterations reported in prostate cancer.

No.	Gene	Application	Significance in Prostate Cancer	References
DNA methylation				
1.	*GSTP1*	Prognostic and therapeutic	Promoter hypermethylation increased PC risk	[60]
2.	*E-Cadherin (CDH1)*	Therapeutic	Methylation mediating outgrowth, differentiation and maturation of the ductal network	[61]
		Prognostic	Hypermethylation with increase in Gleason score and tumor-node-metastasis stage	[62]
3.	*FABP5*	Therapeutic	Promoter hypomethylation increased PC risk	[63]
4.	*EphA5*	Prognostic and therapeutic	Hypermethylation with increase in Gleason score	[64]
5.	*CRMP4*	Diagnostic and prognostic	Predicting tumor progression and metastasis	[65]
		Diagnostic	Methylation predictor of lymph node metastasis	[66]
6.	*KAI1*	Diagnostic and therapeutic	Hypermethylation causing PC progression	[67]
Histone modification				
1.	*PRMT6*	Diagnostic	Predicts for clinically aggressive disease	[68]
2.	*EZH2*	Prognostic	Correlated aggressive tumors	[69]
3.	*PRTM5*	Diagnostic	R761 methylation, a relevant biomarker for AR-mediated proliferation versus arrest in *TMPRSS2:ERG* positive cells	[70]
		Therapeutic	Epigenetic activator of AR in PC via demethylation of H4R3	[71]
4.	*KDM1A*	Therapeutic	Demethylation at K114 controls androgen-dependent gene transcription and TMPRSS2-ERG fusion	[72]
5.	*H2B*	Therapeutic	Hypoacetylated, hypomethylated, and dephosphorylated in PC	[73]

explored as compared to DNA methylation. Important histone modifications are summarized in Table 16.2.

Metabolomics alterations

Metabolomics provide downstream quantification of complete system activity, reflecting contributions from genome, epigenome, transcriptome, and proteome, and their interaction with the environment. During neoplastic transformation, prostate cells undergo metabolic reprogramming to meet the demands of growth and proliferation. It has the dual benefit of identifying metabolic pathways associated with the progression of disease, and also assists in the identification of surrogate biomarkers [74].

Alterations in cysteine and methionine metabolism, nicotinamide adenine dinucleotide metabolism, and hexosamine biosynthesis have been detected. Impaired sphingosine-1-phosphate receptor 2 signaling, downstream of sphingosine, and a potential key oncogenic pathway, have been targeted for therapeutic use in PC [75]. Lucarelli et al. [76] observed the upregulation of androgen-induced protein synthesis, increased amino acid metabolism, and perturbation of nitrogen breakdown pathways, along with high choline-containing compounds and phosphocholine levels. Dysregulation of metabolism of fatty acids, sphingolipids, polyamines in *TMPRSS2-ERG* translocation positive PC specimens, and alterations in cholesterol metabolism, were observed in high-grade tumors, enabling the cells to ensure energy storage [77].

Metabolites pertaining to energy pathways and lipid metabolism, were demonstrated to be of prime importance in defining the degree of disease aggressiveness. These encompass energy metabolism (TCA cycle intermediates, citrate, phosphoenolypyruvate, and adenosine diphosphate), cell growth and proliferation (amino acids, bile acids, glycerol-3-phosphate), and cell membrane constituents (phospholipids, phosphocholine, choline). Similarly, Lucarelli et al. [76] and Kelly et al. [74] reported loss of zinc accumulation capacity, resulting in the inhibition of citrate accumulation ability, an early event in prostate cell transformation. By screening 35 potential biomarkers, Adams et al. [78] observed cholesterol, glycerides, and phospholipids as most promising. The steroid hormone androgen is derived from cholesterol, explaining its elevated amount in rapidly growing cell.

Sarcosine, an N-methyl derivative of glycine, is significantly increased during disease progression, from normal to localized to metastatic PC. Knockdown of glycine-N-methyl transferase enzyme that produces sarcosine from glycine, attenuates PC invasion. Sarcosine, proline, kynurenine, uracil, and glycerol 3-phosphate were also found in elevated concentrations in the urine samples of metastatic PC patients [76,79]. Peptide metabolites namely pyroglutamine, gamma-glutamylphenylalanine, phenylpyruvate, N-acetylcitrulline, and stearoylcarnitine, were also associated with the aggressive disease [80].

Metabolite biomarkers in clinical imaging

Metabolic tracers such as ^{11}C-acetate, ^{11}C/^{18}F-choline, ^{18}F-fluoro-2-deoxyglucose (FDG), [1-^{13}C]-pyruvate, and ratio of [1-^{13}C]-lactate/[1-^{13}C]- pyruvate have been used in positron emission tomography (PET) for PC. Major PET tracers derived from choline are ^{18}F-fluorocholine (FCH), ^{11}C-Choline and ^{18}F-Fluoroethylcholine (FEC). ^{11}C-choline PET has been approved by the Food and Drug Administration (FDA, USA) for PC detection [81,82].

Commercially available molecular biomarkers

The National Cancer Institute defines a biomarker as "a biological molecule found in the blood, other body fluids, or tissues, that is a sign of a normal or abnormal process, or of a condition or disease" [83]. Ideally, a biomarker should be able to establish the presence of a disease and its progression, differentiate between individuals at high or low risk, predict recurrence, and monitor response to treatments. Furthermore, an ideal biomarker assay should be affordable, easily accessible, consistent, noninvasive, and easily quantifiable.

Conventional biomarkers

Prostate-specific antigen (PSA) is a kallikrein type serine protease, produced and secreted by prostate epithelial cells, to maintain liquefaction of seminal fluid. PSA is prostate specific but not cancer specific, as it is found to be elevated in benign prostate hyperplasia (BPH) and prostatitis. Moreover, 15% of PC cases occur in men with very low PSA levels. Another limitation is the inability to distinguish patients with indolent or aggressive disease. This lack of specificity of PSA may lead to overtreatment of indolent tumors, and undertreatment of life threatening PC. Several variants of PSA include free PSA, total PSA, PSA velocity, and PSA density, with little additional diagnostic/prognostic value [84].

Currently, PSA levels, Gleason Score, and tumor stage/volume (TNM classification, International Union Against Cancer), are used in diagnosis and staging of PC; however, these markers are insufficient and only partially effective (Fig. 16.2).

In Table 16.3, we have summarized the relevant information on standard and innovative diagnostic and prognostic biomarkers, currently in widespread clinical use.

Precision medicine

The matching up of a tumor-specific molecular defect, with the targeted therapy designed to treat that particular defect, is the essence of personalized medicine [94]. Prostate cancer is commonly affected by *ERG* rearrangements,

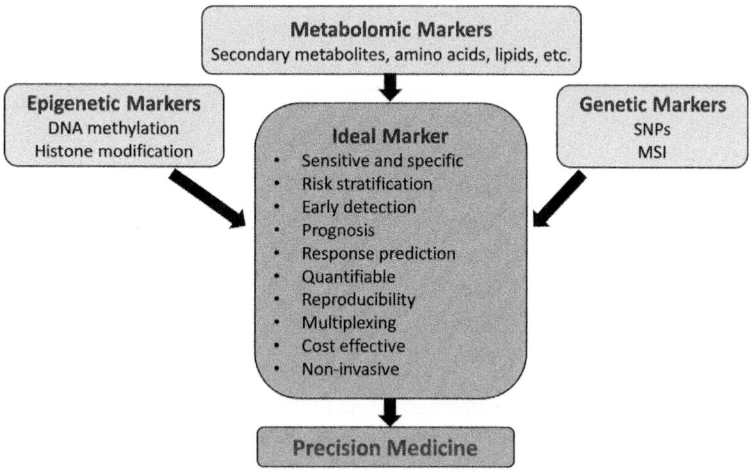

FIGURE 16.2 New biomarkers for prostate cancer.

TABLE 16.3 Diagnostic and prognostic biomarkers in prostate cancer.

No.	Biomarker	Sample	Algorithm	Application	References
Diagnostic biomarkers					
1.	Prostate health index (PHI)	Serum	Combines three PSA subforms [tPSA, %fPSA and [−2]proPSA (p2PSA)] into a single mathematical score	Distinguish PC from BPH	[85]
2.	4K score	Plasma	Includes four different kallikrein markers (tPSA, %fPSA, intact PSA, hK2) and integrates them with age and DRE findings	Calculate risk of high-grade PC and reduce unnecessary biopsies	[86]
3.	SelectMDx	Urine	Measures mRNA levels of a 2-gene panel (*DLX1* and *HOXC6*) in post-DRE urine (*KLK3* expression as internal reference). Results are integrated with risk factors like tPSA, age, prostate biopsy, and family history	Identify increased risk of high-grade cancer and reduce unwanted biopsies	[87]
4.	ConfirmMDx®	Biopsy tissue	Analyzes methylation status of *GSTP1, APC* and *RASSF1*	Prevent repetition of biopsy	[88]
5.	PC Antigen 3	Post-DREurine	Measures *PCA3* mRNA in first-void and post-DRE urine	Prevent unnecessary biopsies	[89]
Prognostic biomarkers					
1.	OncotypeDX GPS	FPE PBx tissue	Multigene assay consisting of 12 genes related to androgen metabolism, cellular organization, proliferation, and stromal response, and 5 reference genes	Predictor of recurrence,PC death and adverse pathology after radical prostatectomy	[90]
2.	Prolaris	FFPE PBx or RP tissue	Evaluates expression of 31 cell-cycle progression genes and 15 housekeeper genes to generate a mathematical score (CCP-score)	Distinguish aggressive from indolent tumors	[91]
3.	ProMark	Tissue	Eight biomarkers used to quantitate multiplex proteomics *in-situ* imaging system	Postbiopsy active surveillance.	[92]
4.	Decipher	RP	22 gene panel including cell proliferation, cell structure, immune system modulation, cell cycle progression and androgen signaling. The score ranges from 0 to 1, levels of >0.6 considered high risk for disease progression	Follow up after radical prostectomy with positive margins	[93]

and is known to harbor a unique fusion between prostate-specific androgen-regulated *TMPRSS2* gene and *ETS* genes *ERG*, *ETV1*, or *ETV4*, in 50% of PC cases [95]. Currently, there are no known therapies that directly target this gene fusion. Nevertheless, future efforts are likely to focus on it.

The major pathways affecting development of PC include AR signaling, PI3K/PTEN/AKT, and DNA damage pathways. AR signaling is an essential target for pharmacological intervention in patients who become resistant to androgen deprivation therapy (ADT). The drugs abiraterone acetate, prednisolone, enzalutamide, apalutamide, and darolutamide are in clinical trial phase 3. These drugs are prescribed in combination with ADT or alone [96]. PI3K/AKT/mTOR pathway plays an essential role in PC initiation and progression. The loss of *PTEN* gene causes hyperactivation of P13K signaling. The drugs, Buparlisib, AZD8186, ipatasertib, bicalutamide, AZD5363, sirolimus, carboplatin, ridaforolimus, etc. are in clinical trials, targeting this signaling axis [97]. DNA damage defects are majorly seen in the advanced state of PC. They are addressed using poly ADP ribose polymerase (PARP) family, which plays a role in the detection of DNA damage and assist in repairing. Olaparib, in clinical trial phase 2, has been evaluated in mCR-PC patients [93].

Epigenetic modifications provide targets for both diagnostic and prognostic markers, and for therapeutics, since they are reversible in nature. *EZH2* gene is overexpressed in metastatic PC, therefore, its inhibitor GSK126 and derivatives are likely to play a therapeutic role. Similarly, SP-2509, an inhibitor of LSD1 (histone demethylase) and MK-8628, ZEN003694, INCB057643 and ODM-207, inhibitors of BET family members that recognize histone acetylation marks may prove useful for personalized treatment [98].

Despite recent advances in genome or epigenome-based studies aiming to find novel biomarkers, availability of ideal markers and their clinical use is still in their infancy, and cancer research has to go a long way to design marker-based individualized treatments.

Conclusions

Early diagnosis of any disease is pivotal and can increase the survival rate of the affected individuals. In PC, prostate-specific antigen (PSA) is regarded as a best conventional biomarker available but it has many limitations. Other PC markers, Gleason Score and Tumor Stage/Volume (TNM Staging), have also proved to be inefficient. Thus, recent information from genetic, epigenetic, and metabolomic profiling has provided immense scope for improvement in current diagnosis and prognosis tools to ensure better management of prostate cancer. PC being a heterogeneous disease in nature, poses difficulties in correct diagnosis and treatment. However, our understanding and knowledge of different aspects of prostate cancer has improved dramatically over the decade, owing to advances in the molecular characterization technologies and in-depth exploration of the genetic and epigenetic basis of PC. The molecular profiles (tumor maps) provided by TCGA data set, gene targets unraveled by GWAS, including both epigenetic alterations and SNPs, MSI, and metabolome profiling, have provided the insight now being translated to aid clinicians in the early diagnosis and designing of treatment regimens. In this chapter, we have provided the overview of the information generated from high-throughput technologies and its utility in the development of potential biomarkers for diagnosis, prognosis, and therapeutic applications in PC. These efforts would further contribute to the field of precision medicine to cater to the need of individual patients, and ultimately, in achieving a decrease in cancer induced mortality.

References

[1] Bray F, Ferlay J, Soerjomataram I, Siegel RL, Torre LA, Jemal A. Global cancer statistics 2018: GLOBOCAN estimates of incidence and mortality worldwide for 36 cancers in 185 countries. CA Cancer J. Clin. 2018;68(6):394–424.

[2] Koochekpour S. Genetic and epigenetic changes in human prostate cancer. Iran. Red Crescent Med. J. 2011;13(2):80–5.

[3] Boyd LK, Mao X, Lu YJ. The complexity of prostate cancer: genomic alterations and heterogeneity. Nat. Rev. Urol. 2012;9(11):652–64.

[4] Inamura K. Prostatic cancers: understanding their molecular pathology and the 2016 WHO classification. Oncotarget 2018;9(18):14723–37.

[5] Abeshouse A, Ahn J, Akbani R, Ally A, Amin S, Andry CD, et al. The molecular taxonomy of primary prostate cancer. Cell 2015;163(4):1011–25.

[6] Kaffenberger SD, Barbieri CE. Molecular subtyping of prostate cancer. Curr. Opin. Urol. 2016;26(3):213–8.

[7] Barbieri CE, Baca SC, Lawrence MS, Demichelis F, Blattner M, Theurillat JP, et al. Exome sequencing identifies recurrent SPOP, FOXA1 and MED12 mutations in prostate cancer. Nat. Genet. 2012;44(6):685–9.

[8] Abeshouse A, Ahn J, Akbani R, Ally A, Amin S, Andry CD, et al. The molecular taxonomy of primary prostate cancer. Cell 2015;163(4):1011–25.

[9] Yadav SS, Li J, Lavery HJ, Yadav KK, Tewari AK. Next-generation sequencing technology in prostate cancer diagnosis, prognosis, and personalized treatment. In: Urologic oncology: seminars and original investigations, vol. 33; 2015. https://doi.org/10.1016/j.urolonc.2015.02.009 (6).

[10] Salvi S, Conteduca V, Gurioli G, Calistri D, Casadio V, De Giorgi U. Impact of candidate genetic polymorphisms in prostate cancer: an overview. Mol. Diagn. Ther. 2016;20(1):1–2.

[11] Ahmed M, Eeles R. Germline genetic profiling in prostate cancer: latest developments and potential clinical applications. Future Sci. OA 2016;2(1). https://doi.org/10.4155/fso.15.87.

[12] Chung S, Nakagawa H, Uemura M, Piao L, Ashikawa K, Hosono N, et al. Association of a novel long non-coding RNA in 8q24 with prostate cancer susceptibility. Cancer Sci. 2011;102(1):245–52.

[13] Eeles RA, Al Olama AA, Benlloch S, Saunders EJ, Leongamornlert DA, Tymrakiewicz M, et al. Identification of 23 new prostate cancer susceptibility loci using the iCOGS custom genotyping array. Nat. Genet. 2013;45(4):385–95.

[14] Cuzick J, Thorat MA. Prostate cancer prevention. Recent results in cancer research, vol. 202. Springer-Verlag Berlin Heidelberg; 2016.

[15] Ahn J, Berndt SI, Wacholder S, Kraft P, Kibel AS, Yeager M, et al. Variation in KLK genes, prostate-specific antigen and risk of prostate cancer. Nat. Genet. 2008;40(9):1032–7.

[16] Vickers AJ, Cronin AM, Roobol MJ, Hugosson J, Jones JS, Kattan MW, et al. The relationship between prostate-specific antigen and prostate cancer risk: the prostate biopsy collaborative group. Clin. Cancer Res. 2010;16(17):4374–81.

[17] Stevens VL, Ahn J, Sun J, Jacobs EJ, Moore SC, Patel AV, et al. HNF1B and JAZF1 genes, diabetes, and prostate cancer risk. Prostate 2010;70(6):601–7.

[18] Srinivasan S, Clements JA, Batra J. Single nucleotide polymorphisms in clinics: fantasy or reality for cancer? Crit. Rev. Clin. Lab. Sci 2016;53(1):29–39.

[19] Kote-Jarai Z, Mikropoulos C, Leongamornlert DA, Dadaev T, Tymrakiewicz M, Saunders EJ, et al. Prevalence of the HOXB13 G84E germline mutation in British men and correlation with prostate cancer risk, tumour characteristics and clinical outcomes. Ann. Oncol. 2015;26(4):756–61.

[20] Sobti RC, Berhane N, Melese S, Mahdi SA, Gupta L, Thakur H, Singh N. Impact of XPD gene polymorphism on risk of prostate cancer on north Indian population. Mol. Cell. Biochem. 2012;362(2):263–8.

[21] Zhou C, Xie LP, Lin YW, Yang K, Mao QQ, Cheng Y. Susceptibility of XPD and hOGG1 genetic variants to prostate cancer. Biomed. Rep 2013;1(4):679–83.

[22] Zhang J, Dhakal IB, Greene G, Lang NP, Kadlubar FF. Polymorphisms in hOGG1 and XRCC1 and risk of prostate cancer: effects modified by plasma antioxidants. Urology 2010;75(4):779–85.

[23] Karyadi DM, Zhao S, He Q, McIntosh L, Wright JL, Ostrander EA, et al. Confirmation of genetic variants associated with lethal prostate cancer in a cohort of men from hereditary prostate cancer families. Int. J. Cancer 2015;136(9):2166–71.

[24] Schoenfeld JD, Margalit DN, Kasperzyk JL, Shui IM, Rider JR, Epstein MM, et al. A single nucleotide polymorphism in inflammatory gene RNASEL predicts outcome after radiation therapy for localized prostate cancer. Clin. Cancer Res. 2013;19(6):1–8.

[25] Chen CH, Gong M, Yi QT, Guo JH. Role of interleukin-6 gene polymorphisms in the development of prostate cancer. Genet. Mol. Res. 2015;14(4):13370–4.

[26] Zhou L, Zheng Y, Tian T, Liu K, Wang M, Lin S, et al. Associations of interleukin-6 gene polymorphisms with cancer risk: evidence based on 49,408 cancer cases and 61,790 controls. Gene 2018;670(11):136–47.

[27] Jurečeková J, Drobková H, Sarlinová M, Babušíková E, Sivoňová MK, Matáková T, et al. The role of interleukin-6 polymorphism (rs1800795) in prostate cancer development and progression. Anticancer Res. 2018;38(6):3663–7.

[28] Liu JM, Cheng SH, Xia C, Deng T, Zhu YC, Wei X, et al. Association between single nucleotide polymorphisms in AKT1 and the risk of prostate cancer in the Chinese Han population. Genet. Mol. Res. 2017;16(1). https://doi.org/10.4238/gmr16019469.

[29] FitzGerald LM, Zhao S, Leonardson A, Geybels MS, Kolb S, Lin DW, et al. Germline variants in IL4, MGMT and AKT1 are associated with prostate cancer-specific mortality: an analysis of 12,082 prostate cancer cases. Prostate Cancer Prostatic Dis. 2018;21(2):228–35.

[30] Henríquez-Hernández LA, Valenciano A, Foro-Arnalot P, Álvarez-Cubero MJ, Cozar JM, Suárez-Novo JF, et al. Intraethnic variation in steroid-5-alpha-reductase polymorphisms in prostate cancer patients: a potential factor implicated in 5-alpha-reductase inhibitor treatment. J. Genet. 2015;94(2):335–41.

[31] Price DK, Chau CH, Till C, Goodman PJ, Leach RJ, Johnson-Pais TL, et al. Association of androgen metabolism gene polymorphisms with prostate cancer risk and androgen concentrations: results from the prostate cancer prevention trial. Cancer 2016;122(15):2332–40.

[32] Choi SY, Kim HJ, Cheong HS, Myung SC. The association of 5-alpha reductase type 2 (SRD5A2) gene polymorphisms with prostate cancer in a Korean population. Korean J. Urol 2015;56(1):19–30.

[33] Han JH, Lee YS, Kim HJ, Lee SY, Myung SC. Association between cytochrome CYP17A1, CYP3A4, and CYP3A43 polymorphisms and prostate cancer risk and aggressiveness in a Korean study population. Asian J. Androl. 2015;17(2):285–93.

[34] Poniah P, Mohamed Z, Apalasamy YD, Zain SM, Kuppusamy S, Razack AH. Genetic polymorphisms in the androgen metabolism pathway and risk of prostate cancer in low incidence Malaysian ethnic groups. Int. J. Clin. Exp. Med. 2015;8(10):19232–8.

[35] Gu CY, Qin XJ, Qu YY, Zhu Y, Wan FN, Zhang GM, et al. Genetic variants of the CYP1B1 gene as predictors of biochemical recurrence after radical prostatectomy in localized prostate cancer patients. Medicine (Baltim.) July 2016;95(27). 10.1097/MD.0000000000004066.

[36] Jingwi EY, Abbas M, Ricks-Santi L, Winchester D, Beyene D, Day A, et al. Vitamin D receptor genetic polymorphisms are associated with PSA level, Gleason score and prostate cancer risk in African-American men. Anticancer Res. 2015;35(3):1549–58.

[37] Nunes SB, de Matos Oliveira F, Neves AF, Araujo GR, Marangoni K, Goulart LR, Araújo TG. Association of vitamin D receptor variants with clinical parameters in prostate cancer. SpringerPlus 2016;5(1). https://doi.org/10.1186/s40064-016-2009-8.

[38] Oh JJ, Byun SS, Lee SE, Hong SK, Jeong CW, Choi WS, et al. Genetic variants in the CYP24A1 gene are associated with prostate cancer risk and aggressiveness in a Korean study population. Prostate Cancer Prostatic Dis. 2014;17(2):149–56.

[39] Bhargava A, Fuentes FF. Mutational dynamics of microsatellites. Mol. Biotechnol. 2010;44(3):250–66.

[40] Schlötterer C. Evolutionary dynamics of microsatellite DNA. Chromosoma 2000;109(6):365–71.

[41] Suzuki H, Komiya A, Aida S, Akimoto S, Shiraishi T, Yatani R, et al. Microsatellite instability and other molecular abnormalities in human prostate cancer. Jpn. J. Cancer Res. 1995;86(10):956–61.

[42] dos Santos ML, Sarkis ÁS, Nishimoto IN, Nagai MA. Androgen receptor CAG repeat polymorphism in prostate cancer from a Brazilian population. Cancer Detect. Prev. 2003;27(5):321–6.

[43] Paz-y-Miño C, Robles P, Salazar C, Leone PE, García-Cárdenas JM, Naranjo M, López-Cortés A. Positive association of the androgen receptor CAG repeat length polymorphism with the risk of prostate cancer. Mol. Med. Rep. 2016;14(2):1791–8.

[44] Gómez R, Torres-Sánchez L, Camacho-Mejorado R, Burguete-García AI, Vázquez-Salas RA, Martínez-Nava GA, et al. Androgen receptor CAG polymorphism and sporadic and early-onset prostate cancer among Mexican men. J. Hum. Genet. 2016;61(9):781–6.

[45] Hakimi JM, Schoenberg MP, Rondinelli RH, Piantadosi S, Barrack ER. Androgen receptor variants with short glutamine or glycine repeats may identify unique subpopulations of men with prostate cancer. Clin. Cancer Res. 1997;3(9):1599–608.

[46] Strom SS, Gu Y, Zhang H, Troncoso P, Babaian RJ, Pettaway CA, et al. Androgen receptor polymorphisms and risk of biochemical failure among prostatectomy patients. Prostate 2004;60(4): 343–51.

[47] Vijayalakshmi K, Thangaraj K, Rajender S, Vettriselvi V, Venkatesan P, Shroff S, et al. GGN repeat length and GGN/CAG haplotype variations in the androgen receptor gene and prostate cancer risk in south Indian men. J. Hum. Genet. 2006;51(11):998–1004.

[48] Rajender S, Vijayalakshmi K, Pooja S, Madhavi S, Paul SF, Vettriselvi V, et al. Longer (TA)n repeat but not A49T and V89L polymorphisms in SRD5A2 gene may confer prostate cancer risk in south Indian men. J. Androl. 2009;30(6):703−10.

[49] El Ezzi AA, Baker MT, Zaidan WR, Hraiki KM, El Saidi MA, Kuddus RH. Association of polymorphisms in the VDR, CYP17 and SRD5A2 genes and prostate cancer among Lebanese men. Asian Pac. J. Cancer Prev. APJCP 2017;18(1):93−8.

[50] Kumazawa T, Tsuchiya N, Wang L, Sato K, Kamoto T, Ogawa O, et al. Microsatellite polymorphism of steroid hormone synthesis gene CYP11A1 is associated with advanced prostate cancer. Int. J. Cancer 2004;110(1):140−4.

[51] Celhar T, Gersak K, Ovcak Z, Sedmak B, Mlinaric-Rascan I. The presence of the CYP11A1 (TTTTA)6 allele increases the risk of biochemical relapse in organ confined and low-grade prostate cancer. Cancer Genet. Cytogenet. 2008;187(1):28−33.

[52] Strand S, Orntoft T, Sorensen K. Prognostic DNA methylation markers for prostate cancer. Int. J. Mol. Sci. 2014;15(9):16544−76.

[53] Hoque MO. DNA methylation changes in prostate cancer: current developments and future clinical implementation. Expert Rev. Mol. Diag 2009;9(3):243−57.

[54] Geybels MS, Zhao S, Wong CJ, Bibikova M, Klotzle B, Wu M, et al. Epigenomic profiling of DNA methylation in paired prostate cancer versus adjacent benign tissue. Prostate 2015;75(16):1941−50.

[55] Zhao S, Geybels MS, Leonardson A, Rubicz R, Kolb S, Yan Q, et al. Epigenome-wide tumor DNA methylation profiling identifies novel prognostic biomarkers of metastatic-lethal progression in men diagnosed with clinically localized prostate cancer. Clin. Cancer Res. 2017;23(1):311−9.

[56] Wu Y, Davison J, Qu X, Morrissey C, Storer B, Brown L, Vessella R, Nelson P, Fang M. Methylation profiling identified novel differentially methylated markers including OPCML and FLRT2 in prostate cancer. Epigenetics 2016;11(4):247−58.

[57] Geybels MS, Alumkal JJ, Luedeke M, Rinckleb A, Zhao S, Shui IM, et al. Epigenomic profiling of prostate cancer identifies differentially methylated genes in TMPRSS2: ERG fusion-positive versus fusion-negative tumors. Clin. Epigenet. 2015;7(1). https://doi.org/10.1186/s13148-015-0161-6.

[58] Devaney JM, Wang S, Funda S, Long J, Taghipour DJ, Tbaishat R, et al. Identification of novel DNA-methylated genes that correlate with human prostate cancer and high-grade prostatic intraepithelial neoplasia. Prostate Cancer Prostatic Dis. 2013;16(4):292−6.

[59] Ngollo M, Dagdemir A, Karsli-Ceppioglu S, Judes G, Pajon A, Penault-Llorca F, et al. Epigenetic modifications in prostate cancer. Epigenomics 2014;6(4):415−26.

[60] Mahon KL, Qu W, Devaney J, Paul C, Castillo L, Wykes RJ, et al. Methylated glutathione S-transferase 1 (mGSTP1) is a potential plasma free DNA epigenetic marker of prognosis and response to chemotherapy in castrate-resistant prostate cancer. Br. J. Canc. 2014;111(9):1802−9.

[61] Keil KP, Abler LL, Mehta V, Altmann HM, Laporta J, Plisch EH, et al. DNA methylation of E-cadherin is a priming mechanism for prostate development. Dev. Biol. 2014;387(2):142−53.

[62] Zhang SQ, Zhang GQ, Zhang L. Correlation between methylation of the E-Cadherin gene and malignancy of prostate cancer. Int. J. Clin. Exp. Pathol. 2016;9(3):3642−7.

[63] Kawaguchi K, Kinameri A, Suzuki S, Senga S, Ke Y, Fujii H. The cancer-promoting gene fatty acid-binding protein 5 (FABP5) is epigenetically regulated during human prostate carcinogenesis. Biochem. J. 2016;473(4):449−61.

[64] Li S, Zhu Y, Ma C, Qiu Z, Zhang X, Kang Z, et al. Downregulation of EphA5 by promoter methylation in human prostate cancer. BMC Canc. 2015;15(1):18−23.

[65] Huang QX, Xiao CT, Chen Z, Lu MH, Pang J, Di JM, et al. Combined analysis of CRMP4 methylation levels and CAPRA-S score predicts metastasis and outcomes in prostate cancer patients. Asian J. Androl. 2018;20(1):56−61.

[66] Gao X, Li LY, Rassler J, Pang J, Chen MK, Liu WP, et al. Prospective study of CRMP4 promoter methylation in prostate biopsies as a predictor for lymph node metastases. J. Natl. Cancer Inst. 2017;109(6). https://doi.org/10.1093/jnci/djw282.

[67] Lee J, Lee MS, Jeoung DI, Kim YM, Lee H. Promoter CpG-Site methylation of the KAI1 metastasis suppressor gene contributes to its epigenetic repression in prostate cancer. Prostate 2017;77(4):350−60.

[68] Almeida-Rios D, Graça I, Vieira FQ, Ramalho-Carvalho J, Pereira-Silva E, Martins AT, et al. Histone methyltransferase PRMT6 plays an oncogenic role of in prostate cancer. Oncotarget 2016;7(33):53018−28.

[69] Xu K, Wu ZJ, Groner AC, He HH, Cai C, Lis RT, et al. EZH2 oncogenic activity in castration-resistant prostate cancer cells is Polycomb-independent. Science 2012;338(6113):1465−9.

[70] Mounir Z, Korn JM, Westerling T, Lin F, Kirby CA, Schirle M, et al. ERG signaling in prostate cancer is driven through PRMT5-dependent methylation of the androgen receptor. Elife 2016;5. https://doi.org/10.7554/elife.13964.

[71] Deng X, Shao G, Zhang HT, Li C, Zhang D, Cheng L, et al. Protein arginine methyltransferase 5 functions as an epigenetic activator of the androgen receptor to promote prostate cancer cell growth. Oncogene 2017;36(9):1223−33.

[72] Metzger E, Willmann D, McMillan J, Forne I, Metzger P, Gerhardt S, et al. Assembly of methylated KDM1A and CHD1 drives androgen receptor−dependent transcription and translocation. Nat. Struct. Mol. Biol. 2016;23(2):132−9.

[73] Cang S, Xu X, Ma Y, Liu D, Chiao J. Hypoacetylation, hypomethylation, and dephosphorylation of H2B histones and excessive histone deacetylase activity in DU-145 prostate cancer cells. J. Hematol. Oncol. 2016;9(1):3−7.

[74] Kelly RS, Vander Heiden MG, Giovannucci E, Mucci LA. Metabolomic biomarkers of prostate cancer: prediction, diagnosis, progression, prognosis, and recurrence. Cancer Epidemiol. Prev. Biomarkers 2016;25(6):887−906.

[75] Ren S, Shao Y, Zhao X, Hong CS, Wang F, Lu X, et al. Integration of metabolomics and transcriptomics reveals major metabolic pathways and potential biomarker involved in prostate cancer. Mol. Cell. Proteom. 2016;15(1):154−63.

[76] Lucarelli G, Rutigliano M, Galleggiante V, Giglio A, Palazzo S, Ferro M, et al. Expert Rev. Mol. Diag 2015;15(9):1211−24.

[77] Meller S, Meyer HA, Bethan B, Dietrich D, Maldonado SG, Lein M, et al. Integration of tissue metabolomics, transcriptomics and immunohistochemistry reveals ERG-and Gleason score-specific metabolomic alterations in prostate cancer. Oncotarget 2016;7(2):1421−33.

[78] Adams CD, Richmond R, Ferreira DL, Spiller W, Tan V, Zheng J, et al. Circulating metabolic biomarkers of screen-detected prostate cancer in the ProtecT study. Cancer Epidemiol. Prev. Biomarkers 2019;28(1):208–16.
[79] Zhang A, Yan G, Han Y, Wang X. Metabolomics approaches and applications in prostate cancer research. Appl. Biochem. Biotechnol. 2014;174(1):6–12.
[80] Huang J, Mondul AM, Weinstein SJ, Koutros S, Derkach A, Karoly E, et al. Serum metabolomic profiling of prostate cancer risk in the prostate, lung, colorectal, and ovarian cancer screening trial. Br. J. Canc. 2016;115(9):1087–97.
[81] Thapar RA, Titus M. Recent advances in metabolic profiling and imaging of prostate cancer. Curr. Metabolom 2014;2(1):53–69.
[82] Kumar D, Gupta A, Mandhani A, Sankhwar SN. NMR spectroscopy of filtered serum of prostate cancer: a new frontier in metabolomics. Prostate 2016;76(12):1106–19.
[83] Batra JS, Girdhani S, Hlatky L. A quest to identify prostate cancer circulating biomarkers with a bench-to-bedside potential. J. Biom. 2014;2014:321680. https://doi.org/10.1155/2014/321680.
[84] Hall IJ, Rim SH, Massetti GM, Thomas CC, Li J, Richardson LC. Prostate-specific antigen screening: an update of physician beliefs and practices. Prev. Med. 2017;103:66–9.
[85] Le BV, Griffin CR, Loeb S, Carvalhal GF, Kan D, Baumann NA, Catalona WJ. Proenzyme prostate specific antigen is more accurate than total and free prostate specific antigen in differentiating prostate cancer from benign disease in a prospective prostate cancer screening study. J. Urol. 2010;183(4):1355–9.
[86] Zappala SM, Scardino PT, Okrongly D, Linder V, Dong Y. Clinical performance of the 4Kscore test to predict high-grade prostate cancer at biopsy: a meta-analysis of US and European clinical validation study results. Rev. Urol. 2017;19(3):149–65.
[87] Leyten GH, Hessels D, Smit FP, Jannink SA, de Jong H, Melchers WJ, et al. Identification of a candidate gene panel for the early diagnosis of prostate cancer. Clin. Cancer Res. 2015;21(13):3061–70.
[88] Kretschmer A, Tilki D. Biomarkers in prostate cancer–current clinical utility and future perspectives. Crit. Rev. Oncol.-Hematol. 2017;120:180–93.
[89] Marks LS, Fradet Y, Deras IL, Blase A, Mathis J, Aubin SM, et al. PCA3 molecular urine assay for prostate cancer in men undergoing repeat biopsy. Urology 2007;69(3):532–5.
[90] Klein ME, Dabbs DJ, Shuai Y, Brufsky AM, Jankowitz R, Puhalla SL, Bhargava R. Prediction of the oncotype DX recurrence score: use of pathology-generated equations derived by linear regression analysis. Mod. Pathol. 2013;26(5):658–65.
[91] Cuzick J, Swanson GP, Fisher G, Brothman AR, Berney DM, Reid JE, et al. Prognostic value of an RNA expression signature derived from cell cycle proliferation genes in patients with prostate cancer: a retrospective study. Lancet Oncol. 2011;12(3):245–55.
[92] Blume-Jensen P, Berman DM, Rimm DL, Shipitsin M, Putzi M, Nifong TP, Small C, Choudhury S, Capela T, Coupal L, Ernst C. Development and clinical validation of an *in situ* biopsy-based multimarker assay for risk stratification in prostate cancer. Clin. Cancer Res. 2015;21(11):2591–600.
[93] Dalela D, Löppenberg B, Sood A, Sammon J, Abdollah F. Contemporary role of the Decipher® test in prostate cancer management: current practice and future perspectives. Rev. Urol. 2016;18(1):1–7.
[94] Cetnar JP, Beer TM. Personalizing prostate cancer therapy: the way forward. Drug Discov. Today 2014;19(9):1483–7.
[95] Yadav SS, Li J, Lavery HJ, Yadav KK, Tewari AK. Next-generation sequencing technology in prostate cancer diagnosis, prognosis, and personalized treatment. In: Urologic oncology: seminars and original investigations, vol. 33; 2015. https://doi.org/10.1016/j.urolonc.2015.02.009 (6).
[96] Aoun F, Rassy EE, Assi T, Kattan J. Personalized treatment of prostate cancer: better knowledge of the patient, the disease and the medicine. Future Oncol. 2016;12(20):2359–61.
[97] Saura C, Roda D, Roselló S, Oliveira M, Macarulla T, Pérez-Fidalgo JA, et al. A first-in-human phase I study of the ATP-competitive AKT inhibitor ipatasertib demonstrates robust and safe targeting of AKT in patients with solid tumors. Cancer Discov. 2017;7(1):102–13.
[98] Nevedomskaya E, Baumgart S, Haendler B. Recent advances in prostate cancer treatment and drug discovery. Int. J. Mol. Sci. 2018;19(5):E1359. https://doi.org/10.3390/ijms19051359.

Chapter 17

MicroRNAs and inflammation biomarkers in obesity

Bruna Jardim Quintanilha[1,2], Bruna Zavarize Reis[3], Telma A. Faraldo Corrêa[2,3], Graziela Biude da Silva Duarte[3] and Marcelo Macedo Rogero[1,2]

[1]*Department of Nutrition, School of Public Health, University of São Paulo, São Paulo, Brazil;* [2]*Food Research Center (FoRC), CEPID-FAPESP, Research Innovation and Dissemination Centers São Paulo Research Foundation, São Paulo, Brazil;* [3]*Department of Food and Experimental Nutrition, Faculty of Pharmaceutical Science, University of São Paulo, São Paulo, Brazil*

Introduction

Obesity, a multifactorial and polygenic condition, is a public health issue of great concern in both developed and developing countries. It is estimated that in 2016 alone there were more than 1.9 billion overweight adults worldwide, that is, 39% of the world's adult population; of these, more than 650 million (approximately 13% of the world's adult population) were obese. Obesity also affects children: an estimated 41 million people under 5 years of age, as well as 340 million people aged 5–19 years, were either overweight or obese in 2016 [1].

Obesity results from the interaction of a series of genetic, metabolic, behavioral, and environmental factors. It leads to metabolic inflammation or meta-inflammation, a chronic form of low-grade inflammation in white adipose tissue (WAT) that differs from a classic inflammatory response [2]. Two proteins play a key role in metainflammation: inhibitor of nuclear factor kappa-B kinase (IKK-β), and c-Jun N-terminal kinase 1 (JNK-1), whose activation in turn activates nuclear factor kappa B (NF-kB) and activator protein 1 (AP-1), respectively. NF-kB and AP-1 then translocate to the cell nucleus and activate the transcription of genes related to inflammation, such as tumor necrosis factor alpha (TNF-α), interleukin 6 (IL-6), and interleukin 1 beta (IL-1β) [3].

Adipocytes and infiltrating inflammatory cells—mainly macrophages—secrete cytokines such as TNF-α, IL-6, and IL-1β, and inflammatory modulators such as leptin, resistin, and adiponectin [2,4]. Obesity is strongly associated with metabolic diseases such as type 2 diabetes (T2D), atherosclerosis, cardiovascular disease (CVD), nonalcoholic fatty liver disease (NAFLD), and some cancers [5].

Micro RNAs

MicroRNAs (miRNAs) play a key role in obesity, since they can posttranscriptionally regulate a large number of genes. They may also be important in maintaining metabolic homeostasis; if so, their regulation could serve as potential therapeutics for metabolic diseases [6]. MicroRNAs are noncoding RNAs that modulate gene expression. Mature miRNAs regulate the expression of proteins, by cleaving mRNA or repressing its translation, depending on the level of complementarity between the miRNA and the target mRNA [7].

Changes in miRNA levels have been shown to affect gene expression and thereby cell function, in several pathophysiological disorders related to obesity, including inflammation, oxidative stress, impaired adipogenesis, insulin signaling, apoptosis, and angiogenesis [8–12].

MicroRNAs are present in tissues or body fluids in stable form, protected from endogenous RNAse activity [6]. They may play a role in communication between adipocytes, as well as between adipose tissue and other tissues. Notably, miRNAs can act as potential diagnostic biomarkers, as they fulfill most of the necessary criteria: they can be rapidly and accurately detected through noninvasive methods, have high sensitivity and specificity to the disease in question, allow for early detection, and have a long half-life in the sample [13–15].

The identification and characterization of miRNAs associated with obesity could yield a new generation of therapeutic targets for antiobesity treatments. Additionally, identifying miRNAs that are dysregulated during the development of obesity could provide early obesity biomarkers for clinical diagnosis.

Precision Medicine for Investigators, Practitioners and Providers. https://doi.org/10.1016/B978-0-12-819178-1.00017-4

MicroRNAs concepts

MicroRNAs are a class of small, noncoding endogenous RNA molecules (~18–25 nucleotides) involved in post-transcriptional gene regulation, by binding to a target messenger RNA (mRNA), which results in the degradation or inhibition of translation [16]. To date, 2654 mature human miRNAs have been identified (MiRBase, release 16, December 2018; http://www.mirbase.org). MicroRNAs can act directly on the target genes or indirectly, by regulating transcription factors that in turn control the expression of genes [4]. One miRNA can regulate several mRNAs, but not all miRNAs incur translational repression. Some of them have the ability to activate the translation of proteins, alter the structure of chromatin by regulating histone modification, and even target genes with low DNA methylation directly [3,16].

Biogenesis

Biogenesis of miRNAs is a sequential process involving a variety of enzymes and proteins [16]. MicroRNAs are usually transcribed by RNA polymerase II from miRNA genes, first forming the primary miRNA transcript (pri-miRNA). This transcript is then cleaved by the DROSHA-DiGeorge syndrome critical region gene 8 (DGCR8) microprocessor complex, creating a shorter sequence called the miRNA precursor (pre-miRNA) that displays a hairpinlike secondary structure. The pre-miRNA is exported to the cytoplasm and processed by DICER, a ribonuclease III enzyme that produces mature miRNA, which is incorporated into an RNA-protein complex (i.e., the RNA-induced silencing complex, RISC). Under most conditions, mature RISC represses gene expression post-transcriptionally, by binding the three prime untranslated regions (3′-UTRs) of specific mRNAs, and mediating mRNA degradation, destabilization, or translational inhibition, according to sequence complementarity to the target [3,17–19].

Evidence shows that, besides intracellular function, miRNAs are present in extracellular fluids in the human body, including plasma, serum, urine, and saliva; recently, they have also been associated with diseases such as obesity, cancer, and cardiovascular disease (CVD) [20–22]. MicroRNAs also play an important role in cell-to-cell communication in peripheral blood, either through membrane-enclosed vesicles such as exosomes (extracellular vesicles of endosomal origin), or by binding to lipoproteins (LDL or HDL), proteins, apoptotic bodies, and ribonucleoprotein complexes (linked to Argonaut) [21,22]. In addition to their stability, it should be noted that circulating miRNAs are conserved across species, have expression patterns that are tissue- and biological-stage specific, and are easily determined through real-time polymerase chain reaction (RT-PCR) [16]. Thus, these molecules are promising noninvasive biomarkers of certain diseases, and even of nutritional status [23].

Inflammation and miRNAs: a role in chronic diseases

Obesity is related to endocrine and metabolic changes in WAT, which is the main source of systemic inflammatory response. White adipose tissue produces a variety of proinflammatory cytokines and chemokines known as adipokines, which include TNF-á, IL-1â, IL-6, and monocyte chemotactic protein (MCP-1). Consequently, obese individuals present an increase in proinflammatory biomarkers such as TNF-á, IL-1â, IL-6, and MCP-1, as well as a decrease in antiinflammatory adipokines such as adiponectin [24–26]. In addition, activation of endothelial cells and oxidative stress exacerbate the inflammatory response [24,27,28].

Two important signaling inflammatory pathways are involved in these processes: the NF-êB and JNK pathways. The former can be activated by TNF-á, lipopolysaccharides (LPSs), and saturated fatty acids; it involves the enzymatic complex IKK, which induces the phosphorylation of inhibitor-êB (IêB). This phosphorylation results in IêB polyubiquitination, which, in turn, leads to IêB degradation, as mediated by the 26S proteasome. This degradation allows NF-êB to translocate to the nucleus, and activate the transcription of several êB-dependent genes such as those encoding proinflammatory cytokines, adhesion molecules, and chemokines.

The JNK signaling pathway can be activated by cytokines, fatty acids, and reactive oxygen species (ROS), among others. The active JNK pathway promotes the activation of transcription factor AP-1, which is related to gene expression of proinflammatory cytokines, and may act directly on the insulin signaling pathway. In this context, an increase in inflammatory response may trigger resistance to insulin action due to an increase in Ser^{307} phosphorylation of insulin receptor substrate 1 (IRS-1), and lower activity of phosphatidylinositol-4,5-bisphosphate-3-kinase (PI3K) in skeletal muscle.

It should be noted that Ser^{307} phosphorylation of IRS-1 is associated with activation of JNK and IKK-â. In addition, activation of IL-6 signaling can increase the expression of cytokine signaling suppressor proteins SOCS-1 and SOCS-3, in the three major insulin-sensitive peripheral tissues: WAT, liver tissue, and muscle tissue. SOCS-1 and SOCS-3 impair insulin action, by binding to IRS-1 and IRS-2, which leads to IRS-1 and IRS-2 ubiquitination and degradation [2,29–32].

There is substantial evidence of miRNAs' transcriptional-level regulation of genes encoding proteins related to the inflammatory process, in metabolic diseases [33]. Regulation may occur on many levels. Inflammation may induce the transcriptional process; if so, miRNAs will become rapidly active, since they do not need to be translated or translocated back into the nucleus, to repress their target. The expression of miRNAs occurs in different cell types with different functions, and some target proteins— for example, proteins regulating miRNA processing— may be linked to the inflammatory response. MicroRNAs can also regulate several mechanisms related to inflammation, such as adipogenesis, macrophage activation, and oxidative stress [8].

MicroRNAs as biomarkers in obesity

MicroRNAs found in human body fluids such as urine and plasma, are called circulating miRNAs; they act not only within cells but also in distant tissues, as hormones controlling gene expression. The physiological function of circulating miRNAs is mostly unknown, but different studies have shown that these molecules have essential roles, including immune cell modulation [34–36] (Fig. 17.1).

The miR-221/222 family

The miR-221/222 gene cluster in humans is located in chromosome Xp11.3; genes in this cluster share an almost identical seed sequence. In terms of biogenesis, they are first transcribed as a single long noncoding RNA precursor; then, in the nucleus, the common pri-miR-221/222 transcript is spliced and split by the Drosha-DGCR8 complex, resulting in the formation of two individual precursors: pre-miR-221 and pre-miR-222. After the final processing steps, mature transcripts are formed: in particular, miR-221-5p or miR-221-3p (23 nucleotides), and possibly also miR-222-5p and miR-222-3p (21 nucleotides) [37,38]. These miRNAs are upregulated in obese individuals, and related to metabolic processes involved in obesity [39].

Both miR-221 and miR-222 are involved in adipogenesis, a process through which preadipocytes differentiate into mature adipocytes. In primary human bone marrow stromal cells (BMSCs), these miRNAs downregulate adipogenesis, by reducing expression of peroxisome proliferator-activated receptor gamma (PPARã) and CCAAT/enhancer-binding protein alpha (CEBPá) [40].

Expression of miR-221 in WAT is different in obese and lean people [41]. miR-221 expression evaluated in abdominal subcutaneous adipose tissue was shown to positively correlate with body mass index (BMI), in a nondiabetic Indian population. This miRNA was found to be highly expressed in abdominal subcutaneous adipose tissue, and upregulated in individuals with obesity (BMI >37 kg/m^2) [42]. In addition, miR-221 expression was found to be upregulated in samples of human subcutaneous fat, from obese women with T2D [43].

Both miR-221 and miR-222 play an important role in the process of chronic low-grade inflammation, through inhibition of endothelial cell proliferation and angiogenesis, which

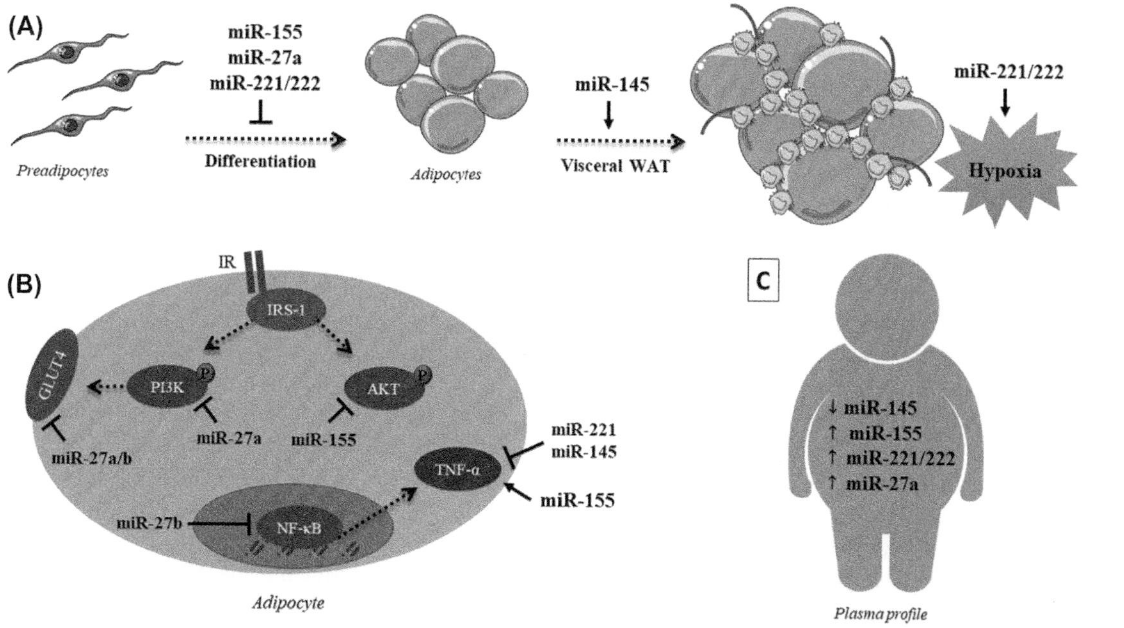

FIGURE 17.1 Mechanisms of some microRNAs in (A) adipogenesis, visceral WAT expansion, and hypoxia; (B) molecular pathways of insulin signaling and NF-κB activation. (C) Plasma profile of miRNAs in obese people: ↓ low expression, ↑ high expression. *AKT*, Protein kinase B; *GLUT4*, glucose transporter member 4; *IR*, insulin receptor; *IRS-1*, insulin receptor substrate 1; *NF-êB*, nuclear factor-kappa B; *PI3K*, activation of phosphoinositide 3-kinase; *TNF-á*, tumor necrosis factor.

contributes to tissue hypoxia. This can impact the production of proinflammatory adipokines and other miRNAs such as miR-27, involved in insulin resistance (IR) [39].

It has been demonstrated that miR-221 plays a relevant role in TNF-á signaling pathway, since it can bind directly to the cytokine's 3′-UTR, and result in its mRNA degradation. In human adipose-derived stem cells (hASCs), hASC-differentiated adipocytes transfected with pre-miR-221 showed 50% reduction in mRNA TNF-á. In addition, pre-miR-221 treatment resulted in a small reduction (~ 10%) in MCP-1 mRNA levels, and had no effect on IL-6 mRNA levels. These results indicate that if miR-221 is overexpressed, TNF-á expression in adipocytes decreases. miR-221 might be helpful in attenuating chronic inflammation in obese women [44].

Some evidence suggests that miR-222 can also affect glucose metabolism. Plasma miR-222 levels have been shown to be higher in patients with T2D, than in those with normal glucose tolerance. When only obese individuals were evaluated, miR-222 plasma concentration increased in the group with T2D. Plasma miR-222 levels were also found to positively correlate with fasting glycemia, and glycated hemoglobin (HBA_{1c}) levels [45].

In women diagnosed with metabolic syndrome, high miR-221 serum levels were observed, and no correlation was found between them and cardiometabolic risk factors. However, upregulation of miR-221/222 in blood vessels of individuals with metabolic syndrome, obesity, or T2D, for example, increases cardiovascular risk, and contributes to the development of atherosclerosis by endothelial dysfunction and neointimal hyperplasia [37].

In samples of human atherosclerotic vessels obtained from patients undergoing coronary bypass graft procedure, miR-221/222 levels were elevated in the intima, and posed a risk of endothelial dysfunction, by suppressing peroxisome proliferator-activated receptor gamma coactivator 1-alpha (PGC-1á), thus contributing to progression of atherosclerosis [46].

Multifunctional miR-155

miR-155 is a multifunctional miRNA: it has been associated with the regulation of different immune-related processes, such as hematopoiesis [47], innate immunity [48], B-cell and T-cell differentiation [49], and cancer [50]. It is one of the most studied miRNAs involved in obesity, since it plays a role in adipogenesis, adipocyte function, and inflammation [51–54]. The induction of miR-155 expression is mediated by TNF-α in adipocytes and in WAT, which explains the role of this miRNA in obesity-mediated inflammation [52]. It was recently demonstrated that adipose tissue macrophages can secrete exosomal miR-155 molecules, which are then efficiently transported into adipocytes [55]. Similarly, in vivo study has shown that transgenic mice overexpressing miR-155 in the B cell lineage produced more TNF-α when challenged with LPSs [56].

In a culture of human adipocytes, miR-155 was found to be substantially upregulated by TNF-α and induced inflammation, chemokine expression, and macrophage migration [52]. These results corroborate findings, whereby miR-155 levels in the adipose tissue of TNF-α knockout mice were lower, than in wild-type mice [52].

Overexpression of miR-155 has been found to significantly reduce insulin-stimulated glucose uptake in 3T3-L1 adipocytes and L6 muscle cells, as well as expression of miR-155 target gene PPARγ. Furthermore, miR-155 overexpression appears to lead to a decrease in insulin-induced phosphorylation of protein kinase B (AKT) in adipocytes, myocytes, and hepatocytes [55]. In addition, miR-155 knockout mice on high-fat diet, exhibit significant weight gain compared to wild-type mice, suggesting that there is a potential miRNA-based mechanism, contributing to the development of diet-induced obesity [57].

Obese subjects display upregulation of miR-155 in adipose tissue, presenting a significant positive correlation between miR-155 and BMI, and between miR-155 and mRNA coding for TNF-α [52]. miR-155 deletion may be promising in protecting against obesity, since it may regulate the development and persistence of obesity via several signaling pathways, including adipogenesis and inflammation. miR-155 has been implicated in adipocyte differentiation toward a white, rather than a brown/beige, phenotype. Brown adipocytes are key sites of energy expenditure; therefore, a brown adipocytelike phenotype may increase energetic efficiency in mammals. Finally, miR-155 may influence adipose tissue accumulation, by activating proinflammatory pathways [58].

miR-145

miR-145 is a member of the miR-143/145 cluster, and the most abundant miRNA in the vascular wall. Due to its positive correlation with high-sensitive C-reactive protein (hs-CRP) levels in acute ischemic stroke (AIS) patients, it is suggested that this miRNA might be a possible biomarker of the beginning and severity of AIS [59,60]. Furthermore, miR-145 is involved in lipid metabolism and inflammatory pathways in a tissue-specific manner.

In WAT, higher expression of both TNF-α and IL-6 was observed, when this miR was downregulated. Decrease in miR-145 levels upregulated expression of ADP-ribosylation factor 6 (ARF6), a small GTPase responsible for activating the NF-kB-mediated inflammatory pathway in macrophages [61]. However, certain studies have yielded different results: in particular, overexpression of miR-145 has been found to increase proinflammatory mediators such as TNF-alpha, IL1-beta,

CCL-2, CCL-3, and CCL-4 in macrophages, by upregulating NF-kB expression and activity [62].

A negative correlation between miR-145 and leptin receptor (LEPR) expression has been found in morbidly obese patients [63]. Obese individuals have lower miR-145 levels than lean individuals, which could be related to the seemingly paradoxical increase in plasma concentration of leptin (an anorexigenic hormone), and LEPR resistance in this phenotype [64].

miR-27a/b

The miR-27 family includes two isoforms: miR-27a, an intergenic miRNA, and miR27b, an intronic miRNA, located within the 14th intron of the human C9orf3 host gene. These miRNAs are homologous and conserved in mammals during evolution [65]. miR-27 has been found to be upregulated in the omental adipose multipotent stem cells of obese human adults [66].

Evidence has shown that miR-27 is a negative regulator of adipocyte differentiation and obesity [67,68]. miR-27a is highly expressed in the stromal vascular fraction (SVF) of murine adipose tissues, and it is involved in adipocyte differentiation, by targeting PPARã 3′-UTR. Overexpression of miR-27a has been found to suppress PPARã expression, and adipocyte differentiation in 3T3-L1 preadipocyte cells [69].

miR-27a and miR-27b are downregulated in response to hydrogen peroxide (H_2O_2) [65], a reactive oxygen species that contributes to oxidative stress, and could also contribute to persistence of inflammation, and development of atherosclerosis. In RAW 264.7 cells transfected with miR-27b, translocation of the p65 subunit of NF-êB appears to be affected. However, miR-27b overexpression does not appear to modify mRNA expression of its target proteins, such as IL-1â, IL-6, and TNF-á. A reduction in CCl-2 mRNA expression has been observed in cells, transfected with miR-27b mimics [65,70].

A study performed in vivo with adipose tissue samples of hyperglycemic rats, and in vitro with 3T3-L1 adipocyte cells exposed to high glucose concentrations, verified an upregulation of miR-27a. Results suggest that this miRNA plays an important role in pathogenesis of T2D [71]. Moreover, miR-27a/b has been related to lipid metabolism through the repression of certain genes, such as sterol regulatory element-binding transcription factor 1c (SREBP1c), retinoid X receptor alpha (RXR-á), adiponectin, PPARã, glucose transporter member 4 (GLUT-4), fatty acid-binding protein 4 (FABP-4), and FASN [68].

miR-27b also targets peroxisome proliferator-activated receptor α (PPARá), which is involved in lipid metabolism through the regulation of genes such as ATP-binding cassette transporter A1 (ABCA1) and ATP-binding cassette subfamily G member 1 (ABCG1). Overexpression of miR-27 has been shown to result in downregulation of PPARá [65]. miR-27b expression in human hepatic Huh7 cells can affect ABCA1 protein levels and cholesterol efflux to apolipoprotein A1 (ApoA1). In vivo, overexpression of pre-miR-27b in mice liver samples was found to reduce hepatic expression of ABCA1 by 50%, and low-density lipoprotein receptor (LDLR) by 20% without changing blood cholesterol and triglyceride levels [72].

References

[1] WHO. Obesity and overweight. 2017 [cited 2017 24 October]; Available from: http://www.who.int/mediacentre/factsheets/fs311/en/.
[2] Rogero M, Calder P. Obesity, inflammation, toll-like receptor 4 and fatty acids. Nutrients 2018;10(4):432.
[3] Quintanilha B, et al. Nutrimiromics: role of microRNAs and nutrition in modulating inflammation and chronic diseases. Nutrients 2017;9(11):1168.
[4] Arner E, et al. Adipose tissue microRNAs as regulators of CCL2 production in human obesity. Diabetes 2012;61(8):1986–93. DB_111508.
[5] González-Muniesa P, et al. Obesity. Nat. Rev. Dis. Primers 2017;3:17034.
[6] Pescador N, et al. Serum circulating microRNA profiling for identification of potential type 2 diabetes and obesity biomarkers. PLoS One 2013;8(10):e77251.
[7] Min P-K, Chan SY. The biology of circulating MicroRNAs in cardiovascular disease. Eur. J. Clin. Investig. 2015;45(8):860–74.
[8] Hulsmans M, De Keyzer D, Holvoet P. MicroRNAs regulating oxidative stress and inflammation in relation to obesity and atherosclerosis. FASEB J. 2011:2515–27.
[9] Ma E, Fu Y, Garvey WT. Relationship of circulating miRNAs with insulin sensitivity and associated metabolic risk factors in humans. Metab. Syndrome Relat. Disord. 2018;16(2):82–9.
[10] Aryal B, et al. MicroRNAs and lipid metabolism. Curr. Opin. Lipidol. 2017;28(3):273–80.
[11] Shah R, et al. Extracellular RNAs are associated with insulin resistance and metabolic phenotypes. Diabetes Care 2017;40(4):546.
[12] Cruz KJC, et al. The role of microRNAs on adipogenesis, chronic low grade inflammation and insulin resistance in obesity. Nutrition 2017;35:28–35.
[13] Creemers EE, Tijsen AJ, Pinto YM. Circulating microRNAs: novel biomarkers and extracellular communicators in cardiovascular disease? Circ. Res. 2012;110(3):483–95.
[14] Zampetaki A, et al. Profiling of circulating microRNAs: from single biomarkers to re-wired networks. Cardiovasc. Res. 2012;93(4):555–62.
[15] Tijsen AJ, Pinto YM, Creemers EE. Circulating microRNAs as diagnostic biomarkers for cardiovascular diseases. Am. J. Physiol. 2012;303(9):H1085–95.
[16] Ross SA, Davis CD. The emerging role of microRNAs and nutrition in modulating health and disease. Annu. Rev. Nutr. 2014;34:305–36.
[17] Cloonan N, et al. MicroRNAs and their isomiRs function cooperatively to target common biological pathways. Genome Biol. 2011;12(12):R126.

[18] Huntzinger E, Izaurralde E. Gene silencing by microRNAs: contributions of translational repression and mRNA decay. Nat. Rev. Genet. 2011;12(2):99–110.

[19] Lin S, Gregory RI. MicroRNA biogenesis pathways in cancer. Nat. Rev. Canc. 2015;15(6):321–33.

[20] El-Khoury V, et al. Assessing cellular and circulating miRNA recovery: the impact of the RNA isolation method and the quantity of input material. Sci. Rep. 2016;6:19529.

[21] Xu J, et al. Circulating microRNAs: novel biomarkers for cardiovascular diseases. J. Mol. Med. (Berl.) 2012;90(8):865–75.

[22] Kawaguchi T, et al. Circulating microRNAs: a next-generation clinical biomarker for digestive system cancers. Int. J. Mol. Sci. 17(9) 2016.

[23] Beckett EL, et al. The role of vitamins and minerals in modulating the expression of microRNA. Nutr. Res. Rev. 2014;27(1):94–106.

[24] Galic S, Oakhill JS, Steinberg GR. Adipose tissue as an endocrine organ. Mol. Cell. Endocrinol. 2010;316(2):129–39.

[25] Greenberg AS, Obin MS. Obesity and the role of adipose tissue in inflammation and metabolism. Am. J. Clin. Nutr. 2006:461S–5S.

[26] Tilg H, Moschen AR. Adipocytokines: mediators linking adipose tissue, inflammation and immunity. Nat. Rev. Immunol. 2006;6(10):772–83.

[27] Elks CM, Francis J. Central adiposity, systemic inflammation, and the metabolic syndrome. Curr. Hypertens. Rep. 2010;12(2):99–104.

[28] Maury E, Brichard SM. Adipokine dysregulation, adipose tissue inflammation and metabolic syndrome. Mol. Cell. Endocrinol. 2010;314(1):1–16.

[29] Baud V, Karin M. Is NF-κB a good target for cancer therapy? Hopes and pitfalls. Nat. Rev. Drug Discov. 2009;8(1):33.

[30] Czech MP. Insulin action and resistance in obesity and type 2 diabetes. Nat. Med. 2017;23(7):804.

[31] Pal M, Febbraio MA, Lancaster GI. The roles of c-Jun NH2-terminal kinases (JNKs) in obesity and insulin resistance. J. Physiol. 2016;594(2):267–79.

[32] Vallabhapurapu S, Karin M. Regulation and function of NF-κB transcription factors in the immune system. Annu. Rev. Immunol. 2009;27:693–733.

[33] Contreras J, Rao DS. MicroRNAs in inflammation and immune responses. Leukemia 2012;26(3):404.

[34] Barile L, Moccetti T, Marbán E, Vassalli G. Roles of exosomes in cardioprotection. Eur. Heart J. 2017;38(18):1372–9.

[35] Guo M, et al. miR-146a in PBMCs modulates Th1 function in patients with acute coronary syndrome. Immunol. Cell Biol. 2010:555–64.

[36] He S, et al. Endothelial extracellular vesicles modulate the macrophage phenotype: potential implications in atherosclerosis. Scand. J. Immunol. 2018;87(4):e12648.

[37] Chistiakov DA, et al. Human miR-221/222 in physiological and atherosclerotic vascular remodeling. BioMed Res. Int. 2015;2015:354517.

[38] Chen WX, et al. miR-221/222: promising biomarkers for breast cancer. Tumour Biol 2013;34(3):1361–70.

[39] Hulsmans M, De Keyzer D, Holvoet P. MicroRNAs regulating oxidative stress and inflammation in relation to obesity and atherosclerosis. FASEB J. 2011;25(8):2515–27.

[40] Skarn M, et al. Adipocyte differentiation of human bone marrow-derived stromal cells is modulated by microRNA-155, microRNA-221, and microRNA-222. Stem Cell. Dev. 2012;21(6):873–83.

[41] Arner P, Kulyté A. MicroRNA regulatory networks in human adipose tissue and obesity. Nat. Rev. Endocrinol. 2015;11(5):276–88.

[42] Meerson A, et al. Human adipose microRNA-221 is upregulated in obesity and affects fat metabolism downstream of leptin and TNF-alpha. Diabetologia 2013;56(9):1971–9.

[43] Ortega FJ, et al. MiRNA expression profile of human subcutaneous adipose and during adipocyte differentiation. PLoS One 2010;5(2):e9022.

[44] Chou WW, et al. Decreased microRNA-221 is associated with high levels of TNF-alpha in human adipose tissue-derived mesenchymal stem cells from obese woman. Cell. Physiol. Biochem. 2013;32(1):127–37.

[45] Ortega FJ, et al. Profiling of circulating microRNAs reveals common microRNAs linked to type 2 diabetes that change with insulin sensitization. Diabetes Care 2014;37(5):1375–83.

[46] Xue Y, et al. MicroRNA-19b/221/222 induces endothelial cell dysfunction via suppression of PGC-1alpha in the progression of atherosclerosis. Atherosclerosis 2015;241(2):671–81.

[47] O'Connell RM, et al. Inositol phosphatase SHIP1 is a primary target of miR-155. Proc. Natl. Acad. Sci. USA 2009;106(17):7113–8.

[48] O'Connell RM, et al. MicroRNA-155 promotes autoimmune inflammation by enhancing inflammatory T cell development. Immunity 2010;33(4):607–19.

[49] Kluiver J, et al. BIC and miR-155 are highly expressed in Hodgkin, primary mediastinal and diffuse large B cell lymphomas. J. Pathol. 2005;207(2):243–9.

[50] Mattiske S, et al. The oncogenic role of miR-155 in breast cancer. Cancer Epidemiol. Prev. Biomark 2012;21(8):1236–43.

[51] Zhang Y, et al. Adipocyte-derived microvesicles from obese mice induce M1 macrophage phenotype through secreted miR-155. J. Mol. Cell Biol. 2016;8(6):505–17.

[52] Karkeni E, et al. Obesity-associated inflammation induces microRNA-155 expression in adipocytes and adipose tissue: outcome on adipocyte function. J. Clin. Endocrinol. Metab. 2016;101(4):1615–26.

[53] Liu S, Yang Y, Wu J. TNFα-induced up-regulation of miR-155 inhibits adipogenesis by down-regulating early adipogenic transcription factors. Biochem. Biophys. Res. Commun. 2011;414(3):618–24.

[54] Mashima R. Physiological roles of miR-155. Immunology 2015;145(3):323–33.

[55] Ying W, et al. Adipose tissue macrophage-derived exosomal miRNAs can modulate in vivo and in vitro insulin sensitivity. Cell 2017;171(2):372–84.

[56] Tili E, et al. Modulation of miR-155 and miR-125b levels following lipopolysaccharide/TNF-α stimulation and their possible roles in regulating the response to endotoxin shock. J. Immunol. 2007;179(8):5082–9.

[57] Maldonado-Avilés JG, et al. Down-regulation of miRNAs in the brain and development of diet-induced obesity. Int. J. Dev. Neurosci. 2018;64:2–7.

[58] Gaudet AD, et al. miR-155 deletion in female mice prevents diet-induced obesity. Sci. Rep. 2016;6:22862.

[59] Boettger T, et al. Acquisition of the contractile phenotype by murine arterial smooth muscle cells depends on the Mir143/145 gene cluster. J. Clin. Investig. 2009;119(9):2634–47.

[60] Jia L, et al. Circulating miR-145 is associated with plasma high-sensitivity C-reactive protein in acute ischemic stroke patients. Cell Biochem. Funct. 2015;33(5):314–9.

[61] Li R, et al. MiR-145 improves macrophage-mediated inflammation through targeting Arf6. Endocrine 2018;60(1):73–82.

[62] Li S, et al. Microrna-145 accelerates the inflammatory reaction through activation of NF-κB signaling in atherosclerosis cells and mice. Biomed. Pharmacother. 2018;103:851–7.

[63] Collares RVA, et al. The expression of LEP, LEPR, IGF1 and IL10 in obesity and the relationship with microRNAs. PLoS One 2014;9(4):e93512.

[64] Mehta R, et al. Circulating miRNA in patients with non-alcoholic fatty liver disease and coronary artery disease. BMJ Open Gastroenterol. 2016;3(1):e000096.

[65] Chen WJ, et al. The magic and mystery of microRNA-27 in atherosclerosis. Atherosclerosis 2012;222(2):314–23.

[66] Roldan M, et al. Obesity short-circuits stemness gene network in human adipose multipotent stem cells. FASEB J. 2011;25(12):4111–26.

[67] Lin Q, et al. A role of miR-27 in the regulation of adipogenesis. FEBS J. 2009;276(8):2348–58.

[68] Zaiou M, El Amri H, Bakillah A. The clinical potential of adipogenesis and obesity-related microRNAs. Nutr. Metabol. Cardiovasc. Dis. 2018;28(2):91–111.

[69] Kim SY, et al. miR-27a is a negative regulator of adipocyte differentiation via suppressing PPARgamma expression. Biochem. Biophys. Res. Commun. 2010;392(3):323–8.

[70] Thulasingam S, et al. miR-27b*, an oxidative stress-responsive microRNA modulates nuclear factor-kB pathway in RAW 264.7 cells. Mol. Cell. Biochem. 2011;352(1–2):181–8.

[71] Herrera BM, et al. Global microRNA expression profiles in insulin target tissues in a spontaneous rat model of type 2 diabetes. Diabetologia 2010:1099–109.

[72] Goedeke L, et al. miR-27b inhibits LDLR and ABCA1 expression but does not influence plasma and hepatic lipid levels in mice. Atherosclerosis 2015;243(2):499–509.

Chapter 18

Micro RNA sequencing for myocardial infarction screening

Sri Harsha Kanuri[1] and Rolf P. Kreutz[2]

[1]Department of Clinical Pharmacology, Institute of Personalized Medicine (IIPM), IU School of Medicine, Indianapolis, IN, United States; [2]Krannert Institute of Cardiology, Indiana University School of Medicine, Indianapolis, IN, United States

Context

Approximately 16.5 million persons in United States suffer from coronary artery disease (CAD) [1]. The lifetime risk of developing CAD in patients greater than 40 years is 32% and 49%, in men and women, respectively [2]. The risk of silent and unrecognized myocardial infarction is approximately 33% and 54%, in men and women, respectively [3]. CAD-specific mortality increase, by 29% and 48%, respectively, in men and women, is estimated to be between 1990 and 2020 in developed countries [4]. The overall death rate by myocardial infarction in the year 2015 was around 223 per 100,000 population [5]. In the United States, the healthcare costs of CAD due to hospitalizations and medications is around 108.9 billion annually [6].

Clinical profile

Patient-related risk factors such as diabetes, hypertension, hyperlipidemia, smoking, and genetics can promote the process of atherosclerosis in the coronary artery intimal wall [7]. Rapid progression of atherosclerosis along with plaque rupture, plaque hemorrhage, and occlusive thrombus can lead to narrowing of coronary artery lumen, acute myocardial infarction (AMI), and sudden cardiac death (SCD) [7]. Patients typically present with ischemic chest pain, ECG abnormality, and elevated biomarkers [8]. Acute coronary syndrome with or without ST elevation myocardial infarction is usually treated with percutaneous coronary intervention (PCI), antithrombotic agents, statins, angiotensin converting enzyme inhibitors, and beta-blockers [8,9].

The presence of concomitant risk factors such as diabetes, hypertension, smoking, and obesity, increases the risk of developing recurrent events after initial CAD presentation [10]. Clopidogrel, a commonly used P2Y12 inhibitor, in addition to aspirin, lowered the incidence of myocardial infarction, stroke, cardiovascular deaths as compared to aspirin alone in patients with acute coronary syndrome [11]. Ticagrelor was shown to be more effective than clopidogrel, in attenuating the risk of combined endpoint of death, myocardial infarction, and stroke, in patients with acute coronary syndrome with or without ST segment elevation [12]. Prasugrel, another $P2Y_{12}$ receptor antagonist, has been shown to attenuate the risk of cardiovascular death, myocardial infarction, and stroke, as compared with clopidogrel in patients with ACS treated with PCI [13].

Although dual antiplatelet therapy is effective in reducing the risk of secondary coronary events, there are limitations to the undifferentiated use of this pharmacologic treatment strategy [12]. Long-term dual antiplatelet therapy for >12 months has been shown to reduce the risk of recurrent coronary events, but at the cost of increased risk of bleeding and noncardiovascular deaths [14]. Drug-gene and drug-drug interactions of antiplatelet drugs can lead to variable treatment outcomes, in secondary prevention of recurrent coronary events [15]. In addition, recently low dose rivaroxaban was approved, for secondary prevention in patients with cardiovascular disease, in addition to low dose aspirin, further increasing antithrombotic treatment options for patients with CAD. In light of persistence of ischemic risk in some patients, and increased risk of bleeding in others, there remains a strong need for

Precision Medicine for Investigators, Practitioners and Providers. https://doi.org/10.1016/B978-0-12-819178-1.00018-6

improved tools, at predicting subjects most likely to benefit from specific pharmacologic treatment strategies.

MiRNA as biomarkers of coronary artery disease

While biomarkers of myocardial necrosis, such as cardiac troponins or creatinine-kinase-muscle brain (MB), have been used clinically in diagnosis of myocardial infarction for a long time, there currently do not exist any validated and reliable blood markers, for diagnosis of subclinical coronary artery disease. MiRNAs are increasingly being considered for early detection and treatment of stable angina, unstable angina, and myocardial infarction. Numerous studies have identified sometimes overlapping miRNA that have been associated with angiographic presence of CAD or angina pectoris.

Diagnosis of coronary artery disease

Whole genome transcriptional profiling of 3 ST-elevation myocardial infarction (STEMI) patients, at baseline and 3 months follow-up, demonstrated that miR-486-3p was a potential biomarker that distinguishes patients with STEMI from stable IHD patients [16].

Analysis of 30 stable angina (SA) patients, 39 unstable angina (UA) patients, 19 AMI, and 16 healthy controls, with miRNA microarray and binary regression model demonstrated upregulation of miR-486 and MiR-92a, and their association with HDL components (HDL_2, HDL_3) in plasma of UA and AMI patients. The authors concluded that these miRNA may be able to differentiate between stable and unstable CAD population [17].

In an explorative analysis (TaqMan microRNA assay) of 367 miRNAs in 34 SA, 19 UA, and 20 control patients, D′ Alessandra et al. found that miR-1, mi-R122, miR-126, miR-133a, miR-133b, miR-337-5p, and miR-433 were positively modulated in UA and SA patients, and suggested as potential biomarkers in the future [18].

Measurement of miR-10 and miR-144 levels in 29 UA, 17 non-ST elevation myocardial infarction (NSTEMI), 14 STEMI, and 20 control patients demonstrated that miR-10a levels were upregulated, and miR-144 levels were downregulated in CAD patients, particularly those with STEMI and higher SYNTAX scores [19,20]. Microarray analysis and PCR validation in 5 patients with stable angina, 5 patients with STEMI, 5 patients with NSTEMI, and 5 controls, revealed that miR-941 was significantly elevated in NSTEMI or STEMI patients [21].

In a cross-sectional study consisting of 69 CAD patients and 30 control subjects, plasma samples quantified with microarray analysis and validated with RT-PCR showed that let-7c, miR-145, and miR-155 were potential biomarkers for detecting CAD in elderly patients [22].

Small RNA sequencing, RT-PCR, and bioinformatic analysis of serum from 6 UA patients, 6 STEMI, and 6 controls patients demonstrated that there are at least 38 dysregulated miRNA in UA and STEMI group. Furthermore, receiver operating characteristic (ROC) analysis revealed that miR-142-3p and miR-17-5p may be useful in diagnosis of UA and STEMI [23].

Assessment of miRNA expression by RT-PCR in 65 CAD patients, 20 UA patients, and 32 control patients of middle age (40–60 years old) demonstrated that miR-149 and miR-424 were reduced 4.49 fold and 3.6 fold, respectively, in stable CAD patients, and reduced 5.09 and 5 fold, respectively, in unstable CAD patients, as compared to non-CAD group. In contrast, miR-765 was elevated 3.98 fold and 5.33 fold, in stable and unstable CAD patients, respectively [24]. Thus, miR-149, miR424, and miR-765 were identified as potential noninvasive biomarkers for coronary artery disease patients in middle-aged patients [24].

miR-765 and miR-149 were significantly associated with coronary artery disease, in an elderly population. In 37 stable CAD patients (72.9 ± 4.2 years), 32 unstable CAD patients (72.03 ± 4.3 years), and 20 healthy volunteers (71.7 ± 5.2 years) followed by real-time PCR analysis revealed that miR-765 levels were increased by 2.9 fold (stable CAD) and 5.8 fold (unstable CAD), whereas miR-149 levels were decreased by 3.5 fold (stable CAD) and 4.2 fold (unstable CAD), respectively [25].

Gene polymorphisms of miRNA

Matrix assisted laser desorption/ionization time-of-flight mass spectrometry (MALDI-TOF), and Sequenom MassARRAY system analysis of 356 CAD patients and 368 control population showed that T-allele of rs2431697 was associated with increased, and G allele of rs2910164 of miR-146a with decreased CAD risk, in a Chinese population [26]. Metaanalysis of 1565 CAD cases and 1541 controls in PubMed, EMBASE, and Chinese National Knowledge Infrastructure (CKNI) databases showed that miR-146a gene polymorphism rs2910164 was associated with increased CAD risk, in Asians and older population [27]. Real-time PCR and restriction fragment length polymorphism (RLFP) performed in 272 CAD patients and 149 control patients demonstrated that TT genotype of miR-146a SNP rs2292382 was more frequently associated with CAD risk in Iranian population [28].

Myocardial infarction

TaqMan microarrays were used to measure the expression of miR-423-5p, miR-208, and miR-1 in 17 AMI, 4 SA, and 3 controls just before percutaneous coronary intervention (PCI), as well as 6, 12, and 24 h after the procedure [29].

Nabialek et al. reported that miR-423-5p was the earliest biomarker that became upregulated with myocardial necrosis, and normalized quickly within 6 h, and thus could be useful in diagnosis of early AMI [29].

Measuring miRNA expression in an initial training cohort (35 atypical CAD and 20 controls), and a validation cohort (122 atypical CAD and 68 controls) with TaqMan Low Density Array and RT-qPCR demonstrated that upregulation of miR-208, miR-215, miR-485, and miR-502, and downregulation of miR-29b was suggested as an important miRNA fingerprint that could signal the onset of atypical angina or silent myocardial infarction [30]. It was suggested that these miRNA could be studied as potential therapeutic targets, for arresting underlying pathophysiologic events such as atherosclerosis, inflammation, fibrosis, and myocardial ischemia [30].

Blood collected from 127 control patients, 176 angina pectoris patients, and 13 AMI was analyzed with ELISA and PCR, to measure miR-133a and troponin I (cTnl) levels. miR-133a levels increased by 72.1 fold within 21.6 h of onset of AMI, which closely resembles the expression pattern of cTnl [31]. miR-133a levels were associated with severity of coronary artery lesions, in CAD patients [31]. 17 STEMI patients and 7 control patients were investigated at the time of admission and 6 months following AMI, and miRNA expression profile was assessed and validated with RT-PCR. The study reported miR-22-5p as a novel and important biomarker that can be useful for diagnosis of AMI [32].

Cardiovascular risk factors and coronary artery disease

In 110 CAD patients undergoing coronary angiography, real-time PCR showed that a lower level of miR-155 was associated with CAD, severity of coronary artery lesions, and proatherogenic metabolic risk factors such as hypertension, total cholesterol, high density lipoprotein (HDL), low density lipoprotein (LDL), and C-reactive protein levels [33]. Also in 30 angiographically confirmed CAD patients and 30 age matched controls, real-time PCR revealed that miR-33a level was 2.9 fold higher in CAD patients, and negatively correlated with total cholesterol, HDL, triglycerides, and VLDL [34].

Real-time PCR of blood samples from 54 patients with diabetes, 46 patients with CAD and diabetes, and 20 healthy volunteers showed that miR-126 and miR-200 were significantly upregulated in patients with CAD and diabetes, and correlated with glycemic and lipid profiles [35]. Interestingly, quantitative real-time PCR analysis of type-2 diabetes (T2D) alone, and T2D patients with CAD, showed that miR-126 was positively correlated with CAD in T2D, and negatively associated with LDL in CAD patients [36].

miR-9 and miR-370 were significantly upregulated in T2D and CAD & T2D patient groups [37]. This study highlights the diagnostic as well as prognostic significance of miR-9 and miR-370, in patients with T2D [37]. Isolation of monocytes from patients with smoking history and nonsmoking group, in a cohort of 76 premature CAD and healthy controls, demonstrated increased expression of miR-124-3p in smoking group [38]. Furthermore, RT-PCR of blood from smokers revealed that miR-124-3p is significantly upregulated in patients with subclinical and advanced atherosclerosis [38].

Coronary plaques and pathogenesis

In 78 CAD patients and 65 controls, it was revealed that miR-206 levels were elevated in diseased endothelial progenitor cells (EPCs) and plasma of CAD patients, and suppressed angiogenesis through downregulation of vascular endothelial growth factor (VEGF) [39]. Similarly miR-23a in 13 AMI patients, 176 angina pectoris patients, and 127 control subjects with PCR revealed that miR-23a is elevated in EPCs and CAD patients, and that it attenuates epidermal growth factor receptor expression [40].

Blood and coronary artery plaques collected from CAD patients, showed that miR-365 level was decreased in plaques, and that miR-365 might be involved in mounting an immune response in CAD patients, through IL-6 production [41]. In another protocol, stepwise multivariate regression analysis demonstrated that miR-17-5p may be a useful biomarker that can reflect the severity of coronary atherosclerosis [42]. miR-21, mi-92a, and miR-99a were significantly elevated in advanced human coronary artery plaques, and implicated in various underlying pathways responsible for coronary atherosclerosis [43].

In another experience, miR-221 was elevated in coronary artery atherosclerotic plaques. It was significantly elevated in patients with risk factors such as hypertension, hypercholesterolemia, and family history of CAD [44]. Integrated backscatter intravascular ultrasound (IB-IVUS) was selected to determine the percentage of lipid volume and fibrous volume in coronary arteries [45]. Subsequently, blood collected from aorta and coronary sinus was analyzed for muscle-specific and vascular enriched miRNA [45]. miR-100 was more concentrated in the coronary sinus than in the aorta, and transcoronary gradient was positively correlated with % lipid volume, and negatively correlated with % fibrous volume of coronary plaques. Coronary artery plaque rupture results in release of miR-100 into the plasma circulation, thereby emphasizing its possible role in plaque stability in symptomatic CAD patients [45].

Ventricular arrhythmias

In a rat model of myocardial infarction and ovarian deficiency, increased risk of ventricular arrhythmias was associated with downregulation of miR-151-5p, induced upregulation of PLM (phospholemman) gene, downregulation of Kir2.1 potassium channel expression, and increased calcium overload [46]. These findings suggest that decreased expression of miR-151-5p might contribute to increased susceptibility to ventricular arrhythmias, and may have significance in biomarker discovery and development of novel therapeutic interventions against ventricular arrhythmias in CAD.

Pilot microarray analysis of pooled plasma from apolipoprotein E (apoE) knockout mice showed that miR-21, miR-23a, and miR-30a were differentially expressed in apoE-mice. This was later confirmed in 32 CAD patients with angiographic stenosis greater than 70% [47].

Nitric oxide pathways

In a mouse ischemic hindlimb model, injection of endothelial progenitor cells (EPCs) transfected with anti-miR-206 resulted in activation of PIK3C2α, AKT, and eNOS [48]. Thus, it has been proposed that miR-206 induced downregulation of phospho-AKT/nitric oxide synthase pathway contributes to decreased angiogenesis, decreased EPC migration, and increased risk of coronary artery disease [48].

Antiangiogenic effects and other mechanisms

Interestingly, miR-361-5p induced downregulation of vascular endothelial growth factor (VEGF), by binding to 3′UTR of VEGF [49]. Additionally, miR-221, miR-222, and miR-92a also exert antiangiogenic effects, thereby resulting in attenuated regenerative capacity of EPCs in CAD patients [50]. Previous reports suggest that miRNAs can modulate atherosclerosis through smooth muscle injury, endothelial damage, and plaque formation [51].

Some miRNAs can induce death of vascular endothelial cells, by promoting early senescence through expression of SIRT1 (miR-217 and miR-34), or by disorganization of cell cycle replication (miR-503), by targeting CCNE1 and CDC25A [52]. They were shown to regulate VEGF (miR361-5p) and epidermal growth factor receptor (EGFR) (miR-23a) expression, which may affect plaque stability (miR-100), vasculopathy, and severity of coronary artery stenosis (miR-17-5p) in CAD patients [40,42,45,49].

MiR-22 expression in peripheral blood mononuclear cells (PBMC) was determined in 21 SA patients, 17 NSTEMI patients, 14 STEMI patients, and 20 control patients with the help of RT-PCR. Furthermore monocyte chemoattractant protein-1 (MCP-1) mRNA and protein levels were measured with RT-PCR and ELISA, respectively [53]. Downregulation of miR-22 in PBMC of CAD patients may participate in inflammatory response, by increasing the levels of MCP-1 protein [53].

Arterial calcification

miR-32 could be involved in calcification of vascular smooth muscle through bone morphogenetic protein-1, runt related transcription factor-2 (RUNX2), osteopontin, and bone-specific phosphoprotein matrix GLA protein, in calcified mice vascular smooth muscle cells in in vitro studies [54]. miR-32 is upregulated in CAD patients with coronary artery calcification [54].

Platelets

Upregulation of miR-384 and miR-624 in platelets of patients with coronary artery disease was documented, as compared to healthy controls [55]. Twenty-one NSTEMI patients receiving dual antiplatelet therapy, including newer P2Y12 antagonists, were analyzed with PCR and multiple electrode aggregometry, to measure miR-223 levels and platelet aggregation, respectively [56]. Higher miR-223 levels were correlated with more potent platelet inhibition, and better responsiveness with newer P2Y12 antagonists [56].

Prognosis of coronary artery disease

miR-1, miR-133a, miR-133b, and miR-208b were associated with myocardial injury and onset of myocardial infarction. This study also reported that miR-133a and miR-208b levels provide prognostic information by predicting all-cause mortality, 6 months from the onset of CAD [57]. Lowered expression of miR-145 was associated with AMI, and low circulating levels of miR-145 was predictive of impending heart failure in patients with AMI [58].

Large series

In 4160 patients with AMI enrolled in Osaka Acute Coronary Insufficiency Study, 11 miRNAs were differentially expressed in serum of patients susceptible for cardiovascular death in the future. Particularly, in patients with a primary myocardial infarction event, elevated levels of miR-155 (fourfold) and miR-380 (threefold) were associated with cardiovascular death within 1 year [59]. Serum samples of 873 CAD patients revealed that upregulation of miR-197 and miR-223 might be associated with cardiovascular death, during a 4-year follow-up period, in association with acute coronary syndrome and stable angina pectoris [60,61].

Presence of coronary collaterals

In aortic blood from 41 patients undergoing coronary angiography, who were classified according to their collateral circulation, miR-423-5p, miR-10b, miR-30d, and miR-126 were significantly elevated in those with low (<0.39) collateral flow index (CFI), as compared to patients with high (>0.39) CFI [62]. Additionally, these biomarkers could also be useful for predicting patients who are at risk of low collateral capacity, which may impact the clinical outcomes of coronary artery disease patients [62].

Fingerprints for recurrent coronary events

Recurrent cardiovascular events occur more frequently in patients with multiple risk factors, including diabetes mellitus [63]. In patients younger than 65 years, smoking, hypertension, and dyslipidemia are important risk factors [64]. The risk of recurrent cardiovascular events and associated mortality in patients with previously diagnosed acute coronary syndrome remains elevated despite treatment with high intensity statins [65]. N-terminal prohormone brain natriuretic peptide, cystatin C, albuminuria, C-reactive protein (CRP), lipoprotein associated phospholipaseA2, and secretory phospholipaseA2 can predict increased risk of recurrent cardiovascular events; however, sensitivity and specificity of these biomarkers remain modest [66,67]. Advantages of miRNAs for usage as biomarkers include tissue-specific expression, reproducibility, small size, stability in blood, as well as effective quantification with PCR and next generation sequencing [68].

We recently demonstrated 70 miRNA that were differentially expressed between CAD patients with recurrent events, as compared to CAD patients with no events [69]. A significant number of these 70 miRNA were associated with clinical presentation of unstable angina and myocardial infarction [69]. Interestingly, some of these differentially expressed miRNAs in CAD patients with recurrent events, were associated with underlying pathogenic mechanisms that promote coronary artery thrombosis, such as platelet activation (miR-340-3p, miR-451a, miR-1976, and miR-6734), endothelial dysfunction (miR-19b-3p, miR-106-3p, miR-185-3p, and miR-589-5p), vascular smooth muscle proliferation (miR-29a-3p, miR-143-3p, miR-152-3p, and miR-589-5p), angiogenesis (miR-485-3p, and miR-18a-3p), coronary artery calcification (miR-27a-3p, miR-29a-3p, miR-223, and miR-4745) and atherosclerosis (miR-10a-5p, miR-27a-3p, miR-331-3p, and miR-106b-3p). Additionally, we found that three miRNAs, namely miR-6087, miR-3653, and miR-551a, were differentially expressed between subjects with recurrent events as compared with controls [69]. In coronary artery disease patients who undergo percutaneous coronary intervention, miR-483, miR-555, and miR-155 may serve as biomarkers that can indicate early phase of coronary artery plaque injury and rupture [70].

Downregulation of miR-199a, and upregulation of SIRT1 protein in myocardial tissue, is associated with increased risk of major adverse cardiac and cerebrovascular events (MACCEs), during 3.2 years follow-up in patients undergoing coronary artery bypass graft (CABG) [71]. In patients with coronary artery disease, upregulation of miRNA-126 is negatively associated with inflammatory markers such as C-reactive protein, tumor necrosis factor-alpha (TNF-alpha), and intracellular adhesion molecule (ICAM-1), thereby resulting in decreased risk and severity of recurrent coronary events in the future [72].

miR-126 is an important biomarker that can predict the occurrence of major adverse cardiovascular events, in patients undergoing primary coronary intervention [73]. Only microvesicle associated miR-126 and miR-199, but not freely circulating miRNAs, can predict the occurrence of recurrent coronary events in stable CAD patients [74]. Upregulation of miR-133a in the coronary circulation might reflect the activation of underlying pathophysiological mechanisms leading to CAD progression and associated complications, and it can be a potential biomarker for predicting coronary artery in-stent restenosis [75].

Elevated miR-223 has been associated with increased platelet responsiveness to newer $P2Y_{12}$ antagonist antiplatelet therapy, and thus might influence the occurrence of recurrent coronary events after primary CAD [56]. Upregulation of miRNA-329, miR-494, and miR-495 has been associated with increased intimal hyperplasia and atherosclerosis, thus favoring coronary artery in-stent restenosis in patients [76]. Vulnerable and unstable coronary artery plaques release miR-21, miR-100, miR-155-5p, miR-483-5p, and miR451a, which can be important harbingers and predictors of impending coronary thrombosis, in patients with CAD [45,70,77]. Downregulation of miR-125a-5p, miR-155, and miR-199a/b-3p and elevation of endothelin-1 in the coronary circulation can accentuate coronary atherosclerosis, and might signal the occurrence of coronary thrombosis after a primary event [77].

Current studies with miRNA in cardiovascular disease and metabolic function

Increased miR-1 and decreased miR-133 lead to myocardial apoptosis, whereas decreased miR-1 and increased miR-133 promote myocardial survival [78]. miR-1 regulates Hsp-60 posttranscriptionally, and leads to cardiomyocyte apoptosis [79]. A recent study reported that miR-24 affords protection against myocardial apoptosis, via inhibiting proapoptotic gene Bcl-2 [80]. miR-199 plays an important role in regulation of size of myocytes and cardiac hypertrophy, through hypoxia-inducible factor 1 alpha [81]. MiR-378 has been shown to offer protection against hypoxic myocardial apoptosis, through downregulation of caspase-3 expression in cardiomyocytes [82].

miR-378 overexpression can protect against doxorubicin induced myocardial apoptosis, through increased calumenin expression, and decreased endoplasmic reticulum stress response [83]. Overexpression of miR-21 has been associated with reduced myocardial apoptosis, attenuated infarct size, and improved ventricular remodeling, through programmed cell death 4 and activator 1 pathway [84]. More specifically, miR-21 induced protection against hypoxia-induced cardiomyocyte injury is mediated through inhibition of excessive autophagy and PTEN/AKT/mTOR signaling pathway [85,86]. Interestingly, the antiischemic agent trimetazine (TMZ) protected against hypoxia-reperfusion induced cardiomyocyte apoptosis, by upregulating miR-21 expression [87].

miRNA inhibition

In rat model of AMI, administration of anti-miR-214 resulted in attenuated apoptosis, myocardial infarct size, and improved left ventricular remodeling through PTEN (phosphatase and tense homolog) and Bim1 expression. This demonstrates the role of miR-214 on left ventricular remodeling in AMI [88,89]. Li et al. reported that, miR-23 accentuates myocardial apoptosis by increasing superoxide levels, decreasing Bax/Bcl2 protein expression ratio, caspase-3 activity, and P53 expression via PTEN dependent manner, in H9C2 cell culture model, thus endorsing its role in AMI [90]. In H9c2 cell culture model, overexpression of miR-122 results in hypoxia-induced cardiomyocyte apoptosis, whereas knockdown of miR-122 results in improved cardiomyocyte survival through PTEN/P13K/AKT, and activation of cellular autophagy [91].

In a cell culture model utilizing human cardiac microvascular endothelial cells, miR-126, miR-130a, and miR-138 afforded protection against inflammatory response and hypoxia induced injury, via P13K/AKT/eNOS signaling pathway [92–94]. Overexpression of miR-106b and miR-495 leads to protection of coronary endothelial cells from hypoxia/reperfusion induced injury, through NF-kB and NLRP3 signaling mechanisms, respectively [95,96]. Transplantation of endothelial stem cells resulted in attenuation of myocardial apoptosis, via downregulation of miR-146a in a rat model of AMI [97]. Platelet-derived miR-4306 can limit the migration of human monocyte–derived macrophages into cardiac tissue, in myocardial infarction mice, thereby preventing cardiac damage through left ventricular remodeling and dysfunction postinfarction [98,99].

miR-22 can lower the production of NLRP3 inflammasome induced cytokine production, and affords protection against coronary artery endothelial cells apoptosis, during hypoxic stress [100]. Downregulation of miR-499 leads to protection of coronary endothelial cells, from inflammation mediated damage, through increased PDCD4 expression and decreased NF-kB signaling pathway [101]. In vitro studies with vascular smooth muscle cells (VSMCs) demonstrate that miR-574 leads to proliferation of VSMCs and decreased apoptosis, making it a viable therapeutic target for treatment of coronary artery disease [102]. Alternatively, miR-362-3p might play a critical role, in atherosclerosis and progression of coronary artery disease, by downregulating the proliferation and migration of VSMCs via ADAMST1 [103].

Hypoxia reperfusion injury

Transgenic mice overexpressing miR-125b, subjected to ischemia and reperfusion injury, demonstrated decreased myocardial infarct size, attenuated caspase 3/7/8 activity, and reduced myocardial apoptosis, via decreased TNF receptor associated factor six levels and NF-kB activation [104]. In rat H9c2 cardiomyocytes, miR-30b mediated protection against hypoxia-reperfusion induced cardiomyocyte injury via increased Bcl2, decreased Bax, downregulated caspase-3, increased AKT and decreased KRAS [105,106]. Both miR-19a and miR-93 overexpression lead to protection, of hypoxia and reoxygenation induced cardiomyocyte injury, via PTEN/P13K/p-AKT pathway, in cell culture model of ischemic heart disease [107,108].

In a cell culture model utilizing neonatal rat ventricular myocytes and H9c2 cells, miR-449a protected against hypoxia induced myocyte injury, through notch-1 signaling pathway [109]. In rat cardiomyocytes subjected to hypoxic and reperfusion injury, upregulation of miR-302 leads to inhibition of antiapoptotic myeloid leukemia cell differentiation protein (Mcl-1), and activation of proapoptotic mechanisms, resulting in cardiomyocyte apoptosis [110]. Accordingly, Fang et al. reported that miR-302 antagonists can be employed as a therapeutic intervention to rescue cardiomyocytes from hypoxia-reperfusion injury induced apoptosis [110].

Downregulation of miR-320 and miR-103 can attenuate myocardial apoptosis, associated with ischemia and reperfusion injury, and thus improve myocardial function [111]. MiR-188-3p and miR-145 can attenuate hypoxia induced myocardial necrosis, by activating autophagic pathways [111]. MiR-509, miR-199a, and miR-204 can function through activation of cardiac regenerative processes and cardiomyocyte proliferation, thereby promoting cardiac repair and regeneration following myocardial infarction [111].

Collagen and myocardial fibrosis

Overexpression of miR-181a leads to deposition of extracellular matrix (collagen-1 and fibronectin) and promotion of myocardial fibrosis, following myocardial infarction in a rat model of myocardial infarction [112]. MiR-328, miR-21, and miR-223 play important roles in cardiac fibrosis, by regulating deposition of fibrous tissue following myocardial infarction in animal models, via TGF-beta, Smad, RASA signaling pathways [113–115]. Interestingly, both miR-22 and miR-101a exert negative influence on cardiac fibrosis, in mouse models of myocardial infarction, by modulating TGF-beta signaling pathway [116,117].

Diagnosis and prognosis

Early diagnosis of recurrent coronary events such as unstable angina or myocardial infarction, with reliable biomarkers, will be important for delivery of personalized clinical interventions in a time-dependent manner, to further improve prognosis of CAD after initial diagnosis [118]. Important features that would establish the utility of miRNAs as biomarkers are cardiac tissue specificity, rapid release kinetics, and stability in blood [57,118]. Droplet digital PCR was recently shown to have a better performance, technically and diagnostically, in quantifying miRNA levels in blood of ST-elevation MI patients, as compared to qRT-PCR in large multicenter trials [119]. miR-6090 and miR-4516 can be regarded as reference genes (optimal endogenous controls) that might be used to normalize the RT-PCR data, for quantification of miRNA expression in the plasma of CAD patients [120].

Robust diagnostic signatures

miRNA-1, miRNA-133, miRNA-208b, and miRNA-499 demonstrated high sensitivity and specificity for diagnosis of acute myocardial infarction [118]. A miRNA signature consisting of miRNA-19b-3p, miR134-5p, and miRNA186-5p was upregulated within 4–72 h of onset of chest pain in acute myocardial infarction, and was also positively correlated with troponin levels [121].

miR-1, miR-133a, miR-133b, miR-208a, miR-208b, and miR-499 were elevated in patients with ST-elevation MI [57,122–124]. These six miRNAs were positively correlated with troponin (hSTnT), and negatively associated with left ventricular ejection fraction [57,125]. However, previous studies indicate that these cardio-specific miRNAs, are not superior to high sensitivity troponin, which is routinely used for diagnosis of early stage of acute myocardial infarction [126]. miR-1, miR-133a, and miR208a are elevated as early as 120 min, and return to baseline within 12 h following an episode of acute myocardial infarction [125]. In contrast, miR-499-5p becomes elevated within 150 min, and returns to baseline with 2–3 days after myocardial infarction [125]. Interestingly, miR-1, miR-133a, and miR-208b were more elevated in patients with myocardial infarction (STEMI or NSTEMI), as compared to patients with unstable angina [31,57]. A miRNA signature consisting of miR196-5p, miR-3163-3p, miR-143-3p, and miR-190a-5p may be useful as novel diagnostic markers for diagnosis of very early onset CAD, in patients with ages around 40 years [127].

miR-21-5p, miR-19b-3p, miR-30d-5p, miR-122-5p, miR-125b-5p, miR134-5p, miR-146a, miR-186-5p, miR-221-3p, miR-320a, miR-328, miR361-5p, miR-375, miR-423, and miR-519e-5p can also be considered as potential and promising new biomarkers for early diagnosis of acute myocardial infarction [121,128–132]. Following an AMI, miR-133, miR-208b, and miR-499b can also provide prognostic significance, by predicting heart failure and all-cause mortality within 6 months [57,122]. Overexpression of miR-23a leads to attenuated expression of telomeric repeat binding factor (TRF2), lower telomere length, and accelerated coronary atherosclerosis, which is associated with poor clinical prognosis [133].

The genotype CC/CT of hsa-miR-196a2 rs11614913, along with diabetes, smoking, age, and pathological changes in coronary arteries was associated with severe prognosis in CAD in Chinese patients [134]. miR-155 and miR-503 are associated with formation of coronary collateral circulation, and thus may affect prognosis in coronary artery disease patients [135,136]. miRNA-related polymorphisms such as miR-4513 (rs2168518) and miR-499 (rs3746444) might be potential biomarkers that can indicate clinical prognosis of CAD patients [137]. miR-126 is positively associated with coronary collateral circulation, and its upregulation may be used to predict collateral formation in ischemic myocardium, supplied by severely stenosed coronary arteries [138]. According to Wang et al., endothelial cell associated miRNAs, such as miR-31 and miR-720, have the potential to be used as reliable biomarkers, for both earlier diagnosis and clinical prognosis of CAD [139].

Challenges and pitfalls

Factors that can influence plasma miRNA levels encompass heparin, statins, ACE inhibitors, antiplatelet therapy, kidney disease, cancer, and cytomegalovirus infection [140]. Additionally, other disadvantages of miRNA that need consideration include the absence of reference threshold values, as well as standardized methods of RNA preparation and normalization methods [140]. Next generation sequencing, qPCR, and microarrays are most commonly used [140]. These methods are very time-consuming, and cannot be used for bedside diagnosis of unstable angina and

STEMI [140]. In the future, programmable oligonucleotide probes that can detect single molecules of miRNA in plasma may need to be developed [140].

Future research

There are around 20 miRNAs that can be considered as potential biomarkers for diagnosis of acute myocardial infarction, yet most of these were demonstrated in small study cohorts with small sample sizes, leading to wide variance in findings. Larger study cohorts with validation are advised [141]. Further research is particularly warranted in areas such as origin, regulation, function, and end-targets of candidate miRNAs [142]. Currently, it takes around 2–3 h for quantification of specific miRNA in blood with PCR [143]. Automated work flow systems that can rapidly, specifically, and sensitively detect very low levels of miRNA in plasma at the bedside should be developed [142,143]. Furthermore, research is needed to examine how drugs (e.g., heparin, statin, ACE inhibitors) and other clinical variables may influence miRNA levels in plasma [142].

The role of endothelial, platelet, coronary, and vascular smooth muscle cell–specific miRNA, and the role they play in the pathogenesis of coronary artery disease require further characterization. miRNA role in pathophysiology of CAD disease progression and complications using in vitro and animal studies are necessary [144]. Biomaterial standardization (serum, plasma, whole blood), and RNA normalization procedures for accurate quantification of miRNA levels are relevant expectations [144]. Lastly, large clinical studies, enrolling subjects from multiple racial and ethnic groups, utilizing standardized procedures for RNA extraction, will need to confirm the utility of candidate miRNA biomarkers for detecting stable angina, unstable angina, and acute myocardial infarction [145].

References

[1] Benjamin EJ, Blaha MJ, Chiuve SE, et al. Heart disease and stroke statistics-2017 update: a report from the American heart association. Circulation 2017;135(10):e146–603.

[2] Lerner DJ, Kannel WB. Patterns of coronary heart disease morbidity and mortality in the sexes: a 26-year follow-up of the Framingham population. Am. Heart J. 1986;111(2):383–90.

[3] de Torbal A, Boersma E, Kors JA, et al. Incidence of recognized and unrecognized myocardial infarction in men and women aged 55 and older: the Rotterdam study. Eur. Heart J. 2006;27(6):729–36.

[4] Sanchis-Gomar F, Perez-Quilis C, Leischik R, Lucia A. Epidemiology of coronary heart disease and acute coronary syndrome. Ann. Transl. Med. 2016;4(13):256.

[5] Benjamin EJ, Virani SS, Callaway CW, et al. Heart disease and stroke statistics-2018 update: a report from the American heart association. Circulation 2018;137(12):e67–492.

[6] Million hearts: strategies to reduce the prevalence of leading cardiovascular disease risk factors—United States, 2011. MMWR Morb. Mortal. Wkly. Rep. 2011;60(36):1248–51.

[7] Ambrose J, Singh M. Pathophysiology of coronary artery disease leading to acute coronary syndromes. 2015.

[8] Fox KAA. Acute coronary syndromes: presentation—clinical spectrum and management. Heart 2000;84(1):93.

[9] API expert consensus document on management of ischemic heart disease. J. Assoc. Physicians India 2006;54:469–80.

[10] Hajar R. Risk factors for coronary artery disease: historical perspectives. Heart Views 2017;18(3):109–14.

[11] Knight CJ. Antiplatelet treatment in stable coronary artery disease. Heart 2003;89(10):1273–8.

[12] Wallentin L, Becker RC, Budaj A, et al. Ticagrelor versus clopidogrel in patients with acute coronary syndromes. N. Engl. J. Med. 2009;361(11):1045–57.

[13] Smith JN, Negrelli JM, Manek MB, Hawes EM, Viera AJ. Diagnosis and management of acute coronary syndrome: an evidence-based update. J. Am. Board Fam. Med. 2015;28(2):283–93.

[14] Wilson SJ, Newby DE, Dawson D, Irving J, Berry C. Duration of dual antiplatelet therapy in acute coronary syndrome. Heart 2017;103(8):573.

[15] Eltayeb H, Nelson A, Willoughby S, Worthley M. Antiplatelet therapy in acute coronary syndromes: current agents and impact on patient outcomes. 2011.

[16] Wei T, Folkersen L, Ehrenborg E, Gabrielsen A. MicroRNA 486-3P as a stability marker in acute coronary syndrome. Biosci. Rep. 2016;36(3).

[17] Niculescu LS, Simionescu N, Sanda GM, et al. MiR-486 and miR-92a identified in circulating HDL discriminate between stable and vulnerable coronary artery disease patients. PLoS One 2015;10(10):e0140958.

[18] D'Alessandra Y, Carena MC, Spazzafumo L, et al. Diagnostic potential of plasmatic MicroRNA signatures in stable and unstable angina. PLoS One 2013;8(11):e80345.

[19] Chen B, Luo L, Wei X, Gong D, Jin L. Altered plasma miR-144 as a novel biomarker for coronary artery disease. Ann. Clin. Lab. Sci. 2018;48(4):440–5.

[20] Luo L, Chen B, Li S, et al. Plasma miR-10a: a potential biomarker for coronary artery disease. Dis. Mark. 2016;2016:3841927.

[21] Bai R, Yang Q, Xi R, Li L, Shi D, Chen K. miR-941 as a promising biomarker for acute coronary syndrome. BMC Cardiovasc. Disord. 2017;17(1):227.

[22] Faccini J, Ruidavets J-B, Cordelier P, et al. Circulating miR-155, miR-145 and let-7c as diagnostic biomarkers of the coronary artery disease. Sci. Rep. 2017;7:42916.

[23] Zhong Z, Hou J, Zhang Q, et al. Circulating microRNA expression profiling and bioinformatics analysis of dysregulated microRNAs of patients with coronary artery disease. Medicine 2018;97(27):e11428.

[24] Sayed AS, Xia K, Li F, et al. The diagnostic value of circulating microRNAs for middle-aged (40-60-year-old) coronary artery disease patients. Clinics 2015;70(4):257–63.

[25] Ali Sheikh MS, Xia K, Li F, et al. Circulating miR-765 and miR-149: potential noninvasive diagnostic biomarkers for geriatric coronary artery disease patients. BioMed Res. Int. 2015;2015: 740301.

[26] Wang Y, Wang X, Li Z, et al. Two single nucleotide polymorphisms (rs2431697 and rs2910164) of miR-146a are associated with risk of coronary artery disease. Int. J. Environ. Res. Public Health 2017;14(5).

[27] Zhou HY, Wei Q, Shi XD, Cao HY, Qin L. miR-146a rs2910164 polymorphism might be associated with coronary artery disease risk in Asians. Cell. Mol. Biol. 2017;63(8):27–9.

[28] Ghaffarzadeh M, Ghaedi H, Alipoor B, et al. Association of MiR-149 (RS2292832) variant with the risk of coronary artery disease. J. Med. Biochem. 2017;36(3):251–8.

[29] Nabialek E, Wanha W, Kula D, et al. Circulating microRNAs (miR-423-5p, miR-208a and miR-1) in acute myocardial infarction and stable coronary heart disease. Minerva Cardioangiol. 2013;61(6):627–37.

[30] Wang J, Pei Y, Zhong Y, Jiang S, Shao J, Gong J. Altered serum microRNAs as novel diagnostic biomarkers for atypical coronary artery disease. PLoS One 2014;9(9):e107012.

[31] Wang F, Long G, Zhao C, et al. Plasma microRNA-133a is a new marker for both acute myocardial infarction and underlying coronary artery stenosis. J. Transl. Med. 2013;11:222.

[32] Maciejak A, Kiliszek M, Opolski G, et al. miR-22-5p revealed as a potential biomarker involved in the acute phase of myocardial infarction via profiling of circulating microRNAs. Mol. Med. Rep. 2016;14(3):2867–75.

[33] Zhu GF, Yang LX, Guo RW, et al. microRNA-155 is inversely associated with severity of coronary stenotic lesions calculated by the Gensini score. Coron. Artery Dis. 2014;25(4):304–10.

[34] Reddy LL, Shah SA, Ponde CK, Rajani RM, Ashavaid TF. Circulating miRNA-33: a potential biomarker in patients with Coronary Artery Disease (CAD). Biomarkers: biochemical indicators of exposure, response, and susceptibility to chemicals. 2018. p. 1–27.

[35] Amr KS, Abdelmawgoud H, Ali ZY, Shehata S, Raslan HM. Potential value of circulating microRNA-126 and microRNA-210 as biomarkers for type 2 diabetes with coronary artery disease. Br. J. Biomed. Sci. 2018;75(2):82–7.

[36] Al-Kafaji G, Al-Mahroos G, Abdulla Al-Muhtaresh H, Sabry MA, Abdul Razzak R, Salem AH. Circulating endothelium-enriched microRNA-126 as a potential biomarker for coronary artery disease in type 2 diabetes mellitus patients. Biomarkers 2017;22(3–4):268–78.

[37] Motawae TM, Ismail MF, Shabayek MI, Seleem MM. MicroRNAs 9 and 370 association with biochemical markers in T2D and CAD complication of T2D. PLoS One 2015;10(5):e0126957.

[38] de Ronde MWJ, Kok MGM, Moerland PD, et al. High miR-124-3p expression identifies smoking individuals susceptible to atherosclerosis. Atherosclerosis 2017;263:377–84.

[39] Wang M, Ji Y, Cai S, Ding W. MiR-206 suppresses the progression of coronary artery disease by modulating vascular endothelial growth factor (VEGF) expression. Med. Sci. Monit. 2016;22:5011–20.

[40] Wang S, He W, Wang C. MiR-23a regulates the vasculogenesis of coronary artery disease by targeting epidermal growth factor receptor. Cardiovasc. Ther. 2016;34(4):199–208.

[41] Lin B, Feng DG, Wang F, et al. MiR-365 participates in coronary atherosclerosis through regulating IL-6. Eur. Rev. Med. Pharmacol. Sci. 2016;20(24):5186–92.

[42] Chen J, Xu L, Hu Q, Yang S, Zhang B, Jiang H. MiR-17-5p as circulating biomarkers for the severity of coronary atherosclerosis in coronary artery disease. Int. J. Cardiol. 2015;197:123–4.

[43] Parahuleva MS, Lipps C, Parviz B, et al. MicroRNA expression profile of human advanced coronary atherosclerotic plaques. Sci. Rep. 2018;8(1):7823.

[44] Bildirici AE, Arslan S, Ozbilum Sahin N, Berkan O, Beton O, Yilmaz MB. MicroRNA-221/222 expression in atherosclerotic coronary artery plaque versus internal mammarian artery and in peripheral blood samples. Biomarkers 2018;23(7):670–5.

[45] Soeki T, Yamaguchi K, Niki T, et al. Plasma microRNA-100 is associated with coronary plaque vulnerability. Circ. J. 2015;79(2):413–8.

[46] Zhang Y, Wang R, Du W, et al. Downregulation of miR-151-5p contributes to increased susceptibility to arrhythmogenesis during myocardial infarction with estrogen deprivation. PLoS One 2013;8(9):e72985.

[47] Han H, Qu G, Han C, et al. MiR-34a, miR-21 and miR-23a as potential biomarkers for coronary artery disease: a pilot microarray study and confirmation in a 32 patient cohort. Exp. Mol. Med. 2015;47:e138.

[48] Tang Y, Zhang Y, Chen Y, Xiang Y, Xie Y. Role of the microRNA, miR-206, and its target PIK3C2alpha in endothelial progenitor cell function - potential link with coronary artery disease. FEBS J. 2015;282(19):3758–72.

[49] Wang HW, Lo HH, Chiu YL, et al. Dysregulated miR-361-5p/VEGF axis in the plasma and endothelial progenitor cells of patients with coronary artery disease. PLoS One 2014;9(5):e98070.

[50] Zhang Q, Kandic I, Kutryk MJ. Dysregulation of angiogenesis-related microRNAs in endothelial progenitor cells from patients with coronary artery disease. Biochem. Biophys. Res. Commun. 2011;405(1):42–6.

[51] Wang R, Dong LD, Meng XB, Shi Q, Sun WY. Unique microRNA signatures associated with early coronary atherosclerotic plaques. Biochem. Biophys. Res. Commun. 2015;464(2):574–9.

[52] Caporali A, Emanueli C. MicroRNA regulation in angiogenesis. Vasc. Pharmacol. 2011;55(4):79–86.

[53] Chen B, Luo L, Zhu W, et al. miR-22 contributes to the pathogenesis of patients with coronary artery disease by targeting MCP-1: an observational study. Medicine 2016;95(33):e4418.

[54] Liu J, Xiao X, Shen Y, et al. MicroRNA-32 promotes calcification in vascular smooth muscle cells: implications as a novel marker for coronary artery calcification. PLoS One 2017;12(3):e0174138.

[55] Sondermeijer BM, Bakker A, Halliani A, et al. Platelets in patients with premature coronary artery disease exhibit upregulation of miRNA340* and miRNA624*. PLoS One 2011;6(10):e25946.

[56] Chyrchel B, Toton-Zuranska J, Kruszelnicka O, et al. Association of plasma miR-223 and platelet reactivity in patients with coronary artery disease on dual antiplatelet therapy: a preliminary report. Platelets 2015;26(6):593–7.

[57] Widera C, Gupta SK, Lorenzen JM, et al. Diagnostic and prognostic impact of six circulating microRNAs in acute coronary syndrome. J. Mol. Cell. Cardiol. 2011;51(5):872–5.

[58] Zhang M, Cheng YJ, Sara JD, et al. Circulating MicroRNA-145 is associated with acute myocardial infarction and heart failure. Chin. Med. J. 2017;130(1):51–6.

[59] Matsumoto S, Sakata Y, Nakatani D, et al. A subset of circulating microRNAs are predictive for cardiac death after discharge for acute myocardial infarction. Biochem. Biophys. Res. Commun. 2012;427(2):280–4.
[60] Orenes-Piñero E, Marín F, Lip GYH. miRNA-197 and miRNA-223 and cardiovascular death in coronary artery disease patients. Ann. Transl. Med. 2016;4(10):200.
[61] Schulte C, Molz S, Appelbaum S, et al. miRNA-197 and miRNA-223 predict cardiovascular death in a cohort of patients with symptomatic coronary artery disease. PLoS One 2016;10(12):e0145930.
[62] Hakimzadeh N, Nossent AY, van der Laan AM, et al. Circulating MicroRNAs characterizing patients with insufficient coronary collateral artery function. PLoS One 2015;10(9):e0137035.
[63] van der Heijden AAWA, Van't Riet E, Bot SDM, et al. Risk of a recurrent cardiovascular event in individuals with type 2 diabetes or intermediate hyperglycemia: the Hoorn study. Diabetes Care 2013;36(11):3498–502.
[64] van den Berg MJ, Westerink J, van der Graaf Y, et al. Risk factors for recurrent cardiovascular events before age 65 years or within 2.5 years of a recent first cardiovascular event. Am. J. Cardiol. 2017;120(2):167–73.
[65] Rockberg J, Jorgensen L, Taylor B, Sobocki P, Johansson G. Risk of mortality and recurrent cardiovascular events in patients with acute coronary syndromes on high intensity statin treatment. Prev. Med. Rep. 2017;6:203–9.
[66] Shlipak MG, Ix JH, Bibbins-Domingo K, Lin F, Whooley MA. Biomarkers to predict recurrent cardiovascular disease: the Heart and Soul Study. Am. J. Med. 2008;121(1):50–7.
[67] Wang J, Tan G-J, Han L-N, Bai Y-Y, He M, Liu H-B. Novel biomarkers for cardiovascular risk prediction. J. Geriatr. Cardiol. 2017;14(2):135–50.
[68] Kaudewitz D, Zampetaki A, Mayr M. MicroRNA biomarkers for coronary artery disease? Curr. Atheroscler. Rep. 2015;17(12):70.
[69] Kanuri SH, Ipe J, Kassab K, et al. Next generation MicroRNA sequencing to identify coronary artery disease patients at risk of recurrent myocardial infarction. Atherosclerosis 2018;278:232–9.
[70] Li S, Lee C, Song J, et al. Circulating microRNAs as potential biomarkers for coronary plaque rupture. Oncotarget 2017;8(29):48145–56.
[71] Yamac AH, Huyut MA, Yilmaz E, et al. MicroRNA 199a is downregulated in patients after coronary artery bypass graft surgery and is associated with increased levels of sirtuin 1 (SIRT 1) protein and major adverse cardiovascular events at 3-year follow-up. Med. Sci. Monit. 2018;24:6245–54.
[72] Wu H, Zhang J. miR-126 in peripheral blood mononuclear cells negatively correlates with risk and severity and is associated with inflammatory cytokines as well as intercellular adhesion molecule-1 in patients with coronary artery disease. Cardiology 2018;139(2):110–8.
[73] Yu XY, Chen JY, Zheng ZW, et al. Plasma miR-126 as a potential marker predicting major adverse cardiac events in dual antiplatelet-treated patients after percutaneous coronary intervention. Euro-Intervention 2013;9(5):546–54.
[74] Jansen F, Yang X, Proebsting S, et al. MicroRNA expression in circulating microvesicles predicts cardiovascular events in patients with coronary artery disease. J. Am. Heart Assoc. 2014;3(6):e001249.
[75] De Rosa R, De Rosa S, Leistner D, et al. Transcoronary concentration gradient of microRNA-133a and outcome in patients with coronary artery disease. Am. J. Cardiol. 2017;120(1):15–24.
[76] Welten SMJ, de Jong RCM, Wezel A, et al. Inhibition of 14q32 microRNA miR-495 reduces lesion formation, intimal hyperplasia and plasma cholesterol levels in experimental restenosis. Atherosclerosis 2017;261:26–36.
[77] Fan X, Wang E, Wang X, Cong X, Chen X. MicroRNA-21 is a unique signature associated with coronary plaque instability in humans by regulating matrix metalloproteinase-9 via reversion-inducing cysteine-rich protein with Kazal motifs. Exp. Mol. Pathol. 2014;96(2):242–9.
[78] Xu C, Lu Y, Pan Z, et al. The muscle-specific microRNAs miR-1 and miR-133 produce opposing effects on apoptosis by targeting HSP60, HSP70 and caspase-9 in cardiomyocytes. J. Cell Sci. 2007;120(Pt 17):3045–52.
[79] Shan ZX, Lin QX, Deng CY, et al. miR-1/miR-206 regulate Hsp60 expression contributing to glucose-mediated apoptosis in cardiomyocytes. FEBS Lett. 2010;584(16):3592–600.
[80] Li DF, Tian J, Guo X, et al. Induction of microRNA-24 by HIF-1 protects against ischemic injury in rat cardiomyocytes. Physiol. Res. 2012;61(6):555–65.
[81] Song XW, Li Q, Lin L, et al. MicroRNAs are dynamically regulated in hypertrophic hearts, and miR-199a is essential for the maintenance of cell size in cardiomyocytes. J. Cell. Physiol. 2010;225(2):437–43.
[82] Fang J, Song XW, Tian J, et al. Overexpression of microRNA-378 attenuates ischemia-induced apoptosis by inhibiting caspase-3 expression in cardiac myocytes. Apoptosis 2012;17(4):410–23.
[83] Wang Y, Cui X, Wang Y, et al. Protective effect of miR378* on doxorubicin-induced cardiomyocyte injury via calumenin. J. Cell. Physiol. 2018;233(10):6344–51.
[84] Dong S, Cheng Y, Yang J, et al. MicroRNA expression signature and the role of microRNA-21 in the early phase of acute myocardial infarction. J. Biol. Chem. 2009;284(43):29514–25.
[85] Huang Z, Wu S, Kong F, et al. MicroRNA-21 protects against cardiac hypoxia/reoxygenation injury by inhibiting excessive autophagy in H9c2 cells via the Akt/mTOR pathway. J. Cell Mol. Med. 2017;21(3):467–74.
[86] Yang Q, Yang K, Li A. microRNA-21 protects against ischemia-reperfusion and hypoxia-reperfusion-induced cardiocyte apoptosis via the phosphatase and tensin homolog/Akt-dependent mechanism. Mol. Med. Rep. 2014;9(6):2213–20.
[87] Yang Q, Yang K, Li AY. Trimetazidine protects against hypoxia-reperfusion-induced cardiomyocyte apoptosis by increasing microRNA-21 expression. Int. J. Clin. Exp. Pathol. 2015;8(4):3735–41.
[88] Yang X, Qin Y, Shao S, et al. MicroRNA-214 inhibits left ventricular remodeling in an acute myocardial infarction rat model by suppressing cellular apoptosis via the phosphatase and tensin homolog (PTEN). Int. Heart J. 2016;57(2):247–50.
[89] Wang X, Ha T, Hu Y, et al. MicroRNA-214 protects against hypoxia/reoxygenation induced cell damage and myocardial ischemia/reperfusion injury via suppression of PTEN and Bim1 expression. Oncotarget 2016;7(52):86926–36.
[90] Li S, Ren J, Sun Q. The expression of microRNA-23a regulates acute myocardial infarction in patients and in vitro through targeting PTEN. Mol. Med. Rep. 2018;17(5):6866–72.

[91] Zhang Z, Li H, Chen S, Li Y, Cui Z, Ma J. Knockdown of MicroRNA-122 protects H9c2 cardiomyocytes from hypoxia-induced apoptosis and promotes autophagy. Med. Sci. Monit. 2017;23:4284–90.
[92] Yang HH, Chen Y, Gao CY, Cui ZT, Yao JM. Protective effects of MicroRNA-126 on human cardiac microvascular endothelial cells against hypoxia/reoxygenation-induced injury and inflammatory response by activating PI3K/Akt/eNOS signaling pathway. Cell. Physiol. Biochem. 2017;42(2):506–18.
[93] Song CL, Liu B, Shi YF, et al. MicroRNA-130a alleviates human coronary artery endothelial cell injury and inflammatory responses by targeting PTEN via activating PI3K/Akt/eNOS signaling pathway. Oncotarget 2016;7(44):71922–36.
[94] Li JB, Wang HY, Yao Y, et al. Overexpression of microRNA-138 alleviates human coronary artery endothelial cell injury and inflammatory response by inhibiting the PI3K/Akt/eNOS pathway. J. Cell Mol. Med. 2017;21(8):1482–91.
[95] An Z, Yang G, Nie W, Ren J, Wang D. MicroRNA-106b over-expression alleviates inflammation injury of cardiac endothelial cells by targeting BLNK via the NF-kappaB signaling pathway. J. Cell. Biochem. 2018;119(4):3451–63.
[96] Zhou T, Xiang DK, Li SN, Yang LH, Gao LF, Feng C. MicroRNA-495 ameliorates cardiac microvascular endothelial cell injury and inflammatory reaction by suppressing the NLRP3 inflammasome signaling pathway. Cell. Physiol. Biochem. 2018;49(2):798–815.
[97] Fang Y, Chen S, Liu Z, et al. Endothelial stem cells attenuate cardiac apoptosis via downregulating cardiac microRNA-146a in a rat model of coronary heart disease. Exp. Ther. Med 2018;16(5):4246–52.
[98] Yang Y, Luo H, Liu S, et al. Platelet microparticles-containing miR-4306 inhibits human monocyte-derived macrophages migration through VEGFA/ERK1/2/NF-kappaB signaling pathways. Clin. Exp. Hypertens. 2018:1–11.
[99] White DA, Su Y, Kanellakis P, et al. Differential roles of cardiac and leukocyte derived macrophage migration inhibitory factor in inflammatory responses and cardiac remodelling post myocardial infarction. J. Mol. Cell. Cardiol. 2014;69:32–42.
[100] Huang WQ, Wei P, Lin RQ, Huang F. Protective effects of microrna-22 against endothelial cell injury by targeting NLRP3 through suppression of the inflammasome signaling pathway in a rat model of coronary heart disease. Cell. Physiol. Biochem. 2017;43(4):1346–58.
[101] Zhang YH, He K, Shi G. Effects of MicroRNA-499 on the inflammatory damage of endothelial cells during coronary artery disease via the targeting of PDCD4 through the NF-Kappabeta/TNF-alpha signaling pathway. Cell. Physiol. Biochem. 2017;44(1):110–24.
[102] Lai Z, Lin P, Weng X, et al. MicroRNA-574-5p promotes cell growth of vascular smooth muscle cells in the progression of coronary artery disease. Biomed. Pharmacother. 2018;97:162–7.
[103] Li M, Liu Q, Lei J, Wang X, Chen X, Ding Y. MiR-362-3p inhibits the proliferation and migration of vascular smooth muscle cells in atherosclerosis by targeting ADAMTS1. Biochem. Biophys. Res. Commun. 2017;493(1):270–6.
[104] Wang X, Ha T, Zou J, et al. MicroRNA-125b protects against myocardial ischaemia/reperfusion injury via targeting p53-mediated apoptotic signalling and TRAF6. Cardiovasc. Res. 2014;102(3):385–95.
[105] Li T, Sun ZL, Xie QY. Protective effect of microRNA-30b on hypoxia/reoxygenation-induced apoptosis in H9C2 cardiomyocytes. Gene 2015;561(2):268–75.
[106] Song CL, Liu B, Wang JP, et al. Anti-apoptotic effect of microRNA-30b in early phase of rat myocardial ischemia-reperfusion injury model. J. Cell. Biochem. 2015;116(11):2610–9.
[107] Sun G, Lu Y, Li Y, et al. miR-19a protects cardiomyocytes from hypoxia/reoxygenation-induced apoptosis via PTEN/PI3K/p-Akt pathway. Biosci. Rep. 2017;37(6).
[108] Ke ZP, Xu P, Shi Y, Gao AM. MicroRNA-93 inhibits ischemia-reperfusion induced cardiomyocyte apoptosis by targeting PTEN. Oncotarget 2016;7(20):28796–805.
[109] Cheng J, Wu Q, Lv R, et al. MicroRNA-449a inhibition protects H9C2 cells against hypoxia/reoxygenation-induced injury by targeting the notch-1 signaling pathway. Cell. Physiol. Biochem. 2018;46(6):2587–600.
[110] Fang YC, Yeh CH. Inhibition of miR-302 suppresses hypoxia-reoxygenation-induced H9c2 cardiomyocyte death by regulating Mcl-1 expression. Oxid. Med. Cell. Longev. 2017;2017:7968905.
[111] Sun T, Dong Y-H, Du W, et al. The role of MicroRNAs in myocardial infarction: from molecular mechanism to clinical application. Int. J. Mol. Sci. 2017;18(4).
[112] Chen P, Pan J, Zhang X, Shi Z, Yang X. The role of MicroRNA-181a in myocardial fibrosis following myocardial infarction in a rat model. Med. Sci. Monit. 2018;24:4121–7.
[113] Zhao D, Li C, Yan H, et al. Cardiomyocyte derived miR-328 promotes cardiac fibrosis by paracrinely regulating adjacent fibroblasts. Cell. Physiol. Biochem. 2018;46(4):1555–65.
[114] Yuan J, Chen H, Ge D, et al. Mir-21 promotes cardiac fibrosis after myocardial infarction via targeting Smad7. Cell. Physiol. Biochem. 2017;42(6):2207–19.
[115] Liu X, Xu Y, Deng Y, Li H. MicroRNA-223 regulates cardiac fibrosis after myocardial infarction by targeting RASA1. Cell. Physiol. Biochem. 2018;46(4):1439–54.
[116] Hong Y, Cao H, Wang Q, et al. MiR-22 may suppress fibrogenesis by targeting TGFbetaR I in cardiac fibroblasts. Cell. Physiol. Biochem. 2016;40(6):1345–53.
[117] Zhao X, Wang K, Liao Y, et al. MicroRNA-101a inhibits cardiac fibrosis induced by hypoxia via targeting TGFbetaRI on cardiac fibroblasts. Cell. Physiol. Biochem. 2015;35(1):213–26.
[118] Kränkel N, Blankenberg S, Zeller T. Early detection of myocardial infarction-microRNAs right at the time? Ann. Transl. Med. 2016;4(24):502.
[119] Robinson S, Follo M, Haenel D, et al. Droplet digital PCR as a novel detection method for quantifying microRNAs in acute myocardial infarction. Int. J. Cardiol. 2018;257:247–54.
[120] Zhang Y, Tang W, Peng L, Tang J, Yuan Z. Identification and validation of microRNAs as endogenous controls for quantitative polymerase chain reaction in plasma for stable coronary artery disease. Cardiol. J. 2016;23(6):694–703.
[121] Wang KJ, Zhao X, Liu YZ, et al. Circulating MiR-19b-3p, MiR-134-5p and MiR-186-5p are promising novel biomarkers for early diagnosis of acute myocardial infarction. Cell. Physiol. Biochem. 2016;38(3):1015–29.
[122] Gidlof O, Smith JG, Miyazu K, et al. Circulating cardio-enriched microRNAs are associated with long-term prognosis following myocardial infarction. BMC Cardiovasc. Disord. 2013;13:12.

[123] Xiao J, Shen B, Li J, et al. Serum microRNA-499 and microRNA-208a as biomarkers of acute myocardial infarction. Int. J. Clin. Exp. Med. 2014;7(1):136–41.

[124] Wang Q, Ma J, Jiang Z, Wu F, Ping J, Ming L. Identification of microRNAs as diagnostic biomarkers for acute myocardial infarction in Asian populations: a systematic review and meta-analysis. Medicine 2017;96(24):e7173.

[125] Gidlof O, Andersson P, van der Pals J, Gotberg M, Erlinge D. Cardiospecific microRNA plasma levels correlate with troponin and cardiac function in patients with ST elevation myocardial infarction, are selectively dependent on renal elimination, and can be detected in urine samples. Cardiology 2011;118(4):217–26.

[126] Li YQ, Zhang MF, Wen HY, et al. Comparing the diagnostic values of circulating microRNAs and cardiac troponin T in patients with acute myocardial infarction. Clinics (Sao Paulo, Brazil) 2013;68(1):75–80.

[127] Du Y, Yang SH, Li S, et al. Circulating MicroRNAs as novel diagnostic biomarkers for very early-onset (</=40 years) coronary artery disease. Biomed. Environ. Sci. 2016;29(8):545–54.

[128] He F, Lv P, Zhao X, et al. Predictive value of circulating miR-328 and miR-134 for acute myocardial infarction. Mol. Cell. Biochem. 2014;394(1–2):137–44.

[129] Yao XL, Lu XL, Yan CY, Wan QL, Cheng GC, Li YM. Circulating miR-122-5p as a potential novel biomarker for diagnosis of acute myocardial infarction. Int. J. Clin. Exp. Pathol. 2015;8(12):16014–9.

[130] Jia K, Shi P, Han X, Chen T, Tang H, Wang J. Diagnostic value of miR-30d-5p and miR-125b-5p in acute myocardial infarction. Mol. Med. Rep. 2016;14(1):184–94.

[131] Wang F, Long G, Zhao C, et al. Atherosclerosis-related circulating miRNAs as novel and sensitive predictors for acute myocardial infarction. PLoS One 2014;9(9):e105734.

[132] Sun T, Dong YH, Du W, et al. The role of MicroRNAs in myocardial infarction: from molecular mechanism to clinical application. Int. J. Mol. Sci. 2017;18(4).

[133] Satoh M, Nasu T, Takahashi Y, et al. Expression of miR-23a induces telomere shortening and is associated with poor clinical outcomes in patients with coronary artery disease. Clin. Sci.(London, England 1979) 2017;131(15):2007–17.

[134] Zhi H, Wang L, Ma G, et al. Polymorphisms of miRNAs genes are associated with the risk and prognosis of coronary artery disease. Clin. Res. Cardiol. 2012;101(4):289–96.

[135] Fei Y, Hou J, Xuan W, Zhang C, Meng X. The relationship of plasma miR-503 and coronary collateral circulation in patients with coronary artery disease. Life Sci. 2018;207:145–51.

[136] Wang J, Yan Y, Song D, Liu L, Liu B. The association of plasma miR-155 and VCAM-1 levels with coronary collateral circulation. Biomark. Med. 2017;11(2):125–31.

[137] Li Q, Chen L, Chen D, Wu X, Chen M. Influence of microRNA-related polymorphisms on clinical outcomes in coronary artery disease. Am. J. Tourism Res. 2015;7(2):393–400.

[138] Nie X, Su L, Zhou Y, et al. [Association between plasma levels of microRNA-126 and coronary collaterals in patients with coronary artery disease]. Zhonghua Xin Xue Guan Bing Za Zhi 2014;42(7):561–5.

[139] Wang HW, Huang TS, Lo HH, et al. Deficiency of the microRNA-31-microRNA-720 pathway in the plasma and endothelial progenitor cells from patients with coronary artery disease. Arterioscler. Thromb. Vasc. Biol. 2014;34(4):857–69.

[140] Paiva S, Agbulut O. MiRroring the multiple potentials of MicroRNAs in acute myocardial infarction. Front. Cardiovasc. Med. 2017;4(73).

[141] Sayed ASM, Xia K, Yang T-L, Peng J. Circulating microRNAs: a potential role in diagnosis and prognosis of acute myocardial infarction. Dis. Mark. 2013;35(5):561–6.

[142] Paiva S, Agbulut O. MiRroring the multiple potentials of MicroRNAs in acute myocardial infarction. Front. Cardiovasc. Med. 2017;4:73.

[143] Li C, Pei F, Zhu X, Duan DD, Zeng C. Circulating microRNAs as novel and sensitive biomarkers of acute myocardial Infarction. Clin. Biochem. 2012;45(10–11):727–32.

[144] Schulte C, Zeller T. microRNA-based diagnostics and therapy in cardiovascular disease-Summing up the facts. Cardiovasc. Diagn. Ther. 2015;5(1):17–36.

[145] Contu R, Latronico MV, Condorelli G. Circulating microRNAs as potential biomarkers of coronary artery disease: a promise to be fulfilled? Circ. Res. 2010;107(5):573–4.

Chapter 19

Cell-free DNA in hepatocellular carcinoma

Mairene Coto-Llerena[1], Luigi M. Terracciano[1], Markus von Flüe[2], Charlotte K.Y. Ng[3] and Salvatore Piscuoglio[1,2]

[1]*Institute of Pathology, University Hospital Basel, Basel, Switzerland;* [2]*Visceral Surgery Research Laboratory, Clarunis, Department of Biomedicine, University of Basel, Basel, Switzerland;* [3]*Department of BioMedical Research, University of Bern, Bern, Switzerland*

Introduction

Hepatocellular carcinoma (HCC) represents more than 90% of primary liver cancers, with >600,000 deaths each year, and increasing incidence [1]. The main risk factors are well described and include viral hepatitis, alcohol abuse, ingestion of the fungal metabolite aflatoxin B1, and nonalcoholic fatty liver disease (NAFLD) [2]. HCC is usually preceded by liver cirrhosis, induced by chronic necroinflammation, and a maladaptive wound healing response of the liver [3]. The transformation of hepatocytes toward HCC is a multistep process, involving the accumulation of epi/genetic alterations in the hepatocytes [4].

HCCs are variably diagnosed during active surveillance of at-risk patients (i.e., those with cirrhosis), or as incidental finding or symptomatic disease (e.g., abdominal discomfort, hepatic decompensation) [5]. In developed countries, surveillance programs lead to early HCC diagnosis in almost 40% of patients [6]. Barcelona Clinic Liver Cancer (BCLC) staging is the most widely recognized clinical algorithm for HCC patient stratification and treatment allocation [7,8]. Patients with early-stage HCC are usually treated with curative intent by resection, liver transplantation, or local ablation and have a median overall survival of 5 years [9]. Unfortunately, most patients treated by resection or ablation have disease recurrence within the first 5 years.

Patients with intermediate-stage HCC may benefit from transarterial chemoembolization (TACE), and have a median overall survival of <2 years [10]. For patients with late-stage unresectable disease, or patients ineligible for liver transplantation, the median overall survival is 1 year [11,12]. The multi-kinase inhibitor sorafenib, with antiproliferative, antiangiogenic, and proapoptotic properties, is the main first-line systemic agent for advanced HCC. However, sorafenib only extends survival by 2.8 months in late-stage patients, and nearly all patients develop resistance to it [13,14].

New kinase inhibitors—first-line lenvatinib [15], and second-line regorafenib [16], cabozantinib [17], and ramucirumab [18]—have been demonstrated to improve clinical outcome, although the median overall survival remains ~1 year. Nivolumab (immune checkpoint PD-1 inhibitor) is another second-line option for patients who have progressed on sorafenib.

The lack of tumor material for molecular profiling hampers the identification of targetable genetic alterations, as HCC is frequently diagnosed by liver imaging with alpha-fetoprotein (AFP) levels in blood [19], without the requirement for histopathologic analysis of the tumor. Alternative sources of tumor materials include surgical resection or liver transplantation. However, both procedures are restricted to patients with early-stage HCC.

In some cases, liver tumor biopsies are performed and may serve as the source of molecular information for late-stage HCC, but the procedure is invasive, may produce insufficient tissue, and is not informative about tumor heterogeneity, or epi/genetic relation to metastatic lesions. To overcome these challenges, liquid biopsies have recently been proposed as an alternative source of tumor DNA [20,21].

Liquid biopsy

Liquid biopsy is the sampling and analysis of nonsolid biological tissue, most commonly blood, but also saliva, urine, cerebrospinal fluid, and other body fluids. In HCC patients, liquid biopsies are usually obtained from blood,

Precision Medicine for Investigators, Practitioners and Providers. https://doi.org/10.1016/B978-0-12-819178-1.00019-8

and used for the extraction of circulating cell-free DNA (cfDNA), circulating tumor cells (CTCs), and exosomes (Fig. 19.1).

CTCs are intact tumor cells, shed into the bloodstream from primary and metastatic lesions. They can be detected via techniques that take advantage of their physical and biological properties, such as the expression of epithelial cell adhesion molecule (EpCAM), N-cadherin, cytokeratin, cancer-specific antigens, or cell size [22]. Isolating CTCs based on size has limited efficiency, since tumor cells in different stages of cell cycle, undergoing apoptosis, or in epithelial-to-mesenchymal transition, may differ in size [23]. The majority of clinical studies on epithelial cancer types has relied on EpCAM expression for CTC isolation, since the CellSearch system (Menarini Silicon Biosystems, Bryn Athyn, PA, USA) is currently the only Food and Drug Administration-approved device in the clinical setting [24,25].

The low frequency of CTCs isolated among blood cells (0–10 CTCs/mL of blood) challenges their use in diagnostics. Nonetheless, CTCs offer the opportunity to perform in vitro assessment of the likelihood of a patient responding to a given therapy [26,27]. Moreover, CTCs can be expanded in vitro or as xenografts, to generate clinically relevant models to study how metastasis occurs at the molecular level [26].

Exosomes are small membrane vesicles (50–140 nm in diameter), secreted by multiple cell types, including cancer cells [28]. They contain small molecules such as DNA, RNA, and proteins, reflective of the function and condition of their originating cells. However, uptake of tumor-derived exosomes from the circulation, by nontumoral cells, can cause dysregulation in their protein expression, proliferation, and differentiation, as well as their antiviral responses [29]. Therefore, a better understanding of the role of exosomes for liquid biopsy in HCC is still necessary.

Circulating tumor DNA (ctDNA)

Circulating cfDNA are double-stranded 180-200-nucleotide DNA fragments, present in plasma, serum, and other bodily fluids [30]. In healthy individuals, white blood cells are the major source of cfDNA in the blood [31], and the average cfDNA concentration in peripheral blood is 30 ng/mL (0–100 ng/mL) [32]. In cancer patients, cfDNA levels are elevated [33–35], and the tumor-derived fraction of cfDNA (i.e., circulating tumor DNA or ctDNA) represents $<0.1\%$ -50% of the total cfDNA detectable in the bloodstream [36,37]. The half-life of ctDNA in the blood circulation ranges from 16 min to 2.5 h [38].

Interestingly, HCC patients have been reported to have aberrantly shorter and longer cfDNA fragments, with the shorter fragments likely to have derived from tumor cells, while the longer fragments are hypothesized to have derived from necrotic cells [39]. ctDNA harbor the same epi/genetic alterations, as the originating tumor cells [40].

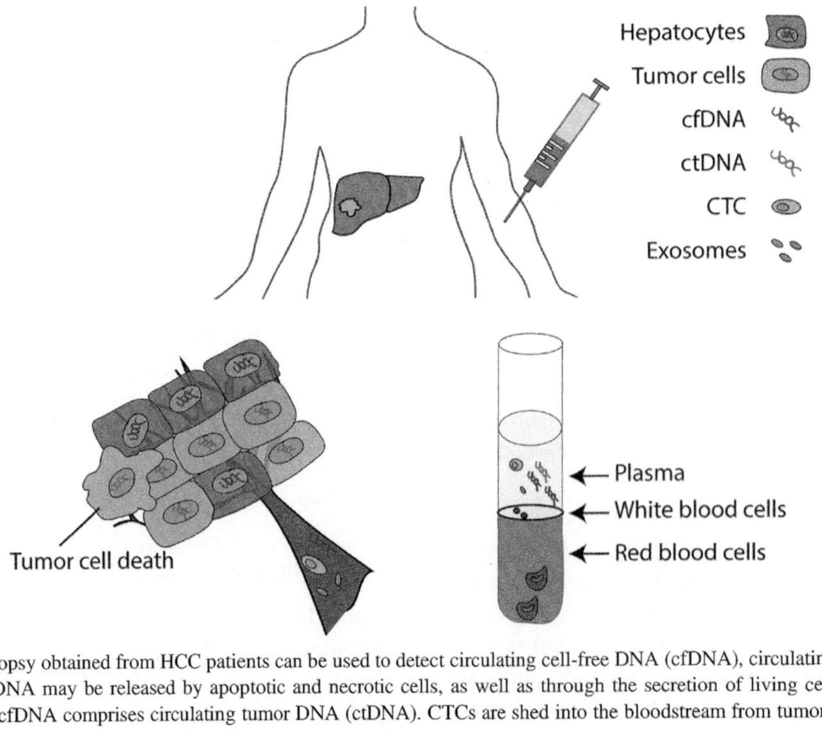

FIGURE 19.1 Liquid biopsy obtained from HCC patients can be used to detect circulating cell-free DNA (cfDNA), circulating tumor cells (CTCs), and exosomes. Circulating cfDNA may be released by apoptotic and necrotic cells, as well as through the secretion of living cells. In cancer patients, the fraction of tumor-derived cfDNA comprises circulating tumor DNA (ctDNA). CTCs are shed into the bloodstream from tumor and/or metastatic lesions. Exosomes, small microvesicles, are released from viable tumor cells to the bloodstream.

Therefore, ctDNA profiling has become an area of increasing clinical relevance in HCC, allowing the identification and tracking of tumor mutations [41,42] without the associated risks of invasive procedures.

cfDNA processing and ctDNA enrichment

Assay sensitivity owing to the low abundance of ctDNA is a clinical challenge. The development of next generation sequencing (NGS), and more recently molecular barcoding (boldsystems.org), has helped to improve sensitivity and specificity of ctDNA detection. However, there is still room for improvement, not only in ctDNA detection, but also in blood processing and DNA extraction.

Sample collection

Optimizing plasma processing will help improve ctDNA detection, by reducing cfDNA released from lyzed hematological cells [43,44]. EDTA collection tubes are usually recommended for the isolation of cfDNA/ctDNA, if the blood is to be processed within a few hours. However, if the blood needs to be stored for longer before processing, Streck (Omaha, NE, USA) or CellSave (Janssen Diagnostics, Raritan, NJ, USA) tubes may be more appropriate [45]. Regarding cfDNA extraction, there is no consensus on the most appropriate method and a wide variety has been used in published studies (Table 19.1). It is possible that the extraction protocols differ not only in their efficiency but also in how they influence downstream analyses. Several recent studies comparing commercially available kits have consistently identified Qiagen QIAamp Circulating Nucleic Acids kit (Germantown, MD, USA), as one that showed high accuracy and reproducibility [46–48], performed best in purifying both short and long exogenous DNA fragments, and gave highest DNA yield [46].

Sample enrichment

Enrichment strategies have been explored to increase detection sensitivity [49,50]. With the shorter length of ctDNA [39], preselecting 90–150 bp DNA fragments provided a twofold enrichment of ctDNA, with increased sensitivity to identify tumor-associated mutations and copy number alterations [49]. Moreover, cfDNA fragmentation is a nonrandom process, demonstrated by the observation that DNA molecules derived from fetal tissues, contained ends that differed from those derived from maternal tissues in pregnant women [51]. Similar observations have been made in HCC, in which ctDNA fragments displayed a tumor-associated preferred plasma DNA end signature [50], which may be used to enhance the sensitivity and specificity of the liquid biopsy approaches.

Quantitative analysis of cfDNA in HCC

In HCC, several studies have reported three to four fold higher concentration of cfDNA in the plasma or serum, compared with chronic hepatitis [52–55], and nearly 20-fold higher than healthy individuals [54]. Quantitative analyses of cfDNA have also been shown to reflect tumor dynamics [38], since cfDNA quantity varies with tumor size [41,56,57] and grade [41,56]. Higher cfDNA concentration is associated with shorter overall survival [53,57], and may serve as predictive biomarker for distant metastasis after curative surgery [53]. Despite the promising results, cfDNA quantity alone lacked robustness as the sole diagnostic tool for HCC [55], but its performance could be enhanced in conjunction with serum AFP levels [55]. Additionally, biological characteristics and potential molecular targets of HCC cannot be derived from quantitative analysis of cfDNA alone.

Hypermethylation pattern

DNA methylation is a heritable epigenetic mark, involving the covalent transfer of a methyl group to the C-5 position of the cytosine ring, by DNA methyltransferases [58]. While in normal cells, DNA methylation assures the proper regulation of gene expression, in cancer cells, it serves as an important mechanism to inactivate tumor suppressor genes. Hypermethylation of tumor suppressor genes is an early event in many tumors, suggesting that altered DNA methylation patterns could be one of the first detectable neoplastic changes [59,60].

Several studies have reported hypermethylation of p15 [61], p16 [62], glutathione S-transferase P1 (*GSTP1*) [63], Ras association domain family 1A (*RASSF1A*), hypomethylation of LINE-1 repetitive sequence [64,65], *TGS10*, *ST8SIA6*, *RUNX2*, and *VIM* [66], as ctDNA markers in HCC patients (Table 19.1). The first studies to use methylation status as a tumor-specific marker to detect ctDNA in HCC patients investigated p15 and p16 [61,62], both of which are frequently abrogated in human neoplasms [67,68]. The authors found that p15 and p16 were methylated in 16% (4/25) and 59% (13/22), respectively, of the plasma/serum DNA of HCC patients [61,62]. Importantly, all cases with evidence of p15/16 methylation in the plasma/serum DNA also showed evidence of methylation in the corresponding tumors [61,62].

Concordant methylation pattern between plasma/serum DNA and HCC tumor is not always observed, underscoring the expectation that ctDNA only accounts for a subset of cfDNA [69]. While *GSTP1* and LINE-1 hypermethylation were detected in 50% (16/32) and 67% (70/105) of the cfDNA [64], *RASSF1A* hypermethylation was a more sensitive marker, detected in 70%–93% of the sera of HCC patients [70,71], and it may identify AFP-negative HCC

TABLE 19.1 List of studies reporting on the analysis of circulating cfDNA in HCC patients.

Study	Blood Collection Tube	Method of cfDNA isolation	Method of analysis	cfDNA level	Methylation of ctDNA	Genetic alterations
Wong et al. [62]	EDTA-Tube	QIAmp blood kit (Qiagen)	Methylation-specific PCR of p16		Yes	
Wong et al. [61]	EDTA-Tube	QIAmp blood kit (Qiagen)	Methylation-specific PCR of p15 and p16		Yes	
Yeo et al. [72]	EDTA-Tube	QIAmp blood mini kit (Qiagen)	Methylation-specific PCR of *RASSF1A*		Yes	
Iizuka et al. [56]	CLOTED BLOOD	DNA extractor SP kit for serum and plasma (Wako Pure Chemical Industries)	Real-time PCR	Yes		
Ren et al. [57]	EDTA-Tube	QIAmp DNA blood midi kit (Qiagen)	cfDNA quantification and allelic imbalance of 8p	Yes		
Wang et al. [98]	EDTA-Tube	Glassmilk DNA isolation system	Methylation-specific PCR of *GSTP1*		Yes	
Tangkijvanich et al. [65]	ND	QIAmp blood kit (Qiagen)	Combined bisulfite restriction PCR for methylation of LINE-1 repetitive sequences		Yes	
Tokuhisa et al. [53]	CLOTTED BLOOD	DNA extractor SP kit for serum and plasma (Wako Pure Chemical Industries)	Real-time PCR of *GSTP1*	Yes		
Zhang et al. [70]	ND	QIAmp UltraSens virus kits (Qiagen)	Methylation-specific PCR of *RASSF1A*, p15 and p16		Yes	
Chan et al. [71]	ND	QIAmp DNA mini kit (Qiagen)	Methylation-specific PCR of *RASSF1A*		Yes	
El-Shazly et al. [99]	Serum separator tube	QIAmp DNA Blood mini kit	Quantitative real-time PCR of Alu repeats	Yes		
Huang et al. [54]	K3-EDTA	TIANamp micro DNA kit (Tiangen)	Quantitative real-time PCR of cfDNA			
Chan et al. [88]	EDTA-Tube	QIAmp DSP DNA blood mini kit (Qiagen)	Whole-genome sequencing			Yes
Chan et al. [75]	ND	QIAmp DSP DNA blood mini kit (Qiagen)	Genome-wide bisulfite sequencing		Yes	Yes
Bettegowda et al. [85]	ND	QIAmp circulating nucleic acid kit (Qiagen)	Safe-SeqS of tumor-specific somatic mutations			Yes
Jiang et al. [39]	EDTA-Tube	QIAmp DSP DNA blood mini kit (Qiagen)	Whole-genome sequencing			Yes
Ono et al. [87]	EDTA-2K	SMI-TEST (Genome Science Laboratories)	Real-time PCR of tumor-specific structural variations			Yes
Ono et al. [87]	EDTA-2K	QIAmp circulating nucleic acid kit (Qiagen)	Whole-exome sequencing			Yes

Continued

TABLE 19.1 List of studies reporting on the analysis of circulating cfDNA in HCC patients.—cont'd

Study	Blood Collection Tube	Method of cfDNA isolation	Method of analysis	cfDNA level	Methylation of ctDNA	Genetic alterations
Sun et al. [69]	EDTA-Tube	ND	Genome-wide bisulfite sequencing		Yes	Yes
Xu et al. [86]	EDTA-Tube	BloodGen mini kit (CoWin Inc.)	Whole-genome sequencing			Yes
Wen et al. [66]	EDTA-Tube	QIAamp DNA blood midi kit (Qiagen)	MCTA-Seq		Yes	
Huang et al. [84]	EDTA-Tube	QIAamp circulating nucleic acid kit (Qiagen)	ddPCR of four mutation hotspots			Yes
Liao et al. [83]	EDTA-Tube	QIAamp DNA mini kit (Qiagen)	Targeted sequencing of *TERT, CTNNB1,* and *TP53* mutation hotspots			Yes
Huang et al. [89]	EDTA-Tube	QIAamp circulating nucleic acid kit (Qiagen)	Whole-exome multiregion sequencing and targeted deep sequencing			
Xu et al. [73]	ND	EliteHealth cfDNA extraction kit (EliteHealth)	Molecular inversion probes for aberrant methylation		Yes	
Ng et al. [41]	Cell-Free DNA Blood Collection Tube (BCT, Streck)	QIAamp circulating nucleic acid kit (Qiagen)	Targeted sequencing of 46 genes frequently altered in HCC	Yes		Yes
Labgaa et al. [42]	EDTA-Tube	QIAamp circulating nucleic acid kit (Qiagen)	Targeted sequencing of 58 genes			Yes
Kirk et al. [82]	EDTA-Tube	QIAamp tissue kit (Qiagen)	PCR-specific Ser-249 *TP*			Yes

cfDNA, cell-free DNA; *ddPCR*, digital droplet PCR; *MCTA-Seq*, Methylated CpG tandems amplification and sequencing; *ND*, Not described in the original manuscript; *PCR*, Polymerase chain reaction; *Safe-SeqS*, Safe-sequencing system.

[71]. Similar to cfDNA quantification, studies have found association between aberrant methylation with increased risk of metastasis or recurrence [61], larger tumors [65,72], and worse prognosis [65,71]. *RASSF1A* methylation in cfDNA, showed a positive correlation with tumor size [64], while LINE-1 hypomethylation was associated with tumor progression and poor prognosis [64]. Combining LINE-1 and *RASSF1A* methylation status in cfDNA was also predictive of disease recurrence following tumor resection [64].

By detecting hypermethylated CpG islands in cfDNA, a recent study identified *TGS10*, *ST8SIA6*, *RUNX2*, and *VIM* as the best hypermethylated markers for the detection of small HCC (<3 cm) [66]. Another study used the methylation profiles of 1098 HCC patients and 835 controls, to construct a diagnostic prediction model using 10 methylation markers [73]. Once applied to cfDNA, the model in the first study achieved 94% sensitivity and 89% specificity in distinguishing HCC patients from cirrhotic or normal controls [66], while in the second study it showed >83% sensitivity and >90% specificity in distinguishing HCC patients from healthy individuals [73]. Both studies detected aberrant methylation in a subset of or all AFP-negative HCC [66,73]. Interestingly, another study showed not only identification of HCC patients, but also differentiation from HBV/HCV infection and fatty liver disease [73].

While regional hypermethylation is associated with the silencing of tumor suppressor genes, global DNA hypomethylation is a genome-wide phenomenon, present in most cancer types [74], and affects the structural-nuclear function by promoting chromosomal and genomic instability. In healthy individuals, plasma cfDNA methylation profiles are fairly stable, without evidence of global hypomethylation [75]. By contrast, the cfDNA of HCC patients showed significant hypomethylation [75].

Studies of methylation profiles suggest that while the liver contributes to ~10% of cfDNA in healthy individuals, the fraction rises to 24% (IQR 19.0%–44.0%) in HCC patients [69]. This variability in liver-derived cfDNA fraction among HCC patients is in agreement with the similarly substantial inter-individual variability, in the proportion of hypomethylated genome [75], since one would expect reduced sensitivity to detect hypomethylation in HCC patients with low fraction of ctDNA. Rising serum concentrations of methylated *RASSF1A* has also been reported in HBV carriers progressing to HCC [71].

Genetic analysis of ctDNA in HCC

Genomic characterization has revealed that each HCC harbors a unique combination of somatic alterations [76–81]. However, most of those alterations occur in genes not directly implicated in neoplasia, and only a few of the genes are considered drivers, in activating key signaling pathways for hepatocarcinogenesis [79]. The key pathways include p53 signaling (e.g., *TP53* and *CDKN2A*), Wnt/β-catenin pathway (e.g., *CTNNB1* and *AXIN1*), chromatin remodeling (e.g., *ARID1A*, *ARID1B*, *ARID2*, *BAP1*, *KMT2D*, *KMT2C*, and *PBRM1*), response to oxidative stress (e.g., *KEAP1* and *NFE2L2*), and telomerase maintenance (e.g., *TERT* promoter mutation, HBV integration into *TERT*, *TERT* fusion genes, and *TERT* copy number amplification) [40].

The fraction of patients with detectable somatic genetic alterations in cfDNA varies substantially between studies. The hotspot *TP53* R249S mutation, for instance, could be detected in the plasma of 36% (19/53) of HCC patients, in an African cohort using direct sequencing [82]. Using the more sensitive NGS, tumor-associated hotspot mutations in the *TERT* promoter, *CTNNB1*, and *TP53* could be detected in 20% (8/41) of plasma samples, in a predominantly HBV-positive cohort [83]. Targeting mutation hotspots by the yet more sensitive droplet digital PCR (ddPCR), detected *TP53* c.747G>T (p.R249S), *CTNNB1* c.121A>G (p.T41A) and c.133T>C (p.S45A), and *TERT* c.-124C>T in the ctDNA of 56% (27/48), of predominantly HBV-positive and BCLC stage A patients [84].

Using digital PCR-based technology, Bettegowda et al. [85] reported that somatic mutations were detectable in the plasma, in 75% (3/4) of advanced-stage HCC patients. Similarly, Labgaa et al. [42] detected at least 1 tumor-specific mutation, in the ctDNA of 67% (4/6) of patients, using ultra-deep sequencing. In addition to somatic mutations, somatic copy number alterations have also been observed, in the cfDNA in 42% (13/31) of patients [86].

Factors such as cohort composition, geographical locations, as well as the sensitivity of the methodologies, appear to influence the proportion of HCC patients with detectable ctDNA, on the basis of somatic genetic alterations. Despite the variations, detectable tumor-specific genetic alterations in the ctDNA are associated with well-established clinicopathologic parameters, such as tumor size, AFP, and vascular invasion [83,86,87], as well as recurrence and extrahepatic metastasis [83,87], consistent with cfDNA quantification and methylation studies. Indeed, our own study showed that somatic mutations in genes frequently altered in HCC were detected in 27% (8/30) of HCC patients, and in 86% (6/7) of those with large tumor (≥5 cm diameter) or metastatic disease [41]. We also observed a positive correlation of detectable tumor-specific mutations in ctDNA, with tumor size and Edmondson grade, although not with BCLC stage or AFP levels [41].

Mutation frequency

The proportion of ctDNA in cfDNA, as estimated by the fraction of DNA sequences harboring tumor-associated mutations, is also highly variable [85]. Among three patients with advanced stage HCC and detectable ctDNA, there was huge variability in the amount of mutant fragments (7.2, 15, and 7910 mutant fragments/5 mL of plasma) [85]. Similarly, ddPCR of four hotspot mutations in HCC driver genes found that mutant allele fractions ranged from 0.33% to 23.7% among patients with detectable ctDNA [84]. Our own NGS-based study identified mutant allele fractions, from 0.06% to 45% in the cfDNA [41]. Based on copy number alterations and/or methylation, it was estimated that a median of 24.0% (range: 4.3%–71.4%) of cfDNA was tumor-derived [69,88]. The observed variability is also consistent with the estimated liver-derived cfDNA fraction, among HCC patients by methylation studies.

Similarity with primary tumor

Analysis of tumor-specific genetic alterations, has allowed us to determine how well ctDNA mirrors the genetic composition of the tumors [41,42]. Earlier studies have demonstrated that both copy number profiles, and detectable hotspot mutations of plasma DNA, highly resemble those of the matched primary tumor [39,83,88]. Beyond the hotspot mutations, more recent ultradeep sequencing analysis, of paired plasma/serum and tumor in a small cohort of eight HCC patients observed that 43% (9/21) of the tumor-associated mutations were detectable in cfDNA [42]. In our own experience, by comparing paired tumors and cfDNA, we found that 87% (80/92) of the somatic mutations were captured in the cfDNA, among seven patients in whom the largest tumor was ≥5 cm or was associated with metastasis [41].

Importantly, we found that the proportion of mutations detected in the matched primary tumors was similar (95%, 87/92) [41]. Our results showed that ctDNA accurately reflects the genetics of the primary tumors in patients with high disease burden [41]. These findings are particularly important in the context of HCC, given that patients with advanced disease are the least likely to undergo surgical resection.

Cancer heterogeneity

Another exciting prospect is the potential for ctDNA to overcome intra-tumor genetic heterogeneity in HCC [89,90]. The extent of tumor heterogeneity was illustrated in a targeted sequencing study of nine tissue samples, including portal vein tumor thrombus, from three patients, showing that only 47.4%–79.4% of subclonal mutations were shared between primary tumor and portal vein tumor thrombus, in each patient [91]. Interestingly, however, >98% of the subclonal mutations were captured in the plasma ctDNA [91]. We also found instances where mutations with low variant allele fraction in the primary tumors were readily detectable only in the cfDNA41, which suggests that ctDNA may also circumvent the issue of intratumor genetic heterogeneity posed by single needle biopsies.

However, there remain limitations. In one of the most comprehensive studies, multi-region whole-exome sequencing of five HCC patients was conducted, and found that cfDNA captured most of the mutations that were heterogeneous between tumor regions, but only given the knowledge of the mutations present in the tumor [89]. Such limitation will hopefully be resolved or reduced with the adoption of molecular barcoding in NGS.

Clinical implications of ctDNA

In HCC, ctDNA has been studied for its potential for risk prediction or early cancer detection in a screening setting, for response to treatment and recurrence monitoring, and as a surrogate for tumor molecular profiling (Fig. 19.2). The identification of genetic alterations associated with drug resistance or drug sensitivity is also crucial, especially considering that the emergence of acquired drug resistance is believed to be the cause of treatment failure, in 90% of patients with metastatic disease [92].

Screening interest

Despite the increased sensitivity in detection methods, ctDNA concentration is typically lower at early stages of the disease, when screening is crucial. Thus far, a study of 50 HCC patients with repeated serum DNA sampling, suggested that aberrant methylation could be detected up to 9 years prior to the diagnosis of HCC [70]. Developing a ctDNA test for early cancer detection would involve the selection of an appropriate marker, broadly applicable for a given cohort. Highly recurrent hotspot mutations such as *TP53* R249S, *CTNNB1* amino acids D32, S33, S37, T41, and S45, and *TERT* c.-124C>T promoter mutations [76,78–81] may serve as potential markers for HCC detection. However, the discovery of frequent *TERT* promoter mutation in cirrhotic preneoplastic lesions [93] suggests that it may lack specificity in a screening setting. Another possibility would be to detect ctDNA on the basis of shorter fragment sizes [39,49], although it is still unclear whether shorter fragment size is a universal characteristic of ctDNA.

Disease follow-up

Response and recurrence monitoring using cfDNA is closer to realistic clinical application. In other cancer types, ctDNA has been used to detect minimal residual disease [94], and track the emergence of drug resistance clones [95]. For HCC, patients who undergo rounds of therapeutic interventions such as TACE and ablation often have recurrence several years later, mandating careful monitoring over time. The correlation between ctDNA fraction and disease burden [41,91,96] supports the use of ctDNA in real-time monitoring of HCC.

In fact, it has been shown that ctDNA level reflects response to treatment and disease progression, with its level falling after resection, and rising prior to recurrence or metastasis [75,87,88]. By contrast, ctDNA remain undetectable in patients with long-term remission [75,87]. ctDNA profiling would also make longitudinal monitoring of clonal evolution possible, and may help to detect micrometastatic disease [97]. However, it should be noted that the timing of sample collection is an important consideration, as serum ctDNA peaks 4 weeks after TACE [87].

New therapeutic targets

In a cohort of 66 HCC patients, 39% (26/66) of the primary tumors harbored potentially targetable alterations, including *TSC1*/*TSC2* mutations (mTOR inhibitors), *EGFR* mutations (gefitinib and erlotinib), *CCND1* amplifications and *CDKN2A* deletions or mutations (palbociclib), *ATM* mutation (olaparib), and *MET* amplification (tivantinib) [89]. The detection of such targetable genetic alterations in ctDNA, such as *VEGFR3* amplification (sorafenib) [89], *TSC2* frameshift mutation (mTOR inhibitors) [41], and *JAK1* hotspot mutation (ruxolitinib) [77], has been reported in just a handful of HCC patients.

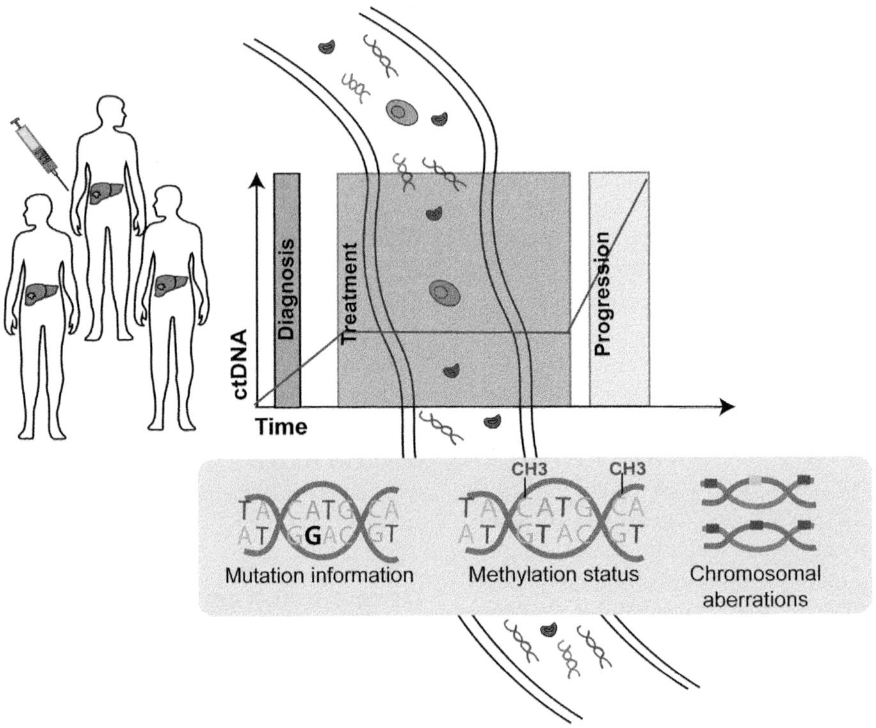

FIGURE 19.2 cfDNA may be used to study tumor methylation patterns, chromosomal aberrations, or mutation profile. ctDNA is being investigated as a marker for the detection of early hepatocellular carcinoma, as well as for assessment of response to treatment, and detection of minimal residual disease.

Conclusions

Through the analysis of cfDNA quantity, methylation, and genomic of ctDNA, ctDNA evidently holds the potential to address several outstanding questions in the clinical management of HCC. Studies carried out thus far have laid the foundation for development of a new generation of reliable, mechanism-based disease biomarkers using ctDNA. However, larger and independent studies, as well as development and standardization of protocol and methodologies in the isolation, detection, and analysis of ctDNA are still needed.

References

[1] Sia D, Llovet JM. Liver cancer: translating '-omics' results into precision medicine for hepatocellular carcinoma. Nat. Rev. Gastroenterol. Hepatol. 2017;14:571–2.

[2] Llovet JM, Zucman-Rossi J, Pikarsky E, et al. Hepatocellular carcinoma. Nat. Rev. Disease Primers 2016;2:16018.

[3] El–Serag HB, Lenhard Rudolph K. Hepatocellular carcinoma: epidemiology and molecular carcinogenesis. Gastroenterology 2007;132:2557–76.

[4] Farazi PA, DePinho RA. Hepatocellular carcinoma pathogenesis: from genes to environment. Nat. Rev. Cancer 2006;6:674–87.

[5] Singal AG, Nehra M, Adams-Huet B, et al. Detection of hepatocellular carcinoma at advanced stages among patients in the HALT-C trial: where did surveillance fail? Am. J. Gastroenterol. 2013;108:425–32.

[6] Singal AG, Conjeevaram HS, Volk ML, et al. Effectiveness of hepatocellular carcinoma surveillance in patients with cirrhosis. Cancer Epidemiol. Biomark. Prev. 2012;21:793–9.

[7] Llovet J, Brú C, Bruix J. Prognosis of hepatocellular carcinoma: the BCLC staging classification. Semin. Liver Dis. 1999;19:329–38.

[8] European Association for the Study of the Liver, Electronic address: easloffice@easloffice.eu, European association for the study of the liver. EASL clinical practice guidelines: management of hepatocellular carcinoma. J. Hepatol. 2018;69:182–236.

[9] Belghiti J, Regimbeau JM, Durand F, et al. Resection of hepatocellular carcinoma: a European experience on 328 cases. Hepatogastroenterol. 2002;49:41–6.

[10] Llovet JM, Bruix J. Systematic review of randomized trials for unresectable hepatocellular carcinoma: chemoembolization improves survival. Hepatology 2003;37:429–42.

[11] El-Serag HB. Hepatocellular carcinoma. N. Engl. J. Med. 2011;365:1118–27.

[12] Llovet JM, Burroughs A, Bruix J. Hepatocellular carcinoma. Lancet 2003;362:1907–17.

[13] Llovet JM, Ricci S, Mazzaferro V, et al. Sorafenib in advanced hepatocellular carcinoma. N. Engl. J. Med. 2008;359:378–90.

[14] Lim HY, Merle P, Weiss KH, et al. Phase II studies with refametinib or refametinib plus sorafenib in patients with RAS-mutated hepatocellular carcinoma. Clin. Cancer Res. 2018;24:4650–61.

[15] Kudo M, Finn RS, Qin S, et al. Lenvatinib versus sorafenib in first-line treatment of patients with unresectable hepatocellular carcinoma: a randomised phase 3 non-inferiority trial. Lancet 2018;391:1163–73.

[16] Bruix J. Regorafenib and the RESORCE trial: a new second-line option for hepatocellular carcinoma patients. Hepatic Oncol. 2016;3:187–9.
[17] Abou-Alfa GK, Meyer T, Cheng A-L, et al. Cabozantinib in patients with advanced and progressing hepatocellular carcinoma. N. Engl. J. Med. 2018;379:54–63.
[18] Zhu AX, Baron AD, Malfertheiner P, et al. Ramucirumab as second-line treatment in patients with advanced hepatocellular carcinoma: analysis of REACH trial results by child-pugh score. JAMA Oncol. 2016. https://doi.org/10.1001/jamaoncol.2016.4115. Published online September 22.
[19] Heimbach JK, Kulik LM, Finn RS, et al. AASLD guidelines for the treatment of hepatocellular carcinoma. Hepatology 2018;67:358–80.
[20] Perakis S, Speicher MR. Emerging concepts in liquid biopsies. BMC Med. 2017;15. https://doi.org/10.1186/s12916-017-0840-6.
[21] De Rubis G, Krishnan SR, Bebawy M. Circulating tumor DNA – current state of play and future perspectives. Pharmacol. Res. 2018;136:35–44.
[22] Alix-Panabières C, Pantel K. Real-time liquid biopsy: circulating tumor cells versus circulating tumor DNA. Ann. Transl. Med. 2013;1:18.
[23] Harouaka R, Kang Z, Zheng S-Y, Cao L. Circulating tumor cells: advances in isolation and analysis, and challenges for clinical applications. Pharmacol. Ther. 2014;141:209–21.
[24] Beije N, Jager A, Sleijfer S. Circulating tumor cell enumeration by the CellSearch system: the clinician's guide to breast cancer treatment? Cancer Treat Rev. 2015;41:144–50.
[25] Miller MC, Doyle GV, Terstappen LWMM. Significance of circulating tumor cells detected by the CellSearch system in patients with metastatic breast colorectal and prostate cancer. J. Oncol. 2010;2010:617421.
[26] Yu M, Bardia A, Aceto N, et al. Cancer therapy. Ex vivo culture of circulating breast tumor cells for individualized testing of drug susceptibility. Science 2014;345:216–20.
[27] Castro-Giner F, Gkountela S, Donato C, et al. Cancer diagnosis using a liquid biopsy: challenges and expectations. Diagnostics (Basel) 2018;8. https://doi.org/10.3390/diagnostics8020031.
[28] Théry C, Zitvogel L, Amigorena S. Exosomes: composition, biogenesis and function. Nat. Rev. Immunol. 2002;2:569–79.
[29] Sato K, Meng F, Glaser S, Alpini G. Exosomes in liver pathology. J. Hepatol. 2016;65:213–21.
[30] Jung K, Fleischhacker M, Rabien A. Cell-free DNA in the blood as a solid tumor biomarker–a critical appraisal of the literature. Clin. Chim. Acta 2010;411:1611–24.
[31] Lui YYN, Chik K-W, Chiu RWK, Ho C-Y, Lam CWK, Lo YMD. Predominant hematopoietic origin of cell-free DNA in plasma and serum after sex-mismatched bone marrow transplantation. Clin. Chem. 2002;48:421–7.
[32] Fleischhacker M, Schmidt B. Circulating nucleic acids (CNAs) and cancer—a survey. Biochim. Biophys. Acta Rev. Cancer 2007;1775:181–232.
[33] Anker P, Mulcahy H, Chen XQ, Stroun M. Detection of circulating tumour DNA in the blood (plasma/serum) of cancer patients. Cancer Metastasis Rev. 1999;18:65–73.
[34] Leon SA, Green A, Yaros MJ, Shapiro B. Radioimmunoassay for nanogram quantities of DNA. J. Immunol. Methods 1975;9:157–64.
[35] Stroun M, Anker P, Lyautey J, Lederrey C, Maurice PA. Isolation and characterization of DNA from the plasma of cancer patients. Eur. J. Cancer Clin. Oncol. 1987;23:707–12.
[36] Vandekerkhove G, Todenhöfer T, Annala M, et al. Circulating tumor DNA reveals clinically actionable somatic genome of metastatic bladder cancer. Clin. Cancer Res. 2017;23:6487–97.
[37] Fiala C, Diamandis EP. Utility of circulating tumor DNA in cancer diagnostics with emphasis on early detection. BMC Med. 2018;16:166.
[38] Diehl F, Schmidt K, Choti MA, et al. Circulating mutant DNA to assess tumor dynamics. Nat. Med. 2008;14:985–90.
[39] Jiang P, Chan CWM, Chan KCA, et al. Lengthening and shortening of plasma DNA in hepatocellular carcinoma patients. Proc. Natl. Acad. Sci. U.S.A. 2015;112:E1317–25.
[40] Ng CKY, Di Costanzo GG, Terracciano LM, Piscuoglio S. Circulating cell-free DNA in hepatocellular carcinoma: current insights and outlook. Front. Med. 2018;5. https://doi.org/10.3389/fmed.2018.00078.
[41] Ng CKY, Di Costanzo GG, Tosti N, et al. Genetic profiling using plasma-derived cell-free DNA in therapy-naïve hepatocellular carcinoma patients: a pilot study. Ann. Oncol. 2018;29:1286–91.
[42] Labgaa I, Villacorta-Martin C, D'Avola D, et al. A pilot study of ultra-deep targeted sequencing of plasma DNA identifies driver mutations in hepatocellular carcinoma. Oncogene 2018;37:3740–52.
[43] Norton SE, Lechner JM, Williams T, Fernando MR. A stabilizing reagent prevents cell-free DNA contamination by cellular DNA in plasma during blood sample storage and shipping as determined by digital PCR. Clin. Biochem. 2013;46:1561–5.
[44] Xue X, Teare MD, Holen I, Zhu YM, Woll PJ. Optimizing the yield and utility of circulating cell-free DNA from plasma and serum. Clin. Chim. Acta 2009;404:100–4.
[45] van Dessel LF, Beije N, Helmijr JCA, et al. Application of circulating tumor DNA in prospective clinical oncology trials - standardization of preanalytical conditions. Mol. Oncol. 2017;11:295–304.
[46] Warton K, Graham L-J, Yuwono N, Samimi G. Comparison of 4 commercial kits for the extraction of circulating DNA from plasma. Cancer Genet. 2018;228–229:143–50.
[47] Page K, Guttery DS, Zahra N, et al. Influence of plasma processing on recovery and analysis of circulating nucleic acids. PLoS One 2013;8:e77963.
[48] Mauger F, Dulary C, Daviaud C, Deleuze J-F, Tost J. Comprehensive evaluation of methods to isolate, quantify, and characterize circulating cell-free DNA from small volumes of plasma. Anal. Bioanal. Chem. 2015;407:6873–8.
[49] Mouliere F, Chandrananda D, Piskorz AM, et al. Enhanced detection of circulating tumor DNA by fragment size analysis. Sci. Transl. Med. 2018;10. https://doi.org/10.1126/scitranslmed.aat4921.
[50] Jiang P, Sun K, Tong YK, et al. Preferred end coordinates and somatic variants as signatures of circulating tumor DNA associated with hepatocellular carcinoma. Proc. Natl. Acad. Sci. U.S.A. 2018;115:E10925–33.
[51] Chan KCA, Jiang P, Sun K, et al. Second generation noninvasive fetal genome analysis reveals de novo mutations, single-base parental inheritance, and preferred DNA ends. Proc. Natl. Acad. Sci. U.S.A. 2016;113:E8159–68.

[52] Leon SA, Ehrlich GE, Shapiro B, Labbate VA. Free DNA in the serum of rheumatoid arthritis patients. J. Rheumatol. 1977;4:139–43.
[53] Tokuhisa Y, Iizuka N, Sakaida I, et al. Circulating cell-free DNA as a predictive marker for distant metastasis of hepatitis C virus-related hepatocellular carcinoma. Br. J. Cancer 2007;97:1399–403.
[54] Huang Z, Hua D, Hu Y, et al. Quantitation of plasma circulating DNA using quantitative PCR for the detection of hepatocellular carcinoma. Pathol. Oncol. Res. 2011;18:271–6.
[55] Liao W, Mao Y, Ge P, et al. Value of quantitative and qualitative analyses of circulating cell-free DNA as diagnostic tools for hepatocellular carcinoma: a meta-analysis. Medicine 2015;94:e722.
[56] Iizuka N, Sakaida I, Moribe T, et al. Elevated levels of circulating cell-free DNA in the blood of patients with hepatitis C virus-associated hepatocellular carcinoma. Anticancer Res. 2006;26:4713–9.
[57] Ren N, Qin L-X, Tu H, Liu Y-K, Zhang B-H, Tang Z-Y. The prognostic value of circulating plasma DNA level and its allelic imbalance on chromosome 8p in patients with hepatocellular carcinoma. J. Cancer Res. Clin. Oncol. 2006;132:399–407.
[58] Robertson KD. DNA methylation and human disease. Nat. Rev. Genet. 2005;6:597–610.
[59] Baylin SB, Jones PA. Epigenetic determinants of cancer. Cold Spring Harb. Perspect. Biol. 2016;8. https://doi.org/10.1101/cshperspect.a019505.
[60] Baylin SB, Jones PA. A decade of exploring the cancer epigenome - biological and translational implications. Nat. Rev. Cancer 2011;11:726–34.
[61] Wong IH, Lo YM, Yeo W, Lau WY, Johnson PJ. Frequent p15 promoter methylation in tumor and peripheral blood from hepatocellular carcinoma patients. Clin. Cancer Res. 2000;6:3516–21.
[62] Wong IH, Lo YM, Zhang J, et al. Detection of aberrant p16 methylation in the plasma and serum of liver cancer patients. Cancer Res. 1999;59:71–3.
[63] Liu M, Cui L-H, Li C-C, Zhang L. Association of APC, GSTP1 and SOCS1 promoter methylation with the risk of hepatocellular carcinoma: a meta-analysis. Eur. J. Cancer Prev. 2015;24:470–83.
[64] Liu ZJ, Huang Y, Wei L, et al. Combination of LINE-1 hypomethylation and RASSF1A promoter hypermethylation in serum DNA is a non-invasion prognostic biomarker for early recurrence of hepatocellular carcinoma after curative resection. Neoplasma 2017;64:795–802.
[65] Tangkijvanich P, Hourpai N, Rattanatanyong P, Wisedopas N, Mahachai V, Mutirangura A. Serum LINE-1 hypomethylation as a potential prognostic marker for hepatocellular carcinoma. Clin. Chim. Acta 2007;379:127–33.
[66] Wen L, Li J, Guo H, et al. Genome-scale detection of hypermethylated CpG islands in circulating cell-free DNA of hepatocellular carcinoma patients. Cell Res. 2015;25:1376.
[67] Cairns P, Mao L, Merlo A, et al. Rates of p16 (MTS1) mutations in primary tumors with 9p loss. Science 1994;265:415–7.
[68] Nobori T, Miura K, Wu DJ, Lois A, Takabayashi K, Carson DA. Deletions of the cyclin-dependent kinase-4 inhibitor gene in multiple human cancers. Nature 1994;368:753–6.
[69] Sun K, Jiang P, Chan KCA, et al. Plasma DNA tissue mapping by genome-wide methylation sequencing for noninvasive prenatal, cancer, and transplantation assessments. Proc. Natl. Acad. Sci. U.S.A. 2015;112:E5503–12.
[70] Zhang Y-J, Wu H-C, Shen J, et al. Predicting hepatocellular carcinoma by detection of aberrant promoter methylation in serum DNA. Clin. Cancer Res. 2007;13:2378–84.
[71] Chan KCA, Lai PBS, Mok TSK, et al. Quantitative analysis of circulating methylated DNA as a biomarker for hepatocellular carcinoma. Clin. Chem. 2008;54:1528–36.
[72] Yeo W, Wong N, Wong W-L, Lai PBS, Zhong S, Johnson PJ. High frequency of promoter hypermethylation of RASSF1A in tumor and plasma of patients with hepatocellular carcinoma. Liver Int. 2005;25:266–72.
[73] Xu R-H, Wei W, Krawczyk M, et al. Circulating tumour DNA methylation markers for diagnosis and prognosis of hepatocellular carcinoma. Nat. Mater. 2017;16:1155–61.
[74] Ehrlich M. DNA hypomethylation in cancer cells. Epigenomics 2009;1:239–59.
[75] Chan KCA, Jiang P, Chan CWM, et al. Noninvasive detection of cancer-associated genome-wide hypomethylation and copy number aberrations by plasma DNA bisulfite sequencing. Proc. Natl. Acad. Sci. U.S.A. 2013;110:18761–8.
[76] Guichard C, Amaddeo G, Imbeaud S, et al. Integrated analysis of somatic mutations and focal copy-number changes identifies key genes and pathways in hepatocellular carcinoma. Nat. Genet. 2012;44:694–8.
[77] Kan Z, Zheng H, Liu X, et al. Whole-genome sequencing identifies recurrent mutations in hepatocellular carcinoma. Genome Res. 2013;23:1422–33.
[78] Ahn S-M, Jang SJ, Shim JH, et al. Genomic portrait of resectable hepatocellular carcinomas: implications of RB1 and FGF19 aberrations for patient stratification. Hepatology 2014;60: 1972–82.
[79] Schulze K, Imbeaud S, Letouzé E, et al. Exome sequencing of hepatocellular carcinomas identifies new mutational signatures and potential therapeutic targets. Nat. Genet. 2015;47:505–11.
[80] Fujimoto A, Furuta M, Totoki Y, et al. Whole-genome mutational landscape and characterization of noncoding and structural mutations in liver cancer. Nat. Genet. 2016;48:500–9.
[81] Cancer Genome Atlas Research Network, Electronic address: wheeler@bcm.edu, Cancer Genome Atlas Research Network. Comprehensive and integrative genomic characterization of hepatocellular carcinoma. Cell 2017;169. 1327–41.e23.
[82] Kirk GD, Camus-Randon AM, Mendy M, et al. Ser-249 p53 mutations in plasma DNA of patients with hepatocellular carcinoma from the Gambia. J. Natl. Cancer Inst. 2000;92:148–53.
[83] Liao W, Yang H, Xu H, et al. Noninvasive detection of tumor-associated mutations from circulating cell-free DNA in hepatocellular carcinoma patients by targeted deep sequencing. Oncotarget 2016;7:40481–90.
[84] Huang A, Zhang X, Zhou S-L, et al. Detecting circulating tumor DNA in hepatocellular carcinoma patients using droplet digital PCR is feasible and reflects intratumoral heterogeneity. J. Cancer 2016;7:1907–14.
[85] Bettegowda C, Sausen M, Leary RJ, et al. Detection of circulating tumor DNA in early- and late-stage human malignancies. Sci. Transl. Med. 2014;6:224ra24.
[86] Xu H, Zhu X, Xu Z, et al. Non-invasive analysis of genomic copy number variation in patients with hepatocellular carcinoma by next generation DNA sequencing. J. Cancer 2015;6:247–53.

[87] Ono A, Fujimoto A, Yamamoto Y, et al. Circulating tumor DNA analysis for liver cancers and its usefulness as a liquid biopsy. Cell Mol. Gastroenterol. Hepatol. 2015;1:516–34.
[88] Chan KCA, Jiang P, Zheng YWL, et al. Cancer genome scanning in plasma: detection of tumor-associated copy number aberrations, single-nucleotide variants, and tumoral heterogeneity by massively parallel sequencing. Clin. Chem. 2013;59:211–24.
[89] Huang A, Zhao X, Yang X-R, et al. Circumventing intratumoral heterogeneity to identify potential therapeutic targets in hepatocellular carcinoma. J. Hepatol. 2017;67:293–301.
[90] Zhai W, Lim TK-H, Zhang T, et al. The spatial organization of intra-tumour heterogeneity and evolutionary trajectories of metastases in hepatocellular carcinoma. Nat. Commun. 2017;8:4565.
[91] Cai Z-X, Chen G, Zeng Y-Y, et al. Circulating tumor DNA profiling reveals clonal evolution and real-time disease progression in advanced hepatocellular carcinoma. Int. J. Cancer 2017;141: 977–85.
[92] Longley DB, Johnston PG. Molecular mechanisms of drug resistance. J. Pathol. 2005;205:275–92.
[93] Nault JC, Mallet M, Pilati C, et al. High frequency of telomerase reverse-transcriptase promoter somatic mutations in hepatocellular carcinoma and preneoplastic lesions. Nat. Commun. 2013;4:2218.
[94] Garcia-Murillas I, Schiavon G, Weigelt B, et al. Mutation tracking in circulating tumor DNA predicts relapse in early breast cancer. Sci. Transl. Med. 2015;7:302ra133.
[95] Murtaza M, Dawson S-J, Tsui DWY, et al. Non-invasive analysis of acquired resistance to cancer therapy by sequencing of plasma DNA. Nature 2013;497:108–12.
[96] Kowalik A, Kowalewska M, Góźdź S. Current approaches for avoiding the limitations of circulating tumor cells detection methods-implications for diagnosis and treatment of patients with solid tumors. Transl. Res. 2017;185:58–84.e15.
[97] Kelley RK, Magbanua MJM, Butler TM, et al. Circulating tumor cells in hepatocellular carcinoma: a pilot study of detection, enumeration, and next-generation sequencing in cases and controls. BMC Cancer 2015;15:206.
[98] Wang J, Qin Y, Li B, Sun Z, Yang B. Detection of aberrant promoter methylation of GSTP1 in the tumor and serum of Chinese human primary hepatocellular carcinoma patients. Clin. Biochem. 2006;39:344–8.
[99] El-Shazly SF, Eid MA, El-Sourogy HA, Attia GF, Ezzat SA. Evaluation of serum DNA integrity as a screening and prognostic tool in patients with hepatitis C virus-related hepatocellular carcinoma. Int. J. Biol. Mark. 2010;25:79–86.

Chapter 20

Non-coding RNA therapy in cancer

Kamila Souckova[1], Tina Catela Ivkovic[1] and Ondrej Slaby[1,2]
[1]Masaryk University, Brno, Czech Republic; [2]Masaryk Memorial Cancer Institute, Brno, Czech Republic

Non-coding RNAs in cancer

MicroRNAs

MicroRNAs (miRNAs) are the most frequently studied group of small noncoding RNAs (ncRNAs), functionally involved in almost all biological processes including cell cycle control, DNA repair, angiogenesis, and inflammation. To date, almost 2700 human mature miRNA sequences have been annotated in miRBASE database [1]. Hundreds of miRNAs have been identified as oncogenic drivers and tumor suppressors in every major cancer type [2,3].

MiRNAs are evolutionarily conserved single-stranded molecules, around 22 nucleotides long, and their expression is tissue and cell type-specific. Their linear canonical biogenesis pathway has been well defined (Fig. 20.1). The main mode of miRNA action is directing RNA-induced silencing complex (RISC) toward target genes and downregulating gene expression by either of two post-transcriptional mechanisms: mRNA degradation or translational repression [4]. The most frequent site of miRNA interaction is the 3′-untranscribed ultraconserved region (3′-UTR) of the target mRNA. Other possible effector pathways of miRNAs are summarized in Fig. 20.1.

MiRNA binding

MiRNA binding is defined by a six to eight nucleotides long seed sequence in the miRNAs 5′end, and further regulated by weaker binding of the rest of the sequence. Low sequence complementarity requirement and different modes of action make miRNAs functionally extremely versatile. Each miRNA has the potential to regulate the expression of hundreds of different genes at the same time. Conversely, more than half of all annotated mRNAs have one or more evolutionarily conserved sequence, predicted to interact with miRNAs [5,6]. Since miRNAs repress translation of tumor suppressors and oncogenes, they can be both oncogenic and tumor suppressive.

Some miRNAs may exhibit both features, depending on the cancer type and/or cellular context [7]. Genes coding for miRNAs are frequently located at fragile sites of the genome, the common break-point regions in human cancer [8]. Silencing of miRNA genes by DNA promoter hypermethylation or histone hypoacetylation has also been described [9,10]. Extensive research has shown complex networks of interactions and feedback circuitries composed of miRNAs, other ncRNAs and protein-coding genes, that can be involved in the pathogenesis of cancer [11,12]. MiRNAs also play an important role in the interplay between tumor cells and the microenvironment, participating in driving tumor development and progression. Most frequently studied tumor suppressive and oncogenic miRNAs are summarized in Table 20.1. Targeting miRNAs could be an opportunity to design more effective therapeutic treatment, because therapeutics focused only on protein-coding genes may not be sufficient for controlling cancer progression (Table 20.2).

Long noncoding RNAs

Long noncoding RNAs (lncRNAs) represent the broadest and most diverse class of noncoding transcripts. According to several large-scale genomic screens, approximately one-third of annotated human genes encode for lncRNAs [13–15]. The majority of lncRNAs have high tissue and cell type-specific expression patterns. They exhibit poor sequence conservation across species, and their expression levels appear to be one order of magnitude lower than of protein-coding genes. Several lncRNAs have been shown to play a crucial role in specific aspects of cancer progression. Many lncRNAs are up- or down-regulated in cancer tissue [16].

LncRNAs are defined by the length (more than 200 nt), and transcribed by RNA polymerase II. They can be categorized according to structural characteristics and modes of action into multiple classes. Two main classes are represented by natural antisense transcripts (NATs) and long

Precision Medicine for Investigators, Practitioners and Providers. https://doi.org/10.1016/B978-0-12-819178-1.00020-4

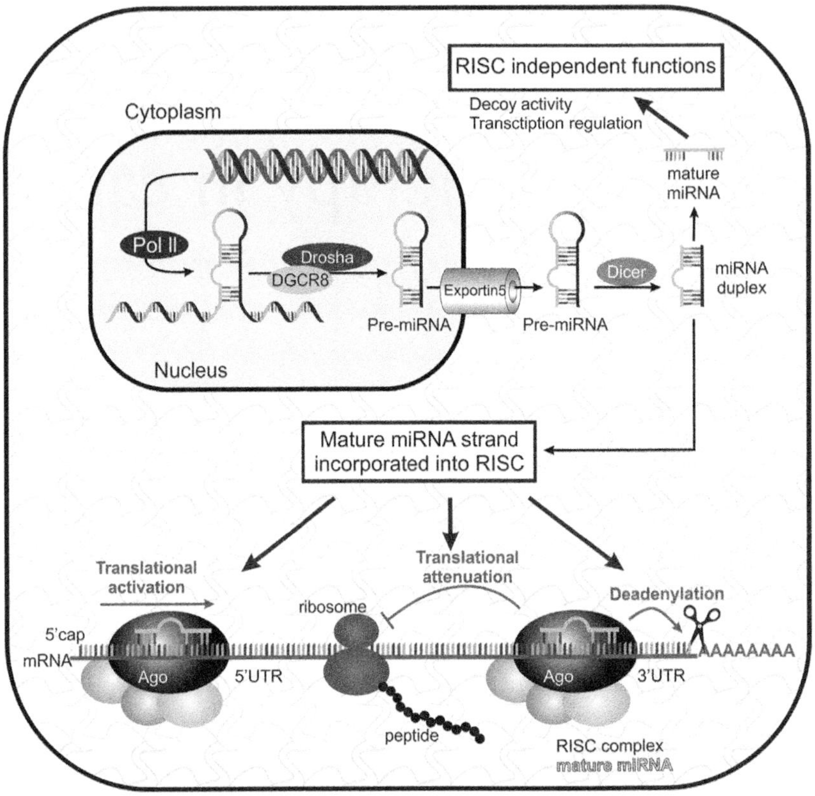

FIGURE 20.1 Biogenetic pathway of miRNA processing and common effector pathways. MiRNAs frequently have their own promoters and are transcribed as autonomous transcription units. They can be transcribed from inter- or intragenic locations, or oriented antisense to neighboring genes. In the nucleus, miRNAs are transcribed by RNA polymerase II (Pol II) into long primary transcripts (pri-miRNA), which are recognized and cleaved by Microprocessor, which includes Drosha and DiGeorge syndrome critical region 8 (DGCR8) to produce 70–90 nt nucleotide hairpin precursor miRNAs (pre-miRNAs). Pre-miRNAs are exported from the nucleus to the cytoplasm, where they are converted to transient mature miRNA duplexes (19–24 nt) by interaction with endoribonuclease Dicer. One strand of the duplex (leading strand) gets integrated into a large protein RISC, whereas the second (passenger strand) is released and degraded. Mature miRNA leads RISC complex to 3′-untranslated region (3′-UTR) of protein-coding genes to induce translational repression or mRNA degradation. MiRNA can also interact with the 5′-untranslated region (5′-UTR) or coding regions of protein-coding genes and cause translational repression or activation. Moreover, miRNAs can act independently on RISC complex by decoy activities, or interact directly with regulatory protein complexes. *Ago*, Argonaute protein.

intergenic ncRNAs (lincRNAs). Other smaller groups are covered, for instance, by pseudogenes, transcribed ultraconserved regions (T-UCRs), circular RNAs (circRNAs), and competing endogenous RNAs (ceRNAs).

NATs are lncRNAs with sequence complementarity to other RNA transcripts. The most known NATs repress transcription of their overlapping genes, thus acting *in cis*. Regulation *in trans* is largely observed with NATs transcribed from pseudogenes that share a high degree of sequence homology with the original gene. NATs mechanism of action is degradation of the sense transcript through small interfering RNA (siRNA)-like mechanism, but also through chromatin structure modification. Importantly, NATs have been detected near key tumor-suppressor genes, such as CDKN2B (ANRIL) [17] and CDKN1A (P21-AS) [18].

LincRNAs are transcribed from DNA sequences between protein-coding genes. Some of them regulate the transcriptional activity of protein-coding genes by guiding the histone methyltransferase polycomb repressive complex 2 (PRC2) to target genomic loci [19]. Ultraconserved regions (UCRs) are a subset of highly conserved sequences, which are localized in both intra- and intergenic regions, and might function as distant enhancers. Some UCRs, the transcribed UCRs (T-UCRs), are expressed in a highly ubiquitous and tissue-specific manner. Expression of many T-UCRs has been detected in different types of cancer, for example, in chronic lymphocytic leukemia, colorectal and hepatocellular carcinomas, and neuroblastomas, often at fragile regions of chromosomes [20]. Many circRNAs overlap-coding genes and arise from head-to-tail splicing of one or more exons. Although circRNAs are generally expressed at low levels, they regulate transcription and interfere with splicing [21].

TABLE 20.1 Examples of significant miRNA oncogenes and tumor suppressors.

MiRNA	Cancer type	Function	Deregulation mechanism/targets	References
cluster miR-17~92	Lung, breast, colorectal and gastric cancer, myeloma and AML	Oncogene	Amplification; positive regulation by E2F and MYC/BIM, PTEN, CDKN1A	[61]
miR-21	Pancreatic, breast, lung, prostate and gastric cancer, CLL, AML, myeloma and glioblastoma	Oncogene	Positive regulation by IL-6 and GF1a/DCD4, PTEN, TPM1	[62]
miR-155	CLL, AML, colorectal, lung, breast cancer and lymphomas	Oncogene	Positive regulation by NF-κB/SHIP1, EBPB	[45]
cluster miR-106b~25	Various cancers	Oncogene	Transcriptional and posttranscriptional regulation/p21, PTEN, CDKN1A, CDKN1C, MDM2, SMAD7	[63]
miR-15a/miR-16	CLL, prostate cancer and pituitary adenomas	Tumor suppressor	Genomic loss; mutations; positive regulation by p53/BCL-2, MCL1	[64]
Let-7 family	Lung, colorectal breast, gastric and ovarian cancer	Tumor suppressor	Negative regulation by MYC/KRAS, NRAS, CDK6, DC25A, MGA2, MYC	[65]
miR-29 family	AML, CLL, lung and breast cancer, cholangiocarcinoma, lymphoma, hepatocellular carcinoma and rhabdomyosarcoma	Tumor suppressor	Genomic loss; negative regulation by MYC; positive regulation by p53/MCL1, CDK6, TCL1, DNMT1, DNMT3a, DNMT3b	[66]
miR-34 family	Colorectal, lung, breast, kidney and bladder cancer, neuroblastoma and melanoma cell lines	Tumor suppressor	Methylation regulation; positive regulation by p53; deletion/CDK4, CDK6, CCNE2, CND1, MET, MYC, CREB, E2F3, BCL-2	[67]
miR-145	Colorectal cancer, various cancers	Tumor suppressor	Genomic loss, methylation regulation, regulation by p53/MUC1, FSCN1, Vimentin, Cadherin, SMAD3, MMP11, Snail1, ZEB1/2, HIF-1, Rock-1	[68]
miR-126	Digestive system cancers	Tumor suppressor	Genomic loss, methylation regulation/KRAS, ADAM9, DNMT1, IRS-1, GOLPH3, SOX3, PI3KR2, VEGF-A, CRK	[69]
miR-31	Various cancers	Tumor suppressor or oncogene	Genomic loss; methylation regulation/SATB2, RASA1, Tima1, Smad3, Smad4, FIH, MET, STK40, SPRY	[70]
miR-708-5p	Various solid and hematological malignancies	Tumor suppressor or oncogene	Posttranscriptional regulation/CD38, Caspase-2, CDKN2B, NNAT, AKT2, ZEB2, p21	[71]

Adjusted and updated from Garzon R, Marcucci G, Croce CM. Targeting microRNAs in cancer: rationale, strategies and challenges. Nat. Rev. Drug Discov. October 2010;9(10):775–89.

TABLE 20.2 Summary of miRNA-based therapeutic agents tested in cancer clinical trials.

Name (company)	Therapeutic agent	Delivery system	Diseases	Trial details	Clinical Trials. gov. identifier
MRG-106 (miRagen Therapeutics)	Anti-miR-155	LNA-modified antisense inhibitor (local intratumoral, subcutaneous, and intravenous injection)	Multiple lymphomas and leukemias, mycosis fungoides	Multicentre phase I	NCT02580552
MesomiR 1 (EnGeneIC)	miR-16 mimic	EnGeneIC delivery vehicle - nonliving bacterial minicells (intravenous injection)	Malignant pleural mesothelioma, nonsmall cell lung cancer	Multicentre phase I	NCT02369198

Adjusted from Rupaimoole R, Slack FJ. MicroRNA therapeutics: towards a new era for the management of cancer and other diseases. Nat. Rev. Drug Discov. March 2017;16(3):203–22.

Pathophysiology in cancer modalities

Besides playing key roles in transcriptional and post-transcriptional regulation of gene expression, some lncRNAs are suggested to act in more complex manners, such as molecular scaffolds that recruit RNA binding proteins, architectural RNA, modulation of protein activity, epigenetics, and gene splicing [12,22]. LncRNAs also act as decoys to compete for or disrupt protein-binding interactions, or as sponges for miRNAs or transcription factors. The act of transcription, rather than RNA itself, could be functionally important for many lncRNAs [23].

Metastasis-associated lung adenocarcinoma transcript 1 (MALAT1) influences gene expression by post-transcriptional mechanisms, mainly through regulation of pre-mRNA processing [24,25]. Another example, and one of the most studied lincRNA so far, is HOX antisense intergenic RNA (HOTAIR), transcribed from the *HOXC* locus during normal development. HOTAIR was one of the first lncRNAs described to have a fundamental role in cancer, mainly in breast tumors [26]. HOTAIR can promote tumor invasion and metastasis, and can also act as an independent predictor of patient survival rates [27,28].

Another lncRNA HULC (highly upregulated in liver cancer), was identified as the most upregulated transcript in hepatocellular carcinoma, and one of the upregulated transcripts in liver metastasis of colorectal cancer [29]. Furthermore, HULC may act as an endogenous sponge, which downregulates activities of miRNAs, including miR-372 [30]. Moreover, lncRNA expression is under genetic control, thus effectively expanding the complexity of eukaryotic transcriptomes and the mechanisms of their regulation. Individual lncRNAs are transcriptionally regulated by important transcription factors, which are deregulated in cancer, such as p53, NF-κB, Sox2, Oct4, and Nanog [31].

Noncoding RNA-based therapeutic strategies

General strategies and delivery systems

Nucleic acid-based therapeutics are emerging as a promising site-specific approach to target pathogenic ncRNAs in a variety of human diseases including cancer. To achieve safe and effective transport of genetic material to tumor cells, and to maintain therapeutic nucleic acids at a constant level over sufficient time, systemically administered therapeutics must cope with many extra- and intracellular barriers. Delivery poses a substantial obstacle because RNA must resist exposure to omnipresent nucleases and innate immune system components. After reaching the target cell, the release of therapeutics from endosomes into the cytoplasm represents one of the key steps of delivery strategies.

Chemical modifications that improve stability, binding affinity, interference capacity, and cellular uptake, are necessary to achieve satisfactory *in vivo* efficacy of RNA-based therapeutics (summarized in Fig. 20.2). Although these modifications have significantly improved biological properties of RNA-based therapeutics, the ability of free RNA to cross cytoplasmic membrane is still rather low. Thus, various nanocarriers have been engineered and implemented.

Nanocarriers for delivery of RNA-based therapeutics

Viral vectors belong to the first genetically modified nanoparticles for targeted delivery of nucleic acids. These vectors can very efficiently deliver genetic material. On the other hand, virus-based carriers have limited vector production and restricted DNA packaging capability. Moreover, relatively high immunogenicity and possible systemic toxicity limit their use in the clinic [32]. Each type of vector derived from lentiviruses, adenoviruses, and adeno-associated viruses, has unique advantages, limitations, and applications [33,34].

Nonviral delivery systems, represented by polymeric vectors, lipid-based vectors, and inorganic particle-based vectors, have been developed as well. Combination of relatively inexpensive synthesis and large-scale production makes these carries extremely convenient for *in vivo* applications, although their transfection capabilities do not usually reach levels, comparable with virus-based vehicles [35].

Polymeric vectors have high structural and composition variability, inducing negligible immunogenicity and low acute toxicity. These features provide great potential for wide clinical use. They can be roughly divided into polycations, hydrophilic or amphiphilic neutral polymers, and neutral hydrophobic polymers. The most frequently used polymeric nanoparticles are cationic. Self-assembled complexes of these positively charged homopolymers or copolymers can form polyplexes with negatively charged nucleic acids. The surface of these particles can be supplied with neutral hydrophilic polymer blocks, which cover their surface and shield their positive charge, leading to lower toxicity and immunogenic response [36].

Experimental and clinical alternatives

Poly(ethylenimine) (PEI) and its conjugates have been extensively used for gene delivery purposes. Compared to low transfection efficacy of low-molecular-weight PEI, high-molecular-weight PEI results in great transfection efficacy; however, it is poorly biodegradable. This leads to

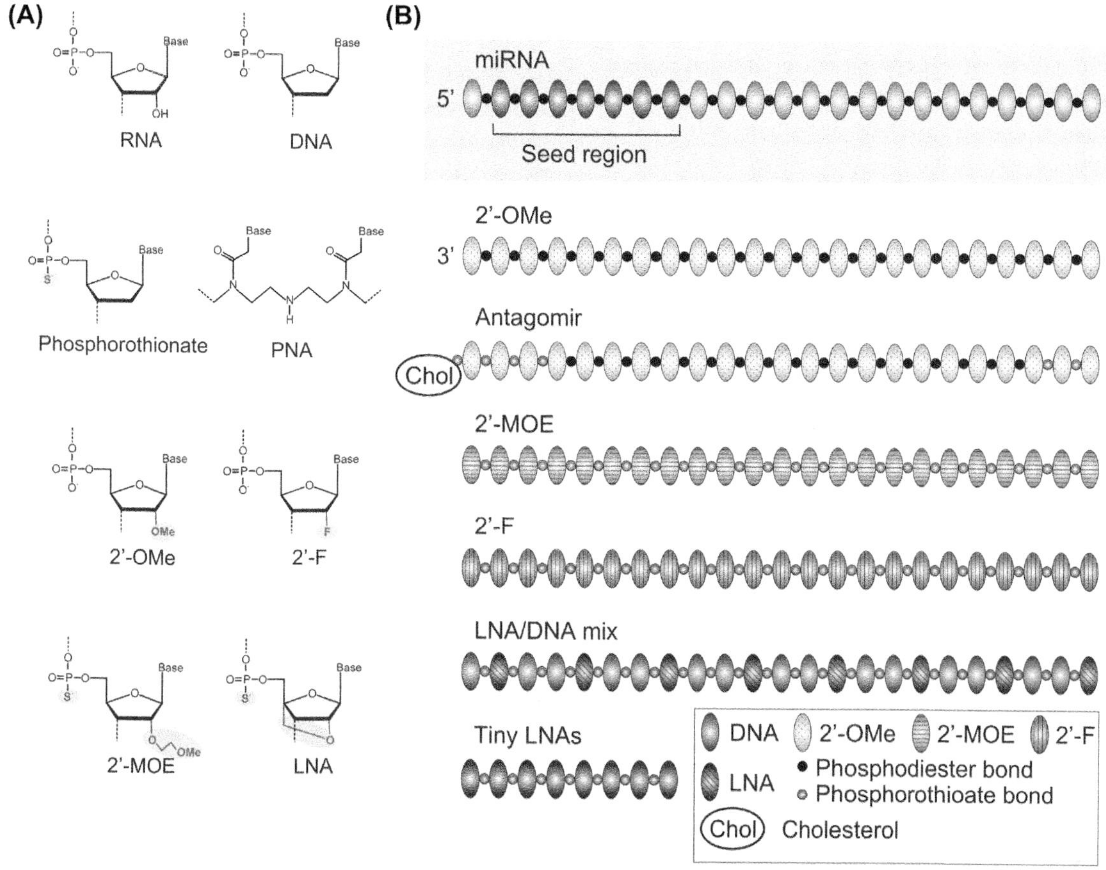

FIGURE 20.2 Chemistry of RNA-based therapeutics. (A) Chemical modifications of nucleosides and sugar-phosphate backbone: replacement of the phosphodiester linkage with phosphorothioate; peptide nucleic acid (PNA); substitutions of 2′-OH of ribose sugar, such as 2′-O-methyl (2′-OMe), 2′-fluoro (2′-F), and 2′-O-methoxyethyl (2′-MOE); locking of the conformation of the backbone with a methylene bridge using locked nucleic acid (LNA); (B) examples of current backbone and nucleoside modifications and their combinations used in antimiRNA therapeutic strategies. *Adapted from Li Z, Rana TM. Therapeutic targeting of microRNAs: current status and future challenges. Nat. Rev. Drug Discov. August 2014;13(8):622–38.*

accumulation in the cell and subsequent cytotoxicity. For overcoming these problems, PEI molecules coated with hydrophilic polymers [36,37], and in combination with iron oxide magnetic nanoparticles [38], have been developed.

Poly(L-lysine) (PLL) is a cationic polypeptide bearing primary amino groups along the linear polymer chain. PLL conjugated with nucleic acids electrostatically interacts with cell membranes, which facilitates cellular uptake [39]. Unfortunately, effective endosomal escape of these polyplexes prevents their wider use in clinical practice.

Lipid-based vectors are another group of broadly used delivery particles. This is a group of hydrophobic polymers, represented by biocompatible and biodegradable liposomes, lipid nanoparticles, and lipid nanoemulsions, which can mimic phospholipidic cell membranes. While cationic liposomes form with negatively charged RNA molecules electrostatic complexes (lipoplexes), neutral liposomes can encapsulate RNA molecule within the vesicle to form stable nucleic acid lipid particles. The major advantages of these delivery systems are biocompatibility, biodegradability, low toxicity, and the possibility to be modified with tissue-specific ligands. Nevertheless, their ability to penetrate the cell membranes is very limited [40].

MiRNA-based oncologic strategies

Research in the field of miRNA therapeutics is currently based on targeting or emulating specific miRNAs involved in the regulation of key cancer processes including cancer onset, progression, invasion, and metastasis. Therapeutic miRNAs may perform oncogenic or tumor suppressive functions, based on pathways and cell types targeted. The main advantage of miRNA-based therapeutic approaches is the ability to affect multiple genes simultaneously, making them extremely efficient in regulating cellular processes and pathways relevant to cell malignancy [41]. As mentioned earlier, in miRNA-based therapy two major strategies have been adopted: inhibition/blocking of oncogenic miRNAs, and restoring the function of tumor suppressor miRNAs.

Antagonism against oncogenic miRNAs

Inhibition strategies are mainly based on antisense oligonucleotides (ASOs) and miRNA sponges. ASOs include a wide range of different backbone and nucleoside modifications and their combinations. Antagomirs were the first approach to inhibit miRNA functions *in vivo*. These anti-miRNA oligonucleotides, are fully complementary to mature miRNAs and are chemically modified with 2′-O methyl protection (2′-OMe) and phosphorothioate bonds to increase stability. Moreover, conjugation with cholesterol increases cellular uptake (Fig. 20.2). Antagomirs can prevent miRNAs of interest from binding to mRNAs, and introduce them into the RNA-induced silencing complex (RISC) [42]. They were developed by Krützfeldt et al. as a proof of concept, used for targeting miR-122 highly abundant in hepatocytes [43]. Antagomirs require high dosages for effective miRNA blocking. That is the reason why they are routinely used only in experimental research.

Most miRNA-based therapeutics in the developmental stage comprise other modifications, such as 2′-F substitutions or locked nucleic acids (LNAs). LNA miRNA inhibitor molecules contain an additional methylene bridge that locks ribose into more thermodynamically stable conformation. Several LNA antimiR-based approaches are currently undergoing preclinical investigation in cancer. One ongoing clinical trial addresses antimiR-155, as miR-155 plays an important role in the activation of the immune system and hematopoiesis [44,45]. Ultrashort LNA oligonucleotides, so-called *tiny LNAs*, were developed for the inhibition of miRNA seed families. These 8-mer oligonucleotides are, compared to longer LNA antimiRs, fully LNA-modified. This modification leads to higher duplex melting temperature [46].

Sponges, zippers, and mimics

MiRNA sponges have been designed to contain multiple tandem binding sites complementary to the heptamer in the seed sequence of targeted miRNA, which means that a single sponge can block the whole miRNA seed family [47]. MiRNA sponges are usually encoded in either plasmid or viral expression vectors, which are driven by a strong promoter, such as the cytomegalovirus promoter. MiRNA sponges are successfully used mostly as experimental tools for now [48].

Another approach to target miRNAs of interest are miRNA-zippers. These molecules are designed to bind two different miRNAs, generating the space for stabilization and specific binding of these miRNAs. The intervention of miRNA-zippers results in miRNA loss-of-function phenotype [49].

Expression levels of tumor suppressive miRNAs, often downregulated in cancer, could be restored by introducing synthetic oligonucleotides identical with targeted miRNA. MiRNA mimics are double-stranded oligonucleotides with a wide range of chemical modifications. After cellular uptake, miRNA mimics are processed into a single-strand form (guide strand), and function in a miRNA-like manner. The opposite (passenger) strand is less stable and modified to prevent RISC loading or to enhance cellular uptake [50].

LncRNAs-based therapeutic strategies

There is much less available information on therapeutic targeting of lncRNAs compared to miRNAs. However there are several reasons why lncRNAs represent an attractive class of therapeutic agents: low expression of lncRNAs may permit lower doses of therapeutics, thus preventing some types of toxicities observed in oligonucleotide-based therapies; high tissue or cell specificity of lncRNA expression provides a unique opportunity for specific regulation by lncRNA-targeting drugs; lncRNAs are predominantly nuclear and act *in cis* so, by lncRNA targeting, locus-specific regulation can be achieved. In general, lncRNAs regulate transcription by many different mechanisms, and they can be targeted by different approaches (summarized in Fig. 20.3).

The therapeutic approach to lncRNA could be roughly divided in several groups: (i) triggering of post-transcriptional RNA degradation pathways; (ii) steric inhibition of lncRNA-protein interactions or prevention of secondary structure formation resulting in loss of function; (iii) transcriptional block of lncRNA genes.

Historically, the majority of lncRNA-based drugs and research tools were developed to downregulate lncRNAs. Posttranscriptional targeting of lncRNAs provides a straightforward strategy to knock down any RNA with oncogenic function. Currently, there are two major approaches for posttranscriptional RNA degradation: double-stranded RNA-mediated interference (RNAi), and application of single-stranded ASOs.

RNAi, represented primarily by siRNAs, is extensively used to inhibit lncRNAs in cancer cells. Many lncRNAs are predominantly located in the nucleus, and thus may be less accessible for siRNA targeting [51]. On the other hand, ASOs are single-stranded DNA sequences designed to be complementary to target lncRNAs. Upon binding to their target RNA, ASOs trigger cleavage of the RNA moiety of DNA:RNA complex by endogenous RNAse H1 activity, independently of RISC machinery. Compared to siRNAs, ASOs-based approach shows higher specificity and fewer off-target effects. The newer generation of ASOs consist of 15–20 nt, and their backbone is modified by phosphorothioate linkages or 2′-OMe. These modifications improve the binding affinity and keep a favorable pharmacokinetic profile for these molecules [52].

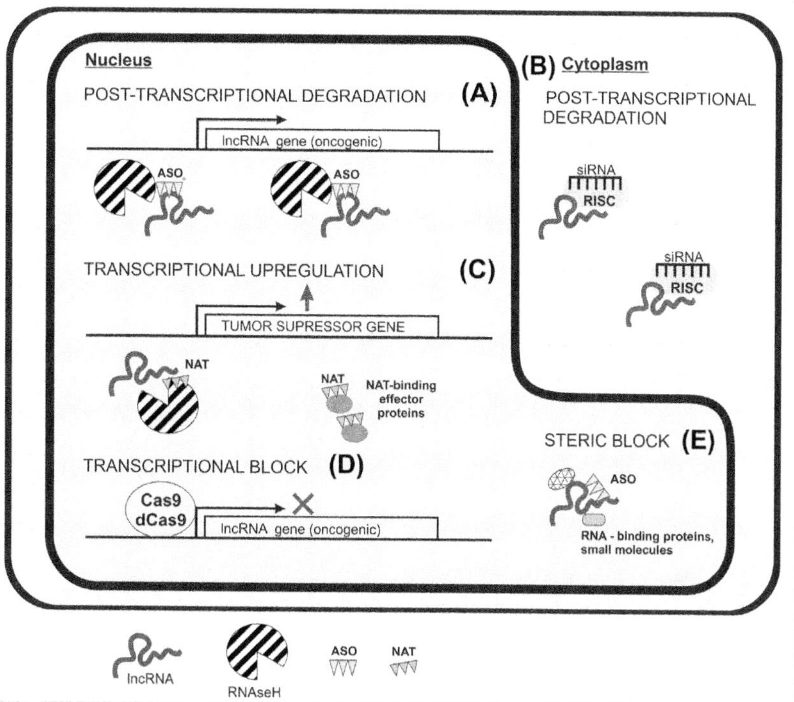

FIGURE 20.3 Examples of different lncRNAs therapeutic targeting. (A, B, C) Regulation of posttranscriptional gene expression by degradation of oncogenic lncRNA, (A) using antisense oligonucleotides (ASOs), (B) small-interfering RNAs (siRNAs), (C) antagoNATs. (D) Transcriptional inhibition of oncogenic lncRNA by dead-Cas9, fused to repressor complex, or through deletion of regions of interest in target lncRNAs. (E) Steric inhibition of lncRNA-protein interactions using RNA binding proteins, small molecules, and modified ASOs, that cannot stimulate RNA degradation pathway. *CRISPR*, clustered regularly interspaced short palindromic repeats; *NAT*, natural antisense transcript; *RISC*, RNA-induced silencing complex.

ASOs can also alter gene expression via steric hindrance and splicing alterations. For instance, ASOs with uniformly modified sugars in the backbone are resistant to RNase H1 cleavage. However, they could be used to modulate splicing patterns of target RNAs by blocking splicing enhancers or repressor binding sites. To achieve RNAse H1-mediated knockdown of target RNAs, gapmer ASOs (ASOs with a long block of deoxynucleotide monomers) are used. These chimeric ASOs are RNA-DNA-RNA hybrids, where RNA residues contain 2′-O-methoxyethyl-modified sugar backbone [53].

Tumor suppressor genes and CRISPR genome-editing

Transcriptional upregulation (derepression) of tumor suppressor genes can be achieved by knockdown of corresponding NATs by single-stranded chemically modified LNAs or ASOs named antagoNATs. AntagoNATs interfere with the function of NATs by blocking the interactions of NATs with effector proteins, or by causing RNAase H1-mediated degradation [54]. The other approach for modulating the expression of lncRNAs is the steric blockade of the promoter by genome-editing techniques. Besides deleting regions of interest by classical CRISPR/Cas9 technique, transcriptional silencing of lncRNAs using new CRISPR-based approaches is appearing feasible [55]. For example, CRISPR interference (CRISPRi) approach using dead-Cas9 [56], or RNA editing using CRISPR/Cas13 [57,58].

LncRNA expression is under genetic control. Highly selective tissue or cell-specific expression of several lncRNAs, combined with therapeutic responses in early stage clinical trials, provide an opportunity for using lncRNA promoters (for example H19), to drive the expression of toxins in a cancer-specific manner [59].

List of Abbreviations

2′-F 2′-fluoro
2′-MOE 2′-methoxyethoxy
2′-OMe 2′-methoxy
AML acute myeloid leukemia
ASO antisense oligonucleotide
Cas9 CRISPR associated protein 9
circRNA circular RNA
CLL chronic lymphocytic leukemia
CRISPR clustered regularly interspaced short palindromic repeats
lincRNA long intergenic ncRNA
LNA locked nucleic acid
lncRNA long noncoding RNA

miRNA microRNA
NAT natural antisense transcript
ncRNA noncoding RNA
NF-κB nuclear factor-κB
PEI poly(ethyleneimine)
PLL poly(L-lysine)
RISC RNA-induced silencing complex
siRNAs small-interfering RNAs
T-UCR transcribed ultraconserved region
UCR ultraconserved region
UTR untranslated region

Acknowledgments

Supported by Ministry of Health of the Czech Republic, grant No. 16-31314A, 16-31765A, 16-31997A, 16-33209A.

References

[1] Griffiths-Jones S, Saini HK, van Dongen S, Enright AJ. miRBase: tools for microRNA genomics. Nucleic Acids Res January 2008;36(Database issue):D154–158.
[2] Munker R, Calin GA. MicroRNA profiling in cancer. Clin. Sci. Lond. Engl. 1979 August 2011;121(4):141–58.
[3] Ji W, Sun B, Su C. Targeting microRNAs in cancer gene therapy. Genes January 9, 2017;8(1).
[4] Bartel DP. MicroRNAs: genomics, biogenesis, mechanism, and function. Cell January 23, 2004;116(2):281–97.
[5] Bartel DP. MicroRNAs: target recognition and regulatory functions. Cell January 23, 2009;136(2):215–33.
[6] Garzon R, Marcucci G, Croce CM. Targeting microRNAs in cancer: rationale, strategies and challenges. Nat. Rev. Drug Discov. October 2010;9(10):775–89.
[7] Pekarsky Y, Croce CM. Is miR-29 an oncogene or tumor suppressor in CLL? Oncotarget July 2010;1(3):224–7.
[8] Calin GA, Sevignani C, Dumitru CD, Hyslop T, Noch E, Yendamuri S, et al. Human microRNA genes are frequently located at fragile sites and genomic regions involved in cancers. Proc. Natl. Acad. Sci. U. S. A. March 2, 2004;101(9):2999–3004.
[9] Oltra SS, Peña-Chilet M, Vidal-Tomas V, Flower K, Martinez MT, Alonso E, et al. Methylation deregulation of miRNA promoters identifies miR124-2 as a survival biomarker in breast cancer in very young women. Sci. Rep. September 26, 2018;8(1):14373.
[10] Baer C, Claus R, Plass C. Genome-wide epigenetic regulation of miRNAs in cancer. Cancer Res January 15, 2013;73(2):473–7.
[11] Anastasiadou E, Faggioni A, Trivedi P, Slack FJ. The nefarious nexus of noncoding RNAs in cancer [Internet] Int. J. Mol. Sci. July 17, 2018 [cited 2018 Dec 5];19(7). Available from: https://www.ncbi.nlm.nih.gov/pmc/articles/PMC6073630/.
[12] Ling H, Fabbri M, Calin GA. MicroRNAs and other non-coding RNAs as targets for anticancer drug development. Nat. Rev. Drug Discov. November 2013;12(11):847–65.
[13] Iyer MK, Niknafs YS, Malik R, Singhal U, Sahu A, Hosono Y, et al. The landscape of long noncoding RNAs in the human transcriptome. Nat. Genet. March 2015;47(3):199–208.
[14] Nik-Zainal S, Davies H, Staaf J, Ramakrishna M, Glodzik D, Zou X, et al. Landscape of somatic mutations in 560 breast cancer whole-genome sequences. Nature 2016 02;534(7605):47–54.
[15] Khurana E, Fu Y, Colonna V, Mu XJ, Kang HM, Lappalainen T, et al. Integrative annotation of variants from 1092 humans: application to cancer genomics. Science October 4, 2013; 342(6154):1235587.
[16] Huarte M. The emerging role of lncRNAs in cancer. Nat. Med. November 5, 2015;21:1253.
[17] Yap KL, Li S, Muñoz-Cabello AM, Raguz S, Zeng L, Mujtaba S, et al. Molecular interplay of the noncoding RNA ANRIL and methylated histone H3 lysine 27 by polycomb CBX7 in transcriptional silencing of INK4a. Mol Cell June 11, 2010;38(5):662–74.
[18] Dimitrova N, Zamudio JR, Jong RM, Soukup D, Resnick R, Sarma K, et al. LincRNA-p21 activates p21 in cis to promote Polycomb target gene expression and to enforce the G1/S checkpoint. Mol Cell June 5, 2014;54(5):777–90.
[19] Wang KC, Chang HY. Molecular mechanisms of long noncoding RNAs. Mol Cell September 16, 2011;43(6):904–14.
[20] Calin GA, Liu C, Ferracin M, Hyslop T, Spizzo R, Sevignani C, et al. Ultraconserved regions encoding ncRNAs are altered in human leukemias and carcinomas. Cancer Cell September 2007;12(3): 215–29.
[21] Memczak S, Jens M, Elefsinioti A, Torti F, Krueger J, Rybak A, et al. Circular RNAs are a large class of animal RNAs with regulatory potency. Nature March 21, 2013;495(7441):333–8.
[22] Arun G, Diermeier SD, Spector DL. Therapeutic targeting of long non-coding RNAs in cancer. Trends Mol. Med. 2018;24(3): 257–77.
[23] Ard R, Allshire RC, Marquardt S. Emerging properties and functional consequences of noncoding transcription. Genetics 2017;207(2):357–67.
[24] Tano K, Mizuno R, Okada T, Rakwal R, Shibato J, Masuo Y, et al. MALAT-1 enhances cell motility of lung adenocarcinoma cells by influencing the expression of motility-related genes. FEBS Lett November 19, 2010;584(22):4575–80.
[25] Tian X, Xu G. Clinical value of lncRNA MALAT1 as a prognostic marker in human cancer: systematic review and meta-analysis. BMJ Open September 30, 2015;5(9):e008653.
[26] Wu Y, Zhang L, Wang Y, Li H, Ren X, Wei F, et al. Long non-coding RNA HOTAIR involvement in cancer. Tumour Biol. J. Int. Soc. Oncodev. Biol. Med October 2014;35(10):9531–8.
[27] Endo H, Shiroki T, Nakagawa T, Yokoyama M, Tamai K, Yamanami H, et al. Enhanced expression of long non-coding RNA HOTAIR is associated with the development of gastric cancer. PLoS One 2013;8(10):e77070.
[28] Kogo R, Shimamura T, Mimori K, Kawahara K, Imoto S, Sudo T, et al. Long noncoding RNA HOTAIR regulates polycomb-dependent chromatin modification and is associated with poor prognosis in colorectal cancers. Cancer Res October 15, 2011;71(20):6320–6.
[29] Yu X, Zheng H, Chan MTV, Wu WKK. HULC: an oncogenic long non-coding RNA in human cancer. J. Cell Mol. Med. 2017;21(2):410–7.
[30] Wang J, Liu X, Wu H, Ni P, Gu Z, Qiao Y, et al. CREB up-regulates long non-coding RNA, HULC expression through interaction with microRNA-372 in liver cancer. Nucleic Acids Res September 2010;38(16):5366–83.
[31] Guttman M, Donaghey J, Carey BW, Garber M, Grenier JK, Munson G, et al. lincRNAs act in the circuitry controlling pluripotency and differentiation. Nature August 28, 2011;477(7364): 295–300.

[32] Nayak S, Herzog RW. Progress and prospects: immune responses to viral vectors. Gene Ther March 2010;17(3):295–304.
[33] Schott JW, Morgan M, Galla M, Schambach A. Viral and synthetic RNA vector technologies and applications. Mol. Ther. J. Am. Soc. Gene. Ther. 2016;24(9):1513–27.
[34] Chira S, Jackson CS, Oprea I, Ozturk F, Pepper MS, Diaconu I, et al. Progresses towards safe and efficient gene therapy vectors. Oncotarget October 13, 2015;6(31):30675–703.
[35] Fernandez-Piñeiro I, Badiola I, Sanchez A. Nanocarriers for microRNA delivery in cancer medicine. Biotechnol. Adv. June 2017;35(3):350–60.
[36] Zhang T, Xue X, He D, Hsieh J-T. A prostate cancer-targeted polyarginine-disulfide linked PEI nanocarrier for delivery of microRNA. Cancer Lett September 1, 2015;365(2):156–65.
[37] Ibrahim AF, Weirauch U, Thomas M, Grünweller A, Hartmann RK, Aigner A. MicroRNA replacement therapy for miR-145 and miR-33a is efficacious in a model of colon carcinoma. Cancer Res August 1, 2011;71(15):5214–24.
[38] Schade A, Delyagina E, Scharfenberg D, Skorska A, Lux C, David R, et al. Innovative strategy for microRNA delivery in human mesenchymal stem cells via magnetic nanoparticles. Int. J. Mol. Sci. May 23, 2013;14(6):10710–26.
[39] Gao S, Tian H, Guo Y, Li Y, Guo Z, Zhu X, et al. miRNA oligonucleotide and sponge for miRNA-21 inhibition mediated by PEI-PLL in breast cancer therapy. Acta Biomater October 2015;25:184–93.
[40] Xue HY, Guo P, Wen W-C, Wong HL. Lipid-based nanocarriers for RNA delivery. Curr. Pharmaceut. Des. 2015;21(22):3140–7.
[41] Slaby O, Laga R, Sedlacek O. Therapeutic targeting of non-coding RNAs in cancer. Biochem. J. December 14, 2017;474(24):4219–51.
[42] Shah MY, Ferrajoli A, Sood AK, Lopez-Berestein G, Calin GA. microRNA therapeutics in cancer – an emerging concept. EBioMedicine October 2016;12:34–42.
[43] Krützfeldt J, Kuwajima S, Braich R, Rajeev KG, Pena J, Tuschl T, et al. Specificity, duplex degradation and subcellular localization of antagomirs. Nucleic Acids Res 2007;35(9):2885–92.
[44] Vigorito E, Kohlhaas S, Lu D, Leyland R. miR-155: an ancient regulator of the immune system. Immunol. Rev. May 2013;253(1):146–57.
[45] Higgs G, Slack F. The multiple roles of microRNA-155 in oncogenesis. J. Clin. Bioinform. September 28, 2013;3(1):17.
[46] Obad S, dos Santos CO, Petri A, Heidenblad M, Broom O, Ruse C, et al. Silencing of microRNA families by seed-targeting tiny LNAs. Nat. Genet. March 20, 2011;43(4):371–8.
[47] Tay FC, Lim JK, Zhu H, Hin LC, Wang S. Using artificial microRNA sponges to achieve microRNA loss-of-function in cancer cells. Adv. Drug Deliv. Rev. January 2015;81:117–27.
[48] Ebert MS, Sharp PA. MicroRNA sponges: progress and possibilities. RNA N Y N November 2010;16(11):2043–50.
[49] Meng S, Zhou H, Feng Z, Xu Z, Tang Y, Li P, et al. CircRNA: functions and properties of a novel potential biomarker for cancer [Internet] Mol Cancer May 23, 2017 [cited 2018 Dec 5];16. Available from: https://www.ncbi.nlm.nih.gov/pmc/articles/PMC5440908/.
[50] van Rooij E, Kauppinen S. Development of microRNA therapeutics is coming of age. EMBO Mol. Med. July 2014;6(7):851–64.
[51] Nobrega MA, Ovcharenko I, Afzal V, Rubin EM. Scanning human gene deserts for long-range enhancers. Science October 17, 2003;302(5644):413.
[52] Crooke ST. Molecular mechanisms of antisense oligonucleotides. Nucleic Acid Therapeut. 2017;27(2):70–7.
[53] Monia BP, Lesnik EA, Gonzalez C, Lima WF, McGee D, Guinosso CJ, et al. Evaluation of 2'-modified oligonucleotides containing 2'-deoxy gaps as antisense inhibitors of gene expression. J. Biol. Chem. July 5, 1993;268(19):14514–22.
[54] Wahlestedt C. Targeting long non-coding RNA to therapeutically upregulate gene expression. Nat. Rev. Drug Discov. June 2013;12(6):433–46.
[55] Zhu S, Li W, Liu J, Chen C-H, Liao Q, Xu P, et al. Genome-scale deletion screening of human long non-coding RNAs using a paired-guide RNA CRISPR-Cas9 library. Nat. Biotechnol. December 2016;34(12):1279–86.
[56] Thakore PI, D'Ippolito AM, Song L, Safi A, Shivakumar NK, Kabadi AM, et al. Highly specific epigenome editing by CRISPR-Cas9 repressors for silencing of distal regulatory elements. Nat Methods December 2015;12(12):1143–9.
[57] Abudayyeh OO, Gootenberg JS, Essletzbichler P, Han S, Joung J, Belanto JJ, et al. RNA targeting with CRISPR-Cas13. Nature 2017 12;550(7675):280–4.
[58] Cox DBT, Gootenberg JS, Abudayyeh OO, Franklin B, Kellner MJ, Joung J, et al. RNA editing with CRISPR-Cas13. Science 2017 24;358(6366):1019–27.
[59] Amit D, Hochberg A. Development of targeted therapy for bladder cancer mediated by a double promoter plasmid expressing diphtheria toxin under the control of H19 and IGF2-P4 regulatory sequences. J. Transl. Med. December 16, 2010;8:134.
[60] Li Z, Rana TM. Therapeutic targeting of microRNAs: current status and future challenges. Nat. Rev. Drug Discov. August 2014;13(8):622–38.
[61] Mogilyansky E, Rigoutsos I. The miR-17/92 cluster: a comprehensive update on its genomics, genetics, functions and increasingly important and numerous roles in health and disease. Cell Death Differ December 2013;20(12):1603–14.
[62] Pfeffer SR, Yang CH, Pfeffer LM. The role of miR-21 in cancer. Drug Dev. Res. September 2015;76(6):270–7.
[63] Mehlich D, Garbicz F, Włodarski PK. The emerging roles of the polycistronic miR-106b∼25 cluster in cancer – a comprehensive review. Biomed. Pharmacother. Biomedecine Pharmacother November 2018;107:1183–95.
[64] Lin K, Farahani M, Yang Y, Johnson GG, Oates M, Atherton M, et al. Loss of MIR15A and MIR16-1 at 13q14 is associated with increased TP53 mRNA, de-repression of BCL2 and adverse outcome in chronic lymphocytic leukaemia. Br. J. Haematol. November 2014;167(3):346–55.
[65] Roush S, Slack FJ. The let-7 family of microRNAs. Trends Cell Biol October 2008;18(10):505–16.
[66] Jiang H, Zhang G, Wu J-H, Jiang C-P. Diverse roles of miR-29 in cancer (review). Oncol. Rep. April 2014;31(4):1509–16.

[67] Saito Y, Nakaoka T, Saito H. microRNA-34a as a therapeutic agent against human cancer. J. Clin. Med. November 16, 2015;4(11):1951–9.
[68] Zeinali T, Mansoori B, Mohammadi A, Baradaran B. Regulatory mechanisms of miR-145 expression and the importance of its function in cancer metastasis. Biomed. Pharmacother. Biomedecine Pharmacother January 2019;109:195–207.
[69] Hu M, Xiong S, Chen Q, Zhu S, Zhou X. Novel role of microRNA-126 in digestive system cancers: from bench to bedside. Oncol. Lett. January 2019;17(1):31–41.
[70] Yu T, Ma P, Wu D, Shu Y, Gao W. Functions and mechanisms of microRNA-31 in human cancers. Biomed. Pharmacother. Biomedecine Pharmacother December 2018;108:1162–9.
[71] Monteleone NJ, Lutz CS. miR-708-5p: a microRNA with emerging roles in cancer. Oncotarget August 1, 2017;8(41):71292–316.
[72] Rupaimoole R, Slack FJ. MicroRNA therapeutics: towards a new era for the management of cancer and other diseases. Nat. Rev. Drug Discov. March 2017;16(3):203–22.

Chapter 21

Cancer-predisposing germline variants and childhood cancer

D.E. Sylvester[1], Y. Chen[1], R.V. Jamieson[2], L. Dalla-Pozza[3] and J.A. Byrne[1]

[1]*Children's Cancer Research Unit, Kids Research and Discipline of Child and Adolescent Health, Faculty of Medicine and Health, University of Sydney, Westmead, NSW, Australia;* [2]*The Children's Hospital at Westmead & Children's Medical Research Institute, Eye & Developmental Genetics Research Group, Westmead, NSW, Australia;* [3]*The Cancer Centre for Children, The Children's Hospital at Westmead, Westmead, NSW, Australia*

Context

Cancer is a genetic disease initiated by the accumulation of deleterious gene variants, altering key cellular pathways and causing oncogenic transformation [1]. This process usually occurs over years, and as such cancer is typically a disease that manifests in the older adult [2]. Cancer in children and younger adults has generally been considered to be a rare sporadic event. However, there is recognition that younger cancer patients may harbor a germline cancer-predisposing genetic alteration that is responsible for, or contributes to, early onset malignancy [3].

Germline genetic variants, which are constitutional and typically present in every cell of the body, are known to predispose to specific cancer types in children [4]. Germline variants that have a deleterious impact on cell function are referred to as pathogenic. For example, carriers of pathogenic germline variants in *RB1* are at high risk of developing an eye cancer in early infancy called retinoblastoma [5]. In contrast, other cancer types more commonly diagnosed in children, such as acute lymphoblastic leukemia, are infrequently associated with a known genetic predisposition [6].

Traditionally, genetic testing for pathogenic germline variants has been arduous, resource intensive, and consequently expensive. In childhood cancer patients, referrals for genetic testing have generally been restricted to those with recognized genotype–phenotype associations, where diagnosis of the molecular genetic variant could be of clinical benefit [4]. With the increasing availability and affordability of next-generation sequencing techniques that allow multiple genes to be sequenced in parallel, a greater proportion of childhood cancer patients can now be considered for germline sequencing to detect pathogenic germline variants.

Precision medicine in pediatric oncology can include germline genomic analysis, as an underlying genetic predisposition to cancer may influence treatment responses and long-term preventative care [7]. Genomic analyses in cancer patients primarily focused on molecular profiling of the tumor to detect therapeutically targetable somatic variants in the cancer genome [8], and matched patient germline samples were initially sequenced to distinguish targetable somatic events from germline variants [9]. Interestingly, the presence of some germline variants can be inferred from tumor analyses, even without direct sequencing of the germline DNA [10]. For example, chromothripsis of the tumor genome, where a large number of structural rearrangements occur after chromosomal 'shattering', could indicate a germline *TP53* variant responsible for Li–Fraumeni syndrome [11]. Similarly a hypermutated tumor, with many somatic variants demonstrating defective DNA mismatch repair, could indicate an underlying syndrome resulting in constitutional mismatch repair deficiency (CMMRD) [12]. The importance of specific germline sequencing and analysis in childhood cancer patients is increasingly apparent and is becoming incorporated into precision medicine pediatric oncology trials [13].

Genetic predisposition and childhood cancer

It has long been recognized that specific cancer types diagnosed in children can be attributed to an underlying genetic susceptibility and that children with particular genetic disorders are at greater risk of developing cancer [14]. In 1971, the likely inheritance of a pathogenic germline variant was described to contribute to the onset of retinoblastoma in infancy [15], with the associated *RB1* gene identified in 1986 [16]. In 1991, Narod et al. estimated that 4% of childhood cancer patients were likely to have a genetic predisposition

Precision Medicine for Investigators, Practitioners and Providers. https://doi.org/10.1016/B978-0-12-819178-1.00021-6

to cancer, based on the number of childhood cancer cases with known genetic conditions [17]. Till date, many cancer predisposition genes have been identified, with more cancer-predisposing genes being expected to be discovered through germline genomic analysis of a growing number of childhood cancer patients [18]. With the advent of next-generation sequencing, it is now considered that up to 10% of children with cancer (age range of 0–19 years) will carry a pathogenic germline variant in a cancer predisposition gene [19,20]. The most common childhood cancer predisposition syndromes with their associated cancer predisposition genes and cancer types are listed in Table 21.1.

Childhood cancer patients known to have a genetic predisposition to cancer either carry a heterozygous pathogenic germline variant in one copy of an autosomal dominant cancer predisposition gene or homozygous or biallelic pathogenic germline variants affecting both copies of an autosomal recessive cancer predisposition gene (Fig. 21.1, Table 21.1). Heterozygous pathogenic variants dysregulate cellular function by altering one of the two copies of a gene, causing a dosage effect that affects a proportion of the translated protein. In contrast, homozygous or biallelic pathogenic variants disrupt both copies of the gene, producing significant or complete protein dysregulation [21]. Onset of malignancy occurs when a predisposed cell acquires a sufficient number of somatic pathogenic gene variants [22]. It is estimated that the acquisition of 2–8 pathogenic gene variants is required for tumorigenesis in humans [2].

Clinical recommendations surrounding therapy, surveillance, and genetic counseling have been developed for childhood cancer patients with pathogenic germline variants where there is a known genotype–phenotype association [18]. Approximately half of the childhood cancer patients (age range 0–26 years) who carried a pathogenic germline variant in a cancer predisposition gene showed a clear association between their cancer type and the germline finding [23]. Most of the remaining childhood cancer patients carried a heterozygous pathogenic germline variant in a gene typically associated with onset of cancer in adulthood [23]. For childhood cancer patients with discordant genotype–phenotype associations, more research is required to understand whether the germline finding is clinically significant or not associated with the development of cancer in childhood.

Indications for next-generation germline sequencing

The suspicion of genetic predisposition to childhood cancer is typically based on the diagnosed cancer type having a known association with pathogenic germline variants and/or the phenotypic features of the child being indicative of an underlying genetic disorder (Table 21.1) [24]. Children diagnosed with cancer may also be suspected of a genetic predisposition based on personal and/or family history and/or other clinical features (Fig. 21.2) [24,25]. For example, children diagnosed with the genetic syndrome ataxia telangiectasia (AT), caused by biallelic pathogenic variants in the gene *ATM*, have a progressive neurodegenerative disorder typically diagnosed early in childhood, and children diagnosed with AT are at high risk of developing hematological malignancies (Table 21.1) [26]. Germline analysis of 40 childhood cancer patients selected for features suggestive of an underlying genetic predisposition detected 8 (20%) with pathogenic germline variants [27], indicating that it may be clinically useful to select childhood cancer patients suspected of genetic predisposition for germline sequencing analysis.

Although features indicative of genetic predisposition in a childhood cancer patient may lead to germline sequencing, it is important to note that an underlying pathogenic germline variant can also be found in children without indicative features [19]. For some childhood carriers of pathogenic germline variants in cancer predisposition genes, there may be no non–cancer-related phenotype, and thus the onset of cancer can be the first indication of an underlying genetic aberration in the child. As a prominent example, at least 50% of children diagnosed with adrenocortical carcinoma (ACC) are found to carry a pathogenic germline variant in the gene *TP53,* and this may consequently be the first evidence of Li–Fraumeni syndrome (Table 21.1) [28]. Pathogenic germline variants can also be de novo in origin, where the gene variant is not inherited from a parent but occurs within a germ cell or during embryogenesis. For example, at least 14% of patients diagnosed with Li–Fraumeni syndrome are expected to carry a de novo *TP53* pathogenic germline variant and thus have no family history [29]. Pathogenic germline variants that are inherited can also exhibit incomplete levels of penetrance and variable phenotypes in familial carriers [30]. While 9 out of 10 childhood carriers of highly penetrant pathogenic germline variants in the *RB1* gene will develop retinoblastoma [31], approximately only 3 out of 10 childhood carriers of pathogenic *WT1* germline variants will develop Wilms tumor (Table 21.1) [30]. Despite the evidence that childhood cancer patients who carry a pathogenic germline variant do not necessarily exhibit features indicative of a genetic predisposition, the often limited availability of resources in the clinical setting may prioritize children with suspected cancer predisposition for genetic testing [27].

Pathogenic germline variants

Cancer predisposition genes have been associated with a range of different cellular functions [32]. The majority of cancer-predisposing variants are in tumor suppressor genes, which are normally responsible for controlling cell proliferation. For example, pathogenic variants in *NF1* lead to reduced neurofibromin function, causing increased RAS activity and subsequent proliferation, with carriers of

TABLE 21.1 Childhood cancer predisposition syndromes with associated cancer types, associated genes, inheritance patterns, and phenotypic syndrom features.

Childhood cancer predisposition syndrome	Associated childhood cancer(s)	Gene(s) or genetic features	Inheritance patterns	Other major phenotypic syndrom feature(s)
Ataxia telangiectasia	Leukemia, lymphoma	*ATM*	AR[a]	Ataxia, telangiectasias, immunodeficiency
Beckwith–Wiedemann syndrome	Wilms, hepatoblastoma	*CDKN1C*, *H19*, *IGF2*, 11p15.5 imprinting	AD[b], complex	Overgrowth, macroglossia
Bloom syndrome	Leukemia, lymphoma, gastrointestinal	*BLM*	AR	Short stature, immunodeficiency
CBL syndrome	Leukemia	*CBL*	AD	Dysmorphism, short neck
Constitutional mismatch repair deficiency	Leukemia, lymphoma, brain, gastrointestinal	*EPCAM*, *MLH1*, *MSH2*, *MSH6*, *PMS2*	AR	Café au lait spots
Costello and Noonan syndromes	Leukemia	*HRAS*, *KRAS*, *NRAS*, *PTPN11*	AD	Dysmorphism, short stature
Denys–Drash syndrome	Wilms, gonadoblastoma	*WT1*	AD	Renal disease, disorders of sex development
Diamond–Blackfan anemia	Leukemia, sarcoma	*RPL5*, *RPL11*, *RPS7*, *RPS19*	AD	Macrocytic anemia, congenital abnormalities
Down syndrome	Leukemia	Trisomy 21	N/A[c]	Developmental delay, dysmorphism, congenital heart defect
Dyskeratosis congenita	Leukemia	*CTC1*, *DKC1*, *NOLA3*, *TERC*, *TERT*	X-linked, AD, AR	Nail dystrophy, abnormal skin pigmentation
Familial adenomatous polyposis syndrome	Gastrointestinal, brain	*APC*	AD	Polyposis
Familial leukemia syndromes	Leukemia	*CEBPA*, *ETV6*, *GATA2*, *PAX5*, *RUNX1*, *SAMD9*	AD, AR	Thrombocytopenia
Familial neuroblastoma	Neuroblastoma	*ALK*, *PHOX2B*	AD	None
Fanconi anemia	Leukemia, brain	*BRCA2*, *FANCA-E*, *RAD51D*	AR	Bone marrow failure, absent radius
Gorlin syndrome	Brain, skin	*PTCH1*, *PTCH2*, *SUFU*	AD	Macrocephaly
Hereditary paraganglioma/pheochromocytoma syndrome	Paraganglioma, pheochromocytoma	*SDHA*, *SDHB*, *SDHC*, *SDHD*	AD	None
Hereditary retinoblastoma	Retinoblastoma, sarcoma	*RB1*	AD	None
Juvenile polyposis syndrome	Gastrointestinal	*BMPR1A*, *SMAD4*	AD	Juvenile polyps
Li–Fraumeni syndrome	Leukemia, brain, sarcoma, adrenocortical carcinoma	*TP53*	AD	None
Lymphoproliferative syndrome (IL2-inducible T-cell kinase deficiency)	Lymphoma	*ITK*	AR	Immunodeficiency
Multiple endocrine neoplasia type 1/2	Thyroid	*MEN1*, *RET*	AD	None

Continued

TABLE 21.1 Childhood cancer predisposition syndromes with associated cancer types, associated genes, inheritance patterns, and phenotypic syndrom features.—cont'd

Childhood cancer predisposition syndrome	Associated childhood cancer(s)	Gene(s) or genetic features	Inheritance patterns	Other major phenotypic syndrom feature(s)
Neurofibromatosis type 1/2	Leukemia, brain, schwannoma	*NF1,NF2*	AD	Café au lait spots, Lisch nodules
Nijmegen breakage syndrome	Lymphoma, brain	*NBN (NBS1)*	AR	Microcephaly, bone marrow failure
Peutz–Jeghers syndrome	Gastrointestinal	*STK11*	AD	Polyps, oral hyperpigmentation
Pleuropulmonary blastoma familial tumor predisposition syndrome	Pleuropulmonary blastoma, sarcoma	*DICER1*	AD	Pulmonary cysts
Rhabdoid tumor predisposition syndrome	Rhabdoid tumors	*SMARCA4, SMARCB1*	AD	None
Rubinstein–Taybi syndrome	Leukemia, lymphoma, brain, sarcoma	*CREBBP*	AD	Short stature, developmental delay
Shwachman–Diamond syndrome	Leukemia	*SBDS*	AR	Pancytopenia
Severe congenital neutropenia	Leukemia	*ELANE, HAX1*	AD, AR	Neutropenia
WAGR (Wilms tumor, aniridia, genitourinary anomalies, retardation) syndrome	Wilms, gonadoblastoma	*WT1*	AD	Aniridia, developmental delay, genitourinary abnormalities
Wiskott–Aldrich syndrome	Lymphoma, leukemia	*WAS*	X-linked	Immunodeficiency
X-linked lymphoproliferative syndrome	Lymphoma	*SH2D1A, XIAP*	X-linked	Dysgammaglobulinemia

[a]*AR, autosomal recessive.*
[b]*AD, autosomal dominant.*
[c]*N/A, not applicable.*

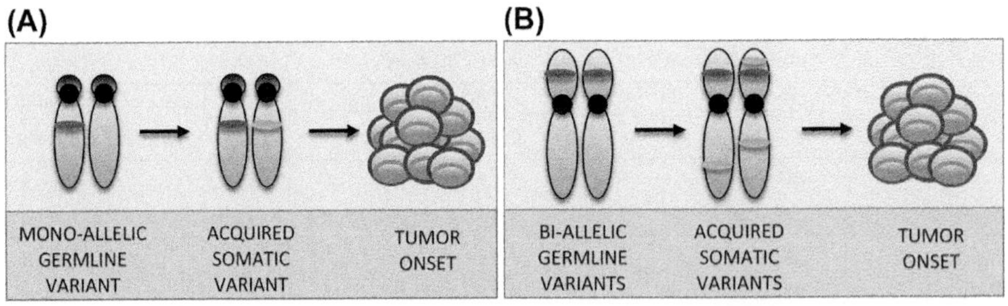

FIGURE 21.1 Pathogenic germline (blue) (dark gray in print version) variants predisposing to childhood cancer occur either (A) in one allele (monoallelic) or (B) in both alleles (biallelic). An accumulation of acquired somatic (yellow) (gray in print version) variants is then required for tumorigenesis. (A) For example, a cell with a pathogenic germline variant in the autosomal dominant gene *RB1* acquires a pathogenic somatic *RB1* variant on the second allele, leading to onset of retinoblastoma. (B) For example, a cell with pathogenic variants in both alleles of the autosomal recessive gene *MSH2* acquires somatic pathogenic variants in other cancer genes (not necessarily on the same chromosome), leading to onset of lymphoma.

pathogenic *NF1* germline variants being at increased risk of developing optic pathway glioma (Table 21.1) [33]. Gain of function germline variants can occur in oncogenes that typically stimulate cell division or prevent cellular differentiation. For example, pathogenic *ALK* variants lead to ALK kinase activation, and carriers of germline pathogenic *ALK* variants are at increased risk of developing neuroblastoma (Table 21.1) [34,35].

Pathogenic germline variants detected in childhood cancer patients can vary in the types of genetic aberrations and their inheritance patterns. Structurally, pathogenic germline variants can range from single nucleotide base substitutions, deletions, or insertions that lead to alternate coding of amino acids and/or premature stop codons to copy number changes or rearrangements of chromosomes, where many genes may be affected [36]. Pathogenic germline

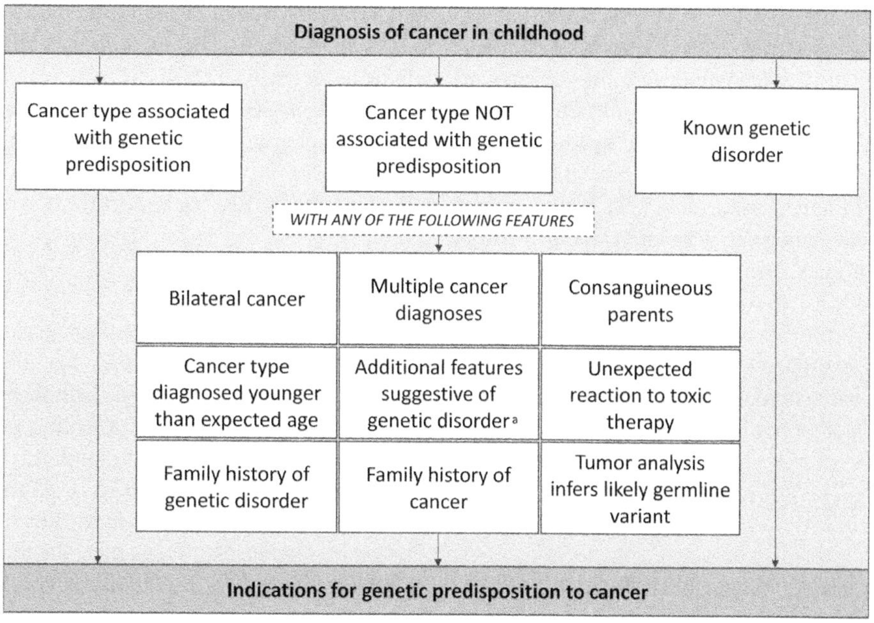

FIGURE 21.2 Features in childhood cancer patients may indicate an underlying genetic predisposition to cancer. [a]For example, aberrant skin pigmentation or congenital anomalies.

variants can be inherited in an autosomal dominant or autosomal recessive pattern, or variants can be X-linked (Table 21.1) [32].

Identifying pathogenic germline variants with next-generation sequencing relies on complex bioinformatic processing and analyses to detect the wide spectrum of possible genomic aberrations [37]. Variants must undergo strict quality control and filtering criteria, with comparison to a reference genome and comprehensive genomic databases [37]. Pathogenicity interpretation of the detected variants is crucial to assessing the possible clinical relevance of the genotype–phenotype associations [38]. Functional consequence and conservation scores of the variants can predict impact on protein function, yet further biological and pedigree evidence may be required to prove pathogenicity or a causal role in cancer development.

Relevance for solid and hematological malignancies

The proportion of childhood cancer patients carrying germline variants in cancer predisposition genes will vary according to cancer type. Some cancer types such as retinoblastoma and ACC can have high rates of detection of an underlying pathogenic variant [5,28]. In contrast, in the majority of children who develop acute lymphoblastic leukemia, the disease is not yet attributable to an underlying germline aberration [39]. Overall, analysis of six studies employing next-generation germline sequencing techniques and investigating a broad spectrum of cancer predisposition genes found that pathogenic germline variants are more frequently detected in childhood cancer patients diagnosed with solid tumors as opposed to hematological malignancies [23]. Germline pathogenic variants were detected in 15.5% of 534 childhood cancer patients diagnosed with non–central nervous system (CNS) solid tumors and in 10.9% of 358 childhood cancer patients diagnosed with CNS tumors [23]. In contrast, only 4.5% of 649 childhood cancer patients diagnosed with hematological malignancies carried pathogenic germline variants [23].

Although childhood cancer patients most commonly develop acute leukemia [40], childhood leukemia patients also have the lowest proportion of identified pathogenic germline variants [23]. Hematopoietic stem cells undergo a high lifetime rate of cell division compared with other tissues [41]. Hematopoietic cells are therefore prone to acquiring somatic genetic variants that can give rise to premalignant clones that are increasingly detected with age [42]. Preleukemic clones can be detected in approximately 1% of newborn cord blood samples, but only 1% of infants with detectable preleukemic cell populations will go on to develop childhood leukemia [43,44]. The sequential acquisition of somatic pathogenic variants in hematopoietic cells is therefore likely to drive tumorigenesis in the majority of childhood leukemia patients.

Pathogenic germline variants have been detected most frequently in children diagnosed with non-CNS solid tumors [23]. Some of the most commonly diagnosed non-CNS solid tumor diagnoses in childhood are sarcomas (for example, osteosarcoma and rhabdomyosarcoma), Wilms tumor, and neuroblastoma [19]. Genetic predisposition to osteosarcoma is mainly associated with Li–Fraumeni and hereditary retinoblastoma syndromes [31,45], and predisposition to rhabdomyosarcoma is also associated with Li–Fraumeni syndrome (Table 21.1) [45]. Pathogenic germline variants in *ALK* and *PHOX2B* have been associated with familial neuroblastoma (Table 21.1) [31]. Other genetic syndromes are known to predispose to neuroblastoma such as Beckwith–Wiedemann and Li–Fraumeni syndromes [31]. Wilms tumor onset is associated with pathogenic germline variants in *WT1* and with other syndromes such as Beckwith–Wiedemann syndrome (Table 21.1) [46]. In contrast to hematopoietic stem cells, tumor-initiating cells associated with solid tumors of childhood may be less prone to acquiring somatic variants. Pathogenic germline variants could therefore contribute to the development of a greater proportion of pediatric solid tumors.

Germline variants associated with adult onset cancer

Next-generation sequencing has enabled childhood cancer patients to be investigated for germline variants in a broad spectrum of cancer predisposition genes. Research studies have identified pathogenic germline variants not only in concordant genotype–phenotype associations but also in genes not previously associated with the cancer diagnosis [23]. The broadening of cancer predisposition syndromes and associated cancer risks is likely to occur as more childhood cancer patients undergo germline sequencing analysis.

Genes in which pathogenic heterozygous germline variants predispose to adult cancer can be considered for germline analysis of childhood cancer patients, as biallelic pathogenic germline variants can predispose to childhood cancer [21]. For example, heterozygous pathogenic germline *BRCA2* variants are associated with the development of breast and ovarian cancers, whereas biallelic *BRCA2* variants cause the genetic syndrome Fanconi anemia, which is associated with a high risk of developing hematological malignancies [47]. Another example is *MSH6*, one of the genes responsible for Lynch syndrome. Heterozygous carriers of *MSH6* pathogenic variants are at increased risk of developing colorectal cancer in adulthood, whereas biallelic pathogenic *MSH6* germline variants cause CMMRD, which confers a risk of malignancies in childhood, particularly brain and colorectal tumors [48]. Other adult cancer-associated autosomal dominant gene variants, such as in *CHEK2*, have been detected when a broader spectrum of cancer predisposition genes have been considered for investigation in childhood cancer patients [13,27]. However, the significance of pathogenic heterozygous germline variants associated with the onset of adult cancer is yet to be understood in the context of childhood cancer. The incidence of childhood cancer is increased both in families known to have a predisposition syndrome associated with cancer in adulthood and in families with a strong family history of cancer, compared with the incidence in families without a history of cancer [49,50]. Nonetheless, cancer samples from children with germline pathogenic *BRCA2* variants have not demonstrated loss of heterozygosity at the *BRCA2* locus, as might be expected if BRCA2 was playing a causal role in tumorigenesis according to the "two-hit" hypothesis [51]. More complex genetic processes may instead be involved, such as polygenic determinants of cancer risk as described in young adult sarcoma patients [52].

Clinical implications

Identification of a predisposing germline variant in a childhood cancer patient allows the clinical team to make decisions regarding therapy. Childhood cancer patients diagnosed with cancer predisposition syndromes, such as Nijmegen breakage syndrome and AT, may be at risk of severe treatment toxicities [53,54]. At the same time, other cancer predisposition syndromes may be associated with favorable responses to specific therapies. For example, hypermutant tumors in patients with CMMRD have been found to respond to immune checkpoint inhibitors [55]. Pathogenic germline variants may also have implications beyond cancer diagnosis. The majority of cancer predisposition syndromes are associated with clinical features unrelated to the cancer diagnosis (Table 21.1) [32]. The spectrum of non–cancer-related manifestations can range from mild features such as skin pigmentations to more significant clinical phenotypes. For example, diagnosis of Wilms tumor with subsequent detection of an associated pathogenic germline *WT1* variant may lead to diagnosis of Denys–Drash syndrome with associated renal disease (Table 21.1) [56].

Ongoing surveillance can also be important in childhood cancer patients with cancer predisposition syndromes. Surveillance protocols are utilized in children diagnosed with genetic syndromes, where there is an identified risk of developing cancer. For example, children diagnosed with Beckwith–Wiedemann syndrome are at an increased risk of developing cancer, usually Wilms tumor or hepatoblastoma, and appropriate screening protocols are recommended [57]. Long-term surveillance of individuals with Li–Fraumeni syndrome can lead to early detection of tumors while the patient is still asymptomatic [58]. Identifying children with pathogenic germline variants at risk of developing cancer could lead to the improved utilization of surveillance and preventative measures [18]. Clinical surveillance recommendations for childhood patients carrying a pathogenic germline variant have been summarized for the major cancer types in Table 21.2. Penetrance and clinical presentation will vary between patients and syndromes; hence, the implementation and timing of clinical surveillance recommendations will depend on the individual and their circumstances [18].

With successful modified treatment regimens, there has been a reduction in the mortality of childhood cancer patients, resulting in a growing number of childhood cancer

TABLE 21.2 Clinical surveillance recommendations for childhood cancer types associated with childhood cancer predisposition syndromes.

Cancer types	Major clinical surveillance recommendations
Brain tumors [45,79–81]	Brain MRI[a] minimum annually Ophthalmologic assessment annually (for neurofibromatosis 1)
Gastrointestinal tumors [79,82]	Colonoscopy and endoscopy annually
Hepatoblastoma [46]	Abdominal ultrasound 3 monthly Serum alpha protein 3 monthly
Kidney tumors [46,81]	Renal ultrasound 3 monthly
Leukemia [83,84]	Baseline bone marrow aspirate Complete blood count analysis minimum annually Annual physical examination Complete metabolic profile annually
Lymphoma [79]	Abdominal ultrasound 6 monthly Complete blood count analysis 6 monthly
Neuroblastoma [31,85]	Abdominal and pelvic ultrasound minimum 6 monthly Chest X-ray minimum 6 monthly Assessment of urinary catecholamine metabolites minimum 6 monthly
Paraganglioma and pheochromocytoma [86]	Plasma and/or urine markers annually Complete blood count annually Whole body MRI biennial
Pleuropulmonary blastoma [87]	Chest CT[b] minimum annually
Retinoblastoma [31]	Ophthalmic examination minimum 3 monthly
Sarcomas [31,45]	Whole body MRI annually Abdominal and pelvic ultrasound annually Annual physical examination
Thyroid cancer [87,88]	Thyroid ultrasound annually Serum calcitonin level screening annually Thyroidectomy (high risk)

[a]*MRI, magnetic resonance imaging.*
[b]*CT, computed tomography.*

survivors [59]. The increasing number of childhood cancer survivors may have substantial chronic health burdens, including the development of second cancers [60]. Childhood cancer survivors carrying pathogenic germline variants have been found to be at significantly increased risk of developing a second cancer [61]. By identifying childhood cancer patients carrying pathogenic germline variants, cancer risk estimates and surveillance strategies may be applied to enable earlier detection and treatment of subsequent cancers.

Diagnosis of a germline variant in a childhood cancer patient can also be important from the family's perspective. Knowledge of a germline variant explaining the cancer diagnosis is often valued by parents, even where this information does not affect therapy or prognosis [62]. Furthermore, detection of a pathogenic germline variant in a child requires testing for parental status, which may lead to interventions for other carriers in the family. For example, a parent found to carry a *BRCA2* variant can be referred to a familial cancer clinic, where appropriate early detection screening strategies or other preventative measures for breast and ovarian cancer can be implemented [51]. Detection of a pathogenic germline variant in a child can also have implications for the parents who may be carriers, with possible associated reproductive or cancer risks, requiring genetic counseling [63].

Disease prevention

The incidence of childhood cancer continues to increase worldwide, with the highest prevalence of childhood cancer being reported within higher socioeconomic communities [40]. While positive lifestyle choices have reduced the incidence of cancers diagnosed in adulthood [64], no comparable high risk exposures or controllable environmental lifestyle factors have been identified for most cancer diagnoses in children [65]. Ionizing radiation can cause childhood cancer, with evidence of radiation-induced incidence of leukemia in atomic bomb survivors [66] and specific viruses such as Epstein–Barr virus being linked with lymphoma [67,68]. It has been proposed that an abnormal or immature immune system, possibly due to lack of infection in combination with underlying genetic susceptibilities, is associated with an increased risk of developing acute childhood leukemia [44].

Although some other risk factors have been linked with the onset of childhood cancer, the identification of childhood carriers of highly penetrant pathogenic germline variants could be the most effective strategy to prevent cancer in children. Implementing surveillance strategies in children identified to be at high risk of developing cancer has proven to be effective for early detection in children with genetic cancer predisposition syndromes, such as Beckwith–Wiedemann syndrome and hereditary retinoblastoma [57,69]. Research is required to determine the costs and benefits of identifying a broader spectrum of children at risk of developing cancer and to measure the implementation and success of surveillance techniques in reducing the incidence, mortality, and/or adverse long-term outcomes of childhood cancer. Ongoing research into novel preventative measures, or strategies for early detection of disease, will ideally result in the prevention, or improved outcomes, in an increasing proportion of childhood cancers.

Importance of multidisciplinary approaches

As germline genomic analysis becomes routine care in pediatric oncology, the integration of multidisciplinary expertise will be vital. A multidisciplinary team will be required to consent, analyze, interpret, and return clinically relevant genomic information to patients and their families (Fig. 21.3). Treating clinicians require a standardized screening tool for personal and family history to identify childhood cancer patients with phenotypic features indicative of a genetic susceptibility [70]. Geneticists and genetic counselors specialized in cancer genetics, in cooperation with oncologists, will also be required to introduce germline genomic testing to childhood cancer patients and their families, to obtain informed consent for all childhood patients undergoing germline analysis and to consider the clinically relevant genes to be prioritized for analysis based on the phenotype in individual cases [63]. Interpretation of genetic data by experienced molecular geneticists is also crucial, often with extended evidence required from parental samples and matched tumor genomic and pathological analyses. For example, in suspected CMMRD, immunohistochemical analysis of patient tissue could identify loss of mismatch repair proteins, and parental sequencing can confirm carrier status of a pathogenic germline variant. Research scientists are required to perform functional studies on variants of uncertain significance (VUS) and novel candidate germline variants and to explore possible

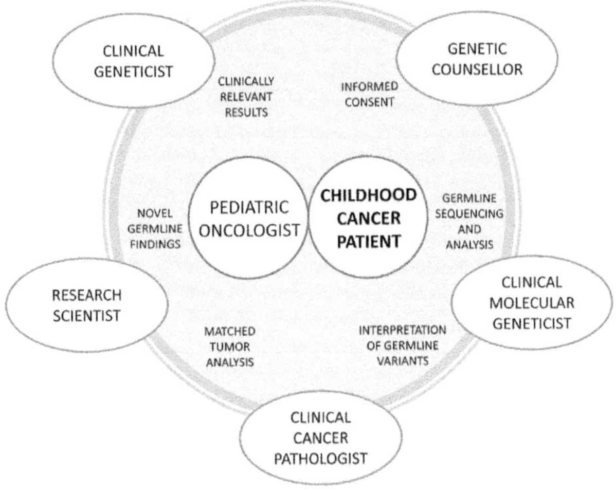

FIGURE 21.3 A multidisciplinary approach is required for the integration and ongoing investigation of germline variants predisposing children to cancer.

genotype–phenotype associations using databases reporting germline variants detected in childhood cancer patients. The return of genomic results in the pediatric oncology setting needs to be approached with sensitivity and will frequently require genetic testing and/or referral of immediate and possibly extended family members to a familial cancer clinic. In summary, a cohesive approach from all disciplines is critical for the successful integration of germline genetic testing into the pediatric oncology clinic.

Germline genetic testing in childhood patients also raises psychological and ethical issues [71]. Performing germline testing in children for a genetic predisposition to cancer raises concerns of psychological harm through distress and anxiety caused by relevant, uncertain, or secondary findings [72]. Traditionally, children with cancer have been offered single gene germline testing associated with their cancer diagnosis where a clear clinical benefit was predicted [4]. The introduction of next-generation sequencing has meant that germline analysis could reveal pathogenic gene variants that may be discordant with and/or completely unrelated to the clinical diagnosis. From an ethical standpoint, predictive or predisposition testing in children for adult onset diseases should not be performed where there is no medical benefit in testing at a young age and there are concerns of psychological harm [72]. Selecting only clinically relevant genes to investigate in germline genomic analyses of childhood cancer patients to avoid predictive testing is complicated by the fact that biallelic loss of genes can lead to onset of childhood cancer, whereas the heterozygous variant is associated with adult cancer. However, research studies investigating the integration of genetic testing within the pediatric oncology clinic have reported that the majority of families are supportive of germline analysis and the return of concordant and discordant results [62,73].

Ongoing lines of investigation and research opportunities

Current whole genome and whole exome sequencing techniques that are typically employed for the germline analysis of childhood cancer patients may not detect the full spectrum of genomic variations, because of the complexity of bioinformatically aligning short genomic reads to a reference genome [74]. It is anticipated that the generation of longer sequencing reads will increase capability to consistently detect a broader spectrum of genomic variations [75]. Long-range sequencing will be more accurate, identify haplotype information, improve the detection of large or complex structural variation, and allow de novo assembly of sequencing reads, with less reliance on a reference genome [76].

Improved genomic classification of VUS is also required. Rigorous assessment and functional studies of VUS detected in well-defined cancer predisposition genes, ideally in combination with detailed family history, could lead to the reclassification of many VUS as either likely pathogenic or likely benign [38]. Furthermore, functional studies and pedigree information of variants detected in novel candidate cancer predisposition genes could lead to the identification of new childhood cancer predisposition genetic syndromes. For example, the recently identified pathogenic germline *ETV6* variants have been studied in multiple pedigrees and found to segregate with thrombocytopenic disease and to predispose children to leukemia [77]. Functional studies of pathogenic germline *ETV6* variants demonstrate reduced or absent ETV6 nuclear localization, interfering with the normal function of nuclear transcriptional repression [77].

Although germline genomic analysis has detected childhood cancer patients who are carriers of pathogenic variants in cancer predisposition genes, no likely predisposing pathogenic germline variants are identified in the majority of patients. A proportion of these patients may carry a deleterious germline variant classified as a VUS or a deleterious variant in a gene yet to be identified as a cancer predisposition gene. The next generation of long-range sequencing techniques may offer increased accuracy and detection of pathogenic variants. Further research into RNA sequencing and expression, epigenetics, and/or polygenic interactions may also uncover novel germline variants in either known or new candidate childhood cancer predisposition genes. For example, aberrant methylation causes gene dysregulation, and methylation in the germline may be an underrecognized cancer predisposing feature, such as the recently identified constitutional *RB1* epimutations detected in retinoblastoma patients [78].

Finally, aggregation of data at an international level is required to determine clear definitions surrounding the relationship between childhood cancer genotypes and phenotypes. Aggregation and sharing of these data would allow research into novel genotype–phenotype associations, which would be expected to broaden the definitions of clinically relevant cancer predisposition genes in childhood cancer patients.

Summary and future directions

Historically, only childhood cancer patients with specific tumors and/or an underlying genetic syndrome were considered to have a genetic predisposition to cancer. New sequencing technologies have allowed a growing number of childhood cancer patients to be investigated for germline aberrations. As a result, pathogenic germline variants are detected in a growing number of childhood cancer patients, both in cancer predisposition genes with a concordant phenotype and also where there is genotype–phenotype discordance.

Expansion of germline genomic testing in pediatric oncology to all patients, with aggregation of data on an international scale, will better define the proportion of children with pathogenic germline variants and broaden the depth and spectrum of cancer predisposition syndromes. The integration of germline genomic analyses into the pediatric oncology clinic requires a cohesive multidisciplinary approach but is likely to be accepted by patient families. Analysis of the outcomes from surveillance and screening techniques employed in children with pathogenic germline variants should help to define the appropriate resources and protocols that are required to best care for these patients and their families.

Acknowledgments

We gratefully acknowledge financial support from Australian Lions Childhood Cancer Research Fund, Cancer Institute NSW through Kids Cancer Alliance Translational Research Center, Perpetual Foundation through the support of The Kids' Cancer Project, Balance Foundation, and donations to the Children's Cancer Research Unit.

References

[1] Hahn WC, Weinberg RA. Rules for making human tumor cells. N. Engl. J. Med. 2002;347(20):1593–603.

[2] Vogelstein B, Papadopoulos N, Velculescu VE, Zhou S, Diaz Jr LA, Kinzler KW. Cancer genome landscapes. Science 2013;339(6127):1546–58.

[3] Plon SE, Nathanson K. Inherited susceptibility for pediatric cancer. Cancer J. 2005;11(4):255–67.

[4] Strahm B, Malkin D. Hereditary cancer predisposition in children: genetic basis and clinical implications. Int. J. Cancer 2006;119(9):2001–6.

[5] Rodriguez-Galindo C, Orbach DB, VanderVeen D. Retinoblastoma. Pediatr. Clin. N. Am. 2015;62(1):201–23.

[6] Kratz CP, Stanulla M, Cave H. Genetic predisposition to acute lymphoblastic leukemia: overview on behalf of the I-BFM ALL host genetic variation working group. Eur. J. Med. Genet. 2016;59(3):111–5.

[7] Tran TH, Shah AT, Loh ML. Precision medicine in pediatric oncology: translating genomic discoveries into optimized therapies. Clin. Cancer Res. 2017;23(18):5329–38.

[8] Meyerson M, Gabriel S, Getz G. Advances in understanding cancer genomes through second-generation sequencing. Nat. Rev. Genet. 2010;11(10):685–96.

[9] Downing JR, Wilson RK, Zhang J, Mardis ER, Pui CH, Ding L, et al. The pediatric cancer genome project. Nat. Genet. 2012;44(6):619–22.

[10] Raymond VM, Gray SW, Roychowdhury S, Joffe S, Chinnaiyan AM, Parsons DW, et al. Germline findings in tumor-only sequencing: points to consider for clinicians and laboratories. J. Natl. Cancer Inst. 2016;108(4):djv351.

[11] Stephens PJ, Greenman CD, Fu B, Yang F, Bignell GR, Mudie LJ, et al. Massive genomic rearrangement acquired in a single catastrophic event during cancer development. Cell 2011;144(1):27–40.

[12] Shlien A, Campbell BB, de Borja R, Alexandrov LB. Combined hereditary and somatic mutations of replication error repair genes result in rapid onset of ultra-hypermutated cancers. Nat. Genet. 2015;47(3):257–62.

[13] Grobner SN, Worst BC, Weischenfeldt J, Buchhalter I, Kleinheinz K, Rudneva VA, et al. The landscape of genomic alterations across childhood cancers. Nature 2018;555(7696):321–7.

[14] Saletta F, Dalla Pozza L, Byrne JA. Genetic causes of cancer predisposition in children and adolescents. Transl. Pediatr. 2015;4(2):67–75.

[15] Knudson Jr AG. Mutation and cancer: statistical study of retinoblastoma. Proc. Natl. Acad. Sci. USA 1971;68(4):820–3.

[16] Friend SH, Bernards R, Rogelj S, Weinberg RA, Rapaport JM, Albert DM, et al. A human DNA segment with properties of the gene that predisposes to retinoblastoma and osteosarcoma. Nature 1986;323(6089):643–6.

[17] Narod SA, Stiller C, Lenoir GM. An estimate of the heritable fraction of childhood cancer. Br. J. Cancer 1991;63(6):993–9.

[18] Ripperger T, Bielack SS, Borkhardt A, Brecht IB, Burkhardt B, Calaminus G, et al. Childhood cancer predisposition syndromes-A concise review and recommendations by the cancer predisposition working group of the society for pediatric oncology and hematology. Am. J. Med. Genet. 2017;173(4):1017–37.

[19] Zhang J, Walsh MF, Wu G, Edmonson MN, Gruber TA, Easton J, et al. Germline mutations in predisposition genes in pediatric cancer. N. Engl. J. Med. 2015;373(24):2336–46.

[20] Parsons DW, Roy A, Yang Y, Wang T, Scollon S, Bergstrom K, et al. Diagnostic yield of clinical tumor and germline whole-exome sequencing for children with solid tumors. JAMA Oncol. 2016;2(5):616–24.

[21] Rahman N, Scott RH. Cancer genes associated with phenotypes in monoallelic and biallelic mutation carriers: new lessons from old players. Hum. Mol. Genet. 2007;16(Spec No 1):R60–6.

[22] Stratton MR, Campbell PJ, Futreal PA. The cancer genome. Nature 2009;458(7239):719–24.

[23] Sylvester DE, Chen Y, Jamieson RV, Dalla-Pozza L, Byrne JA. Investigation of clinically relevant germline variants detected by next-generation sequencing in patients with childhood cancer: a review of the literature. J. Med. Genet. 2018;55(12):785–93.

[24] Knapke S, Nagarajan R, Correll J, Kent D, Burns K. Hereditary cancer risk assessment in a pediatric oncology follow-up clinic. Pediatr. Blood Cancer 2012;58(1):85–9.

[25] Kuhlen M, Borkhardt A. Cancer susceptibility syndromes in children in the area of broad clinical use of massive parallel sequencing. Eur. J. Pediatr. 2015;174(8):987–97.

[26] McKinnon PJ. ATM and the molecular pathogenesis of ataxia telangiectasia. Annu. Rev. Pathol. 2012;7:303–21.

[27] Diets IJ, Waanders E, Ligtenberg MJ, van Bladel DAG, Kamping EJ, Hoogerbrugge PM, et al. High yield of pathogenic germline mutations causative or likely causative of the cancer phenotype in selected children with cancer. Clin. Cancer Res. 2018;24(7):1594–603.

[28] Wasserman JD, Novokmet A, Eichler-Jonsson C, Ribeiro RC, Rodriguez-Galindo C, Zambetti GP, et al. Prevalence and functional consequence of TP53 mutations in pediatric adrenocortical carcinoma: a children's oncology group study. J. Clin. Oncol. 2015;33(6):602–9.

[29] Renaux-Petel M, Charbonnier F, Thery JC, Fermey P, Lienard G, Bou J, et al. Contribution of de novo and mosaic TP53 mutations to Li-Fraumeni syndrome. J. Med. Genet. 2018;55(3):173–80.

[30] Taeubner J, Wieczorek D, Yasin L, Brozou T, Borkhardt A, Kuhlen M. Penetrance and expressivity in inherited cancer predisposing syndromes. Trends Cancer 2018;4(11):718–28.

[31] Kamihara J, Bourdeaut F, Foulkes WD, Molenaar JJ, Mosse YP, Nakagawara A, et al. Retinoblastoma and neuroblastoma predisposition and surveillance. Clin. Cancer Res. 2017;23(13):e98–106.

[32] Rahman N. Realizing the promise of cancer predisposition genes. Nature 2014;505(7483):302–8.

[33] Listernick R, Ferner RE, Liu GT, Gutmann DH. Optic pathway gliomas in neurofibromatosis-1: controversies and recommendations. Ann. Neurol. 2007;61(3):189–98.

[34] Mosse YP, Laudenslager M, Longo L, Cole KA, Wood A, Attiyeh EF, et al. Identification of ALK as a major familial neuroblastoma predisposition gene. Nature 2008;455(7215):930–5.

[35] Chen Y, Takita J, Choi YL, Kato M, Ohira M, Sanada M, et al. Oncogenic mutations of ALK kinase in neuroblastoma. Nature 2008;455(7215):971–4.

[36] den Dunnen JT. Describing sequence variants using HGVS nomenclature. Methods Mol. Biol. 2017;1492:243–51.

[37] Li MM, Datto M, Duncavage EJ, Kulkarni S, Lindeman NI, Roy S, et al. Standards and guidelines for the interpretation and reporting of sequence variants in cancer: a joint consensus recommendation of the association for molecular pathology, American society of clinical oncology, and college of American pathologists. J. Mol. Diagn. 2017;19(1):4–23.

[38] MacArthur DG, Manolio TA, Dimmock DP, Rehm HL, Shendure J, Abecasis GR, et al. Guidelines for investigating causality of sequence variants in human disease. Nature 2014;508(7497):469–76.

[39] Inaba H, Greaves M, Mullighan CG. Acute lymphoblastic leukaemia. Lancet 2013;381(9881):1943–55.

[40] Steliarova-Foucher E, Colombet M, Ries LAG, Moreno F, Dolya A, Bray F, et al. International incidence of childhood cancer, 2001–10: a population-based registry study. Lancet Oncol. 2017;18(6):719–31.

[41] Tomasetti C, Vogelstein B. Cancer etiology. Variation in cancer risk among tissues can be explained by the number of stem cell divisions. Science 2015;347(6217):78–81.

[42] Jaiswal S, Fontanillas P, Flannick J, Manning A, Grauman PV, Mar BG, et al. Age-related clonal hematopoiesis associated with adverse outcomes. N. Engl. J. Med. 2014;371(26):2488–98.

[43] Mori H, Colman SM, Xiao Z, Ford AM, Healy LE, Donaldson C, et al. Chromosome translocations and covert leukemic clones are generated during normal fetal development. Proc. Natl. Acad. Sci. USA 2002;99(12):8242–7.

[44] Greaves M. A causal mechanism for childhood acute lymphoblastic leukaemia. Nat. Rev. Cancer 2018;18(8):471–84.

[45] Kratz CP, Achatz MI, Brugieres L, Frebourg T, Garber JE, Greer MC, et al. Cancer screening recommendations for individuals with Li-Fraumeni syndrome. Clin. Cancer Res. 2017;23(11):e38–45.

[46] Kalish JM, Doros L, Helman LJ, Hennekam RC, Kuiper RP, Maas SM, et al. Surveillance recommendations for children with overgrowth syndromes and predisposition to Wilms tumors and hepatoblastoma. Clin. Cancer Res. 2017;23(13):e115–22.

[47] Howlett NG, Taniguchi T, Olson S, Cox B, Waisfisz Q, De Die-Smulders C, et al. Biallelic inactivation of BRCA2 in Fanconi anemia. Science 2002;297(5581):606–9.

[48] Menko FH, Kaspers GL, Meijer GA, Claes K, van Hagen JM, Gille JJ. A homozygous MSH6 mutation in a child with cafe-au-lait spots, oligodendroglioma and rectal cancer. Fam. Cancer 2004;3(2):123–7.

[49] Brooks GA, Stopfer JE, Erlichman J, Davidson R, Nathanson KL, Domchek SM. Childhood cancer in families with and without BRCA1 or BRCA2 mutations ascertained at a high-risk breast cancer clinic. Cancer Biol. Ther. 2006;5(9):1098–102.

[50] Magnusson S, Borg A, Kristoffersson U, Nilbert M, Wiebe T, Olsson H. Higher occurrence of childhood cancer in families with germline mutations in BRCA2, MMR and CDKN2A genes. Fam. Cancer 2008;7(4):331–7.

[51] Walsh MF, Kennedy J, Harlan M, Kentsis A, Shukla N, Musinsky J, et al. Germline BRCA2 mutations detected in pediatric sequencing studies impact parents' evaluation and care. Cold Spring Harb. Mol. Case Stud. 2017;3(6):a001925.

[52] Ballinger ML, Goode DL, Ray-Coquard I, James PA, Mitchell G, Niedermayr E, et al. Monogenic and polygenic determinants of sarcoma risk: an international genetic study. Lancet Oncol. 2016;17(9):1261–71.

[53] Distel L, Neubauer S, Varon R, Holter W, Grabenbauer G. Fatal toxicity following radio- and chemotherapy of medulloblastoma in a child with unrecognized Nijmegen breakage syndrome. Med. Pediatr. Oncol. 2003;41(1):44–8.

[54] Schoenaker MH, Suarez F, Szczepanski T, Mahlaoui N, Loeffen JL. Treatment of acute leukemia in children with ataxia telangiectasia (A-T). Eur. J. Med. Genet. 2016;59(12):641–6.

[55] Bouffet E, Larouche V, Campbell BB, Merico D, de Borja R, Aronson M, et al. Immune checkpoint inhibition for hypermutant glioblastoma multiforme resulting from germline biallelic mismatch repair deficiency. J. Clin. Oncol. 2016;34(19):2206–11.

[56] Niaudet P, Gubler MC. WT1 and glomerular diseases. Pediatr. Nephrol. 2006;21(11):1653–60.

[57] Maas SM, Vansenne F, Kadouch DJ, Ibrahim A, Bliek J, Hopman S, et al. Phenotype, cancer risk, and surveillance in Beckwith-Wiedemann syndrome depending on molecular genetic subgroups. Am. J. Med. Genet. 2016;170(9):2248–60.

[58] Villani A, Shore A, Wasserman JD, Stephens D, Kim RH, Druker H, et al. Biochemical and imaging surveillance in germline TP53 mutation carriers with Li-Fraumeni syndrome: 11 year follow-up of a prospective observational study. Lancet Oncol. 2016;17(9):1295–305.

[59] Armstrong GT, Chen Y, Yasui Y, Leisenring W, Gibson TM, Mertens AC, et al. Reduction in late mortality among 5-year survivors of childhood cancer. N. Engl. J. Med. 2016;374(9):833–42.

[60] Bhakta N, Liu Q, Ness KK, Baassiri M, Eissa H, Yeo F, et al. The cumulative burden of surviving childhood cancer: an initial report from the St Jude Lifetime Cohort Study (SJLIFE). Lancet 2017;390(10112):2569–82.

[61] Wang Z, Wilson CL, Easton J, Thrasher A, Mulder H, Liu Q, et al. Genetic risk for subsequent neoplasms among long-term survivors of childhood cancer. J. Clin. Oncol. 2018;36(20):2078–87.

[62] Malek J, Slashinski MJ, Robinson JO, Gutierrez AM, Parsons DW, Plon SE, et al. Parental perspectives on whole-exome sequencing in pediatric cancer: a typology of perceived utility. JCO Precis. Oncol. 2017;1(1):1–10.

[63] Druker H, Zelley K, McGee RB, Scollon SR, Kohlmann WK, Schneider KA, et al. Genetic counselor recommendations for cancer predisposition evaluation and surveillance in the pediatric oncology patient. Clin. Cancer Res. 2017;23(13):e91–7.

[64] Torre LA, Siegel RL, Ward EM, Jemal A. Global cancer incidence and mortality rates and trends—an update. Cancer Epidemiol. Biomarkers Prev. 2016;25(1):16–27.

[65] Cogliano VJ, Baan R, Straif K, Grosse Y, Lauby-Secretan B, El Ghissassi F, et al. Preventable exposures associated with human cancers. J. Natl. Cancer Inst. 2011;103(24):1827–39.

[66] Preston DL, Kusumi S, Tomonaga M, Izumi S, Ron E, Kuramoto A, et al. Cancer incidence in atomic bomb survivors. Part III. Leukemia, lymphoma and multiple myeloma, 1950-1987. Radiat. Res. 1994;137(2 Suppl):S68–97.

[67] Pallesen G, Hamilton-Dutoit SJ, Rowe M, Young LS. Expression of Epstein-Barr virus latent gene products in tumour cells of Hodgkin's disease. Lancet 1991;337(8737):320–2.

[68] Vockerodt M, Cader FZ, Shannon-Lowe C, Murray P. Epstein-Barr virus and the origin of Hodgkin lymphoma. Chin. J. Cancer 2014;33(12):591–7.

[69] Imhof SM, Moll AC, Schouten-van Meeteren AY. Stage of presentation and visual outcome of patients screened for familial retinoblastoma: nationwide registration in The Netherlands. Br. J. Ophthalmol. 2006;90(7):875–8.

[70] Jongmans MC, Loeffen JL, Waanders E, Hoogerbrugge PM, Ligtenberg MJ, Kuiper RP, et al. Recognition of genetic predisposition in pediatric cancer patients: an easy-to-use selection tool. Eur. J. Med. Genet. 2016;59(3):116–25.

[71] Wakefield CE, Hanlon LV, Tucker KM, Patenaude AF, Signorelli C, McLoone JK, et al. The psychological impact of genetic information on children: a systematic review. Genet. Med. 2016;18(8):755–62.

[72] Johnson LM, Hamilton KV, Valdez JM, Knapp E, Baker JN, Nichols KE. Ethical considerations surrounding germline next-generation sequencing of children with cancer. Expert Rev. Mol. Diagn. 2017;17(5):523–34.

[73] Brozou T, Taeubner J, Velleuer E, Dugas M, Wieczorek D, Borkhardt A, et al. Genetic predisposition in children with cancer – affected families' acceptance of Trio-WES. Eur. J. Pediatr. 2018;177(1):53–60.

[74] Kamps R, Brandao RD, Bosch BJ, Paulussen AD, Xanthoulea S, Blok MJ, et al. Next-generation sequencing in oncology: genetic diagnosis, risk prediction and cancer classification. Int. J. Mol. Sci. 2017;18(2):308.

[75] Shendure J, Balasubramanian S, Church GM, Gilbert W, Rogers J, Schloss JA, et al. DNA sequencing at 40: past, present and future. Nature 2017;550(7676):345–53.

[76] Pollard MO, Gurdasani D, Mentzer AJ, Porter T, Sandhu MS. Long reads: their purpose and place. Hum. Mol. Genet. 2018;27(R2):R234–41.

[77] Feurstein S, Godley LA. Germline ETV6 mutations and predisposition to hematological malignancies. Int. J. Hematol. 2017;106(2):189–95.

[78] Gelli E, Pinto AM, Somma S, Imperatore V, Cannone MG, Hadjistilianou T, et al. Evidence of predisposing epimutation in retinoblastoma. Hum. Mutat. 2019;40(2):201–6.

[79] Tabori U, Hansford JR, Achatz MI, Kratz CP, Plon SE, Frebourg T, et al. Clinical management and tumor surveillance recommendations of inherited mismatch repair deficiency in childhood. Clin. Cancer Res. 2017;23(11):e32–7.

[80] Evans DGR, Salvador H, Chang VY, Erez A, Voss SD, Schneider KW, et al. Cancer and central nervous system tumor surveillance in pediatric neurofibromatosis 1. Clin. Cancer Res. 2017;23(12):e46–53.

[81] Foulkes WD, Kamihara J, Evans DGR, Brugieres L, Bourdeaut F, Molenaar JJ, et al. Cancer surveillance in gorlin syndrome and rhabdoid tumor predisposition syndrome. Clin. Cancer Res. 2017;23(12):e62–7.

[82] Achatz MI, Porter CC, Brugieres L, Druker H, Frebourg T, Foulkes WD, et al. Cancer screening recommendations and clinical management of inherited gastrointestinal cancer syndromes in childhood. Clin. Cancer Res. 2017;23(13):e107–14.

[83] Porter CC, Druley TE, Erez A, Kuiper RP, Onel K, Schiffman JD, et al. Recommendations for surveillance for children with leukemia-predisposing conditions. Clin. Cancer Res. 2017;23(11):e14–22.

[84] Walsh MF, Chang VY, Kohlmann WK, Scott HS, Cunniff C, Bourdeaut F, et al. Recommendations for childhood cancer screening and surveillance in DNA repair disorders. Clin. Cancer Res. 2017;23(11):e23–31.

[85] Villani A, Greer MC, Kalish JM, Nakagawara A, Nathanson KL, Pajtler KW, et al. Recommendations for cancer surveillance in individuals with RASopathies and other rare genetic conditions with increased cancer risk. Clin. Cancer Res. 2017;23(12):e83–90.

[86] Rednam SP, Erez A, Druker H, Janeway KA, Kamihara J, Kohlmann WK, et al. Von Hippel-Lindau and hereditary pheochromocytoma/paraganglioma syndromes: clinical features, genetics, and surveillance recommendations in childhood. Clin. Cancer Res. 2017;23(12):e68–75.

[87] Schultz KAP, Rednam SP, Kamihara J, Doros L, Achatz MI, Wasserman JD, et al. PTEN, DICER1, FH, and their associated tumor susceptibility syndromes: clinical features, genetics, and surveillance recommendations in childhood. Clin. Cancer Res. 2017;23(12):e76–82.

[88] Wasserman JD, Tomlinson GE, Druker H, Kamihara J, Kohlmann WK, Kratz CP, et al. Multiple endocrine neoplasia and hyperparathyroid-jaw tumor syndromes: clinical features, genetics, and surveillance recommendations in childhood. Clin. Cancer Res. 2017;23(13):e123–32.

Chapter 22

Current status of cancer pharmacogenomics

Juan P. Cayún and Luis A. Quiñones

Laboratory of Chemical Carcinogenesis and Pharmacogenetics (CQF), Department of Basic and Clinical Oncology, Faculty of Medicine, University of Chile, Latin American Society of Pharmacogenomics and Personalized Medicine (SOLFAGEM) & Latin American Network for Implementation and Validation of Pharmacogenomic Clinical Guidelines (RELIVAF), Quinta Normal, Santiago, Chile

Introduction

The development of cancer genomics has significantly increased in recent years, providing insights into genome variations in both normal and tumor cells. Recent findings in cancer pharmacogenomics have provided tumor and host information for optimizing cancer treatment (Fig. 22.1). The 1000 Genomes Project has described human variations among the populations around the world. Phase 3 considered 2504 individuals from 26 populations, and showed over 88 million variants (84.7 million single-nucleotide polymorphisms [SNPs], 3.6 million insertions/deletions, and 60,000 structural variants) [1]. One person has approximately 4.1–5.0 million sites of variation across the genome, and >99.9% of these variations are SNPs and short indels [1].

Traditionally, pharmacogenomics of anticancer drugs has focused the correct dosage regimen, because of drug narrow therapeutic range, which is near the transition between efficacious and toxic dosages. Inter- and intra-individual variability in cancer therapy, and tumor specificity with sufficient efficacy in cancer, and few toxicity effects in normal cells are the other concerns. This scenario has required finding appropriate biomarkers for defining personalized treatments, comprising the right dose in the right patient at the right time [2–4]. The Clinical Pharmacogenetics Implementation Consortium (CPIC) clinical guidelines include recommendations for five anticancer drugs, namely thioguanine, mercaptopurine, tamoxifen, and 5-fluorouracil/capecitabine [5]. In addition, the Royal Dutch Association for the Advancement of Pharmacy – Pharmacogenetics Working Group (DPWG) clinical guidelines, has added tegafur and irinotecan [6]. The Food and Drug Administration (FDA) has listed more than 60 approved drugs with pharmacogenomic biomarkers in labeling information, for efficacy and toxicity [7] (Table 22.1).

Concepts related to cancer pharmacogenomics

The interindividual variability in drug responses has been a constant challenge in medicine. Frederick Vogel first sought to address this problem in 1959. He coined the term "pharmacogenetics" to describe genetic implications in drug response [10]. After the publication of the Human Genome Project, the term "pharmacogenomics" emerged in scientific reports. Pharmacogenomics refers to a wide genomic approach, as opposed to individual genes (pharmacogenetics) [11]; in fact, these terms are interchangeable. The identification of genes related to drug response includes the relevant genes from a pharmacokinetic (absorption, distribution, metabolization, and elimination), or pharmacodynamic perspective (effects on therapeutic target). The pharmacogenomic promise is optimizing drug therapies, providing minimal adverse effects, and maximum efficacy in each person [12,13].

Pharmacogenomics has been defined as a subset of precision medicine [12,14,15], including somatic and germline variations related to drug response [13]. Somatic mutations represent the genetic variations of tumor cells (tumorigenesis, microenvironment, drug resistance, etc.), and germline variations represent the inheritance variations of normal cells (susceptibility) [13,16].

Precision medicine in drug response requires finding appropriate biomarkers, to perform stratification and select the best therapeutic option. Biomarkers are variables (e.g., presence of a mutation, protein level) that are associated with a disease outcome (e.g., disease progression, death). Prognostic biomarkers are associated with a disease

Precision Medicine for Investigators, Practitioners and Providers. https://doi.org/10.1016/B978-0-12-819178-1.00022-8

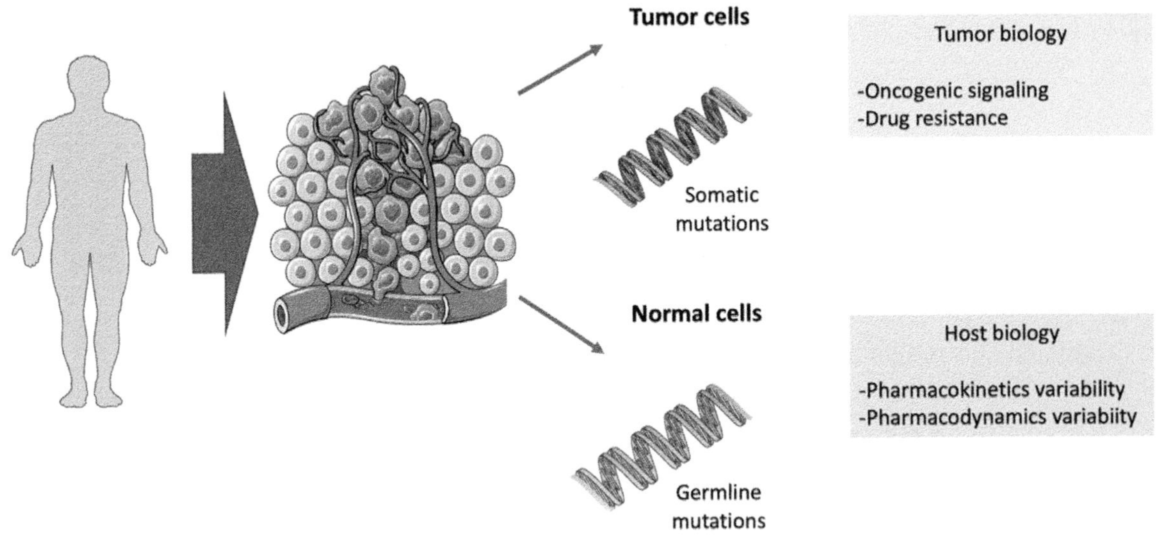

FIGURE 22.1 Overview of cancer pharmacogenomics.

outcome, independently of the treatment. In contrast, predictive biomarkers are variables related to the clinical effect of drugs, where this effect depends on the presence of the biomarker [17]. The gold standard setting comprises randomized clinical trials to demonstrate the existence of predictive biomarkers; however, a first approach could be the definition of prognostic biomarkers.

Germline mutations in cancer therapy

The clearest evidence of the usefulness of the germline variations is the therapeutic group of thiopurines. These anticancer drugs have an antileukemic effect after the conversion into thioguanine nucleotides (TGNs) and incorporation into DNA. For example, 6-mercaptopurine (6-MP) is metabolized to thioguanine nucleotides by hypoxanthine phosphoribosyl transferase (HPRT; Fig. 22.2), and it is used to treat acute lymphoblastic leukemia (ALL). However, 6-MP generates severe myelosuppression, along with the therapeutic effect. This toxicity is explained by the accumulation of TGNs in the cells, caused by low metabolization of 6-MP by the enzyme thiopurine methyltransferase (TPMT). Three polymorphisms in TPMT account for over 90% of relevant clinical variations. Ten percent of the population is heterozygous for these polymorphisms and requires dosage reductions [13,18]. CPIC recommends starting 6-MP treatment with 50% and 10% of the conventional dose, for TPMT intermediate and poor metabolizers, respectively. A similar recommendation has been published by the DPWG [6,19,20].

Another good example is 5-fluorouracil, a pyrimidine analog used mainly in breast and colorectal cancer, that is biotransformed by the enzyme dihydropyridine dehydrogenase (DPD) (>80%) into the inactive metabolite 5,6-dihydro-5-fluorouracil. DPD is found primarily in the liver and gastrointestinal tissue, and it has been identified as the major source of interpatient variability in 5-fluorouracil pharmacokinetics. The metabolite 5-fluoro-2-deoxyuridine monophosphate (5-FdUMP) causes inhibition of thymidylate synthase (TS), leading to low levels of pyrimidines, and exercising antitumor effects. DPD activity is highly variable between patients, and it significantly explains the differential toxicity (Fig. 22.3).

The main source of variability in DPD activity is the presence of genetic polymorphism in *DPYD* gene [13]. This variability is mainly explained by variants c.1905+1G>A, c.1679T>G, c.1236G>A/HapB3, c.1601G>A, and c.2846A>T. DPD phenotypes have been described in the literature, based on the presence of genetic polymorphisms. Normal DPD metabolizers are individuals carrying two normal function alleles (e.g., c.85T>C, *9A, rs1801265, p.C29R; c.1627A>G, *5, rs1801159, p.I543V; c.2194G>A, *6, rs1801160, p.V732I). Intermediate DPD metabolizers are individuals carrying one normal and one decreased function allele (c.2846A>T and c.1129−5923C>G), or two decreased function alleles. Poor DPD metabolizers are individuals carrying two nonfunctioning alleles (c.190511G>A and c.1679T>G) or one nonfunctioning allele plus one decreased function allele. CPIC and DPWG recommend the initial reduction of 50% of 5-fluorouracil for intermediate metabolizers, switching to an alternative drug for poor metabolizers [6,20,22].

The variability of irinotecan-related toxicity can be explained by germline variations. Irinotecan is a prodrug used mainly to treat lung and colorectal cancer, that requires to be converted to SN-38 through carboxylesterase (Fig. 22.4). SN-38 causes anticancer effects via inhibition

TABLE 22.1 Pharmacogenomic biomarkers in FDA approved targeted drugs.

Type of cancer	Biomarker	Detection methodology	Drug
Breast	ER+, PR+	IHC, LBA	Fulvestrant
	HER2+	ISH, IHC, NGS	Trastuzumab, lapatinib, pertuzumab, trastuzumab emtansine
Colorectal	KRAS wt /EGFR+	RT-PCR, IHC	Cetuximab, panitumumab
Head and neck	EGFR+	IHC	Cetuximab
NSCLC	ALK fusion, ALK+	FISH, IHC, NGS	Ceritinib, alectinib, crizotinib
	EGFR exon 19 del, EGFR exon 21 (L858R)	RT-PCR, NGS	Erlotinib, afatinib, gefitinib
	EGFR T790M	NGS	Osimertinib
	ROS1 +	NGS, IHC	Crizotinib
Ovarian	BRCA1/BRCA2 mutations	NGS, PCR	Olaparib, rucaparib
Melanoma	BRAFV600E, V600K	RT-PCR, NGS	Vemurafenib, dabrafenib, trametinib, cobimetinib
Gastrointestinal stromal tumor	KIT+	IHC	Imatinib
Gastric /Gastroesophageal junction	HER2 amplification +	ISH, IHC	Trastuzumab
Acute lymphoblastic leukemia/ lymphoblastic lymphoma	BCR-ABL1	Cytogenetics, FISH, RT-PCR	Imatinib, dasatinib, nilotinib
	BCR-ABL1, BCR-ABL1 T315I	Cytogenetics, FISH, RT-PCR	Ponatinib
Acute myeloid leukemia	FLT3_mutation	NGS	Midostaurin
Aggressive systemic mastocytosis	KIT D816V negative	PCR	Imatinib
Chronic lymphocytic leukemia	CD20 +	IHC	Rituximab
	Chromosome 17p deletion	FISH	Venetoclax
Chronic myelogenous leukemia	BCR-ABL1	Cytogenetics, FISH, RT-PCR	Bosutinib
	BCR-ABL1	Cytogenetics, FISH, RT-PCR	Dasatinib
	BCR-ABL1, BCR-ABL1 T315I	Cytogenetics, FISH, RT-PCR	Ponatinib
	FIP1L1-PDFGRA	NGS, FISH	Imatinib
Myelodysplastic/myeloproliferative neoplasm	PDGFR fusion	FISH	Imatinib
Myelodysplastic syndrome	Chromosome 5q deletion	FISH	Lenalidomide
Non-Hodgkin's lymphoma	CD-20 +	IHC	Rituximab
Small lymphocytic lymphoma	Chromosome 17p deletion	FISH	Ibrutinib
Waldenström macroglobulinemia	Chromosome 17p deletion	FISH	Ibrutinib
dMMR or MSI-H advanced cancer	dMMR, MSI-H	IHC, PCR	Pembrolizumab

dMMR, deficient mismatch repair system; *ER*, Estrogen receptor; *FDA*, Food and Drug Administration; *IHC*, Immunohistochemistry; *ISH*, In situ hybridization; *LBA*, Ligand-binding assay; *MSI-H*, microsatellite instability high; *NGS*, Next-generation sequencing; *NSCLC*, Non-small cancer lung cells; *PR*, Progesterone receptor; *RT-PCR*, reverse transcription polymerase chain reaction. Ref. [8,9].

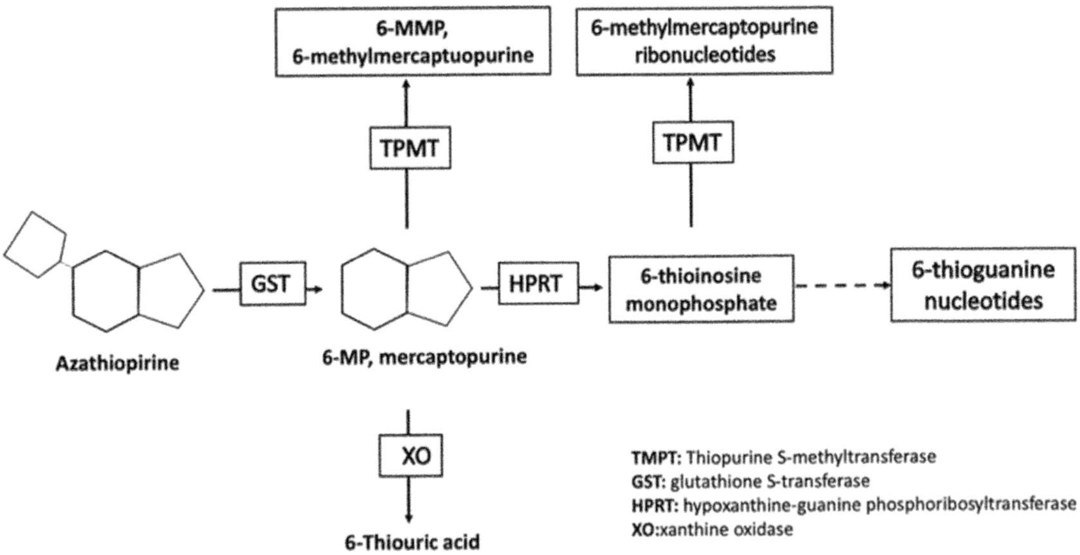

FIGURE 22.2 Simplified scheme for thiopurine metabolism. *Modified from Chouchana L, Roche D, Jian R, Beaune P, Loriot M-A. Poor response to thiopurine in inflammatory bowel disease: how to overcome therapeutic resistance? [Internet] Clin. Chem. July 1, 2013;59(7):1023–1026. Available from: http://www.clinchem.org/cgi/doi/10.1373/clinchem.2012.195750.*

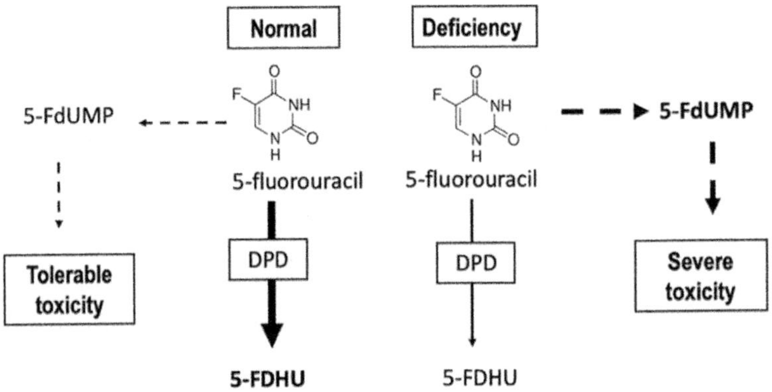

FIGURE 22.3 DPD-dependent inactivation of 5-FU and effects of DPD deficiency. *5-FU*, 5-fluorouracil; *5-FDHU*, 5-fluoro-5,6–dihydrouracil; *5-FdUMP*, 5-fluoro-deoxyuridine monophosphate; *DPD*, dihydropyrimidine dehydrogenase; *TP*, thymidine phosphorylase; *TS*, thymidylate synthase. *Modified from Del Re M, Di Paolo A, van Schaik RH, Bocci G, Simi P, Falcone A, et al. Dihydropyrimidine dehydrogenase polymorphisms and fluoropyrimidine toxicity: ready for routine clinical application within personalized medicine? [Internet] EPMA J. September 25, 2010;1(3):495–502. Available from: http://link.springer.com/10.1007/s13167-010-0041-2.*

of topoisomerase I, causing it to be inactivated by glucuronidation through UDP (uridine diphosphate glucose pyrophosphorylase)-glucuronosyltransferase 1A1 (UGT1A1), and eliminated in bile and urine. Interindividual variations in the UGT1A1 system can explain the increase of severe toxicity, diarrhea, and leukopenia, with the accumulation of SN-38. The UGT1A1 system includes at least 12 enzymes encoded by the *UGT1* locus on chromosome 2. The promoter region of *UGT1A1* contains a number of highly variable TA repeats, affecting the gene expression. Seven TA repeats (*UGT1A1**28 allele) affect the expression level, compared with 6 TA repeats in the wild-type genotype (*UGT1A1**1 allele) [13]. DPWG recommends reducing the initial dose by 30% in *UGT1A1**28/*28 patients, treated with doses higher than 250 mg/m^2, and then increasing the dose in response to neutrophil count [6].

CYP2D6 is an important pharmacogene with broad clinical applications in several drug regimens. Tamoxifen, a selective estrogen receptor modulator (SERM), is biotransformed by CYP2D6 to 4-hydroxytamoxifen and 4-hydroxy N-desmethyltamoxifen (endoxifen; Fig. 22.5). Patients with certain polymorphisms in CYP2D6 have low levels of endoxifen, and as a result, its efficacy is decreased. The *CYP2D6* genotype explains 34%–52% of the variability in absolute endoxifen concentrations. The alleles in *CYP2D6* have been categorized as follows: normal function (*1 and *2), decreased function (*9, *10, *17, and *41),

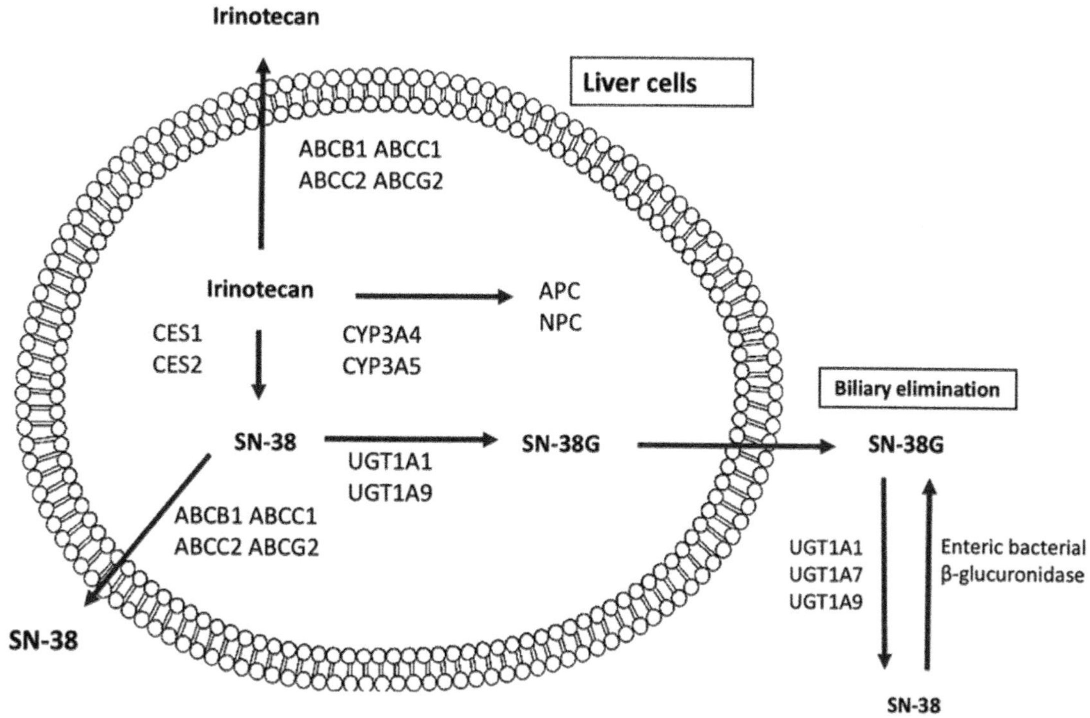

FIGURE 22.4 Irinotecan metabolism. SN-38: 7-ethyl-10-hydroxycamptothecin; SN-38G: glucuronide form of SN-38; CES: carboxylesterases 1 and 2; UGT1A1 and UGT 1A9: uridine diphosphate glucuronosyltransferase 1 and 9, respectively; ABCB1, ABCC1, ABCC2 and ABCG2: ATP-binding cassette subfamily B member 1, subfamily C members 1 and 2, and subfamily G member 2. *Modified from Hahn KK, Wolff JJ, Kolesar JM. Pharmacogenetics and irinotecan therapy. Am. J. Health Syst. Pharm. U. S November 2006;63(22):2211–7.*

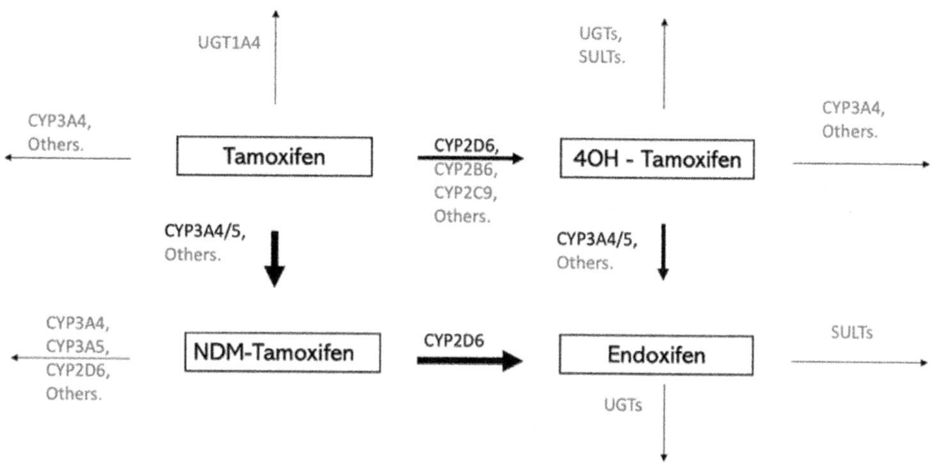

FIGURE 22.5 Main human biotransformation routes of tamoxifen. *CYP*, cytochrome P450; *NDM*, N-desmethyltamoxifen; *UGT*, uridine-5′-diphospho-glucuronosyltransferases; *SULTs*, sulfonyl-transferases. *Modified from Dickschen K, Willmann S, Thelen K, Lippert J, Hempel G, Eissing T. Physiologically based pharmacokinetic modeling of tamoxifen and its metabolites in women of different CYP2D6 phenotypes provides new insight into the tamoxifen mass balance [Internet] Front. Pharmacol. 2012;3:92. Available from: http://journal.frontiersin.org/article/10.3389/fphar.2012.00092/abstract.*

and no function (*3, *4, *5, and *6). In addition, *CYP2D6* can exhibit duplications (*1xN) and deletions.

Alleles with increased function are alleles with two or more normal function gene copies. In accordance with the presence of alleles, patients are classified as ultrarapid metabolizers (UMs), normal metabolizers (NMs), normal or intermediate metabolizers (IMs), and poor metabolizers (PMs) of CYP2D6 [25,26]. CPIC recommends that UMs and NMs should follow the standard of care dosage. For PMs and IMs, the strong recommendation is providing an alternative drug, such as aromatase inhibitor for postmenopausal women, or an aromatase inhibitor plus ovarian function suppressor for premenopausal women. Otherwise, a higher risk of recurrence and worse event-free survival is anticipated. Increasing tamoxifen dose is recommended only when an aromatase inhibitor is contraindicated because the endoxifen levels are not completely compensated. For CYP2D6 IMs and *CYP2D6**10/*10 or *CYP2D6**10/decreased function alleles, alternative hormonal therapy is a moderate recommendation for postmenopausal women; for premenopausal women, an aromatase inhibitor plus ovarian function suppressor can be used [6,25].

ATP-binding cassette subfamily B member 1 (ABCB1) is a transmembrane active efflux pump and regulates the transport of several drugs with pharmacokinetic effects [28]. In methotrexate-based treatments, the A/A genotype for *ABCB1* (rs1045642, C3435T) increases the risk of liver and hematological toxicity in lymphoma and leukemia patients [29,30]. Glutathione-s-transferase (GST) enzymes are implicated in the elimination of several drugs and oxidative stress/reactive oxygen species. GSTs have been found to be highly polymorphic in the population [31].

In breast cancer, *GSTP1* (rs1695) ile105val polymorphism has been related to toxicity and efficacy, in treatments with cyclophosphamide, epirubicin, and fluorouracil [32,33]. This polymorphism has shown effects on the efficacy and toxicity of platinum compounds in colorectal cancer, ovarian cancer, and medulloblastoma [34–36]. In 5-fluorouracil-based chemotherapy, TTAAAG/TTAAAG genotype in thymidylate synthase *(TYMS)* gene, a target enzyme of 5-fluorouracil, has been associated with toxicity and disease progression [37,38].

Somatic mutations in cancer therapy

Several efforts to decipher the cancer genomics and pathways activated in each cancer have been going on [39]. Contrary to passenger mutations, driver somatic mutations promote tumorigenesis by providing selective advantages [40]. Only about 10% of cancers have germline mutations, 10% show germline, and somatic mutations, and 80% only exhibit somatic mutations [41]. Examples of driver mutations in the germline lineage are breast cancer type 1 susceptibility protein (*BRCA1*) and breast cancer type 2 susceptibility protein (*BRCA2*) genes, in familial breast cancer, and adenomatous polyposis coli protein (*APC*) in familial adenomatous polyposis [40].

The International Cancer Genome Consortium (www.icgc.org), The Cancer Genome Atlas (TCGA) project (www.cancergenome.nih.gov), and Cancer Cell Line Encyclopedia (https://portals.broadinstitute.org/ccle) are projects that include molecular analysis of normal and tumor samples, across most types of cancer, integrating genomic profiles with therapeutic uses [39]. The cBioportal platform (http://www.cbioportal.org/) [42] and GDSCTools (https://www.cancerrxgene.org/) [43] provide the opportunity to visualize and download data, on human genomics and cell lines, respectively. The clinical applications of somatic mutations as predictive biomarkers are represented by several approved targeted drugs (Table 22.1).

Breast cancer

For almost 20 years, human epidermal growth factor receptor 2 (HER2) overexpression has been a criterion for selecting breast cancer patients for treatment with trastuzumab, an antiHER2 monoclonal antibody that binds to the extracellular domain, inhibiting downstream signaling [44]. Dual blockade of HER2 has been shown with the pertuzumab antibody. An 11-year follow up, indicated that adjuvant treatment for 1 year with trastuzumab in early breast cancer, provides a clinical benefit, with a hazard ratio of 0.76 (95% confidence interval [CI] 0.68–0.86), even with the crossover of the control arm [45]. Currently, neoadjuvant treatment with trastuzumab and pertuzumab, and adjuvant treatment with trastuzumab for 1 year, is the standard of care in HER2+ early breast cancer. In addition, taxane chemotherapy in combination with trastuzumab and pertuzumab is the standard of care in advanced stages [46].

Non-small cell lung cancer

Epidermal growth factor receptor (*EGFR*) mutations, small in-frame deletions in exon 19, and L858R substitution in exon 21 are predictive biomarkers of the response to tyrosine kinase inhibitors (TKIs; erlotinib, afatinib, and gefitinib), in non-small cell lung cancer (NSCLC). TKIs are the best first-line therapy in EGFR-mutated patients; however, resistance through mutations in exon 20 (p.Thr790Met) has been observed. Currently, clinical trials are going on for evaluating the efficacy of osimertinib, a TKI effective in exon 20 resistance tumors, in resistant NSCLC [47,48]. Anaplastic lymphoma kinase (ALK) is a tyrosine kinase receptor that can suffer rearrangements, generating dysregulated signaling in NSCLC patients. Alectinib is a TKI that inhibits *ALK* and *RET* (Ret proto-oncogene) arrangements in NSCLC patients. Furthermore, lorlatinib is a TKI that targets *ALK* and *ROS1* (ROS proto-oncogene 1) rearrangements, reserved after the progression with alectinib, crizotinib, and ceritinib [48].

Melanoma

In melanoma, mitogen-activated protein kinase (MAPK) pathway is the main signaling pathway involved in tumorigenesis and prognosis. The main component is *BRAF* mutation, which is activated in 40%–60% of cases, followed by *NRAS* mutation in 15%–30% of cases. The main gene hotspots in melanoma altered genes are V600 codon, a *BRAF* hotspot (more than 97% of mutations), and Q61 codon, an *NRAS* hotspot. Selective inhibitors of *BRAF*

V600-mutated kinase (vemurafenib and dabrafenib) and *MEK* kinase (trametinib and cobimetinib) have been approved for therapeutic use, in advanced stages [49,50].

Colorectal cancer

In colorectal cancer, constant activation of the EGFR pathway is associated with increased malignity, especially in Kirsten rat sarcoma viral proto-oncogene (*KRAS*) mutation tumors. *KRAS* wild-type status has been a criterium for patient selection to panitumumab and cetuximab [8], which are both antiEGFR antibodies. After the first results in colorectal cancer, retrospective reanalysis demonstrated the predictive value of *KRAS* status [51]. Currently, the mutation status has been included in Harvey-Ras (*H-RAS*) and neuroblastoma-Ras (*N-RAS*) [52]. Activation of RAS signaling affects interactions with downstream transducers (GTPase-activating proteins, GAPs), and guanine-exchanging/releasing factors (GEFs/GRFs), promoting the release of GTP (guanosine triphosphate) and ERK (extracellular-signal-regulated kinase) translocation into the nucleus, for activating the transcription factors that induce genes like c-Myc (MYC proto-oncogene), c-fos (Fos proto-oncogene), and c-Jun (Jun proto-oncogene) [52,53].

Clinical study designs

A major purpose in pharmacogenomics is finding the sources of variability in the drug responses across the population. The main evidence for classical drugs has been compiled from postapproval uses, focusing on germline variations in genes, related to metabolized enzymes and drug targets. Case-control and retrospective cohort studies have been performed; however, the gold standard is the randomized controlled trial. In contrast, the candidate gene approach has mainly been used in cancer pharmacogenomics, although there have been increasing genome-wide association studies (GWASs). A weakness of clinical studies has been replication and validation, caused in part by genotyping errors, confounders, and small sample sizes [54].

Currently, clinical trial designs include biomarker information to identify differences through stratification. Treatment benefit is based on predictive biomarkers. Enrichment trials have been proposed for treatment with predictive biomarkers and selective and sensitive testing. The main aim is selecting patients with positive biomarkers before the treatment; however, this also increases the probability of false negatives. The success of this type of design has been illustrated in the approvals of imatinib, vemurafenib, and crizotinib [39,55–57].

In contrast, an adaptive-enrichment design has been developed, using interim analysis to reallocate patients in relation to the treatment benefit [58]. This reallocation provides more control over sample size. A key problem is an institutional variability among biomarker assays, motivating the development of umbrella trials. Here, patients are assigned to cohorts based on biomarker and type of cancer; then, the cohorts are randomized into experimental or standard of care. If patients have a negative biomarker, they are randomized to standard of care or nontargeted therapy. The results of each cohort are useful for describing treatment benefit, based on each biomarker. Similarly, basket trials have been developed, to include one target therapy for each cohort, but more than one tumor type [57].

Liquid biopsy and pharmacogenomics

Liquid biopsy has shown that mutations from the primary tumor site, can be detected in nucleic acid analysis from blood samples. Murtaza et al. [16] described drug resistance by monitoring clonal evolution in a case series, using exome sequencing. Importantly, Douillard et al. [17] showed that cell-free DNA (cf-DNA), can determine *EGRF* mutational status in NSCLC patients, with high accuracy and specificity. In colorectal cancer, Yumada et al. [18] described nonresponder patients, with *KRAS* mutation in cf-DNA and wild-type *KRAS* in the primary tumor (3 of 10 patients). This discordance can partially explain the lack of efficacy of EGFR blockade therapy in certain patients.

In fact, Siravegna et al. [19] showed *KRAS* mutation in cf-DNA of colorectal cancer patients, with either primary resistance or acquired resistance to EGFR blockade. After withdrawing of EGRF blockade treatment, *KRAS* mutation levels decreased drastically in the bloodstream, indicating that cf-DNA is a potential tool for monitoring tumor evolution across the treatment. Although several efforts related to liquid biopsy have been developed, the clinical significance of cf-DNA/circulating tumor DNA has not been well established. Further studies are needed to increase the knowledge of the biological concordance with the true primary tumor site and decrease the false-negative rate [20].

Integrative precision medicine in cancer

Across time, precision medicine has combined individual patient characteristics with tumor genomics, including clonal evolution and inter- and intratumor heterogeneity. The understanding of clonal aberration and its dynamics will provide better therapeutic opportunities against tumor resistance [59]. Integrative genomics of cancer treatment has included whole-genomic alterations in tumors, describing the interaction among activated pathways and prognostic biomarkers. In the future, cancer pharmacogenomics in combination with novel predictive biomarkers may increase the knowledge of inherent factors involved in cancer treatment in the postgenomic era [60]. In fact, the tumor mutational burden (TMB), is associated with the improved survival in immune checkpoint inhibitor treatments, with antiCTLA-4 and antiPD-L1/PD-1 [60].

A higher amount of mutations increases the number of neoantigens presented on the major histocompatibility complex, and subsequent recognition by the immune system. These results are consistent with the efficacy of pembrolizumab, an antiPD-1 antibody with approval for cancers with high microsatellite instability (MSI-H), or deficient mismatch repair (dMMR). Mismatch repair system deficiency indicates accumulation of mutations, altering repetitive DNA microsatellite sequences, and providing high rates of mutations and neoantigens [61].

Interestingly, colorectal patients with dMMR treated with fluoropyrimidine chemotherapy plus oxaliplatin, show an increase in survival time, compared with those receiving fluoropyrimidine chemotherapy alone [62]. Conversely, MSI-H esophagogastric patients have poor survival, compared with nonMSI-H patients after cytotoxic chemotherapy [63]. In muscle-invasive bladder cancer patients, ERCC2-mutated patients have more sensitivity to cisplatin treatments, than *ERCC2* wild-type patients do [64]. These differences can be explained in part by the chemotherapy mechanism, cell type, and tumor microenvironment.

Germline-somatic interface

Germline variations can have implications for somatic events in certain types of cancer. Carter et al. [65] showed that 19p13.3, increases somatic mutations in the *PTEN* gene. Lu et al. [66] reported the frequency of rare germline truncations in TCGA tumors, associated with higher somatic mutation frequencies. Other analyses of breast and lung cancer have been reported in the last few years [67,68]. Insights into this interaction could increase the comprehensive knowledge about somatic variations and tumor evolution [65]. In addition, the parallel genotyping of somatic and germline variations could increase therapeutic options to noneligible patients, allowing targeted therapy and a better understanding of disease pathogenesis [69]. Importantly, germline *BRCA* mutations are predictive biomarkers in ovarian and HER2− breast cancer patients, treated with olaparib, a poly(adenosine diphosphate-ribose) polymerase inhibitor [70].

Conclusions

Important advances have been achieved with germline mutations; the toxicity and efficacy recommendations include genetic polymorphisms in *DPYD*, *TPMT*, *CYP2D6*, and *UGT1A1*. In addition, somatic mutations have guided the development and clinical use of targeted therapy, including *BRAF* mutations in melanoma, *RAS* mutations in lung and colorectal cancer, and *HER2* in breast cancer. Clinical validation of liquid biopsy and increasing the knowledge of tumor dynamics and biomarkers to explain drug resistance variability are important goals. Although current genomics information can define the responder population, and increase treatment benefit, insights about tumor microenvironment, including tumor mutational burden and immune response, will provide a broader understanding of inter- and intratumor heterogeneity.

References

[1] 1000 Genomes Project Consortium, Auton A, Brooks LD, Durbin RM, Garrison EP, Kang HM, et al. A global reference for human genetic variation [Internet] Nature October 1, 2015;526(7571):68–74. Available from: http://www.ncbi.nlm.nih.gov/pubmed/26432245.

[2] Paugh SW, Stocco G, McCorkle JR, Diouf B, Crews KR, Evans WE. Cancer pharmacogenomics [Internet]. Nature Publishing Group Clin. Pharmacol. Ther. 2011;90(3):461–6. Available from: https://doi.org/10.1038/clpt.2011.126/nature06264.

[3] Crews KR, Hicks JK, Pui C-H, Relling MV, Evans WE. Pharmacogenomics and individualized medicine: translating science into practice [Internet] Clin. Pharmacol. Ther. October 2012;92(4):467–75. Available from: http://www.pubmedcentral.nih.gov/articlerender.fcgi?artid=3589526&tool=pmcentrez&rendertype=abstract.

[4] McLeod HL. Cancer pharmacogenomics: early promise, but concerted effort needed [Internet] Science March 29, 2013;339(6127):1563–6. Available from: https://www.pharmgkb.org/page/cancerPgx.

[5] Relling MV, Klein TE. CPIC: clinical pharmacogenetics implementation consortium of the pharmacogenomics research network [Internet]. Nature Publishing Group Clin. Pharmacol. Ther. 2009;89(3):464–7. Available from: https://doi.org/10.1038/clpt.2010.279/nature06264.

[6] Swen JJ, Nijenhuis M, de Boer A, Grandia L, Maitland-van der Zee AH, Mulder H, et al. Pharmacogenetics: from bench to byte— an update of guidelines [Internet]. Nature Publishing Group Clin. Pharmacol. Ther. May 16, 2011;89(5):662–73. Available from: https://doi.org/10.1038/clpt.2011.34/nature06264.

[7] Food and Drug Administration. Table of pharmacogenomic biomarker in drug labeling [Internet]. 2015. Available from: http://www.fda.gov/drugs/scienceresearch/researchareas/pharmacogenetics/ucm083378.htm. [Accessed 20 May 2016].

[8] Sánchez NS, Mills GB, Shaw KRM. Precision oncology: neither a silver bullet nor a dream. Pharmacogenomics 2017;18(16):1525–39.

[9] Food and drug administration (FDA). Table of pharmacogenomic biomarkers in drug labeling [Internet] Center for Drug Evaluation and Research 2019. Available from: https://www.fda.gov/Drugs/ScienceResearch/ucm572698.htm.

[10] Vogel F. Moderne problem der humangenetik. Ergeb Inn Med U Kinderheilk 1959;12:52–125.

[11] Relling MV, Evans WE. Pharmacogenomics in the clinic [Internet] Nature October 15, 2015;526(7573):343–50. Available from: http://www.nature.com/articles/nature15817.

[12] Zhang G, Nebert DW. Personalized medicine: genetic risk prediction of drug response [Internet] Pharmacol. Ther. July 2017;175(4):75–90. Available from: https://linkinghub.elsevier.com/retrieve/pii/S0163725817300505.
[13] Relling MV, Dervieux T. Pharmacogenetics and cancer therapy [Internet] Nat. Rev. Canc. 2001;1(2):99–108. Available from: http://www.ncbi.nlm.nih.gov/pubmed/11905809%5Cnwww.ncbi.nlm.nih.gov/pubmed/11905809.
[14] Burki TK. Defining precision medicine [Internet]. Elsevier Ltd Lancet Oncol 2017;18(12):e719. Available from: http://linkinghub.elsevier.com/retrieve/pii/S1470204517308653.
[15] Jameson JL, Longo DL. Precision medicine — personalized, problematic, and promising [Internet] N. Engl. J. Med. June 4, 2015;372(23):2229–34. Available from: http://www.nejm.org/doi/abs/10.1056/NEJMsb0903885.
[16] Filipski KK, Mechanic LE, Long R, Freedman AN. Pharma. Oncol Care April 2014;5:1–5.
[17] Ballman KV. Biomarker: predictive or prognostic? J. Clin. Oncol. 2015;33(33):3968–71.
[18] Filipski KK, Mechanic LE, Long R, Freedman AN. Pharmacogenomics in oncology care [Internet] Front. Genet. January 2014;5(73). Available from: http://www.pubmedcentral.nih.gov/articlerender.fcgi?artid=3986526&tool=pmcentrez&rendertype=abstract. [Accessed 20 May 2016].
[19] Relling MV, Gardner EE, Sandborn WJ, Schmiegelow K, Pui C-H, Yee SW, et al. Clinical pharmacogenetics implementation consortium guidelines for thiopurine methyltransferase genotype and thiopurine dosing: 2013 update [Internet] Clin. Pharmacol. Ther. April 17, 2013;93(4):324–5. Available from: http://doi.wiley.com/10.1038/clpt.2013.4.
[20] Bank P, Caudle K, Swen J, Gammal R, Whirl-Carrillo M, Klein T, et al. Comparison of the guidelines of the clinical pharmacogenetics implementation consortium and the Dutch pharmacogenetics working group [Internet] Clin. Pharmacol. Ther. April 10, 2018;103(4):599–618. Available from: https://onlinelibrary.wiley.com/doi/abs/10.1002/cpt.762.
[21] Chouchana L, Roche D, Jian R, Beaune P, Loriot M-A. Poor response to thiopurine in inflammatory bowel disease: how to overcome therapeutic resistance? [Internet] Clin. Chem. July 1, 2013;59(7):1023–6. Available from: http://www.clinchem.org/cgi/doi/10.1373/clinchem.2012.195750.
[22] Amstutz U, Henricks LM, Offer SM, Barbarino J, Schellens JHM, Swen JJ, et al. Clinical pharmacogenetics Implementation Consortium (CPIC) guideline for dihydropyrimidine dehydrogenase genotype and fluoropyrimidine dosing: 2017 update. Clin. Pharmacol. Ther. 2018;103(2):210–6.
[23] Del Re M, Di Paolo A, van Schaik RH, Bocci G, Simi P, Falcone A, et al. Dihydropyrimidine dehydrogenase polymorphisms and fluoropyrimidine toxicity: ready for routine clinical application within personalized medicine? [Internet] EPMA J. September 25, 2010;1(3):495–502. Available from: http://link.springer.com/10.1007/s13167-010-0041-2. [Accessed 30 January 2019].
[24] Hahn KK, Wolff JJ, Kolesar JM. Pharmacogenetics and irinotecan therapy. Am. J. Health Syst. Pharm. U. S November 2006;63(22):2211–7.
[25] Goetz MP, Sangkuhl K, Guchelaar HJ, Schwab M, Province M, Whirl-Carrillo M, et al. Clinical pharmacogenetics Implementation Consortium (CPIC) guideline for CYP2D6 and tamoxifen therapy. Clin. Pharmacol. Ther. 2018;103(5):770–7.
[26] Varela N, Quiñones LA, Stojanova J, Garay J, Cáceres D, Cespedes S, et al. Characterization of the CYP2D6 drug metabolizing phenotypes of the Chilean mestizo population through polymorphism analyses [Internet] Pharmacol. Res. November 2015;101:124–9. Available from: http://www.ncbi.nlm.nih.gov/pubmed/26211952. [Accessed 20 May 2016].
[27] Dickschen K, Willmann S, Thelen K, Lippert J, Hempel G, Eissing T. Physiologically based pharmacokinetic modeling of tamoxifen and its metabolites in women of different CYP2D6 phenotypes provides new insight into the tamoxifen mass balance [Internet] Front. Pharmacol. 2012;3:92. Available from: http://journal.frontiersin.org/article/10.3389/fphar.2012.00092/abstract. [Accessed 30 January 2019].
[28] Leschziner GD, Andrew T, Pirmohamed M, Johnson MR. ABCB1 genotype and PGP expression, function and therapeutic drug response: a critical review and recommendations for future research [Internet] Pharmacogenomics J June 12, 2007;7(3):154–79. Available from: http://www.nature.com/articles/6500413. [Accessed 31 January 2019].
[29] Suthandiram S, Gan G-G, Zain SM, Bee P-C, Lian L-H, Chang K-M, et al. Effect of polymorphisms within methotrexate pathway genes on methotrexate toxicity and plasma levels in adults with hematological malignancies [Internet] Pharmacogenomics August 2014;15(11):1479–94. Available from: http://www.ncbi.nlm.nih.gov/pubmed/25303299. [Accessed 30 January 2019].
[30] Medicine G, Cho JH, Gregersen PK. Genomics and the multifactorial nature of human autoimmune disease. 2011.
[31] Economopoulos KP, Sergentanis TN. GSTM1, GSTT1, GSTP1, GSTA1 and colorectal cancer risk: a comprehensive meta-analysis [Internet]. Elsevier Ltd Eur. J. Cancer 2010;46(9):1617–31. Available from: https://doi.org/10.1016/j.ejca.2010.02.009.
[32] Oliveira AL, Rodrigues FFO, Santos RE, Aoki T, Rocha MN, Longui CA, et al. GSTT1, GSTM1, and GSTP1 polymorphisms and chemotherapy response in locally advanced breast cancer [Internet] Genet. Mol. Res. June 11, 2010;9(2):1045–53. Available from: http://www.ncbi.nlm.nih.gov/pubmed/20568049. [Accessed 28 February 2017].
[33] Sugishita M, Imai T, Kikumori T, Mitsuma A, Shimokata T, Shibata T, et al. Pharmacogenetic association between GSTP1 genetic polymorphism and febrile neutropenia in Japanese patients with early breast cancer [Internet] Breast Canc. 2016;23(2):195–201. Available from: http://www.ncbi.nlm.nih.gov/pubmed/25008867. [Accessed 30 January 2019].
[34] Rednam S, Scheurer ME, Adesina A. Glutathione S-transferase P1 single nucleotide polymorphism predicts permanent ototoxicity in children with medulloblastoma. 2013. p. 593–8. October 2012.
[35] Kim HS, Kim M-K, Chung HH, Kim JW, Park NH, Song YS, et al. Genetic polymorphisms affecting clinical outcomes in epithelial ovarian cancer patients treated with taxanes and platinum compounds: a Korean population-based study [Internet] Gynecol. Oncol. May 2009;113(2):264–9. Available from: http://www.ncbi.nlm.nih.gov/pubmed/19203783. [Accessed 30 April 2014].

[36] Stoehlmacher J, Park DJ, Zhang W, Yang D, Groshen S, Zahedy S, et al. A multivariate analysis of genomic polymorphisms: prediction of clinical outcome to 5-FU/oxaliplatin combination chemotherapy in refractory colorectal cancer. Br. J. Canc. 2004;91(2):344–54.

[37] Ruzzo a, Graziano F, Loupakis F, Santini D, Catalano V, Bisonni R, et al. Pharmacogenetic profiling in patients with advanced colorectal cancer treated with first-line FOLFIRI chemotherapy [Internet] Pharm. J 2008;8(4):278–88. Available from: http://www.ncbi.nlm.nih.gov/pubmed/17549067.

[38] Rosmarin D, Palles C, Church D, Domingo E, Jones A, Johnstone E, et al. Genetic markers of toxicity from capecitabine and other fluorouracil-based regimens: investigation in the QUASAR2 study, systematic review, and meta-analysis. J. Clin. Oncol. 2014;32:1031–9.

[39] The Cancer Genome Atlas. Cancer genome – TCGA [Internet]. Available from: https://cancergenome.nih.gov/. [Accessed 22 January 2019].

[40] Pon JR, Marra MA. Driver and passenger mutations in cancer [Internet] Annu. Rev. Pathol. 2015;10(1):25–50. Available from: http://www.annualreviews.org/doi/10.1146/annurev-pathol-012414-040312.

[41] Futreal PA, Coin L, Marshall M, Down T, Hubbard T, Wooster R, et al. A census of human cancer genes. Nat. Rev. Canc. 2004;4(3):177–83.

[42] cBioPortal. cBioPortal for cancer genomics [Internet]. Available from: http://www.cbioportal.org/.

[43] Cokelaer T, Chen E, Iorio F, Menden MP, Lightfoot H, Saez-Rodriguez J, et al. GDSCTools for mining pharmacogenomic interactions in cancer. Bioinformatics 2018;34(7):1226–8.

[44] Goldenberg MM. Trastuzumab, a recombinant DNA-derived humanized monoclonal antibody, a novel agent for the treatment of metastatic breast cancer [Internet]. Elsevier Clin. Ther. February 1, 1999;21(2):309–18. Available from: http://linkinghub.elsevier.com/retrieve/pii/S0149291800882880. [Accessed 30 January 2019].

[45] Cameron D, Piccart-Gebhart MJ, Gelber RD, Procter M, Goldhirsch A, de Azambuja E, et al. 11 years' follow-up of trastuzumab after adjuvant chemotherapy in HER2-positive early breast cancer: final analysis of the HERceptin Adjuvant (HERA) trial [Internet] Lancet 2017;389(10075):1195–205. Available from: http://www.ncbi.nlm.nih.gov/pubmed/28215665.

[46] Harbeck N. Advances in targeting HER2-positive breast cancer. Curr. Opin. Obstet. Gynecol. 2018;30(1):55–9.

[47] Garinet S, Laurent-Puig P, Blons H, Oudart J-B. Current and future molecular testing in NSCLC, what can we expect from new sequencing technologies? [Internet] J. Clin. Med. 2018;7(6):144. Available from: http://www.mdpi.com/2077-0383/7/6/144.

[48] 3.2019 NGV. Non-small cell lung cancer. NCCN clin pract guidel oncol (NCCN guidel) [Internet]. 2019. Available from: http://0-bestpractice.bmj.com.lib.exeter.ac.uk/topics/en-gb/1082/pdf/1082.pdf.

[49] Cheng L, Lopez-Beltran A, Massari F, MacLennan GT, Montironi R. Molecular testing for BRAF mutations to inform melanoma treatment decisions: a move toward precision medicine [Internet]. Nature Publishing Group Mod. Pathol. 2018;31(1):24–38. Available from: http://www.ncbi.nlm.nih.gov/pubmed/29148538. [Accessed 30 January 2019].

[50] Cancer Genome Atlas Network. Genomic classification of cutaneous melanoma [Internet] Cell June 18, 2015;161(7):1681–96. Available from: http://www.ncbi.nlm.nih.gov/pubmed/26091043.

[51] Van Cutsem E, Cervantes A, Adam R, Sobrero A, Van Krieken JH, Aderka D, et al. ESMO consensus guidelines for the management of patients with metastatic colorectal cancer. Ann. Oncol. 2016;27(8):1386–422.

[52] Hajdúch M, Jančík S, Drábek J, Radzioch D. Clinical relevance of KRAS in human cancers. J. Biomed. Biotechnol. 2010;2010.

[53] Jeong W-J, Ro EJ, Choi K-Y. Interaction between Wnt/β-catenin and RAS-ERK pathways and an anti-cancer strategy via degradations of β-catenin and RAS by targeting the Wnt/β-catenin pathway [Internet]. Springer US NPJ Precis Oncol 2018;2(1):5. Available from: http://www.nature.com/articles/s41698-018-0049-y.

[54] Wheeler HE, Maitland ML, Dolan ME, Cox NJ, Ratain MJ. Cancer pharmacogenomics: strategies and challenges [Internet] Nat. Rev. Genet 2013;14:23–34. https://doi.org/10.1038/nrg3352. Nature Publishing Group. Available from:.

[55] Iorio F, Knijnenburg TA, Vis DJ, Bignell GR, Menden MP, Schubert M, et al. A landscape of pharmacogenomic interactions in cancer. Cell 2016;166(3):740–54.

[56] López-Cortés A, Guerrero S, Redal MA, Alvarado AT, Quiñones LA. State of art of cancer pharmacogenomics in Latin American populations. Int. J. Mol. Sci. 2017;18(6).

[57] Renfro LA, Mallick H, An M-W, Sargent DJ, Mandrekar SJ. Clinical trial designs incorporating predictive biomarkers [Internet] Cancer Treat Rev February 2016;43(5):74–82. Available from: http://www.ncbi.nlm.nih.gov/pubmed/25936845.

[58] Simon N, Simon R. Adaptive enrichment designs for clinical trials. Biostatistics 2013;14(4):613–25.

[59] Collins DC, Sundar R, Lim JSJ, Yap TA. Towards precision medicine in the clinic: from biomarker discovery to novel therapeutics [Internet]. Elsevier Ltd Trends Pharmacol. Sci. 2017;38(1):25–40. Available from: https://doi.org/10.1016/j.tips.2016.10.012.

[60] Samstein RM, Lee C-H, Shoushtari AN, Hellmann MD, Shen R, Janjigian YY, et al. Tumor mutational load predicts survival after immunotherapy across multiple cancer types [Internet]. Nature Publishing Group Nat. Genet. January 14, 2019:1. Available from: http://www.nature.com/articles/s41588-018-0312-8. [Accessed 30 January 2019].

[61] Boyiadzis MM, Kirkwood JM, Marshall JL, Pritchard CC, Azad NS, Gulley JL. Significance and implications of FDA approval of pembrolizumab for biomarker-defined disease. J. Immunother. Cancer. J. ImmunoTher. Cancer 2018;6(1):1–7.

[62] Tougeron D, Mouillet G, Trouilloud I, Lecomte T, Coriat R, Aparicio T, et al. Efficacy of adjuvant chemotherapy in colon cancer with microsatellite instability: a large multicenter AGEO study [Internet]. Oxford University Press J. Natl. Cancer Inst. July 1, 2016;108(7):djv438. Available from: https://academic.oup.com/jnci/article-lookup/doi/10.1093/jnci/djv438. [Accessed 31 January 2019].

[63] Janjigian YY, Sanchez-Vega F, Jonsson P, Chatila WK, Hechtman JF, Ku GY, et al. Genetic predictors of response to systemic therapy in esophagogastric cancer. Cancer Discov 2018;8(1):49–58.

[64] Li Q, Damish A, Frazier ZJ, Liu D, Reznichenko E, Kamburov A, et al. ERCC2 helicase domain mutations confer nucleotide excision repair deficiency and drive cisplatin sensitivity in muscle-invasive bladder cancer [Internet]. 2018;clincanres.1001 Clin. Cancer Res. 2018 Available from: http://clincancerres.aacrjournals.org/lookup/doi/10.1158/1078-0432.CCR-18-1001.

[65] Carter H, Marty R, Hofree M, Gross AM, Jensen J, Fisch KM, et al. Interaction landscape of inherited polymorphisms with somatic events in cancer. Cancer Discov 2017;7(4):410–23.

[66] Lu C, Xie M, Wendl MC, Wang J, McLellan MD, Leiserson MDM, et al. Patterns and functional implications of rare germline variants across 12 cancer types [Internet]. Nature Publishing Group Nat. Commun. 2015;6:10086.

[67] Liu W, He L, Ramírez J, Krishnaswamy S, Kanteti R, Wang Y-C, et al. Functional EGFR germline polymorphisms may confer risk for EGFR somatic mutations in non-small cell lung cancer, with a predominant effect on exon 19 microdeletions [Internet] Cancer Res April 1, 2011;71(7):2423–7.

[68] Li Q, Seo J-H, Stranger B, McKenna A, Pe'er I, LaFramboise T, et al. Integrative eQTL-based analyses reveal the biology of breast cancer risk loci. Cell 2013;152(3):633–41.

[69] Mandelker D. Toward concurrent testing for somatic and germline variants in cancer patients. Clin. Cancer Res. 2016;22(16):3987–8.

[70] Robson M, Im S-A, Senkus E, Xu B, Domchek SM, Masuda N, et al. Olaparib for metastatic breast cancer in patients with a germline BRCA mutation [Internet] N. Engl. J. Med. 2017;377(17):1700.

B—Proteomics

Chapter 23

Proteomic biomarkers in vitreoretinal disease

Jose Ronaldo Lima de Carvalho, Jr.[1,2,3,4], Karen Sophia Park[1,2], Fábio P. Saraiva[5,6], Stephen H. Tsang[1,2,7], Vinit B. Mahajan[8,9] and Thiago Cabral[4,5,6]

[1]Department of Ophthalmology, Columbia University, New York, NY, United States; [2]Jonas Children's Vision Care and Bernard & Shirlee Brown Glaucoma Laboratory, New York, NY, United States; [3]Department of Ophthalmology, Hospital das Clinicas de Pernambuco (HCPE) – Empresa Brasileira de Servicos Hospitalares (EBSERH), Federal University of Pernambuco (UFPE), Recife, Pernambuco, Brazil; [4]Department of Ophthalmology, Federal University of São Paulo (UNIFESP), São Paulo, Brazil; [5]Adjunct Professor in Ophthalmology, Department of Specialized Medicine, CCS – Federal University of Espirito Santo (UFES), Vitoria, Espirito Santo, Brazil; [6]Vision Center Unit, Ophthalmology, Empresa Brasileira de Servicos Hospitalares (EBSERH), HUCAM-UFES, Vitoria, Espirito Santo, Brazil; [7]Department of Pathology & Cell Biology, Stem Cell Initiative (CSCI), Institute of Human Nutrition, Vagelos College of Physicians and Surgeons, New York, NY, United States; [8]Omics Laboratory, Byers Eye Institute, Department of Ophthalmology, Stanford University, Palo Alto, CA, United States; [9]Veterans Affairs Palo Alto Health Care System, Palo Alto, CA, United States

Context

Proteomics, which is the study of an organism's protein library, has come to the forefront of precision medicine as the field that enables researchers to learn how differences in an individual's gene expression levels and metabolic profiles alter key proteins directly involved in disease pathology. While genomics can provide precise diagnostic information for inherited diseases and those with underlying genetic risk factors, proteomics offers insight into the causal molecular mechanisms by revealing protein expression levels, posttranslational modifications, and metabolic interactions, along with an array of other information. Moreover, knowledge of abnormally up- or downregulated proteins in diseased patients provides a fruitful opportunity to discover biomarkers of disease, which allow for higher specificity in diagnosis and more refined potential therapeutic targets.

Proteomics has been particularly useful for ailments such as myocardial infarction (troponin assays) and specific cancers (distinct protein receptors) [1]. Proteomics is also extended to ophthalmology through molecular assays of liquid biopsies derived from the vitreous and aqueous humor—fluid-filled compartments of the eye that harbor a diverse array of proteins and metabolites. The analytical techniques used to screen protein profiles in such tissues range from small-scale (e.g., multiplex immunoassays) to large-scale (e.g., mass spectrometry), depending on the quantity of proteins and metabolites to be analyzed.

Vitreoretinal diseases, which currently rely heavily on findings from clinical examination rather than molecular assays for diagnosis, may significantly benefit. One avenue is the potential for drug repositioning, which occurs through the identification of previously unknown therapeutic targets that overlap with targets of already-approved drugs [2,3]. The application of such strategies not only holds promise for the treatment of complex orphan diseases, but also offers economic value in the form of time, capital, and efficiency [4].

Sampling options and liquid biopsy techniques

The retina is composed of sensory neural tissue that, when perturbed by retinal biopsy procedures, results in high visual morbidity rates [5]. Only in rare cases such as severe inflammation and suspicion of malignancy do the benefits of retinal biopsy outweigh the risks [6]. A safer alternative is to sample vitreous gel, a liquid that is easier to access.

The vitreous humor occupies the posterior cavity of the eye and is composed of water, collagen, and hyaluronic acid. Water is the main component of the vitreous, comprising approximately 98% of its total mass and giving

Precision Medicine for Investigators, Practitioners and Providers. https://doi.org/10.1016/B978-0-12-819178-1.00023-X

it transparency and optical inertia [7]. Anatomically, the vitreous cavity is divided into the core, cortex, and base. Its position adjacent to the retina makes it an ideal tissue for sampling cytokines and proteins released by the retina under nonhomeostatic disease conditions. For ethical reasons, the analysis of the healthy eye vitreous is not allowed.

To understand the protein profile of a normal vitreous, proteomic studies have been done using postmortem eyes [8]. However, controversy exists over whether the molecular composition of postmortem eye tissue accurately reflects that of living tissue. Many researchers utilize the vitreous from patients with idiopathic macular hole (IMH) as a more suitable proxy for "normal" controls given the benign and idiopathic condition of the disease [9]. Eyes that underwent vitrectomy for epiretinal membrane (ERM) have also been used as a control, despite the abnormal posterior vitreous detachment that may be implicated in its pathology.

Some studies have already demonstrated that these conditions can alter the molecular composition of the vitreous. In cases of IMH, an overexpression of the complement pathway and α-2-macroglobulin was noticed, while abnormally high levels of transthyretin, apolipoprotein-A1, and α-1-antitrypsin were found in cases of ERM [10,11].

The most common method of performing vitreous biopsies is fine needle aspiration (FNA). In this technique, a small amount of the vitreous humor (100–200 μL) is aspirated by a 23-gauge needle that is introduced through the pars plana into the vitreous cavity under local anesthesia (Fig. 23.1 A). A more invasive method, pars plana vitrectomy (PPV), is mainly used in cases in which retinal surgery is indicated or a large volume of vitreous is required for proteomic analysis. In this technique, the procedure is done within the operating room under local anesthesia. Intravenous sedation or, in selective cases, general anesthesia may be used to provide more comfort to the patient. The vitrector, a vitrectomy probe, penetrates through the pars plana to reach the vitreous cavity to cut and aspirate the vitreous gel.

Studies comparing samples from 23-gauge PPV and 23-gauge FNA revealed that these techniques are equally successful, except in the analysis of insoluble proteins, for which PPV was shown to be more effective [12]. 25-gauge and 27-gauge PPV have recently become available and, due to their less invasive nature, may be a safer option for biopsy. However, studies comparing the quality of proteomic data obtained from 23-gauge PPV and FNA versus 25- and 27-gauge PPV have yet to be reported.

A second liquid source that is easily accessible in the eye is the aqueous humor (AH). In spite of its location in the anterior chamber and relatively large distance from the retina, vitreoretinal diseases can alter the protein content of the AH [13–16]. Biopsy of the AH may occur either via the introduction of a 25-gauge needle at the limbus

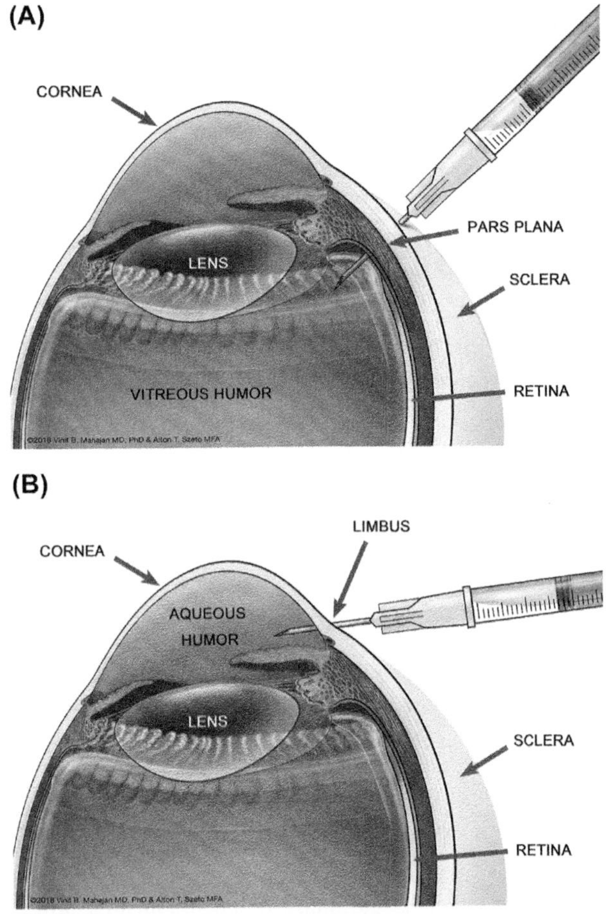

FIGURE 23.1 Eye anatomy and liquid biopsy. (A) Vitreous biopsy. A needle or a vitrector (not shown) penetrates the sclera via the pars plana to reach the vitreous cavity and aspirates the vitreous humor. (B) Aqueous biopsy. A 25-gauge needle is inserted into the anterior chamber through the limbus and collects aqueous humor. Graphical illustrations by Alton Szeto and Vinit Mahajan. *Reproduced with permission from copyright holder.*

(Fig. 23.1 B), the transition zone between the cornea and the sclera, followed by aspiration of AH or through the direct aspiration of AH from the surgical incision site, prior to cataract extraction. Around 100 μL can be aspirated from the anterior chamber. One limitation, however, is that cytokine profiles of the vitreous humor differ considerably from those of the aqueous humor. Thus, this method is reserved for cases in which PPV or FNA may have a high risk and are not indicated. Table 23.1 summarizes the various methods that can be used for proteomic analysis of vitreoretinal diseases.

Available analytical methods

Mass spectrometry (MS) can be used to analyze samples for idiopathic diseases, or those that have not yet been well characterized (untargeted proteomics). Its technique involves the ionization and sorting of specimens based on

TABLE 23.1 Liquid biopsy techniques for vitreoretinal diseases.

Method	Sample	Volume	Anesthesia	Disadvantages
Fine needle aspiration	Vitreous humor	100–200 μL	Local	Inadequate volume; risk of hypotony
Pars plana vitrectomy	Vitreous humor	>200 μL	Local ± sedation or general	More invasive; higher morbidity
Aqueous humor aspiration	Aqueous humor	~100 μL	Local	Inadequate volume; different pool of cytokines

their mass-to-charge ratio (*m/z*), which consequently allows for the profiling and quantification of proteins found in the sample. Before ionization, other procedures may be performed in order to simplify and better characterize the sample, namely (i) liquid chromatography (LC), which involves the isolation of peptides by hydrophobicity and size, (ii) strong-cation exchange chromatography (SCX), which fractionates peptides, and (iii) isoelectric focusing (IEF), which is an additional method of fractionating peptides [17]. Nevertheless, the fractionation of peptides using these techniques prior to MS has become unnecessary with the advent of ultrahigh-pressure liquid chromatography (UHPLC) [18]. UHPLC has become the cornerstone of liquid chromatography, as it allows for a more accurate analysis of complex samples and requires a shorter amount of time for analysis due to its highresolution and high output capabilities [19].

Upon fractionation and separation of the peptides, the sample must be ionized before entering the mass spectrometer, ideally using soft ionization methods that prevent damage to large molecules. Matrix-assisted laser desorption/ionization (MALDI) and electrospray ionization (ESI) are the two most applicable techniques [18,20]. In MALDI, the analyte is irradiated by an ultraviolet (UV) laser, after being dissolved and cocrystallized with a matrix [21]. Upon transformation into the gas phase, the analyte enters the mass spectrometer and is separated by time-of-flight (TOF), or the time spent to travel a known distance inside the mass analyzer. In ESI on the other hand, a liquid solvent is used instead of a matrix to dissolve the analyte [22]. A very strong electric field ionizes the peptide mixture from a spray and generates gas-phase ions. Like in MALDI, the charged ions are delivered to the mass analyzer and separated by time (TOF) or space (trap devices) [20].

The next step after further separation and identification of the peptides is quantification. Unlabeled (e.g., spectral counting or data-independent acquisition—DIA) or labeled (e.g., isotope-coded affinity tags—ICAT; isobaric tags for relative and absolute quantification—iTRAQ; and multiple reaction monitoring—MRM) methods can be used. The use of prespecified peptide-precursor ions in MRM allows for highly sensitive and reproducible quantification of proteins. MRM may also be used to detect short nucleotide polymorphisms (SNPs) and posttranslational modifications (PTMs) that are missed by techniques such as enzyme-linked immunosorbent assay (ELISA).

False-negative results might be expected for proteins that have a high affinity to albumin. Given that the vitreous and aqueous humor exhibit elevated levels of albumin and immunoglobulins, these proteins are depleted from the sample prior to its analysis in MS so as to increase the detectability of less abundant proteins. Advances in MS technology may solve this problem in the near future, improving the results of the analysis.

Bioinformatics and statistical analyses are challenging. A more thorough understanding of the molecular pathways and gene ontology, with the use of bioinformatics, may elucidate the rationale behind disease process, timing, severity, and response to therapy.

Proteomic biomarkers

According to the Biomarkers Definitions Working Group (NIH, Bethesda, MD, USA), the term "biomarker" is defined as "a characteristic that is objectively measured and evaluated as an indicator of normal biological processes, pathogenic processes, or pharmacologic responses to a therapeutic intervention"[23]. Proteomic biomarkers are thus protein indicators of biological processes that are typically discovered through the analysis of a cell protein library in a diseased and/or nondiseased state.

As disease alters cell function and gene expression, resulting changes in the protein profile at the level of transcription, translation, and posttranslational modifications can be detected [18,24]. Given its unbiased approach, LC-MS is often the ideal method for identifying new biomarkers, particularly in circumstances in which there is no prior knowledge of a disease and its causes [9].

While proteomic biomarkers can be derived from the cornea, lens, ciliary body, and retina, the vitreous and aqueous humor are relatively accessible and rich with endogenous proteins [25]. Among the most frequently investigated diseases are age-related macular degeneration (AMD), diabetic retinopathy (DR), retinal detachment (RD), uveitis, and ocular cancers [9]. Table 23.2 provides a summary of protein biomarkers for vitreoretinal diseases that have been identified through LC-MS/MS, multiplex ELISA, and electrophoresis, among other techniques.

Proteomics for drug repositioning

Drug repositioning is the process of applying an existing, approved drug for the treatment of a disease other than that indicated [2]. The existence of documented side-effect profiles and established therapeutic doses that can be reapplied to new indications reduces the number of practical obstacles.

Drug repositioning can be used for rare orphan diseases with small market capitalization, such as inherited retinal degenerations and chronic inflammatory eye diseases. Currently, most prediction methods utilize genomic analyses, retrospective computational methods, and genome-wide association studies (GWAS) to identify candidate drugs that share the same disease pathways as other

TABLE 23.2 Summary of select protein biomarkers for vitreoretinal diseases.

Disease	Method of detection	Biomarkers	References
Vitreous			
ERM	Electrophoresis, MS	SERPINA1, APOA1, TTR	[11]
	iTRAQ, LC-MS/MS	KNG1, FGA, CTSD, CPE, FSTL1, CDHR1, DKK3, B3GNT1, LYPD3, ENO1, MIF, CRYGD	[34]
IMH	LC-MS/MS	A2M, C3, C4A, CFB, CFH	[10]
nvAMD	Electrophoresis, LC-MS/MS, ELISA	SERPINA1, APOA1, TTR, RBP3, TF	[35]
		CLU, OPTC, PEDF, PTGDS	[36]
ROP	Electrophoresis, LC-MS/MS	RPB3, TF, ALB, PEDF, HPGDS, TTR, A2M, CP, AFP, A1BG, HPX	[37]
Retinoblastoma	iTRAQ, LC-MS/MS	GFAP, CRABP1, MMP2, TNC	[38]
DR	Electrophoresis, MS	PDEF, SERPINA1, AHSG, C4	[16]
PDR	Electrophoresis, MS	TF, SERPINA1, SERPINA3, AHSG, HPX, SERPINC1, APOA1, APOJ, FGG, HP	[39]
	Electrophoresis, MS	PEDF, SERPINA5, APOA-IV, PTGDS, SERPINA1, ANKRD15, AHSG, SPTBN5	[40]
	DIGE	ZAG, APOA1, APOH, FGA, C3, C4b, C9, CFB	[41]
	Electrophoresis, LC-MS/MS	AGT, C3, CFI, F2, SERPINA1, SERPINC1	[42]
RRD, DR, PDR, PVR	Electrophoresis, MS	AAT, APOA4, ALB, TF	[43]
RVO	LC-MS/MS	CLU, C3, IGLL5, OPTC, VTN	[44]
BRVO	LC-MS/MS	CLU, C3, PTGDS, VTN	[45]
Posterior uveitis	Multiplex ELISA	IL-23, PDGFRb, SCF, TIMP-1, TIMP-2, BMP-4, NGF, IGFBP-2, IL-17R, IL-1RI	[31]
Aqueous			
ERM	iTRAQ, LC-MS/MS	KNG1, FGA, CTSD, CPE, FSTL1, CDHR1, DKK3, B3GNT1, LYPD3, ENO1, MIF, CRYGD	[34]
nvAMD	Electrophoresis, LC-MS/MS, ELISA	LCN1, CRYAA	[14]
	LC-MS/MS	CTSD and KRT8	[46]
	MRM-MS	CEP55, ACT, DSG1, SOD3, FLG2, 26S proteasome, SERPINA5, C3, PKP2, LRRC15	[13]

Continued

TABLE 23.2 Summary of select protein biomarkers for vitreoretinal diseases.—cont'd

Disease	Method of detection	Biomarkers	References
AMD	LC-MRM-MS	CTDS, KRT8, KRT14, MYO9, HSP70, ACTA	[46]
Coat disease	iTRAQ, LC-MS/MS	HP and APOC-I	[47]
RRD, elevated IOP	LC-MS/MS	AAT, APOA4, ALB, TF	[48]
RRD, AMD, PVRL, INIU	Multiplex ELISA	IL-10, IL-21, ACE	[49]
Other			
Macular schisis (schisis fluid)	LC-MS/MS	OPTC, CRYBB2, CRYAB	[50]
XLRS (schisis fluid)	Electrophoresis, MS	RDH14, SGCE, STK26, TENM1, ALMS1, ZFP90, GRIN1, QSER1, ESCO1, KIF4A, CAPN1	[51]
Retinoblastoma (tumor)	DIGE, LC-MS/MS	FGB, CFH, FN1, ITIH4, FGG, HP, IGHA1, AFM, IGHM, LUM, SERPINA7, VTN	[52]
DR and PDR (tears)	iTRAQ, LC-MS/MS	LCN1, LTF, LACRT, LYZ, SCGB1D1	[53]
RRD, PDR, PVR, MH-RD (silicon oil fluid)	Multiplex ELISA	FGF2, IL-10, IL-12p40, IL-8, VEGF, TGFB	[54]

BRVO, branch retinal vein occlusion; *DR*, diabetic retinopathy; *ERM*, epiretinal membrane; *IMH*, idiopathic macular hole; *INIU*, idiopathic noninfectious uveitis; *IOP*, intraocular pressure; *MH-RD*, macular hole–related retinal detachment; *nvAMD*, neovascular AMD; *PDR*, proliferative diabetic retinopathy; *PVR*, proliferative vitreoretinopathy; *PVRL*, primary vitreoretinal lymphoma; *ROP*, retinopathy of prematurity; *RRD*, rhegmatogenous retinal detachment; *RVO*, retinal vein occlusion; *XLRS*, X-linked retinoschisis.

diseases [3]. Proteomics may be more advantageous in repurposing therapies, as it can directly reveal targetable biomarkers, unlike personalized genetic analysis which typically only reveals genetic risk factors. Proteomics may therefore be particularly useful for vitreoretinal diseases, where the first-line therapy is nonspecific immunosuppressants.

PVR is an example of a disease where ineffective corticosteroids and nonspecific immunosuppressive medications are the primary pharmacologic treatment. PVR, which most commonly develops as a complication of retinal detachment surgery, is characterized by the proliferation of glial and retinal pigment epithelium cells on either side of the retina, forming fibrotic membranes that result in retinal redetachment and the initiation of new retinal tears [26]. While surgery is the mainstay of PVR treatment, the process is complicated and often leads to poor visual function.

Proteomics showed that there were clear molecular differences in the protein profiles of patients with early versus advanced disease. In a study by Roybal et al., patients with early PVR demonstrated preferentially elevated levels of proteins involved in T-cell recruitment, and pathway analysis revealed higher levels of 16 downstream effectors of the mTOR signaling pathway [27]. However, patients with advanced PVR showed upregulated monocyte chemoattractants, suggesting elevated monocyte activity was caused by chronic inflammation.

Together, these results suggest that mTOR inhibitors may be more effective in treating early PVR, preventing progression to advanced disease stages. Further pathway analysis showed elevated levels of IL-13 in PVR vitreous; given that IL-13 is known to produce monocytes that are resistant to glucocorticoids, this may also explain why patients with PVR may not always respond to glucocorticoid treatments [28].

A more effective alternative treatment was also discovered through drug repositioning for NIV, a progressive inflammatory ocular disease that culminates in severe blindness [3]. The natural history of NIV is predominantly characterized by neovascularization, retinal pigmentation, presence of cystoid macular edema, and intraocular inflammation [29]. Traditionally, NIV is treated with nonspecific immunosuppressive medications such as infliximab, which suppresses TNF-α activity, and corticosteroids. Velez et al.'s study, however, suggests that such treatments may have been ineffective due to normal TNF levels in patients with NIV.

Cytokine array analysis also showed that pathways associated with corticosteroids were nonexistent in fibrotic NIV, further attesting to the likely inefficacy of such conventional drugs. Patients with NIV exhibit high levels of VEGF, T-cell proliferative markers, and IL-6, all of which later decreased in concentration when targeted by repurposed bevacizumab (i.e., anti-VEGF therapy), methotrexate, and tocilizumab (i.e., anti-IL-6 therapy), respectively.

Importance for healthcare providers and institutions

Whole exome and whole genome sequencing are incapable of quantifying levels of transcription and protein expression and turnover, which are perhaps the main reasons for the differences seen between two patients of varying disease severity and response to therapy [30].

Randomized clinical trials, for example, guide patient care based on the approach that best suits the large majority of subjects. This "one size fits all" approach to medicine, however, can lead to surprisingly high rates of failure and increased costs[18]. Proteomics may provide valuable information regarding protein expression levels, which can be used to predict disease severity and guide treatment approaches according to each individual's protein profile [31].

The wide establishment of proteomics within clinics and the development of new pointed clinical therapies may necessitate the stratification of patients into small subgroups according to their specific response to therapy. Obtaining insurance approval or reimbursement for treatments may rely more specifically on the methods used for personalized evaluation and the resulting clinical and cost-effectiveness of the procedures [32].

Conclusion

Future proteomics studies should begin to place emphasis on translating mass biomarker data from the bench into practical applications in the clinic. Other noninvasive methods within ophthalmology, such as the sampling of tear fluid, may serve as valuable means of obtaining proteomic biomarker data. Despite the small volume of tear fluid typically obtained, initial proteomic studies have already identified almost 2000 unique proteins and several biomarkers for lacrimal gland diseases [33].

Financial support

The Jonas Children's Vision Care and Bernard & Shirlee Brown Glaucoma Laboratory are supported by the National Institutes of Health [P30EY019007, R01EY018213, R01EY024698, R01EY026682, R21AG050437], National Cancer Institute Core [5P30CA013696], Foundation Fighting Blindness [TA-NMT-0116-0692-COLU], the Research to Prevent Blindness (RPB) Physician-Scientist Award, and unrestricted funds from RPB, New York, NY, USA. J.R.L.C. is supported by the Global Ophthalmology Awards Program (GOAP), a Bayer-sponsored initiative committed to supporting ophthalmic research across the world.

References

[1] Daubert MA, Jeremias A. The utility of troponin measurement to detect myocardial infarction: review of the current findings. Vasc. Health Risk Manag. September 7, 2010;6:691–9. PMC2941782.

[2] Ashburn TT, Thor KB. Drug repositioning: identifying and developing new uses for existing drugs. Nat. Rev. Drug Discov. August 2004;3(8):673–83.

[3] Velez G, Bassuk AG, Colgan D, Tsang SH, Mahajan VB. Therapeutic drug repositioning using personalized proteomics of liquid biopsies. JCI Insight December 21, 2017;2(24):PMC5752263.

[4] Novac N. Challenges and opportunities of drug repositioning. Trends Pharmacol. Sci. May 2013;34(5):267–72.

[5] Cole CJ, Kwan AS, Laidlaw DA, Aylward GW. A new technique of combined retinal and choroidal biopsy. Br. J. Ophthalmol. October 2008;92(10):1357–60.

[6] Lim LL, Suhler EB, Rosenbaum JT, Wilson DJ. The role of choroidal and retinal biopsies in the diagnosis and management of atypical presentations of uveitis. Trans. Am. Ophthalmol. Soc. 2005;103:84–91. Discussion 91–82. PMC1447562.

[7] Le Goff MM, Bishop PN. Adult vitreous structure and postnatal changes. Eye (Lond.) October 2008;22(10):1214–22.

[8] Skeie JM, Mahajan VB. Dissection of human vitreous body elements for proteomic analysis. J. Vis. Exp. January 23, 2011;(47) PMC3182653.

[9] Velez G, Tang PH, Cabral T, Cho GY, Machlab DA, Tsang SH, Bassuk AG, Mahajan VB. Personalized proteomics for precision health: identifying biomarkers of vitreoretinal disease. Transl. Vis. Sci. Technol. September 2018;7(5):12. PMC6159735.

[10] Zhang P, Zhu M, Zhao Y, Qian J, Dufresne C, Turner R, Semba RD, Solomon SD. A proteomic approach to understanding the pathogenesis of idiopathic macular hole formation. Clin. Proteomics 2017;14:37. PMC5688700.

[11] Mandal N, Kofod M, Vorum H, Villumsen J, Eriksen J, Heegaard S, Prause JU, Ahuja S, Honore B, la Cour M. Proteomic analysis of human vitreous associated with idiopathic epiretinal membrane. Acta Ophthalmol. June 2013;91(4):e333–4.

[12] Skeie JM, Brown EN, Martinez HD, Russell SR, Birkholz ES, Folk JC, Boldt HC, Gehrs KM, Stone EM, Wright ME, Mahajan VB. Proteomic analysis of vitreous biopsy techniques. Retina 2012 November–December;32(10):2141–9. PMC3637028.

[13] Kim TW, Kang JW, Ahn J, Lee EK, Cho KC, Han BN, Hong NY, Park J, Kim KP. Proteomic analysis of the aqueous humor in age-related macular degeneration (AMD) patients. J. Proteome Res. August 3, 2012;11(8):4034–43.

[14] Yao J, Liu X, Yang Q, Zhuang M, Wang F, Chen X, Hang H, Zhang W, Liu Q. Proteomic analysis of the aqueous humor in patients with wet age-related macular degeneration. Proteonomics Clin. Appl. August 2013;7(7–8):550–60.

[15] Chiang SY, Tsai ML, Wang CY, Chen A, Chou YC, Hsia CW, Wu YF, Chen HM, Huang TH, Chen PH, Liu HT, Shui HA. Proteomic analysis and identification of aqueous humor proteins with a pathophysiological role in diabetic retinopathy. J. Proteomics June 6, 2012;75(10):2950–9.

[16] Nakanishi T, Koyama R, Ikeda T, Shimizu A. Catalogue of soluble proteins in the human vitreous humor: comparison between diabetic retinopathy and macular hole. J. Chromatogr. B Analyt. Technol. Biomed. Life Sci. August 25, 2002;776(1):89–100.

[17] Zhang Y, Fonslow BR, Shan B, Baek MC, Yates 3rd JR. Protein analysis by shotgun/bottom-up proteomics. Chem. Rev. April 10, 2013;113(4):2343–94. PMC3751594.

[18] Duarte TT, Spencer CT. Personalized proteomics: the future of precision medicine. Proteomes 2016;4(4). PMC5117667.

[19] Dong MW, Zhang K. Ultra-high-pressure liquid chromatography (UHPLC) in method development. Trac. Trends Anal. Chem. December 2014;63:21–30.

[20] Murthy KR, Goel R, Subbannayya Y, Jacob HK, Murthy PR, Manda SS, Patil AH, Sharma R, Sahasrabuddhe NA, Parashar A, Nair BG, Krishna V, Prasad TK, Gowda H, Pandey A. Proteomic analysis of human vitreous humor. Clin. Proteomics 2014;11(1):29. PMC4106660.

[21] Karas M, Hillenkamp F. Laser desorption ionization of proteins with molecular masses exceeding 10,000 daltons. Anal. Chem. October 15, 1988;60(20):2299–301.

[22] Fenn JB, Mann M, Meng CK, Wong SF, Whitehouse CM. Electrospray ionization for mass spectrometry of large biomolecules. Science (New York N.Y.) October 6, 1989;246(4926):64–71.

[23] Strimbu K, Tavel JA. What are biomarkers? Curr. Opin. HIV AIDS November 2010;5(6):463–6. PMC3078627.

[24] Kavallaris M, Marshall GM. Proteomics and disease: opportunities and challenges. Med. J. Aust. June 6, 2005;182(11):575–9.

[25] Steely Jr HT, Clark AF. The use of proteomics in ophthalmic research. Pharmacogenomics August 2000;1(3):267–80.

[26] Pastor JC, Rojas J, Pastor-Idoate S, Di Lauro S, Gonzalez-Buendia L, Delgado-Tirado S. Proliferative vitreoretinopathy: a new concept of disease pathogenesis and practical consequences. Prog. Retin. Eye Res. March 2016;51:125–55.

[27] Roybal CN, Velez G, Toral MA, Tsang SH, Bassuk AG, Mahajan VB. Personalized proteomics in proliferative vitreoretinopathy implicate hematopoietic cell recruitment and mTOR as a therapeutic target. Am. J. Ophthalmol. February 2018;186:152–63. PMC5805631.

[28] Spahn JD, Szefler SJ, Surs W, Doherty DE, Nimmagadda SR, Leung DY. A novel action of IL-13: induction of diminished monocyte glucocorticoid receptor-binding affinity. J. Immunol. September 15, 1996;157(6):2654–9.

[29] Bennett SR, Folk JC, Kimura AE, Russell SR, Stone EM, Raphtis EM. Autosomal dominant neovascular inflammatory vitreoretinopathy. Ophthalmology September 1990;97(9):1125–35. Discussion 1135–1126.

[30] Maier T, Guell M, Serrano L. Correlation of mRNA and protein in complex biological samples. FEBS Lett. December 17, 2009;583(24):3966–73.

[31] Velez G, Roybal CN, Colgan D, Tsang SH, Bassuk AG, Mahajan VB. Precision medicine: personalized proteomics for the diagnosis and treatment of idiopathic inflammatory disease. JAMA Ophthalmol. April 2016;134(4):444–8. PMC4833518.

[32] Doble B, Harris A, Thomas DM, Fox S, Lorgelly P. Multiomics medicine in oncology: assessing effectiveness, cost-effectiveness and future research priorities for the molecularly unique individual. Pharmacogenomics September 2013;14(12):1405–17.

[33] Chen L, Zhou L, Chan EC, Neo J, Beuerman RW. Characterization of the human tear metabolome by LC-MS/MS. J. Proteome Res. October 7, 2011;10(10):4876–82.

[34] Pollreisz A, Funk M, Breitwieser FP, Parapatics K, Sacu S, Georgopoulos M, Dunavoelgyi R, Zlabinger GJ, Colinge J, Bennett KL, Schmidt-Erfurth U. Quantitative proteomics of aqueous and vitreous fluid from patients with idiopathic epiretinal membranes. Exp. Eye Res. March 2013;108:48–58.

[35] Koss MJ, Hoffmann J, Nguyen N, Pfister M, Mischak H, Mullen W, Husi H, Rejdak R, Koch F, Jankowski J, Krueger K, Bertelmann T, Klein J, Schanstra JP, Siwy J. Proteomics of vitreous humor of patients with exudative age-related macular degeneration. PLoS One 2014;9(5):e96895. PMC4020801.

[36] Nobl M, Reich M, Dacheva I, Siwy J, Mullen W, Schanstra JP, Choi CY, Kopitz J, Kretz FT, Auffarth GU, Koch F, Koss MJ. Proteomics of vitreous in neovascular age-related macular degeneration. Exp. Eye Res. May 2016;146:107–17.

[37] Sugioka K, Saito A, Kusaka S, Kuniyoshi K, Shimomura Y. Identification of vitreous proteins in retinopathy of prematurity. Biochem. Biophys. Res. Commun. July 1, 2017;488(3):483–8.

[38] Naru J, Aggarwal R, Singh U, Mohanty AK, Bansal D, Mangat N, Kakkar N, Agnihotri N. Proteomic analysis of differentially expressed proteins in vitreous humor of patients with retinoblastoma using iTRAQ-coupled ESI-MS/MS approach. Tumour Biol. October 2016;37(10):13915–26.

[39] Yamane K, Minamoto A, Yamashita H, Takamura H, Miyamoto-Myoken Y, Yoshizato K, Nabetani T, Tsugita A, Mishima HK. Proteome analysis of human vitreous proteins. Mol. Cell. Proteom. November 2003;2(11):1177–87.

[40] Kim SJ, Kim S, Park J, Lee HK, Park KS, Yu HG, Kim Y. Differential expression of vitreous proteins in proliferative diabetic retinopathy. Curr. Eye Res. March 2006;31(3):231–40.

[41] Garcia-Ramirez M, Canals F, Hernandez C, Colome N, Ferrer C, Carrasco E, Garcia-Arumi J, Simo R. Proteomic analysis of human vitreous fluid by fluorescence-based difference gel electrophoresis (DIGE): a new strategy for identifying potential candidates in the pathogenesis of proliferative diabetic retinopathy. Diabetologia June 2007;50(6):1294–303.

[42] Gao BB, Chen X, Timothy N, Aiello LP, Feener EP. Characterization of the vitreous proteome in diabetes without diabetic retinopathy and diabetes with proliferative diabetic retinopathy. J. Proteome Res. June 2008;7(6):2516–25.

[43] Shitama T, Hayashi H, Noge S, Uchio E, Oshima K, Haniu H, Takemori N, Komori N, Matsumoto H. Proteome profiling of vitreoretinal diseases by cluster analysis. Proteonomics Clin. Appl. September 2008;2(9):1265–80. PMC2600457.

[44] Reich M, Dacheva I, Nobl M, Siwy J, Schanstra JP, Mullen W, Koch FH, Kopitz J, Kretz FT, Auffarth GU, Koss MJ. Proteomic analysis of vitreous humor in retinal vein occlusion. PLoS One 2016;11(6):e0158001. PMC4928959.

[45] Dacheva I, Reich M, Nobl M, Ceglowska K, Wasiak J, Siwy J, Zurbig P, Mischak H, Koch FHJ, Kopitz J, Kretz FTA, Tandogan T, Auffarth GU, Koss MJ. [Proteome analysis of undiluted vitreous humor in patients with branch retinal vein occlusion]. Ophthalmologe March 2018;115(3):203–15.

[46] Kang GY, Bang JY, Choi AJ, Yoon J, Lee WC, Choi S, Yoon S, Kim HC, Baek JH, Park HS, Lim HJ, Chung H. Exosomal proteins in the aqueous humor as novel biomarkers in patients with neovascular age-related macular degeneration. J. Proteome Res. February 7, 2014;13(2):581–95.

[47] Yang Q, Lu H, Song X, Li S, Wei W. iTRAQ-based proteomics investigation of aqueous humor from patients with coats' disease. PLoS One 2016;11(7):e0158611. PMC4944970.

[48] Velez G, Roybal CN, Binkley E, Bassuk AG, Tsang SH, Mahajan VB. Proteomic analysis of elevated intraocular pressure with retinal detachment. Am. J. Ophthalmol. Case Rep. April 2017;5:107–10. PMC5560621.

[49] Kuiper JJ, Beretta L, Nierkens S, van Leeuwen R, Ten Dam-van Loon NH, Ossewaarde-van Norel J, Bartels MC, de Groot-Mijnes JD, Schellekens P, de Boer JH, Radstake TR. An ocular protein triad can classify four complex retinal diseases. Sci. Rep. January 27, 2017;7:41595. PMC5269719.

[50] Patel S, Ling J, Kim SJ, Schey KL, Rose K, Kuchtey RW. Proteomic analysis of macular fluid associated with advanced glaucomatous excavation. JAMA Ophthalmol. January 2016;134(1):108–10. PMC5523109.

[51] Sudha D, Kohansal-Nodehi M, Kovuri P, Manda SS, Neriyanuri S, Gopal L, Bhende P, Chidambaram S, Arunachalam JP. Proteomic profiling of human intraschisis cavity fluid. Clin. Proteomics 2017;14:13. PMC5404285.

[52] Naru J, Aggarwal R, Mohanty AK, Singh U, Bansal D, Kakkar N, Agnihotri N. Identification of differentially expressed proteins in retinoblastoma tumors using mass spectrometry-based comparative proteomic approach. J. Proteomics April 21, 2017;159:77–91.

[53] Csosz E, Boross P, Csutak A, Berta A, Toth F, Poliska S, Torok Z, Tozser J. Quantitative analysis of proteins in the tear fluid of patients with diabetic retinopathy. J. Proteomics April 3, 2012;75(7): 2196–204.

[54] Kaneko H, Takayama K, Asami T, Ito Y, Tsunekawa T, Iwase T, Funahashi Y, Ueno S, Nonobe N, Yasuda S, Suzumura A, Shimizu H, Kimoto R, Hwang SJ, Terasaki H. Cytokine profiling in the sub-silicone oil fluid after vitrectomy surgeries for refractory retinal diseases. Sci. Rep. May 25, 2017;7(1):2640. PMC5454016.

Chapter 24

Role of respiratory proteomics in precision medicine

V.S. Priyadharshini[1,2] and Luis M. Teran[1]

[1]Instituto Nacional de Enfermedades Respiratorias, Delegación Tlalpan, Mexico; [2]Escuela Superior de Medicina del Instituto Politecnico Nacional, Plan de San Luis y Díaz Miron, Mexico

Context

Asthma affects over 300 million people in the world, and chronic obstructive pulmonary disease (COPD) ranks fourth among the deadliest diseases. Idiopathic pulmonary fibrosis (IPF) can similarly cause deaths, and all fibrotic lung diseases are marked by progressive interstitial fibrosis, blocking gaseous exchange and causing rapidly diminishing quality of life. These patients usually only live between three to five years, after diagnosis. These conditions involve intricate molecular interactions, and diverge with diverse molecular pathways, making it difficult for scientists to understand the precise pathogenesis. In respiratory diseases, most patients do not present early symptoms.

Current tests such as chest X-ray, lung function tests, and histopathological techniques are useful; however, they are not specific. Spirometry is more valuable in cases with obstructed airways such as COPD, restrictive pulmonary diseases, and asthma. Nevertheless, there are examples of both overdiagnosis and under diagnosis [1,2]. Genomics has proved useful, yet posttranslational modifications of proteins such as methylation, acetylation, glycosylation, and protein folding alter the phenotype. A gene can also generate different peptide products, through alternative splicing events besides posttranslational modifications, producing several types of proteins in each cell.

Proteomics enables an analysis of complete protein expression, rather than using a single biomarker approach, and advances with new technologies have begun to explain the complexity of the proteome. Indeed, while DNA carries genetic information, it is the proteins that control all biological events both in health and disease.

The human proteome map was constructed using mass spectrometry [3,4]. Respiratory proteomics use biological samples including bronchoalveolar lavage fluid (BAL), plasma, serum, blood cells, nasal lavage fluid (NLF), sputum, exhaled breath condensate, biopsies of the lung and nasal polyps. While analysis of these biological specimens has provided beneficial information, proteomics findings often differ depending on the sample type.

Specialized proteomic equipment and techniques—gel electrophoresis

Dysregulation of proteins is a function of conditions such as disease state, time point, and drug response. Gel-based proteomics remains the most effective way to resolve a complex mixture of proteins. Two-dimensional polyacrylamide gel electrophoresis (2D PAGE) proteins, are separated in isoelectric focusing strips, according to their net charge (first dimension), followed by molecular mass separation in PAGE (second dimension) [5]. Two-dimensional difference gel electrophoresis method (2D DIGE) was invented to overcome intrinsic gel to gel variability of 2D electrophoresis. Samples can be labeled at the protein or peptide level. In 2D DIGE, the protein level is adopted, as size and charge matched samples comigrate within the gel.

A protein with a fluorescent tag covalently binds lysine residues in proteins [6]. Western blotting is performed to detect low abundance proteins, involving the separation of proteins using electrophoresis, transfer onto nitrocellulose membrane, and the precise detection of a target protein by enzyme-conjugated antibodies [7].

Mass spectrometry

Sophisticated mass spectrometry and data analysis software can identify and quantify several thousands of proteins from complex biological mixtures [8]. Mass spectrometers mainly measure mass. In proteomics, this provides

Precision Medicine for Investigators, Practitioners and Providers. https://doi.org/10.1016/B978-0-12-819178-1.00024-1

information about protein identity, chemical modifications, and structure. MS offers sensitivity to detect either digested peptides or intact proteins, and recent advancement allows targeting specific protein modifications. Main components of MS include an ionization source that converts neutral phase samples into mass phase ions. Then a mass analyzer separates and analyzes the mass of the species.

There are several types of mass analyzers, and a detector measures and amplifies the ion current of the mass resolved ions, at each mass/charge (m/z) value. Briefly, ionization sources can be of three different phases, each involving different models: (i) Gas phase—electron ionization, chemical ionization, photoionization, (ii) Solution phase—electrospray(ES), atmospheric pressure photo ionization (PI), atmospheric pressure chemical ionization (CI), (iii) Solid phase-matrix-assisted laser desorption (MALDI), plasma desorption (PD), fast atom bombardment (FAB). Electrospray ionization (ESI) and matrix-assisted laser desorption/ionization (MALDI) are routinely used in the mass spectrometric analysis. ESI ionizes the analytes out of a solution at a high voltage, and is coupled to high pressure liquid chromatography (HPLC), liquid chromatography (LC), or gel electrophoresis, followed by MS/MS analysis. It produces multicharged ions.

MALDI ionizes the samples into the gas phase in vacuum and high voltage; the crystalline matrix is targeted with laser pulses, and rapid sublimation causes the conversion of ions. MALDI-MS is typically used to analyze relatively simple peptide mixtures, whereas integrated liquid-chromatography ESI-MS systems (LC-MS) are favored for the analysis of complex samples.

Other analytical architectures

In the time-of-flight (TOF) technique the ions are accelerated in a flight tube, and the time required to fly to the detector is measured coupled with MALDI. Quadrupole is an ion beam MS ion trap that traps ions using electrical field, and measures by selectively ejecting them to the detector. Fourier transform ion cyclotron resonance (FTICR) is like an ion trap; however, it uses cyclotron motion to resolve ions. LC-MS is a powerful technique that is oriented toward the identification of proteins in complex mixtures. In this strategy, proteins are enzyme digested before liquid chromatography. Then, the mass spectra generated by MS analysis are compared with the theoretical MS/MS spectra from the databases.

Targeted proteomics

Proteomics can also be branched into "shotgun" and targeted proteomics. The shotgun proteomics is a data-dependent acquisition. The most important advantage of this approach is the ability to analyze thousands of proteins in one run. However, "shotgun" methods can often produce low reproducibility of peptide identification, which may be conquered by using the targeted approach. Targeted proteomics are primarily based on multiple-, selected- or parallel reaction monitoring (MRM, SRM, and PRM, respectively), which produces precise and repeatable quantification of predefined proteins [9].

In the recent past, another method of acquisition has been created: data-independent acquisition (DIA). In DIA analysis, precursor ions from determined m/z isolation window are deterministically fragmented. Currently, sequential window acquisition of all theoretical mass spectra (SWATH) (illustrated by DIA) has become standard for biomarker discovery, based on targeted data analysis and spectral libraries [10].

Bottom-up and top-down

Bottom-up proteomics achieves protein identification by analysis of peptide fragments, generated by proteolytic digestion of intact proteins. The proteins can first be separated by GE or chromatography, in which case the sample will contain only one or a few proteins. Alternatively, a complex protein mixture can initially be digested to the peptide level, then separated by online chromatography, coupled to electrospray mass spectrometry (ESI–MS). In the latter case, the digest can contain thousands to hundreds of thousands of peptides, and require the separation in two or more chromatographic dimensions before MS analysis. The identity of the original protein is determined by comparison of the peptide mass spectra, with theoretical peptide masses calculated from a proteomic or genomic database.

In top-down proteomics, intact protein molecular ions generated by ESI are introduced into the mass analzser, and are subjected to gas-phase fragmentation. An obstacle to this approach is the determination of product ion masses, from multiple charged product ions. These can vary in charge state, up to that of the multiple charged protein precursor ion. This can introduce ambiguity in the interpretation of top-down MS–MS spectra. Two approaches have been used to circumvent this limitation. The first is charge state manipulation through gas phase ion-ion interactions, and the second is the use of instruments with high mass measurement accuracy (MMA).

Interfaces with genome, transcriptome, and metabolome

Genomics is the study of the whole genome, whereas the transcriptome is the total complement of ribonucleic acid (RNA) transcripts in a cell. The analysis of mRNAs provides direct insight into the cell- and tissue-specific gene expression features, such as (i) differential gene expression (ii) evaluation of alternative/differential splicing, to assess or predict protein isoforms and (iii) quantitative assessment of genotype influence on gene expression, using allele-specific expression (ASE). This information is integral for a better understanding of the dynamics of cellular and tissue metabolism, and to understand whether and how changes in the transcriptome profiles affect health and disease.

Metabolites are the products and substrates of metabolism that drive essential cellular functions, such as energy production and storage, signal transduction, and apoptosis. These large-scale omics technologies (genomics, transcriptomics, proteomics, metabolomics) [9–13] have revolutionized biology to advance our understanding of biological processes. However, integrating data from omics studies is challenging.

Multiomics integration

Currently available tools include web-based datasets requiring no computational experience, as well as more versatile tools for those with computational experience. User-friendly, web-based tools requiring no computational experience include Paintomics, 3 Omics, and Galaxy (P, M) [14].

To identify single nucleotide polymorphisms (SNPs), associated with measurement of 88 blood proteins (protein quantitative trait loci; pQTLs), Sun et al. studied two large cohorts of current and former smokers, with and without COPD [SPIROMICS (N = 750); COPDGene (N = 590)] [15]. They identified 527 highly significant pQTLs in 38 (43%) of blood proteins tested. The pQTL SNPs explained >10% of measured variation in 13 protein biomarkers, with a single SNP (rs7041; $P = 10-392$) explaining 71%–75% of the measured variation in vitamin D binding protein (gene = GC). Some of these pQTLs (e.g., pQTLs for VDBP, sRAGE, surfactant protein D, and TNFRSF10C) have been previously associated with COPD.

Findings in pulmonary diseases

We will only discuss proteomic studies in noninfectious type of pulmonary diseases in view that present authors have already extensively reviewed other respiratory diseases [16,17]. Proteomics techniques were used to classify asthma into eosinophilic endotype [18], characterized by the significant role eosinophils play in the pathophysiology of the condition. Consistently elevated levels in sputum and/or in blood eosinophils and a considerable response to treatments that suppress eosinophilia are the features. Airway eosinophil activity may be more important than their numbers, and this needs to be investigated.

Proteomic profiles have been studied in biopsies of omalizumab responder (OR) versus non-omalizumab responder (NOR) phenotypes, after taking 36 months of treatment. Baseline galectin-3 expression was present in OR patients but was not detected in NOR. Galectin-3 detection was related to improved respiratory activity in OR, and it could be considered as a potential biomarker of long-term response to omalizumab [19].

Chronic obstructive pulmonary disease

Currently, the most promising blood COPD biomarker is the soluble receptor for advanced glycosylation end products (sRAGE) [20]. The AGER gene codes RAGE, and single nucleotide polymorphisms (SNP) in AGER are associated with COPD and emphysema, in targeted and genome-wide association studies. Three different RAGE variants have been identified (FLRAGE, cRAGE, esRAGE) using 2-DE, MS, and Western blot in lung tissues: FL-RAGE (full-length RAGE) and cRAGE (c-terminal processed full-length RAGE), but not esRAGE (spliced endogenous secretory RAGE) were declined in COPD lungs. cRAGE expression correlated with COPD progression (2.1 fold lower and 3.4 fold lower in mild and very severe COPD, respectively) [21]. In contrast, the three RANGE variants were decreased in IPF (esRAGE expression changed in IPF, but not in COPD). Thus, cRAGE and esRAGE could be used as markers of COPD progression and IPF, respectively.

Plasma sRAGE is a predictor of emphysema progression [22], and sRAGE will be the first blood biomarker of emphysema, to be submitted to the FDA (USA) and European Medicines Agency in the blood biomarker qualification program [23]. It has also been proposed that a combination of biomarkers can be a better predictor of COPD. In a study by Zemans et al. using 1465 COPD gene subjects and 2746 ECLIPSE subjects, combination of five plasma biomarkers (C-reactive protein/CRP, clara cell secretory protein/CC16, fibrinogen, surfactant protein D/SPD, and sRAGE), was associated with 13% of the variance of FEV1 in COPD Gene, and 24% of the variance in ECLIPSE, which was considerably higher than individual biomarkers [24].

Idiopathic pulmonary fibrosis

Tian et al. [25] employed the isobaric tag for relative and absolute quantitation (iTRAQ) combined liquid chromatography-tandem mass spectrometry (LC-MS/MS) method, to study protein expression in lung tissues from IPF patients. They identified a total of 662 proteins with differential expression (455 upregulated and 207 downregulated). KEGG pathway enrichment analysis (www.genome.jp/keg) demonstrated that the differentially expressed proteins in lung tissue mainly belonged to the PI3K-Akt signaling, focal adhesion, extracellular matrix/ECM-receptor interaction, and carbon metabolism pathways. According to the bioinformatic definition of the matrisome, 229 matrisome proteins were identified in lung tissue. These proteins comprised the ECM of the lung, of which 104 were core matrisome proteins, and 125 were matrisome-associated proteins.

A total of 229 ECM proteins were quantified, out of which 56 proteins were differentially expressed (19 upregulated and 37 downregulated). In addition to proteins with well-explored functions such as SCGB1A1, COL1A1, TAGLN, PSEN2, CTSB, TSPAN1, AGR2, CSPG2, and SERPINB3, Tian et al. identified novel ECM proteins with unknown function deposited in IPF lung tissue including, ASPN, LGALS7, HSP90AA1, and HSP90AB1. A few of these differentially expressed proteins were further verified using Western blot analysis and immunohistochemistry.

Similar technology was used by Zhang et al. [26] using serum samples from IPF patients, and profiled using iTRAQ coupled with two-dimensional liquid chromatography/tandem mass spectrometry (2D-LC-MS/MS), and ELISA was used to validate candidate biomarkers. Among 394 proteins, 97 were associated with IPF. Four biomarker candidates generated from iTRAQ experiments— CRP, fibrinogen-α chain, haptoglobin, and kininogen-1— were successfully verified using ELISA. CRP and fibrinogen-a were higher in IPF and haptoglobin and kininogen-1 lower in IPF.

Niu et al. [27] employed a coupled isobaric tag for relative and absolute quantitation, using iTRAQ-LC-MS/MS to analyze protein expression in patients with IPF. A total of 97 differentially expressed proteins were identified in serum samples, 38 upregulated and 59 downregulated. Applying STRING software, a regulatory network containing 87 nodes and 244 edges was built. Functional enrichment showed that differentially expressed proteins were principally involved in protein activation cascade, regulation of response to wounding, and extracellular components. Significantly hemoglobin beta/HBB, CRP, and SERPINA1 were upregulated, and APOA2, AHSG, KNG1, and AMBP downregulated.

The results confirmed the iTRAQ profiling results and AHSG, AMBP, CRP, and KNG1 were inferred as specific IPF biomarkers. Extracellular matrix remodeling is a common feature in lung diseases such as chronic obstructive pulmonary disease (COPD) and idiopathic pulmonary fibrosis (IPF). Åhrman E et al. [28] utilized a sequential tissue extraction strategy to describe disease-specific remodeling in end-stages of COPD and IPF patients. This research was based on a quantitative comparison of the disease proteomes, with a particular focus on the matrisome, using targeted data analysis (SWATH-MS) and data-independent acquisition.

Besides, Åhrman et al. showed significant qualitative and quantitative effects of the solubility of matrisome proteins. COPD was characterized by a disease-specific increase in ECM regulators, metalloproteinase inhibitor 3 (TIMP3), and matrix metalloproteinase 28 (MMP-28), whereas for IPF, impairment in cell adhesion proteins, such as collagen VI and laminins, was most prominent. For both diseases, Åhrman et al. identified an increase in protein levels involved in the regulation of endopeptidase activity, with many proteins from the serpin family.

Ongoing lines of investigation

Eight hours of measurement in the MS were required for the identification of approximately 5000 proteins only 5 years ago, whereas contemporary studies achieved similar results in just 90 min. The next step in clinical proteomics is to apply MS-based protein measurements, for diagnosis in the clinic. Proteins are routinely measured in the clinic using antibody-based techniques, primarily immunohistochemistry (IHC) and enzyme-linked immunosorbent assays (ELISA). Still, posttranslational modifications may affect the experimental result. For example, a comparison between IHC and MS-based analysis of the checkpoint proteins programmed death ligand 1 (PDL1), and programmed death ligand 2 (PDL2) showed that PDL1 staining is reduced when the protein is glycosylated. Therefore, lack of PDL1 staining does not indicate lack of protein expression and activity. Also, MS-based absolute quantification of these proteins showed similar expression levels, whereas PDL2 could hardly be detected by IHC in most samples [29].

Multiple reaction monitoring

The most reliable approach successfully integrated into clinical practice is multiple reaction monitoring (MRM) or selected reaction monitoring (SRM) of targeted MS analysis. The MRM approach overcomes many of the limitations of ELISA and provides the required specificity, multiplexing, and quantification accuracy. MRM analysis is based on rapid selection and fragmentation of proteotypic peptides, and monitoring of three to five predefined fragment ions. To surmount the challenge of the identification of optimal proteotypic peptides, which provide unparalleled protein mapping, large efforts were invested in creating databases such as the PeptideAtlas [30]. Monitoring of just a limited number of preselected ions allows highly sensitive detection of nanogram per milliliter concentrations of peptides in biological samples. However, sensitivity cannot yet compete with immunoassays for hormonal and cytokine blood markers (Fig. 24.1).

MRM assays have been coupled with an immune assay, to enrich for low abundance peptides, overcoming the sensitivity limitation. Such combination is known as Stable Isotope Standards and Capture by Anti-Peptide Antibodies

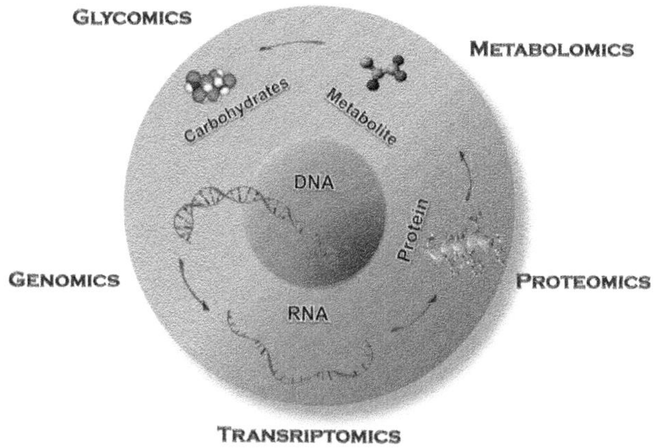

FIGURE 24.1 Multiomics technologies including genomics, transcriptomics, proteomics, metabolomics, and glycomics. These technologies are revolutionary to respiratory medicine.

(SISCAPA) [31]. It uses peptide-specific antibodies to enrich the native peptides of interest, and spiked stable isotope-labeled peptides, which serve as standards for absolute quantification. These peptides are then analyzed by MRM assays, to provide high specificity and quantitative information.

Precision Medicine

In the recent decade, precision medicine has developed as a medical care procedure that employs novel technology in the prevention of diseases striving to tailor treatments according to the precise needs of every patient—initial investigations on precision medicine utilising genomic approaches that have eventually led to important discoveries. More recently, proteomics, which has the power to analyse the expression of proteins globally rather than a single biomarker, has also begun to play a significant role in the development of new systems supporting health care. Lung cancer (LC) is one of the diseases where proteomic-based approaches have made tangible improvement in identifying biomarkers that have the potential to be used in precision respiratory medicine [32–34]. More recently, Veristrat has been used to evaluate good and poor classifications, outcomes, and circulating biomarkers potentially involved in tumour progression and treatment resistance in advanced non–small cell lung cancer (SCLC) patients treated with second-line cytotoxic chemotherapy or erlotinib [35]. In this study, it was demonstrated that VeriStrat good patients benefit from both EGFR TKI and single-agent chemotherapy in the EGFR wild-type population. Thrombospondin-2, C-reactive protein, TNF-receptor 1, and placental growth factor were the analytes most highly associated with overall survival. A biochemical proteomic strategy using kinobeads and quantitative mass spectrometry for the identification of kinase inhibitor resistance mechanisms in cancer cells found MET and ephrin type-A receptor 2 (EPHA2) in gefitinib-resistant HCC827 cells inferring a possible role in developing resistance [36]. siRNA-mediated EPHA2 knockdown or treating cells with the multikinase inhibitor dasatinib restored sensitivity to gefitinib. Using liquid chromatography-tandem mass spectrometry, Tripathi et al. [37] studied mechanisms of chemoresistance of SCLC cell lines and identified five cell surface receptors which differentiated between chemoresistant and chemosensitive cells—among them, upregulation of cell surface glycoprotein MUC18 (MCAM) in chemoresistant tumors. These observations altogether demonstrate the usefulness of proteomics to deliver precision care in respiratory medicine. Angelidis et al. integrated single-cell signatures of ageing in mice to investigate ageing-associated molecular and cellular alterations in the lung, using a combination of transcriptomics and proteomics [38]. While basement membrane collagen IV genes were all downregulated on the mRNA level, they were upregulated at the protein level explaining that protein posttranscriptional regulation takes place. Indeed, proteomic profiling revealed extracellular matrix remodelling in old mice, including increased collagen IV and XVI and lowered Fraser syndrome complex proteins and collagen XIV. Computational integration of the ageing proteome with the single-cell transcriptome predicted the cellular source of regulated proteins and generated an unbiased reference map of the ageing lung.

References

[1] Heffler E, Crimi C, Mancuso S, Campisi R, Puggioni F, Brussino L, Crimi N. Misdiagnosis of asthma and COPD and underuse of spirometry in primary care unselected patients. Respir. Med. September 2018;142:48–52. https://doi.org/10.1016/j.rmed.2018.07.015.

[2] Spero K, Bayasi G, Beaudry L, Barber KR, Khorfan F. Overdiagnosis of COPD in hospitalized patients. Int. J. Chronic Obstr. Pulm. Dis. August 11, 2017;12:2417–23. https://doi.org/10.2147/COPD.S139919.

[3] Reardon S. Precision-medicine plan raises hopes. Nature January 29, 2015;517(7536):540.

[4] Kim MS, Pinto SM, Getnet D, Nirujogi RS, Manda SS, Chaerkady R, et al. A draft map of the human proteome. Nature 2014;509:575–81.

[5] Carrette O, Burkhard PR, Sanchez JC, Hochstrasser DF. State-of-the-art two-dimensional gel electrophoresis: a key tool of proteomics research. Nat. Protoc. 2006;1:812–23.

[6] Beckett P. The basics of 2D DIGE. Methods Mol. Biol. 2012;854:9–19. https://doi.org/10.1007/978-1-61779-573-2_2. PubMed PMID: 22311750.

[7] Kurien BT, Scofield RH. Western blotting: an introduction. Methods Mol. Biol. 2015;1312:17–30. https://doi.org/10.1007/978-1-4939-2694-7_5. Review. PubMed PMID: 26043986.

[8] Fujii K, Nakamura H, Nishimura T. Recent mass spectrometry-based proteomics for biomarker discovery in lung cancer, COPD, and asthma. Expert Rev. Proteomics April 2017;14(4):373–86. https://doi.org/10.1080/14789450.2017.1304215. Epub 2017 Mar 17. Review. PubMed PMID: 28271730.

[9] Nagalakshmi U, Waern K, Snyder M. RNA-Seq: a method for comprehensive transcriptome analysis. Curr. Protoc. Mol. Biol. 2010 Chapter 4: Unit 4.11.1–3.

[10] Kukurba KR, Montgomery SB. RNA sequencing and analysis. Cold Spring Harb. Protoc. 2015;2015:951–69.

[11] Fonseca NA, Marioni J, Brazma A. RNA-seq gene profiling—a systematic empirical comparison. PLoS One 2014;9:e107026.

[12] Wishart DS, Tzur D, Knox C, et al. HMDB: the human metabolome database. Nucleic Acids Res. January 2007;35(Database issue):D5216. https://doi.org/10.1093/nar/gkl923.

[13] Clish CB. Metabolomics: an emerging but powerful tool for precision medicine. Cold Spring Harb. Mol. Case Stud. October 2015;1(1):a000588. https://doi.org/10.1101/mcs. a000588.

[14] Misra BB, Langefeld CD, Olivier M, Cox LA. Integrated omics: tools, advances, and future approaches. J. Mol. Endocrinol. July 13, 2018 pii: JME-18-0055.

[15] Sun W, Kechris K, Jacobson S, Drummond MB, Hawkins GA, Yang J, Chen TH, Quibrera PM, Anderson W, Barr RG, Basta PV, Bleecker ER, Beaty T, Casaburi R, Castaldi P, Cho MH, Comellas A, Crapo JD, Criner G, Demeo D, Christenson SA, Couper DJ, Curtis JL, Doerschuk CM, Freeman CM, Gouskova NA, Han MK, Hanania NA, Hansel NN, Hersh CP, Hoffman EA, Kaner RJ, Kanner RE, Kleerup EC, Lutz S, Martinez FJ, Meyers DA, Peters SP, Regan EA, Rennard SI, Scholand MB, Silverman EK, Woodruff PG, O'Neal WK, Bowler RP, Group SR, Investigators CO. Common genetic polymorphisms influence blood biomarker measurements in COPD. PLoS Genet. 2016;12(8):e1006011. https://doi.org/10.1371/journal.pgen.1006011.

[16] Priyadharshini VS, Teran LM. Personalized medicine in respiratory disease: role of proteomics. Adv. Protein Chem. Struct. Biol. 2016;102:115–46. https://doi.org/10.1016/bs.apcsb.2015.11.008. Epub 2015 Dec 31. Review.

[17] Teran LM, Montes-Vizuet R, Li X, Franz T. Respiratory proteomics: from descriptive studies to personalized medicine. J. Proteome Res. January 2, 2015;14(1):38–50. https://doi.org/10.1021/pr500935s. Epub 2014 Nov 25. Review.

[18] Svenningsen S, Nair P. Asthma endotypes and an overview of targeted therapy for asthma. Front. Med. (Lausanne) September 26, 2017;4:158. https://doi.org/10.3389/fmed.2017.00158. eCollection 2017. Review. PubMed PMID: 29018800; PubMed Central PMCID: PMC5622943.

[19] Riccio AM, Mauri P, De Ferrari L, et al. Galectin-3: an early predictive biomarker of modulation of airway remodeling in patients with severe asthma treated with omalizumab for 36 months. Clin. Transl. Allergy 2017;7(1).

[20] Regan EA, Hersh CP, Castaldi PJ, DeMeo DL, Silverman EK, Crapo JD, Bowler RP. Omics and the search for blood biomarkers in COPD: insights from COPDGene. Am. J. Respir. Cell Mol. Biol. March 15, 2019. https://doi.org/10.1165/rcmb.2018-0245PS.

[21] Ohlmeier S, Mazur W, Salmenkivi K, Myllärniemi M, Bergmann U, Kinnula VL. Proteomic studies on receptor for advanced glycation end product variants in idiopathic pulmonary fibrosis and chronic obstructive pulmonary disease. Proteomics Clin. Appl. 2010;4:97–105.

[22] Coxson HO, Dirksen A, Edwards LD, Yates JC, Agusti A, Bakke P, Calverley PM, Celli B, Crim C, Duvoix A, Fauerbach PN, Lomas DA, Macnee W, Mayer RJ, Miller BE, Muller NL, Rennard SI, Silverman EK, Tal-Singer R, Wouters EF, Vestbo J, Evaluation of CLtIPSEI. The presence and progression of emphysema in COPD as determined by CT scanning and biomarker expression: a prospective analysis from the ECLIPSE study. Lancet Respir. Med. 2013;1(2):129–36. https://doi.org/10.1016/S2213-2600(13)70006-7.

[23] Casaburi R, Celli B, Crapo J, Criner G, Croxton T, Gaw A, Jones P, Kline-Leidy N, Lomas DA, MerrillmD, Polkey M, Rennard S, Sciurba F, Tal-Singer R, Stockley R, Turino G, Vestbo J, Walsh J. The COPD biomarker qualification consortium (CBQC). COPD 2013;10(3):367–77. https://doi.org/10.3109/15412555.2012.752807.

[24] Zemans RL, Jacobson S, Keene J, Kechris K, Miller BE, Tal-Singer R, Bowler RP. Multiple biomarkers predict disease severity, progression and mortality in COPD. Respir. Res. 2017;18(1):117. https://doi.org/10.1186/s12931-017-0597-7.

[25] Tian Y, Li H, Gao Y, Liu C, Qiu T, Wu H, Cao M, Zhang Y, Ding H, Chen J, Cai H. Quantitative proteomic characterization of lung tissue in idiopathic pulmonary fibrosis. Clin. Proteomics February 6, 2019;16:6. https://doi.org/10.1186/s12014-019-9226-4. eCollection 2019. PubMed PMID: 30774578; PubMed Central PMCID: PMC6364390.

[26] Zhang Y, Xin Q, Wu Z, Wang C, Wang Y, Wu Q, Niu R. Application of isobaric tags for relative and absolute quantification (iTRAQ) coupled with two-dimensional liquid chromatography/tandem mass spectrometry in quantitative proteomic analysis for discovery of serum biomarkers for idiopathic pulmonary fibrosis. Med. Sci. Monit. June 17, 2018;24:4146–53. https://doi.org/10.12659/MSM.908702. PubMed PMID: 29909421; PubMed Central PMCID: PMC6036962.

[27] Niu R, Liu Y, Zhang Y, Zhang Y, Wang H, Wang Y, Wang W, Li X. iTRAQ-based proteomics reveals novel biomarkers for idiopathic pulmonary fibrosis. PLoS One January 25, 2017;12(1):e0170741. https://doi.org/10.1371/journal.-pone.0170741. eCollection 2017. PubMed PMID: 28122020; PubMed Central PMCID: PMC5266322.

[28] Åhrman E, Hallgren O, Malmström L, Hedström U, Malmström A, Bjermer L, Zhou XH, Westergren-Thorsson G, Malmström J. Quantitative proteomic characterization of the lung extracellular matrix in chronic obstructive pulmonary disease and idiopathic pulmonary fibrosis. J. Proteomics October 30, 2018;189:23–33. https://doi.org/10.1016/j.jprot.2018.02.027. Epub 2018 Mar 1. PubMed PMID: 29501846.

[29] Morales-Betanzos CA, Lee H, Gonzalez-Ericsson PI, Balko JM, Johnson DB, Zimmerman LJ, Liebler DC. Quantitative mass spectrometry analysis of PD-L1 protein expression, N-glycosylation and expression stoichiometry with PD-1 and PD-L2 in human melanoma. Mol. Cell. Proteomics 2017. https://doi.org/10.1074/mcp.RA117.000037 (Epub ahead of print).

[30] Kusebauch U, Campbell DS, Deutsch EW, Chu CS, Spicer DA, Brusniak MY, Slagel J, Sun Z, Stevens J, Grimes B, Shteynberg D, Hoopmann MR, Blattmann P, Ratushny AV, Rinner O, Picotti P, Carapito C, Huang CY, Kapousouz M, Lam H, Tran T, Demir E, Aitchison JD, Sander C, Hood L, Aebersold R, Moritz RL. Human SRMAtlas: a resource of targeted assays to quantify the complete human proteome. Cell 2016;166:766–78.

[31] Anderson NL, Anderson NG, Haines LR, Hardie DB, Olafson RW, Pearson TW. Mass spectrometric quantitation of peptides and proteins using stable isotope standards and capture by anti-peptide antibodies (SISCAPA). J. Proteome Res. 2004;3:235–44.
[32] Amann JM, Lee JW, Roder H, Brahmer J, Gonzalez A, Schiller JH, Carbone DP. Genetic and proteomic features associated with survival after treatment with erlotinib in first-line therapy of non-small cell lung cancer in Eastern Cooperative Oncology Group 3503. J Thorac Oncol 2010;5:169–78.
[33] Lazzari C, Spreafico A, Bachi A, Roder H, Floriani I, Garavaglia D. Changes in plasma mass spectral profile in course of treatment of non-small cell lung cancer patients with epidermal growth factor receptor tyrosine kinase inhibitors. J Thorac Oncol 2012;7:40–8.
[34] Carbone DP, Ding K, Roder H, Grigorieva J, Roder J, Tsao MS, et al. Prognostic and predictive role of the VeriStrat plasma test in patients with advanced non-small cell lung cancer treated with erlotinib or placebo in the NCIC Clinical Trials Group BR21 trial. J Thorac Oncol 2012;7:1653–60.
[35] Fidler MJ, Fhied CL, Roder J, Basu S, Sayidine S, Fughhi I. The serum-based VeriStrat® test is associated with proinflammatory reactants and clinical outcome in non-small cell lung cancer patients. BMC Cancer 2018 Mar 20;18:310.
[36] Koch H, Busto ME, Kramer K, Médard G, Kuster B. Chemical Proteomics Uncovers EPHA2 as a Mechanism of Acquired Resistance to Small Molecule EGFR Kinase Inhibition. J Proteome Res 2015 Jun 5;14(6):2617–25.
[37] Tripathi SC, Fahrmann JF, Celiktas M, Aguilar M, Marini KD, Jolly MK, et al. MCAM Mediates Chemoresistance in Small-Cell Lung Cancer via the PI3K/AKT/SOX2 Signaling Pathway. Cancer Res 2017 Aug 15;77(16):4414–25.
[38] Angelidis I, Simon LM, Fernandez IE, Strunz M, Mayr CH, Greiffo FR, et al. An atlas of the aging lung mapped by single cell transcriptomics and deep tissue proteomics. Nat Commun 2019 Feb 27;10(1):963. https://doi.org/10.1038/s41467-019-08831-9.

Chapter 25

Cardiovascular proteomics

Sheon Mary[1], Gemma Currie[1], Aletta E. Schutte[2] and Christian Delles[1]

[1]*Institute of Cardiovascular and Medical Sciences, University of Glasgow, Glasgow, United Kingdom;* [2]*Hypertension in Africa Research Team (HART), MRC Unit for Hypertension and Cardiovascular Disease, North-West University, Potchefstroom, South Africa*

Introduction

Aging societies, urbanization, and changes in diet are among the main reasons for the rising prevalence of cardiovascular disease (CVD) worldwide. CVD is the major cause of morbidity and mortality, in high-, low-, and middle income countries [1,2]. CVD often occurs together with other chronic diseases such as obesity, diabetes, and chronic kidney disease, where CVD drives the pathogenesis of these conditions and vice versa, leading to a vicious cycle that further increases the global burden of CVD. CVD has a profound impact on quality of life, and often leads to severe disability due to pain, dyspnoea, neurological deficits, and dementia. Together with the high prevalence of CVD, these factors are responsible for the financial burden associated with CVD. Reduction of the incidence of CVD could lead to significant cost savings, for societies and healthcare systems. In the United States, the health economic benefit associated with a reduction in the incidence of heart disease or hypertension would each be by far greater than that of similar reductions in the incidence of cancer or lung disease [3]. One way of achieving better prevention and therapy of people with CVD is through development of precision medicine strategies [4].

CVD can be driven by well-defined pathogenetic principles. Monogenic forms of hypertension, inherited cardiac arrhythmias, and chemotherapy-related cardiomyopathies are examples of conditions with a clear-cut relationship between a pathogenetic factor and development of CVD. While such conditions can still be modified by environmental and other factors, including age, diet, and drug therapy, the association between causative factor and cardiovascular phenotype is strong. In contrast, the vast majority of CVDs are multifactorial in origin, develop over decades, and are characterized by complex interactions between genetic factors, risk factors, and environmental exposures.

Primary hypertension, chronic kidney disease, ischemic heart disease, heart failure, and stroke are among the most prevalent CVDs, and are prime examples of complex, multifactorial diseases. Dzau and Braunwald proposed that such conditions develop from risk factors to subclinical, functional, and structural changes, before symptoms clinically manifest and disease appears [5]. Each stage in this "cardiovascular continuum" is associated with characteristic pathogenetic principles, such as inflammation and atherogenesis, and interacts with environmental factors, thereby offering the opportunity for targeted preventative and therapeutic strategies (Fig. 25.1).

One of the major challenges is that the relationship between genetic factors and disease phenotype is less robust in most CVDs compared to, for example, forms of breast cancer, where mutations in specific genes are associated with a very high risk of developing disease. In addition, the high prevalence of comorbidities in patients with CVD needs to be taken into account: precise treatment of CVD should ideally also have beneficial effects on comorbid conditions, or at least not lead to adverse effects in other organ systems. Finally, in contrast to many other clinical conditions, tissue-based diagnosis is not available in most patients with CVD. CVDs are therefore not commonly defined and described at the molecular level, which leads to challenges, when clinicians aim to develop genetic or molecular precision medicine approaches.

Genetic risk scores based on robust signals, from genome-wide association studies, have been developed and could inform cardiovascular risk assessment, similar to traditional risk factors such as diabetes or dyslipidemia [6,7]. However, current clinical practice in the prevention and treatment of CVDs is mainly based on risk factors, imaging, functional studies, and biomarkers, with the latter often being proteins such as cardiac troponin I or natriuretic peptides.

Precision Medicine for Investigators, Practitioners and Providers. https://doi.org/10.1016/B978-0-12-819178-1.00025-3

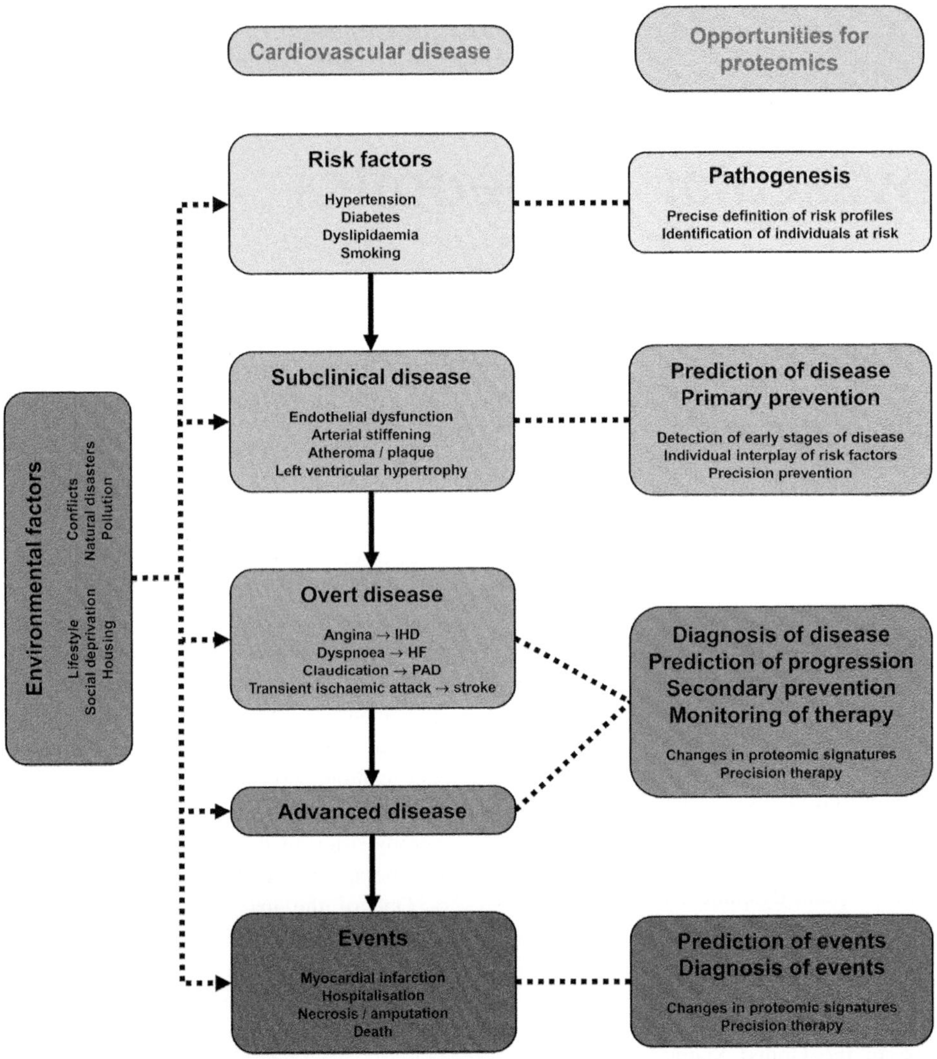

FIGURE 25.1 **Opportunities for cardiovascular proteomics.** The development of cardiovascular diseases, from risk factors to early and more advanced phenotypes, and ultimately to cardiovascular events, is illustrated in the red (dark gray in print version) boxes. Each step of this continuum can be modified by interactions with environmental factors. Opportunities for proteomics to inform precision medicine approaches are illustrated in the blue (gray in print version) boxes and span from precise definition of a disease phenotype to prediction of events and monitoring of therapeutic approaches. Where examples are given in smaller font size, these are not meant to be a complete representation of all risk factors, diseases, environmental factors, and opportunities. *HF*, heart failure; *IHD*, ischemic heart disease; *PAD*, peripheral artery disease.

Proteomics

Proteomics allows simultaneous quantification of multiple protein markers in a biosample [8,9]. In targeted proteomic approaches, these markers are predefined, whereas untargeted approaches assess all proteins in a given sample, in an unbiased fashion. It should be noted, however, that the number of detectable and identifiable features is limited, as a result of methodological constraints. Proteomic approaches describe the current state of a cell, tissue, or organism, as they detect the proteins that are actually expressed and present in a biological sample. In complex disease processes that develop over decades and are driven by genetic and environmental factors and their interaction, proteomics offers a precise definition of the current state of an organism or specific organs.

Mass spectrometry–based proteomic methods

Traditionally, gel-based methods such as sodium dodecyl sulfate (SDS) polyacrylamide gel electrophoresis (SDS-PAGE), and 2D-difference gel electrophoresis (2D-DIGE) have been used for unbiased proteomic studies. Subsequently, proteomics has moved from a gel approach to solution-based analysis. An initial separation step based, for example, on liquid chromatography, gas chromatography, or capillary electrophoresis is followed by mass spectrometry (MS)–based techniques [10,11]. The most common analytical technique used today is nanoflow liquid chromatography coupled to a mass spectrometer (nLC-MS).

Technical issues, however, restrict MS approaches to a relatively narrow range of ions in a scan, depending on abundance. This means that in complex samples containing a wide range of different proteins of different masses and concentrations some proteins of interest may be missed. It should be noted that no proteomic technique can analyze the entire proteome and that different techniques will study different subsets of the proteome. A comparison between different studies, and integration of data, can therefore be challenging.

MS-based methods can be targeted and nontargeted. A shotgun approach, where samples are digested by a proteolytic enzyme, usually trypsin, followed by nLC-MS/MS, provides optimal peptide separation and chromatographic peak resolution. In contrast, an approach where the MS analysis is targeted to the mass:charge ratio of a known compound or peptide allows the monitoring of specific known analytes. Protein quantitation is possible in vitro by methods such as isobaric tags for relative and absolute quantitation (iTRAQ), and tandem mass tags (TMT), and in vivo by Stable Isotope Labeling with Amino acids in Cell culture (SILAC) [12].

Array-based proteomics

Mass spectrometers are expensive pieces of equipment, and require qualified staff with expert knowledge, to oversee the analyses. Service contracts, licenses for analytical software, consumables, and other running costs therefore limit MS to large laboratories and core facilities.

A number of platforms allow multiplexing of antibody-based protein detection, so that up to several dozen proteins can be measured in a small sample volume, at the same time [13]. More recently the development of nucleic acid–based aptamers provides an alternative to antibody-based techniques; these nucleotides directly interact with protein surfaces, and are then measured and quantified [14]. Other novel platforms use antibody-based detection, coupled to nucleotides that can be quantified, using a polymerase chain reaction [15].

All of these techniques are in the first instance suited for the analysis of blood (serum or plasma) samples, but can also be used for cell culture supernatants or tissue homogenates. They are, however, exclusively targeted approaches, based on preselected proteins. This limits their potential to discover novel pathways and molecular targets, but provides robust and quantitative data on selected proteins.

Sample types for proteomic studies

Tissue samples can provide deep insight into the pathological processes within an organ. Precise diagnosis in many conditions such as renal, liver, or bowel disease therefore relies on biopsy material. Where such biopsy material is available, proteomic approaches have indeed shown characteristic signatures of disease processes [16–18]. Precise localization of differentially expressed proteins is possible, with newer proteomic techniques such as matrix assisted Laser desorption/ionization (MALDI)-imaging mass spectrometry [19]. In CVD, however, tissue-based diagnosis is not typically available. Cardiac biopsies are technically demanding, and the often patchy vascular component of CVD does not lend itself to routine tissue sampling. Proteomic studies in cardiovascular tissues are largely restricted to advanced diseases, where samples from explanted hearts or vessels from bypass surgery have been used [17,20].

Biofluids such as blood and urine can display protein signatures that derive from specific organs including the heart and the vasculature, and also reflect generalized processes such as vascular, myocardial, and renal fibrosis. Urine is of interest in the study of renal diseases, but as it is a filtrate of plasma, urine is also able to display extrarenal conditions. In any type of biofluid, small amounts of specific proteins can be masked by more abundant nonspecific proteins such as albumin or immunoglobulins, which leads to analytical challenges.

Extracellular vesicles derived from various cells in the body, contain cell-specific information that can be transferred into target cells. Their membrane composition and content (cargo) can provide insights into their origin, alterations in the molecular (proteomic) make-up of source cells, and information that source cells aim to transfer to other cells [21]. Extracellular vesicles isolated from plasma have been subjected to proteomic analyses, which showed, for example, differential expression of proteins between patients with myocardial infarction, and patients with stable angina [22]. They may have the potential to support risk stratification [23]. Preparation of extracellular vesicles must adhere to agreed protocols, to deliver reliable and reproducible results [23–25].

Proteomics to unravel pathogenetic principles of cardiovascular diseases

Translational studies: Translation from preclinical models to human disease is not straightforward. While some of the major physiological differences between experimental models and humans are well known, the precise molecular differences that determine how diseases develop, and how therapeutic principles counteract the disease processes, are often incompletely understood. Proteomic techniques allow the detection of thousands of proteins in a sample, and therefore not only describe differences between strains of animals, and between animals and humans, but also show conserved proteins and pathways. Species-specific proteomic profiles can thereby guide the choice of models, to interrogate a specific target that is conserved between the model and humans. For example, comparison of urinary proteomic profiles between the Zucker diabetic fatty (ZDF) rat and humans with diabetes has shown similarities in some protein signatures associated with cardiovascular complications of diabetes, whereas protein signatures associated with renal complications were found to be less robustly conserved [26]. Such information is crucial, where animal models are being used to evaluate drug toxicity [27,28] or therapeutic interventions [29], in order to inform translatability to humans.

Insights from studies in human CVD: Studies using cardiac tissue are limited to procedures that require an invasive approach, during which biopsy samples can be obtained, such as implantation of a left ventricular assist device [30], or to samples that are obtained from deceased patients or explanted hearts [17,20]. The majority of human proteomic studies have therefore been conducted in biofluids. We have recently studied the plasma proteome of patients with various stages of CVD, compared to healthy controls, and described 100 differentially expressed proteins including apolipoprotein B, alpha-2-macroglobulin, and low density lipoprotein receptor related protein 2 [31]. We have previously shown that a urinary proteomic signature is associated with presence [32] and severity of coronary artery disease, and that other urinary peptides can predict future acute coronary syndromes [33]. Others have shown that a proteomic signature of diastolic dysfunction can differentiate between patients with and without heart failure [34], thereby linking early pathogenetic stages to more advanced disease.

Extracellular matrix: Changes in extracellular matrix and development of fibrosis are a hallmark of advanced CVD, and unsurprisingly, changes in urinary collagen fragments are one of the commonly found features in human CVDs [32,34]. Changes in cardiac extracellular matrix proteins have been observed, with left ventricular assist device therapy [30], pointing toward reversibility of these processes, and the potential use of proteomic fibrosis markers to monitor response to therapy. Deposition of extracellular matrix also depends on protease activity, and studies into naturally occurring peptides in biofluids can provide information about intrinsic protease activity [35,36].

Sex, gender, and ethnicity: While development and prognosis of CVD is different across ethnicities, and between men and women, only few studies have systematically assessed corresponding differences in protein expression profiles in humans [37–39]. The ongoing African Prospective study on the Early Detection and Identification of Cardiovascular disease and Hypertension (African-PREDICT) will provide opportunities to study the effect of sex and ethnicity on proteomic signatures and other biomarkers, in greater detail [40].

Disease signatures rather than individual biomarkers: As with other omics techniques, the large number of molecular features that can be detected in proteomic studies allows the generation of disease-specific fingerprints that are composed of hundreds of proteins, and thereby comprise a large number of pathogenetic principles. Such signatures can provide more robust information than individual biomarkers. A number of statistical methods such as principal component analysis and support vector machine approaches allow combination of multiple individual protein data into unidimensional classifiers; the above mentioned urinary protein signatures for coronary artery disease [32] and heart failure [34] are examples of this approach. While distilling information facilitates statistic analysis, and application to larger cohorts and ultimately clinical practice, the loss of some information (e.g., proteins that are not consistently expressed in the majority of patients, and thereby will not be entered into the classifier) could prevent identification of rarer disease pathways that play a role in individual patients.

Interfaces with genomics and other omics

The basic paradigm that DNA is transcribed to messenger RNA, and translated into protein, suggests a close relationship between the proteome and the genome. However, while the genome remains relatively static throughout life, the genes that are actually transcribed and translated change as part of normal development, through interaction with the environment, with aging and disease processes. Taking the potential that lies in the genome, together with the current state of an organism, organ, tissue, or cell that is described by the proteome, will therefore provide complementary information.

A number of studies have looked at the genetic determinants of protein expression. An overview of genome-wide association studies, with (multiplexed) proteomics data, has been published by Suhre et al. [41] and an updated version is available at http://www.metabolomix.com/a-table-of-all-published-gwas-with-proteomics/. For example, Zhernakova et al. [42] studied 92 cardiovascular-disease-related proteins in the LifeLines-DEEP cohort, and found that genetic factors contribute most to concentrations of immune-related proteins, whereas the gut microbiome contributes most to proteins involved in metabolism and intestinal health. In the Framingham cohort, Benson et al. [43] identified locus-protein associations in genome-wide and exome array analyses, which explained up to 66% of interindividual plasma protein-level variation. Such data show that there is a strong genetic component to protein expression, but that other, nongenetic factors such as sex, age, disease status, and the gut microbiome also contribute to the variability of protein expression. Similarly, proteomic approaches can be used to identify novel targets of micro-RNAs, and thereby provide information on regulation of gene expression and translation [44,45]; direct applications of such principles to CVD are still awaited.

It should also be noted that there are profound technical differences, between proteomic techniques and genetic/genomic techniques, as summarized by Gerszten et al. [11]. We have already mentioned that no single proteomic technique can assess the complete proteome, whereas genomic techniques especially through next generation sequencing can provide complete or near-complete information on the genetic make-up. Marked differences in the proteome between cells as opposed to relative stability of the genome across all cells in the body, and the wide range of protein concentrations in any given sample, add further to the complexity of comparing proteomic and genomic data. All these factors have to be taken into account, when proteomic and genomic data are being compared and integrated [46,47].

Proteomics and precision medicine in cardiovascular disease

Precision in disease prediction

Primary prevention of CVD is currently based on assessment of risk factors. This approach works well on the population level, but has the potential to misclassify individual patients. Assessing early, subclinical cardiovascular damage is an alternative strategy, to identify those patients who are more likely to progress to overt CVD. Proteomic strategies can contribute to very early disease detection, and facilitate primary prevention and potentially, reversibility of early disease processes.

Development of early disease markers is not an easy task, especially if they are supposed to recognize disease before any other established technique. As in other biomarker approaches, proteomic screens therefore commonly start with established disease, and from there propose molecular changes that could play a role in early detection of disease. This approach has been used for the development of a urinary proteomic signature to predict chronic kidney disease, where differences between patients and healthy controls were first described [48], and then applied to prospective evaluation of this signature's potential, to predict development of chronic kidney disease and particularly diabetic kidney disease (DKD), earlier than established clinical markers such as microalbuminuria [49]. The above mentioned study that established a proteomic signature of diastolic dysfunction, and then applied it to overt heart failure, is an example for this approach [34].

Some CVDs develop over much shorter periods of time than for example ischemic heart disease. A prime example is preeclampsia that develops during the 9 month of pregnancy. Preeclampsia is associated with significant maternal and fetal morbidity and mortality, and is difficult to predict by use of clinical characteristics and biomarkers [50]. The relatively short timeframe of its development means that studies into preeclampsia can be conducted, without the need to store samples from an asymptomatic stage for decades, or to analyze large numbers of samples at enormous costs, in order to find out many years later which of the participants have developed overt CVD. We and others have conducted proteomic studies into the prediction of preeclampsia [51,52], and the description of the placental proteome of preeclampsia [16]. While initial results are promising, confirmation in independent and larger samples is either absent or not yet sufficiently convincing, to translate these findings into clinical practice.

Despite all challenges it is worth considering that the cost savings, associated with reduction of the incidence of CVD, i.e., with primary prevention, are enormous [3]. We believe that proteomic approaches should indeed focus on early disease detection, as this will come with the largest benefit for patients and healthcare systems, and thereby larger populations. The concept of "precision public health," "using the best available data, to target more effectively and efficiently interventions of all kinds, to those most in need," is based on this principle, but subject to ongoing discussions [53,54].

Precision in diagnosis, prognostication, and therapeutic response

The previously proposed concept that DKD develops progressively, from normoalbuminuric to micro and macroalbuminuric stages has been challenged recently by the description of nonalbuminuric DKD [55], and by some patients reverting from microalbuminuria to normoalbuminuria [56]. An albuminuria-based diagnosis of DKD is therefore not sufficiently precise in all patients, and multidimensional proteomic markers, based on multiple pathogenetic principles, has shown potential to improve diagnostic accuracy, and prediction of disease progression [49].

The same principles apply to CVD, where precise diagnosis is often difficult, and based on clinical parameters and longitudinal observation of patients and their symptoms. We have recently described that the response of patients with diastolic heart failure (heart failure with preserved ejection fraction) to spironolactone, depends on markers of collagen cross-linking that essentially describe cardiac fibrosis [57]. Within the same clinical phenotype, there are groups that differ in a biomarker that then determines response to treatment. It is plausible to propose, that even more precise prediction of therapeutic response can be achieved, by more detailed assessment of collagen deposition and degradation, for example by urinary proteomic approaches [34].

The potential of proteomics to simultaneously detect multiple dysregulated pathways can help to identify individual patients' subtypes of a given CVD. We have previously proposed that proteomic approaches could help to arrive at a molecular diagnosis of CVDs such as hypertension [58], and thereby facilitate targeted and precise treatment.

Precision therapy and stratified trials

Most of the studies and data mentioned in this chapter have been generated retrospectively, from existing sample collections, or prospectively to describe disease progression. However, we are missing prospectively applied proteomic data, to inform therapeutic decisions.

A large trial into the use of a urinary proteomic signature to predict patients at high risk of developing DKD has been completed, and is currently awaiting data analysis [59]. In this trial (Proteomic prediction and Renin angiotensin aldosterone system Inhibition prevention Of early diabetic nephRopathy In TYpe 2 diabetic patients with normoalbuminuria; PRIORITY), only patients who are in the high-risk category have been randomized to preventative treatment with spironolactone, whereas the risk in other patients is too low to justify preventative treatment. While we are not aware of ongoing proteomic studies of similar scale in CVD, there are initiatives to comprehensively describe proteomic and other signatures in patients progressing to heart failure, for example, in the Heart OMics in Aging (HOMAGE) project (http://www.homage-hf.eu/).

Summary and conclusions

Despite the potential of cardiovascular proteomics, the available data are often not of sufficient quality and quantity to inform clinical practice. The high costs, especially of MS-based proteomics, are one of the reasons why studies are often underpowered, and data not replicated in independent cohorts. We therefore believe that concerted action, using standardized protocols in international collaboration, is required to help proteomics enter the clinical stage. We should not be afraid of taking a step back and doing the groundwork. Description of the proteome of different CVDs at different stages, in the form of a disease atlas, will be an important milestone in the implementation of cardiovascular proteomics.

Acknowledgments

Our work is supported by grants from the European Commission ("PRIORITY", grant agreement 101813; "HOMAGE" grant agreement 305507); the South African Research Chairs Initiative (SARChI) of the Department of Science and Technology and National Research Foundation (NRF) of South Africa (GUN 86895); South African Medical Research Council, GlaxoSmithKline R&D (Africa Non-Communicable Disease Open Lab grant), the UK Medical Research Council, and with funds from the UK Government's Newton Fund; and the British Heart Foundation (Center of Research Excellence Award; reference number RE/13/5/30177). SM is supported by a Newton International Fellowship from the Academy of Medical Sciences.

References

[1] Patil RR. Urbanization as a determinant of health: a socio-epidemiological perspective. Soc. Work. Publ. Health 2014;29(4):335−41.

[2] Yeates K, Lohfeld L, Sleeth J, Morales F, Rajkotia Y, Ogedegbe O. A global perspective on cardiovascular disease in vulnerable populations. Can. J. Cardiol. 2015;31(9):1081−93.

[3] Dzau VJ, Ginsburg GS, Van Nuys K, Agus D, Goldman D. Aligning incentives to fulfil the promise of personalised medicine. Lancet 2015;385(9982):2118−9.

[4] Currie G, Delles C. Precision medicine and personalized medicine in cardiovascular disease. Adv. Exp. Med. Biol. 2018;1065: 589−605.

[5] Dzau V, Braunwald E. Resolved and unresolved issues in the prevention and treatment of coronary artery disease: a workshop consensus statement. Am. Heart J. 1991;121(4 Pt 1):1244−63.

[6] Thanassoulis G, Peloso GM, Pencina MJ, Hoffmann U, Fox CS, Cupples LA, et al. A genetic risk score is associated with incident cardiovascular disease and coronary artery calcium: the Framingham heart study. Circ. Cardiovasc. Genet 2012;5(1):113−21.

[7] Khera AV, Chaffin M, Aragam KG, Haas ME, Roselli C, Choi SH, et al. Genome-wide polygenic scores for common diseases identify individuals with risk equivalent to monogenic mutations. Nat. Genet. 2018;50(9):1219−24.

[8] Wilkins MR, Pasquali C, Appel RD, Ou K, Golaz O, Sanchez JC, et al. From proteins to proteomes: large scale protein identification by two-dimensional electrophoresis and amino acid analysis. Biotechnology 1996;14(1):61−5.

[9] Lindsey ML, Mayr M, Gomes AV, Delles C, Arrell DK, Murphy AM, et al. Transformative impact of proteomics on cardiovascular health and disease: a scientific statement from the American heart association. Circulation 2015;132(9):852−72.

[10] Carty DM, Schiffer E, Delles C. Proteomics in hypertension. J. Hum. Hypertens. 2013;27(4):211−6.

[11] Gerszten RE, Asnani A, Carr SA. Status and prospects for discovery and verification of new biomarkers of cardiovascular disease by proteomics. Circ. Res. 2011;109(4):463−74.

[12] Terzi F, Cambridge S. An overview of advanced SILAC-labeling strategies for quantitative proteomics. Methods Enzymol 2017;585:29−47.

[13] Ellington AA, Kullo IJ, Bailey KR, Klee GG. Antibody-based protein multiplex platforms: technical and operational challenges. Clin. Chem. 2010;56(2):186−93.

[14] Rohloff JC, Gelinas AD, Jarvis TC, Ochsner UA, Schneider DJ, Gold L, et al. Nucleic acid ligands with protein-like side chains: modified aptamers and their use as diagnostic and therapeutic agents. Mol. Ther. Nucleic Acids 2014;3:e201.

[15] Assarsson E, Lundberg M, Holmquist G, Bjorkesten J, Thorsen SB, Ekman D, et al. Homogenous 96-plex PEA immunoassay exhibiting high sensitivity, specificity, and excellent scalability. PLoS One 2014;9(4):e95192.

[16] Mary S, Kulkarni MJ, Malakar D, Joshi SR, Mehendale SS, Giri AP. Placental proteomics provides insights into pathophysiology of pre-eclampsia and predicts possible markers in plasma. J. Proteome Res. 2017;16(2):1050–60.

[17] Liu S, Xia Y, Liu X, Wang Y, Chen Z, Xie J, et al. In-depth proteomic profiling of left ventricular tissues in human end-stage dilated cardiomyopathy. Oncotarget 2017;8(29):48321–32.

[18] Serhatli M, Baysal K, Acilan C, Tuncer E, Bekpinar S, Baykal AT. Proteomic study of the microdissected aortic media in human thoracic aortic aneurysms. J. Proteome Res. 2014;13(11):5071–80.

[19] Schulz S, Becker M, Groseclose MR, Schadt S, Hopf C. Advanced MALDI mass spectrometry imaging in pharmaceutical research and drug development. Curr. Opin. Biotechnol. 2018;55:51–9.

[20] Chen L, Yang F, Chen X, Rao M, Zhang NN, Chen K, et al. Comprehensive myocardial proteogenomics profiling reveals C/EBPalpha as the key factor in the lipid storage of ARVC. J. Proteome Res. 2017;16(8):2863–76.

[21] Cosme J, Liu PP, Gramolini AO. The cardiovascular exosome: current perspectives and potential. Proteomics 2013;13(10–11):1654–9.

[22] Cheow ES, Cheng WC, Lee CN, de Kleijn D, Sorokin V, Sze SK. Plasma-derived extracellular vesicles contain predictive biomarkers and potential therapeutic targets for myocardial ischemic (MI) injury. Mol. Cell. Proteom. 2016;15(8):2628–40.

[23] Baldan-Martin M, de la Cuesta F, Alvarez-Llamas G, Ruiz-Hurtado G, Ruilope LM, Barderas MG. Proteomic analysis of blood extracellular vesicles in cardiovascular disease by LC-MS/MS analysis. Methods Mol. Biol. 2017;1619:141–9.

[24] Barrachina MN, Calderon-Cruz B, Fernandez-Rocca L, Garcia A. Application of extracellular vesicles proteomics to cardiovascular disease: guidelines, data analysis, and future perspectives. Proteomics 2019;19(1–2):e1800247.

[25] Thery C, Witwer KW, Aikawa E, Alcaraz MJ, Anderson JD, Andriantsitohaina R, et al. Minimal information for studies of extracellular vesicles 2018 (MISEV2018): a position statement of the International society for extracellular vesicles and update of the MISEV2014 guidelines. J. Extracell. Vesicles 2018;7(1):1535750.

[26] Siwy J, Zoja C, Klein J, Benigni A, Mullen W, Mayer B, et al. Evaluation of the Zucker Diabetic Fatty (ZDF) rat as a model for human disease based on urinary peptidomic profiles. PLoS One 2012;7(12):e51334.

[27] Mischak H, Espandiari P, Sadrieh N, Hanig J. Profiling of rat urinary proteomic patterns associated with drug-induced nephrotoxicity using CE coupled with MS as a potential model for detection of drug-induced adverse effects. Proteomics Clin. Appl 2009;3(9):1062–71.

[28] Rouse R, Siwy J, Mullen W, Mischak H, Metzger J, Hanig J. Proteomic candidate biomarkers of drug-induced nephrotoxicity in the rat. PLoS One 2012;7(4):e34606.

[29] Martin-Pozuelo G, Gonzalez-Barrio R, Barbera GG, Albalat A, Garcia-Alonso J, Mullen W, et al. Tomato juice consumption modifies the urinary peptide profile in sprague-dawley rats with induced hepatic steatosis. Int. J. Mol. Sci. 2016;17(11).

[30] Shahinian JH, Mayer B, Tholen S, Brehm K, Biniossek ML, Fullgraf H, et al. Proteomics highlights decrease of matricellular proteins in left ventricular assist device therapydagger. Eur. J. Cardiothorac. Surg. 2017;51(6):1063–71.

[31] Lygirou V, Latosinska A, Makridakis M, Mullen W, Delles C, Schanstra JP, et al. Plasma proteomic analysis reveals altered protein abundances in cardiovascular disease. J. Transl. Med. 2018;16(1):104.

[32] Delles C, Schiffer E, von Zur Muhlen C, Peter K, Rossing P, Parving HH, et al. Urinary proteomic diagnosis of coronary artery disease: identification and clinical validation in 623 individuals. J. Hypertens. 2010;28(11):2316–22.

[33] Neisius U, Koeck T, Mischak H, Rossi SH, Olson E, Carty DM, et al. Urine proteomics in the diagnosis of stable angina. BMC Cardiovasc. Disord. 2016;16:70.

[34] Kuznetsova T, Mischak H, Mullen W, Staessen JA. Urinary proteome analysis in hypertensive patients with left ventricular diastolic dysfunction. Eur. Heart J. 2012;33(18):2342–50.

[35] Casteleiro MA, Stevens R, Klein J. Prediction of proteases involved in peptide generation. Methods Mol. Biol. 2017;1574:205–13.

[36] Magalhaes B, Trindade F, Barros AS, Klein J, Amado F, Ferreira R, et al. Reviewing mechanistic peptidomics in body fluids focusing on proteases. Proteomics 2018;18(18):e1800187.

[37] Baetta R, Pontremoli M, Martinez Fernandez A, Spickett CM, Banfi C. Proteomics in cardiovascular diseases: unveiling sex and gender differences in the era of precision medicine. J. Proteomics 2017;173:62–76.

[38] Dworatzek E, Baczko I, Kararigas G. Effects of aging on cardiac extracellular matrix in men and women. Proteonomics Clin. Appl. 2016;10(1):84–91.

[39] Kim CX, Bailey KR, Klee GG, Ellington AA, Liu G, Mosley Jr TH, et al. Sex and ethnic differences in 47 candidate proteomic markers of cardiovascular disease: the Mayo clinic proteomic markers of arteriosclerosis study. PLoS One 2010;5(2):e9065.

[40] Schutte AE, Gona PN, Delles C, Uys AS, Burger A, Mels CM, et al. The African prospective study on the early detection and identification of cardiovascular disease and hypertension (African-PREDICT): design, recruitment and initial examination. Eur. J. Prev. Cardiol 2019;26(5):458–70.

[41] Suhre K, Arnold M, Bhagwat AM, Cotton RJ, Engelke R, Raffler J, et al. Connecting genetic risk to disease end points through the human blood plasma proteome. Nat. Commun. 2017;8:14357.

[42] Zhernakova DV, Le TH, Kurilshikov A, Atanasovska B, Bonder MJ, Sanna S, et al. Individual variations in cardiovascular-disease-related protein levels are driven by genetics and gut microbiome. Nat. Genet. 2018;50(11):1524–32.

[43] Benson MD, Yang Q, Ngo D, Zhu Y, Shen D, Farrell LA, et al. Genetic architecture of the cardiovascular risk proteome. Circulation 2018;137(11):1158–72.

[44] Ljepoja B, Garcia-Roman J, Sommer AK, Frohlich T, Arnold GJ, Wagner E, et al. A proteomic analysis of an in vitro knock-out of miR-200c. Sci. Rep. 2018;8(1):6927.

[45] Wu Y, Lou QY, Ge F, Xiong Q. Quantitative proteomics analysis reveals novel targets of miR-21 in Zebrafish embryos. Sci. Rep. 2017;7(1):4022.

[46] Delles C, Husi H. Systems biology approach in hypertension research. Methods Mol. Biol. 2017;1527:69–79.

[47] Wanichthanarak K, Fahrmann JF, Grapov D. Genomic, proteomic, and metabolomic data integration strategies. Biomark Insights 2015;10(Suppl. 4):1–6.

[48] Good DM, Zurbig P, Argiles A, Bauer HW, Behrens G, Coon JJ, et al. Naturally occurring human urinary peptides for use in diagnosis of chronic kidney disease. Mol. Cell. Proteom. 2010;9(11):2424–37.
[49] Zurbig P, Mischak H, Menne J, Haller H. CKD273 enables efficient prediction of diabetic nephropathy in nonalbuminuric patients. Diabetes Care 2019;42(1):e4–5.
[50] Carty DM, Delles C, Dominiczak AF. Novel biomarkers for predicting preeclampsia. Trends Cardiovasc. Med. 2008;18(5):186–94.
[51] Carty DM, Siwy J, Brennand JE, Zurbig P, Mullen W, Franke J, et al. Urinary proteomics for prediction of preeclampsia. Hypertension 2011;57(3):561–9.
[52] Bramham K, Mistry H, Lynham S, Ward M, Fiseha D, Poston L, et al. PP046. A unique urinary proteome profile at 15 weeks' gestation in low-risk women with subsequent pre-eclampsia. Pregnancy Hypertens 2013;3(2):83–4.
[53] Chowkwanyun M, Bayer R, Galea S. "Precision" public health – between novelty and hype. N. Engl. J. Med. 2018;379(15):1398–400.
[54] Horton R. Offline: in defence of precision public health. Lancet 2018;392(10157):1504.
[55] Dwyer JP, Parving HH, Hunsicker LG, Ravid M, Remuzzi G, Lewis JB. Renal dysfunction in the presence of normoalbuminuria in type 2 diabetes: results from the DEMAND study. Cardiorenal. Med 2012;2(1):1–10.
[56] Perkins BA, Ficociello LH, Silva KH, Finkelstein DM, Warram JH, Krolewski AS. Regression of microalbuminuria in type 1 diabetes. N. Engl. J. Med. 2003;348(23):2285–93.
[57] Ravassa S, Trippel T, Bach D, Bachran D, Gonzalez A, Lopez B, et al. Biomarker-based phenotyping of myocardial fibrosis identifies patients with heart failure with preserved ejection fraction resistant to the beneficial effects of spironolactone: results from the Aldo-DHF trial. Eur. J. Heart Fail. 2018;20(9):1290–9.
[58] Delles C, Carrick E, Graham D, Nicklin SA. Utilizing proteomics to understand and define hypertension: where are we and where do we go? Expert Rev Proteomics 2018;15(7):581–92.
[59] Lindhardt M, Persson F, Currie G, Pontillo C, Beige J, Delles C, et al. Proteomic prediction and renin angiotensin aldosterone system inhibition prevention of early diabetic nephRopathy in TYpe 2 diabetic patients with normoalbuminuria (PRIORITY): essential study design and rationale of a randomised clinical multicentre trial. BMJ Open 2016;6(3):e010310.

C—Other omics and miscellaneous biomarkers

Chapter 26

Sampling, analyzing, and integrating microbiome 'omics data in a translational clinical setting

Christopher Staley[1,2], Thomas Kaiser[1,2] and Zhigang Zhu[1,2]

[1]Department of Surgery, University of Minnesota, Minneapolis, MN, United States; [2]BioTechnology Institute, University of Minnesota, St. Paul, MN, United States

Context

The microbiome is a community of microorganisms, consisting of trillions of cells, existing throughout a single human [1]. In recent years, the intestinal microbiome, specifically, has been identified as an organ-like assemblage that is critical for human development, functioning, and health [2]. However, a high degree of interpersonal variability in microbiome compositions makes identifying a discrete "healthy" microbiome nearly impossible [3]. Several patterns of microbial imbalances, or dysbioses, ranging from decreased diversity to the presence of specific pathobionts [4]—commensal microorganisms that can promote a disease state during periods of dysbiosis— have been associated with diseases, ranging from inflammatory bowel diseases and obesity to cancers [2]. Thus, characterization of the microbiome has become a new frontier in the pursuit of personalized, precision medicine [5].

In recent years, a wealth of precision medicine initiatives have sought to leverage new 'omics sciences, to increase the speed and accuracy of diagnoses, and to develop personalized treatment options [6]. Use of even a single type of 'omics data can potentially result in the identification of useful biomarkers that show differential abundances in healthy or diseased states [7]. However, due to the high dimensionality of these datasets, which contain hundreds to thousands of features, untangling more complex relationships between the microbiome and clinical outcomes, beyond simple correlative analyses, remains a challenge. Moreover, robust integration of various types of 'omics data, using systems biology approaches, remains very much on the forefront of computational research.

Metagenomics was the first and most widely utilized 'omics field that involves the sequencing of total DNA from a given environment (e.g., stool samples or mucosal biopsies), for direct characterization of the genetic (DNA) content [8–10]. Contemporary metagenomics techniques involve either the amplification and sequencing (amplicon sequencing) of taxonomic marker genes (i.e., the 16S rRNA gene for bacteria and archaea, or internal transcribed spacer region between the 18S and 5.8S rRNA genes for fungi) or untargeted, whole genome shotgun sequencing of all DNA isolated from a sample [11]. Other popular areas of study have included transcriptomics [12], in which RNA genes are sequenced to assess likely cellular functions; metabolomics [13], including the characterization of small molecules, often using analytical chemistry approaches; and proteomics [14], which characterizes proteins present, often using microarrays, mass spectrometry, and other cell biology approaches.

Sampling strategies for 'omics studies

Clinically, the sampling strategy for 'omics studies represents the first, and perhaps most pragmatic, logistical concern. In addition to ensuring an adequate sample size for statistically powered analyses (discussed below), critical concerns include (i) the type of sample, which is directly related to the amount of sample that can be collected, (ii) frequency of sampling, which can be limited by the invasiveness of the sampling procedure, and (iii) downstream applications or analyses that will be performed, which may be limited by sample storage requirements. Studies of the intestinal microbiota, which are associated

Precision Medicine for Investigators, Practitioners and Providers. https://doi.org/10.1016/B978-0-12-819178-1.00026-5

with a variety of health outcomes [2,15], as a primary target for personalized medical interventions have received the most attention. However, a growing number of contemporary sequencing technologies and applications are emerging, to leverage 'omics in the study of solid and nonsolid cancers, from biopsies and blood samples. These two types of applications (intestinal microbiome vs. microenvironment/ circulating microbiome) reflect dichotomous sampling strategies, where the former relies on relatively easily obtainable samples of ample size for any downstream analyses, while the latter reflects a more invasive collection procedure, to obtain a small sample for a targeted application.

Characterization of the intestinal microbiome

Studies of the microbiome often rely on relatively simple sample collections from stool or swabs. Mucosal biopsies are also relatively common, and can be collected as part of standard care. Considerations for fecal sample collection and storage have been recently reviewed [16]. Generally, freezing fecal samples allows for characterization of the microbiome, using metagenomics-based approaches, as well as downstream metabolomics analyses. However, it is important to note that the sample type, and location of collection within the colon, will affect the microbiome characterized [17]. Fecal samples are more representative of the distal luminal contents and are distinct from the mucosal microbiome. Proximal luminal and mucosal communities show a lower degree of dissimilarity [17].

Current investigations aim at the mechanism(s) by which the intestinal microbiome affects host health [18]. Others emphasize the limitations of single time point samples, as simply comparative "snap shots" that cannot be accurately associated with host pathology or phenotype [19]. Longitudinal and cross-sectional study designs have been suggested, to account for interindividual variability [20]. Estimations of effect size for robust power calculations were, until recently, relatively undetermined [21]; however, novel methods employing Dirichlet multinomial [22] and beta diversity [23] calculations are now readily available.

Characterization of the "sterile" microbiome

Many compartments of the body previously thought to be sterile, in fact, contain complex microbial communities that may play crucial roles in sepsis, inflammation, immunity, and cancer progression. Characterization of the circulating microbiome from blood using metagenomics methods has recently been done [24], but factors such as lower bacterial concentrations, polymerase chain reaction (PCR) inhibitors in the blood, storage reagents, and bacterial background in molecular biology reagents make this a more difficult analysis. For example, heparin is a known PCR inhibitor [25]; however, it is a common storage agent for many clinical analyses. In addition, much of the bacterial component found in the blood cannot be cultured in the laboratory [26], making it difficult to determine if microbiomes characterized reflect actual "atopobiosis," or translocation of microbes from oral or intestinal origins, or merely contamination.

A recent report found a greater bacterial burden in samples from pancreatic ductal adenocarcinomas than that found in healthy control biopsies [27], suggesting that other organ tissues may also contain microbes. Thus, care should be taken to consider potential downstream analyses when collecting these more difficult, lower-microbial-burden samples, both to prevent contamination and to ensure that sufficient material is collected.

Analytical approaches for microbiome data

Different types of sequencing and their tradeoffs in terms of resolution, the types of data generated, and cost have been recently reviewed for microbiome studies [20]. In addition, statistical approaches to analyze typical microbiome datasets have now been well described [20,28]. As a result, several features ranging from diversity to composition, and the functional repertoire of the microbiome are now commonly reported in association with clinical studies. While obtaining compositional-based data has become relatively inexpensive and straightforward, linking these data with microbial functions remains a limitation, resulting from costs of analyses and uncertainty in the data obtained. Moreover, due to the large number of features obtained using 'omics methods, new and emerging statistical analyses are necessary to minimize false discovery rates and resolve biologically meaningful conclusions.

Features of microbial communities

Simple descriptions of the microbiome generated from metagenomics-based studies characterize four major features of bacterial communities: (i) alpha diversity, or the number and distribution of taxa within a single sample; (ii) beta diversity, or the degree of difference between two samples; (iii) community composition, or the taxonomic or phylogenetic classification of the species, within and between samples; and (iv) functional characterization of genes, transcripts, metabolites, or proteins being used or produced by the community.

Alpha diversity provides the most widely understood metric to assess patient health, and a decline in diversity is typically associated with poorer health outcomes [3]. Traditionally, alpha diversity has been measured in terms of richness— the number of unique species present— and evenness— how well distributed each species is throughout the community. The Shannon-Weaver index [29] has traditionally been used to characterize both features, while the Simpson index [30] serves as a measure of species dominance. However, organisms in microbiome data are differentiated as taxa [31], defined by a sequence similarity typically >97% [32], with 100% similarity quickly becoming the standard. While conceptually analogous to species, taxa may lack a discrete classification due to incomplete representation in commonly used databases. Moreover, microbiome datasets are typically characterized by a large number of sparse taxa, representing the majority of microbial diversity [33]. To account for this, richness estimates including the Chao1 index [34] and abundance-based coverage estimated (ACE) [35] approximate the total taxonomic diversity, by considering the distribution of taxa that occur at least twice in the data, or fewer than 10 times, respectively.

Beta diversity describes the difference between two or more samples, and can be defined qualitatively or quantitatively based on presence/absence of taxa or based on their relative abundances [36]. The Jaccard index [37] provides a simple measure of the qualitative differences in the numbers of shared and unique species between communities, while the Bray-Curtis distance [38] compares the differences in relative abundance between communities. The UniFrac distance [39] is also commonly used to describe beta diversity and can be considered unweighted (qualitatively) or weighted (quantitatively). Unlike Jaccard and Bray-Curtis measures, however, which account for discrete presence/absence or abundance of taxa, UniFrac measures phylogenetic distances among communities to quantify differences in relatedness. Thus, Jaccard and Bray-Curtis can be used to evaluate empiric differences in community composition, while UniFrac may provide a better view of evolutionary divergence and a measure of potential functional differences, since closely related species often share function [40].

The composition of microbial communities is typically of interest, particularly in clinical settings in which the germ theory of disease has guided microbiological diagnosis for over a century [41]. However, recent studies of the microbiome have revealed that more complex ecological features, including cooccurrence of species, are often of greater relevance in the study of noncommunicable diseases [42]. Owing to the high degree of interpersonal variability, it is not possible to discern discrete compositional patterns among healthy individuals [3]. Furthermore, while many cohort studies are able to identify biomarker taxa associated with disease states, differences in geography and other patient demographics have thus far confounded the identification of universally applicable taxonomic signatures for disease [18–20]. An important consideration when classifying taxa (sequences) is that most amplicon sequencing datasets rely on only short DNA fragments (typically <500 nucleotides), so highly reliable classification at high resolution taxonomic levels (e.g., species and strains) is often not possible. As a result, taxonomy is often binned to less resolved levels (e.g., family or genus). In addition, comparing community composition at low resolution (e.g., phylum) can result in an overestimation of compositional similarity, since a single phylum may contain dozens-to-hundreds of genera.

Functional data is of greater interest than compositional data; however, these data are less abundant in the literature due to greater expense associated with the whole genome shotgun sequencing, transcriptomics, metabolomics, and proteomics methods necessary to generate them [43]. Functional profiling from marker gene sequences has been suggested on the basis that related species share similar functional profiles [40,44,45]. Incomplete species and functional gene representation in databases, as well as similarity of marker genes in highly divergent species, make these methods imperfect tools that should, at best, be used to inform future confirmatory functional studies [19]. Shotgun sequencing provides taxonomic and functional information, but because the approach is untargeted, only highly abundant features are likely to be uncovered at shallow sequencing depth, increasing the cost of identifying informative taxa and functional genes, which are more likely to be of clinical relevance. Furthermore, knowing that a gene is present is not indicative of transcription or translation necessary to produce a functional protein. Metatranscriptomics-based approaches can inform which genes are being actively transcribed but does not inform translation and posttranslational modification necessary for protein synthesis [46]. Metabolomic and proteomic approaches do characterize metabolites and proteins that are being completely synthesized, but these techniques are still very expensive, suffer from the need to annotate chemical formulae against databases and pathways to arrive at translational data, and it is difficult to separate host and microbial products by these methods [47].

Statistical analysis of microbiome data

The primary challenges in analyzing microbiome data are that they are: (1) highly dimensional, often with hundreds-to-thousands of features per sample, and (2) overdispersed, with zero-inflated values for most features [28]. While traditional F-based statistics can be used to evaluate simple features of microbiome data such as alpha diversity or the abundances of a few taxa, microbiome data generally fall outside the assumptions of these tests. To address these issues, new statistical tests and approaches have been developed (Table 26.1). Testing for differences in compositionality (e.g., differentially abundant taxa or functional

TABLE 26.1 Statistical tools for microbiome data analysis.

Tool	Description	References
LEfSe	Linear discriminant analysis of effect sizes—employs a Kruskal-Wallis and Wilcoxon test to identify differentially abundant biomarkers. Galaxy interface available from (http://huttenhower.sph.harvard.edu/galaxy/).	[48]
PERMANOVA	Permutational multivariate analysis of variance— determines the significance of separation of multivariate data across an ordination space. Implementable in the vegan package in R [75].	[49]
ANOSIM	Analysis of similarity— evaluates differences in community composition. Implementable in QIIME2 (https://qiime2.org/) [76], mothur (www.mothur.org/) [31], and vegan.	[50]
Mantel test	Determines correlations between multivariate data. Implementable in vegan.	[51]
ANCOM	Analysis of composition of microbiomes— evaluates differences in community composition. Implementable in QIIME2.	[52]
EMPeror	Interactive tool for 3D visualization of microbiome data (https://omictools.com/emperor-tool).	[55]
SourceTracker	Determine contributions of multiple source communities to sink samples (https://github.com/danknights/sourcetracker/releases).	[56]
SplinectomeR	R Package for longitudinal analysis of microbiome data (https://github.com/RRShieldsCutler/splinectomeR).	[60]

genes) is also highly prone to type 1 error due to the high dimensionality of the data. Linear discriminant analysis of effect sizes (LEfSe) [48] has become a widely used tool for biomarker discovery and employs a sequential Kruskal-Wallis and Wilcoxon test in an algorithm that accounts for the biological relevance of differentially abundant features. A simple solution to determine differences between groups of samples has been to decrease the dimensionality of the data by reducing the community composition to a matrix of beta diversities. Then, tests such as permutational multivariate analysis of variance (PERMANOVA) [49], analysis of similarity (ANOSIM) [50], and Mantel's test [51] can be used to determine differences or relationships between groups of samples. Analysis of composition of microbiomes (ANCOM) has also recently been proposed to determine differences in community compositions, with the advantage that it works without assuming an underlying distribution of features [52].

Visualizing differences in microbiome data is typically done by principal coordinate analysis (PCoA) [53] or nonmetric multidimensional scaling (NMDS) [54], which provides plots of community data in two- or three-dimensional space. The axes on these plots have no units but explain a proportion of variation in the data and, typically, the proximity of points is indicative of community similarity, where closer data points have more similar community composition. EMPeror [55] has become a popular tool for interactive manipulation of these plots.

Beyond simple statistics for hypothesis testing, newer algorithms have been developed to mine microbiome data to identify patterns that cannot be resolved using even the recently adopted statistical approaches. SourceTracker [56] employs a Bayesian algorithm to determine the percentage of the community in user-defined sink samples that are likely to originate from multiple sources. The software can further differentiate which taxa contribute to the source signature. While originally designed to detect contamination in sequence data, the program has since been used to evaluate fecal pollution in recreational waters [57], as well as to determine microbial invasion during mouse cohousing [58] and evaluate the success of microbial engraftment following fecal microbiota transplantation [59]. SplinectomeR [60] is an R package designed to analyze longitudinal 'omics data and helps to overcome issues associated with such studies, including data collection irregularity, subject loss, and period effects. The benefit of this method is that is allows for comparison of changes over time and overall, without losing information from averaging.

Integrating multi-omics data

Among the challenges associated with integrating multi-omics data are that the data are often very sparse and may reflect differences in timescales [61]. In addition, many of the current methods lack robust biological validation. Machine learning is a powerful tool to identify distinguishing features between health and disease and predict future states [20]; however, models constructed in this way may suffer from overfitting and the use of biologically irrelevant features [62]. Thus, considerations of sample size and cross-validation approaches to ensure biological relevance are of paramount concern outside of the computational environment.

There are now several cloud-based or freely available computational methods to integrate multiomics data (Table 26.2), and specific considerations regarding experimental design and statistical analyses have been recently reviewed [43]. These new systems biology tools remain works in progress, with new functionalities added continually to address user demands and new data formats. XCMS Online utilizes robust metabolomics mapping to fit untargeted metabolomics data into metabolic pathway models, and it allows for further integration of genomic and proteomic data to determine biological mechanisms [63,64]. Similarly, MetaboAnalyst provides a user-friendly interface to analyze metabolomics data as well as to incorporate gene expression data [65].

Statistical approaches are being leveraged to understand the interrelationships among datasets [66,67]. Procrustes analysis [68] addresses the problem of data integration by reducing highly dimensional data to lower-dimensional distances for each type of 'omics data, which are then compared to determine if similar patterns exist between data types. Canonical correlation analysis (CCA) attempts to find combinations of variables with a high degree of linear correlation, and robust spar CCA has been adapted for 'omics data that have an abundance of zero-inflated features [69]. In addition, network science approaches have proven useful in attempting to decipher temporal and causal associations in 'omics data [70–72]. Local similarity analysis (LSA) and extended LSA (eLSA) [70,71] allow the incorporation of time-delayed effects in 'omics analyses, which may be useful in integrating 'omics data occurring at different time scales. Bayesian networks have also been applied frequently as predictive models in 'omics data that can dynamically account for unknowns in the data by considering prior probabilities [67,72–74].

TABLE 26.2 Currently available resources to integrate multiomics data.

Tool	URL	Description	References
XCMS Online	http://XCMSOnline.scripps.edu/	Incorporates metabolomics data into metabolic pathways and finds associations to transcriptomic and proteomic data	[63]
MetaboAnalyst	www.metaboanalyst.ca/	Integration of metabolomics and transcriptomic data	[65]
eLSA	http://meta.usc.edu/softs/lsa/	Evaluation of time-dependent associations in 'omics data	[71]
BANJO	https://users.cs.duke.edu/~amink/software/banjo/	Generation of static and dynamic Bayesian inference networks	[72]
BioBayes	www.dcs.gla.ac.uk/BioBayes/	Bayesian parameter estimation for biochemical systems	[74]

Future perspectives for multiomics data and precision medicine

Similar to fingerprints, the microbiome is coming to be recognized as an assemblage that is truly unique to an individual, which may play dynamic and critical roles in disease development, recovery from insults such as antibiotic exposure or surgeries, and even the efficacy of chemotherapeutic agents. As more clinical studies seek to leverage advances in sequencing technologies and 'omics fields to better understand this vital organ, it is important to recognize the potential pitfalls and limitations regarding both study designs and possible conclusions. While the 'omics technologies have now been sufficiently developed to allow for mechanistic investigations into the relationships between the microbiome and host, it is important to acknowledge that data acquisition still vastly outpaces our ability to integrate these data. Tests and algorithms now exist that allow meaningful and robust exploration of single datasets. However, integration of datasets to fully capture the essence of systems biology remains in its infancy. Network-based approaches offer exciting new avenues for hypothesis generation, and machine learning can suggest statistical significance of features or interactions, yet discrepancies between high-level mathematical estimations and biological reality must still be considered.

References

[1] Ursell LK, Metcalf JL, Parfrey LW, Knight R. Defining the human microbiome. Nutr. Rev. 2012;70(Suppl. 1):S38–44.

[2] Marchesi JR, Adams DH, Fava F, et al. The gut microbiota and host health: a new clinical frontier. Gut 2016;65(2):330–9.

[3] Lozupone CA, Stombaugh JI, Gordon JI, Jansson JK, Knight R. Diversity, stability and resilience of the human gut microbiota. Nature 2012;489(7415):220–30.
[4] Packey CD, Sartor RB. Commensal bacteria, traditional and opportunistic pathogens, dysbiosis and bacterial killing in inflammatory bowel diseases. Curr. Opin. Infect. Dis. 2009;3:292–301.
[5] Shukla SK, Murali NS, Brilliant MH. Personalized medicine going precise: from genomics to microbiomics. Trends Mol. Med. 2015;21(8):461–2.
[6] Goodwin S, McPherson JD, McCombie WR. Coming of age: ten years of next-generation sequencing technologies. Nat. Rev. Genet. 2016;17(6):333–51.
[7] Hasin Y, Seldin M, Lusis A. Multi-omics approaches to disease. Genome Biol. 2017;18(1):83.
[8] Handelsman J, Rondon MR, Brady SF, Clardy J, Goodman RM. Molecular biological access to the chemistry of unknown soil microbes: a new frontier for natural products. Chem. Biol. 1998;5(10):R245–9.
[9] Venter JC, Remington K, Heidelberg JF, et al. Environmental genome shotgun sequencing of the Sargasso Sea. Science 2004;304(5667):66–74.
[10] Tyson GW, Chapman J, Hugenholtz P, et al. Community structure and metabolism through reconstruction of microbial genomes from the environment. Nature 2004;428(6978):37–43.
[11] Sadowsky MJ, Staley C, Heiner C, et al. Analysis of gut microbiota – an ever changing landscape. Gut Microb. 2017;8(3):268–75.
[12] Lowe R, Shirley N, Bleackley M, Dolan S, Shafee T. Transcriptomics technologies. PLoS Comput. Biol. 2017;13(5):1–23.
[13] Wishart DS. Emerging applications of metabolomics in drug discovery and precision medicine. Nat. Rev. Drug Discov. 2016;15(7):473–84.
[14] Aslam B, Basit M, Nisar MA, Khurshid M, Rasool MH. Proteomics: technologies and their applications. J. Chromatogr. Sci. 2017;55(2):182–96.
[15] Cho I, Blaser MJ. The human microbiome: at the interface of health and disease. Nat. Rev. Genet. 2012;13(4):260–70.
[16] Vandeputte D, Tito RY, Vanleeuwen R, Falony G, Raes J. Practical considerations for large-scale gut microbiome studies. FEMS Microbiol. Rev. 2017;41(Suppl. 1):S154–67.
[17] Flynn KJ, Ruffin MT, Turgeon DK, Schloss PD. Spatial variation in the native colon microbiota in healthy adults. Cancer Prev. Res. 2018;11(7):393–402.
[18] Gilbert JA, Blaser MJ, Caporaso JG, Jansson JK, Lynch SV, Knight R. Current understanding of the human microbiome. Nat. Med. 2018;24(4):392–400.
[19] Staley C, Kaiser T, Khoruts A. Clinician guide to microbiome testing. Dig. Dis. Sci. 2018;63(12):3167–77.
[20] Knight R, Vrbanac A, Taylor BC, et al. Best practices for analysing microbiomes. Nat. Rev. Microbiol. 2018;16(7):410–22.
[21] Debelius J, Song SJ, Vazquez-Baeza Y, Xu ZZ, Gonzalez A, Knight R. Tiny microbes, enormous impacts: what matters in gut microbiome studies? Genome Biol. 2016;17(1):217.
[22] la Rosa PS, Brooks JP, Deych E, et al. Hypothesis testing and power calculations for taxonomic-based human microbiome data. PLoS One 2012;7(12):1–13.
[23] Kelly BJ, Gross R, Bittinger K, et al. Power and sample-size estimation for microbiome studies using pairwise distances and PERMANOVA. Bioinformatics 2015;31(15):2461–8.
[24] Païssé S, Valle C, Servant F, et al. Comprehensive description of blood microbiome from healthy donors assessed by 16S targeted metagenomic sequencing. Transfusion 2016;56(5):1138–47.
[25] Yokota M, Tatsumi N, Nathalang O, Yamada T, Tsuda I. Effects of heparin on polymerase chain reaction for blood white cells. J. Clin. Lab. Anal. 1999;13(3):133–40.
[26] Potgieter M, Bester J, Kell DB, Pretorius E. The dormant blood microbiome in chronic, inflammatory diseases. FEMS Microbiol. Rev. 2015;39(4):567–91.
[27] Geller LT, Barzily-Rokni M, Danino T, et al. Potential role of intratumor bacteria in mediating tumor resistance to the chemotherapeutic drug gemcitabine. Science 2017;357(6356):1156–60.
[28] Xia Y, Sun J. Hypothesis testing and statistical analysis of microbiome. Genes Dis. 2017;4(3):138–48.
[29] Shannon CE, Weaver W. The mathematical theory of communication. Urbana (IL): The University of Illinois Press; 1949.
[30] Simpson EH. Measurement of diversity. Nature 1949;163(4148):688.
[31] Schloss PD, Westcott SL, Ryabin T, et al. Introducing mothur: open-source, platform-independent, community-supported software for describing and comparing microbial communities. Appl. Environ. Microbiol. 2009;75(23):7537–41.
[32] Martin AP. Phylogenetic approaches for describing and comparing the diversity of microbial communities. Appl. Environ. Microbiol. 2002;68(8):3673–82.
[33] Holmes I, Harris K, Quince C, McInnes P, Wang L. Dirichlet multinomial mixtures: generative models for microbial metagenomics. PLoS One 2012;7(2):e30126.
[34] Chao A. Non-parametric estimation of the number of classes in a population. Scand. J. Stat. 1984;11(4):265–70.
[35] Chao A, Lee S-M. Estimating the number of classes via sample coverage. J. Am. Stat. Assoc. 1992;87(417):210–7.
[36] Ellison AM. Partitioning diversity. Ecology 2010;91(7):1962–3.
[37] Chao A, Chazdon RL, Colwell RK, Shen T-J. A new statistical approach for assessing similarity of species composition with incidence and abundance data. Ecol. Lett. 2004;8(2):148–59.
[38] Bray JR, Curtis JT. An ordination of the upland forest communities of southern Wisconsin. Ecol. Monogr. 1957;27:325–49.
[39] Lozupone C, Knight R. UniFrac: a new phylogenetic method for comparing microbial communities. Appl. Environ. Microbiol. 2005;71(12):8228–35.
[40] Martiny AC, Treseder K, Pusch G. Phylogenetic conservatism of functional traits in microorganisms. ISME J. 2013;7(4):830–8.
[41] Bantock GG. The modern doctrine of bacteriology, or the germ theory of disease. Br. Med. J. 1899;1(1997):846–8.
[42] Byndloss MX, Bäumler AJ. The germ-organ theory of non-communicable diseases. Nat. Rev. Microbiol. 2018;16(2):103–10.
[43] Franzosa EA, Hsu T, Sirota-Madi A, et al. Sequencing and beyond: integrating molecular "omics" for microbial community profiling. Nat. Rev. Microbiol. 2015;13(6):360–72.
[44] Langille MGI, Zaneveld J, Caporaso JG, et al. Predictive functional profiling of microbial communities using 16S rRNA marker gene sequences. Nat. Biotechnol. 2013;31(9):814–21.

[45] Aßhauer KP, Wemheuer B, Daniel R, Meinicke P. Tax4Fun: predicting functional profiles from metagenomic 16S rRNA data. Bioinformatics 2015;31(17):2882–4.

[46] Moran MA, Satinsky B, Gifford SM, et al. Sizing up metatranscriptomics. ISME J. 2013;7(2):237–43.

[47] Ellermann M, Carr JS, Fodor AA, Arthur JC, Carroll IM. Characterizing and functionally defining the gut microbiota: methodology and implications. In: Floch MH, Ringel Y, Walker WA, editors. The microbiota in gastrointestinal pathophysiology. London: Elsevier; 2017. p. 15–25.

[48] Segata N, Izard J, Waldron L, et al. Metagenomic biomarker discovery and explanation. Genome Biol. 2011;12(6):R60.

[49] Anderson MJ. A new method for non-parametric multivariate analysis of variance. Austral Ecol. 2001;26(1):32–46.

[50] Clarke KR. Non-parametric multivariate analyses of changes in community structure. Aust. J. Ecol. 1993;18(1):117–43.

[51] Mantel N. The detection of disease clustering and a generalized regression approach. Cancer Res. 1967;27(2):209–20.

[52] Mandal S, Van Treuren W, White RA, Eggesbø M, Knight R, Peddada SD. Analysis of composition of microbiomes: a novel method for studying microbial composition. Microb. Ecol. Health Dis. 2015;26.

[53] Anderson MJ, Willis TJ. Canonical analysis of principal coordinates: a useful method of constrained ordination for ecology. Ecology 2003;84(2):511–25.

[54] Kenkel NC, Orloci L. Applying metric and nonmetric multidimensional scaling to ecological studies - some new results. Ecology 1986;67(4):919–28.

[55] Vázquez-Baeza Y, Pirrung M, Gonzalez A, Knight R. EMPeror: a tool for visualizing high-throughput microbial community data. GigaScience 2013;2(1):16.

[56] Knights D, Kuczynski J, Charlson ES, et al. Bayesian community-wide culture-independent microbial source tracking. Nat. Methods 2011;8(9). 761-U107.

[57] Staley C, Kaiser T, Lobos A, et al. Application of SourceTracker for accurate identification of fecal pollution in recreational freshwater: a double-blinded study. Environ. Sci. Technol. 2018;52(7):4207–17.

[58] Ridaura VK, Faith JJ, Rey FE, et al. Gut microbiota from twins discordant for obesity modulate metabolism in mice. Science 2013;341(6150):1241214.

[59] Staley C, Vaughn BP, Graiziger CTCT, et al. Community dynamics drive punctuated engraftment of the fecal microbiome following transplantation using freeze-dried, encapsulated fecal microbiota. Gut Microb. 2017;8(3):276–88.

[60] Shields-Cutler RR, Al-Ghalith GA, Yassour M, Knights D. SplinectomeR enables group comparisons in longitudinal microbiome studies. Front. Microbiol. 2018;9:785.

[61] Nicholson JK, Lindon JC. Systems biology: metabonomics. Nature 2008;455(7216):1054–6.

[62] Camacho DM, Collins KM, Powers RK, Costello JC, Collins JJ. Next-generation machine learning for biological networks. Cell 2018;173(7):1581–92.

[63] Gowda H, Ivanisevic J, Johnson CH, et al. Interactive XCMS Online: simplifying advanced metabolomic data processing and subsequent statistical analyses. Anal. Chem. 2014;86(14):6931–9.

[64] Huan T, Forsberg EM, Rinehart D, et al. Systems biology guided by XCMS Online metabolomics. Nat. Methods 2017;14(5):461–2.

[65] Xia J, Sinelnikov IV, Han B, Wishart DS. MetaboAnalyst 3.0—making metabolomics more meaningful. Nucleic Acids Res. 2015;43(W1):W251–7.

[66] McHardy IH, Goudarzi M, Tong M, et al. Integrative analysis of the microbiome and metabolome of the human intestinal mucosal surface reveals exquisite inter-relationships. Microbiome 2013;1(1):17.

[67] Alyass A, Turcotte M, Meyre D. From big data analysis to personalized medicine for all: challenges and opportunities. BMC Med. Genom. 2015;8(1):33.

[68] Hurley JR, Cattell RB. The procrustes program: producing direct rotation to test a hypothesized factor structure. Behav. Sci. 1962;7(2):258–62.

[69] Wilms I, Croux C. Robust sparse canonical correlation analysis. BMC Syst. Biol. 2016;10(1):72.

[70] Ruan Q, Dutta D, Schwalbach MS, Steele JA, Fuhrman JA, Sun F. Local similarity analysis reveals unique associations among marine bacterioplankton species and environmental factors. Bioinformatics 2006;22(20):2532–8.

[71] Xia LC, Steele JA, Cram JA, et al. Extended local similarity analysis (eLSA) of microbial community and other time series data with replicates. BMC Syst. Biol. 2011;5(Suppl. 2):S15.

[72] Hartemink AJ. Reverse engineering gene regulatory networks. Nat. Biotechnol. 2005;23(5):554–5.

[73] Huang SS, Septimus E, Kleinman K, et al. Targeted versus universal decolonization to prevent ICU infection. N. Engl. J. Med. 2013;368(24):2255–65.

[74] Vyshemirsky V, Girolami M. BioBayes: a software package for Bayesian inference in systems biology. Bioinformatics 2008;24(17):1933–4.

[75] Oksanen J, Blanchet FG, Kindt R, et al. Vegan: commity ecology package. 2015.

[76] Hall M, Beiko RG. 16S rRNA gene analysis with QIIME2. In: Beiko R, Hsiao W, Parkinson J, editors. Methods in molecular biology. New York (NY): Humana Press; 2018. p. 113–29.

Chapter 27

Sepsis: future role of omics in diagnosis and therapy

Hermes Vieira Barbeiro, Denise Frediani Barbeiro and Francisco Garcia Soriano
Emergências Clínicas/LIM 51, Faculdade de Medicina da Universidade de São Paulo, São Paulo, Brazil

Introduction to sepsis

The sepsis scenario has a very rapid evolution, and the inflammatory profile changes every hour. Sepsis-aiming precision medicine has to be fast. In a matter of hours, a patient may have died from an infection that has turned into sepsis. In this way, the Surviving Sepsis Campaign and intensive care associations proposed a therapeutic intervention to be performed during the first 4 h [1]. The protocol is similar for all patients, and improves global results.

Despite the significant advances in our understanding of the pathobiology of sepsis, sepsis management remains largely supportive. Tailoring treatment according to clinical and laboratory findings will add further benefit. It is not clearly known why under similar circumstances some patients eliminate more easily an invading microorganism, whereas others develop sepsis and septic shock. A variable to consider is the host genetic, epigenetic, proteomic, metabolomic, and microbiomic factors (Fig. 27.1), which have been related to sepsis outcome. Sepsis can be an area of conflict between standardized, protocoled routines as currently adopted, and precision-based custom handling of each patient.

Genomics—genomic variants

The host activates production and release of many proteins and substances to fight bacteria [2,3]. Human genome contains genes that code proteins in response to infection. There are approximately 22,000 genes that code proteins and 5−10 times more proteins are formed (Fig. 27.1) as the final product [4]. The difference in genes from every person can determine an advantage against infection and less sepsis development [2,5].

Relation between genomics and infection was raised first in 1988. It was demonstrated that individuals with biological parents who died from infectious diseases had a fivefold risk of dying from sepsis [6]. As a general rule, each person presents all the regular genes of the inflammatory cascade. What can change is polymorphisms (genomic variation). Several studies have attempted to associate polymorphisms with incidence or outcome of infectious disease, and its complications in critically ill patients [5,7,8].

Cytokines and other mediators

Tumor Necrosis Factor (TNF) was the first cytokine demonstrated to be involved in sepsis. TNF gene polymorphisms showed association with adverse outcomes, as well as increased incidence of severe sepsis and septic shock [7]. Pro- and antiinflammatory cytokines like the interleukin-1 (IL-1) gene family, interleukin-6 (IL-6) and interleukin-10 (IL-10), can display different inflammatory profiles [9]. Cell signaling pathways activated in the response to a bacterium triggered the search for polymorphism of new genes candidates. Incidence, severity, and mortality of infectious complications in the critically ill are influenced as well by polymorphism of genes involved in pathogen recognition, and signal transduction of inflammatory pathways, such as cluster of differentiation 14 (CD14), tolllike receptors (TLRs), lipopolysaccharide binding protein (LBP), interleukin-1 receptor-associated kinase (IRAK 4), and IRAK 1 [10−16].

These genes can present polymorphisms with increased or decreased signal transduction, and consequently higher or lower inflammatory response to a pathogen. Tyrosine kinase is an enzyme, which catalyzes phosphorylation from nucleoside triphosphate (ATP) to the amino acid tyrosine in proteins. Nonreceptor tyrosine kinases (NRTKs) function in the cytoplasm, and transfer signals within the cell into the nucleus [17]. Feline sarcoma (Fes) or Fujinami poultry

Precision Medicine for Investigators, Practitioners and Providers. https://doi.org/10.1016/B978-0-12-819178-1.00027-7

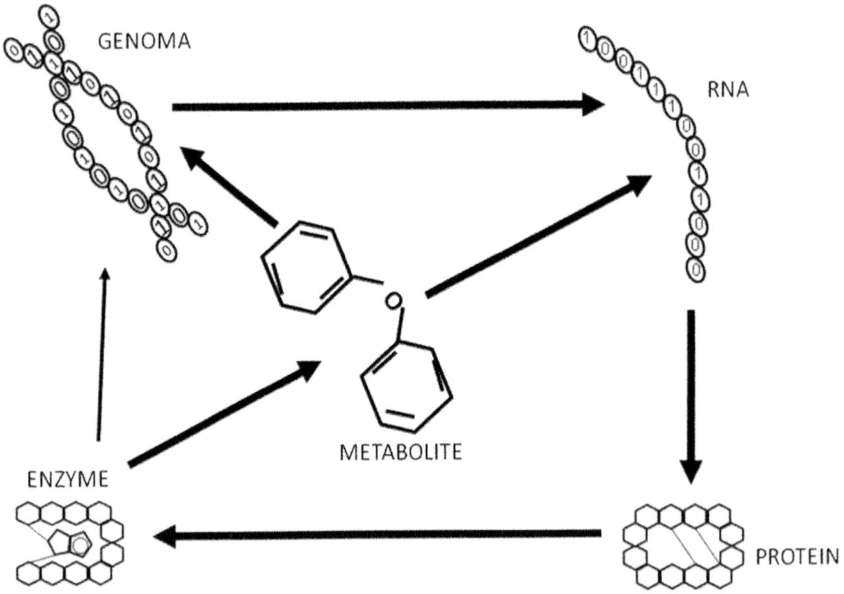

FIGURE 27.1 Schematic description of protein synthesis.

sarcoma protein (Fps), and its homologous-related protein FER, are the two members of a class of NRTKs. Fes signals downstream of receptors in regulation of hematopoietic cell development, survival, migration, inflammatory mediator release, and angiogenesis.

Activated Fes is a potent inducer of myeloid differentiation. The FER gene affects leukocyte recruitment, and intestinal barrier dysfunction caused by lipopolysaccharide (LPS). A study identified a polymorphism of FER, which is associated with a reduced risk of death from sepsis caused by pneumonia [18].

Sepsis causes the activation of hundreds, or almost a 1000, genes. There are several other physiologic pathways involved in sepsis. For example, it has been recognized that protein cascades such as the coagulation cascade, represent important genomic candidate markers [19]. A genetic predisposition to produce high concentrations of plasminogen activator inhibitor-1 (PAI-1) is associated with poor outcome of meningococcal sepsis. This finding suggests that impaired fibrinolysis is an important factor in the pathophysiology of meningococcal sepsis.

Genome-wide association studies

In a genome-wide association study (GWAS) of 740 adult septic patients, 243 autosomal variants were detected clustered in 14 loci, with suggestive evidence for an association with 28-day mortality. Converging evidence for three of them were unveiled in independent data sets. The authors propose vacuolar protein sorting-associated protein 13A (VPS13A), cysteine rich secretory protein LCCL domain containing 2 (CRISPLD2), and the locus of chromosome 13, as foci for future research [2]. The best association signal (rs117983287; $p = 8.16 \times 10^{-8}$) was observed for a missense variant located at chromosome 9q21.2 in the VPS13A gene. VPS13A was further supported by additional GWAS ($p = 0.03$) and sequencing data ($p = 0.04$). Furthermore, CRISPLD2 ($p = 5.99 \times 10^{-6}$) and a region on chromosome 13q21.33 ($p = 3.34 \times 10^{-7}$) were supported by both author data and external biological evidence.

Molecular diagnosis of sepsis

The Molecular Diagnosis and Risk Stratification of Sepsis (MARS) provides a molecular classification of patients with sepsis according to four different endotypes, upon intensive care medicine (ICU) admission. Detection of septic endotypes is in the route for personalized patient management [3]. The authors identified 140 genes that classified the Mars1–4 endotypes. The worst survival outcome at 28 days corresponded to Mars1 endotype (39% did not survive). The same endotype exhibited the worst survival outcome at 1-year follow-up. Sepsis is a multifactorial disease, and molecular diagnosis with risk stratification cannot be performed with single genes, but with interpretation of several genes.

Epigenomic and transcriptomics

Deoxyribonucleic acid (DNA) and histones can be modified by enzymes with epigenetic effect, which causes gene expression to vary without a change in DNA sequence [20]. Histone can be modified by methylation, acetylation, ubiquitination, and phosphorylation [21,22]. Previous epigenetic modifications can be due to habits, environment, and diet, and lead to different clinical scenarios. During an infectious disease, bacterium-induced epigenetic changes can interfere with host cell function, either promoting host defense or allowing pathogen persistence [23,24].

Transcriptomics

Transcriptomics is the measurement of messenger ribonucleic acid (mRNA) levels for genes in specific cells or tissues (Fig. 27.1). This tool points out not only active genes in general, but can also track the effect of epigenomics on genes expression.

Full blood and/or individual blood cell subpopulations are screened, because it is not possible to harvest tissue in septic patients. Whole blood, with all its immunocompetent cells, represents a robust and complex immunological system involved in sepsis pathogenesis [25]. Literature reports different gene expressions, depending on the subpopulation of immunocompetent cells [26]. Practically, no investigation has been performed that evaluated gene expression in individual organs during clinical sepsis, with the exception of experimental models [27].

Interpretation of transcriptomic observations

A number of issues arise in these studies dealing with sepsis. The first one is the complex dynamics of disease progression in sepsis [28–31]. The clinical course of sepsis shows proinflammatory elements are dominant in early stages, later on shifting to functional immunosuppression. Therefore, gene expression differs over time [32].

Sepsis and inflammatory profile

There is much interest to differentiate infectious and noninfectious systemic inflammatory syndrome. Some models evaluated gene expression in volunteers who received an endotoxin dose [33,34]. Target gene groups with their bonds in gene maps were thus identified. Experimental endotoxin injection is of course different from clinical sepsis with a living infectious agent. Clinical trials in sepsis confirmed different gene expression in systemic inflammatory response syndrome (SIRS) of infectious versus noninfectious etiology [4]. The results were confirmed by other studies [35–38]. The authors tested whether profiling transcription is applicable to sepsis diagnosis, using a microarray containing probes for 340 genes related to inflammation. Gene expression pattern was highly homogenous, resulting in 69% of differentially expressed genes. A list of 50 differentially expressed genes showed a positive predictive value of 98%. Regardless of the heterogeneity of the patients, there were a striking correlation between conventional diagnostic classification and the transcription approach [4,39].

Disease etiology and severity

Differences have been found with respect to the septic agent. One report did not reveal different gene expression in gram-positive (G+) and gram-negative (G−) infections, but others did [40,41]. Different expressions of IL-2, IL-10, IL-23, IL-27, interferon-γ (IFN γ), and TNFα have been found when comparing infection to severe sepsis [42]. Gene expression difference between the two groups which predicts postoperative sepsis was observed for IL1β, TNF, CD3D molecule, delta (CD3-TCR complex), and perforin 1 (pore forming protein) (PRF1) [43]. The combination of TNF, IL1β, and CD3D expression had a sensitivity and specificity of 90% and 85%, respectively, with an estimated negative predictive value of 98.1% [43].

Immune dysfunction

A different gene expression was demonstrated in bacterial and virus infections related to immune dysfunction. A T-cell-dominant gene-expression signature was associated with the host response to severe influenza pneumonia. Genes linked to the cell cycle and its regulation, were the main determinants of the host response in influenza infection. The search for an immune response specific pattern to bacterial pneumonia failed [44]. Confirming the abilities of a specific group of genes to differentiate between viral and bacterial infection would influence antibiotic consumption and resistance development, and is of major importance in clinical practice.

Sepsis and organ dysfunction caused by community acquired pneumonia in ICU patients was addressed by gene expression in peripheral blood leukocytes [45]. Transcriptomic analysis defined two sepsis response signatures (SRS). SRS1 found in 41% of patients identified patients with an immunosuppression phenotype (endotoxin tolerance, T-cell exhaustion, and human leukocyte antigen (HLA) class II regulation disorder). This phenotype was associated with higher 14-day mortality [45].

Metaanalysis

A metaanalysis of gene expression in severe sepsis and septic shock, with 45 series, suggested 352 genes as candidates (215 upregulated and 137 downregulated). Examples of upregulated genes were cluster of differentiation 177 (CD177), matrix metalloproteinase-8 (MMP8), haptoglobin (HP), arginase 1 (ARG1), and annexin A3 (ANXA3). The top downregulated genes were Fc fragment of IgE receptor Ia (FCER1A), human ortholog of yeast mitochondrial AAA metalloprotease (YMEI1L1), T cell receptor delta variable 3 (TRDV3), leucine-rich repeat neuronal protein 3 (LRRN3), and MYB proto-oncogene like 1 (MYBL1) [46].

In response to a need for better sepsis diagnosis, several new gene expression classifiers have been recently published, including the 11-gene "Sepsis MetaScore," the "FAIM3-to-PLAC8" ratio, and the Septicyte Lab. The three tools do not show significant differences in overall ability to distinguish noninfectious SIRS from sepsis [47].

Clinical applications for genomics and transcriptomics

Most data of polymorphism and epigenetic changes are useful for prognostic evolution; however, these differences can indicate a different inflammatory and immunologic response. There are genomic and epigenomic patterns that produce a more proinflammatory response, with consequent organ damage. On the other hand, a lower proinflammatory response could mean a difficulty to fight bacteria.

Patients presenting genomic or epigenomic proinflammatory features can be treated more aggressively, with volume expansion, antioxidants, and drugs that inhibit some parts of the inflammatory cascade. In hyperresponsive cases, antiinflammatory therapies can be tested, such as corticoids, anti-TNF drugs. In circumstances of lower response, anti PD-1 or anti PDL-1 pharmacotherapy could be considered, to unblock immune system check points. In addition, IL-7 and interferon-γ are alternatives to improve immune response, in order to fight bacteria. There are ongoing clinical trials for checking the efficiency of these drugs; depending on the outcome, it could be convenient in the future to stratify therapy, depending on observed patterns of inflammation.

Proteomics

Proteins are the final effectors of every function in human being. Many mRNAs are not converted into proteins, so the transcriptome cannot give the true picture of what is occurring in septic inflammation (Fig. 27.1). Maybe proteins can represent a better choice for new biomarkers, which will differentiate infectious systemic inflammation from noninfectious inflammation, and also to find pathogenic mechanisms and pathways that could guide diagnosis and treatment.

Sepsis post liver transplant unveiled new proteins as possible biomarkers. This method was plasma profiling coupling protein chip array, with surface-enhanced laser desorption/ionization with time-of-flight (SELDI-TOF). There were 31 patients with infection and 34 without. Five peaks were differentially expressed and allowed sepsis diagnosis, with a positive likelihood ratio of 5.1, and C-statistics of 0.74 (0.58−0.85) [48]. The new five proteins: CM10 4152.7, CM10 4627.2, CM10 5744.7, CM10 5812.9, and CM10 5912.3 were significantly different, in subjects who developed sepsis at postoperative day 5, in comparison with nonseptic patients. The "septic" protein (CM10 4152) was also differentially expressed in patients with septic shock versus those with nonseptic shock.

Other proteins differentially expressed in patients with systemic inflammatory response syndrome or sepsis encompass complement factor B, haptoglobin, clusterin, 1-B-glycoprotein, complement C4, C reactive protein (CRP) precursor, plasminogen precursor, and transthyretin precursor. After low-dose endotoxin challenge in healthy volunteers, the authors found profound changes of plasma proteome [49]. A new component of 4154 mass units was recognized as the activation peptide of the C1 esterase inhibitor. Resemblance in mass units to 4152 protein peak could reasonably represent the same peptide. Appearance of this peptide could be related to blood proteolytic degradation occurring in septic patients.

Animal models

Experimental studies can standardize bacteria and study proteins in the blood and organs; however, in humans this can only be conducted a posteriori. The model of sepsis with virulent *Streptococcus pyogenes* in mice was used to map proteins in blood plasma and individual organs. Biomarkers grouped into categories can reflect sepsis-related characteristics such as acute phase proteins, coagulation, cytokines/chemokines, vascular endothelial damage, and organ dysfunction. Human biomarkers mapped in the mouse orthologous model and in the protein tissue atlas confirm that the predominant localization of these proteins are plasma, tissues and cells. They are highly tissue and cell specific.

Most biomarkers increase in concentration with disease severity. There was a different behavior for each biomarker group. For example, there was a dose-dependent elevation of acute phase proteins, coagulation proteins, and cytokines/chemokines. However, biomarkers associated with vascular endothelial damage presented smaller increase with vascular damage. Contrasting with previous data, biomarkers linked to organ dysfunction presented only a noticeable increase in the highest infectious dose group. The hypothesis is that organ dysfunction is not a gradual process, but rather confined to some of the most infected animals [50,51].

Another aim of proteomics is to study infectious models and describe protein-protein interactions, using mass spectrometry. With *S. pyogenes* and adhered human plasma protein, a stoichiometric model of protein structure interaction was developed. A stoichiometric surface density model of host-pathogen interactions emerged, using data from proteomic and electron microscopy analysis. The model captured the large differences in network topology, between the surface structure of the wild-type and M1 strain, revealing a dense and a highly organized protein interaction network between *S.s pyogenes* and human plasma proteins [52].

Clinical applications

Proteomics as the final expression of either genomics or transcriptomics is a priority in clinical practice. As not all mRNA is converted to protein, proteomics can convey a more refined and reliable picture of sepsis dynamics.

Metabolomics

Metabolites are small molecules produced by the action of enzymes/proteins (Fig. 27.1). Metabolomics is the large-scale study of such molecules, within cells, fluids, tissues, or organisms, as influenced by both genetic and environmental factors [53]. Metabolomics is the final expression of genes, mRNA, and protein activity; it represents the molecular phenotype, or deep phenotype [54,55]. The Human Metabolome Database contains records for more than 42,000 metabolites, from sugars and other macronutrients, to specific cofactors.

Probably the total amount is higher, and several analytical methods struggle to capture the chemical diversity. The aims of metabolomics are the same as with other "omics" — diagnosis, prognosis, and identification of at-risk patients [54,55]. In a prospective study, blood samples from 65 patients with sepsis and 49 controls were compared using gas chromatography coupled to time-of-flight mass spectrometry. A predictive logistic regression model with six metabolites showed a sensitivity of 0.91 (95% confidence interval/CI 0.69–0.99), and a specificity of 0.84 (95% CI 0.58–0.94), with an area under curve (AUC) of 0.93 (95% CI 0.89–1.01). Myristic acid was the most predictive metabolite, with a sensitivity of 1.00 (95% CI 0.85–1.00) and specificity of 0.95 (95% CI 0.74–0.99) [56].

Changes in the plasma levels of low unsaturated long-chain phosphatidylcholines and kynurenine were also associated with mortality in 20 patients with septic shock [57]. Urine samples from 64 patients with sepsis or septic shock were compared with respect to Sequential Organ Failure Assessment (SOFA) score. Higher amounts of ethanol, glucose, and hippurate, and lower levels of methionine, glutamine, arginine, and phenylalanine were associated to a negative prognosis [58]. The metabolic profile of normal patients and those with SIRS or sepsis was markedly different. Seven metabolites could potentially be used to diagnose sepsis [59].

A significant decrease in the levels of lactitol dehydrate and S-phenyl-D-cysteine and an increase in the levels of S-(3-methylbutanoyl)-dihydrolipoamide-E and N-nonanoyl glycine were observed in sepsis, in comparison to patients with SIRS ($P < 0.05$). Patients with sepsis and septic shock displayed lower levels of glyceryl-phosphorylethanolamine, Ne, Ne dimethyllysine, phenylacetamide, and D-cysteine ($P < 0.05$) in their sera [59].

Volatile metabolome (volatolome)

Volatile organic compounds (VOC) are a promising tool for sepsis screening and diagnosis. VOC can be analyzed in plasma, blood, urine, feces, and respiratory exhaled air. VOC profiles from feces of infants ($N = 843$), up to 3 days prior to clinical late onset sepsis (LOS), were compared with controls. VOC analysis discriminated between infants developing LOS and controls. Early detection of LOS may provide clinicians an opportunity for precocious therapeutic interventions, aimed at prevention of sepsis, possibly improving LOS-related morbidity and mortality [60].

In the emergency department (ED), tools for early screening and diagnosis of sepsis improve outcomes. Common metabolites and protein-mediators were validated as potential biomarkers for a sepsis screening model, to differentiate pediatric intensive care unit (PICU)-sepsis from pediatric emergency department (PED)-sepsis. Metabolomics achieved: sensitivity 0.71, specificity 0.93, and AUROC $= 0.90 \pm 0.03$. The metabolomics-based biomarkers predicted which sepsis patients required care in a PICU versus those that could be safely cared for outside a PICU [61].

Neutropenic febrile patients are a high-risk population, and early diagnosis is crucial. Using metabolomics data, a 5-predictor model had an area under the receiver operating curve of 0.991 (95%CI: 0.972–1.000). Pregnenolone steroids were more abundant in bacteremic patients, and carnitine metabolites in controls. A 3-predictor gene expression model had corresponding results of 0.961 (95% CI: 0.896–1.000), 100%, and 86%. Genes involved in innate immunity were differentially expressed.

Direct breath analysis (exhaled air)

Bacteria have a distinct metabolism, part of which results in the production of bacteria-specific volatile organic compounds (VOCs), which might be used for diagnostic purposes. Isopentanol, formaldehyde, methyl mercaptan, and trimethylamine are VOCs produced by the six most abundant and pathogenic bacteria in sepsis: *Staphylococcus aureus, Streptococcus pneumoniae, Enterococcus faecalis, Pseudomonas aeruginosa, Klebsiella pneumoniae,* and *Escherichia coli.* Humans do not produce these VOCs, so these compounds are credible biomarkers for these pathogens.

Isovaleric acid and 2-methyl-butanal were linked to *St. aureus;* 1-undecene, 2,4-dimethyl-1-heptane, 2-butanone, 4-methyl-quinazoline, hydrogen cyanide, and methyl thiocyanide to *P. aeruginosa*; and methanol, pentanol, ethyl acetate, and indole to *E. coli* [62,63]. Ventilation associated pneumonia (VAP) is consistently underdetected in the hospital. Exhaled breath biomarkers, including measurement of volatile organic compounds, represent an approach [63,64].

Microbiome

Gut

The gut can be a motor of multiple organ dysfunction in critical illness, due to bacterial translocation. It is a potential source of bacteria to precipitate and maintain sepsis. The gut microbial barrier encompasses epithelium, adaptive immune system, and microbiome. Each part plays a crucial role in the maintenance of health, and pathophysiology of critical illness. Serious illness can compromise gut integrity, increasing apoptosis, and permeability. Multiple preclinical studies have demonstrated that targeting gut barrier integrity results in improved survival in critical illness. Enteral nutrition is widely used in the intensive care unit, as a protective tool for gut mass and function.

Microbiota

Bacterial commensals are divided into seven main phyla, which are physiologically dominated by Firmicutes and Bacteroidetes. Nevertheless, substantial inter- and intra-individual variations on genomic, dietary, and environmental factors are identified [65]. The whole gut microbiota, that is, bacteria, fungi, archaea, viruses, and protozoa that colonize the intestinal tract, compose a complex ecosystem. Gut microbiota is estimated to reach in healthy humans a total of 3.10^{13} microorganisms, a population that equals the host cells [66].

Dysbiosis in serious illness

Metabolic diseases, autoimmune disorders, inflammatory bowel disease, neurocognitive impairment, and malignancies exhibit underlying intestinal dysbiosis [67]. A variety of both exogenous and patient-related factors may lead to ICU-acquired dysbiosis, for instance, drugs including antimicrobials, proton pump inhibitors, or depressors of gastrointestinal transit (e.g., opioids), artificial nutrition, sepsis, shock, or bowel ischemia [68]. Loss of diversity is commonly observed, and features deep depletion of potential health-promoting commensal genera like *Faecalibacterium*, *Ruminococcus*, or *Pseudobutyrivibrio*, along with overgrowth of a pathogenic and normally subdominant taxon, namely *Enterobacteriaceae*, and including *Enterococcus* and other pathogens [69–71].

Gut microbial signature

Studies present discrepancies between case-mix, antimicrobial exposure, stool collection time during ICU stay, and analytical methods. ICU-based studies, which adopt routine selective digestive decontamination, further complicate interpretation of the data. Whether intestinal dysbiosis is an independent predictor of poor outcome, or just one more marker of severity, is still speculative. Experimental studies suggest that the intestinal ecosystem modulates the risk of complications such as acute respiratory distress syndrome, ischemia/reperfusion-related acute kidney injury, or sepsis-induced muscle wasting [72–74]. The intestinal barrier may be impaired by dysbiosis and immunosuppression, thereby facilitating the occurrence of sepsis and multiple organ failure [68].

Clinical applications

Preventing bacteria from becoming virulent or reprogramming them to a nonvirulent phenotype is one of the new approaches which may revolutionize the treatment of gut-derived sepsis. There is species-specific colonization resistance: *Clostridium bolteae* and *Blautia producta* act synergistically to prevent vancomycin-resistant enterococci (VRE); *Clostridium scindens* appears to protect from *Clostridium difficile* infection; and *Desulfovibrio*, *Oscillospira*, *Parabacteroides*, and *Coprococcus* genera have been associated with the absence of carriage of extended spectrum beta-lactamase-producing *Escherichia coli* [75–77]. Resident anaerobes have the capacity to prevent intestinal colonization by exogenous microorganisms, via competition for nutrient intake, and induction of a targeted immune response [75,78].

To date, fecal transplant is not typically used in critically ill patients because of ongoing antibiotic use, which is common in the ICU. The most convincing data are in recurrent *C. difficile* infection, where cure rates are three times higher than seen with conventional medical therapy, without apparent side effects [75].

References

[1] Rhodes A, Evans LE, Alhazzani W, Levy MM, Antonelli M, Ferrer R, et al. Surviving sepsis campaign: international guidelines for management of sepsis and septic shock: 2016. Crit. Care Med. 2017;45(3):486–552.

[2] Scherag A, Schöneweck F, Kesselmeier M, Taudien S, Platzer M, Felder M, et al. Genetic factors of the disease course after sepsis: a genome-wide study for 28Day mortality. EBioMedicine 2016;12:239–46.

[3] Scicluna BP, van Vught LA, Zwinderman AH, Wiewel MA, Davenport EE, Burnham KL, et al. Classification of patients with sepsis according to blood genomic endotype: a prospective cohort study. Lancet Respir. Med. 2017;5(10):816–26.

[4] Prucha M, Ruryk A, Boriss H, Möller E, Zazula R, Herold I, et al. Expression profiling: toward an application in sepsis diagnostics. Shock 2004;22(1):29–33.

[5] Sutherland AM, Russell JA. Issues with polymorphism analysis in sepsis. Clin. Infect. Dis. 2005;41(Suppl. 7):S396–402.

[6] Sørensen TI, Nielsen GG, Andersen PK, Teasdale TW. Genetic and environmental influences on premature death in adult adoptees. N. Engl. J. Med. 1988;318(12):727–32.

[7] Stüber F, Petersen M, Bokelmann F, Schade U. A genomic polymorphism within the tumor necrosis factor locus influences plasma tumor necrosis factor-alpha concentrations and outcome of patients with severe sepsis. Crit. Care Med. 1996;24(3):381–4.
[8] Texereau J, Pene F, Chiche JD, Rousseau C, Mira JP. Importance of hemostatic gene polymorphisms for susceptibility to and outcome of severe sepsis. Crit. Care Med. 2004;32(5 Suppl. l):S313–9.
[9] Fang XM, Schröder S, Hoeft A, Stüber F. Comparison of two polymorphisms of the interleukin-1 gene family: interleukin-1 receptor antagonist polymorphism contributes to susceptibility to severe sepsis. Crit. Care Med. 1999;27(7):1330–4.
[10] Hubacek JA, Stüber F, Fröhlich D, Book M, Wetegrove S, Ritter M, et al. Gene variants of the bactericidal/permeability increasing protein and lipopolysaccharide binding protein in sepsis patients: gender-specific genetic predisposition to sepsis. Crit. Care Med. 2001;29(3):557–61.
[11] Feterowski C, Emmanuilidis K, Miethke T, Gerauer K, Rump M, Ulm K, et al. Effects of functional Toll-like receptor-4 mutations on the immune response to human and experimental sepsis. Immunology 2003;109(3):426–31.
[12] Medvedev AE, Lentschat A, Kuhns DB, Blanco JC, Salkowski C, Zhang S, et al. Distinct mutations in IRAK-4 confer hyporesponsiveness to lipopolysaccharide and interleukin-1 in a patient with recurrent bacterial infections. J. Exp. Med. 2003;198(4):521–31.
[13] Sutherland AM, Walley KR, Russell JA. Polymorphisms in CD14, mannose-binding lectin, and Toll-like receptor-2 are associated with increased prevalence of infection in critically ill adults. Crit. Care Med. 2005;33(3):638–44.
[14] Arcaroli J, Silva E, Maloney JP, He Q, Svetkauskaite D, Murphy JR, et al. Variant IRAK-1 haplotype is associated with increased nuclear factor-kappaB activation and worse outcomes in sepsis. Am. J. Respir. Crit. Care Med. 2006;173(12):1335–41.
[15] Khor CC, Vannberg FO, Chapman SJ, Walley A, Aucan C, Loke H, et al. Positive replication and linkage disequilibrium mapping of the chromosome 21q22.1 malaria susceptibility locus. Genes Immun. 2007;8(7):570–6.
[16] Mansur A, Liese B, Steinau M, Ghadimi M, Bergmann I, Tzvetkov M, et al. The CD14 rs2569190 TT genotype is associated with an improved 30-day survival in patients with sepsis: a prospective observational cohort study. PLoS One 2015;10(5):e0127761.
[17] Gocek E, Moulas AN, Studzinski GP. Non-receptor protein tyrosine kinases signaling pathways in normal and cancer cells. Crit. Rev. Clin. Lab. Sci. 2014;51(3):125–37.
[18] Rautanen A, Mills TC, Gordon AC, Hutton P, Steffens M, Nuamah R, et al. Genome-wide association study of survival from sepsis due to pneumonia: an observational cohort study. Lancet Respir. Med. 2015;3(1):53–60.
[19] Hermans PW, Hibberd ML, Booy R, Daramola O, Hazelzet JA, de Groot R, et al. 4G/5G promoter polymorphism in the plasminogen-activator-inhibitor-1 gene and outcome of meningococcal disease. Meningococcal Research Group. Lancet 1999;354(9178):556–60.
[20] Phillips JM, Goodman JI. Identification of genes that may play critical roles in phenobarbital (PB)-induced liver tumorigenesis due to altered DNA methylation. Toxicol. Sci. 2008;104(1):86–99.
[21] Esteller M. Epigenetics provides a new generation of oncogenes and tumour-suppressor genes. Br. J. Cancer 2007;96(Suppl):R26–30.
[22] Dong X, Weng Z. The correlation between histone modifications and gene expression. Epigenomics 2013;5(2):113–6.
[23] Bierne H, Cossart P. When bacteria target the nucleus: the emerging family of nucleomodulins. Cell Microbiol. 2012;14(5):622–33.
[24] Stephens KE, Miaskowski CA, Levine JD, Pullinger CR, Aouizerat BE. Epigenetic regulation and measurement of epigenetic changes. Biol. Res. Nurs. 2013;15(4):373–81.
[25] Leliefeld PH, Wessels CM, Leenen LP, Koenderman L, Pillay J. The role of neutrophils in immune dysfunction during severe inflammation. Crit. Care 2016;20:73.
[26] Peck-Palmer OM, Unsinger J, Chang KC, McDonough JS, Perlman H, McDunn JE, et al. Modulation of the Bcl-2 family blocks sepsis-induced depletion of dendritic cells and macrophages. Shock 2009;31(4):359–66.
[27] Cobb JP, Laramie JM, Stormo GD, Morrissey JJ, Shannon WD, Qiu Y, et al. Sepsis gene expression profiling: murine splenic compared with hepatic responses determined by using complementary DNA microarrays. Crit. Care Med. 2002;30(12):2711–21.
[28] Boomer JS, To K, Chang KC, Takasu O, Osborne DF, Walton AH, et al. Immunosuppression in patients who die of sepsis and multiple organ failure. JAMA 2011;306(23):2594–605.
[29] Xiao W, Mindrinos MN, Seok J, Cuschieri J, Cuenca AG, Gao H, et al. A genomic storm in critically injured humans. J. Exp. Med. 2011;208(13):2581–90.
[30] Hotchkiss RS, Karl IE. The pathophysiology and treatment of sepsis. N. Engl. J. Med. 2003;348(2):138–50.
[31] Cazalis MA, Lepape A, Venet F, Frager F, Mougin B, Vallin H, et al. Early and dynamic changes in gene expression in septic shock patients: a genome-wide approach. Intensive Care Med. Exp. 2014;2(1):20.
[32] Leentjens J, Kox M, van der Hoeven JG, Netea MG, Pickkers P. Immunotherapy for the adjunctive treatment of sepsis: from immunosuppression to immunostimulation. Time for a paradigm change? Am. J. Respir. Crit. Care Med. 2013;187(12):1287–93.
[33] Calvano JE, Bowers DJ, Coyle SM, Macor M, Reddell MT, Kumar A, et al. Response to systemic endotoxemia among humans bearing polymorphisms of the Toll-like receptor 4 (hTLR4). Clin. Immunol. 2006;121(2):186–90.
[34] Talwar S, Munson PJ, Barb J, Fiuza C, Cintron AP, Logun C, et al. Gene expression profiles of peripheral blood leukocytes after endotoxin challenge in humans. Physiol. Genom. 2006;25(2):203–15.
[35] Johnson SB, Lissauer M, Bochicchio GV, Moore R, Cross AS, Scalea TM. Gene expression profiles differentiate between sterile SIRS and early sepsis. Ann. Surg. 2007;245(4):611–21.
[36] Sonna LA, Hawkins L, Lissauer ME, Maldeis P, Towns M, Johnson SB, et al. Core temperature correlates with expression of selected stress and immunomodulatory genes in febrile patients with sepsis and noninfectious SIRS. Cell Stress Chaperones 2010;15(1):55–66.
[37] Seok J, Warren HS, Cuenca AG, Mindrinos MN, Baker HV, Xu W, et al. Genomic responses in mouse models poorly mimic human inflammatory diseases. Proc. Natl. Acad. Sci. U.S.A 2013;110(9):3507–12.
[38] Sweeney TE, Shidham A, Wong HR, Khatri P. A comprehensive time-course-based multicohort analysis of sepsis and sterile inflammation reveals a robust diagnostic gene set. Sci. Transl. Med. 2015;7(287):287ra71.

[39] Průcha M, Zazula R, Russwurm S. Sepsis diagnostics in the era of "omics" technologies. Prague Med. Rep. 2018;119(1):9–29.
[40] Tang F, Liu W, Zhang F, Xin ZT, Wei MT, Zhang PH, et al. IL-12 RB1 genetic variants contribute to human susceptibility to severe acute respiratory syndrome infection among Chinese. PLoS One 2008;3(5):e2183.
[41] Yu SL, Chen HW, Yang PC, Peck K, Tsai MH, Chen JJ, et al. Differential gene expression in gram-negative and gram-positive sepsis. Am. J. Respir. Crit. Care Med. 2004;169(10):1135–43.
[42] Grealy R, White M, Stordeur P, Kelleher D, Doherty DG, McManus R, et al. Characterising cytokine gene expression signatures in patients with severe sepsis. Mediat. Inflamm. 2013;2013:164246.
[43] Hinrichs C, Kotsch K, Buchwald S, Habicher M, Saak N, Gerlach H, et al. Perioperative gene expression analysis for prediction of postoperative sepsis. Clin. Chem. 2010;56(4):613–22.
[44] Parnell GP, McLean AS, Booth DR, Armstrong NJ, Nalos M, Huang SJ, et al. A distinct influenza infection signature in the blood transcriptome of patients with severe community-acquired pneumonia. Crit. Care 2012;16(4):R157.
[45] Davenport EE, Burnham KL, Radhakrishnan J, Humburg P, Hutton P, Mills TC, et al. Genomic landscape of the individual host response and outcomes in sepsis: a prospective cohort study. Lancet Respir. Med. 2016;4(4):259–71.
[46] Sweeney TE, Haynes WA, Vallania F, Ioannidis JP, Khatri P. Methods to increase reproducibility in differential gene expression via meta-analysis. Nucleic Acids Res. 2017;45(1):e1.
[47] Sweeney TE, Khatri P. Benchmarking sepsis gene expression diagnostics using public data. Crit. Care Med. 2017;45(1):1–10.
[48] Paugam-Burtz C, Albuquerque M, Baron G, Bert F, Voitot H, Delefosse D, et al. Plasma proteome to look for diagnostic biomarkers of early bacterial sepsis after liver transplantation: a preliminary study. Anesthesiology 2010;112(4):926–35.
[49] Kasthuri RS, Wroblewski M, Jilma B, Key NS, Nelsestuen GL. Potential biomarkers of an exaggerated response to endotoxemia. Biomarkers 2007;12(3):287–302.
[50] Malmström E, Kilsgård O, Hauri S, Smeds E, Herwald H, Malmström L, et al. Large-scale inference of protein tissue origin in gram-positive sepsis plasma using quantitative targeted proteomics. Nat. Commun. 2016;7:10261.
[51] Huttlin EL, Jedrychowski MP, Elias JE, Goswami T, Rad R, Beausoleil SA, et al. A tissue-specific atlas of mouse protein phosphorylation and expression. Cell 2010;143(7):1174–89.
[52] Sjöholm K, Kilsgård O, Teleman J, Happonen L, Malmström L, Malmström J. Targeted proteomics and absolute protein quantification for the construction of a stoichiometric host-pathogen surface density model. Mol. Cell. Proteom. 2017;16(4 Suppl. 1):S29–41.
[53] Kosmides AK, Kamisoglu K, Calvano SE, Corbett SA, Androulakis IP. Metabolomic fingerprinting: challenges and opportunities. Crit. Rev. Biomed. Eng. 2013;41(3):205–21.
[54] Serkova NJ, Standiford TJ, Stringer KA. The emerging field of quantitative blood metabolomics for biomarker discovery in critical illnesses. Am. J. Respir. Crit. Care Med. 2011;184(6):647–55.
[55] Patti GJ. Separation strategies for untargeted metabolomics. J. Sep. Sci 2011;34(24):3460–9.
[56] Kauppi AM, Edin A, Ziegler I, Mölling P, Sjöstedt A, Gylfe Å, et al. Metabolites in blood for prediction of bacteremic sepsis in the emergency room. PLoS One 2016;11(1):e0147670.
[57] Ferrario M, Cambiaghi A, Brunelli L, Giordano S, Caironi P, Guatteri L, et al. Mortality prediction in patients with severe septic shock: a pilot study using a target metabolomics approach. Sci. Rep. 2016;6:20391.
[58] Garcia-Simon M, Morales JM, Modesto-Alapont V, Gonzalez-Marrachelli V, Vento-Rehues R, Jorda-Miñana A, et al. Prognosis biomarkers of severe sepsis and septic shock by 1H NMR urine metabolomics in the intensive care unit. PLoS One 2015;10(11):e0140993.
[59] Su L, Huang Y, Zhu Y, Xia L, Wang R, Xiao K, et al. Discrimination of sepsis stage metabolic profiles with an LC/MS-MS-based metabolomics approach. BMJ Open Respir. Res. 2014;1(1):e000056.
[60] Berkhout DJC, van Keulen BJ, Niemarkt HJ, Bessem JR, de Boode WP, Cossey V, et al. Late-onset sepsis in preterm infants can be detected preclinically by fecal volatile organic compound analysis: a prospective, multicenter cohort study. Clin. Infect. Dis. 2019;68(1):70–7.
[61] Mickiewicz B, Thompson GC, Blackwood J, Jenne CN, Winston BW, Vogel HJ, et al. Biomarker phenotype for early diagnosis and triage of sepsis to the pediatric intensive care unit. Sci. Rep. 2018;8(1):16606.
[62] Bos LD, Sterk PJ, Schultz MJ. Volatile metabolites of pathogens: a systematic review. PLoS Pathog. 2013;9(5):e1003311.
[63] Bos LD, Martin-Loeches I, Kastelijn JB, Gili G, Espasa M, Povoa P, et al. The volatile metabolic fingerprint of ventilator-associated pneumonia. Intensive Care Med. 2014;40(5):761–2.
[64] Douglas IS. New diagnostic methods for pneumonia in the ICU. Curr. Opin. Infect. Dis. 2016;29(2):197–204.
[65] Li J, Jia H, Cai X, Zhong H, Feng Q, Sunagawa S, et al. An integrated catalog of reference genes in the human gut microbiome. Nat. Biotechnol. 2014;32(8):834–41.
[66] Sender R, Fuchs S, Milo R. Revised estimates for the number of human and bacteria cells in the body. PLoS Biol. 2016;14(8):e1002533.
[67] Feng Q, Chen WD, Wang YD. Gut microbiota: an integral moderator in health and disease. Front. Microbiol. 2018;9:151.
[68] Dickson RP. The microbiome and critical illness. Lancet Respir. Med. 2016;4(1):59–72.
[69] Lankelma JM, van Vught LA, Belzer C, Schultz MJ, van der Poll T, de Vos WM, et al. Critically ill patients demonstrate large interpersonal variation in intestinal microbiota dysregulation: a pilot study. Intensive Care Med. 2017;43(1):59–68.
[70] Yeh A, Rogers MB, Firek B, Neal MD, Zuckerbraun BS, Morowitz MJ. Dysbiosis across multiple body sites in critically ill adult surgical patients. Shock 2016;46(6):649–54.
[71] Ojima M, Motooka D, Shimizu K, Gotoh K, Shintani A, Yoshiya K, et al. Metagenomic analysis reveals dynamic changes of whole gut microbiota in the acute phase of intensive care unit patients. Dig. Dis. Sci. 2016;61(6):1628–34.
[72] Dickson RP, Singer BH, Newstead MW, Falkowski NR, Erb-Downward JR, Standiford TJ, et al. Enrichment of the lung microbiome with gut bacteria in sepsis and the acute respiratory distress syndrome. Nat. Microbiol. 2016;1(10):16113.
[73] Andrade-Oliveira V, Amano MT, Correa-Costa M, Castoldi A, Felizardo RJ, de Almeida DC, et al. Gut bacteria products prevent AKI induced by ischemia-reperfusion. J. Am. Soc. Nephrol. 2015;26(8):1877–88.

[74] Schieber AM, Lee YM, Chang MW, Leblanc M, Collins B, Downes M, et al. Disease tolerance mediated by microbiome *E. coli* involves inflammasome and IGF-1 signaling. Science 2015;350(6260):558–63.
[75] Caballero S, Kim S, Carter RA, Leiner IM, Sušac B, Miller L, et al. Cooperating commensals restore colonization resistance to vancomycin-resistant *Enterococcus faecium*. Cell Host Microbe 2017;21(5):592–602. e4.
[76] Buffie CG, Bucci V, Stein RR, McKenney PT, Ling L, Gobourne A, et al. Precision microbiome reconstitution restores bile acid mediated resistance to *Clostridium difficile*. Nature 2015;517(7533):205–8.
[77] Gosalbes MJ, Vázquez-Castellanos JF, Angebault C, Woerther PL, Ruppé E, Ferrús ML, et al. Carriage of enterobacteria producing extended-spectrum β-Lactamases and composition of the gut microbiota in an amerindian community. Antimicrob. Agents Chemother 2016;60(1):507–14.
[78] Pamer EG. Resurrecting the intestinal microbiota to combat antibiotic-resistant pathogens. Science 2016;352(6285):535–8.

Chapter 28

Precision nutrition to target lipid metabolism alterations in cancer

Marta Gómez de Cedrón and Ana Ramírez de Molina
Precision Nutrition and Cancer Program. Molecular Oncology and Nutritional Genomics of Cancer Group. IMDEA Food Institute, CEI UAM + CSIC, Madrid, Spain

Introduction

Cancer is the second leading cause of mortality and is responsible for one-sixth of deaths worldwide. In the course of this multifactorial condition, a cascade of alterations takes place, modifying the expression of both tumor suppressor genes and oncogenes. Proliferative cells present a distinct metabolism characterized by high rates of glycolysis, lactate production, and increased anabolism. During the last decade, there has been increased interest in understanding lipid metabolic alterations associated with tumorigenesis. In fact, several studies have demonstrated the importance of lipid metabolism regulation, in the promotion of migration, invasion, and angiogenesis in cancer.

Key enzymes involved in lipid-metabolic pathways have been found differentially expressed in normal and tumor tissues. Some of them have been described to be associated with cancer survival, and have been proposed as prognosis markers. In addition, several studies based on meta-analyses have demonstrated a link between obesity, and risk of cancer and disease prognosis.

Precision nutrition aims to design nutritional recommendations to treat or prevent metabolic disorders, including cancer, taking into account not only genetics, but other factors, such as dietary habits, food behavior, physical activity, and the gut microbiota. Herein, we aim to summarize lipid metabolism alterations in cancer, together with modifiable risk factors, such as obesity or gut microbiota dysbiosis, to provide molecular targets for precision nutrition interventions. Finally, we focus on colorectal cancer lipid metabolism alterations as a paradigm for precision nutrition interventions.

Lipid metabolism reprogramming in cancer

Cancer cells alter their metabolism to provide additional energy and anabolic demands of biosynthesis, to sustain cell proliferation and metastasis [1]. Although proliferative metabolism is a hallmark of cancer, there is incredible biological diversity across cancer types and even heterogeneities within a single tumor. In addition, cancer cells display increased metabolic plasticity to continuously adapt to changes in the microenvironment. This indicates that the altered tumor metabolism results from the combination of cell-autonomous genetic alterations, and flexibility to adapt to changes in the microenvironment (oxygenation, pH, and nutrient availability) [2].

The most prominent metabolic alteration in proliferative cancer cells is increased glucose uptake and glycolysis, regardless of the abundance of oxygen (Warburg effect) [3]. Aerobic glycolysis in cancer requires a coordinated adjustment of glutamine metabolism, for the replenishment of TCA cycle intermediates [4]. In the last years, compelling evidence obtained from untargeted/targeted lipidomic studies, cancer preclinical models, and clinical trials, has revealed the crucial role of lipid metabolic pathways to support tumor growth and metastasis. The disruption of cancer associated lipid metabolism, may induce tumor regression and/or inhibit metastatic spread.

One of the most important metabolic hallmarks of cancer cells is the increase of de novo lipogenesis, which not only contributes to cell proliferation but also to maintain redox homeostasis. Thus, enzymes involved in fatty acid (FA) synthesis are considered as therapeutic targets for cancer treatment. High intake of saturated fatty acids from diet, is reported to be a risk factor for the development, and

Precision Medicine for Investigators, Practitioners and Providers. https://doi.org/10.1016/B978-0-12-819178-1.00028-9

a poorer prognosis for certain cancers. In this regard, it has been described that increased lipolysis of fats from the diet, provides an additional source of FAs to sustain cancer cell proliferation, and malignancy in breast cancer.

Lipogenesis and cholesterogenesis

Lipid metabolism alterations in cancer have been shown to affect, not only cancer cell proliferation, but differentiation, malignancy, and motility [5]. The majority of adult mammalian tissues satisfy their lipid requirements, through the uptake of free fatty acids and lipoproteins from the diet, de novo lipogenesis and cholesterogenesis being mainly restricted to liver and adipose tissues. However, neoplastic tissues display aberrant activation of de novo lipogenesis and cholesterogenesis [6]. In fact, overexpression of lipogenic enzymes is reported as a common characteristic of many cancers [7], and the inhibition of different enzymes within the FA biosynthetic pathway, can block cancer cell growth [8]. FAs can be diverted to (i) storage in the form of triacylglycerols (TAGs), (ii) phospholipids for membrane biogenesis, (iii) lipid signaling molecules (iv) energy production by β-oxidation.

Key enzymes involved in de novo lipogenesis and cholesterogenesis are differentially expressed in normal versus tumor tissues, some of them being proposed as prognosis biomarkers. Synthesis of FAs starts with the utilization of citrate to produce acetyl coenzyme-A (AcCoA) and oxaloacetate (OAA) by ATP-citrate lyase (ACL). Then, AcCoA carboxylase (ACC) converts AcCoA to malonyl-CoA. Fatty acid synthase (FASN) sequentially condenses AcCoA molecules, to produce saturated FAs, which are desaturated by stearoyl-CoA desaturase (SCD). FASN is frequently overexpressed in breast, prostate, and other types of cancer [6,9], and ACL has been shown to support tumor formation and transformation [10].

The mevalonate pathway for cholesterol synthesis has also been reported to be upregulated in several cancers such as prostate and breast cancer. Inhibition of HMG-CoA reductase, the rate-limiting enzyme of the mevalonate pathway, induces cell cycle arrest in breast cancer [11]. High expression levels of genes of the mevalonate pathway correlate with poorer prognosis in breast cancer, due to the accumulation of cholesterol, that alters plasmatic membrane fluidity, contributing to the promotion of transformation [12].

Sterol regulatory element-binding proteins (SREBPs), which are the main transcription factors for the expression of the majority of enzymes involved in FA (SREBP1c) and cholesterol synthesis (SREBP2), are controlled by oncogenes which are frequently overexpressed in cancer [6].

Fatty acid oxidation in cancer

Together with de novo lipogenesis and cholesterogenesis in cancer, the mobilization of intracellular FAs to be oxidized in mitochondria (FAO) is starting to be recognized as crucial in cancer cell metabolism. Enhanced mitochondrial β-oxidation of FAs has been linked to tumor promotion in pancreatic cancer [13,14], and highly metastatic breast cancer cells rely partially on FAO-associated reactive oxygen species (ROS) production [15]. Importantly, FAO provides an alternative pathway to support cancer cell survival, when glucose becomes limiting, and/or when there is an increased necessity to face oxidative stress. FAO not only fuels the electron transport chain for ATP production, but it can also be induced to regulate redox homeostasis, allowing adaptation and survival [16]. FAO counteracts oxidative stress in leukemia and glioma cells, by increasing the intracellular pool of NADPH and, on the contrary, the inhibition of β-oxidation induces apoptosis in this scenario. Moreover, FAO is also crucial for the survival of cells from solid tumors, when undergoing a loss of attachment (LOA), which triggers *anoikis* or cell death due to oxidative stress [17,18].

Regulation of FA storage and intracellular FA mobilization (lipolysis and lipophagy)

Increased lipogenesis in cancer cells requires mechanisms to accommodate their lipid content intracellularly and, importantly, there must exist a balance between intracellular lipid storage and lipid mobilization [19]. FAs and their derivatives are thought to be rapidly incorporated into neutral lipids and phospholipids. Lipid droplets (LDs) are complex and dynamic organelles, that store and supply lipids for energy metabolism, membrane synthesis, and the production of essential lipid-derived molecules. LDs consist of a neutral lipid core surrounded by a phospholipid monolayer that is mainly composed of phosphatidylcholine (PC). There is a complex proteome associated with LDs, which includes structural molecules, enzymes, and signaling molecules.

Among the LD-associated proteins are components of the sterol biosynthesis pathway, acyl-CoA metabolism (ACSL3), and TAG biosynthesis. Structural proteins, such as perilipins (PLIN2) or caveolins, are critical for the integrity of LDs, to avoid coalescence, and to protect them from lipolysis. Cancer cells seem to contain a considerably higher amount of lipid droplets than normal cells. Numerous LD proteins, such as PLIN2, hypoxia-inducible protein 2 (HIG2), adipose differentiation-related protein (ADRP) and mannose six phosphate receptor binding protein 1 (Tip47) in several cancer cells, have been shown to promote the accumulation of LDs [20]. These proteins contribute to maintaining endoplasmic reticulum (ER) homeostasis, to protect against ROS, and to increase resistance to therapeutic drugs in cancer cells. It has been demonstrated that LD-depletion, through the inhibition of lipid synthesis or the silencing of PLIN2, significantly attenuates proliferation of colon cancer cells [21].

Increased de novo lipogenesis in cancer cells requires complementary lipolytic mechanisms to remodel the intracellular lipid content, making the stored lipids available for the synthesis of complex lipids and phospholipids, ATP or NADPH production when required. This cooperation is reflected by the fact that several enzymes of lipolysis, such as adipose TAG lipase (ATGL), hormone-sensitive lipase (HSL), or monoacylglycerol lipase (MAGL), have been related to tumorigenesis [22]. In this sense, ATGL knockdown in HCT116 colorectal cancer cells results in reduced proliferation [23], and overexpression of G0/G1 switch protein-2 (G0S2), which diminishes the activity of ATGL, inhibits the growth of K562 leukemia cells both in vitro and in mouse xenografts [24]. Increased levels and enzymatic activity of MAGL, are found in several aggressive cancer types including melanoma, ovarian and breast cancer [22], and inhibition of MAGL suppresses cancer cell migration, invasion, and survival [25].

Recently, a specific function of autophagy in lipid mobilization and regulation of cellular lipid content have been described. Lipophagy has been shown to augment resistance to cell death in cancer [26], by releasing FAs to sustain FAO dependent ATP production, to supply lipids for membrane biogenesis, and to support protumorigenic lipid signaling molecules.

Extracellular lipid uptake

Cancer cells can also incorporate exogenous lipids after inhibition of de novo FA synthesis. The upregulation of cell surface receptors for the uptake of plasma lipids, such as the cluster of differentiation 36 (CD36), correlates with increased metastasis-initiating potential, and poorer prognosis [27,28]. Moreover, the expression of low-density lipoprotein receptor (LDLr) is upregulated in renal cell carcinoma (RCC) cells [29]. Increased fatty acid trafficking, through upregulation of fatty acid binding proteins (FABPs), is another mechanism to promote lipid uptake in cancer cells [30]. A recent study in breast cancer and glioblastoma cell lines revealed that uptake of extracellular FAs during hypoxia is promoted by upregulation of FABP3 and FABP7, and upregulation of FABP5 promotes cell proliferation and growth of prostate cancer [31].

Depending on the source of FAs, de novo lipogenesis or extracellular lipid uptake, the phospholipid composition of membranes will be affected. Thus, increased de novo lipogenesis enriches T cancer cell membranes with saturated and monounsaturated fatty acids [24], which are less prone to lipid peroxidation than polyunsaturated acyl chains, providing resistance to oxidative stress and chemotherapy [32].

Tumor microenvironment and cancer-associated adipocytes (CAA)

Most of the work done in relation to target cancer metabolism has focused identification of oncogenes and tumor suppressor genes, that mediate cancer metabolic reprogramming. However, metabolic reprogramming must be accompanied by the flexibility to adapt to substrate and oxygen availability in the surrounding tumor microenvironment.

Stromal adjacent glycolytic cells support lactate production, to be oxidized by tumor cells (reverse Warburg effect), in lung and colorectal adenocarcinoma. A similar mechanism occurs in ovarian cancer, where adipocytes provide lipids to malignant cells, to be oxidized in mitochondria [33,34].

Cancer microenvironment

A tumor cannot be considered just as a group of proliferative cells, surrounded by an extracellular matrix. Several cell types communicate with the surrounding healthy tissue. The coevolution of malignant cells, cancer-associated fibroblasts (CAFs), immune system cells, vascular endothelial cells, and the extracellular matrix (ECM), impact cancer progression.

One of the reasons why current therapies are not totally efficient is altered tissue microenvironment [35]. In tissues located close to adipose tissue such as breast, prostate, or colon, adipocytes play a critical role in this crosstalk. Cancer-associated adipocytes (CAA) suffer several alterations to support tumor malignancy. These CAAs provide FAs to cancer cells and generate a permissive niche for tumor growth and dissemination. Moreover, rapid expansion and hypertrophy of adipose tissue provoke oxygen deficiency, and compensatory mechanisms to promote angiogenesis favoring tumor spreading [36] (Fig. 28.1).

Precision nutrition targeting lipid metabolism

According to the International Society of Nutrigenetics/Nutrigenomics (ISNN), the future of precision nutrition should be discussed at three levels: (1) stratification of conventional nutritional guidelines into subpopulation by age, gender, and other social determinants (2) individual approaches based on refined phenotyping (3) genetic-directed nutrition based on genetic variants with high penetrance, and response to nutritional interventions.

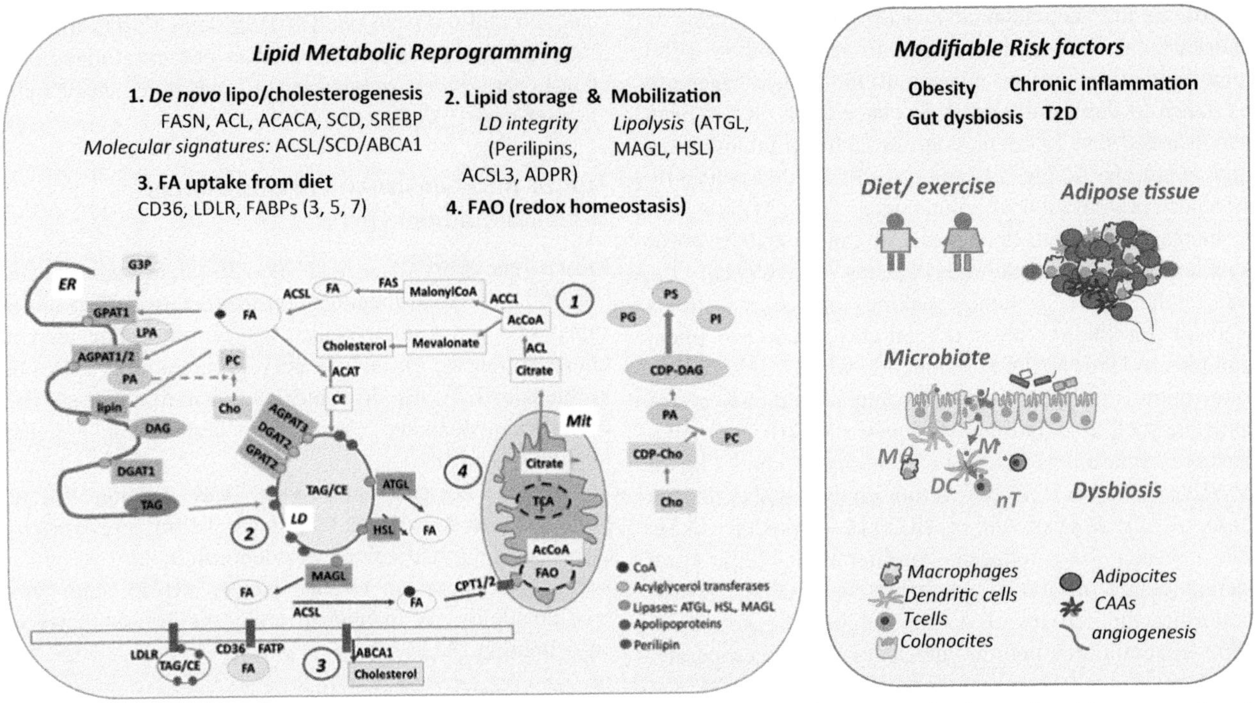

FIGURE 28.1 Summary of major lipid metabolism alterations in cancer. (A) Lipid metabolism interconnected pathways altered in cancer, such as de novo lipogenesis and FAO, as well as processes regulating overall intracellular lipid content, based on LD formation and lipid mobilization by lipolysis. (B) Modifiable risk factors related to lipids (obesity, type 2 diabetes, metabolic syndrome, dysbiosis) that contribute to progression and prognosis in cancer. *Molecules: CE*, cholesterol esters; *Cho*, choline; *DAG*, diacylglycerol; *FA*, fatty acid; *G3P*, glycerol-3-phosphate; *LPA*, lyso-phosphatidic acid; *MAG*, monoacylglycerol; *PA*, phosphatidic acid; *PC*, phosphatidylcholine; *PE*, phosphatidylethanolamine; *PG*, phosphatidylglycerol; *PI*, phosphatidylinositol; *PS*, phosphatidylserine; *TAG*, triacylglycerol. *Enzymes: ACC1*, Acetyl-CoA carboxylase; *ACL*, ATP-citrate lyase; *ACSL*, Long chain acylCoA synthetase; *AGPAT1*, Acylglycerol phosphate acyltransferase; *ATGL*, adipose tissue triacylglycerol lipase; *CD36*, thrombospondin receptor; *CPT1A2*, carnitine palmitoyltransferase 1A2; *DGAT1*, diacylglycerol acyltransferase; *FASN*, fatty acid synthetase; *FATP*, fatty acid transport protein; *GPAT1*, Glycerol phosphate acyl transferase; *HSL*, hormone sensitive lipase; *MAGL*, monoacylglycerol lipase; *SCD*, stearoyl CoA desaturase.

Single nucleotide polymorphisms (SNPs) are the most common forms of sequence variation in the human genome. Additional "omics," including transcriptomics, proteomics, metabolomics, and lipidomics, provide a refined nutritional scenario for personalized nutritional interventions [37,38].

Nutrigenetics

Nutrigenetics studies the effect of genetic variation, on dietary response to disease risk. There are abundant examples of polymorphisms in genes involved in the regulation of food intake, lipid and lipoprotein metabolisms, insulin signaling, glucose, and amino acids metabolism, and chronic inflammation, along with the response to weight loss, weight regain insulin resistance, or serum lipid levels.

SNPs in the cluster of differentiation 36 (*CD36*) gene, have been associated with dyslipidemia in subjects that consume high amounts of fats [39]. SNPs in methylenetetrahydrofolate reductase (*MTHFR*) and methionine synthase (*MTR*) genes, implicated in homocysteine metabolism, increase breast cancer risk associated with low intakes of folate and vitamins B6 and B12 [40]. Vitamin D receptor (*VDR*) gene polymorphisms, elevate predisposition to osteoporosis in postmenopausal women with low calcium intakes [41]. Moreover, SNPs in apolipoprotein C3 (*APOC3*) and apolipoprotein A1 (*APOA1*) genes, lead to increased risk of metabolic syndrome in individuals with Western diets.

Nutrigenomics

Nutrient components can also modulate cancer risk and progression by affecting, directly or indirectly, gene expression. *Nutrigenomics* studies the role of nutrients and bioactive compounds in gene expression and, consequently, proteome and metabolome. The direct association of unhealthy diets with *obesity* and *gut dysbiosis* opens new routes in the understanding of how nutrients could influence cancer prognosis. High intakes of refined grain products and sweets are associated with inflammation and cancer risk, compared to vegetables, fruits, or whole grain. High red meat consumption is associated with gene expression signatures, linked to colon cancer [42].

Abundant saturated fatty acid intake augments the expression levels of genes related to inflammation, insulin resistance, and liver steatosis, contributing to obesity. Low intakes of folate are associated with altered expression of lipid metabolism genes, influencing nonalcoholic fatty liver disease (NAFLD) [43]. In contrast, the Mediterranean diet reduces the expression of genes implicated in inflammation, endoplasmic reticulum (ER) stress or oxidative stress, protecting against insulin resistance [44,45]. Similarly, high intakes of monounsaturated fatty acids (MUFAs), have been associated with a reduction in the expression of genes involved in inflammation and abnormal lipid storage [46]. Moreover, energy-restricted diets supplemented with eicosapentaenoic acid (EPA), and α-lipoic acid, have been associated with upregulation of FAO genes, together with downregulation of lipogenic and proinflammatory genes [47].

Bioactive food compounds from natural sources, such as epigallocatechin-3-gallate (EGCG), curcumin, sulforaphane, and genistein, exert anticancer properties by means of upregulation of tumor suppressor genes, and/or downregulation of oncogenes. Polyphenols contribute to the prevention of obesity, through upregulation of genes implicated in adipogenesis, lipolysis, and FAO [48–51].

Gene expression profiles have also been used to predict responsiveness to nutritional treatments. Expression levels of fatty acid metabolism, citric acid cycle, oxidative phosphorylation, and proinflammation genes, in subcutaneous adipose tissue, are differentially regulated during a low-calorie diet, between weight maintainers and regainers after weight loss [52].

Nutri-epigenetics

The dysregulation of epigenetic marks can alter the phenotype and cell function, leading to the onset and progression of diverse chronic diseases. In this sense, complex interactions among nutritional factors and DNA methylation, histone modifications and noncoding microRNAs (miRNAs), have been implicated in obesity, dyslipidemia, type 2 diabetes, nonalcoholic fatty liver disease, cardiovascular disease, and cancer. High-fat and high sugar consumption have been related to abnormal methylation of neuropeptide genes that control food intake, affecting the development of obesity [53]. Low-protein diets disrupt some histone modifications in genes of glucose homeostasis, insulin signaling, and lipid metabolism [54].

Nutri-epigenetics aims to provide specific nutritional interventions to reverse epigenetic alterations in chronic diseases related to metabolism. PUFAs modulate the expression of microRNAs implicated in suppression of oncogenic and lipogenic genes [55], and fruit polyphenols impact on methylation profile of genes involved in lipid metabolism [56].

Epigenetic marks can also modulate the effect of nutritional treatments on weight loss and changes in metabolic profiles, consequently becoming predictive biomarkers of nutritional interventions [57].

Obesity influences on risk and prognosis in cancer

Environment plays a crucial role in cancer progression and resistance to chemotherapy. Cancer malignancy not only relays on the genetic factors from the patient but also on environmental factors associated with lifestyle [58]. One-third of cancer deaths could be prevented by modifying key risk factors such as high saturated fatty acids intake, smoking, or alcohol consumption. Moreover, physical activity and diet are also considered among the most important environmental risk factors for cancer development, due to the association with obesity.

Obesity is a chronic, multifactorial, and proinflammatory disease, defined as high body weight with excessive adipose tissue accumulation [59]. An estimated 1.5 billion people suffer from overweight, according to the World Health Organization (WHO) [60]. Obesity has been implicated in the development of type-2 diabetes, cardiovascular diseases, nonalcoholic fatty liver disease, and cancer initiation and progression [61,62]. The overall risk of death from cancer is 1.5–1.6 fold higher in individuals with body mass index (BMI) > 40 kg/m^2 [63]. Main types of cancer involved are: prostate [64], postmenstrual endometrial [65], breast [66], ovary [67], bladder [68], liver [69], colon [70], and pancreas [66].

Excess of visceral adiposity triggers alterations in the cellular composition of adipose tissue, which is crucial for the growth and dissemination of tumors that develop close to adipocytes, like breast, ovary or colon tumors [71]. BMI does not provide information about body fat distribution, and additional anthropometric measures such as waist/hip ratio are used. Excess of adipocytes can accumulate in nonclassic locations. Ectopic fat depots are classified as central systemic adipose tissue (CA), or local adipose tissue surrounding tumor microenvironment. CA is responsible for altered levels of sex steroids, insulin resistance, and chronic inflammation [72], along with colorectal [72] and breast cancer [73].

Gut microbiota

Dysbiosis is characterized by a decrease in microbial diversity and increase in proinflammatory species. This imbalanced microbiota is unable to protect from pathogenic organisms, that can trigger inflammation and produce genotoxins or carcinogenic metabolites. Moreover, during obesity and its associated comorbidities, the composition of gut microbiota and intestinal epithelial barrier function are altered. Dysbiosis has been associated with colorectal, gastric, and esophageal cancer [74,75]. Probiotics are considered functional foods and based on epidemiological and experimental data, can support biologically active components in colorectal cancer (CRC) prevention and treatment (Fig. 28.2).

Precision nutrition and lipid metabolism in colorectal cancer

CRC ranks as the third leading cause of cancer-related deaths (WHO). In the course of this multifactorial condition, numerous alterations occur in both tumor suppressor and oncogenic genes. In most cases, early detection allows tumors to be successfully removed by surgery, and can also increase treatment efficiency. Three different chemotherapeutic drugs, together with inhibitors of vascular endothelial growth factor (VEGF) and epithelial growth factor receptor (EGFR), have allowed to reach a median survival of 30 months [76].

The main challenge is to assess inter and intratumor heterogeneity, which can be originated by genomic, epigenomic, transcriptomic, and immune variability. This will lead to stratification for personalized treatment [77]. Experimental and epidemiological studies support the role of lipid metabolism, diet, gene-diet interactions, obesity, and gut microbiota in the etiology and prognosis in CRC.

Key enzymes of lipid metabolism are differentially expressed in normal and tumor CRC tissues. Some of them are associated with cancer survival and have been individually proposed as prognosis markers. Lipid metabolism alterations not only affect the primary tumor but also influence malignancy, migration, and invasion.

A lipid-metabolic signature ColoLipidGene -ACSL1, ABCA1 (ATP-Binding Cassette Subfamily-A Member 1), AGPAT1 (1-Acylglycerol-3-Phosphate O-Acyltransferase 1), and SCD (Stearoyl-CoA-desaturase), is associated with CRC prognosis in stage II patients [78]. A network formed by ACSL1, ACSL4, (Acyl-CoA synthetase) and SCD promotes migration and invasion by CRC cells [79]. Genetic analysis of 57 SNPs located in seven lipid-metabolism locations has shown that rs8086 SNP in ACSL1, was associated with reduced mRNA expression levels of ACSL1, and shorter CRC disease-free survival (DFS) [80].

Analysis of 43 fatty acid metabolism-related genes, and 392 SNPs in 1225 CRC cases and 2032 controls, unveiled associations of HPGD (OH-prostaglandin dehydrogenase 15-(NAD), PLA2G6 (phospholipase A2 group 6), and TRPV3 (transient receptor potential vanilloid 3) to increased risk for CRC, and PTGER2 (prostaglandin E receptor 2) to a lower risk of CRC. It is also highlighted the relevance of genetic variation in fatty acid metabolism genes and CRC risk [81].

Analysis of 30 SNPs in eight fatty acid biosynthesis-related genes, in 1780 CRC cases and 1864 controls, described the association between rs9652472 polymorphism of LIPC (hepatic triglyceride lipase) and increased risk of CRC [82].

Among environmental factors related to lifestyle, one-third of cancer deaths could be prevented by modifying smoking, alcohol consumption, or obesity. Dysbiosis [83], diets rich in red, processed, and grilled meats are strongly associated with colorectal cancer as well [84]. Heterocyclic amines and polycyclic aromatic hydrocarbons are

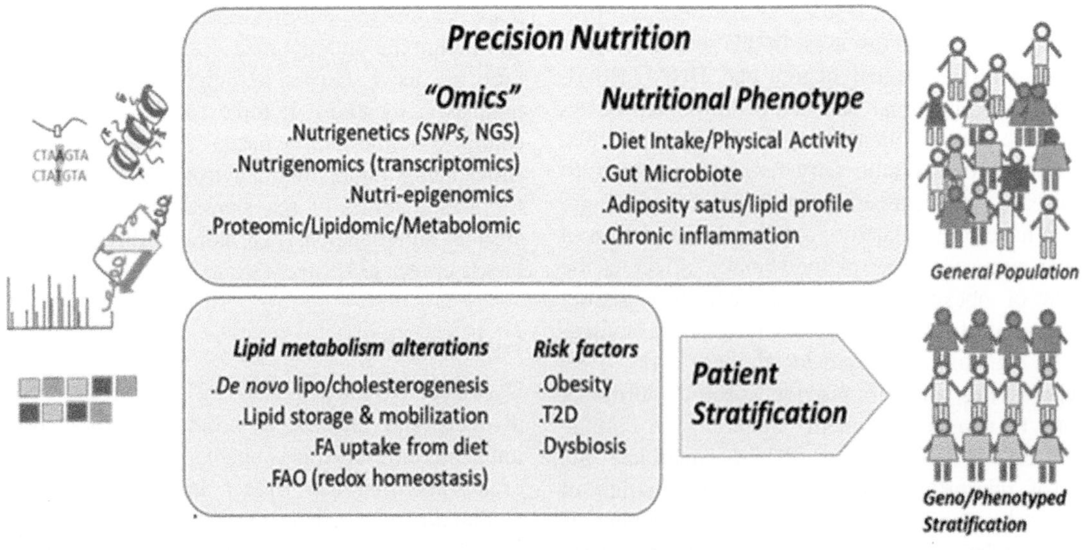

FIGURE 28.2 Precision nutrition approaches for prevention and management of lipid metabolism alterations in cancer, and associated risk factors such as obesity, chronic inflammation and gut dysbiosis.

implicated in the increased of risk, in the context of red meat intake [85].

Fatty acids and lipid nutrients

While ω3 polyunsaturated (PUFAs) fatty acids are associated with protection in CRC [86], ω6 PUFAS display opposite effects [87]. Intake of unsaturated FA (UFA) may be beneficial for health, as the saturated ones are associated with tumorigenesis. Importantly, triglycerides (TG) and LDL were associated with CRC prognosis, with increased levels in patients with distant metastasis. Cholesterol in high-fat diets strongly links with colorectal tumorigenesis [88]. Ceramide sphingolipid can be chemopreventive, and used in combination with tamoxifen induces cell cycle arrest and apoptosis [89]. Phosphatidylcholine (PC) was significantly increased in CRC-derived cells [47].

Other nutrients and lifestyle changes

Folate plasma levels are associated with hypermethylation of several tumor suppressor genes, and with DNA hypomethylation [90]. A large prospective cohort study showed that individuals with the highest folate intake presented a 30% reduction in the risk of developing CRC [91]. Milk, calcium, and vitamin D act as protective agents against CRC development and increase survival [92].

In the presence of excessive accumulation of adipose tissue, because of malignant cell proximity, cancer-associated adipocytes (CAA) suffer delipidation and acquire fibroblast-like features, that will influence malignancy. Increase in systemic ectopic fat shows a positive correlation with CRC, among other malignancies. Epidemiological and experimental data support the positive role of probiotics in CRC prevention and treatment [93].

Genetics variants in diet-related genes have been found associated with CRC, such as SNPs in methylenetetrahydrofolate reductase mutation (MTHFR), implicated in folate biosynthetic pathway, along with levels of folate intake [94]. Genetic variants in transient receptor potential melastatin 7 (TRPM7), have been connected with an elevated risk of both adenomatous and hyperplastic polyps [95]. SNPS in poly ADP ribose polymerase (PARP) gene are associated with CRC risk, depending on high-temperature cooked red meat diets [96].

Acknowledgements

This work has been supported by Ministerio de Economía y Competitividad del Gobierno de España (MINECO, Plan Nacional I + D + i AGL2016-76736-C3), Gobierno regional de la Comunidad de Madrid (P2013/ABI-2728, ALIBIRD-CM), EU Structural Funds.

Author contributions

MGC wrote the paper and ARdM performed a critical revision of the article.

Conflicts of interest

The authors declare no conflict of interest.

References

[1] Hanahan D, Weinberg RA. Hallmarks of cancer: the next generation. Cell 2011;144(5):646–74.

[2] Chen JQ, Russo J. Dysregulation of glucose transport, glycolysis, TCA cycle and glutaminolysis by oncogenes and tumor suppressors in cancer cells. Biochim. Biophys. Acta 2012;1826(2):370–84.

[3] Warburg O. On the origin of cancer cells. Science 1956;123(3191):309–14.

[4] DeBerardinis RJ, et al. The biology of cancer: metabolic reprogramming fuels cell growth and proliferation. Cell Metabol. 2008;7(1):11–20.

[5] Mashima T, et al. Acyl-CoA synthetase as a cancer survival factor: its inhibition enhances the efficacy of etoposide. Cancer Sci. 2009;100(8):1556–62.

[6] Menendez JA, et al. The metastasis inducer CCN1 (CYR61) activates the fatty acid synthase (FASN)-driven lipogenic phenotype in breast cancer cells. Oncoscience 2016;3(7–8):242–57.

[7] Roongta UV, et al. Cancer cell dependence on unsaturated fatty acids implicates stearoyl-CoA desaturase as a target for cancer therapy. Mol. Cancer Res. 2011;9(11):1551–61.

[8] Santos CR, Schulze A. Lipid metabolism in cancer. FEBS J. 2012;279(15):2610–23.

[9] Menendez JA, Lupu R. Fatty acid synthase and the lipogenic phenotype in cancer pathogenesis. Nat. Rev. Canc. 2007;7(10):763–77.

[10] Bauer DE, et al. ATP citrate lyase is an important component of cell growth and transformation. Oncogene 2005;24(41):6314–22.

[11] Rao S, et al. Lovastatin mediated G1 arrest in normal and tumor breast cells is through inhibition of CDK2 activity and redistribution of p21 and p27, independent of p53. Oncogene 1998;17(18):2393–402.

[12] Clendening JW, et al. Dysregulation of the mevalonate pathway promotes transformation. Proc. Natl. Acad. Sci. U. S. A. 2010;107(34):15051–6.

[13] Freigang S, et al. Fatty acid-induced mitochondrial uncoupling elicits inflammasome-independent IL-1alpha and sterile vascular inflammation in atherosclerosis. Nat. Immunol. 2013;14(10):1045–53.

[14] Khasawneh J, et al. Inflammation and mitochondrial fatty acid beta-oxidation link obesity to early tumor promotion. Proc. Natl. Acad. Sci. U. S. A. 2009;106(9):3354–9.

[15] Li S, et al. High metastaticgastric and breast cancer cells consume oleic acid in an AMPK dependent manner. PLoS One 2014;9(5):e97330.

[16] Ludtmann MH, et al. Nrf2 affects the efficiency of mitochondrial fatty acid oxidation. Biochem. J. 2014;457(3):415–24.

[17] Pike LS, et al. Inhibition of fatty acid oxidation by etomoxir impairs NADPH production and increases reactive oxygen species resulting in ATP depletion and cell death in human glioblastoma cells. Biochim. Biophys. Acta 2011;1807(6):726–34.

[18] Ito K, et al. A PML-PPAR-delta pathway for fatty acid oxidation regulates hematopoietic stem cell maintenance. Nat. Med. 2012;18(9):1350–8.
[19] Przybytkowski E, et al. Upregulation of cellular triacylglycerol - free fatty acid cycling by oleate is associated with long-term serum-free survival of human breast cancer cells. Biochem. Cell Biol. 2007;85(3):301–10.
[20] Togashi A, et al. Hypoxia-inducible protein 2 (HIG2), a novel diagnostic marker for renal cell carcinoma and potential target for molecular therapy. Cancer Res. 2005;65(11):4817–26.
[21] Penrose H, et al. Epidermal growth factor receptor mediated proliferation depends on increased lipid droplet density regulated via a negative regulatory loop with FOXO3/Sirtuin6. Biochem. Biophys. Res. Commun. 2016;469(3):370–6.
[22] Nomura DK, et al. Monoacylglycerol lipase regulates a fatty acid network that promotes cancer pathogenesis. Cell 2010;140(1):49–61.
[23] Ou J, et al. Loss of Abhd5 promotes colorectal tumor development and progression by inducing aerobic glycolysis and epithelial-mesenchymal transition. Cell Rep. 2018;24(10):2795–7.
[24] Yue S, et al. Cholesteryl ester accumulation induced by PTEN loss and PI3K/AKT activation underlies human prostate cancer aggressiveness. Cell Metabol. 2014;19(3):393–406.
[25] Nomura DK, et al. Monoacylglycerol lipase exerts dual control over endocannabinoid and fatty acid pathways to support prostate cancer. Chem. Biol. 2011;18(7):846–56.
[26] Amir M, Czaja MJ. Autophagy in nonalcoholic steatohepatitis. Expert Rev. Gastroenterol. Hepatol. 2011;5(2):159–66.
[27] Kuemmerle NB, et al. Lipoprotein lipase links dietary fat to solid tumor cell proliferation. Mol. Cancer Ther. 2011;10(3):427–36.
[28] Pascual G, et al. Targeting metastasis-initiating cells through the fatty acid receptor CD36. Nature 2017;541(7635):41–5.
[29] Sundelin JP, et al. Increased expression of the very low-density lipoprotein receptor mediates lipid accumulation in clear-cell renal cell carcinoma. PLoS One 2012;7(11):e48694.
[30] Koizume S, Miyagi Y. Lipid droplets: a key cellular organelle associated with cancer cell survival under normoxia and hypoxia. Int. J. Mol. Sci. 2016;17(9).
[31] Levi L, et al. Saturated fatty acids regulate retinoic acid signalling and suppress tumorigenesis by targeting fatty acid-binding protein 5. Nat. Commun. 2015;6:8794.
[32] Rysman E, et al. De novo lipogenesis protects cancer cells from free radicals and chemotherapeutics by promoting membrane lipid saturation. Cancer Res. 2010;70(20):8117–26.
[33] Martinez-Outschoorn UE, et al. Energy transfer in "parasitic" cancer metabolism: mitochondria are the powerhouse and Achilles' heel of tumor cells. Cell Cycle 2011;10(24):4208–16.
[34] Nieman KM, et al. Adipocytes promote ovarian cancer metastasis and provide energy for rapid tumor growth. Nat. Med. 2011;17(11):1498–503.
[35] Hanahan D, Coussens LM. Accessories to the crime: functions of cells recruited to the tumor microenvironment. Cancer Cell 2012;21(3):309–22.
[36] Park J, et al. Obesity is associated with higher risk of prostate cancer detection in a biopsy population in Korea. BJU Int. 2014;114(6):891–5.
[37] Kohlmeier M, et al. Guide and position of the international society of nutrigenetics/nutrigenomics on personalized nutrition: Part 2 - ethics, challenges and endeavors of precision nutrition. J. Nutrigenetics Nutrigenomics 2016;9(1):28–46.
[38] Ferguson LR, et al. Guide and position of the international society of nutrigenetics/nutrigenomics on personalised nutrition: Part 1 - fields of precision nutrition. J. Nutrigenetics Nutrigenomics 2016;9(1):12–27.
[39] Ramos-Lopez O, et al. CD36 genetic variation, fat intake and liver fibrosis in chronic hepatitis C virus infection. World J. Hepatol. 2016;8(25):1067–74.
[40] Jiang-Hua Q, et al. Association of methylenetetrahydrofolate reductase and methionine synthase polymorphisms with breast cancer risk and interaction with folate, vitamin B6, and vitamin B 12 intakes. Tumor Biol. 2014;35(12):11895–901.
[41] Stathopoulou MG, et al. The role of vitamin D receptor gene polymorphisms in the bone mineral density of Greek postmenopausal women with low calcium intake. J. Nutr. Biochem. 2011;22(8):752–7.
[42] Pellatt AJ, et al. Dietary intake alters gene expression in colon tissue: possible underlying mechanism for the influence of diet on disease. Pharmacogenetics Genom. 2016;26(6):294–306.
[43] Tryndyak V, et al. Interstrain differences in the severity of liver injury induced by a choline- and folate-deficient diet in mice are associated with dysregulation of genes involved in lipid metabolism. FASEB J. 2012;26(11):4592–602.
[44] Lopez-Moreno J, et al. Mediterranean diet supplemented with coenzyme Q10 modulates the postprandial metabolism of advanced glycation end products in elderly men and women. J. Gerontol. A Biol. Sci. Med. Sci. 2018;73(3):340–6.
[45] Yubero-Serrano EM, et al. Mediterranean diet supplemented with coenzyme Q10 modifies the expression of proinflammatory and endoplasmic reticulum stress-related genes in elderly men and women. J. Gerontol. A Biol. Sci. Med. Sci. 2012;67(1):3–10.
[46] Varela LM, et al. The effects of dietary fatty acids on the postprandial triglyceride-rich lipoprotein/apoB48 receptor axis in human monocyte/macrophage cells. J. Nutr. Biochem. 2013;24(12):2031–9.
[47] Huerta AE, et al. Effects of dietary supplementation with EPA and/or alpha-lipoic acid on adipose tissue transcriptomic profile of healthy overweight/obese women following a hypocaloric diet. Biofactors 2017;43(1):117–31.
[48] Jiao X, et al. Blueberry polyphenols extract as a potential prebiotic with anti-obesity effects on C57BL/6 J mice by modulating the gut microbiota. J. Nutr. Biochem. 2018;64:88–100.
[49] Novotny JA, et al. The effect of obesity and repeated exposure on pharmacokinetic response to grape polyphenols in humans. Mol. Nutr. Food Res. 2017;61(11).
[50] Farhat G, Drummond S, Al-Dujaili EAS. Polyphenols and their role in obesity management: a systematic review of randomized clinical trials. Phytother Res. 2017;31(7):1005–18.
[51] Carpene C, et al. Novel strategies for preventing diabetes and obesity complications with natural polyphenols. Curr. Med. Chem. 2015;22(1):150–64.
[52] Goyenechea E, et al. A nutrigenomic inflammation-related PBMC-based approach to predict the weight-loss regain in obese subjects. Ann. Nutr. Metab. 2009;54(1):43–51.
[53] Duthie SJ. Epigenetic modifications and human pathologies: cancer and CVD. Proc. Nutr. Soc. 2011;70(1):47–56.
[54] Sohi G, et al. Maternal protein restriction elevates cholesterol in adult rat offspring due to repressive changes in histone modifications at the cholesterol 7alpha-hydroxylase promoter. Mol. Endocrinol. 2011;25(5):785–98.

[55] Shah MS, et al. Integrated microRNA and mRNA expression profiling in a rat colon carcinogenesis model: effect of a chemoprotective diet. Physiol. Genom. 2011;43(10):640–54.
[56] Gracia A, et al. Fatty acid synthase methylation levels in adipose tissue: effects of an obesogenic diet and phenol compounds. Genes Nutr. 2014;9(4):411.
[57] Milagro FI, et al. Dietary factors, epigenetic modifications and obesity outcomes: progresses and perspectives. Mol. Asp. Med. 2013;34(4):782–812.
[58] Simonds NI, et al. Review of the gene-environment interaction literature in cancer: what do we know? Genet. Epidemiol. 2016;40(5):356–65.
[59] Gonzalez-Muniesa P, et al. Obesity. Nat. Rev. Dis. Primers. 2017;3:17034.
[60] Overweight, W.H.O.O.a. http://www.who.int/mediacentre/factsheets/fs311/en/; 2017.
[61] Shalitin S, Battelino T, Moreno LA. Obesity, metabolic syndrome and nutrition. World Rev. Nutr. Diet. 2016;114:21–49.
[62] Martin-Timon I, et al. Type 2 diabetes and cardiovascular disease: have all risk factors the same strength? World J. Diabetes 2014;5(4):444–70.
[63] Calle EE, et al. Overweight, obesity, and mortality from cancer in a prospectively studied cohort of U.S. adults. N. Engl. J. Med. 2003;348(17):1625–38.
[64] Zhang X, et al. Impact of obesity upon prostate cancer-associated mortality: a meta-analysis of 17 cohort studies. Oncol. Lett. 2015;9(3):1307–12.
[65] Zhang Y, et al. Overweight, obesity and endometrial cancer risk: results from a systematic review and meta-analysis. Int. J. Biol. Mark. 2014;29(1):e21–9.
[66] Dobbins M, Decorby K, Choi BC. The association between obesity and cancer risk: a meta-analysis of observational studies from 1985 to 2011. ISRN Prev Med 2013;2013:680536.
[67] Liu Z, et al. The association between overweight, obesity and ovarian cancer: a meta-analysis. Jpn. J. Clin. Oncol. 2015;45(12):1107–15.
[68] Qin Q, et al. Obesity and risk of bladder cancer: a meta-analysis of cohort studies. Asian Pac. J. Cancer Prev. APJCP 2013;14(5):3117–21.
[69] Larsson SC, Wolk A. Overweight, obesity and risk of liver cancer: a meta-analysis of cohort studies. Br. J. Canc. 2007;97(7):1005–8.
[70] Moghaddam AA, Woodward M, Huxley R. Obesity and risk of colorectal cancer: a meta-analysis of 31 studies with 70,000 events. Cancer Epidemiol. Biomark. Prev. 2007;16(12):2533–47.
[71] Nieman KM, et al. Adipose tissue and adipocytes support tumorigenesis and metastasis. Biochim. Biophys. Acta 2013;1831(10):1533–41.
[72] Ritchie SA, Connell JM. The link between abdominal obesity, metabolic syndrome and cardiovascular disease. Nutr. Metabol. Cardiovasc. Dis. 2007;17(4):319–26.
[73] Schapira DV, et al. Visceral obesity and breast cancer risk. Cancer 1994;74(2):632–9.
[74] Louis P, Hold GL, Flint HJ. The gut microbiota, bacterial metabolites and colorectal cancer. Nat. Rev. Microbiol. 2014;12(10):661–72.
[75] Sheflin AM, Whitney AK, Weir TL. Cancer-promoting effects of microbial dysbiosis. Curr. Oncol. Rep. 2014;16(10):406.
[76] Cremolini C, et al. First-line chemotherapy for mCRC-a review and evidence-based algorithm. Nat. Rev. Clin. Oncol. 2015;12(10):607–19.
[77] Dienstmann R, et al. Consensus molecular subtypes and the evolution of precision medicine in colorectal cancer. Nat. Rev. Canc. 2017;17(2):79–92.
[78] Vargas T, et al. ColoLipidGene: signature of lipid metabolism-related genes to predict prognosis in stage-II colon cancer patients. Oncotarget 2015;6(9):7348–63.
[79] Sanchez-Martinez R, et al. A link between lipid metabolism and epithelial-mesenchymal transition provides a target for colon cancer therapy. Oncotarget 2015;6(36):38719–36.
[80] Vargas T, et al. 3'UTR polymorphism in ACSL1 gene correlates with expression levels and poor clinical outcome in colon cancer patients. PLoS One 2016;11(12):e0168423.
[81] Hoeft B, et al. Polymorphisms in fatty-acid-metabolism-related genes are associated with colorectal cancer risk. Carcinogenesis 2010;31(3):466–72.
[82] Crous-Bou M, et al. Genetic polymorphisms in fatty acid metabolism genes and colorectal cancer. Mutagenesis 2012;27(2):169–76.
[83] Van Raay T, Allen-Vercoe E. Microbial interactions and interventions in colorectal cancer. Microbiol. Spectr. 2017;5(3).
[84] Jeyakumar A, Dissabandara L, Gopalan V. A critical overview on the biological and molecular features of red and processed meat in colorectal carcinogenesis. J. Gastroenterol. 2017;52(4):407–18.
[85] Helmus DS, et al. Red meat-derived heterocyclic amines increase risk of colon cancer: a population-based case-control study. Nutr. Canc. 2013;65(8):1141–50.
[86] Ma CJ, et al. Prospective double-blind randomized study on the efficacy and safety of an n-3 fatty acid enriched intravenous fat emulsion in postsurgical gastric and colorectal cancer patients. Nutr. J. 2015;14:9.
[87] Whelan J, McEntee MF. Dietary (n-6) PUFA and intestinal tumorigenesis. J. Nutr. 2004;134(12 Suppl. l):3421S–6S.
[88] Bayerdorffer E, et al. Decreased high-density lipoprotein cholesterol and increased low-density cholesterol levels in patients with colorectal adenomas. Ann. Intern. Med. 1993;118(7):481–7.
[89] Morad SA, et al. Tamoxifen magnifies therapeutic impact of ceramide in human colorectal cancer cells independent of p53. Biochem. Pharmacol. 2013;85(8):1057–65.
[90] Bariol C, et al. The relationship between hypomethylation and CpG island methylation in colorectal neoplasia. Am. J. Pathol. 2003;162(4):1361–71.
[91] Gibson TM, et al. Pre- and postfortification intake of folate and risk of colorectal cancer in a large prospective cohort study in the United States. Am. J. Clin. Nutr. 2011;94(4):1053–62.
[92] Huncharek M, Muscat J, Kupelnick B. Colorectal cancer risk and dietary intake of calcium, vitamin D, and dairy products: a meta-analysis of 26,335 cases from 60 observational studies. Nutr. Canc. 2009;61(1):47–69.
[93] Dimitrov DV. The human gutome: nutrigenomics of the host-microbiome interactions. OMICS 2011;15(7–8):419–30.
[94] Eussen SJ, et al. Plasma folate, related genetic variants, and colorectal cancer risk in EPIC. Cancer Epidemiol. Biomark. Prev. 2010;19(5):1328–40.
[95] Dai Q, et al. The relation of magnesium and calcium intakes and a genetic polymorphism in the magnesium transporter to colorectal neoplasia risk. Am. J. Clin. Nutr. 2007;86(3):743–51.
[96] Brevik A, et al. Polymorphisms in base excision repair genes as colorectal cancer risk factors and modifiers of the effect of diets high in red meat. Cancer Epidemiol. Biomark. Prev. 2010;19(12):3167–73.

Chapter 29

The salivary volatome in breast cancer

Jorge A.M. Pereira[1], Ravindra Taware[2], Priscilla Porto-Figueira[1], Srikanth Rapole[2] and José S. Câmara[1,3]

[1]*CQM – Centro de Química da Madeira, Universidade da Madeira, Campus da Penteada, Funchal, Portugal;* [2]*Proteomics Lab, National Centre for Cell Science (NCCS), Ganeshkhind, SPPU Campus, Pune, Maharashtra, India;* [3]*Faculdade de Ciências Exatas e da Engenharia, Universidade da Madeira, Campus da Penteada, Funchal, Portugal*

Introduction

Saliva is a complex body fluid secreted in the oral cavity by three major salivary glands pairs, parotid, submandibular, and sublingual, along with hundreds of minor salivary glands located in the labial, buccal, palatal, lingual, and retromolar regions of the oral mucosa [1]. Its primary role involves the gastrointestinal functions associated with food digestion, such as mastication, swallowing, and taste perception [2]. Additionally, a correct supply of saliva is essential for the protection and preservation of the oral and gastrointestinal tissues against microbial and viral pathogens [3,4]. The articulation of speech is another function to which saliva is certainly important and often forgotten [4]. Biochemically, saliva is a clear, slightly acidic solution composed of approximately 99% of water with soluble analytes, such as nucleic acids, proteins, metabolites, and minerals [5]. It is well established that the salivary constituents interchange with blood and vice versa by various mechanisms, such as active transport, intracellular passive diffusion, and extracellular ultrafiltration [6]. Therefore, saliva shares many constituents with blood, like DNA, RNA, and proteins, and can be an excellent alternative to blood sampling to study the physiology of the body. In fact, despite the large repertoire of molecules it contains, saliva harbors unique molecular constituents that can be discriminatory for the screening and detection of oral and systemic diseases and can provide important clues about the pathophysiological condition of the human body [7]. Furthermore, saliva sampling possesses inherent qualities, such as a noninvasive and painless procedure and easy handling and storage. Together, these conditions make it imperative to profile, identify, and quantitate the salivary metabolites that can help not only in diagnostics but also in prognostics and therapeutic drug monitoring. In this context, different omics approaches, like transcriptomics, proteomics, and metabolomics, are being applied to profile saliva and unveil putative biomarkers [8–15] (Fig. 29.1). Particularly, with the advent of modern mass spectrometry in combination with robust separation techniques, the metabolomics approach is viewed as a valuable resource to pinpoint the role of metabolic dysregulation in many malignant diseases. Many metabolites generated in the cell are volatile in nature, and they are lost during the routine metabolomic preparation and analysis protocols, resulting in an incomplete metabolic landscape. This can be overcome through the specific characterization of the volatile fraction of the metabolome. Such analysis, known as volatomics, deals with the detection, identification, and quantitation of the volatile and semi-volatile organic metabolites (VOMs and semiVOMs) produced by the human body [16,17]. It is an attractive strategy to unveil putative diagnostic biomarkers, particularly when it involves noninvasive sampling procedures. Moreover, VOMs are highly amenable for the integration into sensor devices, often known as eNOSEs, which have a great potential for an easy translation into Point Of Care Testing (POCT) devices [18]. This represents a breakthrough for disease prevention and management, considerably improving the efficiency of the screening programs of high-risk populations or allowing more effective treatments through a more accurate therapeutic drug monitoring.

Volatile organic compounds as disease biomarkers

Interfaces with the human microbiome, genome, metabolome, and volatome

Saliva is being continuously produced along with moisture in the oral cavity and so, to a certain extent, it can be considered a mirror of the human metabolism interaction with the environment, allowing the obtention of near real-time snapshots of such interaction [19]. This is reflected

Precision Medicine for Investigators, Practitioners and Providers. https://doi.org/10.1016/B978-0-12-819178-1.00029-0

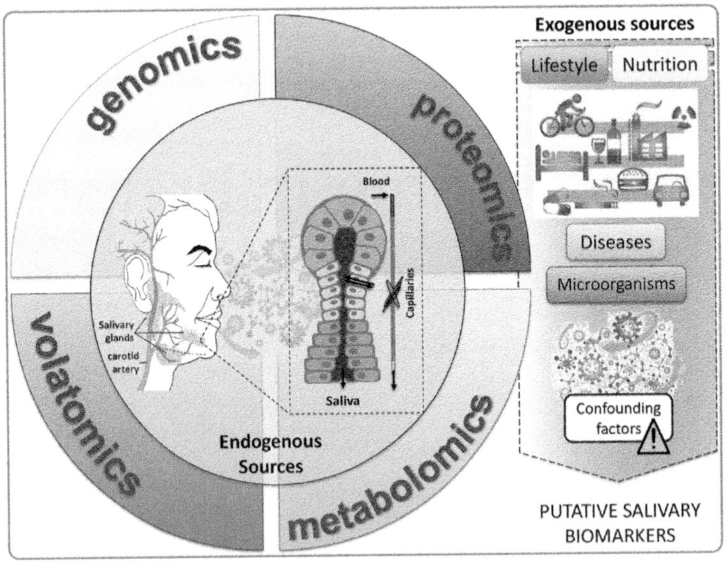

FIGURE 29.1 Overview of the metabolic interplay involving saliva sampling and the definition of putative salivary biomarkers.

in different levels, which include the human microbiome, genome, metabolome, and volatome. Saliva composition will be necessarily modulated by the overall metabolic fitness of each person but also integrates other endogenous and exogenous contributions. The mutual interaction between the individual oral microbiome and saliva composition, for instance, has been extensively studied in the context of oral hygiene. Regarding this [20], showed that chewing was essential to stimulate salivary flow and recover pH levels, thereby inhibiting tooth decay caused by Streptococcus mutants. In a recent study, [21] were able to identify unique VOCs signatures for bacteria species that often colonize our mouth, S. mutants, L. salivarius, and P. acidifaciens. This result shows the potential of a breath test to follow different clinical conditions in oral health. Food intake has an obvious interference in the oral microbiome and saliva composition, but this modulation also involves the genome itself. A recent study points to a correlation between fat mass, genomic composition, and oral microbiome in obese women from Croatia. This result is supported by the observation that there was a higher occurrence of Staphylococcus aureus, a major human pathogen, and a lower presence of Streptococcus oralis, Streptococcus mitis, and Serratia ureilytica in the saliva of these women when compared with the control group [22]. And these three strains are a very tiny part of the microbiome that each person hosts, as over 300 bacterial species have been already identified just in the oral cavity [23,24]. In fact, the volatile composition of saliva includes the VOCs produced by oral bacteria, as aliphatic amines, branched chain fatty acids, indole, phenol, and volatile sulfur-containing compounds [23,24], but also VOCs derived from many other sources, as the VOCs in inhaled and exhaled air, that dissolve differentially in saliva [24]. At a deeper level, the acinar cells that compose the salivary glands are vascularized, allowing VOMs transference from the blood through several mechanisms, as passive diffusion, ultrafiltration, and active diffusion [4,25]. Overall, this shows that the salivary volatile composition includes blood, gingival exudate, nasal cavity, gastrointestinal reflux, food debris, commercial products, and environmental pollution contributions (reviewed in Ref. [26]. In this way, the salivary volatile composition reflects both the oral compositions as biochemical and metabolic blood information, constituting a valuable diagnostic tool for myriad clinical conditions. This includes different local and systemic diseases, malignancy, infections, cardiovascular problems, and genetic disorders, among others, as reviewed by Ref. [27].

The salivary volatome and potential correlations with breast cancer

Previously, different studies have explored the suitability of saliva for cancer diagnosis, unveiling proteins differentially expressed in cancer patients. This includes, for instance, vascular endothelial growth factor (VEGF), epidermal growth factor (EGF), and carcinoembryonic antigen (CEA) that were significantly elevated in the salivary fluid of cancer patients [28]. Specifically, regarding breast cancer, augmented levels of CA15-3, and EGF receptor have been reported to be found in patients with this pathology [29]. In turn, [30] proposed a broader proteomic analysis of saliva to discriminate different phenotypes and stages, and consequent therapeutic interventions, therefore, saving patients at lower risk from unnecessary chemotherapy administration prior to surgery. More recently, significant differences between BC patients at stages I–II and healthy controls were observed in the concentrations of 15 salivary free amino acid [31]. At the genomic level, higher number of detectable mitochondrial RNAs (miRNA) species have been detected in saliva, breast milk, and seminal fluid by comparison with urine, cerebrospinal fluid, and pleural fluid [7] and in fact [32] prevalidated eight mRNA biomarkers and one protein biomarker in saliva for breast cancer detection. Overall, it is becoming clear that different cancer-specific signatures are embedded in saliva [33] and a volatomic fingerprint able to discriminate BC is certainly feasible, as it has been already proposed by Refs. [16,34]. So far, over 350 salivary VOMs have been reported in different studies (reviewed in Ref. [35]. However, this number is very conservative, and the salivary volatome should be much more complex. This statement is greatly supported by the fact that more than 300 different bacterial species have been already identified in just the oral microbiome [23,24]. There is, therefore, a great complexity in the salivary volatome to unveil, but also a great potential to identify putative biomarkers for several diseases, including different forms of cancer and particularly BC.

Diagnostic implications for BC

Although BC diagnosis, screening programs, and treatments clearly improved in the last decade, the fact is that BC continues to figure among the main cause of death in women worldwide [36]. This is certainly a great motivation to continue to develop better and more reliable diagnosis tools, preferentially able to detect the disease in its early stages when the treatments are more efficient and the long-term survival achievable. Additionally, most of current diagnostic tools available are expensive, therefore, representing a severe burden to the developing countries and prohibitive or inexistent in the other countries. In this context, the identification of volatile biomarkers for BC using the patient's biofluids (saliva, urine, exhaled breath) is a breakthrough that ongoing research efforts are trying to achieve [16,34,37,38]. Among the human biofluids, saliva is certainly one of the easiest forms to obtain valuable snapshots of human metabolism in health and disease. Its noninvasive nature and easy to repeatedly collect under controlled parameters, without causing discomfort to patients, make this sample very relevant for volatomic research. Furthermore, saliva sampling is technically much easier than exhaled breath analysis, and its analysis is also facilitated by its lower volatile complexity, when compared with urine, for instance.

Techniques for volatome studies in cells, tissues, and fluids

The identification and quantification of VOMs emitted by biological samples of clinical importance, such as cells, tissues, and biofluids (mainly urine, saliva, exhaled breath) can be carried out using different analytical techniques. Broadly, such techniques can be classified according to the use of gas chromatography (GC) in the analytical layout. This technique, usually combined with mass spectrometry (MS) is the most often reported for volatile studies, allowing a comprehensive characterization of the volatile composition of different samples, but requires expensive and complex equipment (Fig. 29.2). In contrast, approaches involving different sensors sets and architectures, commonly known as *e-noses*, are more suitable for the

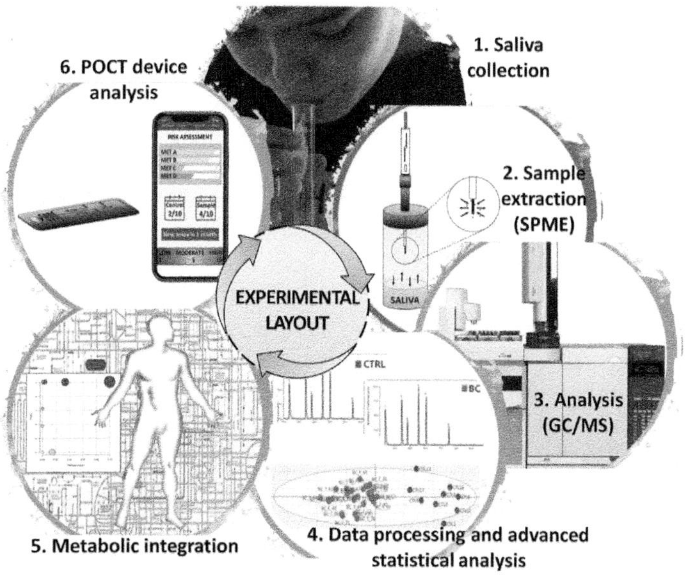

FIGURE 29.2 Experimental layout most often reported for the characterization of the volatile composition of different biofluids, including saliva.

clinical environment, but only target specific groups or family of compounds (reviewed in Ref. [39]).

GC-dependent techniques

GC/MS is the most often reported methodology in the field of volatile analysis. Essentially, the volatile and semi-volatile components of a mixture are injected in a capillary column and separated under a temperature gradient, being finally identified in the detector. In its most simple configuration, the GC apparatus is relatively affordable, in the range of $20.000, but adding a second dimension to the GC separation (GC×GC), as well as robust and powerful mass detection configuration (TOF-MS/MS), can increase the costs up to 20 times or more. Compared with this, the methodologies and equipment to perform sampling are inexpensive (<$100 each SPME fiber). The sampling step is required to trap the sample VOMs and inject them in the GC apparatus, being the solid-phase microextraction (SPME) the most popular sampling format for GC/MS analysis. SPME involves the adsorption/desorption of the sample VOMs in specific sorbents coated in the SPME fiber. These interactions are dependent of the equilibrium constants between the sample headspace and the stationary phase and can be modulated using several parameters, as the time, temperature, and form of exposure of the sample VOMs to the fiber (headspace, static, or dynamic sampling), the type and polarity of the sorbents and thickness of the fiber (reviewed in Refs. [40,41]). Following this sampling procedure, the trapped VOMs are thermally desorbed at high temperature from the SPME fiber (using a carrier gas) to the GC/MS inlet. Then, the VOMs will be separated (retained) along the capillary column and detected and quantified. Overall, SPME is quick, easy to use, is compatible with the preservation of the structure of the biological sample and allows high reproducibility and sensitivity of the following GC/MS analysis (limit of detection up to the ppbv range) [42]. For these reasons, SPME has been widely reported for the analysis of biofluids regarding the identification of putative diseases biomarkers [14,38,40,41,43]. Needle Trap Micro-Extraction (NTME) is a novel approach for the extraction of VOMs in which the sorbent material is packed inside a needle [44]. This format adds to the advantages described to SPME, an exhaustive character that allows an enhanced sensitivity and the ability to do sampling and storage in a single device [37,44,45]. Experimentally, NTME usage is quite simple, just involving the sample headspace pumping through the sorbents packed in the needle trap device by using a disposable 1 mL syringe. The following analytical steps are then performed as the regular SPME, including the same GC/MS configuration [37,45]. This certainly expands and facilitates the use of NTME to field and clinical applications. Thin film microextraction (TFME) can be particularly relevant for many clinical applications. Instead of a fiber, the TFME sorbents adopts a thin sheet format that enables a larger surface area to extraction phase volume ratio [46,47]. Consequently, the extraction efficiency will be largely improved, allowing the analysis of larger amounts of VOMs within shorter periods. TFME can be operated in a similar mode as SPME or NTME, requiring only a modified desorption chamber to transfer the trapped VOMs from the fiber to the GC/MS instrument [48].

GC-independent techniques

The most popular GC-independent techniques for VOCs analysis were essentially designed for real-time measurements of the whole volatile fraction of a given sample (untargeted analysis) or just for specific VOMs (targeted approach). In the first case, proton transfer reaction mass spectrometry (PTR-TOF-MS) is certainly in the forefront of real-time and noninvasive volatomic analysis. PTR-TOF-MS is a direct injection mass spectrometric analytical approach based on chemical ionization, which allows rapid determination of VOMs with high sensitivity and specificity [49]. Most importantly, it does not involve any sample preparation or sample destruction. PTR-MS, different from the conventional MS, use H_3O^+ generated by an external ion source to cause the soft chemical ionization of the sample VOMs [50]. This allows a very low response time and limit of detection (in the range of pg/g), making PTR-MS an excellent alternative to GC-MS. It is nevertheless, much more expensive, which explains why it was not already widely adopted for volatomic analysis. Regarding the targeted volatomic analysis, myriad different platforms using laser spectroscopy or chemical sensors designed for artificial olfaction, often known as *e-noses*, is already available and used in the clinical environment (reviewed in Ref. [51]). This technology offers many advantages over routine analytical platforms, notably its easier usage, not requiring high technical expertise, its lower equipment and operation costs, portability, and real-time measurements [52,53]. Finally, e-noses offer flexibility in choosing different sets of gas sensor arrays for customized applications. However, there are also drawbacks to be considered, namely the inability to identify individual analytes in the complex mixture of VOMs, the moisture interference, lower accuracy, and sensitivity as compared to most of the analytical setups and shorter life span [54].

Challenges and research opportunities in volatomic analysis using biofluids

The identification of putative volatile biomarkers for different diseases using human biofluids collected noninvasively is an exciting strategy that is being pursued by several research groups around the world. In the last decades, major improvements are driving this strategy more closely to the clinical environment and real-time applications. In this context, eventually, one of the most notable works is the artificially intelligent nanoarray for breath analysis developed by Hossam Haick group [55]. This portable sampling device with the size of a small USB drive was used to diagnose and classify 17 different diseases using pattern analysis of the VOMs detected in the exhaled breath of the patients enrolled in the multicenter study [55]. Another approach is applying different forms of ion mobility spectrometry (IMS), notably high-field asymmetric waveform IMS (FAIMS), to gas analysis in medical diagnosis [56]. In this regard, Owlstone Medical Ltd. (Cambridge, UK) is exploring the concept of breath biopsy using FAIMS, and has already developed several solutions able to deliver comprehensive volatomic analysis for the clinical environment. Currently, Owlstone has several clinical trials ongoing on different forms of cancer and asthma, which will challenge the methodologies developed in large-scale studies.

There are still several important challenges to address in volatomic research. Novel extraction procedures and advanced detection systems and methodologies are continuously being developed, thereby producing a steady increase in the quality and coverage of the volatile fingerprints. Surprisingly, this is not being translated into proportional clinical biomarker identifications. There are several reasons that account for this high false discovery rate, with the complexity of human metabolism at the top. So far, over 1700 VOMs have been assigned to the human volatilome, but their origin and dependency from confounding factors as age, gender, diet, genetic background, environmental exposition, etc., has not been satisfactorily elucidated [26]. To address this problem, extensive longitudinal studies enrolling dozens of patients must be implemented. Furthermore, simultaneous characterization of the volatomic composition of different biofluids (saliva, urine, exhaled breath) from the same patient will be crucial to minimize the influence of confounding factors and achieve the full characterization of the metabolic routes of putative biomarkers. This is particularly relevant to discard the exogenous contribution from the volatile fingerprints and improve their utility as a diagnostic tool. Specifically, regarding the contribution of salivary VOMs to BC diagnosis, there are additional points that should be critically considered. As already mentioned, the salivary volatome contains valuable metabolic information about health and diseases, including for BC [16,34]. However, saliva can be easily contaminated with exogenous VOMs from different sources, particularly from the environment during normal breathing, from the diet, and from oral bacteria that colonize our mouth in an extension that depends on the oral hygiene of each subject. Furthermore, saliva flow rate and composition are affected by circadian rhythms. For minimizing these interferences and improve the quality of the volatomic data obtained, the most effective strategies are the design of consistent protocols for saliva collection and the recruitment of large cohorts, preferentially around or above 100 subjects by intervention group.

Acknowledgments

This work was supported by Fundação para a Ciência e a Tecnologia (project PEst-OE/QUI/UI0674/2019, CQM, INNOINDIGO/0001/2015, Portuguese Government funds), and through Madeira 14–20 Program, project PROEQUIPRAM - Reforço do Investimento em Equipamentos e Infraestruturas Científicas na RAM (M1420-01-0145-FEDER-000008), ARDITI-Agência Regional para o Desenvolvimento da Investigação Tecnologia e Inovação, through the project M1420-01-0145-FEDER-000005 - Centro de Química da Madeira – CQM$^+$ (Madeira 14–20) and the Inno-Indigo NCD-CAPomics project, Department of Science & Technology, Govt. of India, grant no. DST/IMRCD/EU/Inno-Indigo/NCDs-CAPomics/2015. Priscilla Porto-Figueira was supported by FCT (PhD fellowship SFRH/BD/129630/2017) and Jorge A. M. Pereira by ARDITI (Post-Doctoral fellowship Project M1420 - 09-5369-FSE-000001). Ravindra Taware acknowledges the Council of Scientific and Industrial Research, New Delhi, India for research associateship.

References

[1] Hand A. Salivary glands. Ten cate's oral histology: development, structure and function. A. Nanci. China: Elsevier; 2008. p. 290–318.
[2] Humphrey SP, Williamson RT. A review of saliva: normal composition, flow, and function. J. Prosthet. Dent. 2001;85(2):162–9.
[3] Moss SJ. Clinical implications of recent advances in salivary research. J. Esthetic Dent. 1995;7(5):197–203.
[4] Pedersen AML, Sorensen CE, Proctor GB, Carpenter GH, Ekstrom J. Salivary secretion in health and disease. J. Oral Rehabil. 2018;45(9):730–46.
[5] Schipper RG, Silletti E, Vingerhoeds MH. Saliva as research material: biochemical, physicochemical and practical aspects. Arch. Oral Biol. 2007;52(12):1114–35.
[6] Wong DT. Salivary diagnostics powered by nanotechnologies, proteomics and genomics. J. Am. Dent. Assoc. 2006;137(3):313–21.
[7] Wong DT. Salivary extracellular noncoding RNA: emerging biomarkers for molecular diagnostics. Clin. Ther. 2015;37(3):540–51.
[8] Carvalho AL, Henrique R, Jeronimo C, Nayak CS, Reddy AN, Hoque MO, Chang S, Brait M, Jiang WW, Kim MM, Claybourne Q, Goldenberg D, Khan Z, Khan T, Westra WH, Sidransky D, Koch W, Califano JA. Detection of promoter hypermethylation in salivary rinses as a biomarker for head and neck squamous cell carcinoma surveillance. Clin. Cancer Res. 2011;17(14):4782–9.
[9] Ishikawa S, Sugimoto M, Kitabatake K, Sugano A. Identification of salivary metabolomic biomarkers for oral cancer screening. Sci. Rep. 2016;6:31520.
[10] Jessie K, Jayapalan JJ, Ong K-C, Abdul Rahim ZH, Zain RM, Wong K-T, Hashim OH. Aberrant proteins in the saliva of patients with oral squamous cell carcinoma. Electrophoresis 2013;34(17):2495–502.

[11] Kawahara R, Bollinger JG, Rivera C, Ribeiro ACP, Brandão TB, Leme AFP, Maccoss MJ. A targeted proteomic strategy for the measurement of oral cancer candidate biomarkers in human saliva. Proteomics 2016;16(1):159–73.

[12] Li Y, John M A R St, Zhou X, Kim Y, Sinha U, Jordan RCK, Eisele D, Abemayor E, Elashoff D, Park NH, Wong DT. Salivary transcriptome diagnostics for oral cancer detection. Clin. Cancer Res. 2004;10(24):8442–50.

[13] Righini CA, De Fraipont F, Timsit JF, Faure C, Brambilla E, Reyt E, Favrot MC. Tumor-specific methylation in saliva: a promising biomarker for early detection of head and neck cancer recurrence. Clin. Cancer Res. 2007;13(4):1179–85.

[14] Taware R, Taunk K, Pereira JAM, Shirolkar A, Soneji D, Camara JS, Nagarajaram HA, Rapole S. Volatilomic insight of head and neck cancer via the effects observed on saliva metabolites. Sci. Rep. 2018;8(1):17725.

[15] Zimmermann BG, Wong DT. Salivary mRNA targets for cancer diagnostics. Oral Oncol. 2008;44:425–9.

[16] Cavaco C, Pereira JAM, Taunk K, Taware R, Rapole S, Nagarajaram H, Câmara JS. Screening of salivary volatiles for putative breast cancer discrimination: an exploratory study involving geographically distant populations. Anal. Bioanal. Chem. 2018;410(18):4459–68.

[17] Silva CL, Passos M, Câmara JS. Investigation of urinary volatile organic metabolites as potential cancer biomarkers by solid-phase microextraction in combination with gas chromatography-mass spectrometry. Br. J. Canc. 2011;105(12):1894–904.

[18] Wilson AD. Advances in electronic-nose technologies for the detection of volatile biomarker metabolites in the human breath. Metabolites 2015;5(1):140–63.

[19] Streckfus CF, Brown RE, Bull JM. Proteomics, morphoproteomics, saliva and breast cancer: an emerging approach to guide the delivery of individualised thermal therapy, thermochemotherapy and monitor therapy response. Int. J. Hyperth. 2010;26(7):649–61.

[20] Ribelles Llop M, Guinot Jimeno F, Mayne Acien R, Bellet Dalmau LJ. Effects of xylitol chewing gum on salivary flow rate, pH, buffering capacity and presence of *Streptococcus mutans* in saliva. Eur. J. Paediatr. Dent. 2010;11(1):9–14.

[21] Hertel M, Preissner R, Gillissen B, Schmidt-Westhausen AM, Paris S, Preissner S. Detection of signature volatiles for cariogenic microorganisms. Eur. J. Clin. Microbiol. Infect. Dis. 2016;35(2):235–44.

[22] Hudek A, Skara L, Smolkovic B, Kazazic S, Ravlic S, Nanic L, Osvatic MM, Jelcic J, Rubelj I, Bacun-Druzina V. Higher prevalence of FTO gene risk genotypes AA rs9939609, CC rs1421085, and GG rs17817449 and saliva containing *Staphylococcus aureus* in obese women in Croatia. Nutr. Res. 2018;50:94–103.

[23] Kakoei S, Barkhori F, Mirzazadeh A, Mohammadi M, Gholamhoseinian A. Influence of menstrual cycle and salivary ß-estradiol on volatile sulfur compound 2012;1(1):5.

[24] Soini H, Klouckova I, Wiesler D, Oberzaucher E, Grammer K, Dixon S, Xu Y, Brereton R, Penn D, Novotny M. Analysis of volatile organic compounds in human saliva by a static sorptive extraction method and gas chromatography-mass spectrometry. J. Chem. Ecol. 2010;36(9):1035–42.

[25] Pfaffe T, Cooper-White J, Beyerlein P, Kostner K, Punyadeera C. Diagnostic potential of saliva: current state and future applications. Clin. Chem. 2011;57(5):675–87.

[26] Amann A, de Lacy Costello B, Miekisch W, Schubert J, Buszewski B, Pleil J, Ratcliffe N, Risby T. The human volatilome: volatile organic compounds (VOCs) in exhaled breath, skin emanations, urine, feces and saliva. J. Breath Res. 2014;8(3):034001.

[27] Malathi N, Mythili S, Vasanthi HR. Salivary diagnostics: a brief review. ISRN Dent. 2014;2014. 158786-158786.

[28] Brooks MN, Wang J, Li Y, Zhang R, Elashoff D, Wong DT. Salivary protein factors are elevated in breast cancer patients. Mol. Med. Rep. 2008;1(3):375–8.

[29] Pink R, Simek J, Vondrakova J, Faber E, Michl P, Pazdera J, Indrak K. Saliva as a diagnostic medium. Biomed. Pap. Med. Fac. Univ. Palacky Olomouc. Czech. Repub. 2009;153(2):103–10.

[30] Streckfus CF, Storthz KA, Bigler L, Dubinsky WP. A comparison of the proteomic expression in pooled saliva specimens from individuals diagnosed with ductal carcinoma of the breast with and without lymph node involvement. J. Oncol. 2009:737619.

[31] Cheng F, Wang Z, Huang Y, Duan Y, Wang X. Investigation of salivary free amino acid profile for early diagnosis of breast cancer with ultra performance liquid chromatography-mass spectrometry. Clin. Chim. Acta 2015;447:23–31.

[32] Zhang L, Xiao H, Karlan S, Zhou H, Gross J, Elashoff D, Akin D, Yan X, Chia D, Karlan B, Wong DT. Discovery and preclinical validation of salivary transcriptomic and proteomic biomarkers for the non-invasive detection of breast cancer. PLoS One 2010;5(12):e15573.

[33] Sugimoto M, Wong DT, Hirayama A, Soga T, Tomita M. Capillary electrophoresis mass spectrometry-based saliva metabolomics identified oral, breast and pancreatic cancer-specific profiles. Metabolomics 2010;6(1):78–95.

[34] Cavaco C, Perestrelo R, Silva C, Aveiro F, Pereira J, Câmara J. Establishment of the saliva volatomic profile as an exploratory and non-invasive strategy to find potential breast cancer biomarkers. Int. Labmate Ltd. 2014:4–5.

[35] de Lacy Costello B, Amann A, Al-Kateb H, Flynn C, Filipiak W, Khalid T, Osborne D, Ratcliffe NM. A review of the volatiles from the healthy human body. J. Breath Res. 2014;8(1):014001.

[36] Bray F, Ferlay J, Soerjomataram I, Siegel RL, Torre LA, Jemal A. Global cancer statistics 2018: GLOBOCAN estimates of incidence and mortality worldwide for 36 cancers in 185 countries. CA Cancer J. Clin. 2018;68(6):394–424.

[37] Porto-Figueira P, Pereira J, Miekisch W, Camara JS. Exploring the potential of NTME/GC-MS, in the establishment of urinary volatomic profiles. Lung cancer patients as case study. Sci. Rep. 2018;8(1):13113.

[38] Taunk K, Taware R, More TH, Porto-figueira P, Pereira JAM, Mohapatra R, Soneji D, Camara JS, Nagrajaram HA, Rapole, Srikanth. A non-invasive approach to explore the discriminatory potential of the urinary volatilome of invasive ductal carcinoma of the breast. RSC Adv. 2018;8(44):25040–50.

[39] Pereira J, Porto-Figueira P, Cavaco C, Taunk K, Rapole S, Dhakne R, Nagarajaram H, Camara JS. Breath analysis as a potential and non-invasive frontier in disease diagnosis: an overview. Metabolites 2015;5(1):3–55.

[40] Pereira J, Silva CL, Perestrelo R, Goncalves J, Alves V, Camara JS. Re-exploring the high-throughput potential of microextraction techniques, SPME and MEPS, as powerful strategies for medical diagnostic purposes. Innovative approaches, recent applications and future trends. Anal. Bioanal. Chem. 2014;406(8):2101–22.

[41] Silva C, Cavaco C, Perestrelo R, Pereira J, Câmara J. Microextraction by packed sorbent (MEPS) and solid-phase microextraction (SPME) as sample preparation procedures for the metabolomic profiling of urine. Metabolites 2014;4(1):71–97.

[42] Biniecka M, Caroli S. Analytical methods for the quantification of volatile aromatic compounds. Trac. Trends Anal. Chem. 2011;30(11):1756–70.

[43] Taware R, Taunk K, Pereira JAM, Dhakne R, Kannan N, Soneji D, Câmara JS, Nagarajaram HA, Rapole S. Investigation of urinary volatomic alterations in head and neck cancer: a non-invasive approach towards diagnosis and prognosis. Metabolomics 2017;13(10):111.

[44] Trefz P, Kischkel S, Hein D, James ES, Schubert JK, Miekisch W. Needle trap micro-extraction for VOC analysis: effects of packing materials and desorption parameters. J. Chromatogr. A 2012;1219:29–38.

[45] Porto-Figueira P, Pereira JAM, Câmara JS. Exploring the potential of needle trap microextraction combined with chromatographic and statistical data to discriminate different types of cancer based on urinary volatomic biosignature. Anal. Chim. Acta 2018;1023:53–63.

[46] Bruheim I, Liu X, Pawliszyn J. Thin-film microextraction. Anal. Chem. 2003;75(4):1002–10.

[47] Shigeyama H, Wang T, Ichinose M, Ansai T, Lee S-W. Identification of volatile metabolites in human saliva from patients with oral squamous cell carcinoma via zeolite-based thin-film microextraction coupled with GC–MS. J. Chromatogr. B 2019;1104:49–58.

[48] Vuckovic D, Cudjoe E, Musteata FM, Pawliszyn J. Automated solid-phase microextraction and thin-film microextraction for high-throughput analysis of biological fluids and ligand-receptor binding studies. Nat. Protoc. 2010;5(1):140–61.

[49] Capozzi V, Yener S, Khomenko I, Farneti B, Cappellin L, Gasperi F, Scampicchio M, Biasioli F. PTR-ToF-MS coupled with an automated sampling system and tailored data analysis for food studies: bioprocess monitoring, screening and nose-space analysis. J. Vis. Exp. 2017;(123).

[50] Romano A, Hanna GB. Identification and quantification of VOCs by proton transfer reaction time of flight mass spectrometry: an experimental workflow for the optimization of specificity, sensitivity, and accuracy. J. Mass Spectrom. 2018;53(4):287–95.

[51] Capelli L, Taverna G, Bellini A, Eusebio L, Buffi N, Lazzeri M, Guazzoni G, Bozzini G, Seveso M, Mandressi A, Tidu L, Grizzi F, Sardella P, Latorre G, Hurle R, Lughezzani G, Casale P, Meregali S, Sironi S. Application and uses of electronic noses for clinical diagnosis on urine samples: a review. Sensors (Basel) 2016;16(10).

[52] Wilson AD. Diverse applications of electronic-nose technologies in agriculture and forestry. Sensors (Switzerland) 2013;13(2):2295–348.

[53] Zhang X, Lv Y, Wang H, Tao X, Zhang X. Classification identification of abalone flavoring liquids based on metal sensor array. Carpathian J. Food Sci. Tech. 2016;8(2):107–12.

[54] Wilson AD, Baietto M. Applications and advances in electronic-nose technologies. Sensors 2009;9(7):5099–148.

[55] Nakhleh MK, Amal H, Jeries R, Broza YY, Aboud M, Gharra A, Ivgi H, Khatib S, Badarneh S, Har-Shai L, Glass-Marmor L, Lejbkowicz I, Miller A, Badarny S, Winer R, Finberg J, Cohen-Kaminsky S, Perros F, Montani D, Girerd B, Garcia G, Simonneau G, Nakhoul F, Baram S, Salim R, Hakim M, Gruber M, Ronen O, Marshak T, Doweck I, Nativ O, Bahouth Z, Shi D-y, Zhang W, Hua Q-l, Pan Y-y, Tao L, Liu H, Karban A, Koifman E, Rainis T, Skapars R, Sivins A, Ancans G, Liepniece-Karele I, Kikuste I, Lasina I, Tolmanis I, Johnson D, Millstone SZ, Fulton J, Wells JW, Wilf LH, Humbert M, Leja M, Peled N, Haick H. Diagnosis and classification of 17 diseases from 1404 subjects via pattern analysis of exhaled molecules. ACS Nano 2017;11(1):112–25.

[56] Covington JA, van der Schee MP, Edge AS, Boyle B, Savage RS, Arasaradnam RP. The application of FAIMS gas analysis in medical diagnostics. Analyst 2015;140(20):6775–81.

Chapter 30

Engineered biomarkers for immunodiagnosis of leprosy

Bruno Araujo Soares[1,2], João Paulo Stanke Scandelari[2], Bruna Mayumi Wagatuma Bottolo[2] and Juliana de Moura[1,2]

[1]Department of Bioprocesses and Biotechnology Engineering, Federal University of Paraná, Curitiba, Paraná, Brazil; [2]Department of Basic Pathology, Federal University of Paraná, Curitiba, Paraná, Brazil

The disease: leprosy

Leprosy or Hansen's disease is one of the oldest bacterial diseases of mankind. Due to the low virulence of the etiologic agent, *Mycobacterium leprae*, and the variation in susceptibility among individuals, infection leads to disease in about 5% of those infected [1].

The disease is characterized by a chronic infection of the skin and peripheral nerves, and may also affect eyes, mucous membranes, bones, and testicles [2].

The long incubation period, from 5 to 20 years, is due to the 14-day bacillus replication, occurring between 30 and 35°C [3], i.e., the face, elbows, and extremities temperatures. The bacillus presents tropism for Schwann cells. Once within the cell, it can induce dedifferentiation into mesenchymal cells. This process results in the demyelination of the peripheral nerves, clinically identified as anesthesia or hypoesthesia. As a result, skin and sensitivity changes are developed [2]. Without treatment, the disease evolves resulting in leonine facies, simian hand, lagophthalmos, and destruction of the cartilaginous nasal septum [2,3].

Studies correlated several genes to disease development, such as *SLC11A1* and *LGALS3*, related to expression of nucleotide-binding oligomerization domain 2 (NOD2) and galectin-3, respectively. Other genes have also been correlated, such as Parkin coregulator gene (*PARK2*), natural resistance-associated macrophage protein (*NRAMP1*), transporter associated with antigen processing 1 and 2 (*TAP1* and *TAP2*); vitamin D3 receptor (*VDR*), and toll-like receptor (*TLR*) genes, as well as *CCDC122* and *C13orf31*, related to expression of coiled-coil domain-containing protein 122 and Laccase domain-containing protein 1, respectively. Tumor necrosis factor superfamily member 15 (*TNFSF15*), and receptor-interacting serine–threonine kinase 2 (*RIPK2*), as well as human leukocyte antigen (HLA-DR) polymorphisms were also investigated [2,4,5].

Epidemiology

In 2000, the World Health Organization's (WHO) goal of reducing the prevalence of leprosy to less than 1 case per 10,000 population, was achieved almost everywhere except in Southeast Asia and the Americas [6].

Of the 153,487 new cases detected in 2017 in Southeast Asia, 92.6% belonged to India and Indonesia (126,164 and 15,910, respectively). In the Americas, of the 29,101 new cases in 2017, 92.3% were reported in Brazil (26,875). Summing up, India, Brazil, and Indonesia accounted for 80.2% of newly registered cases of leprosy worldwide [7].

The detection of individuals with visible deformities (G2D, grade 2 disability), and the detection of leprosy in children indicate the ability of the health system to diagnose leprosy early, along with community knowledge about the disease, and the extension of health services to the population. In 2017, 12,189 new cases with G2D were diagnosed worldwide, with India, Brazil, and Indonesia accounting for 4552, 1949, and 1116 cases, respectively [7].

For endemic countries that have not been able to achieve the goal, proposed pillars of intervention are strengthening coordination, partnerships, and governmental dominance, eliminating leprosy and its complications, eliminating discrimination (stigma), and promoting the social inclusion of individuals with leprosy. The indicators for these objectives are (I) no children diagnosed with leprosy and with G2D, (II) less than one new case per million inhabitants, and (III) no country with laws stigmatizing leprosy [6,7].

Precision Medicine for Investigators, Practitioners and Providers. https://doi.org/10.1016/B978-0-12-819178-1.00030-7

Current diagnosis

Symptoms and clinical signs are scarce early in the disease, delaying diagnosis, which is essentially clinical and epidemiological. Cardinal points are (I) lesions of the skin with changes in sensitivity, (II) involvement of peripheral nerves, thickened or not, associated with sensory, motor, or autonomic alterations, and (III) positive result in bacilloscopy in intradermal smears. When these three signs are present, it is possible to have 97% sensitivity in diagnosis [8].

During inspection, the lesions have the following characteristics: achromic, hypochromic, erythematous, erythematous-hypochromic, or erythematous-violet plaques or patches, and infiltrates or nodules can be seen. At palpation, these lesions are associated with decreased tactile, thermal, and painful sensitivity (hypoesthesia or anesthesia), especially in the hands and feet. Other alterations, such as loss of motor function, supraciliary and ciliary madarosis, alopecia, involvement of the mucous membranes and cartilaginous nasal septum, loss of corneal reflex, limb paresthesia, lagophthalmos, infiltration of the face, testicular involvement, and lymphadenopathy are possible [3].

The classification of the patient reflects their bacilliferous burden, their potential for dissemination of the disease, and indicates the appropriate therapy. WHO admits two groups (I) paucibacillary (PB), when the patient presents with up to five skin lesions, and negative results during smear microscopy, and (II) multibacillary (MB), when more than five skin lesions are unveiled, with positive results during bacilloscopy. If it is not possible to perform bacilloscopy, only the number of lesions shall be considered [2,9]. The flowchart (Fig. 30.1) suggests how to approach a patient with suspected leprosy [9].

Immune response in leprosy

The clinical phenotype of leprosy is remarkably influenced by the ability of the host immune system, to organize a vigorous cell-mediated response to *M. leprae* bacillus.

Based on the host-pathogen interaction, clinical features of the disease have been dichotomized according to individual response, like type 1 helper T lymphocyte/type 2 helper T lymphocyte (Th1/Th2) paradigm, with different cytokine secretion patterns. Protection against leprosy is marked by an efficient Th1 cell-mediated response, and secretion of proinflammatory mediators such as interferon-γ (IFN-γ), tumor necrosis factor-α (TNF-α), interleukin-1β (IL-1β), and IL-2, usually accompanied by low titer anti-*M. leprae* antibodies. On the opposite, a predominant Th2-based immunological profile resulting in ineffective response against *M. leprae,* driven by antiinflammatory cytokines such as IL-4 and IL-10, along with transforming growth factor-β (TGF-β), followed by the production of antibodies, is observed in susceptible subjects [10].

Disease modalities

Immunologically stable tuberculoid versus lepromatous forms described above respectively, are situated in each pole, separated by unstable borderline variations [11].

Leprosy spectrum is divided into five groups – lepromatous leprosy (LL), tuberculoid leprosy (TT), borderline lepromatous (BL), borderline tuberculoid (BT) and borderline borderline (BB) disease, as first proposed by Ridley and Jopling considering histopathological, bacterial, and clinical observations [12].

PB spectrum involves a pronounced Th1 response, infection control, and host resistance to the pathogen. On the other hand, MB individuals are more prone to mycobacterial propagation due to a predominant humoral response mediated by Th2 lymphocytes [13].

Immune pathophysiology

The infection starts with the microorganism entering vacuoles of host macrophages, dendritic cells, and Schwann cells. During infection, TLR, expressed in phagocytic cells (innate immunity), are responsible for pathogen-associated molecular pattern (PAMP) recognition, resulting in antigen presenting cell (APC) activation, especially macrophages, and adaptive immune response initiation [13]. Macrophage activation is essential to model the adaptive immunity in TT and LL, as they can differentiate into M1 or M2 phenotypes.

Naïve macrophages that mature into M1 lead to the secretion of IFN-γ, TNF-α, IL-1, IL-6, and IL-12 inflammatory cytokines and trigger a cell-mediated immune (CMI) response [14]. This results in macrophage expression of induced nitric oxide synthase (iNOS), nitric oxide (NO) formation, and pathogen destruction, through damages caused by reactive oxygen species. In contrast, M2 profile generates IL-10, IL-13, TGF-β, and basic fibroblast growth factor (FGF-b), cytokines that downregulate immune response and promote tissue repair. Therefore, the preferential focus is the generation of M2 biomarkers in LL patients, rather than TT [15].

Although Th1 and Th2 role in leprosy is extensively described, this mechanism alone cannot explain the polarized immune system behavior [16]. Survival aspects of *M. leprae* are also related to other T $CD4^+$ lymphocyte subpopulations, such as Th17 [16–19] and regulatory T cells (Treg) [16,18]. Furthermore, the occurrence and extent of type 1 and 2 hypersensitivity reactions in leprosy may be linked to the proportion between Treg and Th17 cells [18].

Inflammation has been associated with Th17 cells, as IL-17 recruits neutrophils and is involved in tissue repair and synthesis stimulation of bactericidal molecules [20]. This may explain why greater levels of IL-17 have been

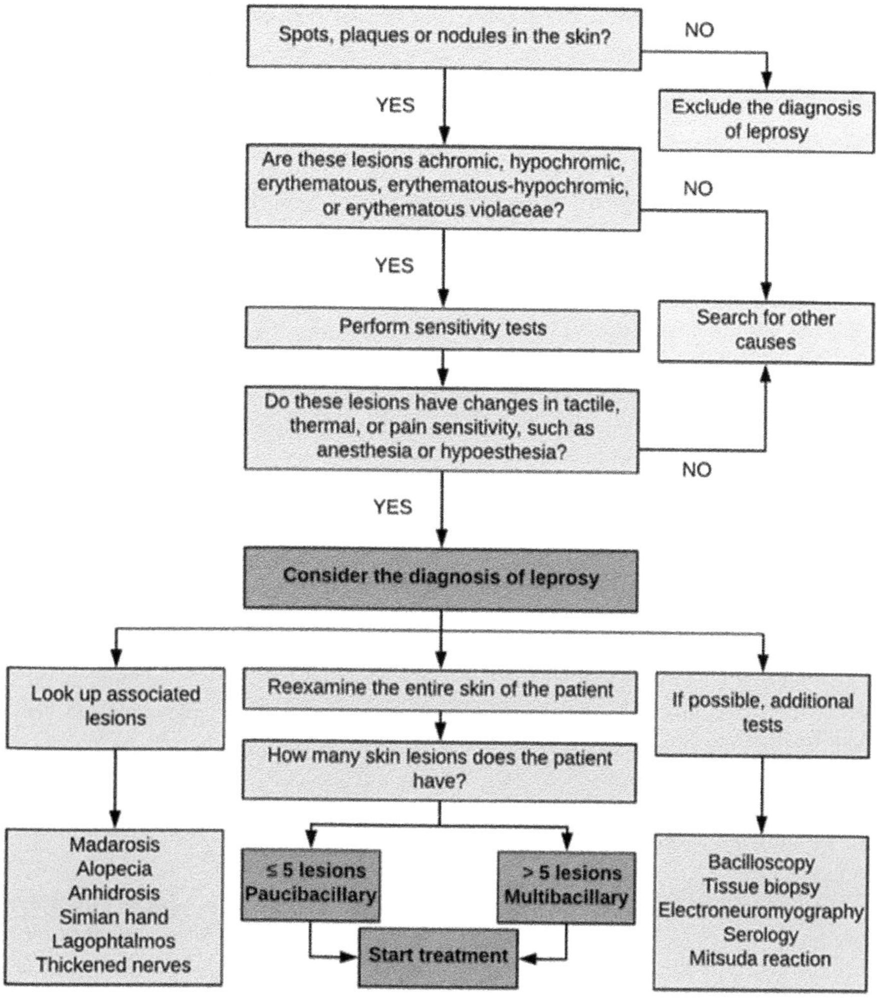

FIGURE 30.1 Approach to the patient with suspicion of leprosy: a step-by-step process for the clinical diagnosis. Flowchart shows schematic illustration of approach used in patients with suspected leprosy according to the WHO criteria [9].

identified in peripheral blood cells, and skin biopsies of PB than in MB patients [17,19]. However, neural tissue inflammation and reduced production of nerve growth factor (NGF), induced by IL-17, have been reported and correlated to severe nerve injury [21].

Contrastingly, Treg subset has been ascribed as inflammation and effector T $CD4^+$ suppressors. Transcription factor forkhead box 3 (FoxP3) is the common transcriptional regulator, responsible for managing Treg differentiation in natural regulatory T cells (nTreg), or in the two types of induced regulatory T cells (iTreg): TGF-β producing T cells (Th3) and IL-10 producing (Tr1) [22]. Increased levels of these cells in leprosy skin lesions have been observed in LL, demonstrating the role of Treg in Th1 unresponsiveness [23,24].

The complex immunology response against *M. leprae* makes it difficult to develop high sensitivity and specificity tests, capable of diagnosing all variations of the disease. This issue may be overcome by the adoption of biotechnological approaches aiming at discovering novel antigens or engineering them to increase antigenicity. The ability to precisely identify infected individuals, including those with low-titer antibody, could be improved with the employment of more sensitive serological analysis [25].

Development of engineered biomarkers for serological diagnosis of leprosy

Immune features and antigenicity of recombinant proteins and their engineered biomolecules have been addressed over the last decades. Despite all success in generating biological information, these biomarkers and the immunochemical techniques employed, were not effective enough for detecting asymptomatic *Mycobacterium* infection, or for diagnosing PB leprosy [26].

Natural, synthetic, and recombinant antigens are discussed, as well as features of serological methods, emphasizing more accurate and affordable biotechnological platforms.

PGL-I

The phenolic glycolipid I (PGL-I) was first identified during immunogenicity assays with extracts of *M. leprae* bacillus, from infected armadillo livers. This molecule was classified as a major component in the bacillus membrane, representing around 2% of mycobacteria dry weight [27].

During the process of host-pathogen interaction, PGL-I seems to be responsible for inhibiting the secretion of proinflammatory cytokines [28], *M. leprae* adhesion and invasion of Schwann cells [29], and for promoting nerve injuries by induction of iNOS in infected macrophages. Most of these pathogenic properties were associated with the saccharide region of PGL- I [30].

Chemical analysis demonstrated that PGL-I structure is composed of three carbohydrates known as 3-O-methyl-rhamnose, 2,3-di-O-methyl-rhamnose, and 3,6- di-O-methyl-glucose, all of them glycosidically linked to a phenyl group, along with a phthiocerol portion with mycocerosic acids [31]. Besides importance in the infection process, PGL-I saccharide fraction was considered restricted to *M. leprae*, and a highly immunogenic antigen, promptly recognized by immunoglobulin M (IgM) antibodies present in leprosy serum [32].

The expense of obtaining natural glycolipid directly from armadillo tissues, and its low solubility in aqueous solutions, created difficulties [33]. Engineering strategies were proposed to improve water-solubility, conserving PGL-I antigenicity. Many PGL-I analogs were designed, by coupling the synthesized di- or trisaccharide haptens to hydrophilic carriers, such as bovine serum albumin (BSA) or human serum albumin (HAS) [32].

The most utilized synthetic neoglycoconjugates differ mainly on the nature of the linker arm between the hapten and the carrier. PGL-I analogs, in which the saccharide antigenic core connected to BSA, through an 8-methoxycarbonyl octyl glycoside linker, is called ND-O-BSA (natural disaccharide octyl-BSA) [33,34] or NT-O-BSA (natural trisaccharide octyl-BSA) [35]. In the other class of trisaccharide conjugates, chemical modifications enabled p-(2- methoxycarbonylethyl) phenyl linkage, resulting in the PGL-I analog NT-P-BSA (natural trisaccharide propyl-BSA) [36].

As can be seen in Table 30.1, *M. leprae* PGL-I has presented elevated detection rates in MB patients. In Philippine population, a semi-synthetic analog of PGL-I, composed by the disaccharide conjugated to BSA (DBSA), was able to identify 74.1% and 79.6% of the MB IgM antibodies by using, as techniques, enzyme-linked immunosorbent assay (ELISA) and dipstick assay, respectively [37]. In Brazil, the specificities of three PGL-I conjugates were compared. By using ELISA, the reactivities were 86.36% and 76.36% for native PGL-I and ND-O-HAS respectively, while NT-P-BSA was able to detect 90% of patients, by using the lateral flow assay (ML Flow) [38].

Similar high diagnostic results were achieved in four endemic countries (Brazil, Ghana, Indonesia, and the Philippines) and one nonendemic (The Netherlands), with anti-PGL-I antibodies, resulting in 97.4% of MB detection in ML flow test. In contrast, a maximum of 40% of the PB was identified on the same research, corroborating failure of PGL-I in distinguishing PB patients from nonleprosy individuals [39].

A meta-analysis of eight trials concluded that the risk of healthy household contacts for developing leprosy is 3 times higher if they display anti-PGL-I positive response. However, even among contacts with anti-PGL-I positive titers, more than 50% never proceeded to clinical leprosy [40].

An analogous conclusion was reached after a study with 28,092 contacts of leprosy patients, from an endemic area in Bangladesh. After 6 years, only 239 of these individuals developed the disease. No significant difference was observed regarding anti-PGL-I titer between healthy and nonhealthy groups. In addition, out of 25 leprosy contacts, only 6 had significant reactivity against synthetic PGL-I, more precisely ND-O-HSA. Curiously, for the patients who developed leprosy, titers were even lower at diagnosis than at baseline. Hence, the relationship between anti-PLG-1 antibody levels and disease onset could not be established [41].

PGL-I seems to be a valuable tool for evaluating the efficiency of multidrug therapy (MDT). After initiating treatment, anti-PGL-I IgM titers were reduced to one half in 2 weeks, and almost undetectable levels after 6 months, in most previously seropositive individuals [42].

Although anti-PGL-I antibody titer correlates with bacterial load and can be a useful indicator for MB infection or treatment monitoring [43], this antigen, regardless of carrying protein or serological technique, has limited value in identifying PB patients. Consequently, the use of data obtained with PGL-I serology should be carefully considered and interpreted together with clinical evidence, and/or with analysis of indicators of cell-mediated immunity.

In this manner, IL-10, a regulatory cytokine that limits Th1 cell function and the effectiveness of immune response against *M. leprae*, IFN-γ-inducible protein 10 (IP-10), associated with *M. leprae* risk of infection and transmission, and chemokine ligand 4 (CCL4), increased in leprosy patients, were combined with anti-PGL-I IgM analysis, in a lateral flow assay with up-converting phosphor reporter (UCP-LFA), excited with infrared light. Even using whole blood as a sample, this field-friendly device was able to discern leprosy patients from contacts and healthy volunteers, and MB from PB patients [44].

TABLE 30.1 PGL-I-based assay performances for detection of antibodies in MB and PB leprosy patients from different regions of the world.

	Antigen	Serum reactivity		Method	Studied population
		MB	PB		
PGL-I	DBSA	74.1%	13.6%	ELISA [37]	Philippines
		79.6%	13.6%	Dipstick assay [37]	
	NT-P-BSA	97.4%	40%	ML flow [39]	Brazil
					Ghana
					Indonesia
					Philippines
	Native PGL-I	86.36%	22.73%	ELISA [38]	Brazil
	ND-O-HSA	76.36%	31.82%		
	NT-P-BSA	90%	0%	ML flow [40]	
	ND-O-BSA	73.1%	39.5%	ELISA [43]	China

MB, multibacillary; *PB*, paucinacillary; *PGL-1*, phenolic glycolipid.

Regarding the importance of PGL-I and its conjugates, such as ND-O-BSA or -HSA and ND-P-BSA in the leprosy serology, intensive efforts have been made aiming at improving antigenicity. Combination with recombinant proteins to react with immunoglobulin G (IgG) antibodies, glycolipid-protein conjugates, or association of antigens, can be followed in Table 30.1.

Recombinant *M. leprae* proteins

Major immunochemical methods employed to address antigen-antibody interaction, are designed for recombinant antigenic proteins (Table 30.2).

Comparing *M. leprae* and *M. tuberculosis* genome sequencing [45], hypothetical and putative proteins could be targeted. Meanwhile, advances obtained with recombinant protein-based assays have hardly been superior to results with anti-PGL-I antibodies, i.e., there was no great advantage in the diagnosis of low-titer antibody patients, such as PB leprosy ones.

Based on the similarity to early secreted antigenic target-6 (ESAT-6) from *Mycobacterium tuberculosis, M. leprae* ESAT-6, after recombinant expression, was employed as coating antigen in ELISA. This protein detected anti-ESAT-6 antibodies in 82.4% and 19.4% of sera from the bacilloscopic index (BI)-positive and negative patients, respectively. Besides antigenicity, the interest in *M. leprae* ESAT-6 stems from the fact of being a secreted protein, whose antibodies may be detected in the early phase of infection [46].

M. leprae serine-rich 45 kDa protein (ML0041) was assayed in India by ELISA. All MB and 47% of PB leprosy cases were identified, achieving high specificity as no tuberculosis patient or endemic healthy volunteers scored positive [47].

Due to high homology between conserved proteins from *M. tuberculosis* and *M. leprae*, and their immunogenicity in stimulating T and/or B cell responses, six recombinant proteins were assayed by ELISA and Western blotting: elongation factor EF-Tu (ML1877), the major membrane protein, MMP-I (ML0841) and II (ML2038), 10-kDa culture filtrate protein, CFP-10 (ML0050), chaperone protein GroES (ML0380), and antigen 85B, Ag85B (ML2028). Except for CPF-10 and GroES, all of them were reactive to MB and PB leprosy sera by Western blotting analysis, highlighting the persistent high antigenicity of MMP-II and Ag85B in both groups, although with high specificity (89%), Ag85B was antigenic to 90% and 65% of MB and PB sera, respectively [48].

A serum pool of leprosy patients was employed as probe for screening *M. leprae*-derived antigen expression [49]. Two these antigens, ML0405 and ML2331, were strongly recognized by MB leprosy sera, while reactivity to PB was absent, or as low as in negative controls. Interestingly, undetectable reactivity was achieved by household contacts to both ML0405 and ML2331, indicating these proteins may be effective in leprosy diagnosis even in endemic areas.

ML0405 and ML2331, recognized by leprosy sera from different world regions (Brazil, Philippines, and Japan), were fused in a single recombinant protein (leprosy IDIR diagnostic 1, LID-1). LID-1 was detected in Brazil as well

TABLE 30.2 Immunoassay performances using *M. leprae* recombinant proteins in the detection of antibodies in MB and PB leprosy patients.

Antigen	Strategy	Protein	Reactivity		Assay
			MB	PB	
ML0405 ML2331	*M. leprae* library and anti-patient sera as probe	ML0405	100%	Low or absent	ELISA [49]
		ML2331	79%		
45-kDa protein	Recombinant expression	ML0041	100%	47%	ELISA [47]
ESAT-6	Recombinant expression	ML0049	82%	19%	ELISA [46]
LID-1	Fusion protein	ML0405 ML2331	87%	32%	ELISA [50]
PADL	Mutiepitopic chimeric protein	ML0405 ML2331 ML2055 ML0411 ML0091	80%	7%	ELISA [53]
ML1877 ML0841 ML2038 ML0380 ML0050 ML2028	Recombinant expression	ML1877	100%	100%	Western Blotting [48]
		ML0841			
		ML2038			
		ML0380	60%	0%	
		ML0050	76%	25%	
		ML2028	100%	100%	
			90%	65%	ELISA
MMP-I MMP-II	Mix of recombinant proteins	ML0841 ML2038c	94.4%	39.7%	ELISA [54]
ND-O-LID[a]	NDO fusioned to LID-1	LID-1	78% –99%	45%–47%	Rapid ELISA [51]
ND-O-LID	NDO fusioned to LID-1	LID-1	87%	32%	LFIA [52]

MB, multibacillary; *PB*, paucinacillary.
[a]*Assayed with leprosy sera from Philippines and Colombia, respectively.*

as Japan MB. This antigen was not recognized by a significant number of PB subjects. As concerns household contacts, LID-1 was efficient enough in predicting leprosy six to 8 months before the onset of symptoms, i.e., anti-LID-1 antibodies had its level increased as soon as 1 year prior the clinical diagnosis [50].

Based on the principle that a single fusion complex of two antigenic molecules could amplify the antibody reactivity of leprosy patients, years later, the synthetic mimetic of PGL-I, (ND-O) was chemically conjugated to LID-1 protein. This composite resulted in a single fusion complex (ND-O-LID), able to detect MB patients with a high degree of sensitivity and specificity, in a fast ELISA protocol. The antigenic performance of each antigen was individually inferior to complex ND-O-LID [51].

Cell phone diagnosis

By using the same antigen, an affordable and user-friendly lateral flow immunochromatographic assay (LFIA) was designed, to detect anti-PGL-I IgM and anti-LID-1 IgG antibodies in undiluted serum. The rapid test, coupled to a cell-phone based reader, was objective and allowed direct correlation between seroreactivity and bacillary load, indicating its advantage for treatment monitoring. Although NDO-LID rapid test has demonstrated high specificity (97%) and ability to detect MB serum leprosy (87%), one-third of PB leprosy patients (32%) were positive as well [52].

A chimeric protein compound of epitopic regions of five recombinant proteins of *M. leprae*, PADL polyepitope (Protein Advances Diagnosis of Leprosy), provided accurate detection of antibodies from MB patients, even though most PB patients' sera were unreactive against this fusion protein by ELISA [53].

Subsequently, a blend of MPP-I and MPP-II from *M. leprae* was used as coating antigen, to determine anti-*M. leprae* antibodies by ELISA. Individual antigenicity of each protein was not as effective as the assayed mixture [54].

Serological markers are highly specific, yet lack of sensitivity is still a problem. All immunoassays employed with recombinant proteins were based on ELISA, Western Blotting, or LFIA (Table 30.2). It is likely that innovative and more sensitive label-free devices, such as immunosensors, might be an alternative to identifying *M. leprae* infection, regardless of the type of immune response [25].

Synthetic peptide-based serodiagnosis

The use of synthetic peptide as antigenic determinants, also named epitopes, for the generation immunochemical methods for detecting antibodies has markedly increased over the last decades as an alternative to the use of full-length proteins. An epitopic sequence can be selected according to reactivity to a specific antibody, reducing the possibility of cross-reactivity and, consequently, increasing specificity. Besides that, synthetic peptides may be easily engineered and produced at a low cost [55].

Classically, the sequences recognized by antibodies are structurally known as a linear or conformational epitope. A linear epitope is represented by the primary sequence of amino acids of the antigen, while the conformational one depends on the tertiary and quaternary structure of the protein [55].

Concerning leprosy B-cell epitopes, 63 overlapping synthetic peptides, covering the whole amino acid (AA) sequence of Ag85B, were tested against antibodies from patients by ELISA. Basically, three peptide regions were immunoreactive and had absorbance correlated to bacillary load. Although with high specificity, the 25-mer peptide corresponding to AA 256–280 was recognized by 88% of MB leprosy, while only 15% of PB presented positive serology [56].

Taking into account that the majority of functional antibodies recognize conformational antigenic determinants, specific antibodies from leprosy patients were purified on the basis of the affinity to membrane-blotted *M. leprae* whole antigens and used in the screening of phage-displayed random peptide libraries [57]. Three phage-displayed peptides (Table 30.3) were chemically synthetized and assayed by ELISA. Without being modified or carried with inert proteins, the synthetic peptide pool was able to detect antibodies in 30% of PB and 74% of MB leprosy patients. However, there was high reactivity to tuberculosis and healthy volunteers sera, compromising the assay specificity [58].

One of these peptides, sequence APDDPAWQNIFNLRR, after adding to the C-terminal region the spacer sequence serine-glycine (SGSG), and a C-terminal cysteine [25], was able to induce an immune response against Ag85B without reproducing, *ipsis litteris*, the native epitope sequence. It was characterized as an Ag85B mimotope [59], after epitope mapping using antisynthetic mimotope antibody (Table 30.3).

Antigenic properties of peptides considerably vary, depending on whether they are on the phage surface, as free molecules in solution, or conjugated to an inert carrier [55]. The peptide was, via C-terminal cysteine, conjugated to keyhole limpet hemocyanin (KLH) and assayed in leprosy patients sera, in order to assess the ELISA performance. Given the flexibility of the engineered peptide, by adding an SGSG spacer and more efficient exposure to antibodies after linking to KLH, the Ag85B mimotope sensitivity reached 91.7%, regardless of the immunological profile of 132 patients.

The strategy of using a quartz crystal microbalance (QCM) and an engineered molecule (APDDPAWQNIFNLRR-SGSG-C), coating a crystal sensor surface, was an alternative to increasing the sensitivity without hampering specificity, thus, distinguishing leprosy from tuberculosis and from healthy blood donors [25].

The QCM is a label-free device, in which alterations in the resonant frequency shift (Δf) of a quartz crystal, are proportional to the changes in mass or thickness of the chip surface, that occur due to interaction between two molecules [60], namely the synthetic mimotope and the patient's antibody [25].

Primary health care and the early leprosy diagnosis

Considering that leprosy is characterized by complex clinical manifestations that may appear after an average incubation period of 5 years, and the lack of experienced

TABLE 30.3 Ag85B mimotopes selected from reactivity to leprosy patients antibodies using Phage Display.

Phage display-derived peptides	Ag 85B mimicked sequence	Method for characterizing the mimotope
APD**DPAWQ**NIFNLRR	PPN**DPAWQ**RNDPILQ	Epitope mapping [25]
ICPRDPAWSYCN	–	Western blotting [58]
NNIHAHRYWGTDLNA	–	

health professionals, misdiagnosis can reach more than 40% of cases [61].

Screening programs using immunoassays based on the antigen-specific response could be applied in endemic areas to early diagnose of *M. leprae* infection, as the detection of the disease before the onset of symptoms, and the subsequent initiation of treatment, are crucial elements for avoiding disabilities and reducing mycobacteria transmission. Although the assessment by a qualified physician is essential, innovative laboratory assays that could confirm the suspicion, guide the treatment, classify the clinical form of leprosy, and predict reactional states would also be of great value [26].

While new technologies are being developed, primary healthcare services persist as an important pillar in leprosy detection, mainly in small and medium-sized towns, and rural regions [62]. Although *M. leprae* infection is rare in developed countries, with an increase of population mobility and human migration worldwide, cases of leprosy are likely to appear [63]. Systematic discussion must include undergraduate students in order to arouse interest in neglected diseases such as leprosy [64].

Acknowledgments

The authors gratefully thank the following institutions for their support: Coordenação de Aperfeiçoamento de Pessoal de Nível Superior, CAPES (BAS), Universidade Federal do Paraná – Tesouro Nacional, UFPR-TN (JPSS and BMWB) and Conselho Nacional de Desenvolvimento Científico e Tecnológico, CNPq (JM).

References

[1] Lázaro FP, Werneck RI, Mackert CCO, et al. A major gene controls leprosy susceptibility in a hyperendemic isolated population from North of Brazil. J. Infect. Dis. 2010;201:1598–605.

[2] White C, Franco-Paredes C. Leprosy in the 21st century. Clin. Microbiol. Rev. 2015;28:80–94.

[3] Reibel F, Cambau E, Aubry A. Update on the epidemiology, diagnosis, and treatment of leprosy. Med. Maladies Infect. 2015;45:383–93.

[4] Lastória JC, Abreu MA. Leprosy: review of epidemiological, clinical, and etiopathogenic aspects - part1. An. Bras. Dermatol. 2014;89:205–18.

[5] Cambri G, Mira MT. Genetic susceptibility to leprosy — from classic immune-related candidate genes to hypothesis- free , whole genome approaches. Front. Immunol. 2018;9:1–9.

[6] World Health Organization. Global leprosy update, 2017: reducing the disease burden due to leprosy. WER, vol. 93; 2018. p. 445–56.

[7] World Health Organization Regional Office for South-East Asia. Global leprosy strategy 2016–2020, vol. 1. WHO; 2016. p. 1–20.

[8] Britton WJ, Lockwood DNJ. Leprosy. Lancet 2004;363:1209–19.

[9] World Health Organization. Guide to eliminate leprosy as a public health problem, vol. 14. Geneva: Guidel WHO; 2000. p. 5–38.

[10] Schuenemann VJ, Singh P, Mendum TA, et al. Genome-wide comparison of medieval and modern *Mycobacterium leprae*. Science 2013;341:179–83.

[11] Gaschignard J, Grant AV, Van Thuc N, et al. Pauci- and multibacillary leprosy: two distinct, genetically neglected diseases. PLoS Neglected Trop. Dis. 2016;10:e0004345.

[12] Fonseca AB de L, do Vale Simon M, Cazzaniga RA, et al. The influence of innate and adaptative immune responses on the differential clinical outcomes of leprosy. Infect. Dis. Poverty 2017;6:5.

[13] Yamamura M, Uyemura K, Deans R, et al. Defining protective responses to pathogens: cytokine profiles in leprosy lesions. Science 1991;254:277–9.

[14] Yang D, Shui T, Miranda JW, et al. *Mycobacterium leprae*-infected macrophages preferentially primed regulatory T cell responses and was associated with lepromatous leprosy. PLoS Neglected Trop. Dis. 2016;10:e0004335.

[15] de Sousa JR, Sotto MN, Simões Quaresma JA. Leprosy as a complex infection: breakdown of the Th1 and Th2 immune paradigm in the immunopathogenesis of the disease. Front. Immunol. 2017;8:18–21.

[16] Sadhu S, Khaitan BK, Joshi B, Sengupta U, Nautiyal AK, Mitra DK. Reciprocity between regulatory T cells and Th17 cells: relevance to polarized immunity in leprosy. PLoS Neglected Trop. Dis. 2016;10:e0004338.

[17] Santos MB, de Oliveira DT, Cazzaniga RA, et al. Distinct roles of Th17 and Th1 cells in inflammatory responses associated with the presentation of paucibacillary leprosy and leprosy reactions. Scand. J. Immunol. 2017;86:40–9.

[18] Saini C, Siddiqui A, Ramesh V, Nath I. Leprosy reactions show increased Th17 cell activity and reduced FOXP3+ Tregs with concomitant decrease in TGF-β and increase in IL-6. PLoS Neglected Trop. Dis. 2016;10:e0004592.

[19] de Almeida-Neto FB, Assis Costa VM, Oliveira-Filho AF, et al. Th 17 cells, interleukin-17 and interferon- γ in patients and households contacts of leprosy with multibacillary and paucibacillary forms before and after the start of chemotherapy treatment. J. Eur. Acad. Dermatol. Venereol. 2015;29:1354–61.

[20] Sadhu S, Mitra DK. Emerging concepts of adaptive immunity in leprosy. Front. Immunol. 2018;9:1–7.

[21] Aarão TL de S, de Sousa JR, Botelho BS, Fuzii HT, Quaresma JAS. Correlation between nerve growth factor and tissue expression of IL-17 in leprosy. Microb. Pathog. 2016;90:64–8.

[22] Saini C, Tarique M, Kumar S, Rao D. Role of FoxP3+ Treg cells mediating immune suppression in leprosy. Curr. Immunol. Rev. 2015;11:66–72.

[23] Bobosha K, Wilson L, van Meijgaarden KE, et al. T-cell regulation in lepromatous leprosy. PLoS Neglected Trop. Dis. 2014;8:e2773.

[24] Saini C, Ramesh V, Nath I. Increase in TGF-β secreting CD4+CD25+ FOXP3+ T regulatory cells in anergic lepromatous leprosy patients. PLoS Neglected Trop. Dis. 2014;8:e2639.

[25] de Santana JF, da Silva MRB, Picheth GF, et al. Engineered biomarkers for leprosy diagnosis using labeled and label-free analysis. Talanta 2018;187:165–71.

[26] de Souza VNB, Iyer AM, Lammas DA, Naafs B, Das PK. Advances in leprosy immunology and the field application: a gap to bridge. Clin. Dermatol. 2016;34:82–95.

[27] Hunter SW, Brennan PJ. A novel phenolic glycolipid from *Mycobacterium leprae* possibly involved in immunogenicity and pathogenicity. J. Bacteriol. 1981;147:728–35.

[28] Arbues A, Lugo-Villarino G, Neyrolles O, Guilhot C, Astarie-Dequeker C. Playing hide-and-seek with host macrophages through the use of mycobacterial cell envelope phthiocerol dimycocerosates and phenolic glycolipids. Front. Cell. Infect. Microbiol. 2014;4:1–7.

[29] Díaz Acosta CC, Dias AA, Rosa TLSA, et al. PGL I expression in live bacteria allows activation of a CD206/PPARγ cross-talk that may contribute to successful *Mycobacterium leprae* colonization of peripheral nerves. PLoS Pathog. 2018;14:e1007151.

[30] Madigan CA, Cambier CJ, Kelly-Scumpia KM, et al. A macrophage response to *Mycobacterium leprae* phenolic glycolipid initiates nerve damage in leprosy. Cell 2017;170. 973–985.e10.

[31] Hunter SW, Fujiwara T, Brennan PJ. Structure and antigenicity of the major specific glycolipid antigen of *Mycobacterium leprae*. J. Biol. Chem. 1982;257:15072–8.

[32] Spencer JS, Brennan PJ. The role of *Mycobacterium leprae* phenolic glycolipid I (PGL-I) in serodiagnosis and in the pathogenesis of leprosy. Lepr. Rev. 2011;82:344–57.

[33] Chatterjee D, Cho S-N, Brennan PJ, Aspinall GO. Chemical synthesis and seroreactivity of O-(3,6-di-O-methyl-β-d-glucopyranosyl)-(1→4)-O-(2,3-di-O-methyl-α-lrhamnopyranosyl)-(1→9)-oxynonanoyl-bovine serum albumin—the leprosy-specific, natural disaccharide-octyl-neoglycoprotein. Carbohydr. Res. 1986;156:39–56.

[34] Zhang J, Chatterjee D, Brennan PJ, Spencer JS, Liav A. A modified synthesis and serological evaluation of neoglycoproteins containing the natural disaccharide of PGL-I from *Mycobacterium leprae*. Bioorg. Med. Chem. Lett. 2010;20:3250–3.

[35] Chatterjee D, Cho S-N, Stewart C, Douglas JT, Fujiwara T, Brennan PJ. Synthesis and immunoreactivity of neoglycoproteins containing the trisaccharide unit of phenolic glycolipid I of *Mycobacterium leprae*. Carbohydr. Res. 1988;183:241–60.

[36] Fujiwara T, Izumi S. Synthesis of the neoglycoconjugates of phenolic glycolipid-related trisaccharides for the serodiagnosis of leprosy. Agric. Biol. Chem. 1987;51:2539–47.

[37] Bührer SS, Gussenhoven GC, Klatser PR, Smits HL. A simple dipstick assay for the detection of antibodies to phenolic glycolipid-I of *Mycobacterium leprae*. Am. J. Trop. Med. Hyg. 1998;58:133–6.

[38] Lobato J, Costa MP, Reis EDM, et al. Comparison of three immunological tests for leprosy diagnosis and detection of subclinical infection. Lepr. Rev. 2011;82:389–401.

[39] Buhrer-Sekula S, Smits HL, Gussenhoven GC, et al. Simple and fast lateral flow test for classification of leprosy patients and identification of contacts with high risk of developing leprosy. J. Clin. Microbiol. 2003;41:1991–5.

[40] Penna MLF, Penna GO, Iglesias PC, Natal S, Rodrigues LC. Anti-PGL-1 positivity as a risk marker for the development of leprosy among contacts of leprosy cases: systematic review and meta-analysis. PLoS Neglected Trop. Dis. 2016;10:e0004703.

[41] Richardus RA, van der Zwet K, van Hooij A, et al. Longitudinal assessment of anti-PGL-I serology in contacts of leprosy patients in Bangladesh. PLoS Neglected Trop. Dis. 2017;11:e0006083.

[42] Cho SN, Cellona RV, Villahermosa LG, et al. Detection of phenolic glycolipid I of *Mycobacterium leprae* in sera from leprosy patients before and after start of multidrug therapy. Clin. Diagn. Lab. Immunol. 2001;8:138–42.

[43] Le W, Haiqin J, Danfeng H, et al. Monitoring and detection of leprosy patients in Southwest China: a retrospective study, 2010–2014. Sci. Rep. 2018;8:11407.

[44] van Hooij A, Tjon Kon Fat EM, Richardus R, et al. Quantitative lateral flow strip assays as user-friendly tools to detect biomarker profiles for leprosy. Sci. Rep. 2016;6:34260.

[45] Cole ST, Eiglmeier K, Parkhill J, et al. Massive gene decay in the leprosy bacillus. Nature 2001;409:1007.

[46] Parkash O, Pandey R, Kumar A, Kumar A. Performance of recombinant ESAT-6 antigen (ML0049) for detection of leprosy patients. Lett. Appl. Microbiol. 2007;44:524–30.

[47] Parkash O, Kumar A, Nigam A, Franken KLMC, Ottenhoff THM. Evaluation of recombinant serine-rich 45-kDa antigen (ML0411) for detection of antibodies in leprosy patients. Scand. J. Immunol. 2006;64:450–5.

[48] Spencer JS, Kim HJ, Wheat WH, et al. Analysis of antibody responses to *Mycobacterium leprae* phenolic glycolipid I, lipoarabinomannan, and recombinant proteins to define disease subtype-specific antigenic profiles in leprosy. Clin. Vaccine Immunol. 2011;18:260–7.

[49] Reece ST, Ireton G, Mohamath R, et al. ML0405 and ML2331 are antigens of *Mycobacterium leprae* with potential for diagnosis of leprosy. Clin. Vaccine Immunol. 2006;13:333–40.

[50] Duthie MS, Goto W, Ireton GC, et al. Use of protein antigens for early serological diagnosis of leprosy. Clin. Vaccine Immunol. 2007;14:1400–8.

[51] Duthie MS, Raychaudhuri R, Tutterow YL, et al. A rapid ELISA for the diagnosis of MB leprosy based on complementary detection of antibodies against a novel protein-glycolipid conjugate. Diagn. Microbiol. Infect. Dis. 2014;79:233–9.

[52] Duthie MS, Balagon MF, Maghanoy A, et al. Rapid quantitative serological test for detection of infection with *Mycobacterium leprae*, the causative agent of leprosy. J. Clin. Microbiol. 2014;52:613–9.

[53] Duthie MS, Hay MN, Morales CZ, et al. Rational design and evaluation of a multiepitope chimeric fusion protein with the potential for leprosy diagnosis. Clin. Vaccine Immunol. 2010;17:298–303.

[54] Tsukamoto Y, Maeda Y, Makino M. Evaluation of major membrane protein-I as a serodiagnostic tool of pauci-bacillary leprosy. Diagn. Microbiol. Infect. Dis. 2014;80:62–5.

[55] Van Regenmortel MHV, Muller S. Synthetic peptides as antigens. In: Van Regenmortel MHV, Muller S, editors. Laboratory techniques in biochemistry and molecular biology. 1st ed. Elsevier Science; 1999. p. 380.

[56] Filley E, Thole JER, Rook GAW, et al. Identification of an antigenic domain on *Mycobacterium leprae* protein antigen 85b, which is specifically recognized by antibodies from patients with leprosy. J. Infect. Dis. 1994;169:162–9.

[57] Smith GP. Surface presentation of protein epitopes using bacteriophage expression systems. Curr. Opin. Biotechnol. 1991;2:668–73.

[58] Alban SM, de Moura JF, Thomaz-Soccol V, et al. Phage display and synthetic peptides as promising biotechnological tools for the serological diagnosis of leprosy. PLoS One 2014;9:e106222.

[59] Geysen HM, Rodda SJ, Mason TJ. A priori delineation of a peptide which mimics a discontinuous antigenic determinant. Mol. Immunol. 1986;23:709–15.

[60] Kößlinger C, Drost S, Aberl F, Wolf H, Koch S, Woias P. A quartz crystal biosensor for measurement in liquids. Biosens. Bioelectron. 1992;7:397–404.

[61] Henry M, GalAn N, Teasdale K, et al. Factors contributing to the delay in diagnosis and continued transmission of leprosy in Brazil – an explorative, quantitative, questionnaire based study. PLoS Neglected Trop. Dis. 2016;10:e0004542.

[62] Lucia Fernandes Penna M, Leide Oliveira MW, Hage Carmo E, Oliveira Penna G, Gomes Temporão J. The influence of increased access to basic healthcare on the trends in Hansen's disease detection ratein Brazil from 1980 to 2006. Rev. Soc. Bras. Med. Trop. 2008;41:6–10.

[63] Ebenso J. An overview of training and development needs. Lepr. Rev. 2012;83:2.

[64] Caitano D. Leprosy : educate to control. Hansenol. Int. 2009;34:7–8.

Chapter 31

Decision support systems in breast cancer

Alexandros Vamvakas[1], Katerina Vassiou[2], Dimitra Tsivaka[1,3] and Ioannis Tsougos[1,3]

[1]*Medical Physics Department, Medical School, University of Thessaly, Larisa, Greece;* [2]*Radiology and Anatomy Department, Medical School, University of Thessaly, Larisa, Greece;* [3]*Neuroimaging Department, IoPPN, King's College London, London, United Kingdom*

Introduction

Breast cancer is the second most common diagnosed cancer in both sexes combined, while it is the leading cause of female cancer deaths globally [1]. Breast cancer is a heterogeneous disease, varying with respect to etiology, prognosis, and response to therapy [2]. Cancer is commonly described as a disease of the genes, while the genetic and epigenetic properties of individual cancer cells are highly variable. In general, malignant tumors are biologically complex and exhibit substantial spatial variation in gene expression, biochemistry, histopathology, and macroscopic structure [3].

Indeed, cancerous cells not only undergo clonal evolution from a single progenitor cell, into more aggressive and therapy-resistant cells, but also exhibit branched evolution; thus, every tumor develops and preserves multiple distinct subclonal populations. This genetic heterogeneity, which is also constantly prone to various environmental stressors, leads to temporal and regional regarding stromal architecture, oxygen consumption, glucose metabolism, protein, and growth factor expression. As a result, tumor subregions develop, each with spatially distinct patterns of blood flow, vessel permeability, cell proliferation, cell death, and other features [3].

The intertumor heterogeneity, along with intratumor heterogeneity [3], can lead to the failure of targeted therapies, even with validated targets and drugs, as resistant clones survive and proliferate [4].

Histological and molecular classification of breast lesions

Based on histopathologic characteristics and gene expression profiles, breast tumors can be classified into multiple categories. Prior studies have demonstrated that tumors with distinct pathologic characteristics show unique clinical behavior and should be treated in a personalized therapeutic framework [5]. Standard breast cancer treatment is based on the tumor characteristics, including clinical stage, histopathologic features, and biomarker profile, and is affected by the patient's age, health, and menopausal status. These clinicopathologic variables have a profound impact on survival and account for most of the differences in clinical outcome among patients with breast cancer [6]. Histopathological analysis plays a central role in the diagnosis of breast cancer. The assessment of histological type and the histological grade, guide clinical management [7].

The TNM staging system (American Joint Committee on Cancer (AJCC) / Union for International Cancer Control (UICC)) incorporates tumor size, regional lymph node status, and metastases [6]. Malignant breast lesions are characterized as invasive or noninvasive, depending on tumor corresponding growth pattern. Generally, invasive lesions are more aggressive and spread outside the breast, whereas noninvasive breast lesions do not exceed breast tissue boundaries, and they are also known as carcinomas in situ. Also, malignant breast lesions are divided into ductal or lobular, considering the primary anatomical region of origin, i.e., ducts or lobules. The four major malignant breast categories are Invasive Ductal Carcinomas (IDCs), Invasive Lobular Carcinomas (ILCs), Ductal Carcinomas in situ (DCIS), and Lobular Carcinomas in situ (LCIS). Invasive Ductal Carcinomas (IDCs) are the most common type, with many thousand new cases diagnosed every year [8].

Breast carcinoma grade differentiation between low, intermediate, and high, constitutes a robust prognostic factor, also highlighting the tumor heterogeneity, and is incorporated in clinical decision-making tools. The breast

Precision Medicine for Investigators, Practitioners and Providers. https://doi.org/10.1016/B978-0-12-819178-1.00031-9

tumor grading system is based on the evaluation of three morphologic parameters, which are the percentage of the tumor arranged in glands and tubular structures, the degree of nuclear pleomorphism, and the mitotic rate.

Molecular biomarkers

Expression of established prognostic and predictive biomarkers, such as hormone receptors, and human epidermal growth factor receptor 2 (HER2) oncoprotein, is the basis for breast cancer molecular classification. Molecular subtypes are indicators of genetic tumor heterogeneity, which is probed with multigene assays and can aid targeted treatment by improving the stratification into low- and high-risk groups [6].

The presence immunohistochemical surrogate biomarkers of the estrogen receptor (ER), progesterone receptor (PR), and HER2 are important for responses to specific treatments (e.g., tamoxifen for patients with ER-positive disease), and prognosis (poorer outcomes for those with ER-negative disease) and may define etiologic subtypes. On the other hand, the triple-negative breast cancers (lack of expression of ER, PR, HER2), usually do not present the typical signs of malignancy on standard mammography, are more likely to be detected as interval and high-grade tumors, and have a poor 5-year survival rate [2].

Molecular classification

The commonly used molecular classification system categorizes invasive breast carcinomas into four distinct molecular subtypes, namely luminal A, luminal B, human epidermal growth factor receptor 2 (HER2)-overexpressing, and basal-like, which provide a clinically useful basis for therapeutic selection. The various molecular subtypes have distinct recurrence and survival rates. In particular, the luminal A subtype shows the best survival, while the worst is observed for the HER2 and basal-like subtypes. Patients with the luminal B subtype exhibit higher proliferation, and have a poorer prognosis than those with the luminal A subtype. The basal-like and HER2 subtypes are more sensitive to preoperative chemotherapy, and are more likely to undergo a pathologic complete response, but have a worse prognosis than luminal and normal-like cancers. Finally, the basal-like molecular subtype is significantly associated with nodal involvement, which is important for the prognosis of early-stage breast cancer [9].

Immunohistochemical surrogate biomarkers of estrogen receptor (ER), progesterone receptor (PR), and HER2 status have been used to define molecular subtypes. Therefore, demand exists for alternative means of classifying breast cancers into distinct molecular subtypes [9].

Inter and intratumor heterogeneity occur at the morphologic, genomic, transcriptomic, and proteomic levels, creating diagnostic and therapeutic challenges [6]. Approaches to medical care are based on the large-scale data synthesis from different sources, such as molecular biology tools including genomics (DNA), proteomics (proteins), metabolomics (metabolites) and transcriptomics (RNA), as well as other factors (heredity, lifestyle). The aim is the characterizing the biological heterogeneity of the lesions, by identifying the molecular phenotypes of the mutations of the disease [8].

Risks of small tissue samples

Biopsy methods do not accurately provide a complete characterization of the entire tumor, leading 30% of the cases to either under or overtreatment [10]. Mapping of the biological variations within the whole tumor site is desirable [11]. Standard histopathologic biomarkers may be an undersampling of the heterogeneous tumor [12]. Whole-mount breast biopsy methods have been adapted in the clinical practice; however, there is still a need for utilizing large size tissue samples for technique validation, as well as, optimizing cost-effectiveness for assessing the histopathology in the future [13].

Clinical imaging in breast cancer

Imaging can provide a more comprehensive view of the inter and intratumor heterogeneity. The key advantage of the several imaging methods noninvasiveness is that they can be applied repeatedly during treatment, providing if feasible spatial and temporal information, which facilitates the monitoring of disease progression or its response to therapy. Survival rates have improved tremendously since the 1980s, likely because of the significant uptake of mammographic screening, along with improvements in breast cancer treatment [2].

Imaging variables

Traditional radiographic imaging relies upon largely qualitative features, such as tumor density, the pattern of enhancement, intratumoral cellular composition (including the presence of blood, necrosis, and mineralization), the regularity of tumor margins, and anatomic relationship to the surrounding tissues [2].

Morphologic and structural differences between tumors are sometimes apparent, and classical subjective descriptors of these characteristics (e.g., spiculated margin) are common. Also, the presence of morphological features typical of benign, borderline, or malignant tumors, increased the level of diagnostic confidence of radiologists. BI-RADS (Breast Imaging – Reporting and Data System) has been established to standardize the interinstitutional imaging reporting and data collection. The BI-RADS lexicon includes different morphological features (margin, shape, size, internal pattern, etc.), and provides an estimated risk of malignancy. So, well-defined tumor margins are considered an indicator of less infiltrative behavior and, thus, lower aggressiveness. Besides, a tumor morphologic phenotype may be associated with a pathologic entity [14].

Tumor size and volume are usually important features in tumor evaluation and may have prognostic value. However, different imaging techniques may improve the assessment of tumor structure, by means of advanced tomographic structural and/or functional information [14].

Tumor screening

Several advanced imaging modalities for breast cancer screening prior to treatment are used except for mammography, including ultrasonography (US), digital breast tomosynthesis (DBT), and magnetic resonance imaging (MRI). Particularly, US and DBT are complementarily utilized for early detection and initial evaluation purposes, presenting a decrease in the frequency of false positive findings [15].

On the other hand, MRI holds a leading role in screening a particular group of high-risk women (e.g., suspicious BI-RADS 3 and 4 lesions), due to its advantages regarding spatial resolution, generation of high contrast images for soft tissue, as well as the lack of ionizing radiation., MRI has now evolved from simple imaging modality to an advanced computational system producing a variety of numerical parameters [16]. Advanced imaging analysis aims at the unraveling of tumor heterogeneity, guiding cancer diagnosis, staging, and planning interventions for treating patients, monitoring therapeutic approaches, predicting treatment response, and determining outcomes [17].

Cellular and functional information

The parameters extracted from the various advanced MR imaging techniques such as diffusion-weighted/tensor imaging (DWI/DTI), perfusion-weighted imaging (PWI), MR spectroscopy (MRS) and MR elastography (MRE), provide significant structural and functional information in a microscopic and cellular level, highlighting aspects of the underlying breast pathophysiology, regarding cellularity, neovascularization, and tumor biochemical processes. Usually, the most critical elements in the determination of tumor grade and prognosis are tumor cellularity and vascularity. These elements can be quantified using diffusion and perfusion techniques, respectively.

Studies have reported the value of DWI for differentiation between benign and malignant breast tumors, by means of apparent diffusion coefficient (ADC) parametric maps [18–20]. In addition, the correlation of ADC values is conducted with several prognostic factors and biological markers, including ER, PR, HER2, and Ki-67 statuses [21–23].

Perfusion-weighted MR imaging also assesses tumor angiogenesis, vascularity, and vessel permeability, mainly utilizing dynamic contrast-enhanced (DCE) imaging techniques. The high gadolinium uptake by tumors helps the clinicians in the accurate differentiation of breast lesions, compared to normal tissue. Furthermore, DCE-MRI signal time-series evaluation through empirical or pharmacokinetic models has presented robust results for tumor characterization [24,25], as well as monitoring response to therapy parameters [26,27].

Benign and malignant breast lesions differ in the characteristics of their microvessels, and hence in the behavior of gadolinium uptake in the lesion, which can be measured with the pharmacokinetic parameters of vascular permeability, such as the transfer constant K^{trans}. K^{trans} measures the transit of contrast agent through the vascular bed at the capillary level and reflects qualitative changes of tumor vessels (i.e., increased porosity/permeability), a surrogate of neoangiogenesis. In addition to K^{trans}, several tissue-specific kinetic parameters may be estimated including: the volume fraction of the extravascular extracellular space (v_e) in tissue, the volume fraction of plasma in tissue (v_p), and the rate constant for efflux of gadolinium contrast back into plasma, from the tissue extracellular space (k_{ep}) [9,24].

Proton nuclear magnetic resonance

In vivo 1H-MRS is a noninvasive method for characterizing the cellular biochemistry which underlies breast pathologies, mainly by monitoring the total choline concentration. As choline complexes are believed to be precursors of the phospholipids that compose cell membranes, increases in choline signals are thought to reflect increased membrane synthesis. In addition, other relevant metabolites, including multiple lipids, can be detected and monitored. Breast 1H-MRS is proposed to be used as an adjunct tool to morphologic and dynamic MRI examination, for the improvement of specificity, obviating unnecessary biopsies. It can be considered as a bridge between metabolism and the anatomic and physiological studies available from MRI [28–32].

Multiparametric evaluation

Tumor biological processes are closely correlated, their accurate interpretation is not always straightforward, and becomes difficult on the basis of individual numeric parameters, as similarities may exist between pathologies, and one should be very careful in correctly evaluating all available MR data [33]. Recently, multiparametric approaches are proposed for improving the diagnostic accuracy through the correlation of the results between different MR imaging techniques [34–36]. However, most clinical radiology and research studies only measure tumor size or average parameter values, such as mean ADC and blood flow. In doing so, spatially rich information is discarded [3].

Computer-assisted-diagnosis, radiomics, and decision support

Quantitative imaging analysis follows the significant advancements within the field of medical image processing and pattern recognition [37]. Medical imaging quantification processes are based on two fundamental principles [17,38,39]

(a) Tumor biological processes presented in histological and molecular levels are reflected in the macroscopic imaging phenotypic characteristics, and

(b) Images contain much more “hidden” information than what is usually captured by the human eye

Computer-Aided-Diagnosis (CAD) systems in breast cancer imaging, have been under development for decades, especially applied in breast cancer detection (CADe) and diagnosis (CADx) [39]. Challenging clinical questions regarding lesion delineation from breast’s diffusive parenchyma (e.g., detection of masses, microcalcifications, and architectural distortions), and breast lesion biological characterization (e.g., benign vs. malignant, tumor grading, suspicious BI-RADS lesions), can thus, be addressed.

Standard computer and machine-learning algorithms

CAD algorithms are mainly composed of two stages, i.e., detection and classification of suspicious regions, into cancerous and normal tissue. Initially, detection is performed using basic image enhancement methods, descriptors of the statistical distribution of intensity values, and decomposition of the image through wavelet transforms, in order to investigate differences between tumorous areas and background. Subsequently, based on an initial hypothesis concerning possible connections between cancer progression and specific imaging data, CAD systems use a set of quantitative imaging features describing the geometrical structure, intensity distribution and texture of a Region of Interest (ROI), automatically or manually contoured.

In addition, the above-mentioned systems may include statistical or machine learning classifiers in order to identify imaging patterns able to support breast tumor differential diagnosis [40]. The vast amounts of data and numerical parameters produced, may pose more of a problem rather than a solution, because the endogenous complexity of the sophisticated imaging methods and big data sets, as well as the nature of the imaging features, trouble radiologists in collecting and rationalizing the abundance of these quantitative metrics, as well as accurately evaluating them during the clinical routine.

The exponential growth in computer technology has led to the development of novel methodologies, with which innumerable quantitative features are extracted from digital medical images, and evaluated through high-throughput computing [37]. Termed *Radiomics*, it has been very recently integrated into clinical practice for breast diagnosis and treatment management.

The concept of radiomics

Etymologically, radiomics comes from the union of the terms radio- (radiation or radiology), and -omics, a common suffix used to form nouns relating to the study of the totality of a field (e.g., genomics, proteomics, metabolomics, transcriptomics). Thus, radiomics is the field that studies numerous features, computationally extracted from radiological images, in order to discover hidden features with increased diagnostic impact.

Obviously, there exist many similarities between CAD and radiomics. However, CAD procedures are based on a predefined hypothesis about the specific clinical tasks, presenting limitations in integrating data from different sources that hold important diagnostic value. In contrast to CAD’s simplicity and the ability for answering only elementary clinical questions, radiomics analysis utilize a plethora of quantitative imaging features – potential imaging biomarkers, extracted from digital images [39,41]. It does not require any a priori hypothesis or knowledge, which indeed is not feasible when one investigates myriads of new information, thus exploiting more complex computational data-driven processes for aiding decision support [42].

Radiomics signatures in breast cancer

Radiomics applications in breast imaging include assessing molecular subtypes, prognosis, and therapeutic responses, by yielding predictive image-based phenotypes of breast cancer. Relationship between the macroscopic appearance of the tumor and its environment are also searched. These relationships can be extracted from clinical breast imaging features relating to tumor size, shape, intensity, and texture, providing a comprehensive tumor characterization, when combined with the rest biological indicators of risk [2,37]. Libraries of quantitative individual voxel-based variables are more sensitively associated with clinical endpoints, compared with the more qualitative radiologic, histopathologic, and clinical data [43]. In addition, the identification and standardization of large-scale radiological phenotypic

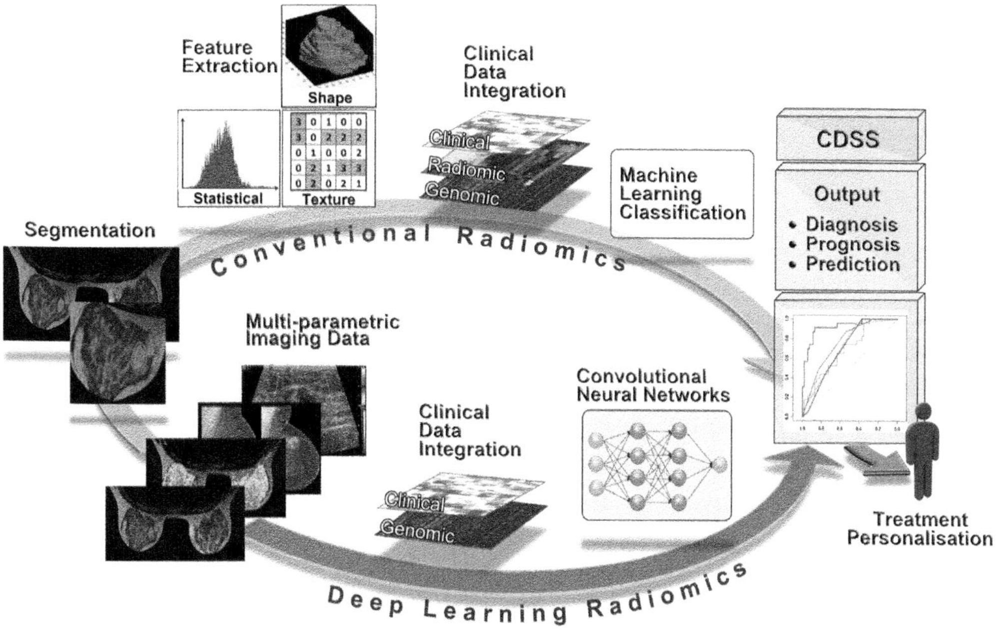

FIGURE 31.1 The workflow of radiomics analysis for aiding clinical decision making. Both the conventional pipeline based on the analysis of engineered imaging features, as well as, a deep learning pathway, providing further automation, are depicted.

characteristics, that provide a strong association with specific tumor heterogeneity profiles, leads to the definition of what it has been called *the radiomics signature* of breast cancer [16,39,40,44].

Radiomics analysis can face complex technical difficulties (medical data collection, development of novel computational methods), and methodological challenges (poor study design, data overfitting, lack of standards for results validating) [38,45]. The basic steps and methodological aspects of a radiomics analysis pipeline are depicted in Fig. 31.1 and analyzed below (Fig. 31.1).

Tumor delineation and habitat imaging

Image segmentation is usually the first step, after data preprocessing (noise reduction, correction of artifacts, normalization). The precise definition of breast lesion boundaries is a very important procedure, as it affects the subsequent qualitative analysis of the radiomics descriptors, extracted from the corresponding regions or volumes of interest (ROI/VOI). In a clinical routine, ROIs are manually segmented by expert radiologists, but besides its time-consuming nature, this approach induces intra/interobserver variability, and reproducibility errors, as many tumors present indistinct and blurring boundaries [46].

Segmentation algorithms

The development and validation of novel semi-automated or automated segmentation algorithms is an open research field, which presents interesting and sophisticated results. Yet, fully automated methods are feasible only when there are strong signal differences between the lesion and the background [47]. In addition, time-cost minimization for segmenting all tumor slices in tomographic imaging modalities, such as MRI, enables the reconstruction of three-dimensional (3D) tumor models, which further facilitate the global assessment of the pathology.

The initial approaches toward automated breast tumor delineation methods for CAD systems, included intensity-based methods, utilizing histogram thresholding for edge enhancement in mammographic data [48]. However, the main disadvantage of the thresholding methods is spatial incoherence (scattering) presented in segmented regions, as this method does not take into account pixel neighborhood information. Region growing methods are an evolution, where segment coherence is obtained from the application of conditions by the user, such as homogeneity criteria between neighboring voxels, and mostly the inclusion of manually induced seed pixels in the final segment [49].

Classification and clustering methods have been developed, for classifying image pixels into different groups of similar intensities, and thus, properties, utilizing sophisticated algorithms such as k-NN, k-means, and fuzzy c-means [50–53]. Furthermore, when this kind of segmentation method is performed on fused MR images (T1, T2, DWI, DCE), it enables tumor separation into subregions, also termed *habitats*, which contributes in the revelation of tumor heterogeneity, and potential selected region-based feature extraction for adaptive analyses [4,54].

Radiomics descriptors

After tumor delineation, radiomics features are extracted from the information contained in the segmented ROIs, that can be used to qualitative assess tumor phenotype, aggressiveness, treatment response, and cancer genetics, as well as differentiate between benign and malignant tumor [17]. The further processing and selection between the variety of radiomics features derived lead to the potential definition of qualitative imaging biomarkers (QIB), that hold prognostic and predictive value for cancer outcome [41]. Therefore, when found to have a significant correlation with tumor biological properties, these parameters may possibly serve as useful endpoints, for the assessment of the severity, degree of change, or status of a cancer lesion, relative to normal [33].

Besides various metrics derived from advanced MR techniques, novel approaches such as texture analysis seem to overcome limitations, regarding diagnostic accuracy and reproducibility [12,55]. Texture features provide more detailed structural and dimensional information of pixel intensity value distribution, which facilitates a more effective intercomparison between images, by means of upgraded quantitative perception of tissue imaging characteristics.

Descriptor classification

Radiomic features may be divided into several categories, depending on their characteristics, such as shape-size-based, histogram-based, textural, and transform-based [11,40]. Shape and size based features provide information about tumor location and different size parameters, like surface, volume, diameter, sphericity, and surface-to-volume ratio. First order histogram parameters, such as mean value, standard deviation, percentiles, skewness, kurtosis, and entropy, enable the rough assessment of pixel intensity global distribution, without considering spatial variations.

Higher order statistics derived features, referred to as textural features, have been widely utilized in breast tumor DCE MRI parametric maps in the past, for improving characterization of breast lesions, and their response to treatment [56–60]. Second order histograms such as gray level co-occurrence matrices (GLCM) [12,41] and gray level run length matrices (GLRLM) [12,44], characterize spatial relationships between pixel intensities in different 2D or 3D directions, and thus, are robust in quantifying tumor structural properties, and various patterns of heterogeneity. GLCM analysis of DCE MR data is robust in differentiating between benign and malignant breast lesions [61]. Finally, additional high-order textural features, such as Gabor textures, temporal kinetics, and fractal-based textures, have been employed for classifying malignant from benign breast tumors, by means of identifying texture related signal variations [62,63].

Multi-omics data

To date, several studies have focused on the correlation and integration of radiomic features with breast genomics and proteomics data, and their results continue to support the notion, that radiomics metrics may perform well for better understanding of molecular and genetic variability, protein expression, prognosis and response to therapy [54,64]. Prediction of tumor subtypes [65], as well as clinical phenotypes, is enhanced through the association of breast tumor MR imaging data with ER, PR, and HER2 statuses [66].

Machine learning classification

For demanding diagnostic problems like pathologies mimicking tumors, or lesions with identical pathophysiological profiles, where data ranges are overlapping, conventional statistically significant correlations might be limited [8]. Thus, innovative computer-assisted diagnostic tools are required to analyze multidimensional data, and to illustrate the above relationships in intelligible and measurable quantities.

In the past, several advanced methods of data analysis have been evaluated, such as logistic regression (LR) [67,68] and Bayesian classifiers [69], aiming at incrementation of MRI diagnostic accuracy. However, the above-mentioned computational processes proved quite demanding and time-consuming, and moreover present limitations in evaluating the big amount of radiomics descriptors generated.

Machine learning classifiers are progressively replacing such alternatives, like support vector machine (SVM), and K nearest neighbor (KNN) classifiers [69–72]. Based on their ability to learn information from the provided training datasets, these sophisticated algorithms have demonstrated a superior efficiency, in making accurate selection and classification of the features extracted from radiological images, upgrading diagnostic overall accuracy of multiparametric imaging [10,73]. Deep learning classification techniques, such as artificial neural networks (ANN), may analogously be used as an automated computer analysis tool, providing further process automation [74].

Deep learning, a class of machine learning algorithms, is an emerging field of Data Science, It excels at learning a hierarchy of increasingly complex imaging features, directly from raw data. Hence, it can be considered the alternative to the quite demanding and time-consuming conventional machine learning methods, which involve segmentation, feature extraction, and classification steps [75].

Automatic feature extraction

On the top of computational cost minimization, the main advantage of implementing deep learning schemes over radiomics workflow is their ability for executing data-driven segmentation and extraction of many self-taught features, in a totally automated way, from raw imaging data sources. Dalmis et al. [76,77] have shown the value of convolutional neural networks (CNN) and U-net, for automated segmenting of breast lesions, and whole breast and fibroglandular tissue respectively.

Deep learning is advantageous in performing multi-parametric clinical big data associations, which facilitates the provision of pathology predictive outcomes, and the development of intelligent clinical decision support systems (CDSS), for implementing individualized patient-specific diagnosis and prognosis.

Current limitations of radiomics

Quantifiable radiomics features are very sensitive to data acquisition parameters (interscanner sampling, image artifacts, reconstruction methods) and the great variability of the feature extraction methods [78]. Also, several limitations are related to lesion segmentation methods in terms of user-dependency, and thus the subjective selection of initial criteria finally affect accuracy, stability, and reproducibility of the proposed methods. International collaborations have been established, e.g., the Image Biomarker Standardization Initiative [79], which work toward standardizing the parameters used in imaging biomarker extraction. Machine learning classification procedure suffers from another pitfall, as classification modeling of the immense numbers of radiomics features is based on relatively small medical image datasets, thus implying potential bias and inaccurate results.

The link between the imaged properties of tumors and tumor biology is not straightforward, and even though most radiomics studies show a statistical correlation between radiomic features and biological phenotype, this correlation does not imply causation. Hence there is a need for further systematic studies, with the proper design for results validation and establishment of standards.

Conflicts of interest

The authors declare that there are no conflicts of interest regarding the publication of this work.

References

[1] Bray F, Ferlay J, Soerjomataram I, Siegel RL, Torre LA, Jemal A. Global cancer statistics 2018: GLOBOCAN estimates of incidence and mortality worldwide for 36 cancers in 185 countries. CA Cancer J. Clin November 2018;68(6):394–424.

[2] Bi WL, Hosny A, Schabath MB, Giger ML, Birkbak NJ, Mehrtash A, Allison T, Arnaout O, Abbosh C, Dunn IF, Mak RH, Tamimi RM, Tempany CM, Swanton C, Hoffmann U, Schwartz LH, Gillies RJ, Huang RY, Aerts HJWL. Artificial intelligence in cancer imaging: clinical challenges and applications. CA Cancer J. Clin March 2019;69(2):127–57.

[3] O'Connor JP, Rose CJ, Waterton JC, Carano RA, Parker GJ, Jackson A. Imaging intratumor heterogeneity: role in therapy response, resistance, and clinical outcome. Clin. Cancer Res January 15, 2015;21(2):249–57.

[4] Napel S, Mu W, Jardim-Perassi BV, Aerts HJWL, Gillies RJ. Quantitative imaging of cancer in the postgenomic era: radio(geno)mics, deep learning, and habitats. Cancer December 15, 2018;124(24):4633–49.

[5] Saha A, Harowicz MR, Cain EH, Hall AH, Hwang ES, Marks JR, Marcom PK, Mazurowski MA. Intra-tumor molecular heterogeneity in breast cancer: definitions of measures and association with distant recurrence-free survival. Breast Cancer Res. Treat November 2018;172(1):123–32.

[6] Turashvili G, Brogi E. Tumor heterogeneity in breast cancer. Front. Med December 8, 2017;4:227.

[7] Heng YJ1, Lester SC, Tse GM, Factor RE, Allison KH, Collins LC, Chen YY, Jensen KC, Johnson NB, Jeong JC, Punjabi R, Shin SJ, Singh K, Krings G, Eberhard DA, Tan PH, Korski K, Waldman FM, Gutman DA, Sanders M, Reis-Filho JS, Flanagan SR, Gendoo DM, Chen GM, Haibe-Kains B, Ciriello G, Hoadley KA, Perou CM, Beck AH. The molecular basis of breast cancer pathological phenotypes. J. Pathol February 2017;241(3):375–91.

[8] Tsougos I, Vamvakas A, Kappas C, Fezoulidis I, Vassiou K. Application of radiomics and decision support systems for breast MR differential diagnosis. Comput. Math. Methods Med September 23, 2018;2018:7417126.

[9] Fan M, Li H, Wang S, Zheng B, Zhang J, Li L. Radiomic analysis reveals DCE-MRI features for prediction of molecular subtypes of breast cancer. PLoS One February 6, 2017;12(2):e0171683.

[10] Park H, Lim Y, Ko ES, Cho H-Ho, Lee JE, Han B-K, Ko EY, Choi JS, Park KW. Radiomics signature on magnetic resonance imaging: association with disease-free survival in patients with invasive breast cancer. Clin. Cancer Res 2018,Oct 1;24(19):4705–14.

[11] Yip SSF, Aerts HJWL. Applications and limitations of radiomics. Phys. Med. Biol. 2016 Jul 7;61(13):R150–66.

[12] Chitalia RD, Kontos D. Role of texture analysis in breast MRI as a cancer biomarker: a review. J. Magn. Reson. Imaging April 2019;49(4):927–38.

[13] Clarke GM, Peressotti C, Constantinou P, Hosseinzadeh D, Martel A, Yaffe MJ. Increasing specimen coverage using digital whole-mount breast pathology: implementation, clinical feasibility and application in research. Comput. Med. Imag. Graph 2011 ;35(7–8):531–41.

[14] García-Figueiras R, Baleato-González S, Padhani AR, Luna-Alcalá A, Vallejo-Casas JA, Sala E, Vilanova JC, Koh DM, Herranz-Carnero M, Vargas HA. How clinical imaging can assess cancer biology. Insights Imag March 4, 2019;10(1):28.

[15] Rahbar H, McDonald ES, Lee JM, Partridge SC, Lee CI. How can advanced imaging be used to mitigate potential breast cancer overdiagnosis? Acad. Radiol June 2016;23(6):768–73.

[16] Leithner D, Horvat JV, Ochoa-Albiztegui RE, Thakur S, Wengert G, Morris EA, Helbich TH, Pinker K. Imaging and the completion of the omics paradigm in breast cancer. Radiologe 2018 Jun 8;58(Suppl 1):7–13.

[17] Abrol S, Kotrotsou A, Salem A, Zinn PO, Colen RR. Radiomic phenotyping in brain cancer to unravel hidden information in medical images. Top. Magn. Reson. Imag February 2017;26(1):43–53.

[18] Woodhams R, Matsunaga K, Kan S. ADC mapping of benign and malignant breast tumors. Magn. Reson. Med. Sci. 2005;4:35–42.

[19] Partridge SC, Nissan N, Rahbar H, Kitsch AE, Sigmund EE. Diffusion-weighted breast MRI: clinical applications and emerging techniques. J. Magn. Reson. Imag February 2017;45(2):337–55.

[20] Woodhams R, Ramadan S, Stanwell P, Sakamoto S, Hata H, Ozaki M, Kan S, Inoue Y. Diffusion-weighted imaging of the breast: principles and clinical applications. Radiographics July–August 2011;31(4):1059–84.

[21] Amornsiripanitch N, Nguyen VT, Rahbar H, Hippe DS, Gadi VK, Rendi MH, Partridge SC. Diffusion-weighted MRI characteristics associated with prognostic pathological factors and recurrence risk in invasive ER+/HER2- breast cancers. J. Magn. Reson. Imag 2018 Jul;48(1):226–36.

[22] Partridge SC1, Amornsiripanitch N. DWI in the assessment of breast lesions. Top. Magn. Reson. Imag October 2017;26(5):201–9.

[23] Choi BB, Kim SH, Park CS, Jung NY. Chonnam, correlation of prognostic factors of invasive lobular carcinoma with ADC value of DWI and SUVMax of FDG-PET. Med. J May 2017;53(2):133–9.

[24] Sorace AG, Partridge SC, Li X, Virostko J, Barnes SL, Hippe DS, Huang W, Yankeelov TE. Distinguishing benign and malignant breast tumors: preliminary comparison of kinetic modeling approaches using multi-institutional dynamic contrast-enhanced MRI data from the International breast MR consortium 6883 trial. J. Med. Imag January 2018;5(1):011019.

[25] Bhooshan N, Giger ML, Jansen SA, Li H, Lan L, Newstead GM. Cancerous breast lesions on dynamic contrast-enhanced MR images: computerized characterization for image-based prognostic markers. Radiology 2010;254:680–90.

[26] Luo J, Johnston BS, Kitsch AE, Hippe DS, Korde LA, Javid S, Lee JM, Peacock S, Lehman CD, Partridge SC, Rahbar H. Ductal carcinoma in situ: quantitative preoperative breast MR imaging features associated with recurrence after treatment. Radiology December 2017;285(3):788–97.

[27] Yang Q, Li L, Zhang J, Shao G, Zheng B. A new quantitative image analysis method for improving breast cancer diagnosis using DCE-MRI examinations. Med. Phys January 2015;42(1):103–9.

[28] Stanwell P, Mountford C. In vivo proton MR spectroscopy of the breast. Radiographics 2007;27(Suppl. 1):S253–66.

[29] Vassiou K, Tsougos I, Kousi E, Vlychou M, Athanasiou E, Theodorou K, Arvanitis DL, Fezoulidis IV. Application value of 3T 1H-magnetic resonance spectroscopy in diagnosing breast tumors. Acta Radiol May 2014;55(4):416–7.

[30] Tsougos I, Svolos P, Kousi E, Athanassiou E, Theodorou K, Arvanitis D, Fezoulidis I, Vassiou K. The contribution of diffusion tensor imaging and magnetic resonance spectroscopy for the differentiation of breast lesions at 3T. Acta Radiol February 2014;55(1):14–23.

[31] Bolan PJ, Kim E, Herman BA, Newstead GM, Rosen MA, Schnall MD, Pisano ED, Weatherall PT, Morris EA, Lehman CD, Garwood M, Nelson MT, Yee D, Polin SM, Esserman LJ, Gatsonis CA, Metzger GJ, Newitt DC, Partridge SC, Hylton NM. ACRIN trial team ISPY-1 investigators., MR spectroscopy of breast cancer for assessing early treatment response: results from the ACRIN 6657 MRS trial. J. Magn. Reson. Imag July 2017;46(1):290–302.

[32] Fardanesh R, Marino MA, Avendano D, Leithner D, Pinker K, Thakur SB. Proton MR spectroscopy in the breast: technical innovations and clinical applications. J. Magn. Reson. Imag 2019. https://doi.org/10.1002/jmri.26700.

[33] Tsougos I. Advanced MR neuroimaging: from theory to clinical practice. CRC Press; 2017. ISBN: 9781498755238.

[34] Fan WX, Chen XF, Cheng FY, Cheng YB, Xu T, Zhu WB, Zhu XL, Li GJ, Li S. Retrospective analysis of the utility of multiparametric MRI for differentiating between benign and malignant breast lesions in women in China. Medicine (Baltimore) January 2018;97(4):e9666.

[35] Mirka H, Tupy R, Narsanska A, Hes O, Ferda J. Pre-surgical multiparametric assessment of breast lesions using 3-Tesla magnetic resonance. Anticancer Res December 2017;37(12):6965–70.

[36] Pinker K, Helbich TH, Morris EA. The potential of multiparametric MRI of the breast. Br. J. Radiol January 2017;90(1069):20160715.

[37] Crivelli P, Ledda RE, Parascandolo N, Fara A, Soro D, Conti M. A new challenge for radiologists: radiomics in breast cancer. BioMed Res. Int. 2018:10. Article ID 6120703.

[38] Valdora F, Houssami N, Rossi F, Calabrese M, Tagliafico AS. Rapid review: radiomics and breast cancer. Breast Cancer Res Treat.; 2018.

[39] Gillies RJ, Kinahan PE, Hricak H. Radiomics: images are more than pictures, they are data. Radiology February 2016;278(2):563–77.

[40] Avanzo M, Stancanello J, El Naqa I. Beyond imaging: the promise of radiomics. Phys. Med. June 2017;38:122–39.

[41] Weaver O, Leung JWT. Biomarkers and imaging of breast cancer. AJR Am. J. Roentgenol February 2018;210(2):271–8.

[42] Florez E, Fatemi A, Claudio PP, Howard CM. Emergence of radiomics: novel methodology identifying imaging biomarkers of disease in diagnosis, response, and progression. SM J. Clin. Med. Imag 2018;4(1):1019.

[43] Verma V, Simone 2nd CB, Krishnan S, Lin SH, Yang J, Hahn SM. The rise of radiomics and implications for oncologic management. J. Natl. Cancer Inst July 1, 2017;109(7).

[44] Sutton EJ, Huang EP, Drukker K, Burnside ES, Li H, Net JM, Rao A, Whitman GJ, Zuley M, Ganott M, Bonaccio E, Giger ML, Morris EA. Breast MRI radiomics: comparison of computer- and human-extracted imaging phenotypes. Eur. Radiol. Exp 2017;1(1):22.

[45] van Griethuysen JJM, Fedorov A, Parmar C, Hosny A, Aucoin N, Nar ayan V, Beets-Tan RGH, Fillion-Robin J-C, Pieper S, Aerts HJWL. Computational radiomics system to decode the radiographic phenotype. Cancer Res November 1, 2017;77(21):e104–7.

[46] Larue RT, Defraene G, De Ruysscher D, Lambin P, van Elmpt W. Quantitative radiomics studies for tissue characterization: a review of technology and methodological procedures. Br. J. Radiol February 2017;90(1070):20160665.

[47] Limkin EJ, Sun R, Dercle L, Zacharaki EI, Robert C, Reuzé S, Schernberg A, Paragios N, Deutsch E, Ferté C. Promises and challenges for the implementation of computational medical imaging (radiomics) in oncology. Ann. Oncol June 1, 2017;28(6):1191–206.

[48] Koenig M, Laue H, Boehler T, Peitgen H-O. Automatic segmentation of relevant structures in DCE MR mammograms. In: Proceedings on SPIE medical imaging; 2007. 65141S–65141S–6.

[49] Al-Faris AQ, Ngah UK, Mat Isa NA, Shuaib IL. Computer-aided segmentation system for breast MRI tumour using modified automatic seeded region growing (BMRI-MASRG). J. Digit. Imag. February 2014;27(1):133–44.

[50] Nie K, Chen JH, Chan S, et al. Development of a quantitative method for analysis of breast density based on three-dimensional breast MRI. Med. Phys. 2008;35:5253–62.

[51] Ertas G, Doran SJ, Leach MO. A computerized volumetric segmentation method applicable to multi-centre MRI data to support computer-aided breast tissue analysis, density assessment and lesion localization. Med. Biol. Eng. Comput January 2017;55(1):57–68.

[52] Niukkanen A, Arponen O, Nykänen A, Masarwah A, Sutela A, Liimatainen T, Vanninen R, Sudah M. Quantitative volumetric K-means cluster segmentation of fibroglandular tissue and skin in breast MRI. J. Digit. Imag 2018 Aug;31(4):425–34.

[53] Chen W, Giger ML, Bick U. A fuzzy c-means (FCM)-based approach for computerized segmentation of breast lesions in dynamic contrast-enhanced MR images. Acad. Radiol January 2006;13(1):63–72.

[54] Sala E, Mema E, Himoto Y, Veeraraghavan H, Brenton JD, Snyder A, Weigelt B, Vargas HA. Unravelling tumour heterogeneity using next-generation imaging: radiomics, radiogenomics, and habitat imaging. Clin. Radiol January 2017;72(1):3–10.

[55] Parekh V, Jacobs MA. Radiomics: a new application from established techniques. Expert Rev. Precis. Med. Drug Dev 2016;1(2):207–26.

[56] Fan M, Wu G, Cheng H, Zhang J, Shao G, Li L. Radiomic analysis of DCE-MRI for prediction of response to neoadjuvant chemotherapy in breast cancer patients. Eur. J. Radiol September 2017;94:140–7.

[57] Amano Y, Woo J, Amano M, Yanagisawa F, Yamamoto H, Tani M. MRI texture analysis of background parenchymal enhancement of the breast. BioMed Res. Int 2017;2017:4845909.

[58] Parekh VS, Jacobs MA. Integrated radiomic framework for breast cancer and tumor biology using advanced machine learning and multiparametric MRI. NPJ Breast Cancer November 14, 2017;3:43.

[59] Braman NM, Etesami M, Prasanna P, Dubchuk C, Gilmore H, Tiwari P, Plecha D, Madabhushi A. Intratumoral and peritumoral radiomics for the pretreatment prediction of pathological complete response to neoadjuvant chemotherapy based on breast DCE-MRI. Breast Cancer Res May 18, 2017;19(1):57.

[60] Chen W, Giger ML, Li H, Bick U, Newstead GM. Volumetric texture analysis of breast lesions on contrast-enhanced magnetic resonance images. Magn. Reson. Med September 2007;58(3):562–71.

[61] Ahmed A, Gibbs P, Pickles M, Turnbull L. Texture analysis in assessment and prediction of chemotherapy response in breast cancer. J. Magn. Reson. Imag 2013;38(1):89–101.

[62] Zheng Y, Englander S, Baloch S, et al. STEP: spatiotemporal enhancement pattern for MR-based breast tumor diagnosis. Med. Phys 2009;36:3192–204.

[63] Soares F, Janela F, Pereira M, Seabra J, Freire MM. 3D lacunarity in multifractal analysis of breast tumor lesions in dynamic contrast-enhanced magnetic resonance imaging. IEEE Trans. Image Process 2013;22:4422–35.

[64] Sutton EJ, Oh JH, Dashevsky BZ, Veeraraghavan H, Apte AP, Thakur SB, Deasy JO, Morris EA. Breast cancer subtype intertumor heterogeneity: MRI-based features predict results of a genomic assay. J. Magn. Reson. Imag November 2015;42(5):1398–406.

[65] Li H, Zhu Y, Burnside ES, Huang E, Drukker K, Hoadley KA, Fan C, Conzen SD, Zuley M, Net JM, Sutton E, Whitman GJ, Morris E, Perou CM, Ji Y, Giger ML. Quantitative MRI radiomics in the prediction of molecular classifications of breast cancer subtypes in the TCGA/TCIA data set. NPJ Breast Cancer 2016;2:16012.

[66] Guo W, Li H, Zhu Y, Lan L, Yang S, Drukker K, Morris E, Burnside E, Whitman G, Giger ML, Ji Y. Tcga breast phenotype research group., prediction of clinical phenotypes in invasive breast carcinomas from the integration of radiomics and genomics data. J. Med. Imag October 2015;2(4):041007.

[67] McLaren CE, Chen WP, Nie K, Su MY. Prediction of malignant breast lesions from MRI features: a comparison of artificial neural network and logistic regression techniques. Acad. Radiol July 2009;16(7):842–51.

[68] Wu S, Berg WA, Zuley ML, Kurland BF, Jankowitz RC, Nishikawa R, Gur D, Sumkin JH. Breast MRI contrast enhancement kinetics of normal parenchyma correlate with presence of breast cancer. Breast Cancer Res 2016;18:76.

[69] Fusco R, Sansone M, Filice S, Carone G, Amato DM, Sansone C, Petrillo A. Pattern recognition approaches for breast cancer DCE-MRI classification: a systematic review. J. Med. Biol. Eng 2016;36(4):449–59.

[70] Vidić I, Egnell L, Jerome NP, Teruel JR, Sjøbakk TE, Østlie A, Fjøsne HE, Bathen TF, Goa PE1. Support vector machine for breast cancer classification using diffusion- weighted MRI histogram features: preliminary study. J. Magn. Reson. Imag 2018 May;47(5):1205–16.

[71] Johansen R, Jensen LR, Rydland J, Goa PE, Kvistad KA, Bathen TF, Axelson DE, Lundgren S, Gribbestad IS. Predicting survival and early clinical response to primary chemotherapy for patients with locally advanced breast cancer using DCE-MRI. J. Magn. Reson. Imag June 2009;29(6):1300–7.

[72] Nahid A-A, Kong Y. Involvement of machine learning for breast cancer image classification: a survey. Comput. Math. Methods Med 2017;2017.

[73] Acharya UR, Hagiwara Y, Sudarshan VK, Chan WY, Ng KH. Towards precision medicine: from quantitative imaging to radiomics. J. Zhejiang Univ. Sci. B January 2018;19(1):6–24.

[74] Antropova N, Abe H, Giger ML. Use of clinical MRI maximum intensity projections for improved breast lesion classification with deep convolutional neural networks. J. Med. Imag January 2018;5(1):014503.

[75] Shen D, Wu G, Suk H-I. Deep learning in medical image analysis. Annu. Rev. Biomed. Eng June 21, 2017;19:221–48.

[76] Dalmış MU, Vreemann S, Kooi T, Mann RM, Karssemeijer N, Gubern-Mérida A. Fully automated detection of breast cancer in screening MRI using convolutional neural networks. J. Med. Imag January 2018;5(1):014502.

[77] Dalmış MU, Litjens G, Holland K, Setio A, Mann R, Karssemeijer N, Gubern-Mérida A. Using deep learning to segment breast and fibroglandular tissue in MRI volumes. Med. Phys February 2017;44(2):533–46.

[78] Saha A, Harowicz MR, Mazurowski MA. Breast cancer MRI radiomics: an overview of algorithmic features and impact of inter-reader variability in annotating tumors. Med. Phys July 2018;45(7):3076–85.

[79] Zwanenburg A, et al. Image biomarker standardisation initiative. 2016. arXiv preprint arXiv:1612.07003.

D—Big data, artificial intelligence, and deep phenotyping

Chapter 32

Precision medicine in the assessment of suicide risk

Thiago Henrique Roza[1,2], Lucas Mohr Patusco[1,2], Aline Zimerman[1,2], Pedro Ballester[3,4] and Ives Cavalcante Passos[1,2]

[1]*Molecular Psychiatry Laboratory, Hospital de Clínicas de Porto Alegre, Porto Alegre, Brazil;* [2]*Programa de Pós-Graduação em Psiquiatria e Ciências do Comportamento, Universidade Federal do Rio Grande do Sul, Porto Alegre, Brazil;* [3]*Machine Intelligence Research Group, PUCRS, Porto Alegre, Brazil;* [4]*Programa de Pós-Graduação em Ciência da Computação, Pontifícia Universidade Católica do Rio Grande do Sul, Porto Alegre, Brazil*

Introduction

Unlike most leading causes of death in the United States, suicide rates have not declined over the past 50 years [1]. Approximately 800,000 deaths by suicide occur each year worldwide, which accounts for one death every 40 s [2]. Such numbers make suicide the 18th leading cause of death globally, being the second among young people aged 15–29 years [2]. Estimates state that almost 80% of suicides took place in low and middle income countries [2]. On the one hand, it is known that suicide is more common in men, with a global rate of suicide deaths of 15 per 100,000 people per year, whereas the rate for women is 8 per 100,000 [3]. On the other hand, studies have found that women are at higher risk for suicide ideation and attempts than men [4,5]. Additionally, it is likely that the number of suicides is underestimated, as a reasonable percentage of them is classified as undetermined intent, and accidental deaths [6].

Glossary

"Completed suicide" is defined as a self-injurious behavior that resulted in fatality and was associated with at least some intent to die [7]; whereas, "suicide attempt" is a potentially self-injurious behavior, in which intention to die as consequence of the act can be inferred [7]. "Suicidal ideation" is the occurrence of thoughts about killing him/herself, or passive thoughts in relation to wanting to be dead, not accompanied by preparatory actions [7].

Predictive models and suicide prevention

Nearly half of people who died by suicide visited a primary care physician within 1 month of their deaths [8]. In addition, another study pointed out that nearly 90% of suicide cases had a psychiatric diagnosis [9]. These findings show an important window of opportunity for preventive strategies; however, the first step is to recognize subjects at higher likelihood to attempt suicide. In this sense, the use of machine learning techniques, to build predictive algorithms to identify who will attempt suicide, is gaining traction in psychiatric research, and may provide predictive models for both clinical practice and public health systems.

Risk and protective factors

There are many factors that can be considered as risk for suicide, and some that may account as protective. Among the protective factors, we have religious involvement, optimism, effective coping strategies, high levels of extroversion, social support, and strong reasons for living [10]. The role of religiosity, however, is not constant across different cultures. While a meta-analysis has suggested lower suicide risk for religious people, its findings show that the effect was only significant for Western and not for Eastern cultures [11].

The risk factors for suicide can be challenging to understand and apply in patient evaluations, as each passing day novel potential suicide risk factors are highlighted in publications. Therefore, there are some models that try to

Precision Medicine for Investigators, Practitioners and Providers. https://doi.org/10.1016/B978-0-12-819178-1.00032-0

organize and synthesize important risk factors and the association among them, as depicted in Fig. 32.1 [10]. In this model, one can see the integration of different levels of risk factors, namely population and individual-level risk factors. Population-level risk factors can be further divided into environmental factors and lack of social cohesion, whereas individual risk factors can be divided into distal (or predisposing), developmental (or mediating), and proximal (or precipitating) factors [10].

Precision medicine in the evaluation of suicide risk

When a patient with a depressive episode, and no known previous history of suicide attempts, first presents for treatment, what is the likelihood that this individual patient will attempt suicide? This question is a major unmet challenge in psychiatric disorder treatment, and also an important question for society at large, as suicide is a tragic but highly preventable event. There was little awareness in medical practice on objective stratification of suicide risk until recently, which has led suicide to be referred to as "the quiet epidemic" [12]. Suicide is usually viewed as an extreme response to a catastrophic event, such as the loss of a close relative [13]. However, many individuals, including patients with mood disorders, go through these kinds of stressors, and yet they do not attempt suicide.

Efforts have largely reported average group-level differences between suicide attempters and nonattempters. However, what is still not known is how to integrate these

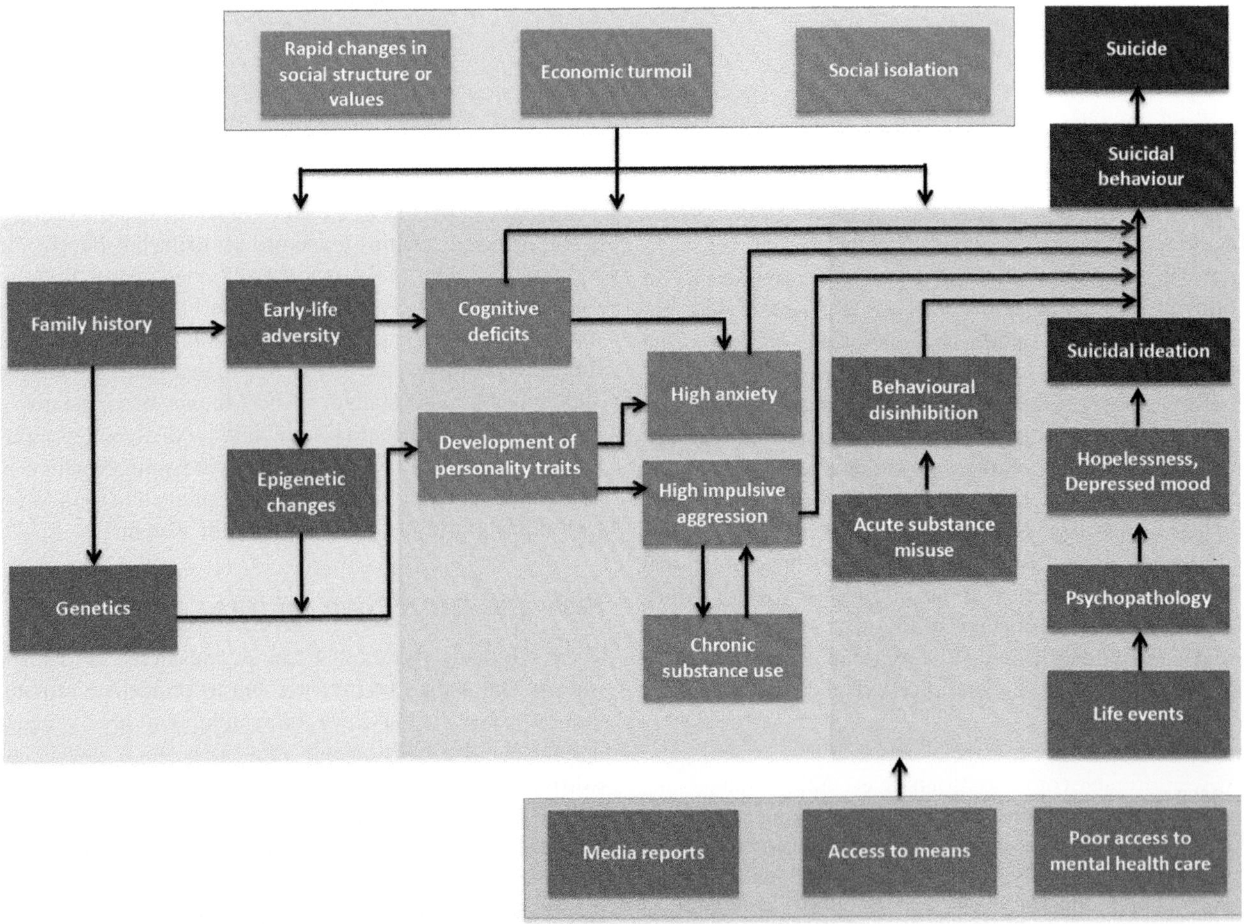

FIGURE 32.1 Model for suicide risk. Suicide is affected by population (lack of social cohesion factors, marked in purple [black in print version]; and environmental factors, highlighted in orange [gray in print version]), and individual risk factors. In addition, individual factors can be distal, or predisposing (in green [light gray in print version]); developmental or mediating (in blue [dark gray in print version]); and proximal, or precipitating (in red [light black in print version]). All these factors interact with each other, increasing the risk of suicidal behavior, which is the outcome to be prevented. As shown in the figure, predisposing factors lead to long-term effects on genetic function, not leading directly to suicidal events, but increasing its risk, by affecting mediating factors. Mediating factors, however, may increase suicide risk by promoting traits (such as anxiety, and impulsive aggression traits), associated to suicidal behavior; such traits increase the likelihood of acting in cases of suicidal ideation, and may be consequences of genetic modifications caused by predisposing factors, but also can be associated to other factors such as chronic substance abuse. Proximal risk factors, such as psychopathology (any mental illness associated with suicide risk such as bipolar disorder, major depressive disorder, schizophrenia, personality disorders, or a combination of mental illness), and acute substance abuse, are also associated to epigenetic and genetic factors, as well as life events which ultimately contribute to suicidal behavior. *Adapted from Turecki G, Brent DA. Suicide and suicidal behaviour. Lancet 2016;387:1227–39. https://doi.org/10.1016/S0140-6736(15)00234-2.*

variables to build a signature of suicide attempt. The use of big data and machine learning models that could leverage each person's unique biological profile may help to develop multimodal risk stratification tools. *Big data* is a broad term used to describe volumes of large and complex amounts of data, which can vary from the molecular level, including genomics, proteomics, and metabolomics, to clinical, sociodemographic, administrative, environmental, and even social media information [14]. Such huge amounts of data cannot be handled without the use of sophisticated computational techniques [15], hence the use of *machine learning*, which is, in general terms, the reception of data and the fitting of mathematical functions to generate a model explaining the data received [16].

Machine learning allows objective predictions of previously *unseen* observations, such as independent datasets not used to generate the prediction models [17]. Typically, these algorithms are implemented in three key stages. First, subjects' data are separated into *training* and *test* sets, with the former being used to fit a machine learning model. Second, model tuning and feature selection routines could be implemented, to improve model performance. Third, the predictive accuracy of the *final* model is evaluated using the *test* set, along with the reporting of algorithm accuracy, area under the curve (AUC), sensitivity, and specificity. It is essential to adequately select the predictor variables that will be used in training an algorithm [18]. Moreover, there are two types of machine learning approaches: hypothesis-driven and data-driven. When using a hypothesis-driven approach, we use models that instantiate prior knowledge on the underlying mechanisms of the variables largely from previously published literature or expert opinion. In data-driven approaches, however, models are usually agnostic to such mechanisms.

There are several advantages concerning the use of machine learning tools for suicide risk assessment. For instance, machine learning may produce more accurate relationship estimations among multivariate data, and can find complex nonlinear patterns between predictors and outcomes [19]. Such characteristics contrast with common statistical methods, which usually yield group-level results, and have difficulty coping with nonlinear multivariate data [20]. Social media, internet-based or smartphone data could contribute to the creation of new digital biomarkers [21]. In addition, clinical practice and doctor-patient relationship may benefit from machine learning–based tools, in which the patient would be more aware of his own healthcare, being able to adequately monitor eventual changes; therefore, playing a more active role in the traditional doctor-patient relationship [17].

Most major psychiatric disorders have been explored in machine learning studies, such as bipolar disorder [17,22–26], major depressive disorder [27], schizophrenia [28], substance use disorder [29], posttraumatic stress disorder [30], and suicide [18]. Moreover, the variables included in these studies ranged from clinical and sociodemographic, to biological, neuroimaging, and neuropsychological markers.

In clinical terms, an individual identified by a machine learning algorithm, as having high suicide risk, could benefit from suicide preventive measures and pharmacological treatment. Such algorithm could be used even in primary care settings, or as a web-based suicide risk calculator [14,31].

Overview of machine learning strategies for the evaluation of suicide risk

A systematic review has evaluated the use of machine learning techniques in the prediction of self-injurious thoughts and behavior, including suicide [32]. However, an overall conclusion about performance across studies could not be reached due to the heterogeneity among them. Even though accuracy was not generally high for the models (most studies reported an accuracy smaller than 0.90), authors pointed out that machine learning strategies present higher predictive accuracy, in comparison with traditional statistical methods. Such methods were able to go beyond known risk factors for self-injurious thoughts and behavior, finding novel variables and boosting model performance; in addition, some studies were even able to identify subgroups among patients, which was important to suggest clinical cutoffs [32].

A study applied machine learning to predict suicide attempts and suicide ideation in patients with fibromyalgia, using electronic health records [33]. In their model, the AUC was 0.80 for suicidal ideation, and 0.82 for suicide attempts [33]. Nonfatal suicide attempts in adolescents in comparison with multiple control groups, such as adolescents with other self-injury, depressed adolescents, and general hospital adolescent controls, were recently addressed. Their discriminatory performance was good, presenting an AUC above 0.8 regardless of the control group. Performance increased in the temporal window of the imminence of a suicide attempt, with AUCs bigger than 0.9 when comparing with general controls [34].

Regular nondeployed US male army soldiers were the focus of another protocol which used administrative data as variables for the prediction of suicide deaths among psychiatric outpatients, in a time of 26 weeks after the mental health visit. Four machine learning algorithms (naïve Bayes, random forests, support vector regression, elastic net penalized regression) were selected. Using the elastic net classifier, 45.6% of suicides occurred after the 15% of visits with the highest predicted risk. In addition, the 5% of visits with the highest predicted risk included only 0.1% of soldiers. Considering all 26 weeks, AUC

(0.72) was higher for soldiers with prior psychiatric hospitalizations [35]. Another research devised three machine learning algorithms for mood disorders, and subsequently used these trained algorithms in a new group of subjects, in order to classify them as suicide attempters and nonattempters. Accuracy of the prediction was 65%–72% ($P < .05$), with the highest accuracy in predictions from the relevance vector machine (RVM) method (72.1% sensitivity and 71.3% specificity; AUC of 0.77). Among the most relevant predictor variables were previous hospitalizations for depression, a history of psychosis, cocaine dependence, and posttraumatic stress disorder (PTSD) [18].

Suicide deaths in the 12 months after inpatient psychiatric treatment were searched in a population of US army soldiers, by also using administrative data as variables for machine learning methods. Interestingly, 52.9% of posthospital suicides actually occurred in the group with the highest predicted suicide risk, which was composed of 5% of hospitalizations [36]. Several of their models presented good performance, with AUCs around 0.85 [36].

Predictor variables from standardized questionnaires were used to predict suicide ideation among treatment-seeking military veterans. Accuracy of 84.4% in the prediction of suicide ideation was achieved, through the use of 25 variables; whereas, by reducing the number of variables to the 10 top indicators of suicide ideation, almost all originary from the Patient Health Questionnaire (PHQ), accuracy was around 81% [37].

Multiple data sources

The combination of multiple sources of data has also been explored. A study implemented a machine learning model, using sociodemographic and questionnaire-based variables, for the prediction of suicidal ideation in primary care patients. One of their best classifiers achieved AUC of 0.856, based on only four variables from the patient health questionnaire (PHQ); whereas other more complex machine learning classifiers reached performance with an AUC around 0.87, with only PHQ variables or a combination of PHQ and sociodemographic variables [38]. In mental health patients, variables based on specific questionnaires, as well as sociodemographic and clinical information, were explored in a model which presented an accuracy of 0.78 [39]. Using clinical scales and sociodemographic data in outpatient psychiatric patients, another model had an overall accuracy of 93.7% for the detection of past month suicide attempt, 90.8% for past year suicide attempt, and 87.4% for lifetime suicide attempt. AUC was 0.93, 0.87, and 0.89, respectively [40].

Suicide ideation in the general population is another area of interest. Using data from a nationwide public survey on health and nutrition, one model presented good performance, with an AUC for the ROC curve of 0.85 [41].

Unstructured text analysis

Text mining strategies (natural language processing) were options to identify suicide ideation and suicide attempts in free-texts in a psychiatric clinical database. Classifier performance was good for suicide ideation, with a sensitivity of 87.8% and a PPV of 91.7%. For suicide attempt, the performance was also good, with a sensitivity of 98.2% and a PPV of 82.8% using a hybrid of machine learning and rule-based approach [42]. A Spanish longitudinal study [43] evaluated patients recently discharged from psychiatric inpatient and emergency room settings over 12 months, and also tried to predict suicidal ideation, using machine learning for natural language processing, with the single question: "How are you feeling today?" PPV was 0.61, sensitivity 0.56, and specificity 0.57 [43].

Another study used unstructured clinical notes from medical records of US veterans [44]. They applied a genetic programming framework in word counts of clinical notes, free text written by health professionals and semiautomatic texts and tables (for instance, list of medications) from patients who committed suicide, in a case control model. Initially, they converted the free-texts into phrases and single words datasets, which were later used in the models. Accuracy was up to 69% [44].

Virginia Woolf's diaries and letters, within and outside the period of 2 months before her death by suicide, were also investigated, in order to evaluate if a text classification machine learning model could identify suicide associated patterns as a text signature. The accuracy was 80% [45].

Internet posts and other machine learning strategies

Suicide risk and emotional distress were screened among users of a Chinese platform of social media, named Weibo [46]. Suicide risk factors and emotional distress parameters were measured through specific scales. Simplified Chinese-Linguistic Inquiry and Word Count (SC-LIWC) categories were fitted. People with higher suicide risk presented higher use of pronouns, prepend and multifunction words, greater total word count, and lower use of verbs. The use of second-person plural was positively associated with stress and severe depression; and work-related words and third-person plural were negatively associated with Weibo Suicide Communication, which was measured by a single-item question in which the participant answered if he/she had told other Weibo users about suicidal thoughts in the past year. Even though the classifier did not have a good performance, among those with Weibo Suicide Communication the AUC for suicide chance was 0.61 ($P = .04$). Therefore, language markers were revealed as culturally sensible, being different for each cultural group [46].

Among Twitter users, a different model, which used Twitter posts as data, presented a sensitivity of 0.53, a specificity of 0.97, a positive predictive value of 0.75, and a negative predictive value of 0.93, for the prediction of suicide risk [47]. Recording the conversation of suicidal and nonsuicidal adolescents at the emergency department and analyzing this data with natural language processing was shown to be a feasible potential alternative in the discrimination of both groups. Accuracy of 96.67% has been announced [48].

Another strategy that could be used in machine learning–based suicide risk assessment is the use of vocal features. For instance, a group of researchers used vocal features (nonverbal features) and words (verbal features) of participants as variables to classify them into three different groups: suicidal, mentally ill but not suicidal, and controls. In order to obtain the verbal and nonverbal features, the subjects were video and audio-recorded as they answered a semistructured interview with five open-ended questions. The feature extraction was made with the transcription and audio signals of the answers to the semistructured questions. The performance of the algorithms combining acoustic and linguistic features was good, presenting AUC higher than 0.8 for most classifications, surpassing 0.9 for the classification between suicidal and controls; moreover, the addition of acoustic features to verbal features was most helpful in the classification between suicidal and mentally ill patients [49], demonstrating another interesting source of variables for machine learning–based assessment of patients with suicide risk.

Another source of data for machine learning is neuroimaging. One study, even with a small sample, explored this potential and showed interesting results [50]. They used fMRI neural representation of concepts related to death and life as variables for machine learning–based discrimination of suicide ideators from controls; their model presented good performance, with an accuracy of 91% [50].

Not only the variables for machine learning–based interventions may vary across studies, but also the strategy for prediction and classification. An interesting strategy for suicide risk assessment tools in clinical practice could be the development of a tool based on more than one stage of assessment. For instance, a group of researchers applied an innovative two-step approach on a large sample of US army soldiers firstly identified as nonideators, because they denied lifetime suicide ideation at a specific survey taken initially [51]. They implemented this two-step machine learning approach in order to predict nonfatal suicide attempts in this group in a follow-up of 45 months. In their approach, they first used administrative data to generate a model to predict those, among the nonideators, at high risk for suicide attempt; by identifying those at high risk, they could exclude the low risk group from further assessment in the second step. In the second step, they used predictors from the survey taken initially, the same one conducted for identifying nonideators. Then, the researchers created a model to predict whether subjects were at high risk for suicide attempt. After the application of the two steps, they identified the 3% of the total nonideators with the highest risk, and this group accounted for 45% of suicide attempts in the whole sample. For the first step the AUC was 0.82, whereas for the second step it was 0.83 for their best model, showing good performance for both models. This approach not only identified those at high risk, but also discriminated those at low risk, which could have the potential to diminish the burden of further suicide assessments in such low-risk groups [51].

Gender differences in machine learning prediction models have also been explored in the literature. A previous study investigated the gender-specific associations with suicide ideation, by using variables from different domains (psychopathology, predeployment factors, demographic variables, military characteristics, deployment experiences) in a national sample of veterans deployed during the Afghanistan and Iraq wars [52]. Among their findings, they pointed out that the highest probability of suicidal ideation was among women with probable depression, probable PTSD, and who had been sexually harassed during deployment; among men, probable depression, probable anxiety, deployment to Iraq and Afghanistan, and more than mild somatic symptoms were variables that combined were associated with increased suicide ideation. Their models presented good classification performance for the prediction of suicidal ideation, with an ROC AUC of 0.91 for men, and 0.92 for women [52]. Such study highlights the difference in prediction models between genders, and the varying importance of different variables for each gender.

Even considering all the potential benefits and interesting results of recent machine learning studies for prediction and classification of suicide risk, one must bear in mind the existing limitations of current scientific literature on the subject. Importantly, there are many challenges regarding the translation of machine learning–based interventions from research settings into clinical practice. It is essential to consider limitations such as reproducibility, datasets of small sizes and insufficient number of details, data protection and privacy concerns, data of poor quality, small number of longitudinal studies, and the eventual presence of confounding factors [53]; such limitations may arise from the fact that most studies are pilots, with small sample size, cross-sectional design, lacking external validity [21]. In addition, one must consider the potential ethical aspects that may arise when applying machine learning and artificial intelligence in the screening of patients with suicide risk, mainly regarding autonomy and liberty of the individuals, as it can be seen in detail in

Mckernan et al. [54]. Furthermore, it is essential to be responsible concerning the legal and ethical implications surrounding the protection of patients' data; such large-scale data could be used by insurance companies (against their own patients), and machine learning algorithms may create some stratifying criteria that are not acceptable in society [55].

Prevention of suicide

Many prevention strategies have been studied lately and can be implemented as tools for suicide prevention. One good example is means restriction, which can be defined as the restriction of more lethal means of suicide [56]. It has been shown that means restriction strategies such as restriction of access to firearms, building of physical barriers at suicide hotspots, pesticide regulation, the withdrawal of specific toxic analgesics from the market, limiting analgesic package size, among others, reduce the rates of suicide by the restricted method and overall suicide rates; to some extent, there is the replacement of one method by another, usually less lethal, which was shown to be not as important as imagined [56,57]. In more specific terms, the use of barriers at suicide hotspots is strong in terms of evidence and has been shown to be highly effective in reducing suicide rates [58]. In addition, firearms are very lethal means of suicide; such information is supported by research that has shown that people who commit suicide using firearms are more likely to die at their first suicide attempt in comparison with people who died using other means [59]. For example, only in the United States, every year more than half of suicide deaths are consequence of firearms, even though less than 2% of nonfatal self-harm attempts utilize firearms [60]. The restriction of access to firearms and other gun control laws, in Australia, were shown to be associated with significant decreases in firearm suicide rates [61]. In South Korea, the prohibition of the herbicide Paraquat had significant impact on suicide rates by all methods in the whole country [62].

There are other examples of preventive measures that have been shown effective in previous research. Community-based, workplace and school-based interventions may have an important impact on suicidality; in addition, appropriate access to care, such as increased access to mental health services for people at risk is likely to have a positive impact on this population [10]. Interventions that aim to strengthen social support and reduce loneliness among at-risk groups, such as elders, may be an interesting prevention tool, as has been shown by a team of investigators as having significant impact on suicidal behavior and rates [63]. Another important strategy is school-based awareness programs, which were demonstrated to reduce rates of suicide ideation and attempts [57]. Educational workshops, about suicide and on the identification of people with suicide risk, for influential people in the community, and suicide hotlines are other preventive measures that might work [64]. The strategy of maintaining follow-up after a suicide attempt is likely to reduce the rates of new suicide attempts, at least in the first 12 months of the suicide attempt, as shown by a meta-analysis [65]. In addition, another quite important strategy for suicide prevention is the establishment of guidelines for responsible media reporting on suicide [57], as the ones developed by the WHO [66]. In this booklet guide they state, for instance, the importance of avoiding the description of suicide methods used, and to be cautious when reporting celebrity suicides. The guide contains many other recommendations for responsible media reporting on suicide, which can be seen in WHO [66].

Interventions in the patient with risk of suicide

Among all pharmacological interventions used to treat risk of suicide, the use of lithium in mood disorders is certainly the most consistent. It has been over 40 years since the first studies suggested that lithium had this effect, and since then we have growing body of evidence supporting it. Summarizing all that, a systematic review of placebo controlled RCTs published in 2013 included data from 48 studies, with a total of 6,674 participants [67]. This study found lithium to be superior to placebo, reducing the risk of death by suicide in 87%, and the risk of death by any cause in 62%. The magnitude of lithium's impact in suicidality is even more significant considering that mood disorders is the most common diagnostic category in suicide.

The use of antidepressants in suicide is a controversial topic. Antidepressant treatment has been shown to decrease suicide ideation and behavior in adult and geriatric populations, being associated with lower suicide risk for such patients [57,68]. In addition, some previous studies pointed out that impact on suicide ideation reduction is greater for SSRI in comparison with other antidepressants such as bupropion and venlafaxine [69,70]. However, in 2004, the FDA issued a warning about the possibility of antidepressants increasing suicide risk in young people [10]. Moreover, research has found evidence of increased risk of suicide in children and adolescents, but not in adults, when using antidepressants; even though such harm cannot be estimated accurately, with the need for more research [71]. Nevertheless, another investigation has shown that prescriptions and sales of SSRI were inversely correlated to suicide rates in several countries, even for young people [72]. In addition, even though it is a controversial subject, current scientific literature does not support the absence of antidepressant treatment for adolescents and children, mainly because untreated depression has the potential to increase suicide risk [57,73]. Furthermore, there is evidence that, in populational terms, prescriptions of SSRI and lithium were inversely associated with suicide rates for people aged 19 and under [74].

Patients with schizophrenia are at a high risk for suicide, with a lifetime risk of approximately 50% for attempting and 10% for completing suicide [75]. For this population, clozapine, the choice for treatment-resistant schizophrenia, has also been demonstrated to decrease the risk of suicide attempts. In the InterSePT study, 980 patients with a diagnosis of schizophrenia or schizoaffective disorder at a high risk for suicide were randomized to receive clozapine or olanzapine in an open label design [76]. After a 2-year follow-up, the clozapine treated group had significantly lower rates of suicidal behavior, suicide attempts, and rescue interventions. Moreover, in a population-based cohort study from Sweden of 26,046 patients treated for schizophrenia, those receiving clozapine had lower chance of attempting (OR 0.45) and completing (OR 0.44) suicide when compared to haloperidol users [77].

Ketamine is another pharmacological option for treating suicidal behavior. Studies using ketamine for treatment of unipolar and bipolar depression have demonstrated rapid antidepressive response afterward the drug administration [78]. In addition, Ketamine has been shown to rapidly reduce suicide ideation, with a dose of 0.5 mg/kg 40 min i.v. infusion, with initial impact on suicidal thoughts soon after administration of the medication, with benefits lasting for up to 1 week [79,80]; the effects of ketamine, regarding suicidal ideation, were noted in MDD, in bipolar depression, and even in cancer [80,81]. Another option is intranasal esketamine, which has also been shown to be effective on depressive and suicidal symptoms [82]. Therefore, ketamine could work as a crisis treatment of people with high suicide risk in the psychiatric emergency; however, the durability and safety of long-term treatment has not been shown yet [83–85].

Current evidence supports efficacy of psychotherapy, mainly cognitive behavior therapy (CBT) and dialectical behavior therapy (DBT) [57]. DBT combines CBT, mindfulness, and systemic characteristics and aims to promote the modification of behavior, through emotional regulation, skills improvement, stress tolerance, and interpersonal effectiveness. In addition, it has been pointed out to diminish rates of recurrent suicidal behavior in patients with borderline personality disorder [86]. Cognitive behavior therapy also has been shown to have impact on suicidal behavior [87]. Another psychotherapeutic approach, namely mentalization-based therapy, which is derived from psychodynamic psychotherapy, and aims to enhance the mentalizing and reflective capacity of the patient, has also been shown to be effective on suicidal behavior [86].

Implications for public health

Groundbreaking discoveries and changes at the population and public level will involve data integration focused on a person-centered approach. Therefore, public health authorities will need to lead the field forward by investing in precision medicine. In such a manner, information will be translated into better care for mental health. "Big data" approaches will be needed to measure precisely our phenome, genome, and exposome; including data from imaging, insurance, pharmacy, social media, as well as omic data (genomics, proteomics, and metabolomics). We also need precise clinical predictions. Integration of data from multiple levels will need to be translated into clinical practice to predict suicide by both the generation of homogeneous groups of patients and the use of calculators to accurately predict at an individual level and facilitate important clinical decisions. Public health authorities will need to be leading the translation of big data into care of consumers, and transforming prediction and information into a greater understanding of risk assessment and, ultimately, into better care.

Machine learning has great potential in the field of suicide risk evaluation in clinical practice settings. However, as this is a recent field of study, which is rapidly growing, but is still quite limited, most authors do not recommend public health and policies to be based on such promising tools yet [32]. Nevertheless, in the future, many aspects of public health system and the general population may benefit from it for identifying patients with suicide risk [32]. The improvement of machine learning models, the overcoming of current limitations, and the translation of the models into practical clinical tools are essential new steps for machine learning to fulfill its promise in the field of suicide assessment in practical terms [88]. In order to do so, it will be fundamental for the machine learning model to be adequate to the setting and population it will be applied to. In addition, it will be essential to overcome current beliefs surrounding such technology not only in healthcare providers, but for the users too. There will be the need for training healthcare professionals to handle such tools, and the modification of existing clinical workflows. Such challenges are big and difficult to overcome [88].

In addition, it will be necessary to pair the accurate prediction tools to accurate interventions for the identified individuals, and this connection of prediction and optimized treatments needs to be explored in research. Moreover, it may be useful to use different intervention strategies for varying levels of suicide risk [88]. One important aspect to consider for future applications of machine learning–based suicide evaluation tools is how to handle eventual false positives and negatives that may persist [88].

The implementation of evidence-based public health interventions, mainly those aiming to prevent suicide, such as means restriction, can have a significant impact on suicide rates and the consequent mortality [57]. In addition, public health policy with interventions aiming air pollution limitation and housing improvement could have a significant impact on suicide rates [89].

In terms of mental healthcare (in the level of the health system), some interventions were shown to have an impact on suicide rates. In England, for instance, the implementation of depression guidelines, the increase in availability of specialist community services, concurrent treatment of mental disorders and substance abuse, formal transfer care from youth to adult services, and review of care after suicide deaths were shown to be associated to reduced suicide rates [90]. Other important public health intervention would be the provision of mental health services for population, and follow-up of people who attempted suicide, as such strategies were shown to have significant impact of suicide rates [57].

Future directions

New areas of study may focus on the investigation of technology and internet and its relationship with suicidality. The potential effect of social media and internet on suicidal ideation and behavior is yet to be explored, as well as the effect of technology and internet-based interventions on suicide prevention [57]. The use of large datasets on suicide evaluation is another interesting field for research, along with the identification of novel suicide methods and the subsequent strategy of restriction for them [57,89]. Furthermore, the control of the spread of such methods through the media is important for study and intervention [89].

The use of machine learning in the prediction of suicidality has yet to grow, but shows promising results. Machine learning is certainly one of the most important and promising tools in the advance of the scientific field of suicide prediction and classification [88]. In addition, most of the research in the use of machine learning in suicide has been published in the past few years, which shows the area is growing rapidly in research and suggests it could benefit from different strategies in order to reach its full potential in the future [32].

Many studies are focused on military populations and mental health patients; therefore, it would be important for further studies to focus on broad populations of civilians, without psychiatric diagnoses [32]. Another limitation lies on the fact that most studies were based on adults, highlighting the need for more studies with adolescents and elderly [88]. Such limitations hold back the use of machine learning in broader populations and settings [88].

Another area for improvement would be to use the massive amount of data offered by electronic communication such as social media and communication channels between patients and healthcare providers, as well as other sources of data, including GPS, web search histories, call logs, and public legal and financial databases [32].

Furthermore, the use of unsupervised learning in this field of research could find novel groups and clusters for individuals in the evaluation of suicide [32]. Unsupervised learning, in contrast with supervised learning, is related to data in which the outcome variable is not given to the machine [88]. Even though, for most research questions in the study field of suicide, supervised machine learning will be the most common approach to be used, as it has been to date in literature [88], the use of unsupervised learning could advance the field even more [32].

Other important limitations that need to be overcome concern study design. Most studies to date implemented a cross-sectional design, which demonstrates the need for more longitudinal studies, with larger sample sizes [32]. In addition, the further exploratory efforts for diagnosis, treatment planning and risk classification should also be explored in future endeavors in machine learning studies [32]. The majority of longitudinal studies in the field use electronic health records as data, and such data include many variables that are not specific for suicide prediction [88]. Also, many studies published to date lack external validation [88]. Lastly, the performance of current models in most cases still needs improvement in order to be used in clinical contexts [88].

Conclusion

Suicide is a major cause of death and its rates have not declined over the past years. One of the major challenges in the assessment of patients with suicide risk is the identification of these patients, besides the subsequent choice of the most appropriate intervention in the form of preventive and treatment measures. Such patients may benefit greatly from precision medicine tools, such as machine learning techniques that may predict and classify patients with high suicide risk. Even though this field is still in its infancy, each year a rising number of scientific papers are published, demonstrating this as a rapidly growing field. If the field overcomes its current limitations, mainly concerning the translation to clinical practice, such tools will be able to be implemented into different complexity levels of current health systems and will have the potential to change the current unfortunate scenario of deaths by suicide.

References

[1] Curtin SC, Warner M, Hedegaard H. Increase in suicide in the United States, 1999-2014. NCHS Data Brief; 2016. p. 1–8.

[2] World Health Organization. WHO | Suicide data. WHO; 2018. https://www.who.int/mental_health/prevention/suicide/suicideprevent/en/.

[3] Saxena S, Krug EG, Chestnov O. World Health Organization. Department of Mental Health and Substance Abuse. Preventing suicide: a global imperative. 2014.

[4] Borges G, Nock MK, Haro Abad JM, Hwang I, Sampson NA, Alonso J, et al. Twelve-month prevalence of and risk factors for suicide attempts in the World Health Organization World Mental Health Surveys. J. Clin. Psychiatry 2010;71:1617–28. https://doi.org/10.4088/JCP.08m04967blu.
[5] Nock MK, Borges G, Bromet EJ, Alonso J, Angermeyer M, Beautrais A, et al. Cross-national prevalence and risk factors for suicidal ideation, plans and attempts. Br. J. Psychiatry 2008;192:98–105. https://doi.org/10.1192/bjp.bp.107.040113.
[6] Katz C, Bolton J, Sareen J. The prevalence rates of suicide are likely underestimated worldwide: why it matters. Soc. Psychiatr. Psychiatr. Epidemiol. 2016;51:125–7. https://doi.org/10.1007/s00127-015-1158-3.
[7] Posner K, Oquendo MA, Gould M, Stanley B, Davies M. Columbia classification algorithm of suicide assessment (C-CASA): classification of suicidal events in the FDA's pediatric suicidal risk analysis of antidepressants. Am. J. Psychiatry 2007;164:1035–43. https://doi.org/10.1176/ajp.2007.164.7.1035.
[8] Ahmedani BK, Simon GE, Stewart C, Beck A, Waitzfelder BE, Rossom R, et al. Health care contacts in the year before suicide death. J. Gen. Intern. Med. 2014;29:870–7. https://doi.org/10.1007/s11606-014-2767-3.
[9] Arsenault-Lapierre G, Kim C, Turecki G. Psychiatric diagnoses in 3275 suicides: a meta-analysis. BMC Psychiatry 2004;4:1–11. https://doi.org/10.1186/1471-244X-4-37.
[10] Turecki G, Brent DA. Suicide and suicidal behaviour. Lancet 2016;387:1227–39. https://doi.org/10.1016/S0140-6736(15)00234-2.
[11] Wu A, Wang J-Y, Jia C-X. Religion and completed suicide: a meta-analysis. PLoS One 2015;10:e0131715. https://doi.org/10.1371/journal.pone.0131715.
[12] Turecki G. The molecular bases of the suicidal brain. Nat. Rev. Neurosci. 2014;15:802–16. https://doi.org/10.1038/nrn3839.
[13] Oquendo MA, Perez-Rodriguez MM, Poh E, Sullivan G, Burke AK, Sublette ME, et al. Life events: a complex role in the timing of suicidal behavior among depressed patients. Mol. Psychiatry 2014;19:902–9. https://doi.org/10.1038/mp.2013.128.
[14] Passos IC, Mwangi B, Kapczinski F. Big data analytics and machine learning: 2015 and beyond. Lancet Psychiatry 2016;3:13–5. https://doi.org/10.1016/S2215-0366(15)00549-0.
[15] Klous S, Wielaard N. We are big data. 1st ed. Paris: Atlantis Press; 2016. https://doi.org/10.2991/978-94-6239-183-3.
[16] Bishop CM. Pattern recognition and machine learning. 1st ed. Springer; 2006.
[17] Librenza-Garcia D, Kotzian BJ, Yang J, Mwangi B, Cao B, Pereira Lima LN, et al. The impact of machine learning techniques in the study of bipolar disorder: a systematic review. Neurosci. Biobehav. Rev. 2017;80:538–54. https://doi.org/10.1016/j.neubiorev.2017.07.004.
[18] Passos IC, Mwangi B, Cao B, Hamilton JE, Wu MJ, Zhang XY, et al. Identifying a clinical signature of suicidality among patients with mood disorders: a pilot study using a machine learning approach. J. Affect. Disord. 2016;193:109–16. https://doi.org/10.1016/j.jad.2015.12.066.
[19] Obermeyer Z, Emanuel EJ. Predicting the future - big data, machine learning, and clinical medicine. N. Engl. J. Med. 2016;375:1216–9. https://doi.org/10.1056/NEJMp1606181.
[20] Passos IC, Mwangi B. Machine learning-guided intervention trials to predict treatment response at an individual patient level: an important second step following randomized clinical trials. Mol. Psychiatry 2018. https://doi.org/10.1038/s41380-018-0250-y.
[21] Passos IC, Mwangi B, Kapczinski F. Personalized psychiatry: big data analytics in mental health. 1st ed. Springer International Publishing; 2019.
[22] Sartori JM, Reckziegel R, Passos IC, Czepielewski LS, Fijtman A, Sodre LA, et al. Volumetric brain magnetic resonance imaging predicts functioning in bipolar disorder: a machine learning approach. J. Psychiatr. Res. 2018;103:237–43. https://doi.org/10.1016/j.jpsychires.2018.05.023.
[23] Wu M-J, Mwangi B, Passos IC, Bauer IE, Cao B, Frazier TW, et al. Prediction of vulnerability to bipolar disorder using multivariate neurocognitive patterns: a pilot study. Int. J. Bipolar Disord. 2017;5:32. https://doi.org/10.1186/s40345-017-0101-9.
[24] Mwangi B, Wu M-J, Cao B, Passos IC, Lavagnino L, Keser Z, et al. Individualized prediction and clinical staging of bipolar disorders using neuroanatomical biomarkers. Biol. Psychiatry Cognit. Neurosci. Neuroimaging 2016;1:186–94. https://doi.org/10.1016/j.bpsc.2016.01.001.
[25] Wu M-J, Mwangi B, Bauer IE, Passos IC, Sanches M, Zunta-Soares GB, et al. Identification and individualized prediction of clinical phenotypes in bipolar disorders using neurocognitive data, neuroimaging scans and machine learning. Neuroimage 2017;145:254–64. https://doi.org/10.1016/j.neuroimage.2016.02.016.
[26] Wu M-J, Passos IC, Bauer IE, Lavagnino L, Cao B, Zunta-Soares GB, et al. Individualized identification of euthymic bipolar disorder using the Cambridge neuropsychological test automated battery (CANTAB) and machine learning. J. Affect. Disord. 2016;192:219–25. https://doi.org/10.1016/j.jad.2015.12.053.
[27] Kessler RC, van Loo HM, Wardenaar KJ, Bossarte RM, Brenner LA, Cai T, et al. Testing a machine-learning algorithm to predict the persistence and severity of major depressive disorder from baseline self-reports. Mol. Psychiatry 2016;21:1366–71. https://doi.org/10.1038/mp.2015.198.
[28] Pinto JV, Passos IC, Gomes F, Reckziegel R, Kapczinski F, Mwangi B, et al. Peripheral biomarker signatures of bipolar disorder and schizophrenia: a machine learning approach. Schizophr. Res. 2017;188:182–4. https://doi.org/10.1016/j.schres.2017.01.018.
[29] Squeglia LM, Ball TM, Jacobus J, Brumback T, McKenna BS, Nguyen-Louie TT, et al. Neural predictors of initiating alcohol use during adolescence. Am. J. Psychiatry 2017;174:172–85. https://doi.org/10.1176/appi.ajp.2016.15121587.
[30] Kessler RC, Rose S, Koenen KC, Karam EG, Stang PE, Stein DJ, et al. How well can post-traumatic stress disorder be predicted from pre-trauma risk factors? An exploratory study in the WHO World Mental Health Surveys. World Psychiatry 2014;13:265–74. https://doi.org/10.1002/wps.20150.
[31] Kapczinski NS, Mwangi B, Cassidy RM, Librenza-Garcia D, Bermudez MB, Kauer-Sant'anna M, et al. Neuroprogression and illness trajectories in bipolar disorder. Expert Rev. Neurother. 2017;17:277–85. https://doi.org/10.1080/14737175.2017.1240615.
[32] Burke TA, Ammerman BA, Jacobucci R. The use of machine learning in the study of suicidal and non-suicidal self-injurious thoughts and behaviors: a systematic review. J. Affect. Disord. 2019;245:869–84. https://doi.org/10.1016/j.jad.2018.11.073.

[33] McKernan LC, Lenert MC, Crofford LJ, Walsh CG. Outpatient engagement lowers predicted risk of suicide attempts in fibromyalgia. Arthritis Care Res. (Hoboken) 2018. https://doi.org/10.1002/acr.23748.

[34] Walsh CG, Ribeiro JD, Franklin JC. Predicting suicide attempts in adolescents with longitudinal clinical data and machine learning. J. Child Psychol. Psychiatry 2018;59:1261–70. https://doi.org/10.1111/jcpp.12916.

[35] Kessler RC, Stein MB, Petukhova MV, Bliese P, Bossarte RM, Bromet EJ, et al. Predicting suicides after outpatient mental health visits in the army study to assess risk and resilience in servicemembers (army STARRS). Mol. Psychiatry 2017;22:544–51. https://doi.org/10.1038/mp.2016.110.

[36] Kessler RC, Warner CH, Ivany C, Petukhova MV, Rose S, Bromet EJ, et al. Predicting suicides after psychiatric hospitalization in US army soldiers: the army study to assess risk and rEsilience in servicemembers (army STARRS). JAMA Psychiatry 2015;72:49–57. https://doi.org/10.1001/jamapsychiatry.2014.1754.

[37] Colic S,J, Richardson D, James Reilly P, Gary Hasey M. Using machine learning algorithms to enhance the management of suicide ideation. In: Conf Proc. Annu Int Conf IEEE Eng Med Biol Soc IEEE Eng Med Biol Soc Annu Conf 2018; 2018. p. 4936–9. https://doi.org/10.1109/EMBC.2018.8513200.

[38] Jordan P, Shedden-Mora MC, Lowe B. Predicting suicidal ideation in primary care: an approach to identify easily assessable key variables. Gen. Hosp. Psychiatry 2018;51:106–11. https://doi.org/10.1016/j.genhosppsych.2018.02.002.

[39] Barros J, Morales S, Echavarri O, Garcia A, Ortega J, Asahi T, et al. Suicide detection in Chile: proposing a predictive model for suicide risk in a clinical sample of patients with mood disorders. Rev. Bras. Psiquiatr. 2017;39:1. https://doi.org/10.1590/1516-4446-2015-1877.

[40] Oh J, Yun K, Hwang J-H, Chae J-H. Classification of suicide attempts through a machine learning algorithm based on multiple systemic psychiatric scales. Front. Psychiatry 2017;8:192. https://doi.org/10.3389/fpsyt.2017.00192.

[41] Ryu S, Lee H, Lee D-K, Park K. Use of a machine learning algorithm to predict individuals with suicide ideation in the general population. Psychiatry Investig. 2018;15:1030–6. https://doi.org/10.30773/pi.2018.08.27.

[42] Fernandes AC, Dutta R, Velupillai S, Sanyal J, Stewart R, Chandran D. Identifying suicide ideation and suicidal attempts in a psychiatric clinical research database using natural language processing. Sci. Rep. 2018;8:7426. https://doi.org/10.1038/s41598-018-25773-2.

[43] Cook BL, Progovac AM, Chen P, Mullin B, Hou S, Baca-Garcia E. Novel use of natural language processing (NLP) to predict suicidal ideation and psychiatric symptoms in a text-based mental health intervention in Madrid. Comput. Math. Methods Med. 2016;2016:8708434. https://doi.org/10.1155/2016/8708434.

[44] Poulin C, Shiner B, Thompson P, Vepstas L, Young-Xu Y, Goertzel B, et al. Predicting the risk of suicide by analyzing the text of clinical notes. PLoS One 2014;9:e85733. https://doi.org/10.1371/journal.pone.0085733.

[45] de Avila Berni G, Rabelo-da-Ponte FD, Librenza-Garcia D, V Boeira M, Kauer-Sant'Anna M, Passos IC, et al. Potential use of text classification tools as signatures of suicidal behavior: a proof-of-concept study using Virginia Woolf's personal writings. PLoS One 2018;13:e0204820. https://doi.org/10.1371/journal.pone.0204820.

[46] Cheng Q, Li TM, Kwok CL, Zhu T, Yip PS. Assessing suicide risk and emotional distress in Chinese social media: a text mining and machine learning study. J. Med. Internet Res. 2017;19:1–10. https://doi.org/10.2196/jmir.7276.

[47] Braithwaite SR, Giraud-Carrier C, West J, Barnes MD, Hanson CL. Validating machine learning algorithms for twitter data against established measures of suicidality. JMIR Ment. Health 2016;3:e21. https://doi.org/10.2196/mental.4822.

[48] Pestian JP, Grupp-Phelan J, Bretonnel Cohen K, Meyers G, Richey LA, Matykiewicz P, et al. A controlled trial using natural language processing to examine the language of suicidal adolescents in the emergency department. Suicide Life Threatening Behav. 2016;46:154–9. https://doi.org/10.1111/sltb.12180.

[49] Pestian JP, Sorter M, Connolly B, Bretonnel Cohen K, McCullumsmith C, Gee JT, et al. A machine learning approach to identifying the thought markers of suicidal subjects: a prospective multicenter trial. Suicide Life Threatening Behav. 2017;47:112–21. https://doi.org/10.1111/sltb.12312.

[50] Just MA, Pan L, Cherkassky VL, McMakin D, Cha C, Nock MK, et al. Machine learning of neural representations of suicide and emotion concepts identifies suicidal youth. Nat. Hum. Behav. 2017;1:911–9. https://doi.org/10.1038/s41562-017-0234-y.

[51] Bernecker SL, Zuromski KL, Gutierrez PM, Joiner TE, King AJ, Liu H, et al. Predicting suicide attempts among soldiers who deny suicidal ideation in the army study to assess risk and resilience in servicemembers (Army STARRS). Behav. Res. Ther. 2018. https://doi.org/10.1016/j.brat.2018.11.018.

[52] Gradus JL, King MW, Galatzer-Levy I, Street AE. Gender differences in machine learning models of trauma and suicidal ideation in veterans of the Iraq and Afghanistan wars. J. Trauma. Stress 2017;30:362–71. https://doi.org/10.1002/jts.22210.

[53] Bzdok D, Meyer-Lindenberg A. Machine learning for precision psychiatry: opportunities and challenges. Biol. Psychiatry Cognit. Neurosci. Neuroimaging 2018;3:223–30. https://doi.org/10.1016/j.bpsc.2017.11.007.

[54] McKernan LC, Clayton EW, Walsh CG. Protecting life while preserving liberty: ethical recommendations for suicide prevention with artificial intelligence. Front. Psychiatry 2018;9:650. https://doi.org/10.3389/fpsyt.2018.00650.

[55] Adkins DE. Machine learning and electronic health records: a paradigm shift. Am. J. Psychiatry 2017;174:93–4. https://doi.org/10.1176/appi.ajp.2016.16101169.

[56] Yip PSF, Caine E, Yousuf S, Chang S-S, Wu KC-C, Chen Y-Y. Means restriction for suicide prevention. Lancet (Lond. Engl.) 2012;379:2393–9. https://doi.org/10.1016/S0140-6736(12)60521-2.

[57] Zalsman G, Hawton K, Wasserman D, van Heeringen K, Arensman E, Sarchiapone M, et al. Suicide prevention strategies revisited: 10-year systematic review. Lancet Psychiatry 2016;3:646–59. https://doi.org/10.1016/S2215-0366(16)30030-X.

[58] Pirkis J, Too LS, Spittal MJ, Krysinska K, Robinson J, Cheung YTD. Interventions to reduce suicides at suicide hotspots: a systematic review and meta-analysis. Lancet Psychiatry 2015;2:994–1001. https://doi.org/10.1016/S2215-0366(15)00266-7.

[59] Anestis MD. Prior suicide attempts are less common in suicide decedents who died by firearms relative to those who died by other means. J. Affect. Disord. 2016;189:106–9. https://doi.org/10.1016/j.jad.2015.09.007.

[60] CDC - Centers for Disease Control and Prevention. WISQARS (web-based injury statistics query and reporting system)|Injury Center|CDC. 2017. https://www.cdc.gov/injury/wisqars/index.html.

[61] Chapman S, Alpers P, Jones M. Association between gun law reforms and intentional firearm deaths in Australia, 1979-2013. J. Am. Med. Assoc. 2016;316:291–9. https://doi.org/10.1001/jama.2016.8752.

[62] Myung W, Lee G-H, Won H-H, Fava M, Mischoulon D, Nyer M, et al. Paraquat prohibition and change in the suicide rate and methods in South Korea. PLoS One 2015;10:e0128980. https://doi.org/10.1371/journal.pone.0128980.

[63] Lapierre S, Erlangsen A, Waern M, De Leo D, Oyama H, Scocco P, et al. A systematic review of elderly suicide prevention programs. Crisis 2011;32:88–98. https://doi.org/10.1027/0227-5910/a000076.

[64] Szekely A, Konkoly Thege B, Mergl R, Birkas E, Rozsa S, Purebl G, et al. How to decrease suicide rates in both genders? An effectiveness study of a community-based intervention (EAAD). PLoS One 2013;8:e75081. https://doi.org/10.1371/journal.pone.0075081.

[65] Inagaki M, Kawashima Y, Kawanishi C, Yonemoto N, Sugimoto T, Furuno T, et al. Interventions to prevent repeat suicidal behavior in patients admitted to an emergency department for a suicide attempt: a meta-analysis. J. Affect. Disord. 2015;175:66–78. https://doi.org/10.1016/j.jad.2014.12.048.

[66] World Health Organization. WHO | Preventing suicide: a resource for media professionals - update 2017. World Health Organization; 2017.

[67] Cipriani A, Hawton K, Stockton S, Geddes JR. Lithium in the prevention of suicide in mood disorders: updated systematic review and meta-analysis. BMJ 2013;346:f3646. https://doi.org/10.1136/bmj.f3646.

[68] Gibbons RD, Brown CH, Hur K, Davis J, Mann JJ. Suicidal thoughts and behavior with antidepressant treatment: reanalysis of the randomized placebo-controlled studies of fluoxetine and venlafaxine. Arch. Gen. Psychiatry 2012;69:580–7. https://doi.org/10.1001/archgenpsychiatry.2011.2048.

[69] Vitiello B, Emslie G, Clarke G, Wagner KD, Asarnow JR, Keller MB, et al. Long-term outcome of adolescent depression initially resistant to selective serotonin reuptake inhibitor treatment: a follow-up study of the TORDIA sample. J. Clin. Psychiatry 2011;72:388–96. https://doi.org/10.4088/JCP.09m05885blu.

[70] Grunebaum MF, Ellis SP, Duan N, Burke AK, Oquendo MA, John Mann J. Pilot randomized clinical trial of an SSRI vs bupropion: effects on suicidal behavior, ideation, and mood in major depression. Neuropsychopharmacology 2012;37:697–706. https://doi.org/10.1038/npp.2011.247.

[71] Sharma T, Guski LS, Freund N, Gotzsche PC. Suicidality and aggression during antidepressant treatment: systematic review and meta-analyses based on clinical study reports. BMJ 2016;352:i65. https://doi.org/10.1136/bmj.i65.

[72] Ludwig J, Marcotte DE, Norberg K. Anti-depressants and suicide. J. Health Econ. 2009;28:659–76. https://doi.org/10.1016/j.jhealeco.2009.02.002.

[73] Moller H-J. Is there evidence for negative effects of antidepressants on suicidality in depressive patients? A systematic review. Eur. Arch. Psychiatry Clin. Neurosci. 2006;256:476–96. https://doi.org/10.1007/s00406-006-0689-8.

[74] Otuyelu E, Foldvari A, Szabo E, Sipos V, Edafiogho P, Szucs M, et al. Antidepressant drugs and teenage suicide in Hungary: time trend and seasonality analysis. Int. J. Psychiatry Clin. Pract. 2015;19:221–5. https://doi.org/10.3109/13651501.2015.1061017.

[75] Nyman AK, Jonsson H. Patterns of self-destructive behaviour in schizophrenia. Acta Psychiatr. Scand. 1986;73:252–62.

[76] Meltzer HY. Clozapine treatment for suicidality in schizophrenia international suicide prevention trial (InterSePT). Arch. Gen. Psychiatry 2003;60:82. https://doi.org/10.1001/archpsyc.60.1.82.

[77] Ringbäck Weitoft G, Berglund M, Lindström EA, Nilsson M, Salmi P, Rosén M. Mortality, attempted suicide, re-hospitalisation and prescription refill for clozapine and other antipsychotics in Sweden-a register-based study. Pharmacoepidemiol. Drug Saf. 2014;23:290–8. https://doi.org/10.1002/pds.3567.

[78] Fond G, Loundou A, Rabu C, Macgregor A, Lancon C, Brittner M, et al. Ketamine administration in depressive disorders: a systematic review and meta-analysis. Psychopharmacology (Berl.) 2014;231:3663–76. https://doi.org/10.1007/s00213-014-3664-5.

[79] Zarate CAJ, Brutsche NE, Ibrahim L, Franco-Chaves J, Diazgranados N, Cravchik A, et al. Replication of ketamine's antidepressant efficacy in bipolar depression: a randomized controlled add-on trial. Biol. Psychiatry 2012;71:939–46. https://doi.org/10.1016/j.biopsych.2011.12.010.

[80] De Berardis D, Fornaro M, Valchera A, Cavuto M, Perna G, Di Nicola M, et al. Eradicating suicide at its roots: preclinical bases and clinical evidence of the efficacy of ketamine in the treatment of suicidal behaviors. Int. J. Mol. Sci. 2018;19. https://doi.org/10.3390/ijms19102888.

[81] Fan W, Yang H, Sun Y, Zhang J, Li G, Zheng Y, et al. Ketamine rapidly relieves acute suicidal ideation in cancer patients: a randomized controlled clinical trial. Oncotarget 2017;8:2356–60. https://doi.org/10.18632/oncotarget.13743.

[82] Canuso CM, Singh JB, Fedgchin M, Alphs L, Lane R, Lim P, et al. Efficacy and safety of intranasal esketamine for the rapid reduction of symptoms of depression and suicidality in patients at imminent risk for suicide: results of a double-blind, randomized, placebo-controlled study. Am. J. Psychiatry 2018;175:620–30. https://doi.org/10.1176/appi.ajp.2018.17060720.

[83] Rajkumar R, Fam J, Yeo EYM, Dawe GS. Ketamine and suicidal ideation in depression: jumping the gun? Pharmacol. Res. 2015;99:23–35. https://doi.org/10.1016/j.phrs.2015.05.003.

[84] Andrade C. Ketamine for depression, 6: effects on suicidal ideation and possible use as crisis intervention in patients at suicide risk. J. Clin. Psychiatry 2018;79. https://doi.org/10.4088/JCP.18f12242.

[85] Sanacora G, Heimer H, Hartman D, Mathew SJ, Frye M, Nemeroff C, et al. Balancing the promise and risks of ketamine treatment for mood disorders. Neuropsychopharmacology 2017;42:1179–81. https://doi.org/10.1038/npp.2016.193.

[86] Stoffers JM, Vollm BA, Rucker G, Timmer A, Huband N, Lieb K. Psychological therapies for people with borderline personality disorder. Cochrane Database Syst. Rev. 2012:CD005652. https://doi.org/10.1002/14651858.CD005652.pub2.

[87] Tarrier N, Taylor K, Gooding P. Cognitive-behavioral interventions to reduce suicide behavior: a systematic review and meta-analysis. Behav. Modif. 2008;32:77–108. https://doi.org/10.1177/0145445507304728.

[88] Linthicum KP, Schafer KM, Ribeiro JD. Machine learning in suicide science: applications and ethics. Behav. Sci. Law 2019. https://doi.org/10.1002/bsl.2392.

[89] Sinyor M, Tse R, Pirkis J. Global trends in suicide epidemiology. Curr. Opin. Psychiatry 2017;30:1–6. https://doi.org/10.1097/YCO.0000000000000296.

[90] Kapur N, Ibrahim S, While D, Baird A, Rodway C, Hunt IM, et al. Mental health service changes, organisational factors, and patient suicide in England in 1997-2012: a before-and-after study. Lancet Psychiatry 2016;3:526–34. https://doi.org/10.1016/S2215-0366(16)00063-8.

Chapter 33

Artificial intelligence in gastroenterology

James K. Ruffle[1,2], Adam D. Farmer[1,3] and Qasim Aziz[1]

[1]*Centre for Neuroscience and Trauma, Blizard Institute, Wingate Institute of Neurogastroenterology, Barts and the London School of Medicine & Dentistry, Queen Mary University of London, London, United Kingdom;* [2]*Department of Radiology, University College London NHS Foundation Trust, London, United Kingdom;* [3]*Department of Gastroenterology, University Hospitals of North Midlands NHS Trust, Stoke on Trent, United Kingdom*

Introduction

Artificial intelligence (AI) frequents mainstream media news, from such areas as driver-less cars, recommendation systems of television streaming services, or advanced robotics [1,2]. Moreover, it has a number of potential applications in science and technology. Nevertheless, this enthusiasm and positivity have also been tempered with concerns, including the notion of "robots" replacing workforces, or even the end of mankind, as the late Stephen Hawking hypothesized back in 2014 [3].

Furthermore, as the discussion of AI becomes ever-more common, the novelty and excitement it affords generates unsustainable hype, referred to by Gartner as the "*peak of inflated expectation*," a preceding state before this overexaggerated excitement falls to the "trough of disillusionment," when the ideas and hype generated simply do not live up to reality [4]. Needless to say, the prospective utility of AI in healthcare is an enticing and rapidly developing field [5]. Will it revolutionize healthcare, damage it, or perhaps have minimal effect at all?

Artificial intelligence and data-driven decision making

Artificial intelligence

AI, in essence, refers to a computer performing a task associated with an intelligent being. Included are "cognitive" functions not dissimilar to the human mind, such as the ability to "learn" [6]. Despite interest in AI seeming relatively recent by virtue of media attention, its concept is in-fact far more developed, such as the works of Alan Turing (1912–54), a pioneer of computer science, or even ancient such as in Aristotle's study of logic (384–322 BC) [6]. One could consider AI as the subject field with many subfields, in the same light that gastroenterology engenders inflammatory bowel diseases, endoscopy, nutrition, *etc*. These areas of gastroenterology all interlink in multiple ways, to which AI subfields are no different (Fig. 33.1).

Machine learning(ML)

ML is defined as an area of computer science, wherein a system can develop an ability to "learn" with data, without any explicit programming. Historically, it was coined by Arthur Samuel in 1959, a pioneer of computer games, while at IBM. Machine learning originates from computational learning theory with pattern recognition, where a computer can develop an algorithm to make a prediction, based upon recognition of patterns or features in the data. The developed predictive *model* can then adapt further to new circumstances, such as improvement in accuracy with more practice or time.

As a simple example in gastroenterology, Kinar et al. trained a machine learning model to detect the presence or absence of colorectal cancer (the target), by complete blood count data (the input predictors), with an overall accuracy of 82% [7]. This particular paradigm is that of a binary classifier, disease positive or disease negative, and used a specific model known as a decision tree ensemble. In training a classification model, there are a number of algorithmic methods: decision trees, an ensemble of trees, support vector machines, K-nearest neighbors, to name but a few. The details on the "types of learning" is beyond the scope of this chapter, but we would recommend Dey's "*Machine learning algorithms: a review*" for further reading [8].

There is often some uncertainty as to the best or "gold standard" methodological approach, to be adopted for a given research question, and thus, some machine learning toolboxes now automatically train a model using a variety

Precision Medicine for Investigators, Practitioners and Providers. https://doi.org/10.1016/B978-0-12-819178-1.00033-2

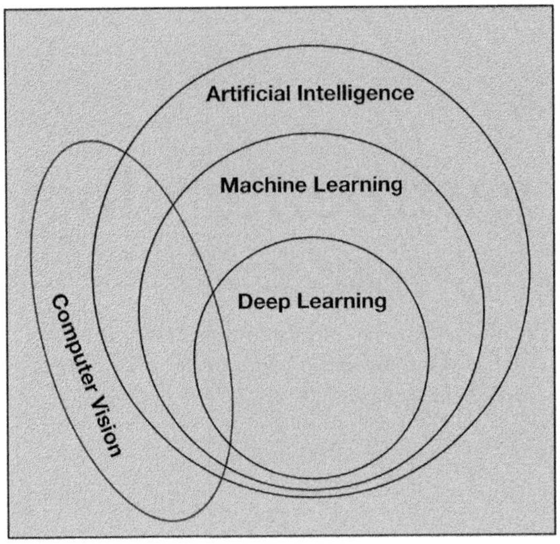

FIGURE 33.1 Nomenclature in artificial intelligence. *Adapted from Ruffle JK, Farmer AD, Aziz Q. Artificial intelligence assisted gastroenterology — promises and pitfalls. Am. J. Gastroenterol 2019;114(3):422–8. https://doi.org/10.1038/s41395-018-0268-4.*

of learning techniques, and depict the receiver operator characteristic or performance for each, enabling the investigator to determine the most effective model to use [9].

Data partitioning: training, validation, testing

Data partitioning is essential when developing a predictive model. The essential elements to partition in the data are referred to as *training, validation,* and *testing*. This follows a simple rationale, namely that if one is to develop a predictive model upon a specific dataset, how could it be subsequently tested on unseen data or samples? For achieving this, it is common to use randomized partitioning of data, to an initial training set to train the model, frequently in the region of 70% of the available samples. The remaining samples would then be allocated for model validation and testing. The function of the validation partition, is to ensure that a model does not "overfit" to a training dataset—a phenomenon which means that the machine will not develop an algorithm to make predictions too close to that of the training data, and then does not generalize well, performing poorly on the never before seen testing samples. The testing dataset is used to test the performance of the model.

Deep learning

Deep learning is an area of AI that uses multiple layers of nonlinear processing to extract features of the data and transform them in order to predict a desirable target [10]. *Feature extraction* is the automated selection of variables, that have some degree of predictive power to predict a target, while *transformation* is the augmentation of data, into a means more effective in building the model. Simply put, physicians intrinsically perform feature extraction on a regular basis, such as in taking a medical history and "extracting" the salient points, in predicting a patient's diagnosis. The difference in deep learning, however, is that this feature extraction is automated computationally.

Using the previously discussed colorectal cancer example [7], by providing complete blood count data, feature extraction can select measures of the full blood count, that yield the most "predictive power" in predicting a diagnosis of colorectal cancer, and furthermore assign weights to each of the variables, dependent on their usefulness. For instance, in this model, hemoglobin, mean corpuscular hemoglobin, mean cell volume, and hematocrit were found to have the highest predictive powers, while other measures (such as monocytes) held a lesser contribution. Deep learning has notably formed a key methodology in the research of medical image analysis, and a well-known framework to develop with here is "TensorFlow" [11].

Neural networks

In recent years, *Neural networks* are also frequently discussed and utilized in machine learning. It is named as of its similarity to the brain's neural network. Neural networks incorporate a network of elements called *neurons*, which receive inputs (such as patient data) and subsequently change their internal state (such as activating), depending on the input, which then produces an output contingent on this (such as predicting a diagnosis). By using the data provided as input vectors, the neural network acts as an interconnected group of nodes, to predict a response. In healthcare research, much of the current AI models seemingly use neural networks [12,13].

Below, we provide a hypothetical example using a neural network, in predicting both a diagnosis from a differential (binary classifier) and also a predicted treatment efficacy (continuous variable) in (Fig. 33.2), using inflammatory bowel disease (IBD) as the example. Similar to the aforementioned classification models, the neural network is a partly all-encompassing terminology, which incorporates a variety of different learning types, examples of which are Scaled-Conjugate Gradient, Levenberg–Marquardt and Bayesian Regularization, to name but a few. Further reading on learning types can be found here [6,8].

Patterns and interactions in data

An additional use that machine learning could provide in developing predictive models is the automated analysis of interaction factors in predicting a target. By and large, a large proportion of epidemiological studies in predicting an outcome, rely on the principal analysis of one variable and relating that to an outcome, for instance, the risk of lung

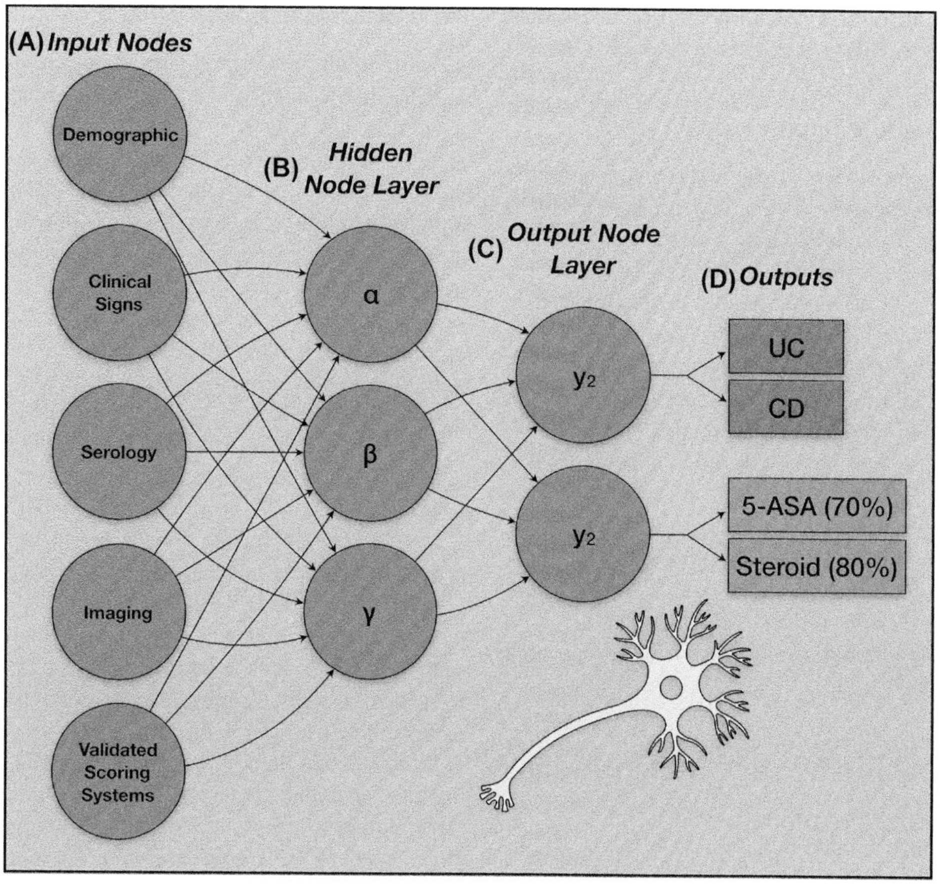

FIGURE 33.2 Example of a neural network to predict the diagnosis of ulcerative colitis (UC) or Crohn's disease (CD), and moreover whether treatment with a 5-aminosalicylic acid (5-ASA) or a corticosteroid, would be more effective. (A) Example input vectors about the patient, to be inputted to the neural network include history, examination findings, an investigation, and so on. (B, C) In neural network design, the utility of these parameters in (A) are combined, weighted and biased, depending on their predictive power at a hidden nodal layer, to provide an output layer. (D) After the internal steps of (B, C), the algorithm determines a response – such as a diagnosis from a differential (output layer y_1) – a binary classifier, or a predicted response to a treatment (output y_2) – a continuous variable. In this hypothetical example, this data could then be taken into consideration between doctor and patient, to determine what may be best for the patient. For instance, the doctor/patient may elect to choose the corticosteroid, as of higher predicted symptom relief, given this particular patient's input data. However, while the steroid in this example displays a 10% greater symptomatic response, the doctor/patient may feel this 10% is "not worth" the adverse effect profile, that accompanies treatment with steroids. Either way, this provides a powerful tool to make a data-driven decision in healthcare, at the individual level. *Figure from Ruffle JK, Farmer AD, Aziz Q. Artificial intelligence assisted gastroenterology – promises and pitfalls. Am. J. Gastroenterol 2019;114(3):422–8. https://doi.org/10.1038/s41395-018-0268-4.*

cancer in smokers. A scientist can then introduce other variables, and interrogate their relationship to the predictor (smoking), or the target (lung cancer), either methodologically, such as with the Bradford Hill causality criteria (a set of viewpoints to determine if an observed epidemiological association is causal) [14,15], or even statistically, such as with logistic regression.

In machine learning, however, this process can be a highly efficient process of recognizing interactions between variables, and how they relate to the target predictor, in an automated way. For example, let us say, a hepatologist wanted to develop a model to predict alcoholic liver disease, by using serological liver function tests results. The presence or absence of alcoholic liver disease would be the target to predict (a binary classifier), and the liver function test values would be the input vectors, on which the model would base its prediction.

In feature extraction, the model would assign particular weights to predict a target. For example, aspartate transaminase (AST) would likely have more predictive power than lactate dehydrogenase (LDH). However, the model would also *likely* determine that, in fact, the interaction of AST with alanine transaminase (ALT) also holds predictive power. Notably, as this is similar to how a physician may calculate the AST/ALT ratio in alcoholic liver disease [16,17]. However, what is also important to appreciate here is our stipulating what the model would *likely* do. There exists an inherent limitation in current AI technology where, despite a model being hugely effective at a prediction, its means of how it achieves this precisely often remains unknown, a "black box" technology of sorts.

Image analysis and computer vision

Computer vision refers to the AI subfield of using computer systems in the processing of visual data, images or videos, for instance how a computer may ascertain information from this (and make a prediction accordingly). This is a frequently discussed area in the media, for example, driverless cars, which has featured prominently on Gartner's recent predictions of technological advances [4,18].

Historically, radiology has been a specialty of humans undertaking pattern-recognition of images to predict a diagnosis, such as spotting polyps on CT colonography [19]. The use of computer vision in radiological investigations is a hot topic, as to how AI could augment and improve the field. Moreover, a key message from radiological academic meetings has been that "*Rads [Radiologists] who use AI will replace rads who don't*" [20]. Perhaps the question could also be asked for certain areas of gastroenterology?

Guideline specific treatment algorithms

AI assisting treatment allocation

A central facet of contemporaneous medical practice (not just in gastroenterology), is the use of treatment guideline pathways or algorithms, typically determined by evidence and expert consensus. The physician identifies the diagnosis and follows the specific investigative and treatment algorithm for that particular disorder, regardless of the individual patient. For instance, in upper gastrointestinal (GI) bleeding, physicians follow a sequence of treatment algorithm of steps X, Y, and Z, which were all determined to be best practice.

In the upper GI bleeding example, one point of outstanding controversy in management is the use of proton pump inhibitors (PPI) prior to endoscopy [21]. There is without question a plethora of research studies investigating upper GI bleeding over recent decades (see Cochrane), and with that considerable patient data from clinical trials. The nature of this data might be demographics, patient comorbidity, GI-bleed score systems, *etc.*, before intervention but also that of patient outcome. This serves as a perfect example of how this data could be used with machine learning, wherein this preinterventional data could be used for predictive vectors, and the predictive outcome being a clinical outcome with or without PPI. Furthermore, this could then be developed further to build a model to ascertain whether a specific patient would benefit from preendoscopic PPI or not.

AI assisting treatment choice

Often enough, there might be a handful of possible treatments, and it might not be a simple "black or white" decision, in choosing what a patient would best benefit from. The quantification of a patient's response to treatment has an important role here, whether that is by the quality of life, symptom relief (e.g., pain, nausea, bowel habit), or other markers of improvement, such as inflammatory serological markers, endoscopic, or radiological appearances. Outside of AI, the use of quality-adjusted life years (QALY) forms a vital area in health outcomes and decisions in treatments, including with an economic role in the UK's National Health Service [22].

Let us consider irritable bowel syndrome (IBS), an enigmatic disorder which has frequently been reclassified, and with that its recommended treatment algorithms [23–25]. The choice of X or Y might not be clear cut, it might frankly feel more like a "gut feeling," or perhaps the *a priori* knowledge that the physician might have about the specific patient, and the condition might lead to choosing one treatment over the other.

There is without question vast amounts of patient data from clinical trials, of different treatment options in IBS, with associated demographics, comorbidities, investigation results, *etc.*, but also response to treatments (be that an objective measure or a patient-quantified response), such as "*on a scale of 1-10, how much improvement did you feel?.*" Machine learning models can be trained to fit continuous variable data (such as symptomatic response following treatment). An example here is the neural network "fitting" algorithms. Could machine learning algorithms be used to predict a percentage improvement to multiple therapies? This could enable the gastroenterologist to make a data-driven decision at the individual patient level, i.e., *personalized medicine* (Fig. 33.3).

Computer vision in endoscopy

The majority of current AI research in gastroenterology focuses on computer vision in endoscopy [26,27], though capsule endoscopy has also recently been studied, such as automatic detection of GI angioectasia [28]. For example, in polyp surveillance in endoscopy, a real-time video is depicted on a screen for the gastroenterologist to interpret, while simultaneously facing the difficulty of navigating the mobilizing colon, through a two-dimensional representation of what is actually a three-dimensional structure.

In addition to these difficulties, there are the distractions from the procedure, such as if the patient is in discomfort or pain, or if the bowel preparation is suboptimal . This could result in a polyp being missed. Perhaps AI computer vision could aid the gastroenterologist here, identifying suspicious lesions from the video feed, in *real-time*. Notably, this

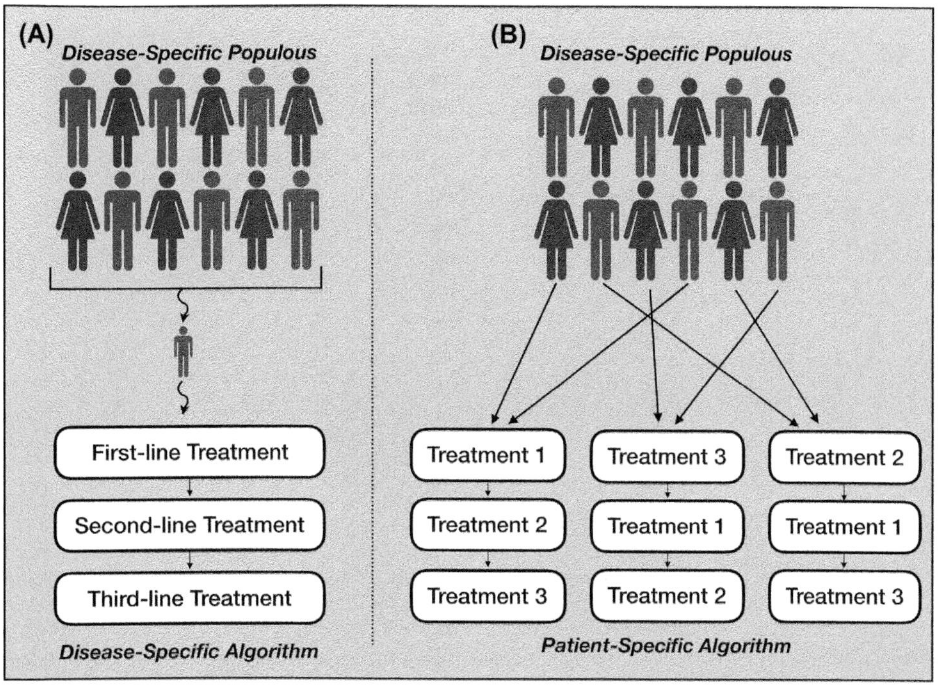

FIGURE 33.3 From disease-specific algorithms to patient-specific with machine learning. (A) Conventional healthcare provision of a given disease, focuses on the definition of a treatment pathway determined by best clinical practice, cost efficiency, scientific literature, *etc.* (B) With machine learning, perhaps models could be developed to stratify patients to treatments, most likely to benefit them specifically, by personalizing medicine. *Adapted from Ruffle JK, Farmer AD, Aziz Q. Artificial intelligence assisted gastroenterology – promises and pitfalls. Am. J. Gastroenterol 2019;114(3):422–8. https://doi.org/10.1038/s41395-018-0268-4.*

would not be dissimilar to if there were a second gastroenterologist watching the endoscopy video-feed, highlighting areas they thought were of concern. We provide a hypothetical example of this below (Fig. 33.4).

There are a number of promising recent studies, developing machine learning models for polyp detection [29–32], or even a step further, the differentiation between adenomatous or hyperplastic [33]. Referring to two studies in particular, Hirasawa et al., developed a neural network to predict gastric cancer, utilizing a *training* set of more than 13,000 endoscopic images, and reported an overall sensitivity of 92%, when *testing* it on a further 2200 images [29]. Meanwhile, Byrne et al. built a neural network using a *training* set of 223 videos (split into >60,000 image frames), and, when tested on a further 125 endoscopy videos, report a 94% accuracy rate, in differentiating diminutive adenomas from hyperplastic polyps [33]. A key learning point to note is the large sample size required to produce robust algorithms.

Aiding in differential diagnoses

Thus far, we have described AI approaches that form a category referred to as *supervised learning*, a terminology which is defined by the human investigator supplying the definitively correct answer for a machine to build a model to predict, such as a diagnosis of bowel cancer, or the known outcome to a particular treatment. This correct answer is often referred to as the *ground truth*. In contrast to supervised learning is *unsupervised learning*, which is when the ground truth is not known (or not supplied to the model).

This area involves providing a dataset with data vectors (such as patient data, symptoms, blood results, imaging), and employing the machine to identify patterns, classifiers or clusters based upon this, without specifying what groups are expected. For instance, this could be useful in gastroenterology, wherein there is an arguably vague and homogenous disease label, for what is perhaps much more heterogeneous, such as functional constipation and IBS with constipation. Perhaps an unsupervised algorithm could identify patterns in patients with constipation, to provide more useful clusters. Functional dyspepsia would, perhaps, also be another poignant example here.

Perspectives for disease prevention and temporality

Machine learning can also predict the temporal evolution of data and trends. Key examples of temporal AI predictions in the commercial sector is financial forecasting, or developing more accurate weather prediction. As a healthcare example, perhaps machine learning could be used to more accurately predict temporal changes, treatment outcomes,

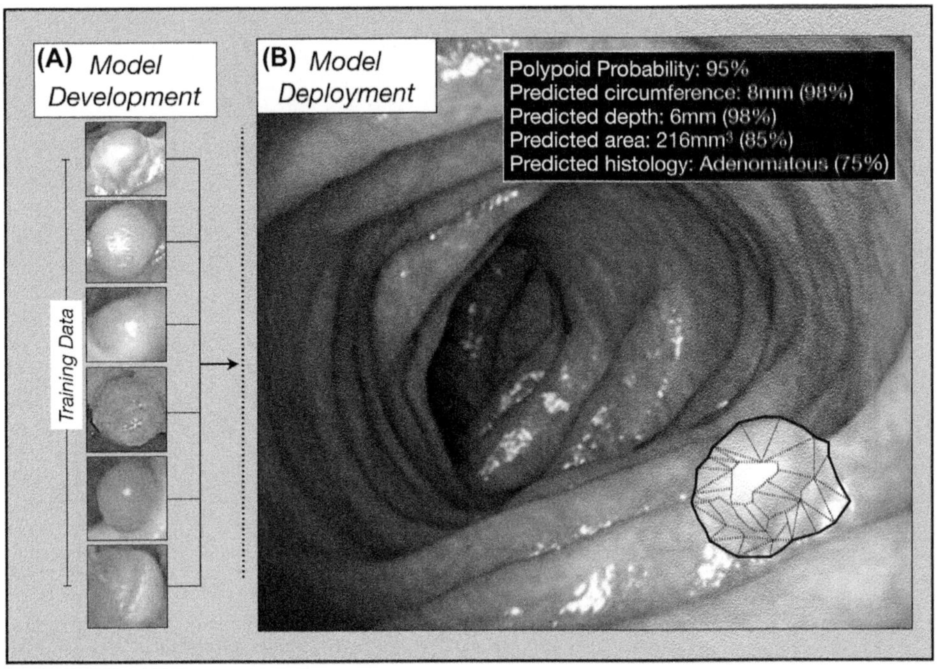

FIGURE 33.4 Computer-vision aided endoscopy. (A) The model is first trained by providing the machine with a large number of training data, images of polyps in this case. (B) A hypothetical example of how the deployment of such a system might take place, whereby in *real-time*, areas of an image that closely relate to the ground truth training data are highlighted to the investigator, and its 3-dimensional geometry recognized, with the probability of it being a polyp illustrated, as well as other metrics. Notably, this is similar to how algorithms of human-testing on the Internet functions: "select all the images with street signs." *Figure from Ruffle JK, Farmer AD, Aziz Q. Artificial intelligence assisted gastroenterology — promises and pitfalls. Am. J. Gastroenterol 2019;114(3):422–8. https://doi.org/10.1038/s41395-018-0268-4.*

or even survival at an individual patient level, as opposed to say the conventional use of a Kaplan-Meir curve, to apply to all patients. For instance, in inflammatory bowel disease (IBD), the physician has to make a decision whether or not to escalate an individual patient to biological therapy. Perhaps temporal AI-analytics could predict the time point in the patient's disease, at which they are likely to necessitate it, benefit from it most, weighted against side effects that accompany it (Fig. 33.5)?

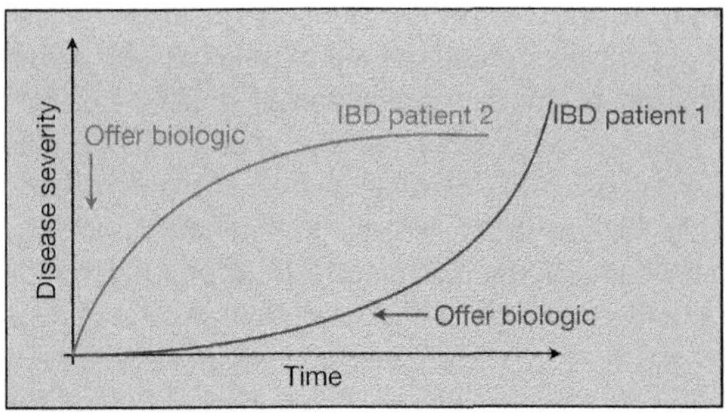

FIGURE 33.5 Predicting disease course temporally to ascertain the effective time-point for treatment intervention. A hypothetical example of two patients with IBD. It is possible that with preexisting patient data, a temporal predictive model could be devised from their disease severity and trajectory, and with that the suggested intervention point with a new therapy. Consider in this example how the gradient of worsening disease severity, differs in both patients significantly, wherein an accurately trained model could purport the point at which an intervention (here, a biological therapy), be introduced to attenuate this gradient.

Research opportunities, limitations, and future directions

Research opportunity

Much of contemporaneous research requires a prospective, expensive/large grant-funded collection of data, over several years, to ascertain a research question as to whether patients benefit most from treatment X or Y. However, data-science and AI methodology offer the utility of using previously acquired large datasets retrospectively to build a model. In fact, they typically rely on this, as the ground truth is therefore established. A salient point in machine learning research is the necessity for large sample sizes, to ensure reproducibility and prevent an overfit; consider the large sample sizes (in the thousands), of the aforementioned computer-vision endoscopy papers as an example.

However, much of this data is archived by research groups, governments, pharmaceutical companies, data repositories, or even made freely available with author's manuscripts. Wherein a dataset was used to determine one clinical question, the dataset can be used for machine learning, to teach a model in a cost-effective, data-effective manner.

Thousands of algorithms to predict thousands of diseases, or one algorithm to predict a thousand diseases?

An "AI-marketplace" has been suggested to become commonplace, by multiple commercial/industrial AI-leaders. Indeed, one for radiology already exists [34]. While this might seem enticing, there lies an inherent limitation to this, which is the differential diagnosis.

If a gastroenterologist reviews a patient, he may conclude with a diagnostic "impression," X, and a series of differentials, Y and Z. In this case, the gastroenterologist needs a test to confirm the impression, rule out differentials, or perhaps even reveal a different diagnosis entirely. These are measures to determine the *ground truth*. Meanwhile, what the gastroenterologist almost certainly does *not* say is that "it's either disorder X or they are completely healthy."

Yet, to date, this is how the vast majority of machine learning models have been developed (and continue to be), where the machine is provided with an array of diseased and healthy controls, and the model needs to identify which is which. Considering the endoscopy-computer vision examples – near all models published thus far function by identifying a singular pathology amongst healthy tissue (e.g., a polyp), rather than identifying *any* pathology, let alone differentiate into distinct pathological diagnoses.

With this in mind, it could be questionable how clinically useful many of these machine learning models actually would be, if deployed in healthcare at present. Rather, we would recommend that models need to be developed that aim to tease out a difficult differential diagnosis in a more effective manner than a physician could.

Algorithm ownership and responsibility

Will hospitals, governments, insurance companies, or even individual physicians, purchase or license the use of algorithms? Legally, who would take responsibility for any erroneous predictions, missed or incorrect diagnoses, for instance, the algorithm purchaser or the designer?

Artificial intelligence does not equate to artificial humanity

The view of "robots replacing the workforce" is not uncommon, forming many a media headline, not least in Hollywood dystopian-themed films. However, while this *may* be a genuine concern in some occupations, we argue the physician/gastroenterologist is not one of them. A principal reason for which is that artificial intelligence is not equivalent to artificial humanity. A machine could be trained to spot a suspicious lesion among healthy tissue on endoscopic images with relative "ease;" however, there are certain areas the machine could not perform or outperform the physician, not least in communication, emotional awareness or empathy.

An AI algorithm may decide that, for a given patient with bowel cancer, he or she would benefit from treatment X over Y, based upon the probability of therapeutic success or survival rate. However, the algorithm would lack an ability to be dynamic, and accommodate the ideas and concerns of the patient (or physician for that matter). Perhaps the agenda of the patient's treatment preference would be vastly different, determined by side effect burden rather than the predicted treatment response on which the model was trained. To that end, it is important to appreciate that while in many ways AI could revolutionize healthcare, the means by which it did so would need to be through AI-assisted practice, not AI-driven.

Stewardship with artificial intelligence in healthcare

As described, supervised or unsupervised learning represent approaches in AI, depending on whether the ground truth is known, or indeed provided. Further to this is the concept of feature extraction and transformation, wherein if an array of different data/variables is provided for a patient, then the model will train, automatically extract and weight the most important or predictive features. While this approach is hugely enticing, perhaps even sounding somewhat "hands-off", this is not the case, and in-fact, the machine can be misled and is vulnerable to garbage or misleading data.

One could train a model to predict a diagnosis of IBS in a cohort of individuals, compared to healthy controls, using demographics, symptom scoring, or psychophysiological data. If for instance, all the IBS patients used in training the model by happenstance were female, the model would *likely* learn this feature, and erroneously only then label female patients as having IBS, namely that a patient being male would be negatively weighted in the model.

References

[1] Baruch J. Steer driverless cars towards full automation. Nature 2016;536(7615):127.

[2] Research N. Netflix research articles. 2018. https://research.netflix.com/articles.

[3] Cellan-Jones R, Hawking S. Warns artificial intelligence could end mankind. BBC News; 2014.

[4] Panetta K. 5 trends emerge in the gartner hype cycle for emerging technologies. 2018. https://www.gartner.com/smarterwithgartner/5-trends-emerge-in-gartner-hype-cycle-for-emerging-technologies-2018/.

[5] Ruffle JK, Farmer AD, Aziz Q. Artificial intelligence assisted gastroenterology – promises and pitfalls. Am. J. Gastroenterol 2019 Mar;114(3):422–8. https://doi.org/10.1038/s41395-018-0268-4. ePub ahead of print.
[6] Russell SJ, Norvig P. Artificial inteliigence, a modern approach. Pearson Education; 2009.
[7] Kinar Y, Kalkstein N, Akiva P, Levin B, Half EE, Goldshtein I, et al. Development and validation of a predictive model for detection of colorectal cancer in primary care by analysis of complete blood counts: a binational retrospective study. J. Am. Med. Inform. Assoc 2016;23(5):879–90.
[8] Dey A. Machine learning algorithms: a review. Int. J. Comput. Sci. Inf. Technol 2016;7(3):1174–9.
[9] MathWorks. MathWorks. Classification learner. 2018. https://uk.mathworks.com/help/stats/classificationlearner-app.html.
[10] Schmidhuber J. Deep learning in neural networks: an overview. Neural Network 2015;61:85–117.
[11] Tensorflow. An open source machine learning framework for everyone. 2018. https://www.tensorflow.org/.
[12] Nigam VP, Graupe D. A neural-network-based detection of epilepsy. Neurol. Res. 2004;26(1):55–60.
[13] Waxman JA, Graupe D, Carley DW. Automated prediction of apnea and hypopnea, using a LAMSTAR artificial neural network. Am. J. Respir. Crit. Care Med 2010;181(7):727–33.
[14] Hill AB. The environment and disease: association or causation? Proc. Roy. Soc. Med 1965;58:295–300.
[15] Fedak KM, Bernal A, Capshaw ZA, Gross S. Applying the Bradford Hill criteria in the 21st century: how data integration has changed causal inference in molecular epidemiology. Emerg. Themes Epidemiol 2015;12:14.
[16] Carey WD. How should a patient with an isolated GGT elevation be evaluated? Clevel. Clin. J. Med 2000;67(5):315–6.
[17] Limdi JK, Hyde GM. Evaluation of abnormal liver function tests. Postgrad. Med. J 2003;79(932):307.
[18] Panetta K. Top trends in the gartner hype cycle for emerging technologies. Gartner; 2017. p. 2017.
[19] Huang A, Li J, Summers RM, Petrick N, Hara AK. Improving polyp detection algorithms for CT colonography: pareto front approach. Pattern Recogn. Lett. 2010;31(11):1461–9.
[20] RSNA AIMI. 2017: Rads who use AI will replace rads who don't Stanford University: artificial intelligence in medicine & imaging. 2017. https://aimi.stanford.edu/about/news/rsna-2017-rads-who-use-ai-will-replace-rads-who-don-t.
[21] Sreedharan A, Martin J, Leontiadis GI, Dorward S, Howden CW, Forman D, et al. Proton pump inhibitor treatment initiated prior to endoscopic diagnosis in upper gastrointestinal bleeding. Cochrane Database Syst. Rev 2010;7.
[22] Whitehead SJ, Ali S. Health outcomes in economic evaluation: the QALY and utilities. Br. Med. Bull 2010;96:5–21.
[23] Farmer AD, Ruffle JK. Irritable bowel syndrome. Hamdan Med. J 2015;8(3).
[24] Farmer A, Ruffle J. Irritable bowel syndrome: enigmatic for doctors, problematic for patients. Trends Urol. Men's Health 2017;8(2):13–6.
[25] Drossman DA, Hasler WL. Rome IV-functional GI disorders: disorders of gut-brain interaction. Gastroenterology 2016;150(6):1257–61.
[26] Misawa M, Kudo SE, Mori Y, Cho T, Kataoka S, Yamauchi A, et al. Artificial intelligence-assisted polyp detection for colonoscopy: initial experience. Gastroenterology 2018;154(8):2027–2029 e3.
[27] Alagappan M, Brown JRG, Mori Y, Berzin TM. Artificial intelligence in gastrointestinal endoscopy: the future is almost here. World J. Gastrointest. Endosc 2018;10(10):239–49.
[28] Leenhardt R, Vasseur P, Li C, Saurin JC, Rahmi G, Cholet F, et al. A neural network algorithm for detection of GI angiectasia during small-bowel capsule endoscopy. Gastrointest. Endosc 2019;89(1):189–94.
[29] Hirasawa T, Aoyama K, Tanimoto T, Ishihara S, Shichijo S, Ozawa T, et al. Application of artificial intelligence using a convolutional neural network for detecting gastric cancer in endoscopic images. Gastric Cancer 2018;21(4):653–60.
[30] East JE, Rees CJ. Making optical biopsy a clinical reality in colonoscopy. Lancet Gastroenterol Hepatol 2018;3(1):10–2.
[31] Yuan Y, Meng MQ. Deep learning for polyp recognition in wireless capsule endoscopy images. Med. Phys 2017;44(4):1379–89.
[32] Zhou T, Han G, Li BN, Lin Z, Ciaccio EJ, Green PH, et al. Quantitative analysis of patients with celiac disease by video capsule endoscopy: a deep learning method. Comput. Biol. Med 2017;85:1–6.
[33] Byrne MF, Chapados N, Soudan F, Oertel C, Linares Perez M, Kelly R, et al. Real-time differentiation of adenomatous and hyperplastic diminutive colorectal polyps during analysis of unaltered videos of standard colonoscopy using a deep learning model. Gut 2017.
[34] Holzberger K. AI marketplace is open for business! developers create deep learning algorithms for radiology. Nuance. 2018. https://whatsnext.nuance.com/healthcare/ai-marketplace-is-open-for-business-developers-create-deep-learning-algorithms-for-radiology/.

Chapter 34

Algorithms for epilepsy monitoring

Giampaolo Vatti[1], Paolo Detti[2], Zabalo Manrique de Lara[2], Lucia Billeci[3], Alessandro Tonacci[3] and Maurizio Varanini[3]

[1]*Department of Neurological and Sensory Sciencese, Azienda Ospedaliera Universitaria Senese, Siena, Italy;* [2]*Department of Information Engineering and Mathematics, University of Siena, Siena, Italy;* [3]*Institute of Clinical Physiology, National Research Council of Italy (IFC-CNR), Pisa, Italy*

Epilepsy and epileptic seizure prediction

Epilepsy is a neurological disorder [1], whose main clinical sign are the so-called "epileptic seizures," having a significant social and sanitary impact, both due to its quite high incidence and chronicity. In fact, it is estimated that the lifetime prevalence of epilepsy is around 7.60 per 1000 persons, whereas its incidence rate appears to be of 61.44 per 100,000 person years [2]. Seizures are unpredictable events so disability caused by epilepsy can be significant, due to uncertainty and consequences of the fits.

Drugs for controlling seizures are effective in about two-thirds of the patients, whereas the remainder one-third is affected by drug resistance [3]. Drug-resistant cases could benefit from resective surgical procedures, but this approach is still largely underutilized and cannot be applied in all cases.

Diagnosis and monitoring

Electroencephalogram (EEG) is the most widely used methodology to support diagnosis, because it is able to detect changes in neuronal activity, occurring during an epileptic seizure. Such changes normally occur during the seizure, but in some instances, they precede the seizure for a few seconds [4]. Therefore, EEG was started to be used as a predictor of epileptic seizures [5]. Indeed, the early identification of epileptic seizures could contribute to the early interruption of an epileptic event, or of potentially harmful ongoing activities (e.g., driving a car, eating, walking, etc.). In general, preictal patterns are patient specific, therefore no predetermined set of features will be able to capture all possible preictal signatures. In this regard, all data should be considered potentially relevant for recognizing preictal patterns, instead of restricting algorithm analysis to specific spatiotemporal features.

Autonomic regulation and the electrocardiogram

Epileptic modifications were observed in the activation of the autonomic nervous system (ANS). The rationale for this association is linked to the fact that modulation of the ANS activity often conducts to variations of the neuronal microenvironment, in turn producing a change of the activation state of neuronal populations [6]. Patients sometimes report previous feelings consistent with tachycardia, bradycardia, asystole, and modification of ventricular repolarization. In addition, about one-third of epileptic subjects present an altered interictal electrocardiographic activity [7]. Cardiac involvement could be responsible for sudden unexpected death in epilepsy. Such cardiac changes precede the onset of the epileptic seizure in 70% of cases, so that it has been proposed to modulate the heart—brain interaction, for the prevention and control of focal crises [8]. Reviews are available on the heart—brain axis [9,10].

Electroencephalogram synchronization

Ictal and preictal states can be related to synchronization patterns in EEG [11—14]. Andrzejak et al. applied the mean phase coherence, as a measure of phase synchronization in EEG, finding a decrease before seizure onset [11]. Mirowski et al. [12] combined nonlinear classifiers with phase-synchrony measures. Kuhlmann [13] addressed the mean phase coherence, too. Seizure prediction performance was evaluated for all possible channel pairs, and for different prediction methods. A network model has been used for representing the recurrence pattern of EEG signals, in

Precision Medicine for Investigators, Practitioners and Providers. https://doi.org/10.1016/B978-0-12-819178-1.00034-4

which temporal synchronization patterns are quantified using spectral graph theoretic features [14].

Other studies have analyzed the typical parameters of heart rate variability (HRV), comparing them between phases (e.g., preictal vs. interictal), and finding higher sympathetic activity during the preictal phase, as much as 5 min before seizure onset [9–11], a sufficient amount of time for interrupting risky activities. Behabani and colleagues [15] employed the typical HRV features in an adaptive decision threshold method, obtaining an average sensitivity in seizure prediction of 78.59%, within an intervention time of 110s, and a false positive (FP) rate of 0.21 h-1. On the other hand, Fujiwara et al. [16] used HRV features, in both time- and frequency domain, within a multivariate statistical process, obtaining a sensitivity of 91% and an FP rate of 0.7 h-1 in predicting a crisis. A recent research applied the support vector machine (SVM) approach, using HRV features for the early detection or prediction of epileptic seizures [17]. Seizure was anticipated within 5 min from onset, with a sensitivity of 94.1% and an FP rate of 0.49 h-1, more robust results if compared with other studies [15,16].

Machine learning tools

Deep learning, a machine learning technique, allows automatic feature extraction from big datasets [17]. The approach has been tested on an intracranial EEG dataset, using a convolutional neural network algorithm, showing mean sensitivity of 69% with a prediction interval ranging from 1 min to several hours. The system offers the ability to the patient to tune the alarm sensitivity. The developed system can be deployed onto low power chips, and that can be implemented in wearable devices. In Tsiouris et al. [18], another deep learning algorithm has been presented, the long short-term memory model, an evolution of recurrent neural networks. The algorithm showed promising results, due to its ability to control more precisely what information needs to be in memory and what can be discarded, in analyzing highly dynamic nonlinear large dataset such EEG.

The approach, tested on the CHB-MIT Scalp EEG database system, could be able to predict all the seizures, with a false prediction rate (FPR) of 0.11–0.02, depending on the duration of the preictal window. In Samiee et al., 2017, an extraction feature method based on sparse rational decomposition and local Gabor binary patterns is proposed. EEG recordings of the CHB-MIT Scalp EEG database show a sensitivity of 70.4% and the overall specificity of 99.1%, in patient-specific detection of epileptic EEG epochs [19]. Chu et al., 2107, analyzed the transition process from normal to seizure attractor state and investigated a precursor phenomenon. Closer to the bifurcation point that triggers the transition from normal to seizure state, the power spectral density of low frequency bands, of the perturbation of an attractor in the EEG, showed a relative increase. Seizure prediction algorithm based on this feature reached sensitivity of 86.67%, with a FPR of 0.367 h−1 and average prediction time of 45.3 min [20].

In Fan and Chou, 2019, a network model for representing the recurrence pattern of EEG signals is presented. In the model, the temporal synchronization patterns are quantified, using the spectral graph theoretic features. Different approaches have been proposed. A maximum sensitivity of 98% and a maximum specificity of 92% have been attained [21].

Patient-specific profiles—PANACEE protocol

Relevant differences exist between patients' physiological reactions and prodromal signs, before seizure occurrence. It is important to employ a tailored solution for covering the large variability. This was the aim of project "PANACEE" (Prevision and analysis of brain activity in transitions: epilepsy and sleep) of Tuscany Region, Italy—PAR FAS 2007–13 1.1.a.1.1.2 B22I14000770002 [22–24]. Overall, the largely patient-specific PANACEE protocol analysis is based on computation of features extracted from EEG/ECG, together with the application of a classification procedure, for the identification of preictal segments in the signals.

Dataset

It consisted of seven patients admitted to the Unit of Neurology and Neurophysiology, Department of Neurological and Neurosensorial Sciences, University of Siena. Only patients with simultaneous EEG and ECG recording and at least three seizures were selected. Patients were monitored with video-EEG with a sampling rate of 512 Hz and electrodes arranged on the basis of the international 10–20 system. The onset of the seizure was identified as the earliest seizure-related change in behavior (as observed on the video recording) or EEG (as recognized by two expert epileptologists).

Diagnosis and classification of seizures according to the criteria of the International League Against Epilepsy [25] was performed by an expert clinician (Table 34.1). Note that for some patients, Seiz. ECG is lower than Seiz. EEG, due to technical problems in the recording of ECG signal.

Epileptic seizure prediction through analysis of electroencephalogram synchronization

Electroencephalogram signal analysis procedure

Recently introduced phase synchronization measures, namely phase lag index (PLI) and weighted phase lag index

TABLE 34.1 Data of the patients of the unit of neurology and neurophysiology at university of Siena.

Age (years)	Type of seiz.	Localization/ Lateralization	Seiz. EEG	Seiz. ECG	Train Seiz. EEG	Test Seiz. EEG	Train EEG (h)	Test EEG(h)
55 M	IAS	T/R	5	0	3	2	0.86	0.54
10 F	IAS	T/L	3	3	2	1	2.93	1.96
24 M	IAS	T/L	5	5	3	2	5.45	4.05
16 M	FBTC	All	8	3	4	4	6.69	4.44
9 M	IAS	T/L	4	3	2	2	3.15	2.17
34 F	IAS	T/L	3	3	2	1	4.08	3.09
19 M	WIAS	T/L	4	3	2	2	8.84	6.92

ECG, Electrocardiogram; *EEG*, Electroencephalogram; *F*, Female; *FBTC*, bilateral tonic–clonic; *IAS*, focal onset impaired awareness; *L*, left; *M*, Male; *R*, right; *Seiz*, Seizure; *T*, temporal; *Test Seiz*, Testing seizures; *Train Seiz*, Training seizures; *WIAS*, focal onset without impaired awareness.

(WPLI), were utilized [26,27]. The method has been first presented [22] and compared with other approaches from the literature. PLI has been introduced [26]d for the assessment of functional connectivity in the brain, and WPLI [27] is an extension of PLI.

Mathematical basis

Given two time series xh(t) and xk(t) (e.g., related to two EEG channels) and a time window Δt containing N time instants, PLI and WPLI are defined as follows:

$$PLI_{h,k,\Delta_t} = \left| \frac{1}{N} \sum_{p=1}^{N} sign(\phi_h(p) - \phi_k(p)) \right| \quad (34.1)$$

$$WPLI_{h,k,\Delta_t} = \frac{\left| \frac{1}{N} \sum_{p=1}^{N} \sin(\phi_h(p) - \phi_k(p)) \right|}{\frac{1}{N} \sum_{p=1}^{N} |\sin(\phi_h(p) - \phi_k(p))|} \quad (34.2)$$

where $\varphi_h(p)$ and $\varphi_k(p)$ are the phases at time instant p, of signals xh(t) and xk(t), respectively, which are determined by the Hilbert transformation [22].

PLI is based on the idea of discarding the phase differences that center around 0 (mod π) and quantify the asymmetry of the distribution of phase differences around zero. In WPLI, each phase difference is weighted according to the magnitude of the lag. As a consequence, phase differences around zero only marginally contribute to the calculation of the WPLI. Both PLI and WPLI take values in [0,1], where 0 indicates that no phase locking occurs and 1 is a perfect phase locking.

In our approach, the synchronization indices PLI and WPLI introduced above are not directly computed on the raw EEG signals, but on the signals resulting of a number of manipulations. Namely, each EEG signal is first filtered on prespecified frequency bands and then differentiated (i.e., the absolute value of the time-derivative of the signal is computed) [23]. As shown [23,24], differentiating makes the basic noise nearly flat and sharpens the regions where the signal exhibits its peaks, which, as shown below, are most likely to be where seizures occur.

Fig. 34.1 shows the behaviors of PLI and WPLI computed over filtered (in the Hz band) [8,13] and differentiated signals. The lower parts of Fig. 34.1A and B show PLI and WPLI on a time span of 300 s, preceding the first seizures of patients PN00 and PN05 (seizure starts/ends at the first/second vertical dotted line). The upper parts of the figures show the EEG data, related to the pairs of EEG channels on which PLI and WPLI are computed, on the same time span. In all the figures, PLI and WPLI are computed using a time window Δt of 6 s (with an overlap of 1 s).

Seizure prediction and classification procedure

A threshold-based classifier, called ThAlgo, is employed. The features used in the classifier are obtained by applying a moving average procedure [22], to the synchronization measures PLI and WPLI. An input parameter of ThAlgo is the prediction interval, defined as a time span immediately preceding each seizure (not univocally defined in the literature), that constitutes the time period in which we would like to obtain the alarm for the incoming crisis. Records inside (outside) the prediction interval constitute the positive (negative) class. The positive classification of at least one record inside a given prediction interval means that the corresponding seizure has been successfully predicted. The prediction time of a seizure is computed, as the distance between the first positive record inside the prediction interval

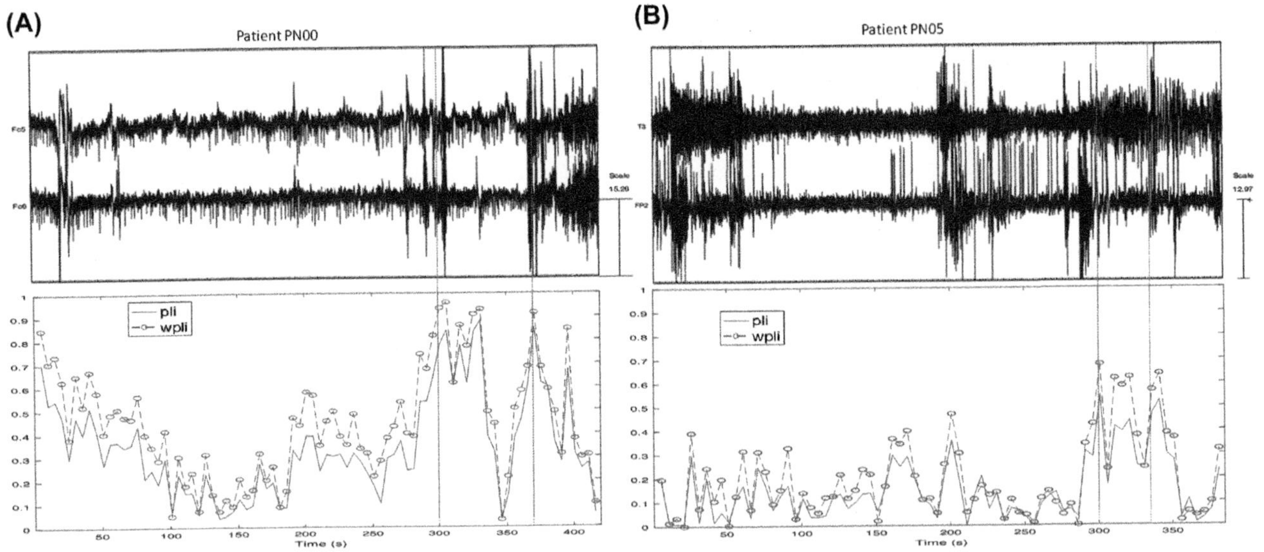

FIGURE 34.1 (A) Electroencephalogram (EEG) on channels h = {Fc5} and k = {Fc6}; $PLI_{h,k}$ and $WPLI_{h,k}$ on the first seizure of patient PN00. (B) EEG on channels h = {T3} and k = {FP2}; phase lag index/$PLI_{h,k}$ and weighted phase lag index/$WPLI_{h,k}$ of the last seizure of patient PN05.

of a seizure and its onset. The positive classification of a record outside the defined prediction intervals is counted as a FP.

ThAlgo employs a set of thresholds (built during the training phase), to classify the records in the test phase. During the training phase, a feature selection phase is carried out, to identify subsets of features and to build thresholds, which are the most relevant for the classification task. In the subsequent test phase, the selected features and the thresholds are linearly combined, to obtain a single feature and a single threshold, which are employed to classify the unlabeled records used for testing.

Experimental results

All patients' datasets have been subsampled at 256 Hz and segmented into seizure (i.e., containing a seizure occurrence) and nonseizure. During this process, all ictal periods and a (postictal) period of 1000 s after the end of each seizure are discarded.

A q-fold cross-validation procedure was included, with $q = 5$ (with $q = 3$) on datasets with at least four (with three) seizure instances. As a consequence, in the experimental campaign, we consider datasets of patients containing at least three seizures. Table 34.1 reports the number of seizure instances used in the training phase and in the test phase. Average number of EEG hours over the q rounds of cross validation, used in the training phase and in the test phase, is also informed.

EEG signals are filtered using a band-pass FIR filter with band [8–13] Hz. The synchronization measures have been computed for each channel pair, on a time window $\Delta t = 6$ s with an overlap of 1 s. In the training and test phase, we consider prediction intervals of length T, with $T \in \{150,200,300\}$ seconds.

Model results

Results are computed choosing for each patient the combination of the parameters of the classifier, giving the best performance. The performance is measured in terms of sensitivity, i.e., number of correctly predicted seizures (Sens%) and specificity (Spec%). We also report the average prediction time tp of a seizure. The results are reported in Table 34.2.

Algorithm ThAlgo correctly predicts 95.7% of the seizures with a specificity equal to 97.8% on average.

TABLE 34.2 Performances of ThAlgo on the data of Table 34.1.

Patient	Sens%	Spec%	t_p (sec)
PN00	100	100	142.80
PN05	100	98.20	29.67
PN06	100	95.46	152.50
PN10	100	96.28	94.90
PN12	90	96.63	221.67
PN13	100	100	67
PN14	80	98.06	131.50
Average	95.71	97.81	120

Sens, Sensitivity; *Spec*, Specificity; *tp*, Time preonset.

Furthermore, all seizures are predicted in five of seven patients, and no false alarm is generated in two patients' data. In our experimental settings, a specificity of 99.87% corresponds to less than 1 FP per hour. Prediction times are 2 min, on average. Because in the experiments the prediction interval is not bigger than 300 s, prediction times bigger than 5 min cannot be achieved.

Prediction via autonomic nervous system by means of ECG signals

ECG analysis procedure

Signal preprocessing is aimed at removing artifacts and improving the signal-to-noise ratio for a clearer, easier analysis. Then, ECG signal was interpolated at 1024 Hz, so that the successive step of QRS detection provided a precise time location of the QRS complexes, resulting in an RR series with reduced quantization. The RR series was processed by an algorithm, for the recognition and correction of values due to artifacts and nonsinusoidal beats, to obtain an RR series that only contains variations generated by the sinus node, and thus reflects the activity of the ANS.

This corrected RR series was analyzed, by using a sequence of successive windows of 180s with 60s of overlap between each other. From each window, the features were extracted, which are informative with respect to the ANS activation. In addition to the classical descriptors, derived in time- and frequency domain, features characterizing recurrences, a basic property of dynamical systems, have been considered. These features were extracted by the recurrence quantification analysis (RQA) [26], which results particularly suitable for the nonstationarity of the signals, and the absence of mathematical assumptions regarding the generating systems [27].

Algorithm features

Feature selection process was performed using the information provided by previous seizures of the same subject, to identify a patient-specific predictive subset. They included time- and frequency-domain features, as well as characteristics related to the RQA. Concerning time domain, mean of normal RR intervals (MeanNN), root mean square of the successive differences of RR intervals (RMSSD), and standard deviation of normal RR intervals (SDNN) were selected. The number of pairs of successive NN (R-R) intervals that differed by more than 50 ms (NN50), the proportion of NN50 divided by the total number of NN intervals (PNN50), variance, dispersion (standard deviation) of points perpendicular to the axis of line-of-identity in the Poincarè plot (SD1), dispersion (standard deviation) of points along the axis of line-of-identity in the Poincarè plot (SD2), the cardiac sympathetic index, the cardiac vagal index, the coefficient of sample entropy, and the Katz fractal dimension were also included.

Features extracted using RQA were recurrence rate, determinism, laminarity, entropy, the maximum length of the diagonal, and the trapping time. On the other hand, frequency-domain features selected were the normalized power of the low frequency band (0.04–0.15 Hz) (LFn), the normalized power of the high frequency band (0.15–0.40 Hz) (HFn), and the ratio of LF to HF (LF/HF) [18].

Fig. 34.2 shows the dynamics of some of the selected features, from 25 min before seizure onset to 5 min afterward. Each of the feature vectors was segmented according to the epileptic phase. Segments classified as ictal included the period from seizure onset to the end of the seizure; preictal included the period from x minutes before seizure onset till onset; and interictal were nonseizure data, apart from at least 50 min from the seizure. The value of x, describing the duration, in minutes, of the preictal phase, was tested in this project as 5, 15, and 30 min, related to different prediction horizons (PHs).

Classification procedure based on support vector machine

Using stepwise regression analysis [18], after the feature selection step was applied, the SVM classification with a kernel function was used, to classify the RR segments as either preictal or interictal. More in depth, a double cross validation was employed, to predict a seizure block (defined as consisting of preictal + interictal phases), based on the other blocks.

According to double cross-validation method, data were partitioned into two subsets: training set and test set. The training set was separated into subdivisions: learning set and validation set [28]. In a patient with N seizures and I-hour–long interictal recordings (at least 50 min far from the seizure), we separated the dataset into N block, each containing (I/N) hour of interictal recording and 15 min of preictal. Then, we used one seizure block for testing set and the other (N−1) blocks for training set. To establish an optimal SVM classifier in training, we perform fivefold cross validation, by selecting 80% of the training set (learning set) and validating the model on the remaining 20% of the training set (validation set), to check if the model is well fit.

Once the SVM model was trained, the performance was evaluated by testing the model, on the fold that was

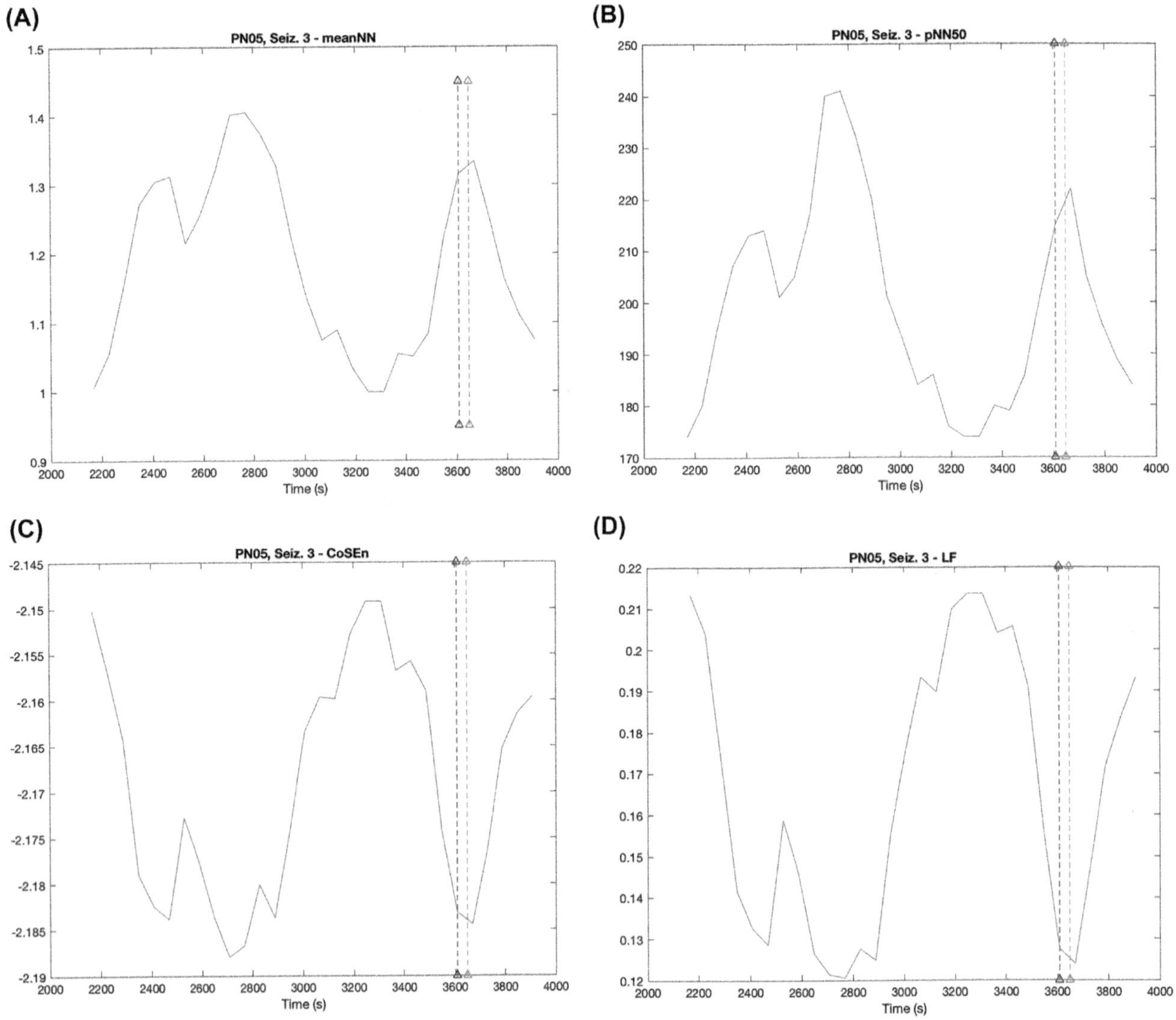

FIGURE 34.2 Each panel presents dynamic changes in one particular heart rate variability parameter from 25 min before seizure onset to 5 min after the end of the seizure. Seizure onset and seizure end are represented by the vertical red and green (gray and light gray in print version) lines, respectively. A representative seizure from one patient is reported (PN05, Seizure 3). (A) meanNN, (B) pNN50, (C) CoSEn, (D) Low frequency.

reserved for testing. Thus, this process is repeated N times and the average prediction rate is reported. The prediction time was also calculated as the time interval from the first vector of feature that was classified as a preseizure and the seizure onset. Concerning sensitivity, its maximum value (100%) was given when at least one preseizure window was correctly classified.

Experimental results

Optimal performances corresponded to PH = 30 min, with both higher sensitivity and specificity highlighted (Fig. 34.3). This evidence is linked to the overall performance of the approach used; however, being the personalization of the seizure recognition and treatment the main clinical target of the present work, the visualization of the performances achieved for each of the patients, taking into account the "optimal" PH (30 min), is shown in Table 34.3.

Integration of EEG and ECG for better seizure prediction

Preprocessing of EEG and ECG

Previous studies have shown the capacity of improving seizure detection by the combination of EEG and ECG features [29–32]. Features from the EEG were obtained for windows of 6 s (with an overlap of 1 s), while features from ECG were extracted in windows of 180s, with 60s of overlap.

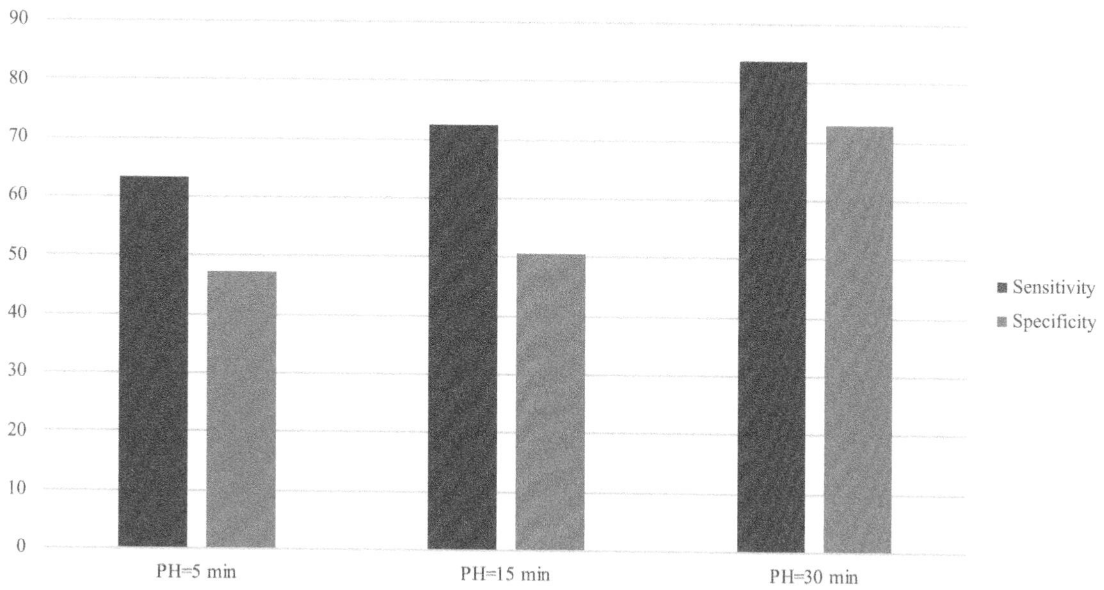

FIGURE 34.3 Average sensitivity and specificity of the support vector machine classifier for the three prediction horizons (PHs) tested.

TABLE 34.3 Patient-specific (and overall average) performances of the support vector machine classifier with PH = 30 min.

Pat.	Sens%	Spec%	FP/h	tp (min)
PN05	100.0	62.5	20.2	23.0
PN06	80.0	87.1	7.4	20.4
PN10	75.0	66.6	19.0	19.5
PN12	66.7	100.0	0.0	4.3
PN13	100.0	58.6	23.6	21.0
PN14	66.7	75.1	13.0	18.0
Average	81.4	74.9	13.8	17.7

FP/h, False positive/hour; *Pat*, Patient; *Sens*, Sensitivity; *Spec*, Specificity; *tp*, Prediction time.

TABLE 34.4 Performance of the support vector machine classifier with the integration of electroencephalogram and electrocardiogram features for patient PN06.

Seizure	Sens%	Spec%	FP/h	tp (min)
1	100	72.32	16	27
2	100	77.02	13.45	24
3	100	98.86	0.65	23
4	100	82.35	10.28	27
5	100	67.44	15	13
Average	100	79.598	11.076	22.8

FP/h, False positive/hour; *Sens*, Sensitivity; *Spec*, Specificity; *tp*, Prediction time.

(1) Averaging of EEG features on 180 s windows. This step was applied given the different dynamics of the two signals, so that we get both EEG and ECG features in 180 s windows;
(2) Application of stepwise regression analysis to the whole set of EEG and ECG features for a patient-specific feature selection;
(3) Segmentation of the set of features in ictal, preictal, and ictal segments;
(4) Classification of the feature segments by the application of SVM with a leave-one-out approach.

Table 34.4 reports the results of the classifier for each seizure and the average in patient PN06.

Final considerations

The current approach relies on the fact that EEG and ANS response can depend on several factors, the most important of which are the location of the seizure, the lateralization, and the use of drugs [33,34]. Therefore, a universal approach, defined for all patients independently from their disease's characteristics, would fail to effectively classify a seizure block, thus potentially affecting the positive outcome of the preventive strategy and, in some instances, of the treatment.

For the EEG, a computationally viable threshold-based approach detected the changes in the synchronization corresponding to the preictal state, a few minutes before seizure onset, with good sensitivity and specificity. ANS modifications, monitored through the ECG signal, were able to predict the onset of epileptic seizures even 30 min in advance, with sensitivity average of 83.7%. The main issue of this approach is represented by the high value of FPs/h (14.9), which overestimates the risk of a seizure, possibly triggering unnecessary alarms to the patient. The patient-specific methodology could be more efficient, in those patients featuring repeatable and stereotypical seizures.

The integration between ECG and EEG signals is essential for improving the performance of the epileptic seizure system. Importantly, the prediction time obtained with our algorithm is enough to drive the patient stopping potentially at-risk activities and taking a rapid, effective pharmacological treatment (in case of drug-sensitive epilepsy). With the aim of developing portable and wearable devices for real-life use, future research directions include (i) testing our approach on larger datasets; (ii) combining EEG and ECG features in the classification procedure, to improve effectiveness and reliability of the classification phase; and (iii) real-world data acquisition and real-time application of the proposed classifiers.

References

[1] Engel J, Pedley TA. Epilepsy – a comprehensive textbook, II edition, vol. 3. Lippincott Williams & Wilkins; 2008.

[2] Fiest KM, Sauro KM, Wiebe S, Patten SB, Kwon CS, Dykeman J, et al. Prevalence and incidence of epilepsy: a systematic review and meta-analysis of international studies. Neurology 2017;88:296–303. Erratum in: Neurology. 2017;89:642.

[3] Jacoby A, Austin JK. Social stigma for adults and children with epilepsy. Epilepsia 2007;48(Suppl. 9):6–9.

[4] Noachtar S, Rémi J. The role of EEG in epilepsy: a critical review. Epilepsy Behav. 2009;15:22–33.

[5] Park Y, Luo L, Parhi KK, Netoff T. Seizure prediction with spectral power of eeg using cost-sensitive support vector machines. Epilepsia 2011;52:1761–70.

[6] Ponnusamy A, Marques JL, Reuber M. Heart rate variability measures as biomarkers in patients with psychogenic nonepileptic seizures: potential and limitations. Epilepsy Behav. 2011;22:685–91.

[7] Moghimi N, Lhatoo SD. Sudden unexpected death in epilepsy or voodoo heart: analysis of heart/brain connections. Curr. Cardiol. Rep. 2013;15:424.

[8] Jansen K, Varon C, Van Hu el S, Lagae L. Peri-ictal ECG changes in childhood epilepsy: implications for detection systems. Epilepsy Behav 2013;29:72–6.

[9] Bou Assi E, Nguyen DK, Rihana S, Sawan M. Towards accurate prediction of epileptic seizures: a review. Biomed. Signal Process. Control 2017;34:144–57.

[10] Gadhoumi K, Lina JM, Mormann F, Gotman J. Seizure prediction for therapeutic devices: a review. J. Neurosci. Methods 2016;260:270–82.

[11] Andrzejak RG, David P, Lehnertz K, Elger CE, Mormann F, Kreuz T. Epileptic seizures are preceded by a decrease in synchronization. Epilepsy Res 2003;53:173–85.

[12] Mirowski P, Madhavan D, LeCun Y, Kuzniecky R. Classification of patterns of EEG synchronization for seizure prediction. Clin. Neurophysiol. 2009;120:1927–40.

[13] (a) Kuhlmann L, et al. Patient-specific bivariate-synchrony-based seizure prediction for short prediction horizons. Epilepsy Res. 2010;91(2–3):214–31.(b) Fan M, Chou C. Detecting abnormal pattern of epileptic seizures via temporal synchronization of eeg signals. IEEE Trans. Biomed. Eng. 2018;66(3):601–8.

[14] Behbahani S, Dabanloo NJ, Nasrabadi AM, Dourado A. Prediction of epileptic seizures based on heart rate variability. Technol. Health Care 2016;24(6):795–810. https://doi.org/10.3233/THC-161225 PMID:27315150.

[15] Fujiwara K, Miyajima M, Yamakawa T, Abe E, Suzuki Y, Sawada Y, et al. Epileptic seizure prediction based on multivariate statistical process control of heart rate variability features. IEEE Trans. Biomed. Eng. 2016;63:1321–32. https://doi.org/10.1109/TBME.2015.2512276 PMID:26841385.

[16] Pavei J, Heinzen RG, Novakova B, Walz R, Serra AJ, Reuber M, et al. Early seizure detection based on cardiac autonomic regulation dynamics. Front. Physiol. 2017;8:765. PMID: 290[16] 51738, https://doi.org/10.3389/fphys.2017.00765.

[17] Kiral-Kornek I, Roy S, Nurse E, Mashford B, Karoly P, Carroll T, et al. Epileptic seizure prediction using big data and deep learning: towards a mobile system. EBioMedicine 2018;27:103–11.

[18] Tsiours KM, Pezoulas VC, Zervakis M, Konitsiotis S, Koutsouris DD, Fotiadis DI. A long short-term memory deep learning network for the prediction of epileptic seizures using EEG signals. Comput. Biol. Med. 2018;99:24–37.

[19] Samiee K, Kov acs P, Gabbouj M. Epileptic seizure detection in long-term eeg records using sparse rational decomposition and local gabor binary patterns feature extraction. Knowl. Based Syst. 2017;118:228–40.

[20] Fan M, Chou C. Detecting abnormal pattern of epileptic seizures via temporal synchronization of EEG signals. IEEE Trans. Biomed. Eng. 2019;66(3):601–8.

[21] Chu H, Chung CK, Jeong W, Cho K-H. Predicting epileptic seizures from scalp eeg based on attractor state analysis. Comput. Methods Progr. Biomed. 2017;143:75–87.

[22] Detti P, Zabalo Manrique de Lara G, Bruni R, Pranzo M, Sarnari F, Vatti G. A patient-specific approach for short-term epileptic seizures prediction through the analysis of EEG synchronization. IEEE Trans. Biomed. Eng. 2019;66(6):1494–504.

[23] Billeci L, Marino D, Insana L, Vatti G, Varanini M. Patient-specific seizure prediction based on heart rate variability and recurrence quantification analysis. PLoS One 2018;13(9):e0204339.

[24] Billeci L, Tonacci A, Marino D, Insana L, Vatti G, Varanini M. A machine learning approach for epileptic seizure prediction and early intervention. Biosyst. Biorobotics 2019;21:972–6.

[25] Fisher RS, Cross JH, French JA, Higurashi N, Hirsch E, Jansen FE, et al. Operational classification of seizure types by the international League against epilepsy: position paper of the ILAE commission for classification and terminology. Epilepsia 2017;58:522–30.

[26] Cornelis JS, et al. Phase lag index: assessment of functional connectivity from multi channel eeg and meg with diminished bias from common sources. Hum. Brain Mapp. 2007;28(11):1178–93.

[27] Vinck M, et al. An improved index of phase-synchronization for electrophysiological data in the presence of volume-conduction, noise and sample size bias. Neuroimage 2011;55(4):1548–65.

[28] Majumdar KK, Vardhan P. Automatic seizure detection in ecog by differential operator and windowed variance. IEEE Trans. Neural Syst. Rehabil. Eng. 2011;19(4):356–65.

[29] Parvez MZ, Paul M. Epileptic seizure prediction by exploiting spatiotemporal relationship of eeg signals using phase correlation. IEEE Trans. Neural Syst. Rehabil. Eng. 2016;24(1):158–68.

[30] Eckmann JP, Kamphorst SO, Ruelle D. Recurrence plots of dynamical systems. Europhys. Lett. 1987;5:973–7.

[31] Marwan N, Wessel N, Meyerfeldt U, Schirdewan A, Kurths J. Recurrence plot based measures of complexity and its application to heart rate variability data. Phys. Rev. E 2002;66:026702.

[32] Rongrong S, Yuanyuan W. Predicting termination of atrial fibrillation based on the structure and quantification of the recurrence plot. Med. Eng. Phys. 2008;30:1105–11. https://doi.org/10.1016/j.medengphy.2008.01.008 PMID:18343707.

[33] Friedman JH. An overview of predictive learning and function approximation. NATO ASI Series F Comp. Syst. Sci. 1994;136:1–61.

[34] Mormann F, Andrzejak RG, Elger CE, Lehnertz K. Seizure prediction: the longand winding road. Brain 2006;130:314–33.

Chapter 35

Precision medicine in ophthalmology: an evolving revolution in diagnostic and therapeutic tools

Marina Roizenblatt[1,2,3], Jaime Roizenblatt[4], Kim Jiramongkolchai[1], Peter Louis Gehlbach[1], Rodrigo Brant[2], Mauricio Maia[2,3] and Rubens Belfort, Jr. [2,3]

[1]*Department of Ophthalmology, Wilmer Eye Institute, Johns Hopkins University School of Medicine, Baltimore, MD, United States;* [2]*Department of Ophthalmology, Federal University of São Paulo, São Paulo, Brazil;* [3]*Vision Institute, IPEPO, Paulista Medical School, Federal University of São Paulo, São Paulo, Brazil;* [4]*Division of Ophthalmology, São Paulo University School of Medicine, São Paulo, Brazil*

Context

At present, precision medicine is an emerging concept that stratifies the patient and their disease using a wide array of individual data, including clinical information, lifestyle, genetics, modern imaging techniques, psychosocial characteristics, and further biomarker information, thus going beyond the classical and observable "signs-and-symptoms" approach [1] (Table 35.1). These characteristics are able to distinguish a given patient from others with similar clinical presentations. The main goal is therefore to enable more precise targeting of disease subgroups, with preventive efforts and the best available care for each individual [2]. The ability to understand and act on these variables with equally high precision is driven by advances in genome editing, cellular reprogramming, tissue engineering, and information technologies especially artificial intelligence (AI), and others [3].

Establishing a precise diagnosis is the initial goal of most therapeutic encounters between a physician and a patient. Precision medicine is a natural extension of this step, which integrates research disciplines and clinical practice. Furthermore, the strategy allows modification of disease risk factors, early detection of disease processes, precise and dynamically adjusted interventions, and determines preventative and therapeutic intervention efficacy, based on real-world outcomes [3–5]. The result is improved clinical prognosis and minimization of unnecessary side effects, especially in patients who are less likely to have a response to a particular treatment [6].

Healthcare costs have steadily increased and have not necessarily translated into significant outcome improvements. The economics of the present situation are not sustainable and demand change. Therefore, an opportunity for investigators, practitioners, healthcare providers, payers, regulators, patients, and drug developers, to achieve higher efficacy while minimizing costs, exists. In parallel, significant decreases in the cost of genome sequencing have facilitated the discovery of many new and rare genetic diseases [7].

In the near future, precision medicine will likely disrupt the present healthcare system, by developing a systematic approach for most illness, including eye diseases. Within ophthalmology, current efforts are directed at defining subsets of disease, such that a more accurate prognosis is determined, and therapy is more tailored [5]. In terms of early diagnosis and precise treatment, the stratification of disease, achieved at the phenotypic and genotypic levels, is already being successfully applied. Most notable among these are inherited eye diseases. Phenotypic stratification in genetically heterogeneous monogenic diseases, such as retinitis pigmentosa, enables identification of associations, which lead to gene-specific phenotypes [8]. In addition, the methodological implementations tend to use AI systems, capable of integrating bioinformatics data into clinically relevant knowledge. This in turn allows the development of a system capable of supporting clinical decision-making, in a cost-effective and personalized way [9].

AI is a general term that implies the use of computers to model intelligent behavior, to solve complex problems with

Precision Medicine for Investigators, Practitioners and Providers. https://doi.org/10.1016/B978-0-12-819178-1.00035-6

TABLE 35.1 Comparison between traditional medicine and precision medicine.

	Traditional Medicine	Precision Medicine
Model temporality	Traditional health model	Recent concepts
Genomics	Relative importance in diagnosis and therapeutic decision	Fundamental importance in diagnosis and therapeutic decision
Treatment Approach	Therapies are designed to treat a large group of people with a similar illness	Treatment is based on a wide array of patient's unique clinical profile
Treatment Effect	Patients with the same disease respond differently and have different side effects for the same therapy	The right treatment is given to the right populations, so that there is a higher success rate and fewer side effects
Therapeutic Definition	Therapeutic success is statistically homogeneous, so that an average response represents individuals well	Therapeutic success is statistically heterogeneous, so the average response is not representative of a sample of individuals
Cost-Effectiveness	Acceptable	Still under debate
Accessibility	There are health disparities among different socioeconomic groups	The current financial barriers to large-scale deployment of precision medicine can increase social disparities

low error rate and with minimal human intervention [10]. The emergence of this novel concept is a consequence of continued progress in enhancing computer processing speed, capacity, and software programming. AI is generally accepted as a branch of engineering, applied at its inception to the field of robotics [10]. Relevant applications in healthcare include healthcare costs, surgical innovation, and clinical research outcomes. AI algorithms focus on optimizing treatments, with respect to safety, quality of life, accelerated recovery, and cost [11]. In synthesizing these related technologies, it is logical to use AI as well as emerging software and robotic technology to improve the quality of treatment offered to patients.

In the field of ophthalmology, Vitreoretinal microsurgery is among the most technically challenging techniques. Exceptional hand-eye precision is required to perform delicate maneuvers which are on the order of microns in scale. Moreover, the target tissue (retina) is fragile and does not regenerate, making the risk of blindness a very real potential consequence of inadvertent tissue damage. The unassisted human hand has limitations with regard to dexterity, tremor, and precision in positioning instruments. As a consequence, robotic approaches are logical, strategic, and promising solutions to the numerous challenges. Improvement in surgeon capabilities, enhanced outcomes, a reduction in patient discomfort, shorter hospitalization, and improved safety are all potential outcomes [12].

New technologies and are being utilized in residency training, including incorporation of virtual reality, formal lectures, wet-lab, and live surgical experiences, to shorten the learning curve, improve surgical performance, and diminish complications. Virtual reality simulation training provides a safe environment, in which to practice surgical technique and decision-making without risk to a patient [13,14]. Analogous to aircraft pilots, surgeons can now train on advanced simulators, with increasingly realistic three-dimensional immersive environments. Through the development of patient-specific virtual model surgical planning, decision-making and simulated execution can be performed virtually, in advance of the actual procedure [11].

Retinal genetic diseases: screening and therapeutic options

In 2001, the global scientific community witnessed the completion of the Human Genome Project, a landmark that enabled a deeper understanding of medicine [15]. Sequencing of the genome has become an essential component of an individualized diagnosis and treatment program, when dealing with hereditary diseases [16]. Screening for retinal genetic diseases is a reality today. The expression pattern within retinal diseases can be used to prioritize genes for screening, because a large fraction of these pathologies result from mutations in genes with a small range of functions. Many proteins targeted to the photoreceptor inner and outer segments, for instance, show mitochondrial RNA abnormalities in photoreceptor inner segments. Thus, genes encoding messages with this subcellular localization might be given high priority for screening [17].

Regarding therapeutic approaches, the American Society of Gene and Cell Therapy defines gene therapy as a field of biomedical research, with the goal of influencing the course of various genetic and acquired diseases at the DNA/RNA level, by introducing, removing, or changing the genetic code, to treat a specific pathology. There is a set of strategies that modify the expression of an individual's genes or repair abnormal genes using special carriers of genetic material, called vectors. In brief, gene therapy can be defined as a pharmacologic method by which nucleic acids, and ultimately bioactive proteins, become the therapeutic agents delivered to the target tissues [18].

In addition to gene therapy, there is cell therapy, which targets pathologies at the cellular or organ levels and aims to replace diseased tissues or organs for healthy and functional ones. Cell therapy is already available in the form of blood transfusions, bone marrow, and organ replacement therapy; however, there is a shortage of donors with matching characteristics. With the advent of stem cell–derived tissues, several groups worldwide are looking for ways to develop cells and organs in vitro, with unlimited potential and already matching the host immune system requirements.

Embryonic stem cells are being used in several clinical trials as a source for all kinds of cell types in the human body; however, there are ethical and rejection issues related to their use in research. Because they came from an embryo, religious and governmental sectors in several countries are against their use.

Recently, induced pluripotent stem (IPS) cells were developed using adult human cells, by Takahashi et al. in Yamanaka's laboratory [19]. Those cells, if acquired from the host, can match the immune system and lack all the ethical restrictions embryonic stem cells face, so research using IPS is thriving. With IPS, Yamanaka received the Nobel Prize [20].

The eye represents an ideal target for such therapy. This is in part attributable to the fact that the eye is an immune privileged site, and also that the eye structure allows to directly visualize, access, and locally treat intraocular cells, with a comparatively small dose. Therefore, an increasing number of clinical trials are now going on in this field [21]. As clear evidence of progress, on December 2017, the United States Food and Drug Administration approved the first retinal gene therapy, which targeted diseases caused by mutations in the gene RPE65. A young boy with the diagnosis of inherited blindness due to Leber Congenital Amaurosis was the first to receive this gene therapy with promising functional results [22].

Retinal diseases are a major cause of irreversible blindness. These conditions can be genetically caused, acquired later in life, or both, this last group of patients being described as complex diseases. Current clinical trials include Leber congenital amaurosis type 2, Leber hereditary optic neuropathy, autosomal recessive retinitis pigmentosa, choroideremia, Stargardt disease, achromatopsia, X-linked retinoschisis, and Usher syndrome. Furthermore, genetic therapy is targeting more common ocular diseases, such as age-related macular degeneration and diabetic retinopathy (Fig. 35.1) [23]. Of course each gene and cell therapy approach rests on an enormous foundation of basic and translational research, from a large number of investigators, institutions, and patients [24].

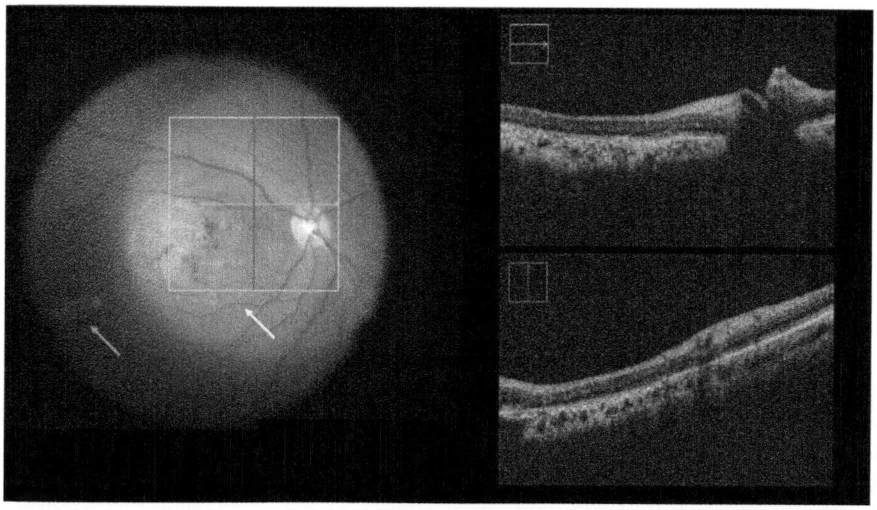

FIGURE 35.1 Intraoperative photograph and optical coherence tomography (OCT), taken before completion of subretinal vector delivery for genetic therapy. The yellow (white in print version) arrow shows the macular lesion, and the green (light gray in print version) arrow identifies the vector delivery cannula. On intraoperative OCT scan, the blue (pale white in print version) arrow represents the horizontal axis, and the pink (black in print version) arrow represents the vertical axis.

Developments in ophthalmological robotic surgery

Over the last two decades, ophthalmic surgery has progressively developed various robotic systems, which have been deployed to increase manual dexterity and expand a surgeon's manipulation capabilities. They also provide more accurate tool positioning, improve precision, cancel physiological hand tremor, and allow sensing information to be acquired and incorporated [25,26].

In the ocular anterior segment, the Da Vinci system has been used under experimental conditions, to suture corneal lacerations [27], to perform penetrating keratoplasties [28], to perform anterior capsulorhexis [29], and to remove pterygia [30]. Vitreoretinal surgical maneuvers also have the potential to be improved by using robotic technology. These include routine maneuvers such as laser photocoagulation and membrane peeling, and those that are rarely performed because they are not within the physiological capabilities of most surgeons, such as retinal vein cannulation, in the setting of retinal vein occlusion [31].

To stabilize the tips of microsurgical instruments, a variety of "accuracy-enhancement" devices are being developed, including Micron (Carnegie Mellon University, Pittsburgh, PA, USA) [32], Steady Hand Eye Robot (SHER) (Johns Hopkins University, Baltimore, MD, USA) [33], smart micromanipulation-aided robotic-surgical tool (SMART) (Johns Hopkins University, Baltimore, MD, USA) [34], force-sensing microforceps (Johns Hopkins University, Baltimore, MD, USA) [35], and the teleoperated system developed with Preceyes (Preceyes, Eindhoven, the Netherlands) [36].

Micron is a handheld actively stabilized tool, designed to increase accuracy for precise manipulation by removing involuntary motion, when actuating the tip to counteract the effect of the undesired handle motion, while still preserving intuitive eye–hand coordination [32]. The SHER works as a comanipulation system ("co-bot"), as the tool is held simultaneously by both the surgeon's hand, and a specially designed actively controlled robotic arm. The robotic device senses forces exerted by the operator on the tool and by the tool on the target tissue. Using this information in various control modes, the robot provides smooth, tremor-free, precise positional control, and force scaling [33]. SMART tools are defined as swept-source optical coherence tomography (OCT)–based, "smart" micromanipulation-aided robotic-surgical tools, which actively counter surgeon hand tremor. Finally, Preceyes is the most clinically studied telemanipulation robotic device. This system is designed to enhance surgical precision in vitreoretinal procedures and consists of a motion controller, for hand motion input by the surgeon, and a table-mounted instrument manipulator holding the surgical instrument [37].

There is no doubt that robotics will contribute to the future of ophthalmic surgeons; however, the exact nature and extent of the contribution will depend on balancing benefits with risks, costs, and time constraints [38]. As a landmark of this new technological era, Edwards et al. [39] have recently published the first-in-human study of remotely controlled robot-assisted retinal surgery, performed through a telemanipulation device. In this chapter, 12 patients were randomly assigned to either robot-assisted surgery or manual retinal surgery for dissecting epiretinal or inner limiting membrane. Surgical outcomes were equally successful in both groups. In addition, Gijbels et al. [40] were the first to report robot-assisted retinal vein cannulation, in four patients diagnosed with retinal vein occlusion. The results showed that it is technically feasible to cannulate and safely inject an anticoagulant into a retinal vein, with the aid of robotic technology and instrumentation.

Virtual reality simulation training

The standardization of the ophthalmologist's educational training program is a well-known challenge, not only because it is difficult to define the minimum competencies, curricula, and surgical outcomes, which the beginning physician should have at the end of the training program, but also because of the absence of tools capable of objectively measuring these abilities or evaluating surgical competence. Traditionally, surgical teaching has been conducted through repetition of exposure. However, numerous factors including patient safety concerns, risk management, increased efficiency demands, and the constant development of new technology and treatment modalities, coupled with an increase in patient numbers, have led to the development of alternative teaching methods [41].

Simulation-based education is already an established tool for building expertise in many specialties worldwide, such as in the high-risk aviation and military industries. In parallel with these industries, the benefits of incorporating specialty-oriented simulation training within medical schools are potentially numerous and span the breadth of medicine [42]. Simulation makes it possible to practice microsurgical decision-making and technique, familiarizing the new resident with the maneuvers performed under a microscope, using both hands and both feet in a safe learning environment, with no risk to patient safety. Simulators now allow for improvement in performance measures of the training fellows, by measuring relevant parameters throughout the simulation. A critical purpose of this device is to enhance resident education and subsequent patient outcomes [43].

Deuchler et al. [44] evaluated to what extent preparation via simulator affects surgeon performance, and whether simulation training impacts actual pars plana vitrectomy performance. A statistically significant relationship between simulator performance and live performance emerged in the operating room. Saleh et al. [45] analyzed the performance of ophthalmic trainees, with 2 hours or less of combined simulation and intraocular surgical experience. In this specific group of students, the simulator was more useful for monitoring performance than evaluating and quantifying overall skills. This was due in part to the great variability in novice performance. Novice variance would be expected to be reduced with both simulator and live surgical experience. Rossi et al. [46] demonstrated potential applications for vitreoretinal surgical simulator trained surgeons, learning new techniques before going into actual patient care.

Currently available simulators for ocular surgery include EYESi (VRmagic, Mannheim, Germany), PhacoVision (Melerit Medical, Linköping, Sweden), and MicroVisTouch (Immersive Touch Inc., Chicago, IL, USA). All three simulators use a stereoscopic microscope and provide a virtual reality environment for training cataract surgery steps. The EYESi simulator is the only one that has been validated in peer-reviewed publications and is available on the market for training in both cataract and vitreoretinal surgery [46,47].

Modern imaging techniques

Because some critical ocular tissues are not fully visible during standard ophthalmologic examination, it is of fundamental importance to develop modern imaging techniques. As assessment of ocular disease is dependent on the ability to visualize pathologic changes, emerging imaging techniques remain highly relevant to ocular assessment and are critical to the personalized medical care of patients. Furthermore, the ability to record, document, and follow interval change is a tremendous advantage of serial imaging, when dealing with progressive disease.

The ocular anterior segment is presently imaged to quantitatively determine curvature, asphericity, thickness, ectatic corneal conditions, anterior segment dimensions, and densitometric analysis of the crystalline lens. Each available technology for characterization of the anterior segment structures has advantages and disadvantages, often related to imaging precision and clinical applicability. Corneal topography, for instance, can be analyzed using keratometer, placido-based systems, scanning-slit systems, and Scheimpflug photography—based systems. Anterior segment analysis can be performed using scanning-slit systems, Scheimpflug photography—based systems, OCT, and very high—frequency ultrasonography. To date, Scheimpflug photography—based systems provide the most complete and modern analysis of the anterior segment in a noninvasive way [48]. Rotating Scheimpflug imaging technology is used by instruments such as Pentacam (Oculus Inc., Dutenhofen, Germany) to measure the anterior and posterior corneal surfaces, as well as other anterior segment structures.

With respect to the ocular posterior segment, OCT imaging systems are the most prevalent technology. Emerging over the last two decades, OCT has revolutionized the diagnosis and therapeutic protocols of retinal diseases. OCT functions to provide a type of optical biopsy. Requiring only seconds, cross-sectional volumetric information on retinal pathology in situ and in real time is provided, allowing the analysis of tissue structure and properties. Clinical OCT resolution approaches that of excisional biopsy and histopathology, but in a noninvasive way [49,50]. Time-domain detection was the first commercially available OCT technique, followed by spectral-domain OCT, and finally by the longer-wavelength swept-source technology [51].

Although in the past all OCT systems had a table top design, present systems are also available coupled to the surgical microscope. The most recent clinically significant innovation has been OCT angiography, which delivers images of microvascular blood flow of all the retinal vascular layers, in a rapid and noninvasive fashion. These images have been unprecedented in resolution, and unlike fluorescein angiography, no dye is required, flow in previously invisible vessels is provided, however leakage is not visualized [52]. The increased scanning speed allows the images to be captured with sufficient resolution, to produce volume maps of the retina and choroids, from which coronal or sagittal sections can be constructed in specific software. The Swept-source OCT has the ability not only to quantify retina and choroid thickness but also to expose the morphology of the intermediate and large vessels [53]. The information provided allows for AI/deep learning algorithms to utilize this imaging data.

Widefield images have also generated a clinical revolution in the ophthalmological field. In the past, traditional fundus cameras offered a 30- to 50-degree field of view. Present technology provides up to 200 degrees, allowing the acquisition of widefield color photography, autofluorescence imaging, and fluorescein angiography for imaging peripheral retinal disease. This has been a relevant advance, as peripheral retina is the site of many vision-threatening eye diseases, and evaluation of the retinal periphery is important for diagnosis, monitoring, and treatment of pathological manifestations [54]. Finally, it is worth noting that spectral imaging able to quantify tissue oxygen saturation rate is emerging [55].

Ongoing lines of investigation and research opportunities in the field

While classical clinical trials generally require the collection of data from many people, the concept of precision medicine focuses on individuals and the genetic and environmental factors that shape that person's response to a particular treatment [56]. According to Feldman et al. [57], there is significant overlap between precision medicine and translational research concepts. Precision medicine strategies take advantage of research tools including genomics and social environment to collect data on patient outcomes and disease progression, along with the use of computational biology and data analysis to collate and interpret large data sets.

The European Society for Translational Medicine defines translational medicine as an interdisciplinary branch of the biomedical field, supported by three main pillars: benchside, bedside, and community [58]. Translational medicine now refers to the bench-to-bedside interface between basic science and clinical medicine, of harnessing knowledge from basic sciences to produce promising new devices and treatment options [59]. The publication of the initial sequence and analysis of the human genome in 2001 [15], associated with subsequent advances in molecular biology, neurosciences, immunology, biomedical engineering, and modern imaging systems, are consistent with the concepts of both precision medicine and translational research [60].

A notable example of research opportunities relates to the potential role of genetic therapies in the preservation and restoration of ocular structures and vision, together with the intensive efforts to characterize embryonic, adult, and IPS cells toward various phenotypes, and data on the way in which injury signals direct stem cell differentiation. Three main approaches to the use of stem cells have been described: sustained drug delivery, immunomodulation, and differentiation into various ocular structures [61].

Further research is also desirable in the use of modern dyes. Vital dyes have been employed in making advanced diagnoses and in assisting surgical technique in various specialties, including ophthalmology. The use of vital dyes to stain preretinal tissues during vitreoretinal surgery, for example, a procedure known as "chromovitrectomy," allows for the visualization of thin and transparent tissues, such as the internal limiting membrane, epiretinal membranes, or the posterior hyaloid. However, there remain unanswered questions about dye properties, chemistry, pharmacology, and indications for use, to achieve the best surgical performance and minimize toxicity [62].

Finally, three-dimensional printing systems merit mention. Their application to cultured cells to potentially fabricate entirely new organs and to treat specified diseases is remarkable. The exponential development of this technology increases the likelihood that it will find new uses in medicine, including ophthalmology [63].

Importance for healthcare providers and institutions

There remains a question regarding the full potential of precision medicine, among the least-advantaged people [64]. There are at present financial barriers to the deployment of large-scale precision medicine efforts. The accessibility of precision medicine will ultimately determine if it becomes a strategy utilized primarily in the privileged social classes or whether financial incentives can drive the technology into broader use. Without major cost reductions, or enhanced efficiencies, precision medicine will result in a massive increase in the already large share of total gross national product, claimed by healthcare in most developed countries [65]. As a consequence, there is reasonable debate about the desirable cost-effectiveness threshold, of a technology that can be impressively effective while being prohibitively expensive [66].

Precision medicine initiatives propose changes to institutional values and frameworks. In addition to the inherent financial issues, there is an emerging tendency for patients to pressure healthcare systems, to supplement and extend traditional health delivery models, and empower patient self-care. This phenomenon is probably due to an increased level of information flow, transparency, customization, collaboration, patient choice, and responsibility-taking, as well as quantitative, predictive, and preventive aspects [67]. Patients are in the best position to demand sharing of both the data and professional credit, which will be necessary to fully capture the value of this new collaborative approach [68]. In this context, care is needed to avoid increasing already existing health disparities among groups. There may be concrete limitations to patient participation, that need to be considered in the process of incorporating precision medicine in a redesigned healthcare system, that forges new multistakeholder partnerships. Such change will require engagement with present organizational barriers in the health system [69].

Conclusion

The integration of precision medicine into current medical practice may well represent an inevitable evolutionary process. Based on this knowledge, researchers, practitioners, and providers should be able to build a knowledge base, which can better guide individualized therapeutic and diagnostic approaches to the needs of each patient.

Funding

MR: Lemann Foundation, Instituto de Visão-IPEPO, São Paulo, Brazil.

PLG: Research to Prevent Blindness, New York, New York, USA, and gifts from the J Willard and Alice S Marriott Foundation, the Gale Trust, Herb Ehlers, Bill Wilbur and Rajandre Shaw, Helen Nassif, Mary Ellen Keck, Don and Maggie Feiner, and Ronald Stiff.

MM and RBJ: CNPq—National Council for Scientific and Technological Development, Brasilia, DF, Brazil.

References

[1] König IR, Fuchs O, Hansen G, von Mutius E, Kopp MV. What is precision medicine? Eur. Respir. J. 2017;50(4):1700391.

[2] Ashley EA. Towards precision medicine. Nat. Rev. Genet. 2016;17(9):507.

[3] Torkamani A, Andersen KG, Steinhubl SR, Topol EJ. High-definition medicine. Cell 2017;170(5):828−43.

[4] Bahcall O. Precision medicine. Nature Publishing Group; 2015.

[5] Rosenbaum JT, Sibley CH, Choi D, Harrington CA, Planck SR. Molecular diagnosis: implications for ophthalmology. Prog. Retin. Eye Res. 2016;50:25−33.

[6] Jameson JL, Longo DL. Precision medicine—personalized, problematic, and promising. Obstet. Gynecol. Surv. 2015;70(10):612−4.

[7] Council NR. Toward precision medicine: building a knowledge network for biomedical research and a new taxonomy of disease. National Academies Press; 2011.

[8] Porter LF, Black GC. Personalized ophthalmology. Clin. Genet. 2014;86(1):1−11.

[9] Castaneda C, Nalley K, Mannion C, Bhattacharyya P, Blake P, Pecora A, et al. Clinical decision support systems for improving diagnostic accuracy and achieving precision medicine. J. Clin. Bioinform. 2015;5(1):4.

[10] Hamet P, Tremblay J. Artificial intelligence in medicine. Metabolism 2017;69:S36−40.

[11] Diana M, Marescaux J. Robotic surgery. Br. J. Surg. 2015;102(2):e15−28.

[12] Roizenblatt M, Edwards TL, Gehlbach PL. Robot-assisted vitreoretinal surgery: current perspectives. Robotic Surg. (Auckland) 2018;5:1−11.

[13] Thomsen ASS, Kiilgaard JF, Kjærbo H, la Cour M, Konge L. Simulation-based certification for cataract surgery. Acta Ophthalmol 2015;93(5):416−21.

[14] Feudner EM, Engel C, Neuhann IM, Petermeier K, Bartz-Schmidt K-U, Szurman P. Virtual reality training improves wet-lab performance of capsulorhexis: results of a randomized, controlled study. Graefe's Arch. Clin. Exp. Ophthalmol. 2009;247(7):955.

[15] Consortium IHGS. Initial sequencing and analysis of the human genome. Nature 2001;409(6822):860.

[16] Carrasco-Ramiro F, Peiro-Pastor R, Aguado B. Human genomics projects and precision medicine. Gene Therapy 2017;24(9):551.

[17] Blackshaw S, Fraioli RE, Furukawa T, Cepko CL. Comprehensive analysis of photoreceptor gene expression and the identification of candidate retinal disease genes. Cell 2001;107(5):579−89.

[18] Gehlbach PL, Roizenblatt M, Mori K. Antiangiogenic gene therapy: prospects for human application. In: Saunders WB, editor. Peyman GA, ed Intravitreal surgery: principles and practice 2018. p. In press.

[19] Takahashi K, Tanabe K, Ohnuki M, Narita M, Ichisaka T, Tomoda K, et al. Induction of pluripotent stem cells from adult human fibroblasts by defined factors. Cell 2007;131(5):861−72.

[20] Maeder ML, Gersbach CA. Genome-editing technologies for gene and cell therapy. Mol. Ther. 2016;24(3):430−46.

[21] Dalkara D, Goureau O, Marazova K, Sahel J-A. Let there be light: gene and cell therapy for blindness. Hum. Gene Ther. 2016;27(2):134−47.

[22] http://med.miami.edu/news/first-u.s.-patient-treated-with-nightstar-gene-therapy-for-x-linked-retinit

[23] Campa C, Gallenga C, Bolletta E, Perri P. The role of gene therapy in the treatment of retinal diseases: a review. Curr. Gene Ther. 2017;17(3):194−213.

[24] MacLaren RE, Bennett J, Schwartz SD. Gene therapy and stem cell transplantation in retinal disease: the new frontier. Ophthalmology 2016;123(10):S98−106.

[25] Ebrahimi A, He C, Roizenblatt M, Niravkumar P, Sefati S, Gehlbach P, et al. Real-time sclera force feedback for enabling safe robot-assisted vitreoretinal surgery. In: Engineering in Medicine and Biology Society (EMBC), 2018 40th annual international conference of the IEEE 2018. IEEE; 2018. In press.

[26] He C, Ebrahimi A, Roizenblatt M, Patel N, Yang Y, Gehlbach PL, et al., editors. User behavior evaluation in robot-assisted retinal surgery. 2018 27th IEEE international symposium on robot and human interactive communication (RO-MAN). IEEE; 2018.

[27] Tsirbas A, Mango C, Dutson E. Robotic ocular surgery. British J. Ophthalmol 2007;91(1):18−21.

[28] Bourges JL, Hubschman JP, Burt B, Culjat M, Schwartz SD. Robotic microsurgery: corneal transplantation. British J. Ophthalmol 2009;93(12):1672−5.

[29] Bourla DH, Hubschman JP, Culjat M, Tsirbas A, Gupta A, Schwartz SD. Feasibility study of intraocular robotic surgery with the da Vinci surgical system. Retina 2008;28(1):154−8.

[30] Bourcier T, Nardin M, Sauer A, Gaucher D, Speeg C, Mutter D, et al. Robot-assisted pterygium surgery: feasibility study in a nonliving porcine model. Transl. Vis. Sci. Technol 2015;4(1):9.

[31] Channa R, Iordachita I, Handa JT. Robotic vitreoretinal surgery. Retina; 2016.

[32] Maclachlan RA, Becker BC, Tabares JC, Podnar GW, Lobes Jr LA, Riviere CN. Micron: an actively stabilized handheld tool for microsurgery. IEEE Trans. Robot. A Publ. IEEE Robot. Autom. Soc 2012;28(1):195−212.

[33] Taylor R, Jensen P, Whitcomb L, Barnes A, Kumar R, Stoianovici D, et al. A steady-hand robotic system for microsurgical augmentation. Int. J. Robot. Res. 1999;18(12):1201−10.

[34] Balicki M, Han JH, Iordachita I, Gehlbach P, Handa J, Taylor R, et al. Single fiber optical coherence tomography microsurgical instruments for computer and robot-assisted retinal surgery. Med Image Comput Comput Assist Interv 2009;12(Pt 1):108−15.

[35] He X, Handa J, Gehlbach P, Taylor R, Iordachita I. A submillimetric 3-DOF force sensing instrument with integrated fiber Bragg grating for retinal microsurgery. IEEE Trans. Bio-Med. Eng 2014;61(2):522−34.

[36] Meenink H, Hendrix R, Naus G, Beelen M, Nijmeijer H, Steinbuch M, et al. Robot-assisted vitreoretinal surgery. Medical robotics. Elsevier; 2012. p. 185−209.

[37] de Smet MD, Meenink TC, Janssens T, Vanheukelom V, Naus GJ, Beelen MJ, et al. Robotic assisted cannulation of occluded retinal veins. PLoS One 2016;11(9):e0162037.

[38] de Smet MD, Naus GJ, Faridpooya K, Mura M. Robotic-assisted surgery in ophthalmology. Curr. Opin. Ophthalmol. 2018.

[39] Edwards T, Xue K, Meenink H, Beelen M, Naus G, Simunovic M, et al. First-in-human study of the safety and viability of intraocular robotic surgery. Nat. Biomed. Eng 2018;1.
[40] Gijbels A, Smits J, Schoevaerdts L, Willekens K, Vander Poorten EB, Stalmans P, et al. In-human robot-assisted retinal vein cannulation, a world first. Ann. Biomed. Eng. 2018:1–10.
[41] Thomsen ASS. Intraocular surgery—assessment and transfer of skills using a virtual-reality simulator. Acta Ophthalmol 2017;95:1–22.
[42] Michael M, Abboudi H, Ker J, Khan MS, Dasgupta P, Ahmed K. Performance of technology-driven simulators for medical students—a systematic review. J. Surg. Res 2014;192(2):531–43.
[43] Ament CS, Henderson BA. Optimizing resident education in cataract surgery. Curr. Opin. Ophthalmol. 2011;22(1):64–7.
[44] Deuchler S, Wagner C, Singh P, Müller M, Al-Dwairi R, Benjilali R, et al. Clinical efficacy of simulated vitreoretinal surgery to prepare surgeons for the upcoming intervention in the operating room. PLoS One 2016;11(3):e0150690.
[45] Saleh G, Theodoraki K, Gillan S, Sullivan P, O'sullivan F, Hussain B, et al. The development of a virtual reality training programme for ophthalmology: repeatability and reproducibility (part of the international forum for ophthalmic simulation studies). Eye 2013;27(11):1269.
[46] Rossi JV, Verma D, Fujii GY, Lakhanpal RR, Wu SL, Humayun MS, et al. Virtual vitreoretinal surgical simulator as a training tool. Retina 2004;24(2):231–6.
[47] Kozak I, Banerjee P, Luo J, Luciano C. Virtual reality simulator for vitreoretinal surgery using integrated OCT data. Clin. Ophthalmol. 2014;8:669.
[48] Piñero DP, editor. Technologies for anatomical and geometric characterization of the corneal structure and anterior segment: a review. Seminars in ophthalmology. Taylor & Francis; 2015.
[49] Drexler W, Fujimoto JG. State-of-the-art retinal optical coherence tomography. Prog. Retin. Eye Res. 2008;27(1):45–88.
[50] Baghaie A, Yu Z, D'Souza RM. State-of-the-art in retinal optical coherence tomography image analysis. Quant. Imag. Med. Surg. 2015;5(4):603.
[51] Adhi M, Duker JS. Optical coherence tomography—current and future applications. Curr. Opin. Ophthalmol. 2013;24(3):213.
[52] Spaide RF, Fujimoto JG, Waheed NK, Sadda SR, Staurenghi G. Optical coherence tomography angiography. Prog. Retin. Eye Res. 2018;64:1–55.
[53] Dansingani KK, Balaratnasingam C, Naysan J, Freund KB. En face imaging of pachychoroid spectrum disorders with swept-source optical coherence tomography. Retina 2016;36(3):499–516.
[54] Witmer MT, Kiss S. Wide-field imaging of the retina. Surv. Ophthalmol. 2013;58(2):143–54.
[55] Keane PA, Sadda SR. Retinal imaging in the twenty-first century: state of the art and future directions. Ophthalmology 2014;121(12):2489–500.
[56] Polanco C. Precision medicine and portable embedded system devices. Nature 2015:609–11.
[57] Feldman AM. Bench-to-bedside; clinical and translational research; personalized medicine; precision medicine—what's in a name? Clin. Trans. Sci 2015;8(3):171–3.
[58] Cohrs RJ, Martin T, Ghahramani P, Bidaut L, Higgins PJ, Shahzad A. Translational medicine definition by the European society for translational medicine. Elsevier; 2015.
[59] Woolf SH. The meaning of translational research and why it matters. JAMA 2008;299(2):211–3.
[60] Rosenberg RN. Translating biomedical research to the bedside: a national crisis and a call to action. Jama 2003;289(10):1305–6.
[61] Aharony I, Michowiz S, Goldenberg-Cohen N. The promise of stem cell-based therapeutics in ophthalmology. Neural Regen. Res 2017;12(2):173.
[62] Badaro E, Novais EA, Penha FM, Maia M, Farah ME, Rodrigues EB. Vital dyes in ophthalmology: a chemical perspective. Curr. Eye Res. 2014;39(7):649–58.
[63] Huang W, Zhang X. 3D printing: print the future of ophthalmology. Investig. Ophthalmol. Vis. Sci. 2014;55(8):5380–1.
[64] Bayer R, Galea S. Public health in the precision-medicine era. N. Engl. J. Med. 2015;373(6):499–501.
[65] Kohane IS. Ten things we have to do to achieve precision medicine. Science (New York, NY) 2015;349(6243):37–8.
[66] Curl P, Vujic I, van't Veer LJ, Ortiz-Urda S, Kahn JG. Cost-effectiveness of treatment strategies for BRAF-mutated metastatic melanoma. PLoS One 2014;9(9):e107255.
[67] Swan M. Emerging patient-driven health care models: an examination of health social networks, consumer personalized medicine and quantified self-tracking. Int. J. Environ. Res. Public Health 2009;6(2):492–525.
[68] Desmond-Hellmann S. Toward precision medicine: a new social contract? American Association for the Advancement of Science; 2012.
[69] Adams SA, Petersen C. Precision medicine: opportunities, possibilities, and challenges for patients and providers. J. Am. Med. Inform. Assoc. 2016;23(4):787–90.

Chapter 36

Phenotyping COPD

José Luis López-Campos[1,2]

[1]*Medical-Surgical Unit of Respiratory Diseases, Biomedicine Institute of Sevilla, (IBIS), University Hospital Virgen del Rocío, University of Sevilla, Sevilla, Spain;* [2]*Research Center in Respiratory Diseases Net (CIBERES), Carlos III Health Institute (ISCIII), Madrid, Spain*

Context

A firm diagnosis can be reached with well-defined criteria, in a majority of patients suffering from airway disease, yet clinical presentation may either vary, and not completely adjust to our definition, or show considerable overlap in real life. Consequently, traditional classification of diseases based on symptoms, signs, and syndromes may not be sufficient to explain and differentiate airway diseases. This phenomenon is especially relevant in chronic obstructive pulmonary disease (COPD), a complex heterogeneous cluster, with several conditions influencing final clinical expression [1].

Identification of airway disease faces major challenges. The symptoms are essentially dyspnea, cough, expectoration, wheezing, or chest tightness. If we add systemic symptoms, which are common to other chronic diseases, we may also mention fatigue or exercise capacity limitation. However, from a global perspective, these symptoms can be present in all airway diseases, differing by complementary aspects such as variability over time, exposure to risk factors, or presence of specific triggers.

Diagnostic criteria may not be straightforward. Each of the consensus documents that defines criteria for main airway diseases, that is, asthma, COPD, and bronchiectasis, has been developed without taking into account the possible overlap with other respiratory diseases. As a result, it is relatively common for a patient to meet diagnostic criteria for more than one airway disorder [2,3].

Even in cases with clear diagnosis, clinical presentation has considerable variability. Accordingly, it is common to see clinical aspects or results in complementary tests that are revealed in some cases but not in others, with traditional medicine being as yet unable to provide an explanation [4,5].

The concept of precision medicine has emerged to address some of these challenges, from a diagnostic and therapeutic point of view [6]. In a condition like COPD with considerably increasing prevalence [7], and notable morbidity and mortality [8,9], it is important to have a diagnosis capable of identifying patient types that allows a more direct therapeutic approach.

Heterogeneity of COPD

It is now acknowledged that patients with COPD have differential clinical behaviors, which can be identified according to symptoms. Beyond classical separation between pink puffer and blue bloater [10], other forms of presentation have been described, some in the 1959 CIBA Symposium [11].

Clinical diagnosis

Clinical criteria for chronic bronchitis [12], pulmonary static and dynamic hyperinflation [13–15], symptoms suggestive of bronchial asthma, bronchial hyperresponsiveness [16], and decline in forced expiratory volume in one second (FEV_1) are relevant markers, associated with clinical outcomes. Evaluation of variability during disease progression is another aspect. Variability has been shown in symptom perception throughout the year, during a month, over a week, and even daily, with a cadence varying over 24 h [17]. Variability applies to both upper and lower airway symptoms [18], with differences between geographical areas [19]. Interestingly, this variability in clinical presentation over time also affects lung function. Considerable variability in the decline of FEV_1 is noticed, with some declines greater than 40 mL/year, while others exhibit less striking worsening or even improved lung function over time, with active treatment [20].

Improvement in lung function over time breaks the preconceived and traditional idea that in COPD, decline in FEV_1 is constant, and cannot be prevented by any intervention [21].

Precision Medicine for Investigators, Practitioners and Providers. https://doi.org/10.1016/B978-0-12-819178-1.00036-8

Lung imaging

Radiological expression changes as well. Regardless of the degree of spirometric involvement, lesions detected in the airway or parenchyma make the patients completely different [22]. Airway patterns, degree of emphysema, or association with bronchiectasis constitutes some of the heterogeneities of COPD from a radiological perspective. Finally, the association of COPD with various comorbidities of the respiratory system establishes a complex relationship that directly affects clinical presentation. Some of the most relevant examples, owing to their frequency or clinical impact, are sleep apnea-hypopnea syndrome, respiratory infections, lung cancer, or pulmonary hypertension.

Association with outcomes

There is no correlation between different clinical variables at the patient level. One interesting idea is that variables associated with COPD may be good markers of disease progression, however very bad markers of each other. The Evaluation of COPD Longitudinally to Identify Predictive Surrogate Endpoints (ECLIPSE) study compared relationship between FEV_1 and clinical outcomes. Cohorts level exhibited correlation between FEV_1 and clinical outcomes [23]. However, a careful analysis of the graphs (Fig. 36.1) revealed enormous dispersion of data. Correlation at the cohort level is not met at the patient level. In epidemiology, this discrepancy is called ecological fallacy [24]. It is necessary to evaluate each of them separately, to establish the impact of the disease on a specific individual, since both provide complementary information.

A number of clinically relevant variables are associated with outcomes. It seems logical to remember that variables related to prognosis should have greater weight in evaluation of patients. These may include diffusing capacity [25], lung hyperinflation [13], exacerbations [26], hyperresponsiveness [27], multidimensional indexes [28], chronic bronchitis [29], physical activity [30], and dyspnea [31].

Interestingly, it has been reported that FEV_1 is more poorly correlated with dyspnea than inspiratory capacity [32], which may be the most important symptom in COPD. Secondary pulmonary function tests (e.g., volumes or diffusing capacity of the lungs for carbon monoxide [DLCO]), may be recommended for staging.

Treatment options

We do have therapies for specific patient types. Core treatment includes nonpharmacological measures, such as tobacco cessation, exercise, and vaccinations, along with use of long-acting bronchodilators. Long-acting bronchodilators have been shown to improve lung function, by decreasing hyperinflation and improving symptoms [33], thus decreasing exacerbations [34]. The rest of the treatments, either oral or inhaled, should be used in patients with specific characteristics [35], for example, inhaled steroids for frequent exacerbators, or roflumilast for those with severe chronic bronchitis and frequent exacerbations.

Biologics in COPD

One promising option is biological therapy [36]. Only mepolizumab has already shown a potential role in eosinophilic COPD, albeit with contradictory results [37]. Probably, one of the key issues regarding biological therapy in COPD is need for advance patient selection, which implies adequate phenotyping.

Intervention endpoints

Clinical studies show their results in terms of average improvement. However, average response is a simplification of a much more complex reality. Just as we mentioned above, here we can apply the same principle of ecological fallacy. Individual evaluation of the clinical presentation, impact on disease perception and progression, and response to therapies are needed in precision medicine.

Classical COPD phenotyping

Although not always formally labeled as phenotypes, during the past decades, two main initiatives have been proposed as means of identifying patient types. In the first version of the *Global Initiative for Obstructive Lung Disease* (GOLD), FEV_1 was the main severity variable [38]. However, FEV_1 was considered insufficient to capture the complexity of this syndrome. Finally, in 2011, GOLD established four types of patients according to three characteristics, that is, lung function, symptoms, and exacerbations [39]. This proposal was maintained with some changes until currently 16 types of patients are identified on the basis of the same variables [40] (Fig. 36.2).

By providing information with two digits, it gives both short- and long-term prognostic information. Additionally, selected variables are clinically relevant and easy to measure. However, more than one variable per axis [41], together with identification of arbitrary cut-off points [42], and lack of a clear progressive prognostic signal [43], are limitations of this approach.

Multidimensional strategies

The first multidimensional features were introduced in the BODE index by Celli et al. [28]. Four variables (body mass

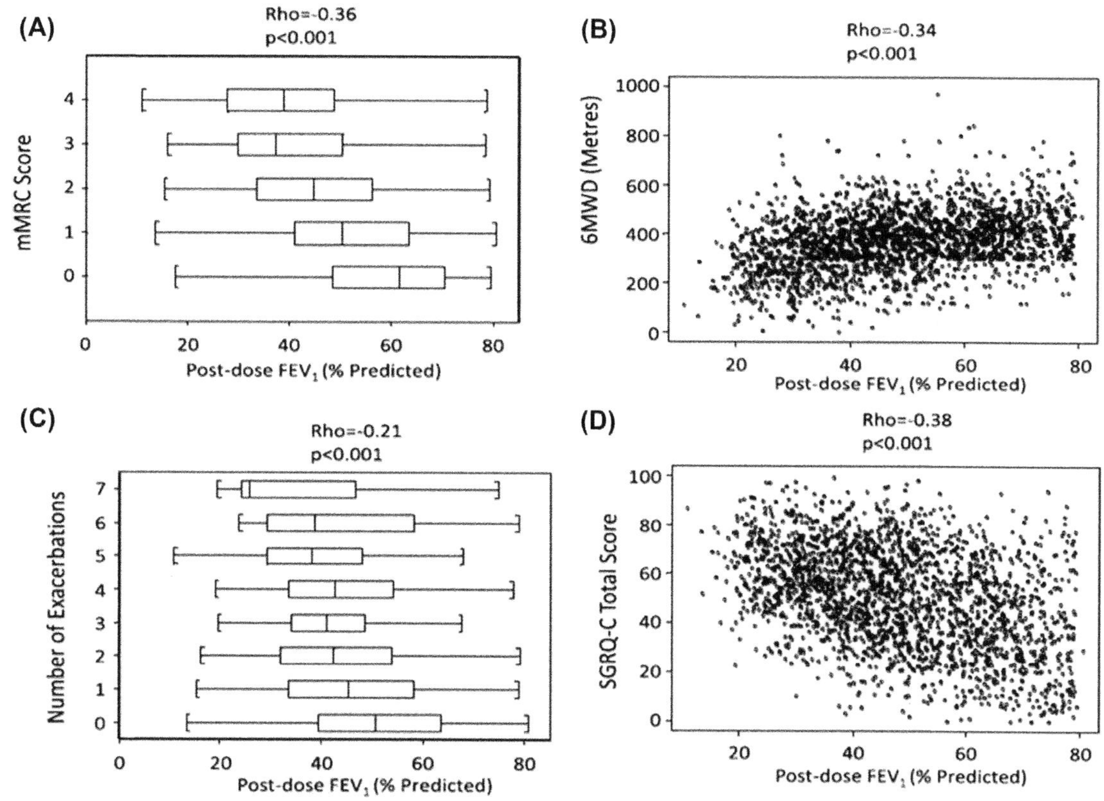

FIGURE 36.1 Relationship between severity of airflow limitation and breathlessness. Assessed using the modified Medical Research Council (mMRC) questionnaire (panel A), 6-minute walking distance (6MWD) (panel B), reported exacerbations in the year before inclusion in the study (panel C), and St George's Respiratory Questionnaire for COPD (SGRQ-C) (panel D). *Reproduced from Agusti A, Calverley PM, Celli B, et al. Characterisation of COPD heterogeneity in the ECLIPSE cohort. Respir. Res 2010;11:122. This figure is from an Open Access article distributed under the terms of the Creative Commons Attribution License (http://creativecommons.org/licenses/by/2.0), which permits unrestricted use, distribution, and reproduction in any medium, provided the original work is properly cited.*

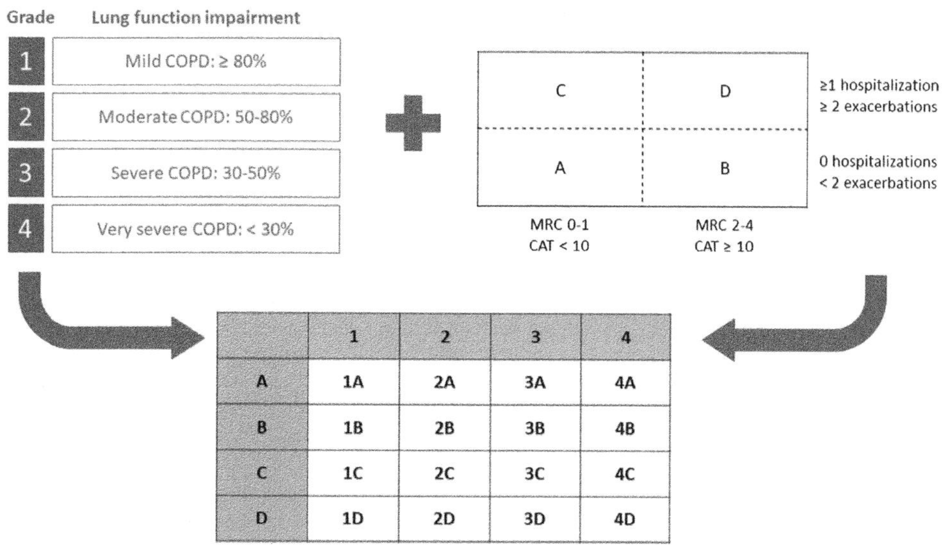

FIGURE 36.2 Different patient types according to GOLD recommendations. *CAT*, Category; *GOLD*, Global Initiative for Obstructive Lung Disease; *MRC*, Medical Research Council questionnaire.

index, obstruction degree, dyspnea perception, and exercise capacity) were combined to identify impairment. Important relationship with various clinical outcomes and aspects of the disease was demonstrated. New indexes not based on BODE appeared, which in turn also had updates [22]. Several protocols display long-term prognostic capacity [44]. These are not intended to evaluate therapeutic response or to choose treatments.

An article published some years ago established a consensus on what should be called clinical phenotype in COPD [45]. It was defined as "single or combination of disease attributes that describe differences between individuals with COPD, as they relate to clinically meaningful outcomes (symptoms, exacerbations, response to therapy, rate of disease progression, or death)." Although intuitive and accepted by the scientific community, this definition resulted in a considerable number of variables that were disease attributes, described differences between individuals, and were related to clinically meaningful outcomes, for example, dyspnea, bronchiectasis, chronic cough and sputum, exacerbations, hyperinflation, bronchial hyperresponsiveness, airway chronic colonization, and emphysema. Hence, the distribution of patients according to phenotypes that began as a relatively simple scheme has become considerably more complicated [46].

The Spanish COPD guidelines (GesEPOC) took the lead, and this was followed by national guidelines from other countries. The latest version classifies patients into high or low risk, according to classical variables (dyspnea, exacerbations, and pulmonary function), recommending the establishment of phenotypes only for high risk patients.

The selected clinical phenotypes are nonexacerbator, asthma-COPD overlap [47], frequent exacerbator with emphysema, and frequent exacerbator with chronic bronchitis (Fig. 36.3). The main strength is tools to identify patient types associated with specific therapies. However, in order to apply this approach to the whole COPD spectrum, phenotypes should be mutually exclusive and collectively exhaustive. In other words, all patients should be represented by one phenotype, being allocated to only one of them. However, this does not occur, for example, bronchiectasis [48], and one patient may fulfill the criteria for more than one phenotype.

Additionally, they are not stable over time. The frequent exacerbator or the nonexacerbator phenotype may cease [49]. Moreover, some phenotypes are just proposals and probably need prospective validation, for example, the asthma-COPD overlap [50].

There is a missing link between clinical phenotypes and biological processes underlying clinical expression, also known as endotypes. Endotypes in COPD have been less explored than they have been in asthma; however, the information we have indicates a lack of correlation between endotypes and phenotypes [51].

Modern COPD phenotyping

Clinical phenotypes are of limited use in clinical practice, which deals with individual patients (i.e., precision medicine) and not with groups (stratified medicine) [52]. Therefore, the concept of treatable traits emerges that may warrant a specific therapeutic approach [53] (Fig. 36.4).

The concept has evolved to be applied to other traits that may not result in direct therapy, but would indicate different disease behavior or biological pathway. Analogous to the concept of phenotypes and endotypes, these include observable traits (phenotraits), namely clinical, morphological, and functional, together with biological traits (endotraits), showing the different biological pathways implicated in a specific individual. An example of this concept is shown in Fig. 36.5, in which each individual patient may show a different degree of each of these traits.

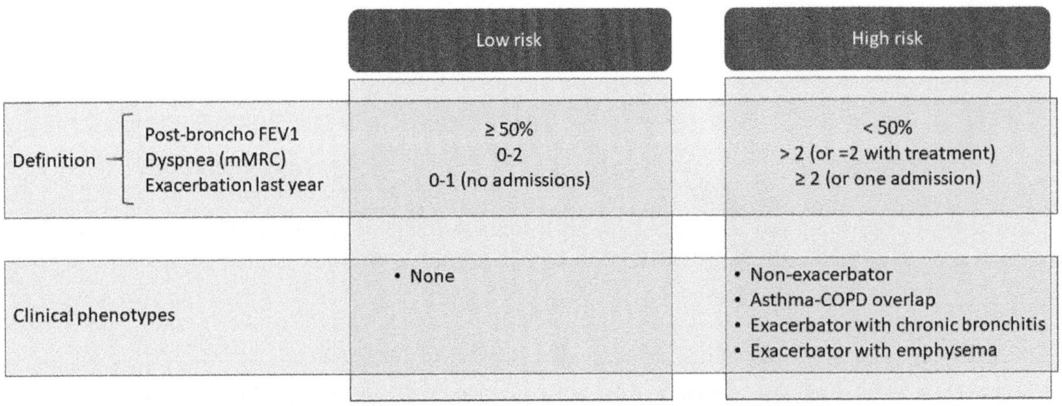

FIGURE 36.3 Concept of clinical phenotypes recorded in the Spanish GesEPOC guidelines [71].

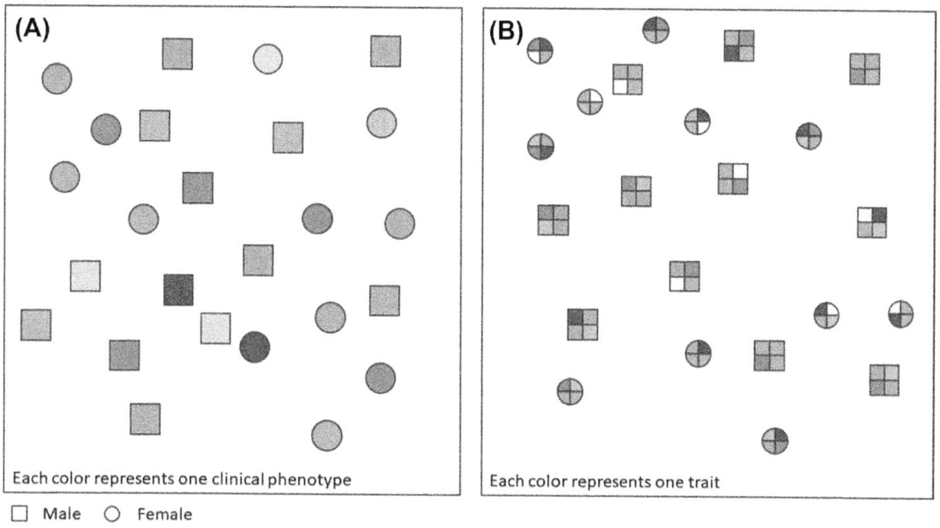

FIGURE 36.4 Diagrammatic representation of two approaches: (A) clinical phenotypes; (B) specific traits.

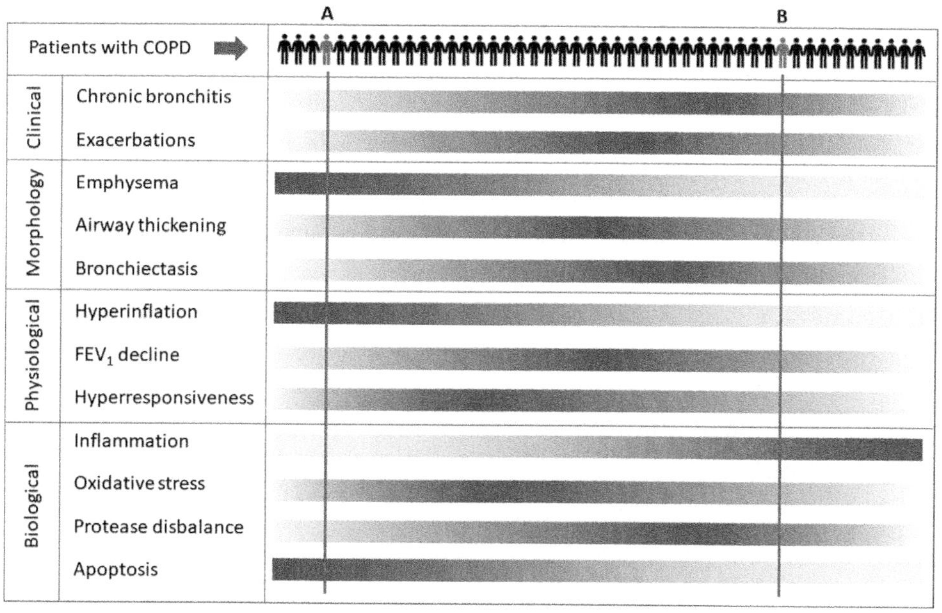

FIGURE 36.5 Diagrammatic representation exemplifying some examples of specific traits and the uneven distribution in a population with COPD. In this example, patient A would represent a case with emphysema, lung hyperinflation, and increased apoptosis, in contrast to patient B with chronic bronchitis, bronchiectasis, and more severe FEV_1 decline and inflammation.

Importance of clinical traits

Four major disease expressions have ignited the interest of researchers in the area of COPD, that is, dyspnea, chronic bronchitis, exacerbations, and symptoms overlapping with asthma. They are important because of prevalence, impact on health-related quality of life, and prognostic implications [29,54,55]. Other relevant clinical outcomes are fatigue or exercise limitation. A systematic clinical approach that includes the evaluation of these traits is needed [56].

Importance of functional traits

There are at least three major functional traits, that is, FEV_1 decline, lung hyperinflation, and airway hyperresponsiveness. Despite FEV_1 decline being the hallmark of COPD, real-life data indicate that this decline behaves differently. Progression of FEV_1 decline over time does not worsen at similar speed [20]. Accordingly, rapid and slow decliners have been described [57]. More strikingly, an improvement in FEV_1 decline has been reported with therapy [20]. The progression of FEV_1 decline is related to the best lung function attained during childhood. Different degrees of lung development are shown to be relevant in subsequent progression in adulthood [58].

The importance of lung development on lung function decline has been reported previously [59], and has been confirmed in recent studies [60,61]. Both can be programmed in early stages of life. The environment during embryonic development and first years of life seems to be essential to acquisition of maximum lung function, with a critical window for subsequent lung health [62,63]. Factors associated with lung growth and aging during embryonic development and early childhood are described in Table 36.1 [64].

FEV_1 decline is also a factor affecting current severity of obstruction. In early stages, it is faster than in more advanced disease. This opens a window of opportunity in early stages, when FEV_1 decline can be reduced with therapy [65,66].

Lung hyperinflation is a phenomenon related to expiratory flow limitation, of equal clinical impact. Regardless of the physiological measurement, lung hyperinflation does not appear in all COPD cases, and it has clinical and prognostic implications [13]. Measures of static lung hyperinflation have been reported to be predictive of respiratory and all-cause mortality [14]. Accordingly, static hyperinflation evaluation is needed. Additionally, dynamic lung hyperinflation, in the setting of increased ventilation, can have important negative consequences for the function of both respiratory and cardiovascular systems, depending on the extent of static lung hyperinflation, and is increasingly implicated as a major cause of dyspnea and exercise limitation [67].

Apart from the potential role in development of COPD [68], bronchial hyperresponsiveness is an independent trait in COPD, and provides important information on phenotype heterogeneity [69]. Another potential use is in evaluation of overlap between COPD and asthma [70], with direct implications in phenotype categorization [71].

Importance of morphological traits

Participation of anatomical lung compartments (i.e., the main airway, small airway, parenchyma, and vasculature) in progression of COPD is recognized. Seminal histological studies have clearly shown that small airway plays a significant role in COPD [72]. Other compartments are also related to specific clinical presentations. Imaging modalities are widely available in clinical care settings, and may contribute to describing COPD heterogeneity, as well as structural and functional effects [73]. The immediate application is in quantification of emphysema, which has shown a good correlation with clinical and lung function parameters [74]. Evaluation of airway branch variation is important. There are two main airway branch

TABLE 36.1 Summary of different early-life factors associated with lung growth and aging during prenatal and early childhood periods.

	Lung growth impairment	Accelerated lung aging
Embryonic development	Increased risk • Maternal lifestyle (smoking) • Maternal malnutrition • Environmental contamination • Intrauterine growth restriction Decreased risk • Multiparous mother	Increased risk • Maternal winter infections • Prenatal malnutrition • Intrauterine growth restriction
Early childhood	Increased risk • Second-hand tobacco smoke exposure (maternal smoking) • Poor nutrition • Pulmonary infections • Premature birth (low birth weight) • Elderly mothers • Bronchopulmonary dysplasia and invasive associated treatments (mechanical ventilation and oxygen therapy)	Increased risk • Second-hand tobacco smoke exposure (maternal smoking) • Pulmonary infections • Premature birth (low birth weight) • Cohabitation with younger siblings • Birth in cold seasons • Elderly mothers • Born in winter months Decreased risk • Cohabitation with pets • Cohabitation with older siblings

Adapted from Delgado Pecellin I, Moreno Valera MJ, Moreno Ortega M, Quintana Gallego E, Carrasco Hernandez L, Lopez-Campos JL. Lung growth and aging: a complex and increasingly confounding network. Arch. Bronconeumol 2019., Feb 20. https://doi.org/10.1016/j.arbres. 2019.01.005. pii: S0300-2896(19)30012-2.

abnormalities, namely greater central airway bifurcation density throughout the lung, and narrower airway lumens in all the lobes [75].

Computed tomography is mostly recommended for lung cancer, bronchiectasis, emphysema-associated lung fibrosis, or preinterventional approaches; yet with more accessible quantification methods it may contribute to disease phenotyping [73], including emphysema-dominant and airway-dominant disease [76]. Fissure integrity for lung volume reduction, or early detection of lung cancer, is another potential application of radiology in COPD, beyond disease phenotyping.

Importance of biological traits

Endotrait assessment can be complex, in part because there is a dissociation between airway inflammation and phenotraits previously discussed. Despite attempts to characterize inflammatory loads in COPD [77], it was not until high-throughput technology and bioinformatics became available that we had an opportunity to advance our understanding. Searched biomarkers include exhaled breath condensate [78], lung biopsies [79], induced sputum [80], plasma variables [81], and bronchoalveolar lavage fluid [82].

Role of proteomics

Proteomic profile of peripheral neutrophils, in patients with stable COPD and healthy age-matched controls [83], identified two groups of patients with COPD, on the basis of differentially regulated protein clusters, yet without difference in lung function. Therefore, systemic inflammatory signals do not necessarily correlate with the GOLD classification.

Breath analysis

Volatile organic compounds in exhaled air, identified using gas chromatography mass spectrometry, have been associated with inflammatory profiles in COPD [84]. The identification of individual molecular compounds is not always possible, for phenotyping in the clinical setting. Yet the combined signal response of the portable electronic nose (eNose) is an option that reveals a characteristic profile. This technology is fast and easy to use, as a point-of-care tool. In patients with asthma and COPD, five validated clusters determined by the clinical/inflammatory characteristics were announced [85].

Interest of metabolomics

Metabolomic profiling of peripheral blood and exhaled breath condensate was recently attempted [86]. While metabolomics is good for identifying biomarkers and clarifying the mechanism of disease, no independent clinical-metabolic phenotypes could be recognized so far.

Molecular phenotyping of COPD represents an essential first step in identification and classification, before diagnostic/prognostic tools and treatment options can be established. Large-scale profiling based on omics technology will provide the means of elucidating global alterations in COPD. Dysregulation at different molecular levels and anatomical locations can dominate in different disease subgroups; consequently, multiomics integration may be necessary [87].

Pilot studies with multiomics

Integrating five to seven omics data blocks has enabled correct classification of COPD diagnosis in small cohorts. These results can serve as guidelines for the design of future systems-based multiomics investigations.

Treatment plan

New P4 medicine (predictive, preventive, personalized, and participatory) guidelines are currently advised. Translational bioinformatics and integrative modeling will continue to open new avenues of scientific research, for addressing challenges in P4 medicine [88]. In COPD, identification of biomarkers (endotypes) specific for different clinical presentations, traditionally termed as clinical phenotypes, is a priority. One or a combination of biomarkers to predict behavior of disease will help select treatment strategies that are able to improve disease impact and prognosis, for example, rapid decliners, those prone to bronchial colonization, frequent exacerbators, or those with different degrees of hyperinflation. Unfortunately, the relationship between endotypes and phenotypes is far from being fully understood. A recent study proposed early endotyping, as a possible means for intervention in COPD [51].

Studying COPD once it had occurred, does not allow tracing back the origins of disease. Susceptible individuals may show distinct pathobiological mechanisms ("COPD endotypes") that ultimately manifest as specific phenotypes via different biological mechanisms. Therefore, grading these endotypes according to severity may allow better diagnosis, and intervention can be designed on the basis of the involved mechanisms [51].

COPD theratypes

As initiatives thus far have not achieved clinical translation, including omics protocols [83,89], complementary approximation to phenotyping, such as analysis according to therapeutic response, could be an alternative (theratypes).

Patient responses considerably vary [90]. Only a few studies attempted to identify responders. One of them was published by Donohue et al. [91]. Functional response to double bronchodilation (umeclidinium/vilanterol) versus single bronchodilation was compared by using an increase of >12% and >200 mL in trough FEV_1 as a marker. One-third of the patients responded positively to both long-acting bronchodilators, another third to one agent, and the remaining third did not respond. Interestingly, the former third of patients also had stronger responses to double bronchodilation. It is hypothesized that response to single bronchodilation may be used to identify candidates for double bronchodilation. Unfortunately, tests to predict response to long-acting bronchodilators have reported conflicting results [92–95].

Responses to double bronchodilation according to sex [96] or baseline disease severity, as measured by COPD Assessment Test [97], were also addressed. A greater response to double bronchodilation corresponded to those with more severe disease impact, thus supporting the GOLD recommendation to administer double-bronchodilator therapies, to more symptomatic patients.

To make things more complex, comparison of two double-bronchodilation combinations highlighted the fact that patients may respond differently to different bronchodilators [98], thereby confirming the individualized response observed by the others [91]. Future trials should evaluate long-term impact of double bronchodilation in different patient types, as polymorphisms have been proposed [99].

References

[1] Barnes PJ, Burney PG, Silverman EK, et al. Chronic obstructive pulmonary disease. Nat. Rev. Disease Primers 2015;1:15076.

[2] Turner RM, DePietro M, Ding B. Overlap of asthma and chronic obstructive pulmonary disease in patients in the United States: analysis of prevalence, features, and subtypes. JMIR Public Health Surveill 2018;4:e60.

[3] Polverino E, Dimakou K, Hurst J, et al. The overlap between bronchiectasis and chronic airway diseases: state of the art and future directions. Eur. Respir. J 2018;52.

[4] Kania A, Krenke R, Kuziemski K, et al. Distribution and characteristics of COPD phenotypes – results from the polish sub-cohort of the POPE study. Int. J. Chronic Obstr. Pulm. Dis. 2018;13:1613–21.

[5] Alcazar-Navarrete B, Trigueros JA, Riesco JA, Campuzano A, Perez J. Geographic variations of the prevalence and distribution of COPD phenotypes in Spain: "the ESPIRAL-ES study". Int. J. Chronic Obstr. Pulm. Dis. 2018;13:1115–24.

[6] Psaty BM, Dekkers OM, Cooper RS. Comparison of 2 treatment models: precision medicine and preventive medicine. JAMA: J. Am. Med. Assoc 2018;320:751–2.

[7] Echazarreta AL, Arias SJ, Del Olmo R, et al. Prevalence of COPD in 6 urban clusters in Argentina: the EPOC.AR study. Archivos de bronconeumologia; 2017.

[8] Lopez-Campos JL, Ruiz-Ramos M, Soriano JB. Mortality trends in chronic obstructive pulmonary disease in Europe, 1994–2010: a joinpoint regression analysis. Lancet Res. Med 2014;2:54–62.

[9] Collaborators GBDCRD. Global, regional, and national deaths, prevalence, disability-adjusted life years, and years lived with disability for chronic obstructive pulmonary disease and asthma, 1990–2015: a systematic analysis for the global burden of disease study 2015. Lancet Res. Med 2017;5:691–706.

[10] Burrows B, Fletcher CM, Heard BE, Jones NL, Wootliff JS. The emphysematous and bronchial types of chronic airways obstruction. A clinicopathological study of patients in London and Chicago. Lancet 1966;1:830–5.

[11] Terminology, definitions, and classification of chronic pulmonary emphysema and related conditions: a report of the conclusions of a CIBA guest symposium. Thorax 1959;14:286–99.

[12] Definition and classification of chronic bronchitis for clinical and epidemiological purposes. A report to the Medical Research Council by their Committee on the Aetiology of Chronic Bronchitis. Lancet 1965;1:775–9.

[13] Casanova C, Cote C, de Torres JP, et al. Inspiratory-to-total lung capacity ratio predicts mortality in patients with chronic obstructive pulmonary disease. Am. J. Respir. Crit. Care Med. 2005;171:591–7.

[14] Tantucci C, Donati P, Nicosia F, et al. Inspiratory capacity predicts mortality in patients with chronic obstructive pulmonary disease. Respir. Med. 2008;102:613–9.

[15] Parshall MB, Schwartzstein RM, Adams L, et al. An official American Thoracic Society statement: update on the mechanisms, assessment, and management of dyspnea. Am. J. Respir. Crit. Care Med. 2012;185:435–52.

[16] Cockcroft DW, Wenzel S. Airway hyperresponsiveness and chronic obstructive pulmonary disease outcomes. J. Allergy Clin. Immunol. 2016;138:1580–1.

[17] Soler-Cataluna JJ, Sauleda J, Valdes L, et al. Prevalence and perception of 24-hour symptom patterns in patients with stable chronic obstructive pulmonary disease in Spain. Arch. Bronconeumol. 2016;52:308–15.

[18] Kumar A, Kunal S, Shah A. Incidence and impact of upper airway symptoms in patients with chronic obstructive pulmonary disease. Arch. Bronconeumol. 2017;53:647–9.

[19] Anzueto A, Calverley PMA, Mueller A, et al. Demographic characteristics and clinical outcomes in patients from Latin America versus the rest of the world: a TIOSPIR((R)) Post-Hoc analysis. Arch. Bronconeumol. 2018;54:140–8.

[20] Vestbo J, Edwards LD, Scanlon PD, et al. Changes in forced expiratory volume in 1 second over time in COPD. N. Engl. J. Med. 2011;365:1184–92.

[21] Hansel TT, Barnes PJ. New drugs for exacerbations of chronic obstructive pulmonary disease. Lancet 2009;374:744–55.

[22] Lopez-Campos JL, Centanni S. Current approaches for phenotyping as a target for precision medicine in COPD management. COPD 2018;15:108–17.

[23] Agusti A, Calverley PM, Celli B, et al. Characterisation of COPD heterogeneity in the ECLIPSE cohort. Respir. Res. 2010;11:122.

[24] Hofstede SN, van Bodegom-Vos L, Kringos DS, Steyerberg E, Marang-van de Mheen PJ. Mortality, readmission and length of stay have different relationships using hospital-level versus patient-level data: an example of the ecological fallacy affecting hospital performance indicators. BMJ Qual. Safety 2018;27:474–83.

[25] Boutou AK, Shrikrishna D, Tanner RJ, et al. Lung function indices for predicting mortality in COPD. Eur. Respir. J 2013;42:616–25.
[26] Soler-Cataluna JJ, Martinez-Garcia MA, Roman Sanchez P, Salcedo E, Navarro M, Ochando R. Severe acute exacerbations and mortality in patients with chronic obstructive pulmonary disease. Thorax 2005;60:925–31.
[27] Hospers JJ, Postma DS, Rijcken B, Weiss ST, Schouten JP. Histamine airway hyper-responsiveness and mortality from chronic obstructive pulmonary disease: a cohort study. Lancet 2000;356:1313–7.
[28] Celli BR, Cote CG, Marin JM, et al. The body-mass index, airflow obstruction, dyspnea, and exercise capacity index in chronic obstructive pulmonary disease. N. Engl. J. Med. 2004;350:1005–12.
[29] Lahousse L, Seys LJM, Joos GF, Franco OH, Stricker BH, Brusselle GG. Epidemiology and impact of chronic bronchitis in chronic obstructive pulmonary disease. Eur. Respir. J 2017;50.
[30] Moy ML, Gould MK, Liu IA, Lee JS, Nguyen HQ. Physical activity assessed in routine care predicts mortality after a COPD hospitalisation. ERJ Open Res 2016;2.
[31] Nishimura K, Izumi T, Tsukino M, Oga T. Dyspnea is a better predictor of 5-year survival than airway obstruction in patients with COPD. Chest 2002;121:1434–40.
[32] Di Marco F, Milic-Emili J, Boveri B, et al. Effect of inhaled bronchodilators on inspiratory capacity and dyspnoea at rest in COPD. Eur. Respir. J 2003;21:86–94.
[33] Di Marco F, Sotgiu G, Santus P, et al. Long-acting bronchodilators improve exercise capacity in COPD patients: a systematic review and meta-analysis. Respir. Res. 2018;19:18.
[34] Beeh KM, Burgel PR, Franssen FME, et al. How do dual long-acting bronchodilators prevent exacerbations of chronic obstructive pulmonary disease? Am. J. Respir. Crit. Care Med. 2017;196:139–49.
[35] Toledo-Pons N, Cosio BG. Is there room for theophylline in COPD? Arch. Bronconeumol. 2017;53:539–40.
[36] Gea J. The future of biological therapies in COPD. Arch. Bronconeumol. 2018;54:185–6.
[37] Pavord ID, Chanez P, Criner GJ, et al. Mepolizumab for eosinophilic chronic obstructive pulmonary disease. N. Engl. J. Med. 2017;377:1613–29.
[38] Pauwels RA, Buist AS, Calverley PM, Jenkins CR, Hurd SS. Global strategy for the diagnosis, management, and prevention of chronic obstructive pulmonary disease. NHLBI/WHO Global Initiative for Chronic Obstructive Lung Disease (GOLD) workshop summary. Am. J. Respir. Crit. Care Med. 2001;163:1256–76.
[39] Rodriguez-Roisin R, Agusti A. The GOLD initiative 2011: a change of paradigm? Arch. Bronconeumol. 2012;48:286–9.
[40] Vogelmeier CF, Criner GJ, Martinez FJ, et al. Global strategy for the diagnosis, management, and prevention of chronic obstructive lung disease 2017 report: GOLD executive summary. Arch. Bronconeumol. 2017;53:128–49.
[41] Jones PW, Adamek L, Nadeau G, Banik N. Comparisons of health status scores with MRC grades in COPD: implications for the GOLD 2011 classification. Eur. Respir. J 2013;42:647–54.
[42] Smid DE, Franssen FME, Gonik M, et al. Redefining cut-points for high symptom burden of the global initiative for chronic obstructive lung disease classification in 18,577 patients with chronic obstructive pulmonary disease. J. Am. Med. Dir. Assoc. 2017;18:1097.e11–24.
[43] Cabrera Lopez C, Casanova Macario C, Marin Trigo JM, et al. Comparison of the 2017 and 2015 global initiative for chronic obstructive lung disease reports. Impact on grouping and outcomes. Am. J. Respir. Crit. Care Med. 2018;197:463–9.
[44] Marin JM, Alfageme I, Almagro P, et al. Multicomponent indices to predict survival in COPD: the COCOMICS study. Eur. Respir. J 2013;42:323–32.
[45] Han MK, Agusti A, Calverley PM, et al. Chronic obstructive pulmonary disease phenotypes: the future of COPD. Am. J. Respir. Crit. Care Med. 2010;182:598–604.
[46] Shirtcliffe P, Weatherall M, Travers J, Beasley R. The multiple dimensions of airways disease: targeting treatment to clinical phenotypes. Curr. Opin. Pulm. Med. 2011;17:72–8.
[47] Plaza V, Alvarez F, Calle M, et al. Consensus on the asthma-COPD overlap syndrome (ACOS) between the Spanish COPD guidelines (GesEPOC) and the Spanish guidelines on the management of asthma (GEMA). Arch. Bronconeumol. 2017;53:443–9.
[48] Martinez-Garcia MA, Maiz L, Olveira C, et al. Spanish guidelines on the evaluation and diagnosis of bronchiectasis in adults. Arch. Bronconeumol. 2018;54:79–87.
[49] Hurst JR, Vestbo J, Anzueto A, et al. Susceptibility to exacerbation in chronic obstructive pulmonary disease. N. Engl. J. Med. 2010;363:1128–38.
[50] Perez de Llano L, Cosio BG, Miravitlles M, Plaza V, group Cs. Accuracy of a new algorithm to identify asthma-COPD overlap (ACO) patients in a cohort of patients with chronic obstructive airway disease. Arch. Bronconeumol. 2018;54:198–204.
[51] Petersen H, Vazquez Guillamet R, Meek P, Sood A, Tesfaigzi Y. Early endotyping: a chance for intervention in chronic obstructive pulmonary disease. Am. J. Res. Cell Mol. Biol 2018;59:13–7.
[52] Agusti A. The path to personalised medicine in COPD. Thorax 2014;69:857–64.
[53] Agusti A, Bel E, Thomas M, et al. Treatable traits: toward precision medicine of chronic airway diseases. Eur. Respir. J 2016;47:410–9.
[54] Mullerova H, Lu C, Li H, Tabberer M. Prevalence and burden of breathlessness in patients with chronic obstructive pulmonary disease managed in primary care. PLoS One 2014;9:e85540.
[55] Han MK, Quibrera PM, Carretta EE, et al. Frequency of exacerbations in patients with chronic obstructive pulmonary disease: an analysis of the SPIROMICS cohort. Lancet Res. Med 2017;5:619–26.
[56] Burgel PR, Paillasseur JL, Janssens W, et al. A simple algorithm for the identification of clinical COPD phenotypes. Eur. Respir. J 2017;50.
[57] Nishimura M, Makita H, Nagai K, et al. Annual change in pulmonary function and clinical phenotype in chronic obstructive pulmonary disease. Am. J. Respir. Crit. Care Med. 2012;185:44–52.
[58] Lange P, Celli B, Agusti A, et al. Lung-function trajectories leading to chronic obstructive pulmonary disease. N. Engl. J. Med. 2015;373:111–22.
[59] Fletcher C, Peto R. The natural history of chronic airflow obstruction. Br. Med. J. 1977;1:1645–8.
[60] Bui DS, Lodge CJ, Burgess JA, et al. Childhood predictors of lung function trajectories and future COPD risk: a prospective cohort study from the first to the sixth decade of life. Lancet Res. Med 2018;6:535–44.

[61] Belgrave DCM, Granell R, Turner SW, et al. Lung function trajectories from pre-school age to adulthood and their associations with early life factors: a retrospective analysis of three population-based birth cohort studies. Lancet Res. Med 2018;6:526–34.
[62] Dratva J, Zemp E, Dharmage SC, et al. Early life origins of lung ageing: early life exposures and lung function decline in adulthood in two European cohorts aged 28–73 years. PLoS One 2016;11:e0145127.
[63] Weiss ST, Ware JH. Overview of issues in the longitudinal analysis of respiratory data. Am. J. Respir. Crit. Care Med. 1996;154: S208–11.
[64] Delgado Pecellin I, Moreno Valera MJ, Moreno Ortega M, Quintana Gallego E, Carrasco Hernandez L, Lopez-Campos JL. Lung growth and aging: a complex and increasingly confounding network. Arch. Bronconeumol 2019, Feb 20. https://doi.org/10.1016/j.arbres.2019.01.005. pii: S0300-2896(19)30012-2.
[65] Tashkin DP, Celli BR, Decramer M, Lystig T, Liu D, Kesten S. Efficacy of tiotropium in COPD patients with FEV1 >/= 60% participating in the UPLIFT(R) trial. COPD 2012;9:289–96.
[66] Celli BR, Thomas NE, Anderson JA, et al. Effect of pharmacotherapy on rate of decline of lung function in chronic obstructive pulmonary disease: results from the TORCH study. Am. J. Respir. Crit. Care Med. 2008;178:332–8.
[67] O'Donnell DE, Webb KA, Neder JA. Lung hyperinflation in COPD: applying physiology to clinical practice. COPD Res. Pract 2015;1:4.
[68] Marcon A, Locatelli F, Keidel D, et al. Airway responsiveness to methacholine and incidence of COPD: an international prospective cohort study. Thorax 2018;73:825–32.
[69] van den Berge M, Vonk JM, Gosman M, et al. Clinical and inflammatory determinants of bronchial hyperresponsiveness in COPD. Eur. Respir. J 2012;40:1098–105.
[70] Tkacova R, Dai DL, Vonk JM, et al. Airway hyperresponsiveness in chronic obstructive pulmonary disease: a marker of asthma-chronic obstructive pulmonary disease overlap syndrome? J. Allergy Clin. Immunol. 2016;138:1571–1579 e10.
[71] Miravitlles M, Soler-Cataluna JJ, Calle M, et al. Spanish guidelines for management of chronic obstructive pulmonary disease (GesEPOC) 2017. Pharmacological treatment of stable phase. Arch. Bronconeumol. 2017;53:324–35.
[72] Hogg JC, Pare PD, Hackett TL. The contribution of small airway obstruction to the pathogenesis of chronic obstructive pulmonary disease. Physiol. Rev. 2017;97:529–52.
[73] Labaki WW, Martinez CH, Martinez FJ, et al. The role of chest computed tomography in the evaluation and management of the patient with chronic obstructive pulmonary disease. Am. J. Respir. Crit. Care Med. 2017;196:1372–9.
[74] Kaya L, Ozel D, Ozel BD. Evaluating qualitative and quantitative computerized tomography indicators of chronic obstructive pulmonary disease and their correlation with pulmonary function tests. Pol. J. Radiol. 2017;82:511–5.
[75] Smith BM, Traboulsi H, Austin JHM, et al. Human airway branch variation and chronic obstructive pulmonary disease. Proc. Natl. Acad. Sci. U.S.A 2018;115:E974–81.
[76] Subramanian DR, Gupta S, Burggraf D, et al. Emphysema- and airway-dominant COPD phenotypes defined by standardised quantitative computed tomography. Eur. Respir. J 2016;48: 92–103.
[77] Roy K, Smith J, Kolsum U, Borrill Z, Vestbo J, Singh D. COPD phenotype description using principal components analysis. Respir. Res. 2009;10:41.
[78] Borrill ZL, Roy K, Singh D. Exhaled breath condensate biomarkers in COPD. Eur. Respir. J 2008;32:472–86.
[79] Bhattacharya S, Srisuma S, Demeo DL, et al. Molecular biomarkers for quantitative and discrete COPD phenotypes. Am. J. Res. Cell Mol. Biol. 2009;40:359–67.
[80] Gray RD, MacGregor G, Noble D, et al. Sputum proteomics in inflammatory and suppurative respiratory diseases. Am. J. Respir. Crit. Care Med. 2008;178:444–52.
[81] Bowler RP, Canham ME, Ellison MC. Surface enhanced laser desorption/ionization (SELDI) time-of-flight mass spectrometry to identify patients with chronic obstructive pulmonary disease. COPD 2006;3:41–50.
[82] Merkel D, Rist W, Seither P, Weith A, Lenter MC. Proteomic study of human bronchoalveolar lavage fluids from smokers with chronic obstructive pulmonary disease by combining surface-enhanced laser desorption/ionization-mass spectrometry profiling with mass spectrometric protein identification. Proteomics 2005;5:2972–80.
[83] Loi ALT, Hoonhorst S, van Aalst C, et al. Proteomic profiling of peripheral blood neutrophils identifies two inflammatory phenotypes in stable COPD patients. Respir. Res. 2017;18:100.
[84] Fens N, de Nijs SB, Peters S, et al. Exhaled air molecular profiling in relation to inflammatory subtype and activity in COPD. Eur. Respir. J 2011;38:1301–9.
[85] de Vries R, Dagelet YWF, Spoor P, et al. Clinical and inflammatory phenotyping by breathomics in chronic airway diseases irrespective of the diagnostic label. Eur. Respir. J 2018;51.
[86] Kilk K, Aug A, Ottas A, Soomets U, Altraja S, Altraja A. Phenotyping of chronic obstructive pulmonary disease based on the integration of metabolomes and clinical characteristics. Int. J. Mol. Sci. 2018;19.
[87] Li CX, Wheelock CE, Skold CM, Wheelock AM. Integration of multi-omics datasets enables molecular classification of COPD. Eur. Respir. J 2018;51.
[88] Younesi E, Hofmann-Apitius M. From integrative disease modeling to predictive, preventive, personalized and participatory (P4) medicine. EPMA J 2013;4:23.
[89] Baralla A, Fois AG, Sotgiu E, et al. Plasma proteomic signatures in early chronic obstructive pulmonary disease. Proteonomics Clin. Appl. 2018;12:e1700088.
[90] Lopez-Campos JL, Calero-Acuna C, Marquez-Martin E, et al. Double bronchodilation in chronic obstructive pulmonary disease: a crude analysis from a systematic review. Int. J. Chronic Obstr. Pulm. Dis. 2017;12:1867–76.
[91] Donohue JF, Singh D, Munzu C, Kilbride S, Church A. Magnitude of umeclidinium/vilanterol lung function effect depends on monotherapy responses: results from two randomised controlled trials. Respir. Med. 2016;112:65–74.
[92] Burgel PR, Le Gros V, Decuypere L, Bourdeix I, Perez T, Deslee G. Immediate salbutamol responsiveness does not predict long-term benefits of indacaterol in patients with chronic obstructive pulmonary disease. BMC Pulm. Med 2017;17:25.
[93] Tashkin DP, Li N, Kleerup EC, et al. Acute bronchodilator responses decline progressively over 4 years in patients with moderate to very severe COPD. Respir. Res. 2014;15:102.

[94] Pascoe S, Wu W, Zhu CQ, Singh D. Bronchodilator reversibility in patients with COPD revisited: short-term reproducibility. Int. J. Chronic Obstr. Pulm. Dis. 2016;11:2035–40.

[95] Konno S, Makita H, Suzuki M, et al. Acute bronchodilator responses to beta2-agonist and anticholinergic agent in COPD: their different associations with exacerbation. Respir. Med. 2017;127:14–20.

[96] Tsiligianni I, Mezzi K, Fucile S, et al. Response to indacaterol/glycopyrronium (IND/GLY) by sex in patients with COPD: a pooled analysis from the ignite program. COPD 2017;14:375–81.

[97] Martinez FJ, Fabbri LM, Ferguson GT, et al. Baseline symptom score impact on benefits of glycopyrrolate/formoterol metered dose inhaler in COPD. Chest 2017;152:1169–78.

[98] Feldman GJ, Sousa AR, Lipson DA, et al. Comparative efficacy of once-daily umeclidinium/vilanterol and tiotropium/olodaterol therapy in symptomatic chronic obstructive pulmonary disease: a randomized study. Adv. Ther. 2017;34:2518–33.

[99] Kang J, Kim KT, Lee JH, et al. Predicting treatable traits for long-acting bronchodilators in patients with stable COPD. Int. J. Chronic Obstr. Pulm. Dis. 2017;12:3557–65.

E—Lifestyle, age-related, and environmental agents

Chapter 37

Lifestyle medicine

Patrice Couzigou
Professor Emeritus of Medicine, University of Bordeaux, Bordeaux, France

Lifestyle medicine needs to be promoted

Lifestyle and behavioral medicine target, particularly in the singular caring relationship, nutrition (both qualitative and quantitative), physical inactivity and sedentary lifestyle, smoking cessation, control of alcohol consumption, and sleep. This text aims to show the importance, the need to specify, to individualize, to personalize the prescription of behavioral modifications, and to propose some ways of accomplishing these objectives.

What observations and what shortcomings?

Noncommunicable diseases are the leading causes of chronic disease and death worldwide. They are responsible for 87% of the causes of death in France (WHO 2014 report). They are mainly dependent on tobacco and alcohol consumption, qualitative and quantitative malnutrition, physical inactivity, sedentary lifestyle (not to be confused with physical inactivity), and sleep disorders. In France around three-fourth of deaths are related to cancer and cardiovascular disease, mainly linked with these risk factors. To this list must be added ultraviolet irradiation (UV), pollution (environmental or professional exposures, food contaminants), and also pesticides, and neuroendocrine disruptors.

Curative versus preventive medicine

The WHO definition of health as "state of complete mental, social, and physical well-being" (not simply an absence of disease or infirmity) could make us forget the asymptomatic phase of diseases linked with these risk factors, during many years. How many people are considered "healthy," just a few days before a cancer diagnosis or the occurrence of a heart attack or stroke? Vascular lesions, tumors were present and had been developing quietly for years. The progression of the disease was probably close to clinical manifestation, due to its natural evolution, possibly favored by a stress. After a new round of cell multiplication, the tumor becomes clinically obvious. These examples reinforce the failures of strictly curative medicine.

Lifestyle medicine is too often derided as mere prevention and supplementation to curative medicine. However the prevention has a major potential impact: it is estimated that 90% of type 2 diabetes, 82% of cardiovascular diseases, 70% of cancers [1], and 35% of dementias [2] could be avoided, by changing our way of life. As a representative illustration, the potentially interested French population is considerable. More than seven million people with obesity, more than 15 million overweight people (not to mention those with poor quality nutrition), more than 12 million people regularly smoking, and at least five million alcohol consumers, for a total population around 67 million people.

About 12 million people not reaching the recommendations concerning physical activity should be added, and 20 million with sedentary lifestyle [3]. Sedentary lifestyle is a risk factor different from physical inactivity, and physical inactivity is globally implicated in mortality, as much as or more than tobacco [4]. Despite these observations, the behavior of the French people, regarding their health, is mainly technical: complementary examinations and drugs, with little direct personal implication.

This is in apparent contradiction with the desire for a healthier, less technical, caring relationship, illustrating the ambivalence of the human being. The internal ecology of the person is not privileged, and is even neglected. The technical response is still predominant, both at the individual level (requests by sick people) and at the collective level, with poor remuneration of the caring act, intellectual and relational, in comparison to the technical act [5].

Precision Medicine for Investigators, Practitioners and Providers. https://doi.org/10.1016/B978-0-12-819178-1.00037-X

Reliance on chemicals

In France a medication is prescribed to almost 90% of consulting patients, compared to only 40% in the Netherlands. Dietary advice in pathologies such as hypertension, type 2 diabetes, and dyslipidemias is undervalued. Drug prevention for hypercholesterolemia , when indicated, is preferred over more simple and effective lifestyle changes. And prevention by drug instead of lifestyle modification tends to aggravate drug dependence.

Neglected complementary attention

A very preferentially technical medical response is one of the reasons for reinforcing the importance of complementary care. The place of complementary approaches is somewhere between general well-being and chronic care for sick people. They are often and falsely called parallel medicine: in the absence of additional evidence , they are not medicine. However, some of the complementary healthcare practices which directly or indirectly involve the sick person, through touch and/or personal involvement, could be part of an internal ecology approach.

On the contrary, external techniques such as homeopathy are very problematic. Beyond the unproven effectiveness, apart from the placebo effect and the caring relationship, the argument of sparing allopathic drugs is absolutely not demonstrated [6]. The media unfortunately contributes to fueling magical thinking, in conflict with the promotion of a better way of life. These unconventional approaches can, however, be useful crutches, especially if they favor a personal involvement to improve the internal ecology.

A better understanding of prevention and screening

Behavioral medicine evokes prevention. This is often poorly understood. It can expand life expectancy more than drug therapies. For example, an adult aged around 50 with chronic hepatitis C, and weak hepatic fibrosis, without antiviral treatment, statistically loses (but a person is not a statistic) less than 2 years of his life expectancy. Remember that cure is at a cost of 30,000 euros or more. Statistically also, and depending on lifestyle and the way of hepatitis contamination, the same person is likely to suffer comorbidities linked to tobacco, alcohol, or obesity. Treatment of these comorbidities, concomitantly or post-antiviral treatment, would improve life expectancy (on average 5 years or more and at a lower cost) than virus C eradication. Prevention can do better than cure.

Obviously it is totally different in case of severe liver fibrosis or cirrhosis. Weak fibrosis in cases of hepatitis C is observed in at least one-third of cases. Actually both treatments, preventive and curative, and not only antiviral prescriptions, should be provided.

The false security of normal laboratory tests

The inactive patient, static, overweight, reassures himself with results of biochemical data or medical imaging. Yet a "normal cholesterol" does not exclude significant risk factors. Absence of imaging abnormalities don't preclude problems during the next months or years. Body mass index, waist circumference, amount of physical activity, and avoidance of sedentary lifestyle should be systematically evaluated by the healthcare professional. From the vantage point of the patient, a prescribed medication, even in primary prevention, means treatment and protection, even when it confers little prevention.

France invests only 2% of health expenditure for prevention, against 3% on average in European countries (OCDE report). In addition, too often prevention is confused with screening, which is not the same concept. When the risk factor is identified, a change in the lifestyle should be considered, like an investor does after a negative balance sheet of his assets. Of course, this raises the problem of motivation; however, working on the motivation is part of the role of caregiver.

Relevance of lifestyle guidance and intervention

Approximately 1.2 million people undergo a colonoscopy every year in France. It would be very useful to inform them, or remind them, of the risk factors behind polyps and colorectal cancers, that is, alcohol, diet, excess body weight, and physical inactivity, which are detected in more than 40% of the cases [7].

A brief intervention has demonstrated efficacy for at-risk alcohol users [8]. Something analogous could be beneficial for colorectal pathology, without forgetting the benefits of this lifestyle for the whole body. At the collective level, a similar comment is to be made concerning mass screenings of cancer, such as for breast or colorectal cancer.

Risk factors and disease prevention

Within most prevention initiatives, emphasis is very insufficiently placed on the risk factors: alcohol, excess weight, physical inactivity, tobacco. Certainly, risk factor does not mean disease. Even communicable illnesses do not always lead to a disease. Hepatitis C causes cirrhosis in about 20% of the cases, which shows that other factors, genetic and environmental, are necessary for its genesis.

The difference between risk factors and pathology can be quite subtle. In the near future, when specific dysbiosis of the intestinal microbiota will be well characterized, in conditions such as obesity, anxiety, depression, will it be interpreted as disease? Will drugs be recommended for dysbiosis?

Lifestyle medicine and interest to medical practice

General medicine isn't the first choice at the end of the medical course. In France it is the very last. Little recognition of lifestyle medicine is one of the causes. Segmentation, hyperspecialization (although necessary) contribute to a reductionist approach by organ or disease, forgetful of human complexity: "*we cannot treat a complex problem as a sum of simple, isolable problems*" (Edgar Morin, 1921).

Compatibility of personalized medicine with evidence-based medicine

Evidence-based medicine (EBM) includes not only collective scientific data , and in particular well-neglected behavioral data, but also, what is often omitted, the caregiver's experience, and the uniqueness of the sick person [9] This faulty understanding does not favor the caregiver's personal involvement in the therapeutic relationship, which is particularly necessary in the behavioral approach [5]. As a result cohesion, consistency, and bonding are words which are insufficiently present in the minds of caregivers who, if more convinced, would deliver more effective medical assistance.

Even when this approach is envisaged, the actual role and responsibility of the general practitioner, is mainly to refer the case to specialists. In French national meetings, including ministerial, the central role, importance, and responsibility of general practitioners is emphasized, yet usually without a single liberal general practitioner in the room. The difficulty of their practice is not addressed, especially in terms of spent time, and corresponding recognition (including financial compensation).

Gaps in medical education and training

An unpublished survey, to which the author contributed a lot, with student unions and young doctors was carried out in 2015 by questionnaire, among 2732 medical students (upper and third cycle general medicine) or doctors, concerning their general medicine course. Undergraduate students (n = 1203) testified that they did not participate in consultations on the topic of smoking (61.2%), alcoholism (57.2%), nutritional monitoring (47.8%), and promotion of physical activity and guidance against inactivity (40.1%). They similarly failed to be instructed in motivational interviewing (54.2%), and to participate in them (47.7%).

Results with postgraduate students were not different. It is true that internship mentors are overworked, and cannot do everything. The inadequacies of the clinical internship concerning liberal and lifestyle medicine have been recently addressed [10] Medical corporations don't appreciate to look at these data. But in fact, medical students do not see enough practical lifestyle medicine. It's a reality during hospital internship and also outpatient training. It is probably one of the explanations for the so called "medical desertification" in France.

Additional hurdles

Another aspect of the holistic approach is that lifestyle medicine is often criticized for wanting to normalize behavior. However, accompanying a person is not wanting to normalize him or her. "*To accompany is to help the person to want to change*" (Myriam Razafindratsima, nongovernmental organization Asmae). The demand of the person must be constantly present. After primary needs are ensured (hunger, thirst, absence of pain, security), the human being gives priority to psychic balance, rather than to biological balance. Lifestyle changes are not easy to accomplish.

Fundamental guidelines

Both at the collective and individual level, promoting physical activity (*"move"*) and good nutrition, and fighting against sedentarism and addictions are to be promoted. Plant-based foods are essential assets in the prevention of diseases. Eating is a unique cultural act that testifies to the special relationship between man and nature. Health inequalities are obviously a major difficulty, and need to be taken in account. Healthcare professionals require more exposure to lifestyle problems, encompassing training in behavioral medicine [11]. The students desire such a training. The author's experience with the optional module Lifestyle Medicine, at the second medical cycle at the University of Bordeaux, argues in favor of this approach. Better consideration (including financial aspect) for relational and intellectual caregiving acts, in comparison to purely technical interventions, is crucial , not only at the individual level but also on a collective level (media).

Green prescription: a first step and nudge

The practice of behavioral medicine can now benefit from the promotion of the green prescription [12]. Several countries are considering this alternative especially with regard to physical exercise. This means integrating motivational interviewing into the caregiving relationship. Clinical consultation should not be limited to ordering tests and images, but include more dialoguing, thus favoring the self-determination and the internal ecology of the person. The healthcare professional is expected to help the person, to choose the action of which he or she feels capable, even minimal or below the optimal goal. Simple, personalized actions of everyday life are to be the initial target, such as walking close to home (in a park, around a lake, or toward a commercial area).

Personalized recommendations

Choosing an activity with a cultural dimension or in contact with nature is preferable. If not possible, at least one should regularly use the stairs, not remaining seated for long either. Getting up and walking a few minutes every hour, avoiding nibbles between the meals and repeated servings at the table, as well as registering in a sports association are essential suggestions. Nonrecreational physical activities such as domestic, travel, and professional ones are to be privileged, and their effectiveness is advocated [13,14].

The written contract of the green prescription

Sport activities or sport "on prescription" is also possible and desirable, focusing such settings as severe obesity, diabetes, cardiovascular disease, and cancer. The person should be asked to commit in writing, based on a visual scale of 0—10, the chances of achieving the recommended goal, till the next consultation. This is important for compliance, and also a source of feedback for the caregiver, about his convincing powers.

The patient should sign the contract, as a moral obligation. If chances are low (less than 50%), a more realistic goal should be negotiated. A journey of a 1000 leagues begins with a single step (Lao Tzu, 4th century BCE). Few goals should be envisaged, as an example, one of physical activity and fighting sedentary lifestyle, one about eating habits.

Finally, the caregiver textually writes on the prescription paper the objective chosen by the patient (and not imposed by the caregiver, even if it helps), along with the personal contract. The prescription (which does not order) should also include the frequency of the planned actions, and strategies to keep them in mind. If there is an associated drug prescription, it will be made on the same form, after the written contract.

The nudge technique of behavioral science

Such a presentation is not symbolically neutral. For skeptics, it is good to remember that it fits the positive reinforcement steps of the Nudge Technique in behavioral science, also known as "boost."

It encourages people to change their behavior, or to make certain choices without constraint, obligations, or fear of punishment. Incidentally this method was transposed to behavioral economics by Richard Thaler (Nobel Prize in Economics 2017). The green prescription could be summarized and illustrated by the aphorism "*the door of change opens from the inside.*" It is not externally imposed as traditional interventions, but arranged together with the patient, with a large input of his or her inner strengths and desires.

Integrated team effort

The prescription is given to the patient, but also forwarded to the general practitioner, if the contract was made with a specialist. Within the context of a multidisciplinary system, it should be accessible to other health professionals, such as the nurse and the pharmacist. They will enhance motivation by recalling the contract, whenever it's their turn to be involved.

Follow-up

The green prescription should be evaluable and verifiable. At the next visit (constancy is essential), the contract is to be rediscussed and if successful, a more ambitious goal will be planned. In difficult circumstances (handicaps, disability, severe diabetes, morbid obesity, stroke, myocardial infarction, cancer treatment), a specialized opinion may be requested, and therapeutic education envisaged.

Challenges and criticisms

The effectiveness of green prescription has been scientifically questioned in the Anglo-Saxon literature [15], even if the practical modalities were slightly different, without clearly adopting the technique of motivational interview. Nevertheless, writing a contract between two people is more effective than establishing a goal. The recommendations of the Canadian sports society, cosigned by many international scientific societies, emphasize the importance of writing [16]. The length of consultation is one of the objections. In the author's practice, it generally takes less than a quarter of an hour for the green prescription. It was quantified in the English literature at 7 min, in case of conventional medical prescription, and 13 min if the green prescription was made by a nurse [15].

Perspectives

Randomized trials about lifestyle medicine will be welcome. They would be helpful to confirm the clinical interest of green prescription, with the original management described above. Different focuses, more attuned to the precision medicine framework, should not be overlooked, such as the potential contribution of internet connectivity, including smartphones, wearable devices, and the internet of things.

The main idea is to use the motivational interview and brief intervention as a first step, even if minimal.

Economic studies would be of great interest. They could confirm preliminary data, pointing toward favorable cost/effectiveness [17] and cost/utility ratio, when compared to pharmaceutical interventions [18].

In France, financial compensation for Public Health Objectives (ROSP), within the context of the green prescription, can now be requested, given the frequency of behavioral problems. The compensation can be based on the number and quality of targets of the green prescription over a year. Ideally these ROSP initiatives should be readable, dynamic, correlated, coordinated with other health professionals, articulated with public health policies, and evaluable.

Exchange between drugs and lifestyle

Are patients ready to give up drug prescriptions at the end of the medical consultation? Population surveys plead, on the contrary, for behavioral medicine. Polls show that formal prescription by the doctor is the first reason for practicing physical activity. The green prescription would help rebalance the caregiving relationship, against excessive technicality. The caring relationship does not have to fade away, in face of current explosive therapeutic advances. On the contrary, it must be part of a gradual and complementary approach, toward caring of the sick person.

References

[1] Allen L, Williams J, Townsend N, Mikkelsen M, Roberts N, Foster C, Wickramasinghe K. Socioeconomic status and non-communicable disease behavioural risk factors in low-income and lower-middle-income countries: a systematic review. Lancet Glob. Health March 2017;5(3):e277–89. https://doi.org/10.1016/S2214-109X(17)30058-X.

[2] Livingston G, Sommerlad A, Orgeta V, et al. Dementia prevention, intervention, and care. Lancet July 19, 2017. https://doi.org/10.1016/S0140-6736(17)31363-6. pii: S0140-6736(17)31363-31366. [Epub ahead of print].

[3] Studies and surveys (ESTEBAN 2014–2016) nutrition section. Chapter physical activity and sedentary lifestyle public health France. September 2017 [in French].

[4] Wen PC, Wu X. Stressing harms of physical inactivity to promote exercise. Lancet 2012;380:192–3.

[5] Couzigou P. Practitioner blindness? Lifestyle medicine is a reality! Screening without prevention, rational? Presse Med. 2013;42:1551–4.

[6] Piolot M, Fagot JP, Riviere S, et al. Homeopathy in France in 2011–2012 according to reimbursements in the French national health insurance database (SNIIRAM). Fam. Pract. August 2015;32(4):442–8.

[7] Gingras D, Beliveau R. Colorectal cancer prevention through dietary and lifestyle modifications. Cancer Microenviron August 2011;4(2):133–9.

[8] www.niaaa.nih.gov/.../alcohol-screening-and-brief-intervention.

[9] Sackett DL, Rosenberg WM, Gray JA, et al. Evidence based medicine: what it is and what it isn't. BMJ 1996;312(7023):71–2.

[10] Phillips J, Charnley I. Third- and fourth-year medical students' changing views of family medicine. Fam. Med. 2016;48:54–60.

[11] Phillips E, Pojednic R, Polak r, et al. Including lifestyle medicine in undergraduate medical curricula. Med. Educ. Online January 2015;20(1):26150. https://doi.org/10.3402/meo.v20.26150.

[12] Couzigou P. Lifestyle medicine need to be promoted – green prescription. Presse Med. 2018 Jul – Aug;47(7–8 Pt 1):603–5. https://doi.org/10.1016/j.lpm.2018.07.007 [French].

[13] Lear SA, Hu W, Rangarajan S, et al. The effect of physical activity on mortality and cardiovascular disease in 130,000 people from 17 high-income,middle-income, and low-income countries: the PURE study. Lancet September 21, 2017. https://doi.org/10.1016/S0140-6736(17)31634-3. pii: S0140-6736(17)31634-3. [Epub ahead of print].

[14] Lear SA, Yusuf S. Physical activity to prevent cardiovascular disease: a simple, low-cost, and widely applicable approach for all populations. JAMA Cardiol 2017 Dec 1;2(12):1358–60. https://doi.org/10.1001/jamacardio.2017.4070. PMID:29117280.

[15] Elley R, Kerse N, Arroll B, et al. Effectiveness of counselling patients on physical activity in general practice: cluster randomised controlled trial. BMJ 2003;326:1–6.

[16] Thornton JS, Frémont P, Khan K, et al. Physical activity prescription: a critical opportunity to address a modifiable risk factor for the prevention and management of chronic disease:a position statement by the Canadian Academy of Sport and Exercise Medicine. Br. J. Sports Med. 2016;50:1109–14.

[17] Garrett S, Elley CR, Rose SB, et al. Are physical activity interventions in primary care and the community cost-effective? A systematic review of the evidence. Br. J. Gen. Pract. March 2011;61(584):e125–33. https://doi.org/10.3399/bjgp11X561249.

[18] He T, Lopez-Olivo MA, Hur C, et al. Systematic review: cost-effectiveness of direct-acting antivirals for treatment of hepatitis C genotypes 2–6 Aliment. Pharmacol. Ther. October 2017;46(8):711–21. https://doi.org/10.1111/apt.14271 [Epub 2017 Aug 24].

Chapter 38

Precision medicine: will technology be leveraged to improve population health?

Juan Pablo Rey-Lopez[1], Blanca Lumbreras[2], Jose J. Ponce-Lorenzo[3], Carlos Campillo-Artero[4] and Maria Pastor-Valero[2]

[1]*University of Sydney, School of Public Health, Sydney, NSW, Australia;* [2]*Department of Public Health, University Miguel Hernández, Alicante, the Valencian Community, Spain; and CIBER of Epidemiology and Public health (CIBERESP);* [3]*Department of Medical Oncology, University General Hospital of Alicante, Alicante, Spain;* [4]*Balearic Health Service, Majorca, Balearic Islands, Spain; and Center for Research in Health and Economics, Barcelona School of Management, Universitat Pompeu Fabra, Barcelona, Catalonia, Spain*

The epidemic of noncommunicable diseases

It is well established that the spectacular growth in life expectancy observed during the last century, in many countries, is largely explained by advances in public health (sanitation, vaccines, food security, tobacco control, etc.) [1,2]. However, this progress is today threatened by an emerging crisis of noncommunicable diseases (NCDs), in almost all countries. According to the World Health Organization (WHO), 71% of deaths worldwide (41 million) are caused by NCDs, representing approximately 15 million of premature global deaths (between 30 and 70 years) per year. Specifically, cardiovascular disease (44%), cancer (22%), chronic respiratory disease (9%), and diabetes (4%), amount to approximately 82% of deaths by NCDs globally [3].

Although genetic susceptibility can exist, most NCDs risk factors are determined by environmental factors, and are, therefore, largely preventable. These factors include obesity, smoking, alcohol consumption, unhealthy diets, and sedentary lifestyles. It is widely recognized that unhealthy behaviors are largely shaped by commercial determinants of health. For example, for food companies, it is more profitable to favor the production of cheap, high-energy, and low nutritional value foods, compared with healthier food alternatives. Rather than individual level approaches, population-level interventions are essential, to influence and change the health status of future generations [4].

In 2011, the WHO pioneered a United Nations (UN) declaration, in which it announced that the NCD epidemic is a worldwide issue, that affects national development. A Global Action Plan for the Prevention and Control of noncommunicable diseases was launched, whose aim is to reduce premature deaths caused by NCDs by a third before 2030. Most population-level interventions are based on successful experiences; for example, those carried out for the prevention and treatment of cardiovascular diseases. These achievements have led to an increase in life expectancy, and a reduction in mortality rates for chronic diseases in middle-aged people (36–64 years).

However, these important breakthroughs are being threatened by the obesity pandemic, the high prevalence of tobacco consumption worldwide, and the increase in sedentary lifestyles, hence the importance and urgency of implementing cost-effective interventions to reduce the prevalence of modifiable health risk factors [4]. Clearly, efforts must be multisectoral, going beyond the health sector, and require strong political will to dramatically change the environmental and social factors that lead to poor health [5].

Precision medicine, precision public health, and the Emperor's new clothes

Although a precise and universally accepted definition of "Precision Medicine" (PM) remains elusive, PM can be understood as *"the use of diagnosed tools and treatments targeted to the needs of the individual patient, on the basis of genetic, biomarker, and psychosocial characteristics."* [6] In 2011, the National Research Council (US) published a report *"Toward precision medicine: Building a knowledge network*

Precision Medicine for Investigators, Practitioners and Providers. https://doi.org/10.1016/B978-0-12-819178-1.00038-1

for biomedical research and a new taxonomy of disease." [7] This document was produced by an ad-hoc Committee of the National Research Council, comprised of experts in both basic biology and clinical medicine. According to this committee, it was urgent to create a "*framework for integrating the rapidly expanding range and detail of biological, behavioral, and experiential information to facilitate basic discovery, and to drive the development of a more accurate and precise classification of disease, which, in turn, will enable better medicine.*"

To progress toward this revolutionary model of health care, a key requirement was to integrate informational resources between two stakeholder groups which had (according to their authors) unrelated, and distinct interests and goals [7]; on the one hand, biomedical researchers, biotechnology, pharmaceutical industries; and on the other, clinicians, health agencies, and payers. In January 2015, PM gained momentum when US President Barak Obama announced the PM Initiative (renamed in 2016 as the All for Us initiative) [8] and Francis Collins (director of the National Institutes of Health) described PM as "*prevention and treatment strategies that take individual variability into account*" [9]. The goals of PM (initially therapeutic) were extended to preventive health goals, and in 2016, the head of the Office of Genomics and Public Health at the Center for Disease Control (CDC) and Prevention, employed the term "Precision Public Health" (PPH) [10]. More recently, the CDC has suggested that PPH offers significant opportunities to improve the health of the population.

Nevertheless, many of the appealing arguments employed by PM or PPH advocates have received extensive criticisms among several public health scholars, and some clinical researchers [11–15]. For example, Joyner and Paneth [11] questioned many of the medical promises posited by PM, raising doubts about the public health benefit derived from mass use of personal genomics to treat common human diseases. In a similar way, Bayer and Galea [12] downplayed the impact of PM to improve the health of populations, based on the argument that the production of health is largely explained by factors outside the healthcare system.

Recently, we argued that current evidence shows that use of technological devices (i.e., physical activity wearables), or implementation of personalized nutrition plans (including genotype data), do not necessarily lead to larger and sustained changes in healthy behaviors, compared with nonpersonalized interventions [13]. One tenet of PPH is that personalized genetic data may help to identify higher-risk individuals. However, risk stratification in clinical practice is well developed, and the scientific evidence of the added predictive value of genetic markers for cardiovascular disease outcomes is till now limited [14].

Importantly, the prospect that the predictive value of risk factors (at the individual level), will be enhanced in the future is unlikely, due to two reasons. First, most known risk factors in epidemiology have tiny effects [15]. It has been suggested that relative risks of at least 50, may serve as a useful discriminatory tool at the individual level [16]. Second, an inconvenient truth for genetic information commercial companies is that the predictive capacity of genetic variants depends on other risk factors that interact with them [17]. Despite these limitations, it is paradoxical to observe how since 2006, there has been a proliferation of direct to consumer (DTC) personal genetic information companies, in high-income countries (the USA and Europe) [18].

A large proportion of this DTC market remains focused on genetic ancestry testing [18], but also includes nutrigenetic, pharmacogenetic, paternity testing, newborn screening, and physical fitness testing [19]. The advent of DTC genetics reflects a dramatic shift in the governance of genetic testing, from a paradigm of protection to a paradigm of open access, where patients are reframed as consumers, and ultimately, are mainly responsible for their health [18]. In addition, contrary to the positive expectations created by big data to improve health, public health scholars have recently warned about the perils of digital technologies, of widening (rather than decreasing) health inequities. For example, personalized unhealthy food and beverage marketing may undermine the efforts to tackle the obesity epidemic [20]. Some academics have even suggested that rather than being driven by patients' needs, healthcare systems are becoming a profit-fuelled investor bubble [21].

Almost 3 decades ago, Geoffrey Rose's (*The Strategy of Preventive Medicine*) [22], emphasized that the high-risk approach, on one hand, and the population-based approach, on the other hand, are not incompatible goals, and that the focus should always be to discover and intervene upon the causes of incidence at the population level. Today, unfortunately, in many countries, just a small percentage of the budget spent on health is invested for prevention purposes (for example, 5% in England) [23].

Challenges in precision medicine

Precision medicine and overdiagnosis

PM argues that there is individual heterogeneity and that not all patients require the same medical treatment. This is especially important in oncology, where the study of genomics can be fundamental to find precise targets on which to act medically, but it extends to all medicine [24].

The fundamental criticism that PM receives with regard to the excessive diagnosis (overdiagnosis), which can occur when the prevailing focus of medical practice in on an individual's genes and biology and does not sufficiently incorporate the important role of environmental factors in disease etiology and health [25]. Overdiagnosis has been defined as the diagnosis of a condition that would otherwise not cause symptoms or harm to a patient during his or her lifetime. Many factors can lead to overdiagnosis, among others, the use of advanced diagnostic technology, the broadening of disease definitions, a medical culture that encourages greater use of tests and treatments and use of nonbeneficial screening tests [26]. Overdiagnosis has been quantified for many cancers with proportions which vary widely across studies, from 2.9% to 88.1% for prostate cancers, from 1% to 12.4% for breast cancer, from 49% to 83% for thyroid cancer and 11.9% for lung cancer [26].

Although some level of overdiagnosis is unavoidable, limiting screening to only people at the highest risk of disease will minimize the problem of overdiagnosis [24]. However, with the growing interest in PM, there is a risk that a large part of the population will undergo unnecessary screening tests for which the benefits are unclear and may have harmful effects [27,28]. Another important problem is the insufficient sample size when testing PM interventions using a standard clinical trial since the genotypically diverse patients with cancer who will respond to any one drug will only represent a small subset of the population [24]. Although PM is typically articulated in terms of personalized benefits to individual patients, the underlying approach to early detection also permits precision initiatives to target interventions in high-risk groups. Because of the increasing use of genomic testing, physicians will encounter situations of uncertainty in which the individual risk-benefit profile of an intervention for an asymptomatic patient is unknown [29,30].

Problems with the validation of clinical biomarkers

The development of PM has led to an increased demand for specific biomarkers for prognostic assessment, treatment guidance, resistance monitoring, and dosing. Although the concept of providing treatment only to those patients likely to benefit is desirable, the randomized controlled trials developed to date have shown negative results [31]. This has not been related to the concept of PM itself, but with the validation of these new biomarkers.

Clinical biomarkers are based on the "omics" technologies, which permit large-scale parallel measurements for the comprehensive analysis of the complete, or nearly complete, cellular specific constituents, such as RNAs, DNAs, proteins, and intermediary metabolites, including genomics, proteomics, and metabolomics [32]. Recent years have seen remarkable advances in this omics-based research, leading to high expectations for the development of noninvasive molecular biomarkers, which enable clinicians to improve the PM process [33].

However, while biomedical investigators have been quick to use these new technologies in research, the analysis and interpretation of the data present unique challenges, and few of the proposed biomarkers have yet achieved clinical application with clearly defined benefits, despite extensive commercial support [34]. The current challenge confronting omics-based research is no longer the development of equipment, but rather the interpretation and analysis of data, implementation and clinical utility, and ultimately, improved health outcomes. The lack of meta-analysis evaluating data derived from these biomarkers and confirmation of results at different settings affect the reproducibility of the studies [35].

Different organizations have developed strategies for the transformation of omics-based discoveries into biomarkers for clinic-based health care. The Institute of Medicine Roundtable on Translating Genomic-Based Research for Health for instance, brings together leaders from government, academia, industry, foundations, associations, patient communities, and other stakeholder groups to examine global aspects of the introduction of genomics and genetics research findings into medicine, public health, education, and policy [36].

Some authors have designed a risk-benefit framework to facilitate the integration of genomic biomarkers into practice [37]. This framework, which includes the quantitative assessment of risk-benefit and quality-adjusted life-years as a summary measurement of clinical utility, aims to help clinicians and policy-makers to estimate the health outcomes of genomic biomarkers. However, this useful approach has several limitations, related to the lack of available data, and has not yet been applied in practice.

The private sector has also developed evidence-based approaches, to evaluate the clinical application of these technologies, such as the BlueCross BlueShield Association's Technology Evaluation Center (TEC) [38]. However, again, the developed criteria have not yet been incorporated into a real context.

There are other initiatives which focus on the evaluation of genomic test applications in clinical practice, such as the Evaluation of Genomic Applications in Practice and Prevention (EGAPP) Initiative [39], which aims to adapt existing evidence review methods to the systematic evaluation of genomic biomarkers, and to link scientific evidence to recommendations for their clinical use. However, according to previous data [40], there are no associations between the implementation of genomic biomarkers (according to the FDA approval) and the clinical utility evaluations and recommendations made by both guideline developers and professional organizations.

All these initiatives include relevant aspects to improve biomarker identification, validation, and clinical implementation, underlining the different concerns affecting each of the four stages necessary for the implementation of a biomarker: analytical validation, clinical validation, demonstration of clinical value, and regulatory approval [41]. For instance, the barriers related to the regulatory, clinical, and ethical issues in conducting human trials, make it difficult to obtain a sufficient sample size to allow adequate biomarker selection [42].

Moreover, the large number of data generated require complex bioinformatics tools and the confirmation of primary results using a second confirmatory method, to avoid the presence of bias and overfitting [41]. The patients included in biomarker clinical implementation studies usually come from a single setting, with inherent bias, while they should reflect the real context in practice: the biomarkers should be validated in a population like that which will receive the biomarker in practice. Therefore, a clinical implementation study should be multicentred in well-defined populations, and the biomarker should be validated prospectively.

Regulation of biomarkers

As for drugs, medical devices, diagnostic tests, including biomarkers, and surgical innovations, by this citation order, a downwards regulatory ladder or gradient of both strictness of regulatory assessment standards and regulatory enforcement has long been in place [43].

One crux to understanding what has spurred very recent changes in the regulation of biomarkers (and diagnostic tests) is that the rate of clinically useful biomarkers to initially claimed with potential clinical utility is considerably low. This has led to the establishment of three main criteria or regulatory standards for their approval and licensing; analytical validity (scientific validity in the EU terminology, the ability with which the biomarker accurately and reproducibly measures what it intends to), clinical validity (analytical performance, the accuracy with which a biomarker identifies a particular condition or an analyte, estimated by means of known parameters: sensitivity, specificity, predictive values, and likelihood ratios), and clinical utility (clinical performance, whether the biomarker can provide information about diagnosis, treatment, management, or prevention of a disease that will be helpful and safe, with high benefit/risk ratio, to a consumer), the final endpoint [43–47].

A significant fraction of biomarkers are devoid of sufficient methodological rigor: they are plagued with known methodological flaws leading to biased (over)estimates of their clinical validity (a surrogate of clinical utility), and many lack validation in subsequent pivotal studies. Most importantly, from a regulatory standpoint, is that they try to prove *clinical validity* but fail to demonstrate clinical utility [43,48,49].

It is worth noting that in most of these studies it is often ignored, that average measures of association (e.g., odds ratios) between putative risk factors or biomarkers and outcomes, are unsuitable for ascertaining their performance. Several studies have pinpointed that even strong associations (OR or RR > 10), are related to low capacity of both risks factors and biomarkers, to discriminate between cases and noncases in the population. Even exposures (e.g., risk factors used as screening tests, genetic mutations), with very high disease risks, have a low population prevalence and account for a small fraction of cases. This is the reason why average measures of association ought to be interpreted in tandem with measures of discriminatory accuracy, which often goes unheard in these studies. Observed progression of precision medicine is slower than expected [50–52]. When known, the *incremental* discriminatory accuracy of newly developed biomarkers (relative to those used in clinical practice), is very low and even marginal.

Pursuant to the new regulations of medical devices and in vitro diagnostic tests (2012 FDA Safety and Innovation Act, 2016 21st Century Cures Act, Regulation (EU) 2017/746 on in vitro diagnostic medical devices), the regulatory assessment of biomarkers should encompass their analytical validity, clinical validity, and particularly their clinical utility, the final endpoint and ultimate added value (health benefit), deriving from its use in clinical practice [44–48,53–56].

Translating intermediate results into practical clinical knowledge should be an indisputable regulatory requirement. Dodging any of these three standards falls short as an adequate basis for regulatory decision making, and could contribute to regulatory errors: type I, harm resulting from a premature approval (leading to overdiagnosis and overtreatment); type II, withholding authorization when the technology could eventually contribute clinical benefit; and type III, the opportunity costs associated with them [48,57,58].

An additional shortcoming is that studies on biomarkers usually focus on the test in isolation failing to take into account important dimensions of their implementation (the organizational elements that need to be in place in order to assure the quality of the tests supporting system infrastructure) and leading to the overestimation of clinical applicability of study results [43,59–62].

Moreover, using in early drug trials biomarkers insufficiently qualified (i.e., failing to the fit-for-purpose process of linking a biomarker with biological processes and clinical outcomes), results in overestimates of efficacy, harm and inefficiency due to both false positive and negative results, what even worsens if the biomarker reaches Phase III trials and the market [43,44,63].

It has been noted that better regulation should assure, inter alia, that newly qualified biomarkers are implemented safely in trials, directed at high specific fit-for-purpose clinical contexts, their clinical utility is not expanded beyond initial qualification claims, until enough evidence is collated, and those recently qualified are not prematurely deployed into routine clinical practice [43,63]. When it comes to the appraisal of biomarkers, in addition to the abovementioned regulatory standards, the relative (incremental) diagnostic value should substitute for absolute value as well, and incremental cost-effectiveness should be set as the fourth barrier criterion for their coverage and reimbursement similarly to drugs [43,64,65].

As for companion diagnostics, it should be made clear that when a biomarker is marketed along with a drug, what produces value is the combination of both; it is a joint product. Their evidence requirements and reimbursement processes vary across countries. Fostering evidence generation for companion diagnostic begs for much greater harmonization, and setting adequate clinical validity and clinical utility standards, and unequivocal and consistent signals to developers as far as prices are concerned, based on their added diagnostic value. The pricing of the combination is an unresolved issue [43,64,65].

Several countries with established health technology assessment (HTA) programs, do not have specific national programs for regulating and providing guidance on diagnostics. Among those that have them (e.g., the UK), there is substantial variability in the methods, processes, extent, and comprehensiveness, as well as in implementation, the nature of implementation guidance and law enforcement. Dissemination and adoption of HTA agencies recommendations, still face serious difficulties in being translated into policy [43,66,67].

Until the new regulatory models are fully and duly implemented, we will be witnessing long-standing interacting effects of regulation failures such as the negative (health) externalities due to type I and II errors, shortage of information, inconsistencies in regulatory agencies' information and standards, regulatory agencies captured by industries, conflict of interests, and changes in professional self-regulation [43,68].

The success of the new EU regulatory model will also be dependent on the ability to provide sufficient information as for the whole regulatory processes, assure a central registry, and make results publicly available. Some blurred definitions of standards and requirements of these new regulations could hinder its progress. Moreover, since Notified Bodies are private organizations independent of national authorities (with some exceptions, e.g., Spain, France), and have different requirements, harmonization should be accomplished, their performance ought to be scrutinized on a regular basis, and a postmarket surveillance reinforced. Finally, patchy regulatory changes introduced in a piecemeal fashion should be avoided, as well as any avoidable delay in their implementation [43,69].

Treatment of nonsmall cell lung cancer-a new era

Lung cancer is one of the most frequent types of cancer diagnosed in high-income countries and is the leading cause of cancer deaths, with more than 1 million global deaths expected per year [70]. Despite the huge relevance of this cancer in humans, detailed knowledge of the molecular hallmarks of this cancer is very recent [71]. The treatment of lung cancer has now entered in a new era, because of discovery of epidermal growth factor receptor (*EGFR*)-activating mutations, anaplastic lymphoma kinase (*ALK*) gene rearrangements, BRAF mutations, HER2 mutations and ROS1 gene rearrangements, which lead to changes in health outcomes in some patients with lung cancer [72–74]. Moreover, compared with other cancers, lung cancer has one of the highest rates of genetic alterations [74], some of which are actionable.

Molecular biomarkers in nonsmall cell lung cancer

Current guidelines recommend evaluation of *EGFR*, *ALK*, and *ROS1* in all patients with lung cancer patients who have metastatic disease, irrespective of clinical characteristics [75]. Recently the guidelines are including also the evaluation of BRAFpV600E mutation.

Of note, multiplexed genetic sequencing panels are preferred over multiple single-gene tests, to identify other treatment options beyond *EGFR*, *ALK*, and *ROS1*.

For laboratories performing next-generation sequencing (NGS), it is recommended that *BRAF*, *KRAS*, *RET*, *ERBB2*, and *MET* be included. The National Comprehensive Cancer Network (NCCN) guidelines have one or more specific treatment recommendations for all of the genes except *KRAS* [76].

The following targeted therapies are indicated for patients with NSCLC who have metastatic disease, with sensitizing molecular alterations [77]:

Tyrosine kinase inhibitors (TKIs) erlotinib, gefitinib, afatinib, and osimertinib are all approved for the treatment of both *EGFR* exon 19 deletions and *EGFR* L858R mutations, whereas osimertinib is the only TKI approved for the treatment of commonly acquired resistance mutation p.T790M.

Crizotinib is approved for the treatment of NSCLC with an *ALK* or *ROS1* rearrangement. Alectinib, ceritinib, and brigatinib are also approved for *ALK*-rearranged NSCLC.

The combination of BRAF inhibitor dabrafenib and MEK inhibitor trametinib was recently approved for NSCLC with *BRAF* p.V600E.

The NCCN guidelines also recommend consideration of emerging targeted therapeutic options, including crizotinib for *MET* mutations or high-level *MET* amplification, cabozantinib or vandetanib for *RET* rearrangements, and ado-trastuzumab emtansine for *ERBB2* mutations.

The French cooperative thoracic intergroup (IFCT)

The results of the French Cooperative Thoracic Intergroup (IFCT) study show the success of a nationwide program [78]. The molecular screening done in the program, which involved nearly 20,000 patients with advanced NSCLC per year, enabled the detection (with an acceptable turnaround time), of at least one potentially actionable molecular alteration in almost 50% of the analyses, and affected the treatment decisions for 51% of patients. When a genetic alteration was detected, the median overall survival was 4.7 months longer than when a genetic alteration was absent.

Challenges in precision medicine in lung cancer

Successful implementation of PM requires widely accessible tumor molecular profiling in routine practice, across different settings worldwide, as well as the existence of research centers equipped with high-quality testing devices [79]. Nonetheless, targeted gene sequencing, or whole-exome sequencing by next-generation sequencing assays, is gradually being integrated into clinical practice. Their success depends on them being made broadly available to practicing clinicians, applicable to small tumor biopsies, and affordable to patients and/or the health care system with a turnaround time to obtain results that are short.

An emerging option is the use of plasma genotyping with sequencing circulating tumor DNA, which has the logistical advantage of being rapid, noninvasive, cheap, and no onerous for the patient [80]. It is expected to become a new standard in daily clinical practice in the near future but still needs standardization, especially for the use of a large panel of genes. By carrying out increasingly extensive molecular analyses, many uncommon or rare alterations are detected for which clinical significance assessment constitutes a real challenge.

Another major challenge is, despite the high response rates for all targeted therapies to remain effective for a finite period of time. Improvements in outcomes for our patients will build on what we understand about how cancers escape targeted therapies. One strategy is to improve target inhibition. As has been seen in ALK+ lung cancers and EGFR-mutant lung cancers, better inhibitors can be developed [81,82]. PM applied in any kind of metastatic solid tumor refractory to standard of care, has not shown significant benefits in terms of survival, in the SHIVA trial [83]. In fact, its current indication is restricted only to certain tumors.

References

[1] Beaglehole R, Ebrahim S, Reddy S, et al. Prevention of chronic diseases: a call to action. Lancet 2007;370:2152–7.

[2] Bloom DE, Chatterji S, Kowal P, et al. Marcro-economic implications of population ageing and selected policy responses. Lancet 2015;385:649–57.

[3] Global health estimates 2015: disease burden by cause, age, sex, by country and by region, 2000–2015. Geneva: World Health Organization; 2016. http://www.who.int/healthinfo/global_burden_disease/estimates/en/index2.html.

[4] WHO. Global status report on noncommunicable diseases 2014. Geneva: World Health Organization; 2014.

[5] WHO. Noncommunicable diseases progress monitor. Geneva: World Health Organization; 2017.

[6] Ramaswami R, Bayer R, Galea S. Precision medicine from a public health perspective. Annu. Rev. Public Health 2018;39:153–68.

[7] National Research Council. Toward precision medicine: building a knowledge network for biomedical research and a new taxonomy of disease. Washington, DC: Natl. Acad. Press; 2011.

[8] White H. The precision medicine initiative. March 12. Washington, DC: Archived Obama White House; 2016. https://obamawhitehouse.archives.gov/node/333101.

[9] Collins FS, Varmus H. A new initiative on precision medicine. N. Engl. J. Med. 2015;372:793–5.

[10] Khoury MJ, Iademarco MF, Riley WT. Precision public health for the era of precision medicine. Am. J. Prev. Med. 2016;50:398–401.

[11] Joyner MJ, Paneth N. Seven questions for personalized medicine. JAMA 2015;314:999–1000.

[12] Bayer R, Galea S. Public health in the precision-medicine era. N. Engl. J. Med. 2015;373:499–501.

[13] Rey-López JP, Sá TH, Rezende LFM. Why precision medicine is not the best route to a healthier world. Rev. Saude Publica 2018;52:12.

[14] Ioannidis JPA. Prediction of cardiovascular disease outcomes and established cardiovascular risk factors by genome-wide association markers. Circ. Cardiovasc. Genet 2009;2:7–15.

[15] Siontis GC, Ioannidis JP. Risk factors and interventions with statistically significant tiny effects. Int. J. Epidemiol. 2011;40:1292–307.

[16] Rockhill B. Theorizing about causes at the individual level while estimating effects at the population level: implications for prevention. Epidemiology 2005;16:124–9.

[17] Keyes KM, Davey Smith G, Koenen KC, Galea S. The mathematical limits of genetic prediction for complex chronic disease. J. Epidemiol. Community Health 2015;69:574–9.

[18] Soo-Jin Lee S, Consuming DNA. The good citizen in the age of precision medicine. Annu. Rev. Anthropol. 2017;46:33–48.

[19] Spencer EG, Topol EJ. Direct to consumer fitness DNA testing. Clin. Chem. 2019;65:45–7.

[20] Montgomery K, Chester J, Nixon L, et al. Big data and the transformation of food and beverage marketing: undermining efforts to reduce obesity? Crit. Public Health 2017. https://doi.org/10.1080/09581596.2017.1392483.

[21] McKee M, Stuckler D. The crisis of capitalism and the marketisation of health care: the implications for public health professionals. J. Public Health Res 2012;1:236–9.

[22] Rose G. The strategy of the preventive medicine. Oxford: Oxford University Press; 1992.

[23] Taylor-Robinson D, Kee F. Precision public health-the Emperor's new clothes. Int. J. Epidemiol. 2018:1–6.

[24] Hunt S, Jha S. Can precision medicine reduce overdiagnosis? Acad. Radiol. 2015;22:1040–1.

[25] Carlsten C, Brauer M, Brinkman F, et al. Genes, the environment and personalized medicine: we need to harness both environmental and genetic data to maximize personal and population health. EMBO Rep. 2014;15:736–9.

[26] Kale MS, Korenstein D. Overdiagnosis in primary care: framing the problem and finding solutions. BMJ 2018;362:k2820.

[27] Korenstein D, Chimonas S, Barrow B, et al. Development of a conceptual map of negative consequences for patients of overuse of medical tests and treatments. JAMA Intern. Med 2018;178:1401–7.

[28] Singh H, Dickinson JA, Thériault G, et al. Overdiagnosis: causes and consequences in primary health care. Can. Fam. Physician 2018;64:654–9.

[29] Adams SA, Petersen C. Precision medicine: opportunities, possibilities, and challenges for patients and providers. J. Am. Med. Inform. Assoc. 2016;23:787–90.

[30] Meagher KM, Berg JS. Too much of a good thing? Overdiagnosis, or overestimating risk in preventive genomic screening. Per Med 2018;15:343–6.

[31] Lee SM, Falzon M, Blackhall F, et al. Randomized prospective biomarker trial of ERCC1 for comparing platinum and nonplatinum therapy in advanced non-small-cell lung cancer: ERCC1 trial (ET). J. Clin. Oncol. 2017;35:402–11.

[32] Vlaanderen J, Moore LE, Smith MT, et al. Application of OMICS technologies in occupational and environmental health research; current status and projections. Occup. Environ. Med. 2010;67:136–43.

[33] Green ED, Guyer MS, National Human Genome Research Institute. Charting a course for genomic medicine from base pairs to bedside. Nature 2011;470:204–13.

[34] Parker LA, Chilet-Rosell E, Hernández-Aguado I, et al. Diagnostic biomarkers: are we moving from discovery to clinical application? Clin. Chem. 2018;64:1657–67.

[35] Voskuil J. How difficult is the validation of clinical biomarkers? F1000Res 2015;28:101.

[36] David SP, Johnson SG, Berger AC, et al. Making personalized health care even more personalized: insights from activities of the IOM genomics roundtable. Ann. Fam. Med. 2015;13:373–80.

[37] Veenstra DL, Roth JA, Garrison LP, et al. A formal risk-benefit framework for genomic tests: facilitating the appropriate translation of genomics into clinical practice. Genet. Med. 2010;12:686–93.

[38] Sullivan SD, Watkins J, Sweet B, et al. Health technology assessment in health-care decisions in the United States. Value Health 2009;12:S39–44.

[39] Veenstra DL, Piper M, Haddow JE, et al. Improving the efficiency and relevance of evidence-based recommendations in the era of whole-genome sequencing: an EGAPP methods update. Genet. Med. 2013;15:14–24.

[40] Chang CQ, Tingle SR, Filipski KK, et al. An overview of recommendations and translational milestones for genomic tests in cancer. Genet. Med. 2015;17:431–40.

[41] Duffy MJ, Sturgeon CM, Sölétormos G, et al. Validation of new cancer biomarkers: a position statement from the European group on tumor markers. Clin. Chem. 2015;61:809–20.

[42] Abel G. Current status and future prospects of point-of care testing around the globe. Expert Rev. Mol. Diagn 2015;15:853–5.

[43] Campillo-Artero C. Regaining HTA from oblivion: improving and integrating the regulation of drugs, devices, diagnostics and surgical innovations. In: Del Llano-Señarís J, Campillo-Artero C, editors. Health technology assessment and health policy today: a multifaceted view of their unstable crossroads. Barcelona: Springer Verlag Health Care; 2014.

[44] Institute of Medicine. Genome-based diagnostics: clarifying pathways to clinical use: workshop report. Washington, DC: The National Academies Press; 2012.

[45] Food and Drug Administration. Food and Drug Administration Safety and Innovation Act. Bethesda: FDA; 2012.

[46] Food and Drug Administration. 23rd Century Cures Act. Bethesda: FDA; 2016.

[47] European Union. Regulation (EU) 2017/746 of the European Parliament and of the council of 5 April 2017 on in vitro diagnostic medical devices and repealing directive 98/79/EC and commission decision 2010/227/EU. 2017. Brussels.

[48] Beckman RA, Clark J, Chen C. Integrating predictive biomarkers and classifiers into oncology clinical development programs. Nat. Rev. Drug Discov. 2011;10:735–48.

[49] Lumbreras B, Ibern P. La regulación de los biomarcadores y su papel en la medicina estratificada. Gest. Clín. Sanit 2010;12:122–5.

[50] Merlo J. Multilevel analysis of individual heterogeneity—a fundamental critique of the current probabilistic risk factor epidemiology. Am. J. Epidemiol. 2014;180:208–12.

[51] Pepe MS, Janes H, Longton G, et al. Limitations of the odds ratio in gauging the performance of a diagnostic, prognostic, or screening marker. Am. J. Epidemiol. 2004;159:882–90.

[52] Merlo J. Multilevel analysis of individual heterogeneity and discriminatory accuracy (MAIHDA) within an intersectional framework. Soc. Sci. Med. 2018;203:74–80.

[53] Helfand M. Web exclusive white paper series on diagnostic test evaluation. Med. Decis. Mak. 2009;29:634–5.

[54] Lord SJ, Irwig L, Bossuyt PMM. Using the principles of randomized controlled trial design to guide test evaluation. Med. Decis. Mak. 2009;29:E1–12.

[55] Lijmer JG, Leeflang M, Bossuyt PMM. Proposals for a phased evaluation of medical tests. Med. Decis. Mak. 2009;29:E13–21.

[56] Annes JP, Giovanni MA, Murray MF. Risks of presymptomatic direct-to-consumer genetic testing. N. Engl. J. Med. 2011;363:1100–1.

[57] Eichler HG, Bloechl-Daum B, Brasseur D, et al. The risks of risk aversion in drug regulation. Nat. Rev. Drug Discov. 2013;12:907–15.

[58] Bauer P, Köning F. The risks of methodology aversion in drug regulation. Nat. Rev. Drug Discov. 2014;13:317–8.

[59] Beastall GH. Adding value to laboratory medicine: a professional responsibility. Clin. Chem. Lab. Med. 2013;51:221–7.

[60] Bossuyt PMM, McCaffery K. Additional patient outcomes and pathways in evaluations of testing. Med. Decis. Mak. 2009;29:E30–8.

[61] Korf BR, Rehm HL. New approaches to molecular diagnosis. JAMA 2013;309:1511–21.

[62] Trikalinos TA, Siebert U, Lau J. Decision-analytic modelling to evaluate benefits and harms of medical tests: uses and limitations. Med. Decis. Mak. 2009;29:E22–9.

[63] Sistare FD, Dieterle F, Troth S, et al. Towards consensus practices to qualify safety biomarkers for use in early drug development. Nat. Biotechnol. 2010;28:446–54.

[64] Towse A, Ossa D, Veenstra D, Carlson J, et al. Understanding the economic value of molecular diagnostic tests: case studies and lessons learned. J. Personalized Med. 2013;3:288–305.

[65] Towse A, Garrison L. Economic incentives for evidence generation: promoting an efficient path to personalized medicine. Value Health 2013;16:539–43.

[66] Plum J, Campbell B, Lyratzopoulos G. How guidance on the use of interventional procedures is produced in different countries: an international survey. Int. J. Technol. Assess. Health Care 2009;25:124–33.

[67] Banta D. Dissemination of health technology assessment. In: del Llano-Senarís J. Campillo-Artero C, editors. (Ortún V, dir). Health technology assessment and health policy today: a multifaceted view of their unstable crossroads. Barcelona: Springer Healthcare; 2014.

[68] Campillo-Artero C, Ibern P. Framing an integrated and adaptive regulatory overhaul of medical technologies: a regulatory science and health economics perspective. Barcelona: Center for Research in Health and Economics; 2015. CRES Working Paper #201512-87.

[69] Fraser AG, Butchart EG, Szymański P, et al. The need for transparency of clinical evidence for medical devices in Europe. Lancet 2018;392:521–30.

[70] Siegel R, Naishadham D, Jemal A. Cancer statistics, 2013. CA Cancer J. Clin 2013;63:11–30.

[71] Hanahan D, Weinberg RA. Hallmarks of cancer: the next generation. Cell 2011;144:646–74.

[72] Lynch TJ, Bell DW, Sordella R, et al. Activating mutations in the epidermal growth factor receptor underlying responsiveness of non-small-cell lung cancer to gefitinib. N. Engl. J. Med. 2004;350:2129–39.

[73] Shaw AT, Kim DW, Nakagawa K, et al. Crizotinib versus chemotherapy in advanced ALK-positive lung cancer. N. Engl. J. Med. 2013;368:2385–94.

[74] Alexandrov LB, Nik-Zainal S, Wedge DC, et al. Signatures of mutational processes in human cancer. Nature 2013;500:415–21.

[75] Lindeman NI, Cagle PT, Aisner DL, et al. Updated molecular testing guideline for the selection of lung cancer patients for treatment with targeted tyrosine kinase inhibitors: guideline from the College of American Pathologists, the International Association for the Study of Lung Cancer, and the Association for Molecular Pathology. J. Mol. Diagn. 2018;20:129–59.

[76] Etinger DSWD, Airsner DL. NCCN clinical practice guidelines in oncology non-small cell lung cancer version 2. 2018. https://www.nccn.org/professionals/physician_gls/pdf/nscl.pdf.

[77] Brown NA, Aisner DL, Oxnard GR. Precision medicine in non-small cell lung cancer: current standards in pathology and biomarker interpretation. Am. Soc. Clin. Oncol. Educ. Book 2018;23:708–15.

[78] Barlesi F, Mazieres J, Merlio JP, et al. Routine molecular profiling of patients with advanced non-small-cell lung cancer: results of a 1-year nationwide programme of the French Cooperative Thoracic Intergroup (IFCT). Lancet 2016;387:1415–26.

[79] Yu HA, Planchard D, Lovly CM. Sequencing therapy for genetically defined subgroups of non-small cell lung cancer. Am. Soc. Clin. Oncol. Educ. Book 2018;38:726–39.

[80] Oxnard GR, Thress KS, Alden RS, et al. Association between plasma genotyping and outcomes of treatment with osimertinib (AZD9291) in advanced non-small-cell lung cancer. J. Clin. Oncol. 2016;34:3375–82.

[81] Soria JC, Ohe Y, Vansteenkiste J, et al., FLAURA Investigators. Osimertinib in untreated EGFR-mutated advanced non-small-cell lung cancer. N. Engl. J. Med. 2018;378:113–25.

[82] Hida T, Nokihara H, Kondo M, et al. Alectinib versus crizotinib in patients with ALK-positive non-small-cell lung cancer (J-ALEX): an open-label, randomised phase 3 trial. Lancet 2017;390:29–39.

[83] Le Tourneau C, Delord JP, Gonçalves A, et al. Molecularly targeted therapy based on tumour molecular profiling versus conventional therapy for advanced cancer (SHIVA): a multicentre, open-label, proof-of-concept, randomised, controlled phase 2 trial. Lancet Oncol. 2015;16:1324–34.

Further reading

[1] Garau M, Towse A, Garrison L, Housman L, Ossa D. Can and should value based pricing be applied to molecular diagnostics? London: Office of Health Economics; 2012.

[2] Fang C, Otero HJ, Greenberg D, Neumann PJ. Cost-utility analyses of diagnostic laboratory tests: a systematic review. Value Health 2011;14:1010–8.

Chapter 39

Network analysis of neuropsychiatry disorders

Grover Enrique Castro Guzman[1], Joana Bisol Balardin[2], Claudinei Eduardo Biazoli, Jr.[3], João Ricardo Sato[3] and Andre Fujita[1]

[1]*University of São Paulo, Department of Computer Science, São Paulo, São Paulo, Brazil;* [2]*Instituto do Cérebro, Hospital Israelita Albert Einstein, São Paulo, SP, Brazil;* [3]*Universidade Federal do ABC, Center for Mathematics, Computing and Cognition, Santo André, SP, Brazil*

Introduction

Precision medicine applied to psychiatry, or precision psychiatry, has been emerging as a quite promising approach to psychiatric care [1,2]. Given the intensive research on the structural and neural correlates of neuropsychiatric disorders over the last 20 years, neuroimaging bears excellent promise for the development of precision psychiatry. Indeed, the current massive availability of neuroimaging data, alongside with behavior, genetic, and environmental information, provides unprecedented research opportunities to improve diagnosis accuracy and prediction of critical clinical outcomes, especially the response to individualized treatments. However, among the many challenges in moving neuroimaging from a primary scientific tool to an integral part of the personalized medicine approach, the needs for reliable and valid proxy measures of brain anatomy, function, and connectivity, as well as for adequate analytical techniques are not minor ones (for a comprehensive review on this topic, see Ref. [3]). Therefore, interdisciplinary efforts involving neuroscience, bioinformatics, and computational psychiatry are required, for an adequate clinical translation of neuroimaging results in mental health.

The contribution of neuroimaging to personalized neuropsychiatric diagnosis and treatment has been primarily driven by (i) the current dominant view of neuropsychiatric disorders as neural systems disorders, formalized in the US National Institute of Mental Health Research Domain Criteria (RDoC) framework [4] and (ii) the rapid methodological advances in measuring, mapping, and modeling neuroimaging-derived neural signals at the systems level [5]. In this context, the human brain connectomic concept has been gaining prominence. Such idea takes into account the whole set of anatomically and/or functionally interconnected brain areas, forming networks and subnetworks, and it is opposed to previous prevailing views that emphasized the existence of distinct and somewhat isolated functional units in the brain [6]. Though most functional magnetic resonance imaging (fMRI) protocols in psychiatric research published to date are aimed at mapping brain functional or structural differences between healthy controls and patients diagnosed with mental disorders, the first usages of the connectomics framework to precision psychiatry are being proposed and tested. An interesting example of an attainable goal comes from the recent study of Drysdale et al. [7], in which measures of brain connectivity derived from resting-state fMRI were used to identify subtypes of depression, in a sample of 711 individuals and to predict treatment responsiveness [7].

In this chapter, we aim at providing some concepts, methods, and discussions on the potential of characterizing individual brain functional networks and its application to precision psychiatry. Importantly, no tool or approach derived from the precision psychiatry emerging paradigm has reached a sufficient level of evidence and significance to be incorporated in current clinical practice. However, with the remarkable pace of innovation in both technological tools and cutting-edge neurobiological knowledge, we envision a promising future application of functional connectomics to the individualized diagnosis and treatment of neurological and psychiatric disorders. First, we briefly describe some methods for functional data acquisition, and brain functional networks modeling and analyses. Then, we present the current state-of-the-art literature on personalized medicine based on individual profiles of brain networks. Finally, we discuss the main challenges, perspectives, technologies, and research opportunities in this field.

Precision Medicine for Investigators, Practitioners and Providers. https://doi.org/10.1016/B978-0-12-819178-1.00039-3

Methods for functional data acquisition

Electroencephalography

Electroencephalography (EEG) is one of the oldest and most well-established ways to infer brain electrical activity (see Ref. [8] for a review), having many applications in the neurological practice, and especially in epileptology. EEG captures the difference of electric potential between two pairs of electrodes (or a standard reference), positioned in distinct regions of the scalp. Although presenting a reduced spatial resolution, this technique has a high (up to the order of hundreds of Hz) temporal resolution. Thus, EEG can be used to infer brain rhythms (spectral analysis, usually from 1 to 100 Hz) and electric potentials evoked by specific events.

Functional magnetic resonance imaging

fMRI is currently one of the most potent primary imaging tools in neuroscience research. The physical basis of the effect captured by fMRI relies on the attenuation of MRI signal by deoxyhemoglobin, given its paramagnetic properties, and thus acting as an endogenous contrast. Most fMRI studies are based on blood oxygenation level dependent (BOLD, [9]) signal, which results from a complex interplay of local hemodynamic and metabolic changes. A phenomenon called hemodynamic coupling is the neurophysiological process linking local neural activity, with variations in hemodynamic states. Hence, the BOLD signal is an indirect measure of local neural activity, i.e., a proxy variable. Given the nature of the BOLD signal and the constraints of MRI acquisition, the temporal resolution of fMRI is quite low concerning relevant neural phenomena (usually of the order of seconds). On the other hand, MRI signal is captured with a relatively high spatial resolution and allows measuring signals not only from superficial cortical regions but also of subcortical structures. The seminal study of Logothetis et al. [10] in primates established the association between brain electrical activity and the BOLD signal. One of the challenges of fMRI data acquisition is the restriction to the magnetic resonance environment, which compromises the collection of data in out of the laboratory experiments and real-life situations.

Functional near-infrared spectroscopy

Functional near-infrared spectroscopy (fNIRS) is a method also based on the hemodynamic coupling process (see Ref. [11] for a review). However, data acquisition is not based on the MRI signal, but on scalp montages combining light emitters and detectors (optodes). In fNIRS, the differences in light absorption curves between oxy and deoxyhemoglobin are used to infer the variations on their concentrations, in superficial small cortical vessels. In contrast to fMRI, fNIRS presents a relatively low spatial resolution and only allows inferences of the cortical surface. On the other hand, fNIRS is relatively robust against motion artifacts and can be used in more naturalistic settings [12,13].

Construction of functional brain networks

Brain functional connectivity is defined as the activity correlation between spatially remote regions. Functional connectivity may be extracted, using many noninvasive brain signal–acquisition approaches in humans. Here, we briefly present the most commonly used methods.

There are several methods to characterize functional connectivity among brain regions-of-interest (ROIs) and then construct functional brain networks (FBNs). To build an FBN, suppose that the vertices of the network are the ROIs. For fMRI data, ROIs represent sets of voxels defined by an atlas (e.g., the CC200 and CC400 atlas). For EEG and fNIRS data, ROIs represent the channels. To identify the edges between the ROIs, one may use one of the methods described in the next sections. Fig. 39.1 illustrates the construction process of FBNs.

Correlation

Pearson's product–moment correlation coefficient

Pearson's correlation [14,15] is a measure of linear dependence between two random variables. It can also be

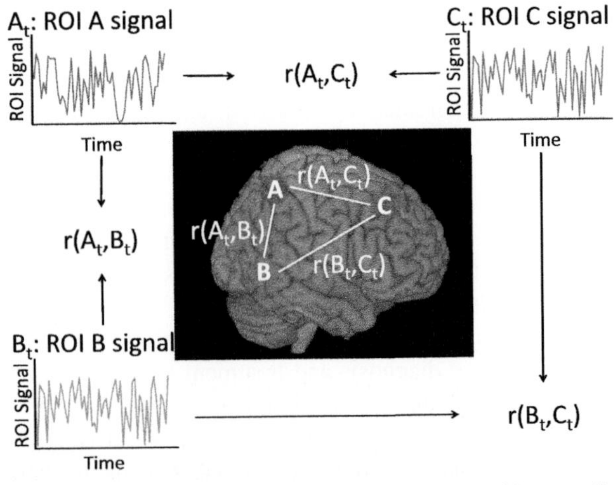

FIGURE 39.1 Construction of functional brain networks. First, coupling measures (e.g., Pearson or Spearman correlation, Granger causality) among the brain signals are calculated in a pairwise manner. In the example, the $r(A_t, C_t)$ represents the coupling measure between regions A and C. These measures determine the functional connectivity to be used to model the graph at the subject level (e.g., to define distances between vertices).

useful to identify functional connectivity between the activities of ROIs X and Y collected over time.

Let T be the number of observations (time series length) of X and Y and $\bar{x} = \frac{\sum_{i=1}^{T} x_i}{T}$ and $\bar{y} = \frac{\sum_{i=1}^{T} y_i}{T}$ be the means of X and Y, *respectively. Then, the Pearson's correlation coefficient* ρ_{Pearson} *between two ROIs* X *and* Y *is defined as*

$$\rho_{\text{Pearson}}(X, Y) = \frac{\sum_{i=1}^{T}(x_i - \bar{x})(y_i - \bar{y})}{\sqrt{\sum_{i=1}^{T}(x_i - \bar{x})^2 \sum_{i=1}^{T}(y_i - \bar{y})^2}} \quad (39.1)$$

The Pearson correlation is greater than zero in the case of a positive (increasing) linear relationship and lower than zero in the case of a negative (decreasing) linear relationship. In the case of linearly independent ROIs, $\rho_{\text{Pearson}}(X, Y) = 0$.

The statistical test for $H_0 : \rho_{\text{Pearson}} = 0$ versus $H_1 : \rho_{\text{Pearson}} \neq 0$ cannot be carried out analytically as it is usually done, because neuronal (time series) data are not independent. In this case, we use the block bootstrap procedure described in Algorithm 1.

The presence of Pearson's correlation between a pair or ROIs means that there is a linear relationship between the times series. Thus, to construct an FBN by using the Pearson's correlation, one may use Algorithm 1 between all pairs of ROIs and add an edge between the ROIs that present a P-value lower than a certain threshold (e.g., 0.05). The disadvantage is that it does not provide the direction of information flow between ROIs, i.e., the edge is undirected. For a better review, refer to Ref. [16].

The R (https://www.r-project.org) function to estimate the Pearson's correlation coefficient is cor with parameter method= "pearson" (package stats).

Spearman's rank correlation coefficient

Unlike the Pearson's correlation coefficient, Spearman's rank correlation [17] does not require assumptions of linearity in the relationship between ROIs.

The Spearman's correlation coefficient ρ_{Spearman} can be easily calculated by applying the Pearson's correlation in the data converted into ranks. Thus, Spearman's rank correlation can capture monotonic nonlinear relationships, i.e., if values of ROI Y tend to increase (or decrease) when values of ROI X increase.

The interpretation of the Spearman's correlation coefficient is similar to Pearson's correlation coefficient: ρ_{Spearman} assumes values between -1 and $+1$, where Spearman's correlation is greater than zero in the case of a monotonically increasing relationship (for all x_1 and x_2 such that $x_1 < x_2$, we have $y_1 < y_2$) and lower than zero in the case of a monotonically decreasing relationship (for all x_1 and x_2 such that $x_1 < x_2$, we have $y_1 > y_2$). In the case of monotonically independent ROIs, $\rho_{\text{Spearman}}(X, Y) = 0$.

The statistical test for $H_0 : \rho_{\text{Spearman}} = 0$ versus $H_1 : \rho_{\text{Spearman}} \neq 0$ can be carried out by using Algorithm 1 and replacing ρ_{pearson} and ρ^*_{Pearson} by ρ_{Spearman} and ρ^*_{Spearman}, respectively.

Both advantage and disadvantage of the Spearman's correlation are similar to Pearson's correlation, but the interpretation of coefficients differs. The presence of Spearman's correlation between a pair or ROIs means that there is a monotonic, which includes the linear (as in Pearson's correlation) relationship between the times series. To construct an FBN based on the Spearman's correlation, one may use Algorithm 1 between all pairs of ROIs and add an edge between ROIs that present a P-value lower than a certain threshold (e.g., 0.05).

For a comprehensive review, please refer to Ref. [16].

ALGORITHM 1 Block bootstrap to test the correlation between ROIs *X* and *Y*.

Input: the two time series X and Y.
Output: the P-value for H_0: ρ_{Spearman} versus H_1: $\rho_{\text{Pearman}} \neq 0$.

1. Calculate ρ_{Pearson} between ROIs X and Y using Eq. (39.1).
2. Let T be the time series length of X and Y and $l = T^{\frac{1}{3}}$, for instance, be the block's size. Divide the data into $T - l + 1$ blocks (the blocks may be overlapping). Resample the blocks with replacement and generate bootstrap time series X^* and Y^* independently.
3. Calculate the Pearson's correlation coefficient $(\rho^*_{\text{Pearson}})$ between X^* and Y^* using Eq. (39.1).
4. Repeat Steps 2 and 3 until the desired number of bootstrap replicates is obtained.
5. The P-value is the fraction of replicates of $|\rho^*_{Pearson}|$ that are at least as large as the observed statistic on the original dataset $(|\rho_{Pearson}|)$.

The R function to estimate the Spearman's correlation coefficient is cor with parameter method = "spearman' (package stats).

Granger causality

The main drawback of methods based on correlation is the fact that we can only identify which ROI is associated with which other ROI, but we cannot identify which one "causes" the other. When temporal data are available, it is possible to infer the direction of the association based on the intuitive idea that the cause never comes after its effect. The concept of Granger causality is based on this idea. If past values of a time series x_t are useful for predicting future values of a time series y_t, then we may suggest that the former may "cause" the latter.

To identify Granger causality in the time domain, we usually use the vector autoregressive (VAR) model (Ongoing lines of investigation and research opportunities section).

Vector autoregressive model

Granger causality among ROIs may be identified by using a VAR model. Let k be the number of ROIs (time series); p be the order of the model (number of time points in the past to be analyzed); T be the time series length; $y_{i,t}$ be the ith time series, and $\varepsilon_{i,t}$ be the vector of random variables for the ith time series, with

zero mean and covariance matrix

$$\sum = \begin{pmatrix} \sigma_{1,1}^2 & \sigma_{2,1} & \cdots & \sigma_{k,1} \\ \sigma_{1,2} & \sigma_{2,1}^2 & \cdots & \sigma_{k,2} \\ \vdots & \vdots & \ddots & \vdots \\ \sigma_{1,k} & \sigma_{2,k} & \cdots & \sigma_{k,k}^2 \end{pmatrix}$$

The disturbances ε_t are serially uncorrelated but may be contemporaneously correlated, i.e., $\sum$ may not necessarily be an identity matrix.

Then, the equations system of a k-dimensional VAR model of order p is as follows:

$$\begin{cases} y_{1,t} = v_1 + a_{1,1}^1 y_{1,t-1} + \cdots a_{1,1}^p y_{1,t-p} + \cdots a_{k,1}^1 y_{k,t-1} + \cdots \\ \quad + a_{k,1}^p y_{k,t-p} + \varepsilon_{1,t} \\ y_{2,t} = v_2 + a_{1,2}^1 y_{1,t-1} + \cdots a_{1,2}^p y_{1,t-p} + \cdots a_{k,2}^1 y_{k,t-1} + \cdots \\ \quad + a_{k,2}^p y_{k,t-p} + \varepsilon_{2,t} \\ \vdots \\ y_{k,t} = v_k + a_{1,k}^1 y_{1,t-1} + \cdots a_{1,k}^p y_{1,t-p} + \cdots a_{k,k}^1 y_{k,t-1} + \cdots \\ \quad + a_{k,k}^p y_{k,t-p} + \varepsilon_{k,t} \end{cases}$$

This equations system can be rewritten in a matricial form, such as follows. Let

$$\mathbf{Y} = \begin{pmatrix} y_{1,p+1} & y_{2,p+1} & \cdots & y_{k,p+1} \\ y_{1,p+2} & y_{2,p+2} & \cdots & y_{k,p+2} \\ \vdots & \vdots & \ddots & \vdots \\ y_{1,T} & y_{2,T} & \cdots & y_{k,T} \end{pmatrix},$$

$$\mathbf{Z} = \begin{pmatrix} y_{1,p} & y_{1,p-1} & \cdots & y_{1,1} & \cdots & y_{k,p} & y_{k,p-1} & \cdots & y_{k,1} \\ y_{1,p+1} & y_{1,p} & \cdots & y_{1,2} & \cdots & y_{k,p+1} & y_{k,p} & \cdots & y_{k,2} \\ \vdots & \vdots & \ddots & \vdots & \ddots & \vdots & \vdots & \ddots & \vdots \\ y_{1,T-1} & y_{1,T-2} & \cdots & y_{1,T-p} & \cdots & y_{k,T-1} & y_{k,T-2} & \cdots & y_{k,T-p} \end{pmatrix}$$

and

$$\beta = \begin{pmatrix} a_{1,1}^1 & a_{1,2}^1 & \cdots & a_{1,k}^1 \\ \vdots & \vdots & \ddots & \vdots \\ a_{1,1}^p & a_{1,2}^p & \cdots & a_{1,k}^p \\ \vdots & \vdots & \ddots & \vdots \\ a_{k,1}^1 & a_{k,2}^1 & \cdots & a_{k,k}^1 \\ \vdots & \vdots & \ddots & \vdots \\ a_{k,1}^p & a_{k,2}^p & \cdots & a_{k,k}^p \end{pmatrix}$$

Then, the VAR model can be described as $\mathbf{Y} = \mathbf{Z}\beta + \mathbf{u}$. The coefficients of this model ($a_{i,j}^l$, with $i,j = 1,\ldots,k$ and $l = 1,\ldots,p$) can be estimated by using the ordinary least squares as $\widehat{\beta} = (\mathbf{Z'Z})^{-1}\mathbf{Z'Y}$. The $((T-p)\times k)$ matrix of residues $\mathbf{u}$ can be estimated as $\widehat{\mathbf{u}} = \mathbf{Y} - \mathbf{Z}\widehat{\beta}$, and the $(k \times k)$ covariance matrix $\sum$ as $\widehat{\sum} = \frac{\widehat{\mathbf{u}}'\widehat{\mathbf{u}}}{(T-p)-(kp)}$.

A necessary and sufficient condition for an ROI $y_{i,t}$ being not Granger causal for ROI $y_{j,t}$ is if and only if $a_{i,j}^l = 0$. Thus, Granger noncausality may be identified by testing the significance of the elements of the autoregressive coefficients (β), i.e., $H_0 : \mathbf{C}\beta = 0$ versus $H_1 : \mathbf{C}\beta \neq 0$, where $\mathbf{C}$ is a matrix of contrasts of the parameters we wish to test. This test can be achieved by applying the Wald's test [18].

For example, suppose we are interested in testing if $y_{i,t}$ Granger causes $y_{j,t}$. Let $\mathbf{c}$ be a $(1 \times k)$ matrix with one in the ith position and $\mathbf{0}$ be a $(1 \times k)$ matrix of zeros. Then, the $(p \times (kp))$ matrix of contrasts $\mathbf{C}$ is defined as follows:

$$\mathbf{C} = \begin{pmatrix} \mathbf{c} & \mathbf{0} & \cdots & \mathbf{0} \\ \mathbf{0} & \mathbf{c} & \cdots & \mathbf{0} \\ \vdots & \vdots & \ddots & \vdots \\ \mathbf{0} & \mathbf{0} & \cdots & \mathbf{c} \end{pmatrix}.$$

Then, the Wald's test statistic is given by $W = \frac{(\mathbf{C}\hat{\beta}_j)'(\mathbf{C}(\mathbf{Z}'\mathbf{Z})^{-1}\mathbf{C}')^{-1}(\mathbf{C}\hat{\beta}_j)}{\hat{\sum}_{j,j}}$. The Wald's test statistic W follows χ^2 distribution with rank$(\mathbf{C})$ degrees of freedom.

The main advantage of the VAR model is the capacity to identify Granger causality, i.e., directionality at the edges of the FBNs.

The R function to identify Granger causality by estimating the coefficients of the VAR model is VAR (package vars).

Methods to analyze functional brain networks

Once an FBN is constructed (for example, by using one of the methods described in Construction of functional brain networks section), the next steps usually consist in analyzing, comparing, and identifying structural differences between the individual presenting a disease versus healthy controls. To this end, one possible approach is to model FBNs as graphs. A graph $G = (V, E)$ is composed of a set of n vertices $V = \{v_1, v_2, \ldots, v_n\}$ and a set of edges E that connect pairs of vertices of V. The vertices represent the brain ROI while the edges represent the presence of correlation or Granger causality between ROIs. Graphs can be subclassified in undirected (without orientation at the edges, i.e., when edges represent correlation, for example) and directed (with orientation at the edges, i.e., when edges represent Granger causality, for instance) [19,20]. Moreover, the edges may present weights, which represent the strength of functional interaction between two ROIs [21].

There are several measures and approaches for FBNs structural analysis. In the next sections, we describe some of them.

Graph measures of functional integration

Functional integration is the ability to combine and process specialized information spread over several brain regions. The most commonly used measure of functional integration is the average shortest path length between all pairs of ROIs (vertices of the graph) in the FBN. Another measure related to the average shortest path length is the global efficiency, which is simply the average inverse shortest path length [22]. The main advantage of calculating global efficiency instead of the average shortest path length is the fact that it can be computed on disconnected networks. Paths between disconnected vertices are defined to have infinite length, consequently zero efficiencies. Notice that the larger is the average shortest path length, the greater is the distance between ROIs; thus, the less integrated and efficient is the communication among different ROIs.

Paths can be easily generalized for directed and weighted networks. While an unweighted path length is equal to the number of edges in the path, a weighted path length is equal to the total sum of individual edge lengths. Edge lengths are inversely related to edge weights, i.e., the larger weights represent the stronger associations and consequently "closer" distances.

For a complete review, refer to Ref. [20].

The R function to calculate the length of all the shortest paths among ROIs is shortest.paths (package igraph).

Graph measures of functional segregation and clustering analyses

Another measure closely related and complementary to functional integration is the functional segregation. *Functional segregation* is the ability to process specialized information within densely interconnected ROIs. Intuitively, functional segregation is related to the structures of clusters or modules in a graph. Thus, measures of segregation quantify the presence of groups within the FBN, and the presence of groups suggests a segregated neural processing architecture.

One of the most basic measures of segregation is based on the number of triangles in the FBN (the number of triangles around the ith ROI is given by $t_i = \frac{1}{2}\sum_{j,h\in N} a_{ij}a_{ih}a_{jh}$). The number of triangles is directly proportional to segregation. Locally, the fraction of triangles around an individual ROI is known as the clustering coefficient (the clustering coefficient of vertex v_i is given by $C_i = \frac{1}{n}\sum_{i\in N}\frac{2_{t_i}}{k_i(k_i)-1}$) and is equivalent to the fraction of the ROI's neighbors that are also neighbors of each other [23]. Thus, the average clustering coefficient of the entire FBN reflects the prevalence of clustered connectivity around individual ROIs. Another measure similar to the clustering coefficient is the transitivity $\left(T = \frac{\sum_{i\in N} 2t_i}{\sum_{i\in N} k_i(k_i-1)}\right)$ [24]. Both the clustering coefficient and the transitivity have been generalized for weighted [25] and directed [26] FBNs.

Other measures of segregation are based on clusters (also known as network's modular structure or community structure). A cluster can be identified by maximizing the number of within-group edges and minimizing the number of between-group edges [27]. The degree to which the network may be subdivided into such delineated and nonoverlapping groups is quantified by the *modularity*. Suppose that a network can be fully subdivided into a set of nonoverlapping modules M. Let e_{uv} be the proportion of all edges that connect ROIs in module u with vertices in module v. Then, the modularity can be defined as $Q = \sum_{u\in M}\left[e_{uu} - \left(\sum_{v\in M} e_{uv}\right)^2\right]$ [28]. Another related concept is the *participation coefficient*. Let $k_i(m)$ be the number of edges between ROI i and all ROIs in module $m \in M$, then the participation coefficient of ROI i is given by Ref. $y_i = 1 - \sum_{m\in M}\left(\frac{k_i(m)}{k_i}\right)^2$ [29]. For a complete review, refer to Ref. [20].

To test whether two or more sets of FBNs are equally clustered, one may use the nonparametric statistical test called analysis of cluster structure variability (ANOCVA) [30]. ANOCVA compares the clustering structures of multiple groups simultaneously and also identifies ROIs that contribute to the differential clustering. It is based on the silhouette statistic [31] and analysis of variance. Essentially, the silhouette statistic is used to measure the "variability" of the clustering structure in each population. Next, the method compares the silhouette among populations. The intuitive idea behind this approach is that we assume that populations with the same clustering structures also have the same "variability." The ANOCVA algorithm and the bootstrap-based test are described in Algorithms 2 and 3, respectively [32].

The R functions to calculate the transitivity and modularity are the transitivity and modularity, respectively (package igraph). The implementation of ANOCVA is available in the R package anocva.

ALGORITHM 2 ANOCVA

Input: the k types of populations $T_1, \ldots, T_k$, a dissimilarity metric, and a clustering algorithm.

Output: the statistics $s_q^{(\bar{\bar{\mathbf{A}}}_j, \mathbf{l}_{\bar{\bar{\mathbf{A}}}})}$ and $s_q^{(\bar{\bar{\mathbf{A}}}_j, \mathbf{l}_{\bar{\bar{\mathbf{A}}}})}$.

1. Let N be the number of ROIs and n_j be the number of subjects collected for the jth population ($j = 1, \ldots, k$). The ROIs of the ith subject ($i = 1, \ldots, n_j$) taken from the jth population are represented by the matrix $\mathbf{X}_{i,j} = (\mathbf{x}_{i,j,1}, \ldots, \mathbf{x}_{i,j,N})$, where each item $\mathbf{x}_{i,j,q}(q = 1, \ldots, k)$ is a vector containing the features. Define the (N x N) matrix of dissimilarities among ROIs of each matrix $\mathbf{X}_{i,j}$ by $\mathbf{A}_{i,j} = \{d(\mathbf{x}_{i,j,q}, \mathbf{x}_{i,j,q'})\}$.
2. Let $n = \sum_{j=1}^{k} n_j$, then define the following average matrices of dissimilarities: $\bar{\mathbf{A}}_j = \frac{1}{n_j}\sum_{j=1}^{n_j} \mathbf{A}_{i,j} = \frac{1}{n_j}\sum_{i=1}^{n_j}\{d(\mathbf{x}_{i,j,q'}, \mathbf{x}_{i,j,q'})\}$ and $\bar{\bar{\mathbf{A}}} = \frac{1}{n}\sum_{j=1}^{k} n_j \bar{\mathbf{A}}_j$.
3. To determine the clustering labels $\mathbf{l}_{\bar{\bar{\mathbf{A}}}}$, apply the clustering algorithm to the matrix of dissimilarities $\bar{\bar{\mathbf{A}}}$.
4. Estimate the silhouette statistic [31] of the qth ROI based on the dissimilarity matrix $\bar{\bar{\mathbf{A}}}$ and the vector of labels $\mathbf{l}_{\bar{\bar{\mathbf{A}}}}$, i.e., $s_q^{(\bar{\bar{\mathbf{A}}}_j, \mathbf{l}_{\bar{\bar{\mathbf{A}}}})}$ for $q = 1, \ldots, N$.
5. Estimate the silhouette statistic [31] of the qth ROI based on the dissimilarity matrix $\bar{\mathbf{A}}_j$ and the vector of labels $\mathbf{l}_{\bar{\bar{\mathbf{A}}}}$, i.e., $s_q^{(\bar{\mathbf{A}}_j, \mathbf{l}_{\bar{\bar{\mathbf{A}}}})}$ for $q = 1, \ldots, N$.

Motifs

Motifs are distinct structural patterns in the network, for example, triangles. Triangles may represent feed-forward loops, feedback loops, and bidirectional loops. One way to discriminate between individuals with a disorder is by the differences in the frequencies of motifs.

The significance of a motif h in the FBN is determined by its frequency of occurrence (J_h) in all subsets of the network (subnetworks), usually normalized as the motif z-score by comparison with ensembles of random null-hypothesis networks. Let $< J_{rand,h} >$ and $\sigma^J_{rand,h}$ be the mean and standard deviation for the number of occurrences of h in an ensemble of random networks, respectively. Then, the z-score of motif h is defined as $z_h = \frac{J_h - <J_{rand,h}>}{\sigma^J_{rand,h}}$ [33].

The frequency of occurrence of different motifs around an individual vertex is known as the motif fingerprint of that vertex and it reflects the functional role of the corresponding ROI (let h' be any n_h vertex motif, $F_{nh,i}$ be the n_h vertex motif fingerprint for vertex v_i, and $J'_{h,i}$ be the number of occurrences of motif h' around vertex i. Then, the n_h vertex motif fingerprint of the FBN is given by $F_{nh}(h') = \sum_{i\in N} F_{n_h,i}(h') = \sum_{i\in N} J_{h',i})$ [34]. An FBN can be characterized by the frequency of occurrence of different motifs, also known as the *motif profile* of the FBN. For a complete review regarding motifs, refer to Ref. [20].

The R function to search for motifs is motifs (package igraph).

ALGORITHM 3 Bootstrap

Input: the dataset $\{T_1, T_2, \ldots, T_k\}$, the dissimilarity metric, the clustering algorithm, and the clustering labels for $\bar{\bar{\mathbf{A}}}$, i.e., $\mathbf{l}_{\bar{\bar{\mathbf{A}}}}$.

Output: the P-values for $\Delta\widehat{S}$ and $\Delta\widehat{S}_q$.

1. To construct bootstrap samples T_j^*, for $j = 1, \ldots, k$, resample with replacement n_j subjects from the entire dataset $\{T_1, T_2, \ldots, T_k\}$.
2. Use Algorithm 2 with T_j^* as input to compute $\bar{\mathbf{A}}_j^*, \bar{\bar{\mathbf{A}}}^*, \mathbf{l}^*_{\bar{\bar{\mathbf{A}}}}, s_q^{(\bar{\bar{\mathbf{A}}}^*, \mathbf{l}^*_{\bar{\bar{\mathbf{A}}}})*}$, and $s_q^{(\bar{\bar{\mathbf{A}}}^*, \mathbf{l}^*_{\bar{\bar{\mathbf{A}}}})*}$ for $j = 1, \ldots, k$ and $q = 1, \ldots, N$.
3. Calculate $\Delta\mathbf{S} = \sum_{j=1}^{k} \delta\mathbf{S}_j^T \delta\mathbf{S}_j$ and $\Delta s_q = s_q^{(\bar{\bar{\mathbf{A}}}, \mathbf{l}_{\bar{\bar{\mathbf{A}}}})} - \frac{1}{k}\sum_{j=1}^{k}\left(s_q^{(\bar{\bar{\mathbf{A}}}_j, \mathbf{l}_{\bar{\bar{\mathbf{A}}}})}\right)$.
4. Repeat Steps 1 to 3 until the desired number of bootstrap replicates is obtained.
5. The P-values for the bootstrap tests based on the observed statistics $\Delta\widehat{S}$ and $\Delta\widehat{S}_q$ are the fraction of replicates of $\Delta\widehat{S}^*$ and $\Delta\widehat{s}_q^*$ on the bootstrap dataset T_j^*, respectively, that are at least as large as the observed statistics on the original dataset.

Centrality

Important ROIs (hubs) often interact with many other ROIs and are associated with functional integration, segregation, and communication. The concept of an important ROI (vertex) in the FBN is called *centrality*. There are several measures of centrality, basically because it depends on what one considers as "important." In this section, we describe the more commonly used ones.

The *degree* is the most common measure of centrality. The degree of a vertex v is defined as the number of vertices connected to it. The interpretation is simple: it represents how much an ROI interacts with other ROIs of the brain. Thus, it is supposed that ROIs that highly interact with other ROIs are essential (Fig. 39.2A).

A natural extension of degree centrality is the *eigenvector centrality*. Eigenvector centrality takes into account that an ROI v_i is important if it is connected to other important ROIs. Let $\mathbf{A}$ be the adjacency matrix of graph $G(\mathbf{A}_{ij} = 1$ if ROIs v_i and v_j are connected and $\mathbf{A}_{ij} = 0$, otherwise). Then the eigenvector centrality of ROI v_i is given by the ith element of the eigenvector associated with the largest eigenvalue of $\mathbf{A}$ (Fig. 39.2B).

The *closeness centrality* is an estimate of the number of necessary steps to access every other ROI from ROI v. The closeness centrality of ROI v is defined by the inverse of the average length of the shortest paths from/to all the other ROIs in the FBN. Let $d(v, v_i)$ be the shortest path length between vertices$v$ and v_i, then the closeness centrality of a vertex v is defined as $\frac{1}{\sum d(v,v_i)}$ for all $v \neq v_i$. In other words, a high closeness centrality means that ROI is "close" to several other ROIs (Fig. 39.2C).

A related and often more sensitive measure is the *betweenness centrality*. Betweenness centrality quantifies the number of times a vertex acts as a bridge along the shortest path between two other ROIs. Let σ_{jk} be the total number of shortest paths from ROI v_j to ROI v_k and $\sigma_{jk}(i)$ be the number of those paths that pass through ROI v_i, then the betweenness centrality of vertex v_i is given by$\sum_{j \neq i \neq k} \frac{\sigma_{jk}(i)}{\sigma_{jk}}$. The notion of betweenness centrality can be straightforwardly extended to edges and could be used to detect important functional connections. Because a high betweenness centrality is related to how much an ROI or connection is used to connect two other ROIs, once they are deleted, the FBN becomes more prone to be disconnected (Fig. 39.2D).

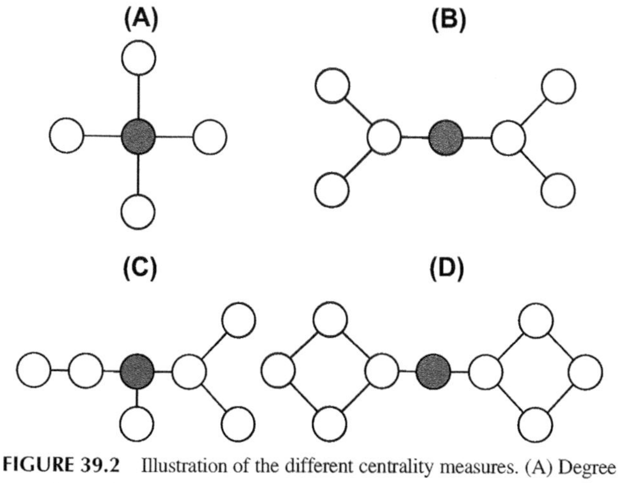

FIGURE 39.2 Illustration of the different centrality measures. (A) Degree centrality. (B) Eigenvector centrality. (C) Closeness centrality. (D) Betweenness centrality. *Red vertices [dark gray in print version]* represent the vertex with the highest centrality measure.

The degree, eigenvector, closeness, and betweenness centrality measures can be easily calculated by using the functions degree, eigencentrality, closeness, and betweenness, respectively, of the R package igraph.

Entropy

Another descriptive measure used to analyze FBNs is the concept of *graph spectral entropy*, introduced by Takahashi et al. [35]. The graph spectral entropy quantifies the "uncertainty" of the FBN. Thus, it can be used as a complementary approach to the integration, segregation, motifs, and centrality measures to characterize the organization of FBNs.

The spectrum of an undirected graph G is the set of eigenvalues of its adjacency matrix $\mathbf{A}$. A graph with n vertices has n real eigenvalues $\lambda_1 \geq \lambda_2 \geq \ldots \geq \lambda_n$. Given a family of random graphs g generated by some probability law, the eigenvalues are random vectors for which we can take the expectation with respect to the law of the graph's family. Takahashi et al. [35] defined the spectral density of a general graph family g as $\rho_g(\lambda) = \lim_{n \to \infty} < \frac{1}{n}\sum_{j=1}^{n} \delta(\lambda - \lambda_j)/\sqrt{n} >$, where δ is the Dirac delta function and the brackets "<>" indicate the expectation with respect to the law of the random graph. Let ρ_g be the spectra of the adjacency matrix of a random graph g. The spectral entropy $H(\rho_g)$ is defined by Ref. [35] as $H(\rho_g) = -\int_{-\infty}^{+\infty} \rho_g(\lambda) \log \rho_g(\lambda) d\lambda$, where we assume $0 \log 0 = 0$. The higher is the entropy, the higher is the uncertainty, and vice-versa. This uncertainty can be empirically interpreted as how difficult is to predict whether a pair of ROIs has a functional connectivity. For example, suppose a graph generated by an Erdös-Rényi random graph model with probability p. When $P = 0$, i.e., a graph without edges (empty graph), or $p = 1$, i.e., a graph with all possible edges (complete graph), the graph present very low entropy, because all edges are totally predictable. However, when$p = 0.5$, this graph presents the highest entropy because the presence/absence of an edge is totally unpredictable. Moreover, notice that the entropy is equal for graphs with parameters p and $1 - p$ (Fig. 39.3).

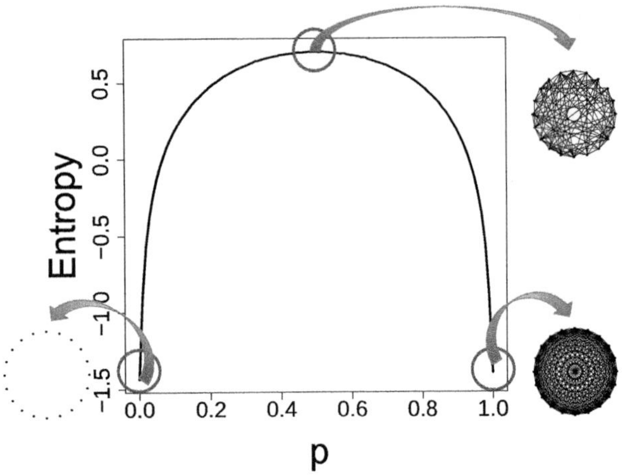

FIGURE 39.3 Erdös-Rényi random graph entropy. Notice that for $p = 0$ and $p = 1$, we obtain the lowest graph entropy, because the graph is totally predictable. For $P = 0.5$, we obtain the highest entropy, which is also the case that the presence/absence of an edge is the most unpredictable.

Illustrations of the application of graph spectral entropy on neuroscience can be found in Refs. [36,37], for example. In the former, they identified that the somatomotor subnetwork presents higher entropy in people diagnosed with attention-deficit/hyperactivity disorder (ADHD) than controls. In the latter, they identified higher entropy in the cerebellum of autistic people than controls.

For an excellent review regarding graph spectral entropy, refer to Ref. [38].

The spectral graph entropy can be calculated by using function graph, entropy of the R package statGraph.

Ongoing lines of investigation and research opportunities

Diagnosis in psychiatry

Currently, the diagnosis of psychiatric disorders is based on patient symptoms and the psychopathological examination performed by a properly trained clinician. Extensive phenotypic heterogeneity and superposition of mental health conditions have been pointed out as challenges for developing improved diagnostic procedures and treatments [39]. Neuroimaging-derived network-based markers could stratify patients into more homogeneous subgroups. Potentially, such subgroups would be formed by individuals who are more likely to respond to a given treatment or intervention. In the most influential work applying this rationale, fMRI functional connectivity was used to define subtypes of major depressive disorder in a fully data-driven manner [7]. The authors developed a method for identifying features in the whole-brain patterns of functional connectivity in resting-state networks that correlate to specific symptom combinations. Initially [19], they created individual brain network maps, depicting how each brain region of a parcellation scheme is functionally correlated to all other brain regions. Then, they used the canonical correlation analysis to define a low dimensional representation of a subset of those connectivity features (analogous to principal components), which were predictive of a linear combination of clinical symptoms. This data-driven approach identified a first connectivity component (canonical variate), comprising a combination of predominantly frontostriatal and orbitofrontal connectivity features, that were correlated with anhedonia and psychomotor retardation; and a second component associated with anxiety and insomnia, which is composed of predominantly limbic connectivities involving the amygdala, ventral hippocampus, ventral striatum, subgenual cingulate, and lateral prefrontal control areas. Then, using these features and hierarchical clustering [40,41], Drysdale et al. [7] tested whether those features tend to cluster in patient subgroups. This analysis revealed four patient clusters, defined by distinct and relatively homogeneous patterns of connectivity, along with the first and second canonical variates, then suggesting four putative subtypes of depression. Finally, the authors investigated whether they could develop classifiers for diagnosis at the individual level. To assess classification performance of patients ($n = 333$) versus controls ($n = 378$), as well as statistical significance, they carried out a leave-one-out cross-validation and a permutation test, respectively. Support-vector machine (SVM) classifiers (using linear kernel functions) presented the best results, yielding overall accuracy rates of up to 89.2%. This result was replicated in an independent sample. Responsiveness to transcranial magnetic stimulation therapy was better for a subsample of patients which were classified in one of the connectivity-defined subtypes.

Later [42], other authors extended the approach presented by the former ones [7] to identify brain-based dimensions of psychopathology, by using sparse canonical correlation. While the latter focused on discovering depression subtypes within categories of psychopathology based solely on resting-state fMRI functional connectivity features, the former aimed at linking a broad range of symptoms that are present across categories to individual differences in FBNs. The authors [43] studied 999 individuals of ages 8–22 who completed both functional neuroimaging, and a comprehensive evaluation of psychiatric symptoms. The sample was divided into discovery ($n = 663$) and replication datasets ($n = 336$), that were matched on age, gender, race, and overall psychopathology. Next, they constructed subject-level functional networks using a 264-node parcellation system that includes an a priori assignment of nodes to canonical resting-state networks. The brain connectivity feature selection was limited to the top 10% of the most variable connections, because features that do not vary across subjects cannot be predictive of individual differences. Characteristics of clinical symptoms included items from a structured psychiatric interview [43] that covers a diverse range of psychopathological domains, including mood and anxiety disorders, psychosis-spectrum symptoms, ADHD, among others. A sparse canonical correlation model, using elastic net regularization and parameter tuning, over both clinical and connectivity features, identified specific patterns of functional connectivity, that were linked to distinct combinations of psychiatric symptoms. Overall results (replicated in an independent dataset) indicated that all four dimensions of psychopathology (mood, psychosis, fear, and externalizing behavior) were characterized by decreased segregation of the default-mode and executive networks (frontoparietal and salience). Specific patterns of connectivity, linked to mood and psychosis, became more prominent with increasing age, while sex differences were present for connectivity related to mood.

Therapeutics

An example of treatment response prediction based on network biomarkers is a study [44] in social anxiety disorder (SAD). Based on the evidence that the amygdala is the most common locus of dysfunction in SAD, the authors adopted a seed-driven approach to create maps of functional connectivity and test whether brain connectomics could predict response to cognitive-behavioral therapy better than conventional clinical measures. The authors divided patients into two categories, defined according to scores of a clinical measure of symptoms: those with a good response to therapy ($n = 19$) and those with no sufficient response ($n = 19$). They used logistic regression of initial severity (clinical scores) and amygdala connectomic measures combined with leave-one-out cross-validation. As a result, they were able to categorize patients according to the response to therapy with 81% accuracy, 84% sensitivity, and 78% specificity. All these pioneering studies provide examples of attainable goals, including the potential of machine learning methods, to translate complex patterns discovered in large neuroimaging datasets into clinical practice (for a complete point of view, see Ref. [45]).

From a therapeutic and rehabilitation perspective, one of the current challenges is the lack of knowledge on how to decrease/increase the strength of specific connections. Although techniques such as TDCS and TMS are promising neuromodulatory technologies, the present state-of-the-art cannot assure the probably needed levels of specificity. Medicines are still very unspecific regarding brain areas and connections. Surgical procedures (e.g., gamma-knife) might perhaps be improved, by interrupting or modulating specific connections in an individualized manner.

Intrasubject variability and behavior

The work of Finn et al. [46] provides an innovative approach to the individual variation in whole-brain functional connectivity. The authors chose a set of brain regions and extracted the representative BOLD signal of each area. Then, for each subject, they calculated temporal correlations among these signals and modeled a weighted graph. The authors defined the graph of the individual as the functional connectome profile. Functional connectomes were found to be relatively stable over time and unique for each subject, and this graph could be considered as a brain functional "fingerprint" to identify the subjects. Similarities of these connectomes between subjects were associated with a cognitive measure (e.g., fluid intelligence). Sato et al. [47] have also replicated the connectome fingerprinting results and showed that they are stable even when considering an average of 3 months. The authors also observed an association between the functional connectome profiles and cognition and psychiatric symptoms [48,49].

The intrasubject reliability of the functional connectome assures that it is suitable to be considered for individual characterization. Correlation of individual connectomes with cognitive and behavioral profiles suggests that individual differences in brain organization may explain clinical relevant behavioral variability [50,51]. By modulating specific brain connections and networks, it could be possible to impact more particular aspects of cognition and behavior (e.g., psychiatric symptoms). Remarkably, Kaufmann et al. [52] have shown that delays on the individualization of the connectome, during childhood and adolescence, was associated with an increased incidence of mental disorders. As such, it might be a candidate marker in prevention strategies.

Developmental trajectories and prevention

Another research topic is the susceptibility to psychiatric disorder related to neurodevelopmental trajectories, and possible ways to modulate such trajectories. Kessler et al. [53] have reported that most patients with psychiatric illnesses have the first onset during childhood and adolescence. Most psychiatric subclinical and clinical disorders or early manifestations of symptoms, begin during the so-called developmental period. During this period, the brain is under intense processes of structural and functional changes [54,55]. Given these profound changes, childhood and adolescence are sensitive windows for the impact of environmental factors in the developmental trajectory. Some investigators [56] reported that the quality of family interaction regarding cohesion and conflict is associated with spontaneous brain activity at insula and orbitofrontal cortex, two critical regions for emotional regulation.

The characterization of typical developmental trajectories of brain networks is still an open question. However, this step might be crucial to early diagnosis, prevention, and treatment of mental disorders. An interesting hypothesis states that at least some psychiatric disorders are related to atypical neurodevelopmental trajectories, i.e., deviations from typical ones. These possible deviations are supposedly dependent on both genetic and environmental factors, and their interaction with the timing of developmental processes.

Deviations from typical trajectories may occur in many different ways, and adequately taking such heterogeneity into account is arguably one of the main challenges for precision medicine in mental health. The present state-of-the-art on this topic points toward the combination of longitudinal network analyses and machine learning tools. One of the main difficulties is that brain networks data are highly multidimensional, particularly when considering neuroimaging data. Thus, to control the false positives rate, avoid high collinearity among variables, and separate "the noise from the signal," modern machine learning methods are needed. Furthermore, inferences at the single subject level (e.g., healthy subject vs. patient) are straightforward outputs from these methods. Sato et al. [57] used the one-class support-vector machine (OC-SVM) to demonstrate an association between ADHD and maturational delay in the precuneus-dorsal anterior cingulate functional connectivity. Similarly, Sato et al. (2018) [48] applied OC-SVM to show that children with maturational delays of the default-mode network are more prone to psychiatric symptoms than typical and precocious ones. However, the methodological combination of functional connectivity analyses and machine learning, to determine the most appropriate treatments to a given subject, remains mostly untested.

Wearable diagnostic tools

The development of low-cost wearable devices is of great importance. The use of a Holter monitor, for instance, is a gold-standard exam to the diagnosis of a series of cardiologic conditions. A similar device might be applied to better characterize the brain signal, about daily activities and the expression of symptoms. However, one of the main current limitations for neuropsychiatric applications is the time-scale of interest: 1 day may be a short time interval to extract information about low-frequency variability, slower than one circadian cycle. Current wearable gadgets (e.g., actigraphy, heart-rate monitors) and internet-of-things (e.g., appliance or vehicle devices sending data to a cloud database) are an attempt to overcome this limitation [58]. Validation of the signal quality of OpenBCI [59], Emotiv [60], and Muse EEG [61], which are low-cost devices, makes them attractive to further developments. Advances in fNIRS instrumentation have achieved notable milestones regarding potential in real-life situations [12,62].

Acknowledgments

This work was partially supported by São Paulo Research Foundation (2015/01587-0, 2018/17996-5, 2018/04654-9), CNPq (304876/2016-0), CAPES (Finance Code 001), Alexander von Humboldt Foundation, Newton Fund/The Academy of Medical Sciences, and European Research Council.

References

[1] Fernandes BS, Williams LM, Steiner J, Leboyer M, Carvalho AF, Berk M. The new field of precision psychiatry. BMC Med. 2017;15(1):80.
[2] Hahn T, Nierenberg A, Whitfield-Gabrieli S. Predictive analytics in mental health: applications, guidelines, challenges and perspectives. Mol. Psychiatry 2017;22(1):37.
[3] Ecker C, Murphy D. Neuroimaging in autismfrom basic science to translational research. Nat. Rev. Neurol. 2014;10(2):82.
[4] Insel TR. The nimh research domain criteria (rdoc) project: precision medicine for psychiatry. Am. J. Psychiatry 2014;171(4):395–7.
[5] Milham MP, Craddock RC, Klein A. Clinically useful brain imaging for neuropsychiatry: how can we get there? Depress. Anxiety 2017;34(7):578–87.
[6] Fornito A, Bullmore ET, Zalesky A. Opportunities and challenges for psychiatry in the connectomic era. Biol. Psychiatry Cogn. Neurosci. Neuroimaging 2017;2(1):9–19.
[7] Drysdale AT, Grosenick L, Downar J, Dunlop K, Mansouri F, Meng Y, Fetcho RN, Zebley B, Oathes DJ, Etkin A, et al. Resting-state connectivity biomarkers define neurophysiological subtypes of depression. Nat. Med. 2017;23(1):28.
[8] Jackson AF, Bolger DJ. The neurophysiological bases of eeg and eeg measurement: a review for the rest of us. Psychophysiology 2014;51(11):1061–71.
[9] Ogawa S, Lee TM, Kay AR, Tank DW. Brain magnetic resonance imaging with contrast dependent on blood oxygenation. Proc. Natl. Acad. Sci. 1990;87(24):9868–72.
[10] Logothetis NK, Pauls J, Augath M, Trinath T, Oeltermann A. Neurophysiological investigation of the basis of the fmri signal. Nature 2001;412(6843):150.
[11] Scholkmann F, Kleiser S, Metz AJ, Zimmermann R, Pavia JM, Wolf U, Wolf M. A review on continuous wave functional near-infrared spectroscopy and imaging instrumentation and methodology. Neuroimage 2014;85:6–27.
[12] Balardin JB, Zimeo Morais GA, Furucho RA, Trambaiolli L, Vanzella P, Biazoli Jr C, Sato JR. Imaging brain function with functional near-infrared spectroscopy in unconstrained environments. Front. Hum. Neurosci. 2017;11:258.
[13] Pinti P, Aichelburg C, Gilbert S, Hamilton A, Hirsch J, Burgess P, Tachtsidis I. A review on the use of wearable functional near-infrared spectroscopy in naturalistic environments. Jpn. Psychol. Res. 2018;60(4):347–73.
[14] Galton F. I. co-relations and their measurement, chiefly from anthropometric data. Proc. R. Soc. Lond. 1889;45(273–279):135–45.

[15] Pearson K. Notes on the history of correlation. Biometrika 1920;13(1):25–45.

[16] de Siqueira Santos S, Takahashi DY, Nakata A, Fujita A. A comparative study of statistical methods used to identify dependencies between gene expression signals. Briefings Bioinf. 2013;15(6):906–18.

[17] Spearman C. "General intelligence," objectively determined and measured. Am. J. Psychol. 1904;15(2):201–92.

[18] Graybill FA. Theory and application of the linear model. 1976. 04; QA279, G7.

[19] Power JD, Cohen AL, Nelson SM, Wig GS, Barnes KA, Church JA, Vogel AC, Laumann TO, Miezin FM, Schlaggar BL, et al. Functional network organization of the human brain. Neuron 2011;72(4):665–78.

[20] Rubinov M, Sporns O. Complex network measures of brain connectivity: uses and interpretations. Neuroimage 2010;52(3): 1059–69.

[21] Supekar K, Menon V, Rubin D, Musen M, Greicius MD. Network analysis of intrinsic functional brain connectivity in alzheimer's disease. PLoS Comput. Biol. 2008;4(6):e1000100.

[22] Latora V, Marchiori M. Efficient behavior of small-world networks. Phys. Rev. Lett. 2001;87(19):198701.

[23] Watts DJ, Strogatz SH. Collective dynamics of small-worldnetworks. Nature 1998;393(6684):440.

[24] Newman ME. The structure and function of complex networks. SIAM Rev. 2003;45(2):167–256.

[25] Onnela JP, Saramäki J, Kertész J, Kaski K. Intensity and coherence of motifs in weighted complex networks. Phys. Rev. E 2005;71(6):065103.

[26] Fagiolo G. Clustering in complex directed networks. Phys. Rev. E 2007;76(2):026107.

[27] Girvan M, Newman ME. Community structure in social and biological networks. Proc. Natl. Acad. Sci. 2002;99(12):7821–6.

[28] Newman ME. Fast algorithm for detecting community structure in networks. Phys. Rev. 2004;69(6):066133.

[29] Guimera R, Amaral LAN. Cartography of complex networks: modules and universal roles. J. Stat. Mech. Theory Exp. 2005;2005(02):P02001.

[30] Fujita A, Takahashi DY, Patriota AG. A non-parametric method to estimate the number of clusters. Comput. Stat. Data Anal. 2014;73:27–39.

[31] Rousseeuw PJ. Silhouettes: a graphical aid to the interpretation and validation of cluster analysis. J. Comput. Appl. Math. 1987;20:53–65. https://doi.org/10.1016/0377-0427(87) 90125-7.

[32] Patriota AG, Vidal MC, de Jesus DAC, Fujita A, Anocva. A nonparametric statistical test to compare clustering structures. In: Theoretical and applied aspects of systems biology. publisher; 2018. p. 113–25.

[33] Milo R, Shen-Orr S, Itzkovitz S, Kashtan N, Chklovskii D, Alon U. Network motifs: simple building blocks of complex networks. Science 2002;298(5594):824–7.

[34] Sporns O, Kötter R. Motifs in brain networks. PLoS Biol. 2004;2(11):e369.

[35] Takahashi DY, Sato JR, Ferreira CE, Fujita A. Discriminating different classes of biological networks by analyzing the graphs spectra distribution. PLoS One 2012;7(12):e49949.

[36] Sato JR, Takahashi DY, Hoexter MQ, Massirer KB, Fujita A. Measuring network's entropy in adhd: a new approach to investigate neuropsychiatric disorders. Neuroimage 2013;77:44–51.

[37] Sato JR, Calebe Vidal M, de Siqueira Santos S, Brauer Massirer K, Fujita A. Complex network measures in autism spectrum disorders. IEEE ACM Trans. Comput. Biol. Bioinform 2018;15(2):581–7.

[38] de Siqueira Santos S, Takahashi DY, Sato JR, Ferreira CE, Fujita A. Statistical methods in graphs: parameter estimation, model selection, and hypothesis test. In: Mathematical Foundations and Applications of Graph Entropy. 6; 2016. p. 183–202.

[39] Ecker C, Spooren W, Murphy D. Translational approaches to the biology of autism: false dawn or a new era? Mol. Psychiatry 2013;18(4):435.

[40] Ravasz E, Somera AL, Mongru DA, Oltvai ZN, Barabási AL. Hierarchical organization of modularity in metabolic networks. Science 2002;297(5586):1551–5.

[41] Rihel J, Prober DA, Arvanites A, Lam K, Zimmerman S, Jang S, Haggarty SJ, Kokel D, Rubin LL, Peterson RT, et al. Zebrafish behavioral profiling links drugs to biological targets and rest/wake regulation. Science 2010;327(5963):348–51.

[42] Xia M, Womer FY, Chang M, Zhu Y, Zhou Q, Edmiston EK, Jiang X, Wei S, Duan J, Xu K, et al. Shared and distinct functional architectures of brain networks across psychiatric disorders. Schizophr. Bull. 2018;45(2):450–63.

[43] Calkins ME, Merikangas KR, Moore TM, Burstein M, Behr MA, Satterthwaite TD, Ruparel K, Wolf DH, Roalf DR, Mentch FD, et al. The philadelphia neurodevelopmental cohort: constructing a deep phenotyping collaborative. J. Child Psychol. Psychiatry 2015;56(12):1356–69.

[44] Whitfield-Gabrieli S, Ghosh S, Nieto-Castanon A, Saygin Z, Doehrmann O, Chai X, Reynolds G, Hofmann S, Pollack M, Gabrieli J. Brain connectomics predict response to treatment in social anxiety disorder. Mol. Psychiatry 2016;21(5):680.

[45] Bzdok D, Meyer-Lindenberg A. Machine learning for precision psychiatry: opportunities and challenges. Biol. Psychiatry Cogn. Neurosci. Neuroimaging 2018;3(3):223–30.

[46] Finn ES, Shen X, Scheinost D, Rosenberg MD, Huang J, Chun MM, Papademetris X, Constable RT. Functional connectome fingerprinting: identifying individuals using patterns of brain connectivity. Nat. Neurosci. 2015;18(11):1664.

[47] Sato JR, White TP, Biazoli Jr CE. Commentary: a test-retest dataset for assessing long-term reliability of brain morphology and resting-state brain activity. Front. Neurosci. 2017;11:85.

[48] Sato JR, Biazoli Jr CE, Salum GA, Gadelha A, Crossley N, Vieira G, Zugman A, Picon FA, Pan PM, Hoexter MQ, et al. Association between abnormal brain functional connectivity in children and psychopathology: a study based on graph theory and machine learning. World J. Biol. Psychiatry 2018;19(2):119–29.

[49] Biazoli Jr CE, Salum GA, Pan PM, Zugman A, Amaro Jr E, Rohde LA, Miguel EC, Jackowski AP, Bressan RA, Sato JR, Commentary. Functional connectome fingerprint: identifying individuals using patterns of brain connectivity. Front. Hum. Neurosci. 2017;11:47.

[50] Finn ES, Constable RT. Individual variation in functional brain connectivity: implications for personalized approaches to psychiatric disease. Dialogues Clin. Neurosci. 2016;18(3):277.

[51] Shen X, Finn ES, Scheinost D, Rosenberg MD, Chun MM, Papademetris X, Constable RT. Using connectome-based predictive modeling to predict individual behavior from brain connectivity. Nat. Protoc. 2017;12(3):506.

[52] Kaufmann T, Alnss D, Doan NT, Brandt CL, Andreassen OA, Westlye LT. Delayed stabilization and individualization in connectome development are related to psychiatric disorders. Nat. Neurosci. 2017;20(4):513.

[53] Kessler RC, Amminger GP, Aguilar-Gaxiola S, Alonso J, Lee S, Ustun TB. Age of onset of mental disorders: a review of recent literature. Curr. Opin. Psychiatr. 2007;20(4):359.

[54] Sato JR, Salum GA, Gadelha A, Picon FA, Pan PM, Vieira G, Zugman A, Hoexter MQ, Anés M, Moura LM, et al. Age effects on the default mode and control networks in typically developing children. J. Psychiatr. Res. 2014;58:89–95.

[55] Sato JR, Salum GA, Gadelha A, Vieira G, Zugman A, Picon FA, Pan PM, Hoexter MQ, Anes M, Moura LM, et al. Decreased centrality of subcortical regions during the transition to adolescence: a functional connectivity study. Neuroimage 2015;104:44–51.

[56] Sato JR, Biazoli CE, Salum GA, Gadelha A, Crossley N, Vieira G, Zugman A, Picon FA, Pan PM, Hoexter MQ, et al. Associations between childrens family environment, spontaneous brain oscillations, and emotional and behavioral problems. Eur. Child Adolesc. Psychiatry 2018:1–11.

[57] Sato JR, Hoexter MQ, Castellanos XF, Rohde LA. Abnormal brain connectivity patterns in adults with adhd: a coherence study. PLoS One 2012;7(9):e45671.

[58] Li X, Dunn J, Salins D, Zhou G, Zhou W, Rose SMSF, Perelman D, Colbert E, Runge R, Rego S, et al. Digital health: tracking physiomes and activity using wearable biosensors reveals useful health-related information. PLoS Biol. 2017;15(1):e2001402.

[59] Rashid U, Niazi I, Signal N, Taylor D. An eeg experimental study evaluating the performance of Texas instruments ads1299. Sensors 2018;18(11):3721.

[60] Badcock NA, Preece KA, de Wit B, Glenn K, Fieder N, Thie J, McArthur G. Validation of the emotiv epoc eeg system for research quality auditory event-related potentials in children. Peer J. 2015;3:e907.

[61] Krigolson OE, Williams CC, Norton A, Hassall CD, Colino FL. Choosing muse: validation of a low-cost, portable eeg system for erp research. Front. Neurosci. 2017;11:109.

[62] Brockington G, Balardin JB, Morais GAZ, Malheiros A, Lent R, Moura LM, Sato JR. From the laboratory to the classroom: the potential of functional near-infrared spectroscopy in educational neuroscience. Front. Psychol. 2018;9.

Chapter 40

Nutrigenetic approaches in obesity and weight loss

Omar Ramos-Lopez[1,2] and J. Alfredo Martinez[1,3,4,5]

[1]*Department of Nutrition, Food Science and Physiology, University of Navarra, and Center for Nutrition Research, University of Navarra, Pamplona, Spain;* [2]*Faculty of Medicine and Psychology, Autonomous University of Baja California, Tijuana, BC, Mexico;* [3]*CIBERobn, Physiopathology of Obesity, Carlos III Institute, Madrid, Spain;* [4]*Navarra Institute for Health Research (IdiSNA), Pamplona, Spain;* [5]*Madrid Institute of Advanced Studies (IMDEA Food), Madrid, Spain*

Introduction

Obesity epidemic is one of the most important health challenges worldwide, as denoted by a recent report of the World Health Organization, revealing that more than 39% of adults presented overweight or obesity (about 1.9 billion people) [1]. According to the Global Burden of Disease study, excessive body weight accounted for 4 million deaths, among adults in 195 Countries over 25 Years [2].

Genetic factors may contribute to the onset and development of excessive adiposity and accompanying comorbidities, by affecting energy homeostasis and body weight regulation [3]. Nutrigenetic studies are enabling to clarify the involvement of gene-diet interactions, in determining specific adiposity phenotypes, and modulating therapy outcomes [4].

Herein, we review genetic variants and biomarkers related to obesity and weight loss, which may serve to understand disease etiology and envisage future therapeutic targets and innovative treatments (Fig. 40.1).

Gene-diet interactions and obesity predisposition

Single nucleotide polymorphisms (SNPs) are the most studied genetic variants in the field of precision nutrition [5]. Multiple SNPs are associated with obesity predisposition through interactions with dietary factors (Table 40.1). Relevant interactions between polymorphisms located at lipid-metabolism genes (*PPARG*, *APOA5*, *APOA2*, and *APOB*) and high-fat diets, were found in relation to greater adiposity markers. Interestingly, SNP rs2301241 in *TXN* gene (acting as an antioxidant), was associated with higher waist circumference (WC) values in subjects with low vitamin E intakes. Obesity risk was significantly higher in T-allele carriers of $-13{,}910\,C > T$ polymorphism (rs4988235), upstream *LCT* gene, only among subjects consuming moderate or high amounts of lactose (Table 40.1).

Because the magnitude of associations between individual SNPs and adiposity traits is generally modest, studies using genetic risk scores (GRS), have examined the additive effect of multiple loci and diet interactions [6]. Thus, a validated GRS for obesity was associated with higher BMI and WC values among individuals consuming high amounts of fats, compared to those with low-fat intakes [7]. Also, higher intakes of animal protein, saturated fat, and carbohydrates were positively associated with greater percentages of body fat mass, among individuals carrying a high genetic risk group for obesity [8].

Nominally significant interactions were detected between BMI-associated GRS and protein intake, on obesity and fat mass among women within the Malmö Diet and Cancer Study [9]. Nevertheless, a longitudinal analysis evidenced no relationships between adiposity-associated GRS and dietary protein, in relation to subsequent change in body weight and waist circumference in three different Danish cohorts [10].

The role of micronutrient status in this field has also been explored. For instance, a diet with a high content of ascorbic acid was associated with higher WC gain, among people genetically predisposed to abdominal obesity [11]. On the contrary, a significant interaction between a GRS from six WC-related SNPs and dietary calcium was reported concerning WC changes, where each risk allele was associated with greater WC reductions per 1000 mg of calcium intake [12].

Precision Medicine for Investigators, Practitioners and Providers. https://doi.org/10.1016/B978-0-12-819178-1.00040-X

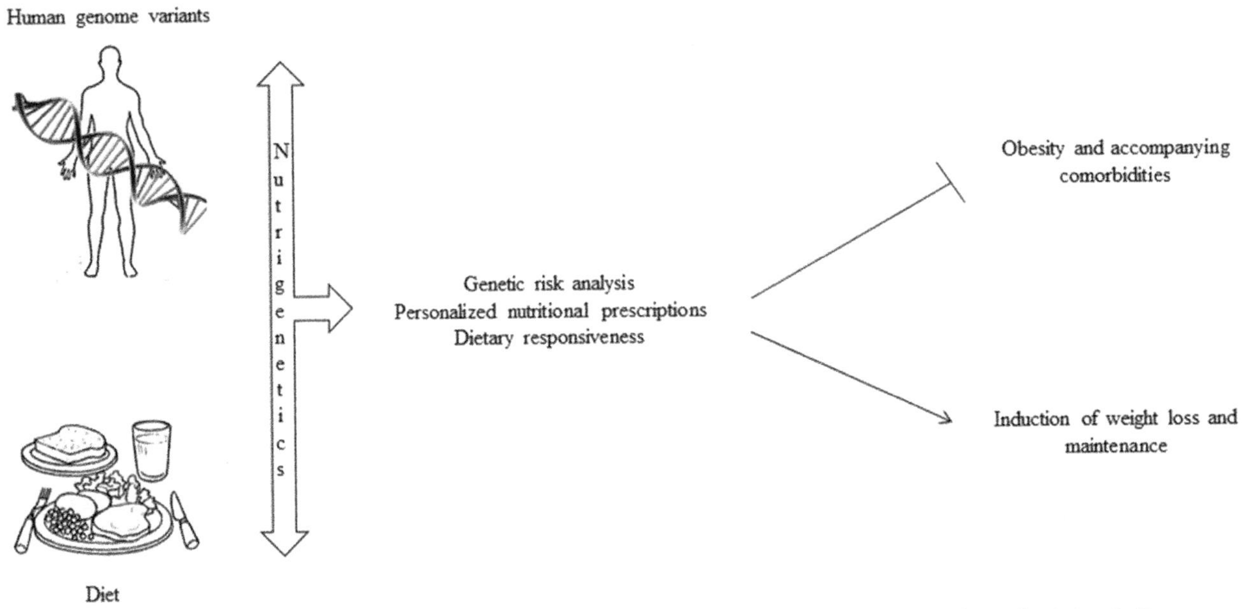

FIGURE 40.1 Nutrigenetic approaches in precision nutrition, for prevention and management of obesity and associated chronic diseases.

TABLE 40.1 Nutrigenetic examples of SNPs-diet interactions involved in obesity predisposition.

Gene	Polymorphism	Allele	Diet interaction	Main findings	Reference
FTO	rs8050136	A	High carbohydrate	Increased obesity risk	54
LCT	rs4988235	T	High lactose	Increased obesity risk	55
PPARG	rs1801282	G	High fat	Higher BMI	56
TXN	rs2301241	T	Low vitamin E	Higher WC	57
ADAM17	rs10495563	A	Low n-6 PUFA	Increased obesity risk	58
TNFA	rs1800629	A	High fat	Increased obesity risk	59
APOA5	rs662799	T	High fat	Higher adiposity markers	60
LEPR	rs1137101	G	High SFA/High fat	Increased obesity risk	61
APOB	rs1469513	G	High fat	Increased obesity risk	62
APOA2	rs5082	C	High fat dairy foods	Higher BMI	63

ADAM17, ADAM metallopeptidase domain 17; *APOA2*, apolipoprotein A2; *APOA5*, apolipoprotein A5; *APOB*, apolipoprotein B; *BMI*, body mass index; *FTO*, fat mass and obesity associated; *LCT*, lactase; *LEPR*, leptin receptor; *PPARG*, peroxisome proliferator activated receptor gamma; *PUFA*, polyunsaturated fatty acids; *SFA*, saturated fatty acids; *TNFA*, tumor necrosis factor a; *TXN*, thioredoxin; *WC*, waist circumference.

The association of a GRS constructed from 32 BMI-associated variants with adiposity, was strengthened with greater intake of fried foods and sugar-sweetened beverages, in three American cohort studies [13,14]. Similarly, an increased risk for obesity was found among individuals with low habitual coffee consumption, who were genetically predisposed to obesity in the basis of a GRS calculated from 77 BMI-related loci [15].

Under the assumption that the assessment of dietary patterns provides more reliable information regarding real food intake, compared to particular macronutrient consumption, higher Mediterranean dietary pattern adherence was associated with decreased obesity risk, in subjects carrying high GRS according to FTO polymorphisms [16]. Associations between genetic predisposition and obesity traits were stronger with a healthier diet (based on habitual consumption of whole grains, fish, fruits, vegetables, and nuts/seeds), in 18 cohorts of European ancestry [17].

The increment in BMI was smaller among individuals with a strong genetic predisposition to obesity, categorized in the highest tertiles of three diet quality scores (Alternative Healthy Eating Index 2010 [AHEI-2010]; Alternative Mediterranean Diet score [AMED]; and the Dietary Approach to Stop Hypertension [DASH]) [18]. Genetic association with weight gain was also significantly attenuated, with increasing adherence to the AHEI-2010 dietary score in two prospective cohort studies [19].

Gene-diet interactions involving weight loss and adiposity outcomes

Potential gene-diet interactions have also been reported to influence the heterogeneity of adiposity outcomes (Table 40.2). Investigations include SNPs mapped to genes involved in the regulation of critical physiological processes, such as circadian rhythm, inflammatory response, lipid metabolism, insulin signaling, amino acid breakdown, and blood glucose homeostasis (Table 40.2).

Greater body fat losses were reported in highly sensitive carriers of obesity GRS in response to dietary therapy, in a large Korean population [20]. Also, highest diabetes-susceptibility loci were associated with greater WC reductions, after 1-year of intensive lifestyle advice within the Look AHEAD (Action for Health in Diabetes) clinical trial [21]. Likewise, potentially modest benefits in weight loss and physical activity were found among participants with higher GRS for coronary artery disease, who received risk-reducing strategies based on diet and exercise [22]. Furthermore, obese and overweight Spanish adolescents with lower obesity GRS, evidenced greater benefits on weight loss and metabolic profile improvements, after 3 months of a multidisciplinary intervention program [23].

On the other hand, changes in body weight over a 5-year lifestyle intervention were not influenced by BMI-related GRS in a middle-aged Danish cohort [24]. Moreover, diabetes genetic risk counseling did not significantly modify mean weight loss, among overweight individuals under a validated diabetes prevention program, designed to improve diet quality and physical activity level [25]. Meanwhile, GRS comprising 15 SNP previously associated with childhood BMI, did not influence changes in BMI or cardiometabolic traits in Danish Children, following a lifestyle intervention [26].

The impact of genetic information disclosure on obesity management

The Food4Me European randomized controlled trial revealed that disclosure of information concerning fat-mass and obesity-associated (*FTO*) genotype risk, had a greater effect on changes in adiposity markers, compared with nonpersonalized intervention group [27]. Similarly, the Coriell Personalized Medicine Collaborative trial reported that individuals receiving their *FTO* genotype alone had greater intentions to lose body weight at follow-up than those who received no risk genetic advice [28]. This effect was enhanced in participants carrying a high genetic risk.

Genetic information has also been included in nutrigenetic tests, aimed at assessing the effect on changing certain obesity-related dietary behaviors. Apolipoprotein E (*APOE*)-based personalized nutrition resulted in a greater reduction in saturated fat intake (percentage of total energy) than did standard dietary advice, although no differences by *APOE* genotypes (E4+ vs. E4−) were found [29]. Greater

TABLE 40.2 Nutrigenetic trials analyzing SNPs-diet interactions involved in adiposity outcomes in response to nutritional interventions.

Gene	Polymorphism	Allele	Diet interaction	Main outcomes	Reference
FTO	rs1558902	A	High protein	Greater weight loss	64
TFAP2B	rs987237	G	High protein	Higher weight regains	65
MTNR1B	rs10830963	G	High protein	Lower weight loss in women	66
IL6	rs2069827	C	Mediterranean diet	Lower weight gains	67
IRS1	rs2943641	C	High carbohydrate	Greater weight loss	68
PPM1K	rs1440581	C	High fat	Less weight loss	69
TCF7L2	rs7903146	C	High fiber	Greater weight loss	70
PPARG	rs1801282	G	High fat	Lower weight loss	71
ADCY3	rs10182181	G	Moderately high-protein	Lower decrease of fat mass, trunk and android fat	72
FGF21	rs838147	C	Low-carbohydrate/high-fat	Less reduction of total fat mass and trunk fat	73

ADCY3, adenylate cyclase 3; *FGF21*, fibroblast growth factor 21; *FTO*, fat mass and obesity associated; *IL6*, interleukin 6; *IRS1*, insulin receptor substrate 1; *MTNR1B*, melatonin receptor 1B; *PPARG*, peroxisome proliferator activated receptor gamma; *PPM1K*, protein phosphatase, Mg2+/Mn2+ dependent 1K; *TCF7L2*, transcription factor 7 like 2; *TFAP2B*, transcription factor AP-2 beta.

improvements in Mediterranean diet scores were reported among subjects receiving genotypic feedback, targeting specific variants in five nutrient-responsive genes, after 6 months of follow-up [30]. On the other hand, there was no evidence that including phenotypic (anthropometry and blood biomarkers) and genotypic information (5 diet-responsive genetic variants), enhanced the effectiveness of a customized nutritional program based on individual baseline diet [31].

Genotype-based nutritional advice appears to be a useful strategy for the prevention and treatment of obesity by favoring some stable dietary changes and improving adiposity outcomes. These findings could be related to individuals perceiving gene-tailored counseling, as more understandable and useful than general dietary recommendations [32], with counseling more positively scored toward the end of personalized intervention [33].

Other genetic variants

Two of the most studied are highly polymorphic copy number variants (CNVs), encompassing salivary (*AMY1A*) and pancreatic (*AMY2A*) amylase genes. A pioneering study reported the association between reduced copy number of *AMY1A* gene and increased BMI [34]. Afterward, significant and positive contributions of *AMY1A* copy number, to lower obesity risks in Mexican and French children were found [35,36]. However, an independent study involving two East Asian populations of Chinese and Malay ethnicity, was not able to replicate the association between *AMY1A* gene and obesity or BMI [37].

A significant interaction between *AMY1A* copy number and starch intake on BMI and body fat percentage was detected, in which BMI tended to decrease with increasing *AMY1A* copy numbers in the low-starch intake group, and tended to increase with increasing *AMY1* copy numbers in the high-starch intake group [38]. Furthermore, a randomized trial revealed greater reductions in body weight and WC, among individuals carrying the A allele of the *AMY1A-AMY2A* rs11185098 genotype (indicating higher amylase amount and activity), compared to those without the A allele [39].

Possible contributions of CNVs in five genes (*LEPR*, *NEGR1*, *ARHGEF4*, and *CPXCR1*), and four intergenic regions (12q15c, 15q21.1a, and 22q11.21d), to development of obesity, particularly abdominal obesity, were recently reported in Mexican children [40]. Meanwhile, a significant interaction between *APOB* Ins/Del polymorphism and dietary n-3 polyunsaturated fatty acid (PUFA) intake, regarding obesity risk in type 2 diabetic patients, was found [41]. Thus, a higher general obesity risk was detected in carriers of the Del allele, than Ins/Ins homozygotes, when dietary n-3 PUFA intake was low [41]. Instead, a variable number of tandem repeats (VNTR) polymorphism, within the exon III of the *DRD4* gene, was not related to success in weight loss in obese children, after 1-year lifestyle intervention [42].

Host genetics, microbiota composition, and obesity risk: potential interactions

Emerging evidence suggests complex interactions between host genetic background and gut microbiome, concerning the risk of developing obesity [43]. SNPs located within intronic and untranslated regions of *PLD1* gene were associated with abundant levels of genus *Akkermansia*, which has been shown to affect obesity susceptibility [44]. On the other hand, differences in the abundance of *Prevotella* genus were related to a human variant adjacent to *LYPLAL1*, a gene, reported to be associated with body fat distribution, waist-hip ratio, and insulin sensitivity in some populations [45].

Also, analyses of the gut microbial community composition in twins unveiled reduced abundances of *Actinobacteria* and *Bifidobacterium* that were significantly linked to the minor allele at the *APOA5* SNP rs651821, which is known to be associated with metabolic syndrome [46]. Moreover, elevated levels of the beneficial *Bifidobacteria* have been found among carriers of the lactase nonpersistence genotype [47], whose protective effect against obesity traits has been documented in some populations of European descent [48,49].

This gut-microbiome-related *LCT* gene profile was associated with long-term improvements of body fat composition and distribution, among subjects eating a low-calorie, high-protein diet [50].

In genetically distinct inbred mouse strains, susceptibility to obesity and metabolic syndrome was modulated by interactions between gut microbiota, host genetics, and diet [51]. In the same way, microbiota transplant experiments in animal models revealed that alterations of the gut microbiota composition modified the strain-specific susceptibility to diet-induced metabolic disease [52]. Furthermore, a significant enrichment of *Enterobacteriaceae* (phylum Proteobacteria) was linked to chromosome 3 locus (rs29982345) in mice [53]. Interestingly, this genomic region containing three amylase genes was subsequently associated with body fat growth during high-fat/high-sucrose feeding [53].

Future directions

Genomic studies have identified a number of genetic variants associated with obesity predisposition. This knowledge has contributed to the design of genotype-based nutritional strategies to induce long-term weight loss. Nevertheless, recent investigations have revealed the involvement of epigenetic marks and gut microbiota composition in body weight regulation. Furthermore, metabolomic analyses have allowed metabolically categorizing individuals in different groups, based on food consumption or in response to dietary prescriptions. The integration of these scientific insights is needed for the implementation of tailored nutritional interventions.

References

[1] World Health Organization. Obesity and overweight. 2018. Available from: https://www.who.int/news-room/fact-sheets/detail/obesity-and-overweight.

[2] GBD 2015 Obesity Collaborators, Afshin A, Forouzanfar MH, et al. Health effects of overweight and obesity in 195 Countries over 25 years. N. Engl. J. Med. 2017;377:13–27.

[3] Goni L, Milagro FI, Cuervo M, Martínez JA. Single-nucleotide polymorphisms and DNA methylation markers associated with central obesity and regulation of body weight. Nutr. Rev. 2014;72:673–90.

[4] Heianza Y, Qi L. Gene-diet interaction and precision nutrition in obesity. Int. J. Mol. Sci. 2017. https://doi.org/10.3390/ijms18040787.

[5] Ramos-Lopez O, Milagro FI, Allayee H, et al. Guide for current nutrigenetic, nutrigenomic, and nutriepigenetic approaches for precision nutrition involving the prevention and management of chronic diseases associated with obesity. J. Nutrigenet. Nutrigenomics 2017;10:43–62.

[6] Moonesinghe R, Liu T, Khoury MJ. Evaluation of the discriminative accuracy of genomic profiling in the prediction of common complex diseases. Eur. J. Hum. Genet. 2010;18:485–9.

[7] Celis-Morales CA, Lyall DM, Gray SR, et al. Dietary fat and total energy intake modifies the association of genetic profile risk score on obesity: evidence from 48 170 UK biobank participants. Int. J. Obes. 2017;41:1761–8.

[8] Goni L, Cuervo M, Milagro FI, Martínez JA. A genetic risk tool for obesity predisposition assessment and personalized nutrition implementation based on macronutrient intake. Genes Nutr 2015;10:445.

[9] Rukh G, Sonestedt E, Melander O, et al. Genetic susceptibility to obesity and diet intakes: association and interaction analyses in the Malmö diet and cancer study. Genes Nutr 2013;8:535–47.

[10] Ankarfeldt MZ, Larsen SC, Ängquist L, et al. Interaction between genetic predisposition to adiposity and dietary protein in relation to subsequent change in body weight and waist circumference. PLoS One 2014;9:e110890.

[11] Larsen SC, Angquist L, Ahluwalia TS, et al. Dietary ascorbic acid and subsequent change in body weight and waist circumference: associations may depend on genetic predisposition to obesity—a prospective study of three independent cohorts. Nutr. J. 2014;13:43.

[12] Larsen SC, Ängquist L, Ahluwalia TS, et al. Interaction between genetic predisposition to obesity and dietary calcium in relation to subsequent change in body weight and waist circumference. Am. J. Clin. Nutr. 2014;99:957–65.

[13] Qi Q, Chu AY, Kang JH, et al. Fried food consumption, genetic risk, and body mass index: gene-diet interaction analysis in three US cohort studies. BMJ 2014;348:g1610.

[14] Qi Q, Chu AY, Kang JH, et al. Sugar-sweetened beverages and genetic risk of obesity. N. Engl. J. Med. 2012;367:1387–96.

[15] Wang T, Huang T, Kang JH, et al. Habitual coffee consumption and genetic predisposition to obesity: gene-diet interaction analyses in three US prospective studies. BMC Med 2017;15:97.

[16] Hosseini-Esfahani F, Koochakpoor G, Daneshpour MS, Sedaghati-Khayat B, Mirmiran P, Azizi F. Mediterranean dietary pattern adherence modify the association between FTO genetic variations and obesity phenotypes. Nutrients 2017. https://doi.org/10.3390/nu9101064.

[17] Nettleton JA, Follis JL, Ngwa JS, et al. Gene × dietary pattern interactions in obesity: analysis of up to 68 317 adults of European ancestry. Hum. Mol. Genet. 2015;24:4728–38.

[18] Ding M, Ellervik C, Huang T, et al. Diet quality and genetic association with body mass index: results from 3 observational studies. Am. J. Clin. Nutr. 2018;108:1291–300.

[19] Wang T, Heianza Y, Sun D, et al. Improving adherence to healthy dietary patterns, genetic risk, and long term weight gain: gene-diet interaction analysis in two prospective cohort studies. BMJ 2018. https://doi.org/10.1136/bmj.j5644.

[20] Cha S, Kang JH, Lee JH, et al. Impact of genetic variants on the individual potential for body fat loss. Nutrients 2018. https://doi.org/10.3390/nu10030266.

[21] Peter I, McCaffery JM, Kelley-Hedgepeth A, et al. Association of type 2 diabetes susceptibility loci with one-year weight loss in the look AHEAD clinical trial. Obesity (Silver Spring) 2012;20:1675–82.

[22] Knowles JW, Zarafshar S, Pavlovic A, et al. Impact of a genetic risk score for coronary artery disease on reducing cardiovascular risk: a pilot randomized controlled study. Front. Cardiovasc. Med 2017. https://doi.org/10.3389/fcvm.2017.00053.

[23] Moleres A, Rendo-Urteaga T, Zulet MA, et al. Obesity susceptibility loci on body mass index and weight loss in Spanish adolescents after a lifestyle intervention. J. Pediatr. 2012;161:466–470.e2.

[24] Sandholt CH, Allin KH, Toft U, et al. The effect of GWAS identified BMI loci on changes in body weight among middle-aged Danes during a five-year period. Obesity (Silver Spring) 2014;22:901–8.

[25] Grant RW, O'Brien KE, Waxler JL, et al. Personalized genetic risk counseling to motivate diabetes prevention: a randomized trial. Diabetes Care 2013;36:13–9.

[26] Hollensted M, Fogh M, Schnurr TM, et al. Genetic susceptibility for childhood BMI has no impact on weight loss following lifestyle intervention in Danish children. Obesity 2018;26:1915–22.

[27] Celis-Morales C, Marsaux CF, Livingstone KM, et al. Can genetic-based advice help you lose weight? Findings from the Food4Me European randomized controlled trial. Am. J. Clin. Nutr. 2017;105:1204–13.

[28] Wang C, Gordon ES, Norkunas T, et al. A randomized trial examining the impact of communicating genetic and lifestyle risks for obesity. Obesity 2016;24:2481–90.

[29] Fallaize R, Celis-Morales C, Macready AL, et al. The effect of the apolipoprotein E genotype on response to personalized dietary advice intervention: findings from the Food4Me randomized controlled trial. Am. J. Clin. Nutr. 2016;104:827–36.

[30] Livingstone KM, Celis-Morales C, Navas-Carretero S, et al. Effect of an internet-based, personalized nutrition randomized trial on dietary changes associated with the Mediterranean diet: the Food4Me study. Am. J. Clin. Nutr. 2016;104:288–97.

[31] Celis-Morales C, Livingstone KM, Marsaux CF, et al. Effect of personalized nutrition on health-related behaviour change: evidence from the Food4Me European randomized controlled trial. Int. J. Epidemiol. 2017;46:578–88.

[32] Nielsen DE, El-Sohemy A. A randomized trial of genetic information for personalized nutrition. Genes Nutr 2012;7:559–66.

[33] Nielsen DE, Shih S, El-Sohemy A. Perceptions of genetic testing for personalized nutrition: a randomized trial of DNA-based dietary advice. J. Nutrigenet. Nutrigenomics 2014;7:94–104.

[34] Falchi M, El-Sayed Moustafa JS, Takousis P, et al. Low copy number of the salivary amylase gene predisposes to obesity. Nat. Genet. 2014;46:492–7.

[35] Mejía-Benítez MA, Bonnefond A, Yengo L, et al. Beneficial effect of a high number of copies of salivary amylase AMY1 gene on obesity risk in Mexican children. Diabetologia 2015;58:290–4.

[36] Bonnefond A, Yengo L, Dechaume A, et al. Relationship between salivary/pancreatic amylase and body mass index: a systems biology approach. BMC Med 2017;15:37.

[37] Yong RY, Mustaffa SB, Wasan PS, et al. Complex copy number variation of AMY1 does not associate with obesity in two East Asian cohorts. Hum. Mutat. 2016;37:669–78.

[38] Rukh G, Ericson U, Andersson-Assarsson J, Orho-Melander M, Sonestedt E. Dietary starch intake modifies the relation between copy number variation in the salivary amylase gene and BMI. Am. J. Clin. Nutr. 2017;106:256–62.

[39] Heianza Y, Sun D, Wang T, et al. Starch digestion-related amylase genetic variant affects 2-year changes in adiposity in response to weight-loss diets: the POUNDS lost trial. Diabetes 2017;66:2416–23.

[40] Antúnez-Ortiz DL, Flores-Alfaro E, Burguete-García AI, et al. Copy number variations in candidate genes and intergenic regions affect body mass index and abdominal obesity in Mexican children. BioMed Res. Int. 2017. https://doi.org/10.1155/2017/2432957.

[41] Rafiee M, Sotoudeh G, Djalali M, et al. Dietary ω-3 polyunsaturated fatty acid intake modulates impact of insertion/deletion polymorphism of ApoB gene on obesity risk in type 2 diabetic patients. Nutrition 2016;32:1110–5.

[42] Roth CL, Hinney A, Schur EA, Elfers CT, Reinehr T. Association analyses for dopamine receptor gene polymorphisms and weight status in a longitudinal analysis in obese children before and after lifestyle intervention. BMC Pediatr 2013;13:197.

[43] Cuevas-Sierra A, Ramos-Lopez O, Riezu-Boj JI, Milagro FI, Martinez JA. Diet, gut microbiota and obesity: links with host genetics and epigenetics and potential applications. Adv. Nutr 2019. https://doi.org/10.1093/advances/nmy078.

[44] Davenport ER, Cusanovich DA, Michelini K, Barreiro LB, Ober C, Gilad Y. Genome-wide association studies of the human gut microbiota. PLoS One 2015;10:e0140301.

[45] Li J, Fu R, Yang Y, et al. A metagenomic approach to dissect the genetic composition of enterotypes in Han Chinese and two Muslim groups. Syst. Appl. Microbiol. 2018;41:1–12.

[46] Lim MY, You HJ, Yoon HS, et al. The effect of heritability and host genetics on the gut microbiota and metabolic syndrome. Gut 2017;66:1031–8.

[47] Goodrich JK, Davenport ER, Clark AG, Ley RE. The relationship between the human genome and microbiome comes into view. Annu. Rev. Genet. 2017;51:413–33.

[48] Almon R, Álvarez-León EE, Serra-Majem L. Association of the European lactase persistence variant (LCT-13910 C>T polymorphism) with obesity in the Canary Islands. PLoS One 2012;7:e43978.

[49] Manco L, Dias H, Muc M, Padez C. The lactase -13910C>T polymorphism (rs4988235) is associated with overweight/obesity and obesity-related variables in a population sample of Portuguese young adults. Eur. J. Clin. Nutr. 2017;71:21–4.

[50] Heianza Y, Sun D, Ma W, et al. Gut-microbiome-related LCT genotype and 2-year changes in body composition and fat distribution: the POUNDS lost trial. Int. J. Obes. 2018;42:1565–73.

[51] Ussar S, Griffin NW, Bezy O, et al. Interactions between gut microbiota, host genetics and diet modulate the predisposition to obesity and metabolic syndrome. Cell Metab 2015;22:516–30.

[52] Kreznar JH, Keller MP, Traeger LL, et al. Host genotype and gut microbiome modulate insulin secretion and diet-induced metabolic phenotypes. Cell Rep 2017;18:1739–50.

[53] Parks BW, Nam E, Org E, et al. Genetic control of obesity and gut microbiota composition in response to high-fat, high-sucrose diet in mice. Cell Metab 2013;17:141–52.

[54] Vimaleswaran KS, Bodhini D, Lakshmipriya N, et al. Interaction between FTO gene variants and lifestyle factors on metabolic traits in an Asian Indian population. Nutr. Metab. 2016;13:39.

[55] Corella D, Arregui M, Coltell O, et al. Association of the LCT-13910C>T polymorphism with obesity and its modulation by dairy products in a Mediterranean population. Obesity 2011;19:1707–14.

[56] Memisoglu A, Hu FB, Hankinson SE, et al. Interaction between a peroxisome proliferator-activated receptor gamma gene polymorphism and dietary fat intake in relation to body mass. Hum. Mol. Genet. 2003;12:2923–9.

[57] Mansego ML, De Marco G, Ivorra C, et al. The nutrigenetic influence of the interaction between dietary vitamin E and TXN and COMT gene polymorphisms on waist circumference: a case control study. J. Transl. Med. 2015;13:286.

[58] Junyent M, Parnell LD, Lai CQ, et al. ADAM17_i33708A>G polymorphism interacts with dietary n-6 polyunsaturated fatty acids to modulate obesity risk in the genetics of lipid lowering drugs and diet network study. Nutr. Metabol. Cardiovasc. Dis. 2010;20:698–705.

[59] Joffe YT, van der Merwe L, Carstens M, et al. Tumor necrosis factor-alpha gene -308 G/A polymorphism modulates the relationship between dietary fat intake, serum lipids, and obesity risk in black South African women. J. Nutr. 2010;140:901–7.

[60] Sánchez-Moreno C, Ordovás JM, Smith CE, Baraza JC, Lee YC, Garaulet M. APOA5 gene variation interacts with dietary fat intake to modulate obesity and circulating triglycerides in a Mediterranean population. J. Nutr. 2011;141:380–5.

[61] Domínguez-Reyes T, Astudillo-López CC, Salgado-Goytia L, et al. Interaction of dietary fat intake with APOA2, APOA5 and LEPR polymorphisms and its relationship with obesity and dyslipidemia in young subjects. Lipids Health Dis 2015;14:106.

[62] Doo M, Won S, Kim Y. Association between the APOB rs1469513 polymorphism and obesity is modified by dietary fat intake in Koreans. Nutrition 2015;31:653–8.

[63] Smith CE, Tucker KL, Arnett DK, et al. Apolipoprotein A2 polymorphism interacts with intakes of dairy foods to influence body weight in 2 U.S. populations. J. Nutr. 2013;143:1865–71.

[64] Zhang X, Qi Q, Zhang C, et al. FTO genotype and 2-year change in body composition and fat distribution in response to weight-loss diets: the POUNDS LOST trial. Diabetes 2012;61:3005–11.

[65] Stocks T, Ängquist L, Hager J, et al. TFAP2B -dietary protein and glycemic index interactions and weight maintenance after weight loss in the DiOGenes trial. Hum. Hered. 2013;75:213–9.

[66] Goni L, Cuervo M, Milagro FI, Martínez JA. Gene-gene interplay and gene-diet interactions involving the MTNR1B rs10830963 variant with body weight loss. J. Nutrigenet. Nutrigenomics 2014;7:232–42.

[67] Razquin C, Martinez JA, Martinez-Gonzalez MA, Fernández-Crehuet J, Santos JM, Marti A. A Mediterranean diet rich in virgin olive oil may reverse the effects of the -174G/C IL6 gene variant on 3-year body weight change. Mol. Nutr. Food Res. 2010;54(Suppl. 1):S75–82.

[68] Qi Q, Bray GA, Smith SR, Hu FB, Sacks FM, Qi L. Insulin receptor substrate 1 gene variation modifies insulin resistance response to weight-loss diets in a 2-year randomized trial: the Preventing Overweight Using Novel Dietary Strategies (POUNDS LOST) trial. Circulation 2011;124:563–71.

[69] Xu M, Qi Q, Liang J, et al. Genetic determinant for amino acid metabolites and changes in body weight and insulin resistance in response to weight-loss diets: the Preventing Overweight Using Novel Dietary Strategies (POUNDS LOST) trial. Circulation 2013;127:1283–9.

[70] Heni M, Herzberg-Schäfer S, Machicao F, Häring HU, Fritsche A. Dietary fiber intake modulates the association between variants in TCF7L2 and weight loss during a lifestyle intervention. Diabetes Care 2012;35:e24.

[71] Garaulet M, Smith CE, Hernández-González T, Lee YC, Ordovás JM. PPARγ Pro12Ala interacts with fat intake for obesity and weight loss in a behavioural treatment based on the Mediterranean diet. Mol. Nutr. Food Res. 2011;55:1771–9.

[72] Goni L, Riezu-Boj JI, Milagro FI, et al. Interaction between an ADCY3 genetic variant and two weight-lowering diets affecting body fatness and body composition outcomes depending on macronutrient distribution: a randomized trial. Nutrients 2018. https://doi.org/10.3390/nu10060789.

[73] Heianza Y, Ma W, Huang T, et al. Macronutrient intake-associated FGF21 genotype modifies effects of weight-loss diets on 2-year changes of central adiposity and body composition: the POUNDS lost trial. Diabetes Care 2016;39:1909–14.

F—Precision interventions

Chapter 41

Reproductive medicine involving genome editing: its clinical and social conundrums

Tetsuya Ishii

Office of Health and Safety, Hokkaido University, Sapporo, Hokkaido, Japan

Introduction

Germ cells and somatic cells are the two fundamental cell types that constitute multicellular organisms, including humans. Since the dawn of genetic engineering in the 1960s, the potential of genetic modification in human germ cells and embryos (collectively termed the germline) has been suggested [1]. In 1982, the US President's Commission for the study of ethical issues, related to genetic engineering with human beings, advised that

> *it would be wise to have engaged in careful prior thought, about steps such as treatments that can lead to heritable changes in human beings, or those intended to enhance human abilities, rather than simply correct deficiencies caused by well-defined genetic disorders in patients [2].*

However, conventional genetic engineering methodologies, including transgenesis, are generally inefficient and imprecise, thus rendering germline genetic modification unrealistic. Moreover, in terms of the reproductive use of germline genetic modification, controversy has arisen due to the potential transfer of genetic modifications to future generations. The major points in dispute thus include concerns over the safety and welfare of subsequent generations, changes in the nature of human reproduction and parent–child relationships, exacerbation of prejudice against persons with disabilities, and the potential misuse of genetic enhancement [3]. Against this backdrop, some countries have legally prohibited the clinical use of germline genetic modification [4].

Compared with the older methods of genetic engineering, the newer genome editing techniques being adopted worldwide are more efficient, precise, and versatile. Major genome editing techniques involve the use of zinc finger nucleases, transcription activator-like effector nucleases, and clustered regularly interspaced short palindromic repeat (CRISPR)-Cas9 nucleases, which utilize designable bacterial DNA cutting enzymes (nucleases) [5,6]. To develop novel modalities for patients with cancers or genetic diseases, clinical trials of genome editing in somatic cells are ongoing, primarily in the United States and China [7].

Regarding germline genome editing (GGE), the first report in 2015 demonstrated the potential of CRISPR-Cas9 to correct a pathogenic mutation in human one cell–stage embryos (zygotes). However, this report also revealed technical issues: failed gene correction, mosaicism, and off-target mutations [8]. Despite the rapid advances, subsequent GGE reports also revealed some limitations and risks [9–16]. Such basic research on GGE has been focused on the medically beneficent goal, of preventing genetic disease at the prenatal stage. However, it has also sparked tremendous debate worldwide, as observed in the previous controversies surrounding germline genetic modification.

At least 61 academies and other organizations, such as the US National Academies of Sciences, Engineering, and Medicine (NASEM), the UK Nuffield Council of Bioethics (NCB), and the Science Council of Japan, have released relevant statements, primarily regarding whether it is acceptable to modify the germline genome, so that the genetic modification is inheritable by future generations [17].

Meanwhile, fertility clinics in nine countries have performed several techniques that modify the composition of the mitochondrial genome in oocytes or zygotes, by transferring cytoplasm (including mitochondria) from a donor egg (oocyte) [18]. Such cases were primarily intended to help specific infertile patients have genetically related children, without the direct use of oocytes donated from a third party. However, such practices led to several

Precision Medicine for Investigators, Practitioners and Providers. https://doi.org/10.1016/B978-0-12-819178-1.00041-1

adverse events, inciting new regulations in the United States and China.

On the other hand, the United Kingdom legalized a form of germline mitochondrial genome modification (called mitochondrial donation) in 2015, which led to nothing worse than genetically related children free from serious mitochondrial disease [19]. In November 2018, a Chinese researcher claimed that his laboratory had applied CRISPR technology to human embryos, which were later delivered as healthy twin girls with resistance to HIV infection [20]. Eventually, this first case drew stringent criticism, for threatening the offspring's health, by using experimental reproductive techniques for such a minor reason [21].

Importantly, the reproductive use of GGE does not depend on gamete donation. The current regulatory landscape suggests that GGE will likely be practiced in some countries other than China [4]. Still, a number of clinical and social conundrums remain.

Genome editing

Among genome editing tools, CRISPR-Cas9 has distinct advantages in terms of the ease of preparing single guide RNA (sgRNA), and the greater utility for multiplex genome editing, in which several sites across the genome are simultaneously modified by using separate sgRNAs [6]. As such, CRISPR-Cas9 has spread rapidly among laboratories worldwide. Below, we explain the general genome editing procedures by taking CRISPR-Cas9 as an example.

The CRISPR-Cas9 genome system uses a nuclease derived from *Streptococcus pyogenes* that binds to a target DNA sequence through RNA-DNA interaction and cuts the DNA between the target sequence and a trinucleotide motif (NGG). To improve the utility of genetic engineering, the CRISPR RNA (crRNA) and trans-activating crRNA of bacterial origin, together with a targeting RNA sequence, are assembled into an sgRNA. To realize a specific gene modification, several target sequences (20 bp) whose 3′ terminal leads to a trinucleotide motif (NGG) are meticulously selected within a gene of interest, while minimizing any high homology with other sequences in a reference genome.

After the target sequences are determined, sgRNAs that include an RNA sequence to bind a target DNA sequence are produced. Then, each sgRNA (several different sgRNAs are used when conducting multiplex genome editing) is introduced, along with Cas9 nucleases, into cells. Inside cells, the Cas9 nucleases and sgRNA form a complex, then bind to a target DNA sequence via the targeting RNA included in the sgRNA. Then, DNA double strands are broken specifically at the target site. Although DNA double-strand breaks (DSBs) are toxic to the cells, they are repaired in some surviving cells.

In other cells, an error-prone repair process efficiently leaves insertion or deletion (indel) mutations, which can disrupt a gene of interest through frameshift mutagenesis. This genetic modification process is called nonhomologous end joining (NHEJ). By introducing a DNA template, along with the designed nucleases, a gene can be inserted or a mutation can be corrected at target sites. This process is called homology-directed repair (HDR). HDR is less efficient than NHEJ, and unsuccessful HDR leaves indel mutations.

Recently, base editing technology was developed by integrating DNA base-converting enzymes into a CRISPR system, with deactivated Cas9 (dCas9) [22]. The base editing can efficiently convert a specific DNA base into another base, without breaking DNA double strands. In addition to systems that target gene modifications in the nuclear genome, a genome editing system targeting mitochondria is available [23], despite numerous mitochondrial DNA (mtDNA) copies existing in the cells (200,000−300,000 mtDNA copies per one human oocyte) [24].

Although such genome editing techniques allow efficient and versatile genetic modifications, there remain some technical issues. Chief among them is the potential for off-target effects, on nontarget sites in the genome (Fig. 41.1) [25]. Cas9 nucleases could induce DSBs at nontarget sites, for reasons such as the improper selection of a target sequence. Subsequently, off-target DSBs could induce large-scale genomic alternations such as translocations, inversions, and large deletions, in addition to small indels of various lengths, including point mutations.

CRISPR-Cas9 technology and p53

Recently, another issue has arisen in relation to P53-dependent responses to DSBs. Wild-type p53 inhibits gene modification using CRISPR-Cas9 in human pluripotent stem cells (hPSCs) [26]. However, because hPSCs can acquire de novo P53 mutations, some hPSC lines in which a target gene was modified via CRISPR-Cas9 treatment may have originally had p53 mutations, and these lines could be used for stem cell therapy. Therefore, unintended genetic modifications are possible, which could lead to safety issues in a clinical setting.

Basic research on human germline genome editing

To date, basic research on human GGE has been conducted primarily in China (Table 41.1). CRISPR-Cas9 is frequently used in basic research to test the feasibility of GGE. At this level of research, such modified embryos are subject to genetic analysis, then disposed. They are never used to achieve a pregnancy through embryo transfer.

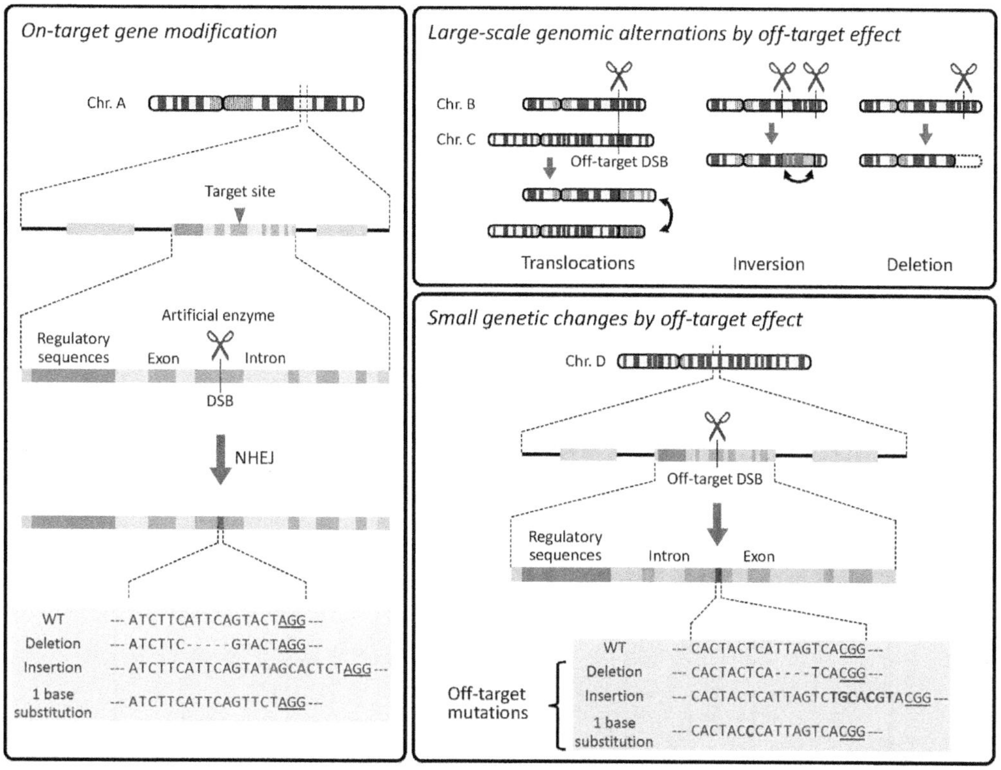

FIGURE 41.1 On-target gene modification and off-target effects in genome editing. Genome editing can efficiently modify a DNA sequence at a target site. Simultaneously, the nucleases may cause off-target effects across the genome. Specifically, the nucleases can create DSBs at nontarget sites. Although off-target DSBs could induce chromosome translocations (by creating concurrent DSBs at two loci), inversions and large deletions, such large-scale genomic alterations, can be readily detected. In contrast, off-target mutations could lead to small insertions or deletions (indels) of various lengths, including point mutations, which may be found in the exons, introns, regulatory regions, or at other locations. *Reuse with permission from Araki M, Ishii T. Trends Biotechnol February 2016;34(2):86–90.*

Except for a report from the United Kingdom that investigated the developmental role of *OCT4* in human embryos [27], the remaining nine assessed the feasibility of using GGE ultimately to prevent genetic disease, by treating individuals at the prenatal stage. The targeted genes were *HBB*, which is responsible for beta thalassemia (autosomal recessive disease) [8,10,12,14], *CCR5*, which is associated with resistance to HIV infection (biallelic mutations confer the resistance) [9], *G6PD*, which is responsible for glucose-6-phosphate dehydrogenase deficiency (X-linked recessive disease) [10], *MYBPC3*, which is responsible for hypertrophic cardiomyopathy (autosomal dominant disease) [11], *FANCF*, which is responsible for Fanconi anemia (a largely autosomal recessive disease) [14], *DNMT3B*, which is responsible for immunodeficiency, centromere instability, and facial anomalies (ICF) syndrome (autosomal recessive disease) [14], *RNF2*, which is associated with such conditions as Angelman syndrome (imprinted mode of maternal inheritance of mutations) [15], and *FBN1*, which is responsible for Marfan syndrome (an autosomal dominant disease) [16]. All these genes are located in the nuclear genome.

In the 2005 report that was the first to use genome editing in human embryos [8], Cas9 mRNA and gRNA were, along with a repair template DNA, microinjected into abnormally fertilized zygotes (3 pronuclei: 3PN), that were in vitro fertilization (IVF) by-products destined for disposal. The research results showed three defects. First, the efficiency of *HBB* modification via HDR was low (5.6%). Second, the resultant embryos were genetically mosaic: embryonic cells, with the corrected gene coexisting with those with the uncorrected gene. In a clinical setting, those two defects could result in genetic diseases (particularly autosomal dominant diseases). Third, off-target indel mutations were observed in the modified embryos. This defect could lead to the birth of children affected with unexpected conditions, or to the later development of diseases in adulthood.

The aforementioned poor results might have arisen due to the use of 3PN zygotes. In subsequent studies, therefore, Cas9 protein was injected into 2PN zygotes or oocytes along with a spermatozoon (10–12). Those efforts improved the issues of off-target mutations and/or mosaicism (Table 41.1). Notably, a report by Ma et al. described

TABLE 41.1 Reports of basic research on genome editing in the human germline (as of October 29, 2018).

Report by	From	Purpose	Targeted gene	Injection of	Into	Off-target mutations	Mosaicism
Liang et al. 2015.	China	Disease prevention	*HBB*	Cas9 mRNA + sgRNA + DNA template	Discarded zygotes (3PN)	Yes	Yes
Kang et al. 2016.	China	Disease prevention	*CCR5*	Cas9 mRNA + sgRNA + DNA template	Discarded zygotes (3PN)	Not detected	Yes
Tang et al. 2017.	China	Disease prevention	*HBB, G6PD*	Cas9 protein + sgRNA + DNA template	Newly created zygotes (2PN)	Not detected	Yes
Ma et al. 2017.	United States, South Korea, China	Disease prevention	*MYBPC3*	Cas9 protein + sgRNA + DNA template + spermatozoon	Oocytes	Not detected	Not detected
Tang et al. 2018.	China	Disease prevention (For reproducibility of Ma et al. 2017)	*MYBPC3, HBB*	Cas9 protein + sgRNA + DNA template	Discarded zygotes (3PN)	Yes	N.D.
Liang et al. 2017.	China	Disease prevention	*HBB*	YEE-BE3 mRNA + sgRNA	Enucleated oocytes, nuclear transfer zygotes	Not detected	Yes
Zhou et al. 2017.	China	Disease prevention	*HBB, FANCF, DNMT3B*	BE3 or SaKKH-BE3 mRNA + sgRNA	Discarded zygotes (3PN)	Yes	N.D.
Li et al. 2017.	China	Disease prevention	*RNF2*	BE3 mRNA + sgRNA	Discarded zygotes (3PN)	Yes	N.D.
Zheng et al. 2018.	China	Disease prevention	*FBN1*	BE3 mRNA + sgRNA	Newly created zygotes	Not detected	N.D.
Forgaty et al. 2017.	United Kingdom	Embryology	*OCT4*	Cas9 protein + sgRNA	Surplus zygotes (2PN)	Not detected	Yes

BE, base editor; *N.D.*, not determined; *PN*, pronuclei.

successful gene correction, without detectable mosaicism or off-target mutations, through simultaneous injection of a spermatozoon with MYBPC3 mutation, and only Cas9 protein and sgRNA (no DNA templates), into an oocyte without the *MYBPC3* mutation [11].

The authors claimed that some blastomeres from Cas9-treated embryos were genetically corrected, using a wild-type maternal allele. However, two other groups subsequently rebutted the claim [28,29]. The rebuttals asserted that blastomeres in which both alleles matched the wild-type should have, in fact, only the maternal allele, due to the large-scale deletion including the paternal allele. In turn, Ma et al. made counterarguments, based on their extended testing of the remaining blastomere samples, and suggested that human embryos have the potential for nonmeiotic homologous chromosome–based DNA repair [30]. Meanwhile, Tang et al. confirmed the DNA template-mediated gene correction using 3PN zygotes [12]. Further research will be needed to settle these disputes.

Non-CRISPR-Cas9 methods

Base editor (BE) systems have also been tested to correct a point mutation in 3PN and 2PN embryos (Table 41.1). Of note, BE systems that include rat cytidine deaminase induce a base conversion (cytidine to thymidine), without breaking double-strand DNA, and this mechanism differs from that of genome editing tools such as CRISPR/Cas9 [22]. In the three relevant reports, mRNA of BE3 or modified BE3 was injected in nuclear transfer zygotes, as well as 3PN and 2PN zygotes [13–15]. However, off-target mutations or mosaicism were observed in those reports (Table 41.1).

In 2018, Zheng et al. reported a highly efficient correction (89%) of $FBN1^{T7498C}$, a point mutation responsible for Marfan syndrome. Moreover, they showed that no off-target base conversions or indels were detected, in any of the tested sites of cell samples, by deep sequencing combined with whole-genome sequencing analysis [16]. Nonetheless, the possibility that only biopsied cell samples contained undetectable off-target mutations could not be ruled out.

In vitro spermatogenesis

With regard to genome editing of the male germline, some rodent experiments demonstrated that specifically disrupted or corrected genes of spermatogonial stem cells (SSCs) are heritable by the resultant neonates [31]. However, they require transplantation into the seminiferous tubules of animal or human testes, which may be clinically risky. In vitro spermatogenesis is still premature even in animal experiments [32], and no comparable works have been reported in human SSCs. A preliminary experiment in which the CRISPR-Cas9 plasmid was directly transfected into human spermatozoa was recently reported in an academic meeting [33]. In general, target-gene modifications in the zygotes or oocytes have been focused, with improved conditions of genome editing or changes in genome editing systems. There has been no consensus regarding the potential for off-target effects in the medical applications of genome editing [34].

The first reproduction using germline genome editing

The major procedure used in human GGE is currently microinjection of artificial DNA-modifying enzyme mRNA or protein, into zygotes or oocytes (Table 41.1). This procedure is feasible at most fertility clinics that use micromanipulators for a common technique, intracytoplasmic sperm injection (ICSI).

In 2018, a Chinese researcher claimed that twin girls were born healthy via GGE, despite the absence of relevant accounts in peer-reviewed journals [20]. At the Second International Summit on Human Genome Editing, Jiankui He of the Southern University of Science and Technology of China presented a lecture entitled "CCR5 gene editing in mouse, monkey and human embryos using CRISPR/Cas9." This GGE study sought to confer HIV resistance to the resultant progeny, by intentionally disrupting the CCR5 genes via NHEJ, not by correcting a pathogenic mutation via HDR.

He pointed out that children with HIV-positive parents are at a high risk of HIV infection and can often expect to be discriminated against. He also stressed the scientific rationale of his reproductive study, explaining that some people in Northern European populations have a naturally occurring CCR5 mutation (CCR5Δ32), which confers resistance to HIV infection [35]. He created CCR5-disrupted mice, by microinjecting CCR5-targeting CRISPR-Cas9 into mouse zygotes. These mice were dissected and confirmed to have no apparent abnormalities in several tissues.

Subsequently, he examined the proximity to Δ32 in the CCR5 gene and identified one promising sequence of 20nt that did not exhibit high homology with any other regions in a reference genome. The fidelity of the pair of a Cas9 and a sgRNA that binds to the selected sequence was confirmed, using monkey zygotes and human cell lines. To achieve a high efficiency of CCR5-targeted modification and reduce mosaicism, the conditions including the forms of Cas9 (mRNA and protein), injection timing, and dose of enzymes were optimized using human zygotes donated by infertile patients. To assess the potential risks of off-target effects, Cas9-sgRNA was injected into human zygotes, and the resultant blastocysts were subject to two different analyses, using whole genome sequencing. One was an off-target analysis of several cells, biopsied from some of the modified blastocysts (so-called preimplantation genetic diagnosis, PGD), and the other was an off-target analysis of embryonic stem cells, established from the remaining modified blastocysts.

One off-target deletion mutation (6 kbp) was found. However, it was considered not to affect other gene functions. Although another Chinese group also reported that they detected no off-target mutations in CCR5-modified human 3PN embryos, they did observe mosaicism in the embryos (Table 41.1) [9]. His data appeared to be more convincing, because of the meticulously optimized GGE conditions for reducing the risks of off-target mutations and mosaicism.

Couples were preliminarily recruited, each of them (eight couples) consisting of an HIV-positive husband and HIV-negative wife. An outline of the genomic analyses, of the couple who had twins via GGE, is shown below. After removing HIV by sperm washing, one spermatozoon, along with the CCR5-targeting Cas9 protein-sgRNA, was microinjected into one oocyte. Using Sanger sequencing, whole genome sequencing, MiSeq (sequencing for targeted genes), and deep sequencing, he rigorously analyzed many samples that were obtained from modified preimplantation embryos, developing fetuses, and born offspring, in addition to the eight enrolled couples (Table 41.2).

He sought to obtain genomic data of genetically modified subjects as precisely as possible, by comparing sequencing data from the subjects with that of parents. Consequently, one twin (fictitious name, Lulu) had CCR5 WT/-15bp, whereas the other twin (Nana) had CCR5-14bp/+1bp, which was consistent with the genomic data at PGD and prenatal testing. Moreover, both twins had no "detectable" off-target mutations in "some tested tissues." Follow-up of the twins for 17 years, and experiments to assess resistance to HIV infection in immune cells, was planned. On January 2019, the Chinese government confirmed that CCR-5-modified children were born in this GGE research and deemed that Jiankui He had violated Chinese legislation [36]. It was reported that other medical institutes will perform the follow-up of the twins because he can no longer perform GGE research in China.

Clinical conundrums

Three key points can be highlighted in the clinical implementation of GGE: assessment of genetic modifications before embryo transfer, prenatal testing, and long-term follow-up. Those points are all vital to ensure the birth of children, for whom prospective parents desire GGE, and their subsequent health. There are clinical conundrums in each stage.

Embryo testing

Before the transfer of embryos to the uterus, modified via genome editing, PGD should be performed to determine that a target gene was modified as expected, and no significant off-target mutations were induced [37]. A sufficient number of modified embryos—possibly 10 according to the relevant basic research—must be subjected to PGD, to allow for the meticulous selection of one or more modified embryos to transfer. However, this depends on the ovarian response or reserve of the prospective mother, after hormonal stimulation for oocyte retrieval. Some embryos are likely to die due to cytotoxicity of DNA-modifying enzymes, which reduces the number of embryos for transfer. Indeed, the first GGE basic research showed that 17.4% of injected 3PN zygotes died [8].

As Jiankui He demonstrated, all cells of modified embryos can be subjected to PGD, to rigorously confirm on-target and off-target gene modifications and whether modified embryos are genetically mosaic. However, it would be impossible to conduct such exhaustive testing in a clinical setting, because the embryos would be destroyed in the process. Recently, fertility clinics performing PGD have shown a preference for blastocyst biopsy (sampling several cells from the trophectoderm of a blastocyst, at day 5 postfertilization), over blastomere biopsy (sampling one cell taken from a cleavage-stage embryo at day 3 post-fertilization), because the former technique provides more DNA, and inflicts less damage on the embryos.

As in the experiments by He, blastocyst biopsy is likely to be adopted in the clinical implementation of GGE, to effectively confirm the correction of a pathogenic mutation at a target gene. Notably, it is common for the genome of human early embryos to be unstable [38]. This may lead to mosaicism in the embryos, potentially affecting the accuracy of PGD. Blastocysts can be chromosomally mosaic among trophectoderm cells, and between trophectoderm cells and inner cells that develop into a fetus body post-implantation [37]. The use of inappropriately designed nucleases can induce several mutated bases (indel mutations and point mutations), in addition to large-scale genomic alternations (Fig. 41.1).

Investigating "potential" off-target sites is one of the possible approaches; however, this approach addresses only hot spots of off-target mutations in the genome. Although whole genome or exome sequencing is applicable, it cannot be error-prone due to the amplification of DNA samples. In addition, it is likely to be difficult to distinguish such small mutations, from single nucleotide polymorphisms (SNPs). As demonstrated by He in the first GGE case (Table 41.2), a comparison of the modified embryo's genome with the

TABLE 41.2 The outline of genome analyses in the first clinical application of germline genome editing to disrupt CCR5 genes.

Subject	An HIV-positive husband and an HIV-negative wife (fictitious names: Mark and Grace, respectively).	Preimplantation blastocysts after genome editing.	Twin fetuses.	Twin female infants (Lulu and Nana).
Sample	Peripheral blood genomes.	Genomes of 3–5 cells biopsied from the trophectoderm.	Fetal cell-free DNA in maternal blood in 12, 19, and 24 weeks of pregnancy.	Genomes from cord blood, umbilical cord tissue, and placenta.
Genetic analysis	SS for CCR5 and 30×WGS.	SS for CCR5 and 30×WGS.	MiSeq targeted sequencing, and ultra deep (40,000×) sequencing on 609 cancer-associated genes.	SS for CCR5, MiSeq-targeted sequencing, and 100×WGS.
Results	Both parents had CCR5 WT/WT. With parental genomes, de novo indel mutations, haplotypes for higher sensitivity, and personalized off-target hot spots were gained.	Data suggested: two of four blastocysts had CCR5 WT/-15bp and -4bp/+1bp, and one of the two embryos had one potential intergenic off-target deletion mutation.	Data on CCR5 genes were identical to those of preimplantation embryos. No detectable off-target mutations found. No detectable novel oncogene variants.	CCR5 gene: WT/-15bp (Lulu) and -4bp/+1bp (Nana). No large deletions and detectable off-target mutations found in tested tissue samples.

SS, Sanger sequencing; *WGS*, whole genome sequencing. For further details, see presentation on 28 November 2018 by He Jiankui at second International Summit on Human Genome Editing. http://nationalacademies.org/gene-editing/2nd_summit/second_day/index.htm (accessed 21 January 2019).

parental genomes could aid in identifying small off-target mutations and SNPs in the genome of modified embryos. However, this approach cannot rule out the presence of small off-target mutations that went undetected but could nevertheless affect the health of the offspring. Currently, there is no consensus regarding how to comprehensively assess the risk of off-target effects on the germline genome. Thus, the transfer of blastocysts modified via genome editing, inevitably involves some reproductive risks.

Fetal testing

When embryos modified via genome editing are successfully implanted, prospective parents may encounter unexpected consequences: miscarriages, stillbirths, and fetal abnormalities that prenatal genetic testing can find. In the first GGE case (Table 41.2), Jiankui He performed three separate analyses of fetal cell-free DNA, which was present in the maternal blood. The data given by He at his presentation, which was purportedly obtained by CCR5-targeted sequencing and deep sequencing at 609 cancer-associated genes, appeared to be reliable, despite the fact that it was never published.

However, He's analysis of off-target mutations was focused on cancer-associated genes, and thus left room for further investigation of off-target mutations during pregnancy. Fetal cell biopsy, such as amniocentesis and chorionic villus sampling, entails substantial risks of miscarriages (approximately 1/300). Even if further genetic testing using biopsied fetal cells is performed, the risks of overlooking small off-target mutations could remain to some extent. When prenatal testing finds adversely affected fetuses, they cannot be treated due to the presence of harmful mutations in all or most cells in the body. In such cases, parents who consented to GGE to protect their future children from serious disease may feel conflicted about undergoing an elective abortion, because aborting the affected fetus would contradict their intention to safeguard their child's health.

Follow-up of resultant children

Even for children born healthy after GGE, it would be necessary to perform health monitoring for a relatively long period of time. Small off-target mutations that PGD and subsequent prenatal testing overlooked might later impose side effects on the resultant children. Although such follow-up might be clinically needed for the entire life of an individual or, for an entire generation of individuals, decades-long prospective clinical trials are uncommon [39]. Moreover, if such long-term follow-up is deemed mandatory, it may infringe on the dignity, welfare, and privacy of families [39]. Investigators would thus need to perform post-GGE follow-up, while respecting the family's autonomy.

Unfortunately, this could result in parents withdrawing their consent to the follow-up of their children born via GGE.

There has been only one follow-up survey for the previous germline mitochondrial genome modifications; however, the power of this survey on resultant teenagers was limited, as it was an inquiry-based investigation, with no medical examinations [40]. In retrospect, some follow-up investigations post-ICSI have suggested the importance of birth defects, developmental outcomes, medical health, and reproductive health [41]. If a GGE study were performed to treat the infertility of parents, and included a follow-up to investigate the reproductive health of the resultant youth, the minimum required period for clinical follow-up post-GGE would be approximately two or three decades, which is longer than the follow-up Jiankui He planned. On the other hand, if GGE were performed to prevent the onset of genetic disease in offspring, the minimum required period for follow-up could be until the presumed timing of disease onset.

If GGE is permitted in some countries, physicians and parents should make a coordinated effort to ensure the health of the GGE-conceived offspring, while informing the children of the details of their conception, so that they understand the importance of follow-up. On becoming legally competent, the GGE-conceived youth should be entitled to refuse follow-up, if their conditions have been adequately examined. However, it is necessary to clarify whether the conditions include developmental outcomes, medical health, reproductive health, or all of the above.

Social conundrums

Two social surveys showed that approximately 60% of respondents accept the use of GGE for any medical purposes, whereas less than 30% of respondents accept its use for nonmedical purposes (genetic enhancement) [42,43]. In this context, it is important to note that the direct subjects of experimental GGE are unborn children, who cannot provide consent. Therefore, if the medical use of GGE were permitted, the prospective parents would need to consent to GGE, in the best interests of their future children. More specifically, in such a situation, the prospective parents could use GGE to bear genetically related children, with or without a trait that they desire, although they can, to some extent, fulfill his or her reproductive needs, by using gametes provided by a specific donor.

At the same time, gamete donation and donor conception are themselves controversial, even if those are legal. Oocyte donation in particular has raised ethical issues, including the potential for female exploitation, the commodification of eggs, and harm to female donors [44]. For the resultant families, because one intended parent is not genetically related to the donor-conceived child, there will likely be emotional conflict, regarding whether or not to disclose the fact of donor conception to the child [45], and a "resemblance talk," about the lack of physical similarity between the intended parent and the child [46].

Acceptable medical uses

The US NASEM report 2017 asserts that clinical use of GGE might be permitted only for compelling medical reasons, in the absence of reasonable alternatives, after the potential risks and benefits are clarified [47]. The UK NCB report 2018 concluded that GGE could be acceptable if it is intended to secure and be consistent with the welfare of the future person. It should not increase disadvantage, discrimination, or division in society [48]. In addition to those statements, the acceptance of mitochondrial donation in the United Kingdom suggests that reproductive use of GGE could be acceptable, for the purpose of having genetically related children free from a serious disease.

Suggested GGE-applicable cases include compelling cases, of high penetrance of a serious or life-threatening autosomal genetic disorder—causing mutation, where PGD is inapplicable, such as autosomal dominant diseases in which one or both parents are homozygous (e.g., Huntington's disease) and autosomal recessive diseases where both parents are homozygous (e.g., cystic fibrosis) [31]. However, prospective parents with these genetic backgrounds are vanishingly rare [49] and may be unwilling to build a family if they are seriously affected.

On the other hand, does "serious disease" include AIDS or HIV infection, which was the subject of the first GGE? Jiankui He's use of GGE for prenatally preventing HIV infection substantially differs from the use of GGE for prenatally preventing genetic disease, in that AIDS develops after infection by external HIV, not internal genetic mutations. That is, HIV infection is largely preventable through the exercise of safety precautions in sexual intercourse, and the avoidance of careless contact with the blood of HIV carriers. Nonetheless, the Chinese parents who underwent the first GGE presumably felt compelling need, to provide HIV resistance to their genetically related children, and considered that GGE use would secure the welfare of their children.

The first IVF was performed in 1978, for an infertile female suffering from complications of blocked fallopian tubes. Thus, this experimental reproductive technique was performed in a clinical setting, for a carefully selected infertility case with a clear etiology [50]. To date, it has been suggested that at least 18 genes in males and 29 genes in females are associated with infertility [51]. For treating infertility cases associated with a mutated gene, the reproductive use of GGE could be permissible, at least in Mexico [52]. However, it is a truism that there is no medicine without risks. Given that experimental GGE can

TABLE 41.3 Situations of regulation on the reproductive use of germline genetic modification in 39 select countries.[a]

Legal prohibition [24]	United States[b], Canada, Costa Rica, Brazil, Finland, Sweden, Lithuania, Bulgaria, Czech Republic, Germany, Denmark, the Netherlands, Belgium, Austria, Switzerland, Italy, France, Spain, Portugal, Australia, New Zealand, Republic of Korea, Singapore, Israel[c]
Legal prohibition, except mitochondrial donation [1]	United Kingdom[d]
Prohibition by guidelines [4]	Japan, China[e], India, Ireland
Ambiguous [10]	Russia, Iceland, Slovakia, Greece, South Africa, Chile, Mexico[f], Argentina, Peru, Colombia

[a]*For details, see Ishii T. Brief Funct Genomics January 2017;16(1):46–56. https://doi.org/10.1093/bfgp/elv053. For recent updates or issues, see below.*
[b]*Sec. 734, Consolidated Appropriations Act 2018 forbids FDA to spend federal budget for reviewing the clinical trial. This annual law has been renewed since 2016.*
[c]*Israel Amendment Law No. Three Prohibition on Genetic Intervention (Human Cloning and Genetic Change in Reproductive Cells), 5776/2016 (valid till May 23, 2020).*
[d]*Human Embryology and Fertilisation Act 2008, in principle, prohibits it. Human Fertilisation and Embryology (Mitochondrial Donation) Regulations 2015 permits mitochondrial donation.*
[e]*Ministry of Health: Technical Standards and Ethical Principles of Assisted Reproductive Technologies and Sperm Banks (2003). See Ishii T., Hibino Y. Repro Biomed Soc February 28, 2018;5:93–109.*
[f]*In Mexico, research on infertility treatment involving germline genetic modification is implicitly legal under Article 56, General Health Law 1984. See Ishii T. J Law Biosci August 2017;4(2):384–390.*

potentially subject offspring to a substantial risk of off-target effects, its use for infertility treatment, which will likely promote its widespread use, might be expected to ruin the welfare of some of the resultant children [37].

Necessity of clear regulation

The United Kingdom became the first country to permit the use of mitochondrial donation for disease prevention. Other germline genetic modifications and other uses of mitochondrial donation remain unlawful. A legal study has shown that 24 of 39 countries legally prohibit the clinical use of germline genetic modification, whereas 9 are ambiguous (Table 41.3) [4].

In China, germline genetic modification is prohibited by public guidelines for medical practitioners, not by laws defining social norms (Table 41.3). On the other hand, 24 countries categorically prohibit the clinical use of germline genetic modification, and thus the reproductive use of GGE would be impermissible in these countries (Table 41.3). Perhaps the most convincing reason for such a prohibitive approach is the irreversible impact of germline genetic modification on the health of resultant children and their descendants. However, some citizens of these countries who disagree will likely go abroad to seek GGE, as so-called reproductive tourists [53].

To respond to the shocking GGE report from China, the World Health Organization (WHO) is considering an expert panel, to discuss the implications of human GGE [54]. In the discussions at WHO and in individual countries, the acceptability of human GGE will ultimately depend on whether parental autonomy, over the GGE-based pursuit of genetically related children, outweighs the irreversible impact on the health, of the next and subsequent generations.

References

[1] Sinsheimer RL. The prospect for designed genetic change. Am. Sci 1969;57(1):134–42.

[2] US_President's_Commission_for_the_Study_of_Ethical_Problems_in_Medicine_and_Biomedical_and_Behavioral_Research. Splicing life: a report on the social and ethical issues of genetic engineering with human beings. 1982.

[3] Frankel MS, Chapman AR. Human inheritable genetic modifications. Assessing scientific, ethical, religious, and policy issues. American Association for the Advancement of Science; 2000. https://wwwaaasorg/sites/default/files/migrate/uploads/germline2pdf.

[4] Ishii T. Germ line genome editing in clinics: the approaches, objectives and global society. Brief. Funct. Genom 2017;16(1):46–56.

[5] Gaj T, Gersbach CA, Barbas 3rd CF. ZFN, TALEN, and CRISPR/Cas-based methods for genome engineering. Trends Biotechnol. 2013;31(7):397–405.

[6] Barrangou R, Doudna JA. Applications of CRISPR technologies in research and beyond. Nat. Biotechnol 2016;34(9):933–41.

[7] Cornu TI, Mussolino C, Cathomen T. Refining strategies to translate genome editing to the clinic. Nat. Med 2017;23(4):415–23.

[8] Liang P, Xu Y, Zhang X, Ding C, Huang R, Zhang Z, et al. CRISPR/Cas9-mediated gene editing in human tripronuclear zygotes. Protein Cell 2015;6(5):363–72.

[9] Kang X, He W, Huang Y, Yu Q, Chen Y, Gao X, et al. Introducing precise genetic modifications into human 3PN embryos by CRISPR/Cas-mediated genome editing. J. Assist. Reprod. Genet 2016;33(5):581–8.

[10] Tang L, Zeng Y, Du H, Gong M, Peng J, Zhang B, et al. CRISPR/Cas9-mediated gene editing in human zygotes using Cas9 protein. Mol. Genet. Genom 2017;292(3):525−33.

[11] Ma H, Marti-Gutierrez N, Park SW, Wu J, Lee Y, Suzuki K, et al. Correction of a pathogenic gene mutation in human embryos. Nature 2017;548(7668):413−9.

[12] Tang L, Zeng Y, Zhou X, Du H, Li C, Liu J, et al. Highly efficient ssODN-mediated homology-directed repair of DSBs generated by CRISPR/Cas9 in human 3PN zygotes. Mol. Reprod. Dev 2018;85(6):461−3.

[13] Liang P, Ding C, Sun H, Xie X, Xu Y, Zhang X, et al. Correction of beta-thalassemia mutant by base editor in human embryos. Protein Cell 2017;8(11):811−22.

[14] Zhou C, Zhang M, Wei Y, Sun Y, Sun Y, Pan H, et al. Highly efficient base editing in human tripronuclear zygotes. Protein Cell 2017;8(10):772−5.

[15] Li G, Liu Y, Zeng Y, Li J, Wang L, Yang G, et al. Highly efficient and precise base editing in discarded human tripronuclear embryos. Protein Cell 2017;8(10):776−9.

[16] Zeng Y, Li J, Li G, Huang S, Yu W, Zhang Y, et al. Correction of the Marfan syndrome pathogenic FBN1 mutation by base editing in human cells and heterozygous embryos. Mol. Ther. J. Am. Soc Gene Ther 2018;26(11):2631−7.

[17] Carolyn B. Do CRISPR germline ethics statements cut it? CRISPR J 2018;1(2):115−25.

[18] Ishii T. Reproductive medicine involving mitochondrial DNA modification: evolution, legality, and ethics. EMJ Repro. Health 2018;4(1):88−99.

[19] The human fertilisation and embryology (mitochondrial donation) regulations 2015. 2015. p. 572.

[20] Jiankui H. CCR5 gene editing in mouse, monkey and human embryos using CRISPR/Cas9. In: Session 3 human embryo editing at second international summit on human genome editing; 2018. http://nationalacademies.org/gene-editing/2nd_summit/second_day/index.htm.

[21] Shen A. Chinese scientists condemn 'crazy' and 'unethical' gene-editing experiment. South China Morning Post; 26 November 2018. www.scmp.com/news/china/society/article/2175105/chinese-scientists-condemn-crazy-and-unethical-gene-editing.

[22] Hess GT, Tycko J, Yao D, Bassik MC. Methods and applications of CRISPR-mediated base editing in eukaryotic genomes. Mol. Cell 2017;68(1):26−43.

[23] Gammage PA, Viscomi C, Simard ML, Costa ASH, Gaude E, Powell CA, et al. Genome editing in mitochondria corrects a pathogenic mtDNA mutation in vivo. Nat. Med. 2018;24(11):1691−5.

[24] Ishii T. Mitochondrial manipulation for infertility treatment and disease prevention. In: Schatten H, editor. Human reproduction: updates and new horizons. New Jersey: Wiley Blackwell; 2016. p. 205−30.

[25] Araki M, Ishii T. Providing appropriate risk information on genome editing for patients. Trends Biotechnol. 2016;34(2):86−90.

[26] Ihry RJ, Worringer KA, Salick MR, Frias E, Ho D, Theriault K, et al. p53 inhibits CRISPR-Cas9 engineering in human pluripotent stem cells. Nat. Med 2018;24(7):939−46.

[27] Fogarty NME, McCarthy A, Snijders KE, Powell BE, Kubikova N, Blakeley P, et al. Genome editing reveals a role for OCT4 in human embryogenesis. Nature 2017;550(7674):67−73.

[28] Egli D, Zuccaro MV, Kosicki M, Church GM, Bradley A, Jasin M. Inter-homologue repair in fertilized human eggs? Nature 2018;560(7717):E5−7.

[29] Adikusuma F, Piltz S, Corbett MA, Turvey M, McColl SR, Helbig KJ, et al. Large deletions induced by Cas9 cleavage. Nature 2018;560(7717):E8−e9.

[30] Ma H, Marti-Gutierrez N, Park S-W, Wu J, Hayama T, Darby H, et al. Ma et al. reply. Nature 2018;560(7717):E10−23.

[31] Ishii T. Germline genome-editing research and its socioethical implications. Trends Mol. Med 2015;21(8):473−81.

[32] Ishii T, Saitou M. Promoting in vitro gametogenesis research with a social understanding. Trends Mol. Med 2017;23(11):985−8.

[33] Spicer C. Scientists use CRISPR in human sperm cells. BioNews 957, 9 July 2018. www.bionews.org.uk/page_137024.

[34] Joung JK. Unwanted mutations: standards needed for gene-editing errors. Nature 2015;523(7559):158.

[35] Smolen-Dzirba J, Rosinska M, Janiec J, Beniowski M, Cycon M, Bratosiewicz-Wasik J, et al. HIV-1 infection in persons homozygous for CCR5-Delta32 allele: the next case and the review. AIDS Rev 2017;19(4):219−30.

[36] Xinhua A. Guangdong releases preliminary investigation result of gene-edited babies. January 2019. at http://www.xinhuanet.com/english/2019-01/21/c_137762633.htm.

[37] Ishii T. Reproductive medicine involving genome editing: clinical uncertainties and embryological needs. Reprod. Biomed. Online 2017;34(1):27−31.

[38] Vanneste E, Voet T, Le Caignec C, Ampe M, Konings P, Melotte C, et al. Chromosome instability is common in human cleavage-stage embryos. Nat. Med. 2009;15(5):577−83.

[39] Cwik B. Designing ethical trials of germline gene editing. N. Engl. J. Med. 2017;377(20):1911−3.

[40] Chen SH, Pascale C, Jackson M, Szvetecz MA, Cohen J. A limited survey-based uncontrolled follow-up study of children born after ooplasmic transplantation in a single centre. Reprod. Biomed. Online 2016;33(6):737−44.

[41] Pereira N, O'Neill C, Lu V, Rosenwaks Z, Palermo GD. The safety of intracytoplasmic sperm injection and long-term outcomes. Reproduction 2017;154(6):F61−f70.

[42] Scheufele DA, Xenos MA, Howell EL, Rose KM, Brossard D, Hardy BWUS. Attitudes on human genome editing. Science 2017;357(6351):553−4.

[43] McCaughey T, Sanfilippo PG, Gooden GE, Budden DM, Fan L, Fenwick E, et al. A global social media survey of attitudes to human genome editing. Cell Stem Cell 2016;18(5):569−72.

[44] Baylis F. The ethics of creating children with three genetic parents. Reprod. Biomed. Online 2013;26(6):531−4.

[45] Frith L, Blyth E, Crawshaw M, van den Akker O. Secrets and disclosure in donor conception. Sociol. Health Illness 2018;40(1):188−203.

[46] Becker G, Butler A, Nachtigall RD. Resemblance talk: a challenge for parents whose children were conceived with donor gametes in the US. Soc. Sci. Med. 2005;61(6):1300−9.

[47] NASEM. National academies of science, engineering, and medicine (NASEM). In: Human genome editing: science, ethics, and governance. The National Academies Press; 2017. https://www.nap.edu/catalog/24623/human-genome-editing-science-ethics-and-governance.

[48] Nuffield NCB. Council on bioethics (NCB), "Genome editing and human reproduction: social and ethical issues". at http://nuffieldbioethics.org/project/genome-editing-human-reproduction, 2018.

[49] Lander ES. Brave new genome. N. Engl. J. Med. 2015;373(1):5–8.

[50] Johnson MH, Elder K. The Oldham notebooks: an analysis of the development of IVF 1969-1978. IV. Ethical aspects. Reprod. Biomed. Soc. Online 2015;1(1):34–45.

[51] Harper JC, Aittomaki K, Borry P, Cornel MC, de Wert G, Dondorp W, et al. Recent developments in genetics and medically assisted reproduction: from research to clinical applications. Eur. J. Hum. Genet. 2018;26(1):12–33.

[52] Ishii T. Peer commentary: mitochondrial replacement techniques and Mexico's rule of law: on the legality of the first maternal spindle transfer case. J. Law Biosci. 2017;4:384–90.

[53] Ishii T, Hibino Y. Mitochondrial manipulation in fertility clinics: regulation and responsibility. Reprod. Biomed. Soc. Online 2018;5:93–109.

[54] WHO. WHO establishing expert panel to develop global standards for governance and oversight of human gene editing. 14 December 2018. www.who.int/ethics/topics/gene-editing/en/.

Chapter 42

Safety and efficacy of guided biopsy

Miguel A. Bergero[1] and Pablo F. Martinez[2]

[1]*Urology, Sanatorio Privado San Geronimo, Santa Fe, Argentina;* [2]*Urology, Hospital Italiano de Buenos Aires, Buenos Aires, Argentina*

Introduction

Prostate cancer (PC) is the most common cancer in men [1]. The introduction of prostate-specific antigen (PSA) testing led to an increase in diagnosis, with the disadvantage of overdetection and overtreatment of clinically insignificant prostate cancer (ISPC) [2]. Transrectal ultrasound–guided (TRUS) biopsy suffers from low sensitivity and specificity for prostate cancer detection (PCD) or clinically significant prostate cancer (SPC) detection [3]. Increased number of biopsies (e.g., saturation biopsies), or combined biopsies with new markers (e.g., PCA 3), have not solved these problems either [4,5].

Multiparametric magnetic resonance imaging (MP-MRI) look more reliable for PCD [6,7]. Several strategies have been developed for targeted biopsy (TB) of prostate lesions identified on MRI, and PC and SPC detection rates are higher. Furthermore, morbidity is lower with these procedures than with TRUS. MRI-guided biopsy (MRGB) includes (1) in-bore MRI-targeted biopsy (MRI-TB), which is performed during MRI scan, using real-time image guidance; (2) MRI-TRUS fusion–targeted biopsy (FUS-TB), where software is used to perform MRI and TRUS image fusion, improving accuracy; and (3) cognitive registration TRUS-targeted biopsy (COG-TB), where an MRI lesion is cognitively targeted using TRUS guidance [8,9].

Standard prostate biopsy

TRUS is the mainstay of PC diagnosis [3]. Although it allows real-time visualization of the prostate, it is considered to be unreliable, due to the inability of gray-scale ultrasonography to distinguish PC from benign tissue [10]. Thus, approximately 30% of PC is missed, principally located either in the lateroanterior part of the peripheral zone (PZ), in the anterior part of the transitional zone (TZ), or in the anterior fibromuscular stroma (AFMS) [11,12]. Sextant biopsy schemes have been all but abandoned and extended 12-core biopsy protocols are currently recommended. Although these biopsy schemes have led to increased PCD, they have also resulted in increased indolent cancer detection; consequently, up to 50% of the tumors detected are small and well differentiated [8].

Overdiagnosis has led to the overtreatment, as suggested by the results of the European Randomized Study of Screening for Prostate Cancer [2]. In addition, the number of unnecessary biopsies has increased, along with the morbidity associated with the procedure [13]. Boccon-Gibod et al. [14] compared prostate biopsy with radical prostatectomy (RP) specimens and observed that 42% of the patients had a PC volume <0.5 mL, and 29% exhibited ISPC (Gleason score [GS] ≤6 or volume <0.5 mL). In contrast, GS 4 was ignored by prostate biopsy in 50% of the cases, thereby demonstrating that biopsy also underdiagnoses SPC. These results were correlated with the 36% GS upgrading, observed by Epstein in RP specimens [15]. RP specimens from patients who underwent surgery during active surveillance also showed 20%–45% GS upgrading [2,16,17].

Saturation biopsy

Jones et al. [18] observed no statistically significant difference between saturation and standard biopsies (44.5% vs. 52%). Descazeaud et al. [19] also reported that PCD was similar. Lane et al. [20] reported a high PCD rate on a second biopsy, despite initial saturation biopsy strategy. Up to one in four patients showed evidence of PC on a second biopsy. Delongchamps et al. [21] evaluated a saturation biopsy scheme on autopsied prostate glands, showing no increase in PCD rate over a less extensive scheme. Saturation biopsies might not only miss PC but also overestimate final GS on whole-mount analysis.

Precision Medicine for Investigators, Practitioners and Providers. https://doi.org/10.1016/B978-0-12-819178-1.00042-3

Transperineal biopsy

In a prospective study conducted by Emiliozzi et al. [22], PC was detected in 43 of 107 patients; 38% of the patients were assessed using transperineal (TP) 6-core approach and 32% using TRUS 6-core approach. Of the 43 PC diagnosed, 95% were found with TP and 79% with TRUS ($P = 0.012$) [22]. In a different study, Kawakami et al. [23] showed that there was no statistical difference in PCD rate (86% vs. 82%, $P = 0.51$) or in the characteristics of PC detected using the TP 14-core approach and the TRUS 12-core approach. A recent metaanalysis by Xue et al. [24] confirmed no significant difference in PCD between these approaches. However, combining the two techniques so as to saturate the whole prostate may optimize PCD, as demonstrated by Kawakami et al. [23].

MRI-guided versus standard prostate biopsy in naïve patients

Accumulated data regarding TB in overall PCD has been dissimilar [25]. In most publications, overall PCD with MRGB was similar to TRUS [26–32]. A metaanalysis by Wegelin et al. [33], comparing three different techniques for MRI-targeted prostate biopsies, found no difference between MRGB and TRUS for overall PCD. However, in a randomized prospective study, Porpiglia et al. [34] reported superiority for MRGB, compared with TRUS, in overall PCD (50.5% vs. 29.5% [$P = 0.002$]). Pokorny et al. [8] observed a marked difference in PCD with MRGB (70%) versus TRUS (56.5%), with excellent sensitivity (SN [92%]), specificity (SP [97%]), positive predictive value (PPV [92%]), and negative predictive value (NPV [97%]). Panebianco et al. [11] also observed that MRGB performed better in detecting PC (73%) than TRUS (38%). However, when the percentage of positive cores was reviewed in a different series, it was observed that TB allowed an equal or higher number of PC to be diagnosed, with fewer biopsies. The same was seen when assessing the mean maximum cancer core length. This was recently confirmed by Kasivisvanathan et al. [35], in a well-designed study.

MRGB versus standard prostate biopsy in naïve patients

The definition of SPC varies widely, but most use either Harnden or Epstein criteria [36,37]. Haffner's study demonstrated that MRGB was more accurate in detecting SPC than TRUS ($P = <0.001$), with a diagnostic accuracy of 0.98 under the ROC curve. The study also found that PC that was not detected by TB was primarily indolent (80%) [26]. Likewise, Mendhiratta et al. [30] observed that 83% of PC detected by systematic biopsy, but not by TB, was clinically indolent.

Another randomized trial was reported as a negative study, showing no difference in the detection of SPC between MRGB and TRUS. Nevertheless, it did demonstrate that MRI can play a role in the visualization of SPC. Of all the biopsy-proven SPC occurring in the MRI group, 87% were detected by TB. Of the 29 MRI-detected lesions with a PI-RADS score of 4 or 5, 28 (97%) were SPC confirmed by TB [31]. However, in a recent prospective, multicenter, paired-cohort study, that used a template prostate mapping biopsy as a reference test, Ahmed et al. [6] showed that for the correct diagnosis of SPC, the best-case scenario might lead to 102 (18%) more cases, of clinically significant cancer being detected in 576 men, compared with the standard pathway of TRUS biopsy.

Additionally, Mozer et al. [28] observed that the proportion of cores positive for SPC was significantly higher, with the targeted-core protocol (31%) than with the extended 12-core protocol (7.5%), and the median number of cores taken per diagnosis of SPC was significantly higher for the extended 12-core protocol ($X = 12$ vs. $X = 2$) (Table 42.1).

MRGB versus standard prostate biopsy in patients with repeat negative biopsies

In most studies involving extended 12-core schemes, PC was detected in around 10%–23% of patients on a second biopsy. Furthermore, repeat random biopsies (RBs), in patients with persistently suspicious PC, showed gradually decreasing PCD as the number of biopsies increased [38]. The prostate saturation biopsy schemes were initially recommended for patients who showed a rising level of PSA, despite previous negative prostate biopsies, but new results based on the true prevalence suggest that approximately 40% of PC might remain undetected, regardless of the number of cores taken. In addition, saturation biopsies can be associated not only with increased patient morbidity but also with the risk of indolent cancer detection [39].

In a prospective, randomized, single-center study, Sciarra et al. [38] observed that MRGB (45.5%) detected more PC than TRUS (24.5%), in patients undergoing a second biopsy. Using a matched multisession TRUS population from his institution's database for comparison, Hambrock et al. [12] reported that cancer detection rate was significantly higher with MRGB (59%) than with TRUS (22%). Just as in these studies, Schoots et al. [40] found in a metaanalysis that MRGB improved overall PC detection, in men with a previous negative biopsy, with a relative SN of 1.62.

TABLE 42.1 MRI-guided biopsy (MRGB) in naïve patients.

Author	Patients and methods	Disease significance	MRI	TB modalities	MRI lesions	PCD characteristics												Comments
						PCDΔ	PCD-RB			PCD-TB			PCD-MRI lesions	NPC		PPC		
							PC	SPC	ISPC	PC	SPC	ISPC		RB	TB	RB	TB	
Haffner 2011	N:555 RB versus TB Nonrandomized trial	ISPC Epstein criteria	1.5T WEC T2W- T1W- DCE No PI- RADS	COG-TB	351	54%	52%	Diagnostic accuracy 0.88		44%	Diagnostic accuracy 0.98					4.7 MCCL	5.6 MCCL	Detection ac- curacy of SPC was higher with TB TB detected 16% more 4/ 5-grade cases PPC > TB**
Delongchamps 2013	N:133 RB versus TB Nonrandomized trial	ISPC GS ≤ 6 MCCL <5 mm	1.5T EC T2W- DWI- DCE No PI- RADS	FUS-TB Urostation®	82		33%	14%	7%	47%	20%	3%				10.5%	30%	SPC > TB**
Pokorny 2014	N:223 RB versus TB Nonrandomized trial	ISPC GS ≤ 6	3T WCE T2W- DWI- DCE PI- RADS	MRI-TB	142		56.5%	63%	37.3%	70%	94%	6%	Highly suspicious MRI 86% PCD	15%	56.5%	33%	61%	SPC > TB** TB decreased the diagnosis of ISPC by 89.5%
Quentin 2014	N:128 RB versus TB Nonrandomized trial	SPC GS ≥ 6 MCCL >5 mm	3T WCE T2W- DWI- DCE PI- RADS	MRI-TB	329	61%	53%	42%	11%	53%	45.5.%	8%		14.4%	35.5%	25%	65%	NPC > TB** PPC > TB**
Mozer 2015	N:152 RB versus TB Nonrandomized trial	SPC ≥1 NPC: GS 3 + 4 or GS 6 ≥ 4 mm	1.5T WCE T2W- DWI- DCE No PI- RADS	FUS-TB Urostation®			56.5%	37%	20%	54%	43.5%	10.5%	Highly suspicious MRI 89% PCD	7.5% SPC**	31% SPC**	4 MCCL**	8 MCCL**	SPCD > TB NPP > TB** PPC > TB**
Peltier 2015	N:110 RB versus TB Nonrandomized trial	SPC GS ≥ 6 MCCL >6 mm >2 NPC	3T EC T2W- DWI- DCE No PI- RADS	FUS-TB Urostation®	175		54.5%	29%	16%	48%	46.5%	5.5%				10% SPC	29% SPC	SPCD > TB** PPC > TB**
Mendhiratta 2015	N:382 RB versus TB Nonrandomized trial	SPC GS ≥ 7	3TEC T2W- DWI- DCE No PI- RADS	FUS-TB Artermis®	382	54%	49%	27%	22.5%	43.5%	31%	13%	Highly suspicious MRI 86% PCD					SPCD > TB ISPCD < TB** MRGB missed 83% of ISPC

Continued

TABLE 42.1 MRI-guided biopsy (MRGB) in naïve patients.—cont'd

						PCD characteristics												
							PCD-RB			PCD-TB				NPC		PPC		
Author	Patients and methods	Disease significance	MRI	TB modalities	MRI lesions	PCDΔ	PC	SPC	ISPC	PC	SPC	ISPC	PCD-MRI lesions	RB	TB	RB	TB	Comments
Panebianco 2014	N: 1140 RB versus TB Nonmulticentric randomized trial	Only Gleason informed	3T EC T2W, DWI-DCE PI-RADS	COG-TB	440		38%	37%	1%	73%	72%	1%	T2W + DCE + DWI 97% accuracy of PCD					PCD > TB SPCD > TB MP-MRI was unable to detect 96% of ISPC smaller than 0.5 cm^3
Baco 2015	N:175 RB versus TB Nonmulticentric randomized trial	SPC GS 6 ≥ 5 mm GS 7	1.5T WEC T2W-DWI PI-RADS	FUS-TB Urostation®	63		54%	49.5%	4.5%	56%	38.5%	21%	Highly suspicious MRI 97% PCD					No difference in PCD and SPCD**
Tonttila 2015	N:103 RB versus TB Nonmulticentric randomized trial	Only Gleason informed	3T WEC T2W-DWI-DCEDC No PI-RADS	COG-TB	40		57%	45%	15%	51%	42.5%	9.5%	Highly suspicious MRI 95% PCD			4 mm	5 mm	No difference in PCD and SPCD
Kasivisvanathan 2018 PRECISION	N:500 RB versus TB Multicentric randomized trial	SPC GS ≥ 7	1.5T/3T WEC/ WERC T2W-DWI-DCE PI-RADS	FUS-TB	175		48%	26%	22%	47%	38%	9%	Highly suspicious MRI 83% SPCD	18%	44%	7.8 mm	6.5 mm	SPCD > TB** ISPCD < TB**

Abbreviations: **, statistical significance; *CB*, combined biopsy (RB + TB); *EC*, endorectal coil; *EM*, electromagnetic tracking; *MCCL*, maximum cancer core length; *NPC*, number of positive cores; *PCDΔ*, overall PCD (RB + TB); *PPC*, proportion of positive cores; *RB*, random biopsy; *TB*, targeted biopsy; *TP*, transperineal biopsy; *WEC*, without EC.

Most studies show a significantly higher detection rate of SPC and a lower detection rate of ISPC, by MRGB over TRUS. Thus, Sonn et al. [41] found that MRGB detected 91% (21/23) of SPC, while TRUS detected 54% (15/28). Hambrock et al. [12] observed that 37 of 40 (93%) patients who were diagnosed by MRGB had SPC, and all patients who underwent RP had SPC that was previously diagnosed by an MRGB. Salami et al. [42] detected 17% more SPC with MRGB than with TRUS, and the rate of SPC missed with TRUS was higher than with MRGB (21% vs. 4.5%). This was also observed in men undergoing repeat biopsy, in which detection rate for SPC was lower with TRUS (66%) than with MRGB (93%) [12]. This evidence was ratified by Schoots's metaanalysis [40]. Miyagawa et al. [43] found that the number of cores taken per diagnosis of PC was higher for the extended 12-core protocol than for the targeted-core protocol. This was also observed by Lee et al. [44] and Salami et al. [42] (Table 42.2).

MRI-TRUS–targeted biopsy versus MRI-targeted transperineal prostate biopsy

Radtke et al. [45] showed that systematic template biopsies missed 21% of SPC and MRGB missed 20%. They concluded that, using transperineal prostate biopsy, MRGB detected at least as much SPC as systematic template biopsies. Pepe et al. [46] compared the accuracy of TRUS-MRGB versus TP-MRGB, with lower detection rate for SPC with TRUS-MRGB (66.7%) than with TP-MRGB (93.3%). SN, SP, PPV, and diagnostic accuracy of TP-MRGB versus TRUS-MRGB were 97.2% versus 66.7%, 78.2% versus 71.2%, 59% versus 42.1%, 97.2% versus 87.5%, and 70% versus 57.5%, respectively. Recently, MP-MRI in men who required further biopsies, using a template prostate mapping biopsy, could be used to safely avoid a repeat biopsy in 14% of the cases, while detecting 97% of the SPC. SN, SP, NPV, and PPV were 80.6%, 68.5%, 83.3%, and 64.3%, respectively, when the 5-point Likert scale response for SPC is "likely" or "highly likely."(6).

Standard prostate biopsy plus MRGB

Siddiqui et al. [47] showed that adding standard biopsy to TB led to 103 more cases of PC (22%), 83% of which were low risk, while only 5% were high risk. MRGB diagnosed 30% more SPC (173 vs. 122 [P= <0.001]) and 17% less ISPC (213 vs. 258 [P= <0.001]) than TRUS. By contrast, Filson et al. [48] reported that, in patients with a suspicious MRI, MRGB plus TRUS detected not only more PC but also more SPC than each one separately (378 vs. 303 vs. 250 [P= <0.001]). MRGB alone detected more SPC than TRUS, and less ISPC (131) than CB (204) and TRUS (208).

Comparing three different techniques for MRGB

Puech et al. [49] published the first paper that compared FUS-TB with COG-TB, finding no difference in PCD and SPCD rates. Wysock et al. [50] also compared both techniques and observed that PCD and SPCD rates were higher with FUS-TB, mainly in lesions that were highly suspicious for cancer.

Delongchamps et al. [51] used conditional logistic regression for FUS-TB, COG-TB, and TRUS to detect PC. FUS-TB performed better than standard biopsy in overall cancer detection, while COG-TB showed a poor performance. Probability of detecting PC that had not been detected with TRUS was significantly higher with FUS-TB than with COG-TB. Finally, FUS-TB detected more Gleason pattern >6 than TRUS, while the detection rate with COG-TB was lower.

Three MRGB techniques, in patients with prior negative systematic biopsy, revealed comparable detection rates of overall PC (FUS-TB, 49%; COG-TB, 44%; and MRI-TB, 55% [$P = .4$]); also no significant differences were seen in the detection rates of SPC (FUS-TB, 34%; COG-TB, 33%; and MRI-TB, 33% [P = > 0.9]). The main limitation of the study was a low rate of PI-RADS ≥ 3 lesions on MP-MRI, causing underpowering for primary outcome [52].

T2-weighted imaging (T2W), dynamic contrast-enhanced (DCE) imaging, magnetic resonance spectroscopic imaging, diffusion weighted imaging, and apparent diffusion coefficient have increased the accuracy of MRI in PC diagnosis. Two or more of these parameters improve PCD (multiparameter MRI/MP-MRI) [53,54].

In 2012, the European Society of Urogenital Radiology (ESUR) developed the Prostate Imaging Reporting and Data System (PI-RADS) classification, which is a scoring system used to standardize prostate MRI reporting. This classification allowed the method to be replicated, and some publications have shown a clinical correlation between a low score and a high probability of benignity. In 2015, ESUR and the American College of Radiology launched the updated PI-RADS V2 (V2: version 2) [55].

In 2003, the Standards for Reporting of Diagnostic Accuracy (STARD) were published, which is a checklist of items aimed at improving the quality of reports. In 2013, a group of experts adapted STARD for MRGB, to improve the quality of reporting as well as facilitate the comparison between TRUS and MRGB [56].

Other studies used a 5-point Likert scale to report the likelihood of the presence of a clinically significant disease. This scoring system was based on the outputs of a consensus group that was convened before the publishing of the PI-RADS MP-MRI reporting consensus, although the two systems have subsequently been found to be similar [6].

TABLE 42.2 MRI-guided biopsy in patients with prior negative biopsy.

						PCD characteristics													
							PCD-RB			PCD-TB				NPC		PPC			
Author	Patients and methods	Disease significance	MRI	TB modalities	MRI lesions	PCDΔ	PC	SPC	ISPC	PC	SPC	ISPC	PCD-MRI lesions	RB	TB	RB	TB	APC	Comments
Sciarra 2010	N:180 RB versus TB Randomized trial	Only Gleason informed	1.5T EC MRSI-DCE No PI-RADS	COG-TB	45		24%	14.5%	10%	49%	28%	18%	MRSI + DCE 91% PCD						PCD > TB SPCD > TB
Hambrock 2010	N:68 RB versus TB Nonrandomized trial	ISPC Epstein criteria	3T EC T2W-DWI-DCE No PI-RADS	MRI-TB	114		22%			59%	54%	4.5%			18%			68%	PCD > TB**
Miyagawa 2010	N:85 RB versus TB Nonradomized trial		1.5T WEC T2W-DWI-DCE No PI-RADS	FUS-TB Real-time virtual sonography		61%	40%			53%				9%	32%				PCD > TB NPB > TB**
Lee 2012	N:87 RB versus TB Nonrandomized trial	ISPC Epstein criteria	3 T WEC T2W-DWI No PI-RADS	COG-TB	82	52.9%	10%			53%				4%	29%			39%	PCD > TB
Vouragnti 2012	N:195 RB versus TB No randomized trial	Only Gleason informed	3 T EC T2W-DWI-DCE No PI-RADS	FUS-TB EM tracking sensor	195	37.5%	23%	5%	18%	29%	11%	18%	Highly suspicious MRI 67% PCD					59%	SPCD > TB PSAD was a significant predictor of PCD**
Sonn 2013	N:105 RB versus TB Nonradomized trial	ISPC Epstein criteria	3T WEC T2W-DWI-DCE No PI-RADS	FUS-TB Artemis®	164	34%	27%	14%	12%	24%	20%	2%	Highly suspicious MRI 88% PCD		16.5%			44.5%	PCD > TB SPCD > TB PSAD was a significant predictor of SPCD**
Salami 2015	N:140 RB versus TB Nonrandomized trial	ISPC Epstein criteria	3T EC T2W-DWI-DCE No PI-RADS	FUS-TB UroNav®	140	65%	48%	31%	18%	52%	48%	4%	Highly suspicious MRI 92% PCD	14%	55%	4.15	6.21	26%	PCD > TB SPCD > TB**

Abbreviations: **, statistical significance; *CB*, combined biopsy (RB + TB); *EC*, endorectal coil; *EM*, electromagnetic tracking; *MRSI*, magnetic resonance spectroscopy imaging; *NPC*, number of positive cores; *PCDΔ*, overall PCD (RB + TB); *PPC*, proportion of positive cores; *RB*, random biopsy; *TB*, targeted biopsy; *TP*, transperineal biopsy; *WEC*, without EC.

Suspicious lesions as predictors of prostate cancer detection with MRGB

Hadaschik et al. [57] noted that the PCD rate was 96% (23/24) in those patients who had a highly suspicious lesion. Filson et al. [48] observed PC in 79% of 825 patients who had a suspicious MRI, compared with 21% of 217 patients with a normal MRI. In addition, 80% of patients with a highly suspicious lesion on MRI had PC with a Gleason pattern ≥7, compared with 24% with a grade-3 region of interest ($P < .001$). Likewise, Baco et al. [31] observed 87% of PCD in patients with suspicious lesions on MRI, whereas SPC was detected in 97% of a subgroup of patients with PI-RADS 4 or 5.

Rais-Bahrami et al. [58] studied the correlation between MP-MRI suspicion score and the presence of PC, informing marked correlation with high-risk disease, with an SP of 0.89 and an NPV of 0.91 under the ROC curve. Also Vourganti et al. [59] observed correlation between the level of suspicion in images and the detection of PC ($P = 0.0004$) and SPC ($P = 0.033$). Kasivisvanathan et al. [35] noted that SPC was highest among participants with PI-RADS V2 score of 5 (83%), followed by 4 (60%) and 3 (12%). Conversely, the percentage of men without cancer was highest among participants with a nonsuspicious PI-RADS V2.

PSA as predictor of prostate cancer detection with MRGB

Vourganti et al. [59] showed that prostate volume was larger (54 vs. 71 mL, p: 0.0006), PSA was higher (18.7 vs. 11.2 ng/mL [$P = 0.0005$]), and PSA density (PSAD) was greater (0.38 vs. 0.16 ng/ml^2 [P= <0.001]), in SPC; these variables also had a correlation with SPC detection. In multivariate analysis, only PSAD remained a significant predictor ($P = 0.0026$). SPC cut-off value was less than 0.15 ng/ml^2, as suggested in the National Comprehensive Cancer Network guidelines (nccn.org/professionals/physician_gls/default/aspx).

Panebianco et al. [60] evaluated the probability of any PC during follow-up of negative MRI, either in naïve patients or with previous biopsy, and found that PSA and PSAD were both independent predictors of subsequent SPC diagnosis, with PSAD being the strongest (hazard ratio 7.57). In agreement with this report, Washino et al. [61] demonstrated that equivocal PI-RADS V2 score ≤3 and PSAD <0.15 ng/ml^2 yielded no SPC, and no additional detection of PC on further biopsies.

Prostate cancer index lesion with MRGB

The index lesion was defined as the highest Gleason score, or the largest tumor volume if Gleason were the same, in order of priority. MRGB currently allows clinicians to direct biopsies to the region of interest, rather than to a random area. Baco et al. [62] noted a 95% (128/135) concordance of the index lesion location between TB and prostatectomy specimens, and a 90% concordance of primary Gleason pattern between TB and prostatectomy specimens. In a study of similar characteristics, Russo [63] observed that 104 of 115 index lesions were correctly diagnosed by MRI, with a sensitivity of 90.4%, including 98/105 clinically significant index lesions with a sensitivity of 93.3%.

Porpiglia et al. [64] noted that, in experienced hands, two cores in the middle of the index lesion allow for a 92.5% accuracy. Additionally, Gleason heterogeneity was observed in 12.6% versus 26.4%, for ≤8 or ≥8 mm index lesions, respectively, with a prevalence of Gleason pattern ≥4 in the center of the target. Similarly, Radtke et al. [65] showed that MRI detected 110 of 120 (92%) index lesions, and that 107 (89%) index lesions harbored SPC. However, this author noted that TB (two cores per lesion) alone diagnosed 96 of 120 (80%) index lesions, standard biopsy alone diagnosed 110 of 120 (92%), and combined standard biopsy and TB detected 115 of 120 (96%). Combined MRGB and standard biopsy detected 97% of all SPC lesions, yielding better results than MRGB or standard biopsy separately.

Prostate cancer detection with MRGB outside of the peripheral zone

The mean percentage of anteriorly located cancer in an unselected population is about 20%. Nontargeted strategies to reach these anterior lesions have been less successful, discovering only 2% of PC, as shown by Chon et al. [66]. Hambrock et al. [12] evaluated MRI-TB in 71 patients with a persistently elevated PSA and prior negative TRUS biopsies, underlining that the principal tumor site was the most ventral aspect of TZ in 26 of 46 cases (57%), followed by PZ paramedian region in 9 (20%) and PZ anterior horns in 5 (11%). Salami et al. [42] observed that the anatomical locations of cancer missed by TRUS biopsies but detected by FUS-TB were 4 of 23 (17.5%) in AFMS, 12 of 23 (52.2%) in TZ, and 3 of 23 (13%) in anterior PZ. Lee et al. [44] noted that 32 patients had suspicious anterior lesions on MRI, and TB-detected PC in 19 (59.5%) of them. The same was shown by Vourganti et al. [59].

Pepe et al. [46] observed that TP-MRGB diagnosed not only more PC (93.7%) located in the anterior zone of the prostate, compared with the TRUS-MRGB approach (25%), but also a greater number of SPC ($P = 0.001$). Radtke et al. [67] showed no difference in the detection of PC and SPC for FUS-TB versus TRUS in the initial biopsy subgroup ($P = 0.181$ for all PC, $P = 0.195$ for SPC defined as GS $\geq 3 + 4$, and $P = 0.661$ for GS $\geq 4 + 3$). However, in the repeat-biopsy subgroup, FUS-TB performed significantly better than TRUS ($P = 0.004$ for GS $\geq 3 + 4$, and $P = 0.02$ for GS $\geq 4 + 3$).

Prostate cancer upgrading with MRGB

Siddiqui et al. [68] remarked that MRGB contributed to improving the Gleason score in 32% of cases. In addition, MRGB detected 67% more Gleason score 4 + 3, and missed 36% of cases of Gleason score 3 + 4 than TRUS, thus reducing the detection of low-risk PC. Vourganti et al. [59] also showed that MRGB had improved Gleason score in 38% of cases detected by TRUS. According to Porpiglia et al. [69], primary and secondary GSs were determined accurately in 85 (72.6%) and 111 (86% [$P = 0.018$]) of the cases for the TRUS group, and in 56 (47.9%) and 103 (79.8% [P= <0.001]) for the FUS-TB group. The primary GS 4 rates were quite comparable between TRUS and FUS-TB (86.7% vs. 86.5%); conversely, for secondary GS 4, FUS-TB showed better performance (59.6% vs. 84.5%). The rate of correctly classified GS was significantly lower for TRUS than for FUS-TB (P= <0.001). The rates of pathological GS upgrading were significantly higher for TRUS (39.3%) compared with FUS-TB (7.8% [P= <0.001]).

Margel et al. [70] evaluated the impact of MP-MRI on PC reclassification among candidates for active surveillance in a prospective cohort study and revealed that approximately one-third of patients on active surveillance improved their GS with a confirmatory MRGB. He also observed a GS reclassification in only 3.5% of cases with a normal MRI. In line with these findings, Walton-Diaz reported [71] not only that 22.4% of the patients had GS ≥ 7 on confirmatory MRGB but also that stable findings on MP-MRI were associated with GS stability in patients with GS 6 who were on active surveillance, which could potentially reduce the number of unnecessary biopsies in men undergoing active surveillance. Recently, a randomized trial was reported as a negative study showing no difference in the upgrading rate between TB and RB. However, significant differences in upgrading on TB were seen between sites, likely reflecting different levels of expertise with the TB technique, as highlighted in the limitations of the study [72].

Negative predictive value of MRI and follow-up of negative MRGB

Several studies attempted to validate the clinical usefulness of negative MRIs. Villers [73] compared the histopathology results of RP specimens with MRI findings in men with a suspicious area detected by an MRI biopsy and reported an NPV of 85% for foci >0.2 cm3 and an NPV of 95% for foci <0.5 cm3. Girometti [74] reported that MRIs had an NPV of 100%. Similarly, in a prospective cohort study in which patients underwent an MP-MRI, Perdona [75] noted an NPV of 91%. According to Filson [76], MP-MRIs do not have a very high NPV. This author reported the results of RB among men with negative MP-MRIs, observing an NPV of only 54%. In addition, a GS ≥ 7 cancer was found in 16% of the cases on biopsy. In a recent metaanalysis that evaluated 48 studies, the median MP-MRI NPV was 82.4% (IQR 69%–92.5%) for overall PC and 88.1% (IQR 85.7% –92.3%) for SPC [77]. Most recently, PROMIS, a paired validating confirmatory multicenter study that assessed the diagnostic accuracy of MP-MRI in PCD, showed an NPV of 89% (83%–94%) [6].

Few studies evaluated the follow-up of patients with a negative biopsy. Itatani [78] performed a retrospective single-institution study, observing an NPV of 89.6% for SPC during a 5-year follow-up. Similar results were noted by Venderink [79], who observed 1.7% (5/300) of SPC detection after a median 41-month follow-up of patients. Panebianco [60] performed a Kaplan–Meier analysis to assess any-grade PC and SPC diagnosis-free survival probabilities in patients with a negative MRI; the results showed that, after a 2-year follow-up, any-grade PC and SPC diagnosis-free survival probabilities were 94% for naïve patients and 95% for patients with a previous negative biopsy, respectively. After a 4-year follow-up, any-grade PC diagnosis-free survival probabilities were 84% in naïve patients and 96% in patients with a previous negative biopsy (long-rank P= < 0.001). Moreover, SPC diagnosis-free survival after a 4-year follow-up remained unchanged (Table 42.3).

Quality of life and safety of MRGB

Kasivisvanathan [35] informed that health-related quality of life 24 h and 30 days after the intervention did not differ significantly between the MRGB group and the TRUS group. The intervention was associated with similar results regarding immediate postintervention discomfort and pain in the two groups. By contrast, Egbers [80] observed that the intensity of pain was significantly shorter and milder after an MRGB compared with TRUS ($P = 0.005$). The discomfort may have been associated with the use of anesthesia, as Arsov [81] established that patients with a periprostatic nerve block with 2% mepivacaine before

TABLE 42.3 MRI-guided biopsy (MRGB) in naïve patients or with prior negative biopsy or active surveillance.

Author	Patients and methods	MRI	TB modalities	Disease significance	MRI lesions	PCD characteristics													Comments
						PCDΔ	PCD-RB			PCD-TB			PCD-MRI lesions	NPC		PPC			
							PC	SPC	ISPC	PC	SPC	ISPC		RB	TB	RB	TB		
Pinto 2011	N = 101 RB versus TB Nonrandomized trial	3T WEC T2W- DWI- DCE No PI- RADS	FUS-TB EM		101	54.5%							Highly suspicious MRI 89.5% PCD**	12%	21%				NPC > TB
Siddiqui 2013	N = 582 RB versus TB Nonrandomized trial	1.5T WCE T2W- DCE- MRSI No PI- RADS	FUS-TB EM	SPC GS $\geq$ 7	582	54%	44%	22.5%	77.5%	43.5%	35%	65%		9.2%	47.5%				SPCD > TB NPC > TB** 32% of GS upgrading with TB GS upgrading was associated with MRI suspicion score**
Sonn 2013	N = 151 RB versus TB Nonrandomized trial	3T WEC T2W- DWI- DCE No PI- RADS	FUS-TB Artemis®	SPC GS $\geq$ 7	279	56%	45%	21.5%	78.5%	35%	24%	76%	Highly suspicious MRI 94% SPC	7%	21%	3.3 MCCL	5.1 MCCL		SPCD > TB NPCD > TB** PPC > TB**
Radtke 2015	N = 294 TP versus TB Nonrandomized trial	3T WEC T2W- DWI- DCE PI- RADS	FUS-TB BiopSee®	SPC GS $\geq$ 7	225	51%	44%	79%		38%	87%		Highly suspicious MRI 76% PCD	7.5%	46%				SPCD > TB TB missed 44% of the ISPC
Maxeiner 2015	N = 169 RB versus TB Nonrandomized trial	3T WEC T2W- DWI PI- RADS	FUS-TB Aplio 500®	SPC GS $\geq$ 7	316	42%	16%			26%	88%			12%	21%				SPCD > TB TB detected 88% of the SPC
Siddiqui 2015	N = 1003 CB versus RB versus TB Nonrandomized trial	3T EC T2W- DCE- MRSI No PI- RADS	FUS-TB UroNav®	Low risk GS 6 or GS 3 + 4 (<50% PPC or < 33% NPC)	1003	69% 25% SPCD	47%	21%	26%	46%	25%	21%							SPCD > TB** CB detected 22% more PC (83% ISPC) TB diagnosed 30% more high-risk PC and 17% fewer low-risk PC than RB
Filson 2016	N = 825 CB versus RB versus TB Nonrandomized trial	3T WEC T2W- DWI- DCE No PI- RADS	FUS-TB Artemis®	SPC GS $\geq$ 7	825	70.5% 46% SPCD**	55.5%	30%	25.5%	52.6%	36.5%	16%	Highly suspicious MRI 80% SPC**						SPCD > CB** CB detected 20% more SPC. PCD was associated with MRI suspicion score **

Abbreviations: **, statistical significance; *CB*, combined biopsy (RB + MRGB); *EC*, endorectal coil; *EM*, electromagnetic tracking sensor; *MCCL*, maximum cancer core length; *N*, number of patients; *NPC*, number of cores positive; *PCDΔ*, overall PCD (RB + MRGB); *PPC*, proportion of cores positive; *RB*, random biopsy; *TB*, targeted biopsy; *TP*, transperineal biopsy; *WEC*, without EC.

MRGB have significantly lower pain levels compared with those with intrarectal instillation of a 2% lidocaine gel (IB-GB) before TRUS. However, the addition of TRUS significantly increases the number of biopsy cores, and MRGB still requires significantly less time in comparison with IB-GB.

One of the most frequent complications of TRUS is bleeding. In a systematic review of complications arising from prostate biopsies, hematospermia was the one most frequently (10%–90%) reported after a TRUS [13]. Some authors associated this condition with the number of biopsies performed. For instance, Berger [82] reported hematospermia in 31.8% of cases after 6-core biopsies, 37.4% after 10-core biopsies, and 38.4% after 15-core biopsies (P= < 0.001). Egbers [80] observed no difference when comparing MRI-TB with TRUS (36% vs. 33% [P= > 0.05]). Another frequent event associated with TRUS is hematuria (20%–80%). Egbers [80] reported hematuria in 51% of cases after an MRGB compared with 79% of cases after a TRUS biopsy ($P = 0.006$); although minor hematuria is common after a prostate biopsy, significant bleeding requiring hospitalization occurred in less than 1% of cases. Finally, hematochezia had a low incidence after TRUS biopsies and was not related to the number of biopsies performed [83]. In addition, MRGB showed no significant differences with TRUS in terms of incidence and duration. Recently, a PRECISION study noted that bleeding complications were less frequent in the MRGB group than in the TRUS one, including events of hematuria (30% vs. 63%), hematospermia (32% vs. 60%), pain in the site of the procedure (13% vs. 23%), and hematochezia (14% vs. 22%) [35].

Urinary infection is a frequent comorbidity associated with TRUS and hospitalization. According to the Global Prevalence Study of Infections in Urology, 3.1% of patients required hospitalization after a biopsy. But the most potentially life-threatening infectious complication is sepsis [84]. However, other series from North America and Brazil reported lower sepsis rates of 0.6% and 1.7%, respectively [85].

Most reported infectious complications result from *Escherichia coli*, with high rates of resistance to fluoroquinolones, ampicillin, and sulfamethoxazole-trimethoprim. Fluoroquinolone resistance has increased globally. Various strategies have been explored to reduce infectious complications, such as cleansing the rectum with povidone-iodine before a TRUS biopsy, switching or expanding the antimicrobial regimen, performing rectal swab cultures, using targeted prophylaxis, or applying different biopsy techniques. Based on the data currently available, TP appears to be associated with a lower risk of infectious diseases and hospitalization. While MRGB allows as few as one or two cores to be taken, the most common approach for TB continues to be the transrectal one as there is slight evidence that it reduces the risk of infections. In a systematic review, only one study reported possible benefits of the transrectal approach, but they were not statistically significant [83].

There is a low risk of acute urinary retention after a standard TRUS biopsy, ranging from 0.2% to 1.7%. Urinary retention is usually transient, and most patients do not require surgery. A risk also exists for short-term worsening of voiding complaints after a TRUS biopsy. Reported rates of dysuria typically range from 6% to 25% [13]. Raaijmakers [86] observed that prostate volume and higher voiding symptoms were associated with risk of urinary retention after a prostate biopsy. Similarly, Zaytoun [87] showed that a larger prostate predicted urinary retention after biopsy (OR: 4.45 [P=< 0.001]). There is no convincing evidence suggesting that the number of biopsy cores increases the risk of urinary retention; but, based on the data currently available, the reported incidence of acute urinary retention after MRGB is sporadic, from 0% to 1% [83].

Mortality after prostate biopsy is extremely rare, and most deaths reported are the result of a septic shock. However, men who were hospitalized due to infectious complications had a 12-fold higher 30-d mortality rate in comparison to those who were not (95% CI, 8.59–16.80 [P= < 0.0001]). Lethal Fournier's gangrene has also been reported. Bleeding after the procedure is usually self-limiting and rarely a life-threatening complication [83].

References

[1] Torre LA, Bray F, Siegel RL, Ferlay J, Lortet-Tieulent J, Jemal A. Global cancer statistics, 2012. CA A Cancer J. Clin. 2015;65:87–108.

[2] Schröder FH, Hugosson J, Roobol MJ, Tammela TLJ, Ciatto S, Nelen V, et al. Screening and prostate-cancer mortality in a randomized european study. N. Engl. J. Med. 2009;360(13):1320–8.

[3] Heidenreich A, Bastian PJ, Bellmunt J, Bolla M, Joniau S, Kwast T Van Der, et al. EAU guidelines on prostate cancer . Part 1 : screening , diagnosis , and local treatment with curative intent — update 2013. Eur. Assoc. Urol. 2013;65(1):124–37.

[4] Loeb S, Partin AW. Review of the literature: PCA3 for prostate cancer risk assessment and prognostication. Rev. Urol. 2011;13(4):e191–5.

[5] Numao N, Kawakami S, Sakura M, Yoshida S, Koga F, Saito K, et al. Characteristics and clinical significance of prostate cancers missed by initial transrectal 12-core biopsy. BJU Int. 2012;109(5):665–71.

[6] Ahmed HU, El-Shater Bosaily A, Brown LC, Gabe R, Kaplan R, Parmar MK, et al. Diagnostic accuracy of multi-parametric MRI and TRUS biopsy in prostate cancer (PROMIS): a paired validating confirmatory study. Lancet 2017;389(10071):815–22.

[7] Simmons LAM, Kanthabalan A, Arya M, Briggs T, Barratt D, Charman SC, et al. The PICTURE study: diagnostic accuracy of multiparametric MRI in men requiring a repeat prostate biopsy. Br. J. Canc. 2017;116(9):1159–65.

[8] Pokorny MR, De Rooij M, Duncan E, Schröder FH, Parkinson R, Barentsz JO, et al. Prospective study of diagnostic accuracy comparing prostate cancer detection by transrectal ultrasound-guided biopsy versus magnetic resonance (MR) imaging with subsequent mr-guided biopsy in men without previous prostate biopsies. Eur. Urol. 2014;66(1):22–9.
[9] Delongchamps NB, Peyromaure M, Schull A, Beuvon F, Bouazza N, Flam T, et al. Prebiopsy magnetic resonance imaging and prostate cancer detection: comparison of random and targeted biopsies. J. Urol. 2013;189(2):493–9.
[10] Gosselaar C, Roobol MJ, Roemeling S, Wolters T, Van Leenders GJLH, Schröder FH. The value of an additional hypoechoic lesion-directed biopsy core for detecting prostate cancer. BJU Int. 2008;101(6):685–90.
[11] Panebianco V, Barchetti F, Sciarra A, Ciardi A, Indino EL, Papalia R, et al. Multiparametric magnetic resonance imaging vs. standard care in men being evaluated for prostate cancer: a randomized study. Urol. Oncol. Semin. Orig. Investig. 2015;33(1):17.e1–7.
[12] Hambrock T, Somford DM, Hoeks C, Bouwense SAW, Huisman H, Yakar D, et al. Magnetic resonance imaging guided prostate biopsy in men with repeat negative biopsies and increased prostate specific antigen. J. Urol. 2010;183(2):520–8.
[13] Loeb S, Vellekoop A, Ahmed HU, Catto J, Emberton M, Nam R, et al. Systematic review of complications of prostate biopsy. Eur. Urol. 2013;64(6):876–92.
[14] Boccon-Gibod LM, Dumonceau O, Toublanc M, Ravery V, Boccon-Gibod LA. Micro-focal prostate cancer: a comparison of biopsy and radical prostatectomy specimen features. Eur. Urol. 2005;48(6):895–9.
[15] Epstein JI, Feng Z, Trock BJ, Pierorazio PM. Upgrading and downgrading of prostate cancer from biopsy to radical prostatectomy: incidence and predictive factors using the modified gleason grading system and factoring in tertiary grades. Eur. Urol. 2012;61(5):1019–24.
[16] van As NJ, Norman AR, Thomas K, Khoo VS, Thompson A, Huddart RA, et al. Predicting the probability of deferred radical treatment for localised prostate cancer managed by active surveillance. Eur. Urol. 2008;54(6):1297–305.
[17] Reese AC, Feng Z, Landis P, Trock BJ, Epstein JI, Carter HB. Predictors of adverse pathology in men undergoing radical prostatectomy following initial active surveillance. Urology 2015;86(5):991–7.
[18] Jones JS, Patel A, Schoenfield L, Rabets JC, Zippe CD, Magi-Galluzzi C. Saturation technique does not improve cancer detection as an initial prostate biopsy strategy. J. Urol. 2006;175(2):485–8.
[19] Descazeaud A, Rubin M, Chemama S, Larré S, Salomon L, Allory Y, et al. Saturation biopsy protocol enhances prediction of pT3 and surgical margin status on prostatectomy specimen. World J. Urol. 2006;24(6):676–80.
[20] Lane BR, Zippe CD, Abouassaly R, Schoenfield L, Magi-Galluzzi C, Jones JS. Saturation technique does not decrease cancer detection during followup after initial prostate biopsy. J. Urol. 2008;179(5):1746–50.
[21] Delongchamps NB, De La Roza G, Jones R, Jumbelic M, Haas GP. Saturation biopsies on autopsied prostates for detecting and characterizing prostate cancer. BJU Int. 2009;103(1):49–54.
[22] Emiliozzi P, Corsetti A, Tassi B, Federico G, Martini M, Pansadoro V. Best approach for prostate cancer detection: a prospective study on transperineal versus transrectal six-core prostate biopsy. Urology 2003;61(5):961–6.
[23] Kawakami S, Yamamoto S, Numao N, Ishikawa Y, Kihara K, Fukui I. Direct comparison between transrectal and transperineal extended prostate biopsy for the detection of cancer. Int. J. Urol. 2007;14(8):719–24.
[24] Xue J, Qin Z, Cai H, Zhang C, Li X, Xu W, et al. Comparison between transrectal and transperineal prostate biopsy for detection of prostate cancer: a meta-analysis and trial sequential analysis. Oncotarget 2017;8(14):23322–36.
[25] Bergero MA, Martinez PF, Radtke JP, Hadaschick B. Multiparametric-mri-guided biopsy in the era of precision medicine. Arch. Esp. Urol. 2017;70(10):833–44.
[26] Haffner J, Lemaitre L, Puech P, Haber GP, Leroy X, Jones JSVA. Role of magnetic resonance imaging before initial biopsy: comparison of magnetic resonance imaging-targeted and systematic biopsy for significant prostate cancer detection. BJU Int. 2011;108(8):E171–8.
[27] Quentin M, Blondin D, Arsov C, Schimmöller L, Hiester A, Godehardt E, et al. Prospective evaluation of magnetic resonance imaging guided in-bore prostate biopsy versus systematic transrectal ultrasound guided prostate biopsy in biopsy naïve men with elevated prostate specific antigen. J. Urol. 2014;192(5):1374–9.
[28] Mozer P, Rouprêt M, Le Cossec C, Granger B, Comperat E, De Gorski A, et al. First round of targeted biopsies using magnetic resonance imaging/ultrasonography fusion compared with conventional transrectal ultrasonography-guided biopsies for the diagnosis of localised prostate cancer. BJU Int. 2015;115(1):50–7.
[29] Peltier A, Aoun F, Lemort M, Kwizera F, Paesmans M, Van Velthoven R. MRI-targeted biopsies versus systematic transrectal ultrasound guided biopsies for the diagnosis of localized prostate cancer in biopsy naïve men. BioMed Res. Int. 2015;2015:571708.
[30] Mendhiratta N, Rosenkrantz AB, Meng X, Wysock JS, Fenstermaker M, Huang R, et al. Magnetic resonance imaging-ultrasound fusion targeted prostate biopsy in a consecutive cohort of men with no previous biopsy: reduction of over detection through improved risk stratification. J. Urol. 2015;194(6):1601–6.
[31] Baco E, Rud E, Eri LM, Moen G, Vlatkovic L, Svindland A, et al. A randomized controlled trial to assess and compare the outcomes of two-core prostate biopsy guided by fused magnetic resonance and transrectal ultrasound images and traditional 12-core systematic biopsy. Eur. Urol. 2016;69(1):149–56.
[32] Tonttila PP, Lantto J, Pääkkö E, Piippo U, Kauppila S, Lammentausta E, et al. Prebiopsy multiparametric magnetic resonance imaging for prostate cancer diagnosis in biopsy-naive men with suspected prostate cancer based on elevated prostate-specific antigen values: results from a randomized prospective blinded controlled trial. Eur. Urol. 2016;69(3):419–25.
[33] Wegelin O, van Melick HHE, Hooft L, Bosch JLHR, Reitsma HB, Barentsz JO, et al. Comparing three different techniques for magnetic resonance imaging-targeted prostate biopsies: a systematic review of in-bore versus magnetic resonance imaging-transrectal ultrasound fusion versus cognitive registration. Is there a preferred technique? Eur. Urol. 2017;71(4):517–31.

[34] Porpiglia F, Manfredi M, Mele F, Cossu M, Bollito E, Veltri A, et al. Diagnostic pathway with multiparametric magnetic resonance imaging versus standard pathway: results from a randomized prospective study in biopsy-naïve patients with suspected prostate cancer. Eur. Urol. 2017;72(2):282–8.

[35] Kasivisvanathan V, Rannikko AS, Borghi M, Panebianco V, Mynderse LA, Vaarala MH, Briganti A, Budäus L, Hellawell G, Hindley RG, Roobol MJ, Eggener S, Ghei M, Villers A, Bladou F, GM1 V, Virdi J, Boxler S, Robert G, Singh PB, Venderink W, Hadaschik MCPSGC. MRI-targeted or standard biopsy for prostate-cancer diagnosis. N. Engl. J. Med. 2018;378(19):1767–77.

[36] Epstein JI, Walsh PC, Carmichael M, Brendler CB. Pathologic and clinical findings to predict tumor extent of nonpalpable (stage T1 c) prostate cancer. JAMA 1994;271(5):368–74.

[37] Harnden P, Naylor B, Shelley MD, Clements H, Coles B, Mason MD. The clinical management of patients with a small volume of prostatic cancer on biopsy: what are the risks of progression? A systematic review and meta-analysis. Cancer 2008;112(5):971–81.

[38] Sciarra A, Panebianco V, Ciccariello M, Salciccia S, Cattarino S, Lisi D, et al. Value of magnetic resonance spectroscopy imaging and dynamic contrast-enhanced imaging for detecting prostate cancer foci in men with prior negative biopsy. Clin. Cancer Res. 2010;16(6):1875–83.

[39] Delongchamps NB, Haas GP. Saturation biopsies for prostate cancer: current uses and future prospects. Nat. Rev. Urol. 2009;6(12):645–52.

[40] Schoots IG, Nieboer D, Giganti F, Moore CM, Bangma CH, Roobol MJ. Is magnetic resonance imaging-targeted biopsy a useful addition to systematic confirmatory biopsy in men on active surveillance for low-risk prostate cancer? A systematic review and meta-analysis. BJU Int. 2018;122(6):946–58.

[41] Sonn GA, Chang E, Natarajan S, Margolis DJ, MacAiran M, Lieu P, et al. Value of targeted prostate biopsy using magnetic resonance-ultrasound fusion in men with prior negative biopsy and elevated prostate-specific antigen. Eur. Urol. 2014;65(4):809–15.

[42] Salami SS, Ben-Levi E, Yaskiv O, Ryniker L, Turkbey B, Kavoussi LR, et al. In patients with a previous negative prostate biopsy and a suspicious lesion on magnetic resonance imaging, is a 12-core biopsy still necessary in addition to a targeted biopsy? BJU Int. 2015;115(4):562–70.

[43] Miyagawa T, Ishikawa S, Kimura T, Suetomi T, Tsutsumi M, Irie T, et al. Real-time virtual sonography for navigation during targeted prostate biopsy using magnetic resonance imaging data. Int. J. Urol. 2010;17(10):855–61.

[44] Lee SH, Chung MS, Kim JH, Oh YT, Rha KH, Chung BH. Magnetic resonance imaging targeted biopsy in men with previously negative prostate biopsy results. J. Endourol. 2012;26(7):787–91.

[45] Radtke JP, Kuru TH, Boxler S, Alt CD, Popeneciu IV, Huettenbrink C, et al. Comparative analysis of transperineal template saturation prostate biopsy versus magnetic resonance imaging targeted biopsy with magnetic resonance imaging-ultrasound fusion guidance. J. Urol. 2015;193(1):87–94.

[46] Pepe P, Garufi A, Priolo G, Pennisi M. Transperineal versus transrectal MRI/TRUS fusion targeted biopsy: detection rate of clinically significant prostate cancer. Clin. Genitourin. Cancer 2017;15(1):e33–6.

[47] Siddiqui MM, Rais-Bahrami S, Turkbey B, George AK, Rothwax J, Shakir N, et al. Comparison of MR/ultrasound fusion-guided biopsy with ultrasound-guided biopsy for the diagnosis of prostate cancer. JAMA, J. Am. Med. Assoc 2015;313(4):390–7.

[48] Filson CP, Natarajan S, Margolis DJA, Huang J, Lieu P, Dorey FJ, et al. Prostate cancer detection with magnetic resonance-ultrasound fusion biopsy: the role of systematic and targeted biopsies. Cancer 2016;122(6):884–92.

[49] Puech P, Rouvière O, Renard-Penna R. Prostate cancer diagnosis: multiparametric MR-targeted biopsy with cognitive and transrectal US–MR fusion guidance versus systematic biopsy—prospective. Radiology 2013;268(2):461–9.

[50] Wysock JS, Rosenkrantz AB, Huang WC, Stifelman MD, Lepor H, Deng FM, et al. A prospective, blinded comparison of magnetic resonance (MR) imaging-ultrasound fusion and visual estimation in the performance of MR-targeted prostate biopsy: the profus trial. Eur. Urol. 2014;66(2):343–51.

[51] Delongchamps NB, Zerbib M. Re: role of magnetic resonance imaging before initial biopsy: comparison of magnetic resonance imaging-targeted and systematic biopsy for significant prostate cancer detection. Eur. Urol. 2012;61(3):622–3.

[52] Wegelin O, Exterkate L, van der Leest M, Kummer JA, Vreuls W, de Bruin PC, et al. The future trial: a multicenter randomised controlled trial on target biopsy techniques based on magnetic resonance imaging in the diagnosis of prostate cancer in patients with prior negative biopsies. Eur. Urol. 2019;75(4):582–90. pii: S0302-2838(18)30939-4.

[53] Puech P, Potiron E, Lemaitre L, Leroy X, Haber GP, Crouzet S, et al. Dynamic contrast-enhanced-magnetic resonance imaging evaluation of intraprostatic prostate cancer: correlation with radical prostatectomy specimens. Urology 2009;74(5):1094–9.

[54] Vilanova JC, Luna-Alcalá A, Boada MBJ. Multiparametric MRI. The role of MRI techniques in the diagnosis, staging and follow up of prostate cancer. Arch. Esp. Urol. 2015;68(3):316–33.

[55] MR Prostate Imaging Reporting and Data System vesion 2.0. American College of Radiology. Accessed June 2015. Available online: http//www.acr.org/Quality-Sa-fety/Resources/PIRADS/.

[56] Moore CM, Kasivisvanathan V, Eggener S, Emberton M, Fütterer JJ, Gill IS, et al. Standards of reporting for MRI-targeted biopsy studies (START) of the prostate: recommendations from an international working group. Eur. Urol. 2013;64(4):544–52.

[57] Hadaschik BA, Kuru TH, Tulea C, Rieker P, Popeneciu IV, Simpfendörfer T, et al. A novel stereotactic prostate biopsy system integrating pre-interventional magnetic resonance imaging and live ultrasound fusion. J. Urol. 2011;186(6):2214–20.

[58] Rais-Bahrami S, Siddiqui MM, Turkbey B, Stamatakis L, Logan J, Hoang AN, et al. Utility of multiparametric magnetic resonance imaging suspicion levels for detecting prostate cancer. J. Urol. 2013;190(5):1721–7.

[59] Vourganti S, Rastinehad A, Yerram NK, Nix J, Volkin D, Hoang A, et al. Multiparametric magnetic resonance imaging and ultrasound fusion biopsy detect prostate cancer in patients with prior negative transrectal ultrasound biopsies. J. Urol. 2012;188(6):2152–7.

[60] Panebianco V, Barchetti G, Simone G, Del Monte M, Ciardi A, Grompone MD, et al. Negative multiparametric magnetic resonance imaging for prostate cancer: what's next? [Figure presented]. Eur. Urol. 2018;74(1):48–54.

[61] Washino S, Okochi T, Saito K, Konishi T, Hirai M, Kobayashi Y, et al. Combination of prostate imaging reporting and data system (PI-RADS) score and prostate-specific antigen (PSA) density predicts biopsy outcome in prostate biopsy naïve patients. BJU Int. 2017;119(2):225–33.

[62] Baco E, Ukimura O, Rud E, Vlatkovic L, Svindland A, Aron M, et al. Magnetic resonance imaging-transectal ultrasound image-fusion biopsies accurately characterize the index tumor: correlation with step-sectioned radical prostatectomy specimens in 135 patients. Eur. Urol. 2015;67(4):787–94.

[63] Russo F, Regge D, Armando E, Giannini V, Vignati A, Mazzetti S, et al. Detection of prostate cancer index lesions with multiparametric magnetic resonance imaging (mp-MRI) using whole-mount histological sections as the reference standard. BJU Int. 2016;118(1):84–94.

[64] Porpiglia F, De Luca S, Passera R, De Pascale A, Amparore D, Cattaneo G, et al. Multiparametric magnetic resonance/ultrasound fusion prostate biopsy: number and spatial distribution of cores for better index tumor detection and characterization. J. Urol. 2017;198(1):58–64.

[65] Radtke JP, Schwab C, Wolf MB, Freitag MT, Alt CD, Kesch C, et al. Multiparametric magnetic resonance imaging (MRI) and MRI–transrectal ultrasound fusion biopsy for index tumor detection: correlation with radical prostatectomy specimen. Eur. Urol. 2016;70(5):846–53.

[66] Chon CH, Lai FC, McNeal JE, Presti JC. Use of extended systematic sampling in patients with a prior negative prostate needle biopsy. J. Urol. 2002;167(6):2457–60.

[67] Radtke JP, Boxler S, Kuru TH, Wolf MB, Alt CD, Popeneciu IV, et al. Improved detection of anterior fibromuscular stroma and transition zone prostate cancer using biparametric and multiparametric MRI with MRI-targeted biopsy and MRI-US fusion guidance. Prostate Cancer Prostatic Dis. 2015;18(3): 288–96.

[68] Siddiqui MM, Rais-Bahrami S, Truong H, Stamatakis L, Vourganti S, Nix J, Hoang AN, Walton-Diaz A, Shuch B, Weintraub M, Kruecker J2, Amalou H, Turkbey B, Merino MJ, Choyke PL, Wood BJPP. Magnetic resonance imaging/ultrasound-fusion biopsy significantly upgrades prostate cancer versus systematic 12-core transrectal ultrasound biopsy. Eur. Urol. 2013;64(5):713–9.

[69] Porpiglia F, De Luca S, Passera R, Manfredi M, Mele F, Bollito E, et al. Multiparametric-magnetic resonance/ultrasound fusion targeted prostate biopsy improves agreement between biopsy and radical prostatectomy gleason score. Anticancer Res. 2016;36(9):4833–40.

[70] Margel D, Yap SA, Lawrentschuk N, Klotz L, Haider M, Hersey K, et al. Impact of multiparametric endorectal coil prostate magnetic resonance imaging on disease reclassification among active surveillance candidates: a prospective cohort study. J. Urol. 2012;187(4):1247–52.

[71] Walton Diaz A, Shakir NA, George AK, Rais-Bahrami S, Turkbey B, Rothwax JT, et al. Use of serial multiparametric magnetic resonance imaging in the management of patients with prostate cancer on active surveillance. Urol. Oncol. Semin. Orig. Investig. 2015;33(5):202e1–7.

[72] Klotz L, Loblaw A, Sugar L, Moussa M, Berman DM, Van der Kwast T, et al. Active surveillance magnetic resonance imaging study (ASIST): results of a randomized multicenter prospective trial. Eur. Urol. 2019;75(2):300–9.

[73] Villers A, Puech P, Mouton D, Leroy X, Ballereau C, Lemaitre L. Dynamic contrast enhanced, pelvic phased array magnetic resonance imaging of localized prostate cancer for predicting tumor volume: correlation with radical prostatectomy findings. J. Urol. 2006;176(6):2432–7.

[74] Girometti R, Bazzocchi M, Como G, Brondani G, Del Pin M, Frea B, et al. Negative predictive value for cancer in patients with "gray-Zone" PSA level and prior negative biopsy: preliminary results with multiparametric 3.0 tesla MR. J. Magn. Reson. Imag. 2012;36(4):943–50.

[75] Perdonà S, Di Lorenzo G, Autorino R, Buonerba C, De Sio M, Setola SV, et al. Combined magnetic resonance spectroscopy and dynamic contrast-enhanced imaging for prostate cancer detection. Urol. Oncol. Semin. Orig. Investig. 2013;31(6):761–5.

[76] Filson C, Natarajan S, Margolis D, Huang J, Lieu P, Frederick J, et al. Prostate cancer detection with magnetic resonance ultrasound fusion biopsy: the role of systematic and targeted biopsies. Cancer 2016;122(6):884–92.

[77] Moldovan PC, Van den Broeck T, Sylvester R, Marconi L, Bellmunt J, van den Bergh RCN, et al. What is the negative predictive value of multiparametric magnetic resonance imaging in excluding prostate cancer at biopsy? A systematic review and meta-analysis from the european association of Urology prostate cancer guidelines panel. Eur. Urol. 2017;72(2):250–66.

[78] Itatani R, Namimoto T, Atsuji S, Katahira K, Morishita S, Kitani K, et al. Negative predictive value of multiparametric MRI for prostate cancer detection: outcome of 5-year follow-up in men with negative findings on initial MRI studies. Eur. J. Radiol. 2014;83(10):1740–5.

[79] Venderink W, van Luijtelaar A, Bomers JG, van der Leest M, Hulsbergen-van de Kaa C, Barentsz JO, Sedelaar JPFJ. Results of targeted biopsy in men with magnetic resonance imaging lesions classified equivocal, likely or highly likely to Be clinically significant prostate cancer. Eur. Urol. 2018;73(3):353–60. pii: S0302(17): 30110–0.

[80] Egbers N, Schwenke C, Maxeiner A, Teichgräber U, Franie T. MRI-guided core needle biopsy of the prostate: acceptance and side effects. Diagnostic Interv. Radiol. 2015;21(3):215–21.

[81] Arsov C, Rabenalt R, Quentin M, Hiester A, Blondin D, Albers P, et al. Comparison of patient comfort between MR-guided in-bore and MRI/ultrasound fusion-guided prostate biopsies within a prospective randomized trial. World J. Urol. 2016;34(2):215–20.

[82] Berger AP, Gozzi C, Steiner H, Frauscher F, Varkarakis J, Rogatsch H, et al. Complication rate of transrectal ultrasound guided prostate biopsy: a comparison among 3 protocols with 6, 10 and 15 cores. J. Urol. 2004;171(4):1478–80.
[83] Borghesi M, Ahmed H, Nam R, Schaeffer E, Schiavina R, Taneja S, et al. Complications after systematic, random, and image-guided prostate biopsy [figure presented]. Eur. Urol. 2017;71(3):353–65.
[84] Wagenlehner F, et al. Infective complications after prostate biopsy: outcome of the global prevalence study of infections in Urology (GPIU) 2010 and 2011, a prospective multinational multicentre prostate biopsy study. Eur. Urol. 2013;63:521–7.
[85] Nam RK, Saskin R, Lee Y, Liu Y, Law C, Klotz LH, et al. Increasing hospital admission rates for urological complications after transrectal ultrasound guided prostate biopsy. J. Urol. 2010;183(3):963–9.
[86] Raaijmakers R, Kirkels WJ, Roobol MJ, Wildhagen MFSF. Complication rates and risk factors of 5802 transrectal ultrasound-guided sextant biopsies of the prostate within a population-based screening program. Urology 2002;60(5):826–30.
[87] Zaytoun OM, Anil T, Moussa AS, Jianbo L, Fareed K, Jones JS. Morbidity of prostate biopsy after simplified versus complex preparation protocols: assessment of risk factors. Urology 2011;77(4):910–4.

Chapter 43

Diet and the microbiome in precision medicine

Miguel Toribio-Mateas[1,2] and Adri Bester[1]

[1]*Bowels and Brains Lab, School of Applied Science, London South Bank University, London, United Kingdom;* [2]*School of Health and Education, Faculty of Transdisciplinary Practice, Middlesex University, London, United Kingdom*

Interactions between microbiome, health, and disease

The role of the gut flora in health and disease is as crucial as it is complex. Ample evidence from both human and animal studies describes the impact of dietary interventions on the idiosyncrasies of the gut microbiota, the tens of trillions of microorganisms [1,2] that inhabit our gastrointestinal system, in quantities, and diversity increasing from stomach to small intestine to colon [3,4].

Under natural conditions, intestinal bacteria share their habitat with a dynamic community of viruses, protozoa, helminths, and fungi [5–7], many of which exhibit parasitic behavior [8]. Every unique community of microorganisms interacts with their human host, through immune, neuroendocrine, and neural pathways [9], thereby casting local as well as systemic effects on the host's health, as well as disease risks. In fact, alterations in normal commensal gut microbiota result in an increase in pathogenic microbes, which deranges both microbial and host homeostasis.

Consumption of ultraprocessed foods is considered to lead to the microbial imbalance known as dysbiosis, that has been widely reported as a key contributor to the multiple system dysregulation, observed in the pathogenesis of cardiovascular [10–12], metabolic [13–16], neuroimmune [17–20], and neurodegenerative [21–24] conditions.

A recent review by Valdes et al. [25] provides a succinct yet comprehensive summary, of the role of the gut microbiota in nutrition and health, analyzing various popular food-based interventions such as low FODMAPS, polyphenol-rich diets, etc., and correlates them with microbial diversity and health. Nutrients in food are sources of information, so the cross talk between the host microbiome and their epigenome is also of paramount importance. This is beautifully depicted by Qin and Wade [26] and expanded on by Prescott et al. [27], from an angle that takes into consideration the interconnected web of biological responses. The input is based on the total exposures to factors that interact epigenetically with the microbiome and its host over time, otherwise known as the human exposome.

In practical terms, this means that when practitioners assess an individual's clinical presentation, they should consider how their lifestyle might be contributing to the composition and the functionality of their gut microbiota, and vice-versa. Seeing the person through a wide-angle lens helps clinicians account for environmental as well as psychosocial exposures as forces that, according to Prescott et al. [27], "operate collectively and cumulatively, to increase the burden of chronic disease." As food choices do not exist in isolation, practitioners with a whole-person approach should be mindful of other factors such as trauma or stress, relationships, sleep, and movement, when working with their patients.

Western diets and microbiota composition

Microbiota composition analyses of Western diets tend to report narrower bacterial diversity, along with altered microbial profiles, when compared with a Mediterranean dietary pattern [28]. It is worth remembering that Western diets are not just "high in fat" or "carbohydrate-rich," as they are often described. They feature a variety of what the NOVA food classification [29] defines as "ultraprocessed foods," i.e., hyperpalatable, "energy-dense, high in unhealthy types of fat, refined starches, free sugars and salt, and poor sources of protein, dietary fiber, and micronutrients" [30]. Typical ultraprocessed food items are fizzy drinks, margarines and spreads, cookies, biscuits, breakfast cereals, energy bars, energy drinks, prepared pies, pizzas, meat nuggets, and prepackaged or "ready meals,"

Precision Medicine for Investigators, Practitioners and Providers. https://doi.org/10.1016/B978-0-12-819178-1.00043-5

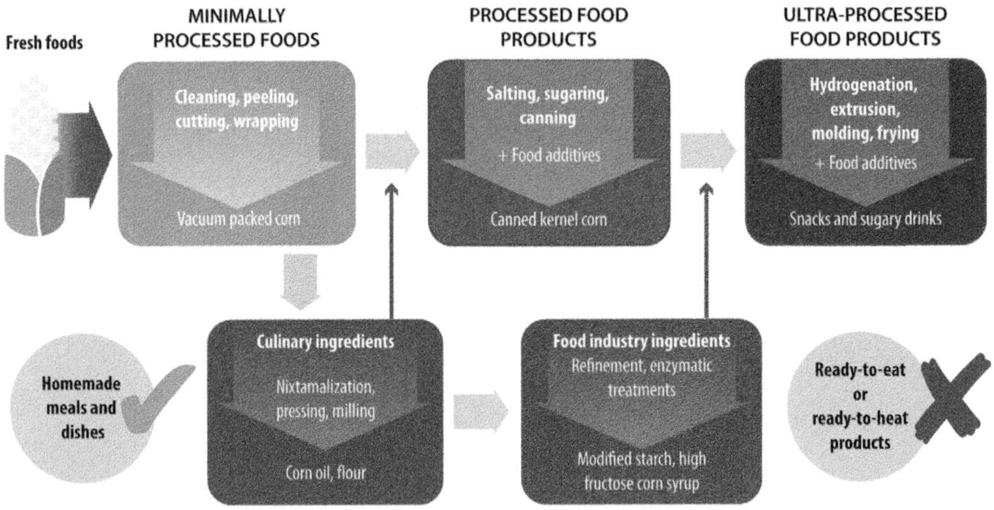

FIGURE 43.1 Industrial processing stages, from fresh food to ultraprocessed food products. *Reproduced with permission Aguayo-Patrón, S.V., Calderón de la Barca A.M. Old fashioned vs. Ultra-processed-based current diets: possible implication in the increased susceptibility to type 1 diabetes and celiac disease in childhood. Foods 2017;6:100. Licence information: Image appears in Foods. 2017 Nov 15;6(11). pii: E100. https://doi.org/10.3390/foods6110100. Licensee MDPI, Basel, Switzerland. This article is an open access article distributed under the terms and conditions of the Creative Commons Attribution (CC BY) license (http://creativecommons.org/licenses/by/4.0/).*

containing large amounts of saturated and transfats along with sucrose, emulsifiers, sweeteners, and high glycemic carbohydrates, e.g., maltodextrin, while offering little micronutrient value, as seen in Fig. 43.1.

Demonizing individual food items is not particularly helpful clinically. However, it is now known that dietary emulsifiers are able to alter the mucus layer thickness, thereby affecting the interactions between gut microbes and host cells [31], and mediating increased levels of proinflammatory lipopolysaccharide (LPS) [32]. Sweeteners have been shown to induce metabolic aberrations mediated by alterations in gut microbiota in animal models [33] and are known to reduce microbial diversity in humans [34], leading to altered glucose homeostasis, decreased satiety, increased caloric consumption, and weight gain [35].

Sweeteners can also affect cognitive processes such as memory, reward learning, and taste perception [36], all of which can affect an individual's relationship with food, as well as food choices [37]. Interestingly, a Western dietary pattern has been seen to affect autonomic regulation and vagal cardiac activity [38], and reduced heart rate variability—a sign of autonomic dysregulation—has been seen in individuals suffering from strong attachment to hyperpalatable foods, which some authors refer to as "food addiction" [39,40].

The mediterranean diet pattern as a source of richness and diversity

A growing number of studies of good methodological quality continue to provide substantiation for the ability of the Mediterranean diet (MD henceforth), to effect positive changes in gut microbiota composition and diversity. On that basis, it is proposed that an MD pattern provides a reliable answer to the question of what "a healthy diet for the microbiome" might be. Gut microbiota and metabolome analysis of 51 vegetarians, 51 vegans, and 51 omnivores, distributed across four geographically distant cities in Italy, was conducted by De Filippis et al. [41]. Adherence to the Mediterranean dietary pattern represented a vital factor, contributing to the wider diversity of their gut microbiota. The 11-unit dietary score by Agnoli et al. [42] was employed, and the diversity and abundance of "healthy foods" such brightly colored vegetables, fruit, nuts, and minimally processed cereals consumed by participants was characterized using the Healthy Food Diversity (HFD) index by Drescher et al. [43]. Participants with the highest adherence level to the MD along with the highest HFD scores presented with highest levels of fecal short-chain fatty acids, irrespective of whether their diets included specific food items such as meat, fish of dairy, etc. [41].

In the authors' own clinical experience, using a practical tool such as the "50-food challenge" data collection chart (please see Fig. 43.2) [9] can help boost patients' creativity around food choices, thereby promoting engagement with the concept of dietary diversity, and optimizing compliance.

As another taster of the ample evidence substantiating the MD's beneficial effect on the gut microbiota, in a transversal study of 31 adults without a previous diagnosis of cancer, autoimmune or digestive diseases, researchers at the Department of Functional Biology of the University of Oviedo in Asturias, Spain, found that participants with the closest adherence to a Mediterranean-style dietary pattern experienced statistically significant changes in a number of

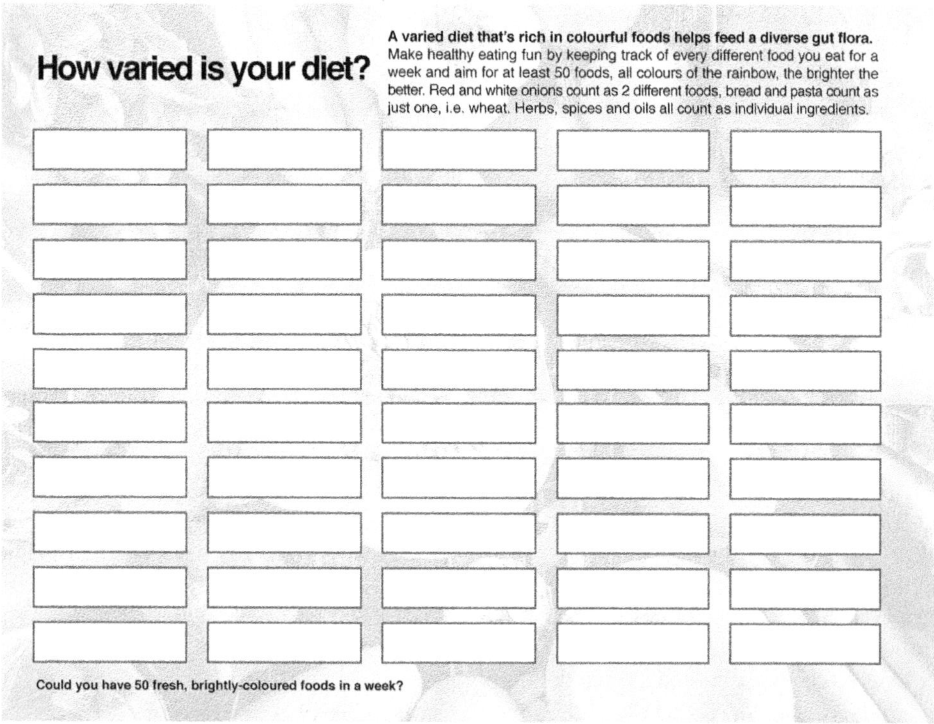

FIGURE 43.2 50-food challenge chart. *Reproduced with permission from Toribio–Mateas, M. Harnessing the power of microbiome assessment tools as part of neuroprotective nutrition and lifestyle medicine interventions. Microorganisms 2018;6:35. The "50-food challenge" chart is an example of a simple but powerful data collection tool used in clinical practice to engage with patients in a light-hearted way so that they report back to their practitioner on their dietary diversity. The rationale is to motivate individuals to vary the foods they have every day, so that they are increasing their micronutrient diversity, thereby feeding different classes of gut microbes. Licence information: Image appears in Microorganisms. 2018 Apr 25;6(2). pii: E35. https://doi.org/10.3390/microorganisms6020035. Licensee MDPI, Basel, Switzerland. This article is an open access article distributed under the terms and conditions of the Creative Commons Attribution (CC BY) license (http://creativecommons.org/licenses/by/4.0/).*

bacterial communities. This included overall higher abundance of Bacteroidetes, Prevotellaceae, and *Prevotella* and a lower concentration of Firmicutes and *Lachnospiraceae*, assessed by 16S rRNA gene sequencing [44].

Adherence to the MD was defined as scores $\geq$4 in the validated Mediterranean Diet Score (MDS) by Trichopoulou et al. [45]. These alterations in gut microbiota richness and spread are consistent with effects of increased dietary fiber from vegetables, legumes, and whole—as in minimally processed—grains, as well as phenolic compounds and carotenoids typically featured in MD foods, such as seasonal and citrus fruits, leafy, pod, and root vegetables, in addition to bulbs, e.g., onions, garlic, leeks, not forgetting red wine and coffee [46,47]. Subjects with an MDS $\geq$4 also had higher concentrations of fecal short-chain fatty acids, butyrate and propionate [44], measured by high-performance liquid chromatography.

The same research team reported a link between dietary bioactive compounds in the MD and the fecal metabolic phenolic profile of 74 healthy volunteers. According to Gutierrez-Diaz et al. [48], participants with median MDS $\geq$ 4 (based on the MDS by [45]), displayed higher levels of commensal Clostridia belonging to the cluster XVIa, particularly *Faecalibacterium prausnitzii*. This bundle of microbes from the Clostridia class falls under the Firmicutes phylum and is known for their ability to colonize the mucin layer of human colon, thereby aiding in the maintenance of gut homeostasis [49]. It includes species such as *Eubacterium rectale*, *Papillibacter cinnamivorans*, *Eubacterium ventriosum*, *Butyrivibrio crossotus*, *Clostridium orbiscindens*, *Coprococcus eutactus*, *Roseburia intestinalis*, and *F. prausnitzii*, known to play a major role in mediating the production of butyrate from fermentable dietary carbohydrates [50,51].

F. prausnitzii is considered to have strong antiinflammatory properties [52]. This is largely mediated by its ability to produce butyrate, thereby protecting the gut mucosa [53], but also by butyrate-independent pathways, which seem to include its ability to block NF-kappaB activation and IL-8 production [54]. Low levels of *F. prausnitzii*, *E. rectale*, and *Eubacterium hallii* have been associated with a peripheral inflammatory state, in patients with cognitive impairment and brain amyloidosis [21].

Low levels of genus *Roseburia* microbes have been seen in patients with primary sclerosing cholangitis and ulcerative colitis [55], as well as in those affected by constipation-predominant irritable bowel syndrome [56]. In the first metagenome-wide study of gut microbiota in type 2 diabetes mellitus (T2D), researchers found that *R. intestinalis* and *F. prausnitzii* concentrations were lower in T2D, compared with healthy individuals [57].

Further thoughts on the Mediterranean dietary pattern

For laypeople, the thought of the MD conjures images of olive oil, vibrantly colorful salads, nuts/seeds, and red wine. For clinicians, the MD is likely to elicit thoughts of large studies like PREDIMED [58]. For researchers, however, the richness in bioactive compounds found in typically Mediterranean foods probably makes the MD as exciting as it is challenging, particularly when trying to match individual dietary components with single clinical endpoints. In fact, the PREDIMED study itself featured myocardial infarction, stroke, or cardiovascular death as a primary composite endpoint [59].

The intervention group consisting of men (aged 55–75) and women (aged 60–80 years), supplementing their diet with extra-virgin olive oil (EVOO) or mixed nuts (walnuts, almonds, and hazelnuts) [60], experienced a 30% reduction in the incidence of cardiovascular events. How much might the gut microbiota of participants have contributed to that outcome? From post hoc analysis of PREDIMED data, Estruch et al. [61] found that those who consumed as much EVOO and as many nuts as they wished (ad libitum) experienced a decrease in bodyweight and less gain in central adiposity, compared with the control diet (featuring no EVOO or nuts) and mention the gut microbiota as a potential mediator of this metabolic change. Was the participants' gut microbiota also a mediator of the beneficial effects on cognitive function [62–64], breast cancer [65], and metabolic syndrome [66] confirmed by further secondary outcome analysis of PREDIMED data? Quite possibly.

In a controlled-feeding, randomized crossover study of 18 healthy men and women (mean age ≅ 53 years), Holscher et al. [67] found that those who consumed up to 42 g walnuts daily for 3 weeks had between 49% and 160% higher relative abundance of butyrate-producing *Clostridium* clusters XIVa and IV Firmicutes species such as *Faecalibacterium*, *Clostridium*, *Dialister*, and *Roseburia*, along with 16%–38% lower relative abundance of *Ruminococcus*, *Dorea*, *Oscillospira*, and *Bifidobacterium*. Walnut consumption was also seen to reduce the proinflammatory secondary bile acids deoxycholic and lithocholic acid, as well as LDL cholesterol. In a larger (n = 194, 134 women and 60 men) and longer randomized, controlled, prospective, crossover study, Bramberger et al. [68] put 96 people on a similar amount of walnuts as Holscher et al. (43 g/day) for 8 weeks, after which time participants moved on to a nut-free diet. Probiotic and butyric acid producing species were enhanced. Noteworthy is that walnut consumption increased the abundance of *Ruminococcaceae* and *Bifidobacteria*, while *Blautia* and *Anaerostipes* (both *Clostridium* cluster XIVa species) were found in significantly lower numbers. The authors believe walnuts can help improve microbial diversity, with positive metabolic ramifications.

PREDIMED nuts also featured almonds and hazelnuts, and a number of studies have looked at both. With the MD being a system made up of a number of individual but interwoven components, it is difficult to unravel the contribution of every one. The systematic review by De Souza et al. [69] provides a thorough appraisal of nuts. Different factors including fatty acid profiles, vegetable protein, fiber, vitamin, and mineral content, as well as levels of phytosterols and phenolics, make each individual nut unique [70], as does the food processing.

Holscher et al. [71] provided 42 g/d of whole (unprocessed), whole roasted, roasted chopped almonds, or almond butter. They found that the relative abundance of *Roseburia*, *Clostridium*, and *Lachnospira* increased significantly, compared to those following a nut-free control diet period. Roasted chopped almonds seemed to have the strongest effect on relative abundance of the genus *Roseburia*, *Lachnospira*, and *Oscillospira*, while whole roasted almonds also increased the relative abundance of *Lachnospira* genus, with no differences for the almond butter group.

Complementing dietary diversity with fermented foods as sources of live microorganisms

Fermented foods, based on both dairy—yoghurt, kefir, cheese—and nondairy—sauerkraut, kimchi, kombucha—substrates are easily accessible dietary tools that can help modulate gut microbiota. A recent protocol supplemented 100 g a day of a probiotic yoghurt versus a high-strength probiotic supplement. The yoghurt contained just two strains of *Lactobacillus acidophilus* LA5 and *Bifidobacterium lactis* BB12 with 10^7 colony forming units (CFU)/g, and the supplement contained *Lactobacillus casei* 3×10^3, *L. acidophilus* 3×10^7, *Lactobacillus rhamnosus* 7×10^9, *Lactobacillus bulgaricus* 5×10^8, *Bifidobacterium breve* 2×10^{10}, *Bifidobacterium longum* 1×10^9, and *Streptococcus thermophilus* 3×10^8 CFU/g, with 100 mg of fructo-oligosaccharide and lactose as carrier substances. The control group was given a conventional yoghurt containing the starter cultures of *S. thermophilus* and *L. bulgaricus* and experienced no statistically significant improvement in mental health markers, which included depression, anxiety, and stress based on validated scales [72]. The fact the study reported similar effectiveness of a supplement and a whole-food intervention is encouraging. Yoghurt is a low cost and versatile food that can be served or blended with a variety of fresh foods and spices, e.g., fruit, cinnamon, ginger, thereby minimizing expense, while improving patient compliance.

In another randomized controlled trial, researchers at the University of Connecticut [73] found that premenopausal women who ate just 339 g of yoghurt for 9 weeks experienced a reduction in biomarkers of chronic inflammation and endotoxin exposure (including LPS, LPS-binding protein, IgM endotoxin-core antibody (IgM EndoCAb), and zonulin). The yoghurt used in the intervention group happened to be low fat. Based on the natural low fat content of yoghurt, it is questionable whether comparable results would have been achieved using full-fat yoghurt instead.

While the endpoints measured were different, a 5-year prospective study of 4545 elderly individuals (55–80 years of age), at high cardiovascular risk, evaluated the association between yoghurt consumption and the reversion of abdominal obesity status. Whole-fat yoghurt had a more beneficial effect than low fat yoghurt varieties [74]. Additional health benefits of regular yoghurt consumption include the support of effective weight management [75], improved lactose digestion [76], and lower risk of developing T2D [77].

It has been difficult to distinguish between yoghurt benefits derived from the nutrients contained in the food matrix, e.g., calcium, vitamin B_{12}, etc., and those associated with the fermenting microorganisms, making it more of a challenge attributing separate effects to fermented foods alone, than as part of a whole dietary pattern.

Final considerations

Diversity of fresh produce is a key contributor to a healthy microbial population, a state known as *eubiosis*, which contrasts with the previously discussed perturbed state of dysbiosis. The authors are both practitioners and clinical researchers, involved in the design and implementation of clinical trials, assessing the impact of foods on the microbiome and on mental health markers. Having carried out extensive, ongoing review of the literature, they propose that polyphenol-rich vegetables, legumes, fruit, herbs, spices, and olive oil, plus the fermented foods typically featured in the Mediterranean dietary pattern, plus the tendency to include fewer ultraprocessed food items than standard British or American diets are the two most significant factors contributing to the characteristics of a eubiotic or healthy gut ecosystem, such as that described in studies delving into the microbial composition and metabolites in the gut of Mediterranean populations. This is illustrated in Fig. 43.3.

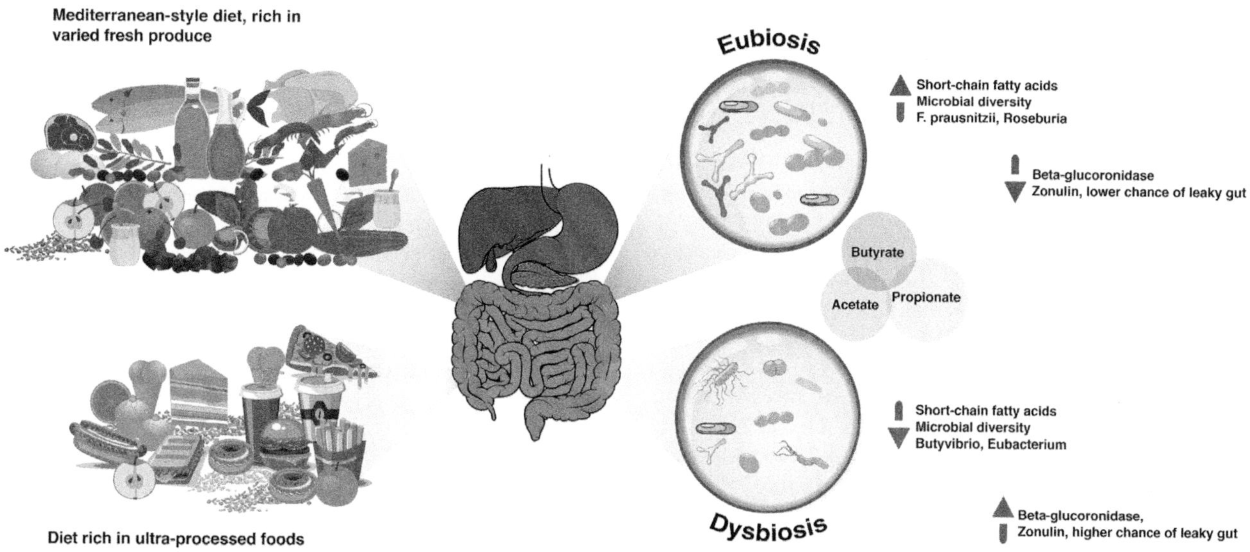

FIGURE 43.3 Achieving and maintaining a healthy gut microbial ecosystem with a Mediterranean-type dietary pattern. Dietary patterns that are rich in fresh produce, including a variety of brightly colored vegetables and fruits, olive oil, nuts, and seeds, e.g., a Mediterranean-style diet, are seen to promote eubiosis, contributing to higher levels of all three short-chain fatty acids and to a wider microbial diversity. Other characteristics of a eubiotic gut ecosystem include lower levels of beta-glucuronidase, documented to help with normal elimination of toxicants, and lower zonulin levels, seen as an indication of reduced susceptibility of damage to the intestinal barrier, i.e., less "leaky gut". On the other hand, dietary patterns rich in ultraprocessed foods and particularly those rich in refined carbohydrates combined with high fat levels are seen to promote gut dysbiosis. Lower microbial diversity and lower levels of short-chain fatty acid levels are seen in patients whose diets consist of mostly of ultraprocessed foods. Other markers are also affected. Beta-glucuronidase may be higher, which could pose issues with toxicant elimination via reduced activity of phase II detoxification pathways. There is also a higher susceptibility for barrier tissue damage. Higher zonulin levels in stool would give practitioners an indication that this is the case. A disrupted intestinal barrier tends to be correlated with a number of negative health outcomes. *Adapted from Toribio–Mateas, M. Harnessing the power of microbiome assessment tools as part of neuroprotective nutrition and lifestyle medicine interventions. Microorganisms 2018;6:35.*

References

[1] Backhed F, Ley RE, Sonnenburg JL, Peterson DA, Gordon JI. Host-bacterial mutualism in the human intestine. Science 2005;307:1915–20.

[2] Backhed F, Roswall J, Peng Y, Feng Q, Jia H, Kovatcheva-Datchary P. Dynamics and stabilization of the human gut microbiome during the first year of life. Cell Host Microbe 2015;17.

[3] Sekirov I, Russell SL, Antunes LC, Finlay BB. Gut microbiota in health and disease. Physiol. Rev. 2010;90:859–904.

[4] Brown EM, Sadarangani M, Finlay BB. The role of the immune system in governing host-microbe interactions in the intestine. Nat. Immunol. 2013;14:660–7.

[5] Chabe M, Lokmer A, Segurel L. Gut protozoa: friends or foes of the human gut microbiota? Trends Parasitol 2017;33:925–34.

[6] Marzano V, Mancinelli L, Bracaglia G, DEL Chierico F, Vernocchi P, DI Girolamo F, Garrone S, Tchidjou Kuekou H, D'argenio P, Dallapiccola B, Urbani A, Putignani L. "Omic" investigations of protozoa and worms for a deeper understanding of the human gut "parasitome". PLoS Negl. Trop. Dis. 2017;11:e0005916.

[7] Jenkins TP, Rathnayaka Y, Perera PK, Peachey LE, Nolan MJ, Krause L, Rajakaruna RS, Cantacessi C. Infections by human gastrointestinal helminths are associated with changes in faecal microbiota diversity and composition. PLoS One 2017;12:e0184719.

[8] Leung JM, Graham AL, Knowles SCL. Parasite-microbiota interactions with the vertebrate gut: synthesis through an ecological lens. Front. Microbiol. 2018;9:843.

[9] Toribio-Mateas M. Harnessing the power of microbiome assessment tools as part of neuroprotective nutrition and lifestyle medicine interventions. Microorganisms 2018;6:35.

[10] Battson ML, Lee DM, Jarrell DK, Hou S, Ecton KE, Weir TL, Gentile CL. Suppression of gut dysbiosis reverses western diet-induced vascular dysfunction. Am. J. Physiol. Endocrinol. Metab. 2018 May 1;314(5):E468–77.

[11] Battson ML, Lee DM, Weir TL, Gentile CL. The gut microbiota as a novel regulator of cardiovascular function and disease. J. Nutr. Biochem. 2017;56:1–15.

[12] Heianza Y, Ma W, Manson JE, Rexrode KM, Qi L. Gut microbiota metabolites and risk of major adverse cardiovascular disease events and death: a systematic review and meta-analysis of prospective studies. J. Am. Heart Assoc 2017;6.

[13] Haro C, Garcia-Carpintero S, Rangel-Zuniga OA, Alcala-Diaz JF, Landa BB, Clemente JC, PEREZ-Martinez P, Lopez-Miranda J, Perez-Jimenez F, Camargo A. Consumption of two healthy dietary patterns restored microbiota dysbiosis in obese patients with metabolic dysfunction. Mol. Nutr. Food Res. 2017;61.

[14] Reimer RA, Willis HJ, Tunnicliffe JM, Park H, Madsen KL, Soto-Vaca A. Inulin-type fructans and whey protein both modulate appetite but only fructans alter gut microbiota in adults with overweight/obesity: a randomized controlled trial. Mol. Nutr. Food Res. 2017;61.

[15] Kassaian N, Aminorroaya A, Feizi A, Jafari P, Amini M. The effects of probiotic and synbiotic supplementation on metabolic syndrome indices in adults at risk of type 2 diabetes: study protocol for a randomized controlled trial. Trials 2017;18:148.

[16] Mitchell CM, Davy BM, Halliday TM, Hulver MW, Neilson AP, Ponder MA, Davy KP. The effect of prebiotic supplementation with inulin on cardiometabolic health: rationale, design, and methods of a controlled feeding efficacy trial in adults at risk of type 2 diabetes. Contemp. Clin. Trials 2015;45:328–37.

[17] Miyake S, Kim S, Suda W, Oshima K, Nakamura M, Matsuoka T, Chihara N, Tomita A, Sato W, Kim SW, Morita H, Hattori M, Yamamura T. Dysbiosis in the gut microbiota of patients with multiple sclerosis, with a striking depletion of species belonging to clostridia XIVa and IV clusters. PLoS One 2015;10:e0137429.

[18] Le Berre L, Rousse J, Gourraud PA, Imbert-Marcille BM, Salama A, Evanno G, Semana G, Nicot A, Dugast E, Guerif P, Adjaoud C, Freour T, Brouard S, Agbalika F, Marignier R, Brassat D, Laplaud DA, Drouet E, Van Pesch V, Soulillou JP. Decrease of blood anti-alpha1,3 Galactose Abs levels in multiple sclerosis (MS) and clinically isolated syndrome (CIS) patients. Clin. Immunol. 2017;180:128–35.

[19] Montassier E, Berthelot L, Soulillou JP. Are the decrease in circulating anti-alpha1,3-Gal IgG and the lower content of galactosyl transferase A1 in the microbiota of patients with multiple sclerosis a novel environmental risk factor for the disease? Mol. Immunol. 2018;93:162–5.

[20] Nagy-Szakal D, Williams BL, Mishra N, Che X, Lee B, Bateman L, Klimas NG, Komaroff AL, Levine S, Montoya JG, Peterson DL, Ramanan D, Jain K, Eddy ML, Hornig M, Lipkin WI. Fecal metagenomic profiles in subgroups of patients with myalgic encephalomyelitis/chronic fatigue syndrome. Microbiome 2017;5:44.

[21] Cattaneo A, Cattane N, Galluzzi S, Provasi S, Lopizzo N, Festari C, Ferrari C, Guerra UP, Paghera B, Muscio C, Bianchetti A, Volta GD, Turla M, Cotelli MS, Gennuso M, Prelle A, Zanetti O, Lussignoli G, Mirabile D, Bellandi D, Gentile S, Belotti G, Villani D, Harach T, Bolmont T, Padovani A, Boccardi M, Frisoni GB. Association of brain amyloidosis with pro-inflammatory gut bacterial taxa and peripheral inflammation markers in cognitively impaired elderly. Neurobiol. Aging 2017;49:60–8.

[22] Chen CH, Lin CL, Kao CH. Irritable bowel syndrome is associated with an increased risk of dementia: a nationwide population-based study. PLoS One 2016;11:e0144589.

[23] Devos D, Lebouvier T, Lardeux B, Biraud M, Rouaud T, Pouclet H, Coron E, Bruley des Varannes S, Naveilhan P, Nguyen J-M, Neunlist M, Derkinderen P. Colonic inflammation in Parkinson's disease. Neurobiol. Dis. 2013;50:42–8.

[24] Tetz G, Brown SM, Hao Y, Tetz V. Parkinson's disease and bacteriophages as its overlooked contributors. Sci. Rep. 2018;8:10812.

[25] Valdes AM, Walter J, Segal E, Spector TD. Role of the gut microbiota in nutrition and health. BMJ 2018;361:k2179.

[26] Qin Y, Wade PA. Crosstalk between the microbiome and epigenome: messages from bugs. J. Biochem. 2018;163:105–12.

[27] Prescott SL, Logan AC. Each meal matters in the exposome: biological and community considerations in fast-food-socioeconomic associations. Econ. Hum. Biol. 2017;27:328–35.

[28] Garcia-Mantrana I, Selma-Royo M, Alcantara C, Collado MC. Shifts on gut microbiota associated to mediterranean diet adherence and specific dietary intakes on general adult population. Front. Microbiol. 2018;9:890.

[29] Monteiro CA, Cannon G, Moubarac JC, Levy RB, Louzada MLC, Jaime PC. Ultra-processing. An odd "appraisal". Publ. Health Nutr. 2018a;21:497–501.

[30] Monteiro CA, Cannon G, Moubarac JC, Levy RB, Louzada MLC, Jaime PC. The UN decade of nutrition, the NOVA food classification and the trouble with ultra-processing. Publ. Health Nutr. 2018b;21:5–17.

[31] Cani PD, Everard A. Keeping gut lining at bay: impact of emulsifiers. Trends Endocrinol. Metabol. 2015;26:273–4.
[32] Chassaing B, Van de Wiele T, DE Bodt J, Marzorati M, Gewirtz AT. Dietary emulsifiers directly alter human microbiota composition and gene expression ex vivo potentiating intestinal inflammation. Gut 2017;66:1414–27.
[33] Suez J, Korem T, Zeevi D, Zilberman-Schapira G, Thaiss CA, Maza O, Israeli D, Zmora N, Gilad S, Weinberger A, Kuperman Y, Harmelin A, Kolodkin-Gal I, Shapiro H, Halpern Z, Segal E, Elinav E. Artificial sweeteners induce glucose intolerance by altering the gut microbiota. Nature 2014;514:181–6.
[34] Frankenfeld CL, Sikaroodi M, Lamb E, Shoemaker S, Gillevet PM. High-intensity sweetener consumption and gut microbiome content and predicted gene function in a cross-sectional study of adults in the United States. Ann. Epidemiol. 2015;25:736–42. e4.
[35] Pearlman M, Obert J, Casey L. The association between artificial sweeteners and obesity. Curr. Gastroenterol. Rep. 2017;19:64.
[36] Burke MV, Small DM. Physiological mechanisms by which non-nutritive sweeteners may impact body weight and metabolism. Physiol. Behav. 2015;152:381–8.
[37] Creze C, Candal L, Cros J, Knebel JF, Seyssel K, Stefanoni N, Schneiter P, Murray MM, Tappy L, Toepel U. The impact of caloric and non-caloric sweeteners on food intake and brain responses to food: a randomized crossover controlled trial in healthy humans. Nutrients 2018;10.
[38] Kim JY, Kang HL, Kim DK, Kang SW, Park YK. Eating habits and food additive intakes are associated with emotional states based on EEG and HRV in healthy Korean children and adolescents. J. Am. Coll. Nutr. 2017;36:335–41.
[39] Meule A, Kubler A. Food cravings in food addiction: the distinct role of positive reinforcement. Eat. Behav. 2012;13:252–5.
[40] Young HA, Benton D. Heart-rate variability: a biomarker to study the influence of nutrition on physiological and psychological health? Behav. Pharmacol. 2018;29:140–51.
[41] de Filippis F, Pellegrini N, Vannini L, Jeffery IB, LA Storia A, Laghi L, Serrazanetti DI, DI Cagno R, Ferrocino I, Lazzi C, Turroni S, Cocolin L, Brigidi P, Neviani E, Gobbetti M, O'toole PW, Ercolini D. High-level adherence to a mediterranean diet beneficially impacts the gut microbiota and associated metabolome. Gut 2016;65:1812–21.
[42] Agnoli C, Krogh V, Grioni S, Sieri S, Palli D, Masala G, Sacerdote C, Vineis P, Tumino R, Frasca G, Pala V, Berrino F, Chiodini P, Mattiello A, Panico S. A priori–defined dietary patterns are associated with reduced risk of stroke in a large Italian cohort. J. Nutr. 2011;141:1552–8.
[43] Drescher LS, Thiele S, Mensink GB. A new index to measure healthy food diversity better reflects a healthy diet than traditional measures. J. Nutr. 2007;137:647–51.
[44] Gutierrez-Diaz I, Fernandez-Navarro T, Sanchez B, Margolles A, Gonzalez S. Mediterranean diet and faecal microbiota: a transversal study. Food Funct 2016;7:2347–56.
[45] Trichopoulou A, Kouris-Blazos A, Wahlqvist ML, Gnardellis C, Lagiou P, Polychronopoulos E, Vassilakou T, Lipworth L, Trichopoulos D. Diet and overall survival in elderly people. BMJ 1995;311:1457–60.
[46] Tap J, Furet JP, Bensaada M, Philippe C, Roth H, Rabot S, Lakhdari O, Lombard V, Henrissat B, Corthier G, Fontaine E, Dore J, Leclerc M. Gut microbiota richness promotes its stability upon increased dietary fibre intake in healthy adults. Environ. Microbiol. 2015;17:4954–64.
[47] Gonzalez S, Fernandez M, Cuervo A, Lasheras C. Dietary intake of polyphenols and major food sources in an institutionalised elderly population. J. Hum. Nutr. Diet. 2014;27:176–83.
[48] Gutierrez-Diaz I, Fernandez-Navarro T, Salazar N, Bartolome B, Moreno-Arribas MV, DE Andres-Galiana EJ, Fernandez-Martinez JL, DE Los Reyes-Gavilan CG, Gueimonde M, Gonzalez S. Adherence to a mediterranean diet influences the fecal metabolic profile of microbial-derived phenolics in a Spanish cohort of middle-age and older people. J. Agric. Food Chem. 2017;65:586–95.
[49] Lopetuso LR, Scaldaferri F, Petito V, Gasbarrini A. Commensal Clostridia: leading players in the maintenance of gut homeostasis. Gut Pathog. 2013;5:23. 23.
[50] Van Den Abbeele P, Belzer C, Goossens M, Kleerebezem M, DE Vos WM, Thas O, DE Weirdt R, Kerckhof FM, VAN DE Wiele T. Butyrate-producing *Clostridium* cluster XIVa species specifically colonize mucins in an in vitro gut model. ISME J. 2013;7:949–61.
[51] El Aidy S, Van Den Abbeele P, Van de Wiele T, Louis P, Kleerebezem M. Intestinal colonization: how key microbial players become established in this dynamic process: microbial metabolic activities and the interplay between the host and microbes. Bioessays 2013;35:913–23.
[52] Lopez-Siles M, Duncan SH, Garcia-Gil LJ, Martinez-Medina M. *Faecalibacterium prausnitzii*: from microbiology to diagnostics and prognostics. ISME J. 2017;11:841–52.
[53] Sokol H, Seksik P, Furet JP, Firmesse O, Nion-Larmurier I, Beaugerie L, Cosnes J, Corthier G, Marteau P, Dore J. Low counts of *Faecalibacterium prausnitzii* in colitis microbiota. Inflamm. Bowel Dis. 2009;15:1183–9.
[54] Sokol H, Pigneur B, Watterlot L, Lakhdari O, Bermudez-Humaran LG, Gratadoux JJ. *Faecalibacterium prausnitzii* is an anti-inflammatory commensal bacterium identified by gut microbiota analysis of Crohn disease patients. Proc. Natl. Acad. Sci. U.S.A. 2008;105.
[55] Bajer L, Kverka M, Kostovcik M, Macinga P, Dvorak J, Stehlikova Z, Brezina J, Wohl P, Spicak J, Drastich P. Distinct gut microbiota profiles in patients with primary sclerosing cholangitis and ulcerative colitis. World J. Gastroenterol. 2017;23:4548–58.
[56] Gobert AP, Sagrestani G, Delmas E, Wilson KT, Verriere TG, Dapoigny M, DEL'homme C, Bernalier-Donadille A. The human intestinal microbiota of constipated-predominant irritable bowel syndrome patients exhibits anti-inflammatory properties. Sci. Rep. 2016;6:39399.
[57] Qin J, Li Y, Cai Z, Li S, Zhu J, Zhang F. A metagenome-wide association study of gut microbiota in type 2 diabetes. Nature 2012;490.
[58] Estruch R, Ros E, Salas-Salvado J, Covas MI, Corella D, Aros F, Gomez-Gracia E, Ruiz-Gutierrez V, Fiol M, Lapetra J, Lamuela-Raventos RM, Serra-Majem L, Pinto X, Basora J, Munoz MA, Sorli JV, Martinez JA, Martinez-Gonzalez MA. Primary prevention of cardiovascular disease with a Mediterranean diet. N. Engl. J. Med. 2013;368:1279–90.
[59] Guasch-Ferre M, Salas-Salvado J, Ros E, Estruch R, Corella D, Fito M, Martinez-Gonzalez MA. The PREDIMED trial, Mediterranean diet and health outcomes: how strong is the evidence? Nutr. Metabol. Cardiovasc. Dis. 2017;27:624–32.
[60] Martínez-González MA, Salas-Salvadó J, Estruch R, Corella D, Fitó M, Ros E. Benefits of the Mediterranean diet: insights from the PREDIMED study. Prog. Cardiovasc. Dis. 2015;58:50–60.

[61] Estruch R, Martínez-González MA, Corella D, Salas-Salvadó J, Fitó M, Chiva-Blanch G, Fiol M, Gómez-Gracia E, Arós F, Lapetra J, Serra-Majem L, Pintó X, BUIL-Cosiales P, Sorlí JV, Muñoz MA, BASORA-Gallisá J, Lamuela-Raventós RM, Serra-Mir M, Ros E. Effect of a high-fat Mediterranean diet on bodyweight and waist circumference: a prespecified secondary outcomes analysis of the PREDIMED randomised controlled trial. Lancet Diabetes Endocrinol 2016;4:666–76.

[62] Martinez-Lapiscina EH, Clavero P, Toledo E, San Julian B, Sanchez-Tainta A, Corella D, Lamuela-Raventos RM, Martinez JA, Martinez-Gonzalez MA. Virgin olive oil supplementation and long-term cognition: the PREDIMED-NAVARRA randomized, trial. J. Nutr. Health Aging 2013;17:544–52.

[63] Martínez-Lapiscina EH, Clavero P, Toledo E, Estruch R, Salas-Salvadó J, San Julián B, Sanchez-Tainta A, Ros E, Valls-Pedret C, Martinez-Gonzalez MÁ. Mediterranean diet improves cognition: the PREDIMED-NAVARRA randomised trial 2013;84:1318–25.

[64] Valls-Pedret C, Sala-Vila A, Serra-Mir M, Corella D, DE LA Torre R, Martinez-Gonzalez MA, Martinez-Lapiscina EH, Fito M, Perez-Heras A, Salas-Salvado J, Estruch R, Ros E. Mediterranean diet and age-related cognitive decline: a randomized clinical trial. JAMA Intern. Med 2015;175:1094–103.

[65] Toledo E, Salas-Salvado J, Donat-Vargas C, BUIL-Cosiales P, Estruch R, Ros E, Corella D, Fito M, Hu FB, Aros F, Gomez-Gracia E, Romaguera D, Ortega-Calvo M, Serra-Majem L, Pinto X, Schroder H, Basora J, Sorli JV, Bullo M, Serra-Mir M, Martinez-Gonzalez MA. Mediterranean diet and invasive breast cancer risk among women at high cardiovascular risk in the PREDIMED trial: a randomized clinical trial. JAMA Intern. Med 2015;175:1752–60.

[66] Babio N, Toledo E, Estruch R, Ros E, Martínez-González MA, Castañer O, Bulló M, Corella D, Arós F, Gómez-Gracia E, Ruiz-Gutiérrez V, Fiol M, Lapetra J, Lamuela-Raventos RM, Serra-Majem L, Pintó X, Basora J, Sorlí JV, SALAS-Salvadó J. Mediterranean diets and metabolic syndrome status in the PREDIMED randomized trial 2014;186:E649–57.

[67] Holscher HD, Guetterman HM, Swanson KS, An R, Matthan NR, Lichtenstein AH, Novotny JA, Baer DJ. Walnut consumption alters the gastrointestinal microbiota, microbially derived secondary bile acids, and health markers in healthy adults: a randomized controlled trial. J. Nutr. 2018a;148:861–7.

[68] Bamberger C, Rossmeier A, Lechner K, Wu L, Waldmann E, Fischer S, Stark RG, Altenhofer J, Henze K, Parhofer KG. A walnut-enriched diet affects gut microbiome in healthy caucasian subjects: a randomized, controlled trial. Nutrients 2018;10.

[69] de Souza RGM, Schincaglia RM, Pimentel GD, Mota JF. Nuts and human health outcomes: a systematic review. Nutrients 2017;9.

[70] Lamuel-Raventos RM, Onge M-PS. Prebiotic nut compounds and human microbiota. Crit. Rev. Food Sci. Nutr. 2017;57:3154–63.

[71] Holscher HD, Taylor AM, Swanson KS, Novotny JA, Baer DJ. Almond consumption and processing affects the composition of the gastrointestinal microbiota of healthy adult men and women: a randomized controlled trial. Nutrients 2018b;10.

[72] Mohammadi AA, Jazayeri S, Khosravi-Darani K, Solati Z, Mohammadpour N, Asemi Z, Adab Z, Djalali M, Tehrani-Doost M, Hosseini M, Eghtesadi S. The effects of probiotics on mental health and hypothalamic-pituitary-adrenal axis: a randomized, double-blind, placebo-controlled trial in petrochemical workers. Nutr. Neurosci. 2016;19:387–95.

[73] Pei R, Dimarco DM, Putt KK, Martin DA, Gu Q, Chitchumroonchokchai C, White HM, Scarlett CO, Bruno RS, Bolling BW. Low-fat yogurt consumption reduces biomarkers of chronic inflammation and inhibits markers of endotoxin exposure in healthy premenopausal women: a randomised controlled trial. Br. J. Nutr. 2017;118:1043–51.

[74] Santiago S, Sayon-Orea C, Babio N, Ruiz-Canela M, Marti A, Corella D, Estruch R, Fito M, Aros F, Ros E, Gomez-Garcia E, Fiol M, Lapetra J, Serra-Majem L, Becerra-Tomas N, Salas-Salvado J, Pinto X, Schroder H, Martinez JA. Yogurt consumption and abdominal obesity reversion in the PREDIMED study. Nutr. Metabol. Cardiovasc. Dis. 2016;26:468–75.

[75] Mozaffarian D, Hao T, Rimm EB, Willett WC, Hu FB. Changes in diet and lifestyle and long-term weight gain in women and men. N. Engl. J. Med. 2011;364:2392–404.

[76] Savaiano DA. Lactose digestion from yogurt: mechanism and relevance. United States: American Society for Clinical Nutrition, Inc; 2014.

[77] Chen M, Sun Q, Giovannucci E, Mozaffarian D, Manson JE, Willett WC, Hu FB. Dairy consumption and risk of type 2 diabetes: 3 cohorts of US adults and an updated meta-analysis. BMC Med. 2014;12:215.

[78] Aguayo-Patrón SV, Calderón de la Barca AM. Old fashioned vs. Ultra-processed-based current diets: possible implication in the increased susceptibility to type 1 diabetes and celiac disease in childhood. Foods 2017;6:100.

Chapter 44

Nanotheranostics in oncology and drug development for imaging and therapy

Bluma Linkowski Faintuch[1] and Salomao Faintuch[2]

[1]*Radiopharmacy Center, Institute of Energy and Nuclear Research, São Paulo, São Paulo, Brazil;* [2]*Department of Radiology, Harvard Medical School, Boston, United States*

Introduction

> *It's far more important to know what person the disease has, than what disease the person has.*
>
> Hippocrates of Cos (460–377 BCE)

Cancer was one of the first, and financially is the largest application, field for molecular pharmacology. The majority of new drugs in the cancer pipeline are extensively investigated from the genomic, metabolomic, and proteomic point of view. Indeed, natural history confirms that malignancies evolve during years or even decades, essentially as molecular aberrations, with phenotypic expression only relatively late in tumorigenesis [1].

If this powerful stimulus for early diagnosis and prevention is not enough, there are imperative reasons for targeted therapy later on. No known condition is genomically as heterogeneous, both spatially and chronologically [2]. Clonal mutations are a feature of cancer, particularly the more aggressive ones. Intratumoral genetic differences are already important in initial lesions, and even more so after metastasization. This could explain why chemotherapy responses often exhibit incomplete or unpredictable patterns, and why ensuing growth and metastases may require ever different prescriptions and strategies [3].

Precision medicine (PM) has been announced as "prevention and treatment strategies, that take in account individual variability in genes, environment, and lifestyle, by integrating detailed information from multiple sources in a holistic manner" [4]. Such approach closely fits the uncertainties and challenges of oncological management, as discussed below.

Cancer nanomedicine and nanotechnologies

Systems biology, digital revolution, and high expectations of demanding and well-informed patients are rapidly conducting to a proactive type of P4 medicine, meaning that it is predictive, preventive, personalized, and participatory [5].

Nanomaterials could help to fill the gap, between ambitious promises and the hard reality. Nanomedicine is the application of materials between 1 and 100 nm, to healthcare modalities [6]. From inorganic particles to proteins and organic structures, this new route is finding significant indications in oncological management, even though its diagnostic and therapeutic spectrum is much wider [7].

Nanomaterials are helpful as delivery vehicles for drugs and/or imaging agents. Liposomes, polymer carriers (micelles, hydrogels, polymersomes, dendrimers, nanofibers), metallic particles (gold, silver, titanium), carbon nanostructures (nanotubes, nanodiamonds, graphene), inorganic particles (silica), and hybrid nanomaterials are representative examples [8].

In the cancer realm, one should not overlook nanotheranostics, with approaches in cancer screening, diagnosis, imaging, therapy, and prevention of toxicity. Nanoparticles are relatively independent of physiological absorption and transportation mechanisms. They can breach cell boundaries, and directly hit subcellular and molecular targets for which they have been designed, even in the case of relatively ischemic and anoxic malignancies.

Within such circumstances, dosage can be substantially reduced, and collateral damage minimized. At the same time, chemical specificity is enhanced, with less likelihood of false positives and negatives during diagnostic procedures, as well as of nonresponders when therapy is the objective [9].

Clinical imaging

Available imaging modalities already visualize anatomical changes, such as X-ray, computed tomography (CT), ultrasound, and magnetic resonance imaging (MRI), along with functional changes, encompassing positron emission

Precision Medicine for Investigators, Practitioners and Providers. https://doi.org/10.1016/B978-0-12-819178-1.00044-7

tomography (PET), single-photon emission computed tomography (SPECT), functional MRI (fMRI), and optical imaging [10]. Nanoparticles are able to enhance many of these results.

Molecular imaging

Molecular imaging is defined by the Society of Nuclear Medicine as the "visualization, characterization, and measurement of biological processes at the molecular and cellular levels, in humans and other living systems" [11].

Molecular changes within cells and tissues occur for years and even decades before detectable cancer emerges [12]. Prolonged dedifferentiation or maldifferentiation, as well as shifts in growth factors, cytokines, and inflammatory mediators can be part of the carcinogenesis pathway; consequently real-time visualization at the molecular level is a priority. Not only a general oncological diagnosis can be anticipated, but also clues about tumor biology and invasiveness.

Probes

Molecular functional imaging (MFI) probes have been developed for assessing metabolism, angiogenesis, hypoxia, receptor imaging, apoptosis, and other phenomena in vivo. High sensitivity, along with temporal and spatial resolution, has been responsible for the growing acceptance of such techniques. Phenotypic, targeted cell-tracking and reporter gene probes exist in the laboratory or practice, although others are envisaged as well. Among the targeting molecules, peptides, aptamers, high-molecular-weight antibodies, engineered protein fragments, and signaling components deserve to be highlighted [10].

Modern probes are designed as a single multifunctional molecule, often containing a target recognition unit, a signal-releasing unit, and a linker unit. High signal-to-noise ratio and low toxicity are of course indispensable features [13]. Multiple radionuclide-bound probes are already adopted for PET, SPECT, or in the case of MRI, paramagnetic chelates [14].

For optical molecular imaging, distribution and kinetics of a fluorescent probe are assessed using a fluorescent imaging instrument. Photoacoustic imaging probes for high-resolution imaging in deep tissues, ultrasound probes, and hybrid approaches are additional options. Skin carcinoma and melanoma; head and neck cancer; thyroid, breast, gynecologic, prostate, colon, liver, and brain malignancies are the most investigated conditions [15].

Nanoparticle theranostics

The concept of theranostics, which is a portmanteau of therapeutics and diagnostics, tries to bridge the gap between customized therapy and targeted diagnosis. In other words, the molecular constructs that allow diagnosis should concomitantly or subsequently be endowed with therapeutic properties. Within such circumstances, an integrated plan can be designed for the patient, theoretically fusing diagnosis and precision treatment [3].

Radiolabeled nanoparticles with imaging qualities and potential cytotoxic abilities are the most widely used oncological theranostic protocols. The technologies are not necessarily overlapping, as isotopes for imaging are different from those emitting cancer-killing particles, yet a combined strategy can be devised. As the tumor is radiotraced, prompt therapy within the same general framework can be planned and executed.

General requirements

High tumor affinity, low toxicity to surrounding tissues, deep tissue penetration, and easy dosage adjustment are logical priorities. Equally indispensable is the possibility of synthesizing the chemical moiety into multifunctional hybrid nanoparticles, as few molecules are naturally endowed with imaging, biodistribution, solubility, cell penetration, or cancer killing abilities [16].

Probe models

Nanoscale probes differ according to imaging modality, cancer vector, and therapeutic cargo. A multicomponent nanoplatform delivery system is designed to encompass and optimize anticancer treatment, with imaging functions and pharmacological properties as alluded to. In certain contexts, this image-guided therapy can include photodynamic and photothermal therapy, radiosensitization, and image-guided surgery [17].

The formatting depends on the theranostic system, be it optical (fluorescent imaging), magnetic (MRI), radiological (computed tomography), nuclear medicine (alpha), beta (−), beta (+), gamma, and Auger electron sources, or photoacoustic (nanotubes, nanorods) [18].

Recent examples are superparamagnetic iron oxide (SPIO) and ultrasmall SPIO (USPIO) for MRI, gold nanoparticles for CT and also radiolabeled methods, quantum dots (QDs) and upconversion nanoparticles (UCNPs) for optical imaging, and carbon or gold nanoparticles for photoacoustic evaluation [19].

Radioactive probes

Prostate-Specific Membrane Antigen (PSMA) ligand probes, which are not nanoparticles however are endowed with theranostic properties, carrying therapeutic loads such as Lutetium-177, Lead-212/203, Actinium-225, Bismuth-223, and other isotopes are strongly envisaged as options for resistant metastatic prostate cancer, with modified ligands for assorted tumors [20].

Tumor hypoxia is a well-known marker of chemotherapy resistance, as well as aggressiveness and poor prognosis. Resistance tracking probes can provide imaging clues about such areas, at the same enabling targeted treatment [21].

Brain malignancies represent a particularly tough diagnostic and therapeutic challenge due to complex anatomy, proximity of essential and vital structures, and especially the blood–brain barrier, which hampers access of large molecules. Customized theranostics could be ideally suited for noninvasive handling of these dangerous lesions [22].

Radioactive multifunctional probes are of obvious interest; however, chemotherapy-loaded paramagnetic particles, or gadolinium nanohybrid dendrimers, for use with MRI, should not be overlooked [23]. Other focused compounds are polymeric and lipid-based nanosystems [24].

Radiomics and radiogenomics

The amalgamation of omics with radiologic techniques (radiomics) is not yet a therapeutic tool; however, it may importantly subsidize the handling of malignancies. Specifically genomic patterns (radiogenomics) could help to tailor therapy in highly heterogeneous or metastatic diseases, by decomposing and digitally mapping each point and voxel of the images.

Metabolomics and proteomics

By the same token, radioproteomics and radiometabolomics are encouraging tools to reveal the molecular subtypes of cancers (deep phenotyping). Virtual biopsy models could be envisaged, in a safe and noninvasive fashion. New diagnostic and therapeutic targets would also be unearthed The National Cancer Institute (NCI) Quantitative Imaging Network (QIN) is one of the major initiatives in this domain [4].

Brain cancer

Brain low grade glioma has undergone radioproteomic investigation, based on combined analysis of datasets accrued in two different cohorts. Reverse-phase protein array (RPPA) expression levels, and radiologist scored image features, were submitted to multiple response regression. Different proteins were associated with each imaging feature, shedding light on the phenotypic aberrations, triggered by metabolomic derangements in low-grade glioma. Deranged signaling pathways encompassed IL8 (interleukin −8), PTEN (phosphatase and tensin homolog), PI3K/Akt (phosphatidylinositol 3-kinase/ protein kinase B), Neuregulin, MAPK/ERK (mitogen activated protein kinase/ extracellular signal regulated kinase), p70S6K (ribosomal protein S6 kinase), and EGF (epidermal growth factor) [25].

Breast tumor

Multiparametric in vivo MRI enriches the morphologic profile collected by conventional MRI imaging, with metabolic information about biochemicals or metabolites involved in metabolic pathways. Altered choline, phospholipid, and energy metabolism has emerged, with proton (^{1}H), phosphorus (^{31}P), and carbon (^{13}C) breast magnetic resonance spectroscopy (MRS).

Also lipid metabolite peaks were visualized by quantitative MRS, with differences not only between benign and malignant masses, but also according to recurrence-free survival. This paves the way for improved therapeutic decisions in this population [26].

Tumor volume, apparent diffusion coefficient, volume transfer coefficient, and extracellular volume ratio have also been measured, providing valuable clues concerning tumor microstructure, blood vessels, and grade [27].

Pharmacodynamic and pharmacotherapic investigation

PET and SPECT are highly sensitive diagnostic tools, detecting as little as 10^{-12} M of tracer. They are also well suited for organ accumulation tests, regardless of tissue depth [7]. The amount of nanomedicine delivered to cancer can be calculated as well, something beyond the scope of most noninvasive protocols. These abilities are especially useful for biodistribution, drug release, and therapeutic efficacy assessment of nanomedicines, in real-time.

PET relies on positron-emitting radionuclides, emitting pairs of γ-rays to generate imaging contrast. The most frequently used are ^{11}C, ^{13}N, ^{15}O, ^{18}F, ^{44}Sc, ^{62}Cu,^{64}Cu, ^{68}Ga, ^{72}As, ^{74}As, ^{76}Br, ^{82}Rb, ^{86}Y, ^{89}Zr, and ^{124}I. SPECT is based on noncoincident γ-rays generated by radionuclides. Commonly used radioisotopes are ^{99m}Tc, ^{111}In, ^{123}I, ^{125}I, and ^{201}Tl [28].

By replacing gamma or positron radionuclides with alpha, beta or auger-electron emitters, a therapeutic effect can be achieved. Actually for more than 60 years ^{123}I/^{131}I "theranostics" were prescribed, and are still recommended in many guidelines, for well-differentiated thyroid cancers, in association with total surgical thyroidectomy.

Radioimmunotheranostics

Monoclonal antibodies (mAbs), which are not nanoparticles, however can be counted among the remarkable advances in theranostics, can be shaped as tumor seeking vectors, as adopted in many settings of oncological therapy. The option for radioimmunotherapy has been successfully introduced for lymphomas and leukemias [29]. Yet, long-lived radionuclides have to be used, given the slow accumulation and excretion of mAbs. Healthy tissue irradiation is also a concern, thus demanding careful design and utilization. Pretargeted radiotherapy is one of the alternatives to increase radiation dose, and attenuate toxicity [7].

Positron-emitting radioisotopes can be combined with a therapeutic radionuclide of the same element (which emits β^- or α-particles, or low-energy Auger/conversion electrons), composing custom-made pairs of elements. By means of PET imaging, dosimetry and quantification of therapy are enabled.

The basis of such a theranostic approach, as alluded to, is the matched-pair of radionuclides. A major gap is thus filled, as single agents are rarely simultaneously suitable for diagnosis and cancer chemotherapy [30].

Oncologic drug development

Targeted therapies generate significant hurdles for drug development laboratories, from preclinical models to custom-designed human trials, encompassing specific biomarkers and end-points.

Nuclear imaging techniques are able to play a role during drug development. They can be applied to investigate biodistribution of nanomedicines, to quantify drug release, to find the optimal clinical dosage schedule, and to assess real-time therapeutic efficacy [7].

In 2015, 13 of the 45 drugs approved by FDA were characterized as precision medicines, by the Personalized Medicine Coalition (PMC). The proportion is steadily increasing, as in 2014 only 9 out of 41 new drugs were analogously classified [30]. Total precision medicine market was over USD 39,726.7 million in 2015 [1].

The pharmaceutical industry is obviously a key stakeholder, for materialization of better and more efficacious drugs and techniques, reducing attrition in the transition between the laboratory stage and the real world, as well as optimizing development costs [31].

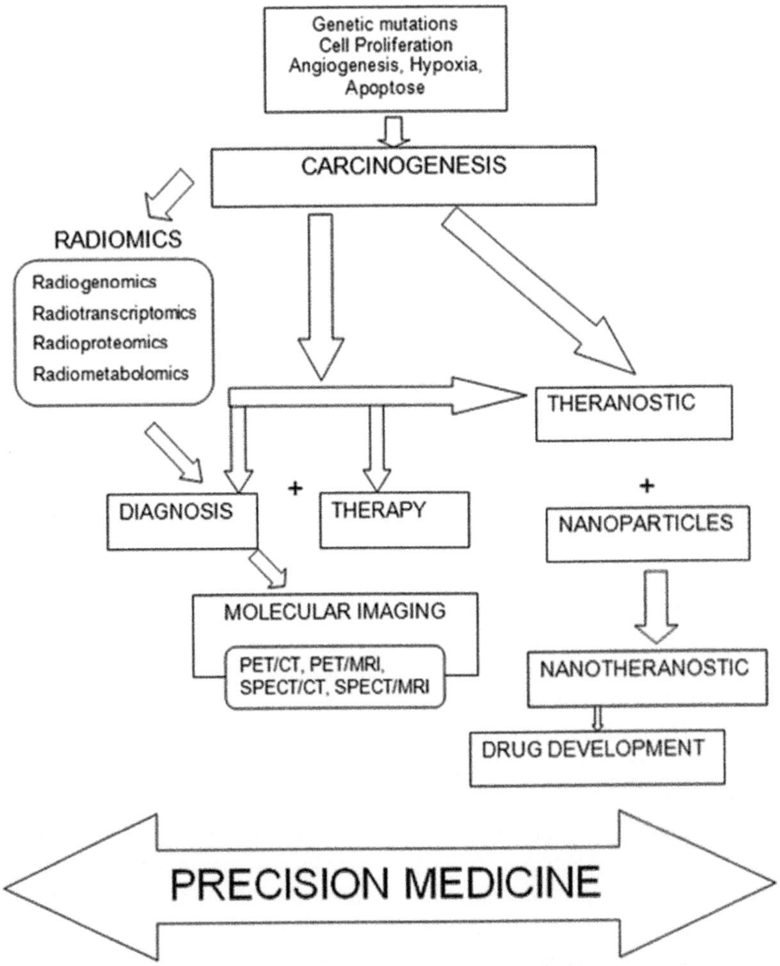

FIGURE 44.1 Oncological nanotheranostics and functional imaging fit the larger picture of Precision Medicine.

Challenges and perspectives

Delivering the right treatment to the right patient at the right time is not easily accomplished. Yet molecular functional imaging entails the visualization of biochemical shifts, since early carcinogenesis stages, thus enabling "real time" snapshots of the natural history of the disease.

Nanotheranostics concurs with visual confirmation of targets, and monitoring of therapeutics. It converts comparatively blind, mass approaches, or "one size fits all" prescriptions, into more focused interventions.

The incorporation of the same techniques into new drug development cycles, will help to pinpoint new targets and biomarkers, reduce false starts, and accelerate the transition toward successful clinical trials.

Barriers must be overcome in the field of cancer theranostic probes, regarding blood circulation time, tumor microenvironment penetration, cell internalization, and efficient drug release. Nanomedicine size is a constraint, as molecules below 6 nm undergo rapid renal excretion, whereas those in the range of 200 nm tend to be phagocytized by the reticuloendothelial system. It has been estimated that as little as 1%–5% of synthesized nanoparticles effectively reach the tumor [7].

As emphasized in another chapter, the best shortcut to this goal is an academic industrial alliance, integrating scientists, clinical professionals, commercial companies, regulators, and other legitimate parties, all sharing the same interest in healthcare progress, within the principles and frameworks of precision diagnosis and therapy (Fig. 44.1).

References

[1] Conde J. Time to empower cancer nanotechnology initiative for precision medicine. Proc. Nat. Res. Soc. 2017;1:01001. https://nrs.org/journal/pnrs/browse-the-journal/volume-1.

[2] Herold CJ, Lewin JS, Wibmer AG, Thrall JH, Krestin GP, Dixon AK, Schoenberg SO, Geckle RJ, Muellner A, Hricak H. Imaging in the age of precision medicine: summary of the proceedings of the 10th biannual symposium of the international society for strategic studies in radiology. Radiology 2016;279(1):226–38.

[3] Ghasemi M, Nabipour I, Omrani A, Alipour Z, Assadi M. Precision medicine and molecular imaging: new targeted approaches toward cancer therapeutic and diagnosis. Am. J. Nucl. Med. Mol. Imaging 2016;6(6):310–27.

[4] Gillies RJ, Kinahan PE, Hricak H. Radiomics: images are more than pictures, they are data. Radiology 2016;278(2):563–77.

[5] Hood L. Systems biology and P4 medicine: past, present, and future. Rambam. Maimonides Med. J. 2013;4(2):e0012.

[6] Tran S, DeGiovanni PJ, Piel B, Rai P. Cancer nanomedicine: a review of recent success in drug delivery. Clin. Transl. Med. 2017;6(44):1–21.

[7] Steen EJL, Edem PE, Nørregaard K, Jørgensen JT, Shalgunov V, Kjaer A, Herth MM. Pretargeting in nuclear imaging and radionuclide therapy: improving efficacy of theranostics and nanomedicines. Biomaterials 2018;179:209–45.

[8] Tong R, Kohane DS. New strategies in cancer nanomedicine. Annu. Rev. Pharmacol. Toxicol. 2016;56:41–57.

[9] Wang EC, Wang AZ. Nanoparticles and their applications in cell and molecular biology. Integr. Biol. (Camb.) 2014;6(1):9–26.

[10] Vaidya T, Agrawal A, Mahajan S, Thakur MH, Mahajan A. The continuing evolution of molecular functional imaging in clinical oncology: radiogenomics part 1. Mol. Diagn. Ther. 2019 November 1;23(1):1–26.

[11] Kurdzeil K, Ravizzini G, Croft B, Tatum J, Choyke P, Kobayashi H. The evolving role of nuclear molecular imaging in cancer. Expert Opin. Med. Diagn. 2008;2(7):829–42.

[12] Diaz- Cano SJ. Tumor heterogeneity: mechanisms and bases for a reliable application of molecular marker design. Int. J. Mol. Sci. 2012;13(2):1951–2011.

[13] Angelakeris M. Magnetic nanoparticles: a multifunctional vehicle for modern theranostics. Biochim. Biophys. Acta Gen. Subj. 2017;1861(6):1642–51.

[14] Lamb J, Holland JP. Advanced methods for radiolabeling multimodality nanomedicines for SPECT/MRI and PET/MRI. J. Nucl. Med. 2018;59(3):382–9.

[15] Saji H. In vivo molecular imaging. Biol. Pharm. Bull. 2017;40(10):1605–15.

[16] Silva CO, Pinho JC, Lopes JM, Almeida AJ, Gaspar MM, Reis C. Current trends in cancer nanotheranostics: metallic, polymeric, and lipid-based systems. Pharmaceutics 2019;11(1):22.

[17] Sikkandhar MG, Nedumaran AM, Ravichandar R, Singh S, Santhakumar I, Goh ZC, Mishra S, Archunan G, Gulyás B, Padmanabhan P. Theranostic probes for targeting tumor microenvironment: an overview. Int. J. Mol. Sci. 2017;18(5). pii:E1036.

[18] Mottaghy FM. Non-invasive molecular imaging and theranostic probes. Methods 2017;130:1–3.

[19] Sisay B, Abrha S, Yilma Z, Assen A, Molla F, Tadese E, Wondimu A, Gebre-Samuel N, Pattnaik G. Cancer nanotheranostics: a new paradigm of simultaneous diagnosis and therapy. J. Drug Deliv. Ther. 2014;4(5):79–86. www.jddtonline.info/index.php/jddt/article/view/967.

[20] Beheshti M, Heinzel A, von Mallek D, Filss C, Mottaghy FM. PSMA radioligand therapy of prostate cancer. Q. J. Nucl. Med. Mol. Imaging 2019;63(1):29–36.

[21] Liu JN, Bu W, Shi J. Chemical design and synthesis of functionalized probes for imaging and treating tumor hypoxia. Chem. Rev. 2017;117(9):6160–224.

[22] Bhatt A, Gurnany E, Modi A, Gulbake A, Jain A. Theranostic potential of targeted nanoparticles for brain cancer. Mini. Rev. Med. Chem. 2017;17(18):1758–77.

[23] Shen C, Wang X, Zheng Z, Gao C, Chen X, Zhao S, Dai Z. Doxorubicin and indocyanine green loaded superparamagnetic iron oxidenanoparticles with PEGylated phospholipid coating for magnetic resonance withfluorescence imaging and chemotherapy of glioma. Int. J. Nanomed. 2018;20(14):101–17.

[24] Silva CO, Pinho JO, Lopes JM, Almeida AJ, Gaspar MM, Reis C. Current trends in cancer nanotheranostics: metallic, polymeric, and lipid-based systems. Pharmaceutics 2019;11(22).

[25] Lehrer M, Bhadra A, Ravikumar V, Chen JY, Wintermark M, Hwang SN, Holder CA, Huang EP, Fevrier-Sullivan B, Freymann JB, Rao A. TCGA Glioma phenotype research group. Multiple-response regression analysis links magnetic resonance imaging features to de-regulated protein expression and pathway activity in lower grade glioma. Oncoscience 2017;23(4):57–66.

[26] Thakur SB, Horvat JV, Hancu I, Sutton OM, Bernard-Davila B, Weber M, Oh JH, Marino MA, Avendano D, Leithner D, Brennan S, Giri D, Manderski E, Morris EA, Pinker K. Quantitative in vivo proton MR spectroscopic assessment of lipid metabolism: value for breast cancer diagnosis and prognosis. J. Magn. Reson. Imaging 2019;50(1):239–49.

[27] Jagannathan NR. Application of in vivo MR methods in the study of breast cancer metabolism. NMR Biomed. 2018 Nov 20 e:4032.

[28] Kazuma SM, Sultan D, Zhao Y, Detering L, Meng Y, Luehmann HP, Abdalla DSP, Liu Y. Recent advances of radionuclide-based molecular imaging of atherosclerosis. Curr. Pharmaceut. Des. 2015;21(36):5267–76.

[29] Larson SM, Carrasquillo JA, Cheung NKV, Press O. Radioimmunotherapy of human tumours. Nat. Rev. Canc. 2015;15(6):347–60.

[30] Qaim SM, Scholten B, Neumaier B. New developments in the production of theranostic pairs of radionuclides. J. Radioanal. Nucl. Chem. 2018;318(3):1493–509.

[31] Hollingsworth SJ. Precision medicine in oncology drug development: a pharma perspective. Drug Discov. Today 2015;20(12):1455–11463.

Section III

Hospital, managed care, and public health applications

Chapter 45

Organoids for cell therapy and drug discovery

Cláudia C. Miranda[1,2] and Joaquim M.S. Cabral[1,2]

[1]*iBB – Institute for Bioengineering and Biosciences and Department of Bioengineering, Instituto Superior Técnico, Universidade de Lisboa, Lisbon, Portugal;* [2]*The Discoveries Centre for Regenerative and Precision Medicine, Lisbon Campus, Instituto Superior Técnico, Universidade de Lisboa, Lisbon, Portugal*

Introduction

Animal models are a very important part of the drug screening process, before initiating human trials. Nevertheless, the differences between species often lead to biological variability that can delay the translation to human studies, as toxicity and efficacy of a drug candidate are predicted in animal testing [1]. During the period of 1969–2002, 75 drugs or drug products were removed from the market, due to unpredicted drug effects on humans [2]. Of all the drugs that enter human clinical trials, only 10% prove high efficacy for the target disease [3]. Currently, there are already some alternatives for in vitro tests with human-derived cells. Still, most of the platforms rely on the use of immortalized or primary cell lines, and two-dimensional (2D) cultures. These often fail to mimic either the three-dimensional (3D) interactions between cells, or the interactions between tissues and organs of the human body as whole, restraining each test to a single tissue response.

Stem cells, and particularly, human pluripotent stem cells (hPSC), hold the promise of providing an assortment of solutions to various medical problems, as they possess a long-term self-renewal capacity, and the ability to differentiate into virtually any type of cell within the human body. These properties impose stem cells as an appealing alternative for therapeutic applications. Moreover, hPSCs offer the possibility to develop cell therapies, tissue regeneration, and better understanding of embryonic development and directed cell differentiation. In the future, cell therapies and disease modeling will be able to provide patient-specific treatments [4].

Stem cells, either originating from primary tissues or derived from hPSCs, have the ability to differentiate and give rise to a 3D tissue that resembles the respective organ in terms of cell type, composition, structure, and function [5]. Organoids have been addressed as a humanized approach, toward the assessment of the efficacy and toxicity of new potential drugs and, at the same time aim to increase the success rate of new drug development, in order to reduce the costs and delays.

Development and utilization of human pluripotent stem cells

Research efforts are still required regarding the establishment of standard culture protocols, including purification steps, to assure that cells applied to the patient are not pluripotent any more, eliminating the possibility of teratoma formation. In addition, scaleup processes under good manufacturing practices (GMPs) should be implemented in order to achieve a clinical relevant number of cells [6,7]. Also, the integrative nature of most reprogramming techniques raised concerns for cellular therapy applications, as the first iPSCs were generated by using retrovirus or lentivirus [8,9]. In order to bypass this problem, several alternative approaches have been developed, including adenovirus [10], plasmid transfection [11], RNA [12], or direct introduction of the four reprogramming proteins [13].

The majority of human embryonic stem cells (hESCs) and human induced (hiPSC)-based products in clinical trials, are directed toward treatment of ophthalmologic diseases, namely age-related macular degeneration, myopic macular degeneration, and Stagardt disease (Table 45.1). However, there are also trials involving severe heart failure, type I diabetes, and spinal cord injury [14,15]. In fact, the first clinical trial using cellular therapy with hPSC was directed toward spinal cord injury, using hESC-derived oligodendrocytes [16]. It was started at Geron Corporation

Precision Medicine for Investigators, Practitioners and Providers. https://doi.org/10.1016/B978-0-12-819178-1.00045-9

TABLE 45.1 Clinical trials based on hPSC-derived products.

Company/Sponsor	Location	Disease	Cell Therapy	PSC source	Patients	Phase
Chabiotech Co. Ltd.	South Korea	Macular degeneration	RPEs	hESC	12	I/II
Ocata Therapeutics	CA, USA	Stargardt macular dystrophy	RPEs	hESC	16	I/II
		Macular degeneration	RPEs	hESC	16	I/II
		Myopic macular degeneration	RPEs	hESC	n/a	I/II
Pfizer	UK	Macular degeneration	RPEs	hESC	10	I
Cell Cure Neurosciences Ltd.	Israel	Macular degeneration	RPEs	hESC	15	I/II
ViaCyte	CA, USA	Type I diabetes mellitus	PECs	hESC	40	I/II
Assistance Publique, Hopitaux de Paris	France	Heart failure	CD15+ Isl-1$^+$ progenitors	hESC	6	I
International Stem Cell Corp.	Australia	Parkinson disease	NSCs	Human parthenogenetic	12	I/II
Asterias Biotherapeutics	CA, USA	Spinal cord injury	OPCs	hESC	13	I/II

NSCs, neural stem cells; *OPCs*, oligodendrocyte progenitor cells; *PECs*, pancreatic progenitor cells; *RPEs*, retinal pigment epithelium cells.

(Menlo Park, CA, USA), and is currently in phase I/II trial by Asterias Biotherapeutics (Fremont, CA, USA).

Cell therapy with stem cell products

Neurodegenerative diseases also represent a target for cell therapy, especially involving cell replacement (e.g., Parkinson disease, Alzheimer disease, amyotrophic lateral sclerosis). A rat model of Parkinson disease has proven that injection of dopaminergic neuron progenitors, derived in vitro, is able to engraft and retain the ability to maturate into region-specific dopaminergic neurons [17]. Moreover, these neurons were able to survive long-term in this model, forming synaptic connections and restoring motor functions [18].

Heart disease is the principal cause of death worldwide, and several preclinical studies are currently testing hPSC-derived therapy, in large animal models of heart disease. Both cell sheets containing hPSC-derived cardiomyocytes [19] and fibrin-based patches containing hPSC-derived endothelial cells and smooth muscle cells [20] have demonstrated promising results after transplantation to pigs with induced myocardial infarction. hPSC-derived cardiomyocytes were also successfully engrafted, upon injection into nonhuman primate hearts [21].

Organoids as a platform for drug development

3D models simulate the effect of drugs, during the first phases of embryonic development. Valproic acid is a widely used antiepileptic drug, which may interfere with the formation of the neural tube of embryos, often leading to birth defects such as microcephaly, a condition associated with reduced cognitive function and impaired motor functions [22]. In order to facilitate the screening of potentially neurotoxic drugs, Miranda et al. developed a platform based on hPSC-derived neural progenitor cells (NPCs) that is able to re-create in vitro the effect of VPA [23]. In cells exposed to VPA during the neural induction of hPSCs, there was a prevalence of neural progenitor structures, such as neural rosettes, over neuronal differentiation, confirming the hypothesis that VPA can induce microcephaly.

Tailored-drugs for a specific type of cancer can be obtained from pancreatic organoids that are derived from biopsies of cancer patients, or from a bank of organoids from cells extracted from colorectal tumors [24,25]. This approach can reduce the time for identification of the most effective drug, according to the subtype of cancer.

When cell therapy is a reliable alternative, organoids may be the best option for cell transplants. Organoids have shown to integrate better upon transplantation to a host, than progenitor cells. In mouse models, cerebral organoids exhibit enhanced survival, while promoting better vascularization compared to transplantation of NPCs [26].

Disease models with stem cells

Lancaster et al. [27] used encapsulated hPSCs to generate neural 3D organoids. This strategy employed an extracellular matrix (ECM), specifically Matrigel, to serve as support for development of a brain organoid that was able to model microcephaly, a disease where the brain fails to properly develop, leading to a decrease in number of neurons and synaptic connections.

Different models have been devised that range from neurological diseases such as Rett-syndrome, to vascular and hematological diseases, and even cancer models. Rett-syndrome is a neurological disease, closely related with motor function impairment and autistic-like behavior, which is commonly caused by mutations in the MECP2 gene of the X chromosome, crucial for the correct functioning of the brain [28]. Using hiPSC technology to reprogram patient-specific cells, it was possible to recapitulate the development of the in vitro human brain, and confirm the existence of features of Rett-syndrome, such as a lower number of neurons and less complex neurites [29].

Recently, a model of vascular disease based on diabetes-induced vessel damage was reported [30]. Endothelial cells were derived from hPSC, using a basement membrane, to create human blood vessel organoids that upon transplantation to mice and exposure to a diabetic environment were able to mimic diabetic vasculopathy features, such as thickening of vascular basement membrane. A model for prediction of drug absorption in the small intestine was also developed, using intestinal epithelial cells derived from hiPSCs [31]. These cells demonstrated characteristics such as tight junctions, metabolic enzymes, and drug transporters that closely mimic the mechanisms of oral drug absorption, making this system a powerful tool toward the prediction of absorption in vivo.

Aleman et al. have developed a model of metastasis from colorectal cancer, which simulates tumor cell integration within different tissues, such as lung, heart, and liver [32]. A microfluidic device with interconnected chambers, each containing cells from the abovementioned tissues, encapsulated in hyaluronic acid, as well as an initial chamber with colorectal tumor encapsulated cells, was able to track the detachment of tumor cells from the tumor, and its integration in the other constructs. Interestingly, there was a preferential migration of tumor cells into the lung and liver constructs, which correspond to the organs where metastasis occurs in colorectal cancer patients.

Bioprocesses for stem cell expansion and differentiation

The use of cells for cell therapy replacement approaches involves the need of a large number of cells. A single severe ischemic event, can lead to loss of 10^9 cardiomyocytes that need to be replaced [33]. In the case of Parkinson disease, one of the most common neurodegenerative conditions, it is estimated that between 30 and 70 million human parthenogenetic-derived neural stem cells per patient would be needed to replace the lost dopaminergic neurons [34]. Spinal cord injury currently affects more than 200,000 patients, with an estimated increase of 17,000 patients per year in the USA alone [35]. In order to treat it, current clinical trials are using up to 20 million oligodendrocyte progenitor cells, per injected dose [36].

Commitment of hPSCs into a desired lineage, in clinically relevant numbers, still faces several technical issues. Commonly, in vitro differentiation protocols are based on adherent cultures that often fail to mimic in vivo development in terms of spatial organization. In addition, 2D cultures have a limited available surface for cell growth and attachment, increasing the cost, labor, and time associated to the culture. Therefore, 3D conditions have emerged as a promising alternative to perform efficient expansion and controlled commitment of hPSCs.

Stem cell cultivation and bioprocessing in bioreactors and spinner flasks have been thoroughly addressed in the past few years (Fig. 45.1) [37,38], as they represent a classical approach for scaling up. Successful expansion of hPSCs as 3D aggregates was attained using different culture media and system parameters, yielding up to 2.85×10^6 cells/mL by using a perfusion method, which represents an increase of 47% in cell yield when compared with batch cultures [39]. In terms of directed differentiation, 3D suspension-based systems have yielded up to 50 million cardiomyocytes in a 100 mL controlled bioreactor scale [40] and an average 90% efficiency of hPSC-derived cardiomyocytes in a 1 L spinner flask [41].

Although the most studied scalable differentiation in bioreactors relies on cardiac induction, neural induction protocols that yield large number of NPCs are starting to emerge. It is already possible to attain an integrated expansion and neural induction protocol that leads to the generation of more than 14×10^6 NPCs in only 6 days at a 30 mL scale, which represents ~80% of differentiation efficiency [42].

Nevertheless, expansion and directed differentiation of hPSCs in spinner flasks as 3D cultures may promote aggregate agglomeration or inflict damages to the culture, due to shear stress derived from agitation. Lei et al. addressed the problem of hPSC aggregate agglomeration

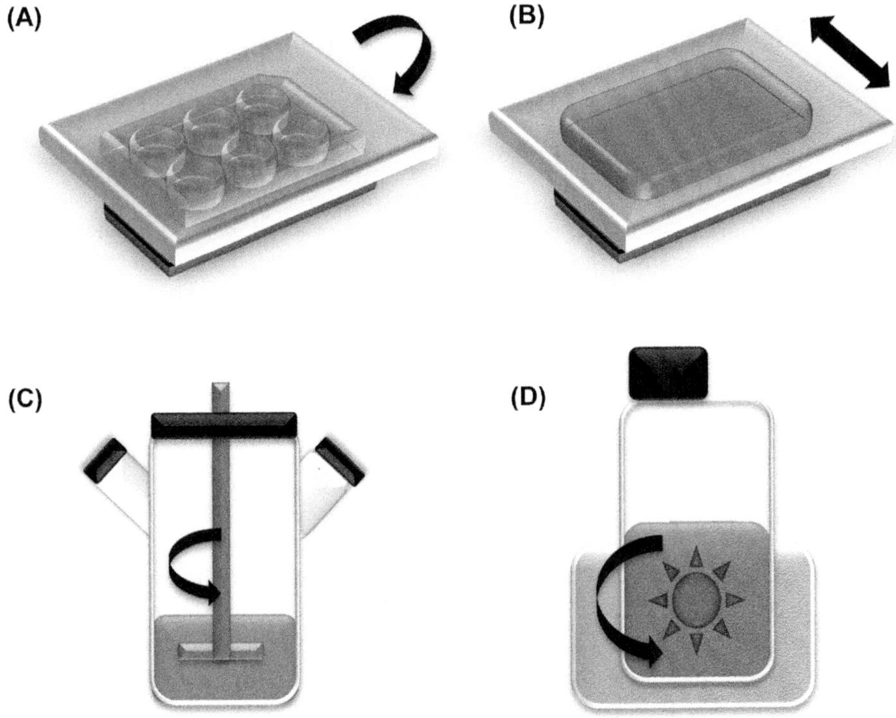

FIGURE 45.1 Examples of platforms for hPSC expansion and differentiation. (A). Rotary agitation shaker. (B). Wave bioreactor. (C). Spinner flask. (D). Vertical-wheel bioreactor.

and scalability, by using a thermoresponsive hydrogel, PNIPAAm-PEG [43]. The embedding of hPSCs within this hydrogel promoted a high expansion rate and, at the same time, enabled their differentiation into dopaminergic progenitor cells, yielding an ~80-fold expansion after 15 days. This system also promoted a scalable derivation of oligodendrocyte progenitor cells (OPCs) [44], and midbrain dopaminergic neurons [45] that upon transplantation were able to efficiently engraft, migrate, and in the case of the OPCs, mature into oligodendrocytes.

An important step of the bioprocessing of stem cells relies on the purification of the final product (Fig. 45.2). Although the efficiencies obtained by directed differentiation protocols are increasing, with the optimization of signaling pathways' modulation, there is still the risk, especially for cell therapy, that a small fraction of hPSCs will fail to differentiate, introducing the problem of teratoma formation upon in vivo transplantation. Even for drug screening applications, there is the risk of contamination of results by a residual population of hPSCs. For that purpose, several purification strategies have been developed, and adapted to enrich the desired cell population [46].

Multiorganoid systems for drug screening

The first organ-on-a-chip system that was released to the market was reported in 2010 (Fig. 45.3) [47]. This perfused system comprised a multiwell plate with an array of 12 wells, which was used to culture hepatocytes on an extracellular matrix (ECM)-coated scaffold. In the past few years, several organ- and multiorgan-chips have been developed (Table 45.2).

One of the major problems in drug development is the effect on the kidneys. Drug-induced kidney injury can be a limiting factor in pharmacotherapy, creating the need to monitor biomarkers that can reflect the in vivo situation. Kidney-on-a-chip platforms often employ cell models that do not fully recapitulate human cell behavior. Models range from nonhuman cells, such as Madin-Darby canine kidney cells, to immortalized renal tubular cell line HK-2. In terms of functionality, primary renal proximal tubule epithelial cells are the most promising. Still, they are derived from a primary source that can be susceptible to donor variability, which can be surpassed by using immortalized PTEC lines.

Kidney organoids

Although the best approach to derive human kidney organoids appears to be by differentiation of hPSCs, the complexity of the nephron architecture makes it difficult to efficiently differentiate kidney organoids that can fully recreate their structure and function. Nevertheless, different approaches have been described in the past years, either using 2D or 3D systems [48]. 3D kidney organoids demonstrated to be more sensitive to monitor drug-induced

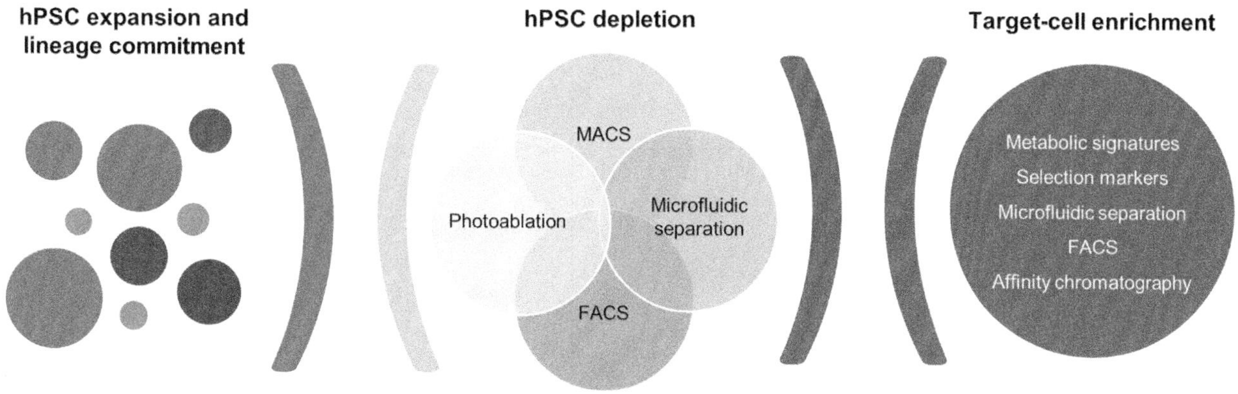

FIGURE 45.2 From hPSC to a target-cell–enriched population. After expansion and lineage commitment of hPSCs, there is the need to deplete the culture from leftover hPSCs, in order to reduce the risk of teratoma formation. For that, methods such as magnetic activated cell sorting (MACS), fluorescent-activated cell sorting (FACS), photoablation or microfluidic separation, could be used. In the end of the differentiation protocols, the target cell population can also be enriched through FACS, metabolic signatures, selection markers, microfluidic separation, or even affinity chromatography.

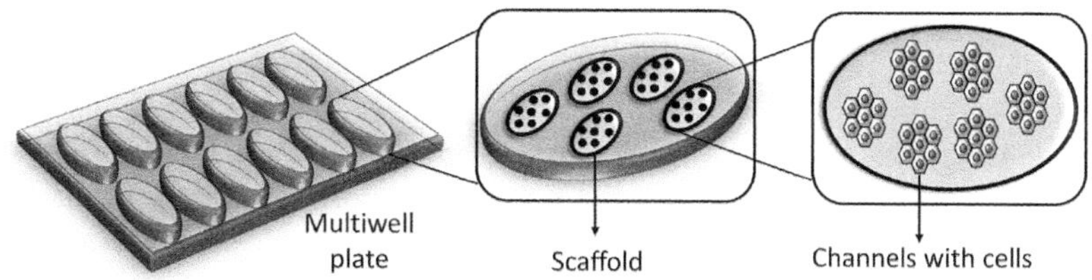

FIGURE 45.3 Schematics of the first commercially available organ-on-a-chip platform. As cell culture medium is perfused within the bioreactor, based on flow rate and time of culture, it is possible to determine oxygen consumption rates of the cells.

toxicity [49]. Exposure of kidney tissue to three different drugs— Cisplatin, Gentamicin, and Doxorubicin— and assessment of acute and chronic toxicity revealed that cells cultured in 2D failed to represent chronic exposure responses, whereas 3D-cultured cells were able to serve as a monitor of chronic toxicity. Moreover, the cells grown in 2D were shown to be less sensitive to lower drug concentrations, than 3D-cultured cells.

Gut organoids

Trietsch et al. recently developed a parallelized organ-o-a-chip platform that mimics intestinal epithelial tubes, and is able to assess barrier integrity on real time [50]. This platform, OrganoPlate (Mimetas, Leiden, The Netherlands), incorporates 40 microfluidic cell culture chambers, each one containing three parallel lanes—one for a polarized cell culture of a human colon adenocarcinoma cell line, the center for ECM gel, and the third one for flow alone—with fluid flow in the outer lanes (Fig. 45.4). If the intestinal barrier is compromised, cells from the first lane will leak the fluid from the tube to the ECM lane. The ability of drugs to compromise the intestinal membrane was assessed, monitoring the passing of a colored agent from the lumen side of the barrier to the "blood vessel." The selected drugs for assessment were staurosporine, an apoptotic inductor, and acetylsalicylic acid which affects tight junctions, the latter one with more pronounced effects.

Neurodegenerative conditions

Amyotrophic lateral sclerosis (ALS) is a neurodegenerative disorder linked to motor neuron loss in the spinal cord and motor cortex, causing paralysis and eventually death [51]. Although patient-specific ALS hiPSCs were derived a few years ago [52], it was not until recently that a platform that integrates disease modeling, with the possibility of drug testing, was developed [53]. This ALS-on-a-chip technology (Massachusetts Institute of Technology, Cambridge, MA, USA) uses a combination of 3D-skeletal muscle bundles, and hiPSC-derived light-sensitive motor neuron spheroids, to analyze muscular contraction based on light activation of the motor neurons. Hence, using ALS-motor neurons instead of healthy motor neurons, there were fewer muscular contractions, as well as motor neuron degradation and death, which could be reversed by treatments with drugs rapamycin and bosutinib that can potentially be used to treat ALS.

Malignancies

Cancer-on-a-chip platforms are a promising tool toward recreating tumor microenvironment, as well as to facilitate the understanding of its behavior that can lead to an improvement in the assessment of drug efficacy. The critical step toward metastasis propagation is the invasion by cancer cells of the bloodstream. For this, Zervantonakis

TABLE 45.2 Current organ- and multiorgan platforms.

Type of Platform	Organ	Cell Source	hPSC-derived	3D	Multiorgan	Type of Culture	Output	References
Liver	Liver	Rat liver cells	No	Yes	No	Perfused microwell array	• Cell viability • Oxygen consumption rate	[47]
Gut	Gut	Human colon adenocarcinoma	No	No	No	Microfluidic	• Fluorescent dye to test gut membrane integrity	[50]
Kidney	Kidney	Human renal cortical epithelial	No	Yes	No	96-Well plate	• Hydrolase activity • Adenylate cyclase • Glucose uptake	[49]
Cancer	Cancer	Human fibrosarcoma	No	Yes	Yes	Microfluidic	• Local diffusive permeability of the endothelial barrier • Tumor cell intravasation	[54]
		Breast carcinoma						
	Endothelial	Primary MVEC						
		Macrovascular endothelial						
Metastasis	Liver	Human hepatoma with RFP	No	Yes	Yes	Microfluidic	• Fluorescent imaging of RFP • Cell viability	[32]
	Lung	Human lung epithelial						
	Cancer	Human colorectal carcinoma						
Gut-liver	Gut	Human colon carcinoma	No	Yes	Yes	Microfluidic	• Confocal imaging • Hepatic enzyme activity • Pharmacokinetic profile of paracetamol	[55]
	Liver	Human hepatoma						
Liver-heart	Liver	Human primary hepatocytes Human hepatoma	No	Yes	Yes	Microfluidic	• Cell viability • Concentration of albumin, α-GST and CK-MB • Beating rate of cardiac organoids	[58]
	Heart	hiPSC-derived CMs	Yes					
Three-organ	Liver	Hepatic stellate cells Kupffer cells Human primary hepatocytes	No No No	Yes	Yes	Microfluidic	• Biochemical assessment of circulating media (albumin, urea, LDH, α-GST) • Cell viability • Cardiac organoid beat rates • Transepithelial resistance and short circuit current electrophysiological sensing	[60]
	Heart	iPSC-derived CMs Human primary cardiac fibroblasts	Yes No					
	Lung	Lung MVECs Airway stromal mesenchymal cells Bronchial epithelial cells	No No No					

Four-organ	Gut	Small intestine barrier model	No	Yes	Yes	Microfluidic	• LDH activity • Glucose and lactate concentration • Immunohistochemical staining	[56]
	Liver	HepaRG Human primary hepatic stellate		Yes				
	Kidney	Human proximal tubule		No				
	Skin	Human juvenile prepuce		No				
Four-organ	Liver	Human hepatoma	No	No	Yes	Microfluidic	• Albumin and urea production • Spontaneous contractile activity of CMs • Skeletal muscle contractility • Electrophysiological action potentials in MTs	[59]
	Heart	hiPSC-derived CMs	Yes					
	Muscle	Human skeletal myofibers	No					
	Brain	hSCSC-derived MTs hiPSC-derived cortical like neurons	No Yes					

α-GST, α-glutathione-S-transferase; *CMs*, cardiomyocytes; *CK-MB*, creatine kinase MB; *hSCSC*, human spinal cord stem cell line; *LDH*, lactate dehydrogenase; *MVEC*, microvascular endothelial cells; *RFP*, red fluorescent protein.

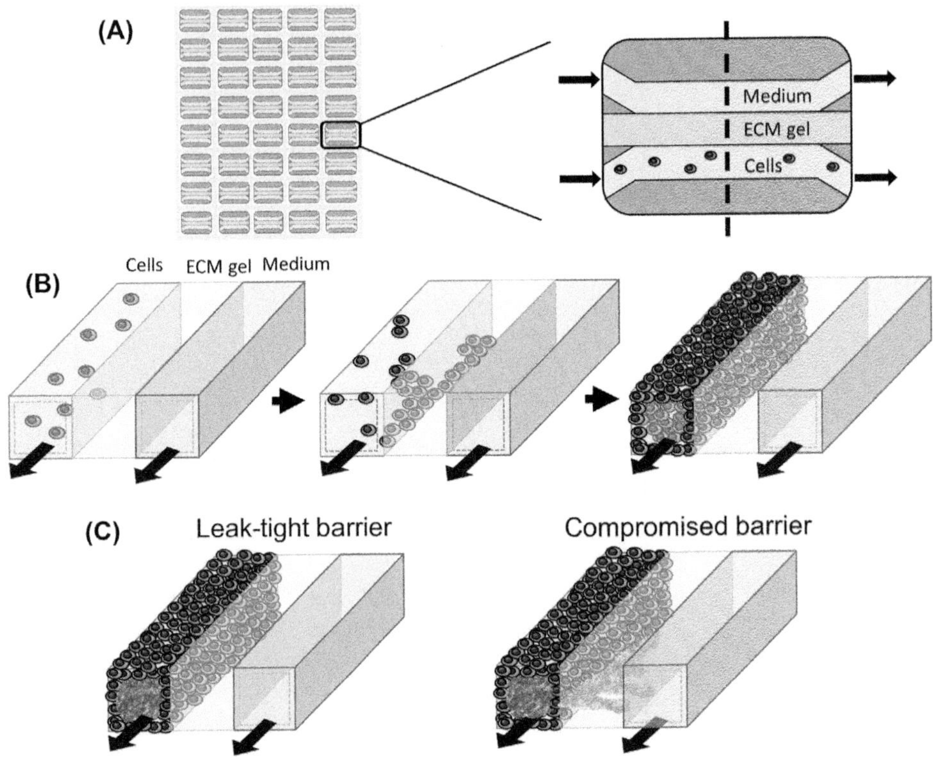

FIGURE 45.4 The OrganoPlate is comprised of 40 microfluidic cell culture chambers (A), contained within a well of a 384-well plate. Each one of these chambers houses three microfluidic tubes, one for circulating media, another for an ECM gel, and the last one containing Caco-2 cells. (B). The seeding and growth of the cells in the tube leads to a polarized cell barrier. (C). Upon injection of the medium in the lumen with a colored agent, the leak-tight cell barrier will prevent passing into the ECM gel. If the barrier is compromised, there will be a leak of the colored agent into the ECM gel, which can be detected by fluorescence analysis.

et al. simulated the 3D interface between a tumor and blood vessels, using a microfluidic-based approach that combines high-resolution live imaging, with control of microenvironment and endothelial permeability, in the presence of macrophages [54].

The use of breast carcinoma cells demonstrated that the intravasation process was accelerated by the presence of macrophage-secreted tumor necrosis factor alpha (TNF-α), due to a decrease in endothelial barrier function. Another model of metastasis from colorectal cancer, as previously mentioned, also employs the use of different tissue organoids [32]. Here, a metastasis-on-a-chip system was developed, using 3D photopatterning, to design separate chambers with flow of culture medium that passes first through a microfluid chamber containing a colorectal cancer (CRC) organoid. Then, the medium continues to multiple downstream chambers, where lung, heart, liver, and endothelial constructs are retained. When cancer cells start to detach from the original location, they flow through other chambers and eventually incorporate in other tissues, creating metastasis that can be identified through immunofluorescence analysis.

Drug metabolization

A pharmacokinetic (PK) profile of several drugs has been determined using microfluidic devices, which started by using cell lines as 3D spheroid cultures of liver and gut cells [55], and are currently being optimized toward the use of organoids. Although organ-on-a-chip is a promising tool for identification of novel drug products and toxicity assessment, it is important to consider that most drugs only start to cause cytotoxic effects, after initial metabolization that occurs in the liver.

A four-organ-on-a-chip device based on liver, intestine, skin, and kidney cells, under a controlled microenvironment, was established to allow prolonged culture periods for up to 28 days [56]. Although most of the modules—intestine, skin, and kidney— represent 2D structures, the liver buds are in a 3D conformation that resembles an in vivo approach. The fact that this microchip uses physiological parameters, in terms of fluid-to-tissue ratios, makes it a great candidate toward future absorption, distribution, metabolism, and excretion studies.

Anticancer drug metabolism

Doxorubicin is a chemotherapeutic agent that is used for a wide range of cancers, mainly metabolized by the liver [57]. As most of the drugs used for chemotherapy, doxorubicin has been largely linked to secondary effects, such as cardio and hepatic toxicity due to production of reactive oxygen species and apoptosis induction [57]. This fact makes doxorubicin a great model for the design of drug discovery platforms with different organoids, in order to correctly predict the effects in the whole system, instead of individually assessing organ-specific responses.

Using primary hepatocytes and hPSC-derived cardiomyocytes to generate micropatterned liver and heart organoids, respectively, Zhang et al. developed a multi-organ system to simulate the metabolism of doxorubicin [58]. This system includes different modules, containing sensors capable of analyzing organoid morphology, biomarker detection, and culture condition monitoring. Not only anticancer efficacy, but also cardiac toxicity related to liver metabolism, is thus enabled.

Multiple drugs

Oleaga et al. developed a microfluidic chamber that highlights the importance of the inclusion of a liver organoid chamber, in the design of a drug testing platform [59]. This system is comprised by four modules, representing liver, cardiac, muscle, and neural tissues, and has the particularity of forcing the culture medium through the liver compartment, before entering the other separate chambers. Here, the authors tested a set of drugs with different properties and targets— doxorubicin, atorvastatin, valproic acid, acetaminophen, and N-acetyl-m-aminophenol— and compared the results with literature reports. In fact, most of the outcomes from the microfluidic chip were similar to previously reported cytotoxic effects, which validate the use of this platform for assessing organ-organ communication, drug toxicity assessment, and predictive studies of novel compounds.

Skardal et al. [60] developed a chip platform, comprised by three different modules that represent liver, heart, and lung tissue, which can be used individually or in an interconnected manner. Initially, this study was performed to test the effect of bleomycin, a drug used against lung cancer, and that has secondary effects in the formation of lung fibrosis and inflammation. There was, in fact, an increase in inflammatory factors secreted by the lung; but, unexpectedly, there was also a detection of another side-effect in the 3-organoid system. Although bleomycin usually does not present cardiotoxic effects, as demonstrated by exposure of the isolated heart organoid to the drug, the cells completely stopped beating in the 3-organ system, indicating that there was lung inflammatory factor–driven cardiotoxicity.

Ongoing lines of research

Besides their use as drug discovery tool, organoids also represent a great advantage in disease modeling applications, acting as a promising tool toward the understanding of disease mechanisms, such as genetic diseases and viral infections [61]. The possibility of having a personalized treatment approach is also possible due to hiPSC technology, enabling an advantage toward the identification of the most suitable drug for that specific person. Using a biopsy of cancer tissue, it is possible to generate an organoid in vitro that will have a specific response for each person, or even allow the creation of a cancer biobank [62].

Body on a chip

The development of a body-on-a-chip system, which incorporates different organ components within a single microfluidic device, aims to complement or even replace the use of animal models, to evaluate the overall effect of a specific drug in different tissues. Still, the main sources are primary cells and immortalized cell lines. Although the first source is able to simulate equivalent levels of function as the original cells, their accessibility and the loss of function in longer periods of time reduces their applications [63]. Immortalized cell lines frequently accumulate mutations that lead to karyotypic anomalies, which can lead to divergent results when compared with same tissue cell types [64].

Although hPSC-derived cells are able to replicate the behavior of the target cells, they still present lower level of enzymatic activity [65]. The majority of organoids consists in only a limited number of cell types [66], and fails to reflect the complete organization of the target tissue. Despite displaying some mature cell types and functions, organoids are still in an early developmental stage [67]. Robust protocols that will allow standardization of the differentiation process, as well as validation parameters for each type of organoid, are also needed. All these barriers notwithstanding organoid technology, and especially hPSC-derived technology, can be considered as a new golden standard for preclinical drug screening applications.

References

[1] Lin JH, Chiba M, Balani SK, et al. Species differences in the pharmacokinetics and metabolism of indinavir, a potent human immunodeficiency virus protease inhibitor. Drug Metab. Dispos. 1996;24(10):1111–20.

[2] Wysowski DK, Swartz L. Adverse drug event surveillance and drug withdrawals in the United States, 1969–2002: the importance of reporting suspected reactions. Arch. Intern. Med. 2005;165(12):1363–9.

[3] Hay M, Thomas DW, Craighead JL, Economides C, Rosenthal J. Clinical development success rates for investigational drugs. Nat. Biotechnol. 2014;32(1):40–51.

[4] Takahashi K, Tanabe K, Ohnuki M, et al. Induction of pluripotent stem cells from adult human fibroblasts by defined factors. Cell 2007;131(5):861–72.

[5] Lancaster MA, Knoblich JA. Organogenesis in a dish: modeling development and disease using organoid technologies. Science 2014;345(6194):1247125.

[6] Azarin SM, Palecek SP. Development of scalable culture systems for human embryonic stem cells. Biochem. Eng. J. 2010;48(3):378.
[7] Carpenter MK, Frey-Vasconcells J, Rao MS. Developing safe therapies from human pluripotent stem cells. Nat. Biotechnol. 2009;27(7):606–13.
[8] Yu J, Vodyanik MA, Smuga-Otto K, et al. Induced pluripotent stem cell lines derived from human somatic cells. Science 2007;318(5858):1917–20.
[9] Takahashi K, Yamanaka S. Induction of pluripotent stem cells from mouse embryonic and adult fibroblast cultures by defined factors. Cell 2006;126(4):663–76.
[10] Stadtfeld M, Nagaya M, Utikal J, Weir G, Hochedlinger K. Induced pluripotent stem cells generated without viral integration. Science 2008;322(5903):945–9.
[11] Okita K, Nakagawa M, Hyenjong H, Ichisaka T, Yamanaka S. Generation of mouse induced pluripotent stem cells without viral vectors. Science 2008;322(5903):949–53.
[12] Yakubov E, Rechavi G, Rozenblatt S, Givol D. Reprogramming of human fibroblasts to pluripotent stem cells using mRNA of four transcription factors. Biochem. Biophys. Res. Commun. 2010;394(1):189–93.
[13] Kim D, Kim CH, Moon JI, et al. Generation of human induced pluripotent stem cells by direct delivery of reprogramming proteins. Cell Stem Cell 2009;4(6):472–6.
[14] Kimbrel EA, Lanza R. Current status of pluripotent stem cells: moving the first therapies to the clinic. Nat. Rev. Drug Discov. 2015;14(10):681–92.
[15] Fox IJ, Daley GQ, Goldman SA, Huard J, Kamp TJ, Trucco M. Stem cell therapy. Use of differentiated pluripotent stem cells as replacement therapy for treating disease. Science 2014;345(6199):1247391.
[16] Keirstead HS, Nistor G, Bernal G, et al. Human embryonic stem cell-derived oligodendrocyte progenitor cell transplants remyelinate and restore locomotion after spinal cord injury. J. Neurosci. 2005;25(19):4694–705.
[17] Kriks S, Shim JW, Piao J, et al. Dopamine neurons derived from human ES cells efficiently engraft in animal models of Parkinson's disease. Nature 2011;480(7378):547–51.
[18] Grealish S, Diguet E, Kirkeby A, et al. Human ESC-derived dopamine neurons show similar preclinical efficacy and potency to fetal neurons when grafted in a rat model of Parkinson's disease. Cell Stem Cell 2014;15(5):653–65.
[19] Kawamura M, Miyagawa S, Miki K, et al. Feasibility, safety, and therapeutic efficacy of human induced pluripotent stem cell-derived cardiomyocyte sheets in a porcine ischemic cardiomyopathy model. Circulation 2012;126(11 Suppl. 1):S29–37.
[20] Xiong Q, Ye L, Zhang P, et al. Functional consequences of human induced pluripotent stem cell therapy: myocardial ATP turnover rate in the in vivo swine heart with postinfarction remodeling. Circulation 2013;127(9):997–1008.
[21] Chong JJ, Yang X, Don CW, et al. Human embryonic-stem-cell-derived cardiomyocytes regenerate non-human primate hearts. Nature 2014;510(7504):273–7.
[22] Ornoy A. Valproic acid in pregnancy: how much are we endangering the embryo and fetus? Reprod. Toxicol. 2009;28(1):1–10.
[23] Miranda CC, Fernandes TG, Pinto SN, Prieto M, Diogo MM, Cabral JMS. A scale out approach towards neural induction of human induced pluripotent stem cells for neurodevelopmental toxicity studies. Toxicol. Lett. 2018;294:51–60.
[24] Boj SF, Hwang CI, Baker LA, et al. Organoid models of human and mouse ductal pancreatic cancer. Cell 2015;160(1–2):324–38.
[25] van de Wetering M, Francies HE, Francis JM, et al. Prospective derivation of a living organoid biobank of colorectal cancer patients. Cell 2015;161(4):933–45.
[26] Daviaud N, Friedel RH, Zou H. Vascularization and engraftment of transplanted human cerebral organoids in mouse cortex. eNeuro 2018;5(6).
[27] Lancaster MA, Renner M, Martin CA, et al. Cerebral organoids model human brain development and microcephaly. Nature 2013;501(7467):373–9.
[28] Marchetto MC, Carromeu C, Acab A, et al. A model for neural development and treatment of Rett syndrome using human induced pluripotent stem cells. Cell 2010;143(4):527–39.
[29] Fernandes TG, Duarte ST, Ghazvini M, et al. Neural commitment of human pluripotent stem cells under defined conditions recapitulates neural development and generates patient-specific neural cells. Biotechnol. J. 2015;10(10):1578–88.
[30] Wimmer RA, Leopoldi A, Aichinger M, et al. Human blood vessel organoids as a model of diabetic vasculopathy. Nature 2019;565(7740):505–10.
[31] Akazawa T, Yoshida S, Ohnishi S, Kanazu T, Kawai M, Takahashi K. Application of intestinal epithelial cells differentiated from human induced pluripotent stem cells for studies of prodrug hydrolysis and drug absorption in the small intestine. Drug Metab. Dispos. 2018;46(11):1497–506.
[32] Aleman J, Skardal A. A multi-site metastasis-on-a-chip microphysiological system for assessing metastatic preference of cancer cells. Biotechnol. Bioeng. 2018;116(4):936–44.
[33] Mummery CL. Cardiology: solace for the broken-hearted? Nature 2005;433(7026):585–7.
[34] Garitaonandia I, Gonzalez R, Sherman G, Semechkin A, Evans A, Kern R. Novel approach to stem cell therapy in Parkinson's disease. Stem Cells Dev. 2018;27(14):951–7.
[35] Singh A, Tetreault L, Kalsi-Ryan S, Nouri A, Fehlings MG. Global prevalence and incidence of traumatic spinal cord injury. Clin. Epidemiol. 2014;6:309–31.
[36] Priest CA, Manley NC, Denham J, Wirth 3rd ED, Lebkowski JS. Preclinical safety of human embryonic stem cell-derived oligodendrocyte progenitors supporting clinical trials in spinal cord injury. Regen. Med. 2015;10(8):939–58.
[37] Rodrigues CA, Fernandes TG, Diogo MM, da Silva CL, Cabral JM. Stem cell cultivation in bioreactors. Biotechnol. Adv. 2011;29(6):815–29.
[38] Fernandes TG, Rodrigues CAV, Diogo MM, Cabral JMS. Stem cell bioprocessing for regenerative medicine. J. Chem. Technol. Biotechnol. 2014;89(1):34–47.
[39] Kropp C, Kempf H, Halloin C, et al. Impact of feeding strategies on the scalable expansion of human pluripotent stem cells in single-use stirred tank bioreactors. Stem Cells Transl. Med. 2016;5(10):1289–301.
[40] Kempf H, Kropp C, Olmer R, Martin U, Zweigerdt R. Cardiac differentiation of human pluripotent stem cells in scalable suspension culture. Nat. Protoc. 2015;10(9):1345–61.
[41] Chen VC, Ye J, Shukla P, et al. Development of a scalable suspension culture for cardiac differentiation from human pluripotent stem cells. Stem Cell Res. 2015;15(2):365–75.

[42] Miranda CC, Fernandes TG, Diogo MM, Cabral JM. Scaling up a chemically-defined aggregate-based suspension culture system for neural commitment of human pluripotent stem cells. Biotechnol. J. 2016;11(12):1628–38.
[43] Lei Y, Schaffer DV. A fully defined and scalable 3D culture system for human pluripotent stem cell expansion and differentiation. Proc. Natl. Acad. Sci. U.S.A. 2013;110(52):E5039–48.
[44] Rodrigues GMC, Gaj T, Adil MM, et al. Defined and scalable differentiation of human oligodendrocyte precursors from pluripotent stem cells in a 3D culture system. Stem Cell Rep. 2017;8(6):1770–83.
[45] Adil MM, Rodrigues GM, Kulkarni RU, et al. Efficient generation of hPSC-derived midbrain dopaminergic neurons in a fully defined, scalable, 3D biomaterial platform. Sci. Rep. 2017;7:40573.
[46] Rodrigues GM, Rodrigues CA, Fernandes TG, Diogo MM, Cabral JM. Clinical-scale purification of pluripotent stem cell derivatives for cell-based therapies. Biotechnol. J. 2015;10(8):1103–14.
[47] Domansky K, Inman W, Serdy J, Dash A, Lim MH, Griffith LG. Perfused multiwell plate for 3D liver tissue engineering. Lab Chip 2010;10(1):51–8.
[48] Taguchi A, Nishinakamura R. Higher-order kidney organogenesis from pluripotent stem cells. Cell Stem Cell 2017;21(6):730–746 e6.
[49] DesRochers TM, Suter L, Roth A, Kaplan DL. Bioengineered 3D human kidney tissue, a platform for the determination of nephrotoxicity. PLoS One 2013;8(3):e59219.
[50] Trietsch SJ, Naumovska E, Kurek D, et al. Membrane-free culture and real-time barrier integrity assessment of perfused intestinal epithelium tubes. Nat. Commun. 2017;8(1):262.
[51] Pasinelli P, Brown RH. Molecular biology of amyotrophic lateral sclerosis: insights from genetics. Nat. Rev. Neurosci. 2006;7(9):710–23.
[52] Dimos JT, Rodolfa KT, Niakan KK, et al. Induced pluripotent stem cells generated from patients with ALS can be differentiated into motor neurons. Science 2008;321(5893):1218–21.
[53] Osaki T, Uzel SGM, Kamm RD. Microphysiological 3D model of amyotrophic lateral sclerosis (ALS) from human iPS-derived muscle cells and optogenetic motor neurons. Sci. Adv. 2018;4(10):eaat5847.
[54] Zervantonakis IK, Hughes-Alford SK, Charest JL, Condeelis JS, Gertler FB, Kamm RD. Three-dimensional microfluidic model for tumor cell intravasation and endothelial barrier function. Proc. Natl. Acad. Sci. U.S.A. 2012;109(34):13515–20.
[55] Lee DW, Ha SK, Choi I, Sung JH. 3D gut-liver chip with a PK model for prediction of first-pass metabolism. Biomed. Microdevices 2017;19(4):100.
[56] Maschmeyer I, Lorenz AK, Schimek K, et al. A four-organ-chip for interconnected long-term co-culture of human intestine, liver, skin and kidney equivalents. Lab Chip 2015;15(12):2688–99.
[57] Carvalho C, Santos RX, Cardoso S, et al. Doxorubicin: the good, the bad and the ugly effect. Curr. Med. Chem. 2009;16(25):3267–85.
[58] Zhang YS, Aleman J, Shin SR, et al. Multisensor-integrated organs-on-chips platform for automated and continual in situ monitoring of organoid behaviors. Proc. Natl. Acad. Sci. U.S.A. 2017;114(12):E2293–302.
[59] Oleaga C, Bernabini C, Smith AS, et al. Multi-organ toxicity demonstration in a functional human in vitro system composed of four organs. Sci. Rep. 2016;6:20030.
[60] Skardal A, Murphy SV, Devarasetty M, et al. Multi-tissue interactions in an integrated three-tissue organ-on-a-chip platform. Sci. Rep. 2017;7(1):8837.
[61] Garcez PP, Loiola EC, Madeiro da Costa R, et al. Zika virus impairs growth in human neurospheres and brain organoids. Science 2016;352(6287):816–8.
[62] Sachs N, de Ligt J, Kopper O, et al. A living biobank of breast cancer organoids captures disease heterogeneity. Cell 2018;172(1–2):373–386 e10.
[63] Gomez-Lechon MJ, Donato MT, Castell JV, Jover R. Human hepatocytes as a tool for studying toxicity and drug metabolism. Curr. Drug Metabol. 2003;4(4):292–312.
[64] Vcelar S, Jadhav V, Melcher M, et al. Karyotype variation of CHO host cell lines over time in culture characterized by chromosome counting and chromosome painting. Biotechnol. Bioeng. 2018;115(1):165–73.
[65] Guguen-Guillouzo C, Corlu A, Guillouzo A. Stem cell-derived hepatocytes and their use in toxicology. Toxicology 2010;270(1):3–9.
[66] Yin X, Mead BE, Safaee H, Langer R, Karp JM, Levy O. Engineering stem cell organoids. Cell Stem Cell 2016;18(1):25–38.
[67] Takasato M, Er PX, Chiu HS, et al. Kidney organoids from human iPS cells contain multiple lineages and model human nephrogenesis. Nature 2015;526(7574):564–8.

Chapter 46

Printing of personalized medication using binder jetting 3D printer

Ziyaur Rahman[1], Naseem A. Charoo[2], Mathew Kuttolamadom[3], Amir Asadi[3] and Mansoor A. Khan[1]
[1]Irma Lerma Rangel College of Pharmacy, Texas A&M Health Science Center, Texas A&M University, College Station, TX, United States; [2]Zeino Pharma FZ LLC, Dubai Science Park, Dubai, United Arab Emirates; [3]Engineering Technology & Industrial Distribution, College of Engineering, Texas A&M University, College Station, TX, United States

Introduction

Most of the commercially available medications cater to the needs of adult patients. The predominant dosage forms are capsules and tablets, which are often not suitable for children. Limited pediatric formulations are available in the USA, which, generally, are liquids or powders for reconstitution [1]. For example, only a single pediatric-friendly product (Purixan) [2], is commercially available for the anticancer drug (mercaptopurine), although cancer is the number one cause of mortality in children after unintended fatalities [3,4]. Because of this, there is a great demand for highly specialized pediatric-friendly products.

The lack of availability of pediatric formulations is primarily due to evolving physiology of children, where fixed doses do not work [1]. Physicians or patient care providers have no choice, but to prescribe the dosage forms, which are intended for adults and unpalatable [5]. Often times, physicians prescribe an extemporaneous preparation, where the drug product is manipulated in a pharmacy with a goal to adjust only the dose, due to unavailability of lower strength; however, scant attention is paid to render the product children-friendly. Additionally, there are efficacy and safety issues related to dose accuracy, stability, sterility, pharmacokinetics, pharmacodynamics, and consistency in the preparation of extemporaneous preparations [5–7].

An ideal pediatric dosage form should offer flexibility to titrate the dose, to cover the diverse pediatric age group 0–17 years. Further, pediatric dosage form should be easy to swallow, dissolve, or disperse in the mouth or a sip of water, palatable, contain minimal, and only FDA approved excipients, exhibit adequate bioavailability and stability at high temperature and humidity, etc. [8,9]. Pharmaceutical companies are reluctant to invest in developing pediatric-friendly innovative formulations, owing to the complexities and high cost of product development.

A series of initiatives have been taken to promote pediatric drug development. For instance, Best Pharmaceuticals for Children Act (BPCA) added incentives, in terms of 6 months market exclusivity extension, and/or patent extension, for companies performing additional clinical studies in the pediatric population [10]. Likewise, Pediatric Research Equity Act (PREA) granted FDA powers, to make pediatric studies mandatory for products that would be used in pediatric patients, and also in certain cases extend the clinical studies data obtained in adults, to approve pediatric medicines [11].

Dose flexibility

The heterogeneity in age, anatomy, physiology, and weight of pediatric patients requires dose adjustment and flexibility [1]. Dose accuracy is also very critical for certain drugs, especially narrow therapeutic index class [12–14]. It can be ensured by dispensing the medication in unit dosage forms. Solid dosage forms can be used to deliver unit dosage forms, and unlike the liquid dosage forms, are more stable and easy to handle, with fewer chances of microbial contamination. However, dose adjustment/flexibility is difficult to achieve in solid-dosage forms, as each strength has to be manufactured separately, which is not a cost-effective business model.

3D printing

Another option would be to manufacture each strength form at the point of dispensing, which is not feasible in a

Precision Medicine for Investigators, Practitioners and Providers. https://doi.org/10.1016/B978-0-12-819178-1.00046-0

pharmacy. Newer manufacturing methods such as 3D printing can be used for tailored/flexible/adjusted solid-dosage forms and can be manufactured in a pharmacy, where resources are limited. It has potential application in personalized medicine to meet the needs of individual patients, such as printing individual dosage forms, dosage forms containing multiple drugs, and tailored drug release rate. It offers many advantages, such as rapid and on-demand pharmacy capability, flexibility in terms of dosage forms, and complexity in terms of shape and geometry, compared to traditional manufacturing methods of pharmaceuticals, such as molding, compacting, and extrusion [1].

Similarly, personalized medications are required for geriatric patients, for the management of diseases due to age-related physiological changes, multiple comorbidities, multiple medications, and differences in pharmacokinetics and pharmacodynamics response to a drug. The presence of several providers of medicinal products further complicates disease management. Geriatric patients are at greater risk of developing adverse effect due to polypharmacy [16,17]. Flexible mono or multidrug dosing capability, and easy to swallow dosage form, could minimize adverse events.

It is possible for a pharmacist to print a once-a-day personalized polypill, instead of dispensing a cocktail of multiple pills based on doctor prescription [18–20]. Polypharmacy is one of the primary factors causing medication nonadherence in elderly and geriatric patients [21], and an emergency room visit is often due to nonadherence of medications [22].

More than two dozen 3D printing methods are reported in the literature, such as fused deposition modeling (FDM), selective laser sintering (SLS), stereolithography (SLA), and binder jetting (BJ) [15]. BJ method was used in the manufacturing of FDA approved drug product. This method produced dosage forms that have a highly porous structure, which causes it to dissolve/disperse in the mouth within seconds, without a sip of water [23,24], making it ideal for patients who have difficulty chewing or swallowing medications especially elderly and children.

3D printing - binder jetting

BJ was developed by Sachs and coworkers of the Massachusetts Institute of Technology in 1986, which was later commercialized by ZCorp Inc (3D Systems, Rock Hill, NC, USA) [25,26]. This technology is also referred to as powder bed inkjet printing. The printing process begins by spreading powder material as a thin layer, over the build platform, using either a roller or powder-jetting reservoir (powder jetting system). In the next step, a binder solution or dispersion is deposited, as small droplets onto defined printing areas of the first layer, using a drop-on-demand printhead.

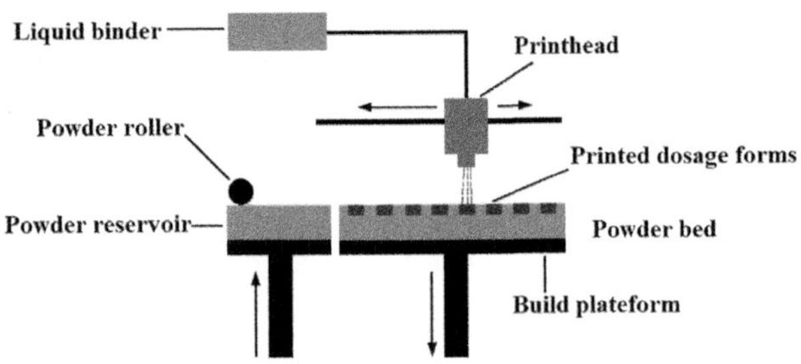

FIGURE 46.1 Structure of the binder jetting 3-D printer.

The binder helps the powder particles to fuse/bond together. After printing the first layer, the build platform moves down to one layer thickness, to enable printing of the next layer. This process is repeated multiple times to form the desired shape of dosage forms. Residual powder on build platform around the built structure provides support during the printing process (Fig. 46.1). The unprinted powder is recovered after postprocessing and can be reused, after ensuring that no significant changes in powder physicochemical properties have occurred. Postprocessing is also required (e.g., drying of the entire powder bed, including printed dosage forms, and support powder bed) to remove the solvent system of ink, and separate the formed object from the loose powder [27].

BJ produces a highly porous structure that has poor resolution, rough surface, high friability, and poor mechanical strength [28,29]. High porosity results in rapid disintegration of the printed dosage forms. Spritam is a first 3D printed drug product employing BJ method to print the dosage forms [23]. The manufacturer labels it ZipDose technology (Aprecia Pharmaceuticals, Langhorne, PA, USA), where the dosage form disintegrates in a few seconds [24]. Spritam contains high dose (250–1000 mg) of the drug and disintegrates very rapidly (about 11 s in mouth), which is ideal for children and geriatric patients [23]. The disintegration time (DT) of orally disintegrating tablets, manufactured using traditional compression method, is 30 s or less [30].

Excipients

The excipients perform various functions such as a diluent, disintegrant, lubricant, flow aid, binder, etc., [31]. One of the primary requirements of excipients is nontoxicity, and free of untoward pharmacological effects. These requirements can be met if the drug product contains FDA approved inactive ingredients [32]. IIG (inactive ingredient) database has information about excipients, their grade, and percentage allowed in various dosage forms (FDA inactive ingredient database). The excipients can also be from generally regarded as safe (GRAS) category, if not listed in the IIG database [33]. FDA frequently updates IIG and GRAS database, as new safety information becomes available, from internal and external studies.

BJ process needs powder material for layering, and binder for bonding particles by wetting and dissolution of drug, polymer, and other excipients present in the formulations. This process is similar to the wet-granulation process, used in tablets and capsule manufacturing [12–14]. Practically, all the approved excipients used in solid dosage forms can be employed in BJ process. For example, FDM requires thermoplastic polymers, SLS requires thermoplastic and laser absorbable polymers, and SLA requires UV polymerizable monomers [15].

Spritam contains the following excipients: colloidal silicon dioxide, glycerin, mannitol, microcrystalline cellulose, Polysorbate 20, povidone, sucralose, butylated hydroxyanisole, and natural and artificial spearmint flavor [23]. These are commonly used excipients in solid dosage forms. Possibly, colloidal silicon dioxide, mannitol, and microcrystalline cellulose are components of the powder bed, and glycerin and Polysorbate 20 are components of the binder. Povidone, sucralose, butylated hydroxyanisole, and natural and artificial spearmint flavor, may be components of powder bed or binder solution/dispersion.

Controlled release preparations

Powder components may contain diluent, disintegrants, sweetener, and glidant. Most commonly used diluents are lactose, mannitol, and microcrystalline cellulose. Disintegrants examples are pregelatinized starch, crospovidone, croscarmellose sodium, and sodium starch glycolate. Silicon dioxide is most commonly used as glidant [34,35]. Polyethylene oxide, polyvinyl pyrollidone (PVP K30), lactose, mannitol, maltitol, maltodextrin, Kollidon SR, Eudragit E100, Eudragit RLPO, ethylcellulose, and hydroxypropyl methylcellulose (E50), have been investigated for designing sustained release, delayed release, and rapidly dispersible tablets by BJ [36–41].

Drug(s) can be incorporated in binder solvent or powder component, depending upon the solubility, stability, and dose of the drug. If the drug is soluble in binder solvent and the dose is low, it can be solubilized in the solvent. It is recommended to incorporate drug in powder material if it is insoluble, thermolabile, and the dose is high. Similar to the drug, polymer binders could be incorporated in powder material or binder solvent, depending upon solvent characteristics, viscosity, and quantity. Polymers investigated as binders are polylactic acid, polycaprolactone, and polyvinylpyrrolidone (K17, K25, and K30 grades) ethylcellulose. Binder solvent could be aqueous, organic (ethanol, acetone, isopropanol), or hydro-organic solvents. Solvent selection is based on the solubility of drug and binder, viscosity, and surface tension. A combination of solvents can be used to optimize the printing process parameters and quality of printed dosage forms. Water, ethanol, acetone, and chloroform are investigated for BJ process [36–42].

Factors affecting binder jetting 3D printing process

Factors related to the printhead, binder, powder, and in-process heating, need to be comprehensively understood before using the process, for manufacturing clinical supplies of a drug product.

Printhead: Printhead could be thermal or piezoelectric type. In thermal printhead, a heating pulse from the microheater (thin film resistor), causes the binder to vaporize and the binder forms vapor bubbles over the resistor. The vapor bubbles expand in binder reservoir, and thus, increase the pressure, which causes the binder ejection through the nozzle [43]. The temperature at the thermal resistor can reach as high as 300°C, but only for a few seconds, and expose around 0.5% of volume [44]. The thermal printer has higher speed, and the cost of fabrication parts is lower than piezoelectric printhead. Disadvantages of the thermal printhead are exposure of the liquid binder to thermal stress, which may critically affect thermally sensitive drug, polymer and other excipients, if they are part of the binder, besides low droplet directionality and nonuniform droplet size [45].

With the piezoelectric option, a direct mechanical pulse is generated by the piezoelectric actuator, and this causes a force to expel the binder from the printhead. In contrast to the thermal printhead, the piezoelectric printhead has the ability to generate and control the size of the droplet, and ejection directionality. In general, the thermal printhead is more suitable for aqueous-based binder, while the piezoelectric printhead is more suited to organic solvents [44]. Clogging of the printhead is a common issue with both printers, which is related to formulation properties, such as viscosity and surface tension, and particle size distribution, if it is a dispersion.

Speed of printhead, frequency of droplets, path of the droplet (raster), droplet spacing, droplet pattern, nozzle diameter, and resonance frequency, affect the performance of the printhead, and thus, the quality of the printed dosage forms [46]. The flow rate of the binder and the fast axis speed of the printhead determine the quantity of binder deposited per unit line length [42]. The printing speed is the forward travel rate of the printhead (in Y-direction). It is generally accepted that higher printing speed, would cause printing inaccuracy and other defects, in the printed dosage forms [47,48].

Binder or solvent system: Binder system could be aqueous, organic or hydro-organic, or a mixture of two or more organic solvents. Selection of solvent systems depends upon the printhead type, solubility, stability, and compatibility of the binder system components. Organic solvents should be preferably from Class 3. The use of Class 2 solvents should be limited, and class I solvents should not be used due to toxicity issues [49].

Ideally, all the components of the binder should be soluble and compatible with each other, and should not change the solubility characteristics of other components. However, it is possible to add insoluble components of formulation as fine particulates. The particle size distribution of insoluble components should be in colloidal range, to avoid sedimentation as well as clogging of the printhead. Printed dosage forms would have content uniformity issue if the dispersed drug sedimented during the printing process. In addition, if solvents are used as a binder and binder polymer is part of the powder bed, the solvents should not cause excessive swelling of the polymer. Excessive swelling of the polymer may increase the thickness of powder layer, and interfere in printing if swollen layer thickness is more than a powder layer. It is recommended to use low viscosity/molecular weight grade of the polymer.

Most important physicochemical properties of the binder system, are viscosity and surface tension. There has to be a balance between these two properties. Surface tension should be high enough to enable the formation of droplets. Furthermore, it should resist leakage of ink, when the printhead is not in operation. High ejection pressure is observed if the binder viscosity is too high, which could lead to uneven jetting, clogging, and consequently raising the temperature of the printhead. Conversely, low viscosity leads to the formation of satellite drops behind main drops. Ideally, the viscosity should be low enough to enable jetting, but prevent the formation of satellite droplets [50,51].

Satellite droplet formation does not affect the formation of primary droplets but can affect the deposition of binder on the powder bed. Satellite drops can lead to nonuniformity in printed dosage forms, as well as changes in the composition of the powder bed, due to deposition of binder on the outside of the printing region [51,52]. Viscosity and surface tension can also affect the filling of binder in printhead [53]. Additionally, high viscosity and surface tension of binder would prevent interaction between binder and powder, which is needed to impart mechanical strength to dosage forms. Literature reports viscosity of 1.6–5.99 mPa.S, and surface tension of 25.7–52 mN/m, for binder solution/suspension for BJ [50,54–59].

These properties of the binder system are affected by the qualitative and quantitative composition of binder solution/suspension. Surface tension can be reduced by the addition of surface-active agents, such as Tween 80 [60]. On the other hand, viscosity can be adjusted by polymer viscosity/molecular weight grade and concentration. Polyols, such as propylene glycol, polyethylene glycol, and glycerol are also used as viscosity modifiers. In addition to their viscosity modifying effects, these also act as humectants, and thus, prevent rapid evaporation of the solvent. Evaporation of solvent changes the viscosity of the binder, and thus, leads to clogging of the printhead. Binder amount determines the mechanical strength and dimensional accuracy of the printed dosage forms. Excess binder would result in dimensional deformity, while insufficient binder would not impart sufficient mechanical strength. Binder level and type also determine product in-vitro and in-vivo performance [44].

Three phenomena may take place, when binder droplets impinge on powder bed namely, spreading, bouncing (i.e., rebounding of impinging droplets on a powder bed), or splashing. These phenomena depend upon the physical properties of droplets, such as velocity and size, powder bed porous structure, and wettability of droplet, and powder bed. The most important factor of determining the flow dynamics of impinging droplets through the powder bed is droplet velocity. Spreading dominates at low droplet velocity, while splashing and/or bouncing dominate at high droplet velocity.

Spreading would increase the initial contact area, between the binder droplets and powder bed, which would subsequently influence droplet penetration and saturated area. Bouncing of droplets causes rebounding from powder droplets upon initial contact, which would potentially offset the permeation zone, along the Y direction from the initial location. Additionally, the splashing of droplets into two or more smaller droplets causes a decrease in droplet permeation area. Splashing and bouncing of droplets lead to dimensional inaccuracy, possibly changing the composition of printed dosage forms and powder, if the drug is present in binder solution/dispersion [61–65].

Powder: Factors related to powder bed, such as particle size, shape, flowability, wettability, saturation level, feed-to-powder ratio, and spreading speed, have been investigated for their effect on the dimensional accuracy, and mechanical strength of the 3D printed structure by BJ process [66–72].

Good flowability of powder is essential, for proper processing of BJ printer. Sufficient powder flowability, allows the roller to spread the powder as a thin layer on the building platform. Poor flowability leads to a decrease in resolution and produces deformity or quality defects in the printed dosage forms. On the other hand, high flowability causes powder bed instability.

Spreading speed of powder, which is performed by a rotating roller, is related to particle size and shape, and flow properties of the powder material. Improper spreading speed produces a nonuniform powder layer. Lower spreading speed is recommended, although at the cost of printing time [71]. Slow speed is recommended for smaller particle powder materials (>5 μm), due to large Van der Waals forces. On the other hand, relatively higher speed can be used for coarse particles [73]. It is suggested that for the desirable spreading of powders in the BJ process, the layer thickness should be at least thicker than the largest particle, and preferably three times the particle size [74].

Particle size affects flowability, pore size distribution, and penetration of binder, which may ultimately affect the quality of the printed dosage forms [75–77]. Changing the particle size of powder material changes the flow characteristics and pore size distribution that will change the binder droplet penetration [78]. Similar to particle size distribution, morphology affects flow and spreadability, packing density, etc. Multifaceted and anisotropic particles stick to each other and create higher friction compared to spherical shaped particles. The irregular shape and wide size distribution powder material produce poor flowability of the powder. The powder that does not flow well does not pack well, thus produces low density printed dosage forms [79].

The powder bed has to be wet sufficiently by the binder solution/dispersion. If the powder layer is not sufficiently wetted, it can cause delamination and poor mechanical strength of printed dosage forms. The wetting of the powder is related to hydrophilicity or hydrophobicity of powder material, and viscosity and surface tension of the binder. In fact, binder components should be selected, based on hydrophilicity or hydrophobicity of powder material. Notwithstanding this, too much wetting of powder decreases resolution, with potential for the excess binder to seep in the lateral and vertical direction of the powder bed, outside the printed area, and resulting in a change in the composition of powder, with a high probability of producing printing defects [80–82].

The saturation level of the powder bed is the percentage of pore volume filled by the binder. The saturation level is primarily linked to the particle size and shape, and layer thickness. Optimal saturation amount of binder needs to be used, to produce the desired mechanical strength and dosage form meeting critical quality attributes. Low and high saturation of a powder bed with a binder, produces deformity in shape, and undesired mechanical properties. Low saturation level would not be adequate to bind particles together and create bonds in successive layers.

In contrast, over saturation would cause the binder to permeate beyond the designed, printed area. Furthermore, it would also cause bleeding, and feathering of the binder in the powder bed. Bleeding is the macroscopic flow of binder within the printed boundary. Binder successively flows from top to bottom, and gradually accumulates at the bottom layers, which results in printing inaccuracy. Feathering is a microscopic flow of binder, resulting in broadening of printed geometry [80,81].

In-process heating: Printed dosage forms can be dried in an oven, after their formation. Printing and drying can also be done simultaneously if the printer is equipped with in-process heating. The power level of heating element and time for drying of the binder should be selected in such a way, that the sprayed binder is dried at an optimal condition. Dimensional accuracy and mechanical strength of the final dosage form, also depend upon in-process drying parameters. Excessive drying prevents bond formation, and thus, leads to poor mechanical properties of the tablets. In contrary, insufficient heating leads to inadequate adhesion strength and distortion of structural integrity [71,80].

Challenges of the 3D printing process

Stability of printed dosage forms: Like other dosage formulations, printed dosage forms should be physically and chemically stable. Unlike FDM or SLS printing methods, BJ does not involve extreme thermal exposure [15]. BJ exposes the dosage forms to mild thermal condition (below 50°C), during printing or postprinting, to remove the solvent from powder [82]. However, for thermolabile drugs, air-drying or vacuum drying at low temperature can be employed. Most likely, the drug would be present in amorphous forms in the printed dosage forms, if the drug is a component of the binder system. On the other hand, the drug may exist as a mixture of amorphous and crystalline forms, in the printed dosage forms, if it is a component of the powder bed.

The ratio of amorphous to crystalline forms in the printed dosage forms would depend upon the solubility of the drug in the binder system, besides printing process parameters. The amorphous form is unstable and tends to revert to stable crystalline forms on exposure to high temperature and humidity. Such conversion may change clinical performance significantly. Therefore, it is critical to maintain the amorphous to crystalline ratio, especially for poorly water-soluble drugs [12–14,83].

Quality control: BJ imparts unique features to dosage forms, which cannot be measured by traditional or pharmacopeial quality control methods. This dictates development and validation of new quality control methods to ensure the quality of the product during manufacturing and in-use. For example, the mechanical strength of BJ printed tablets would be low, compared to traditionally manufactured tablets. Mechanical strength of conventional tablets is measured by hardness and friability tester. The friability limit of 1% is considered acceptable. However, BJ printed tablets would fail if the same criteria are applied to them.

Poor mechanical strength of printed dosage forms is due to porous internal structures, which are essential for quick disintegration. Again, the traditional disintegration test may not measure DT accurately and distinguish between formulations [15,84].

Quality defects: Defects reported in the literature are warping, delamination or cracking, and deformities [15]. It requires a proper understanding of the process and material attributes, to remove or minimize these defects in the printed dosage forms. Potential defects that may be observed in BJ printing include variability in layer thickness, improper layering due to process variations, and compositional variations that may mainly occur due to variability in the quality of recycled material. As noted above, quality control tests appropriate to the requirements of BJ process would have to be applied. For instance, in process test parameters, such as layer thickness, print head speed and height, and the number of times the material can be recycled, would need to be defined and controlled.

Recycling of material: The leftover binder solution/dispersion, after completion of the printing process can be used, provided the solution/dispersion has been demonstrated to be stable. Similar to the binder, powder material obtained after postprocessing of printed dosage forms, can be reused as such or mixed with fresh powder material. Unlike binder system, powder material which is exposed to process conditions, including mild thermal stress, may exhibit compositional changes, owing to binder deposition or migration outside the printed area. Exposure of material to these conditions has potential to change critical material attributes, which may manifest in critical quality attributes of printed dosage forms such as mechanical properties, disintegration, dissolution, chemical assay, and impurity profile.

The critical material attributes that may change include amorphous/crystalline fraction, formation of new polymorphic form, flow properties, particle size distribution, particle shape, drug distribution, content uniformity, and mechanical properties. Moreover, the ratio of reused to new powder can affect thermal and mechanical properties, of in-process, as well as final material. The impact of the number of recycling of used material, the ratio of reuse and new material on the critical attributes of raw material, in-process parameters and finished dosage forms, must be thoroughly evaluated [15].

cGMP: Current good manufacturing practice (cGMP) compliance of the facility and equipment, is required if the drug product is manufactured for mass distribution. cGMP certification has to be submitted to the FDA, along with other modules for a drug product application [85]. The facility and equipment would be inspected by the FDA before approving the drug product. At this point, cGMP compliant BJ printer is not commercially available in the market. It would be challenging for pharma companies to procure cGMP compliant BJ printer.

Opportunity in the personalization of medication

The commercialization of BJ process represents a major challenge. This is due to the simple fact that the process cannot compete with well-established traditional manufacturing methods. Currently, a manufacturing time of approximately an hour would be required, to fabricate 10–50 tablets, depending upon the process and complexity of dosage forms [44]. However, it can be used in pharmacies and hospitals to personalize medications, to meet individual patient needs. For example, printing solid dosage forms of otherwise unstable liquid dosage forms, new strength of the approved drug product, polypill (single pill containing multiple drugs), dosage forms without excipients to which a particular patient is allergic to (such as lactose or dye intolerance), and for ease of administration (geriatric and children patients).

Reportedly, some hospitals in the USA are planning to start manufacturing of their internal supplies of generic drugs, to contain unexpected shortage of drug products, and control healthcare cost [86]. BJ printing provides a great opportunity in that direction. FDA approval is required for generic or brand drug products. Interestingly, FDA approval is not required if compounded drugs are not for mass distribution. Compounded medicines are exempted from many FDA requirements, which are covered under Section 503A of the FD&C Act. Section 503A of the FD&C Act by the Food and Drug Administration Modernization Act of 1997 (Public Law 105–115) (the Modernization Act) exempt requirement of cGMP under section 501(a) (2) (B), labeling of drugs with adequate directions for use under section 502(f) (1) and approval of drugs under new drug applications (NDAs) or abbreviated new drug applications (ANDAs) under section 505 of compounding of nonsterile formulation.

However, there are certain requirements that pharmacies and hospitals have to meet, before employing BJ for personalized medicines, e.g., compliance to USP chapter on pharmacy compounding. Compounding can be done solely for identifiable individual patients, active and inactive ingredients should meet applicable USP or NF monographs, or USP chapters on pharmacy compounding if monograph does not exist [87]. Nevertheless, cGMP compliance is required if pharmacies and hospitals are involved in the compounding of sterile drug products or biologics [88].

Before printing dosage forms by BJ in a hospital and pharmacy, process and formulation variables have to be understood, and the quality and stability of dosage forms should be confirmed in the lab. However, it is not possible to develop a printing formula for all the drugs used in the hospital or dispensed in a pharmacy. It is possible to develop a master printing formula, which should work with most commonly used medicines in a hospital, or which requires minimal tweaking, depending upon the characteristics of the drug. Stability of the printed dosage form should not be an issue since compounded medicines are consumed in a short period.

Conclusion

FDA has approved over 100 medical devices, and only one 3D printed drug product [23,89]. Unlike a medicinal product, there are technical challenges in implementing 3D printing of dosage forms for mass distribution. However, current research has shown tremendous benefits in personalizing medication using 3D printing. Printing of dosage forms in hospital and pharmacy needs limited regulatory oversight, compared to drug products intended for mass distribution. State pharmacy boards may play a crucial role, in ensuring the safe dispensing of personalized medicines.

References

[1] Zajicek A, Fossler MJ, Barrett JS, Worthington JH, Ternik R, Charkoftaki G, Lum S, Breitkreutz J, Baltezor M, Macheras P, Khan M, Agharkar S, MacLaren DD. A report from the pediatric formulations task force: perspectives on the state of child-friendly oral dosage forms. AAPS J. 2013;15(4):1072–81.

[2] Purixan® FDA label. January 22, 2019. https://www.accessdata.fda.gov/drugsatfda_docs/label/2018/205919s003lbl.pdf.

[3] American cancer society: key statistics for childhood cancers. January 22, 2019. http://www.cancer.org/cancer/cancerinchildren/detailedguide/cancer-in-children-key-statistics.

[4] American childhood cancer organization. January 22, 2018. http://www.acco.org/about/.

[5] Schell KH. Compliance issues and extemporaneous preparation of medications for pediatric patients. J. Pharm. Technol. 1992;8(4):158–61.

[6] Liu Q. Clinical pharmacology issues with pediatric formulations, FDA/CDER/OTS/OCP. January 22, 2019. http://www.fda.gov/downloads/AdvisoryCommittees/CommitteesMeetingMaterials/PediatricAdvisoryCommittee/UCM289937.pdf.

[7] Gudeman J, Jozwiakowski M, Chollet J, Randell M. Potential risks of pharmacy compounding. Drugs R D. 2013;13(1):1–8.

[8] Khan MA. Some challenges in the development of pediatric formulations, FDA pediatric advisory committee. 2012. http://www.fda.gov/downloads/AdvisoryCommittees/CommitteesMeetingMaterials/PediatricAdvisoryCommittee/UCM289938.pdf.

[9] Ivanovska V, Rademaker CMA, van Dijk L, Mantel-Teeuwisse AK. Pediatric drug formulations: a review of challenges and progress. Pediatrics 2014;134(2):361–72.

[10] Best pharmaceuticals for children act (BPCA). 2002. https://bpca.nichd.nih.gov/Pages/default.aspx.

[11] Pediatric research equity act. 2007. http://www.fda.gov/downloads/Drugs/DevelopmentApprovalProcess/DevelopmentResources/UCM049870.pdf.

[12] Rahman Z, Korang-Yeboah M, Siddiqui A, Mohammad A, Khan MA. Understanding effect of formulation and manufacturing variables on the critical quality attributes of warfarin sodium product. Int. J. Pharm. 2015;495(1):19–30.

[13] Korang-Yeboah M, Rahman Z, Shah D, Mohammad A, Wu S, Siddiqui A, Khan MA. Impact of formulation and process variables on solid-state stability of theophylline in controlled release formulations. Int. J. Pharm. 2016;499(1–2):20–8.

[14] Rahman Z, Dharani S, Barakh Ali SF, Afrooz H, Reddy IK, Khan MA. Effect of processing parameters and controlled environment storage on the disproportionation and dissolution of extended-release capsule of phenytoin sodium. Int. J. Pharm. 2018;550(1–2):290–9.

[15] Rahman Z, Barakh Ali SF, Ozkan T, Charoo NA, Reddy IK, Khan MA. Additive manufacturing with 3D Printing: progress from bench to bedside. AAPS J. 2018;20(6):101.

[16] Frank C. Challenges and achievements in caring for the elderly. Can. Fam. Physician 2010;56(11):1101–2.

[17] Koronkowski M, Eisenhower C, Marcum Z. An update on geriatric medication safety and challenges specific to the care of older adults. Ann. Long Term Care 2016;24(3):37–40.

[18] Khaled SA, Burley JC, Alexander MR, Yang J, Roberts CJ. 3D printing of five-in-one dose combination polypill with defined immediate and sustained release profiles. J. Control. Release 2015;217:308–14.

[19] Khaled SA, Burley JC, Alexander MR, Yang J, Roberts CJ. 3D printing of tablets containing multiple drugs with defined release profiles. Int. J. Pharm. 2015;494(2):643–50.

[20] Charlesworth CJ, Smit E, Lee DS, Alramadhan F, Odden MC. Polypharmacy among adults aged 65 years and older in the United States: 1988–2010. J. Gerontol. A Biol. Sci. Med. Sci. 2015;70(8):989–95.

[21] Gellad WF, Grenard JL, Marcum ZA. A systematic review of barriers to medication adherence in the elderly: looking beyond cost and regimen complexity. Am. J. Geriatr. Pharmacother. 2011;9(1):11–23.

[22] Cutler RL, Fernandez-Llimos F, Frommer M, Benrimoj C, Garcia-Cardenas V. Economic impact of medication non-adherence by disease groups: a systematic review. BMJ Open 2018;(1):e016982.

[23] Spritam® FDA label. January 22, 2019. https://www.accessdata.fda.gov/drugsatfda_docs/label/2018/207958s008lbl.pdf.

[24] Aprecia. January 09, 2019. https://www.aprecia.com/technology/zipdose.

[25] Sachs EM, Haggerty JS, Cima MJ, Williams PA. Three-dimensional printing techniques. U.S. Patent 5340656. 1994.

[26] Sachs E, Cima M, Williams P, Brancazio D, Cornie J. Three dimensional printing: rapid tooling and prototypes directly from a CAD model. J. Eng. Ind. 1992;114:481–8.

[27] Sachs E, Cima M, Cornie J. Three-dimensional printing: rapid tooling and prototypes directly from a CAD model. CIRP Ann. Manuf. Technol. 1990;39(1):201–4.

[28] Katstra W, Palazzolo R, Rowe C, Giritlioglu B, Teung P, Cima M. Oral dosage forms fabricated by three dimensional Printing™. J. Control. Release 2000;66(1):1–9.

[29] Seitz H, Rieder W, Irsen S, Leukers B, Tille C. Threedimensional printing of porous ceramic scaffolds for bone tissue engineering. J. Biomed. Mater. Res. B Appl. Biomater. 2005;74(2):782–8.

[30] FDA guidance for industry-orally disintegrating tablets. 2008.

[31] Darji MA, Lalge RM, Marathe SP, Mulay TD, Fatima T, Alshammari A, Lee HK, Repka MA, Narasimha Murthy S. Excipient stability in oral solid dosage forms: a review. AAPS PharmSciTech 2018;19(1):12–26.

[32] FDA inactive ingredient database. January 22, 2019. https://www.accessdata.fda.gov/scripts/cder/iig/index.cfm.

[33] FDA data of generally recognized as safe. January 22, 2019. https://www.accessdata.fda.gov/scripts/cder/iig/index.cfmhttps://www.fda.gov/food/ingredientspackaginglabeling/gras/.

[34] Rahman Z, Zidan AS, Khan MA. Risperidone solid dispersion for orally disintegrating tablet: its formulation design and non-destructive methods of evaluation. Int. J. Pharm. 2010;400(1–2):49–58.

[35] Rahman Z, Siddiqui A, Khan MA. Orally disintegrating tablet of novel salt of antiepileptic drug: formulation strategy and evaluation. Eur. J. Pharm. Biopharm. 2013;85(3 Pt B):1300–9.

[36] Lee K-J, Kang A, Delfino JJ, West TG, Chetty D, Monkhouse DC, et al. Evaluation of critical formulation factors in the development of a rapidly dispersing captopril oral dosage form. Drug Dev. Ind. Pharm. 2003;29(9):967–79.

[37] Wang CC, Tejwani Motwani MR, Roach WJ, Kay JL, Yoo J, Surprenant HL, et al. Development of near zero-order release dosage forms using three-dimensional printing (3-DP) technology. Drug Dev. Ind. Pharm. 2006;32(3):367–76.

[38] Wu BM, Borland SW, Giordano RA, Cima LG, Sachs EM, Cima MJ. Solid free-form fabrication of drug delivery devices. J. Control. Release 1996;40(1):77–87.

[39] Yu DG, Yang XL, Huang WD, Liu J, Wang YG, Xu H. Tablets with material gradients fabricated by three-dimensional printing. J. Pharm. Sci. 2007;96(9):2446–56.

[40] Yu D-G, Branford-White C, Yang Y-C, Zhu L-M, Welbeck EW, Yang X-L. A novel fast disintegrating tablet fabricated by three-dimensional printing. Drug Dev. Ind. Pharm. 2009;35(12):1530–6.

[41] Yu DG, Branford-White C, Ma ZH, Zhu LM, Li XY, Yang XL. Novel drug delivery devices for providing linear release profiles fabricated by 3DP. Int. J. Pharm. 2009;370(1–2):160–6.

[42] Huang W, Zheng Q, Sun W, Xu H, Yang X. Levofloxacin implants with predefined microstructure fabricated by three-dimensional printing technique. Int. J. Pharm. 2007;339(1–2):33–8.

[43] Kumar AV, Dutta A, Fay JE. Electrophotographic printing of part and binder powders. Rapid Prototyp. J. 2004;10:13.

[44] Alomari M, Mohamed FH, Basit AW, Gaisford. Personalised dosing: printing a dose of one's own medicine. Int. J. Pharm. 2015 30;494(2):568–77.

[45] Murphy SV, Atala A. 3D bioprinting of tissues and organs. Nat. Biotechnol. 2014;32:773–85.

[46] Rahmati S, Shirazi S, Baghayeri H. Piezo-electric head application in a new 3D printing design. Rapid Prototyp. J. 2009;15:187–91.

[47] Trapp J, Rubenchik AM, Guss G, Matthews MJ. In situ absorptivity measurements of metallic powders during laser powder-bed fusion additive manufacturing. Appl. Mater. Today 2017;9:341–9.

[48] Bidare P, Bitharas I, Ward RM, Attallah MM, Moore AJ. Fluid and particle dynamics in laser powder bed fusion. Acta Mater. 2018;142:107–20.

[49] ICH- impurities:guidline for residual solvents, Q3C(R7). 2018.

[50] Pardeike J, Strohmeier DM, Schrodl N, Voura C, Gruber M, Khinast JG, Zimmer A. Nanosuspensions as advanced printing ink for accurate dosing of poorly soluble drugs in personalized medicines. Int. J. Pharm. 2011;420:93–100.

[51] Hirshfield L, Giridhar A, Taylor LS, Harris MT, Reklaitis GV. Dropwise additive manufacturing of pharmaceutical products for solvent-based dosage forms. J. Pharm. Sci. 2014;103:496–506.

[52] Shimoda J. New bubble jet head technologies used in canon color bubble jet printer BJC-70. Recent Prog. Ink. Jet Technol. 1996:143–6.

[53] Bohórquez JH, Canfield BP, Courian KJ, Drogo F, Hall CAE, Holstun CL, Scandalis AR, Shepard ME. Laser-comparable inkjet text printing. Hewlett Packard J. 1994:9–17.

[54] Buanz ABM, Saunders MH, Basit AW, Gaisford S. Preparation of personalized-dose salbutamol sulphate oral films with thermal ink-jet printing. Pharm. Res. 2011;28:2386–92.

[55] Genina N, Fors D, Palo M, Peltonen J, Sandler N. Behavior of printable formulations of loperamide and caffeine on different substrates-effect of print density in inkjet printing. Int. J. Pharm. 2013;453:488–97.

[56] Genina N, Fors D, Vakili H, Ihalainen P, Pohjala L, Ehlers H, Kassamakov I, Haeggstrom E, Vuorela P, Peltonen J, Sandler N. Tailoring controlled release oral dosage forms by combining inkjet and flexographic printing techniques. Eur. J. Pharm. Sci. 2012;47:615–23.

[57] Lee BK, Yun YH, Choi JS, Choi YC, Kim JD, Cho YW. Fabrication of drug loaded polymer microparticles with arbitrary geometries using a piezoelectric inkjet printing system. Int. J. Pharm. 2012;427:305310.

[58] Raijada D, Genina N, Fors D, Wisaeus E, Peltonen J, Rantanen J, Sandler N. A step toward development of printable dosage forms for poorly soluble drugs. J. Pharm. Sci. 2013;102:3694–704.

[59] Sandler N, Maattanen A, Ihalainen P, Kronberg L, Meierjohann A, Viitala T, Peltonen J. Inkjet printing of drug substances and use of porous substrates-towards individualized dosing. J. Pharm. Sci. 2011;100:3386–95.

[60] Rahman Z, Xu X, Katragadda U, Krishnaiah YS, Yu L, Khan MA. Quality by design approach for understanding the critical quality attributes of cyclosporine ophthalmic emulsion. Mol. Pharm. 2014;11(3):787–99.

[61] Alam P, Toivakka M, Backfolk K, Sirviö P. Impact spreading and absorption of Newtonian droplets on topographically irregular porous materials. Chem. Eng. Sci. 2007;62(12):3142–315.

[62] Chandra S, Avedisian CT. Observations of droplet impingement on a ceramic porous surface. Int. J. Heat Mass Transf. 1992;35(10):2377–88.

[63] Marston JO, Thoroddsen ST, Ng WK, Tan RBH. Experimental study of liquid drop impact onto a powder surface. Powder Technol. 2010;203(2):223–36.

[64] Yu Z, Wang F, Fan LS. Experimental and numerical studies of water droplet impact on a porous surface in the film-boiling regime. Ind. Eng. Chem. Res. 2008;47(23):9174–82.

[65] Yarin AL. Drop impact dynamics: splashing, spreading, receding, bouncing. Annu. Rev. Fluid Mech. 2006;38(1):159–92.

[66] Asadi-Eydivand M, Solati-Hashjin M, Farzad A, Abu Osman NA. Effect of technical parameters on porous structure and strength of 3D printed calcium sulfate prototypes. Robot. Comput. Integr. Manuf. 2016;37:57–67.

[67] Bai Y, Williams CB. An exploration of binder jetting of copper. Rapid Prototyp. J. 2015;21(2):177–85.

[68] Bergmann C, Lindner M, Zhang W, Koczur K, Kirsten A, Telle R, Fischer H. 3D printing of bone substitute implants using calcium phosphate and bioactive glasses. J. Eur. Ceram. Soc. 2010;30(12):2563–7.

[69] Farzadi A, Solati-Hashjin M, Asadi-Eydivand M, Abu-Osman NA. Effect of layer thickness and printing orientation on mechanical properties and dimensional accuracy of 3D printed porous samples for bone tissue engineering. PLoS One 2014;9(9):e108252.

[70] Lu K, Reynolds WT. 3DP process for fine mesh structure printing. Powder Technol. 2008;187(1):11–8.

[71] Miyanaji H, Zhang S, Lassell A, Zandinejad A, Yang L. Process development of porcelain ceramic material with binder jetting process for dental applications. JOM 2016;68(3):831–41.

[72] Vaezi M, Chua CK. Effects of layer thickness and binder saturation level parameters on 3D printing process. Int. J. Adv. Manuf. Technol. 2011;53(1):275–84.
[73] Yang L, Zhang S, Oliveira G, Stucker B. Development of a 3D printing method for production of dental application. Proceedings of the 24th international solid freeform fabrication symposium, Austin, TX, USA. 2013.
[74] Caradonna MA, Cima MJ, Grau J, Moon J, Sachs EM, Saxton PC, Serdy JG, Uhland SA. Jetting layers of powder and the formation of fine powder beds thereby. European patent application EP1009614. 2000.
[75] Derby B. Inkjet printing ceramics: from drops to solid. J. Eur. Ceram. Soc. 2011:312543–50.
[76] Hogekamp S, Pohl M. Methods for characterizing wetting and dispersing of powder. Chem. Ing. Tech. 2004;76:385–90.
[77] Lanzetta M, Sachs E. Improved surface finish in 3D printing using bimodal powder distribution. Rapid Prototyp. J. 2003;9:157–66.
[78] Hapgood KP, Litster JD, Biggs SR, Howes T. Drop penetration into porous powder beds. J. Colloid Interface Sci. 2002;253:353–66.
[79] Sachs EM, Haggerty JS, Cima MJ, Williams PA. Three-dimensional printing techniques. US Patent. 1993. p. 5204055.
[80] Miyanaji H, Zhang S, Lassell A, Zandinejad AA, Yang L. Optimal process parameters for 3d printing of porcelain structures. Procedia Manuf. 2016;5:870–87.
[81] Bredt JF. Binder stability and powder/binder interaction in three dimensional printing, thesis [Ph.D.]. Cambridge: Dept. of Mechanical Engineering, Massachusetts Institute of Technology; 1995. p. 382.
[82] Voura C, Gruber M, Schroedl N, Strohmeier D, Eltzinger B, Bauer W, Brenn G, Khinast J, Zimmer A. Printable medicines: a microdosing device for producing personalised medicines. Pharmaceut. Technol. Eur. 2011;5(1):32–6.
[83] Siddiqui A, Rahman Z, Khan SR, Awotwe-Otoo D, Khan MA. Root cause evaluation of particulates in the lyophilized indomethacin sodium trihydrate plug for parenteral administration. Int. J. Pharm. 2014;473(1–2):545–51.
[84] The United States pharmacopeia–national formulary (USP 40–NF 35). 2017.
[85] FDA guidance for industry- quality systems approach to pharmaceutical CGMP regulations. 2006.
[86] Abelson R, Thomas K. Fed up with drug companies, hospitals decide to start their own. The New York Times. January 18, 2018. https://www.nytimes.com/2018/01/18/health/drug-prices-hospitals.html.
[87] FDA guidance – Pharmacy Compounding of Human Drug Products Under Section 503A of the Federal Food, Drug, and Cosmetic Act. 2016.
[88] FDA guidance – Current Good Manufacturing Practice—Guidance for Human Drug Compounding Outsourcing Facilities Under Section 503B of the FD&C Act. 2018.
[89] Jacobson M. Lessons for medical device manufacturers using 3D printing. Med. Device Online July 18, 2018. https://www.meddeviceonline.com/doc/lessons-for-medical-device-manufacturers-using-d-printing-0001.

Chapter 47

3D printing in orthopedic trauma

Mohit Kumar Patralekh[1] and Hitesh Lal[2]
[1]*Central Institute of Orthopaedics, Safdarjung Hospital and Vardhman Mahavir Medical College, New Delhi, India;* [2]*Sports Injury Centre, Safdarjung Hospital and Vardhman Mahavir Medical College, New Delhi, India*

Introduction

Computed tomography (CT) or magnetic resonance imaging (MRI) data of trauma patients, can be used for manufacturing graspable objects from three-dimensional (3D) reconstructed images. 3D printed patient-specific instrumentation helps to achieve precise implant placement and to achieve improved surgical results. Also, customized implants can be created to match an individual patient's anatomy. The application of 3D printing is not limited to the operating theater, as it also helps in the manufacturing of more individualized prosthetics and orthoses [1–5].

3D printing technology involves the conversion of a computer-generated 3D image into a physical model. Creation of the 3D model is based on 3D DICOM (digital imaging and communications in medicine) format data, derived from CT or MRI. It needs conversion into a file format, which is recognizable by the 3D printer. For accomplishing this, the DICOM file is uploaded into a program (e.g., Mimics from Materialize software for Windows or Osirix(free-open source) for Mac), which allows 3D reconstruction of the image. It is then exported in a file format (stereolithography [STL]), which is readable by the software (computer-aided design [CAD]), which is used for designing 3D objects. Errors or defects or in the STL file are then corrected, before exporting it to the 3D printer.

3D printers create objects, layer by layer. Old manufacturing techniques involved subtraction of layers from the raw material, but 3D printing works by "additive manufacturing," in which the raw material is "added" layer by layer in a predetermined manner, and therefore, it achieves a precise 3D object. Industry grade printers use lasers to accurately sinter granular substrates like metal or plastic powders. The printer keeps on adding a new layer of unfused powder over the previous layer, and the cycle continues until the entire object is generated. These printers have high speeds, have ability to recycle unfused powder, and can use stronger materials with higher melting points, like titanium.

One can manufacture unique patient-specific objects, more cost-effectively than using conventional implant manufacturing techniques. 3D printing technology makes it possible to combine any complex shape, and porous and solid sections can be combined together, for providing optimal strength and performance [6–8].

3D printing can help in the teaching and training of novice surgeons in complex surgical areas, such as pelvi-acetabular trauma. Careful (and actually somewhat easier!) preoperative analysis and practice on the 3D printed model allows the surgeon to select an optimal surgical approach, plan implant placement, visualize screw trajectory, anticipate intraoperative difficulties, and access the need for special equipment. It is also possible to sterilize and review the model intraoperatively if necessary [6,9,10]. Finally, it can help us to evaluate the restoration of individual anatomy after surgery. In some cases, it can also help in making a precise anatomical diagnosis, where it is not otherwise obvious, and for planning suitable management accordingly. 3D printing of artificial cartilage scaffolds individualized for the case and 3D bioprinting are some areas of growing scientific interest. 3D printing, also called as *additive manufacturing* and *rapid prototyping*, has been considered as the "second industrial revolution," and this belief is true, in particular, for orthopedic trauma surgery [1–10].

Pioneering efforts

In 1997, Kacl et al. reported that rapid prototyping might be useful for surgical planning and teaching. His paper did not reveal any difference between stereolithography, and workstation-based 3D reformations, in the management of intraarticular calcaneal fractures [11]. In 1998, Potamianos et al. reported usage of rapid prototyping technique, for the construction of an anatomical model, in a case presenting

Precision Medicine for Investigators, Practitioners and Providers. https://doi.org/10.1016/B978-0-12-819178-1.00047-2

with a complicated shoulder injury. The model thus created provided a definitive interpretation of joint pathoanatomy and enabled a complete appraisal of the degree of injury [12]. Brown et al., in 2003, found that 3D printing helped in surgical planning, and in reducing the exposure to radiation, during 117 complex surgical cases [13]. Guarino et al. in 2007, reported treatment of 10 patients with pediatric scoliosis, and 3 complex pelvic fracture patients, and concluded that 3D printing improved the placement accuracy of the pedicle and pelvic screws, and therefore, decreased the risk of iatrogenic neurovascular injury [14].

Applications of 3D printing in specific anatomical areas

Upper limb

Acromion

Beliën et al. used a 3D model for managing cases of os acromiale and acromion fractures. Initially, a 3D acromial model was manufactured, and then a distal clavicular reconstruction plate was prebent to fit into the individual anatomical shape and curvatures of the acromion. They have published their experience in five cases, including three os acromiale and two acromial fracture. Patients were evaluated using the DASH (disabilities of arm, shoulder, and hand) and Constant–Murley shoulder scores. The fracture or nonunion got united in all cases. If the surgery was performed before the occurrence of secondary damage (like an impingement syndrome), they observed that pain completely disappeared. The surgeon had the luxury of preparing for the surgery in advance, which improved comfort and reduced surgical time. The model could also be used to educate the patient and the operating team, about the planned surgery [15].

Clavicle

Jeong et al. described minimally invasive plating for midshaft clavicular fractures, using intramedullary indirect reduction technique and prebent plates, made using 3D printed models. This allowed for the reduction of the fracture, and accurate plating with minimal soft tissue damage [16]. Kim et al. also utilized a 3D printed clavicle model for preoperative planning, and as a tool during operation for minimally invasive plating, for displaced comminuted midshaft clavicle fractures. CT scan of both clavicles was taken for patients with unilateral comminuted displaced midshaft clavicle fracture. Both clavicles were 3D printed to achieve real-size clavicular models. The uninjured clavicle was then 3D printed into the opposite side model, by mirror imaging technique, to make a preinjury replica of the fractured side clavicle. The 3D-printed uninjured clavicle model served as a template for selecting the precontoured clavicle locking plate, which best fit the model. The plate was inserted via small incisions and fixed with locking screws without exposing the fracture site. Seven comminuted clavicle fractures treated in this manner united nicely [17].

Proximal humerus

You et al. managed 66 elderly cases with complex proximal humeral fractures, who were randomly assigned to two groups. In the test group, 3D printing was used to make 3D model of the fracture, using data acquired from thin-slice CT scan, and processed using Mimics software. It helped to confirm the diagnosis, design individualized surgical plan, simulate operative procedures, and in performing the actual surgery as planned. The 3D model provided a 360 degree visual display, and palpatory sense of the severity and direction of the fracture dislocation, which helped in precise preoperative diagnosis, surgical planning, implant measurement, selection of suitable anatomical locking plate, and simulation of surgery. Shorter surgical duration, lower blood loss, and a reduced number of fluoroscopies were seen, compared to the control group ($P < .05$) [18].

Distal humerus

Kim et al. used 3D-printed plates, for fixing intercondylar humeral fractures. Thirteen patients with intercondylar humeral fractures were randomized, to receive an open reduction and internal fixation with either conventional plates ($n = 7$) or 3D-printed plates ($n = 6$). The 3D-printing group had a significantly shorter mean surgical time. At the last follow-up, no significant difference was found, although the 3D-printing group had a slightly greater rate of good or excellent outcomes (83.1%) in comparison to the conventional plating group (71.4%) [19].

Zheng et al. managed 12 male and 6 female cases with cubitus varus deformities, utilising templates created by rapid prototyping. Osteotomy templates best fitting the angle and range of osteotomy, were "reversely" made from the 3D model by rapid prototyping, and were used for guidance during the corrective surgery. Average postoperative carrying angle in the 18 cases with cubitus varus deformity was 7.3 degrees (range, 5–11 degrees), with a mean correction of 21.9 degrees (range, 12–41 degrees) at 12–24 months follow-up [20].

Zheng et al. used a novel navigation template for osteotomy in cubitus varus, created by computer-assisted design and 3D printing technology. An osteotomy was done with the help of this navigation template, followed by fixation with 2K wires, and immobilization in a long arm plaster with the elbow flexed at 20 degrees. All patients had a good cosmetic appearance. Mean union time was 6.7 weeks (range, 6–8 weeks). An excellent result was achieved in 12 cases and a good result in 2 cases according to Bellemore criteria [21].

Gemalmaz et al. treated an 18 years previously operated patient, having 40 degrees of cubitus varus deformity (with 20 degrees flexion), which was a sequela of an 8-year-old malunited right supracondylar humerus fracture, using a custom 3D printed resection guide during the operation. They obtained an exact osteotomy, accurate correction, and nice functional outcome [22].

Yang et al. treated 40 patients with elbow fractures, randomly divided into a 3D printing surgery group and a conventional surgery group. The 3D printing group showed shorter surgical time, lower blood loss, and higher elbow function score. Polylactic acid (PLA) is environment-friendly, whereas acrylonitrile butadiene styrene (ABS) emits odor during printing. Curling of edges happened during the printing process in four of 10 ABS models but in only a single PLA model. PLA was thus found to be a more appropriate material [23].

Recently, Zaho et al. treated five cases with severe distal humeral bone defects, with customized 3D printing prostheses. The length of the bone defect was 5–12 cm. All had an open fracture of Gustilo type. The MEPS scores and the Enneking scores were all improved, and there was no prosthetic loosening or joint dislocation [24].

Distal radius

Muinck et al. systematically reviewed 3D planned corrective osteotomies, for distal radial malunions. Palmar tilt, radial inclination, and ulnar variance improved significantly. Average flexion–extension, prosupination, and grip strength, also improved. Complications were observed in 11 out of 68 cases (16%). Overall, 3D-planned corrective osteotomies were more useful for the treatment of complex malunited distal radial fractures [25].

Roner et al. developed a case-specific ramp-guide, and compared the accuracy of navigation of 3D planned opening-wedge osteotomies, with state-of-the-art guide relying only on predrilled holes. Surgery of the ramp-guide group was significantly shorter, less rotational and translational residual malalignment error was observed, and significantly fewer distal fragment screws were misaligned, although plate positioning was not much affected [26].

Hand

Zang et al. used 3D printing for planning thumb reconstructions with second toe transfer. Five cases with grade 3 thumb defects were reconstructed with a wrap around the flap and second toe transfer, planned by 3D printing technique using CT scans of hands and feet, in Boholo surgical simulation software. A mirror image of the injured side was created using the uninjured thumb. 3D models of the great toe and the second toe were created, for conceptualizing the donor site dimensions, and also for reconstructing the donor site defect, by planning suitable iliac bone and superficial circumflex iliac artery flaps. Polylactic acid (PLA) models of the reconstructed thumb and donor toe were 3D printed. All reconstructed thumbs survived, but partial flap necrosis occurred in one case, which was managed by dressings. Reconstructed thumbs had an overall good appearance and function [27].

Recently, Fuller et al. made a 3D model of a bone reduction clamp for hand fractures. This prototype was used in the operating room, and met the surgeon's expectations [28].

Miscellaneous

Taylor et al. used 3D printing techniques, as an adjunct in vascularized bone flap transfers to the upper limb. Utilizing open source software and CT data, 3D models were printed in the surgeon's office. Examples included medial femoral trochlea flap, for avascular necrosis and nonunion of the scaphoid, avascular necrosis and nonunion of lunate, medial femoral condyle flap for wrist arthrodesis, free fibular osteocutaneous flap, and distal radial infected nonunion. 3D model-based templates helped in rapid and accurate contouring of vascularized bone flaps in situ, prior to donor pedicle ligation [29].

3D printed implants for eroded glenoids, after total shoulder replacement surgery, have shown results [30]. 3D printed prosthesis, mirrored on the contralateral wrist for replacement of whole scaphoid or lunate, following avascular necrosis, has suitable geometry, mechanical properties, and cytocompatibility properties for in vivo usage [31]. Berg et al. experimentally explored the usage of 3D models of fractured and intact scaphoids, used prebent plates made using the model, and advocated for its use after economic justification [32].

Lower limb

Acetabulum

Role of 3D printing in acetabular fractures, encompasses educational models for understanding fracture pattern, classifying it, preoperative planning, designing precontoured plates as well as drilling guides for screw placement, and also for planning hip arthroplasty for acetabular fracture patients, with acetabular bone defects [33–60]. Hurson et al. reported 12 acetabular fracture cases, classified and planned preoperatively using 3D printing, and proved that these models markedly helped surgeons in understanding the individual fracture anatomy, more so new surgeons [61]. Maini et al. found that patient-specific precontoured plates for acetabular fractures, made using 3D models, were better than intraoperatively contoured plate. Also, real-time 3D pelvis model was found to be an accurate technique for preoperative planning in acetabular fractures [62]. Bagaria et al. found that 3D printing could help surgeons understand complex fractures and achieve near anatomical reduction [63].

According to Kim et al., 3D printed acetabular models helped in understanding complex pathoanatomy of acetabular fracture, and in planning the appropriate positioning of reduction clamps, screws entry sites, and trajectories. Prebending reconstruction plates reduced surgical time. The optimal position of the guide wire, planned during the simulation, was used as a reference during the real surgery for percutaneous posterior column screw fixation, and helped resident training, besides precise positioning. Optimal positioning of anatomical plates was similarly planned using 3D printed clavicular models and gave good results [49].

Pelvis

Cai et al. used 3D printing technology, for minimally invasive cannulated screw fixation of unstable pelvic fractures in 137 patients. Duration of surgery and the mean number of fluoroscopies were significantly more in the control group. However, no significant difference in functional outcomes was noticed between the two groups [64].

Wu et al. assessed 3D printing technology for the operative treatment of old pelvic fractures. Model creation from CT DICOM data needed 7 h (range: 6–9 h). There was a good correlation between preoperative planning and postoperative follow-up radiographs, in all nine patients. No wound problem or nonunion occurred. The result was excellent in two cases, good in five, and poor in two patients, based on the Majeed score[65].

Zeng et al. evaluated the efficacy 3D printing assisted internal fixation. for unstable pelvic fracture using minimally invasive para-rectus approach. Outcomes were 97.37% excellent and good on Matta scoring and 94.4% excellent and good on Majeed assessment. The average surgical time was 110 min, intraoperative blood loss 320 mL, and incision length 6.5 cm. The technique was, thus, feasible, safe, and effective, with advantages of minimal trauma, a little bleeding, rapid healing, and accurate reduction [35]. Interestingly, 3D printed intraoperative guides have also been used in pelvic and hip surgery, for curved peri-acetabular osteotomies [66].

Liu et al. reported use of a 3D printed guiding template, for sacroiliac screw insertion, which could significantly shorten the surgical time, provide a satisfactory outcome of stabilization of the pelvic ring, and protect doctors and patients from X-ray exposure [67].

Distal femur

Lin et al. treated 21 cases of distal femoral fractures, using 3D printing with Mimics software. The positioning of plates and screws were preoperatively rehearsed in the navigation module; 3D assistance of screw entry point was obtained; 21 plates and 180 screws were placed with the help of the femoral module; CT with 3D reconstruction was performed in 21 cases postoperatively. Plate positioning was found consistent with prediction, with no significant differences in spatial location of screws [68].

Arnal-Burró et al. reported usage of 3D printed cutting guides, for open-wedge distal femoral osteotomies. Axial correction accuracy, operative time, fluoroscopy time, and costs were optimum in the 3D guides group [69]. Similarly, Shi et al. reported medial closed-wedge distal femoral osteotomy, performed with assistance by 3D-printed cutting guides and locking guides, to treat valgus knee combined with lateral compartment disease. 3D-printed cutting and locking guides increased the precision, in patients with lateral compartment disease and valgus deformity, made the operation shorter, and reduced fluoroscopy exposure [70]. Similarly, Chen et al. concluded that 3D printed cutting blocks, greatly improved the accuracy of distal femoral osteotomy, for correction of valgus knees with osteoarthritis [71].

Anterior cruciate ligament (ACL) reconstruction

Rankin et al. developed a case-specific, arthroscopic ACL femoral tunnel guide, for anatomical positioning of the ACL graft in the femoral tunnel, based on MRI scan of the patient's uninjured opposite knee, for finding the femoral footprint relative to the femoral articular cartilage borders. Transparent acrylic-based photopolymer, PA220 plastic and 316L stainless steel case-specific ACL femoral tunnel guides, were manufactured by a 3D printing technique. No significant difference was observed, in the size and positioning of the center of the ACL femoral footprint guide, to MRI site [72].

Proximal tibia

Huang et al. applied a 3D printing technique for tibial plateau fractures and assessed fixation outcomes in terms of deviations of screw placement. Accurate, fixation outcomes were achieved. There was no significant difference in the deviations of screw length, entry point, and projection angle, between the ideal and real screw trajectories [73,74].

Giannetti et al. compared the outcomes after minimally invasive reduction and internal fixation, with and without 3D printing, for displaced tibial plateau fractures. Significant reduction in operative time, radiation exposure, and blood loss were seen in 3D printing group. Functional outcomes were equivalent, and no complications were observed [75].

Vaishya et al. managed a 36-year-old male with Schatzker type 2 right proximal tibial fracture, using 3D printed model, for fracture pattern delineation, and identification of exact placement of plate and screws. LISS system was used, besides an extra 7 mm cancellous screw proximally to achieve anatomical reduction with minimal soft tissue damage and blood loss [6].

Yang et al. reported 3D printing assisted osteotomy, for malunited lateral plateau fractures in seven cases. 3D printing technology helped in accurate planning and performing the osteotomy, reduced the risk of postoperative deformity, shortened the operative time, and decreased blood loss [76].

Tibial pilon and malleolar fractures

Chung et al. used 3D printing for understanding complicated fracture patterns, preoperative templating, selection of anatomical plates, and planning screw trajectories for reduction and fixation of distal tibial fractures, and got nice results [77].

Talus

Wu et al. reported 3D printing techniques for achieving correct posterior screw placement, and safe zone geometry for screw fixation of talar neck, using CT data of 15 normal feet. Mimics software was used for 3D reconstruction, and 4 mm screws were simulated from lateral tubercle of the posterior process to talar head. Screw lengths and trajectories at nine locations, which did not breach the cortex, were evaluated. The safe zone was found between the 30 and 60 degrees location; the maximum height of each safe zone was 7.8 ± 1.2 degrees, and the width of each safe zone was 13.6 ± 1.4 degrees. The safe zone of posterior screw fixation was defined, assuming fractures to be reduced [78].

Chiu et al. treated a 30-year-old male, who had sustained a traumatic loss of the whole talus, with an anatomical antibiotic-loaded talus cement spacer, manufactured using 3D printing techniques, 7 weeks after the initial trauma. The external fixator was maintained for another 3 weeks. After 14 months, the patient could walk independently without any pain for 15 min, with the help of a crutch occasionally, and infection was under control. The range of motion of his left ankle was, however, limited to 15 degrees in the flexion-extension arc, and a minimal subtalar motion was there [79].

Calcaneum

Chung et al. used 3D printing to make models of calcaneal fracture and intact ipsilateral calcaneum by mirror imaging from the contralateral side. They also created preshaped calcaneal plates and used these for percutaneous fixation of calcaneal fractures [10].

Wu et al. reported 3D printing assisted percutaneous reduction, and cannulated screw fixation, for intraarticular calcaneal fractures. A thin slice CT scan of bilateral calcanei was taken, and a mirror image of the opposite side (to achieve prefracture anatomy), and fractured side calcaneal models were 3D printed. Bohler and Gissane angles measured on X-ray films demonstrated significant improvement, immediately after surgery, and did not change significantly on follow-up. The AOFAS score ranged from 76 to 100 (mean 88.2), and the results were excellent in 10 feet, good in 7, and fair in 2 [80].

Ankle ligament reconstruction

Sha et al. performed anatomical reconstruction of lateral ankle ligaments, by making fibular channels with the patient-specific navigational template, in 15 cases with chronic lateral ankle instability. By using the 3D template, fibular channels were made easily, and lateral ligaments were precisely reconstructed [81].

Miscellaneous topics

Atypical femoral fracture: bowed femur

Park et al. used preoperative templating and 3D printed model, for studying difficulties encountered with commercially available intramedullary nailing systems, for treatment of atypical femoral fractures with marked bowing. The 3D printed femur model had a mean anterior bow radius of curvature of 772 mm, and a lateral bowing angle of 15.48 degree. Nail position in the medullary canal, perforation of femoral cortex by a distal tip, and site of perforation in relation to the knee joint were studied. On simulated fracture reduction none, of the nails gave adequate fracture reduction. Nail fitting can be improved by using a nailing system with a small radius of curvature, and by applying patient-specific techniques [82].

Validity of "mirroring"

Surgeons usually consider bilateral bones to be symmetrical and use mirror imaging 3D technology. Zhang et al. measured short axis and long axis at the three selected transverse sections of bilateral tibia and femora; at 5, 10 and 15 cm from each end, to find the symmetry on CT images. They printed full-size mirror image of the contralateral long bone, which is considered similar to the affected side, and used it as a reference for fracture reduction. Significant differences were found between the short axes of the left and right femoral condyles, 5 cm above the intercondylar keel, and short axis of distal tibia 15 cm above the talar dome. The "Comparison of long axis and short axis of three equidistant transverse sections," allows one to judge the symmetry of bilateral long bones, and prevents blind preoperative planning with contralateral mirror models [83].

Bagaria et al. did a multicentric study involving 5 surgeons and created 3D printed biomodels for 50 surgical cases including pelvic fractures (11), periarticular fractures (24), complex primary (7), and revision replacement surgeries (8), using CT scan data. Preoperative planning, preoperative implant selection, surgical rehearsal, surgical simulation, intraoperative referencing, navigation, and inventory management were relevant advantages, besides better surgical accuracy and reduced surgical time [84].

External fixation

Qiao et al. used 3D printing and computer-assisted reduction techniques, for developing a custom external fixator, which assisted in fracture reduction. CT data were utilized for reconstituting and reducing the 3D image of the fracture, and on this basis, a Q-Fixator was created by 3D printing techniques. Experiments on three fracture models demonstrated excellent reduction. It allowed easy manipulation, accurate reduction, was minimally invasive, and easy. Stress adjustment and fracture healing optimization may be other possible future applications [85].

3D printed bone clips

Yeon et al. explored the use of 3D printed PLA/hydroxyapatite/silk composite bone clips, in experimental rat models. These clips are relatively noninvasive (drilling of bone is not required), have a patient-specific design, are mechanically stable, and are highly biocompatible, as a possible internal fixation device [86].

3D printing for tissue engineering and regenerative medicine

Bioprinting techniques used in skin tissue engineering, range from laser-induced forward transfer to extrusion-based techniques. Vascularization of new tissues and biological linkage are major challenges. Progress in this multidisciplinary field requires close interaction between material scientists, tissue engineers, and clinicians [87].

Bioactive scaffolding for bone and cartilage

Turnbull et al. reviewed in detail the role of 3D printing, in tissue engineering for the generation of appropriate scaffoldings for bone and cartilage, which may have paradigm-shifting applications in the field of orthopedic trauma. 3D printing can make new alternatives to bone grafts. However, materials like polymers, ceramics, and hydrogels in standard form, are unable to fully demonstrate the properties of the bone. Bioactive composite 3D scaffolds (polymers, hydrogels, metals, ceramics, and bio-glasses) can overcome this limitation [88].

Li et al. attempted to create a multilayer composite scaffold of cartilage, bone, and calcified layers, simulating natural full-thickness bone-cartilage structure. The bone and calcified layers were created using 3D printing. The cartilage layer was manufactured, by an improved temperature-gradient thermally induced crystallization technology. The layers were confirmed by micro CT, and scanning electron microscopy. Biomechanical testing demonstrated superior mechanical properties, compared to scaffolds without the calcified layer. These scaffoldings might find applications in bone and cartilage full-thickness injury repair methods [89].

Exoskeleton and bracing

Saharan et al. used 3D printed, lightweight exoskeletons (iGrab) based on twisted and coiled polymer (TCP) muscles, which were lightweight, provided high power to mass ratio, and enough stroke. Silver coated nylon threads were used for making TCP muscles, which can easily be actuated electrothermally. Hand orthosis manufactured using various actuation technologies were reviewed, and authors presented their design of the tendon-driven exoskeletal prosthesis, with muscles confined to the forearm area [90]. Paterson et al. reported the use of customized wrist splints made by 3D printing [91].

Prosthesis

Patient-specific sockets may be created by 3D printing techniques for appropriately customized rehabilitation solution, after lower limb amputation surgery. They are anatomical and provide higher durability and strength [92–94]. The combination of rapid prototyping and robotic technologies, has led to the advent of functional prosthetic hands [95]. 3D printing permits the creation of customized, lightweight, well-fitting, and affordable prosthesis, especially for growing children.

Xu et al. treated an 8-year-old boy, suffering from traumatic right wrist amputation, as a result of a mincing machine accident. A 3D-printed prosthetic hand was manufactured, and the child was well rehabilitated [96].

Reliability

Zou et al. evaluated the precision and reliability of stereolithography appearance of 3D printed model. CT data for bone/prosthesis and model were collected, and 3D reconstructed. The intraclass correlation coefficient (ICC) was used for evaluating the degree of similarity between the model and real bone/prosthesis regarding several anatomical parameters. No significant difference was found in the anatomical parameters except the maximum height of long bone. All ICCs were greater than 0.990. Overall, usage of 3D printed model for diagnosis and treatment purposes, in complex orthopedic diseases, was precise and reliable [97,98].

References

[1] Wong TM, Jin J, Lau TW, Fang C, Yan CH, Yeung K, To M, Leung F. The use of three-dimensional printing technology in orthopaedic surgery: a review. J. Orthop. Surg. January 31, 2017;25(1). 2309499016684077.

[2] Gadia A, Shah K, Nene A. Emergence of three-dimensional printing technology and its utility in spine surgery. Asian Spine J. April 1, 2018;12(2):365–71.

[3] Garg B, Gupta M, Singh M, Kalyanasundaram D. Outcome and safety analysis of 3D printed patient specific pedicle screw jigs for complex spinal deformities: a comparative study. Spine J. Jan 2019 1;19(1):56–64. May 3.

[4] Vaish A, Vaish R. 3D printing and its applications in orthopedics. J. Clin. Orthop. Trauma March 1, 2018;9:S74–5.

[5] Trauner KB. The emerging role of 3D printing in arthroplasty and orthopedics. J. Arthroplast. Aug 2018 1;33(8):2352–4.

[6] Vaishya R, Vijay V, Vaish A, Agarwal AK. Three-dimensional printing for complex orthopedic cases and trauma: a blessing. Apollo Med. April 1, 2018;15(2):51.

[7] Rengier F, Mehndiratta A, von Tengg-Kobligk H, Zechmann CM, Unterhinninghofen R, Kauczor HU, et al. 3D printing based on imaging data: review of medical applications. Int. J. Comput. Assist. Radiol. Surg. 2010;5:335–41.

[8] Kern R. 3-D printed implants hit the market, pave the way for more personalized devices. Grey Sheet 2013;39:1–3.

[9] Starosolski ZA, Kan JH, Rosenfeld SD, Krishnamurthy R, Annapragada A. Application of 3-D printing (rapid prototyping) for creating physical models of pediatric orthopedic disorders. Pediatr. Radiol. 2014;44:216–21.

[10] Chung KJ, Hong DY, Kim YT, Yang I, Park YW, Kim HN. Preshaping plates for minimally invasive fixation of calcaneal fractures using a real-size 3D-printed model as a preoperative and intraoperative tool. Foot Ankle Int. November 2014;35(11):1231–6.

[11] Kacl GM, Zanetti M, Arngwerd M, et al. Rapid prototyping (stereolithography) in the management of intra-articular calcaneal fractures. Eur. Radiol. 1997;7(2):187–91.

[12] Potamianos P, Amis AA, Forester AJ, McGurk M, Bircher M. Rapid prototyping for orthopaedic surgery. Proc. Inst. Mech. Eng. Part H J. Eng. Med. May 1, 1998;212(5):383–93.

[13] Brown GA, Firoozbakhsh K, DeCoster TA, et al. Rapid prototyping: the future of trauma surgery? J. Bone Joint Surg. Am. 2003;85-A(Suppl. 4):49–55.

[14] Guarino J, Tennyson S, McCain G, et al. Rapid prototyping technology for surgeries of the pediatric spine and pelvis. J. Pediatr. Orthop. 2007;27(8):955–60.

[15] Beliën H, Biesmans H, Steenwerckx A, Bijnens E, Dierickx C. Prebending of osteosynthesis plate using 3D printed models to treat symptomatic os acromiale and acromial fracture. J. Exp. Orthop. December 1, 2017;4(1):34.

[16] Jeong HS, Park KJ, Kil KM, Chong S, Eun HJ, Lee TS, Lee JP. Minimally invasive plate osteosynthesis using 3D printing for shaft fractures of clavicles. Arch. Orthop. Trauma Surg. November 1, 2014;134(11):1551–5.

[17] Kim HN, Liu XN, Noh KC. Use of a real-size 3D-printed model as a preoperative and intraoperative tool for minimally invasive plating of comminuted midshaft clavicle fractures. J. Orthop. Surg. Res. December 2015;10(1):91.

[18] You W, Liu LJ, Chen HX, Xiong JY, Wang DM, Huang JH, Ding JL, Wang DP. Application of 3D printing technology on the treatment of complex proximal humeral fractures (Neer3-part and 4-part) in old people. Orthop. Traumatol. Surg. Res. November 1, 2016;102(7):897–903.

[19] Shuang F, Hu W, Shao Y, Li H, Zou H. Treatment of intercondylar humeral fractures with 3D-printed osteosynthesis plates. Medicine January 2016;95(3).

[20] Zhang YZ, Lu S, Chen B, Zhao JM, Liu R, Pei GX. Application of computer-aided design osteotomy template for treatment of cubitus varus deformity in teenagers: a pilot study. J. Shoulder Elb. Surg. January 1, 2011;20(1):51–6.

[21] Zheng PF, Chen J, Xux XP, Jiang B, Yao QQ, Wang LM. Accurate osteotomy assisted by individualized navigation templates for the treatment of children cubitus varus. China J. Orthop. Traumatol. April 2017;30(4):377–82.

[22] Gemalmaz HC, Sarıyılmaz K, Ozkunt O, Sungur M, Kaya I, Dikici F. A new osteotomy for the prevention of prominent lateral condyle after cubitus varus correctional surgery-made possible by a 3D printed patient specific osteotomy guide: a case report. Int. J. Surg. Case Rep. January 1, 2017;41:438–42.

[23] Yang L, Grottkau B, He Z, Ye C. Three dimensional printing technology and materials for treatment of elbow fractures. Int. Orthop. November 1, 2017;41(11):2381–7.

[24] Zhao X, Yu L, Tao S, Deng L, Wei C, Wang B, Zhu S. Treatment of severe distal humeral bone defects with three-dimensional printing technology. Zhongguo xiu fu chong jian wai ke za zhi Zhongguo xiufu chongjian waike zazhi (Chin. J. Reparative Reconstr. Surg.) December 2018;32(12):1534–9.

[25] de Muinck Keizer RJ, Lechner KM, Mulders MA, Schep NW, Eygendaal D, Goslings JC. Three-dimensional virtual planning of corrective osteotomies of distal radius malunions: a systematic review and meta-analysis. Strategies Trauma Limb Reconstr. August 1, 2017;12(2):77–89.

[26] Roner S, Carrillo F, Vlachopoulos L, Schweizer A, Nagy L, Fuernstahl P. Improving accuracy of opening-wedge osteotomies of distal radius using a patient-specific ramp-guide technique. BMC Musculoskelet. Disord. December 2018;19(1):374. https://doi.org/10.1186/s12891-018-2279-0.

[27] Zang CW, Zhang JL, Meng ZZ, Liu LF, Zhang WZ, Chen YX, Cong R. 3D printing technology in planning thumb reconstructions with second toe transplant. Orthop. Surg. May 2017;9(2):215–20.

[28] Fuller SM, Butz DR, Vevang CB, Makhlouf MV. Application of 3-dimensional printing in hand surgery for production of a novel bone reduction clamp. J. Hand Surg. September 1, 2014;39(9):1840–5.

[29] Taylor EM, Iorio ML. Surgeon-based 3D printing for microvascular bone flaps. J. Reconstr. Microsurg. July 2017;33(06):441–5.

[30] Stoffelen DV, Eraly K, Debeer P. The use of 3D printing technology in reconstruction of a severe glenoid defect: a case report with 2.5 years of follow-up. J. Shoulder Elb. Surg. August 1, 2015;24(8):e218–22.

[31] Gittard SD, Narayan R, Lusk J, Morel P, Stockmans F, Ramsey M, Laverde C, Phillips J, Monteiro-Riviere NA, Ovsianikov A, Chichkov BN. Rapid prototyping of scaphoid and lunate bones. Biotechnol. J. Healthc. Nutr. Technol. January 2009;4(1):129–34.

[32] ten Berg PW, Dobbe JG, Streekstra GJ. Short report letter: three-dimensional printed anatomical models in scaphoid surgery. J. Hand Surg. (Eur. Vol.) January 2018;43(1):101–2.

[33] Mai JG, Gu C, Lin XZ, Li T, Huang WQ, Wang H, Tan XY, Lin H, Wang YM, Yang YQ, Jin DD. Application of three-dimensional printing personalized acetabular wing-plate in treatment of complex acetabular fractures via lateral-rectus approach. Zhonghua wai ke za zhi (Chin. J. Surg.) March 2017;55(3):172–8.

[34] Cimerman M, Kristan A. Preoperative planning in pelvic and acetabular surgery: the value of advanced computerised planning modules. Injury April 1, 2007;38(4):442–9.

[35] Hurson C, Tansey A, O'donnchadha B, Nicholson P, Rice J, McElwain J. Rapid prototyping in the assessment, classification and preoperative planning of acetabular fractures. Injury October 1, 2007;38(10):1158–62.

[36] Taller S, Sram J, Lukáš R, Endrych L, Džupa V. Fixation of acetabular fractures. a novel method of pre-operative Omega plate contouring. Acta Chir. Orthop. Traumatol. Cech. 2014;81(3):212–20.

[37] Chen H, Wang G, Li R, Sun Y, Wang F, Zhao H, Zhang P, Zhang X. A novel navigation template for fixation of acetabular posterior column fractures with antegrade lag screws: design and application. Int. Orthop. April 1, 2016;40(4):827–34.

[38] Zhuang Y, Cao S, Lin Y, Li R, Wang G, Wang Y. Minimally invasive plate osteosynthesis of acetabular anterior column fractures using the two-incision minimally invasive approach and a preshaped three dimension plate. Int. Orthop. October 1, 2016;40(10):2157–62.

[39] Zeng C, Xing W, Wu Z, Huang H, Huang W. A combination of three-dimensional printing and computer-assisted virtual surgical procedure for preoperative planning of acetabular fracture reduction. Injury October 1, 2016;47(10):2223–7.

[40] Li BF, Zhang Y, Tai GL, Dai YY, Xie HB, Guo XZ, Zhu CR, Wang F. Application of 3D digital orthopedic techniques in treatment of acetabular fracture. Nan fang yi ke da xue xue bao (J. South. Med. Univ.) June 2016;36(7):1014–7.

[41] Chana-Rodríguez F, Mañanes RP, Rojo-Manaute J, Gil P, Martínez-Gómiz JM, Vaquero-Martín J. 3D surgical printing and pre contoured plates for acetabular fractures. Injury November 1, 2016;47(11):2507–11.

[42] Upex P, Jouffroy P, Riouallon G. Application of 3D printing for treating fractures of both columns of the acetabulum: benefit of pre-contouring plates on the mirrored healthy pelvis. Orthop. Traumatol. Surg. Res. May 1, 2017;103(3):331–4.

[43] Chen X, Chen X, Zhang G, Lin H, Yu Z, Wu C, Li X, Lin Y, Huang W. Accurate fixation of plates and screws for the treatment of acetabular fractures using 3D-printed guiding templates: an experimental study. Injury June 1, 2017;48(6):1147–54.

[44] Manganaro MS, Morag Y, Weadock WJ, Yablon CM, Gaetke-Udager K, Stein EB. Creating three-dimensional printed models of acetabular fractures for use as educational tools. Radiographics May 11, 2017;37(3):871–80.

[45] Liu ZJ, Jia J, Zhang YG, Tian W, Jin X, Hu YC. Internal fixation of complicated acetabular fractures directed by preoperative surgery with 3D printing models. Orthop. Surg. May 2017;9(2):257–60.

[46] Merema BJ, Kraeima J, Ten Duis K, Wendt KW, Warta R, Vos E, Schepers RH, Witjes MJ, IJpma FF. The design, production and clinical application of 3D patient-specific implants with drilling guides for acetabular surgery. Injury November 1, 2017;48(11):2540–7.

[47] Zeng CJ, Huang WH, Huang HJ, Wu ZL. Laparoscopic acetabular fracture fixation after three-dimensional modelling and printing. Indian J. Orthop. September 2017;51(5):620–3.

[48] Boudissa M, Courvoisier A, Chabanas M, Tonetti J. Computer assisted surgery in preoperative planning of acetabular fracture surgery: state of the art. Expert Rev. Med. Dev. January 2, 2018;15(1):81–9.

[49] Kim JW, Lee Y, Seo J, Park JH, Seo YM, Kim SS, Shon HC. Clinical experience with three-dimensional printing techniques in orthopedic trauma. J. Orthop. Sci. March 1, 2018;23(2):383–8.

[50] Huang Z, Song W, Zhang Y, Zhang Q, Zhou D, Zhou X, He Y. Three-dimensional printing model improves morphological understanding in acetabular fracture learning: a multicenter, randomized, controlled study. PLoS One January 17, 2018;13(1):e0191328.

[51] Maini L, Verma T, Sharma A, Sharma A, Mishra A, Jha S. Evaluation of accuracy of virtual surgical planning for patient-specific pre-contoured plate in acetabular fracture fixation. Arch. Orthop. Trauma Surg. April 1, 2018;138(4):495–504.

[52] Wang YC, Ma Y, Yu WZ, Li YF, Liu YH. Application of the computer-assisted virtual reduction combined with 3D printing technique in acetabular fractures. Zhongguo gu shang (China J. Orthop. Traumatol.) July 2017;30(7):627–32.

[53] Lin X, Xiao X, Wang Y, Gu C, Wang C, Chen J, Liu H, Luo J, Li T, Wang D, Fan S. Biocompatibility of bespoke 3D-printed titanium alloy plates for treating acetabular fractures. BioMed Res. Int. 2018;2018.

[54] Zhou X, Zhang Q, Song W, Zhou D, He Y. Clinical significance of three-dimensional skeleton-arterial model in the management of delayed reconstruction of acetabular fractures. BMC Surg. December 2018;18(1):30.

[55] Lim PK, Stephenson GS, Keown TW, Byrne C, Lin CC, Marecek GS, Scolaro JA. Use of 3D printed models in resident education for the classification of acetabulum fractures. J. Surg. Educ. November 2018;75(6):1679–84. https://doi.org/10.1016/j.jsurg.2018.04.019.

[56] Shon HC, Choi S, Yang JY. Three-dimensional printing-assisted surgical technique with limited operative exposure for both-column acetabular fractures. Ulusal travma ve acil cerrahi dergisi (Turk. J. Trauma Emerg. Surg.) July 1, 2018;24(4):369–75.

[57] Hsu CL, Chou YC, Li YT, Chen JE, Hung CC, Wu CC, Shen HC, Yeh TT. Pre-operative virtual simulation and three-dimensional printing techniques for the surgical management of acetabular fractures. Int. Orthop. August 20, 2018:1–8.

[58] Maini L, Mishra A, Agarwal G, Verma T, Sharma A, Tyagi A. 3D printing in designing of anatomical posterior column plate. J. Clin. Orthop. Trauma July 1, 2018;9(3):236–40.

[59] Sánchez-Pérez C, Rodríguez-Lozano G, Rojo-Manaute J, Vaquero-Martín J, Chana-Rodríguez F. 3D surgical printing for preoperative planning of trabecular augments in acetabular fracture sequel. Injury September 1, 2018;49:S36–43.

[60] Sánchez-Pérez C, Rodríguez-Lozano G, Rojo-Manaute J, Vaquero-Martín J, Chana-Rodríguez F. 3D surgical printing for preoperative planning of trabecular augments in acetabular fracture sequel. Injury September 1, 2018;49:S36–43. https://doi.org/10.1016/j.injury.2018.07.014.

[61] Hurson C, Tansey A, O'Donnchadha B, et al. Rapid prototyping in the assessment, classification and pre-operative planning of acetabular fractures. Injury 2007;38:1158–62.

[62] Maini L, Sharma A, Jha S, Tiwari A. Three-dimensional printing and patient-specific pre-contoured plate: future of acetabulum fracture fixation? Eur. J. Trauma Emerg. Surg. 2016 1-0.
[63] Bagaria V, Deshpande S, Rasalkar DD, et al. Use of rapid prototyping and three-dimensional reconstruction modeling in the management of complex fractures. Eur. J. Radiol. 2011;80(3):814–20.
[64] Cai L, Zhang Y, Chen C, Lou Y, Guo X, Wang J. 3D printing-based minimally invasive cannulated screw treatment of unstable pelvic fracture. J. Orthop. Surg. Res. December 2018;13(1):71. https://doi.org/10.1186/s13018-018-0778-1.
[65] Wu XB, Wang JQ, Zhao CP, Sun X, Shi Y, Zhang ZA, Li YN, Wang MY. Printed three-dimensional anatomic templates for virtual preoperative planning before reconstruction of old pelvic injuries: initial results. Chin. Med. J. February 20, 2015;128(4):477.
[66] Zeng C, Xiao J, Wu Z, Huang W. Evaluation of three-dimensional printing for internal fixation of unstable pelvic fracture from minimal invasive para-rectus abdominis approach: a preliminary report. Int. J. Clin. Exp. Med. 2015;8(8):130–9.
[67] Liu Y, Zhou W, Xia T, Liu J, Mi BB, Hu LC, Shao ZW, Liu GH. Application of the guiding template designed by three-dimensional printing data for the insertion of sacroiliac screws: a new clinical technique. Curr. Med. Sci. December 1, 2018;38(6):1090–5.
[68] Lin H, Huang W, Chen X, Zhang G, Yu Z, Wu X, Wu C. Digital design of internal fixation for distal femoral fractures via 3D printing and standard parts database. Zhonghua yi xue za zhi February 2016;96(5):344–8.
[69] Arnal-Burró J, Pérez-Mañanes R, Gallo-del-Valle E, Igualada-Blazquez C, Cuervas-Mons M, Vaquero-Martín J. Three dimensional-printed patient-specific cutting guides for femoral varization osteotomy: do it yourself. Knee December 1, 2017;24(6):1359–68.
[70] Shi J, Lv W, Wang Y, Ma B, Cui W, Liu Z, Han K. Three dimensional patient-specific printed cutting guides for closing-wedge distal femoral osteotomy. Int. Orthop. June 27, 2018:1–6.
[71] Chen G, Li G, Lin Z, Chen X, Zhang G, You F, Chen J, Zeng Q, Zheng F, Yu Z. Effectiveness of distal femoral osteotomy assisted by three-dimensional printing technology for correction of valgus knee with osteoarthritis. Chin. J. Reparative Reconstr. Surg. February 2017;31(2):134–8.
[72] Rankin I, Rehman H, Frame M. 3D-printed patient-specific ACL femoral tunnel guide from MRI. Open Orthop. J. 2018;12:59.
[73] Huang H, Hsieh MF, Zhang G, Ouyang H, Zeng C, Yan B, Xu J, Yang Y, Wu Z, Huang W. Improved accuracy of 3D-printed navigational template during complicated tibial plateau fracture surgery. Australas. Phys. Eng. Sci. Med. March 1, 2015;38(1):109–17.
[74] Huang H, Zhang G, Ouyang H, Yang Y, Wu Z, Xu J, Xie P, Huang W. Internal fixation surgery planning for complex tibial plateau fracture based on digital design and 3D printing. J. South. Med. Univ. February 2015;35(2):218–22.
[75] Giannetti S, Bizzotto N, Stancati A, Santucci A. Minimally invasive fixation in tibial plateau fractures using an pre-operative and intra-operative real size 3D printing. Injury March 1, 2017;48(3):784–8.
[76] Yang P, Du D, Zhou Z, Lu N, Fu Q, Ma J, Zhao L, Chen A. 3D printing-assisted osteotomy treatment for the malunion of lateral tibial plateau fracture. Injury December 1, 2016;47(12):2816–21.
[77] Chung KJ, Huang B, Choi CH, Park YW, Kim HN. Utility of 3D printing for complex distal tibial fractures and malleolar avulsion fractures: technical tip. Foot Ankle Int. December 2015;36(12):1504–10.
[78] Wu JQ, Ma SH, Liu S, Qin CH, Jin D, Yu B. Safe zone of posterior screw insertion for talar neck fractures on 3-dimensional reconstruction model. Orthop. Surg. February 2017;9(1):28–33.
[79] Chiu SY, Wan KW. Use of three-dimensional printing techniques in the management of a patient suffering from traumatic loss of the talus. J. Foot Ankle Surg. January 1, 2019;58(1):176–83.
[80] Wu M, Guan J, Xiao Y, Wang Z, Chen X, Zhao Z, Zhang K, Zhu J. Application of three-dimensional printing technology for closed reduction and percutaneous cannulated screws fixation of displaced intraarticular calcaneus fractures. Chin. J. Reparative Reconstr. Surg. November 1, 2017;31(11):1316.
[81] Sha Y, Wang H, Ding J, Tang H, Li C, Luo H, Liu J, Xu Y. A novel patient-specific navigational template for anatomical reconstruction of the lateral ankle ligaments. Int. Orthop. January 1, 2016;40(1):59–64.
[82] Park JH, Lee Y, Shon OJ, Shon HC, Kim JW. Surgical tips of intramedullary nailing in severely bowed femurs in atypical femur fractures: simulation with 3D printed model. Injury June 1, 2016;47(6):1318–24.
[83] Zhang W, Ji Y, Wang X, Liu J, Li D. Can the recovery of lower limb fractures be achieved by use of 3D printing mirror model? Injury November 1, 2017;48(11):2485–95.
[84] Bagaria V, Chaudhary K. A paradigm shift in surgical planning and simulation using 3D graphy: experience of first 50 surgeries done using 3D-printed biomodels. Injury November 1, 2017;48(11):2501–8.
[85] Qiao F, Li D, Jin Z, Gao Y, Zhou T, He J, Cheng L. Application of 3D printed customized external fixator in fracture reduction. Injury June 1, 2015;46(6):1150–5.
[86] Yeon YK, Park HS, Lee JM, Lee JS, Lee YJ, Sultan MT, Seo YB, Lee OJ, Kim SH, Park CH. New concept of 3D printed bone clip (polylactic acid/hydroxyapatite/silk composite) for internal fixation of bone fractures. J. Biomater. Sci. Polym. Ed. June 13, 2018;29(7–9):894–906.
[87] Kogelenberg SV, Yue Z, Dinoro JN, Baker CS, Wallace GG. Three-dimensional printing and cell therapy for wound repair. Adv. Wound Care May 1, 2018;7(5):145–56.
[88] Turnbull G, Clarke J, Picard F, Riches P, Jia L, Han F, Li B, Shu W. 3D bioactive composite scaffolds for bone tissue engineering. Bioact. Mater. Sep 2018 1;3(3):278–314.
[89] Li Z, Jia S, Xiong Z, Long Q, Yan S, Hao F, Liu J, Yuan Z. 3D-printed scaffolds with calcified layer for osteochondral tissue engineering. J. Biosci. Bioeng. Sep 2018 1;126(3):389–96.
[90] Saharan L, Sharma A, de Andrade MJ, Baughman RH, Tadesse Y. Design of a 3D printed lightweight orthotic device based on twisted and coiled polymer muscle: iGrab hand orthosis. In: Active and passive smart structures and integrated systems 2017, vol. 10164. International Society for Optics and Photonics; April 2017. p. 1016428.
[91] Paterson AM, Donnison E, Bibb RJ, Ian Campbell R. Computer-aided design to support fabrication of wrist splints using 3D printing: a feasibility study. Hand Ther. December 2014;19(4):102–13.

[92] Herbert N, Simpson D, Spence WD, Ion W. A preliminary investigation into the development of 3-D printing of prosthetic sockets. J. Rehabil. Res. Dev. March 1, 2005;42(2):141.

[93] Rogers B, Bosker GW, Crawford RH, Faustini MC, Neptune RR, Walden G, Gitter AJ. Advanced trans-tibial socket fabrication using selective laser sintering. Prosthet. Orthot. Int. March 2007;31(1):88–100.

[94] Hsu LH, Huang GF, Lu CT, Hong DY, Liu SH. The development of a rapid prototyping prosthetic socket coated with a resin layer for transtibial amputees. Prosthet. Orthot. Int. March 2010;34(1):37–45.

[95] Laurentis KJ, Mavroidis C. Mechanical design of a shape memory alloy actuated prosthetic hand. Technol. Health Care January 1, 2002;10(2):91–106.

[96] Xu G, Gao L, Tao K, Wan S, Lin Y, Xiong A, Kang B, Zeng H. Three-dimensional-printed upper limb prosthesis for a child with traumatic amputation of right wrist. Medicine December 1, 2017;96(52):e9426.

[97] Zou Y, Han Q, Weng X, Zou Y, Yang Y, Zhang K, Yang K, Xu X, Wang C, Qin Y, Wang J. The precision and reliability evaluation of 3-dimensional printed damaged bone and prosthesis models by stereolithography appearance. Medicine February 2018;97(6).

[98] Lal H, Patralekh MK. 3D printing and its applications in orthopaedic trauma: A technological marvel. J. Clin. Orthop. Trauma. 2018;9(3): 260–8. https://doi.org/10.1016/j.jcot.2018.07.022.

Chapter 48

Smartphone-based clinical diagnostics

Shengwei Zhang[1], Taleb Ba Tis[2] and Qingshan Wei[1,3]

[1]Department of Chemical and Biomolecular Engineering, North Carolina State University, Raleigh, NC, United States; [2]Department of Materials Science and Engineering, North Carolina State University, Raleigh, NC, United States; [3]Emerging Plant Disease and Global Food Security Cluster, North Carolina State University, Raleigh, NC, United States

Introduction

Most molecular detection methods are currently designed and developed for laboratory use, which show several drawbacks, such as complicated steps and procedures, and requirement of highly trained professionals. The concept of point-of-care (POC) testing or diagnostics was timely introduced to address these issues. In POC testing, samples are collected and analyzed at the bedside, or in the clinician's office, and the results can be obtained in real time. Due to their cost-effectiveness, fast result turnaround, and small device footprint, POC diagnostic tools have shown great promise for predictive and personalized healthcare.

Potential social, economical, and public-health impact

There is an urgent need for next-generation sensing and measurement tools, for accurate and reliable diagnosis of human diseases in resource-limited settings. In many parts of the world, especially developing countries in Africa, Asia, and Latin America, the access to advanced diagnostic technologies or laboratories is still quite limited. At the same time, many high-risk diseases, such as HIV infection, hepatitis, parasite, and viral infections, are still endemic in these regions. With limited medical professional presence, disease diagnostics could only be performed in a very inefficient way, for example, by visual checking of disease symptoms. As such, the development of portable, cost-effective, and ready-to-use diagnostic devices, is of great importance in addressing the public health issues in developing countries and remote areas.

Mobile phone users worldwide have reached 5 billion in 2017 [1]. Global smartphone users have increased from 1.57 billion in 2014 to 2.53 billion in 2018 [2]. The increase of smartphones is even faster in developing countries, where the ownership rate has increased from 21% in 2013 to 37% in 2015 [3]. Therefore, smartphones are becoming easily accessible and affordable tools, for most people around the world.

Smartphones can be converted into compact, low-cost sensing and imaging tools, for applications in global health, in particular in the field and resource-limited settings. A number of review articles have summarized these applications, including sensing, nanoscale imaging, POC detection, and disease diagnostics [4–13].

Smartphone technologies

Current smartphones work like a portable computer in a pocket. Equipped with built-in sensors, such as camera, GPS module, and accelerometer, smartphones can be used as data acquisition devices or wearable sensors. The cameras installed in smartphones have undergone massive technological advances. In the past decade, the total pixel count of smartphone image sensor chips doubles almost every 2 years, following a Moore's Law trend [7]. High-performance image sensors have been applied to the newest smartphone models, producing much better image quality than before. As of July 2018, smartphone-based CMOS image sensors (e.g., Sony IMX 586), provide pixel counts as high as 48 megapixels, and pixel size as small as 0.8 μm [14]. As a result, the optical resolution of a smartphone microscope can be enhanced theoretically to the sub-micron level.

These advances in imaging hardware, have made it possible to take high-quality images with smartphones, which is closely comparable with those obtained from high-end CMOS/CCD cameras. In addition, powerful processors and adequate memory, have made it possible to process and store data directly on the phone. Almost every smartphone comes with wireless communication like Wi-Fi, cellular network, and Bluetooth. As such, smartphone-acquired data can also be easily transmitted to nearby computers or remote servers, for various telemedicine applications.

Precision Medicine for Investigators, Practitioners and Providers. https://doi.org/10.1016/B978-0-12-819178-1.00048-4

Detection methods in smartphone-based devices

Smartphone-based microscopy

To prepare a smartphone-based microscope, it requires to integrate multiple different optical components with the smartphone camera. Bright-field imaging can be carried out on the smartphone by simply adding an external lens and light source. The external lens functions equivalently to the objective lens in a traditional microscope, and the smartphone camera is analogous to the CCD/CMOS camera, on the benchtop counterpart. The magnification of the smartphone microscope can be estimated from the expression below: $M = f_1/f_2$, where f_1 is the focal length of the built-in lens of the smartphone camera, and f_2 is the focal length of the external lens. The external lens can be either a single-piece lens or a compact lens-module used on a smartphone or Raspberry Pi camera.

Such lens modules are usually cheap, readily available, and provide better image quality than single-piece lenses. Cost-effective, battery-powered white LEDs can be used as light sources for bright-field imaging. A dilemma in bright-field imaging, is the trade-off between spatial resolution and field of view (FOV). Lenses with higher magnification can provide better resolution, but with reduced FOV. In a more balanced example, Switz et al. used a reversed smartphone camera lens as the external lens, and achieved resolution less than 5 μm over a FOV larger than 10 mm^2 [15].

Fluorescence imaging

In fluorescence imaging, light source and optical filters with specific wavelengths are required, to excite the fluorophores and collect emission signals. A compact laser diode module is often used as the light source, when a high signal to noise ratio (SNR) is needed, while color LEDs become a more economical option, in combination with excitation filters. Effective excitation filters could be band-pass or short-pass optical filters, to narrow the spectral bandwidth of LEDs. In smartphone fluorescence microscopy, a strong background due to autofluorescence or nonspecific scattering is often the main limitation of high image contrast. To circumvent that, optical configuration based on either side (i.e., waveguide coupling), or tilted illumination at a high angle, can be adapted [16]. Fluorescence imaging on a smartphone with large FOV has also been reported [16]. With these techniques, high-sensitivity imaging results were reported, including imaging of 100-nm diameter nanoparticles (Fig. 48.1A), single viruses, and single DNA molecules [17,18].

Diffraction-based computational microscopy

Imaging methods based on computational imaging have also been applied in smartphone microscopy, especially for cellular imaging and detection. In smartphone-based holographic imaging (Fig. 48.1B), the conventional lens system is removed. An aperture ($\sim$100 μm) is placed in front of the light source, to generate a partially coherent light illumination. The spatially filtered light is then shed onto the sample, and creates holograms, due to the interference between the scattered light from the samples and uninterrupted background. The hologram is then captured by the image sensor of the smartphone. The geometry and morphology of the sample can be reconstructed from the holograms digitally. This method provides a simple way to perform microscopic imaging on the smartphone without using any optical lenses [19]. Similarly, smartphone-based digital diffraction detection (or "3D") has been reported for cellular imaging. In this method, antibody-labeled microbeads are added to the sample, to label specific antigens on the cell surface. Diffraction patterns are created by the illumination through a pinhole between sample and light source. The diffraction image is captured by the smartphone camera and later processed on a cloud server, to reconstruct the image. Both amplitude and phase information can be reconstructed in the output, which allows the number and position of microbeads bound on the cells to be identified for differentiation of different cell types. [20].

Smartphone-based quantitative sensing

With smartphone camera as the detector, optical information from samples can also be quantitatively analyzed on the phone. In the fluorometric analysis, sample concentration is correlated with fluorescence intensity, measured by the detector. In smartphone-based fluorometric sensors, the fluorescent intensity is proportional to the brightness of the image taken by the camera. Under the same acquisition conditions like exposure time, ISO, and F aperture, a relationship can be found between the pixel brightness and analyte concentration. The setup of the smartphone fluorometric sensor is similar to smartphone-based fluorescence microscopy, where a laser module, an emission filter, and an external lens (if needed) are used. A more cost-effective way in recent work is to use the flashlight on the phone as the light source (Fig. 48.1C), which was filtered by a colored adhesive tape as excitation filter. The fluorescent signals are then separated by a colored piece of glass, as the emission filter [21].

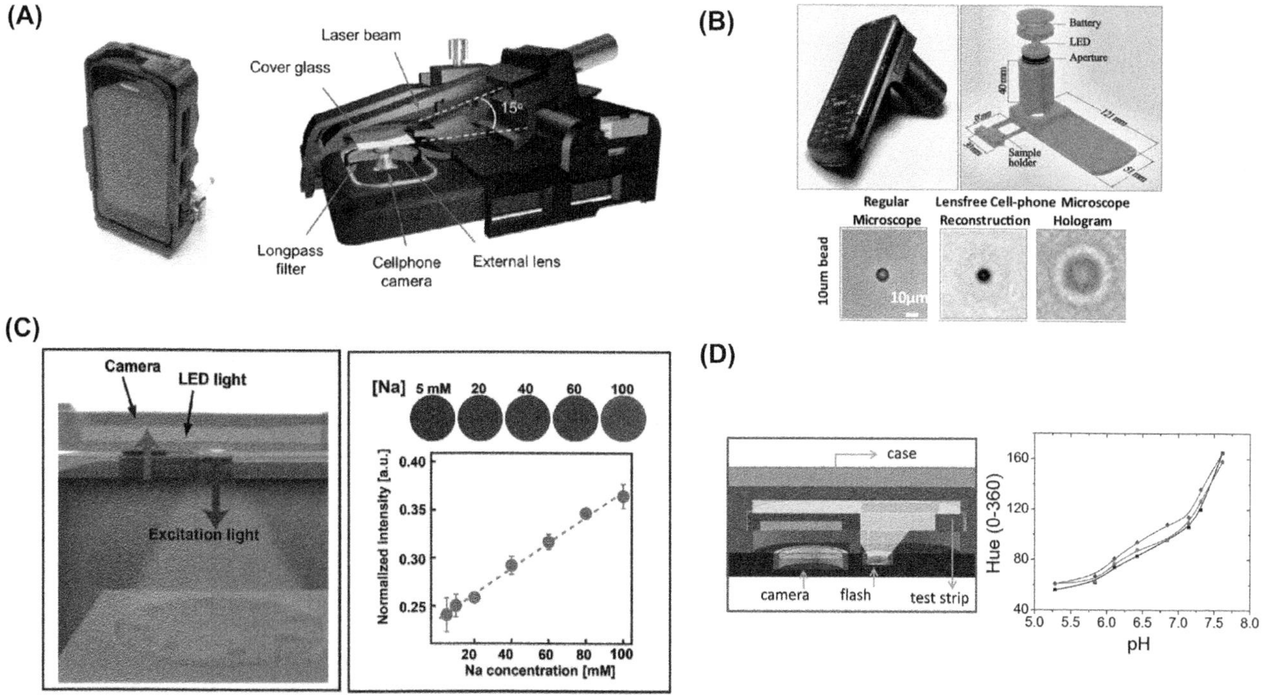

FIGURE 48.1 Representative detection methods in smartphone-based diagnostics. (A) Smartphone-based fluorescence microscopy [17]. (B) Lens-free microscopy on a cellphone [19]. (C) Fluorometric measurement of electrolyte concentrations from sweat [21]. (D) Colorimetric measurement of pH from sweat [23]. *Reproduced with permission from the references mentioned above. Copyright 2013 American Chemical Society, 2010 Royal Society of Chemistry, 2018 Royal Society of Chemistry and 2013 Royal Society of Chemistry.*

Colorimetric sensing

Colorimetric analysis on the phone is a little bit more complicated, as both color and brightness information is involved and analyzed. One approach to quantify the colors from different sample spots is to compare red, green, and blue (RGB) values, extracted from the cellphone images. Either the value of R/G/B component itself or the difference in these color components can quantitatively reflect the color difference between samples [22]. Some studies also converted RGB values into other color space (like hue, saturation, lightness/HSL), to better correlate changes of color with respect to analyte concentrations, such as pH level (Fig. 48.1D) [23]. Imaging setup of colorimetric smartphone device, is similar to those of smartphone-based bright-field microscopes. Typically, a white light based on either the phone flashlight or an external LED is used to illuminate the samples. Reflected or transmitted light is then collected by the lens system of the smartphone camera. External lenses can be used in case higher magnification is needed.

Electrical and electrochemical sensing methods

Smartphone-connected electrochemical biosensors are used in the detection of ions and small molecules from biological samples. Wireless communication modules like WiFi, Bluetooth, and NFC modules, can be installed on the electrochemical biosensor, and the data are transmitted to the smartphone for next-step analysis. Moreover, amperometric [24], potentiometric [24], and impedimetric [25] sensors are able to be connected with smartphone interfaces.

Clinical diagnostic applications of smartphone devices

Ions and small molecules

Colorimetric and fluorometric detection of small analytes has been demonstrated by smartphones. By choosing proper sensing dyes, a variety of ions and molecules can be measured on the smartphones. Table 48.1 shows representative examples of these detections.

Colorimetric measurement of pH from sweat and saliva has been carried out on a smartphone [23]. A saliva sample was collected on a test strip, and the image was taken by the smartphone for colorimetric analysis. Hue values were extracted from RGB images, to better reflect changes of color with respect to pH level. Fluorometric analysis has also been applied to detect common ions, from tear and sweat. Several chemical probes, namely crown ethers, o-acetanisidide, and seminaphtorhodafluor, were immobilized in the detection region, to generate fluorescent signals. Concentrations of Na^+, K^+, Ca^{2+} and H^+ from tear

TABLE 48.1 Smartphone-based detection of ions and small molecules.

Analyte	Sample	Method of detection	Results	References
pH	Saliva	Colorimetric	Range: 5.0–9.0	[23]
Na^+ K^+ Ca^{2+} pH	Tear	Fluorescence	Na^+: Sensitivity = 2.7 mM K^+: Sensitivity = 1.4 mM Ca^{2+}: Sensitivity = 0.02 m pH: Sensitivity = 0.06 pH units	[26]
Cl^- Na^+ Zn^{2+}	Sweat	Fluorometric	Cl^-: Dynamic range = 5–100mM Na^+: Dynamic range = 20–60mM Zn^{2+}: Dynamic range = 1–20μM	[21]
Na^+	Saliva	Fluorometric	–	[27]
Glucose	Human serum	Enzymatic colorimetric	LOD = 0.7 mM (buffer) LOD = 0.3 mM (serum)	[28]
Glucose	Blood	Enzymatic colorimetric	Range = 110 –586 mM	[29]
Lactate	Oral fluids	Enzymatic colorimetric	LOD (oral fluid) = 0.5 mmol/L	[30]
Vitamin B12	Blood	AuNP-based colorimetric	Vitamin B12: Sensitivity = 87% Specificity = 100%	[31]
Vitamin D	Human serum	AuNP-based colorimetric	Vitamin D: Accuracy = 15 nM Precision = 10 nM	[32]
Cholesterol	Blood	Colorimetric	Range: 140mg/dl-400 mg/dL Within 1.8% accuracy	[33]
Pregnanediol glucuronide (PdG)	Urine	Colorimetric, ELISA	Accuracy = 82.20%	[34]
Cortisol	Saliva	Chemiluminescence	LOD = 0.3 ng/mL (saliva) LOD = 0.1 ng/mL (buffer) Range: LOD = 60 ng/mL	[35]

samples were measured on a smartphone, when integrated with a paper-based microfluidic system (Fig. 48.2A). With independent light source and filters, this device can detect four ions simultaneously. This system has the potential to be applied in the diagnosis of dry eye [26].

Electrolytes from sweat can provide information about nutritional health and physical performance. In the work by Sekine et al., sweat sample was collected and mixed with preloaded dyes, in a soft and conformal microfluidic device. Concentrations of Na^+, Cl^-, and Zn^{2+} were read out by using a smartphone imaging device [21]. Fluorometric detection of sodium ion was also conducted on a smartphone device, where excitation and emission lights were separated by transmission grating [27].

Blood glucose

Glucose oxidase (GOx)- horseradish peroxidase (HRP) enzymatic colorimetric assay, is commonly used in the detection of glucose. In this assay, glucose is oxidized and produces H_2O_2. Then a color substance is produced from two colorless precursors, via an enzymatic oxidation process mediated by HRP. Glucose detection has recently been carried out, in test solutions as well as human serum, on a paper-based device, with a smartphone readout [28]. Based on a similar detection method, the glucose level in blood was measured, on an enzyme-immobilized hydrophilic PET film by a smartphone [29].

Lactate, vitamins, and steroids

As the product of anaerobic respiration, lactate can be detected in a similar way of enzymatic colorimetry. The functionalized paper was used as a substrate, to enhance reagents stability and homogeneity of color distribution. The L-lactate in the oral fluid was then analyzed on a portable smartphone reader [30]. Concentrations of Vitamin B12 [31] and Vitamin D (Fig. 48.2B) [32] have also been measured

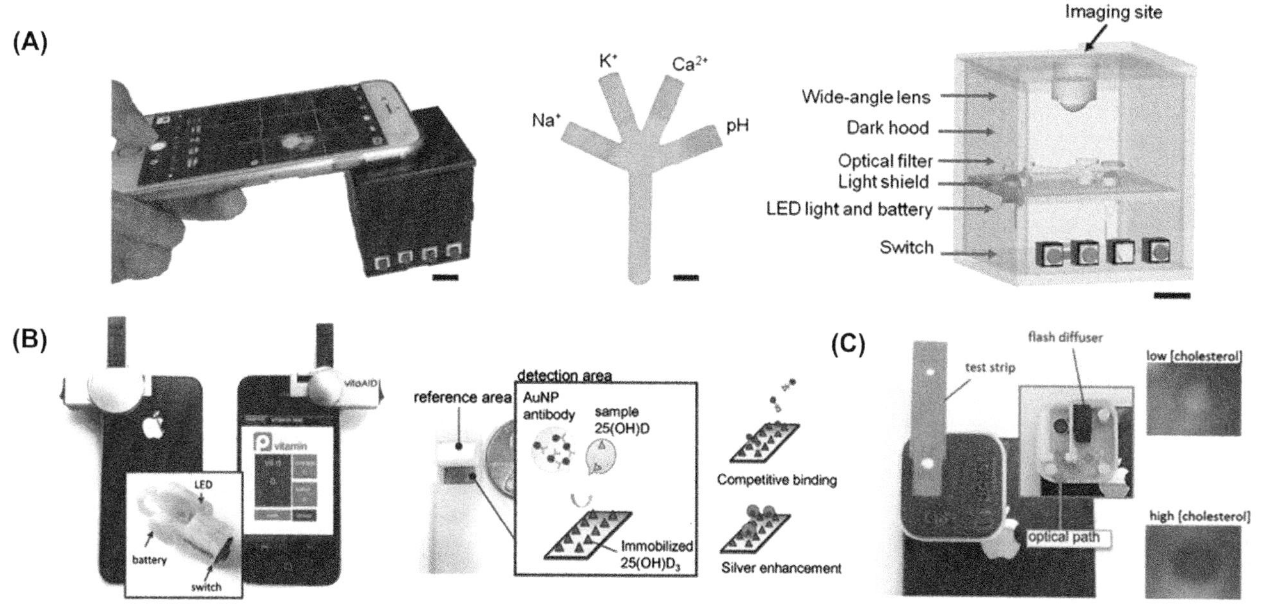

FIGURE 48.2 Detection of ions and small molecules on smartphone platforms. (A) Smartphone-based colorimetric analysis of electrolyte concentrations from the tear [26]. (B) Quantification of vitamin D level on a smartphone platform [32]. (C) Cholesterol testing on a smartphone [33]. *Reproduced with permission from the references mentioned above. Copyright 2014 Royal Society of Chemistry and 2014 Royal Society of Chemistry.*

on the smartphone, in combination with a gold nanoparticle (AuNP)-based colorimetric immunoassay. In these works, antibodies of the vitamins were conjugated on AuNPs. Competitive binding of antibody-modified AuNP was performed on the test line. The color intensities of these lines were then quantified through the colorimetric analysis. Several steroids, including cholesterol (Fig. 48.2C), and steroid hormones, including cortisol, have also been measured on the smartphone by colorimetric or fluorometric methods [33–35]. Representative cases of ions and small molecules detection for diagnostic applications on the smartphone are shown in Fig. 48.2.

Detection of proteins

Traditionally, tests of proteins are performed with sandwich immunoassays, and the assay signals are quantified by benchtop readers. The smartphone-based sensing technologies provide an alternative approach in a much faster and cost-effective way. Table 48.2 summarizes representative results in the quantification of protein biomarkers by smartphones for disease diagnosis.

Petryayeva et al. designed a fluorescence-based assay, to detect thrombin concentration in whole blood and serum. A paper-in-PDMS microfluidic chip was fabricated, containing thrombin-sensitive test spot and insensitive reference spot. Upon the exposure of test spot to the sample, quenched photoluminescence of quantum dot (QD630) was recovered. The intensity of photoluminescence was calibrated to quantify the concentration of thrombin [36]. Using inkjet printing technology, Joh et al. developed a sensitive point-of-care "D4" immunoassay, which is capable of detecting multiple protein targets from a drop of blood. They tested their assay on a compact and cost-effective smartphone reader for leptin detection. The results were comparable with those obtained from a tabletop glass slide scanner [37]. Measurement of CRP concentration was similarly reported by using a colorimetric assay and smartphone readout [38].

Antibody biomarkers of infectious disease

Enzyme-linked immunosorbent assay (ELISA) is the gold standard for antibody detection. A number of novel ELISA assays that operate on the smartphone devices have recently been reported. For example, researchers from Sia's lab designed a smartphone dongle for the POC diagnosis of HIV and syphilis (Fig. 48.3A). In this device, a disposable cassette with microfluidic channels was preloaded with reagents. Blood sample flew through the channels, reacted with the gold-labeled antibodies, and subsequently was washed by the washing buffer. The silver enhancing reagents were then added, which amplified the signals of gold nanolabels by darkening the color and increasing optical density (OD) of test spots. Little power was required to run this device, since the flow of the sample and washing buffers, was driven by a mechanical vacuum pump at the time of assay. The audio jack on the smartphone was used for powering the electronics and data transmission. In the field test, the smartphone dongle met the need of current clinical requirements, with a sensitivity of 92%–100% and specificity of 79%–100% [39].

TABLE 48.2 Examples of protein detection on smartphone platforms for disease diagnosis.

Analyte	Sample	Method of detection	Results	References
Thrombin	Whole blood and serum	Fluorescent	Time ≤ 30 min LOD = 18 NIH units/mL	[36]
Leptin	Serum	Fluorescent	LOD = 0.71 ng/mL	[37]
C-reactive protein	–	Colorimetric	LOD = 0.026 μg/mL	[38]
HIV antibody Syphilis antibody	Whole blood	Gold/silver amplified ELISA	HIV: sensitivity = 100% specificity = 87% Syphilis: sensitivity = 92% specificity = 92%	[39]
HIV-1-p17 hemagglutinin (HA)	Blood plasma	Bioluminescence	LOD = 10 pM	[40]
Highly pathogenic H5N1	Throat swab samples	Fluorescent	Sensitivity = 96.55% Specificity = 98.55%	[41]
Influenza hemagglutinin (HA)	–	Digital diffraction detection	LOD = 0.9 ng	[42]
Mumps IgG Measles IgG Herpes simplex virus (HSV) IgG	–	Colorimetric	Assay time = 1 min Accuracy: Mumps IgG = 99.6% Measles IgG = 98.6% HSV-1 IgG = 99.4% HSV-2 IgG = 99.4%	[43]
HE4	Urine	Colorimetric	Sensitivity = 89.5% Specificity = 90% Assay time = 5 h LOD = 19.5 ng/mL Range = 19.5 −1250 ng/mL	[44]
Prostate-specific antigen (PSA)	Whole blood	Colorimetric fluorescent	Colorimetric detection: Assay time = 13 min LOD = 0.9 ng/mL Fluorescence detection Assay time = 22 min LOD = 0.08 ng/mL	[45]
Antigen of brain natriuretic peptide (BNP), suppression of tumorigenicity 2 (ST2)	Serum	Fluorescent	LOD of BNP = 5 pg/mL LOD of ST2 = 1 ng/mL	[46]

In HIV detection, Arts and his colleagues developed an alternative method, based on bioluminescent sensor protein, and applied the system on the smartphone for the detection of HIV p17 antigen. This platform can be applied in the detection of hemagglutinin (HA), and dengue virus type I as well [40]. Other viral diseases, like avian influenza, have also been detected on smartphone platforms. Yeo et al. reported a fluorescent ELISA platform on the smartphone to detect highly pathogenic H5N1 influenza virus [41].

Smartphone-based digital diffraction detection (or "D3") assay (Fig. 48.3B) has also been applied in molecular diagnostics, including the detection of biomarkers of breast and cervical cancer [20]. By using the D3 assay, molecular diagnostic of avian influenza on the smartphone

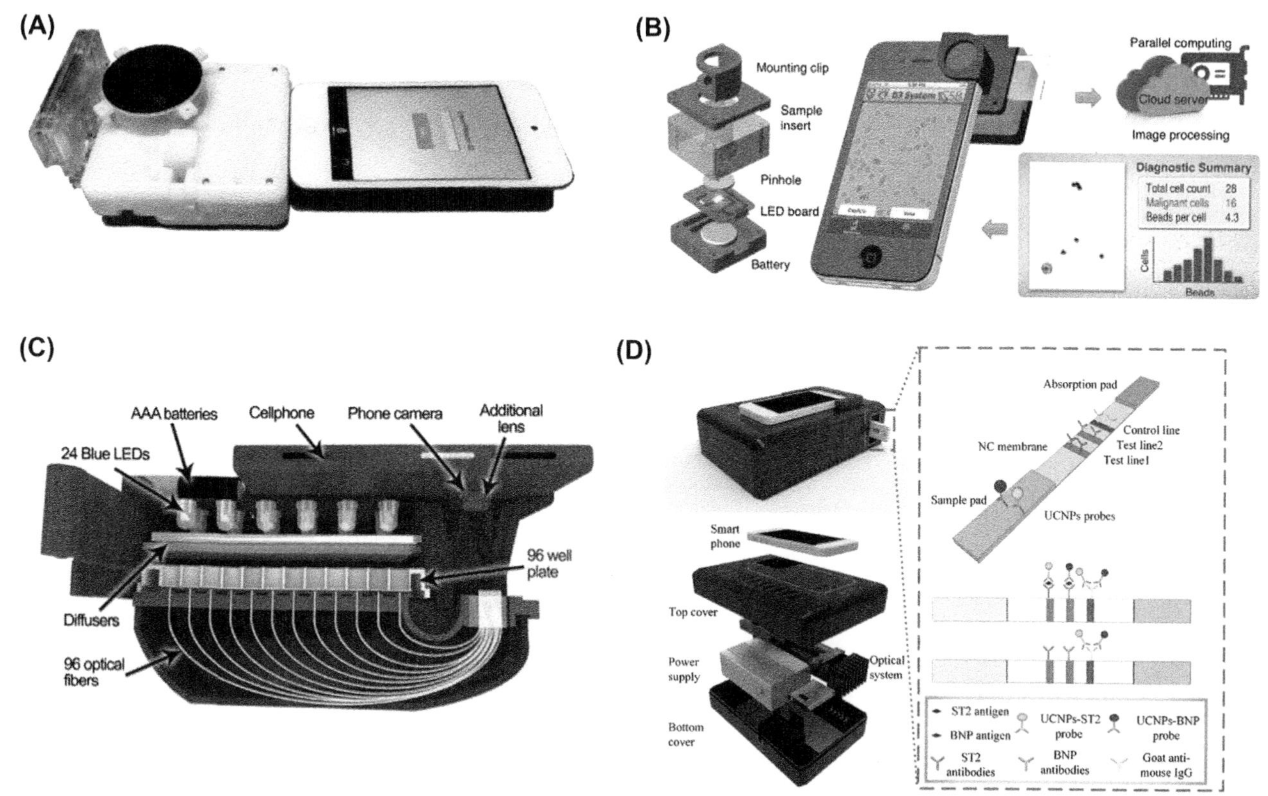

FIGURE 48.3 Protein detection on smartphone platforms. (A) A smartphone dongle for point-of-care diagnosis of HIV and syphilis [39]. (B) Digital diffraction imaging (D3 assay) on a smartphone for protein biomarker detection [20]. (C) Microplate reader on a smartphone for portable ELISA testing [43]. (D) Smartphone-based lateral flow strip reader for biomarker detection associated with heart failure [46]. *Reproduced with permission from the references mentioned above. Copyright 2015 The American Association for the Advancement of Science and 2017 American Chemical Society.*

has been demonstrated [42]. A smartphone-based 96-well plate reader (Fig. 48.3C), for the detection of characteristic antibodies for multiple viral infections, including mumps, measles, and HSV infection, has also been reported. This compact microtiter plate reader builds a bridge between benchtop ELISA techniques and portable diagnostics in field settings [43].

Oncological and nononcological diagnosis

Human epididymis protein 4 (HE4) is a biomarker for ovarian cancer. A colorimetric sandwich ELISA assay with cellphone readout has been reported by Wang et al. in the detection of HE4 [44]. In another example, Barbosa et al. performed ELISA quantification of prostate-specific antigen (PSA) on the smartphone. Colorimetric and fluorometric tests were compared in the detection of PSA from a whole blood sample [45]. Biomarker detection of the noncancerous disease on the smartphone has also been reported recently. Brain natriuretic peptide (BNP), and suppression of tumorigenicity 2 (ST2), are two biomarkers used in the prognosis evaluation of heart failure. You et al. designed a fluorescent lateral flow assay (LFA) platform (Fig. 48.3D), which detects these biomarkers using upconversion fluorescent nanoparticles, and the results were quantified by a smartphone reader. These two biomarkers can be detected at the same time with high sensitivity and specificity [46].

Detection of viral nucleic acids

Polymerase chain reaction (PCR) and loop-mediated isothermal amplification (LAMP) are two commonly used nucleic acid amplification methods. However, limited by complicated assay steps, and the size and cost of the instruments, these methods are seldom adopted in remote areas, or regions with limited healthcare infrastructure. Below is a list of recent examples of nucleic acid detection on smartphone platforms, for diagnosis of major viral diseases (Table 48.3).

Compared with PCR, LAMP is an isothermal DNA amplification method, that is more commonly used on smartphone-based nucleic acid detection. Unlike PCR, which requires temperature cycling, LAMP works at a constant temperature, which significantly simplifies the device design. Combined with the reverse transcription step, LAMP can be applied to detect RNA as well. As LAMP is typically run at 60−65°C, a heat source is needed. The fluorescent signal from dyes can be easily detected with a smartphone camera.

TABLE 48.3 Examples of nucleic acid detection on smartphone platforms.

Analyte	Sample	Method of detection	Results	References
HIV-1 RNA	–	Digital reverse-transcription loop-mediated isothermal amplification (RT-LAMP)	Resolution: twofold change at 10^5 copies/mL	[47]
HIV-1 RNA	Whole blood	RT-LAMP	LOD: 3 viruses/60 nL droplet (670 viruses/μL) Resolution: 10-fold change at 6.7×10^4 μL^{-1}, 100-fold change at 670 μL^{-1}	[48]
Hepatitis C RNA	–	Colorimetric, digital RT-LAMP	Upper limit of quantification (ULQ) = $1.1-1.6 \times 10^7$ copies/mL	[49]
Herpes simplex virus type 2 (HSV-2) DNA	–	LAMP	Sensitivity: 100 copies/reaction	[53]
Hepatitis B viral DNA HIV viral RNA	Clinical	Isothermal amplification	LOD: 10–50 fmol ($6 \times 10^9-3 \times 10^{10}$ copies)/10 μL Resolution: up to 40-fold change	[54]
Zika virus (ZIKV)	Urine Blood Saliva	RT-LAMP	LOD_{95} = 2 PFU/mL LOD_{50} = 4.9 PFU/mL	[55]
Human papillomavirus (HPV) DNA	–	Digital diffraction detection	LOD: ~50 atto-mole	[20]
Kaposi's sarcoma herpesvirus (KSHV) DNA	–	AuNP-colorimetric	LOD = 500 pM	[57]
Single DNA		Fluorescent imaging	Sizing accuracy: <1 kbp for >10 kbp	[18]

Integration of microfluidic digital amplification assay with a smartphone camera detector has been reported recently [47–49]. In a digital analysis, the sample is injected into an array of discrete microchambers, so that each microchamber contains one or zero target molecules of interest. After the reaction, only these chambers containing analytes can produce signals, which are considered positive regardless of signal intensity. The percentage of positive microwells will be used to calculate analyte concentration, based on Poisson statistics. A digital microfluidic chip named "SlipChip" has recently been applied in biomedical detection, including PCR and immunoassay (Talis Biomedical Corporation, Menlo Park, CA, USA) [50–52]. More recently, the SlipChip was tested on smartphone platforms for LAMP amplification [48,49]. Selck et al. tested RT-LAMP on SlipChip to detect HIV-1 RNA (Fig. 48.4A). They demonstrated that digital analysis was more robust against temperature fluctuations and reaction time variations, compared with traditional analog analysis.

Given the limited precision of temperature control and imaging quality of smartphone LAMP platform, this research provided a method to run LAMP reactions with improved robustness, under low hardware requirements [48]. Colorimetric RT-LAMP has also been reported in the detection of hepatitis C virus (HCV) RNA. In this work, a rotational SlipChip device was designed and built as the substrate. Positive and negative reactions generated blue and purple colorimetric signals, respectively, which were captured by the unmodified cellphone camera directly. From the cellphone images, quantification of RNA concentration was achieved, after image processing and digital counting [49].

Damhorst et al. developed an integrated assay with a microfluidic blood lysis module and a microfluidic chip for real-time, quantitative measurement of amplified DNA from reverse transcription. They applied this system in the detection of HIV from blood samples [47]. Similar real-time analysis was also conducted inside a Thermos cup, which holds the heat source, a microfluidic chip, and optical components (Fig. 48.4B) [53]. Quantum dot barcode has also been applied in the detection of RNA of HIV and hepatitis B virus (HBV) on the smartphone (Fig. 48.4C). In this study, different capturing DNA oligonucleotides were conjugated to the barcode, as biorecognition elements for target DNA. In the presence of target DNA, red-fluorescence labeled detection DNA, will bind with target DNA to form a sandwich structure. Five different barcodes loaded with different capture DNAs, were immobilized on the glass slide with microwells. Three were used for DNA detection, and two for the positive and negative control, respectively. The first smartphone image was taken to identify the positions of all five color barcodes. By using different filters for barcodes, five more images were then taken, and used for determining the presence of target DNA. Multiplexed detection of HIV, HBV, and HCV was achieved with this smartphone-supported biosensor [54].

In another example of multiplexed detection of Zika and Chikungunya viruses (Fig. 48.4D), quenching of unincorporated amplification signal reporters (QUASR) technique, was used to produce signals. A Bluetooth microcontroller was used to control LED excitation light and the isothermal hot plate. This platform can handle blood, urine, or saliva sample directly, without lysis procedure [55]. Energy consumption is an important issue in field applications of nucleic acid amplification and detection, especially when a heat plate is needed. Jiang et al. designed a field-deployable PCR system, powered by a solar cell. They also tested the system in the detection of Kaposi's sarcoma herpesvirus (KSHV) [56].

D3 assay has been applied in the detection of human papillomavirus, from cervical specimen [20]. In another example, Mancuso and his colleagues designed a cellphone accessory, to measure the concentration of KSHV from biosamples, using AuNP-based colorimetric method [57]. Imaging and length measurement of fluorescently labeled single DNA molecules has also been achieved on a smartphone, using a lightweight, cost-effective imaging attachment. This fluorescent microscopy device provides a new approach for DNA-associated disease diagnostics, and DNA-protein interaction study, by direct visualization of single DNA strands on a portable platform [18].

Detection and imaging of bacteria and microbes

Bacteria and parasites are pathogens for many infectious diseases. In the developing world, tuberculosis, malaria, and filarial parasite diseases remain to be major public health issues. Optical microscopy is an effective way in the detection of pathogen and diagnosis of these diseases. A portable, easy-to-use imaging platform would be helpful in the disease control and evaluation of treatment in remote areas, given the limited healthcare infrastructure and trained personnel present. Both smartphone-based microscopic and analytical methods have been adopted in the detection of pathogenic bacteria and parasite. Table 48.4 summarizes recent examples of clinical-related detection and identification of pathogenic bacteria and parasites on smartphone-based devices.

Detection and imaging of bacteria on mobile phone devices have been explored in the last few years. As early

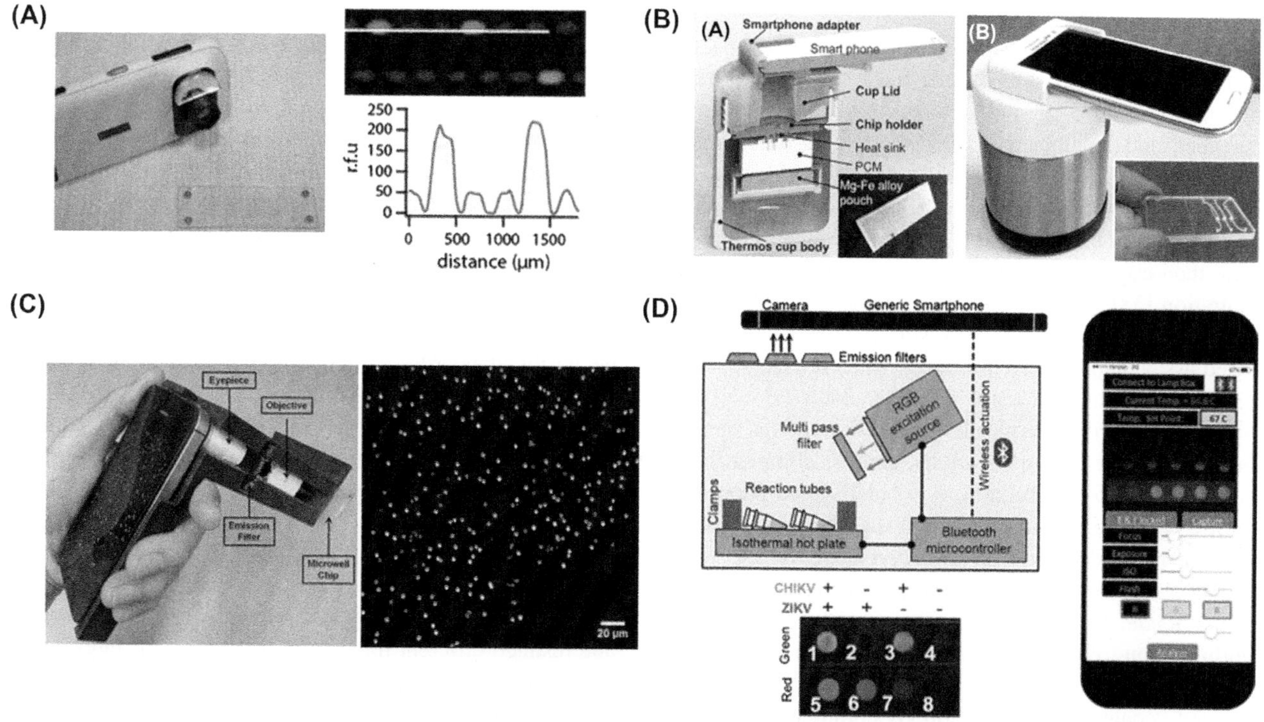

FIGURE 48.4 Nucleic acid detection on smartphones. (A). Digital LAMP assay with smartphone readout [48]. (B) SmartCup for portable nucleic acid amplification and detection [53]. (C) Multiplexed detection of infectious disease DNA on a smartphone with quantum dot barcode [54]. (D) Virus RNA detection on a smartphone platform [55]. *Reproduced with permission from the references mentioned above. Copyright 2013 American Chemical Society, 2016 Elsevier and 2015 American Chemical Society.*

TABLE 48.4 Examples of pathogenic bacteria and parasites detection on smartphone platforms.

Bacteria/Parasite	Sample	Method of detection	Results	References
Plasmodium falciparum *Mycobacterium tuberculosis*	Blood cell, sputum	Bright-field imaging Fluorescent imaging	Resolution = 1.2 μm	[9]
Bacillus anthracis	–	Bright-field imaging	Range: 50–5000 spores in 3–5 h.	[58]
Escherichia coli	Blood serum	AuNP-base colorimetric	LOD = 8 CFU/mL	[59]
M. tuberculosis	–	AuNP-based colorimetric	Time = 65 min LOD = 10 ug/mL	[60]
E. coli *Neisseria gonorrhoeae*	Urine	Immunoagglutination assay	LOD = 10 CFU/mL for both bacteria	[61]
Loa loa, filarial parasites	Blood	Bright-field video	100% sensitivity 94% specificity	[62]
P. falciparum	Blood	Immunoassay	LOD = 20.6 par./μL	[63]
Soil-transmitted helminth infection	Urine	Bright-field imaging	–	[64–66]
Schistosoma haematobium eggs *Schistosoma mansoni* eggs	Stool and urine	Bright-field microscopy	*S. mansoni*: Sensitivity = 50% specificity = 99.5% *S. haematobium*: Sensitivity = 35.6% Specificity = 100%	[79]
Strongyle eggs	Feces	Fluorescent imaging	LOD = 50 EPG (egg per gram of feces)	[67]
Giardia lamblia cysts	Water	Fluorescence	LOD = ~12 cysts/10 mL	[68]

as 2009, bright-field and fluorescent microscopy of bacteria and parasite-infected blood cells have been demonstrated on a Nokia N73 phone equipped with a microscope eyepiece and an objective [9]. Smartphone-based microscopy was combined with microfluidic incubation device to identify spores of *B. anthracis*. Spores were allowed for germination in the incubation chamber. Bacterial filaments were then trapped in a fine filter for optical imaging during aspiration [58]. Paper-based microfluidic assays have been applied in bacterial detection as well. In the cellulose paper-based assay, signals were produced when the nanoparticles aggregated in the existence of target bacteria, causing a change in the color. With this assay, *E. coli* was detected and differentiated from *S. aureus*, another common bacterial pathogen [59]. Similarly, An AuNP-based paper platform for the detection of tuberculosis pathogen was also applied in the detection of *M. tuberculosis* on the smartphone [60]. Cho et al. detected *E. coli* and *Neisseria gonorrhoeae* by bright-field imaging of an immune-agglutination assay. After filtered by the microfluidic paper device, target bacterial antigen in the urine sample induced coagulation of antibody-conjugated particles, which can be captured by the smartphone camera [61].

Detection of parasites was also carried out on smartphones. By capturing videos of blood sample flow in a thin imaging chamber, filarial parasite *Loa loa* was identified on a smartphone microscope (Fig. 48.5A), namely CellScope. Differential images between each recorded video frame and a time-averaged frame were calculated, and parasites were identified in those differential images. A microcontroller was used to control the sample flow and illumination. The results were finally displayed directly in an iOS app [62]. In another example, with an unmodified mobile phone and a commercially available immunoassay kit, malarial parasites were detected on the smartphone. This simple procedure can be easily standardized, allowing rapid training and easy implementation in resource-limited areas [63]. Smartphone devices for the diagnosis of soil-transmitted helminth diseases have been made available. In these studies, bright-field imaging was carried out on smartphones to observe *Schistosoma haematobium* after simple processing of stool or urine samples [64–66]. Parasite eggs from fecal samples could be counted on a smartphone after labeled with fluorescent dyes (Fig. 48.5B). Due to the different size of eggs and resulted different fluorescent intensities, the eggs of two parasites, strongyle and ascarid, can be discriminated on the cellphone, based on the intensity threshold set in the mobile phone app [67]. Detection of water-borne pathogens, especially parasites, is an important way of controlling the spread of some parasite diseases. A smartphone fluorescence

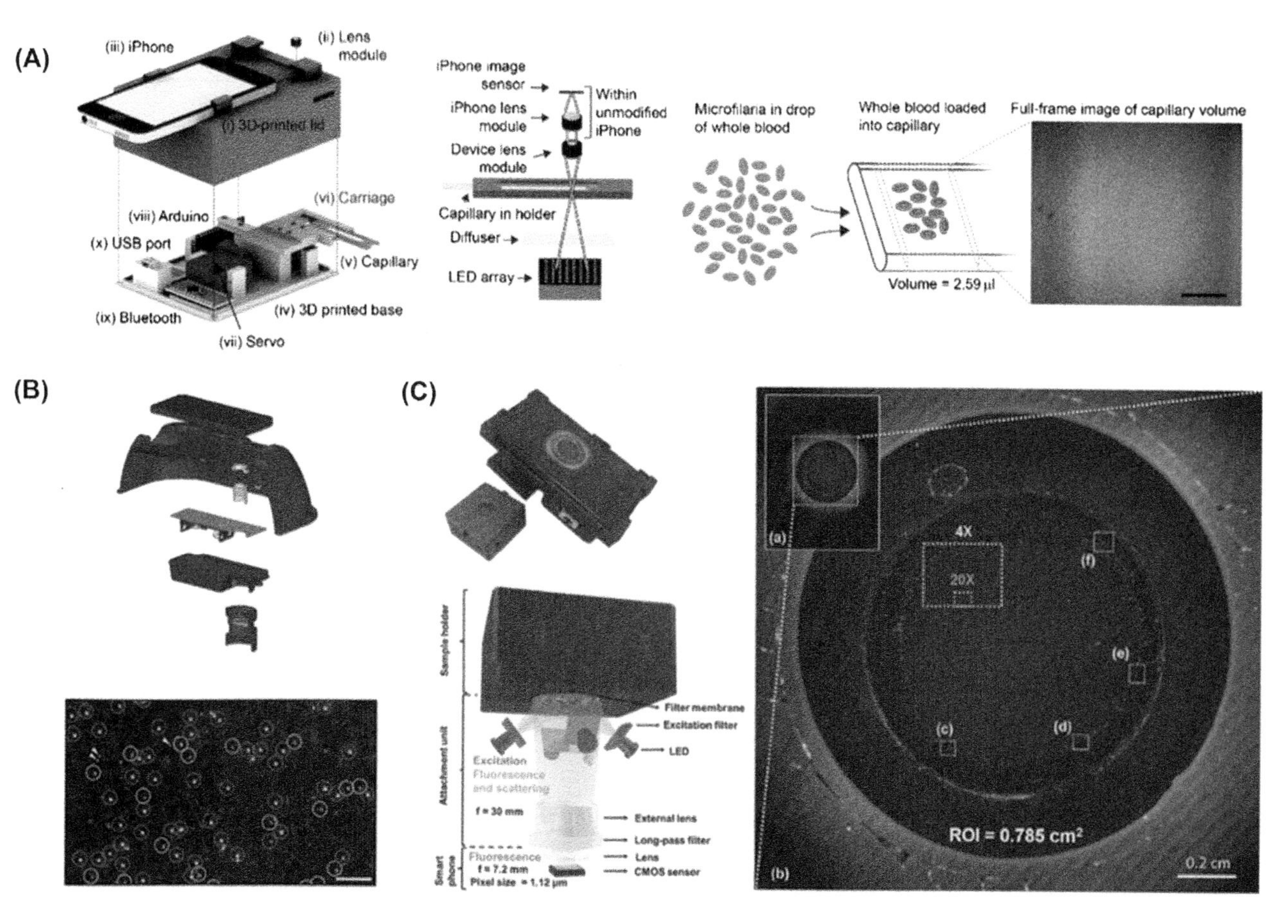

FIGURE 48.5 Parasite and microbe detection on smartphone platforms. (A) Quantification of blood-borne filarial Parasites on a smartphone microscope [62]. (B) Counting of fluorescently labeled parasite eggs on the smartphone [67]. (C) Detection and quantification of *Giardia lamblia* cysts on a smartphone fluorescent microscope [68]. *Reproduced with permission from the references mentioned above. Copyright 2015 American Association for the Advancement of Science, 2016 Australian Society for Parasitology and 2014 Royal Society of Chemistry.*

microscopy was used to detect fluorescently labeled *Giardia lamblia* cysts from water samples (Fig. 48.5C). In this application, the images were sent by the mobile phone to the server and analyzed with a machine-learning algorithm to obtain accurate counts of cysts [68].

Detection and imaging of human cells

Imaging, counting, and analysis of human cells can provide rich information in disease diagnosis. Labeling of specific antigens on the cells can also provide molecular information like protein expression, which is important in identification, risk management, and treatment of disease. A few examples of cell imaging, counting, and analysis on the smartphone are summarized in Table 48.5.

Smartphone microscopes have shown great potential in cancer diagnosis via single-cell imaging or counting. The smartphone-based D3 (digital diffraction detection) platform has been demonstrated for cellular imaging of immunolabeled cancer cells [20]. Another imaging modality, autofluorescence (AF) imaging, was applied in the detection of basal cell carcinoma. From the difference of AF intensity, malignant tissue can be identified from normal healthy tissue. This simple method can be applied in the primary evaluation of suspicious skin tissue [69]. Recently, a fluorescent imaging cytometry platform installed on a smartphone was reported for counting and magnetic separation of cancer cells. Stained breast cancer cells were magnetically levitated, imaged, and counted. By using different staining procedure and filters, different cells can be distinguished. Combination of magnetic levitation and fluorescence imaging also allowed spatial separation and imaging of different cells due to their difference in density, and therefore, levitation height [70].

Imaging and quantitative analysis of blood cells provides valuable information on the morphology and population of blood cells, which is essential in the diagnosis of diseases, such as malaria, sickle cell anemia, and HIV infection. Proof-of-concept work was reported recently, where white blood cells in a microfluidic chamber were counted on a smartphone-based fluorescent imaging cytometer. In this work, a fluorescence imaging device was

TABLE 48.5 Representative cases on cellular detection using mobile phone detectors.

Cell	Sample	Method of detection	Results	References
Breast cancer cell Cervical cancer cell	–	D3 (digital diffraction analysis) platform	–	[20]
Basal cell carcinoma	–	Autofluorescence imaging	–	[69]
Cancer cell imaging	–	Fluorescence imaging	LOD = 680 cells/mL Counting accuracy: $R^2 = 0.95–0.96$	[70]
Blood cell counting	Blood	Fluorescence imaging Bright-field imaging Colorimetric analysis	WBC: $R^2 = 0.98$, -2.9×10^4 cells/μL bias RBS: $R^2 = 0.98$, 230 cells/μL bias Hb: $R^2 = 0.92$ CC, 0.036 g/dL bias	[72]
Sickle cell detection	Blood	Bright-field imaging, magnetic levitation	Statistically significant difference between the control group and sickle cell anemia sample.	[74]
CD4+ T cell count	Blood	Bright-field imaging	Accuracy: 93.3% (threshold = 200 cells/μL), 96.6% (threshold = 500 cells/μL)	[75]
CD4+ T cell count	Blood	Colorimetric ELISA	Accuracy: 97% (threshold = 350 cells/μL)	[76]
Sperm	Semen	Bright-field video	Sensitivity = 87.5% specificity = 90.9%	[77]
Sperm	Semen	Bright-field video	Sperm concentration criterion: sensitivity = 95.83% specificity = 97.10% accuracy = 96.77% Sperm motility criterion: sensitivity = 95.83% specificity = 98.04% accuracy = 97.31%	[78]

built on a smartphone using an LED as a light source and a plastic color film as an inexpensive emission filter. Blood cell sample labeled with fluorescent dyes was continuously injected into a microfluidic chamber using a syringe pump. The microfluidic chip was placed in front of the smartphone imaging system. Video clips were taken to record cell flow through the chamber. The JPEG images were then extracted for processing and cell counting [71].

In another smartphone device, blood cells were analyzed by an imaging cytometry platform (Fig. 48.6A) which is capable of the quantification of white blood cell (WBC) counts, red blood cell (RBC) counts, and hemoglobin (Hb) concentrations [72]. Sickle cell disease or sickle cell anemia remains to be a major public health issue in Africa. As a hematologic disease, this disease can be diagnosed directly by a blood test. A smartphone was used in the imaging and identification of sickle cells from blood smear samples [73]. In a different example, by using magnetic levitation, sickle cell anemia sample can be distinguished from normal RBC samples with a smartphone. A difference in the density between sickle cells and normal RBCs resulted in the different height distribution pattern of sickle cell anemia blood sample and normal blood sample under magnetic levitation. This effect can be further magnified by induced dehydration of RBC. As a result, sickle cell anemia genotype (SS) can be distinguished from normal blood cell with a statistically significant difference [74].

The number of a specific kind of WBC called CD4+ T cells gives important information in the diagnosis of HIV infection. The CD4+ T cell count, or CD4 test, is traditionally carried out on a benchtop flow cytometer. As an alternative approach, CD4 cells were imaged and counted on a smartphone platform (Fig. 48.6B). To do that, antibodies were conjugated onto the chip surface to capture CD4+ cells. The CD4+ cell concentration can then be used to identify HIV-positive and negative samples using the threshold values suggested by WHO [75]. Wang et al. developed a colorimetric ELISA system for CD4+ cell counts on the smartphone. Magnetic beads conjugated with anti-CD4 antibody were used for capturing CD4+ cells. Under the actuation of a magnet, the cells labeled with beads were moved between chambers for washing, secondary antibody targetting, and color development. Both mobile and desktop apps were developed to quantify CD4+ cell counts by measuring the color change of the assay [76].

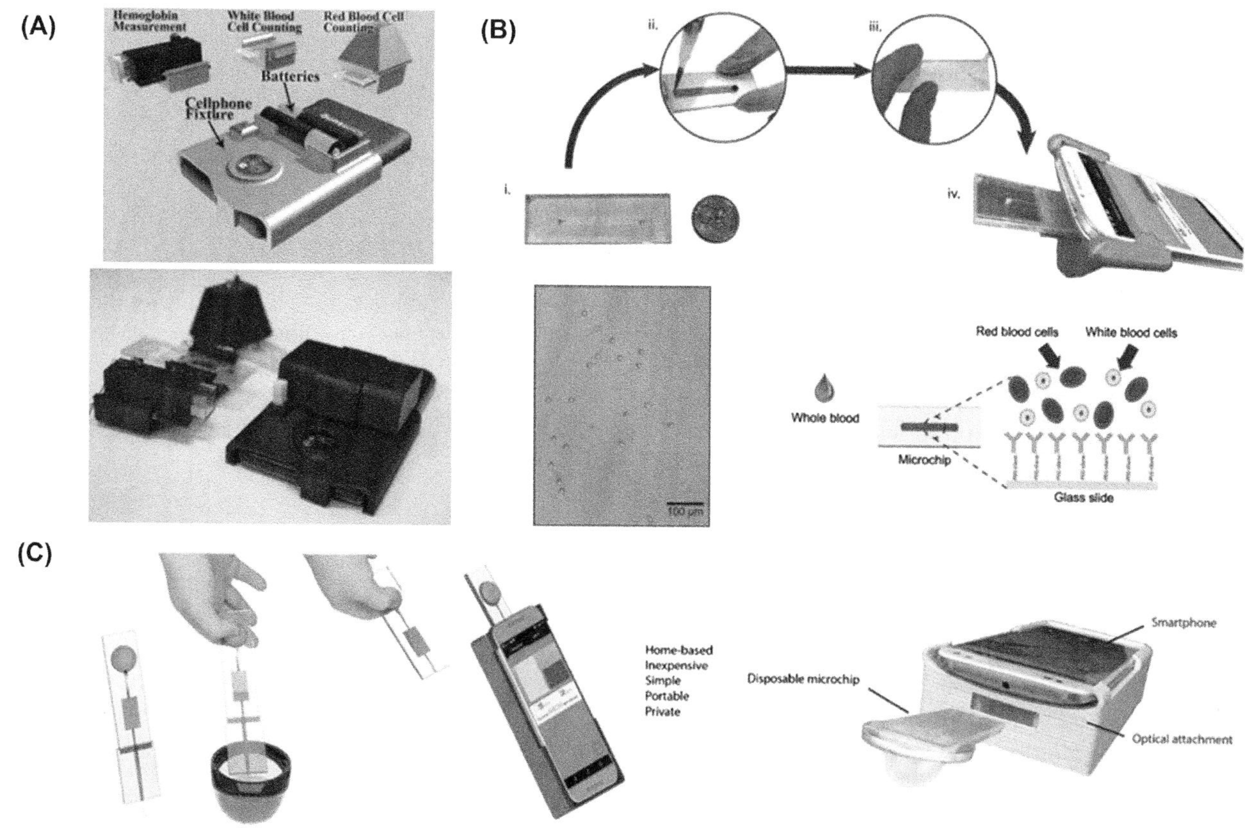

FIGURE 48.6 Cell imaging and analysis on smartphone platforms. (A) A cost-effective blood analysis platform on a smartphone for WBC, RBC, and Hb quantification [72]. (B) CD4+ cell counting on a smartphone device [75]. (C) Smartphone device for POC semen analysis [78]. *Reproduced with permission from the references mentioned above. Copyright 2013 Royal Society of Chemistry, 2017 Royal Society of Chemistry and 2017 American Association for the Advancement of Science.*

Smartphones have also been applied in the analysis of semen and diagnosis of male infertility. A simple ball lens-based smartphone microscope was used in the semen analysis. By analyzing the videos, the numbers and motilities of spermatozoa were calculated on three smartphones with different models for comparison [77]. Similarly, in the work by Kanakasabapathy et al., bright-field videos of spermatozoa were taken under the magnification of two aspheric lenses. Semen samples were drawn into the counting chamber by a manual vacuum pump. From the smartphone-captured videos, the concentration and motility of sperm can be calculated by an Android application on the phone (Fig. 48.6C). Untrained users can then test semen quality at home, conveniently with smartphones [78].

Summary and outlook

The cost-effectiveness and portability of smartphone devices have enabled various diagnostic applications in the resource-limited or field settings. In the context of global health challenges, simple, easy-to-use, and low-cost smartphone devices have found increased applications in both developed and developing countries for personalized health monitoring and disease diagnosis.

Currently, adding optical attachments onto smartphones is still the mainstream approach of designing smartphone-based detection and diagnosis systems. Highly customizable and compact optical attachments provide the opportunity of using a single smartphone in multiple scenarios to perform a series of detection by switching the attachments. In the context of sample preparation and liquid handling, microfluidic devices have been extensively combined with smartphones. With the integration of microfluidic devices, the sample volume can be reduced to microliters, and the sample processing procedures can be significantly simplified. Power-free and vacuum-driven sample handling systems on smartphone diagnostic tools have made another step toward real field use. Subsequent data processing after mobile sensing and imaging can be realized either directly on the phone or a cloud server. Remote data sharing can be easily available between patients and clinicians with connected mobile phone devices. In the meanwhile, parallel processing of high-volume data can help clinicians, and other healthcare specialists make more accurate decisions on diagnostics and therapeutics.

Based on the current achievements in smartphone-based detection and diagnostics, several areas can be envisioned in the future research and development. To move forward toward clinical applications, more closely collaborations between engineers and clinicians will be expected, which allow to better identify the need and validate the POC systems under more rigorous conditions. Multiplexed, high-throughput analysis of multiple biomarkers is another field in smartphone detection diagnostics. Novel designs of assays are needed to achieve this goal, including efforts on the design of high-throughput microfluidic systems for parallel sample preparation and assay reactions. More robust and user-friendly imaging and sensing system would be needed for field applications in those areas with few trained professionals. Big-data produced by the massive smartphone devices can better assist clinicians and policy makers in analyzing and tracking emerging public health issues, such as the outbreak of new infectious diseases. Finally, machine learning and artificial intelligence is a promising area in data processing of smartphone images for faster and more accurate result analysis. All these efforts will be made toward designing more powerful and precise smartphone diagnostic tools for POC and field applications.

References

[1] GSMA. Number of mobile subscribers worldwide hits 5 billion. 2017. Available from: https://www.gsma.com/newsroom/press-release/number-mobile-subscribers-worldwide-hits-5-billion/.

[2] Statista. Number of smartphone users worldwide from 2014 to 2020 (in billions). 2016. Available from: https://www.statista.com/statistics/330695/number-of-smartphone-users-worldwide/.

[3] Poushter J. Smartphone ownership and internet usage continues to climb in emerging economies. 2016. Available from: http://www.pewglobal.org/2016/02/22/smartphone-ownership-and-internet-usage-continues-to-climb-in-emerging-economies/.

[4] Roda A, et al. Smartphone-based biosensors: a critical review and perspectives. Trac. Trends Anal. Chem. 2016;79:317–25.

[5] Zhang D, Liu Q. Biosensors and bioelectronics on smartphone for portable biochemical detection. Biosens. Bioelectron. 2016;75:273–84.

[6] McCracken KE, Yoon J-Y. Recent approaches for optical smartphone sensing in resource-limited settings: a brief review. Anal. Methods 2016;8(36):6591–601.

[7] Ozcan A. Mobile phones democratize and cultivate next-generation imaging, diagnostics and measurement tools. Lab Chip 2014;14(17):3187–94.

[8] McLeod E, Wei Q, Ozcan A. Democratization of nanoscale imaging and sensing tools using photonics. Anal. Chem. 2015;87(13):6434–45.

[9] Contreras-Naranjo JC, Wei Q, Ozcan A. Mobile phone-based microscopy, sensing, and diagnostics. IEEE J. Sel. Top. Quantum Electron. 2016;22(3):1–14.

[10] Liu X, Lin T-Y, Lillehoj PB. Smartphones for cell and biomolecular detection. Ann. Biomed. Eng. 2014;42(11):2205–17.

[11] Xu X, et al. Advances in smartphone-based point-of-care diagnostics. Proc. IEEE 2015;103(2):236–47.

[12] Hernández-Neuta I, et al. Smartphone-based clinical diagnostics: towards democratization of evidence-based health care. J. Intern. Med. 2019;285(1):19–39.

[13] Bates M, Zumla A. Rapid infectious diseases diagnostics using Smartphones. Ann. Transl. Med. 2015;3(15).

[14] Sony Corporation SSSC. Sony releases stacked CMOS image sensor for smartphones with industry's highest*1 48 effective megapixels. 2018. Available from: https://www.sony.net/SonyInfo/News/Press/201807/18-060E/index.html.

[15] Switz NA, D'Ambrosio MV, Fletcher DA. Low-cost mobile phone microscopy with a reversed mobile phone camera lens. PLoS One 2014;9(5):e95330.

[16] Zhu H, et al. Cost-effective and compact wide-field fluorescent imaging on a cell-phone. Lab Chip 2011;11(2):315–22.

[17] Wei Q, et al. Fluorescent imaging of single nanoparticles and viruses on a smart phone. ACS Nano 2013;7(10):9147–55.

[18] Wei Q, et al. Imaging and sizing of single DNA molecules on a mobile phone. ACS Nano 2014;8(12):12725–33.

[19] Tseng D, et al. Lensfree microscopy on a cellphone. Lab Chip 2010;10(14):1787–92.

[20] Im H, et al. Digital diffraction analysis enables low-cost molecular diagnostics on a smartphone. Proc. Natl. Acad. Sci. U.S.A. 2015;112(18):5613–8.

[21] Sekine Y, et al. A fluorometric skin-interfaced microfluidic device and smartphone imaging module for in situ quantitative analysis of sweat chemistry. Lab Chip 2018;18(15):2178–86.

[22] Jung Y, et al. Smartphone-based colorimetric analysis for detection of saliva alcohol concentration. Appl. Opt. 2015;54(31):9183–9.

[23] Oncescu V, O'Dell D, Erickson D. Smartphone based health accessory for colorimetric detection of biomarkers in sweat and saliva. Lab Chip 2013;13(16):3232–8.

[24] Nemiroski A, et al. Universal mobile electrochemical detector designed for use in resource-limited applications. Proc. Natl. Acad. Sci. U.S.A. 2014;111(33):11984–9.

[25] Zhang D, et al. Smartphone-based portable biosensing system using impedance measurement with printed electrodes for 2,4,6-trinitrotoluene (TNT) detection. Biosens. Bioelectron. 2015;70:81–8.

[26] Yetisen AK, et al. Paper-based microfluidic system for tear electrolyte analysis. Lab Chip 2017;17(6):1137–48.

[27] Lipowicz M, Garcia A. Handheld device adapted to smartphone cameras for the measurement of sodium ion concentrations at saliva-relevant levels via fluorescence. Bioengineering (Basel Switz.) 2015;2(2):122–38.

[28] Chun HJ, et al. Paper-based glucose biosensing system utilizing a smartphone as a signal reader. BioChip J. 2014;8(3):218–26.

[29] Devadhasan JP, et al. Whole blood glucose analysis based on smartphone camera module. J. Biomed. Opt. 2015;20(11):117001.

[30] Calabria D, et al. Smartphone–based enzymatic biosensor for oral fluid L-lactate detection in one minute using confined multilayer paper reflectometry. Biosens. Bioelectron. 2017;94:124–30.

[31] Lee S, et al. NutriPhone: a mobile platform for low-cost point-of-care quantification of vitamin B 12 concentrations. Sci. Rep. 2016;6:28237.

[32] Lee S, et al. A smartphone platform for the quantification of vitamin D levels. Lab Chip 2014;14(8):1437–42.
[33] Oncescu V, Mancuso M, Erickson D. Cholesterol testing on a smartphone. Lab Chip 2014;14(4):759–63.
[34] Ogirala T, et al. Smartphone-based colorimetric ELISA implementation for determination of women's reproductive steroid hormone profiles. Med. Biol. Eng. Comput. 2017;55(10):1735–41.
[35] Zangheri M, et al. A simple and compact smartphone accessory for quantitative chemiluminescence-based lateral flow immunoassay for salivary cortisol detection. Biosens. Bioelectron. 2015;64:63–8.
[36] Petryayeva E, Algar WR. Single-step bioassays in serum and whole blood with a smartphone, quantum dots and paper-in-PDMS chips. Analyst 2015;140(12):4037–45.
[37] Joh DY, et al. Inkjet-printed point-of-care immunoassay on a nanoscale polymer brush enables subpicomolar detection of analytes in blood. Proc. Natl. Acad. Sci. U.S.A. 2017;114(34):E7054–62.
[38] McGeough CM, O'Driscoll S. Camera phone-based quantitative analysis of C-reactive protein ELISA. IEEE Transac. Biomed. Circ. Syst. 2013;7(5):655–9.
[39] Laksanasopin T, et al. A smartphone dongle for diagnosis of infectious diseases at the point of care. Sci. Transl. Med. 2015;7(273):273re1.
[40] Arts R, et al. Detection of antibodies in blood plasma using bioluminescent sensor proteins and a smartphone. Anal. Chem. 2016;88(8):4525–32.
[41] Yeo S-J, et al. Smartphone-based fluorescent diagnostic system for highly pathogenic H5N1 viruses. Theranostics 2016;6(2):231.
[42] Im H, et al. Digital diffraction detection of protein markers for avian influenza. Lab Chip 2016;16(8):1340–5.
[43] Berg B, et al. Cellphone-based hand-held microplate reader for point-of-care testing of enzyme-linked immunosorbent assays. ACS Nano 2015;9(8):7857–66.
[44] Wang S, et al. Integration of cell phone imaging with microchip ELISA to detect ovarian cancer HE4 biomarker in urine at the point-of-care. Lab Chip 2011;11(20):3411–8.
[45] Barbosa AI, et al. Portable smartphone quantitation of prostate specific antigen (PSA) in a fluoropolymer microfluidic device. Biosens. Bioelectron. 2015;70:5–14.
[46] You M, et al. Household fluorescent lateral flow strip platform for sensitive and quantitative prognosis of heart failure using dual-color upconversion nanoparticles. ACS Nano 2017;11(6):6261–70.
[47] Damhorst GL, et al. Smartphone-imaged HIV-1 reverse-transcription loop-mediated isothermal amplification (RT-LAMP) on a chip from whole blood. Engineering 2015;1(3):324–35.
[48] Selck DA, et al. Increased robustness of single-molecule counting with microfluidics, digital isothermal amplification, and a mobile phone versus real-time kinetic measurements. Anal. Chem. 2013;85(22):11129–36.
[49] Rodriguez-Manzano J, et al. Reading out single-molecule digital RNA and DNA isothermal amplification in nanoliter volumes with unmodified camera phones. ACS Nano 2016;10(3):3102–13.
[50] Du W, et al. SlipChip. Lab Chip 2009;9(16):2286–92.
[51] Shen F, et al. Digital PCR on a SlipChip. Lab Chip 2010;10(20):2666–72.
[52] Liu W, et al. SlipChip for immunoassays in nanoliter volumes. Anal. Chem. 2010;82(8):3276–82.
[53] Liao S-C, et al. Smart cup: a minimally-instrumented, smartphone-based point-of-care molecular diagnostic device. Sensor. Actuator. B Chem. 2016;229:232–8.
[54] Ming K, et al. Integrated quantum dot barcode smartphone optical device for wireless multiplexed diagnosis of infected patients. ACS Nano 2015;9(3):3060–74.
[55] Priye A, et al. A smartphone-based diagnostic platform for rapid detection of Zika, chikungunya, and dengue viruses. Sci. Rep. 2017;7:44778.
[56] Jiang L, et al. Solar thermal polymerase chain reaction for smartphone-assisted molecular diagnostics. Sci. Rep. 2014;4:4137.
[57] Mancuso M, Cesarman E, Erickson D. Detection of Kaposi's sarcoma associated herpesvirus nucleic acids using a smartphone accessory. Lab Chip 2014;14(19):3809–16.
[58] Hutchison JR, et al. Reagent-free and portable detection of Bacillus anthracis spores using a microfluidic incubator and smartphone microscope. Analyst 2015;140(18):6269–76.
[59] Shafiee H, et al. Paper and flexible substrates as materials for biosensing platforms to detect multiple biotargets. Sci. Rep. 2015;5:8719.
[60] Veigas B, et al. Gold on paper–paper platform for Au-nanoprobe TB detection. Lab Chip 2012;12(22):4802–8.
[61] Cho S, et al. Smartphone-based, sensitive µPAD detection of urinary tract infection and gonorrhea. Biosens. Bioelectron. 2015;74:601–11.
[62] D'ambrosio MV, et al. Point-of-care quantification of blood-borne filarial parasites with a mobile phone microscope. Sci. Transl. Med. 2015;7(286):286re4.
[63] Scherr TF, et al. Mobile phone imaging and cloud-based analysis for standardized malaria detection and reporting. Sci. Rep. 2016;6:28645.
[64] Bogoch II, et al. Mobile phone microscopy for the diagnosis of soil-transmitted helminth infections: a proof-of-concept study. Am. J. Trop. Med. Hyg. 2013;88(4):626–9.
[65] Bogoch II, et al. Evaluation of portable microscopic devices for the diagnosis of Schistosoma and soil-transmitted helminth infection. Parasitology 2014;141(14):1811–8.
[66] Ephraim RK, et al. Diagnosis of Schistosoma haematobium infection with a mobile phone-mounted foldscope and a reversed-lens CellScope in Ghana. Am. J. Trop. Med. Hyg. 2015;92(6):1253–6.
[67] Slusarewicz P, et al. Automated parasite faecal egg counting using fluorescence labelling, smartphone image capture and computational image analysis. Int. J. Parasitol. 2016;46(8):485–93.
[68] Koydemir HC, et al. Rapid imaging, detection and quantification of Giardia lamblia cysts using mobile-phone based fluorescent microscopy and machine learning. Lab Chip 2015;15(5):1284–93.
[69] Lihachev A, et al. Autofluorescence imaging of basal cell carcinoma by smartphone RGB camera. J. Biomed. Opt. 2015;20(12):120502.
[70] Knowlton S, et al. 3D-printed smartphone-based point of care tool for fluorescence-and magnetophoresis-based cytometry. Lab Chip 2017;17(16):2839–51.
[71] Zhu H, et al. Optofluidic fluorescent imaging cytometry on a cell phone. Anal. Chem. 2011;83(17):6641–7.
[72] Zhu H, et al. Cost-effective and rapid blood analysis on a cell-phone. Lab Chip 2013;13(7):1282–8.
[73] Breslauer DN, et al. Mobile phone based clinical microscopy for global health applications. PLoS One 2009;4(7):e6320.

[74] Knowlton S, et al. Sickle cell detection using a smartphone. Sci. Rep. 2015;5:15022.

[75] Kanakasabapathy MK, et al. Rapid, label-free CD4 testing using a smartphone compatible device. Lab Chip 2017;17(17):2910–9.

[76] Wang S, et al. Micro-a-fluidics ELISA for rapid CD4 cell count at the point-of-care. Sci. Rep. 2014;4:3796.

[77] Kobori Y, et al. Novel device for male infertility screening with single-ball lens microscope and smartphone. Fertil. Steril. 2016;106(3):574–8.

[78] Kanakasabapathy MK, et al. An automated smartphone-based diagnostic assay for point-of-care semen analysis. Sci. Transl. Med. 2017;9(382):eaai7863.

[79] Coulibaly JT, et al. Accuracy of mobile phone and handheld light microscopy for the diagnosis of schistosomiasis and intestinal protozoa infections in Côte d'Ivoire. PLoS Neglected Trop. Dis. 2016;10(6):e0004768.

Section IV

Perspectives and challenges

Chapter 49

Information technology and patient protection

Claude J. Pirtle and Jesse M. Ehrenfeld
Vanderbilt University Medical Center, Nashville, TN, United States

If you had walked into a physician's office or clinic about 15–20 years ago, you would have noticed copious amounts of paper located in numerous places around the healthcare setting. Until recently, large filing cabinets held the most critical personal data about patients, including medical histories, medications, lab results, among many other essential pieces to a patient's care. Healthcare is rapidly advancing both in terms of the treatment interventions offered, and the framework with which patient data is processed, stored, and reported. Despite this modernization of medical records, and the plentitude of data available to healthcare providers, there are still many hurdles to be overcome, from appropriate data storage to effective utilization for patient care optimization. With the evolution of a multitude of technologies, we are seemingly becoming closer to arriving at a precision metric of care, versus the older one-size-fits-all approach.

Over four in five of all nonfederal acute care hospitals adopted a basic electronic health record (EHR), according to the Office of the National Coordinator [1]. EHRs are the cornerstone of health information technology in a physician's office or hospital setting, but the use of multiple information systems is embedded in environments all around. Bytes of information, including vital patient data, are being transferred, typically in a secure fashion, from radiological studies, mobile health platforms, laboratory results, among many other data mediums.

Protecting patient data and privacy was a much easier concept when locking a file cabinet was the only thing needed. Today we live in a globally connected world, and the precipitous increase in electronic data stored has led to a growing number of individuals from across the world attempting to hack into hospitals and access data. To help better protect patients and their information, Congress instituted the Health Insurance Portability and Accountability Act of 1996 (HIPAA). This act made it necessary for the U.S. Department of Health and Human Services (HHS), to script regulatory procedures to protect certain health information. HHS was able to accomplish this with the creation of the HIPAA Security and Privacy Rule.

Data storage and manipulation in the healthcare environment

Healthcare organizations must build their storage systems to be flexible and scalable to meet the growing data demand. Healthcare data management has a strict requirement for the privacy, confidentiality, security, and traceability of patient and organizational data. These requirements have made many organizations apprehensive about moving to a cloud-based environment. Traditionally, healthcare systems have avoided cloud state storage in favor of on-premises options because of the immediate control physical that datacenters offered to the IT staff. However, more and more medical centers have recently begun to implement cloud storage, as a part of their overall IT infrastructure approach, because of the reduced maintenance costs, improved backup performance, scalability, and improved HIPAA-compliance. HIPAA compliance is also a concern for organizations looking to migrate to the cloud. Many organizations are still hesitant to trust a third-party vendor, to host EHRs and PHI (Personal Health Information), fearing that lackluster security could lead to a data breach. Most HIPAA-compliant cloud vendors make it evident and are willing to discuss how their solution complies with HIPAA regulations.

Healthcare organizations are demanding more storage space for big data analytics, and the volume of unstructured data needing to be stored, for analytic initiatives. In the past, large health systems managing petabytes of essential healthcare data required that on-premises server rack systems be added to their arsenal if more storage was needed.

Precision Medicine for Investigators, Practitioners and Providers. https://doi.org/10.1016/B978-0-12-819178-1.00049-6

By using the cloud as a vehicle, the increase in cloud storage can be easily managed by a phone call or electronic message to the storage vendor, and in most cases, a committed cloud storage vendor can accommodate the necessary requests (or partial request) within a few hours. The U.S. spending in healthcare cloud computing was estimated at 3.74 billion dollars in 2015, and spending is expected to increase to 9.48 billion dollars by 2020 [2,3].

Big data manipulation

Big data is defined by Gartner as "high volume, high velocity, and/or high variety information assets, that require new forms of processing to enable enhanced decision making, insight discovery and process optimization." [4] (see Fig. 49.1). Healthcare data is continuously created, ranging from patient data and revenue cycle data to many other forms. From 2014, the total amount of healthcare data has been estimated to be about 150 exabytes [5]. Velocity is a term used to describe the amount of data that is being generated or acted upon. IBM estimates that about 2.5 quintillion bytes of data is generated each day[6].

This figure is a few years old, and data creation has likely accelerated, with the closer interconnection of the Internet, personal devices, and continuous connectivity. Just to put things into perspective, Google processes over 3.5 billion searches daily [7]. Variety of information is also a large consideration of the data gathered. Structured text, unstructured text, images, meta-data, among others, are collected daily [5,6].

The data stored has to be sifted through, in an effort to appropriately report external metrics (for example, in the U.S. to the Centers for Medicare and Medicaid Services), by a certain time, and in parallel, this same data has to be appropriately cataloged, for the ultimate reason for its creation—patient care. To extrapolate and develop a type of structure, the data can come from internal services (e.g., the electronic health record, the picture archiving, and communication system/PACS) or from an external service, such as an insurance company replying with claims, an outside laboratory reporting a microbiology culture, or patient wearable devices.

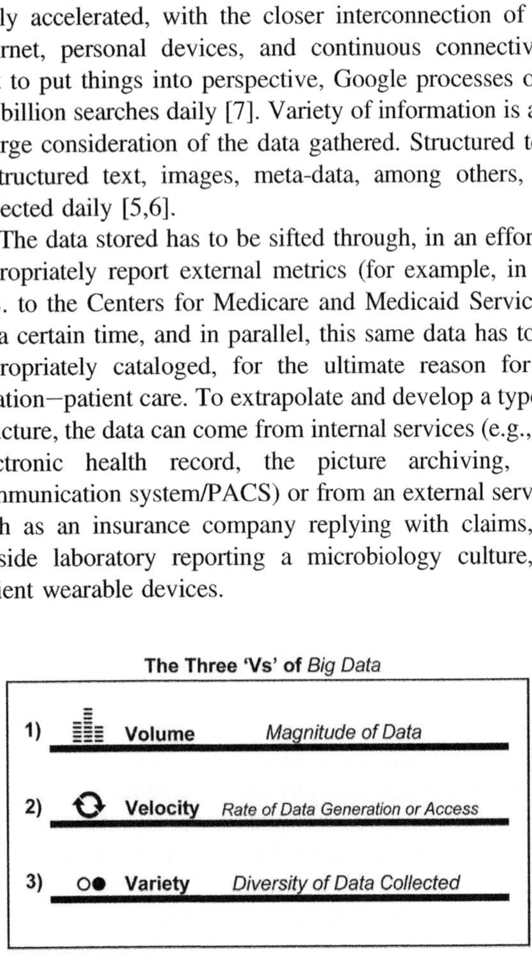

FIGURE 49.1 Big data defined – volume, velocity, and variety.

Data processing platforms

A number of tools are available to help analyze and make sense of big data, and one of the most well-known is Hadoop. Hadoop is an open-source software platform created in the late 1990s, that fabricates a framework for distributed storage and processing, of large amounts with varying structures of data at the same time. Using the MapReduce algorithm, Hadoop distributes the burden of large amounts of data onto commodity hardware. After the processing of the data clusters in parallel is completed, the data is then reliably integrated back together, to form a final result. Hadoop is robust in many aspects, particularly that it allows almost limitless simultaneous tasks to be completed. There are a growing number of centers that have leveraged the power of Hadoop, and other tools like it, to facilitate data processing.

The electronic health record timeline and a promise

EHRs were initially developed and used at academic centers in the early to mid-1960s. Most of the academic institutions with these electronic records developed and maintained the platforms themselves. An early example would be the Massachusetts General Hospital's Computer Stored Ambulatory Record (COSTAR), which came online in the late 1960s. COSTAR was a set of modules that allowed the scheduling and registration of patients, clinical information storage, among other functionality that we take for granted in today's electronic medical records [8]. The U.S. Department of Veterans Affairs began using an EHR in the 1970s, across the Veteran Health System (VHA), allowing it unrestricted access to any veteran's records in the system [9]. Epic was founded in 1979 [10] and also Cerner Systems in 1979 [11]. These two industry juggernauts combined make up almost 50% of the hospital electronic medical records market in the United States, according to the Office of the National Coordinator [12].

In 2008, less than 10% of the hospitals in the United States had an electronic health record [1]. The HITECH Act was a portion of the American Recovery and Reinvestment Act, which totaled 787 billion dollars [13]. The Act formally propelled the adoption of EHRs by healthcare practices by introduced legislation. Of the money allocated, 18 billion dollars was earmarked to allow the Centers for Medicare and Medicaid Services (CMS) to incentivize the transition to, and meaningful use of EHRs, with the goal of increasing structured data, improving the efficiency and quality of care allotted, improve population health, privacy, and security protection of patient health information (PHI) [14].

CMS and the Office of the National Coordinator (ONC) have established standards for electronic health records. Being a Certified EHR Technology (CEHRT) allows the purchaser of that product and the end-users, to understand that the said product meets a defined standard, and contains a certain technological capability and functionality [15]. The HITECH Act was integral in framing the creation of the Certified EHR Technology rule.

The exchange of patient data

A common vernacular allows for an electronic health record to understand, for example, a complete metabolic panel (CBC) ordered and resulted in Miami, Florida, to also mean the same thing in an EHR in Boise, Idaho. We still have an abundance of work to complete, but some standards that have revolutionized the conversation include Logical Observation Identifiers Names and Codes, Health Level 7 (HL7), and Digital Imaging and Communications in Medicine. Logical Observation Identifiers Names and Codes (LOINC), started in 1994, developed a common terminology standard for laboratory and clinical observations. LOINC includes an extensive common standard for laboratory tests, measurements such as vital signs, instruments for surveys, among many other codes [16].

HL7 contains a complete suite of standards specifications, whose members provide a framework for the integration, retrieval, and sharing of electronic health information. The framework allows a standard definition of how information is bundled and communicated from one system to another [17]. Digital Imaging and Communications in Medicine (DICOM) enables the transmission, storing, retrieval, processing, and displaying of medical imaging information in a standardized format. The standard was initially developed in 1985, and it has continued to evolve even as technology has become more robust [18].

Challenges and pitfalls

Most medical systems are a silo of patient data that is not readily exchanged. Communication breakdown often occurs when patients transition between healthcare organizations. The creation and integration of Healthcare Information Exchanges (HIE) with EHRs, could improve care coordination, by providing clinicians with complete information in real-time when meeting a patient. Research has shown that HIE improves access to patient test results, reduces the rate of diagnostic imaging tests [3], and improves the care of patients with chronic conditions. However, HIE adoption has been sparse.

An enormous push for interoperability has been put forth by the Office of the National Coordinator for Information Technology. They have put together a road map entitled "Connecting Health and Care for the Nation – A Shared Nationwide Interoperability Roadmap." Yearly goals are set forth, in an effort to improve the exchange of patient information and achieve nationwide interoperability, in an effort to improve health and lower costs. Some of the most important points include improving technical standards and implementation guidance, consistency in implementation, the innovation of new standards and use of APIs (Application Programming Interfaces), alignment of state and federal requirements that allow interoperability, among many other key initiatives [19].

Priorities for reliable data movement

Technologies are on the horizon that could improve the interoperability, including the innovation and use of APIs (see Fig. 49.2). According to the United States Department of Health and Human Services, an API is a "technology that allows one software program to access the services provided by another software program" [20]. Recently legislation has been passed in the United States, that specifically calls for the development of APIs by EHR vendors that require "no special effort" for a third-party developer, to exchange health information with the application.

An example would be the development of a third-party application, to allow the synchronous exchange of data of a glucometer with an electronic health record. With an API, the glucometer (a blood sugar measuring device), does not have to readily customize their code to exchange information with an EHR, using a standard such as Health Level 7's FHIR (Fast Health Interoperability Resources). This is very important, considering that as of July 17, 2017, 86 vendors were certified by the U.S. Government to provide electronic health records [21]. Without these universal APIs, each electronic health record would require custom coding for each system.

Common electronic standards

FHIR is a standard developed by HL7 to improve the exchange of data [22]. This standard combines ideas of prior HL7 products, and also new technologies, such as

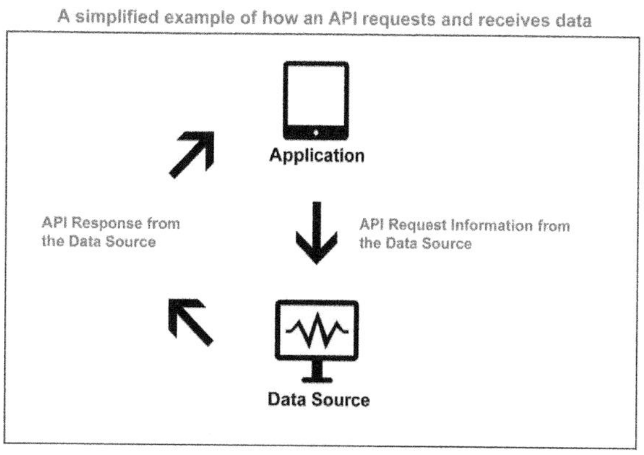

FIGURE 49.2 Data movement using an Application Programming Interface (API).

web standards. The standard allows the seamless communication of electronic health records with mobile applications, essentially connecting Patient-Generated Health Data (PGHD) directly with the EHR, with other electronic health records, among many other opportunities around the web. The goal of FHIR was to build a standardized framework to access data in multiple electronic health record environments—completely agnostic of the electronic health record system.

The ability of some remote monitoring devices to have a two-way conversation already exists in some minor markets. One could imagine an electrophysiology device such as an implantable cardioverter-defibrillator, capable of self-interrogation and rhythm assessment. This device could seamlessly use a wireless connection (via Bluetooth or wireless cell signal), to feed the found rhythm to a cardiologist that resides miles away from the patient's site.

Genomic databases

Warfarin is a blood thinner synthesized in 1948 and approved for human use in 1954. Major bleeding events were found to occur at a rate of 7.2 per 100 patient-years [23]. Over the years, it has been noted that warfarin has a very narrow therapeutic index. Genetic variants can either increase or slow the degradation of the drug. These include VKORC1, CYP2C9, among others [24]. Having a complete database of genetic variants of all patients that reside in a state (or country), could lead to an interconnected scenario, where the EHR of an ambulatory provider, is able to access the continuously updated database, for a patient or group's genetic variance.

Many patient registries have been created in the interim to serve as a centralized repository; however, many of these registries are disease-specific, and few are interconnected. They usually focus on a specific condition or disease and can be sponsored by a nonprofit agency, government agency, or private company [25]. The Cystic Fibrosis (CF) Foundation Patient Registry, for instance, collects information of individuals with CF, in an effort to assist teams providing care, and guide quality improvement initiatives [26].

A large number of strides, including CEHRTs, APIs, FHIR, and patient registries, have been made in the correct direction to improve interoperability and create a cohesive data ecosystem, where the sharing of information is not only effortless, but also the societal norm.

Importance of healthcare provider data and institutional data

Limitless research opportunities are now available, using the data of thousands of electronic health records. Data mining allows access to electronic data, that can be analyzed with some caveats. Structured data, for example, blood pressure or temperature, account for only about 20% of EHR data. For the most part, the EHR contains lines of unstructured bits and bytes, like a physician's discharge summary or a nurse's note. Through advanced analytical techniques such as natural language processing, we are improving the ability to interrogate this type of data.

Electronic health record, stress, and burnout

Healthcare provider burnout has come to light very recently as a serious issue worldwide, with burnout rates over 50% [27]. Burn out has been shown to lead to a high clinician and staff turnover, higher medical error rates, lower patient satisfaction, and care quality, and most devastatingly physician suicide [28]. Although burnout is a contribution of multiple factors, from emotional to social stressors, many studies have shown that the EHR has come between clinical medicine and the physician. Physicians reported that EHRs contributed most to stress and burnout, with over 21% of physicians selecting this over prior authorizations, student debt, regulatory compliance, and a host of other issues[29].

For every hour that a physician provides direct care to patients, almost two additional hours are spent on the electronic health record and other clinical work of that day [30]. A unique opportunity does exist to refine better the human-technology integration, whether it is an improvement in clinical workflows, decrease in electronic health record clerical duties, or mandatory regulatory reporting by physician practices. A number of corporations, including the EHR vendors, are aware and working toward a shared solution.

Patient privacy and protection

Health information is our most precious information, that should only be shared with our complete explicit permission. As with the movement of information from multiple data sources, privacy and security issues become apparent. HIPAA currently protects data and information of consumers. The Privacy Rule, or *Standards for Privacy of Individually Identifiable Health Information*, establishes national standards for the protection of certain health information. The *Security Standards for the Protection of Electronic Protected Health Information* (the Security Rule), establishes a national set of security standards, for protecting certain health information that is held or transferred in electronic form. The Security Rule operationalizes the protections contained in the Privacy Rule by addressing the technical and non-technical safeguards, that organizations called *covered entities* must put in place, to secure individuals' "electronically protected health information" (e-PHI). Within HHS, the Office for Civil Rights (OCR) has responsibility for enforcing the Privacy and Security Rules, with voluntary compliance activities and civil money penalties [31,32]. The most recent updates to these rules occurred in 2009, with the passage of the American Recovery and Reinvestment Act.

Data ownership and disposal

The aggregation of multiple sources of data allows the movement of data from multiple vendors, such as Pharmacy Benefit Managers (PBMs), insurance companies, among a myriad of others. Currently, no time limit exists for the storage and use of a patient's data. This brings into question who exactly owns the data. There are some advocates who believe patients should own their data, whereas most healthcare facilities believe that as the entity that generates the data, they should own and have control over it.

With multiple consumers of patient data, and big data analytics requiring multiple facets of data to be collected, it can quickly be seen how data privacy and security comes into light. As of December 27, 2018, OCR received notification of 351 data healthcare breaches, of 500 or more records. In the event, your medical record was stolen, some experts estimate that your record is worth almost 10 times as much, as your credit card number on the black market [33]. In a 2016 study by the Ponemon Institute, nearly 90% of the institutions represented in the study had a data breach in the past 2 years, with the leading cause of data breaches in healthcare found to be a criminal attack [34].

Discrimination, transparency, and consent

Collected data, particularly concerning genetic risk, could have a rippling effect, not only affecting the individual patient but also family members, who may be perceived to be at an increased genetic predisposition for a certain disease or future symptomatology. Many downstream potential effects could have grounds of discrimination for health insurance, life insurance, employment, and potential termination from an employer. The data privacy laws in the U.S. and across the world continue to lag behind the ongoing technological innovation.

Transparency and fairness are important aspects. Data in most countries, including the United States, currently can be stored for unlimited periods of time [35]. This allows for the data owner to use it for scientific research, or they may enter an agreement with a partner after your transaction has already occurred, to study a segment of your data.

Understanding the full spectrum of consenting to the use of your data allows you to have some control of your data privacy and security. Hundreds of mobile phone medical applications are now available on a multitude of platforms, to help you improve diet, comply with recent hypertension guidelines, or allow the tracking of your steps. However, not being completely informed about the current use and future use of your data in this instance is a disservice. Remotely monitoring your location by using the GPS tracker, could be an essential solution to becoming an avid marathon runner. However, this application creator could be selling your GPS data, diet routine, and weight loss information to a third party without your known consent.

An article in 2016 showed that 86% of randomly selected diabetes applications on a mobile platform, used cookies and other mechanisms, that could allow them to send information about the end-user to third parties. A majority of the applications did not have a privacy policy to protect the end user [36].

Data homogeneity and population selection bias

In a 2016 publication, a Toronto-based startup was building an auditory test, to help identify neurological disease in narrative speech. They ran into an issue because the technology only worked optimally for English speakers of a particular dialect [37]. Looking further into the data, the company had only collected data from a small region of the country, instead of a disparate collection of data to train their artificial intelligence model with. Large corporations have recently come under scrutiny, due to their accuracy and precision in identifying people of Caucasian ancestry, versus other lineages.

A 2016 article divulges that about 81% of our genome-wide association study data, is of European descent. This is an advancement from a prior 2009 study that showed that 96% of the genome sequencing was of European data [38]. Having a biased homogenous data population narrows the scope of research that can be done on specific data sets. This could also allow for the development of erroneous, or partial observations to occur. Slowly the pendulum is moving toward a more heterogeneous data collection strategy, through augmented recruitment and sampling techniques. The further development of these strategies will harmonize the research community data sets to improve health as a whole.

Conclusions

Technology offers a multitude of opportunities to improve the healthcare, of not only a segment of the population but potentially of the world. Innovation continues to improve many different aspects, from the quality of patient care to research opportunities, and the speed in which data is retrieved and accessed. Advanced in how healthcare data are collected, stored, transmitted, and shared will enable the implementation of new approaches to precision medicine. However, it will be critical for developers and clinicians to understand the limitations of the underlying technology, as these tools are brought into the marketplace.

References

[1] Office of the National Coordinator for Health Information Technology. "Non-federal acute care hospital electronic health record adoption," health IT quick-state #47. September 2017. https://dashboard.healthit.gov/quickstats/pages/FIG-Hospital-EHR-Adoption.php.
[2] Venditto G. Exploring the strategic benefits of a move to the cloud. 2018. https://www.healthcareitnews.com/news/exploring-strategic-benefits-move-cloud.
[3] Vest JR, Jasperson S. How are health professionals using health information exchange systems? Measuring usage for evaluation and system improvement. J. Med. Syst. 2011;36(5):3195−204.
[4] Martin-Sanchez F, Verspoor K. Big data in medicine is driving big changes. Yearb. Med. Inform 2014;9(1):14−20. https://doi.org/10.15265/IY-2014-0020. Published 2014 Aug 15.
[5] Bellazzi R. Big data and biomedical informatics: a challenging opportunity. Yearb. Med. Inform 2014;9(1):8−13. https://doi.org/10.15265/IY-2014-0024. Published 2014 May 22.
[6] Bresnick J. Understanding the many V's of healthcare big data analytics. June 5, 2017. https://healthitanalytics.com/news/understanding-the-many-vs-of-healthcare-big-data-analytics.
[7] Google Search Statistics. http://www.internetlivestats.com/google-search-statistics/, 2019.
[8] Committee, on improving the patient record, and of medicine institute. In: Dick RS, et al., editors. The computer-based patient record: an essential technology for health care. Revised Edition. National Academies Press; 1997 ProQuest Ebook Central. http://ebookcentral.proquest.com/lib/vand/detail.action?docID=3376738. Created from vand on 2019-01-21 15:47:20.
[9] Atherton J. Virtual Mentor 2011;13(3):186−9. https://doi.org/10.1001/virtualmentor.2011.13.3.mhst1-1103.
[10] About. 2019. https://www.epic.com/about.
[11] Loria G. What is Cerner EMR? A guide to the Vendor's solutions. 2019. https://www.softwareadvice.com/resources/what-is-cerner-emr/#overview.
[12] Cohen J. Epic, Cerner make up 50% of hospital EHR market share, ONC data shows. July 17, 2018. https://www.beckershospitalreview.com/ehrs/epic-cerner-make-up-50-of-hospital-ehr-market-share-onc-data-shows.html.
[13] HITECH act summary. 2019. https://www.hitechanswers.net/about/about-the-hitech-act-of-2009/.
[14] Meaningful use. 2019. https://www.cdc.gov/ehrmeaningfuluse/introduction.html. Last Reviewed: January 18, 2017.
[15] Certified EHR technology. 2019. https://www.cms.gov/Regulations-and-Guidance/Legislation/EHRIncentivePrograms/Certification.html. Last Reviewed October 25, 2018.
[16] About LOINC. 2019. https://loinc.org/about/.
[17] Introduction to HL7 standards. 2019. http://www.hl7.org/implement/standards/.
[18] History. 2019. https://www.dicomstandard.org/history/.
[19] Connecting health and care for the nation − a shared nationwide interoperability roadmap. 2019. https://www.healthit.gov/sites/default/files/hie-interoperability/nationwide-interoperability-roadmap-final-version-1.0.pdf.
[20] Secure API server showdown winner announced. 2019. https://www.hhs.gov/about/news/2018/05/17/secure-api-server-showdown-winner-announced.html. Last Revised May 16, 2018.
[21] Almedia W. Updated list: 2015 edition certified EHRs. July 21, 2017. https://www.healthcareitnews.com/news/updated-list-2015-edition-certified-ehrs.
[22] Introducing HL7 FHIR. 2019. https://www.hl7.org/fhir/summary.html.
[23] Pirmohamed M. Warfarin: almost 60 years old and still causing problems. Br. J. Clin. Pharmacol. 2006;62(5):509−11.
[24] Dean L. Warfarin therapy and VKORC1 and CYP genotype. 2012 mar 8 [updated 2018 Jun 11]. In: Pratt V, McLeod H, Rubinstein W, et al., editors. Medical genetics summaries. Bethesda (MD): National Center for Biotechnology Information (US); 2012 [Internet]. Available from: https://www.ncbi.nlm.nih.gov/books/NBK84174/.
[25] List of registries. 2019. https://www.nih.gov/health-information/nih-clinical-research-trials-you/list-registries. Last Reviewed November 14th, 2018.
[26] Patient registry. 2019. https://www.cff.org/Research/Researcher-Resources/Patient-Registry/.
[27] Physician burnout. Content last reviewed July 2017. Rockville, MD: Agency for Healthcare Research and Quality; 2017. https://www.ahrq.gov/professionals/clinicians-providers/ahrq-works/burnout/index.html.
[28] Drummond D. Physician burnout: its origin, symptoms, and five main causes. Fam. Pract. Manag. 2015 Sep−Oct;22(5):42−7.
[29] Reaction data. https://www.reactiondata.com/report/physician-burnout/.
[30] Sinsky C, Colligan L, Li L, Prgomet M, Reynolds S, Goeders L, et al. Allocation of physician time in ambulatory practice: a time and motion study in 4 specialties. Ann. Intern. Med. 2016;165:753−60. https://doi.org/10.7326/M16-0961.
[31] The security rule. 2019. https://www.hhs.gov/hipaa/for-professionals/security/index.html. Last Updated May 12, 2017.
[32] The HIPAA privacy rule. 2019. https://www.hhs.gov/hipaa/for-professionals/privacy/index.html. Last Updated April 16, 2015.
[33] Humer C, Finkle J. Your medical record is worth more to hackers than your credit card. September 24th, 2014. https://www.reuters.com/article/us-cybersecurity-hospitals/your-medical-record-is-worth-more-to-hackers-than-your-credit-card-idUSKCN0HJ21I20140924.
[34] Sixth annual benchmark study on privacy & security of healthcare data. May 2016. https://www.ponemon.org/local/upload/file/Sixth%20Annual%20Patient%20Privacy%20%26%20Data%20Security%20Report%20FINAL%206.pdf.
[35] "Big data" in health care raises some ethical concerns. September 1, 2013. https://www.reliasmedia.com/articles/63767-big-data-in-health-care-raises-some-ethical-concerns.
[36] Blenner SR, Köllmer M, Rouse AJ, Daneshvar N, Williams C, Andrews LB. Privacy policies of android diabetes apps and sharing of health information. JAMA 2016;315(10):1051−2. https://doi.org/10.1001/jama.2015.19426.

[37] Gershgorn D. If AI is going to be the world's doctor, it needs better textbooks. September 6th, 2018. https://qz.com/1367177/if-ai-is-going-to-be-the-worlds-doctor-it-needs-better-textbooks/.

[38] Popejoy AB, Fullerton SM. Genomics is failing on diversity. October 12th, 2016. https://www.nature.com/news/genomics-is-failing-on-diversity-1.20759.

Further reading

[1] Largest healthcare data breaches of 2018. December 27, 2018. https://www.hipaajournal.com/largest-healthcare-data-breaches-of-2018/.

Chapter 50

Blockchain solutions for healthcare

Peng Zhang[1] and Maged N. Kamel Boulos[2]
[1]*Vanderbilt University Medical Center, Nashville, TN, United States;* [2]*Sun Yat-sen University, Guangzhou, Guangdong, China*

Introduction

In 2018, Kamel Boulos and colleagues discussed how blockchain technology could secure patient and provider identities, manage medical supply lines, enable public and open geo-tagged data, and much more [1]. Indeed, recent years have seen a growing interest in blockchain technologies among health and healthcare research and practice communities.

In March 2018, a consortium of scholarly publishers, including Springer Nature, launched Phase 1 of Blockchain for Peer Review "*to make the peer review process more transparent, recognizable and trustworthy*" [2]. Earlier, in 2017, the U.S. Centers for Disease Control and Prevention (CDC) started experimenting with blockchain, for sharing public health data, to help public health workers respond faster to a crisis [3].

By 15 January 2019, a PubMed query using the term "blockchain" had retrieved 101 indexed papers, up from 41 on 20 June 2018 [4]. Blockchain solutions are currently being explored in different parts of the world for securing patient and provider identities, for securely storing health records and maintaining a single version of the truth [5], for managing pharmaceutical and medical device supply chains, for medical fraud detection, for medical data sharing among researchers [6], for research data monetization [7], in crisis mapping and recovery scenarios, using blockchain-enabled augmented reality [8], and even for tackling environmental plastic pollution with blockchain rewards, or littercoin (twitter.com/littercoin) [9].

The Internet of things (IoT) [10] is proposed as the foundation of the smart, healthy cities and regions, of today and tomorrow [11]. Geospatially-enabled blockchain solutions exist that use a crypto-spatial coordinate system to add an immutable spatial context that regular blockchains lack [12]. These geospatial blockchains do not just record an entry's specific time, but also require and validate its associated proof of location, allowing accurate spatio-temporal mapping of physical world events, and enabling a myriad of location-based, smart cities, and IoT application possibilities [13].

The market for IoT devices and apps that negotiate with, and pay each other for secure, safe operation and services, is expected to grow in the near future. Examples of these IoT devices include mobile and wearable devices that pay for public transportation [14], and autonomous connected devices and vehicles for smart city emergency/disaster response, such as a drone defibrillator, or a drone for the delivery of ordered medicines and medical supplies [15], or a self-driving ambulance car, or helicopter.

The blockchain-powered, distributed peer-to-peer apps powering these smart devices, drones, and vehicles, would cut out the "middleman" and the dependence on third-party centralized providers, for navigation and other geospatial data, and would mitigate the possibility of an IoT-powered autonomous vehicle, being hijacked and driven to a wrong location [16].

The challenges facing blockchain technologies today include interoperability [17], security [18,19] and privacy, jurisdictional issues [20], as well as the need to find suitable and sustainable business models of implementation. These challenges will be briefly discussed at the end of this chapter.

The rise of blockchain technology and its concepts

Centralized systems and their limitations

Traditional information systems are managed in a centralized manner, meaning that typically one entity or authority is responsible for storing, maintaining, and modifying data and the exchange of data, on behalf of the clients it serves. Given the structure of those systems that authority also has the exclusive rights to access, create, modify, and delete any of the records, at its own discretion. Although this model has worked fairly well across different industries, many incidents of attacks and data loss in recent years, have

Precision Medicine for Investigators, Practitioners and Providers. https://doi.org/10.1016/B978-0-12-819178-1.00050-2

demonstrated the danger of a centralized system—vulnerabilities to a single point of attack and corruption, as well as insider attacks.

Examples of such incidents include the Equifax (one of the three major consumer credit reporting agencies) data breach [21], where about 143 million American consumers' sensitive personal information, including Social Security Numbers (SSN) and driver's license numbers, were compromised. Imaginably, this attack raised deep concerns among consumers, regarding identity theft and threats to personal financial safety. What was worse, Equifax executives were reported to have sold their stocks in the company after the breach was publicly discovered [22], resulting in a trust deficit in the company.

Another unfortunate attack exploited 21.5 million records, from the U.S. Office of Personnel Management (OPM) [23]. The OPM possessed millions of records containing current and former government employees' background checks data, which also included large sets of sensitive personal information, such as SSNs, dates of birth, and physical addresses.

Decentralization using blockchain technology

Blockchain technology is a decentralized platform that removes the dependency and trust upon a single, centralized authority but, at the same time, ensures secure and so-called *trustless* transactions, directly between interacting entities [24]. It distributes information across a number of parties, the majority of which must reconcile, in order for the entire network to reach consensus regarding data stored on-chain. By nature, blockchain technology offers decentralization, to prevent any party from being the single source of truth, immutability in order to protect against data tampering via auditable trails of all transactions on-chain, and consensus as well, which requires the majority of network participants to reach an agreement to ensure information integrity enabled by cryptography. This technology provides the foundations for a number of application domains, including cryptocurrencies and other more advanced decentralized apps (DApps) [25].

Decentralized apps build upon a related concept called *smart contracts*, which are enhancements to the fundamental elements of blockchain technology. As well implemented and supported in the Ethereum Blockchain [25], smart contracts enable programmability of blockchains so that the program code can directly control the exchanges or redistributions of cryptocurrency and other digital assets, such as variants of cryptocurrencies and pieces or representations of data.

These exchanges and redistributions can occur between any users or parties in the blockchain ecosystem, without involving a third party or an escrow. Instead, smart contract code can control and guarantee the transaction, according to a set of predefined agreements, established between the involved participants. Smart contracts can store data of varying formats and define custom operations on the data, and therefore, facilitate the development of DApps. DApps are web-based applications that interact with blockchain ecosystems and provide similar web services to the end-users so that the underlying technology is encapsulated by similar user experience [26].

Potential applications of blockchain technology in healthcare

In the healthcare domain, building secure and efficient information systems to improve clinical workflows and healthcare interoperability, creating provenance tracking systems for safe prescription drug delivery, and also expediting insurance claims adjudication are addressed [27]. These interests primarily focus on the improvement of quality of care and community wellbeing.

Challenges facing state-of-the-art health information systems

The current state of healthcare information systems faces a number of challenges that impede advancements toward an interoperable ecosystem. For instance, the U.S. healthcare law, HIPAA, requires that healthcare practices must comply with security and privacy regulations, to ensure that patients' confidentiality and safety are best protected, as they receive care and generate sensitive health information. Given the fast development pace of the Internet of things and edge devices, people have, however, become much more reliant on portable, smart computing devices, to collect and store information ranging from financial, to social, to health and wellness-related.

Much of the data collected through these smart devices cannot easily integrate with high-fidelity health records, documented by care practitioners during patient visits [28]. This is due to the fact that these instruments, i.e., devices and/or mobile apps, collecting health and wellness data, despite their continuity, do not typically hold the same strict standards as certified medical equipment or license. As a result, the lack of interoperability between traditional healthcare practices and modern healthcare solutions becomes much transparent due to the inadequacies of trust relationships established between the two types of entities [29].

Another challenge that makes achieving interoperability difficult is the heterogeneity in representations of healthcare identities of patients, and often times of care providers. Key electronic health record (HER) vendors in the healthcare space, such as Epic [30] and Cerner [31], provide enterprise solutions to maintain and resolve healthcare identities of both providers and patients, under the condition that they are interacting with healthcare services or personnel within the same network. Although these EHR systems have spread across a significant portion of large-scale healthcare organizations, their adoption is less common in mid-size care facilities, regional clinics, or private practices. Identity matching of patients and care teams across vendors is also not an easy process, as each vendor may encode identifiers differently from other vendors, or the common set of information (usually a combination of first and last names, date of birth, social security number, and address) requested from patients, may change throughout their lifetime [32].

What is the potential role of blockchain technology in healthcare

General cryptographic theories that guarantee information security can be applied to the healthcare domain, except with much more fine-grained and discreet considerations. First of all, the network need not carry out the same level of decentralization as public blockchains that distribute all information to anyone with Internet access [33]. Instead, a consortium of entities or organizations within the same sector, and adhering to similar compliance requirements, may form a relatively small-scale and permissioned blockchain that is sufficient to remove the need for a centralized authority. However, it could serve the healthcare community effectively while following the same codes as if a traditional, standalone healthcare facility.

Smart contracts are also useful in this scenario, to expand the functionality of bitcoin-like blockchains that only support pure cryptocurrency transactional data, to more advanced logic that dictates the transactions of complex data and data structures, and provides non-repudiation of those transactions. Applied to the same scenario of data integration, between certified and regulated medical entities, and popular edge device and app providers as described above, any historical record of data exchange would persist in the blockchain, making it transparent to the network which sources generally provide reputable data, and also which ones may be rogue players in the ecosystem.

Smart contracts can be created to enforce a membership-based ecosystem, thus creating a permissioned blockchain, which only designated entities or organizations can involve in critical decisions, including adding or removing members and validating transactions to include them in the shared ledger [34]. This model does not necessarily create a closed environment by default, because users of such a permissioned blockchain, such as healthcare professionals and health app providers, can engage in exchanging data using the blockchain framework [29]. In other words, only the maintainers of this model of blockchain are permissioned.

Operator identity and asymmetric public keys

One of the key innovative features of blockchain technology is its implementation of blockchain identities that is based on cryptographic models, which have been proven computationally infeasible to break. For instance, users of the bitcoin blockchain can remain pseudo-anonymous during their transactions with other bitcoin users, by communicating only with their uniquely associated addresses, which are also asymmetric public keys in cryptography. A public key is mathematically linked with a private key, and the pair is used in conjunction to protect some piece of data via encryption and to validate the digital signature of a user. Even though public keys can be freely distributed, as long as their paired private keys are kept secure, encrypted data can be protected without compromising any personal information.

Inspired by the way cryptography is employed in blockchain technology, the creation, and management of healthcare identities can be approached similarly, via a decentralized, unified identity system, with public identifiers linked to cryptographic objects, instead of personal information for easy lookups. This model would produce a universal patient index registry, which can be distributed across a wide range of care practices, not limited by geographic or network boundaries [35]. Today, Ethereum-based uPort (Ethereum, Zug, Switzerland) is one of the largest blockchain-based efforts to manage identities with personal mobile devices [36].

It demonstrates how cryptography-based identities, can be uniquely registered in a blockchain system, associating appropriate reputation with an individual as they interact with other trusted authorities, who provide attestations or verifications of claims made by the individual. Obviously, for healthcare uses, alternative means of identity management must also be considered, as personal mobile devices are not accessible to a large population. Nonetheless, the idea of maintaining and exchanging attestations from authorities can be explored further, for the healthcare space. For example, healthcare organizations are generally considered as trusted sources, which can provide authoritative attestations for patients and/or their healthcare staff, such as a digital certificate acknowledging the completion of vaccination, and proof of a patient's healthcare insurance coverage.

Potential value and utility of blockchain technology for healthcare practitioners

This technology also has significant value and utility for healthcare practitioners and payers alike, due to the enablement of a more extensive and far-reaching network, which nationally and potentially universally connects the healthcare workforce. Compared to patients who are consumers of healthcare services, healthcare practitioners have far more legal liabilities and other duties that they must fulfill, before and during their healthcare practice.

Healthcare providers and practitioners must maintain appropriate and unexpired credentials or licenses, and comply with all required immunization and vaccination requirements, within required time periods, to ensure a safe environment for both patients and their colleagues. Just as patient data are oftentimes siloed within a healthcare organization, the licensure and compliance data associated with each provider's employment is also frequently challenged with difficulties of data transfer and exchange, particularly when healthcare professionals move to a new healthcare employer. This does not only elongate the employment process of those practitioners but can also create a lot of financial hurdles for patients and insurance companies, as delayed propagation of information can lead to miscalculations of in—or out-of-network health costs [37].

These issues can be mitigated with blockchain technology that offers a decentralized database as a shared provider registry, with the most up-to-date information being available, and consistent across the entire network [48]. The registry may also record licensure and/or immunization compliance information to alleviate healthcare practitioners' burden of keeping up with employment-related data. The state of Illinois has conducted a pilot study using blockchain technology to create a certification and credential registry of providers [38].

Clinical research protocols

For researchers interested in improving medical outcomes via clinical trials, data sharing is as crucial to them as to patients. Particularly in rare disease studies, where cohort populations are small-scaled and geographically dispersed, patient recruiting, and data sharing across these geographic regions may be crucial to the success of a trial as they must rely on data aggregation from various sources to achieve a statistically significant cohort size [39]. Researchers may benefit from a decentralized ecosystem targeting these special needs, through which they can exchange relevant information, and keep other researchers interested in similar studies, informed about their latest clinical research findings [29].

Data auditing

A very important utility of blockchain technology is the audit trail that ensures the transparency of data. It can represent an extremely valuable asset to healthcare payers and insurance entities, for surfacing healthcare and insurance fraud. According to the National Health Care Anti-Fraud Association (Washington, DC, USA), a conservative estimation of the annual financial losses to healthcare fraud are about USD $68 billion [40]. Although patients suffer from health care fraud that requires larger out-of-pocket payments to a certain degree, payers, unfortunately, bear much more costs collectively, as fraud could happen at every stage of a patient visit, and throughout the billing process [41].

Due to the complexities and variabilities of medical visits and treatments, with the current clinical workflow where information experiences discontinuity, it is particularly hard to identify where fraud would occur. This is where decentralized frameworks may be able to innovate upon and revolutionize the single trusted system, by tracking information every step of the way and distributing summaries at each checkpoint, across multiple stakeholders in the industry.

Blockchain challenges

As any other technology threatening to disintermediate legacy processes and commercial interests, there are challenges of blockchain interoperability, security, and privacy, as well as the need to find suitable and sustainable business models of implementation. Among the challenges facing geospatial blockchain implementation today, there are four particularly pressing ones, requiring careful consideration and innovative solutions, both technical and regulatory.

Interoperability, security, and privacy

Blockchains from different providers and services should seamlessly talk to each other as appropriate [17]. Security should not be overlooked [42]. After all, the whole rationale of using a blockchain is to let people who did not previously know or trust one another, share data in a secure, tamperproof way. But the security of even the best-conceived blockchain can fail in some scenarios (e.g., the so-called *51% attacks* [18], calling for adequate preemptive mechanisms to be put in place, in order to mitigate or prevent blockchain security breaches. Blockchain's promise of transparency needs to be reconciled with the European Union's now much stricter privacy rules under GDPR (General Data Protection Regulation), which require personal data to be erasable on demand [43].

Jurisdictional issues of blockchain-based applications

The computers, or nodes, validating transactions on a blockchain, can conceivably be (and arguably should be) in a variety of jurisdictions, many of which have differing applicable legislation. As each node is playing a role in validating a transaction, it is very plausible that every transaction could be governed by each jurisdiction hosting a node. Regulatory authorities have not yet ruled on these jurisdictional issues, in many parts of the world, but it is easy to envision many contentious or highly litigious scenarios [20].

Future challenges

Other challenges are related to the rise of extremely powerful quantum computing. Just as the aforementioned 51% attacks where once considered improbable, the prospect of quantum computing power being arrayed against blockchain, to crack codes and breach security, has now entered the realm of possibility, albeit remote, with the advent of IBM's newly announced (2019), world's first commercial 20-qubit quantum computer [19,44].

Conclusions

Even though bitcoin is a controversial subject, seen by some as a flawed economic concept [45], it has given us blockchain, with its foundations of decentralization, cryptographic security, and immutability. It is impressive to see how these concepts are being used today in many serious ways, decoupled from their original implementation in bitcoin.

The challenges facing blockchain technologies today are not insurmountable, and we expect blockchain technologies to get increasingly powerful and robust, as they become coupled with artificial intelligence (AI) [46], in various real-world healthcare solutions, involving AI-mediated data exchange on blockchains [47].

References

[1] Kamel Boulos MN, Wilson JT, Clauson KA. Geospatial blockchain: promises, challenges, and scenarios in health and healthcare. Int. J. Health Geogr. 2018;17(1):25. https://doi.org/10.1186/s12942-018-0144-x. Published 2018 Jul 5.

[2] See https://www.blockchainpeerreview.org/.

[3] See https://www.technologyreview.com/s/608959/why-the-cdc-wants-in-on-blockchain/.

[4] See https://www.ncbi.nlm.nih.gov/pubmed/?term=blockchain.

[5] See, for example, https://medicalchain.com/en/.

[6] See, for example, https://www.nature.com/articles/d41586-018-02641-7.

[7] See https://www.theguardian.com/science/2018/feb/18/genetics-how-do-you-make-money-from-your-dna.

[8] See https://medium.com/mimir-blockchain/how-augmented-reality-can-change-how-we-navigate-a-natural-disaster-d7fbde0d735b.

[9] See https://medium.com/@littercoin/say-hello-to-littercoin-a-blockchain-reward-for-producing-open-data-on-plastic-pollution-ff1c29f215b7.

[10] See https://www.slideshare.net/sl.medic/how-the-internet-of-things-and-people-can-help-improve-our-health-wellbeing-and-quality-of-life.

[11] See https://www.biomedcentral.com/collections/smarthealthycities.

[12] See, for example, https://www.foam.space/.

[13] See http://researchly.leobosankic.com/2018/06/19/proof-of-location/.

[14] See http://www.tmleuven.be/project/guide2wear/G2W_D3.1.pdf.

[15] See, for example, http://www.flyzipline.com/.

[16] See https://www.geospatialworld.net/article/blockchain-geospatial-systems/.

[17] See https://www.technologyreview.com/s/611187/how-to-get-blockchains-to-talk-to-each-other/.

[18] Security issues today; see, for example, https://www.technologyreview.com/the-download/611238/self-serving-cryptocurrency-miners-are-attacking-small-blockchain-networks/.

[19] Security issues with the rise of quantum computing in the future. 2019. see, for example, https://www.forbes.com/sites/dantedisparte/2019/01/09/did-ibm-just-break-blockchain/.

[20] See https://www.techopedia.com/who-owns-the-data-in-a-blockchain-application-and-why-it-matters/2/33461.

[21] Gressin S. The Equifax data breach: what to do. Washington, DC: Federal Trade Commission; September 8, 2017.

[22] Owens JC, Cherney MA. Equifax executives sold stock after data breach, before informing public [Internet]. MarketWatch. MarketWatch; 2017 [cited 2018 Dec 18]. Available from: https://www.marketwatch.com/story/equifax-executives-sold-stock-after-data-breach-before-informing-public-2017-09-07.

[23] United States. Office of personnel management. A white paper: a fresh start for federal pay: the case for modernization. US Office of Personnel Management; 2002.

[24] Nakamoto S. Bitcoin: a peer-to-peer electronic cash system. 2019. https://bitcoin.org/bitcoin.pdf.

[25] Buterin V. A next-generation smart contract and decentralized application platform. White paper. January 2014.

[26] Zhang P, White J, Schmidt DC, Lenz G. Applying software patterns to address interoperability in blockchain-based healthcare apps. June 5, 2017. arXiv preprint arXiv:1706.03700.

[27] Zhang P, Schmidt DC, White J, Lenz G. Blockchain technology use cases in healthcare. Adv. Comput. 2018;111:1−41. https://doi.org/10.1016/bs.adcom.2018.03.006. https://www.dre.vanderbilt.edu/~schmidt/PDF/blockchain-bookchapter-2018.pdf.

[28] Zhang P, White J, Schmidt D. Architectures and patterns for leveraging high-frequency, low-fidelity data in the healthcare domain. IEEE International Conference on Healthcare Informatics (ICHI). June 4, 2018. p. 463−4. IEEE.

[29] Zhang P, White J, Schmidt DC, Lenz G, Rosenbloom ST. Fhirchain: applying blockchain to securely and scalably share clinical data. Comput. Struct. Biotechnol. J January 1, 2018;16:267−78.

[30] Epic. Epic community [Internet]. 2019 [cited 2018 Dec 18]. Available from, https://www.epic.com/.

[31] Cerner. Ambulatory solutions | Cerner. Cerner − cerner.com [Internet]. 2019 [cited 2018 Dec 18]. Available from, https://www.cerner.com/.

[32] McCoy AB, Wright A, Kahn MG, Shapiro JS, Bernstam EV, Sittig DF. Matching identifiers in electronic health records: implications for duplicate records and patient safety. BMJ Qual. Saf March 1, 2013;22(3):219−24.

[33] Zhang P, Walker MA, White J, Schmidt DC, Lenz G. Metrics for assessing blockchain-based healthcare decentralized apps. Ine-health networking, applications and services (Healthcom), 2017 IEEE 19th international conference on. October 12, 2017. p. 1−4. IEEE.

[34] Cachin C. Architecture of the hyperledger blockchain fabric. Workshop on distributed cryptocurrencies and consensus ledgers. 310; July 2016.

[35] Krawiec R, Housman D, White M, Filipova M, Quarre F, Barr D, Nesbitt A, Fedosova K, Killmeyer J, Israel A. Blockchain: opportunities for health care. edn. Blockchain: Opportunities for health care; 2016. p. 1−16.

[36] Lundkvist C, Heck R, Torstensson J, Mitton Z, Sena M. Uport: a platform for self-sovereign identity. [Internet]. uport. Uport. 2017 [cited 2018 Dec 18]. Available from: http://blockchainlab.com/pdf/uPort_whitepaper_DRAFT20161020.pdf.

[37] Geer L. How blockchain will transform practitioner credentialing. 2018 [Internet], Hashed%20Health.%20Jorge%20Vieira/wp-content/uploads/2017/01/HH_logo_web_crop.png [cited 2018 Dec 18]. Available from: https://hashedhealth.com/blockchain-will-transform-practitioner-credentialing/.

[38] FriedmanFeb S. Illinois builds momentum for blockchain [Internet]. GCN. 2019. [cited 2018 Dec 18]. Available from: https://gcn.com/articles/2018/02/05/illinois-blockchain.aspx.

[39] Karmen C, Ganzinger M, Kohl CD, Firnkorn D, Knaup-Gregori P. A framework for integrating heterogeneous clinical data for a disease area into a central data warehouse. InMIE 2014:1060−4.

[40] Statistics [Internet]. 2019. Important information | services that need approval. [cited 2018 Dec 18]. Available from: https://www.bcbsm.com/health-care-fraud/fraud-statistics.html.

[41] Healthcare Fraud [Internet]. Healthcare news & insights. 2012 [cited 2018Dec18]. Available from: http://www.healthcarebusinesstech.com/healthcare-fraud/.

[42] See https://www.technologyreview.com/s/610836/how-secure-is-blockchain-really/.

[43] See https://spectrum.ieee.org/tech-talk/computing/networks/3-obstacles-to-moving-social-media-platforms-to-a-blockchain.

[44] See https://www.engadget.com/2019/01/08/ibm-q-system-one-quantum-computer/.

[45] See https://www.cnbc.com/2018/05/01/warren-buffett-bitcoin-isnt-an-investment.html.

[46] See our healthcare AI collection at https://web.archive.org/web/20190331010845/https://plus.google.com/collection/40E-LE.

[47] Mamoshina P, Ojomoko L, Yanovich Y, Ostrovski A, Botezatu A, Prikhodko P, Izumchenko E, Aliper A, Romantsov K, Zhebrak A, Obioma Ogu I, Zhavoronko A. Converging blockchain and next-generation artificial intelligence technologies to decentralize and accelerate biomedical research and healthcare. Oncotarget 2017;9(5):5665−90. http://doi.org/10.18632/oncotarget.22345.

[48] Zhang P. Architectures and patterns for moving towards the use of high-frequency, low-fidelity data in healthcare. PhD diss. Vanderbilt University; 2018.

Chapter 51

Ethical questions in gene therapy

Iñigo de Miguel Beriain[1,2] and Jessica Almqvist[3]

[1]*Chair in Law and the Human Genome Research Group, Department of Public Law, University of the Basque Country, UPV/EHU, Bizkaia, Spain;* [2]*IKERBASQUE, Basque Foundation for Science, Bilbao, Spain;* [3]*Associate Professor of Public International Law. Department of Public Law and Legal Philosophy, Autónoma University in Madrid, Madrid, Spain*

Introduction: gene editing and precision medicine

Gene editing is an outstanding technology, which might bring a totally new scenario in precision medicine. Over the last years, it has come to have multiple applications in biomedical research, such as for the "*creation of disease models with desired genetic mutations, screening in a high-throughput manner for drug resistance genes, and making appropriate editions to genes* in vivo *for disease treatment.*" [1] There are enormous expectations that gene editing will cure genetic diseases by changing the genomic combinations that trigger them [2,3]. If this ever becomes a reality, medicine will never be the same. For the first time in history, it will be possible to stop treating the consequences of genetic diseases and instead concentrate on erasing pathologies from their genetic roots [4]. And, these interventions could be done on an individual basis, with a treatment tailored to each patient's circumstances, since DNA is unique and individualized.

The most promising gene-editing tool, CRISPR-Cas, is still in its infancy [5]. It was described for the first time in 1993 by the Spanish researcher Francis Martinez Mojica [6]. However, it took until 2005 to find that the CRISPR loci, encoded the instructions for an adaptive immune system that protected microbes against specific infections [7]. As late as September 2012, experts continued to be skeptical about the potential of this technology, to become a robust system for editing the human genome [8]. Indeed, there was no definitive agreement on its real potential until 5 years ago, in January 2013, when the group of Doudna and Charpentier published their classic paper on CRISPR in Science [9], and soon after the team of Feng Zhang reported mammalian genome editing [10].

Basic research related to gene editing on human embryos is becoming an increasingly recurrent phenomenon, mostly in China but also in the United States and even in the EU [11,12]. So far, there is not a lot of clinical trials involving testing gene-editing tools on patients. The biotech company Sangamo Therapeutics (Richmond, CA, USA) initiated the first in-vivo clinical trial, using a gene-editing tool called zinc finger nucleases (ZFNs), in November 2018. The aim of the trial was to treat the Hunter syndrome in a small cohort of patients, and its results are still being analyzed [13]. Preliminary results suggest that treatment for rare disease is safe, but its effectiveness is unclear [14].

Only recently, in December 2018, the U.S. Food and Drug Administration (FDA) approved the first in-body CRISPR medicine testing [15]. For the time being, clinical applications of germline genome editing techniques on human embryos remain out of reach. The only exception is the experiment carried out by the Chinese researcher He Jankui, who, in November 2018, shocked the world when he announced that he had brought to life two genetically modified human beings, prompting an outcry from the scientific community [16,17].

The road ahead

Several obstacles are in the way of optimal development. Most of the problems are scientific/technical and might be solved (or not) at some point. Others are ethically driven, and much more difficult to tackle. In order to move forward, the ethical concerns raised by gene editing call for public debate [18] that is not limited beforehand by "unmovable" boundaries but fosters an open discussion of precisely these concerns.

A preliminary question: assumable risk

First of all, we must face the question of assumable risk, i.e., the level of risk we are willing to assume, to bring gene editing of human beings into practice. Indeed, the risks involved are a key concern that cautions scientists from considering clinical applications of gene editing tools [19]. Those who object to gene editing, stress the risks posed to future generations, who cannot decide themselves if they are ready to assume them, but who would be beneficiaries or victims of decisions that may affect their genetic makeup. Since risk assessments are particularly complex, it is difficult to provide a straightforward response to these concerns, especially considering that our control of this technology is still limited. Problems include both off-target and on-target effects, that might cause severe harm, including cancer, to patients and their descendants [20]. Also, mosaicism is a problem that must be taken into account, although recent research shows that both the effects of off-target mutations and mosaicism can be reduced [21].

In any case, the question about the risks involved is not really of an ethical one but of a scientific/technical nature. As Alto Charo notes, "*with regard to the possibility of particular applications causing physical harm to individuals living in the future, one can perform a traditional risk-benefit assessment, albeit with the stakes raised by the large number of future people affected, and with skepticism that we can confidently predict the effects*" [22]. However, "*prediction and management of risks is a common feature of most planning*" [23] and "*rather than allowing uncertainty to slide toward paralysis and prohibition, a better approach has been to develop increasingly sophisticated means of calculating probabilities, and incorporating ongoing risk monitoring and mitigation strategies*" [24].

Recently, the Organizing Committee of the Second International Summit on Human Genome Editing made a statement along similar lines. While acknowledging that the "*the variability of effects produced by genetic changes, makes it difficult to conduct a thorough evaluation of benefits and risks,*" it does not rule out the possibility of addressing these risks [25].

Ethical objections to gene-editing

Not all germline gene editing in clinical contexts gives rise to ethical objections. Only some do. Certainly, any application of this technology, which at the moment is in an experimental phase, will require the free and informed consent of the interested person or his/her legal representatives. This is required not only by bioethics law but also international human rights law. For instance, the Statement on Human Genome Editing II reads: Nevertheless, germline genome editing could become acceptable in the future, if these risks are addressed, and if a number of additional criteria are met. These criteria include strict independent oversight, a compelling medical need, an absence of reasonable alternatives, a plan for long-term follow-up, and attention to societal effects, as stated by article seven of the International Covenant on Civil and Political Rights [26].

Besides the requirement of free and informed consent, it is difficult to find substantive ethical objections to modifications of our somatic line [27]. By contrast, as we shall see, cases where there is evidence that medical treatments will alter the human germline, even if those alterations are unintended, are far more controversial, although not ruled out necessarily. Unintentional human germline modification is what happens, for example, in many cases of chemotherapy. Because of this, patients undergoing this treatment are advised to abstain from reproducing for a period of time to avoid any damage to their offspring.

However, in these cases, it is usually considered that the intended purpose—the cure of a patient who will die unless treated—is important enough to outweigh the negative effects of such treatment, including the alteration of a patient's germline [28].

The most controversial cases are those where the editing is motivated precisely by a wish to change the human germline, to produce consequences that will affect not only the interested person or the patient but also all his or her descendants. However, even if the two sets of cases are somewhat different—in the first one there is no desire to alter the genome of the offspring, and in the second one there is—the ethical problems triggered in both scenarios are practically the same. This is because those who object to human germline modification that will alter the genes of our descendants, do not seem to really care about what a scientist or a doctor intended to do. Instead, they focus on what they believe is really bad or even morally wrong about human germline modification as such.

Criticisms to germline editing

Some object out of a belief in the futility of these tools. Others believe that using these tools on humans or embryos, will damage the integrity of the human genome, which would be contrary to human dignity. Others again object to these applications, out of a belief that using these tools are contrary to descendants' autonomy to decide themselves, if they want to have their germline modified or not. Another objection is captured by the slippery slope argument. From this perspective, even if therapeutic uses of germline modification may be ethically sound, permitting it will open the door to eugenics.

The futility argument

Those who defend that these tools are futile, argue that we do not need these tools to prevent the birth of persons suffering from genetic diseases, since we can achieve the same objective by using existing (and supposedly less controversial) technologies already available in medical practice: that of preimplantation genetic diagnosis (PGD) [29,30]. However, this argument contains many inaccuracies [31]. Even if PGD and in vitro fertilization (IVF) are already used, to avoid passing on some 250 genetic diseases, such as cystic fibrosis, Huntington's disease, hemophilia and phenylketonuria [32], from a scientific point of view, genetic modification has a meaningful use that cannot be achieved by IVF/PGD.

For a start, it is not always possible to obtain mutation free embryos from prospective parents. As noted by G. Cavaliere, "*in cases when one of the prospective parents is homozygous for a dominant genetic disorder, the risk of transmission to offspring is as high as 100%.*" [33] Moreover, even in cases where IVF and PGD may prevent a disease in a couple's offspring, such as Huntington's disease, human germline modification "*may have the advantage of preventing disease in subsequent generations as well. In the case of autosomal recessive disorders, children who are born as the result of PGD, are often carriers of the condition their parents selected against*" [32]. Thus, in some cases, PGD is not useful at all. Gene editing, instead, can perfectly amend the situation by changing the affected gene in the in-vitro embryo and in this way cure the genetic disorder, including for future generations.

Moreover, if genetic modification tools become sufficiently safe to be used in clinical contexts, they will allow not only to prevent the transmission of pathologies as such but also to improve our ability to prevent their appearance in the first place. For example, germline modification can alter BRCAs genes to obtain expressions that reduce the probability of developing cancer. PGD could never do this in practice [34].

To this must be added that PGD has ethical problems of its own. PGD has always been criticized for having eugenic purposes since it selects embryos to be implanted, on the basis of their genetic characteristics to reach perfection [35]. Moreover, it does not actually cure any pathology, but only ensures that human beings free from certain pathologies are born. The nature of gene editing is very different, since it modifies the genes themselves. Therefore, it has curative or perfective purposes but not eugenic ones. The fact that gene editing cures diseases before they evolve generates much fewer ethical problems. To this should be added that gene editing will make it possible to reduce the number of embryos, that are now generated by artificial fertilization, in order to guarantee the success of this reproductive procedure. This, in turn, implies less destruction of embryonic lives. This is a consideration that diminishes the controversy surrounding the genesis of the human being [36]. In sum, it is simply not true that gene editing is futile given other technologies, above all, PGD/IVF.

Respect for human dignity

According to article 1 of the UNESCO Declaration on the Human Genome and Human Rights: "*The human genome underlies the fundamental unity of all members of the human family, as well as the recognition of their inherent dignity and diversity. In a symbolic sense, it is the heritage of humanity.*" Moreover, article 24 of the same Declaration indicates that germline interventions are contrary to human dignity. According to this provision: "*The International Bioethics Committee of UNESCO should contribute to the dissemination of the principles set out in this Declaration and to the further examination of issues raised by their applications and by the evolution of the technologies in question (...) It should make recommendations, in accordance with UNESCO's statutory procedures, addressed to the General Conference and give advice concerning the follow-up of this Declaration, in particular regarding the identification of practices that could be contrary to human dignity, such as germ-line interventions.*"

This UNESCO Declaration, thus holds that we are morally obliged to prevent human germline modifications since man-made changes in the human genome—seen as the symbolic heritage of humankind—are acts that would be contrary to human dignity. However, in our view, there are important reasons for doubting this claim. For a start, human dignity is a deeply contested concept, that is not helpful to sort out what is really at stake [33]. More importantly, scientifically speaking, the human genome is not a stable entity but in continuous change. Considering this reality, the very idea of keeping the current human genome from changes is questionable for the simple reason that it is not possible.

What if the drafters of the Declaration really meant that we must protect the integrity of the human genome from clinical interventions, such as gene editing, but allowing random changes to occur? In our view, this claim is even more problematic. Why would variants introduced by nature in a random way be morally acceptable, while the variants that result from human ingenuity and directed mutations ethically reprehensible? [37] This position is difficult to sustain unless one assumes that God is behind nature, or that nature is itself a kind of self-conscious or intelligent being, whose works we must not interfere with. Indeed, it builds on genetic essentialism, which is a deeply controversial doctrine, that is certainly not shared by everyone [38].

Of course, we may choose to entertain an extremely pessimistic view, of the possible consequences of human interventions into nature [28], a view so pessimistic that all of us would prefer to ban all human interventions altering the human genome in order to prevent an apocalyptic ending. However, is it reasonable to extend our fears up to a point, that we will always consider it better and safer to ban and sanction germline modification, regardless of the benefits that it can bring? Is it not a moral wrong to ignore the circumstances, of people born with genetic predispositions to developing pathological conditions? Is it possible that the deliberate human refusal to intervene in nature, is equally or even more morally reprehensible if the intervention would prevent people from being born with a genetic disease, or high propensities to develop one? [39,40].

Respect for the autonomy of future generations

Decisions must often be taken by parents without taking the autonomy of their offspring into account [41]. Since these are interventions on human embryos or gametes, the decision on whether or not to perform them implies that the future parents will decide who they like their offspring to be, without giving them the possibility of even having a say about it.

However, this argument is fairly weak, since people of one generation always take decisions affecting future generations. For instance, when people decide to procreate, they do not necessarily take the opinion of those who will be born as a result. Neither do parents ask their children, if they want to be vaccinated or not, or if they want to undergo medical interventions that pose a risk to their lives, but which seem to be the better option to preserve or restore their health. These decisions may seem contrary to the autonomy of future generations if we apply the general requirement of free and informed consent, as described by Article 6 of the Universal Declaration on Bioethics and Human Rights, adopted at the General Conference of UNESCO on 19 October 2005, "Any preventive, diagnostic and therapeutic medical intervention is only to be carried out, with the prior, free and informed consent of the person concerned, based on adequate information. The consent should, where appropriate, be express, and may be withdrawn by the person concerned at any time, and for any reason without disadvantage or prejudice".

Nevertheless, in our view, they are not inherently immoral decisions if the motives behind the decisions have sufficient weight. Free and informed consent is not an absolute rule. For example, the UNESCO Declaration on Bioethics and Human Rights acknowledges limits for persons without the capacity to consent. Indeed, its Article 7, paragraph (a) mentions the 'best interest of the child', and according to paragraph seven of the same provision, "research should only be carried out for his or her direct health benefit, subject to the authorization and the protective conditions prescribed by law, and if there is no research alternative of comparable effectiveness, with research participants able to consent."

From this perspective, the key question has to do with the reasons for authorizing a clinical application where this requirement does not hold full force. For example, we accept authorizations, when it is about a child whose health we hope to preserve or restore, even if it is at the expense of his or her personal autonomy, and when it is not possible or reasonable to postpone the decisions [42,43]. Why do we have good reasons for overriding free consent in these cases, but not when it comes to gene editing if it is to the direct health benefit of future generations?

Some may respond that these cases raise very different moral dilemmas since gene editing has much graver consequences than other kinds of interventions. Especially significant is that gene editing, is said to, even alter the identity of the subject, whose germline has been modified. In fact, this could be the most salient concern behind the adoption of legal prohibitions, such as Articl6 90 of the EU regulation on clinical trials [44], according to which "*no gene therapy clinical trials may be carried out, which result in modifications to the subject's germ line genetic identity*" [45].

Nevertheless, this argument is not as convincing as it might seem at first sight. It is difficult to demonstrate that human germline modification, actually changes the identity of the subject, partly because it is not entirely clear what a change of identity means. It is doubtful if a person's identity always can or should be preserved at all costs. For example, certain diseases, such as Alzheimer, can change the identity of people suffering from it without a choice. However, we would hardly seek to preserve the identity of a person with Alzheimer because we believe it is good.

Other conditions, such as deafness or blindness from birth, become part of the identities of those persons who have them. Even so, would we really believe that parents are immoral if they were to decide to modify genetically their offspring to prevent them from suffering any of these conditions? [46].

A slippery slope to eugenics

Even if those who defend this argument in principle does not object to gene editing for therapeutic purposes, they believe it is necessary to ban all gene editing, to prevent eugenics [47]. From their standpoint, the problem is not that human germline modification is morally wrong or really bad. Instead, what they are against is eugenics. However, in their view, it is not possible to permit human germline modification, to prevent or treat diseases, while prohibiting usages for other purposes. Permitting gene-editing tools to prevent and treat diseases, runs the risk of opening the door to eugenic usages with all what that implies, from parents using the tools to make their children smarter, or nations using these tools to create a more powerful military capacity.

This is the controversial slippery slope argument [48]. The ban on genetic enhancement will collapse, due to the strong incentives of nations and parents to defect from it, in order to compete with other nations and parents. So, in order to prevent this from happening, we must ban all applications of gene editing tools, even if it means not being able to use them for therapeutic purposes [49].

As worrying as this development might seem, there are good reasons for taking a cautious stance toward the merits of the slippery slope [50,51]. First of all, the argument is built on a problematic understanding of causality. It assumes that a given action A, in this case, permitting the germline modification for therapeutic purposes, *necessarily* leads to the final result X, i.e. gene editing for eugenic purposes. From this perspective, it is not possible to establish any intermediate points to prevent X from occurring, for example, by imposing a ban on using genetic tools for eugenics.

A somewhat different problem with the slippery slope argument is that it is based on a distinction between interventions to treat or cure a disease, and interventions for the sake of human enhancement. However, this distinction is not clear and can be difficult to draw in practice [52,53]. For example, the attempt to improve the germline of a human being so that he or she has more possibilities to tackle AIDS, eugenics? Is it eugenics to modify genes, such as the BRCAs, to reduce the risk of developing cancer? In principle, we may respond affirmatively in both cases, since the purpose is not to treat diseases, but to prevent them from developing in the first place, thus, eugenics. However, if this is true, how then should Article 13 of the Oviedo Convention [54], adopted in 1997, be interpreted? This provision only allows an intervention that seeks to modify the human genome for preventive, diagnostic, or therapeutic purposes (thus, not for eugenic purposes), as long as it does not alter the human germline. But how can we distinguish between an intervention made for preventive purposes and eugenics, if we consider that, for example, altering the BRCAs genes is eugenics?

Stretching our imagination of possible scenarios a bit further, let us consider the case where the use of alcohol generates considerable proneness to aggressiveness because of a specific genetic configuration. Would an intervention aimed at changing the genetic propensity to avoid this proneness, be done for therapeutic or eugenic purposes? [55].

Finally, even if the modification techniques were to be used by parents to enhance the genetic makeup of their offspring, it is not entirely clear why these kinds of applications actually would be morally objectionable [56,57]. According to the EU Charter of Fundamental Rights, "eugenic practices, in particular those aiming at the selection of persons" is prohibited [58]. However, the Charter does not clarify what "eugenic practices" refer to. The explanatory notes to the Charter indicate that the scope of the prohibition is very limited, and essentially refers to practices that qualify as a grave crime under international criminal law [59], such as the organization and implementation of selection programs for sterilization, forced pregnancy, and compulsory ethnic marriage, among others [60]. To sum up, in our view, the slippery slope argument is disputed and seems quite weak.

Conclusion

In our view, there are important reasons for promoting continued scientific progress of gene editing, above all, because of its potential to erasing genetic disorders, or even genetic predispositions to developing grave diseases of future generations. There is an urgent need for a public debate on the ethics of gene editing, which is not determined beforehand by those who believe that this technology is just wrong or really bad. It should also seriously take what is good or even right about gene editing. If nothing else, we owe it to future generations that will be born with genetic disorders that could have been avoided.

Acknowledgments

Iñigo de Miguel Beriain would like to acknowledge the support by the Basque Government, Grant supporting Research Groups IT-1066-16.

References

[1] Li S, Yang Y, Qiu Y, Chen Y, Xu L, Ding Q. Applications of genome editing tools in precision medicine research. Hereditas (Beijing) 2017;39:177–88. https://doi.org/10.16288/j.yczz.16-395.

[2] Carroll D, Charo RA. The societal opportunities and challenges of genome editing. Genome Biol. 2015;16:242. En, http://doi.org/10.1186/s13059-015-0812-0.

[3] Bellver Capella V. La revolución de la edición genética mediante CRISPR-CAS9 y los desafíos éticos y regulatorios que comporta. Cuadernos de Bioética 2016;27. at: http://aebioetica.org/revistas/2016/27/90/223.pdf.

[4] Doudna JA, Sternberg SH. A crack in creation: gene editing and the unthinkable power to control evolution. Boston: Houghton Mifflin Harcourt; 2017.

[5] Lander ES. The heroes of CRISPR. Cell January 14, 2016;164:18–28.

[6] Mojica FJM, Juez G, Rodríguez-Valera F. Transcription at different salinities of Haloferax mediterranei sequences adjacent to partially modified PstI sites. Mol. Microbiol. 1993;9:613–21.

[7] Mojica FJM, Díez-Villaseñor C, García-Martínez J, Soria E. Intervening sequences of regularly spaced prokaryotic repeats derive from foreign genetic elements. J. Mol. Evol. 2005;60:174–82.

[8] (a) Barrangou R. RNA-mediated programmable DNA cleavage. Nat. Biotechnol. 2012;30:836–8.(b) Carroll D. A CRISPR approach to gene targeting. Mol. Ther. 2012;20:1658–60.

[9] Jinek M, Chylinski K, Hauer M, Douadn JA, Charpentier E. A programable dual-RNA guided ADN endonuclease in adaptative bacterial inmunity. Science 2012;337(6096):816–21.

[10] Cong L, Ran FA, Cox D, Lin S, Barretto R, Habib N, Hsu PD, Wu X, Jiang W, Marraffini LA, Zhang F. Multiplex genome engineering using CRISPR/Cas systems. Science 2013;339:819–23.
[11] Callaway E. Gene-editing research in human embryos gains momentum. Experiments are now approved in Sweden, China and the United Kingdom. Nature April 19, 2016;532(7599). https://www.nature.com/news/gene-editing-research-in-human-embryos-gains-momentum-1.19767.
[12] Cyranoski D. Japan set to allow gene editing in embryos. Nature 2018. at: https://www.nature.com/articles/d41586-018-06847-7.
[13] Kaiser J. New gene-editing treatment might help treat a rare disorder, hints first human test. Sci. Mag. September 5, 2018 at: https://www.sciencemag.org/news/2018/09/new-gene-editing-treatment-might-help-treat-rare-disorder-hints-first-human-test.
[14] Ledford H. First test of in-body gene editing shows promise. Nat. Biotechnol. September 05, 2018 at: https://www.nature.com/articles/d41586-018-06195-6.
[15] Sheridan C. Go-ahead for first in-body CRISPR medicine testing. Nat. Biotechnol. December 14, 2018 https://www.nature.com/articles/d41587-018-00003-2.
[16] Cyranoski D, Ledford H. Genome-edited baby claim provokes international outcry. Nature November 26, 2018. at: https://www.nature.com/articles/d41586-018-07545-0.
[17] Hanock T. Chinese scientist who edited genes likely to face criminal charges. Financial Times; January 21, 2019. at: https://www.ft.com/content/8fa09ff8-1d66-11e9-b126-46fc3ad87c65.
[18] Organizing Committee of the Second International Summit on Human Genome Editing. Statement on human genome editing II. November 29, 2018. at: http://www8.nationalacademies.org/onpinews/newsitem.aspx?RecordID=11282018b.
[19] Baltimore D, et al. A prudent path forward for genomic engineering and germline gene modification. Science 2015;348(6230):36–8.
[20] Lanphier E, Urnov F, Haecker SE, Werner M, Smolenski J. Don't edit the human germ line. Nature 2015;519:410–1.
[21] Li X-J, et al. CRISPR: established editor of human embryos? Cell Stem Cell 2017;21(3):295–6. Available at: https://www.ncbi.nlm.nih.gov/pmc/articles/PMC5819596/.
[22] Charo RA. Human germline engineering and human rights. AJIL Unbound 2018;112:344–9. 345-346. At: https://www.cambridge.org/core/services/aop-cambridge-core/content/view/E9C98F9BE3054B32BC590EDEA988502A/S2398772318000880a.pdf/div-class-title-germline-engineering-and-human-rights-div.pdf.
[23] Nuffield Council. Genome editing and human reproduction. London: Nuffield Council on Bioethics; 2018. Available at. http://nuffieldbioethics.org/wp-content/uploads/Genome-editing-and-human-reproduction-FINAL-website.pdf.
[24] The National Academy of Sciences, Engineering and Medicine. Human genome editing. Science, ethics and governance. Washington D.C.: The National Academies Press; 2017. p. 111–2.
[25] Organizing Committee of the Second International Summit on Human Genome Editing. Statement on human genome editing II. November 29, 2018.
[26] International Covenant on Civil and Political Rights. Adopted and opened for signature, ratification and accession by General Assembly resolution 2200A (XXI) of 16 December 1966 entry into force 23 March 1976. As of 29 January 2019, this treaty has been ratified by 172 states.
[27] This conclusion is in line, for example, with explanatory report to the convention for the protection of human rights and dignity of the human being with regard to the application of biology and medicine: convention on human rights and biomedicine, adopted on 4 April 1997, para. 92.
[28] Constam D. Comment on "human dignity and gene editing". EMBO Rep. 2018;20:e47220.
[29] Lander ES. What we don't know, international summit on gene editing. Washington: Commissioned papers; 2015. December 1–3: 20-27. At: http://nationalacademies.org/cs/groups/pgasite/documents/webpage/pga_170455.pdf.
[30] Darnosky M. Executive Director of the Center for Genetics and Society, BBC News, "US 'will not fund research for modifying embryo DNA'". 2015. At: http://www.bbc.co.uk/news/health-32530334.
[31] De Miguel Beriain I, Marcos del Cano AM. Gene editing in human embryos. A comment on the ethical issues involved. In: Soniewicka M, editor. The ethics of reproductive genetics. Between utility, principles, and virtues. Springer; 2018. p. 173–87.
[32] Gyngell C, Douglas T, Savulescu J. The ethics of germline gene editing. J. Appl. Philos. 2017;34(4):498–513. 499.
[33] Cavaliere G. Genome editing and assisted reproduction: curing embryos, society or prospective parents? Med. Health Care Philos. 2017;21(2):215–25.
[34] Soniewicka M. Failures of imagination: disability and the ethics of selective reproduction. Bioethics 2015;29(8):557–63.
[35] Sandel M. The case against perfection. Cambridge: The Belknap Press of the University of Harvard Press; 2007. 50–7.
[36] Savulescu J, Pugh J, Douglas T, Gyngell C. The moral imperative to continue gene editing research on human embryos. Protein Cell 2015;6(7):476–9.
[37] De Miguel Beriain I. Should human germ line editing be allowed? Some suggestions on the basis of the existing regulatory framework. Bioethics 2018;8(00):1–7. https://doi.org/10.1111/bioe.12492.
[38] Morar N. An empirically informed critique of Habermas' argument from human nature. Sci. Eng. Ethics 2015;21(1):95–113.
[39] De Miguel Beriain I. Human dignity and gene editing: using human dignity as an argument against modifying the human genome and germline is a logical fallacy. EMBO Rep. 2018;19:e46789.
[40] De Miguel Beriain I. Response by the author. EMBO Rep. 2018:e47346. https://doi.org/10.15252/embr.201847346|. Published online 13.12.2018.
[41] Collins F. Statement on NIH funding of research using gene-editing technologies in human embryos. 2015. at: http://www.nih.gov/about/director/04292015_statement_gene_editing_tech- nologies.htm.
[42] Mullin A. Children, parents, and responsibility for children's health. In: Arras JD, Fenton E, Kukla R, editors. The Routledge companion to bioethics, part VI. Reproduction. New York: Routledge; 2014. p. 381–92.
[43] Savulescu J. Bioethics: why philosophy is essential for progress. J. Med. Ethics 2015;41:28–33. at 30.
[44] Regulation (EU) No 536/2014 of the European Parliament and of the Council of 16 April 2014 on clinical trials on medicinal products for human use, and repealing Directive 2001/20/EC.
[45] De Miguel Beriain I. Legal issues regarding gene editing at the beginning of life: an EU perspective. Regen. Med. 2017;12(6):669–79.

[46] Savulescu J. Education and debate: deaf lesbians, "designer disability," and the future of medicine. BMJ 2002;325(7367):771–3.
[47] Knoepfler P. A better baby with gene editing? GMO sapiens: the life-changing science of designer babies. Singapore: World Scientific; 2016.
[48] Holtug N. Human gene therapy: down the slippery slope. Bioethics 1993;7(3):402–19.
[49] Gardner W. Can human genetic enhancement be prohibited? J. Med. Philos. 1995;20:65–84.
[50] Walton D. The slippery slope argument in the ethical debate on genetic engineering of humans. Sci. Eng. Ethics 2016;23(6):1507–28.
[51] Burgess JA. The great slippery-slope argument. J. Med. Ethics 1993;19(3):169–74.
[52] Macintosh KL. Enhanced beings. Human germline modification and the law. New York: Cambridge University Press, esp; 2018. p. 11–29.
[53] Anderson WF. Human gene therapy: why draw a line? J. Med. Philos. 1989;14:681–93.
[54] Convention for the protection of human rights and dignity of the human being with regard to the application of biology and medicine: convention on human rights and biomedicine, opened for signature on 4 April 1997, ETS No. 164 (referred to as the Oviedo convention).
[55] Caplan AL, McGee G, Magnus D. What is immoral about eugenics? BMJ 1999;319(7220):1284–5.
[56] Caplan AL, McGee G, Magnus D. What is immoral about eugenics? West. J. Med. 1999;171(5–6):335–7.
[57] Douglas T. The harms of enhancement and the conclusive reasons. Camb. Q. Healthc. Ethics 2015;24(1):23–36. En http://doi.org/10.1017/S0963180114000218.
[58] Michalowski S. Article 3. In: Peers S, Hervey T, Kenner J, Ward A, editors. EU charter of fundamental rights: a commentary. Oxford: Hart Publishers; 2014. p. 39–102.
[59] Rome statute of the international criminal court, 2187 UNTS 90, entered into force. 1 July 2002.
[60] Explanations relating to the charter of fundamental rights. 2007. OJ C 303/2, 14.12.2007, p. 17–35, at p. 18. At: https://eur-lex.europa.eu/LexUriServ/LexUriServ.do?uri=OJ:C:2007:303:0017:0035:en:PDF.

Further reading

[1] Resnik D. Debunking the slippery slope argument against human germ-line gene therapy. J. Med. Philos. 1994;19(1):23–40.

Chapter 52

Regulatory issues for artificial intelligence in radiology

Filippo Pesapane[1], Matteo B. Suter[2], Marina Codari[3], Francesca Patella[4], Caterina Volonté[5] and Francesco Sardanelli[6,7]

[1]*Postgraduate School in Radiodiagnostics, Università degli Studi di Milano, Milan, Italy;* [2]*Medical Oncology Unit, ASST Sette Laghi, Varese, Italy;* [3]*Department of Electronics, Informatics and Bioengineering, Politecnico di Milano, Milan, Italy;* [4]*Radiology Unit, ASST Santi Paolo e Carlo, Milan, Italy;* [5]*Independent Researcher, London, United Kingdom;* [6]*Unit of Radiology, IRCCS Policlinico San Donato, San Donato Milanese, Italy;* [7]*Department of Biomedical Sciences for Health, Università degli Studi di Milano, San Donato Milanese, Italy*

Artificial intelligence (AI), machine learning (ML) and deep learning (DL)

AI is expected to have a massive impact on the radiologist's daily life: awareness of this trend is a necessity, especially for young generations who will face this revolution [1,2]. Data protection issues, cybersecurity implications, and accountability and responsibility controversies will also be discussed, and the main differences between the policy of the European Union (EU) and that of the United States (US) will be considered.

ML indicates a subfield of AI that allows computers to learn without being explicitly program. It has been extensively applied to medical imaging [3,4]. Among the techniques that fall under the ML umbrella, DL has emerged as one of the most promising [2]. Indeed, AI includes ML that also includes DL (Fig. 52.1). DL methods belong to representation-learning methods with multiple levels of representation, which process raw data to perform classification or detection tasks [5]. The DL approach incorporates computational models and algorithms, such as artificial neural networks (ANN), which are structured in layers composed of interconnected nodes: each node performs a weighted sum of the input data, that is subsequently passed to an activating function [6].

There are three different kind of layers: the input layer, which receives input data; the output layer, which produce the results of data processing; and the hidden layer(s), which extract the patterns within data. A deep ANN differs from the single-hidden layer for its large number of hidden layers, which characterize the depth of the network [7]. Among the different deep ANNs, convolutional neural networks (CNNs) are used to obtain feature maps, where the intensities of each pixel/voxel are calculated, as the sum of each pixel/voxel of the original image to its neighbors, weighted by convolution matrixes. The architecture of deep CNNs allows the composing of complex features (such as shapes), from simpler features (e.g., image intensities), to decode raw image data without the need of detecting specific features [8] (Fig. 52.2).

Radiologists are already familiar with computer-aided detection/diagnosis (CAD) systems, first introduced in the sixties for applications to chest X-ray and mammography [4]. However, the advancement in algorithm development, combined with the ease of access to computational resources, currently allows AI to be applied in radiological decision-making, with a higher functional level [6].

An increasing partnership between radiologists and computer scientists is strategically requested [9]. Creating this kind of multidisciplinary "AI team" will help to ensure that patient safety standards are met, and cause judicial transparency, which will allow legal liability to be assigned to the radiologist component human authority [10].

Machine learning in imaging procedures for improved workflow and communication

AI may find multiple applications, from image acquisition and processing to aided reporting, follow-up planning, data storage, data mining, and many others. After being trained using the vast number of examinations and images, AI algorithms may look at medical images, to identify patterns giving information about abnormal findings [11,12]. This is crucial because not all abnormalities are representative of

Precision Medicine for Investigators, Practitioners and Providers. https://doi.org/10.1016/B978-0-12-819178-1.00052-6

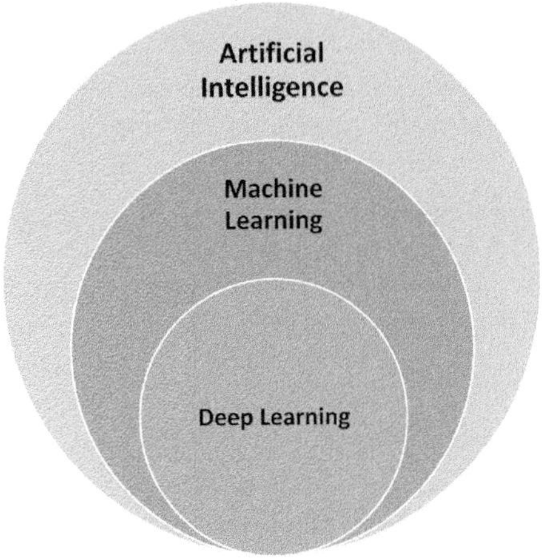

FIGURE 52.1 Deep learning as a subset of machine learning methods, which represent a branch of the existing artificial intelligence techniques. *Reprinted from Pesapane F, Codari M, Sardanelli F. Artificial intelligence in medical imaging: threat or opportunity? Radiologists again at the forefront of innovation in medicine. Eur. Radiol. Exp. 2018;2(1):35. Reprinted with permission.*

disease, and must be actioned. AI systems are educated on a case-by-case base. However, unlike CAD systems, which just highlight the presence or absence of image features known to be associated with a disease state [13,14], AI looks at specific labeled structures and also learns how to extract image features, either visible or hidden to the human eye. This approach mimics human analytical cognition, allowing them to get better performances than those obtained with CAD software [15].

Until a few years ago, the clinical application to ML on medical imaging in terms of detection and characterization produced limited results, such as differentiation of normal from abnormal chest x-ray examination [16–18] or mammograms [19,20]. The AI application to advanced imaging modalities, such as CT and MRI, is now in the relatively early phase. Examples of promising results are differentiation of malignant from benign chest nodules on CT scans [21], diagnosis of neurologic and psychiatric diseases [22,23], and identification of biomarkers in glioblastoma [24]. Interestingly, MRI was shown to predict survival in women with cervical cancer [25,26] and in patients with amyotrophic lateral sclerosis [27].

To summarize, the following tasks could be accomplished by AI with an immediate positive impact [28]:

- *prioritization of reporting*: automatic selection of findings deserving a faster action;
- *comparison of current with previous examinations, especially in oncologic follow-up* (tens of minutes are needed for this currently: AI could do this for us. We will supervise the process, extracting data to be

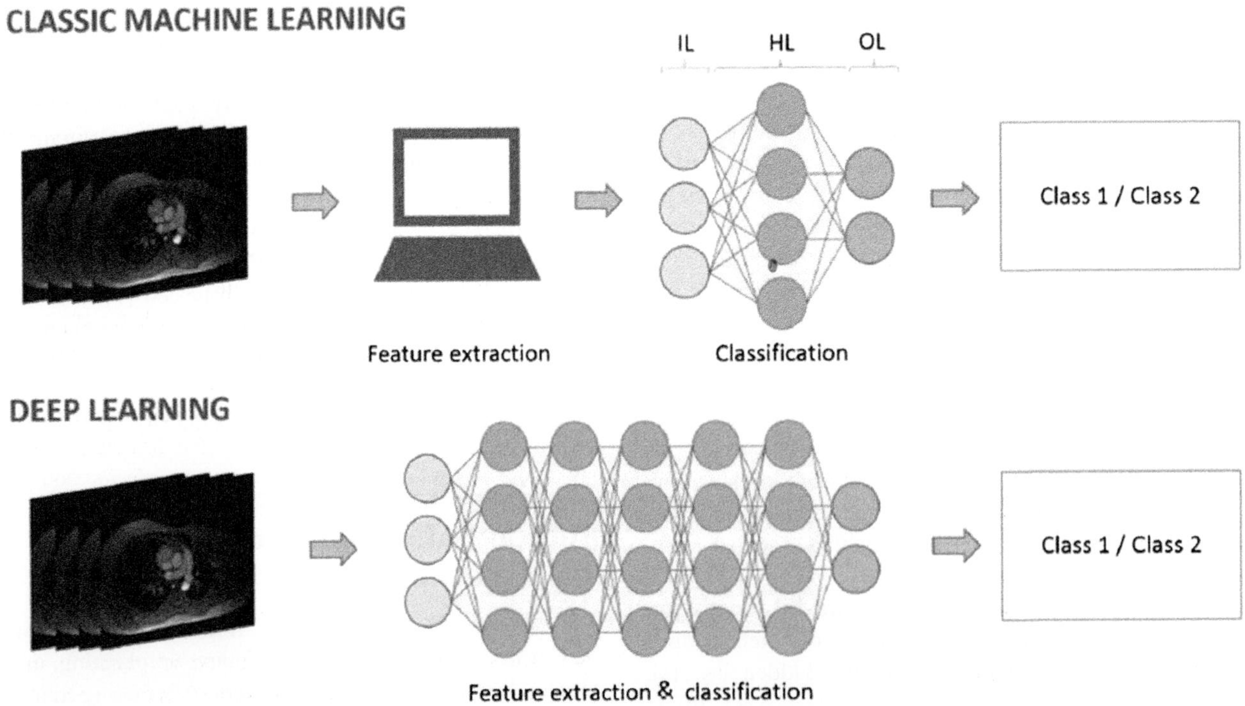

FIGURE 52.2 Comparison between classic machine learning and deep learning approaches applied to a classification task. Both depicted approaches use an artificial neural network organized in different layers (IL input layer, HL hidden layer, OL output layer). The deep learning approach avoids the design of dedicated feature extractors by using a deep neural network that represents complex features as a composition of simpler ones. *Reprinted from Pesapane F, Codari M, Sardanelli F. Artificial intelligence in medical imaging: threat or opportunity? Radiologists again at the forefront of innovation in medicine. Eur. Radiol. Exp. 2018;2(1):35 Reprinted with permission.*

integrated into the report, and drawing the conclusions considering the clinical context and therapy regimens. AI could also take into account the time interval between examinations.

- *quick identification of negative studies*: at least in this first phase, AI will favor sensitivity and negative predictive value, over specificity and positive predictive value, finding the normal studies, and leaving abnormal ones for radiologists [36]. This would be useful in high volume screening settings. This concept of *quick negative* should also represent valuable help for screening programs in underserved countries [10,37].
- *aggregation of electronic medical records*: allowing radiologists to access clinical information, to adapt the protocol or to interpret the exam in full clinical context;
- *automatic recall and rescheduling of patients*: for findings deserving an imaging follow-up;
- *immediate use of clinical decision support systems*: for ordering, interpreting, and defining further patient management;
- *internal peer-review of reports*;
- *tracking of resident training*;
- *quality control of technologists' performances, and tracked communication between radiologists and technologists*;
- *data mining regarding relevant issues*: including radiation dose [29].
- *anticipation of the diagnosis of cancerous lesions in oncologic patients*: using texture analysis and other advanced approaches [30];
- *prediction of treatment response to therapies for tumors*: such as intra-arterial treatment for hepatocellular carcinoma [31].
- *evaluation of the biological relevance of borderline cases*: such as B3 lesions diagnosed at pathology, of needle biopsy of breast imaging findings [32];
- *estimation of functional parameters*: such as the fractional flow reserve from CT coronary angiography using DL [33];
- *detection of perfusion defects and ischemia:* for example in the case of myocardial stress perfusion defects, and induced ischemia [34];
- *segmentation and shape modeling*: such as brain tumor segmentation [35] or, more in general, brain structure segmentation [36];
- *reducing diffusion MRI data processing to a single optimized step*: for example, making microstructure prediction on a voxel-by-voxel basis, as well as automated model-free segmentation from MRI values [37].

Radiomics

Radiomics deals with the high-throughput extraction of peculiar quantitative features from radiological images [38–40]. Data derived from radiomic investigation, such as intensity, shape, texture and wavelength, can be retrieved from medical images [39,41–44] through ML processes. In this way, valuable information for predicting treatment response, differentiating benign and malignant tumours, and assessing cancer genetics in many cancer types van be provided [39,45–47]. Because of the rapid growth of this area, numerous published radiomics investigations lack standardized evaluation, of both the scientific integrity and the clinical relevance [38]. Independent validation datasets are still needed to confirm the diagnostic and prognostic value of radiomic features, yet radiomics has shown several promising applications for precision medicine [38,39].

Radiogenomics

Radiogenomics investigates the relationship between imaging characteristics of a disease (namely, the imaging-phenotype or radiologic phenotype), and its gene expression patterns, gene mutations, and other genome-related features with reference with either the patient genotype or, in the case of tumours, of lesion genotype [48,49]. Actually, radiogenomics is an ambiguous term, as it used to refer to the study of genetic variation, such as single nucleotide polymorphisms, along with a cancer patient's risk of developing toxicity following radiation therapy [50], or to predicting tumor response to radiation therapy [51,52]. The term was only later extended to the relationship between imaging phenotype and patient or lesion genotype. In fact, the last decade, the addition of genomic data to radiology allowed to establish new correlations between cellular genomics and medical imaging, delivering a new branch of imaging named imaging genomics or, more simply, radiogenomics.

By using innovative technology, radiogenomics aims to develop imaging biomarkers that can predict risks and patient outcomes, allowing a better stratification of the patients for more precise management [53]. Examples can be found in breast cancer [54], glioblastoma [55], low-grade glioma [56], and kidney cancer [57]. However, radiogenomics still needs time before playing a significant practical role in cancer care, due to the limitations of the available big-data that, nowadays, often lacks complete characterization of the patients and poor integration of individual datasets [58].

Ethical risks

No governments have legislated about radiogenomics, despite its applications being ready to change several national healthcare systems around the world [58]. One relevant example is the insurance system that might discriminate patients with medical conditions, who are determined to be predominantly genetic and not lifestyle-related [59].

Quantitative imaging as a biomarker

Radiological images data can be the subject of mathematical analysis, in order to extract from voxel properties quantitative functional biomarkers, which may be a surrogate for histology, susceptibility to specific therapies or outcomes [44,45,60,61]. Important functional biomarkers are those that derive from perfusion and diffusion imaging.

Perfusion imaging

Perfusion imaging, either CT- or MRI-based, mostly performed before/after contrast agent (CA) intravenous injection, enables a spatially resolved quantification of the hemodynamic status of a tissue [60]. The first assumption is that CA administration, induces time-dependent changes in tissue signal, that can be monitored by the dynamic acquisition of images before, during, and after CA injection. The second postulation is that pathological tissues have different hemodynamic properties from normal tissues that can be depicted by perfusion biomarkers [60].

There are two methods to analyze perfusion data. One consists of the bare assessment of time-dependent changes, in signal intensity on dynamic scans, without employing any pharmacokinetic model [62]. The resulting biomarkers (such as time to peak or maximum relative enhancement) are called *semi-quantitative*, since they do not reflect the tissue biology consistently, and most of all, they cannot be reliably compared between patients [62]. Despite the lack of reproducibility, this method meets a large consent in everyday practice, because of the relative simplicity to compute semi-quantitative biomarkers [62].

Hemodynamic parameters can also be extracted using the principles of tracer-kinetic theory [60]. This method provides quantitative biomarkers (such as K trans, Vp, Kep), that technically benefit from greater precision and reproducibility than semi-quantitative parameters, and thus, can be used during clinical trials to assess tissue physiology [60,61].

Diffusion imaging

Introduced for the first time by Le Bihan [63], diffusion-weighted imaging (DWI) is an MRI technique, which uses the quantitative estimation of random motion of water molecules (the so-called *Brownian motion or diffusion*), as a surrogate for tissue characterization. According to this theory, water diffusion in biological tissues depends on the presence of impediments, such as cell membranes, fibers, or electric charges at the proteins or cell membrane surfaces. The higher the impediments, the lower the water displacements [64]. Thus, the restriction of water diffusion can be the result of several pathological conditions leading to an increase of obstacles, such as cellular proliferation in tumors, or cell swelling during cytotoxic edema after brain ischemia. Therefore, DWI sequences can be successfully employed in a number of fields, such as oncology or neurology, to detect pathological structural changes [63].

According to the acquisition protocol and the post-processing algorithm, DWI sequences can provide diverse biomarkers. The apparent diffusion coefficient (ADC) is a biomarker, obtained by a mono-exponential approach, whose values directly reflects the freedom of water displacements: the higher the ADC, the higher the diffusion [65].

ADC depicts water diffusion just in interstitial spaces. However, Brownian motion is also present in the vascular compartment [65]. Thus, in 1986 Le Bihan proposed a further algorithm, the intravoxel incoherent motion (IVIM) that, exploiting a biexponential approach (which means multiple DWI scans using different diffusion sensitizing impulses), quantifies the water motion both in the interstitial space, by an ADC-alike biomarker named D, and in the capillary space, by two perfusion parameters D* and *f* [65].

The main advantage of IVIM is the possibility to capture the hemodynamic properties of a tissue, without CA administration [63]. However, although these techniques have been used for more than 2 decades, they are still hampered by measurement error and bias, due to the lack of optimal standardization of acquisition protocol and post-processing metrology [61].

Human expert and computer algorithms: synergies and conflicts

The extensive application of AI-based systems may represent a paradigm shift in radiology, moving it from the visual assessment of medical images, which can be subjective, to the quantitative, automated and computer-based detection of complex patterns within image data [66]. In the last few years, DL has become the methodology of choice for medical image processing [67,68].

The optimists see in AI technological revolution another great opportunity to enhance radiologist role in the healthcare system [2], opposed to those who predict a relatively fast replacement of radiologists by AI systems, which will soon exceed human accuracy and reduce reporting time [17].

Data quality and interpretation rationale

ML is a data-driven approach. It means that algorithm performances are highly sensitive to data quality. In this scenario, radiologists play a key role in the development of successfully AI systems. Indeed, they are the owners of the clinical knowledge, needed to select/generate high-quality data that will be used to train and validate AI applications in supervised learning approaches. However, the accurate selection and labeling of medical imaging is a time consuming and challenging task, especially, when dealing with complex diseases.

For minimizing this problem, an unsupervised learning approach has been used, giving a concrete alternative to learn discriminative features without explicit labeling [2,69]. However, there is another open challenge that radiologists will need to face, namely, to work with black-box systems. Despite their promising performances, it is not easy to understand what drives the AI systems to its conclusions [70]. This lack of interpretability represents the main challenge for radiologists who cetainly want to understand how the algorithm comes to its conclusion, what features have been selected, how to detect possible failure, and most importantly, to communicate to patients the rationale behind their diagnosis or treatment planning [66]. This open issue urges the research community to develop specific tools, that will allow to safely use and trust AI systems in daily clinical practice.

Regulatory issues and policy initiatives

Many systems, e.g., the software that controls an airplane on autopilot or a fully driverless car, exert direct and physical control over objects in the human environment [71]. Other systems, including medical and radiological devices, provide sensitive services, that require training and certification [68,71–74]. These applications raise additional questions concerning the standards to which AI systems are held, and the procedures and techniques available to ensure those standards are being met [58,75].

Legal concerns

Is AI software used in healthcare, a medical device for legislation's purposes? Not all AI programs used in healthcare will be deemed so. There is a distinction between those research programs that enhance medical knowledge and those that promote changes in healthcare [76,77]. The precautionary principle approach, also called *ex-ante regulation*, finds some limits or sometimes outrightly bans certain applications, due to their potential risks. This means that these applications are never tested because of what could happen in the worst-case scenarios. The permissionless innovation approach, also called *ex-post regulation*, allows experimentation to proceed freely, and the issues that do arise are addressed as they emerge [58].

Ex-post regulation is hindered by the autonomous nature of AI systems, which evolve and constantly change according to their experiences and learning, in an unforeseeable way [15]. On the other hand, ex-ante regulation is obstructed by AI applications discreetness (their development requires little physical infrastructure), discreteness (their components can be designed by different subjects, for different purposes, and without actual coordination), diffuseness (these subjects can be dispersed geographically, and yet collaborate on the same project) and opacity (it can be difficult for outside observers, to identify and understand all the features of an AI system) [78].

A further issue for policymakers is time; companies understood the potential of ML, and they are continuously collecting new types of data to analyze and exploit [79]. In this quickly and unpredictably changing the environment, regulations need to be timely to be relevant.

Geographical and political differences

Broadly speaking, while in the United States, the technology sector prospered in a permissionless innovation policy environment, the European decision-makers adopted a different policy for this revolutionary technological branch [77]. Certainly, swifter approval of AI medical devices helps generate revenue for manufacturers, and radiologists may benefit from having more tools at their disposal. But the final goal of bringing new devices to market should be to improve prevention, diagnosis, treatment, and prognosis of diseases, with a potentially positive impact on patient outcome. Therefore, systems for approving new medical devices must provide pathways to market for important innovations, while also ensuring that patients are adequately protected [58].

The EU approach

In the EU, the definition of medical device is provided by Article 1(2) of Directive 93/42/EEC [80]: the term "medical device" is applied to any instrument or other tool, including any kind of software, intended by the manufacturer to be used for human beings, for the purpose, among others, of diagnosis, prevention, monitoring, treatment, or alleviation of disease [81].

The European regulatory regimen currently in force stems from three Directives on medical devices [80,82,83], and it requires manufacturers to ensure that the devices they produce are fit for their intended purpose. This means that they must comply with a number of essential requirements, set out by the Directives. Depending on the risk classification of the device, whether the essential requirements have been met can be assessed either by the manufacturer or by a notified body, which is an independent accredited certification organization, appointed by the competent authorities of EU Member States.

This regulatory framework has been reformed by the new Medical Devices Regulation (MDR) [84], which will apply from 26 May 2020, and the new In Vitro Diagnostic Medical Device Regulation (IVDR) [85], which will apply from 26 May 2022. Because they are Regulations, as opposed to Directives, once they apply, they do so directly, without the need for the governments of EU Member States to pass legislation to implement their scope [86].

The U.S. approach

In the United States, the regulatory approval allowing machines to do the work of trained radiologists is a major obstacle still unsolved [58]. The amount of testing and effort necessary to secure clearance from the Food and Drug Administration (FDA), for allowing machines to provide primary interpretations of imaging studies without a radiologist, would be overwhelming.

At the end of 2016, the 21st Century Cures Act [87] clarified the scope of FDA regulatory jurisdiction, over software used in healthcare, specifying that a medical device is an instrument or other tool, "intended for use in the diagnosis of disease or other conditions, or in the cure, mitigation, treatment, or prevention of disease, in man or other animals, or intended to affect the structure or any function of the body of man or other animals" [88].

Every AI system falling within this definition will be regulated by the FDA, as provided by the Federal Food, Drug, and Cosmetic Act [76]. The FDA categorizes the medical devices into three classes, according to their use and risk, and regulates them accordingly. The higher the risk, the stricter the control.

Independent diagnosis versus diagnosis decision support system

The black box nature and the rapid growth of ML applications will make it difficult for the FDA to approve all the new medical devices, that are continuously developed in a timely fashion, given the volume and the complex nature of testing and verification involved. Therefore, developers often present AI systems as aid tools for radiologists, rather than as a tool that substitutes them [89].

Data protection and cybersecurity implications

The concept of circulating enormous amounts of confidential information, in the vast number of copies, among many unregulated companies is increasingly insane and risky. Therefore, in the last decade, personal data regulation is increasing, and privacy concerns are growing [90].

However, we still need data as an integral part of the technology development, especially for AI. The current lack of well-annotated big datasets, for training AI algorithms, is a key obstacle to the wider introduction of these systems in radiology [4,67,91]. Access to big data of medical images is needed to provide training material to AI devices so that they can learn to recognize imaging abnormalities [4,92]. One of the problems is that sensitive data should either be harvested illicitly or collected from unknown sources [93], because of the lack of unique and clear regulations [94].

In the era of electronic medical records, AI complicates an already complex cybersecurity landscape [95]: the concept of confidentiality requires that a physician withholds information from the medical record, in order to truly keep it confidential [96]. Once a clinical decision based on AI is integrated into clinical care, withholding information from electronic records will become increasingly difficult, since patients whose data are not recorded, cannot benefit from AI analyses [96].

Before using government overregulation, we need to face the data protection and cybersecurity implications technologically. Data protection can no longer rely on current technologies, that allow spreading of personal data and require data sharing at a large and uncontrolled scale [97]. A possible solution could come from blockchain technology (BCT), namely an open-source software that allows the creation of a large, decentralized and secure public databases, containing ordered records arranged in a block structure [98]. Different blocks are stored digitally, in nodes, using the computers of the blockchain network members themselves, who are both users and maintainers of the entire system.

The information on all transactions, present and past, are stored in the nodes [99]. Although the best-known use of BCT is in the field of economics (i.e., cryptocurrencies), its usefulness is extending to other fields, including the health data. Particularly, BCT appeals to radiogenomics due to its emphasis on sharing, distribution, and encryption [99]. Newer BCT efforts, such as smart contracts, second-layer systems, permissioned blockchains, further the potential health care use, and there has been no hype surrounding the potential of the technology in medicine [100]. As the blocks are impossible to change, it is impossible to delete or to modify anything without leaving a trace, and this is critical in the case of sensitive data, such as medical information. Unfortunately, there is another side of the coin: at this moment, to obtain greater security, privacy is lost. The patients should accept to share their sensitive data, without a central authority to decide what is right or wrong.

Data protection in the EU

In Europe, the General Data Protection Regulation (GDPR) applied from 24 May 2018, updated the European legal framework for data protection. According to GDPR, all data processing and use should be opt-in, and consumer consent for data use should be clear. In general, it completely prohibits current data marketing, based on third-party non-opt-in personal data. GDPR is a more suitable instrument to regulate AI because it has an extended territorial scope and wider rights for data subjects. For instance, enhanced notification requirements (e.g., the personal data breaches must be notified to the supervising authority within 72 h), and rights to compensation for material or nonmaterial damage, and additional liability for data controllers and processors [101].

In addition, the EU adopted the Cybersecurity Directive [102], to set out a number of requirements for EU member states, which aim to prevent cyberattacks and keep their consequences under control. Among other things, member states are required to ensure that operators of essential services, take appropriate measures to minimize the impact of incidents, to preserve service continuity, and to ensure that supervisory authorities are notified of incidents without undue delay [102].

Data protection in the United States

In the United States, the Health Insurance Portability and Accountability Act (HIPAA), is a compliance focus for what concerns health information [76]. This Act includes elaborated rules requiring, among other things, the formulation of policies and the setup of training systems, for those who have access to sensitive data [76]. Cybersecurity is dealt with by the FDA, which requires manufacturers to report only a limited number of risks their devices present, and actions taken to minimize the vulnerability [76].

Accountability and responsibility

As soon as AI starts making autonomous decisions about diagnoses and treatments, stopping to be only a support tool, a problem arises as to whether its developer can be held accountable for the decision. The first question is: who will be sued if an AI-based device makes a mistake? While this is a difficult question, similar questions have been posed and resolved, when other technologies were introduced. Regarding AI, errors mainly appear when confounding factors are correlated with pathologic entities in the training datasets, rather than actual signs of disease. When AI programs decide, their decision is based on the collected data (this means on the quality of the collected data) and on their algorithms.

The reason why their decisions are unpredictable is twofold [78]. AI programs, even if they mimic human brain neural network, think differently [15,91,93]: There is a huge number of possibilities in every given situation, and humans are unable to process all of these and consider them in order to make a decision. We only consider what is more obvious for our brain, while AI systems can consider every potential scenario and every consideration [67,73,74,79,91,103,104]. Because of this, when faced with a decision, we do not share a common basis with AI devices.

Therefore, we are unable to predict what they will decide in a given set of circumstances. On the other hand, AI systems are designed to learn from their genuine experiences, and these are by their very nature unpredictable. Because it is not possible to foresee what experiences a system will have, it is not possible to foresee how the system will develop either. When something "goes wrong," it is worth considering whether the device itself or its designer/builder is to be considered at fault. In a broader theoretical framework, we should distinguish between data analysis and decision making [58].

Ongoing trends of investigation and research opportunities

Radiomic data can be used to infer a particular mutation or genotype, defining treatment possibilities, prognosis, or the need to perform further biopsies. In an era where clones marked by specific mutations, can prompt different therapeutic decisions (e.g., the T790M mutation in lung cancer) [105], radiogenomics data, which represent the entire tumor (or tumors), rather than just a histological sample, can be the morphological counterpart to the liquid biopsy and the cell-free DNA [106]. One should well keep in mind that cancer experiences a branch evolutionary growth, in which up to 70% of the somatic mutations are not detectable in every tumor region [107]. Inserting radiomics as a step between imaging and biopsy, could change accuracy in determining specific genotype dramatically.

Radiomics, using a morphological feature, has been able to distinguish cancer from normal tissue, and even define histological grade for certain tumors. A study from Wibimer et al. [108] was able to discriminate prostate cancer from benign prostate tissue, and even add information about aggressiveness through Gleason score [109]. A recent study determined, through radiomic and semantic features, if a meningioma was either high or low grade [110]. Another brilliant example in assessing aggressiveness is the work by Hanania et al., on pancreatic intraductal papillary mucinous neoplasm, using radiomics to stratify patients for surgical resection [111].

Cancer prognosis

A study by Coroller et al. assessed metastatic potential of lung cancer, through 35 radiomic features, combined in radiomic signatures [112]. An analogous effort was carried out by Diehn et al., for high grade central nervous system tumors [113]. Prognostic tools that avoid invasive techniques in delicate regions, such as the central nervous system, could have a great clinical impact. Radiogenomics has already proved its worth for hepatocellular carcinoma [48], identifying subtypes more sensitive to immunotherapy, and breast cancer [114], in a delicate situation, such as that of neoadjuvant chemotherapy for locally advanced disease.

Radiomic and other omics

Radiomic offers information beyond those represented in genomic signatures, as evidenced by a study by Emami-nejad et al., where an image feature-based classifier had higher discriminatory power than a genomic biomarker-based one, in predicting recurrence risk for early-stage lung cancer [115]. Radiomic features could define tumor-infiltrating lymphocytes subsets, and more broadly tumor microenvironment (also named as habitat imaging) [116].

An aspect of radiomics worth exploring is its temporal application, as an instrument of treatment monitoring and optimization. For example, a dynamic analysis could identify early responders and refractory patients, allowing a timely change in therapy. It is also worth mentioning another aspect that could greatly accelerate research: data-sharing. Open access to anonymized medical images, coupled with histology, clinical history, and genomic signatures, could greatly aid in the creation and validation of radiomic models. Programs, such as the Cancer Imaging Archive, supported by the NCI, represent an interesting step in this direction [117].

References

[1] Lakhani P, Prater AB, Hutson RK, Andriole KP, Dreyer KJ, Morey J, et al. Machine learning in radiology: applications beyond image interpretation. J. Am. Coll. Radiol. 2018;15(2):350–9.

[2] Pesapane F, Codari M, Sardanelli F. Artificial intelligence in medical imaging: threat or opportunity? Radiologists again at the forefront of innovation in medicine. Eur. Radiol. Exp. 2018;2(1):35.

[3] Samuel AL. Some studies in machine learning using the game of checkers. IBM J. Res. Dev. 1959;3:210–29.

[4] Lee JG, Jun S, Cho YW, Lee H, Kim GB, Seo JB, et al. Deep learning in medical imaging: general overview. Korean J. Radiol. 2017;18(4):570–84.

[5] LeCun Y, Bengio Y, Hinton G. Deep learning. Nature 2015;521(7553):436–44.

[6] King Jr BF. Guest editorial: discovery and artificial intelligence. AJR Am. J. Roentgenol. 2017;209(6):1189–90.

[7] Erickson BJ, Korfiatis P, Akkus Z, Kline TL. Machine learning for medical imaging. Radiographics 2017;37(2):505–15.

[8] Chartrand G, Cheng PM, Vorontsov E, Drozdzal M, Turcotte S, Pal CJ, et al. Deep learning: a primer for radiologists. Radiographics 2017;37(7):2113–31.

[9] Sardanelli F. Trends in radiology and experimental research. Eur. Radiol. Exp 2017;1(1).

[10] Liew C. The future of radiology augmented with Artificial Intelligence: a strategy for success. Eur. J. Radiol. 2018;102:152–6.

[11] Sardanelli F, Hunink MG, Gilbert FJ, Di Leo G, Krestin GP. Evidence-based radiology: why and how? Eur. Radiol. 2010;20(1):1–15.

[12] Dodd JD. Evidence-based practice in radiology: steps 3 and 4—appraise and apply diagnostic radiology literature. Radiology 2007;242(2):342–54.

[13] Azavedo E, Zackrisson S, Mejare I, Heibert Arnlind M. Is single reading with computer-aided detection (CAD) as good as double reading in mammography screening? A systematic review. BMC Med. Imaging 2012;12:22.

[14] Dheeba J, Albert Singh N, Tamil Selvi S. Computer-aided detection of breast cancer on mammograms: a swarm intelligence optimized wavelet neural network approach. J. Biomed. Inform. 2014;49:45–52.

[15] Kohli M, Prevedello LM, Filice RW, Geis JR. Implementing machine learning in radiology practice and research. AJR Am. J. Roentgenol. 2017;208(4):754–60.

[16] Chockley K, Emanuel E. The end of radiology? Three threats to the future practice of radiology. J. Am. Coll. Radiol. 2016;13(12 Pt A):1415–20.

[17] Obermeyer Z, Emanuel EJ. Predicting the future – big data, machine learning, and clinical medicine. N. Engl. J. Med. 2016;375(13):1216–9.

[18] Becker AS, Bluthgen C, Phi van VD, Sekaggya-Wiltshire C, Castelnuovo B, Kambugu A, et al. Detection of tuberculosis patterns in digital photographs of chest X-ray images using Deep Learning: feasibility study. Int. J. Tuberc. Lung Dis. 2018;22(3):328–35.

[19] Becker AS, Mueller M, Stoffel E, Marcon M, Ghafoor S, Boss A. Classification of breast cancer in ultrasound imaging using a generic deep learning analysis software: a pilot study. Br. J. Radiol. 2018;91(1083):20170576.

[20] Becker AS, Marcon M, Ghafoor S, Wurnig MC, Frauenfelder T, Boss A. Deep learning in mammography: diagnostic accuracy of a multipurpose image analysis software in the detection of breast cancer. Investig. Radiol 2017;52(7):434–40.

[21] Cheng JZ, Ni D, Chou YH, Qin J, Tiu CM, Chang YC, et al. Computer-aided diagnosis with deep learning architecture: applications to breast lesions in US images and pulmonary nodules in CT scans. Sci. Rep. 2016;6:24454.

[22] Lu X, Yang Y, Wu F, Gao M, Xu Y, Zhang Y, et al. Discriminative analysis of schizophrenia using support vector machine and recursive feature elimination on structural MRI images. Medicine (Baltim.) 2016;95(30):e3973.

[23] Li R, Zhang W, Suk HI, Wang L, Li J, Shen D, et al. Deep learning based imaging data completion for improved brain disease diagnosis. Med. Image Comput. Comput. Assist. Interv 2014;17(Pt 3):305–12.

[24] Korfiatis P, Kline TL, Coufalova L, Lachance DH, Parney IF, Carter RE, et al. MRI texture features as biomarkers to predict MGMT methylation status in glioblastomas. Med. Phys. 2016;43(6):2835–44.

[25] Matsuo K, Purushotham S, Moeini A, Li G, Machida H, Liu Y, et al. A pilot study in using deep learning to predict limited life expectancy in women with recurrent cervical cancer. Am. J. Obstet. Gynecol. 2017;217(6):703–5.

[26] Obrzut B, Kusy M, Semczuk A, Obrzut M, Kluska J. Prediction of 5-year overall survival in cervical cancer patients treated with radical hysterectomy using computational intelligence methods. BMC Cancer 2017;17(1):840.

[27] van der Burgh HK, Schmidt R, Westeneng HJ, de Reus MA, van den Berg LH, van den Heuvel MP. Deep learning predictions of survival based on MRI in amyotrophic lateral sclerosis. Neuroimage Clin. 2017;13:361–9.

[28] Nance Jr JW, Meenan C, Nagy PG. The future of the radiology information system. AJR Am. J. Roentgenol. 2013;200(5):1064–70.

[29] Chen H, Zhang Y, Zhang W, Liao P, Li K, Zhou J, et al. Low-dose CT via convolutional neural network. Biomed. Opt. Express 2017;8(2):679–94.

[30] Chaudhary K, Poirion OB, Lu L, Garmire LX. Deep learning-based multi-Omics integration robustly predicts survival in liver cancer. Clin. Cancer Res. 2018;24(6):1248–59.

[31] Abajian A, Murali N, Savic LJ, Laage-Gaupp FM, Nezami N, Duncan JS, et al. Predicting treatment response to intra-arterial therapies for hepatocellular carcinoma with the use of supervised machine learning-an artificial intelligence concept. J. Vasc. Interv. Radiol. 2018,Jun;29(6):850–857.e1.

[32] El-Sayed ME, Rakha EA, Reed J, Lee AH, Evans AJ, Ellis IO. Predictive value of needle core biopsy diagnoses of lesions of uncertain malignant potential (B3) in abnormalities detected by mammographic screening. Histopathology 2008;53(6):650–7.

[33] Itu L, Rapaka S, Passerini T, Georgescu B, Schwemmer C, Schoebinger M, et al. A machine-learning approach for computation of fractional flow reserve from coronary computed tomography. J. Appl. Physiol. 2016;121(1):42–52.

[34] Nakajima K, Okuda K, Watanabe S, Matsuo S, Kinuya S, Toth K, et al. Artificial neural network retrained to detect myocardial ischemia using a Japanese multicenter database. Ann. Nucl. Med. 2018,Jun;32(5):303–10.

[35] Pereira S, Pinto A, Alves V, Silva CA. Brain tumor segmentation using convolutional neural networks in MRI images. IEEE Trans. Med. Imaging 2016;35(5):1240–51.

[36] Moeskops P, Viergever MA, Mendrik AM, de Vries LS, Benders MJ, Isgum I. Automatic segmentation of MR brain images with a convolutional neural network. IEEE Trans. Med. Imaging 2016;35(5):1252–61.

[37] Golkov V, Dosovitskiy A, Sperl JI, Menzel MI, Czisch M, Samann P, et al. Q-space deep learning: twelve-fold shorter and model-free diffusion MRI scans. IEEE Trans. Med. Imaging 2016;35(5):1344–51.

[38] Lambin P, Leijenaar RTH, Deist TM, Peerlings J, de Jong EEC, van Timmeren J, et al. Radiomics: the bridge between medical imaging and personalized medicine. Nat. Rev. Clin. Oncol. 2017;14(12):749–62.

[39] Yip SSF, Parmar C, Kim J, Huynh E, Mak RH, Aerts H. Impact of experimental design on PET radiomics in predicting somatic mutation status. Eur. J. Radiol. 2017;97:8–15.

[40] Parekh VS, Jacobs MA. Integrated radiomic framework for breast cancer and tumor biology using advanced machine learning and multiparametric MRI. NPJ Breast Cancer 2017;3:43.

[41] Davnall F, Yip CS, Ljungqvist G, Selmi M, Ng F, Sanghera B, et al. Assessment of tumor heterogeneity: an emerging imaging tool for clinical practice? Insights Imaging 2012;3(6):573–89.

[42] Aerts HJ, Velazquez ER, Leijenaar RT, Parmar C, Grossmann P, Carvalho S, et al. Decoding tumour phenotype by noninvasive imaging using a quantitative radiomics approach. Nat. Commun. 2014;5:4006.

[43] Rahmim A, Salimpour Y, Jain S, Blinder SA, Klyuzhin IS, Smith GS, et al. Application of texture analysis to DAT SPECT imaging: relationship to clinical assessments. Neuroimage Clin 2016;12:e1–9.

[44] Pesapane F, Patella F, Fumarola EM, Panella S, Ierardi AM, Pompili GG, et al. Intravoxel incoherent motion (IVIM) diffusion weighted imaging (DWI) in the periferic prostate cancer detection and stratification. Med. Oncol. 2017;34(3):35.

[45] Patella F, Franceschelli G, Petrillo M, Sansone M, Fusco R, Pesapane F, et al. A multiparametric analysis combining DCE-MRI- and IVIM -derived parameters to improve differentiation of parotid tumors: a pilot study. Future Oncol. 2018,Dec;14(28):2893–903.

[46] King AD, Chow KK, Yu KH, Mo FK, Yeung DK, Yuan J, et al. Head and neck squamous cell carcinoma: diagnostic performance of diffusion-weighted MR imaging for the prediction of treatment response. Radiology 2013;266(2):531–8.

[47] Peng SL, Chen CF, Liu HL, Lui CC, Huang YJ, Lee TH, et al. Analysis of parametric histogram from dynamic contrast-enhanced MRI: application in evaluating brain tumor response to radiotherapy. NMR Biomed. 2013;26(4):443–50.

[48] Kuo MD, Jamshidi N. Behind the numbers: decoding molecular phenotypes with radiogenomics—guiding principles and technical considerations. Radiology 2014;270(2):320–5.

[49] Rutman AM, Kuo MD. Radiogenomics: creating a link between molecular diagnostics and diagnostic imaging. Eur. J. Radiol. 2009;70(2):232–41.

[50] Barnett GC, Coles CE, Elliott RM, Baynes C, Luccarini C, Conroy D, et al. Independent validation of genes and polymorphisms reported to be associated with radiation toxicity: a prospective analysis study. Lancet Oncol. 2012;13(1):65–77.

[51] Yard BD, Adams DJ, Chie EK, Tamayo P, Battaglia JS, Gopal P, et al. A genetic basis for the variation in the vulnerability of cancer to DNA damage. Nat. Commun. 2016;7:11428.

[52] Radiogenomics Consortium. RGC. 2009. Available from: https://epi.grants.cancer.gov/radiogenomics/.

[53] Pinker K, Chin J, Melsaether AN, Morris EA, Moy L. Precision medicine and radiogenomics in breast cancer: new approaches toward diagnosis and treatment. Radiology 2018;287(3):732–47.

[54] Mazurowski MA, Zhang J, Grimm LJ, Yoon SC, Silber JI. Radiogenomic analysis of breast cancer: luminal B molecular subtype is associated with enhancement dynamics at MR imaging. Radiology 2014;273(2):365–72.

[55] Gutman DA, Cooper LA, Hwang SN, Holder CA, Gao J, Aurora TD, et al. MR imaging predictors of molecular profile and survival: multi-institutional study of the TCGA glioblastoma data set. Radiology 2013;267(2):560–9.

[56] Mazurowski MA, Clark K, Czarnek NM, Shamsesfandabadi P, Peters KB, Saha A. Radiogenomics of lower-grade glioma: algorithmically-assessed tumor shape is associated with tumor genomic subtypes and patient outcomes in a multi-institutional study with the Cancer Genome Atlas data. J. Neuro. Oncol. 2017;133(1):27–35.

[57] Karlo CA, Di Paolo PL, Chaim J, Hakimi AA, Ostrovnaya I, Russo P, et al. Radiogenomics of clear cell renal cell carcinoma: associations between CT imaging features and mutations. Radiology 2014;270(2):464–71.

[58] Pesapane F, Volonte C, Codari M, Sardanelli F. Artificial intelligence as a medical device in radiology: ethical and regulatory issues in Europe and the United States. Insights imaging. 2018.

[59] BBC News. John Hancock adds fitness tracking to all policies. 2018. updated 20-09-2018. Available from: https://www.bbc.co.uk/news/technology-45590293?utm_source=The+Medical+Futurist+Newsletter&utm_campaign=ce21cecbb6-EMAIL_CAMPAIGN_2018_09_25&utm_medium=email&utm_term=0_efd6a3cd08-ce21cecbb6-420636970&mc_cid=ce21cecbb6&mc_eid=ed8a9cd6e8.
[60] Ingrisch M, Sourbron S. Tracer-kinetic modeling of dynamic contrast-enhanced MRI and CT: a primer. J. Pharmacokinet. Pharmacodyn. 2013;40(3):281–300.
[61] Shukla-Dave A, Obuchowski NA, Chenevert TL, Jambawalikar S, Schwartz LH, Malyarenko D, et al. Quantitative imaging biomarkers alliance (QIBA) recommendations for improved precision of DWI and DCE-MRI derived biomarkers in multicenter oncology trials. J. Magn. Reson. Imaging 2019 Jun;49(7):e101–21.
[62] Noij DP, de Jong MC, Mulders LG, Marcus JT, de Bree R, Lavini C, et al. Contrast-enhanced perfusion magnetic resonance imaging for head and neck squamous cell carcinoma: a systematic review. Oral Oncol. 2015;51(2):124–38.
[63] Le Bihan D, Breton E, Lallemand D, Grenier P, Cabanis E, Laval-Jeantet M. MR imaging of intravoxel incoherent motions: application to diffusion and perfusion in neurologic disorders. Radiology 1986;161(2):401–7.
[64] Padhani AR, Liu G, Koh DM, Chenevert TL, Thoeny HC, Takahara T, et al. Diffusion-weighted magnetic resonance imaging as a cancer biomarker: consensus and recommendations. Neoplasia 2009;11(2):102–25.
[65] Le Bihan D, Breton E, Lallemand D, Aubin ML, Vignaud J, Laval-Jeantet M. Separation of diffusion and perfusion in intravoxel incoherent motion MR imaging. Radiology 1988;168(2):497–505.
[66] Hosny A, Parmar C, Quackenbush J, Schwartz LH, Aerts H. Artificial intelligence in radiology. Nat. Rev. Cancer 2018;18(8):500–10.
[67] Miller DD, Brown EW. Artificial intelligence in medical practice: the question to the answer? Am. J. Med. 2018;131(2):129–33.
[68] Thrall JH, Li X, Li Q, Cruz C, Do S, Dreyer K, et al. Artificial intelligence and machine learning in radiology: opportunities, challenges, pitfalls, and criteria for success. J. Am. Coll. Radiol. 2018;15(3 Pt B):504–8.
[69] Kallenberg M, Petersen K, Nielsen M, Ng AY, Pengfei D, Igel C, et al. Unsupervised deep learning applied to breast density segmentation and mammographic risk scoring. IEEE Trans. Med. Imaging 2016;35(5):1322–31.
[70] Seah JCY, Tang JSN, Kitchen A, Gaillard F, Dixon AF. Chest radiographs in congestive heart failure: visualizing neural network learning. Radiology 2019;290(2):514–22.
[71] Calo R. Artificial intelligence policy: a primer and roadmap: social science research network. 2017. Available from: https://lawreview.law.ucdavis.edu/issues/51/2/Symposium/51-2_Calo.pdf.
[72] Yi PH, Hui FK, Ting DSW. Artificial intelligence and radiology: collaboration is key. J. Am. Coll. Radiol. 2018.May;15(5):781–3.
[73] King Jr BF. Artificial intelligence and radiology: what will the future hold? J. Am. Coll. Radiol. 2018;15(3 Pt B):501–3.
[74] Ravi D, Wong C, Deligianni F, Berthelot M, Andreu-Perez J, Lo B, et al. Deep learning for health informatics. IEEE J. Biomed. Health Inform. 2017;21(1):4–21.
[75] Christenses HI, Carnegie C, Krovi V, Smart B. From internet to robotics: Stone et al. Supra note. 2016 [105-9]. Available from: http://jacobsschool.ucsd.edu/contextualrobotics/docs/rm3-final-rs.pdf.
[76] Tsang LKD, Mulryne J, Strom L, Perkins N, Dickinson R, Wallace VM, Jones B. The impact of artificial intelligence on medical innovation in the European Union and United States. 2017. Available from: https://www.arnoldporter.com/~/media/files/perspectives/publications/2017/08/the-impact-of-artificial-inteelligence-on-medical-innovation.pdf.
[77] Thierer AD, O'Sullivan A, Russel R. Artificial intelligence and public policy Mercatus research paper. 2017. Available from: https://www.mercatus.org/system/files/thierer-artificial-intelligence-policy-mr-mercatus-v1.pdf.
[78] Scherer MU. Regulating artificial intelligence systems: risks, challenges, competencies, and strategies. Harv. J. Law Tech 2016;29(2):354–400.
[79] Mitchell T, Brynjolfsson E. Track how technology is transforming work. Nature 2017;544(7650):290–2.
[80] European Economic Community. 93/42/EEC – council directive concerning medical devices official journal of the European communities. 1993. Available from: http://ec.europa.eu/growth/single-market/european-standards/harmonised-standards/medical-devices_en.
[81] European Commission. MDCG 2018-2 future EU medical device nomenclature – description of requirements. 2018. Available from: https://ec.europa.eu/docsroom/documents/28668.
[82] European Economic Community. 90/385/EEC – council directive on the approximation of the laws of the member states relating to active implantable medical devices council directive. 1990. Available from: https://ec.europa.eu/growth/single-market/european-standards/harmonised-standards/implantable-medical-devices_en.
[83] European Commission. Directive 98/79/EC of the European parliament and of the council on in vitro diagnostic medical devices official journal of the European communities. 1998. Available from: https://ec.europa.eu/growth/single-market/european-standards/harmonised-standards/iv-diagnostic-medical-devices_en.
[84] The European Parliament and the Council of The European Union. Regulation (EU) 2017/745 of the European parliament and of the council on medical devices, amending directive 2001/83/EC, regulation (EC) No 178/2002 and regulation (EC) No 1223/2009 and repealing council directives 90/385/EEC and 93/42/EEC official journal of the European communities. 2017. Available from: https://eur-lex.europa.eu/legal-content/EN/TXT/?uri=CELEX%3A32017R0745.
[85] The European Parliament and the Council of The European Union. Regulation (EU) 2017/746 of the European parliament and of the council on in vitro diagnostic medical devices and repealing directive 98/79/EC and commission decision 2010/227/EU official journal of the European communities. 2017. Available from: https://eur-lex.europa.eu/legal-content/EN/TXT/?uri=CELEX%3A32017R0746.
[86] European Union. Regulations, directives and other acts. 2018. updated 23/04/29018. Available from: https://europa.eu/european-union/eu-law/legal-acts_en.
[87] 114th Congress (2015-2016). H.R.34 – 21st century cures act. 2016. updated December 13, 2016. Available from: https://www.congress.gov/bill/114th-congress/house-bill/34.

[88] U.S. Food & Drug Administration. Is the product a medical device? U.S. Department of Health and Human Services; 2018. updated 22/03/2018. Available from: https://www.fda.gov/MedicalDevices/DeviceRegulationandGuidance/Overview/ClassifyYourDevice/ucm051512.htm.

[89] Recht M, Bryan RN. Artificial intelligence: threat or boon to radiologists? J. Am. Coll. Radiol. 2017;14(11):1476–80.

[90] Statista. Online privacy – statistics & facts: statistics portal. 2018. cited 2017. Available from: https://www.statista.com/topics/2476/online-privacy.

[91] Krittanawong C. The rise of artificial intelligence and the uncertain future for physicians. Eur. J. Intern. Med. 2018;48:e13–4.

[92] Kruskal JB, Berkowitz S, Geis JR, Kim W, Nagy P, Dreyer K. Big data and machine learning-strategies for driving this bus: a summary of the 2016 intersociety summer conference. J. Am. Coll. Radiol. 2017;14(6):811–7.

[93] Castelvecchi D. Can we open the black box of AI? Nature 2016;538(7623):20–3.

[94] Dilsizian SE, Siegel EL. Artificial intelligence in medicine and cardiac imaging: harnessing big data and advanced computing to provide personalized medical diagnosis and treatment. Curr. Cardiol. Rep. 2014;16(1):441.

[95] Allen G, Chan T. Artificial intelligence and national security. 2017. Available from: https://www.belfercenter.org/publication/artificial-intelligence-and-national-security.

[96] Char DS, Shah NH, Magnus D. Implementing machine learning in health care – addressing ethical challenges. N. Engl. J. Med. 2018;378(11):981–3.

[97] Helbing D, Frey B, Gigerenzer G, Hafen E, Hagner M, Hofstetter Y, et al. Will democracy survive big data and artificial intelligence? Scientific American. 2017. updated 23 March 2017. Available from: https://www.scientificamerican.com/article/will-democracy-survive-big-data-and-artificial-intelligence/.

[98] Radanovic I, Likic R. Opportunities for use of blockchain technology in medicine. Appl. Health Econ. Health Policy 2018;16(5):583–90.

[99] Funk E, Riddell J, Ankel F, Cabrera D. Blockchain technology: a data framework to improve validity, trust, and accountability of information exchange in health professions education. Acad. Med. 2018,Dec;93(12):1791–4.

[100] Gordon WJ, Catalini C. Blockchain technology for healthcare: facilitating the transition to patient-driven interoperability. Comput. Struct. Biotechnol. J. 2018;16:224–30.

[101] Powles J, Hodson H. Google DeepMind and healthcare in an age of algorithms. Health Technol. 2017;7(4):351–67.

[102] The European Parliament and the Council of The European Union. Directive (EU) 2016/1148 of the European parliament and of the council concerning measures for a high common level of security of network and information systems across the Union official journal of the European communities. 2016. updated 19/07/2016. Available from: https://eur-lex.europa.eu/legal-content/EN/TXT/?toc=OJ:L:2016:194:TOC&uri=uriserv:OJ.L_.2016.194.01.0001.01. ENG.

[103] Mnih V, Kavukcuoglu K, Silver D, Rusu AA, Veness J, Bellemare MG, et al. Human-level control through deep reinforcement learning. Nature 2015;518(7540):529–33.

[104] Russell S, Bohannon J. Artificial intelligence. Fears of an AI pioneer. Science 2015;349(6245):252.

[105] Oxnard GR, Arcila ME, Sima CS, Riely GJ, Chmielecki J, Kris MG, et al. Acquired resistance to EGFR tyrosine kinase inhibitors in EGFR-mutant lung cancer: distinct natural history of patients with tumors harboring the T790M mutation. Clin. Cancer Res. 2011;17(6):1616–22.

[106] Corcoran RB, Chabner BA. Application of cell-free DNA analysis to cancer treatment. N. Engl. J. Med. 2018;379(18):1754–65.

[107] Gerlinger M, Rowan AJ, Horswell S, Math M, Larkin J, Endesfelder D, et al. Intratumor heterogeneity and branched evolution revealed by multiregion sequencing. N. Engl. J. Med. 2012;366(10):883–92.

[108] Wibmer A, Hricak H, Gondo T, Matsumoto K, Veeraraghavan H, Fehr D, et al. Haralick texture analysis of prostate MRI: utility for differentiating non-cancerous prostate from prostate cancer and differentiating prostate cancers with different Gleason scores. Eur. Radiol. 2015;25(10):2840–50.

[109] Fehr D, Veeraraghavan H, Wibmer A, Gondo T, Matsumoto K, Vargas HA, et al. Automatic classification of prostate cancer Gleason scores from multiparametric magnetic resonance images. Proc. Natl. Acad. Sci. U. S. A. 2015;112(46):E6265–73.

[110] Coroller TP, Bi WL, Huynh E, Abedalthagafi M, Aizer AA, Greenwald NF, et al. Radiographic prediction of meningioma grade by semantic and radiomic features. PLoS One 2017;12(11):e0187908.

[111] Hanania AN, Bantis LE, Feng Z, Wang H, Tamm EP, Katz MH, et al. Quantitative imaging to evaluate malignant potential of IPMNs. Oncotarget 2016;7(52):85776–84.

[112] Coroller TP, Grossmann P, Hou Y, Rios Velazquez E, Leijenaar RT, Hermann G, et al. CT-based radiomic signature predicts distant metastasis in lung adenocarcinoma. Radiother. Oncol. 2015;114(3):345–50.

[113] Diehn M, Nardini C, Wang DS, McGovern S, Jayaraman M, Liang Y, et al. Identification of noninvasive imaging surrogates for brain tumor gene-expression modules. Proc. Natl. Acad. Sci. U. S. A. 2008;105(13):5213–8.

[114] Teruel JR, Heldahl MG, Goa PE, Pickles M, Lundgren S, Bathen TF, et al. Dynamic contrast-enhanced MRI texture analysis for pretreatment prediction of clinical and pathological response to neoadjuvant chemotherapy in patients with locally advanced breast cancer. NMR Biomed. 2014;27(8):887–96.

[115] Emaminejad N, Qian W, Guan Y, Tan M, Qiu Y, Liu H, et al. Fusion of quantitative image and genomic biomarkers to improve prognosis assessment of early stage lung cancer patients. IEEE Trans. Biomed. Eng. 2016;63(5):1034–43.

[116] Sun R, Limkin EJ, Vakalopoulou M, Dercle L, Champiat S, Han SR, et al. A radiomics approach to assess tumour-infiltrating CD8 cells and response to anti-PD-1 or anti-PD-L1 immunotherapy: an imaging biomarker, retrospective multicohort study. Lancet Oncol. 2018;19(9):1180–91.

[117] The Cancer Imaging Archive (TCIA). [Available from: http://www.cancerimagingarchive.net.

Chapter 53

Precision medicine at the academic-industry interface

Patrick J. Silva[1,a] and Kenneth S. Ramos[1,2,3,4,b]

[1]*University of Arizona Health Sciences, Office of the Senior Vice President Health Sciences, Tucson, AZ, United States;* [2]*University of Arizona College of Medicine-Phoenix, Tucson, AZ, United States;* [3]*University of Arizona College of Medicine-Tucson, Tucson, AZ, United States;* [4]*University of Arizona Center for Applied Genetics and Genomic Medicine, Tucson, AZ, United States*

Introduction

Precision medicine has emerged as a healthcare delivery platform that emphasizes the individualization of care, through the integration of novel technologies and approaches into the diagnosis, treatment, and clinical management of patients and populations. A major driver for this movement has been the sobering reality that healthcare, as we know it today, will not be sustainable long term, without approaches that help to augment diagnostic accuracy and precision, deliver targeted therapies that improve efficacy and decrease toxicity and help stratify populations into subpopulations.

The notion that precision medicine is not an inherently new approach must also be recognized, as differential diagnosis and blood typing for blood transfusions have been a cornerstone of medicine for over a century. A major facet of the current precision medicine evolution is the test-then-treat paradigm, where a biomarker (often a single molecule or a disease-related variant of a normal molecule), is measured or detected to inform decisions about the presence, absence, or scope of disease. The test-then-treat paradigm is not new, but the tools and throughput with which to test, coupled with novel therapeutics, have expanded exponentially in recent years. Higher content analytic assays and the expanding menu of drugs to target specific pathways have driven demand for clinically actionable decision rules and clinical guidelines, by which the biomedical industry makes investment and product development decisions.

The development of new decision rules largely happens at the academic-industry interface. Translational research in academic medical centers, supported by the private sector, is a cornerstone of the regulatory science, underlying the evolution of the regulatory pathways for new therapeutics, approved to objectively determine molecular defects instead of using the traditional pathological classifications of disease that have driven therapeutic marketing approvals of the FDA, EMEA, and other agencies.

Background

The completion of the Human Genome Project (HGP) in 2001 was a major milestone, which enabled precision medicine with the tools of high-throughput analytics [1]. The reference genome sequence resulting from the HGP provided insights on the *structure* of the genome but was not revealing of *function*, particularly in disease processes. The cost of the project has been estimated at $2.7 billion [2]. By the mid-2000s, it was practical to objectively measure changes in gene expression, and differences in genome sequences of diseased individuals, relative to the reference genome or healthy individuals. There were still major hurdles to leveraging genome sequences and expression profiles in clinical decision-making: (1) the cost of sequencing a genome was orders of magnitude in excess,

a Dr. Silva is Executive Director, Biomedical Corporate Alliances in the Office of the Senior Vice President Health Sciences at the University of Arizona.

b Dr. Ramos is currently Professor of Medicine, Texas A&M College of Medicine, Houston TX, Center for Genomic and Precision Medicine, Alkek Chair of Medical Genetics Houston TX, Executive Director, Texas A&M Institute of Biosciences and Technology Houston TX, Associate Vice President for Research, Texas A&M University Health Science Center College Station, TX, Assistant Vice Chancellor for Health Services, The Texas A&M University System, College Station TX and Professor of Medicine at University of Arizona Health Sciences, Tucson AZ. He maintained an academic appointment at University of Arizona.

Precision Medicine for Investigators, Practitioners and Providers. https://doi.org/10.1016/B978-0-12-819178-1.00053-8

of what would be considered cost-effective for clinical decision making, (2) differences in genetic analytes were generally not clinically actionable, because the functional implications of those differences could not be assumed to be causative, and (3) outputs were not clinically actionable, due to lack of knowledge of structure-function relationships. Monogenic congenital rare diseases were an early exception to this hurdle, and consequently, the treatment of rare Mendelian disorders with protein replacement therapy, or dietary modifications, has been a field of medicine experiencing substantial progress [3].

The HGP had a positive, yet indirect, impact on the diagnostic ecosystem, as it enabled and incentivized rapid acceleration of new high-throughput analytic technologies, that allow broad measurement of sequence differences and gene expression profiles. The HGP was primarily reliant on Sanger sequencing [4], but the reference genome allowed for the construction of microarrays: silicon chips with an array of molecular probes, for most of the protein-encoding mRNAs expressed from the human genome. This technology made measurements of genome-wide changes in gene expression possible and even practical.

More importantly, the HGP provided a template that enabled sequencing by synthesis, or *next-generation sequencing* [5–7] and importantly, in silico reassembly [8]. Much like a picture on a puzzle box that guides the reassembly of a complex puzzle, the reference genome made it practical to sequence an entire genome from small fragments or reads, at low cost, and then reassembly of the pieces using computers [8]. In 2003, the first practical demonstration of NGS and commercialization by Solexa, an Oxford University spinout, proved to be a breakthrough in precision medicine [5]. In 2007, Illumina acquired Solexa and rapidly developed the user base for NGS technology.

The thousand-dollar barrier

These events culminated in a major inflection point in the cost of DNA sequencing, and a rate of technological advancement ($/per unit output), that improved in excess of Moore's law (2-fold improvement every 18 months) [2]. The threshold of sequencing a complete genome at reasonable depth for under $1000 threshold was deemed imminent [9] and achieved by 2017 (Fig. 53.1). The significance of overcoming the $1000 cost barrier is that many clinical uses of NGS have become practical below this price point. It is important to note that the costs generally reflect only the production of the sequence data, and perhaps a report flagging the key features, but not any substantive clinical annotation or analysis, which could easily cost much more in most clinical cases[9].

The reduced cost of NGS, however, has enabled targeted sequencing of somatic and germline specimens to identify known disease-causing and actionable mutations, assess disease-relevant changes in gene expression relative to comparator expression profiles, or even identify variants of unknown significance, in the case of diagnostic odysseys of rare disease. Importantly, the significance of changes in gene expression or differences in genetic sequences can now be explored at a population scale at a feasible cost, by gathering and comparing data sets across populations of healthy and diseased individuals or through the natural history of the disease.

Wide spectrum analysis

It was not until population-level sequencing became financially practical that genome sequence information could augment the understanding of the genetic, epigenetic, and proteomic variation, underlying disease in ways that could be used to augment clinical decision-making [94]. These insights are developed by collecting and retrospectively analyzing, biological specimens from individuals with the disease compared to healthy subjects, and comparing the thousands of variations in various analytes from various tissues, to identify those differences with an underlying biologic rationale. Analytes may include mutations, epigenetic modifications, expression changes, and changes in protein levels.

Molecular classification of cancer

Lung cancer, like most other diseases, has historically been described by morphological, pathological, or anatomical characteristics [10]. More recently, genetic tools and the practicality of widespread use of sequencing, have changed how lung cancer is diagnosed and described [11]. Lung cancer is characterized by mutations in at least 20 different genes, that are therapeutically addressed with a few targeted therapies, approved by regulators or in the clinical stage development [12]. Many more tumor driver mutations have been reported in genes that are currently therapeutically intractable [13,14].

While EGFR mutations can be present in up to 20% of non-small cell lung cancers in Asian populations [15], many of these driver mutations have a prevalence of less than 1% in the general population. Furthermore, tumor heterogeneity expands temporally as a tumor evolves branches of cell lineages with different mutations, as exemplified in adenocarcinoma of the lung [13]. Choice of treatment can drive subclonal expansion, and tumor recurrence in the direction of cell lineages, because of biopsy error and undiscovered driver mutations, that go unchallenged during the course of therapy. In essence, many targeted drugs only treat a fraction of the tumor whose biology is revealed by a biopsy, with many biopsies likely producing false negatives for certain tumor-driving mutations.

In cancer, precision biopsy and computational tools that allow deconvolution of tumor heterogeneity, and clonal

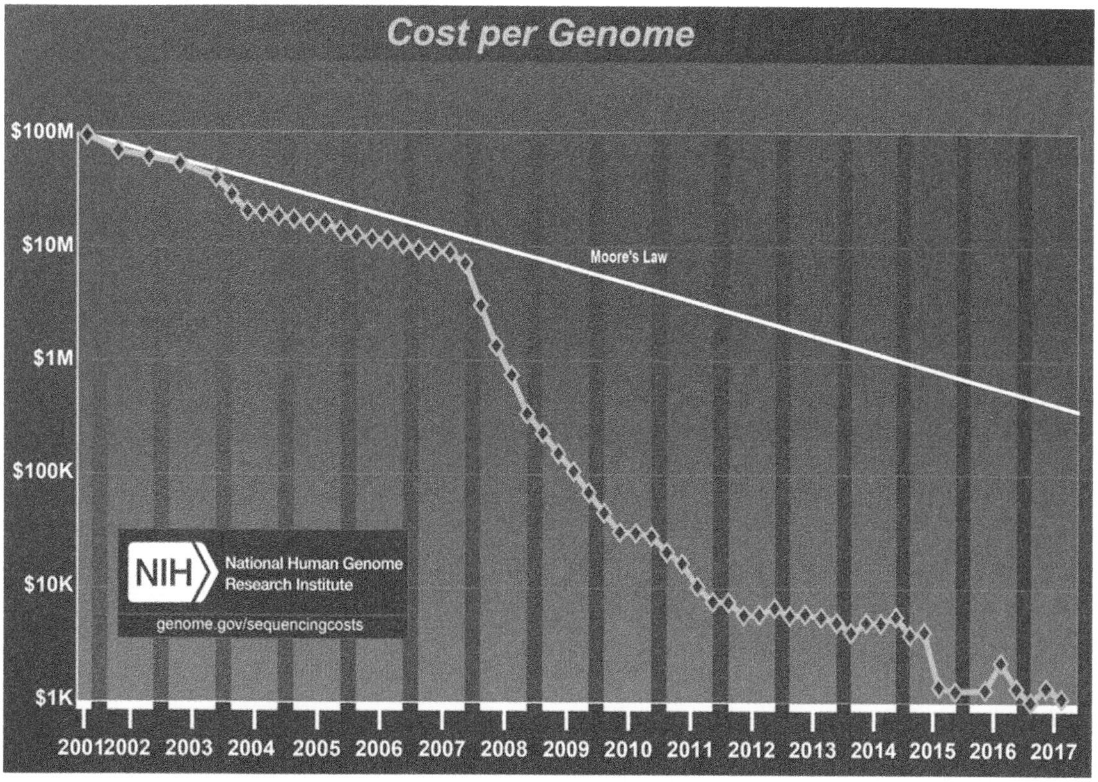

FIGURE 53.1 Cost per megabase to sequence DNA between 2001 and 2017 relative to Moore's Law. Wetterstrand KA. DNA Sequencing Costs. *Data from the NHGRI Genome Sequencing Program (GSP). Available at: www.genome.gov/sequencingcostsdata.*

discrimination of mutations can shed light on the individual origins of disease and could prove powerful in treating the multiple cancer cell lineages that may be driving advanced cancers [16]. However, the clinical and statistical significance of mutations (or gene expression changes), among individual clinical cases, requires population-scale collections of specimens and longitudinal clinical annotation, to achieve statistical significance using traditional statistical tests; the individuality paradox, or the n-of-1 issue.

As molecular disease classification allows more precise discrimination between and within individuals over the natural history of the disease, the disease population gets more segmented, and in a sense, smaller. The dichotomy of precision medicine is that as we continue to characterize disease at the individual level, we must recruit larger populations, to find the similarities necessary to achieve statistical power and clinical relevance. As a corollary, imprecise classification "molecular reductionism" [17] of inherently heterogeneous and complex diseases, has been attributed to pivotal clinical trial failures, and a major driver of the decline in drug approvals[18].

Fundamentally, this means a higher number of private sector-medical center interaction nodes, necessary to run a drug registration trial, and the high transaction costs that accompany any approach, to assemble a precisely defined disease cohort for a multisite clinical trial [18]. Population-scale specimen collections and data commons are key tools to understand and address the personalized medicine—population health paradox, both in devising more precise and reliable clinical decision tools, to enable testing of new drugs in patients that are likely to respond.

Specimen banking

Academic specimen banks are an increasingly important source of population biomarker information for the clinical research community, and especially, biomedical companies trying to understand the prevalence of biomarkers, disease subtypes, or defining the increasingly precise use cases (e.g., label strategy) for early-stage drugs [18—21]. Academic medical centers employ physician—scientists, involved in clinical trials and research and engaged in the collection of blood, tissue, or genetic material, during the course of the provision of standard care. In this setting, longitudinal annotation is often also captured in the electronic health record.

Collectively, these paired longitudinal data and specimens sets provide a valuable resource, to understand the finer aspects of the disease, such as the degree of homogeneity or heterogeneity, the prevalence of the targeted subtypes of disease in populations, and the discovery and validation of diagnostic biomarkers. The collection of a longitudinal specimen bank, with robust clinical annotation, is a valuable resource to industry collaborators attempting to understand a target disease, at a population level for clinical trial planning [21]. Biorepositories are also large capital expenditures, and ongoing commitments by academic institutions, or units within, which might have very specialized uses for the repository[21].

In essence, academic medical centers are developing increasingly robust data ecosystems around their specimen collections, often aligned with their inherent regional clinical centers of excellence, that provide a deep pool of information about select disease populations. These resources are routinely sought out by industry collaborators, primarily large biopharma companies, to help inform their pipeline strategy and drug development portfolio strategy: prioritization of indications for clinical development, label expansion strategies for approved products, program mothballing, and acquisition divestiture decisions[18].

Reference and research biobanks

Specimen banking is not new. The U.S. Armed Forces Institute of Pathology (AFIP) was established in 1862 and was superseded by the National Pathology Tissue Repository (Ragin and Park [22] provide a thorough general review of current biorepository management considerations). Biospecimen banking has accelerated concurrent with the commoditization of high-throughput analytic technologies, enabling smaller samples to be utilized; thus, most banks have been established recently [21]. Holub et al. [95] estimate that over 60 million samples are currently archived worldwide.

Assembling a cohort of specimens (or clinical trials subjects), with highly specified clinical and biological criteria, can be quite challenging, when multiple filtering criteria, such as the presence of a combination of molecular biomarkers or past treatments is applied (the individuality paradox). Striving toward disease homogeneity, by filtering out unlike clinical cases reduces sample size, which has significant implications for research and industry-sponsored clinical trials. There is a practical limit in just how specific (i.e., homogeneous) a cohort can be assembled, to meet common statistical analysis standards, before the transaction costs of engaging many institutions overwhelm the strategic benefits of a more homogenous cohort.

Biobank access and administration

Specimen sharing among nonprofit research institutions is somewhat standardized by NIH guidelines [23], including the use of standardized data use agreements (DUAs) and Uniform Biologic Material Transfer Agreements (UBMTAs). However, significant frictions and transaction costs still occur, whenever industry users wish to access a biorepository under the stewardship of a public entity. Substantive issues of biosafety, scientific merit, ethics, informed consent, and legal risk must usually be analyzed, by individuals with appropriate expertise with each industry request, and aligned with the objectives and mission of the repository and host institution, prior to contracting. Many stakeholders are involved in the decision to share a specimen, and these issues are addressed by governance committees that meet periodically, and so requests often queue for committee review before contracting.

In addition, the matter of fees and cost recovery are critical sustainability considerations, often perceived to be at odds with ethical and legal preferences. Increasingly, companies wish to have data rights extending beyond the scope of the original project, prompting the transfer in order to build out their data ecosystems and to train their artificial intelligence (AI) engines. These are complex considerations in areas where most institutions administrative rank-and-file have minimal expertise. Consequently, a request for specimens takes time to be analyzed, and an outgoing transfer agreement formulated.

The provenance of specimens is increasingly important to industry users compliance goals, so this diligence is mutually beneficial and generally welcomed by industry users. Even after specimens or case-level clinical data are shared, there are residual administrative issues that may require management. For example, if a patient wants to revoke consent, or an institution wishes to ensure this is possible, the practicality of doing so is daunting, in most instances where consent forms are filed on paper, and the chain of communication between the patient and the user is opaque and temporary. As such, the institutions with the best command of these administrative issues, are generally deemed by industry collaborators as partners of choice.

Biobank networks

Biorepositories have begun to coordinate and establish networks, to enable users of those specimens to achieve scale, by being able to assemble a cohort out of a larger population, often from several institutions. Efficiencies of scale can be further augmented by administrative efficiencies. Sometimes these alliances might be formed as consortia at a project level, to examine a specific disease phenotype such at the NIH supported TOPCAT [24] and HALTC [25] cohorts, where multiple endpoints and a large sample size require the development of a cohort, that would be impractical for a drug registration trial. In other instances, multiinstitutional level alliances or consortia are formulated to leverage specimen collection efforts across broad disease areas and large populations [26] (Table 53.1).

TABLE 53.1 Population-scale programs with confirmed cohorts over 100,000 subjects.

Cohort	Traits	Size (# subjects)	Notes	References
Million Women Study	Breast cancer	1,280,296	Key finding: linkages between hormone therapy and breast cancer, alcohol, and breast cancer.	[68]
All of Us	Disease agnostic; lifestyle data, electronic health record, and biospecimens	1,000,000	Expected to collect and store 34 million samples in a facility at Mayo Clinic in Rochester, MN.	[69]
China Kadoorie Biobank	Cause-specific mortality and morbidity and any hospital admission through linkages with registries and health insurance databases	512,891	Joint venture between University of Oxford's Clinical Trial Service Unit and Epidemiological Studies Unit (CTSU) and the Chinese Academy of Medical Sciences (CAMS).	[70]
UK Biobank	Disease agnostic; lifestyle data, electronic health record, and biospecimens	502,649	502,649 was the number of subjects who completed a health questionnaire. 50,000 whole genomes (w/Regenergon) and 450,000 whole exomes planned.	[71,72]
Million Veteran Program	Disease agnostic; lifestyle data, electronic health record, and biospecimens	500,000	Goal: 1,000,000 by 2025. Linked to VA EHR and epidemiological data infrastructure. Periodic survey instruments used to augment annotation of specimens.	[73]
GIANT	Human body shape, size, and adipose status	322,154	Consortium of consortia and cohorts. Over 59 cohorts involved worldwide, including several on this list.	[74,75]
Biobank Japan	47 target diseases	291,274	Annual blood sampling of the cohort, with outcome data on subjects that developed target diseases.	[76]
Heart Failure Molecular Epidemiology for Therapeutic Targets (HERMES)		230,000	~30,000 HF cases/~200,000 HF controls, ongoing.	[77]
Kaiser Permanente Biobank	Disease agnostic; lifestyle data, electronic health record, and biospecimens	200,000	500,000 planned. Deep clinical annotation with EHR linkages throughout their health system.	[78]
International blood pressure consortium	Blood pressure	200,000	A consortium of consortia: Cohorts for Heart and Aging Research in Genomic Epidemiology (blood pressure) and the GBPGEN consortium (Global Blood Pressure Genetics Consortium).	[79]
The CARDIoGRAMplusC4D Consortium	Coronary artery disease (CAD)	194,427	63,746 CAD cases and 130,681 controls. A consortium of consortia.	[80]
Global Lipids Genetics Consortium	Lipids and coronary artery disease	188,577	Composed of over 40 cohorts worldwide.	[81]

Continued

TABLE 53.1 Population-scale programs with confirmed cohorts over 100,000 subjects.—cont'd

Cohort	Traits	Size (# subjects)	Notes	References
LifeLines	Disease agnostic; lifestyle data, electronic health record, and biospecimens	167,729	Longitudinal over a 30-year period. Questionnaire every 1.5 years and examination every 5 years.	[82]
Mexico City Prospective Study	Tobacco and diabetes health effects	159,456	Supported out of Oxford Clinical Trial Service Unit.	[83,84]
Diabetes Genetics Replication And Meta-analysis Consortium (DIAGRAM)	Diabetes	149,821	34,840 diabetes cases/114,981 diabetes controls. Consortium of consortia, meta-analysis.	[85]
Children's Hospital of Philadelphia	Inherited diseases	100,000	World's largest pediatric biobank: autism, intellectual disability, attention deficit-hyperactivity disorder, epilepsy, and diabetes.	[78]

National and transnational consortia

Whether small or large, clear governance issues represent critical determinants of collaborative and impactful repository development. Almost all involve a mixture of charitable and industry support, to augment predominantly government project and core grant support. In the case of the TIES Cancer Research Network, a federated model was employed, that allowed institutions to maintain ownership, possession, and control of specimens and data, while having joint governance for operational management, policies, procedures, and data use terms [26]. The primary benefit of the federated model is that the pooling of clinical cases across institutions enables scale, and increases the capacity to assemble a precisely defined cohort of relevant size.

The federated model is clearly favored in cases where maintaining control of risk management is a priority. The most expansive example of a biobank network is the newly formed Biobanking and Biomolecular Resources Research Infrastructure—European Research Infrastructure Consortium (BBMRI-ERIC), a pan-European agency dedicated to enabling transnational specimen collaborations, and advancing best practices in biobanking, consortia, and bioethics [96].

Multipartner corporate biobanks

Certain biopharmaceutical industry players are taking a different strategy, that involves recruiting clinical cases through alliances with large health systems [27,28], creation of their own clinical trials, or acquisition of population-scale collections, as exemplified by Amgen (1000 Oaks, Ca, USA) acquisition of deCode genetics (Reykjavik, Iceland) [29]. These large collections enable genome-wide association studies (GWAS), a major step toward reverse engineering complicated disease processes, sharing a common genetic link. For example, Nioi et al. [30] described a loss of function mutation in ASGR-1, in the deCode cohort dataset that directly resulted in therapeutic targeting the underlying mechanism of cardiovascular disease.

In this case, Amgen choose to pursue the target (ASGR-1), in lieu of escalating investments into an existing program around LCAT loss-of-function, based on analysis of omics and health records revealing that loss of LCAT function was associated with beneficial lipid profiles, but no meaningful improvement in cardiovascular events [31]. These clinically annotated omic data sets, in essence, allowed a de facto clinical gene knock-out study, to determine how valid a target might be in the most clinically relevant model possible—a human population with and without the disease. This approach to therapeutic development is sometimes referred to as reverse translation, or bedside-to-bench.

Amgen is now developing a monoclonal antibody mimicking the ASGR1 loss of function mutation [31], that does appear to be beneficial for HDL levels, again based on population data implicating ASGR-1 loss-of-function with favorable HDL levels. CRISPR technology is expected to catalyze and accelerate validation of pharmaceutically actionable targets, revealed by interrogation of these large cohorts, and possibly directly used as a therapeutic strategy. Regeneron (Eastview, NY, USA) infrastructure has allowed sequencing of 250,000 samples, from its collaboration with Geisinger Health System (Danville, PA, USA), and leadership in a precompetitive consortium of pharma companies, to substantially sequence all of the specimens in the UK Biobank (Table 53.2).

More recently, consumer genomics companies have sought to leverage their specimen and data collections. 23andMe (Mountainview, CA, USA) claims over 30 pharma collaborations, predicated on access to their annotated and consented sequence databases [32] (Table 53.3), but there is no indication in the literature yet, validating the utility of consumer genomics datasets, as viable drug discovery tools. Quality clinical annotation (or lack thereof) is likely to be a key determinant of their utility in drug and target discovery.

Reconfiguration of biobanks and biorepositories

Biorepositories are among the largest investments an institution can make, so financial sustainability is an important consideration [21]. The recurring costs of specimen repositories create an impetus to optimize utilization, within the constraints of scientific merit, ethical frameworks, and policy applicable to medical centers and academic institutions, in ways that allow sharing the cost of this valuable public resource with industry collaborators. Companies are increasingly vigilant about compliance with patient privacy rules, informed consent, ethical considerations, and statutory standards and requirements, in the use of these specimen collections and their datasets. Thus, the value of these resources as collaborative currency requires that a robust quality and compliance program be in place. Put another way, inadequate informed consent or property rights could render specimens a net liability for any potential user, and as such, a deterrent to the translational research progress.

Clinical data commons

A macro-trend in the biomedical sector, is the construction of population-scale data commons (hundreds of thousands to millions of lives), with overarching strategic goals. These could enable meta-analysis of a living and growing longitudinal data commons, to iteratively improve mechanistic

TABLE 53.2 Pharma population-scale exome or genome sequencing collaboration cohorts and consortia.

Company	Collaborator(s)	Traits	Size (# subjects)	Notes	References
Regeneron	Geisinger Health System	Disease agnostic; Lifestyle data, electronic health record, and biospecimens.	250,000	Informed development of two drugs: (1) monoclonal antibody to angiopoietin-like protein 3 (ANGPTL3), an inhibitor of lipoprotein lipase and endothelial lipase that appears to play a central role in lipoprotein metabolism, (2) label expansion for an approved blockbuster drug for moderate-to-severe atopic dermatitis.	[86]
	UK Biobank; GlaxoSmithKline		50,000		[87]
	Abbvie; Alnylam; AstraZeneca; Biogen; Pfizer		500,000	Sequence exomes of entire UK biobank; expansion of GSK collaboration	[88,89]
deCODE cohort	Icelandic government; deCODE; Amgen	Disease agnostic; Lifestyle data, electronic health record, and biospecimens.	140,000	Initially a public health investment that morphed into a private enterprise to access capital. Unfortunate timing (financial crisis) required sale.	[90]
	Amgen; Somalogic		40,000	Largest clinical cohort for proteomic analysis to date signed December 2018.	[91]
Astra Zeneca	AstraZeneca's Centre for Genomics Research	Not disclosed	500,000	Analysis of 2 million by 2016.	[92]
	Human Longevity		500,000	AZ will share 500,000 from AZ clinical trials and access HLIs health database.	
	Wellcome Trust Sanger Institute			Joint alliance research team will collaborate to discover new biomarkers and drug targets.	
	Institute for Molecular Medicine Finland (FIMM), University of Helsinki, Finland			Leverage Finnish volunteer recall mechanism to expand clinical annotation on rare variants of clinical significance discovered in Finnish populations.	
Genentech	Human Longevity		10,000		[93]

TABLE 53.3 Consumer genetics sequencing collaborations.

Vendor	Collaborator(s)	Content	Estimated size	Notes	References
23andMe	Genentech, 30 undisclosed	SNPs	1,000,000	Highly engaged population with iterative annotation via surveys to subjects who have submitted samples.	[32]
Ancestry	Calico	SNPs	1,000,000	Collaboration with Calico to interrogate potential longevity loci in Ancestry cohort.	[78]
Human Longevity	Astra Zeneca	Whole genome, exome, SNPs	200,000	Health Nucleus provides a battery of diagnostic services that include extensive genome sequencing.	[78]

understanding of personalized treatment responses, personalized disease trajectories, and generalized medical hazards. A convergence of trends is making this possible: (1) implementation of somewhat uniform data quality standards, (2) falling costs of high-throughput analytics (RNAseq, NGS, proteomics), (3) improved patient-reported data collection tools, (4) improved electronic health record infrastructure, and (5) cloud computing. These new sources of increasingly standardized health and outcome data, have the promise to augment randomized clinical trials, as a source of clinical validation of new products, technologies, and medical decision making.

Mining of healthcare big data

Large provider networks, academic health systems, and payer networks are the most common sources of population-level clinical data [33]. Electronic health records represent a relatively new infrastructure in the healthcare system, and standardized data structures are quite new, so data from these sources vary in quality, often depending on how recently they were created. Examples of such data include diagnosis codes [34,35], claims data [36–38], prescription orders, pathology results, and unstructured data, such as physician notes. The challenges in curation [39], use [38], and analysis [37] of these types of data have been reviewed thoroughly elsewhere so we will focus on the use of such data in omic themed collaborations, at the academic-industry interface.

Private sector companies that provide diagnostic decision-making tools (i.e., clinical diagnostic vendors, reference laboratories, and diagnostic medical imaging companies), which are not directly involved in the provision of healthcare, have become increasingly interested in capturing data and insight, on the clinical cases they touch. Inspired by Google's data monetization model, and perhaps Google's investments, 23andMe has pioneered the "data monetization" business model in the annotated –omics space.

Blinded clinical data

Some would argue this is an example of a barter system [40], and there is little doubt that many of the elements, involved in the academic medical center–industry collaboration involves barter, or the exchange of nonfinancial collaborative currency like data (perhaps to the dismay of their respective chief financial officers). Many molecular diagnostic companies are following suit, and devising ways to build data ecosystems to help them identify new diagnostic targets, disease areas, drug classes, and even monetize unique insights, enabled by their analytic platforms and AI engines. Many of these business models leverage insights derived from population-level clinical data sets. These organizations are accelerating demand for de-identified case level clinical data sets, that can be used to augment the informative utility of omic data sets, much in the way clinical outcome and blood chemistry data in the deCode cohort, have assisted Amgen in prioritizing drug targets in HDL homeostasis, and essentially enabling the creation of in silico clinical trials.

At a macro level, organizations possessing clinically annotated -omics data recognize their value, by virtue of accelerating demands for these type of data. However, the value of these data as collaborative currency is complicated, when considering who owns the data (patients), who stores and curates the data (providers), and the challenges of ensuring consent and provenance, for the myriad exchanges and uses that can occur, as such data is replicated, shared, and used across the cloud and the healthcare ecosystem. The chain of custody of case-level data is an important element in managing the legal and ethical considerations inherent in the use of such data to build population-level data commons and feed artificial intelligence engines. Companies who are consumers of these data are increasingly vigilant of the legal risk associated with specimen and data provenance (or lack thereof).

Aggregated datasets involve different considerations, such as case duplication, quality assurance, de-identification, and validation challenges. In summary, optimization of annotation clinical data ecosystems holds many opportunities for quality improvement and disintermediation, and perhaps there is a prominent role for patients in managing their "account ledger" for these data ecosystems, and indirectly providing downstream consent, further annotation, and participation in studies both physical and in silico. A healthcare blockchain has been proposed as a solution to many of these data curation and management challenges, and a team at MIT's Media Lab have begun to explore its development[41].

Collaboration models

Large health systems covering rural populations over large geographies are particularly well positioned in that they have had a longitudinal relationship with a larger percentage of their patients. Data ecosystems and biorepositories tied to large populations are valuable. Longitudinal clinical data sets, with longitudinal biospecimen samples, are exquisitely valuable since the natural history of the disease can be compared and contrasted across a population, at a molecular level where specimens allow.

Diagnostic licensing and ownership

Patenting and licensing of novel reagents, analytic methods, and biomarkers have long been a strategy of academic medical centers, and have accelerated somewhat in the postgenomic era. The fate of this strategy was brought into a question, by a court case involving a generalized biomarker concept out of the University of Colorado: homocysteine as a biomarker for cardiovascular risk. The commercial entities charged with exploiting those patents, Metabolite Labs and Competitive Technologies Inc., brought a case against Laboratory Corporation of America [42], that in essence asserted property rights over the general notion of "measuring homocysteine and correlating it with vitamin deficiency."

The nuances are many and reviewed elsewhere [43], but the case was the first in a series of key cases referred to the Supreme Court of the United States (SCOTUS), to address the natural phenomena doctrine. It was an attempt by the diagnostic industry to have the courts more clearly define, to what extent the measurement and clinical decision use of a naturally occurring molecule could be owned by virtue of patent rights. The Supreme Court changed course, and issued a *writ of certiorari*; basically, a legal maneuver to avoid reviewing the case, and referring it back to a lower court. However, the industry had not given up seeking clarity from the courts on the ownability of biomarkers.

The fate of BRCA sequencing tests, based on discoveries of the clinical significance of hereditary BRCA-1/2 cancers, was eventually more impactful to biomarker patent and license practice. Patents owned by the University of Utah, and licensed by Myriad Genetics (Salt Lake City, UT, USA), were enforced against reference labs offering BRCA-1/2 mutation tests, then challenged by the Association for Molecular Pathology (Rockville, MD, USA), which ultimately the SCOTUS agreed to hear [44]. This case, shaped case law, and involved the submission of amicus briefs by many stakeholders, including the American Medical Association, Association of University Technology Managers (AUTM), and the Biotechnology Industry Organization (BIO), the Pharmaceutical Research Manufacturers Association of America (PhRMA), among other major interest groups in the diagnostic ecosystem.

cDNA versus gene sequence

The parable of the blind men and the elephant was an apt description of the context of the broader socioeconomic context of the amicus briefs, that simultaneously informed and obscured the question of law faced by the SCOTUS. The Myriad case prompted partial resolution of natural phenomena doctrine: a cDNA or other synthetically derived DNA sequence could be patented, a gene sequence (or naturally mutated variant thereof) could not.

Probe versus mutation

A case brought by the Mayo Clinic against Prometheus (San Diego, CA, USA), later clarified to what extent one can patent, and thus own, the process of clinical inference from a biomarker measurement [45]. In essence, it is possible to patent the probe, but not the mutation, and perhaps the method of clinical interpretation, to the extent it involves a transformative step, not a mere mental step [46]. An example of a transformative step would be a "test-then-treat" claim. The "test-then-treat" paradigm was later refined and arguably clarified, by the SCOTUS in the Mayo v. Prometheus case [47,48].

Some have argued that the Mayo decision has rendered personalized medicine discoveries unpatentable, under the interpretation of natural phenomena doctrine, set forth in the decision [45]. We have argued, that patent protection of diagnostic approaches in precision medicine is increasingly narrow, and collaboration models with industry warrant a strategic approach, leveraging the key opinion leaders, unique populations, analytic capabilities, and other precision medicine resources, for emphasizing clinical validation beyond just patent licensing for assay commercialization [18,48].

European patent law

Article 53(c) of the European Patent Convention states that "European patents shall not be granted in respect of... diagnostic methods practiced on the human or animal body...." (ref EPO). However, Rules 26–29 of the European Patent Convention (EPC) and Biotech Directive 98/44/EU (European Parliament and the Council of the European Union of 6 July 1998 on the legal protection of biotechnological inventions), provide some clarity that has eluded the US patent system. Rule 27 states, "biological material which is isolated from its natural environment or produced by means of a technical process, even if it previously occurred in nature" is patentable subject matter. Rule 29 states "An element isolated from the human body or otherwise produced by means of a technical process, including the sequence or partial sequence of a gene, may constitute a patentable invention, even if the structure of that element is identical to that of a natural element."

Article 52(2) and 53(c) allow for patentability of in vitro and ex vivo diagnostic methods utilizing "laws of nature" under the EPC if they fulfill all other patentability criteria [49]. In summary, analytes, probes, and in vitro/ex vivo assay methods, and their relation to clinical classifications and decisions, are generally patentable in EU countries under the EPC and the Biotech Directive, which has been incorporated into law in all EU countries [50].

Collectively, these court cases have severely limited the scope of claims, that can be granted in a biomarker patent, and discouraged academic institutions from patent protecting biomarker discoveries by their faculty. Diagnostic companies are increasingly moving toward multianalyte, platform strategies, as evidenced by Foundation Medicine

(Morrisville, NC, USA) parallel review by CMS and FDA, of their 324-analyte targeted next-generation sequencing CDx assay [51]. The CDx assay illuminates the fallacy in patenting and licensing a single mutation for a single clinical decision, when the trend is toward products that measure hundreds of mutations from 327 genes, for what could be myriad clinical decision use cases, as more permutations of new drugs and cancer targets emerge.

The potential utility of assays like Foundation One CDx will require interrogation of well-annotated data ecosystems, robust biorepositories, and increasingly fragmented clinical cohorts, and their importance to expand clinical use and content of multi-analyte offerings (i.e., market size).

Clinical research and clinical trials

Most biopharma industry-sponsored clinical research at academic medical centers is prospective in nature, and geared toward registration of therapeutic products. These trials are increasingly expensive, due to the number of clinical trial sites required, to overcome increasingly fragmented enrollment criteria. For most drugs, where the anticipated markets are in excess of $1B per year, every day of lost patent life is worth millions of dollars. The major friction in clinical trial design is ascertaining whether a particular clinical site is likely to have a meaningful number, of the desired variety of clinical cases. The electronic health record is an increasingly important tool at the academic-industry interface, to enable site selection and full accrual of a study[52].

For example, TrinetX (trinetx.com) has established a federated network of academic medical centers and health systems, to facilitate clinical data sharing for research and clinical trial site selection, for biopharma companies and CROs [52]. This system is largely reliant on analyzing the most consistently available flavor of deidentified clinical data, from the EHR: ICD9/10 diagnosis codes. This is highly impactful, in that it allows external collaborators to start to understand the demographics of a health center, in a more granular and digital fashion than what is enabled by medical science liaisons, interacting with key opinion leaders, who may not readily be able to channel metrics from their clinical specialties.

In many disease areas, particularly oncology and cardiovascular disease, drug development increasingly involves genotype driven therapeutic mechanisms and enrollment criteria, which further augments the need for a granular understanding of the molecular demographics of disease, in populations across health centers. The quality of the clinical data infrastructure, of an academic medical center or health system, is increasingly a major determinant of the institution's stature as a partner of choice.

Precision medicine as a specialty

A precision medicine specialty can drive deep industry engagement, and consequently, growth in the clinical trials enterprise in that medical specialty. Thought leadership in a specific disease vertical attracts both patients seeking superior care, and biopharma companies seeking access to these high concentrations of highly relevant clinical cases. In leading academic medical centers, a handful of thought leaders in certain disease verticals, are aggressively leveraging the populations they treat to build new tools (assays and biomarkers), gain new insights (clinical decision rules), and develop new therapeutic strategies (drug development and interventional workflows).

These centers of excellence are ultra-fertile for clinical trials, and often cultivate an associated omic-based precision medicine capability, that is attractive to patients and biopharma companies developing targeted therapies. These centers tend to be coordinating centers for multicenter trials, for many drugs and new classes of drugs, lead in large NIH funded clinical trials and registries, and publish the most impactful drug trial papers. They are also fragile, and arguably, portable enterprises, nucleated by one or a very few key clinical opinion leaders.

Market expansion (evidence-based marketing)

Validation of new use cases for developmental stage and marketed decision tools, drive further clinical data consumption, by the companies developing such products. This is particularly important where a company's product, or that of a competitor, might have been used in the clinical workflow such that improvements in outcomes, cost of care delivery, or cost of disease burden might be better understood. Therapeutic medical devices and capital surgical equipment represent two classes of biomedical products sold into increasingly consolidated provider organizations. These sales channels and buying decisions are increasingly complex and require increasing evidence in clinical benefit and cost-benefit justification.

Historically, a common perception has been that commercial launch of a therapeutic medical device, upon confirmation of 510(k) regulatory treatment by the FDA (or a CEA mark by the EMEA), was adequate to drive sales. More recently, reimbursement and consequently, investment by medical device developers has not followed rapid growth in the number of medical device technologies and products. In part, there has not been commensurate growth in healthcare spending, resulting in heightened competition in an increasingly noisy marketplace, more products, and constant healthcare spending [53]. This phenomenon is quite evident in the field of continuous glucose monitoring, where much commercial investment brought new products to market, but the prevalence of diabetes and cost burden for payers, has them looking past regulatory approval to clinical outcome data, confirming clinical benefit [54].

More generally, reliable clinical outcome data, whether prospective or retrospectively obtained, is increasingly important in driving the clinical adoption of new technologies and products. Academic medical centers tend to play a key role in the clinical research ecosystem at the industry-patient interface, and access to the patients and population health economic data are increasingly useful to companies, increasingly compelled to make an evidence-based market argument.

Return on investment validation (justifying the price)

The economic justification of using new technologies is increasingly important for a number of reasons [55], consequently obtaining data with which a company can justify a return on investment is useful. At the front end of the product development cycle, diagnosis, billing, and cost data across the clinical workflow, inform the scope and nature of the healthcare problem being addressed with a particular technology, even where that case-level data may not involve the use of similar technology. At the back end of product development cycles, pharmacoeconomic analysis is increasingly important to justify very expensive enzyme replacement therapies for rare diseases (Fabry) [56], autologous cell therapies (Chimeric Antigen T-cell; CAR-T) [57], or a course of curative chemotherapy for a prevalent disease Hepatitis C [58,59] (hepatitis C).

Cost/benefit ratio

New therapies have increasingly smaller target populations, and expensive costs of goods sold (COGS) and selling price, even though in some instances like hepatitis C, also high prevalence. These are factors driving evidence-based usage and reimbursement decisions. Conversely, therapies that modify disease in a profound way, dramatically alter the cost burden of disease. It is increasingly important for drug companies, to quantify the costs and benefits of therapy, in ways that are relevant across the purchase decision chain (patient, provider, and payer).

Data structures that allow case-level cost analysis would make this type of analysis more feasible, but currently, the relevant pieces of information exist across disparate and disconnected nodes on the healthcare delivery supply chain. For example, aggregate drug purchase and average selling price data might be obscured within pharmacy benefit manager data systems, while clinical outcome and complication data, might be dispersed through payer and provider data systems.

Rationalizing regulated product development strategy (lowering the cost of failure)

Beyond the very direct approach of correcting a congenital molecular defect with enzyme replacement therapy, there are currently many examples, where the molecular basis of disease is defined by gene expression profiling (Ref. [60] Oncotype DX. Genomic Health, Redwood City, CA, USA), mutation inventory (Foundation one CDX, Cambridge, MA, USA), or germline genotyping (Vertex CF, Cambridge, MA, USA). Oncotype Dx Breast Recurrence Score is an example of an expression profiling test with prognostic utility, as well as the provision of correlative information of predictive utility of the response of breast cancer patients to chemotherapy. This therapeutic approach was not designed for the molecular biology of the disease but has proven useful in guiding treatment decisions [60].

Tissue agnostic label approval

Further, the Foundation One CDx, a targeted sequencing assay which not only reveals whether a tumor harbors mutations, known to exist is certain tumor types, but can also reveal novel fusions like neurotrophic receptor tyrosine kinase (NTRK) fusions with novel fusion partners [61], that can be treated with off label use of drugs (larotrectinib) that inhibit the pathogenic pathways, likely activated by these novel defects [62]. The off label use of investigational TRK kinase inhibitors guided by targeted NGS, exemplifies the promise of precision medicine in cancer treatment, and the National Cancer Institute (NCI) has established support mechanisms, for so-called *basket trials*to expand the clinical evaluation of these approaches [63], which in essence treat the molecular basis of disease, instead of the tissue of origin or pathological categorization. Larotrectinib was recently approved by the FDA to treat cancers with confirmed NTRK fusions, notable in that this is among the first FDA approvals for an oncology drug, based on a molecular cancer classification, not a tissue of origin (i.e., a tissue agnostic label approval).

In cystic fibrosis, a variety of gene therapy approaches have failed, but a rationally designed small molecule, ivacaftor, has proven effective as a disease-modifying therapy, for patients with G551D mutations in the cystic fibrosis conductance regulator gene (CFTR) [64]. Ivacaftor was designed to interact with, and open the CFTR chloride ion channel, which addresses known dysfunction associated with a number of mutations. The use of ivacaftor has since been expanded, for use in patients with a number of other CFTR mutations (reviewed by Gentzsch and Mall [65]).

Workflow and clinical decision optimization

Informatics-oriented companies in the healthcare space, particularly those focused on radiology, pathology, genomic medicine, managed care, and accountable care business models, are evolving their collaboration approaches particularly rapidly. The workflows in healthcare delivery have been slower to change than the flow of new tools becoming available to providers. For example, NGS sequencing reports may require in excess of 3 weeks to be delivered to the physician, due to billing considerations (CMS 14 day rule), informatics architecture, and other workflow-related factors. These reports may then be used by the physician for treatment decisions, then attached to an electronic health record as a PDF document, rendering secondary digital queries of embedded data, difficult to impossible for later research.

In other words, the workflows support primary use of precision medicine data, but not the secondary uses that provide such promise, to better understand subsequent care decisions, or advance population health. Many other examples of the interplay of health data and workflows exist, and informatics oriented healthcare companies see opportunity in collaborating, to understand and help solve these issues; either as a direct software-as-a-service business opportunity, or to enable greater throughput of clinical cases, that increase the use of their machines, their services, or their products.

It is worth noting that biopharma companies are attentive to these workflow issues, as they impact the volume use of their products, and potentially clinical outcomes. Health systems without integrated diagnostic enterprises, or independent hospital affiliates, often lose the opportunity to ingest raw data generated in the course of laboratory analysis. For example, when samples are sent externally for NGS analysis, the raw .vcf sequence file format is usually not returned to the physician, and thus, not available for research. Collaborations between these types of companies and academic medical centers are often structured as strategic alliances, involve much barter of data and human resources, and the occasional in-kind contribution to a project or initiative; and commonly, dollars never change hands in these partnerships. The benefits of these kinds of collaborations tend to be much less tangible because they impact the bottom line (cost reduction), more visibly than the top line (sales), particularly for the provider side of the relationship. However, the operational visibility of academic health centers may benefit from such investments.

Policy, ethical, and regulatory considerations

Pharma companies and physician—scientists are often explicitly interested in variants of unknown function and pursue these efforts using extensive genome sequencing. While these efforts are often fruitful academic exercises, they can create significant challenges in healthcare decision making, as variants of unknown significance are often nonactionable targets, that create issues and potential liabilities for the provider and the corresponding institutions. Consequently, providers default to only focusing on actionable genetic findings, as defined by the American College of Medical Genetics [66] or Clinical pharmacogenetics implementation consortium (CPIC) guidelines [67].

In instances where sufficient molecular insight is available, a reasonable compromise is the enrollment in clinical research trials, specifically designed to help define management decisions. This option is only available for certain diseases and select populations and continues to pose a major impediment to widespread utilization of genomic data in clinical practice. Clearly, the collection of broad, untargeted genotyping information in the course of the standard of care, must be done thoughtfully with clearly laid out action plans, and ideally under a research consent which addresses some of the management issues for providers.

Concluding statements

There are significant frictions to be addressed to enable compliant and frictionless flow of specimens and data between patients, providers, researchers, and industry collaborators. Computational tools such as mobile computing and blockchain ledger technology, have potential to substantially impact these frictions, but financial incentives for investment in these technologies is a challenge, requiring use cases and a revenue model, that fits the reimbursement based healthcare system. Deployment of these computational technologies as infrastructure, to mitigate costs in large drug clinical trials, might be a viable first use.

Institutions should optimize all of the currencies for precision medicine collaborations: biorepository quality, maximally useful data commons, maximize the ingestion of omics data for its clinical cases, *minimize* transactional frictions underlying industry collaborations, and streamline relevant administrative workflows. Academic partners of choice for industry collaborators, tend to perform well in most of these criteria. Layering a prudent and practical proprietary strategy, that might involve patenting and licensing, should certainly be facilitated, where protecting the investment in clinical trials and productization is warranted. However, managing and leveraging the currencies of data and specimens, are critically important industry-facing functions of the academic medical center research enterprise, that sometimes get overshadowed by the perception that industry engagement is mostly about patenting and licensing.

References

[1] Venter JC, Adams MD, Myers EW, et al. The sequence of the human genome. Science 2001;291(5507):1304–51.

[2] National Institutes of Health. The cost of sequencing a human genome. July 6, 2016. https://www.genome.gov/27565109/the-cost-of-sequencing-a-human-genome/.

[3] Strande NT, Berg JS. Defining the clinical value of a genomic diagnosis in the era of next-generation sequencing. Annu. Rev. Genom. Hum. Genet. 2016;17(1):303–32.

[4] Sanger F, Nicklen S, Coulson AR. DNA sequencing with chain-terminating inhibitors. Proc. Natl. Acad. Sci. U.S.A. 1977;74:5463–7.

[5] Balasubramanian S, Klenerman D, Barnes C. Arrayed polynucleotides and their use in genome analysis. Patent US20030022207. 2003.

[6] Braslavsky I, Hebert B, Kartalov E, Quake SR. Sequence information can be obtained from single DNA molecules. Proc. Natl. Acad. Sci. U.S.A. 2003;100:3960–4.

[7] Mitr RD, Shendure J, Olejnik J, Edyta-Krzymanska-Olejnik, Church GM. Fluorescent *in situ* sequencing on polymerase colonies. Anal. Biochem. 2003;320:55–65.

[8] Weber JL, Myers EW. Human whole-genome shotgun sequencing. Genome Res. 1997;7:401–9. https://doi.org/10.1101/gr.7.5.401.

[9] Payne K, Gavan SP, Wright SJ, Thompson AJ. Cost-effectiveness analyses of genetic and genomic diagnostic tests. Nat. Rev. Genet. 2018;19(4):235–46. https://doi.org/10.1038/nrg.2017.108.

[10] Travis WD, Brambilla E, Burke AP, Marx A, Nicholson AG. WHO classification of tumours of the lung, pleura, thymus and heart. Lyon: International Agency for Research on Cancer; 2015.

[11] Sharma SV, Bell DW, Settleman J, Haber DA. Epidermal growth factor receptor mutations in lung cancer. Nat. Rev. Cancer 2007;7:169–81.

[12] Masri M, McManus M, Mudad R. Treatment of advanced non-small cell lung cancer in the era of targeted therapy. Curr. Pulmonol. Rep. 2018;7:79. https://doi.org/10.1007/s13665-018-0204-5.

[13] de Bruin EC, McGranahan N, Mitter R, Salm M, Wedge DC, et al. Spatial and temporal diversity in genomic instability processes defines lung cancer evolution. Science 2014;346(6206):251–6.

[14] Hirsch FR, Scagliotti GV, Mulshine JL, et al. Lung cancer: current therapies and new targeted treatments. Lancet Januray 21, 2017;389(10066):299–311. https://doi.org/10.1016/S0140-6736(16)30958-8.

[15] Shigematsu H, Lin L, Takahashi T, et al. Clinical and biological features associated with epidermal growth factor receptor gene mutations in lung cancers. J. Natl. Cancer Inst. 2005;97(5):339–46. https://doi.org/10.1093/jnci/dji055.

[16] Jamal-Hanjani M, Quezada SA, Larkin J, Swanton C. Translational implications of tumor heterogeneity. Clin. Cancer Res. 2015;21(6):1258–66.

[17] Scannell JW, Blanckley A, Boldon H, Warrington B. Diagnosing the decline in pharmaceutical R&D efficiency. Nat. Rev. Drug Discov. 2012;11:191–200.

[18] Silva P, Schiabley V, Ramos KS. Academic medical centers as innovation ecosystems: addressing the population –omics & personalized diagnosis paradox through novel industry partnership models in precision medicine. J. Transl. Med. 2018;16:28. https://doi.org/10.1038/nrd3681.

[19] Paskal W, Paskal AM, Debski T, Gryziak M, Jaworowski J. Aspects of modern biobank activity - comprehensive review. Pathol. Oncol. Res. 2018;24(4):771–85.

[20] McCormick T, Martin K, Hehenberger M. The evolving role of biomarkers, focusing on patients from research to clinical practice. The IBM Institute for Business Value; 2007. p. 4. http://www-935.ibm.com/services/us/gbs/bus/pdf/g510-6640-00_biomarker.pdf.

[21] Vaught J, Rogers J, Carolin T, Compton C. Biobankonomics: developing a sustainable business model approach for the formation of a human tissue biobank. JNCI Monogr. 2011;42:24–31.

[22] Ragin C, Park JY. Biospecimens, biobanking and global cancer research collaborations. Ecancermedicalscience 2014;8:454. https://doi.org/10.3332/ecancer.2014.454. Published 2014 Aug 28.

[23] Office of biorepositories and biospecimen research. NCI best practices for biospecimen resources. 2011. Bethesda (MD), https://biospecimens.cancer.gov/bestpractices/.

[24] Pitt B, Pfeffer MA, Assmann SF, et al. Spironolactone for heart failure with preserved ejection fraction. N. Engl. J. Med. 2014;370(15):1383–92.

[25] Di Bisceglie AM, Shiffman ML, Everson GT, et al. Prolonged therapy of advanced chronic hepatitis C with low-dose peginterferon. N. Engl. J. Med. 2008;359:2429–41.

[26] Jacobson RS, Becich MJ, Bollag RJ, et al. A federated network for translational cancer research using clinical data and biospecimens. Cancer Res. 2015;75(24):5194–201.

[27] Murray MF, Carey DJ, Davis FD, et al. The Geisinger-Regeneron initiative: a new kind of genomics research collaboration. Report, Geisinger Health Systems. 2014.

[28] Ledford H. Drug firm seeks genome bounty. Nature April 22, 2016;532:427. https://doi.org/10.1038/nature.2016.19797.

[29] McCallister E. BioCentury Dec. 24, 2012;20(51):A1–4.

[30] Nioi P, Sigurdsson A, Thorleifsson G, et al. Variant ASGR1 associated with a reduced risk of coronary artery disease. N. Engl. J. Med. 2016;374:2131–41. https://doi.org/10.1056/NEJMoa1508419.

[31] Mullard A. An audience with Sean Harper. Nat. Rev. Drug Discov. 2018;17:10–1. https://doi.org/10.1038/nrd.2017.262.

[32] Servick K. Can 23andMe have it all? Science September 25, 2015;349(6255):1472–4. https://doi.org/10.1126/science.349.6255.1472. 1476-7.

[33] Scott CT, Caulfeild TC, Borgelt T, Illes J. Personal medicine—the new banking crisis. Nat. Biotechnol. 2012;30:141–7.

[34] Weiner MG. POINT: is ICD-10 diagnosis coding important in the era of big data? Yes. Chest 2018;153(5):1095–8.

[35] Liebovitz DM, Fahrenbach J. COUNTERPOINT: is ICD-10 diagnosis coding important in the era of big data? No. Chest 2018;153(5):1095–8.

[36] Tyree PT, Lind BK, Lafferty WE. Challenges of using medical insurance claims data for utilization analysis. Am. J. Med. Qual. 2006;21(4):269–75.

[37] Brady BL, Tkacz J, Meyer R, Bolge SC, Ruetsch C. Assessment of Rheumatoid Arthritis Quality Process Measures and Associated Costs. Popul Health Manag 2017;20(1):31–40. https://doi.org/10.1089/pop.2015.0133.

[38] Bloem BR, Ypinga JHL, Willis A, et al. Using medical claims analyses to understand interventions for Parkinson patients. J. Parkinson's Dis. 2018;8(1):45–58.

[39] Erickson SM, Rockwern B, Koltov M, McLean RM. Medical practice and quality committee of the American College of

Physicians. Putting patients first by reducing administrative tasks in health care: a position paper of the American College of Physicians. Ann. Intern. Med. 2017;166(9):659–61.

[40] Tett G. How Big Tech brought back the barter economy. Financial Times Magazine; April 18, 2018.

[41] Orcutt M. Who will build the health-care blockchain? MIT Tech. Rev. 2017 https://www.technologyreview.com/s/608821/who-will-build-the-health-care-blockchain/.

[42] Supreme Court of the United States. Laboratory Corporation of America Holdings, Dba Labcorp v Metabolite Laboratories. 2006. 548 S Ct 124.

[43] Klein RD, Mahoney MJ. LabCorp v. Metabolite Laboratories: the supreme court listens, but declines to speak. J. Law Med. Ethics 2008;36(1):141–9. https://doi.org/10.1111/j.1748-720X.2008.00243.x.

[44] Supreme Court of the United States. Association for molecular pathology v myriad genetics. 2013. 133 S Ct 2107.

[45] Holman CM. Mayo, Myriad, and the future of innovation in molecular diagnostics and personalized medicine. N.C. J. Law Technol. 2014;15:639–78.

[46] Thomas J. Mayo v. prometheus: implications for patents, biotechnology, and personalized medicine. Congressional Research Service Report for Congress No. R42815. 2012.

[47] Supreme Court of the United States. Mayo Collaborative Services, dba Mayo Medical Laboratories v Prometheus Laboratories, Inc. 2012. 132 S Ct 1289.

[48] Silva P, Ramos KS. Academic medical centers as innovation ecosystems: evolution of industry partnership models beyond the Bayh-Dole Act. Acad. Med. 2018;93:1135–41.

[49] The European Patent Convention. Article 53. Exceptions to patentability. 2018. https://www.epo.org/law-practice/legal-texts/html/epc/2016/e/ar53.html.

[50] Schrell A, Bauser H, Brunner H. Biotechnology patenting policy in the European Union – as exemplified by the development in Germany. In: Fiechter A, Sautter C, editors. Green gene technology. Advances in biochemical engineering/biotechnology. Berlin, Heidelberg: Springer; 2007. p. 107.

[51] Ramamurhty L, Maxwell K, Sawchyn B, Anhorn R. Perspective: FDA/CMS parallel review advances coverage for cancer comprehensive genomic profiling. Am. J. Manag. Care Evid. Based Oncol. 2018;24:6.

[52] Topaloglu U, Palchuk MB. Using a federated network of real-world data to optimize clinical trials operations. J. Clin. Oncol. Clin. Cancer Inf. February 16, 2018 https://doi.org/10.1200/CCI.17.00067Published online on ascopubs.org/journal/cci.

[53] Hines JZ, Lurie P, Yu E, Wolfe S. Left to their own devices: breakdowns in United States medical device premarket review. PLoS Med. 2010;7(7):e1000280. https://doi.org/10.1371/journal.pmed.1000280.

[54] Heinemann L, DeVries JH. Citation for: evidence for continuous glucose monitoring: sufficient for reimbursement? Diabet. Med. 2014;31:122–5.

[55] Antman EM, Creager MA, Houser SR, Warner JJ, Konig M, American Heart Association. American Heart Association principles on the accessibility and affordability of drugs and biologics: a presidential advisory from the American Heart Association. Circulation 2017;12;136(24):e441–7. https://doi.org/10.1161/CIR.0000000000000551. Epub 2017 Nov 9.

[56] Pharmacoeconomic review report: Migalastat (Galafold): (amicus therapeutics): indication: Fabry disease [Internet]. Ottawa (ON): Canadian Agency for Drugs and Technologies in Health; February 2018.

[57] Lin JK, Lerman BJ, Barnes JI, Boursiquot BC, Tan YJ, Robinson AQL, Davis KL, Owens DK, Goldhaber-Fiebert JD. Cost effectiveness of chimeric antigen receptor T-cell therapy in relapsed or refractory pediatric B-cell acute lymphoblastic leukemia. J. Clin. Oncol. September 13, 2018. https://doi.org/10.1200/JCO.2018.79.0642. JCO2018790642. [Epub ahead of print].

[58] Gilead Sciences Canada, Inc. Pharmacoeconomic review report: Sofosbuvir/Velpatasvir/Voxilaprevir (Vosevi). Indication: hepatitis C infection genotype 1 to 6. CADTH common drug review. Ottawa (ON): Canadian Agency for Drugs and Technologies in Health; February 2018.

[59] Pharmacoeconomic review report: Glecaprevir/Pibrentasvir (Maviret): (AbbVie Corporation): indication: hepatitis C genotype 1 to 6 CADTH common drug reviews. Ottawa (ON): Canadian Agency for Drugs and Technologies in Health; February 2018.

[60] Paik S, Shak S, Tang G, et al. A multigene assay to predict recurrence of tamoxifen-treated, node-negative breast cancer. N. Engl. J. Med. 2004;351:2817–26.

[61] Vaishnavi A, Capelletti M, Le AT, et al. Oncogenic and drug-sensitive NTRK1 rearrangements in lung cancer. Nat. Med. 2013;19(11):1469–72.

[62] Doebele RC, Davis LE, Vaishnavi A, et al. An oncogenic NTRK fusion in a patient with soft-tissue sarcoma with response to the tropomyosin-related kinase inhibitor LOXO-101. Cancer Discov. 2015;5(10):1049–57.

[63] Mullard A. NCI-MATCH trial pushes cancer umbrella trial paradigm. Nat. Rev. Drug Discov. August 2015;14(8):513–5. https://doi.org/10.1038/nrd4694.

[64] Ramsey BW, Davies J, McElvaney NG, et al. A CFTR potentiator in patients with cystic fibrosis and the G551D mutation. N. Engl. J. Med. 2011;365(18):1663–72.

[65] Gentzsch M, Mall MA. ion channel modulators in cystic fibrosis. Chest August 2018;154(2):383–93. https://doi.org/10.1016/j.chest.2018.04.036. Epub 2018 May 8.

[66] Richards S, Aziz N, Bale S, et al. Standards and guidelines for the interpretation of sequence variants: a joint consensus recommendation of the American College of Medical Genetics and Genomics and the Association for Molecular Pathology. Genet. Med. 2015;17(5):405–24.

[67] Caudle KE, Dunnenberger HM, Freimuth RR, et al. Standardizing terms for clinical pharmacogenetic test results: consensus terms from the Clinical Pharmacogenetics Implementation Consortium (CPIC). Genet. Med. 2016;19(2):215–23.

[68] Allen NE, Beral AV, Casabonne D, et al. On behalf of the million Women Study Collaborators. Moderate alcohol intake and cancer incidence in women. J. Natl. Cancer Inst. 2009;101(5):296–305. https://doi.org/10.1093/jnci/djn514.

[69] Khoury MJ, Evans JP. A public health perspective on a national precision medicine cohort: balancing long-term knowledge generation with early health benefit. J. Am. Med. Assoc. 2015;313(21):2117–8.

[70] Chen Z, Chen J, Collins R, et al. China Kadoorie Biobank of 0.5 million people: survey methods, baseline characteristics and long-term follow-up. Int. J. Epidemiol. 2011;40(6):1652–66.

[71] Collins R. Lancet March 31, 2012;379(9822):1173–4. https://doi.org/10.1016/S0140-6736(12)60404-8.
[72] Allen N, Sudlow C, Downey P, et al. UK Biobank: current status and what it means for epidemiology. Health Policy Technol. 2012;1:123–6. https://doi.org/10.1016/j.hlpt.2012.07.003.
[73] Gaziano JM, Concato J, Brophy M, et al. Million Veteran Program: a mega-biobank to study genetic influences on health and disease. J. Clin. Epidemiol. February 2016;70:214–23. https://doi.org/10.1016/j.jclinepi.2015.09.016. Epub 2015 Oct 9.
[74] Locke AE, Kahali B, Berndt SI, et al. Genetic studies of body mass index yield new insights for obesity biology. Nature 2015;518(7538):197–206.
[75] Shungin D, Winkler TW, Croteau-Chonka DC, et al. New genetic loci link adipose and insulin biology to body fat distribution. Nature 2015;518(7538):187–96.
[76] Nagai A, Hirata M, Kamatani Y, et al. Overview of the BioBank Japan Project: study design and profile. J. Epidemiol. 2017;27(3S):S2–8.
[77] Small AM, O'Donnell CJ, Damrauer SM. Curr. Cardiol. Rep. 2018;20:22. https://doi.org/10.1007/s11886-018-0969-8Xs.
[78] Kaiser J. Who has your DNA—or who wants it? Science 2015;349:1475.
[79] Ehret GB, Munroe PB, et al. International Consortium for Blood Pressure Genome-Wide Association Studies, genetic variants in novel pathways influence blood pressure and cardiovascular disease risk. Nature 2011;478:103–9.
[80] Deloukas P, Kanoni S, et al. For CARDIoGRAMplusC4D Consortium. Large-scale association analysis identifies new risk loci for coronary artery disease. Nat. Genet. 2012;45(1):25–33.
[81] Willer CJ, Schmidt EM, Sengupta S, Peloso GM, Gustafsson S, Kanoni S, et al. Discovery and refinement of loci associated with lipid levels. Nat. Genet. 2013;45:1274–83.
[82] Scholtens S, Smidt N, Swertz MA, et al. Cohort Profile: LifeLines, a three-generation cohort study and biobank. Int. J. Epidemiol. 2015;44(4):1172–80. https://doi.org/10.1093/ije/dyu229.
[83] Tapia-Conyer R, Kuri-Morales P, Alegre-Diaz J, et al. Int. J. Epidemiol. 2006;35(2):243–9. https://doi.org/10.1093/ije/dyl042.
[84] Alegre-Diaz J, Herrington W, Lopez-Cervantes M, et al. Diabetes and cause-specific mortality in Mexico City. N. Engl. J. Med. 2016;375:1961–71.
[85] Morris AP, Voight BF, Teslovich TM, Ferreira T, Segre AV, Steinthorsdottir V, et al. Large-scale association analysis provides insights into the genetic architecture and pathophysiology of type 2 diabetes. Nat. Genet. 2012;44:981–90.
[86] https://www.genengnews.com/topics/omics/regeneron-genetics-center-surpasses-250k-exomes-sequenced-and-ramping-up/.
[87] U.K. Biobank. Regeneron and GSK Announce largest gene sequencing initiative on World's most detailed health database to improve drug discovery and disease diagnosis. 2018. https://newsroom.regeneron.com/news-releases/news-release-details/uk-biobank-regeneron-and-gsk-announce-largest-gene-sequencing.
[88] Regeneron Partners With AbbVie, Alnylam, AstraZeneca, Biogen, Pfizer to Sequence UK Biobank Samples. 2018. https://www.genomeweb.com/sequencing/regeneron-partners-abbvie-alnylam-astrazeneca-biogen-pfizer-sequence-uk-biobank-samples#.XDVR4FVKiUk.
[89] Regeneron to Lead $50M Exome Sequencing Consortium with UK Biobank. 2018. https://www.genengnews.com/topics/omics/regeneron-to-lead-50m-exome-sequencing-consortium-with-uk-biobank/.
[90] Greely HT. The uneasy ethical and legal underpinnings of large-scale genomic biobanks. Annu. Rev. Genom. Hum. Genet. 2007;8:343–64.
[91] https://www.amgen.com/media/news-releases/2018/12/decode-genetics-an-amgen-subsidiary-and-somalogic-announce-collaboration-to-perform-largescale-protein-analysis-of-up-to-40000-human-samples/.
[92] Ledford H. AstraZeneca launches project to sequence 2 million genomes. Nature 2016;532:427. https://doi.org/10.1038/nature.2016.19797.
[93] AstraZeneca taps gene pioneer Venter for huge drug-hunting sweep. 2018. https://www.reuters.com/article/us-astrazeneca-genomics/astrazeneca-taps-gene-pioneer-venter-for-huge-drug-hunting-sweep-idUSKCN0XI2Z1.
[94] Guidelines for validation of next-generation sequencing-based oncology panels: a joint consensus recommendation of the Association for Molecular Pathology and College of American Pathologists. J. Mol. Diagn. 2017;19(3):341–65.
[95] Holub P, Swertz M, Reihs R, van Enckevort D, Müller H, Litton JE. BBMRI-ERIC directory: 515 Biobanks with over 60 million biological samples. Biopreserv. Biobanking 2016;14(6):559–62.
[96] Simell BA, Törnwall OM2, Hämäläinen I, et al. Transnational access to large prospective cohorts in Europe: current trends and unmet needs. Nat. Biotechnol. March 25, 2019;49:98–103. https://doi.org/10.1016/j.nbt.2018.10.001.

Chapter 54

The future of precision medicine

Reza Mirnezami[1] and Arsalan Wafi[2]

[1]*Colorectal Surgeon and Honorary Lecturer in Surgery, Department of Surgery & Cancer, Imperial College London, London, United Kingdom;*
[2]*Clinical Research Fellow, Department of Cardiovascular Surgery, St George's Hospital, University of London, London, United Kingdom*

Origins of the precision medicine concept and its context in the real world

The Human Genome Project (HGP), launched in 1990, was the result of a growing understanding of the complexity of factors governing human health and disease, coupled with a heightened awareness of the influence of the genetic make-up of an individual, on disease susceptibility. This ambitious venture into the "molecular unknown" represented a monumental scientific effort, academic challenge, and financial burden [1]. In 2003, mapping of the entire human genomic landscape was complete, and since that time we have witnessed rapid growth in "omics" scientific disciplines, spurred on by biotechnological advances and increasing public demand.

The term "omics" in this context, typically refers to a quantitative or semi-quantitative analysis of biomolecular structure, function or activity, within a given specific biological domain (e.g., genome, proteome, metabolome). Translational omics refers to the effective deployment of this derived molecular information to the clinical setting [2]. The requirement for analysis of increasingly large-scale population-based data is burdensome from the chemometric aspect. However, this analysis should guarantee the discovery of a large number of latent associations, between an individual's molecular phenotype, lifestyle behaviors, and environmental exposures.

The vision of precision medicine (PM) in the future, is ultimately that of a health space, within which the biomolecular infrastructure of every individual, through health, disease, and pharmaco-therapeutic intervention, is defined, stored, and readily accessed, from a global data repository or virtual cloud, for the greater well-being of not only that individual, but also others experiencing similar "health events." This in itself, will require secure data analysis, collection, and storage on an unprecedented scale, utilizing hundreds-to-thousands of multilevel and multinational bio-information banks [3].

The road ahead

The concern, however, as demonstrated by the relatively limited number of examples, of genuinely transformative genomic translations to bedside medicine, is that in the real world, such data mining may prove disappointingly ineffective. Here, it is important to bear in mind that biomarker-disease phenotype associations should not be confused with, nor do they guarantee, *precision*, per se. Comprehensive omics-wide evaluation, even at the individual level, will generate data on the terabytes scale, and unfortunately, in this context, the deeper one scratches the surface, the deeper the data quagmire becomes. Most of the anticipated newly discovered associations between disease and molecular phenotype are likely destined to have small effect-size and low predictive validity [4].

Precision medicine will, therefore, need to overcome the apparent irony, that mankind's evolutionary survival mechanism of having a biodiverse gene or omics pool, will in itself serve as a hindrance to efficient and valid translational application. It is, therefore, clear that despite the unquestionable promise of PM, the clinical practice of medicine is unlikely to undergo a change at the rate that was originally foreseen when the HGP was completed. This is typical of one of the fundamental laws of technological advancement; we invariably overestimate the short-term impact of a truly transformational discovery, while under-estimating its longer-term effects [5].

Diagnostic, therapeutic, and preventative potential of precision medicine

Full sequencing of the first human genome cost an estimated $400 million. Today, the cost stands at $9,500, and within the next four to 5 years, it is projected for $1,000, or

Precision Medicine for Investigators, Practitioners and Providers. https://doi.org/10.1016/B978-0-12-819178-1.00054-X

less By mid-2015, costs for such sequencing were estimated to amount to just above $4000 and today, the cost stands at just over $1000 [6]. Correspondingly, the last decade has witnessed an exponential growth in the use of "genetic fingerprinting," and this is perhaps best exemplified in the case of cancer medicine [7]. The need for PM tools in cancer detection, surveillance, and therapeutic monitoring is driven by a host of unique selling points, that tackle the limitations of traditional laboratory-based tests, for example, glycoprotein-based tumor markers, such as carcinoembryonic antigen (CEA) and cancer antigen 125 (CA-125).

First, each individual's "omics" diversity means that conventional point-of-care tests lack specificity and sensitivity. Moreover, they only offer a time-specific single point measurement of the level of a given target entity and compare this value with a reference range. Such linear approaches fail to take account of, and thus fail to capture the fundamentally dynamic nature of molecular fluxes that regulate human health and disease; in addition, these tests, like most in current clinical use, fail to appreciate that "normal reference range" is likely to vary from individual to individual [8].

Wide spectrum cancer tests

CancerSEEK, a multi-analyte blood test, employs combined assays for evaluating genetic alterations, as well as circulating expression of gene-encoded proteins [9]. Cohen and colleagues applied the CancerSEEK test to the evaluation of eight different cancer subtypes (ovary, liver, stomach, pancreas, esophagus, colorectum, lung, and breast), amongst 1005 patients with nonmetastatic malignancy. The authors reported high sensitivity and specificity for detecting a variety of cancer subtypes, including several that currently lack validated biomarker screening for early detection. Other similar examples in ovarian cancer are the ROMA and OVA1 tests, which look at multiple disease-specific markers (including HE4, ApoA1, and transferrin), and which have demonstrated high sensitivity and specificity [10,11].

New drug screening and indication

Two very broad areas of potential for PM drug development and selection are (1) selection of the most effective therapeutic agent for a particular patient and (2) the ability to mitigate one of the key challenges in cancer treatment, which is drug resistant. In terms of drug selection, there has been recent steady growth in the development of companion diagnostics (CDx) tests that are designed to direct the clinician to the optimal therapeutic agent for a given patient. For example, there are currently 10 CDx products available for Trastuzumab (Herceptin), which is approved for the treatment of early-stage breast cancers that are found to overexpress human epidermal growth factor receptor-2 (HER2+) and have spread to the draining lymph node basin, or those which are HER2+, and have not spread to the lymphatics but are deemed to harbor high-risk features [12].

Herceptin is a highly effective inhibitor of cancer cell proliferation in HER2+ tumors; however, the drug only works in approximately one-quarter to one-third of patients that are found to overexpress the HER protein. One of the first and most commonly used CDx for Herceptin is the HercepTest, which boasts specificities of above 93% in determining HER2/neu protein gene expression [13].

Personalized therapy

From the point of view of CDx development, the opportunities for the pharmaceuticals industry are expansive, as one can envisage a nested arrangement, as illustrated in Fig. 54.1. Where originally, CDx would be designed to perform at one level, e.g., to define the patients with HER2 overexpressing tumors, the requirements now are far greater, e.g., to define the patients with HER2 expressing tumors that would be nonresponders, and/or that would experience unacceptable treatment toxicity.

Refractory tumors

Omics approaches offer the opportunity to interrogate resistant cancer subtypes, identify molecular mechanisms underlying drug resistance, and potentially manipulate cancers to "convert" them to a treatment-sensitive state. One of the most challenging aspects of lung cancer care is the ability of tumors to rapidly shift to a multidrug-resistant (MDR) phenotype. An example of such a study is the targetting of a unique biomarker, in a multidrug-resistant form of non-small cell lung cancer, leading to a halt in cancer growth. In this study the screening of a 1000-gene panel in patients with the anaplastic lymphoma kinase-rearranged non-small cell lung cancer (ALK-NSCLC) phenotype, led to the identification of SHP2 (a non-receptor protein tyrosine phosphatase) as a targetable biomarker of therapeutic resistance. In an elegant demonstration, targeting this biomarker with SHP099, a small molecule inhibitor of SHP2, led to a halt in the growth of the treatment resistance cancer cells [14].

Cardiovascular prevention

Disease prevention is one of the central pillars of P4 medicine [15]. The inherited metabolic disorder, familial hypercholesterolemia, which carries an increased risk of early cardiac-related morbidity and mortality, has been targeted recently in screening studies with lipid panel and

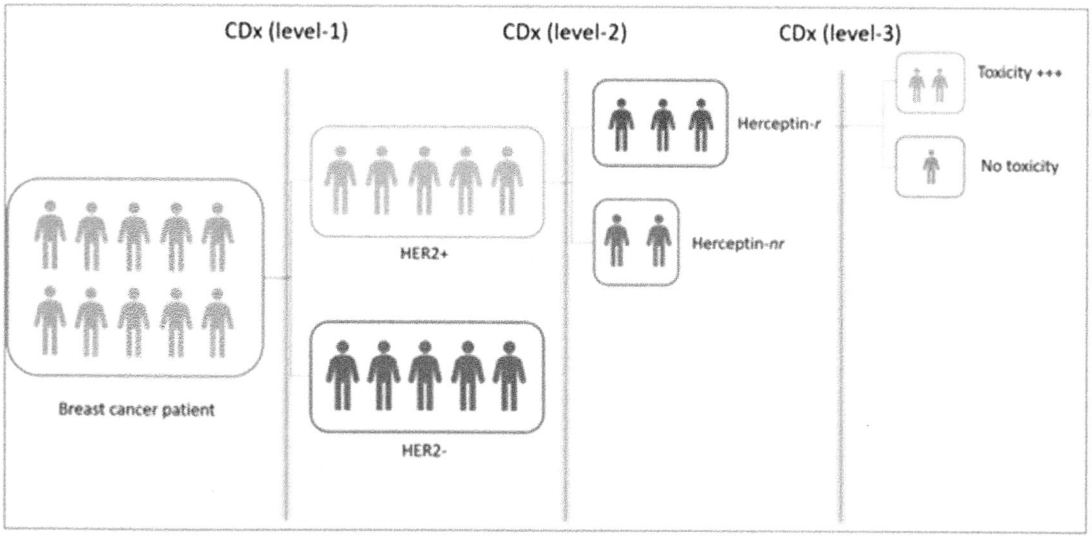

FIGURE 54.1 Figure illustrating the potential for the development of multiple companion diagnostics tests for a given condition. CDx (level-1) predicts HER2 status. CDx (level-2) defines which patients with HER2+ status will be Herceptin responsive (Herceptin-r), and who will be nonresponsive (Herceptin-nr). CDx (level-3), in turn, determines the relative likelihood of drug-associated toxicity, making the decision-making in terms of therapeutic intervention far more precise for a given individual.

genetic testing, with promising results both in terms of cost-effectiveness, and also in terms of potential for the development of personalized cardiovascular protective therapies.

Importance for patients and healthcare providers

Ultimately, PM is envisaged to fulfill a tripartite health manifesto: (1) to promote well-being and prevent the development of illness; (2) to provide patients with the most reliable, safe and well-tolerated diagnostic tests; (3) to provide patients with the most effective treatments tailored to them, with maximal expected therapeutic benefit and minimal unwanted toxicity. These anticipated benefits must be carefully and transparently defined to patients and the wider public to ensure that this most critical stakeholder group is won over, and above all, safeguarded.

Patient participation

Patient and public trust in PM is critical to ensuring support for, and adherence to clinician recommendations. It is also crucial that patient recruitment in omics research is prioritized, by reinforcing the concept of participation being a voluntary act and protecting the values of reciprocity, nonexploitation, and service for the public good [16]. Accessibility and fairness is a well-documented barrier to healthcare, and there remains inconsistent coverage, and the risk of incurring significantly high costs for trials, when reaching further out continues to hamper patient enrollment beyond major cities [17]. Therefore, it is a given fact that for patients to benefit fully, precision medicine must be widely available in all geopolitical climates, for all socioeconomic groups.

Individual attention versus public health needs

Precision medicine will provide clinicians with the necessary tools to make diagnoses and provide treatments, with the reassurance that the patient's state, as well as the disease process, have been comprehensively and exhaustively analyzed. Not only does this encourage improvements in efficacy and reduction in complications, but it also promises to provide a more efficient system of healthcare. At a clinic level, one can envisage more productive consultations and a more effective method of patient follow-up. At the hospital level, it will lead to an overhaul of referral, admission, and discharge pathways, which should result in a reduction of waiting times, improved service utilization, and reduction in length of stay and readmission rates. At a regional and national level, access to a universal cloud of data will streamline the flow of patients to the most appropriate tertiary services, in turn improving the efficiency of the hub and spoke model of healthcare.

Medical education and healthcare personnel training

To effectively translate new knowledge to the bedside, a basic working knowledge of biomarkers, cellular biology, and algorithmic computational science will become a minimal requirement for future clinicians. The upside of such a prerequisite is that it will give rise to new super-specialties, based on omics driven medicine. Medicine will become a more dynamic environment, in which new knowledge of biomarker associations and disease states will be generated at an exponential rate. Staying up-to-date with such an influx of new findings for clinicians presents a great challenge, and it is likely that computer-based decision support systems will be relied upon heavily [18].

Such systems will offer treatment algorithms, which in themselves require extensive validation first. Currently, there are no universally accepted recommendations for validating biomarker associations or treatment algorithms, and widespread utilization of such tools outside of the clinical trial setting will require a great amount of trust in the system from clinicians. The immense time and financial burden of such technology and manpower will, in no doubt, need a thorough justification to gain support.

Role and regulation of academia and industry

An exponential growth in discoveries in terms of markers of disease susceptibility, diagnosis, prognosis, and therapy is expected [19]. It is the responsibility of both investigators and institutions to protect the integrity of their research, and ensure that findings are appropriately validated and tested for reproducibility. It is imperative that academic research organizations avoid poor experimental design and poor quality research. A candidate omics-based test must be adequately designed, to answer a specific, well-defined, and relevant clinical question, while being conducted with adequate statistical and bioinformatics rigor. Validation will not only be undertaken on even larger heterogeneous datasets, but will have to be dynamic and expect to evolve on a regular basis, as further discoveries on upstream, downstream, and parallel biomarker pathways are made.

Practically, this will require complex processing pipelines, integration of novel multiomic analytical tools, with each pipeline module consisting of a large number of data treatment steps (initial alignment of data, data refinement, variant calling, and omic classifying, and finally biostatistical mining) [20]. How feasible this is, remains one of the many unanswered questions. With the advent of machine learning and super-computer technology, multidimensional heterogeneous data processing will be scaled up and will allow massive data sets to be analyzed in a single hub, avoiding the network-speed minimizing latency issues. The huge financial burden of such endeavors on institutions will present a large barrier to the progress of omics-based academia, and it is likely that simultaneous scaling out approaches (linkage of less powerful but more accessible and more numerous processing units), will be undertaken [21].

Translational issues

Studies will need to prioritize biomarkers based on available clinical knowledge and discard futile ideas early, implementing only relevant and beneficial concepts that are reproducible and stand the test of time. Omics discoveries that offer results will be transferred to the bedside, while discarded ideas remain in the pool of omics libraries, for potential future reevaluation. The authors (AW and RM) anticipate that a phase of highly intensive and likely highly costly omics wide investigation can be expected, in the early phases of populating this data library. As the candidate markers with the greatest potential are identified, so a "less is more" stance can be assumed in translational biomarker deployment, leading in turn to both a sharp reduction in costs, as well as analytical workload. This projected relationship is illustrated in Fig. 54.2. Moreover, a comprehensive understanding of pharmaco-therapeutic efficacy should yield cost-effective repurposing of existing therapies, for previously unknown disease associations [8].

Innovative regulatory pathways

Regulatory bodies in place to approve the translation of discoveries to the bedside will likely evolve, to keep up with the expansion of the omics knowledge field. It will not be feasible to implement lengthy and costly trials to test every association. More scientifically robust critical appraisals of literature, including novel emerging techniques for synthesizing evidence, for example, through use of AI-equipped natural language processing software, will more than likely need to be utilized [22]. Regulatory bodies must also deal with a dramatic increase in administrative burden, because of higher numbers of contracts, regulatory processes, and generation of multiple protocols. Overly stringent regulatory frameworks are known to impede the progress of translational omics initiatives and can restrict research to only the more financially commanding and resourceful research centers.

Drug repurposing

Drug repurposing is a new and interesting avenue [23]. A deeper molecular understanding of diseases is likely to demonstrate latent biomolecular commonality between conditions, which in turn could facilitate successful drug repurposing initiatives. Traditionally, computer-based docking screens have been used to discover new ligands, for targets of known structures, and prediction of new substrates for enzymes of unknown function, through which new drugs are created [24]. Recently, the reciprocal approach to this has gained popularity in which a known approved drug (rather than the target), is used to predict

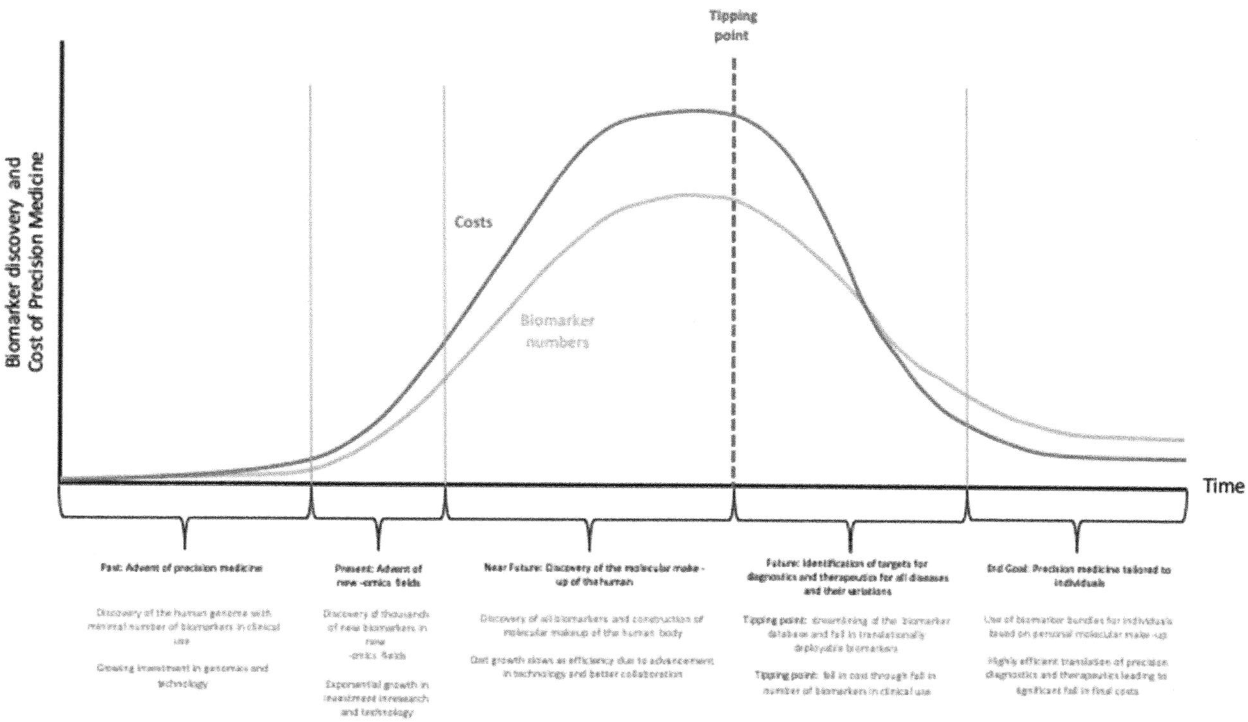

FIGURE 54.2 Past, present and future projected trends of biomarker discovery, utilization, and cost in precision medicine.

new targets, for repurposing use, by analyzing known patterns of the chemistry of the ligands [23]. This approach is paving the way, toward commonplace drugs being used to treat previously unassociated conditions. Such approaches present an opportunity for accelerated industry growth and reduced production costs.

Opportunities for the pharmaceutical industry

In 2016, for the third year in a row, personalized medicines (PMs) accounted for more than 20% of all new molecular entities (NMEs), the Food and Drug Administration (FDA) agency approved in the United States. That ratio is a sharp increase from 2005 when PMs accounted for just 5% of NME approvals [8,25]. However, with the potential for great reward, so comes the potential for significant failure. AstraZeneca's Iressa (Gefitinib), an epidermal growth factor receptor (EGFR) inhibitor, first approved for use in advanced non-small cell lung cancer (NSCLC), was retracted from the market due to poor efficacy. Only later was it discovered that the drug's effect is restricted to the subtypes of advanced NSCLC, with a specific EGFR mutation. The drug was subsequently reapproved for use in this subgroup of patients, but the firm took a huge financial hit and lost time in the market [26]. A more damaging example is that of IMFINZI (durvalumab) use in metastatic NSCLC, as phase 3 trials failed to show a significant progression-free survival, and led to the firm's largest fall in market share in its history [27].

Role of alliances and coalitions

Precision medicine alliances and coalitions are international groups comprising multistakeholder, multisectoral and multiprofessional large-scale collaborations, aimed at answering and optimizing global and expert questions, toward the implementation, democratization, and standardization of precision medicine. Personalized Medicine Connective, a United States-based non-profit organization founded in 2015 by industry leaders, aims to quantitatively describe an integrative framework to generate, optimize, and capture value, in the utilization of personalized/precision medicine [28]. The group recently carried out a cost-benefit analysis using metastatic melanoma as a case example. The organization concluded that improved diagnostics could identify approximately twice the number of patients at the early stages of the disease, and in turn reduce the need for second-line surgery by 72% for late disease, lowering the overall cost of melanoma treatment by more than $1 billion per year in the United States alone [8].

Below is a list of some of the alliances and coalitions at the forefront of PM:

- The Precision Medicine Initiative – The White House (www.whitehouse.gov/precisionmedicine)

- WR Worldwide Innovative Networking In Personalized Cancer Medicine (www.winconsortium.org)
- European Alliance For Personalized Medicine (www.euapm.eu)
- The European Personalized Medicine Association (www.epemed.org)
- PMC Personalized Medicine Coalition (www.personalizedmedicinecoalition.org)
- Global Alliance For Genomics & Health (www.ga4gh.org)
- Genomic Medicine Alliance (www.genomicmedicinealliance.org)
- Cancercommons (www.cancercommons.org)
- Personal Connected Health Alliance (www.pchalliance.org)
- The Precision Medicine Alliance (www.precisionmedicinealliance.org)

Public-health impact

The Precision Medicine Initiative introduced by the Obama administration in 2015, was devised with a clear strategy to take a near-term focus on cancers, and in the longer-term to take strides toward generating knowledge more broadly applicable to the wider spectrum of well-being and disease [29]. The stance taken by this group followed a logical rationale; cancer is an obvious choice for enhancing the reputation of PM in the near-term. Cancer is common and imposes a tremendous physical, emotional, psychological, and economic burden on patients, relatives, and society. Cancers are a leading cause of morbidity and mortality, and they evoke fear, because of their symptoms and because of the perceived consequences of cancer treatment. For these reasons, public enthusiasm for, and engagement with, an initiative demonstrating a firm commitment to beating cancer is likely to be strong, and this represents a powerful advantage in the early phases of PM growth.

Social and ethical challenges

Informed consent

"Untargeted" molecular phenotyping, leads to the generation and storage of vast amounts of person-specific biological data, without a clear aim or understanding of the data's social and medical impact. The technological and analytical tools used are heterogeneous, and owing to an exponential growth in newly devised phenotyping approaches, most remain in relative technological infancy, which makes standardization difficult, and weakens the validity of the biomarker data generated. Specific questions and study methodologies are not a feasible model of consent in omics studies carrying out untargeted multilevel tests, where the hypotheses may be only very broadly definable.

Data reassessment and shifting conclusions

The dynamic nature of knowledge generation, which is expected in PM, as marker validity is confirmed and new knowledge is acquired, means that there will be a moral and legal obligation to recontact patients, should a finding of clinical significance be identified further down the pipeline. Naturally, this will pose considerable logistic challenges, and there are questions regarding who would take overall responsibility of longitudinal data oversight and custodianship. At the other end of the discussion, is the argument that data should not be held hostage by patient concerns regarding data disclosure, or by provider concerns regarding litigation.

Biobanks and serial investigations

That is not to say that privacy and confidentiality concerns are subordinate to the greater good, only that there may be a balance, and that there should be an endorsement of "broad consent" for future use of biospecimens. In the research context, it has been argued that privacy and confidentiality as they have been understood, should perhaps be replaced by some sort of "responsible use" doctrine [30].

Data security and disease stigmata

Governance of information from a virtual cloud will require stringent security measures in place, to protect confidential information. However, the promise of protecting confidential information may have to be broken in specific circumstances, such as imminent threats and public health concerns. This leads to the next issue of social stigma. History teaches us that one can learn vast amounts, about the risk of stigmatizing certain conditions such as HIV, and the social implications this has on patients and communities. Interpretational uncertainties, heterogeneous levels of health literacy, and attitudes toward illnesses, highlight the importance of educating the public, as well as maintaining transparency in PM [31].

Access and disclosure of genetic tests

Policymakers have generated many laws such as the 2008 Genetic Nondiscrimination Act (GINA, USA), and the Affordable Care Act/genetic testing (Obamacare, USA, 2015), to combat discrimination based on genetic tests or family history, in employment and in health insurance. However, fear of discrimination and the consequences of transparent health disclosure, continue to represent significant factors associated with patient nonengagement in molecular phenotyping studies [32,33]. The PM community will have to combat these problems, with new models such as a "trusted governance" system, where patients consent to allow a group consisting of citizens and privacy experts to review data sharing and permit its use in appropriate contexts.

Science credibility and investigator integrity

In 2016, the publishing group, Springer, retracted 107 papers from one journal, after discovering they had been accepted with falsified peer reviews [8]. Moreover, in 2008, the FDA retracted approval status for a diagnostic test for ovarian cancer (Ovasure), which was sold to patients on the market for 4 months, based on findings that some of the supporting data had been falsified [34]. These reports highlight the importance of funders, regulatory bodies, and scientific journals in ensuring rigorous development and scrutinization of omics-based tests.

Funding bodies play a driving role in cultivating a culture of integrity and transparency in science. While funders seek to accelerate progress through discovery, translation, and clinical applications, the priority should be preventing misinformation. The use of independent verification and validation bodies should be encouraged, despite the stereotype that funders do not consider third party validation as innovative science. Without this assurance, the danger remains that debatable hypotheses and claims are encouraged, with the threat of promising ideas being overlooked.

Ongoing lines of investigation and research opportunities in the field

Complex interactions between host, environment, gut microbiome, and diet are now being explored in unprecedented detail. It is anticipated that a key element of PM moving forward, will be the individually tailored diet, comprised of food elements specifically selected for their disease beating, or disease-preventative properties. This is not a new concept but is now starting to gain real scientific credibility. Advances in microbiome science are beginning to enlighten the traditional understanding of nutrition, and are poised for clinical translation. Nutrition and gastrointestinal microflora are synergistically linked, such that the gut microbiome is responsible for nutrient signal transduction for the host, and in turn, the food consumed by the host, dictates and regulates gut microbiome composition.

Microbiome-host cross-talk

We now understand that the composition and diversity of food consumed by an individual will directly influence which microbes will colonize and flourish, and which will be pushed into extinction [35]. This, in turn, is understood to have consequences on host health, host risk of future inflammatory and metabolic disorders, and host therapeutic responsiveness. Thus, next-generation dietary advice will need to take into account, the disease-beating or preventative properties of bioactive compounds in different food products, and match these up with an individual's gut microbial fingerprint and genomic profile, to create a new and largely untapped field in PM: Precision Nutrition. This presents tremendous opportunities in terms of research and also in terms of novel/disruptive collaborations between public health organizations, biomedical science, and the food industry, on a scale not previously witnessed.

Artificial intelligence-driven healthcare

In 2017, Google subsidiary DeepMind announced that its AlphaZero AI application had learned to play the ancient game of chess, in entirely self-taught fashion, in just 4 hours. What was truly staggering to note was that having been primed only with the basic rules of the game, AlphaZero set about learning and staging moves that had not previously been described [36]. This ability for emerging AI platforms to offer "fresh" perspectives on old problems encountered in healthcare has generated considerable excitement. There is a now a growing movement toward high-quality research in these areas, and again, the prospects for collaborative enterprise between computational scientists and physicians is tremendous.

Regional focus, global perspective

The challenge we face is to develop next-generation healthcare models that can be readily deployed in any part of the globe, and that can defy economic disparities. Frugal innovative approaches, such as drug repurposing and dietary manipulation of health and disease, will be of particular importance in this regard.

References

[1] Langmead B, Trapnell C, Pop M, Salzberg S. Ultrafast and memory-efficient alignment of short DNA sequences to the human genome. Genome Biol. 2009;10(3):R25.

[2] Mirnezami R, Nicholson J, Darzi A. Preparing for precision medicine. N. Engl. J. Med. 2012;366(6):489–91.

[3] Kambatla K, Kollias G, Kumar V, Grama A. Trends in big data analytics. J. Parallel Distrib. Comput. 2014;74(7):2561–73.

[4] Barsanti-Innes B, Hey S, Kimmelman J. The challenges of validating in precision medicine: the case of excision repair cross-complement group 1 diagnostic testing. Oncologist 2016;22(1):89–96.

[5] National academies of sciences, engineering, and medicine; division on earth and life studies; institute for laboratory animal research; roundtable on science and welfare in laboratory animal use. Advancing disease modeling in animal-based research in support of precision medicine: proceedings of a workshop. Washington (DC): National Academies Press (US); May 30, 2018. 1, Introduction to Precision Medicine and Animal Models. Available from: https://www.ncbi.nlm.nih.gov/books/NBK507222/.

[6] National Human Genome Research Institute (NHGRI). The cost of sequencing a human genome. 2019 [Online] Available at: https://www.genome.gov/sequencingcosts/.

[7] Whitkus R, Doebley J, Wendel J. Nuclear DNA markers in systematics and evolution. Adv. Cell. Mol. Biol. Plants 1994:116–41.

[8] Wafi A, Mirnezami R. Translational –omics: future potential and current challenges in precision medicine. Methods 2018;151:3–11.

[9] Cohen J, Li L, Wang Y, Thoburn C, Afsari B, Danilova L, Douville C, Javed A, Wong F, Mattox A, Hruban R, Wolfgang C, Goggins M, Dal Molin M, Wang T, Roden R, Klein A, Ptak J, Dobbyn L, Schaefer J, Silliman N, Popoli M, Vogelstein J, Browne J, Schoen R, Brand R, Tie J, Gibbs P, Wong H, Mansfield A, Jen J, Hanash S, Falconi M, Allen P, Zhou S, Bettegowda C, Diaz L, Tomasetti C, Kinzler K, Vogelstein B, Lennon A, Papadopoulos N. Detection and localization of surgically resectable cancers with a multi-analyte blood test. Science 2018;359(6378):926–30.

[10] Montagnana M, Danese E, Ruzzenente O, Bresciani V, Nuzzo T, Gelati M, Salvagno G, Franchi M, Lippi G, Guidi G. The ROMA (Risk of Ovarian Malignancy Algorithm) for estimating the risk of epithelial ovarian cancer in women presenting with pelvic mass: is it really useful? Clin. Chem. Lab. Med. 2011;49(3).

[11] Longoria T, Ueland F, Zhang Z, Chan D, Smith A, Fung E, Munroe D, Bristow R. Clinical performance of a multivariate index assay for detecting early-stage ovarian cancer. Am. J. Obstet. Gynecol. 2014;210(1):78.e1–9.

[12] Rosenbaum J, Weisman P. The evolving role of companion diagnostics for breast cancer in an era of next-generation omics. 2019.

[13] Jacobs T, Gown A, Yaziji H, Barnes M, Schnitt S. Specificity of HercepTest in DeterminingHER-2/neuStatus of breast cancers using the United States food and drug administration–approved scoring system. J. Clin. Oncol. 1999;17(7):1983.

[14] Dardaei L, Wang H, Singh M, Fordjour P, Shaw K, Yoda S, Kerr G, Yu K, Liang J, Cao Y, Chen Y, Lawrence M, Langenbucher A, Gainor J, Friboulet L, Dagogo-Jack I, Myers D, Labrot E, Ruddy D, Parks M, Lee D, DiCecca R, Moody S, Hao H, Mohseni M, LaMarche M, Williams J, Hoffmaster K, Caponigro G, Shaw A, Hata A, Benes C, Li F, Engelman J. SHP2 inhibition restores sensitivity in ALK-rearranged non-small-cell lung cancer resistant to ALK inhibitors. Nat. Med. 2018;24(4):512–7.

[15] Auffray C, Charron D, Hood L. Predictive, preventive, personalized and participatory medicine: back to the future. Genome Med. 2010;2(8):57.

[16] Carter P, Laurie G, Dixon-Woods M. The social licence for research: whycare.dataran into trouble. J. Med. Ethics 2015;41(5):404–9.

[17] Chandra A, Skinner J. Geography and racial health disparities. 2003.

[18] Payne T. Computer decision support systems. Chest 2000;118(2):47S–52S.

[19] Lee S, Celik S, Logsdon B, Lundberg S, Martins T, Oehler V, Estey E, Miller C, Chien S, Dai J, Saxena A, Blau C, Becker P. A machine learning approach to integrate big data for precision medicine in acute myeloid leukemia. Nat. Commun. 2018;9(1).

[20] Diao Y, Roy A, Bloom T. Building highly-optimized, low-latency pipelines for genomic data analysis. In: Proceedings of the Conference on Innovative Data Systems Research (CIDR '15), 2015. ACM SIGMOD International Conference on Management of Data - SIGMOD '15; 2015.

[21] Liu B, Madduri R, Sotomayor B, Chard K, Lacinski L, Dave U, Li J, Liu C, Foster I. Cloud-based bioinformatics workflow platform for large-scale next-generation sequencing analyses. J. Biomed. Inform. 2014;49:119–33.

[22] Wong A, Plasek J, Montecalvo S, Zhou L. Natural language processing and its implications for the future of medication safety: a narrative review of recent advances and challenges. Pharmacotherapy 2018;38(8):822–41.

[23] Roundtable on translating genomic-based research for health; board on health sciences policy; institute of medicine. Drug repurposing and repositioning: workshop summary. Washington (DC): National Academies Press (US); August 8, 2014. https://doi.org/10.17226/18731. Available from: https://www.ncbi.nlm.nih.gov/books/NBK202175/.

[24] Kolb P, Ferreira R, Irwin J, Shoichet B. Docking and chemoinformatic screens for new ligands and targets. Curr. Opin. Biotechnol. 2009;20(4):429–36.

[25] Personalised Medicine at FDA. A 2016 progress report [internet]. Personalized medicine coalition. 2017. Available from: http://www.personalizedmedicinecoalition.org/Userfiles/PMC-Corporate/file/PM-at-FDA.pdf.

[26] Laack E, Sauter G, Bokemeyer C. Lessons learnt from gefitinib and erlotinib: key insights into small-molecule EGFR-targeted kinase inhibitors in non-small cell lung cancer. Lung Cancer 2010;69(3):259–64.

[27] Antonia S, Goldberg S, Balmanoukian A, Chaft J, Sanborn R, Gupta A, Narwal R, Steele K, Gu Y, Karakunnel J, Rizvi N. Safety and antitumour activity of durvalumab plus tremelimumab in non-small cell lung cancer: a multicentre, phase 1b study. Lancet Oncol. 2016;17(3):299–308.

[28] Realising the value of precision medicine: Today [Internet]. Personalized medicine connective. 2017. Available from: http://www.pmconnective.org/docs/articles/keeling-waldron-1.pdf.

[29] Ginsburg G, Phillips K. Precision medicine: from science to value. Health Aff. 2018;37(5):694–701.

[30] Brothers K, Rothstein M. Ethical, legal and social implications of incorporating personalized medicine into healthcare. Pers. Med. 2015;12(1):43–51.

[31] Huston J. Information governance standards for managing e-health information. J. Telemed. Telecare 2005;11(2_Suppl.):56–8.

[32] Laedtke A, O'Neill S, Rubinstein W, Vogel K. Family physicians' awareness and knowledge of the genetic information non-discrimination act (GINA). J. Genet. Couns. 2011;21(2):345–52.

[33] Rosenbaum S. The patient protection and affordable care act: implications for public health policy and practice. Publ. Health Rep. 2011;126(1):130–5.

[34] (a) Buchen L. Cancer: missing the mark. Nature 2011;471(7339):428–32.(b) Baker M. Cancer institute tackles sloppy data [Internet] Nature 2017. https://www.nature.com/news/cancer-institute-tackles-sloppy-data-1.11580.

[35] Zhernakova A, Kurilshikov A, Bonder M, Tigchelaar E, Schirmer M, Vatanen T, Mujagic Z, Vila A, Falony G, Vieira-Silva S, Wang J, Imhann F, Brandsma E, Jankipersadsing S, Joossens M, Cenit M, Deelen P, Swertz M, Weersma R, Feskens E, Netea M, Gevers D, Jonkers D, Franke L, Aulchenko Y, Huttenhower C, Raes J, Hofker M, Xavier R, Wijmenga C, Fu J. Population-based metagenomics analysis reveals markers for gut microbiome composition and diversity. Science 2016;352(6285):565–9.

[36] Mirnezami R, Ahmed A. Surgery 3.0, artificial intelligence and the next-generation surgeon. Br. J. Surg. 2018;105(5):463–5.

Chapter 55

Precision medicine glossary

Joel Faintuch[1] and Jacob J. Faintuch[2]
[1]*Department of Gastroenterology, São Paulo University Medical School, São Paulo, São Paulo, Brazil;* [2]*Department of Internal Medicine, Hospital das Clinicas, São Paulo, São Paulo, Brazil*

16S rRNA genes: The most used Bacteria and Archaea genes for metagenomic analysis. The small ribosomal subunits of these microorganisms (particle size 16 Svedberg) are good markers for taxonomic classification.

16S sRNA sequencing: Most metagenomic studies rely on this modality of sequencing. Taxonomic classification usually proceeds till genus level, although species identification may be possible as well. Fungi, viruses, and other microorganisms which lack 16S rRNA genes are not recognized.

Algorithm: A set of computer-defined steps to achieve diagnosis, therapy, prognosis, or other question.

Allele: Version or variant of a given gene.

Amplicon sequencing: Metagenomic sequencing based on a single RNA or DNA marker gene.

Artificial intelligence: Cognitive abilities, programs, and techniques built into systems and machines, in order to enable them to react, take decisions, and perform jobs.

Artificial neural network: See Neural network.

Autosomal: Chromosomes (22) and their genes, not related to sex determination (X and Y)

Axenic: Animals born in aseptic conditions (cesarian section), and maintained in sterile laboratories, with autoclaved food and water. Also known as Germ-free animals.

Bayesian statistics: Mathematical procedures geared at prior evidence, or data, which can change with new information. It leads to inferences and hypotheses classified as probability or likelihood, but not true or false. See also Boolean statistics.

Big data: Large amounts of numerical and non-numerical information. They demand specific architectures, algorithms, and statistical methods. Unique trends, patterns, and correlations based on network interactions can be extracted by adequate processing.

Biobank: A large collection of biological samples and health information, kept in optimal conditions to enable research. Usually organized by major institutions.

Bioengineered bacteria: See Genetic engineering.

Biorepository: A limited biobank, usually under the responsibility of a single team or company.

Boolean statistics: One of the statistical methods employed in computer science, as well as analysis of big data. Directed toward complex information in which not all variables are fully known. Yet it detects inferences or deterministic dependencies, defined as true or false. See also Bayesian statistics.

cDNA: A complementary copy of a stretch (or gene) of DNA produced by recombinant technique, or reverse transcription of mRNA.

Carrier: Organism with a single recessive pathogenic variant of a gene, usually free from disease (normal phenotype)

DNA technology. It can also be produced from RNA, by the enzyme reverse transcriptase.

DNA (deoxyribonucleic acid): A large polymer of nucleotides typically containing approximately 150 base pairs.

DNA polymorphism: See Polymorphism.

Codon: A series of three nucleotides on the mRNA molecule, which encode one amino acid. There are codons for all usual amino acids.

Commensal: Microorganism that benefits from the gut environment without causing any harm.

Congenital: Present at birth. Maybe genetical or environmental (related to pregnancy conditions).

Deep learning: Machine-based learning with algorithms designed for abstract models and employing multiple layers of processing. Depending on the modeling, it becomes sensitive to progressively more abstract patterns.

Deletion: Absence or removal of a segment (single base or nucleotide) of DNA. May involve part of a gene, or up to several genes.

Diversity: A quality related to the number of microbial taxa and their distribution. Alpha diversity describes the

Precision Medicine for Investigators, Practitioners and Providers. https://doi.org/10.1016/B978-0-12-819178-1.00055-1

multiplicity of taxonomic units within a sample or community (the different trees in a single forest). Beta diversity addresses the diversity structure between a number of samples or communities (the comparison between different forests).

Dominant: Disease or trait that can be expressed (phenotype) by a single copy of the pathogenic gene variant (heterozygotes).

Duplication: Repeated segments of DNA, involving part of a gene, one gene, or several genes.

Dysbiosis: Qualitative or quantitative imbalances in the microbiome, associated with diseases, functional impairments, or health risks. Dysbiosis is often linked to diversity troubles (alpha or beta diversity), as well as to abundance (excessive, insufficient) of bacteria taxa.

Epigenetic: Alteration in gene expression that is independent of the corresponding DNA coding sequence, and which can be inherited. One example is gene methylation, which may be influenced by the microbiome, as well as by other environmental circumstances.

Exon: Nucleotide sequence in the DNA molecule that encodes polypeptide (protein) synthesis.

Exome: The assemblage of all exons in the chromosomes. The human exome encompasses approximately 180,000 exons.

Exposome: The sum total of environmental influences (nongenetic), from birth onwards, which shape human health and disease, together with the genome. This is a wide-reaching concept, including not only absorbed molecules and epigenetics, but also lifestyle, urban environment, climate, and education.

GWAS/Genome-wide association sequencing: Analysis of a genome-wide set of genetic variants, in a defined cohort of individuals, to determine whether any of these are associated with a given trait or disease. Usually, single-nucleotide polymorphisms are focused.

Gene map: See Locus.

Gene promoter: See Promoter.

Genetic engineering: The synthesis, deletion, or insertion in the genome of nucleotide sequences, which may be artificial or exist in natural conditions, aiming to change both the genotype and the phenotype of a microorganism.

Genetic variant: See Variant.

Genome/genomics: The collection of genes and noncoding nucleotides of an organism.

Genome sequencing: See Sequence analysis.

Genotype: The assemblage of sequenced genes of humans or microorganisms, which, together with epigenetic and environmental factors, shape the phenotype.

Genotyping: Genetic tests to detect specific pathogenic variants.

Germ-free: see Axenic;

Germline mutation: An inheritable mutation, transmitted to offspring.

Gene expression: The series of phenomenons including transcription and translation, by which a gene is activated and produces the protein it encodes.

Hypermutability: See Instability.

Indel: A small genetic variation, related to insertion or deletion (*in - del*) of bases in the genome.

Insertion: Inclusion of one or more nucleotide base pairs into the DNA.

Instability: A region of DNA that is highly susceptible to mutations. See also Microsatellite instability.

Intron: Noncoding nucleotide sequence in the DNA molecule. These segments are usually removed during RNA transcription.

Locus (loci): Defined positions on a chromosome, related to genes, alleles (gene variants), traits, and genetic markers. Gene maps list all the loci of a given genome and can identify the locus for a biological trait.

MMR/DNA mismatch repair: An intrinsic protection mechanism to conserve DNA sequence. It removes base mispairs generated by replication or recombination. Many types of DNA damage are corrected, related to insertion, deletion, and misincorporation of bases, including microsatellite formation.

MSI/Microsatellite instability: See Microsatellite DNA.

mtDNA/Mitochondrial DNA: DNA outside the chromosomes, carried in the mitochondria. It is inherited only from the mother and contains as few as 37 genes. This DNA can undergo mutations and be associated with inherited diseases.

Machine learning: The use of algorithms that find patterns in data without explicit instructions. Depending on the inputs, it may discover associations with outputs. See also Pattern recognition.

Metabolite: See Metabolome.

Metabolome: The profile of metabolites (small molecules generated by metabolic activity, with <1500 Da) of a given fluid, tissue, or organism. It reflects physiologic changes, along with those related to eventual diseases, organ failures, or intoxications.

Metabonome: The larger metabolome, or total metabolite pool, encompassing for instance plasma, urine, and fecal samples. It corresponds to the sum-total of the metabolome, or meta-metabolome.

Metagenome: The genes and genomes of the participants in a collected sample. They may encompass many individuals and are usually evaluated by gene sequencing of total DNA.

Metaproteome: The assemblage of all proteins (or just small proteins), collected in a sample.

Metatranscriptome: The set of transcripts of the genes expressed in a sample, potentially related to a phenotype. Both community-wide, and single-organism metatranscriptomics may be addressed.

Microarray: Scanning of the whole genome or targeted regions, to search for copy number variants, or single nucleotide polymorphisms (SNPs).

Microbiome: A community of microorganisms that occupy a specific habitat or environment. Some limit microbiome to the genes or genomes only, the whole microbial collection being designated as microbiota. However, in both circumstances, the taxonomic catalog is the same, typically based on analysis of 16S rRNA gene sequences (metagenomics). As a consequence, many groups employ the terms interchangeably.

Microbiota: See Microbiome.

Microsatellite DNA: Segments of repeated sequences of DNA (two to five nucleotides), anywhere in the genome. They are unstable and highly susceptible to mutations.

Microsatellite instability (MSI): This is a condition of genetic hypermutability (predisposition to mutation), that results from impaired DNA mismatch repair (MMR).

Mosaicism: A nonhomogeneous genetic pattern, in which cells of the same individual or experimental sample, exhibit different numbers and arrangements of genes.

Multivariate analysis: Simultaneous analysis of more than one statistical variable. See also Pattern recognition.

Mutation: A definitive change in the DNA, affecting copy number or intrinsic DNA features. See also Germline mutation and Somatic mutation.

NGS/Next-generation sequencing: High-throughput sequencing technique utilizing massive parallel sequencing, for improved results. The entire human genome can be sequenced with NGS on the same day.

Natural language processing: Computer analysis and processing of uncoded, untrained spoken, or written language. This is a complex challenge due to variations in vocabulary and grammar, often requiring machine learning.

Neural network: An artificial mathematical connectionist system, employing interconnected layers of units forming different layers or levels, sequentially utilized for pattern recognition and problem solving, often in association with machine learning.

Nucleotide: The monomer which, after polymerization, can generate nucleic acids (DNA, RNA). Nucleotides encompass a nitrogenous base, a sugar, and a phosphate group.

Oligonucleotide: A small sequence (polymer) of nucleotides, usually a fragment of DNA or RNA. They are often used for diagnosis and research.

Oncogene: A gene that promotes cellular proliferation, potentially resulting in benign tumors or cancer.

PCR/Polymerase chain reaction: An important gene sequencing tool for genetic, genomic, and microbiological studies. Tiny amounts of cDNA can be amplified to substantial quantities, thus allowing diagnosis and other procedures.

PRR/Pattern recognition receptors: A feature of certain immune cells for recognition of bacteria and other foreign antigens.

Pattern recognition (computer science): An approach for classification of big data and extraction of clinically useful patterns, signals, or biomarkers. May employ algorithms, machine learning, and other tools.

Phage: Virus that parasites bacteria. Can be used as a carrier (vector) of DNA into the bacteria.

Phenotype: Collection of physical, biochemical, and psychological traits or features of an individual. The phenotype is susceptible to genetic, epigenetic, and environmental impacts.

Phylogenetics: Taxonomic classification of genome sequences of microorganism populations, generated by microbiome analysis, into existing phyla, or major microbial divisions.

Plasmid: Self-replicating DNA segment. Naturally occurring or modified plasmids can be used to carry genes into bacteria and other cells.

Polymorphism: DNA variants, including deletions, insertions, or copy number changes. Maybe devoid of danger or signal disease risk.

Promoter: Noncoding DNA sequence, which starts gene transcription.

Proteome: See Metaproteome

RNA (ribonucleic acid): A large nucleotide polymer analogous to DNA, however containing one different base (uracil in RNA, thymine in DNA).

Recessive: Disease or trait only expressed (phenotype) by two copies of the pathogenic gene variant (homozygotes).

Recombinant DNA: A technology to join together different fragments of natural or synthetic DNA, thus creating new sequences (genes).

Reporter gene: Marker of gene transfection to another cell. The product of the reporter gene, which is attached to the main gene, is usually easy to identify, thus confirming success of the procedure.

SNP/Single nucleotide polymorphism: A variation of a single nucleotide, located in a specific position of the genome. It can be associated with both a gene and a noncoding region of DNA.

Sequence analysis: Procedures for estimation of structure, features, and functions of sequenced DNA or RNA strands. This leads to the identification of genes (exomes), and of noncoding nucleotides as well (exons and introns). Alignment, clustering, chimera detection, phylogenetics, assembly, and annotation are typical steps in this process.

Shotgun sequencing: A high-throughput system that randomly breaks down DNA/RNA strands in small pieces, and then reassembles them by computer programs that juxtapose overlapping fragments. This is one of the most modern techniques in use.

Single-nucleotide polymorphism: The most common type of genetic variant.

Somatic mutation: A mutation in somatic cells, not transmitted to subsequent generations. It may be related to cancer or other diseases.

Splicing: The process by which noncoding regions (introns) are removed from the RNA transcript, followed by molecule reconstitution.

Systems biology: A mathematical modeling approach to areas of knowledge based on quantitative models and network interactions. It allows for hypothesis testing and experimental validations.

Taxon (taxa): A taxonomic unit of microorganisms, ranked in the scientific classification. Examples of taxa are phylum, class, order, family, genus, and species.

Telomere: A repeating structure found at the end of chromosomes. Important for genetic integrity and prevention of cancer may diminish with age, free radicals, and other agents.

Transcription: The first step of gene expression. Consists of the copying of a segment of DNA by RNA, usually by messenger RNA (mRNA).

Reverse transcription: Copying of a segment of RNA (RNA template), generating cDNA. Conducted by the enzyme reverse transcriptase.

Translation: Generation of the corresponding protein (or polypeptide) from messenger RNA.

Transcriptome: see Metatranscriptome

Tumor suppressor gene: A gene that inhibits tumor progression.

Variant: Alternative forms of a gene, leading or not to an altered phenotype. When related to mutations, they are sometimes used as synonyms.

Volatile organic compounds/VOC: Small molecules generated by human or bacterial metabolism, as well as contaminants originating from ingested or absorbed foods, drugs, tobacco products, and environmental sources.

They may be identified in the breath (breathome or volatolome), as well as in other samples (feces, urine, secretions).

Volatolome: See Volatile organic compounds.

Vectors: Transduction agents employed in gene therapy, to carry foreign DNA into a foreign cell. Viral agents (nonpathogenic or partially inactivated) can be used, along with plasmids, physical, and other methods.

WES/Whole-exome sequencing: Sequencing of all coding nucleotides.

WGS/Whole-genome sequencing: Sequencing of transcribing and nontranscribing regions.

Wild type: Normal allele or organism. Contrasts with mutant, variant, or genetically engineered.

TABLE 55.1 Helpful dictionary and glossary internet sites.

Description	Location
BioTech life science	www.gate2biotech.com/biotech-life-science-dictionary
Biochemistry and molecular biology	https://themedicalbiochemistrypage.org/glossary.php
Molecular biology	www.genscript.com/molecular-biology-glossary/search
Metabolites and other molecules	www.chem.qmul.ac.uk/iubmb/
Biochemistry	https://biochemden.com/biochemistry-glossary
Genetics and genomics	www.genome.gov/glossary/?textonly=true
Mendelian inheritance in man(human genes, genetic disorders)	www.omim.org
NCI cancer glossary	www.cancer.gov/publications/dictionaries/cancer-terms
Glossary of biostatistics	www.medicalbiostatistics.com/Glossary.pdf
Gene reviews and glossary	www.ncbi.nlm.nih.gov/books/NBK1116/

X-linked disorders (or X linkage): Diseases or traits transmitted by genes in sex-determining chromosome X. Y-linkage is uncommon (Table 55.1).

Further reading

[1] Faintuch J, editor. Microbiome, dysbiosis, probiotics and bacteriotherapy. Sao Paulo (Brazil): Manole; 2017.
[2] Farkas D. Glossary: definitions useful in understanding pharmacogenetics. EJIFCC 2008;19:92–3.
[3] Filipp FV. Precision medicine driven by cancer systems biology. Cancer Metastasis Rev. 2017;36:91–108.
[4] Hutson M. Artificial intelligence, in so many words. Science 2017;357(6346):19.
[5] Kaushansky K. Glossary of molecular biology terminology. Hematol. Am. Soc. Hematol. Educ. Progr. 2001:522–40.
[6] Luketa S. New views on the megaclassification of life. Protistology 2012;7:218–37.
[7] Marchesi JR, Ravel J. The vocabulary of microbiome research: a proposal. Microbiome 2015;3:31.

[8] Precision medicine alliance. Available from: https://precisionmedicinealliance.org.

[9] Shetty SA, Hugenholtz F, Lahti L, Smidt H, de Vos WM. Intestinal microbiome landscaping: insight in community assemblage and implications for microbial modulation strategies. FEMS Microbiol. Rev. 2017;41:182–99.

[10] Ursell LK, Metcalf JL, Parfrey LW, Knight R. Defining the human microbiome. Nutr. Rev. 2012;70(Suppl. 1):S38–44.

[11] Wagenaar DJ, Weissleder R, Hengerer A. Glossary of molecular imaging terminology. Acad. Radiol. 2001;8:409–20.

[12] Yates LR, Seoane J, Le Tourneau C, Siu LL, Marais R, Michiels S, Soria JC, Campbell P, Normanno N, Scarpa A, Reis-Filho JS, Rodon J, Swanton C, Andre F. The European Society for Medical Oncology (ESMO) precision medicine glossary. Ann. Oncol. 2018;29:30–5.

Chapter 56

Useful Internet sites

Joel Faintuch[1] and Jacob J. Faintuch[2]

[1]*Department of Gastroenterology, São Paulo University Medical School, São Paulo, São Paulo, Brazil;* [2]*Department of Internal Medicine, Hospital das Clinicas, São Paulo, São Paulo, Brazil*

TABLE 56.1 General genome and microbiome sources.

Address	Microbiome
www.metahit.eu	METAHIT metagenomics
www.hmpdacc.org/	Human microbiome project-data
www.kegg.jp/kegg	KEGG encyclop. Genes genomes
http://www.mg-rast.org/	Metagenomics server MG-RAST
www.theseed.org/servers	The network-based SEED API
https://press3.mcs.anl.gov/gensc/	Genomic standards consortium
www.mybiosoftware.com/	Functional gene ecological analysis
www.micca.org	Microbial community analysis
https://omictools.com(copyrighter-tool)	Gene copy number correcting tool
https://img.jgi.doe.gov	IMG/M integrated Microbial genomes
https://bioinformatics.ca/	CAMERA Microbial res. Analysis
https://github.com/jcvi/metarep	METAREP Metagenome analyzer
www.metavelvet.dna.bio.keio.ac.jp/MSL.html	de novo metagenomic assembler
https://bitbucket.org/biobakery/metaphlan2	MetaPhlAn2 Metagen.Phylog.Anal.
http://g2im.u-clermont1.fr/phylopdb/	PhylOPDb: 16S rRNA database
http://eggnog-mapper.embl.de	Genome-wide function. Orthology annotation
www.insdc.org	Int. Nucleotide sequence database
http://omim.org	Gene locus information
www.ncbi.nlm.nih.gov/books/NBK1116/	Gene reviews
https://w3id.org/biocompute/1.3.0	Biocompute standard reporting of genome sequencing
www.phenoscanner.medschl.cam.ac.uk/phenoscanner	Genetic variant/phenotype associations

Precision Medicine for Investigators, Practitioners and Providers. https://doi.org/10.1016/B978-0-12-819178-1.00056-3

TABLE 56.2 Sources for metabolome and transcriptome.

Address	Metabolome and omics tools
www.hmdb.ca	Human metabolome data base
http://hmp2.org	Integrative human metabolome data
www.lipidmaps.org	Lipid Maps lipidomic gateway
https://bmcbioinformatics.biomedcentral.com/	Metabolome searcher
https://biocyc.org	BioCyc database collection
http://fiehnlab.ucdavis.edu	SetupX metabolomics base
https://metacyc.org/	Metabolic pathway ddatabase
https://pubchem.ncbi.nlm.nih.gov/	Substance, compound, bioassay
www.metaboanalyst.ca	Metabolomics and biochemical pathways
https://omictools.com	OMIC tools
www.qiagenbioinformatics.com/products	Ingenuity pathway analysis/IPA
gmd.mpimp-golm.mpg.de	Golm metabolome ddatabase
https://omictools.com(transcriptomics)	Bioinformatics for transcriptome
https://github.com/kwanjeeraw/metabox	Metabolomics and omics integration
http://xcmsonline.scripps.edu	Metabolomics and omics integration

TABLE 56.3 Bioinformatics platforms.

Address	Bioinformatics
http://ab.inf.uni-tuebingen.de	MEGAN aalgorithm in bioinformatics
http://comet2.gobics.de	COMET comparative aanal. Metagenomes
http://www.mg-rast.org/	MG RAST Metagen. Rapid aannotation
https://github.com/biocore(Evident)	Evident software package
http://picrust.github.io	PICRUSt phylogenetic iinvestigation
https://github.com/danknights/sourcetracker/releases	Source tracker
https://github.com/RRShieldsCutler/splinectomeR	Splinectome R
www.bioinformatics.babraham.ac.uk	Babraham bioinformatics
http//hannon lab.cshl.edu/	FASTX toolkit hannon laboratory
https://omictools.com(transcriptomics)	Bioinformatics for transcriptome
https://omictools.com(Emperor)	3D visualization of microbiome data
https://qiime2.org	Analysis of similarity
https://www.mothur.org	Analysis of similarity
http://huttenhower.sph.harvard.edu/galaxy	Linear discriminant analysis
http://meta.usc.edu/softs/lsa/	Extended similarity
https://users.cs.duke.edu/~amink/software/banjo/download	Bayesian network inference
http://www.dcs.gla.ac.uk/BioBayes	Bayesian statistics

TABLE 56.4 Germline variant and mutation databases.

Address	Variants and other information
www.internationalgenome.org/	1000 genomes project
https://decipher.sanger.ac.uk/	Genomic variation and phenotype
http://exac.broadinstitute.org/	Exome aggregation consortium
www.ncbi.nlm.nih.gov/clinvar/	Genomic variation and health
esp.gs.washington.edu/drupal/	Exome sequencing project
http://omim.org/	Mendelian inheritance in man

TABLE 56.5 Somatic variant and mutation databases.

Address	Variants and mutations
http://atlasgeneticsoncology.org	Genetics in oncology and hematology
https://candl.osu.edu/	Cancer mutations
cancer.sanger.ac.uk/cosmic	Somatic mutations in cancer
https://civicdb.org/home	Variants in cancer
www.genecardsorg	Multiomics gene information
https://dcc.icgc.org	Cancer genomes and mutations
https://ckb.jax.org/	Cancer hotspot genes
https://pct.mdanderson.org/	Personalized cancer therapy
cgap.nci.nih.gov/chromosomes	Chromosome aberrations
www.mycancergenome.org	Câncer mutations
www.pharmgkb.org	Pharmacogenomics database
https://pmkb.weill.cornell.edu	Clinical cancer variants
http://oncokb.org	Oncology gene database

TABLE 56.6 Prediction of biological effects of somatic and germline mutations.

Address	Variants and mutations
wiki.chasmsoftware.org/index.php/Main_Page	Cancer specific mutations
https://cadd.gs.washington.edu/	Impact of variants
http://fathmm.biocompute.org.uk/	Functional analysis
http://mutationassessor.org	Amino acid substitutions
http://genetics.bwh.harvard.edu/pph2/	Polymorphism phenotyping
http://structure.bmc.lu.se/PON-P2/	Amino acid substitutions
http://pantherdb.org	Functional classification/Gene ontology/Proteome
www.ngrl.org.uk/Manchester/	Diagnostic mutation database/Bioinformatics
provean.jcvi.org/index.php	Protein variation effect analyzer
varianttools.sourceforge.net/Annotation/DbNSFP	Variant annotation database
www.rostlab.org/services/snpdbe/	Database functional effects

TABLE 56.7 Specialized clinical sites.

Address	Diseases and other specialized sites
http://vmc.vcu.edu/momspi	Multi-omic microbiome study: Pregnancy
www.americangut.org	American gut project
http://ibdmdb.org	Inflammatory bowel disease Multi-omic
www.pharmacomicrobiomics.org	The PharmacoMicrobiomics portal
www.brenda-enzymes.org	Braunschweig enzyme database
www.github.com/alifar76	Inflammatory bowel disease prediction
http://www.gastro.org	Am. Gastr.Assoc. Fecal Microbiota registry
http://www.hpmcd.org	Human pan-Microbe communities
www.homd.org	Human oral microbiome database
www.omim.org	Gene catalog and genetic diseases
www.aacr.org/Research/Research/pages/aacr-project-genie.aspx	Cancer genomics database

TABLE 56.8 Metaproteomic platforms.

Address	Site resources
https://omictools.com (peptide-intens)	Pipasic peptide intens. Weighted proteome
https://github.com/compomics/meta	Metaproteomics data analysis and interpretation
https://unipept.ugent.be OK	Biodiversity of complex metaproteomes
https://mikrobiologie.uni-greifswald.de	Amino acid sequences of a given protein database
www.uniprot.org/proteomes	Uniprot metaproteomic databases
https://ab.inf.uni-tuebingen.de	Metaproteome dataset
www.pfam.xfam.org/	Protein families database
www.ebi.ac.uk/pride/	PRIDE proteomics database

TABLE 56.9 Designer bacterial and human genome tools.

Address	Site resources
https://crispr.mit.edu	Optimized CRISPR sgRNA design
www.e-crisp.org/E-CRISPOK	Design of CRISPR constructs
www.multicrispr.net/basic_input.html	RNA target search in input sequences
https://portals.broadinstitute.or/gpp/public/analysis-tools/sgrna-design	Picks candidate sgRNA sequences for targets (CRISPRa and CRISPRi)
https://cm.jefferson.edu/off-spotter	Identifier of all genome hits of gRNAs
https://crispr.cos.uni-heidelberg.de/	CRISPR/Cas9 target online predictor
http://chopchop.cbu.uib.no/index.ht	CHOPCHOP genome editing

TABLE 56.10 Large-scale genomic, phenotypic, and clinical datasets and biobanks

Address	Site resources
www.ukbiobank.ac.uk	Health data on 500,000 volunteers
http://datacompass.lshtm.ac.uk/40/	Cardiometabolic genomics 30,000 people
www.intervalstudy.org.uk	Multiomics 50,000 participants
www.epiccvd.eu	Cardiovascular genomics 520,000 people
www.ckbiobank.org/site	Chronic diseases 510,000 Chinese
www.discovehrshare.com	Whole exome 50,000 adults
www.research.va.gov/mvp	DNA/health databank > 1 million veterans
www.rpgeh.kaiser.org	Genomic/health biobank 500,000 people
http://victr.vanderbilt.edu/pub/biovu	Gene/phenotype bank 250,000 people
https://emerge.mc.vanderbilt.edu/	Rare genetic variants 25,000 people
https://allofus.nih.gov	DNA/health databank >1 million people
www.genius-chd.com	Gene biobank 250,000 coronary cases
www.hermesconsortium.org	Gene biobank 11,000 heart failure cases
www.afgen.org	Exome/genome 40,000 atrial fibrillation
www.caliberresearch.org	Databank 10 million general/cardiac
www.soluciones.si/blog/2013/08/13/instalacion-abucasis-generalitat-valenciana/	Clinical databank 5.1 million patients Valencia, Spain
http://mondriaanfoundation.org	Gene/pharmacy 500,000 cases
www.hic.nihr.ac.uk	Databank 5 disease areas
www.ucr.uu.se/swedeheart	Databank 2 million heart trouble
www.escardio.org/Research/Registries-&-surveys/Observational-research-programme/registry-overview	Multiple databanks cardiac diseases
www.ucl.ac.uk/nicor	Databanks 2 million cardiovascular patients

TABLE 56.11 Clinical trials and general scientific information.

Address	Resources
https//www.thecochranelibrary.com	Systematic reviews for evidence-based medicine
https://www.bmj.com/specialties/clinical-evidence	Practical evidence-based information
https://www.cancer.gov/about-cancer/treatment/clinical-trials	Cancer trials
https://www.fda.gov	Official information on drugs, devices, radiation-emitting products, and others
https://clinicaltrials.gov	Largest database of clinical trials
https://www.centerwatch.com	User-friendly clinical trials information
apps.who.int/trialsearch/	International trial search tool
www.clinicalstudyresults.org	Results of finished trials
www.drugbank.ca	Drugs and drug targets
www.ebi.ac.uk/chembl	Drug compounds
https://reaganudall.org/	Medical evidence database

TABLE 56.12 Dictionary and glossary internet sites.

Address	Resources
www.gate2biotech.com/biotech-life-science-dictionary	BioTech life science
https://themedicalbiochemistrypage.org/glossary.php OK	Biochemistry and molecular biology
www.genscript.com/molecular-biology-glossary/search	Molecular biology
https://www.qmul.ac.uk/sbcs/iupac/	Chemical molecules
https://biochemden.com/biochemistry-glossary	Biochemistry
www.genome.gov/glossary/index.cfm?id=208&textonly=true	Genetics and genomics
www.omim.org	Mendelian inheritance in man gene catalog and genetic diseases
http://www.cancer.gov/publications/dictionaries/cancer-terms	NCI cancer glossary
www.medicalbiostatistics.com/Glossary.pdf	Glossary of biostatistics
www.ncbi.nlm.nih.gov/books/NBK1116/	Gene reviews and glossary

Further reading

[1] Prawira A, Pugh TJ, Stockley TL, Siu LL. Data resources for the identification and interpretation of actionable mutations by clinicians. Ann. Oncol. 2017;28(5):946–57.

[2] Prosperi M, Min JS, Bian J, Modave F. Big data hurdles in precision medicine and precision public health. BMC Med. Inf. Decis. Mak. 2018;18(1):139.

[3] Hemingway H, Asselbergs FW, Danesh J, Dobson R, Maniadakis N, Maggioni A, van Thiel GJM, Cronin M, Brobert G, Vardas P, Anker SD, Grobbee DE, Denaxas S. Big data from electronic health records for early and late translational cardiovascular research: challenges and potential. Eur. Heart J. 2018;39(16):1481–95.

Index

Note: 'Page numbers followed by "f" indicate figures and "t" indicate tables'.

B

E

F

H

I

M

N

Q

R

S

T

Technical Bureau
Dug

CPI Antony Rowe
Eastbourne, UK
November 20, 2019